PROCESS HEAT EXCHANGE

Edited by
Vincent Cavaseno
and the Staff of Chemical Engineering

CHEMICAL ENGINEERING

McGraw-Hill Publications Co., New York, N.Y.

Copyright © 1979 by Chemical Engineering McGraw-Hill Pub. Co.
1221 Avenue of the Americas, New York, New York 10020

All rights reserved. No parts of this work may be reproduced or utilized in any form or by any means, electronic or mechanical, including photocopying, microfilm and recording, or by any information storage and retrieval system without permission in writing from the publisher.

Printed in the United States of America

Library of Congress Cataloging in Publication Data

Main entry under title:

Process heat exchange.

    1. Heat—Transmission. 2. Heat exchangers.
I. Chemical engineering.
TJ260.P76   621.4'022   79-18473
ISBN 0-07-606621-5 (soft cover)
      0-07-010742-4 (hard cover)

CAVASENO, VINCENT
PROCESS HEAT EXCHANGE
000341466

66.045 C37

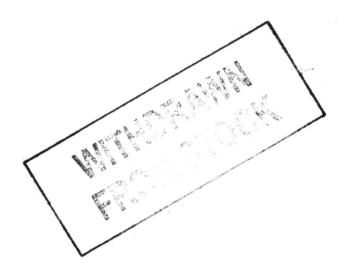

-3 MAY 2001

PROCESS HEAT EXCHANGE

CONTENTS

Introduction ... 1

Section I. HEAT EXCHANGER DESIGN AND SPECIFICATION ... 3

(A) Shell-and-Tube Heat Exchangers ... 5

How to select the optimum shell-and-tube heat exchanger ... 14
Design of heat exchangers ... 37
Design heat exchangers for liquids in laminar flow ... 44
Design parameters for condensers and reboilers ... 52
Designing vertical thermosyphon reboilers ... 56
Double-tubesheet heat-exchanger design stops shell-tube leakage ... 60
Guide to trouble-free heat exchangers ... 68
Horizontal-thermosiphon-reboiler design ... 72
How to design piping for reboiler systems ... 79
How to find the optimum layout for heat exchangers ... 88
Predicting heat-exchanger performance by successive summation ... 92
Preventing vibration in shell-and-tube heat exchangers ... 99
Selecting tubes for CPI heat exchangers—I ... 103
Selecting tubes for CPI heat exchangers—II ... 106
Selecting tubes for CPI heat exchangers—III ... 109
Use of computer programs in heat-exchanger design ... 115

(B) Heat exchangers of Other Designs ... 117

Designing direct-contact coolers/condensers ... 127
Designing spiral-plate heat exchangers ... 137
Designing spiral-tube heat exchangers ... 145
Evaluating plate heat-exchangers ... 150
How heat pipes work ... 153
Uses of inflated-plate heat exchangers ... 156
Where and how to use plate heat exchangers ... 163

(C) Materials of Constructon for Heat Exchangers ... 165

Choosing materials of construction for plate heat exchangers—I ... 168
Choosing materials of construction for plate heat exchangers—II ... 170
Graphite heat exchangers—I ... 173
Graphite heat exchangers—II ... 176
Selecting steel tubing for high-temperature service ... 180
Stainless-steel heat exchangers—Part I ... 185
Stainless-steel heat exchangers—Part II

Section II. HEAT EXCHANGE EQUIPMENT—OPERATION AND MAINTENANCE 189

Design and specification of air-cooled steam condensers — 191
Continuous tube cleaning improves performance of condensers and heat exchangers — 200
Evaluate reboiler fouling — 204
Finding the natural frequency of vibration of exchanger tubes — 208
In-place annealing of high-temperature furnace tubes — 209
Preventing fouling in plate heat exchangers — 211
Operating and maintenance records for heating equipment—I — 215
Operation and maintenance records for heating equipment—II — 217
Operating performance of steam-heated reboilers — 219
Performance of steam heat-exchangers — 223

Section III. HEAT TRANSFER IN REACTION UNITS 229

Controlling heat-transfer systems for glass-lined reactors — 231
Heat more efficiently—with electric immersion heaters — 235
Heat transfer in mechanically agitated units — 239
Heating and cooling in batch processes — 246
Picking the best vessel jacket — 252

Section IV. HEAT TRANSFER IN PIPING SYSTEMS 259

Calculating heat transfer from a buried pipeline — 261
Designing steam tracing — 263
Electric pipe tracing — 270
Steam tracing of pipelines — 274
Heating pipelines with electrical skin current — 282

Section V. FIRED HEATERS 285

Fired Heaters—I Finding the basic design for your application — 287
Fired Heaters—II Construction materials, mechanical features, performance monitoring — 293
Fired Heaters—III How combustion conditions influence design and operation — 303
Fired Heaters—IV How to reduce your fuel bill — 315
Generalized method predicts fired-heater performance — 320
Guide to economics of fired heater design — 328

Section VI. STEAM GENERATION AND TRANSMISSION 337

Balancing boilers against plant loads — 339
Basic data for steam generators—at a glance — 343
Converting boiler horsepower to steam — 345
Converting gas boilers to oil and coal — 346
Designing steam transmission lines without steam traps — 355
Estimating the costs of steam leaks — 358
How to select package boilers — 359
How to size and rate steam traps — 368
How to test steam traps — 373
Improving boiler efficiency — 377
Install steam traps correctly — 380
Select the right steam trap — 386
Short-cut calculation for steam heaters and boilers — 392
Steam traps — 393

Section VII. COOLING SYSTEMS — 399

- Air cooler or water tower—which for heat disposal? — 401
- Cooling-tower basin design — 410
- Design of air-cooled exchangers — 412
- Operation and maintenance of cooling towers — 431
- Proper startup protects cooling-tower systems — 434

Section VIII. HEAT TRANSFER CALCULATIONS — 437

- Calculate enthalpy with a pocket calculator — 439
- Calculating radiant heat transfer — 447
- Charts simplify spiral finned-tube calculations — 448
- Estimating liquid heat capacities—Part I — 454
- Estimating liquid heat capacities—Part II — 458
- New correlation for thermal conductivity — 463
- Quick estimation of gas heat-transfer coefficients — 465
- Steam from flashing condensate — 469
- Relating heat emission to surface temperature — 470
- Relating heat-exchanger fouling factors to coefficients of conductivity — 471

Section IX. HEAT TRANSFER MEDIA — 473

- Cooling-water calculations — 475
- Controlling corrosive microorganisms in cooling-water systems — 486
- Cycle control cuts cooling-tower costs — 489
- Heat-transfer agents for high-temperature systems — 492
- Low-toxicity cooling-water inhibitors—how they stack up — 502
- Organic fluids for high-temperature heat-transfer systems — 504
- Understanding vapor-phase heat-transfer media — 513
- Using wastewater as cooling-system makeup water — 518
- Water that cools but does not pollute — 524

Section X. WASTE-HEAT RECOVERY — 533

- Heat recovery in process plants — 535
- How to avoid problems of waste-heat boilers — 545
- Mystery leaks in a waste-heat boiler — 549
- Rankine-cycle systems for waste heat recovery — 554
- Useful energy from unwanted heat — 559

Section XI. MISCELLANEOUS — 565

- Conserving fuel by heating with hot water instead of steam — 567
- Guide to trouble-free evaporators — 571
- Hot water for process use—storage vs. instantaneous heaters — 577
- Sizing vacuum equipment for evaporative coolers — 581
- Spot heating with portable heating systems — 583
- Low-cost evaporation method saves energy by reusing heat — 586
- New directions in heat transfer — 593

Index — 601

INTRODUCTION

Industry uses 3,500 billion kWh equivalents of energy, or nearly 40% of the total consumed in this country, according to a recent Annual Survey of Manufacturers[1]. Over 80% of industry's share is employed in the chemical process industries (CPI).

It takes only a quick calculaton to show that even a small percentage improvement in energy efficiency holds vast potential for reducing fuel consumption and cutting operating costs. And the price of energy is not going down; in the future the savings will be even greater.

In the CPI, heat is the predominant form of energy used. So, in order for engineers to help meet energy goals, they must have a thorough understanding of heat exchange and its associated equipment in CPI processes.

This necessity was underscored in a recent Department of Energy document[2], which included recommendations for energy conservation measures. Some of these are: improvements in process heaters, boilers, steam systems and waste heat recovery systems; the application of heat exchangers, air coolers and insulation; and housekeeping measures.

In this book, we have presented a range of useful information on the transfer of heat in the CPI. The opening section focuses on heat exchange equipment, with shell-and-tube exchangers in the spotlight. That section then weaves through other heat exchanger designs (such as plate and spiral exchangers) and materials of construction for heat exchangers. Proper operation and regular maintenance of equipment is a must for peak performance; Section II presents a series of articles on that subject.

Sections III and IV outline heat transfer considerations in reaction units and piping systems, respectively. The design and optimization of fired heaters for process fluids in the CPI are thoroughly covered next, in Section V. Steam, which is by and large the most important heat transfer medium, is discussed in Section VI, which includes articles on boiler design and improvement as well as methods to reduce steam losses in transmission lines. Cooling towers are discussed in Section VII. This is followed by a section on heat transfer calculations, and one on water and other media used as cooling or heating agents. Finally, methods of waste-heat recovery are outlined in Section X.

This book was designed to be helpful to engineers in many sectors of industry. For the design engineer, it will be a useful guide for the proper design and specification of heat exchange systems and equipment. For the plant operations engineer, it provides timely tips on saving energy and lowering operating costs through design modifications and correct operating procedures. The plant process engineer will find it a handy troubleshooting guide for tackling operating problems. For the plant maintenance engineer, practical details on programmed and crises maintenance are presented.

All in all, this book provides a combination of theoretical and practical information on process heat exchange that is virtually unavailable elsewhere.

August 1979

---

1. 1975 Bureau of Census survey
2. Industrial Energy Efficiency Improvement Program, Annual Report Support Document, Vol. II, June 1978, Dept. of Energy, Asst. Secretary for Conservation and Solar Applications, Div. of Industrial Convervation.

Section I
HEAT EXCHANGER DESIGN AND SPECIFICATION (A., B., C.)

(A) Shell-and-Tube Heat Exchangers

How to select the optimum shell-and-tube heat exchanger
Design of heat exchangers
Design heat exchangers for liquids in laminar flow
Design parameters for condensers and reboilers
Designing vertical thermosyphon reboilers
Double-tubesheet heat-exchanger design stops shell-tube leakage
Guide to trouble-free heat exchangers
Horizontal-thermosiphon-reboiler design
How to design piping for reboiler systems
How to find the optimum layout for heat exchangers
Predicting heat-exchanger performance by successive summation
Preventing vibration in shell-and-tube heat exchangers
Selecting tubes for CPI heat exchangers—I
Selecting tubes for CPI heat exchangers—II
Selecting tubes for CPI heat exchangers—III
Use of computer programs in heat-exchanger design

How to select the optimum shell-and-tube heat exchanger

Developing an optimum unit can be a complex process because of the many interdependent parameters involved. An experienced engineer is required—either to design the exchanger himself or to evaluate offers from vendors based on his specifications.

John P. Fanaritis and James W. Bevevino, Struthers Wells Corp.

Because a shell-and-tube heat exchanger has no moving parts and exchanges heat between two fluids, an engineer could develop the mistaken impression that the design of this equipment is simple and straightforward. Although this type of exchanger is not usually a sophisticated piece of equipment, many considerations are involved in obtaining the optimum design for a given service.

This article does not provide design formulas or special methods for determining the optimum design, because these vary from process to process. Instead, the parameters involved and the complexities that surround the design are presented.

Included are charts and drawings—taken from the Tubular Exchanger Manufacturers Assn. (TEMA)—that pertain to the construction features of common heat exchangers, as well as to the nomenclature involved in describing them.

Also included are curves and data that provide approximate present-day prices for the several types of heat exchangers. In addition, a curve is provided to estimate the expected surface area for a given shell size, based on alternate tube-lengths. These graphs and charts can be used as guides in determining the approximate size and today's prices for the common variety of shell-and-tube heat exchangers.

Designing such an exchanger is based on the designer's knowledge and experience in heat transfer, mechanical design, utility, maintenance and cost. Inasmuch as the considerations that enter into the selection of the optimum heat exchanger are very difficult to evaluate quantitatively, a designer's experience is of the utmost importance.

Every designer's goal should be to provide a heat exchanger that will meet the specified performance requirements and provide long-term, trouble-free service at a minimum cost to the user. Due to the complexity in the design, and the interrelation between variables, each exchanger application will have as many different designs offered as there are designers or bidders. Under these circumstances, the ultimate choice rests with the purchaser.

Although each vendor or bidder warrants that his proposed design will meet the specified performance requirements, many questions require answers from the user in selecting the design. For example:

- Has the shell side of the heat exchanger been properly evaluated for maximum efficiency in heat-transfer rate, corresponding to the pressure drop?
- Are fluid velocities within reasonable limits, to avoid erosion of components or mechanical failure due to flow-induced vibrations?
- Have vents and drains been included where needed?
- Has differential expansion between the shell and tubes been considered?
- What type of tube-to-tubesheet joint is best?
- Are the specified metals compatible for good mechanical design and weldability?

Questions like the above must be answered to make a

reasonable assessment of a heat-exchanger design.

Usually, the repair costs connected with correcting a deficiency in a heat exchanger are only a small part of the penalty paid for poor design; the resulting loss in plant production for one day will often cost many times the purchase price of the heat exchanger.

Some exchangers are used in critical services involving extremely high pressures and/or temperatures, and handle hazardous fluids such as hydrogen. The purchaser must evaluate whether a particular vendor has the experience to provide the design integrity and the quality in fabricating for such special services.

Performance data specified by user

Selecting a heat-exchanger design must not be done casually; a considerable amount of time may easily be spent in heat-transfer and mechanical-design calculations. A purchaser should establish firm requirements for his heat exchanger at the time he asks for bids. The practice of imposing on a vendor the design and pricing of alternative selections is wasteful and in many cases costly.

To provide an optimum heat-exchanger design, the manufacturer or designer should be furnished by the user with the following minimum service information: (1) total heat load, Btu/h; (2) fluid quantities entering and leaving the exchanger, lb/h; (3) specific heat, thermal conductivity, viscosity, molecular weight or specific gravity of the fluids in appropriate units; (4) heat-exchanger ingoing and outgoing temperatures, °F; (5) operating pressures, psia; (6) allowable pressure drops, psi; (7) fouling factors; (8) design pressures and temperatures, psia and °F; (9) heat-exchanger type; (10) materials of construction; (11) tube-wall thickness for corrosion considerations, in; (12) corrosion allowance; (13) specifications, codes and standards; (14) size or space limitations; and (15) horizontal or vertical installation.

Process conditions

Since the size and resulting cost of a heat exchanger depend greatly on the log-mean-temperature difference (LMTD), a process designer should consider the effect of the operating temperature levels in the early stages of process design.

A high LMTD generally results in a smaller heat exchanger. Therefore, when considering operating temperature levels, a larger LMTD can be achieved by increasing the temperature level of the cooling medium. Close temperature approaches, where small differentials exist between the inlet of one fluid stream and the outlet of the other, will result in very low LMTDs.

There are no specific rules for determining the optimum operating temperatures. These should be selected based on the service and utility of the heat exchanger. Inefficient design and poor heat-exchanger performance can result when the LMTD is too high or too low. For a good design that is to cover many services, the lesser temperature difference between the shellside and the tubeside fluids should be greater than 10°F, the greater temperature difference should exceed 40°F.

Flow quantities—Fluid flowrates (lb/h) on both the shell and tube sides can affect the size and design of an exchanger. A designer may be forced to resort to multiple shells in series when the LMTD and flow quantities are low, and when a large temperature difference exists between the shellside and the tubeside fluids. Under these conditions, a countercurrent flow pattern must be maintained.

Under low-flow conditions, the designer may resort to multiple shells in series, to achieve reasonable fluid velocities and heat-transfer rates. When flowrates are extremely high for the surface requirement, multiple shells in parallel may be needed to achieve reasonable velocities, pressure drops, and an efficient heat-exchanger design.

Fouling factors—Dirt, scale or other deposits formed on the tube—inside and/or outside—which results in resistance to the flow of heat is called "fouling." The size and cost of a heat exchanger are related to specified fouling resistance; haphazard guessing of fouling can be costly.

Inasmuch as fouling factors are difficult to determine, they should be based on experience. Therefore, the user of the heat exchanger has the responsibility of providing the designer with the fouling factors peculiar to his operation. There are very limited data available for accurate assessment of the degree of fouling that should be applied for given service conditions. Fouling varies and depends on the material of construction of the tubes, the types of fluids involved, temperatures, velocities and other operating conditions. Thus, the selection of fouling factors is arbitrary. Many complaints in heat-exchanger operation that cannot be traced to errors in thermal design are generally traced to fouling.

If heavy fouling is anticipated for a particular service, the user should make provisions for periodic chemical or mechanical cleaning of the exchanger. If heavy tubeside fouling is foreseen, a straight-tube heat exchanger with larger-diameter tubes (1-in O.D. at least) should be specified. But when heavy shellside fouling is anticipated, the purchaser should specify a removable-bundle design, with tubes on a square pitch for mechanical cleaning of the bundle.

Allowable pressure drop—Selecting the optimum allowable pressure drop involves consideration of the overall process. However, high pressure drops may result in a smaller size (less costly) heat exchanger for other than isothermal-service requirements. The savings in the initial cost of a heat exchanger must be evaluated against a possible increase in operating costs.

For reasonable designs, the allowable pressure drops should be 5 psi, or higher for operating pressures in excess of 10 psig. In some instances, it is not practical to use all of the available pressure drop, because the resulting high fluid velocities could cause erosion or vibration damage to heat-exchanger components.

Heat-exchanger type and maintenance

Since there are many shell-and-tube heat-exchanger types to choose from, the preferred heat exchanger should be based on desired characteristics for utility and maintenance. The various types and construction features are shown in the "Standards of Tubular Exchanger Manufacturers Assn." (TEMA), and reproduced in this article for convenience (Fig. 1 and 2).

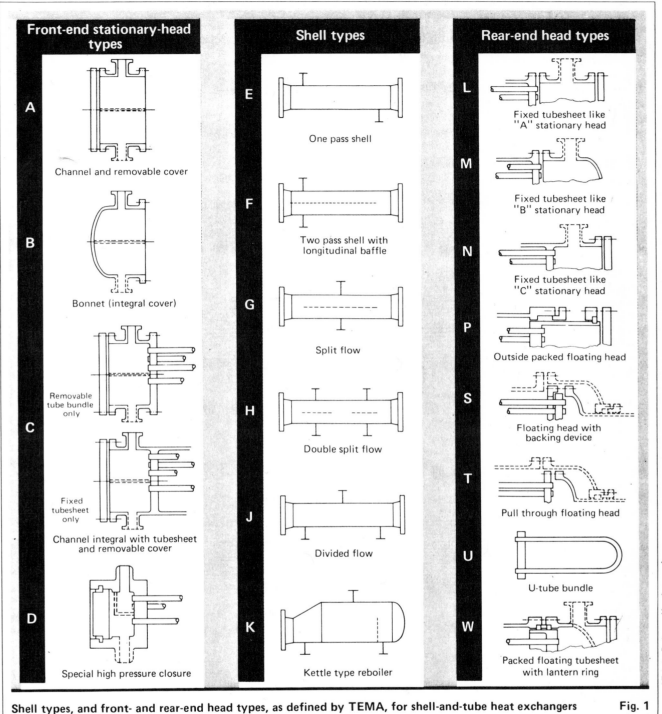

Shell types, and front- and rear-end head types, as defined by TEMA, for shell-and-tube heat exchangers — Fig. 1

Comparative costs between the common types of heat exchangers can be derived from the graphs. Supplemental information is included further on in this article.

Users or purchasers are responsible for specifying design conditions and materials of construction, because these factors, plus the corrosion allowance and specified tube-wall thickness, are relevant to the service life of the equipment. Materials of construction are generally selected based on pressure-temperature requirements; corrosion resistance to the operating fluid streams; and economics, based on anticipated service life versus initial cost.

Careful consideration should be given to the selection of tube material and tube-wall thickness, because heat is transferred through the wall of a tube. There-

HEAT EXCHANGER DESIGN AND SPECIFICATION

Nomenclature of heat-exchanger components

1. Stationary head—channel
2. Stationary head—bonnet
3. Stationary-head flange—channel or bonnet
4. Channel cover
5. Stationary-head nozzle
6. Stationary tubesheet
7. Tubes
8. Shell
9. Shell cover
10. Shell flange—stationary-head end
11. Shell flange—rear-head end
12. Shell nozzle
13. Shell-cover flange
14. Expansion joint
15. Floating tubesheet
16. Floating-head cover
17. Floating-head flange
18. Floating-head backing device
19. Split shear-ring
20. Slip-on backing flange
21. Floating-head cover—external
22. Floating-tubesheet skirt
23. Packing-box flange
24. Packing
25. Packing follower ring
26. Lantern ring
27. Tie rods and spacers
28. Transverse baffles or support plates
29. Impingement baffle
30. Longitudinal baffle
31. Pass partition
32. Vent connection
33. Drain connection
34. Instrument connection
35. Support saddle
36. Lifting lug
37. Support bracket
38. Weir
39. Liquid-level connection

Construction types

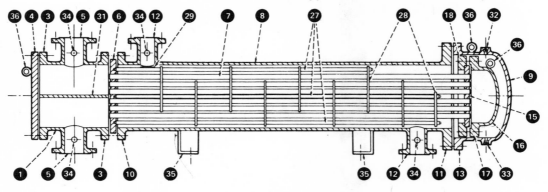

AES

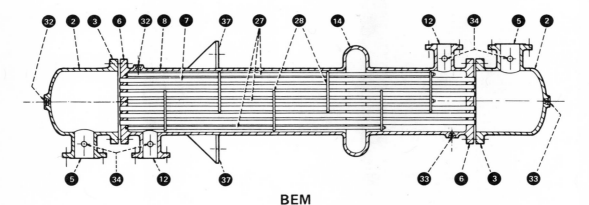

BEM

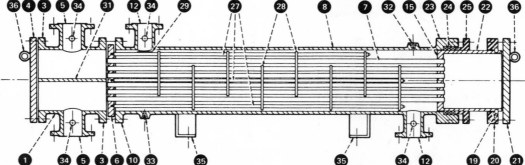

AEP

HOW TO SELECT THE OPTIMUM SHELL-AND-TUBE HEAT EXCHANGER

Construction types (cont'd)

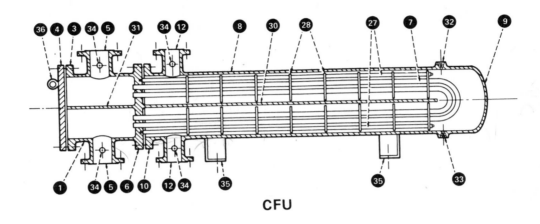

CFU

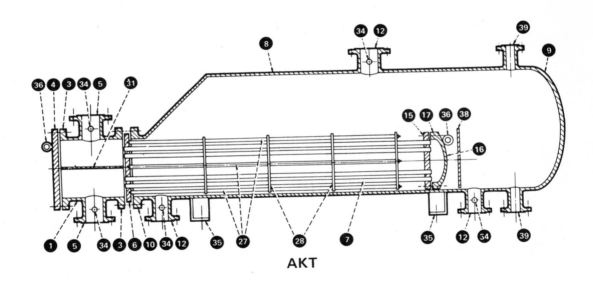

AKT

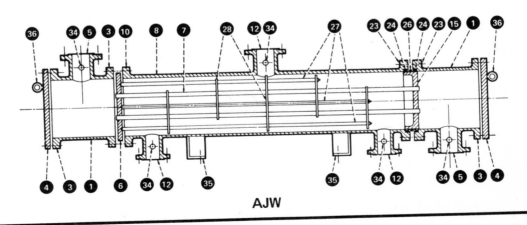

AJW

Fig. 2 Heat-exchanger construction types, showing their components and standard terminology

Courtesy of Tubular Exchanger Manufacturers Assn.

HEAT EXCHANGER DESIGN AND SPECIFICATION

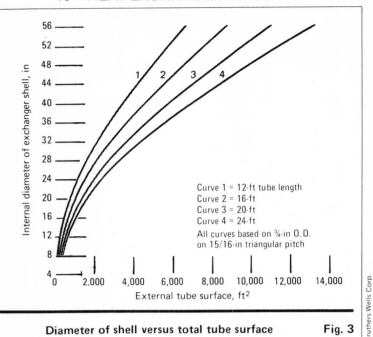

Curve 1 = 12-ft tube length
Curve 2 = 16-ft
Curve 3 = 20-ft
Curve 4 = 24-ft
All curves based on ¾-in O.D. on 15/16-in triangular pitch

Diameter of shell versus total tube surface   Fig. 3

Struthers Wells Corp.

fore, desirable characteristics should include material with high thermal conductivity and a thin wall.

Fluid allocation

Physical-property data should be as accurate as possible. Data for specific heat, density and/or molecular weight, thermal conductivity and viscosity should be provided to the designer.

To determine which fluid should be in the shell and which one in the tube, consider the following factors:

Corrosion—Fewer costly alloy components are needed if the corrosive fluid is inside the tubes.

Fouling—Placing the fouling liquid in the tubes allows better velocity control; increased velocities tend to reduce fouling. Straight tubes allow mechanical cleaning without removing the tube bundle.

Temperature—For high-temperature services requiring special or expensive alloy materials, fewer alloy components are needed when the hot fluid is placed within the tubes.

Pressure—Placing a high-pressure stream in the tubes will require fewer high-pressure components.

Pressure drop—For the same pressure drop, higher heat-transfer coefficients are obtained on the tubeside. A fluid with a low allowable pressure drop should generally be placed inside the tubes.

Viscosity—Higher heat-transfer rates are ordinarily obtained by placing a viscous fluid on the shell side.

Flowrate—Placing the fluid with the lower flowrate on the shellside usually results in a more economical design. Turbulence exists on the shellside at much lower velocities than within the tubes.

Considerations by the designer

A heat-exchanger designer should be informed at the start regarding any size or space limitations for installation of the exchanger. Limited space can exist when the heat exchanger is to be installed in a building or within a structure with other equipment.

Restrictions on the size of a heat exchanger may affect the initial cost, because the designer may not be able to optimize the design. This is particularly true when restrictions are imposed on tube lengths.

In addition to being a specialist in heat transfer, a designer should have a firm grasp of mechanical design, fabrication, and costs of the equipment involved.

He must evaluate the many variables in establishing the following characteristics of the heat exchanger: (1) tube O.D. and length; (2) tube pitch; (3) number of tube passes; (4) number of shell passes; (5) number of baffles and baffle type; (6) number of shells; (7) fluid velocities; (8) actual pressure drops; (9) shell size; (10) fluid distribution at the inlet and outlet of the shell; (11) tube-to-tubesheet attachment; (12) ease of maintenance; and (13) vibration, operating differential-expansion between shell and tubes, and other potential problem areas.

Tube size and length

Heat-exchanger designs with small-diameter tubes (⅜-in to 1-in O.D.) generally are more economical than designs with larger tubes, because the smaller tubes provide for a more compact unit. However, the use of such small tubes may be prohibited by an extremely low allowable tubeside-pressure-drop. Normally, ⅝-in-O.D. tubes are the smallest considered for process heat exchangers, but there are some applications where smaller tubes may be better. Larger-diameter tubes are used when heavy fouling is expected, and when the inside of the tubes is to be cleaned mechanically.

Because tubes in the ⅝-1-in-O.D. range are normally common for shell-and-tube exchangers, tubes in these sizes are more readily available in various materials of construction. Under equal-velocity conditions, smaller tube diameters increase the heat-transfer coefficient, as well as the pressure drop.

Ordinarily, the investment per unit area of heat-transfer surface is less for longer heat exchangers. Therefore, the purchaser should avoid restrictions on length wherever possible. In addition to potential savings in construction through the use of longer tubes, higher heat-transfer rates (less surface) are possible in sensible-heat-transfer service.

Tube pitch or arrangement

In shell-and-tube heat exchangers, tubes are generally arranged on a triangular, square or rotated-square pitch. Although the tube pitch can vary for a given tube size, the designer should limit the center-to-center spacing to the minimum, as outlined in the TEMA Standards, for good mechanical design.

Triangular-tube patterns provide better shellside heat-transfer coefficients in sensible-heat exchange, and provide more surface area for a given shell diameter. Square-pitch tube patterns are generally used when mechanical cleaning of the outside of the tubes is necessary or expected. However, square- and rotated-square-tube patterns provide lower pressure drops, and, therefore, correspondingly lower heat-transfer coefficients in most cases involving sensible heat.

Some designs require widely spaced triangular pitches to facilitate lower pressure drops or reduced shellside velocities. An extended tube pitch may also be used to provide for a special tube-to-tubesheet attachment, or for other special mechanical-design feature.

Number of passes on tubeside and shellside

The number of tubeside or shellside passes provided in the design of an optimum heat exchanger depends on operating temperatures, allowable pressure drops, fluid velocities, relative cost, and the experience of the designer.

One or more passes can be used on the tubeside; multiple passes are used to increase velocity and, therefore, the heat-transfer rate. In selecting the number of tubeside passes, the designer must limit the velocity to maintain an allowable pressure drop, as well as to avoid erosion of the tube material.

The number of passes on the shellside, or the flow configuration of the shell (one pass, two passes, split flow, divided flow, etc.), is primarily a function of the operating temperatures, flowrates and allowable pressure drop.

Inasmuch as the number of shell- or tubeside passes can affect the value of the corrected log-mean-temperature difference, the size and cost of the heat exchanger are closely related to the selection of the flow configurations. There is a distinct advantage to a countercurrent-flow pattern between the shell and tubeside fluids, except when one fluid is isothermal. In cocurrent flow, the hot fluid cannot be cooled below the cold-fluid outlet temperature; thus, the ability to recover heat with cocurrent flow is limited. In multipass heat exchangers—where there is a combination of cocurrent and countercurrent flow in alternate passes—the resultant LMTD is less than that calculated for countercurrent flow, but greater than the one based on cocurrent flow.

Shellside baffling

There are many alternatives available regarding baffle arrangement on the shellside. The baffles may be one of many types; the cuts may be horizontal or vertical, and the window openings may be large or small. The flowpath on the shellside depends on the type and arrangement of the baffles. In some instances, flow patterns greatly affect the performance of the heat exchanger, whereas in other ones, flow patterns are relatively unimportant. For example, flow patterns in the shellside are less important when a liquid is being vaporized, when a vapor is being condensed, or when the shellside heat-transfer coefficient is much higher than that of the tubes.

Baffles are generally used to direct the shellside fluid through a prescribed path, and to support the tubes within the bundle. Sometimes, baffles are called tube supports, if this is their primary purpose. The three most common baffle types are segmental, multisegmental and longitudinal.

Segmental baffles—This baffle type provides a high degree of turbulence and good heat transfer, because it directs the fluid over the tubes primarily in cross-flow. Unless the shellside fluid is being condensed, a horizontal baffle cut should be used to reduce accumulations of deposits at the bottom of the shell, and to prevent stratification of the shellside fluid. Vertically cut baffles are required for horizontal condensers, to allow the condensate to flow freely without covering or flooding an excessive amount of tubes.

Baffle cuts for segmental baffles are expressed as percentage values of the diameter, or net free area. A 20%-diameter cut is considered optimum, because it permits the highest heat transfer for a given pressure drop. As the baffle cut increases, the flow pattern deviates more and more from cross-flow. It is not good practice to apply cross-flow, heat-transfer equations when the distance between adjacent segmental baffles is greater than the shell diameter. For greater spacing, the actual flow pattern should be analyzed.

For large heat exchangers with high flowrates, it is often more economical to omit tubes in the baffle-window area. This provides for better cross-flow, while providing support for all tubes at every baffle. The "no-tubes-in-the-window" baffle design is often necessary to prevent flow-induced tube vibration.

Multisegmental baffles—Characterized by large open areas, various multisegmental baffle types can be used to reduce baffle spacing, or to reduce cross-flow because of pressure-drop limitations. Certain types of multisegmental baffles allow the fluid to flow nearly parallel to the tubes, offering a much lower pressure drop.

Longitudinal baffles—This type of baffle provides for multipass or split-flow, shellside-flow patterns. The longitudinal baffles can be welded to the shell, or sealed against it by a flex seal or other device. Sealing of the long baffle against the shell by means other than welding can be done when the pressure drop is relatively low on the shellside (less than 12 psi).

Number of shells

For an optimum heat exchanger, a designer should strive to provide the fewest number of shells for a given service. There are, however, services that require multiple shells in series and/or in parallel (Fig. 3).

When the service involves a relatively large surface, and the operating temperatures run over a broad range, multiple shells in series at the lower temperatures may be economically beneficial because of less-expensive materials of construction. If service requirements indicate a large temperature cross (outlet temperature of the hot stream being lower than outlet temperature of the cold stream), a countercurrent-flow arrangement is required. Multiple shells may then be the only approach to a reasonable design.

Fluid velocities, pressure drop, shell size

Usually, a designer will strive to use the available pressure drop in arriving at the optimum heat exchanger. The maximum heat-transfer rate and minimum surface-area result when all of the available pressure drop is utilized. However, flow velocities must be limited so as not to cause destructive mechanical damage due to vibration or erosion of heat-exchanger components.

A heat exchanger's shell size is directly related to the number of tubes and the tube pitch. Nevertheless, two

HEAT EXCHANGER DESIGN AND SPECIFICATION

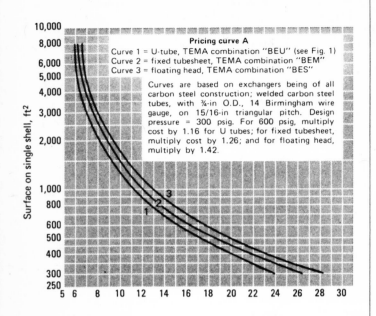

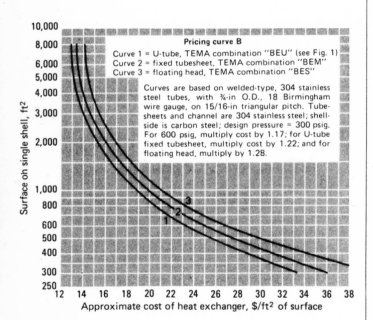

The following multipliers and additives apply to prices derived from Curves A and B:

Multiplier for a kettle-type reboiler over the regular exchanger—with the same size and tube bundle—having a shell diameter of half again the channel size is 1.22 for Curve A, and 1.10 for Curve B.

Multiplier for a removable channel cover (TEMA Type A) over an integral channel cover (TEMA Type B) is 1.075.

Multiplier for a pull-through floating-head (TEMA Type BET) over a conventional floating head (TEMA Type BES) exchanger of the same bundle diameter is 1.18 for Curve A, and 1.14 for Curve B.

Additive for seal-welding tubes to tubesheets is $3/tube, for 3/4-in-O.D. tubes.

Approximate current prices for several types of shell-and-tube heat exchangers — Fig. 4

heat exchangers with the same number of tubes and the same tube pitch may have different shell diameters, because different methods have been used to provide for the entrance and escape of the shell fluid.

TEMA Standards provide design parameters to follow regarding entrance and escape areas for the shell fluid (Fig. 2). A minimum shell size can be maintained by utilizing flow-distributor belts at the shell inlet and outlet nozzles. Distributor belts enable the shell fluid to enter or exit from more than one point around the bundle.

Highly oversized shells are used for special features such as no-tubes-in-the-window shell baffles, vapor escape in reboiler service, and shrouding* within the shell.

Tube-to-tubesheet attachment

In most cases, the tubes are rolled and then expanded into grooved holes in the tubesheet. However, some services require welding of the tube ends to the tubesheet. For severe pressure and/or temperature service, the combination of tube-end-welding and rolled expansion may be the best tube-to-tubesheet attachment for long-term, trouble-free service.

Other considerations

When the designer has been advised that service conditions of the heat exchanger will require maintenance for cleaning or other purposes, he must provide features that will enable easy access or disassembly of the unit. For heat exchangers in high-pressure service, special closures are often required for maintenance. In some of these closures, shear rings or special gaskets are applied, to avoid massive flanges and bolting.

A designer must also consider the effects of differential expansion between the shell and the tubes. This is particularly important with fixed-tubesheet heat exchangers, because with large temperature differences, an overstressed condition can occur in the shell and/or tubes, due to differing coefficients of expansion.

Vibration is another important factor to be considered. Many vibration failures are flow-induced. When the vortex shedding frequency around the tubes just about equals the tubes' natural frequency, failure of the tubes can occur due to fatigue, wear or impact.

Large, low-pressure heat exchangers operating with gas or vapor in the shell are susceptible to acoustic vibration, which can occur when the tube-vortex shedding and the gas-column resonant frequencies coincide. When this results, potentially destructive pressure fluctuation can occur within the exchanger.

Although our present knowledge of the vibration phenomenon is such that a solution cannot be completely explained, a designer should be aware of the potential problem and should use existing correlations to avoid such a problem. If predictive methods indicate the probability of destructive vibrations, the designer should consider the use of shorter, unsupported tube spans, changing the tube pitch, reducing flow velocities and/or revising the shellside baffle arrangement.†

---

*A solid or perforated plate attached to the tube bundle, generally in an inlet-nozzle area, to prevent direct vapor impingement on the tubes.

†An article entitled "Preventing vibration in shell-and-tube heat exchangers" appears on pp. 92-108.

Relevant factors affecting cost

The three most relevant factors that have the greatest effect on size, and therefore on the cost, of a shell-and-tube exchanger are pressure drops, log-mean-temperature difference and fouling factors (Fig. 4):

Pressure drops—The allowable pressure drops imposed by the purchaser can drastically affect the size of an exchanger. If unrealistically low allowable pressure drops are imposed, the designer is forced to use lower fluid velocities to maintain the pressure-drop limitations. Because heat-transfer rates in sensible-heat exchangers are directly related to velocities, the lower velocities can result in a larger exchanger.

To design the optimum heat exchanger, all of the available pressure drops must be used, provided that the resultant fluid velocities are reasonable and do not tend to create erosion and vibration problems.

Log-mean-temperature difference—The size, or surface, of a heat exchanger is inversely proportional to the overall heat-transfer rate and the corrected LMTD. Assuming that reasonable operating temperatures have been specified (as indicated earlier under "Process conditions"), a designer should try to maximize the product of the heat-transfer rate and the LMTD so as to arrive at the optimum unit design.

Fouling factors—The size or surface of a heat exchanger depends, to a large extent, on the specified fouling factors. The surface of an exchanger depends on the overall heat-transfer rate, which is a function of the fouling factors, the heat-transfer coefficients, and the tube-wall resistance. When relatively high heat-transfer coefficients can be achieved on the shell and tube sides, the effect of fouling on the size—and, therefore, on the initial cost—is very significant. Fouling factors should be specified based on experience, because haphazard guessing can be costly.

Intangibles in bid evaluations

There are generally many intangibles to consider in a bid evaluation of a shell-and-tube heat exchanger, which are not normally an integral part of the proposed bid.

The purchaser is usually provided with a specification sheet, and possibly a sketch describing various mechanical and thermal design characteristics of the exchanger. However, the designer may have incorporated, for reliability of service, a number of design features not detailed in the proposal. Such special features are generally the result of the designer's experience. They may include special baffling for the prevention of vibration, special tube-end attachments, etc. For some exchangers in severe service, the experience of the vendor may mean the difference between success or failure either during the initial phases of plant operation or for reliable long-term plant performance.

Another factor that should be considered in the bid evaluation is whether the vendor has the reputation of providing quality in his product. The purchaser must have confidence in the vendor and feel that he is receiving a unit that is well engineered for the specified operating requirements.

The purchaser has the ultimate task of selecting the "best" heat exchanger, based on the proposal received from vendors. The selection should not be based only on price; there have been many sad operating experiences merely because the purchaser bought the lowest-priced equipment, without considering the quality and reliability of the different designs.

Many heat exchangers are purchased on the lowest cost per square foot of surface. This method of purchasing is misleading and generally does not result in the purchase of the optimum exchanger. Although the price per square foot of surface is usually lower for the heat exchanger with the most surface area, the total selling price may be higher. Two heat exchangers may be identical in surface area, but one may have special design features that make it more expensive.

An order should be placed based on the amount of money needed to obtain a unit for specific operating requirements. The manufacturer should be reliable and should stand behind his product when there are complaints or service claims.

An optimum heat-exchanger design is a function of many interacting parameters. It is, therefore, most important that the purchaser be experienced in heat-exchanger design for the proper evaluation of an offering; he should have the ability to evaluate a design and determine whether the offered exchanger is fairly priced. Without this experience, the purchaser is forced to rely heavily on the reputation and experience of the vendor.

References

1. Lord, R. C., Minton, P. E., and Slusser, R. P., *Chem. Eng.*, Jan. 26, 1970, pp. 96-118.
2. "Standards of Tubular Exchanger Manufacturers Assn.," New York.
3. Makris, Arthur J., "Heat Exchanger Design Variables and the Effect of Each on Ultimate Cost," paper presented to the Process Heat Exchanger Soc., Mar. 29, 1961.

The authors

John P. Fanaritis is Executive Vice-President of Struthers Wells Corp., P.O. Box 8, Warren, PA 16365, where he is currently directing engineering projects. He joined Struthers Wells immediately after graduating from the University of Pittsburgh with a B.S. degree in chemical engineering. While with Struthers Wells, he has been actively involved in the design of distillation equipment, direct-fired boilers and heat-recovery equipment. He has also written several articles for various technical publications.

James W. Bevevino is Manager, Heat Exchanger Dept., Struthers Wells Corp., P.O. Box 8, Warren, PA 16365, where he is responsible for thermal design and sales of process heat exchangers. He has also worked at Struthers in various other capacities since he graduated from college. He holds B.S. and M.S. degrees in mechanical engineering. The former he received from Cleveland State University, the latter from Rensselaer Polytechnic Institute. He has written for technical publications before, and is a member of TEMA's Technical Committee and a representative for Struthers Wells on the Board of Directors of the Heat Transfer Research Institute.

Design Of Heat Exchangers

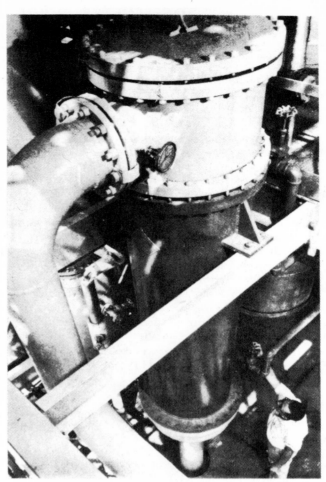

The Patterson-Kelley Co.

Divided into two major sections, the report covers: the classification and application of fluid-fluid heat exchangers, including the newer designs; and functional design of the shell-and-tube type by a shortcut rating technique.

R. C. LORD, P. E. MINTON AND R. P. SLUSSER, Union Carbide Corp.

Changes that have taken place in the chemical and petroleum industries in the past ten years have forced the process heat-exchanger designer into the role of a versatile heat-transfer specialist. He is expected to possess calculation tools and knowhow, so that after considering initial costs, utilities and maintenance he will be able to select the optimum piece of equipment for a given job.

Plants now being constructed are five to ten times the size of those that would have been considered large ten years ago. Often, it is not possible or practical to scale up a heat exchanger by installing ten times as many tubes in the same basic design. Therefore, the designer's experience is of utmost value because the considerations that enter into selecting an optimum exchanger are often difficult to evaluate quantitatively.

Such considerations, for example, might involve the heat sensitivity of the material being handled, or its fouling characteristics as functions of the surface temperature and fluid velocity. Other considerations might involve factors unrelated to performance of the unit itself, such as ease of cleaning and maintenance, and complexity of fabrication.

There are an almost unlimited number of alternatives in selecting heat-tranfer equipment for a process stream, though usually there is only one "best" design. Fortunately, computers and electronic calculators can now assist the designer by performing many of the routine, time-consuming calculations. These tools also make possible sophisticated analyses of special problems.

Although various types of heat-transfer equipment will be mentioned in what follows, the emphasis will be on functional design of shell-and-tube exchangers (Fig. 1), taking into consideration process variables,

Originally published January 26, 1970

DESIGN OF HEAT EXCHANGERS

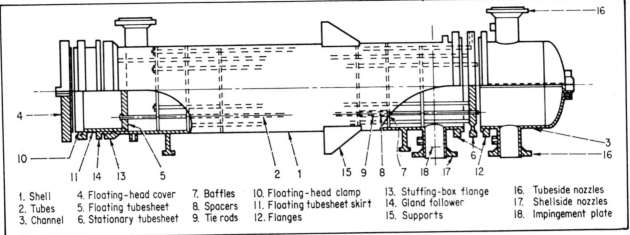

1. Shell
2. Tubes
3. Channel
4. Floating-head cover
5. Floating tubesheet
6. Stationary tubesheet
7. Baffles
8. Spacers
9. Tie rods
10. Floating-head clamp
11. Floating tubesheet skirt
12. Flanges
13. Stuffing-box flange
14. Gland follower
15. Supports
16. Tubeside nozzles
17. Shellside nozzles
18. Impingement plate

COMPONENTS of shell-and-tube exchanger. Model shown is outside-packed stuffing-box unit—Fig. 1

utilities and mechanical design. In future articles, similar treatment will be accorded to spiral and air-cooled exchangers.* Still another article will discuss the factors that influence the design and performance of condensers, vaporizers and reboilers.

The following types of heat-transfer equipment will not be discussed, as most have already been dealt with in other CHEMICAL ENGINEERING reports: direct-contact heat exchangers (including direct-fired heaters, spray and packed devices, baffle decks); indirect-fired heaters (including radiant-section and forced-air stoichiometric heaters); submerged-combustion heat exchangers; regenerative heat exchangers; electric-resistance heaters; and solids heaters and coolers.

Importance of Design Details

Many designers assume they have performed their function adequately if their design meets the duty specified. However, many instances of poor design are not immediately detected. For example, an exchanger with poor shellside design may require more cooling water than anticipated. Or it may be necessary to often remove an exchanger for rerolling or replacing tubes because of inadequate provisions for differential expansion of tubes and shell.

As simple a thing as omitting vents for removing corrosive noncondensable gas can cause failures that may take months to appear. Usually, the repair costs of the exchanger are only part of the penalty paid for poor design; lost production from a single process interruption will often cost many times the purchase price of the heat exchanger.

If poor design causes greater than expected use of utilities, the increment will possibly go unnoticed, and the results will show up as additional plant cost for as long as the unit operates. Often, in the course of expanding a plant or constructing a new one, the original heat-exchanger design is duplicated and its faulty operation passed on to the new installation.

SHELL-AND-TUBE DESIGN CONSIDERATIONS

The two process variables that have the greatest effect on the size (cost) of a shell-and tube heat exchanger are the allowable pressure drops of streams, and the mean temperature difference between the two streams. Other important variables include the physical properties of the streams, the location of fluids in an exchanger, and the piping arrangement of the fluids as they enter and leave the exchanger. (See design features in Table I.)

Pressure Drop

Selection of optimum pressure drops involves consideration of the overall process. While it is true that higher pressure drops result in smaller exchangers, investment savings are realized only at the expense of operating costs. Only by considering the relationship between operating costs and investment can the most economical pressure drop be determined.

Available pressure drops vary from a few millimeters of mercury in vacuum service to hundreds of pounds per square inch in high-pressure processes. In some cases, it is not practical to use all the available pressure drop because resultant high velocities may create erosion problems.

Reasonable pressure drops for various pressure levels are listed below. Designs for smaller pressure drops are often uneconomical because of the large surface area (investment) required.

Pressure Level	Reasonable △P
Sub-atmospheric	1/10 absolute pressure
1 to 10 psig.	1/2 operating gage-pressure
10 psig. and higher	5 psi. or higher

In some instances, velocities of 10 to 15 ft./sec. help to reduce fouling, but at such velocities the pressure drop may have to be from 10 to 30 psi.

---
* Another article, by a different author, will deal with plate-and-frame heat exchangers.—Ed.

Design Features of Shell-and-Tube Heat Exchangers—Table I

Design Features	Fixed Tubesheet	Return Bend (U-Tube)	Outside-Packed Stuffing Box	Outside-Packed Lantern Ring	Pull-Through Bundle	Inside Split Backing Ring
Is tube bundle removable?	No	Yes	Yes	Yes	Yes	Yes
Can spare bundles be used?	No	Yes	Yes	Yes	Yes	Yes
How is differential thermal expansion relieved?	Expansion joint in shell	Individual tubes free to expand	Floating head	Floating head	Floating head	Floating head
Can individual tubes be replaced?	Yes	Only those in outside rows without special designs	Yes	Yes	Yes	Yes
Can tubes be chemically cleaned, both inside and outside?	Yes	Yes	Yes	Yes	Yes	Yes
Can tubes be physically cleaned on inside?	Yes	With special tools	Yes	Yes	Yes	Yes
Can tubes be physically cleaned on outside?	No	With square or wide triangular pitch	With square or wide triangular pitch	With square or wide triangular pitch	With square or wide triangular pitch	With square or wide triangular pitch
Are internal gaskets and bolting required?	No	No	No	No	Yes	Yes
Are double tubesheets practical?	Yes	Yes	Yes	No	No	No
What number of tubeside passes are available?	Number limited by number of tubes	Number limited by number of U-tubes	Number limited by number of tubes	One or two	Number limited by number of tubes. Odd number of passes requires packed joint or expansion joint	Number limited by number of tubes. Odd number of passes requires packed joint or expansion joint
Relative cost in ascending order, least expensive = 1	2	1	4	3	5	6

Mean Temperature Difference

Process-fluid inlet and outlet temperatures, and utility fluid-temperature levels, are usually selected in the early stages of a process design. At this time also, consideration should be given to the effect of mean temperature difference on heat-exchanger investment because the temperature level of the heat-transfer-medium has a great effect on the heat-transfer area required.

When there is a choice of temperature levels, it should be remembered that higher-heating-medium (or lower-cooling-medium) temperatures produce greater mean temperature differences; if these are excessive, however, fouling, product decomposition, or deposition of other materials may occur. Close temperature approaches, i.e. small temperature differences between the inlet of one stream and the outlet of the other, have a pronounced effect on the mean temperature difference.

Although there are no specific rules for determining the best temperature approach, the following recommendations are made regarding terminal temperature differences for various types of heat exchangers; any departure from these general limitations should be economically justified by a study of alternate system-designs:

- The greater temperature difference should be at least 20 C.
- The lesser temperature difference should be at least 5 C. When heat is being exchanged between two process streams, the lesser temperature difference should be at least 20 C.
- In cooling a process stream with water, the outlet-water temperature should not exceed the outlet process-stream temperature if a single body having one shell pass—but more than one tube pass—is used.
- When cooling or condensing a fluid, the inlet coolant temperature should not be less than 5 C. above the freezing point of the highest freezing component of the fluid.
- For cooling reactors, a 10 to 15 C. difference should be maintained between reaction and coolant temperatures to permit better control of the reaction.
- A 20 C. approach to the design air-temperature is the minimum for air-cooled exchangers. Economic justification of units with smaller approaches requires careful study. Trim coolers or evaporative coolers should also be considered.
- When condensing in the presence of inerts, the outlet coolant temperature should be at least 5 C. below the dewpoint of the process stream.

Types of Flow

In an exchanger having one shell pass and one tube pass, where two fluids may transfer heat in either cocurrent or countercurrent flow, the relative direction of the fluids affects the value of the mean temperature difference. This is the log mean in either case, but there is a distinct thermal advantage to counterflow, except when one fluid is isothermal.

In cocurrent flow, the hot fluid cannot be cooled below the cold-fluid outlet temperature; thus, the ability of cocurrent flow to recover heat is limited. Nevertheless, there are instances when cocurrent flow works better, as when cooling viscous fluids, because a higher heat-transfer coefficient may be obtained. Cocurrent flow may also be preferred when there is a possibility that the temperature of the warmer fluid may reach its freezing point.

Fluid Properties, Location

Physical-property data of the fluids should be as accurate as possible, but since most physical properties of mixtures must be calculated or estimated, there is no point in attempting to determine true film temperatures. Physical-property data at the average bulk temperature are sufficient. For fluids in shell-and-tube exchangers, data for specific heat, density, thermal conductivity and viscosity are needed.

To determine which should be the shellside and which the tubeside fluid, consider these factors:

Corrosion—Fewer costly alloy or clad components are needed if the corrosive fluid is inside the tubes.

Fouling—This can be minimized by placing the fouling fluid in the tubes to allow better velocity control; increased velocities tend to reduce fouling. Straight tubes can be physically cleaned without removing the tube bundle; chemical cleaning can usually be done better on the tubeside. Finned tubes on square pitch (Fig. 2) sometimes are easier to clean physically; chemical cleaning is usually not as effective on the shellside because of bypassing.

Temperature—For high-temperature services requiring expensive alloy materials, fewer alloy components are needed when the hot fluid is placed on the tubeside.

Pressure—Placing a high-pressure stream in the tubes will require fewer (more costly) high-pressure components.

Pressure Drop—For the same pressure drop, higher heat-transfer coefficients are obtained on the tubeside. A fluid with a low allowable pressure drop should generally be placed on the tubeside.

Viscosity—Higher heat-transfer rates are generally obtained by placing a viscous fluid on the shellside.

Toxic and Lethal Fluids—Generally, the toxic fluid should be placed on the tubeside, using a double tubesheet to minimize the possibility of leakage. ASME (American Soc. of Mechanical Engineers) code requirements for lethal service must be followed.

Flowrate—Placing the fluid with the lower flowrate on the shellside usually results in a more economical design. Turbulence exists on the shellside at much lower velocities than on the tubeside.

Tube Size and Arrangement

These factors are important in determining the performance of a shell-and-tube exchanger:

Tube Diameter, Length—Designs with small-diameter tubes (⅝ to 1 in.) are more compact and more economical than those with larger-diameter tubes, although the latter may be necessary when the allowable tubeside pressure drop is small. The smallest tube size normally considered for a process heat exchanger is ⅝ in., although there are applications where ½, ⅜, or even ¼-in. tubes are the best selection. Tubes of 1 in. dia. are normally used when fouling is expected because smaller ones are impractical to clean mechanically. Falling-film exchangers and vaporizers generally are supplied with 1½ and 2-in. tubes.

Since the investment per unit area of heat-transfer service is less for long exchangers with relatively small shell diameters, minimum restrictions on length should be observed.

Arrangement—Tubes are arranged in triangular, square, or rotated-square pitch (Fig. 2). Triangular tube-layouts result in better shellside coefficients and provide more surface area in a given shell diameter, whereas square pitch or rotated-square pitch layouts are used when mechanical cleaning of the outside of the tubes is required. Sometimes, widely spaced triangular patterns facilitate cleaning. Both types of square pitches offer lower pressure drops—but lower coefficients—than triangular pitch.

Flow-Induced Vibration

Vibration is usually caused by the shedding of vortices from the downstream side of a tube. As a vortex is shed, the flow pattern (and therefore the pressure distribution) changes, resulting in oscillations in the magnitude and direction of the fluid pressure forces acting on the tube. If the frequency of vortex shedding approaches the tube's natural frequency, the tube will vibrate with a large amplitude and will eventually fail, causing leakage of one fluid into the other. Vibration may be eliminated by reducing velocities, decreasing the tubes' unsupported span, or by altering the method of fixing the ends of the unsupported spans. A future article will discuss vibration prediction as a function of various design parameters.

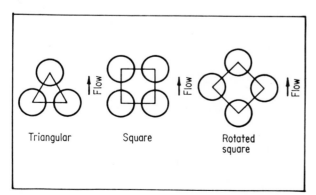

TUBE ARRANGEMENTS that are used for shell-and-tube heat exchangers—Fig. 2

HEAT EXCHANGER DESIGN AND SPECIFICATION

Baffles in the Shell

The path the shellside fluid takes depends on the type and arrangement of baffles. In some cases, flow patterns greatly affect the performance of the exchanger, while in others the patterns are relatively unimportant. Generally, the flow pattern is less important when condensing a vapor or when the coefficient of the shellside is much higher than that of the tubeside.

Most baffles serve two purposes: to direct the shellside fluid through a prescribed path, and to support the tubes and prevent vibration. Baffles are sometimes called tube supports, if such is their primary purpose. The three most common baffle types are: segmental, window and longitudinal.

Segmental Baffles—This type provides a high degree of turbulence and good heat transfer because it directs the fluid over the tubes primarily in crossflow. Unless the shellside fluid is being condensed, a horizontal baffle-cut should be used to reduce accumulation of deposits on the bottom of the shell and to prevent stratification of the shellside fluid. Vertically cut baffles are required for condensers to allow the condensate to flow freely to the outlet without covering an excessive amount of tubes.

Baffle cuts for segmental baffles are expressed as percentages of the diameter. A 20% cut is the optimum,[23] as it affords the highest heat transfer for a given pressure drop. As the baffle cut increases, the flow pattern deviates more and more from crossflow. It is not good practice to apply crossflow heat-transfer equations when the distance between adjacent segmental baffles is greater than the shell diameter. For greater spacing, tubes should be omitted and the actual flow pattern analyzed.

For large exchangers, it is often more economical to omit tubes from the baffle cut. This practice provides better crossflow and still supports all tubes at every baffle. Often, this is necessary to prevent flow-induced tube vibration.

Window baffles—These are considered when crossflow is not practical because of pressure-drop limitations. Window baffles allow the fluid to flow parallel to the tubes, offering much lower pressure drop. Various types of baffles are available to produce parallel flow, but all are characterized by large open areas. Recommended types are shown in Fig. 3.

Longitudinal Baffles—This type divides the shell into two or more sections, providing multipass shellside flow, but this type should not be used unless it is welded to the shell and tubesheet. Although several devices have been tried to seal the baffle and shell, none has been very successful. When multipass shells are required, it is cheaper to use a separate shell for each pass, unless the shell diameter is large enough to easily weld a longitudinal baffle to the shell.

Flow Distribution, Bypass Prevention

For large flows of vapor or liquid, shell fluid-distribution belts are often required. Such belts, which are enlarged sections of the shell that enable fluid to enter at more than one point, commonly provide impingement protection. Sometimes, tubes are omitted from the layout to provide better distribution.

To achieve good shellside heat transfer, bypassing of the fluid must be reduced or eliminated. The clearance between the outermost tubes and the inside of the shell constitutes a bypass lane, as do the lanes provided for the tubeside partition ribs (Fig. 4). Although some bypass also occurs through the tube holes in the baffles, this can be minimized by close tolerances and is not as detrimental to heat transfer as bypasses around or through a bundle. Notches in baffles are also bypasses and should consequently be used with caution. Notches are usually not required for draining because the necessary fabrication tolerances provide ample draining.

Sealing or Blocking Devices—These are strips that prevent bypass around a bundle by "sealing" or blocking the clearance area between the outermost tubes and the inside of the shell (Fig. 4). Some common types include: (1) tie rods and spacers that hold the baffles in place but can be located at the periphery of the baffle to prevent bypassing; (2) sealing strips, which are longitudinal and extend from baffle to baffle; (3) tie rods with "winged" spacers. The wings are extended longitudinal strips attached to the spacers.

Dummy Tubes—These are tubes that do not pass through the tubesheet. Usually closed at one end, they are used to prevent bypassing through lanes parallel to the direction of fluid flow inside the bundle (Fig. 4). In moderate to large exchangers, one dummy tube is as effective in promoting heat transfer as 50 process tubes. Strips or tie rods and spacers can also be used to prevent bypassing through a bundle.

SHORTCUT RATING METHODS

For engineers who do not always have access to computers for designing or evaluating heat exchang-

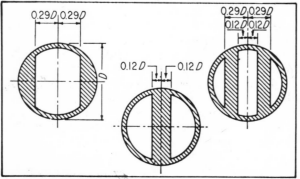

WINDOW BAFFLES that give good results. Note large openings and proportion to diameter—Fig. 3

ers, C. H. Gilmour's[15] shortcut methods are very useful; these techniques are summarized here.

The techniques' main advantages are: uniformity of approach for all heat-transfer mechanisms; convenient grouping of terms for trial-and-error calculations; use of consistent, familiar units that minimize chances for error; and the form of the equations, which is adaptable to rating by means of programmable desk calculators.

Primarily, the method combines into one relationship the classical empirical equations for film heat-transfer coefficients with heat-balance equations and with relationships describing tube geometry, baffles and shell. The resulting overall equation is recast into three separate groups that contain factors relating to: physical properties of the fluid, performance or duty of the exchanger, and mechanical design or arrangement of the heat-transfer surface. These groups are then multiplied together with a numerical factor to obtain a product that is equal to the fraction of the total driving force—or log mean temperature-difference (LMTD or ΔT_M)—that is dissipated across each element of resistance in the heat-flow path.

When the sum of the products for the individual resistance equals one, the trial design may be assumed to be satisfactory for heat transfer. The physical significance is that the sum of the temperature drops across each resistance is equal to the total available LMTD. The pressure drop on both tubeside and shellside must be checked to assure that both are within acceptable limits. As shown in the sample calculation on p. 29, usually several trials are necessary to obtain a satisfactory balance between heat transfer and pressure drop.

Tables II and III respectively summarize the equations used with the method for heat transfer and for pressure drop. The column on the left lists the conditions to which each equation applies. The second column lists the standard form of the correlation for film coefficients that is found in texts. The remaining columns then tabulate the numerical, physical-property, work, and mechanical-design factors, all of which together form the recast dimensional equation. The product of these factors gives the fraction of total temperature drop or driving force ($\Delta T_f / \Delta T_M$) across the resistance.

As described above, the addition of $\Delta T_i / \Delta T_M$, tubeside factor; plus $\Delta T_o / \Delta T_M$, shellside factor; plus $\Delta T_s / \Delta T_M$, fouling factor; plus $\Delta T_w / \Delta T_M$, tube-wall factor, determine the heat-transfer adequacy. Any combination of $\Delta T_i / \Delta T_M$ and $\Delta T_o / \Delta T_M$ may be used, as long as a horizontal orientation on the tubeside is used with a horizontal orientation on the shellside, and a vertical tubeside orientation has a corresponding shellside orientation.

The units in the pressure-drop equations (Table III) are consistent with those used for heat transfer. The pressure drop in psi. is calculated directly. Because the method is a shortcut approach to design,

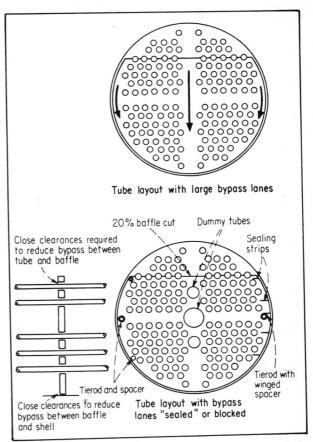

CLEARANCES in tube layout (top sketch) must be eliminated by blocking devices (lower sketch) to prevent shellside bypassing—Fig. 4

certain assumptions pertaining to thermal conductivity, tube pitch and shell diameter are made:

General Assumptions

For many organic liquids, thermal conductivity data are either not available or difficult to obtain. Since molecular weights (M) are known, for most design purposes the Weber equation, which follows, yields thermal conductivities with quite satisfactory accuracies:

$$k = 0.86 \, (cs^{4/3}/M^{1/3})$$

An important compound for which the Weber equation does not work well is water (the calculated thermal conductivity is less than the actual value). Fig. 5 gives the physical-property factor for water (as a function of fluid temperature) that is to be substituted in the equations for sensible-heat transfer with water, or for condensing with steam.

If the thermal conductivity is known, it is best to obtain a pseudo-molecular weight by:

$$M = 0.636 \, (c/k)^3 s^4$$

This value is substituted in the applicable equation to solve for the physical-property factor.

HEAT EXCHANGER DESIGN AND SPECIFICATION

Empirical Heat-Transfer Relationships

Eq. No.	Mechanism or Restriction	Empirical Equation
	Inside the tubes	
(1)	No phase change (liquid), $N_{Re} > 10{,}000$	$\dfrac{h}{cG} = 0.023\,(N_{Re})^{-0.2}(N_{Pr})^{-2/3}$
(2)	No phase change (gas), $N_{Re} > 10{,}000$	$h = 0.0144\,G^{0.8}(D_i)^{-0.2}c_p$
(3)	No phase change (gas), $2{,}100 < N_{Re} < 10{,}000$	$h = 0.0059\,[(N_{Re})^{2/3} - 125]\,[1 + (D/L)^{2/3}]\,(c_p/D_i)\,(\mu_f/\mu_b)^{-0.14}$
(4)	No phase change (liquid), $2{,}100 < N_{Re} < 10{,}000$	$\dfrac{h}{cG} = 0.116\left[\dfrac{(N_{Re})^{2/3} - 125}{N_{Re}}\right][1 + (D/L)^{2/3}](N_{Pr})^{-2/3}(\mu_f/\mu_b)^{-0.14}$
(5)	No phase change (liquid), $N_{Re} < 2{,}100$	$\dfrac{h}{cG} = 1.86\,(N_{Re})^{-2/3}(N_{Pr})^{-2/3}(L/D_i)^{-1/3}(\mu_f/\mu_b)^{-0.14}$
(6)	Condensing vapor, vertical, $N_{Re} < 2{,}100$	$h = 0.925\,k\,(g\rho_i^2/\mu\Gamma)^{1/3}$
(7)	Condensing vapor, horizontal, $N_{Re} < 2{,}100$	$h = 0.76\,k\,(g\rho_i^2/\mu\Gamma)^{1/3}$
(8)	Condensate subcooling, vertical	$h = 1.225\,(k/B)\,(cB\Gamma/kL_B)^{5/6}$
(9)	Nucleate boiling, vertical	$\dfrac{h}{cG} = 4.02\,(N_{Re})^{-0.3}(N_{Pr})^{-0.6}(\rho_L\sigma/P^2)^{-0.425}\Sigma$
	Outside the tubes	
(10)	Nucleate boiling, horizontal	$\dfrac{h}{cG} = 4.02\,(N_{Re})^{-0.3}(N_{Pr})^{-0.6}(\rho_L\sigma/P^2)^{-0.425}\Sigma$
(11)	No phase change (liquid), crossflow	$\dfrac{h}{cG} = 0.33\,(N_{Re})^{-0.4}(N_{Pr})^{-2/3}(0.6)$
(12)	No phase change (gas), crossflow	$h = 0.11\,G^{0.6}D^{-0.4}c_p(0.6)$
(13)	No phase change (gas), parallel flow	$h = 0.0144\,G^{0.8}D^{-0.2}c_p(1.3)$
(14)	No phase change (liquid), parallel flow	$\dfrac{h}{cG} = 0.023\,(N_{Re})^{-0.2}(N_{Pr})^{-2/3}(1.3)$
(15)	Condensing vapor, vertical, $N_{Re} < 2{,}100$	$h = 0.925\,k(g\rho_L^2/\mu\Gamma)^{1/3}$
(16)	Condensing vapor, horizontal, $N_{Re} < 2{,}100$	$h = 0.76\,k(g\rho_L^2/\mu\Gamma)^{1/3}$
	Tube wall	
(17)	Tube wall (sensible-heat transfer)	$h = (24\,k_w)/(d_o - d_i)$
(18)	Tube wall (latent-heat transfer)	$h = (24\,k_w)/(d_o - d_i)$
	Fouling	
(19)	Fouling (sensible-heat transfer)	$h = \text{assumed}$
(20)	Fouling (latent-heat transfer)	$h = \text{assumed}$

Notes:
1. If $W_i/(n s_i d_i^{2.56}) > 0.3$, multiply $\Delta T_i/\Delta T_M$ by 1.3
2. Surface-condition factor (Σ') for copper and steel = 1.0; for stainless steel = 1.7; for polished surfaces = 2.5.
3. For square pitch, numerical factor = 5.42.
4. For square pitch, numerical factor = 9.53.
5. $G = W_o \rho_L/(A \rho_v)$

for Rating Heat Exchangers—Table II

	Numerical Factor	Physical-Property Factor	Work Factor	Mechanical-Design Factor		Eq. No.
$\Delta T_i/\Delta T_M =$	10.43	$\times \dfrac{(Z_i^{0.467} M_i^{0.22})}{s_i^{0.89}}$	$\times \dfrac{W_i^{0.2}(t_H - t_L)}{\Delta T_M}$	$\times \dfrac{d_i^{0.8}}{n^{0.2} L}$		(1)
$\Delta T_i/\Delta T_M =$	9.87		$\times \dfrac{W_i^{0.2}(t_H - t_L)}{\Delta T_M}$	$\times \dfrac{d_i^{0.8}}{n^{0.2} L}$		(2)
$\Delta T_i/\Delta T_M =$	44,700	$\times (Z_f/Z_b)^{0.14}$	$\times \dfrac{W_i(t_H - t_L)}{\Delta T_M}$	$\times \dfrac{1}{[(N_{Re})^{2/3} - 125][1 + (d_i N_{PT}/12L)^{2/3}]nL}$		(3)
$\Delta T_i/\Delta T_M =$	2,260	$\times \left(\dfrac{M_i^{0.22}}{s_i^{0.89} Z_b^{1/3}}\right)\left(\dfrac{Z_f}{Z_b}\right)^{0.14}$	$\times \dfrac{W_i(t_H - t_L)}{\Delta T_M}$	$\times \dfrac{1}{[(N_{Re})^{2/3} - 125][1 + (d_i N_{PT}/12L)^{2/3}]nL}$		(4)
$\Delta T_i/\Delta T_M =$	17.5	$\times \left(\dfrac{M_i^{0.22}}{s_i^{0.89}}\right)\left(\dfrac{Z_f}{Z_b}\right)^{0.14}$	$\times \dfrac{W_i^{2/3}(t_H - t_L)}{\Delta T_M}$	$\times \dfrac{1}{n^{2/3} L^{2/3} (N_{PT})^{1/3}}$		(5)
$\Delta T_i/\Delta T_M =$	4.75	$\times \dfrac{(Z_i M_i)^{0.333}}{s_i^2 c_i}$	$\times \dfrac{W_i^{4/3} \lambda_i}{\Delta T_M}$	$\times \dfrac{1}{n^{4/3} d_i^{4/3} L}$		(6)
$\Delta T_i/\Delta T_M =$	2.92	$\times \dfrac{(Z_i M_i)^{0.333}}{s_i^2 c_i}$	$\times \dfrac{W_i^{4/3} \lambda_i}{\Delta T_M}$	$\times \dfrac{1}{n^{4/3} d_i L^{4/3}}$	(See Note 1)	(7)
$\Delta T_i/\Delta T_M =$	1.22	$\times \left[\dfrac{(Z_i M_i)^{0.333}}{s_i^2}\right]^{1/6}$	$\times \dfrac{W_i^{0.222}(t_H - t_L)}{\Delta T_M}$	$\times \dfrac{1}{(n^{4/3} d_i^{4/3} L)^{1/6}}$		(8)
$\Delta T_i/\Delta T_M =$	0.352	$\times \left(\dfrac{Z_i^{0.3} M_i^{0.2} \sigma_i^{0.425}}{s_i^{1.075} c_i}\right)\left(\dfrac{\rho_v^{0.7}}{P_i^{0.85}}\right)$	$\times \dfrac{W_i^{0.3} \lambda_i}{\Delta T_M}$	$\times \left(\dfrac{1}{n^{0.3} L^{0.3}}\right) \Sigma'$	(See Notes 2 and 5)	(9)
$\Delta T_o/\Delta T_M =$	0.352	$\times \left(\dfrac{Z_o^{0.3} M_o^{0.2} \sigma_o^{0.425}}{s_o^{1.075} c_o}\right)\left(\dfrac{\rho_v^{0.7}}{P_o^{0.85}}\right)$	$\times \dfrac{W_o^{0.3} \lambda_o}{\Delta T_M}$	$\times \left(\dfrac{1}{n^{0.3} L^{0.3}}\right) \Sigma$	(See Notes 2 and 5)	(10)
$\Delta T_o/\Delta T_M =$	4.28	$\times \dfrac{Z_o^{0.267} M_o^{0.222}}{s_o^{0.89}}$	$\times \dfrac{W_o^{0.4}(T_H - T_L)}{\Delta T_M}$	$\times \dfrac{N_{PT}^{0.282} P_y^{0.6}}{n^{0.718} L}$	(See Note 3)	(11)
$\Delta T_o/\Delta T_M =$	7.53		$\times \dfrac{W_o^{0.4}(T_H - T_L)}{\Delta T_M}$	$\times \dfrac{N_{PT}^{0.282} P_y^{0.6}}{n^{0.718} L}$	(See Note 4)	(12)
$\Delta T_o/\Delta T_M =$	21.7		$\times \dfrac{W_o^{0.2}(T_H - T_L)}{\Delta T_M}$	$\times \dfrac{d_o^{0.8} N_{PT}^{0.685}}{n^{0.315} L}$		(13)
$\Delta T_o/\Delta T_M =$	22.9	$\times \dfrac{Z_o^{0.467} M_o^{0.22}}{s_o^{0.89}}$	$\times \dfrac{W_o^{0.2}(T_H - T_L)}{\Delta T_M}$	$\times \dfrac{d_o^{0.8} N_{PT}^{0.685}}{n^{0.315} L}$		(14)
$\Delta T_o/\Delta T_M =$	4.75	$\times \dfrac{(Z_o M_o)^{0.333}}{s_o^2 c_o}$	$\times \dfrac{W_o^{4/3} \lambda_o}{\Delta T_M}$	$\times \dfrac{1}{n^{4/3} d_o^{4/3} N_{PT}^{1/3} L}$		(15)
$\Delta T_o/\Delta T_M =$	2.64	$\times \dfrac{(Z_o M_o)^{0.333}}{s_o^2 c_o}$	$\times \dfrac{W_o^{4/3} \lambda_o}{\Delta T_M}$	$\times \dfrac{N_{PT}^{0.177}}{n^{1.156} L^{4/3} d_o}$		(16)
$\Delta T_w/\Delta T_M =$	159	$\times c/k_w$	$\times \dfrac{W(t_H - t_L)}{\Delta T_M}$	$\times \dfrac{d_o - d_i}{n d_o L}$		(17)
$\Delta T_w/\Delta T_M =$	88	$\times 1/k_w$	$\times \dfrac{W \lambda}{\Delta T_M}$	$\times \dfrac{d_o - d_i}{n d_o L}$		(18)
$\Delta T_s/\Delta T_M =$	3,820	$\times c/h$	$\times \dfrac{W(t_H - t_L)}{\Delta T_M}$	$\times \dfrac{1}{n d_o L}$		(19)
$\Delta T_s/\Delta T_M =$	2,120	$\times 1/h$	$\times \dfrac{W \lambda}{\Delta T_M}$	$\times \dfrac{1}{n d_o L}$		(20)

Empirical Pressure-Drop Relationships for Rating Heat Exchangers—Table III

Eq. No.	Mechanism or Restriction	Empirical Equation
	Inside the tubes	
(21)	No phase change, $N_{Re} > 10,000$	$\Delta P = \dfrac{(Z_i)^{0.2}}{s_i} \left(\dfrac{W_i}{n}\right)^{1.8} \dfrac{N_{PT}[(L_o/d_i) + 25]}{(5.4 d_i)^{3.8}}$ (See note 1)
(22)	No phase change, $2,100 < N_{Re} < 10,000$	$\Delta P = \left(\dfrac{Z_i}{s_i}\right) \left(\dfrac{W_i}{n}\right) \dfrac{N_{PT}[(L_o/d_i) + 25][(N_{Re})^{2/3} - 125]}{(50.2 d_i)^3}$ (See note 1)
(23)	No phase change, $N_{Re} < 2,100$	$\Delta P = \dfrac{(Z_b)^{0.526} (Z_f)^{0.14}}{s_i} \left(\dfrac{W_i}{n}\right)^{4/3} \dfrac{N_{PT}(L_o)^{2/3}}{(5.62 d_i)^4}$
(24)	Condensing	$\Delta P = \dfrac{(Z_i)^{0.2}}{s_i} \left(\dfrac{W_i}{n}\right)^{1.8} \dfrac{N_{PT}[(L_o/d_i) + 25]}{(5.4 d_i)^{3.8}} \times 0.5$ (See note 1)
	Shellside	
(25)	No phase change, crossflow	$\Delta P = \dfrac{0.326}{s_o} (W_o)^2 \dfrac{L_o}{P_B^2 D_o}$
(26)	No phase change, parallel flow	$\Delta P = \dfrac{(Z_o)^{0.2}}{s_o} \left(\dfrac{W_o}{n}\right)^{1.8} \left[\dfrac{n^{0.366} L_o}{(N_{PT})^{1.434}(4.912 d_o)^{4.8}} + \dfrac{0.31 n^{0.0414} (W_o)^{0.2} L_o}{d_o (N_{PT})^{1.76}(4.912 d_o)^4 Z^{0.2} B_o^2}\right]$ (See notes 2 and 3)
(27)	Condensing	$\Delta P = \left(\dfrac{0.081}{s_o}\right) (W_o)^2 \left(\dfrac{L_o}{P_B^2 D_o}\right)$

Notes: 1. For U-bends, use $[(L_o/d_i) + 16]$ instead of $[(L_o/d_i) + 25]$. 2. B_o is equal to fraction of flow area through baffle.
3. Number of baffles $(N_B) = 0.48 (L_o/d_o)$.

Tube pitch for both triangular and square-pitch arrangements is assumed to be 1.25 times the tube diameter. This is a standard pitch used in the majority of shell-and-tube heat exchangers. Slight deviations do not appreciably affect results.

Shell diameter is related to the number of tubes (nN_{PT}) by the empirical equation:

$$D_o = 1.75 \, d_o (nN_{PT})^{0.47}$$

This gives the approximate shell diameter for a packed floating-head exchanger. The diameter will differ slightly for a fixed tubesheet, U-bend, or a multipass shell. For greater accuracy, tube-layout tables can be used to find shell diameters.

Derivation of Dimensional Equations

The following shows how the design equations are developed for a heat exchanger with sensible-heat transfer and Reynolds number 10,000 on the tube side, and with sensible-heat transfer and crossflow (flow perpendicular to the axis of tubes) on the shellside. Equations with other heat-transfer mechanisms are derived similarly.

For the film coefficient or conductance, h, and the heat-balance, these equations apply:

Value of h	Heat-Balance Eq.
$h_i = \dfrac{0.023 \, c_i G_i}{(c_i \mu_i / k_i)^{2/3} (D_i G_i / \mu_i)^{0.2}}$	$W_i c_i (t_H - t_L) = h_i A \Delta T_i$
$h_w = 24 \, k_w / (d_o - d_i)$	$W_i c_i (t_H - t_L) = h_w A \Delta T_w$
$h_o = \dfrac{0.33 \, c_o G_o^{(0.6)}}{(c_o \mu_o / k_o)^{2/3} (D_o G_o / \mu_o)^{0.4}}$	$W_o c_o (T_H - T_L) = h_o A \Delta T_o$
h_s = assumed value	$W_i c_i (t_H - t_L) = h_s A \Delta T_s$

Since the resistances involved in a tube-and-shell exchanger are the tubeside film, the tube wall, the scale caused by fouling, and the shellside film, then:

$$\Delta T_i + \Delta T_w + \Delta T_s + \Delta T_o = \Delta T_M$$

therefore:

$$\Delta T_i / \Delta T_M + \Delta T_w / \Delta T_M + \Delta T_s / \Delta T_M + \Delta T_o / \Delta T_M = 1$$

or:

$$\underbrace{\dfrac{W_i c_i (t_H - t_L)}{h_i A \Delta T_M}}_{\text{Tubeside product}} + \underbrace{\dfrac{W_i c_i (t_H - t_L)}{h_w A \Delta T_M}}_{\text{Tube-wall product}} + \underbrace{\dfrac{W_i c_i (t_H - t_L)}{h_s A \Delta T_M}}_{\text{Fouling product}} + \underbrace{\dfrac{W_o c_o (T_H - T_L)}{h_o A \Delta T_M}}_{\text{Shellside product}} = 1$$

This last equation is obtained by dividing each heat-balance equation by ΔT_M and solving for $\Delta T_f / \Delta T_M$. The design equations are derived by substituting for h the appropriate correlation for the coefficient; for k, the value obtained from the Weber equation; for A, the equivalent of the surface area in terms of the number of tubes, outside diameter and length, according to the relation $A = \pi n (d_o / 12) L$; for mass velocity on the tubeside, $G_i = 183 \, W_i / (d_i^2 n)$; and for mass velocity on the shellside, $G_o = 411.4 \, W_o / (d_o n N_{PT}^{0.47} P)$.

The resulting equation is rearranged to separate the physical-property, work, and mechanical-design parameters into groups. To obtain consistent units, the numerical factor in the equation combines the constants and coefficients. The form of the equations shown in Table II as Eq. (1), (11), (18) and (19) omits dimensionless groups such as Reynolds or Prandtl numbers, but includes single functions of the common design parameters such as number of tubes, tube diameter, tube length, baffle pitch, etc.

The individual products calculated from the four equations are added to give the sum of the products (SOP). A valid design for heat transfer should give

Nomenclature

A	Outside surface area, sq. ft.	t	Temperature on tubeside, deg. C.
B	Film thickness, $[0.00187Z\Gamma/g_c s^2]^{1/3}$, ft.	ΔT_M	Logarithmic mean temperature difference (LMTD), deg. C.
c	Specific heat, Btu./(lb.)(°F.)	U	Overall coefficient of heat transfer, Btu./[(hr.)(sq.ft.)(°F.)]
D_i	Inside tube diameter, ft.		
D_o	Inside shell diameter, in.	W	Flowrate, (lb./hr.)/1,000
d	Tube diameter, in.	Z	Viscosity, cp.
f	Fanning friction factor, dimensionless	Γ	Tube loading, lb./(hr.)(ft.)
G	Mass velocity, lb./(hr.)(sq.ft. cross-sectional area)	λ	Heat of vaporization, Btu./lb.
g_c	Gravitational constant, (4.18×10^8) ft. (hr.)^2	θ	Time, hr.
h	Film coefficient of heat transfer, Btu./[(hr.)(sq.ft.)(°F)]	μ	Viscosity, lb./(hr.)(ft.)
		ρ_v	Vapor density, lb./cu.ft.
k	Thermal conductivity, Btu./[(hr.)(sq.ft.)(°F./ft.)]	ρ_L	Liquid density, lb./cu.ft.
		σ	Surface tension, dynes/cm.
L	Total series length of tubes, ($L_o N_{PT} \times$ number of shells), ft.	Σ, Σ'	Surface condition factor, dimensionless
L_A	Length of condensing zone, ft.	*Subscripts*	
L_B	Length of subcooled zone, ft.	o	Conditions on shellside or outside tubes
L_o	Length of shell, ft.	i	Conditions on tubeside or inside tubes
M	Molecular weight, lb./(lb.-mol)	b	Bulk fluid properties
N_{PT}	Number of tube passes per shell, dimensionless	f	Film fluid properties
		H	High temperature
n	Number of tubes per pass (or in parallel) dimensionless	L	Low temperature
		s	Scale or fouling material
P	Pressure, psia.	w	Wall or tube material
P_B	Baffle spacing, in.	*Dimensionless Groups*	
ΔP	Pressure drop, lb./sq.in.	N_{Re}	Reynolds Number, DG/μ
Q	Heat transferred, Btu.	N_{Pr}	Prandtl Number, $c\mu/k$
s	Specific gravity (referred to water at 20°C.), dimensionless	N_{St}	Stanton Number, h/cG
T	Temperature on shellside, deg. C.	N_{Nu}	Nusselt Number, hD/k

SOP = 1. If SOP comes out to be less or more than one, the products for each resistance are adjusted by the appropriate exponential function of the ratio of the new design parameter to that used previously.

In what follows, each of the equations in Tables II and III is reviewed, giving the conditions where each equation applies and its limitations. In several cases, numerical factors are inserted or assumptions made so as to adapt the empirical relationships to the design of shell-and-tube exchangers. Such modifications have been made to increase the accuracy and broaden the use of the method.

More-sophisticated rating methods are available that make use of complex computer programs; the described method is intended only as a general, shortcut approach to shell-and-tube heat-exchanger selection. Accuracy of the technique is limited by the accuracy with which fouling factors, fluid properties and fabrication tolerances can be predicted. Nevertheless, test data obtained on hundreds of heat exchangers attest to the method's applicability.

Equations for Tubeside Heat Transfer

Eq. (1), No Phase Change (Liquid), $N_{Re} > 10,000$ —For liquids with Reynolds numbers greater than 10,000 and L_o/D greater than 60.

Eq. (2), No Phase Change (Gases), $N_{Re} > 10,000$ and $L_o/D > 60$—Because the Prandtl number of common gases is approximately 0.78 and viscosity enters only as $\mu^{0.2}$, the relationship of physical properties for gases is essentially a constant. This constant is combined with the numerical coefficient in Eq. (1) to eliminate the physical-property factors for gases, resulting in Eq. (2).

Eq. (3) and (4), No Phase Change (Gas and Liquid), $2,100 < N_{Re} < 10,000$—These are for gases (Eq. 3) and liquids (Eq. 4) where the Reynolds number is between laminar and fully developed turbulent-flow

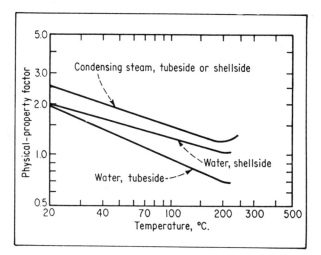

PHYSICAL-PROPERTY FACTORS for water and steam vs. temperature—Fig. 5

regions, and L_o/D is greater than 60. This flow regime should be avoided in tubeside design of shell-and-tube exchangers; experimental data in this range shows wide scatter, and operation is unpredictable.

Eq. (5), No Phase Change (Liquid), $N_{Re} < 2,100$— For liquids in laminar flow in small-diameter tubes, at moderate temperature difference and with large kinematic viscosity (μ_L/ρ). The accuracy of the correlation decreases as the operating conditions—or the geometry of the heat-transfer surface—are changed to increase the effect of natural convection. Where possible, higher-viscosity fluids should generally be placed in crossflow on the shellside of an exchanger, rather than inside the tubes.

Eq. (6), Condensing Vapor, Vertical, $N_{Re} < 2,100$ —For film condensation of vapors inside vertical tubes with a terminal Reynolds number ($4\Gamma/\mu$) less than 2,100. The tube loading (Γ) for vertical tubes (with condensing inside the tube) is equal to $(W_i \times 12)/(n\pi d_i)$. Eq. 6 gives a satisfactory correlation of the data for $N_{Re} < 2,100$. For larger Reynolds numbers, the equation should be adjusted by means of the Dukler plot. To use the dimensional Eq. (6) most conveniently, the constant in the equation should be multiplied by the ratio of the value obtained by the Nusselt equation to the Dukler plot value.

The above applies to condensable vapors only. The presence of noncondensable gases in the vapors decreases the film coefficient (see Ref. 13 for theory), a decrease that depends upon the relative sizes of the gas-cooling load and the total cooling and condensing duty. One model for the analysis of partial condensers assumes that the latent heat of condensation is transferred only through the condensate film, whereas the gas-cooling duty is transferred through both the gas film and condensate films.

Eq. (7), Condensing Vapor, Horizontal, $N_{Re} < 2,100$— For film condensation in horizontal tubes with a terminal Reynolds number ($4\Gamma/\mu$) less than 2,100. The tube loading (Γ) for condensing inside horizontal tubes is $W/(nL2)$, and the Reynolds number in most cases will be less than 2,100. The equation will yield conservative results since it does not include allowances for turbulence due to vapor-liquid shear or splashing of the condensate. At high tube loadings, the liquid condensate on the bottom of the tube blanks off part of the condensing surface. To correct for this, the heat-transfer factor $\Delta T_i/\Delta T_M$ should be multiplied by 1.3, if $W_i/(nsd_i^{2.56}) > 0.3$.

Eq. (8), Condensate Subcooling, Vertical— For laminar films flowing in layer form down vertical tubes, and where the condensate is to be subcooled below the bubble point. In these instances, it is convenient to treat the condenser-subcooler as two separate heat exchangers—the first operating only as a condenser (no subcooling), and the second operating as a liquid cooler only. Fig. 6 shows the assumptions that must be made to determine the length of each section, so as to calculate intermediate temperatures that will permit in turn the calculation of the LMTD.

To estimate the tube length required for the condensing section, Eq. (6) is used in combination with expressions for tube wall, fouling and shellside. The length necessary for subcooling is calculated by using Eq. (8), in combination with the tube wall, fouling, and shellside equations. For the subcooling section only, the arithmetic mean temperature-difference, $[(t_M - T_M) + (t_L - T_L)]/2$, of the tubeside and shellside fluids should be used instead of the log mean temperature-difference. In this equation, t_M and T_M represent, respectively, the tubeside and shellside temperatures at the point where the condensed liquid begins to be subcooled.

Eq. (9), Nucleate Boiling, Vertical— For nucleate boiling inside vertical tubes. The vapors generated by nucleate boiling inside vertical tubes induce flow through the heat exchanger, due to the gravity differences between the effluent two-phase mixture and the liquid feed to the exchanger. The effect is to create two separate zones for heat transfer. In the bottom of the tubes, the liquid is heated to its boiling point (elevated above the normal boiling point of the inlet liquid because of hydrostatic head). In this zone, there is sensible-heat transfer.

In the top part of the tube, the vapors are produced by nucleate boiling. In a rigorous analysis of a thermosyphon reboiler, the calculation of the heat transfer is combined with the hydrodynamics of the system to determine the circulating rate through the reboiler. However, for most design purposes, this calculation is not necessary. For atmospheric pressure and above, the assumption of nucleate boiling over the full tube length will give satisfactory results. On the other hand, the assumption of nucleate boiling over the entire length for the case of reboilers in vacuum service will produce overly optimistic results. Natural-circulation reboilers should not, as a rule, be used in vacuum service.

A surface-condition factor, Σ, appears in the empirical correlations for boiling coefficients. This is not to be confused with a fouling factor but is a measure of the number of nucleation sites for bubble formation on the heated surface. The expression for the film coefficient is adjusted by multiplying by the surface-condition factor.

For a highly polished surface with a relatively small number of sites, the factor is 0.4; for stainless steel and chrome-nickel alloys, 0.6; and for copper and steel, 1.0. Since $\Delta T_i/\Delta T_M$ is inversely proportional to the film coefficient h, various factor equations of Table II should be multiplied by the reciprocal of the surface-condition factor, Σ', or 1.0 for copper and steel, 1.7 for stainless steel or chrome-nickel alloys, and 2.5 for polished chrome.

Equations for Shellside Heat Transfer

Eq. (10), Nucleate Boiling Outside Horizontal— This equation is to be used for rating kettle-type

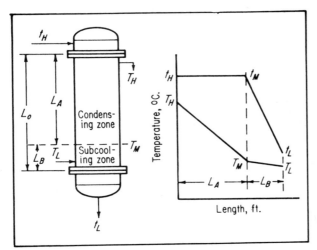

PARAMETERS used for calculating condensate subcooling in vertical tubes, as per Eq. (8)—Fig. 6

reboilers. Because the equation applies only to the nucleate-boiling mechanism, which is in turn a function of the temperature difference across the heating surface, criteria should be set that will establish the maximum heat flux.

Maximum flux is related to the rate at which bubbles can be disengaged from a heated surface, rise through a pool of liquid and become vapor at the liquid surface. If there is insufficient area above the heated surface for vapor disengagement, the bubbles are not released fast enough, the surface becomes vapor-bound and the coefficient decreases sharply.

In industrial reboilers or vaporizers, at the approximate maximum condition for vapor generation with nucleate boiling, the superficial vapor velocity (velocity above the projected area of the bundle) is in the range of 0.3 to 1.5 ft./sec. A guide for estimating this velocity is:

$$V_s = [0.0825 (\rho_L - \rho_g)^{1/4} \sigma^{1/4}]/\rho_g^{1/2} \quad \text{(Ref. 74)}$$

where V_s is the superficial vapor velocity in ft./sec.

To make the disengaging area as large as possible, tubes in kettle-type reboilers should be on rotated square-pitch with wide spacing. This is apart from mechanical considerations that permit easier bundle cleaning with wide spacings. As in boiling inside vertical tubes, a surface-condition factor corrects the empirical correlations for boiling coefficients.

Eq. (11), No Phase Change (Liquid), Crossflow—For liquids flowing perpendicular to axes of tubes through shells with crossflow baffles. Correlations of test results for shellside heat-transfer show wide scatter in data. This is due primarily to variation in the relative area of that portion of the tubes that is in parallel flow (baffle-cut area) to the portion of the tubes in crossflow (crossflow area between baffles) and, secondly, to bypassing of shellside fluid: (1) between the outer tubes and the shell, (2) through baffle-to-shell clearances, (3) through baffle-hole and tube clearances and (4) through lanes formed by channel partitions.

DESIGN OF HEAT EXCHANGERS 25

The equation for crossflow applies only to exchangers of good shellside design. This implies: (a) baffle cuts between 20 and 25% of the shell diameter, (b) blocking with sealing strips any bypass lanes and (c) maintaining minimum fabrication tolerances. To assure a crossflow pattern on the shellside, adjacent baffles should be spaced no further apart than one shell diameter; minimum spacings should not be less than 2 in., or one-fifth of the shell diameter, whichever is greater.

Even with good shellside design, some bypassing is unavoidable. The 0.6 multiplier for the coefficient (h) accounts for the loss in efficiency due to bypassing of the shellside fluid when TEMA (Tubular Exchanger Manufactures Assn.) fabrication tolerances are used, along with good shellside design.

Eq. (12), No Phase Change (Gas), Crossflow—For gases flowing perpendicular to axes of tubes through shells with crossflow baffles. As with gas flowing inside the tubes, the Prandtl number is nearly constant and the viscosity enters only to the 0.27 power. Therefore, the physical-property factor for the gas is included in the numerical factor. The same considerations listed for liquids in crossflow shellside design apply also to gases.

Eq. (13) and (14), No Phase Change (Gas and Liquid), Parallel Flow—Eq. (13), for gases, and Eq. (14), for liquids, apply to flows parallel to the axis of the tubes, the tubes being stayed at intervals by tube supports. Where shellside pressure drop is too high because of crossflow baffles, parallel-flow baffles (Fig. 3) or tube supports can be used. Here, shellside flow is nearly parallel to the tubes' axis.

Since there is little mixing of the shellside fluid in heat exchangers with parallel-flow baffles, provision must be made for uniform fluid distribution, usually by inlet and outlet distribution belts on the shell.

There is a similarity between Eq. (14) and (1), and between Eq. (13) and (2). The essential difference is that the equivalent diameter—(flow area × 4)/(wetted perimeter)—is used for tube diameter in Eq. (13) and (14), whereas for flow inside the tubes—Eq. (1) and (2)—the inside tube diameter is used.

With parallel-flow tube supports (Fig. 3), the shellside flow is not strictly parallel to the tubes' axis, and some crossflow occurs; this increases the film coefficient about 30%. The calculated coefficient for true parallel flow is therefore multiplied by 1.3

Equations for Shellside Condensing

Eq. (15), Condensing Vapor, Vertical, $N_{Re} < 2,100$—For film condensation of vapors outside vertical tubes with a terminal Reynolds number ($4\Gamma/\mu$) less than 2,100. The loading factor (Γ) is ($W_o \times 12$)/($n\pi d_o$). Process condensers are not usually designed for condensing vapors outside the tubes in vertical shells. This is due to the difficulty in removing noncondensable gases that accumulate in the process side and blank off the condensing surface. Therefore, condensing in the shellside of vertical exchangers is usually confined to steam heaters or reboilers.

Unlike the inside of the tubes where the condensed liquid flows in layer form down the length of the tube, the flow of condensate on the outside of the tubes is impeded by the tube supports. Here, most of the condensate layer is removed from each tube, and the bulk of the condensate falls off the edge of the baffle, providing poor contact with the cooling surface. For this reason, vapor-in-shell condensers with cross baffles are not often used where subcooling of the condensate is necessary.

An additional effect, especially where crossflow baffles are used, is the stripping of condensate from the tubes by the vapors flowing through the unit. This is more pronounced in the high-velocity section where the vapors enter. For shells of large diameter, window baffles (Fig. 3), together with inlet vapor distributors, are often used to introduce vapors into the bundle so as to avoid high velocities in entrance areas that can cause tube vibration.

Eq. (16), Condensing Vapor, Horizontal, N_{Re} <2100—For film condensation of vapors outside horizontal tubes with a Reynolds number $(4\Gamma/\mu)$ less than 2,100. The tube loading (Γ) is $W_o/(2n_r \times 2L_o)$, where n_r = number of tube rows across the center of the bundle. In a horizontal heat exchanger where condensing is occurring outside the tubes, the condensate cascades vertically over the tube bank and collects on the bottom of the heat-exchanger shell.

Many of the disadvantages of condensing in vertical shells with crossflow baffles apply also to horizontal condensing in the shell. Noncondensable gases reduce the effectiveness of the condensing surface; and limited contact between the condensate and heat-transfer surface makes condensate subcooling difficult.

As with vertical condensers, the effect of vapor shear on the condensate layer, and the stripping of the condensate by the baffles, yields higher heat-transfer coefficients than are obtained by the Nusselt equation. Where horizontal vapor-in-shell condensers are used, the preferred arrangement is for the vapors to flow in a single pass across the shell, normal to the axis of the tubes.

Heat Conduction Through Tube Wall, Fouling

Eq. (17) and (18), Tube Wall—For calculating the tube-wall factor. The heat flow through a tube with thin walls can be described by the Fourier equation, $Q/\theta = kA(dt/dx)$. In the integrated form, $Q/\theta = (kwA\Delta T_w)/x$, where x is the thickness of the tube. Expressed in the form of heat-transfer coefficient, and using the inside and outside diameters of the tube instead of the tube thickness, $h_w = 24k/(d_o - d_i)$. This is the value for h_w that is substituted into the heat-balance equation in order to obtain $\Delta T_w/\Delta T_M$ in the dimensional equation. Eq. (17) is used whenever

CLASSIFICATION AND APPLICATIONS

Heat exchangers are classified according to their geometry and type of construction. Most metals serve as materials of construction, although for some specific applications such nonmetallic substances as impervious graphite, glass, stoneware and plastics serve even better. Teflon exchangers, for example, are unmatched in corrosion resistance for certain services, and are usually made in bundles of tubes 0.1 to 0.25-in.-dia.;[116, 117] impervious-graphite exchangers,[111, 117] as well as glass units[116] are fabricated in a variety of exchanger types, including the shell-and-tube, which is the most widely used one in the chemical process industries. This last category [10 to 26] is subdivided into three major types: fixed tubesheet, return bend (U-tube) and floating tubesheet.

Shell-and-Tube Fixed Tubesheet, Return Bend

The fixed-tubesheet exchanger design has straight tubes secured at both ends in tubesheets welded to the shell. Usually, the tubesheets extend beyond the shell and serve as flanges for bolting tubeside headers.

Because there are no gaskets or packed joints on the shellside, fixed-tubesheet exchangers provide maximum protection against leakage of shellside fluid to the outside. For the same reasons, the tube bundle cannot be removed for inspection or cleaning. Since clearance between the outermost tubes and the shell is only the minimum required for fabrication, tubes may completely fill the exchanger shell.

When necessary, an expansion joint incorporated in the shell, provides for differential thermal expansion. The need for this joint is determined by considering both the magnitude of the differential expansion and the cycling conditions to be expected during operation. Tubeside headers, channel covers, gaskets, etc., are accessible for maintenance and replacement, and tubes can be replaced and cleaned internally. The shellside can be cleaned only by backwashing or circulating a cleaning fluid.

Fixed-tubesheet exchangers find use primarily in services where the shellside fluids are nonfouling, such as steam, refrigerants, gases, Dowtherm, some cooling waters and clean process streams.

In the return-bend unit, both ends of U-shaped tubes are fastened to a single tubesheet, thus eliminating the problem of differential thermal expansion because the tubes are free to expand and contract. Tube bundles can also be removed for inspection and cleaning, or nonremovable units made with the tubesheets welded to shells.

For removable bundle units, return-bend exchangers provide about the same minimum clearance between the outermost tubes and the inside of the shell as fixed-tubesheet exchangers. The number of tube holes in the tubesheet for any given shell, however, is less than for the fixed-tubesheet kind because of limitations on bending

sensible-heat transfer is involved on either tubeside or shellside; Eq. (18) is used when there is latent-heat transfer on each side as, for example, reboilers heated by condensing steam.

Eq. (19) and (20), Fouling—For conduction of heat through scale or solids deposits. Fouling coefficients are selected by the designer, based upon his experience. As a rule, for most processes with sensible-heat transfer in both tubes and shell, or in condensers where clean fluids are handled, high-fouling coefficients (low-fouling factors) are assumed.

Fouling coefficients of 1,000 to 500 (fouling factors of 0.001 to 0.002) will normally require heat exchangers 10 to 30% larger than needed for a clean-service rating, depending upon the influence of the other resistances to heat transfer. A different approach, applicable to steam-heated reboilers subject to fouling, is to provide 50 to 75% greater area than the clean-service requirement.

Selection of a fouling factor is arbitrary and there is usually little data for accurate assessment of the degree of fouling that should be assumed for a given design. Since fouling varies with material, velocities and temperatures, the extent to which fouling considerations will influence a design depends on operating conditions and the design itself. Eq. (19) is used for sensible-heat transfer on the tubeside or shellside; Eq. (20) where latent heat is transferred on both sides of the heating surface.

Equations for Pressure Drop Inside Tubes

Eq. (21), No Phase Change, $N_{Re} > 10,000$—For pressure drop, an expression of the Fanning equation for noncompressible fluids is used, where the friction factor f in the Fanning equation $= (0.046/N_{Re})^{0.2}$. The equation has been revised to account for pressure losses in the inlet and outlet nozzles, and the inlet and return channels. Based upon tests, these losses for most exchangers are approximately equivalent, for straight tubes, to 300 tube diameters for each tube pass, and 200 tube diameters for each tube pass for return-bend exchangers.

Eq. (22), No Phase Change, $2,000 < N_{Re} < 10,000$—For cases where the Reynolds number lies between the laminar- and turbulent-flow ranges.

Eq. (23), No Phase Change, $N_{Re} < 2,100$—This is an expression of the Fanning equation for noncompressible fluids, where the friction factor f in the Fanning equation $= 16/N_{Re}$. In the laminar-flow range, velocity head losses in the nozzles and channels are negligible and can be neglected.

Eq. (24), Condensing—The equation for calculating the pressure drop for condensing inside tubes is identical to that for no phase change, except for a factor of 0.5 that is used with the condensing equation. For total condensers, the weight rate of flow used in the calculation should be the inlet flowrate. Since for partial condensers the average flow is greater than

---

tubes. The number of tubeside passes must always be an even number; the maximum is limited only by the number of return bends.

Tubeside headers, channels, gaskets, etc., are accessible for maintenance and replacement, and the tube bundle can be removed and cleaned or replaced. Although replacement of tubes in the outside rows presents no problems, the others can be replaced only when special tube supports are used, which allow the U-tubes to be spread apart so as to gain access to tubes inside the bundle. The inside of the tubes may be cleaned only with special tools and then only when the bending radius of the tubes is fairly generous. Because of this, return-bend exchangers are usually found in non-fouling service, or where chemical cleaning is effective. U-bend construction is widely used for high-pressure applications.

Shell-and-Tube Floating Tubesheet

These exchangers have straight tubes secured at both ends in tubesheets. One tubesheet is free to move, thereby providing for differential thermal expansion between the tube bundle and the shell. Tube bundles may be removed for inspection, replacement and external cleaning of the tubes. Likewise, tubeside headers, channel covers, gaskets, etc., are accessible for maintenance and replacement, and tubes may be cleaned internally. The basic types are:

Outside-Packed Stuffing Box (Fig. A-1)—In this exchanger, shellside fluid is sealed by rings of packing compressed within a stuffing box by a packing-follower ring. The packing allows the floating tubesheet to move back and forth. Since the stuffing box only contacts shellside fluid, shellside and tubeside fluids do not mix, should leakage occur through the packing. The number of tubeside passes is limited only by the number of tubes in the bundle. Since the outer tube limit approaches the inside of the floating tubesheet skirt, clearances between outermost tubes and shell are dictated by skirt thickness.

Used for shellside services up to 600 psi. and 600 deg. F., these exchangers are not applicable when leakage of the shellside fluid to the outside cannot be tolerated.

Outside-Packed Lantern Ring (Fig. A-1)—Here, the shellside and tubeside fluids are each sealed by separate rings of packing (or O-rings) separated by a lantern ring provided with weep holes, so that leakage through either packing will be to the outside. The width of the tubesheet must be sufficient to allow for the two packings, the lantern ring and for differential thermal expansion. A small skirt is sometimes attached to the floating tubesheet to provide bearing surface for packings and lantern ring.

Since there can be no pass partition at the floating end, the number of tubeside passes is limited to one or two. Slightly larger than required for return-bend exchangers, the clearance between the outermost tubes and the inside of the shell must prevent tube-hole distortion during tube rolling near the outside edge of the tubesheet.

Outside-packed, lantern-ring units are generally limited to 150 psi. and 500 deg. F. This construction cannot be

for total condensers, the multiplying factor should be 0.7 instead of 0.5.

Because the estimation of the pressure drop for condensing vapors is not clear-cut, the equation should be used only to approximate the pressure drop so as to prevent the design of exchangers with impossible pressure-loss conditions.

Pressure Drop Outside the Tubes

Eq. (25), No Phase Change, Crossflow—This equation is based upon an analysis of pressure drop due to expansion and contraction losses with fluids flowing normal to staggered tubes. Test data on commercial exchangers have been used to adjust the correlation employed to make allowances for normal shellside leakage streams. The equation applies to exchangers with cross baffles having a 20% baffle cut, and with bypass areas normal to the shellside flow (pass partition lanes and tube-to-shell clearances) blocked by sealing strips or dummy tubes.

Eq. (26), No Phase Change, Parallel Flow—For flow of fluids parallel to the axis of the tubes, the supports of which are of the type shown in Fig. 3. The first two terms are identical to those of Eq. (21). The first part of the last term accounts for the effect of frictional pressure loss, and the second part of the last term for velocity head losses through baffle openings. The equation applies only where there is no significant shellside flow normal to the tubes (crossflow).

Eq. (27), Condensing—This equation for pressure drop for condensing vapors in baffled shells is identical to Eq. (25), except that a lower numerical factor is used. For total condensers, the calculation should be based on the inlet flowrate. For partial condensers, the average flowrate should be used with Eq. (25). For pressure drop for condensing outside the tubes, only approximate results should be expected, which themselves should be used only to prevent shellside designs that would result in impossible pressure-loss situations.

Rating With Electronic Calculator

For engineers not having access to computers, the method described above is easily adapted to rating by an electronic calculator that can be programmed. Since the job is thus simplified to entering the design parameters in the keyboard, the time for rating most heat exchangers can be reduced typically to 10 to 15 min.—approximately the time required to enter the data into the machine. The design equations from Tables II and III may be stored on magnetic tape, punched cards or magnetic cards.

Fig. 7 shows the format of a worksheet for rating with an electronic calculator. The format indicates which parameters are to be supplied for each design,

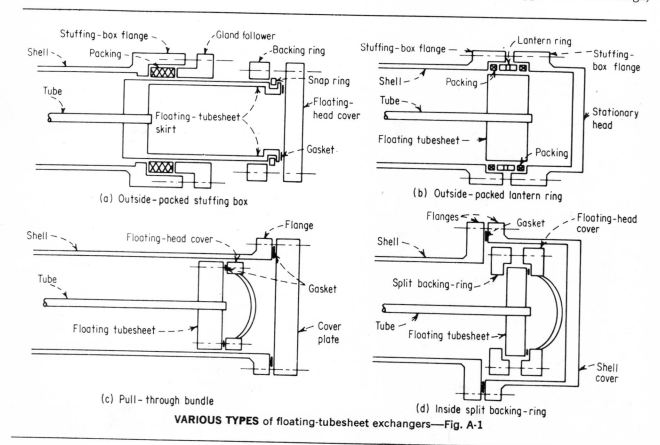

VARIOUS TYPES of floating-tubesheet exchangers—Fig. A-1

and the programs are set up so that the data are always entered into the calculator in the same order. The particular data shown on the worksheet pertain to the sensible-heat shell-and-tube exchanger described in the sample calculation that follows.

SAMPLE CALCULATION

This example applies the rating method to the design of a liquid-liquid heat exchanger under these conditions:

Conditions	Tubeside	Shellside
Flowrate, lb./hr.	307,500	32,800
Inlet temperature, °C.	105	45
Outlet temperature, °C.	unknown	90
Viscosity, cp.	1.7	0.3
Specific heat, Btu./hr./°F.	0.72	0.9
Molecular weight	118	62
Specific gravity with reference to water at 20 °C.	0.85	0.95
Allowable pressure drop, psi.	10	10
Maximum tube length, ft.	12	12
Minimum tube dia., in.	5/8	5/8
Material of construction	steel ($k=26$)	

Preliminary Calculations:

Heat transferred = $32,800 \times (90 - 45) \times 1.8 \times 0.9$ = 2,391,000 Btu./hr.

Temperature decrease of hot fluid = $2,391,000 / (307,500 \times 0.72 \times 1.8)$ = 6 °C.

Shellside outlet temperature = $105 - 6 = 99$ °C.

ΔT_M (or LMTD) = $(54 - 15)/\ln(54/15) = 30.4$ °C.

For a first-trial calculation, the approximate surface can be calculated using an assumed overall heat-transfer-coefficient, U, of 250 Btu./(hr.)(sq.ft.)(°F.). The assumed value of U can be obtained from tabulations in texts and handbooks and is used only to estimate the approximate size for a first trial:

$A = 2,391,000/(250 \times 30.4 \times 1.8) = 175$ sq. ft.

Since the given conditions specify a maximum tube length of 12 ft. and a minimum tube diameter of 5/8 in., the number of tubes required is:

$n = 175/(12 \times 0.1636) = 89$ tubes

and the approximate shell diameter will be:

$D_o = 1.75 \times 0.625 \times 89^{0.47} = 9$ in.

With the exception of baffle spacing, all preliminary calculations have been made for the quantities to be substituted into the dimensional equations. For the first trial, we may start with a baffle spacing equal to about half the shell diameter. After calculating the shellside pressure drop, we may adjust the baffle spacing. Also, it is advisable to check the Reynolds number on the tubeside to confirm that the proper equations are being used.

To find the maximum number of tubes (n_{max}) in

used when leakage of either fluid to the outside is not acceptable, or when possible mixing of tubeside and shellside fluids cannot be tolerated.

Pull-Through Bundle (Fig. A-1)—This type of exchanger has a separate head bolted directly to the floating tubesheet. Both the assembled tubesheet and head are small enough to slide through the shell, and the tube bundle can be removed without breaking any joints at the floating end. Although this feature can reduce shellside maintenance, it increases tubeside maintenance. Clearance requirements (the largest for any type of shell-and-tube exchanger) between the outermost tubes and the inside of the shell must provide for both the gasket and the bolting at the floating tubesheet.

The number of tubeside passes is limited only by the number of tubes. With an odd number of passes, a nozzle must extend from the floating-head cover through the shell cover. Provision for both differential thermal expansion and tube-bundle removal must be made by such methods as packed joints or internal bellows. Since this type of exchanger requires an internal gasket between the floating tubesheet and its head, applications are usually restricted to services where never-visible failures of the internal gasket are not intolerable.

Inside Split Backing-Ring (Fig. A-1)—In this design, the floating cover is secured against the floating tubesheet by bolting to a strong, well-secured split backing-ring. This closure, located beyond the end of the shell, is enclosed by a shell cover of larger diameter. Shell cover, split backing-ring and floating-head cover must be removed for the tube bundle to slide through the shell.

Clearances between the outermost tubes and the inside of the shell—which are about the same as those for outside-packed stuffing-box exchangers—approach the inside diameter of the gasket at the floating tubesheet. This type of construction has the same limitation on the number of tubeside passes as the pull-through bundle, but is more suitable for higher shellside temperatures and pressures.

Air-Cooled Heat Exchangers [27 to 39]

Air-cooled exchangers become especially attractive in locations where water is scarce or expensive to treat. Although an air-cooled unit requires a higher initial investment than its water-cooled counterpart, maintenance and operating costs are usually less.

The air-cooled units have axial-flow fans that force or induce a flow of ambient air across a bank of externally finned tubes. A typical exchanger has a bank of finned tubing, a steel supporting structure with plenum chambers and fan rings, axial-flow fans, drive assemblies and miscellaneous accessories such as louvers, fan guards, fencing, hail screens and vibration switches. Detailed descriptions and design methods of the various types of aircooled exchangers will appear in a forthcoming article.

Extended Surfaces

The ratio of outside to inside surface for bare tubes is usually in the range of 1.1 to 1.5, depending on the tube

WORKSHEET FOR DESIGN and evaluation of heat exchangers, using electronic calculator—Fig. 7

diameter and wall thickness. When the thermal resistance on the outside of the tube is much greater than that on the inside, more heat can be transferred (per unit length) through tubes that have a higher ratio of outside to inside areas. Such tubes, called extended-surface tubes, are made with surface-area ratios of 3 to 40.

Internal extended surfaces (less common) are used when the tubeside thermal resistance is high. Double extended surfaces—tubes with extended surfaces on both the inside and outside—are also available for high thermal resistances on both sides.

Extended surfaces normally have fins longitudinal or transverse to the tube, although other forms such as pegs and spines are available. Longitudinal fins, for parallel-flow designs, are especially suited for services where the allowable pressure drop is low and the finside fluid clean.

Transverse fins are generally for crossflow, although some low-finned tubing is also used in parallel flows. Low-finned tubing can be readily incorporated into shell-and-tube exchangers.

Double-Pipe and Cascade Exchangers

The double-pipe exchanger consists of two concentric pipes separated by mechanical closures. Inexpensive, rugged and easily maintained, they are primarily for low flowrates and are well adapted to high-temperature, high-pressure applications due to their relatively small diameters. Because of the small amount of heat-transfer surface per section, double-pipe exchangers are generally found in small total-surface requirement applications.

Cascade coolers—also referred to as trombone, trickle or serpentine coolers—consist of a bank of tubes in series, one above the other, over which water trickles. Relatively inexpensive to construct and operate, they are often preferred to shell-and-tube exchangers for cooling high-temperature streams where fouling on the outside of the tubes can be significant. The outside is easier to clean, and the design eliminates the potential hazards associated with rapid vaporization. When made of special materials (glass, ceramics, impervious graphite, Teflon), cascade coolers are frequently used to cool very corrosive liquids.

Pipe Coils, Bayonet Heat Exchangers [109 to 113]

Because of the almost unlimited number of possible designs, pipe coils are made to order only. Some of the more common ones are helical coils, spiral pancake coils and serpentine coils. They are used primarily for heating or cooling tanks, or for slurries or viscous materials.

Bayonet heat exchangers consist of a pair of concentric tubes, the outer one having one end sealed. The inner tube is referred to as the bayonet, the outer one as the scabbard. A number of scabbards are attached to a tubesheet, the surface of the scabbard being the heat-transfer area. While the bayonets are unsupported, the scabbards may or may not be supported by baffles.

Since all parts subject to differential thermal expansion move independently of each other, bayonet tubes are used when there are extreme temperature differences be-

parallel that still permits flow in the turbulent region ($N_{Re} = 12,600$), a convenient relationship is $n_{max} = W_i/(2d_iZ_i)$. In this example, $n_{max} = 307.5/(2 \times 0.495 \times 1.7) = 183$. For any number of tubes less than 183 tubes in parallel, we are in the turbulent range and can use Eq. (1).

From Table II, the appropriate expressions for rating are: Eq. (1) for tubeside, Eq. (11) for shellside, Eq. (18) for tube wall, and Eq. (19) for fouling. Eq. (21) and (25) respectively are used for tubeside and shellside pressure drops. If we now substitute values:

Heat-Transfer Calculation

Tubeside, Eq. (1):

$$\frac{\Delta T_i}{\Delta T_M} = 10.43 \left[\frac{1.7^{0.467} \times 118^{0.222}}{0.85^{0.89}}\right] \times$$

$$\left[\frac{307.5^{0.2} \times 6}{30.4}\right]\left[\frac{0.495^{0.8}}{89^{0.2} \times 12}\right]$$

$$= 10.43 \times 4.27 \times 0.621 \times 0.0193 = 0.535$$

Shellside, Eq. (11):

$$\frac{\Delta T_o}{\Delta T_M} = 4.28 \left[\frac{0.3^{0.267} \times 62^{0.222}}{0.95^{0.89}}\right] \times$$

$$\left[\frac{32.8^{0.4} \times 45}{30.4}\right]\left[\frac{1^{0.282} \times 5^{0.6}}{89^{0.718} \times 12}\right]$$

$$= 4.28 \times 1.89 \times 5.98 \times 0.00872 = 0.424$$

Results of Trial Calculations—Table IV

	1st Trial	2nd Trial	3rd Trial	4th Trial
Number of tubes	89	109	109	110
Shell diameter, in.	9	9	9	10
Baffle spacing, in.	5	5	3½	3½
Product of factors:				
Tubeside	0.535	0.514	0.514	0.513
Shellside	0.424	0.367	0.296	0.293
Tube wall	0.052	0.042	0.042	0.042
Fouling	0.250	0.204	0.204	0.202
Total sum of products	1.261	1.127	1.056	1.050
Tubeside ΔP, psi.	14.3	10	10	9.8
Shellside ΔP, psi.	3.9	3.9	11.4	10.1

Tube Wall, Eq. (18):

$$\frac{\Delta T_w}{\Delta T_M} = 159 \left[\frac{0.72}{26}\right]\left[\frac{307.5 \times 6}{30.4}\right]\left[\frac{0.625 - 0.495}{89 \times 0.625 \times 12}\right]$$

$$= 159 \times 0.0277 \times 60.7 \times 0.000195 = 0.052$$

Fouling, Eq. (19):

$$\frac{\Delta T_s}{\Delta T_M} = 3,820 \left[\frac{0.72}{1,000}\right]\left[\frac{307.5 \times 6}{30.4}\right]\left[\frac{1}{89 \times 0.625 \times 12}\right]$$

$$= 3,820 \times 0.00072 \times 60.7 \times 0.00150 = 0.250$$

(SOP) $= 0.535 + 0.424 + 0.052 + 0.250 = 1.261$

Because SOP is greater than 1, the assumed exchanger is inadequate. The surface area must be in-

tween fluids. Since large tubes are normally selected, bayonet tubes are suitable for vacuum applications (low pressure drop). Costs are high because only the outer tube transfers heat to the shellside fluid.

Spiral-Tube, Plate Exchangers [40 to 51]

Spiral-coil exchangers consist of one or more concentric, spirally wound coils that are tightly clamped between a cover plate and a casing, each coil being attached to a manifold at either end. Economical and easy to install, these units come in sizes ranging from 1 to 325 sq. ft. and pressures up to 600 psi. A forthcoming article will describe their design.

In plate devices, heat transfers through plates instead of tubes. Some plates have smooth surfaces, others are corrugated or embossed. The important kinds are: spiral, plate-and-frame, plate-and-fin, and patterned plates.

Spiral Plates—These consist essentially of two long strips of plates wound to form a pair of concentric spiral passages. Fluids are prevented from mixing by alternately welding the two passages so that only one passage is open at either end of the spiral. As one fluid enters at the center of the spiral and flows toward the outside, the other one enters at the periphery and flows toward the center.

Spiral exchangers offer high turbulence at low flow-rates, resistance to fouling, no differential thermal-expansion problems in most services, and relative ease of cleaning. Compactness is another good feature: 2,000 sq. ft. of heat-transfer surface can be provided in a 54-in.-dia. unit with a plate width of 72 in. Spiral units, particularly useful for handling viscous or solids-containing fluids, are well suited as condensers or reboilers.

Plate-and-Frame Exchanger—This unit consists of a number of embossed plates with corner openings mounted between a top carrying-bar and a bottom guide-bar. The plates are gasketed and arranged so that when a series of plates are clamped together, the corrugations on successive plates interlock to form narrow flow passages. Fluids are directed through the adjacent layers between the plates, either in series or parallel.*

Plate-and-Fin Exchangers—These exchangers are made from a stack of layers, each one consisting of corrugated aluminum sheets (fins) between flat aluminum separating plates to form individual fluid passages. Each layer is closed at the edge with solid aluminum bars and the stack is bonded together by a brazing process to yield an integral rigid structure with a series of passages. Either bolt-on or integrally welded headers can be provided. These units are very compact: about nine times more heat-transfer surface is available in 1 cu. ft. of a brazed aluminum plate-fin exchanger than in a shell-and-tube unit. Weight, also, is kept to a minimum. Design pressures go up as high as 700 psi. but temperatures should generally not exceed 150 F.

Patterned Plates—Formed from two plates—one or both of which are embossed, welded or brazed together to

creased by adding tubes or increasing the tube length, or the performance must be improved by decreasing the baffle spacing. Since the maximum tube length is fixed by the conditions given, the alternatives are increasing the number of tubes and/or adjusting the baffle spacing. To estimate assumptions for the next trial, pressure drops are calculated.

Pressure-Drop Calculation

Tubeside, Eq. (21):

$$\Delta P = (1.7^{0.2}/0.85)(307.5/89)^{1.8}[(12/0.495) + 25]/(5.4 \times 0.495)^{3.8} = 14.3 \text{ psi.}$$

Shellside, Eq. (25):

$$\Delta P = (0.326/0.95)(32.8^2)[12/(5^3 \times 9)] = 3.9 \text{ psi.}$$

To decrease the pressure drop on the tubeside to the acceptable limit of 10 psi., the number of tubes must be increased. This will also decrease the SOP. In addition, shellside performance can be improved by decreasing the baffle spacing, since the pressure drop of 3.9 on the shellside is lower than the allowable 10 psi. Before proceeding with successive trials to balance the heat-transfer and pressure-drop restrictions, Table IV is now set up for clarity.

Second Trial

As a first step in adjusting the heat-transfer surface and pressure drop, calculate the number of tubes to give a pressure drop of 10 psi. on the tubeside. The pressure drop varies inversely as $n^{1.8}$. Therefore, $14.3/10 = (n/89)^{1.8}$, and $n = 109$.

Each individual product of the factors is then adjusted in accordance with the applicable exponential function of the number of tubes. Since the tubeside product is inversely proportional to the 0.2 power of the number of tubes, the product from the preceding trial is multiplied by $(n_1/n_2)^{0.2}$, where n_1 is the number of tubes used in the preceding trial, and n_2 is the number to be used in the new one. The shellside product of the preceding trial is multiplied by $(n_1/n_2)^{0.718}$, and the tube-wall and fouling products by n_1/n_2. New adjusted products are then calculated as follows:

Tubeside product	$= (89/109)^{0.2} \times 0.535$	$= 0.514$
Shellside product	$= (89/109)^{0.718} \times 0.424$	$= 0.367$
Tube-wall product	$= (89/109) \times 0.052$	$= 0.042$
Fouling product	$= (89/109) \times 0.250$	$= 0.204$
	SOP	$= 1.127$

Last Trial

For the third trial, baffle spacing is decreased to 3.5 in. from 5 in. Only the shellside product must be adjusted since only it is affected by the baffle spacing. Therefore, the shellside factor of the previous trial is multiplied by the ratio of the baffle spacing to the 0.6 power:

---

form coil-like channels—patterned plates have one fluid circulating through the channels while the other fluid contacts the plates' outer surfaces. These units are inexpensive, lightweight, easy to install or remove and easy to clean on the outside; their main application is for heating or cooling tanks. What is known as a Lamella heat exchanger is formed by welding patterned plates into a tubesheet to form a bundle that is placed inside a shell. Although usually made for pressures up to 250 psi., these plates have been designed for as high as 1,500 psi.

Falling-Film, Jacketed Units [76 to 85, 105 to 108]

Falling-film exchangers are vertical, one-tube pass, shell-and-tube units with devices such as orifice plates, individual ferrules in each tube, or overflow weirs for distributing liquid to each tube. Liquid flows by gravity in a film around the inner tube wall.

Used both for sensible-heat transfer and for vaporization, falling-film units produce greater velocity, more turbulence, and thereby better heat transfer for a given flowrate than could be obtained if the tubes were filled with fluid. Often, these exchangers are more economical for close temperature approaches than multishell installations. Since the tubes can be physically cleaned while the exchanger is in operation, the units are especially useful when poor quality water, or salt water, is to be placed in the tubes.

For agitated and unagitated vessels that require frequent cleaning, and for glass-lined vessels that are difficult to equip with internal coils, jackets are often provided. Any one of several jacket types may be used: patterned plate, dimpled construction, spiral-wound channels, or concentric cyclinders.

Scraped-Surface Exchangers [86 to 92]

Provided with scraper elements that continuously sweep the heat-transfer surface to prevent fouling and to increase heat transfer, these units (low and high speed) are widely used for viscous and rapidly fouling fluids.

Low-speed units usually consist of jacketed pipes connected in series (by special return bends), where the heat-transfer medium flows in the annulus between the two pipes. The scraper assembly consists of a series of blades attached to U-springs (which hold the blades against the inner pipe wall) that in turn are attached to a central shaft driven at 15 to 50 rpm.

High-speed exchangers operate in the range of 200 to 2,000 rpm. Unlike low-speed units, the shaft occupies a major portion of the internal volume, so the process fluid flows in the comparatively small annular space between the outside of the shaft and the heat-transfer surface. The rapidly rotating scraper blades create thin process-side films and violent agitation, resulting in high heat-transfer rates. These machines are applicable where short residence times are essential, to prevent fouling and for crystallization. They are not usually recommended for cooling highly viscous fluids because high power requirements greatly increase the heat that must be rejected.

$$(3.5/5.0)^{0.6} \times 0.367 = 0.296$$

The sum of the products (SOP) for this trial is 1.056. The shellside pressure drop $\Delta P_o = 3.9 \times (5.0/3.5)^3 = 11.4$ psi.

Because we have now reached the point where the assumed design nearly satisfies our conditions, tube-layout tables can be used to find a standard shell-size containing the next increment above 109 tubes. A 10-in.-dia. shell in a fixed-tubesheet design contains 110 tubes.

Again, correcting the products of the heat-transfer factors from the previous trial:

Tubeside product $= (109/110)^{0.2} \times 0.514 = 0.513$
Shellside product $= (109/110)^{0.718} \times 0.296 = 0.293$
Tube-wall product $= (109/110) \times 0.042 = 0.042$
Fouling product $= (109/110) \times 0.204 = 0.202$
$$\text{SOP} = 1.050$$

Tubeside pressure drop:
$$\Delta P_i = 10 \times (109/110)^{1.8} = 9.8 \text{ psi.}$$

The shellside pressure drop is now corrected for the actual shell diameter of 10 in. instead of 9 in.:
$$\Delta P_o = 11.4 \times (9/10) = 10.1 \text{ psi.}$$

Any value of SOP between 0.95 and 1.05 is satisfactory as this gives a result within the accuracy range of the basic equations; unknowns in selecting the fouling factor do not justify further refinement. Therefore, the above is a satisfactory design for heat transfer and is within the pressure-drop restrictions specified. The surface area of the heat exchanger is $A = 110 \times 12 \times 0.1636 = 216$ sq. ft. The design overall coefficient is $U = 2,391,000/(30.4 \times 1.8 \times 216) = 202$ Btu./(hr.)(sq.ft.)(°F.).

The foregoing example shows that the essence of the design procedure is selecting tube configurations and baffle spacings that will satisfy heat-transfer requirements within the pressure-drop limitations of the system.

Procedure for Other Conditions

The preceding example was for rating a heat exchanger of single tube-pass and single shell-pass design, with countercurrent flows of tubeside and shellside fluids. Often, it will be necessary to use two or more passes for the tubeside fluid. In this case, the LMTD is corrected with the Bowman, Mueller and Nagle charts given in heat-transfer texts and the TEMA guide. If the correction factor for LMTD is less than 0.8, multiple shells should be used.

Bear in mind that n in all equations is the number of tubes in parallel through which the tubeside fluid flows; N_{PT} is the number of tubeside passes per shell (total number of tubes per shell $= nN_{PT}$); and L, the total-series length of path, equals shell length $(L_o)(N_{PT}) \times$ (number of shells).

The above procedure can be used for any shell-and-tube heat exchanger with sensible-heat transfer—or with no phase change of fluids—on both sides of the

Thin-Film Heat Exchangers [101 to 104]

Exchangers in this class are all mechanically aided, turbulent film devices. Centrifugal force or rotating blades create extreme agitation and distribute the process fluid. Primarily for handling viscous or heat-sensitive materials, they find wide use for concentrating solutions, evaporation of easily foaming liquids, and distillation under high vacuum.

Since only a small quantity of the process fluid is in the exchanger at any given instant, residence times are small and gases or vapors related are easily disengaged. Exchangers in this group are classified into three types:

Agitated Film—These devices consist of fixed blades rotating in a jacketed vertical or horizontal cylinder or truncated cone. A small clearance exists between the tips of the rotor blades and the heat-transfer wall (differing mechanically from scraped-surface exchangers). The action of the rotor blades forces liquid against the wall and maintains a thin, highly turbulent film on the wall.

Wiped Film—These units are similar to the agitated-film, except that the blades—which float or are hinged, instead of being fixed—ride on a thin film of liquid on the walls. The units are usually installed vertically.

Centrifugal Film—One type of centrifugal-film unit (Fig. A-2) consists of a housing containing a stack of rotating, hollow elements supported on a hollow bottom spindle. Heat is supplied from inside the cone elements by vapors entering through the spindle, and the process liquid is sprayed onto the underside of the cone elements at their upper edges.

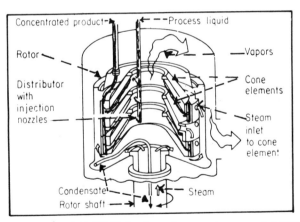

CENTRIFUGAL-FILM heat exchanger is one of several thin-film types—Fig. A-2

As centrifugal force spreads liquid into a thin layer over the heating surface, released vapors move up the center of the cone stack to be discharged at the side of the housing. Concentrated material leaving the individual cone elements is gathered at the periphery of the cone stack. The material is then displaced upward through channels at the cone periphery into a collection trough and then into a paring tube. Heating-medium condensate is collected from the heating side by another paring tube.

Other types of units spray the process fluid on the inside

tubes. Also, N_{Re} on the tubeside must be greater than 10,000, and the viscosity of the fluid on the shellside must be moderate (500 cp. maximum).

As pointed out, the designer should assume as part of his job the specification of tube arrangement that will prevent the flow in the shell from taking bypass paths either around the space between the outermost tubes and the shell, or in vacant lanes of the bundle formed by channel partitions in multipass exchangers. He should insist that exchangers be fabricated in accordance with TEMA tolerances.

By using the appropriate equations from Tables II and III, the technique described for rating heat exchangers with sensible-heat transfer can be used also for rating exchangers that involve boiling or condensing. The method can also be used in the design of partial condensers, or condensers handling mixtures of condensable vapors and noncondensable gases; and in the design of condensers handling vapors that form two liquid phases. However, for partial condensers and for two-phase liquid-condensate systems, a special treatment is required (see original articles describing the method[15]).

In addition to designing exchangers for specified performances, the method is also useful for evaluating the performance of existing exchangers. Here, the mechanical-design parameters are fixed, and the flowrates and temperature conditions (work factor) are the variables that are adjusted.

References

Text Books and Handbooks
1. McAdams, W. H., "Heat Transmission," 3rd ed., McGraw-Hill, New York (1954).
2. Kern, D. Q., "Process Heat Transfer," McGraw-Hill, New York (1950).
3. Jakob, M., "Heat Transfer," Vol. I, Wiley, New York (1949).
4. Jakob, M., "Heat Transfer," Vol. II, Wiley, New York (1957).
5. Knudsen, J. G., and Katz, D. L., "Fluid Dynamics and Heat Transfer," McGraw-Hill, New York (1958).
6. Brown, A. I., and Marco, S. M., "Introduction to Heat Transfer," McGraw-Hill, New York (1958).
7. Schack, A., "Industrial Heat Transfer," Wiley, New York (1965).
8. Eckert, E. R. G., and Drake, R. M., "Heat and Mass Transfer," 2nd ed., McGraw-Hill, New York (1959).
9. Kreith, F., "Principles of Heat Transfer," International Text Book Co., Scranton, Pa. (1965).
10. Kays, W. M., and London, A. L., "Compact Heat Exchangers," 2nd ed., McGraw-Hill, New York (1955).
11. Holman, J. P., "Heat Transfer," 2nd ed., McGraw-Hill, New York (1963).
12. "Chemical Engineers' Handbook," Perry, J. H., ed., 3rd ed., McGraw-Hill, New York (1950).
13. "Chemical Engineers' Handbook," Perry, J. H., et al., ed., 4th ed., McGraw-Hill, New York (1963).
14. Fraas, A. P., and Ozisik, M. N., "Heat Exchanger Design," Wiley, New York (1965).

Design Methods—Shell-and-Tube
15. Gilmour, C. H., Shortcut to Heat Exchanger Design, Parts I through VII, *Chem. Eng.*, Oct. 1952, p. 144; Mar. 1953, p. 226; Apr. 1953, p. 214; Oct. 1953, p. 203; Feb. 1954, p. 190; Mar. 1954, p. 209; Aug. 1954, p. 199.

Shell-and-Tube Exchangers
16. Donahue, D. A., Heat Exchangers, *Petrol. Process.*, Mar. 1956, p. 101.
17. Tinker, T., Shell Side Characteristics of Shell and Tube Heat Exchangers, *Proc. Inst. Mech. Eng.*, 1951, p. 89.
18. Tinker, T., Shell Side Characteristics of Shell and Tube Heat Exchangers, *Trans. ASME*, 80-1, p. 36 (1958).
19. Bell, K. J., Exchanger Design, *Petro./Chem. Eng.*, Oct. 1960, p. C-26.
20. Whitley, D. L., Calculating Heat Exchanger Shell Side Pressure Drop, *Chem. Eng. Progr.*, Sept. 1961, p. 59.
21. "Heat Exchangers," The Patterson-Kelley Co., Manual No. 700-A (1959).
22. "The Engineer's Reference Library," a *Power* Handbook on Heat Exchangers, p. 129, McGraw-Hill, New York.

---

of one or more conical heat-transfer elements where vapors are released over the entire surface and flow to a condenser either internal or external; concentrated liquid is again gathered at the periphery of the cones. Heat can be supplied by condensing vapors or by radiation.

Evaporative Coolers [93 to 98]

Evaporative-cooled units may be described as combinations of shell-and-tube exchangers with cooling towers. A typical cooler consists of a tube bundle, a water basin, fans, recirculating pumps, motors, water distribution system, mist eliminators and controls.

As a recirculating pump draws water from the basin beneath the unit and pumps it through a distributor system onto the tubes, air is forced or induced over the wetted tubes and through the rain of water droplets. Part of the water thus evaporates, cooling both the tube surfaces and the water itself. One disadvantage, however, is that a continued accumulation of impurities from the evaporated water and from the washed air may result in sealing or corrosion.

Used for both sensible-heat transfer and for condensing, evaporative coolers are economical for services between 170 F. down to a 10 F. approach to the design wet-bulb temperature of the ambient air. Above 170 F., dry-air cooling is usually more economical. Between 150 and 170 F., both dry-air and evaporative cooling should be considered.

Some advantages of evaporative coolers: (1) they only use about 5% of the water required for water-cooled units, (2) usually they require less power than dry-cooled units, (3) thermal pollution of surface waters is eliminated, (4) maintenance is generally less than that for water-cooled types, although usually more than for dry-air coolers, (5) all parts are readily accessible and the cooling coil can be cleaned on both sides.

Froth-Contact Heat Exchangers [99, 100]

These consist simply of a bank of tubes immersed in an air-water froth formed above a perforated plate. Heat transfers in two stages: first, the water flows around the tubes in a turbulent, frequently renewed film where heat transfer is high. Then, an actively moving froth surrounds the tubes and heat is transferred to the air; here, heat transfer is low but the area is large. Overall heat-transfer coefficients are about the same as for watercooled equipment. Compared to a heat-exchanger cooling-tower combination, the froth unit would be smaller, thereby saving on installation costs.

Waste-Heat Boilers [70, 71, 72]

This boiler increases the overall efficiency of a plant by generating steam from waste heat that cannot be retained in a system. Although units for low-pressure steam generally present no mechanical-design problems, those for high-pressure steam are difficult to design.

23. Lohrisch, F. W., What Are Optimum Exchanger Conditions?, *Hydrocarbon Process. Petrol. Refiner*, May 1963, p. 177.
24. Gilmour, C. H., No Fooling—No Fouling, *Chem. Eng. Progr.*, July 1965, p. 49.
25. Bergman, D. J., High Temperature Exchanger Problems, *Hydrocarbon Process.*, Oct. 1966, p. 158.
26. Morton, D. S., Heat Exchangers Dominate Process Heat Transfer, *Chem. Eng.*, June 11, 1962, p. 170.

Air-Cooled Exchangers
27. Smith, E. C., Air-Cooled Heat Exchangers, *Chem. Eng.*, Nov. 17, 1958, p. 145.
28. Mathews, R. T., Air Cooling in Chemical Plants, *Chem. Eng. Progr.*, May 1959, p. 68.
29. Perkins, B. G., Which Cooling Medium—Water or Air?, *Petrol. Refiner*, 38-4, p. 99 (1959).
30. Todd, J. R., Field Tests Needed for Air Coolers, *Petrol. Refiner*, 38-4, p. 115 (1959).
31. Campbell, J. C., Field Testing Air-Cooled Heat Exchangers, *Chem. Eng. Progr.*, July 1960, p. 58.
32. Williams, C. L., and Damron, R. D., Which Cools Cheaper: Water or Air?, *Hydrocarbon Process.*, Feb. 1965, p. 139.
33. Young, E. H., and Briggs, D. E., Bond Resistance of Bimetallic Finned Tubes, *Chem. Eng. Progr.*, July 1965, p. 71.
34. Smith, E. G., Gunter, A. Y., and Victory, S. P., Fin Tube Performance, *Chem. Eng. Progr.*, July 1966, p. 57.
35. Kulkarni, M. V., and Young, E. H., Bimetallic Finned Tubes, *Chem. Eng. Progr.*, July 1966, p. 68.
36. Cook, E. M., Comparison of Equipment for Removing Heat From Process Streams, *Chem. Eng.*, May 25, 1964, p. 137.
37. Cook, E. M., Operating Problems of Air-Cooled Units, and Air-Water Combinations, *Chem. Eng.*, July 6, 1964, p. 131.
38. Cook, E. M., Rating Methods for Selection of Air-Cooled Heat Exchangers, *Chem. Eng.*, Aug. 3, 1964, p. 97.
39. Howarth, E., Extended Surfaces: Their Use, Methods of Manufacture, and Properties, *Trans. Inst. Chem. Engrs. (London)*, June 1962, p. A84.

Plate Heat Exchangers
40. Jackson, B. W., and Troupe, R. A., Laminar Flow in a Plate Heat Exchanger, *Chem. Eng. Progr.*, July 1964, p. 62.
41. Hargis, A. M., Beckman, A. T., and Loiacono, J. J., Applications of Spiral Plate Heat Exchangers, *Chem. Eng. Progr.*, July 1967, p. 62.
42. Buonopane, R. A., et al., Heat Transfer Design Method for Plate Heat Exchangers, *Chem. Eng. Progr.*, July 1963, p. 57.
43. Hamberg, M., A User Looks at Plate-Fin Exchangers for Low-Temperature Processes, *Hydrocarbon Process. and Petrol. Refiner*, May 1963, p. 149.
44. Fletcher, W., and Ymse, G., Ramen's Heat Exchanger, *Trans. Inst. Chem. Engrs. (London)*, June 1962, p. A105.
45. Lamb, B. R., The Rosenblad Spiral Heat Exchanger, *Trans. Inst. Chem. Engrs. (London)*, June 1962, p. A108.
46. Performance of Plate Heat Exchangers, *Brit. Chem. Eng.*, Aug. 1960, p. 559.
47. Watson, E. L., et al., Plate Heat Exchanger Flow Characteristics, *Ind. Eng. Chem.*, Sept. 1960, p. 733.
48. Smith, V. C., and Troupe, R. A., Pressure Drop Studies in a Plate Heat Exchanger, *AIChE J.*, May 1965, p. 487.
49. Lawry, F. J., Plate-Type Heat Exchangers, *Chem. Eng.*, June 29, 1959, p. 89.
50. Tangri, N., and Jayaraman, R., Heat Transfer Studies on a Spiral Plate Heat Exchanger, *Trans. Inst. Chem. Engrs. (London)*, 40, p. 161 (1962).
51. Low Temperature Heat Transmission, special report, *Brit. Chem. Eng.*, Jan. 1960, p. 11.

Condensing
52. "Chemical Engineers' Handbook," Perry, J. H., ed., 3rd ed., p. 686, McGraw-Hill, New York (1950).
53. English, K. G., Jones, W. T., Spillers, R. C., and Orr, V., Flooding in a Vertical Updraft Partial Condenser, *Chem. Eng. Progr.*, July 1963, p. 51.
54. Dukler, A. E., Dynamics of Vertical Falling Film Systems, *Chem. Eng. Progr.*, Oct. 1959, p. 62.
55. Thomas, D. G., Enhancement of Film Condensation Heat Transfer Rates on Vertical Tubes by Vertical Wires, *Ind. Eng. Chem. Fundamentals*, Feb. 1967, p. 97.
56. Gregorig, R., An Analysis of Film Condensation on Wavy Surfaces Including Surface Tension Effects, *Z. Angew. Math. Phys.*, 5, p. 36 (1954).
57. Dobratz, C. J., and Oldershaw, C. F., Desuperheating Vapors in Condensers Unnecessary, *Chem. Eng.*, July 31, 1967, p. 146.
58. Leonard, W. K., and Estrin, J., The Effect of Vapor Velocity on Condensation on a Vertical Surface, *AIChE J.*, Mar. 1967, p. 401.
59. Greve, F. W., Measurement of Pipe Flow by the Coordinate Method, Bull. No. 32, Engineering Experiment Station, Purdue University, West Lafayette, Ind., Aug. 1928.
60. Hewitt, G. F., and Wallis, G., "Flooding and Associated Phenomena in Falling Film Flow in a Vertical Tube," *Proc. Am. Soc. Mech. Engrs.*, Multi-Phase Symposium, p. 62, Philadelphia, Pa., Nov. 17, 1963.

Vaporizers
61. Tong, L. S., "Boiling Heat Transfer and Two-Phase Flow," Wiley, New York (1965).
62. "Chemical Engineers' Handbook," Perry, J. H., others, ed., 4th ed., McGraw-Hill, New York (1963).
63. Fair, J. R., Vaporizer and Reboiler Design, *Chem. Eng.*, July 8, 1963, p. 119.
64. Fair, J. R., Vaporizer and Reboiler Design, *Chem. Eng.*, Aug. 5, 1963, p. 101.
65. Parker, N. H., How to Specify Evaporators, *Chem. Eng.*, July 22, 1963, p. 135.
66. Standiford, F. C., Evaporation, a feature report, *Chem. Eng.*, Dec. 9, 1963, p. 157.
67. Mallinson, J. H., Chemical Process Applications for Compression Evaporation, *Chem. Eng.*, Sept. 2, 1963, p. 75.
68. Palen, J. W., and Small, W. M., A New Way to Design Kettle and Internal Reboilers, *Hydrocarbon Process.*, Nov. 1964, p. 199.
69. Hughmark, G. A., Designing Thermosyphon Reboilers, *Chem. Eng. Progr.*, July 1964, p. 59.
70. Csathy, D., Evaluating Boiler Designs for Process-Heat Recovery, *Chem. Eng.*, June 5, 1967, p. 117.
71. Caruana, G., A Review of Evaporators and Their Applications, *Brit. Chem. Eng.*, July 1965, p. 466.
72. Seher, E., Waste Heat Boilers for the Chemical and Metallurgical Industries, *Brit. Chem. Eng.*, May 1965.
73. Gilmour, C. H., Nucleate Boiling—A Correlation, *Chem. Eng. Progr.*, Oct. 1958, p. 77.
74. Kutateladze, S. S., "Heat Transfer in Condensation and Boiling," (1952); translated as U.S. Atomic Energy Commission Report A.E.C.-TR-3770 (1953).
75. Sims, G. E., Akturk, V., and Evans-Lutterodt, K. O., Simulation of Pool Boiling Gas Injection at the Interface, *Intern. J. Heat Mass Transfer*, 6, p. 531 (1963).

Falling-Film Heat Exchangers
76. Sack, M., Falling Film Shell-and-Tube Heat Exchangers, *Chem. Eng. Progr.*, July 1967, p. 55.
77. Hartley, D. E., and Murgatroyd, W., Criteria for the Break-Up of Thin Liquid Layers Flowing Isothermally Over Solid Surfaces, *Intern. J. Heat Mass Transfer*, 7-9, p. 1003 (1964).
78. Bays, G. S., and McAdams, W. H., Heat Transfer Coefficients in Falling Film Heaters, Streamline Flow, *Ind. Eng. Chem.*, Nov. 1937, p. 1240.
79. McAdams, W. H., et al., Heat Transfer to Falling Water Films, *Trans. Am. Soc. Mech. Engrs.*, Oct. 1940, p. 627.
80. Penman, T. O., and Tait, R. W. F., Heat Transfer in Liquid-Film Flow, *Ind. Eng. Chem. Fundamentals*, Nov. 1965, p. 407.
81. Whitt, F. R., Performance of Falling Film Evaporators, *Brit. Chem. Eng.*, Dec. 1966, p. 1523.
82. Portalski, S., Studies of Falling Liquid Film Flow—Film Thickness on a Smooth Vertical Plate, *Chem. Eng. Sci.*, 18, p. 787 (1963).
83. Butterworth, D., The Laminar Flow of Liquid Down the Outside of a Rod Which Is At a Small Angle From the Vertical, *Chem. Eng. Sci.*, 22, p. 911 (1967).
84. Zhivaikin, L., Liquid Film Thickness in Film-Type Units, *Intern. Chem. Eng.*, July, 1962, p. 337.
85. Hewitt, G. F., Lacey, P. M. C., and Nicholls, B., Transitions in Film Flow in a Vertical Tube, United Kingdom Atomic Energy Authority Research Group Report, AERE-R 4614.

Scraped Surface
86. Bachmann, T. H., and Lineberry, D. D., Refrigerant Systems in Scraped Surface Exchangers, *Chem. Eng. Progr.*, July 1967, p. 68.
87. Hoskins, A. P., Votator Heat Exchangers, *Trans. Inst. Chem. Engrs.*, June 1962, p. A97.
88. Skelland, A. H. P., et al., Heat Transfer in a Water-Cooled Scraped-Surface Heat Exchanger, *Brit. Chem. Eng.*, May 1962, p. 346.
89. Skelland, A. H. P., and Seung, L. S., Power Consumption in a Scraped-Surface Heat Exchanger, *Brit. Chem. Eng.*, Apr. 1962, p. 264.
90. Bott, T. R., and Romero, J. J. B., Heat Transfer Across a Scraped Surface, *Can. J. Chem. Eng.*, Oct. 1963, p. 213.
91. Houlton, H. G., Heat Transfer in the Votator, *Ind. Eng. Chem.*, June 1944, p. 522.
92. Bott, T. R., and Sheikh, M. R., Effects of Blade Design in Scraped Surface Heat Transfer, *Brit. Chem. Eng.*, Apr. 1964, p. 229.

Evaporative Coolers
93. Berkeley, F. D., Now Consider Evaporative Cooling, *Petrol. Refiner*, 40-1, p. 169 (1961).
94. Harris, L. S., For Flexibility, the Air Evaporative Cooler, *Chem. Eng.*, Dec. 24, 1962, p. 77.
95. Maze, R. W., Practical Tips on Cooling Tower Sizing, *Hydrocarbon Process.*, Feb. 1967, p. 123.
96. DeMonbrun, J. R., Factors to Consider in Selecting a Cooling Tower, *Chem. Eng.*, Sept. 9, 1968, p. 106.
97. "Cooling Tower Design Trends," HPI Handbook Issue, *Petro/Chem. Eng.*, Mar. 1964, p. 54.
98. "Cooling Tower Fundamentals and Application Principles," The Marley Co., Kansas City, Mo.

Froth-Contact Heat Exchangers
99. Poll, A., and Smith, W., Froth-Contact Heat Exchangers, *Chem. Eng.*, Oct. 26, 1964, p. 111.
100. Poll, A., and Smith, W., The Froth-Contact Heat Exchanger: Preliminary Pilot-Scale Results, *Trans. Inst. Chem. Engrs.*, June, 1962, p. A93.

Thin-Film Evaporators
101. Mutzenberg, A. B., et al., Agitated Thin-Film Evaporators, *Chem. Eng.*, Sept. 13, 1965, p. 175.
102. Carter, A. L., and Kraybill, R. R., Low Pressure Evaporation, *Chem. Eng. Progr.*, Feb. 1966, p. 99.
103. Reay, W. H., Recent Advances in Thin-Film Evaporation, *Ind. Chemist*, June 1963, p. 3.
104. Gudheim, A. R., and Donovan, J., Heat Transfer in

HEAT EXCHANGER DESIGN AND SPECIFICATION

Thin-Film Centrifugal Processing Units, *Chem. Eng. Progr.*, Oct. 1957, p. 476.

Agitated Vessels
105. Chapman, F. S., and Holland, F. A., Heat-Transfer Coefficients for Agitated Liquids in Process Vessels, *Chem. Eng.*, Jan. 18, 1965, p. 153.
106. Chapman, F. S., and Holland, F. A., Heat-Transfer Correlations in Jacketed Vessels, *Chem. Eng.*, Feb. 15, 1965, p. 175.
107. Ackley, E. J., Film Coefficients of Heat Transfer for Agitated Process Vessels, *Chem. Eng.*, Aug. 22, 1960, p. 133.
108. Zakanycz, S., and Salamone, J. J., Nomographs for Unsteady State Heat Transfer, *Ind. Eng. Chem.*, Jan. 1963, p. 27.

Pipe Coils
109. Seban, R. A., and McLaughlin, E. F., Heat Transfer in Tube Coils With Laminar and Turbulent Flow, *Intern. J. Heat Mass Transfer*, 6, p. 387 (1963).
110. Noble, M. A., et al., Heat Transfer in Spiral Coils, *Petrol. Eng.*, Apr. 1952, p. 723.
111. Ito, H., Friction Factors for Turbulent Flow in Curved Pipes, *Trans. Am. Soc. Mech. Engrs.*, 81-2, p. 123 (1959).
112. White, C., Streamline Flow Through Curved Pipes, *Proc. Roy. Soc. (London)*, Series A, 123, p. 645 (1929).
113. Stuhlbarg, D., How to Design Tank Heating Coils, *Petrol. Refiner*, 38-4, p. 143 (1959).

Nonmetallic Exchangers
114. Friedman, S. H., Choosing a Graphite Heat Exchanger, *Chem. Eng.*, July 9, 1962, p. 133.
115. McEwen, C. K., Heat Exchange in Glass, *Chem. Eng.*, Sept. 2, 1963, p. 124.
116. Githens, R. E., et al., Flexible Tube Heat Exchangers, *Chem. Eng. Progr.*, July 1965, p. 55.
117. Hood, R. R., Designing Heat Exchangers in Teflon, *Chem. Eng.*, May 22, 1967, p. 181.
118. Blackburn, D. G., et al., Graphite Block Exchangers, *Trans. Inst. Chem. Engrs.*, June 1962, p. A111.

Improved Boiling Surfaces
119. Czikk, A. M., and O'Neill, P. S., Application of Enhanced Heat Transfer Surface to VTE Process Plant, Paper No. 10 at Symposium on Enhanced Tubes for Desalination Plants, sponsored by Office of Saline Water, Mar. 11-12, 1969, Washington, D.C.
120. Carnavos, T. C., Thin Film Distillation, First International Symposium on Water Desalination, Oct. 5, 1965, Washington, D.C.
121. Young, R. K., and Hummel, R. L., Improved Nucleate Boiling Heat Transfer, *Chem. Eng. Progr.*, July 1964.

Mechanical Design
122. Gardner, K. A., Heat-Exchanger Tube-Sheet Design, *J. Appl. Mech.*, Dec. 1948, p. 377.
123. Gardner, K. A., Heat-Exchanger Tube-Sheet Design—Fixed Tube Sheets, *J. Appl. Mech.*, June 1952, p. 159.
124. Miller, K. A. G., The Design of Tube Plates in Heat Exchangers, *Proc. Inst. Mech. Engrs. (London)*, Series B, 1, p. 215 (1952).
125. Yu, Y. Y., Rational Analysis of Heat-Exchanger Tube-Sheet Stresses, *J. Appl. Mech.*, Sept. 1956, p. 468.
126. Kopp, S., and Sayre, M. F., Expansion Joints for Heat Exchangers, Am. Soc. Mech. Engrs. paper, Vol. 6, No. 211, presented at annual meeting, New York, Nov. 1959.
127. Donohue, D. A., Heat Exchanger Design—Mechanical Design, *Petrol. Refiner*, Jan. 1956, p. 155.
128. Murphy, G., Analysis of Stresses and Displacement in Heat Exchanger Expansion Joints, *Trans. Am. Soc. Mech. Engrs.*, 74, p. 397 (1952).
129. Dahl, N. C., Toroidal-Shell Expansion Joints, *Trans. Am. Soc. Mech. Engrs.*, 75, p. A497 (1953).

Vibration
130. Lentz, G. J., Flow Induced Vibration of Heat Exchanger Tubes Near the Shell-Side Inlet and Exit Nozzles, AIChE Preprint 20, National Heat-Transfer Conference, Minneapolis, Minn., Aug. 1969.
131. Chen, Y. N., Flow-Induced Vibration and Noise in Tube-Bank Heat Exchangers Due to Von Karman Streets, Am. Soc. Mec. Engrs. Paper 67-VIBR-48 (1967).
132. Nelms, H. A., and Segaser, C. L., Survey of Nuclear Reactor Systems Primary Circuit Heat Exchangers, ORNL-4399, UC-38 Engineering and Equipment, Oak Ridge National Laboratory, Apr. 1969.
133. Thompson, H. A., Fatigue Failures Induced in Heat Exchanger Tubes by Vortex Shedding, Am. Soc. Mech. Engrs. Paper 69-PET-6, Sept. 1969.

Piping
134. Kern, R., How to Design Heat Exchanger Piping, *Petrol. Refiner*, 39-2, p. 137 (1960).
135. Volkin, R. A., Economic Piping of Parallel Equipment, *Chem. Eng.*, Mar. 27, 1967, p. 148.
136. Kern, R., Thermosyphon Reboiler Piping Simplified, *Hydrocarbon Process.*, Dec. 1968, p. 118.

Insulation and Tracing
137. Chapman, F. S., and Holland, F. A., Keeping Piping Hot—By Insulation, *Chem. Eng.*, Dec. 20, 1965, p. 79.
138. Chapman, F. S., and Holland, F. A., Keeping Piping Hot—By Heating, *Chem. Eng.*, Jan. 17, 1966, p. 133.
139. Medley, R. G., and Shafer, W. A., Electric Tracing Design Simplified, *Petrol. Refiner*, Feb. 1962, p. 151.
140. Holstein, W. H., Jr., What It Costs to Steam and Electrically Trace Pipelines, *Chem. Eng. Progr.*, Mar. 1966, p. 107.

141. Butz, C. H., When Is Electricity Cheaper Than Steam for Pipe Tracing?, *Chem. Eng.*, Oct. 10, 1966, p. 230.
142. House, F. F., Pipe Tracing and Insulation, *Chem. Eng.*, June 17, 1968, p. 243.
143. Barnhart, J. M., Insulation Saves Heat, Saves Money, *Chem. Eng.*, June 11, 1962, p. 164.

High-Temperature Fluids
144. Voznick, H. P., and Uhl, V. W., Molten Salt for Heat Transfer, *Chem. Eng.*, May 27, 1963, p. 129.
145. High Temperature Heating Media, *Chem. Eng. Progr.*, May 1963, p. 33.
146. Geiringer, P. L., "Handbook of Heat Transfer Media," Reinhold New York (1962).

Maintenance
147. Thomas, J. W., Exchanger Design Tips That Reduce Maintenance Costs, *Hydrocarbon Process.*, Feb. 1965, p. 153.
148. Metz, B. A., and Bird, P. G., Onstream Exchanger Cleaning Works, *Hydrocarbon Process.*, Jan. 1968, p. 133.
149. Small, W. M., Not Enough Fouling? You're Fooling!, *Chem. Eng. Progr.*, Mar. 1968, p. 82.
150. Webber, W. O., Under Fouling Conditions—Finned Tubes Can Save Money, *Chem. Eng.*, Mar. 21, 1960, p. 149.
151. Kern, D. Q., Heat Exchanger Design for Fouling Service, *Chem. Eng. Progr.*, July 1966, p. 51.
152. Kaye, S., and Bird, P. G., Stop Silt Settling in Exchangers, *Hydrocarbon Process.*, Aug. 1965, p. 149.
153. "Guide for Inspection of Refinery Equipment," Am. Petrol. Inst., Chapter VII (1958).

Trouble Shooting
154. Gilmour, C. H., Trouble-Shooting Heat-Exchanger Design, *Chem. Eng.*, June 19, 1967, p. 221.
155. Discrepancies Between Design and Operation of Heat Transfer Equipment, *Chem. Eng. Progr.*, Jan. 1961, p. 71.
156. AIChE Standard Testing Procedures, Sections I and II, AIChE.

Fluid Flow
157. Simpson, L., Sizing Piping for Process Plants, *Chem. Eng.*, June 17, 1968, p. 192.

Acknowledgements

The authors thank Mr. C. H. Gilmour for his instruction and advice through the years, as well as for his review of this report's manuscript. They are also grateful to Mr. J. A. Moore for his review, and to Union Carbide Corp. for permission to publish this report.

Meet the Authors

Richard C. Lord   **Paul E. Minton**   **Robert P. Slusser**

Richard C. Lord is a project engineer in the engineering department at Union Carbide Corp.'s Technical Center, P.O. Box 8361, South Charleston, W. Va., where he is part of the heat-transfer technology group. Formerly, he was involved in activities related to process design and economics. He has a B.S. degree in chemical engineering from the University of Maine and is a member of AIChE.

Paul E. Minton is a project engineer in the engineering department at Union Carbide Corp.'s Technical Center, P.O. Box 8361, South Charleston, W. Va., where he is a part of the heat-transfer technology group. A B.S. graduate in chemical engineering from the Missouri School of Mines and Metallurgy, he is a member of AIChE.

Robert P. Slusser is a project engineer in the engineering department at Union Carbide Corp.'s Technical Center, P.O. Box 8361, South Charleston, W. Va., where for the last 10 years he has been designing equipment as a member of the heat-transfer technology group. His past work includes 12 years as production supervisor in the vinyl-resin facilities. He holds B.S. and M.S. degrees in chemical engineering from Massachusetts Institute of Technology and is a member of AIChE.

Design Heat Exchangers for Liquids in Laminar Flow

Liquids in laminar flow have local heat-transfer coefficients that vary inversely as the distance from the tube entrance increases. When the thermal-entrance length is exceeded, the heat-transfer coefficient becomes constant. Here are techniques for analyzing these design parameters.

NIELS MADSEN, University of Rhode Island

We will derive and show in this article how to handle the most common equations for the design of heat exchangers operating in the laminar-flow regime with liquids such as viscous oils and molten polymers.

These equations will be obtained by (1) using the log mean-temperature difference, (2) averaging the mean overall heat-transfer coefficient with respect to the heat-transfer area, and (3) applying a correction factor for heat transfer if the inside heat-transfer coefficient varies inversely as a fractional power of the distance from the entrance of the tube.

Though thermal analysis of such exchangers is usually done numerically by using finite differences and a computer, the analytical methods, given here, will provide considerable insight into the problem.

The Heat-Transfer Coefficient

Let us begin by developing the heat-transfer coefficient for laminar flow inside a tube. When a fluid enters a tube, a laminar momentum-boundary layer will start to build up at the wall; and if heating takes place simultaneously, a thermal-boundary layer will also build up. Both momentum- and thermal-boundary layers will be initiated at the tubewall, and grow symmetrically toward the centerline of the tube.

If the critical Reynolds number of 2,000 is not exceeded for a fluid having constant properties, the final, developed, velocity profile will be parabolic, and the Fanning friction factor will be $16/N_{Re}$.

Similarly, referring to the thermal-boundary layer, a definite temperature profile will develop, with a fixed value for the heat-transfer coefficient. This coefficient will correspond to a Nusselt number that depends on the boundary conditions assumed for the system, as listed in Table I.

At the entrance of the tube, both the momentum- and thermal-boundary layers will usually approach a thickness of zero, and the heat-transfer coefficient and the boundary layer will approach infinity.

The mathematical analysis of this problem has intrigued mathematicians and engineers. With simplifying assumptions such as constant fluid properties, constant heat-transfer rate or constant wall temperature, the problem is amenable to mathematical analysis. Graetz [1], Nusselt [2], Pohlhausen [3], Lévêque [4], Drew [5], Sellars et al. [6], and Kays [7] have offered such solutions, with various assumptions about the flow and the thermal-boundary conditions. Table II summarizes some of these.

Most of the attempts have found rather limited application in engineering because of the difficulty in specifying:

- Flow conditions at the entrance to the tube.
- Effects of free convection.
- Effects of varying viscosity that are caused by the temperature change with location within the heated fluid.

Originally published August 19, 1974

Consequently, it is often more satisfactory to use purely empirical equations such as the Sieder-Tate equation [8]:

$$N_{Nu_m} = 1.86[N_{Re}N_{Pr}(D/L)]^{1/3}(\mu/\mu_w)^{0.14} \quad (1)$$

or the Hausen equation [9]:

$$N_{Nu_m}\left(\frac{\mu_w}{\mu}\right)^{0.14} = 3.66 + \frac{0.067[N_{Re}N_{Pr}(D/L)]^{1/3}}{1 + 0.04[N_{Re}N_{Pr}(D/L)]^{2/3}} \quad (2)$$

When comparing equations for laminar flow, it is very important to distinguish between the local heat-transfer coefficient, h, and the mean coefficient, h_m, which is the coefficient averaged over the distance from the thermal entrance of the duct, as given by:

$$h_m = (1/L)\int_0^L h\,dL$$

If the local heat-transfer coefficient can be expressed as proportional to the reciprocal of the distance from the entrance, raised to a fractional power, $h = C/L^{1/n}$. Integration yields: $h_m = [n/(n-1)]h$. For example, for the Sieder-Tate equation where $n = 3$, the relation becomes:

$$N_{Nu} = h_i D_i/k = 1.24[N_{Re}N_{Pr}(D_i/L)]^{1/3}(\mu/\mu_w)^{0.14} \quad (3)$$

Eq. (3) is shown in Fig. 1 together with the Lévêque and Graetz equations. The difference in Nusselt numbers calculated with the Sieder-Tate equation for a constant-property fluid is seen to be 15% higher than if the Lévêque equation is used. The Sieder-Tate equation will suit our purpose because it compensates for variation in fluid temperatures.

Analytical Equations for Heat Flow—Table II
(Fluid Properties Are Constant)

Boundary Condition			
Wall Temperature, t_w	Heat-Flux Density, q/A	Velocity Profile	Eq. Reference
Constant	Variable	Uniform	Graetz [1a]
Constant	Variable	Parabolic	Graetz [1b]
Constant	Variable	Linear	Lévêque [4]
Constant	Variable	Langhaar	Kays [7]
	Constant	Langhaar	Kays [7]
	Variable	Analytical	Pohlhausen [3]

As originally proposed, the Sieder-Tate equation was used for predicting a Nusselt number calculated with the arithmetic mean-temperature difference instead of the logarithmic mean-temperature difference. However, the latter is more appropriate. The distinction between the two means is immaterial except when the two temperature differences have a ratio greater than two.

The intersection of the lines representing the Sieder-Tate and the Graetz equations in Fig. 1, assuming that $N_{Pr}(\mu/\mu_w)^{0.52} = 1$, yields:

$$L/D_i = 0.039N_{Re}, \text{ or } L_{hyd}/D_i = 0.039N_{Re} \quad (4)$$

Eq. (4) expresses the hydraulic-entrance length, L_{hyd}, since the thermal- and momentum-boundary layers coincide at a Prandtl number of one. Eq. (4) gives a slightly low value because the intersection of the straight-line segments in Fig. 1 falls below the Graetz equation. The value of 0.039 compares very well with 0.029, 0.050 and 0.058 calculated by Schiller [10], Prandtl and Tietjens [11] and Langhaar [12], respectively.

The thermal-entrance length will always be determined by the intersection of the line segment for the Sieder-Tate equation and the horizontal segment located according to the boundary conditions and the dimensionless group, $N_{Pr}(\mu/\mu_w)^{0.52}$, that is:

$$L_{th}/D_i = 0.039N_{Re}N_{Pr}(\mu/\mu_w)^{0.52} \quad (5)$$

For a constant-property fluid, Gröber, Erk and Grigull [13] give an equation with a dimensionless constant of 0.05.

Heat-Transfer Rate

The thermal analysis of a heat exchanger consists of the integration of a fundamental differential equation, i.e. the heat-transfer rate equation:

$$dq = U(\Delta t)dA \quad (6)$$

For a tubular heat exchanger, the local overall coefficient of heat transfer, U, will depend on (1) inside and outside diameters of the tube, (2) thermal conductance of the material of which the tube is made, (3) the unit resistance for inside and outside dirt and scale, and (4) local individual coefficients of heat transfer.

The heat-transfer coefficient may be defined on the basis of either the outside or inside areas of the tube. By

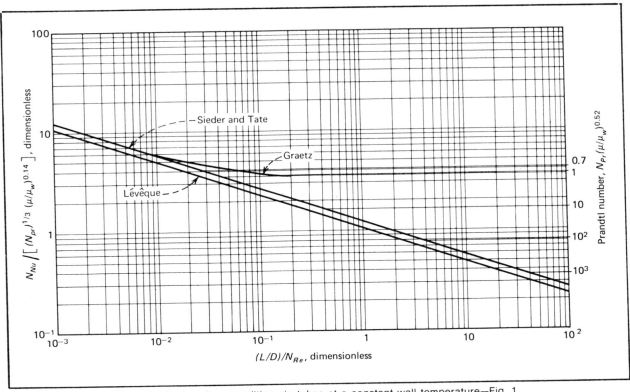

HEAT-TRANSFER relations for laminar-flow conditions in tubes at a constant wall temperature—Fig. 1

convention, it is usually reported on the basis of the outside area. The area, A, in Eq. (6) must always be the one for which the heat-transfer coefficient, U, is reported.

If U is based on the outside area, then:

$$\frac{1}{U_o} = \frac{D_o}{h_i D_i} + R_{d_i}\left(\frac{D_o}{D_i}\right) + \frac{D_o \ln(D_o/D_i)}{2\pi k_w} + R_{d_o} + \frac{1}{h_o}. \quad (7)$$

If U is based on the inside area (as will prove more convenient for designing a heat exchanger with laminar flow on the tubeside), then:

$$\frac{1}{U_i} = \frac{1}{h_i} + R_{d_i} + \frac{D_i \ln(D_o/D_i)}{2\pi k_w} + R_{d_o}\left(\frac{D_i}{D_o}\right) + \frac{D_o}{h_i D_i} \quad (8)$$

The local individual heat-transfer coefficients, h_i and h_o, are usually estimated at the thermal midpoint of the two fluids, and assumed to have this constant value throughout the heat exchanger. This leads to a constant U_m, and Eq. (6) can be integrated by making this assumption, to become:

$$q_T = U_m (\Delta t_{lm}) A \quad (9)$$

As will be shown later, the condition of constant $U = U_m$ is restrictive, and the present treatment allows the heat-transfer rate equation to be solved for many cases of varying local overall coefficients.

At the entrance of a tube, the individual coefficient of heat transfer will approach infinity because the heat-transfer and momentum-boundary layers will be very thin. Only after the fluid has traversed definite distances from the entrance will the heat-transfer and momentum-boundary layers meet at the centerline of the tube, and a developed temperature and velocity profile occur. After the thermal-entrance length has been exceeded, the heat-transfer coefficient will have attained an asymptotic value, and the assumption of a constant local individual heat-transfer coefficient will be satisfactory.

In general, the local coefficient for the entrance region will drop from infinity to the asymptotic value for developed flow and temperature profiles. For a heat exchanger operating in the turbulent regime, the length of the entrance region is usually short compared to the tube length; and since the error will be toward a smaller heat-transfer coefficient than the actual one, it will lead to overdesign and can be tolerated.

Therefore, it has not greatly concerned the heat-transfer engineer except perhaps for very short heat-exchange equipment. When a heat exchanger operates in the laminar regime, the situation is very different. Frequently, entrance conditions prevail throughout the tube—particularly if the fluid has a high viscosity, which is usual when laminar flow is encountered.

Integrating the Heat-Transfer Equation

Eq. (6) cannot be integrated without some further assumptions. First, assume either counterflow or parallel flow. Any modification of these flow geometries—such as crossflow, or simultaneous counterflow and parallel flow—will produce heat-transfer results lying between these two extremes. (An allowance for these can be made by introducing a correction coefficient, usually applied to the mean temperature-difference.) Second, measure the area from the point where the cold fluid enters to the point where it leaves. Third, assume that there is no

Nomenclature

a	$\dfrac{(D_i)^{1-(1/n)}}{C k_f (N_{Re})^{1/n} (N_{Pr})^{1/3} (\mu/\mu_w)^{0.14} \Sigma R}$, $ft^{-1/n}$
A	Heat-transfer surface, ft^2
C	Constant, dimensionless
c_p	Specific heat at constant pressure, $Btu/(lb_m)(°F)$
D	Diameter, ft
h	Local individual coefficient of heat transfer, $Btu/(h)(ft^2)(°F)$
k	Thermal conductivity, $Btu/(h)(ft)(°F)$
L	Distance from entrance of tube, ft
N_{Nu}	Nusselt number $= hD/k$
N_{Pr}	Prandtl number $= c_p\mu/k$
N_{Re}	Reynolds number $= DV\rho/\mu$
n	Integer, dimensionless
q	Rate of heat flow, Btu/h
Q	Rate of heat transfer, Btu/h
R	Unit thermal resistance, $(h)(ft^2)(°F)/Btu$
R_d	Unit dirt-and-scale resistance, $(h)(ft^2)(°F)/Btu$
S	Cross-sectional area, ft^2
t	Bulk temperature, °F
Δt	Temperature difference between hot and cold fluid, °F
U	Local overall heat-transfer coefficient, $Btu/(h)(ft^2)(°F)$
V	Bulk velocity, ft/s
W	Total mass rate of flow, lb_m/h
λ	Enthalpy of phase change, Btu/lb_m
μ	Dynamic viscosity, $lb_m/(ft)(h)$
μ/μ_w	Viscosity ratio

Superscripts

$'$	Cold side
$''$	Hot side

Subscripts

1	At section 1
2	At section 2
I	First integral
II	Second integral
a	At cold end for cold fluid
b	At hot end for cold fluid
f	Referring to fluid
hyd	Hydraulic
i	Inside
lm	Logarithmic mean
m	Mean value of
o	Outside
T	Total
th	Thermal
w	Referring to tubewall

change of phase in either fluid, and that within allowable error the specific heats of the two fluids are constant. We can then write dq in terms of the temperature changes, as:

$$dq = W'c'_p dt' = \pm W''c''_p dt'' \qquad (10)$$

where the positive sign for the extreme-right term is chosen if we are referring to a counterflow heat exchanger, and the negative sign if to a parallel-flow heat exchanger. In Eq. (10), we have made the assumption that the amount of thermal energy is completely transferred from the hot to the cold stream without heat loss to the surroundings.

Since the total mass rates of flow and the specific heats are constant, Eq. (10) is easily integrated from one end of the heat exchanger, a, to the other end, b. Solving for Wc_p yields:

$$W'c'_p = q_T/(t'_b - t'_a) \qquad (11)$$

$$\pm W''c''_p = q_T/(t''_b - t''_a) \qquad (12)$$

Substituting these values into Eq. (10), we can write:

$$dq = q_T\left(\dfrac{dt'}{t'_b - t'_a}\right) = q_T\left(\dfrac{dt''}{t''_b - t''_a}\right) \qquad (13)$$

Anticipating the need for the differential, $d(\Delta t)$, in integrating Eq. (6), Δt is written in terms of its definition as the local driving force and the local temperature difference between the hot and cold streams t'' and t':

$$d(\Delta t) = d(t'' - t') = dt'' - dt' \qquad (14)$$

By substituting the values obtained from Eq. (10) for dt' and dt'', we get:

$$d(\Delta t) = \dfrac{dq}{dT}[(t''_b - t''_a) - (t'_b - t'_a)] = \dfrac{dq}{dT}(\Delta t_b - \Delta t_a) \qquad (15)$$

Solving Eq. (15) for dq, and substituting this value for dq into Eq. (6) yields:

$$\dfrac{q_T}{\Delta t_b - \Delta t_a}\left(\dfrac{d(\Delta t)}{\Delta t}\right) = U dA \qquad (16)$$

Eq. (16) holds for given flowrates, steady-state heat transfer, and the assumptions previously given. The term $q_T/(\Delta t_b - \Delta t_a)$ is a constant. Eq. (16) is a differential equation with separated variables. As such, it may be integrated between limits as indicated by the boundary values at stations (a) and (b):

$$\dfrac{q_T}{\Delta t_b - \Delta t_a}\int_{\Delta t_a}^{\Delta t_b}\dfrac{d(\Delta t)}{\Delta t} = \int_0^A U dA \qquad (17)$$

After integrating and substituting limits and rearranging, we get:

$$q_T = \dfrac{\Delta t_b - \Delta t_a}{\ln(\Delta t_b/\Delta t_a)}\int_0^A U dA \qquad (18)$$

We can further simplify Eq. (18) to:

$$q_T = U_m(\Delta t_{lm})A \qquad (19)$$

in which the following substitutions are made:

$$\Delta t_{lm} \text{ for } (\Delta t_b - \Delta t_a)/\ln(\Delta t_b/\Delta t_a)$$

$$U_m \text{ for } (1/A)\int_0^A U dA$$

Eq. (19) is, of course, the usual integrated form for the heat-transfer rate equation, but it differs in that U_m can be solved for numerous cases where U is known as a function of the area separating the place where U is calculated from the entrance to the tube. Since $A = \pi DL$ for a circular tube, a more general expression is:

$$U_m = (1/L)\int_0^L U dL \qquad (20)$$

Hence, the usual interpretation of U_m as being a mean value of a nearly constant U gives way to the general relation of Eq. (17).

If a phase change occurs in either fluid, Wc_p for the fluid with phase change will be replaced by λdW, where

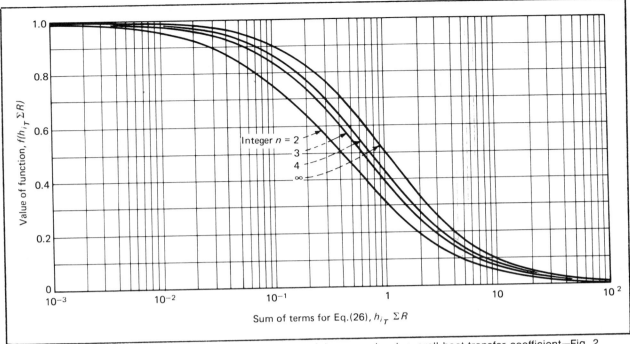

VALUES of the correction function apply to the calculation of the mean, local, overall heat-transfer coefficient—Fig. 2

dW now indicates the mass rate at which the phase is changed, while λ stands for the enthalpy change when unit mass is transformed. A boiling or condensing section must be treated as a unit distinct from a single-phase section. These cannot be handled with an overall Δt_{lm} and U_m for both single-phase and multiphase flow because we assumed constant heat capacity, Wc_p, in deriving Eq. (19). If the heat capacity varies appreciably, we must perform the calculation by finite steps, and the entering and leaving temperatures of both fluids must be calculated for each element. The procedure is described in some detail by Kreith [14].

Computing the Mean Coefficient

The mean overall heat-transfer coefficient that must be used in Eq. (19) is defined by Eq. (18), in which the definite integral may be evaluated as the sum of several successive integrals. This makes it possible to calculate the value of the total integral even though the local overall heat-transfer coefficient, U, becomes a different function of the length from the tube entrance.

For our purpose, let us assume that (1) as long as the distance from the entrance is shorter than the thermal entrance length, the local heat-transfer coefficient, h_i, will vary inversely as the length from the entrance, L, raised to a power $1/n$; and (2) after the value of the coefficient becomes equal to that for developed flow (i.e., after the thermal entrance length, L_{th}, has been exceeded), h_i will remain independent of the tube length. Hence, we may write:

$$U_{i_m} = \frac{1}{L_T}\left[\int_0^{L_{th}} U_{i_I} dL + \int_{L_{th}}^{L_T} U_{i_{II}} dL\right] \quad (21)$$

where the local individual heat-transfer coefficient, h_{i_I},

used in calculating U_{i_I} after the first integral sign may be written as:

$$h_{i_I} = \frac{Ck_f(N_{Re})^{1/n}}{L^{1/n}D^{1-(1/n)}}(N_{Pr})^{1/3}(\mu/\mu_w)^{0.14} \quad (22)$$

For the first integral in Eq. (21), we may then write the following equation, if we assume that the total length of the tube falls short of the thermal entrance length:

$$U_{i_m} = \frac{1}{L_T}\int_0^{L_T} U_i dL$$
$$= \frac{1}{L_T}\int_0^{L_T} \frac{dL}{\frac{1}{h_i} + R_{d_i} + \frac{D_i \ln(D_o/D_i)}{2\pi k_w} + \left(\frac{D_i}{D_o}\right)R_{d_o} + \frac{D_i}{h_o D_o}} \quad (23)$$

After factoring out the sum of the fixed thermal resistances, ΣR, Eq. (23) becomes:

$$U_{i_m} = \frac{1}{L_T \Sigma R}\int_0^{L_T} \frac{dL}{1 + 1/(h_i \Sigma R)} \quad (24)$$

where:

$$\Sigma R = R_{d_i} + \frac{D_i \ln(D_o/D_i)}{2\pi k_w} + \left(\frac{D_i}{D_o}\right)R_{d_o} + \frac{D_i}{h_o D_o}$$

Let us temporarily introduce the quantity a (as defined below) into Eq. (22), and substitute the resulting value of h_{i_I} into Eq. (24), to yield:

$$U_{i_m} = \frac{1}{L_T \Sigma R}\int_0^{L_T} \frac{dL}{1 + aL^{1/n}} \quad (25)$$

where:

$$a = \frac{D_i^{1-1(1/n)}}{Ck_f(N_{Re})^{1/n}(N_{Pr})^{1/3}(\mu/\mu_w)^{0.14}\Sigma R}$$

The integral in Eq. (25) can now be solved for integral values of n by substituting a new variable,

Solutions of the Integral in Eq. (25) for Several Values of the Integer—Table III

Integer, n	$\int_0^{L_T} \dfrac{dL}{1 + aL^{1/n}}$
2	$\dfrac{2(L_T)^{1/2}}{a}\left[1 - \dfrac{\ln[1 + a(L_T)^{1/2}]}{a(L_T)^{1/2}}\right]$
3	$\dfrac{3(L_T)^{2/3}}{2a}\left[1 - \dfrac{2}{a(L_T)^{1/3}} + \dfrac{2\ln[1 + a(L_T)^{1/3}]}{[a(L_T)^{1/3}]^2}\right]$
4	$\dfrac{4(L_T)^{3/4}}{3a}\left[1 - \dfrac{3}{a(L_T)^{1/4}} + \dfrac{3}{[a(L_T)^{1/4}]^2} - \dfrac{3\ln[1 + a(L_T)^{1/4}]}{[a(L_T)^{1/4}]^3}\right]$
∞	$\dfrac{L_T}{a}\left[\dfrac{a}{1+a}\right]$

$y = 1 + aL^{1/n}$. Since $dL = (n/a)y^{n/(n-1)}dy$, we can now find an expression for the integral. The results are shown in Table III for several values of n.

A general solution for the integral may be written:

$$\int_0^{L_T} \frac{dL}{1 + aL^{1/n}} = \frac{n}{n-1}\left(\frac{L_T^{1-(1/n)}}{a}\right)\left[1 - \frac{n-1}{n-2}\left(\frac{1}{aL_T^{1/n}}\right) + \frac{n-1}{n-3}\left(\frac{1}{(aL_T^{1/n})^2}\right) - \frac{n-1}{n-4}\left(\frac{1}{(aL_T^{1/n})^3}\right) \cdots \frac{(n-1)\ln(1 + aL_T^{1/n})}{(aL_T^{1/n})^{n-1}}\right] \quad (26)$$

where n is an integer. The sum of the terms within the brackets will be denoted by $\phi(aL_T^{1/n})$ or by $f(h_{i_T}\Sigma R)$, since $aL_T = 1/(h_{i_T}\Sigma R)$. Values of this function are tabulated in Table IV for various values of $h_{i_T}\Sigma R$ and n, and the function is shown graphically in Fig. 2.

Substituting the general value of the integral from Eq. (26) into Eq. (25), we get:

$$U_{i_m} = n/n(n-1)h_{i_T}k_f(N_{Re}D_i/L)^{1/n}(N_{Pr})^{1/3}(\mu/\mu_w)^{0.14} \quad (27)$$

$$U_{i_m} = n/(n-1)h_{i_T}f(h_{i_T}\Sigma R) = h_{i_{mT}}f(h_{i_T}\Sigma R) \quad (28)$$

where h_{i_T} is the local individual heat-transfer coefficient at the exit of the tube, and $h_{i_{mT}}$ is the mean individual coefficient.

If the total length of the tube exceeds L_{th}, the mean overall heat-transfer coefficient will be:

$$U_{i_m} = \frac{L_{th}}{L_T}\left(\frac{n}{n-1}\right)h_{i_{th}}f(h_{i_{th}}\Sigma R) + \left(\frac{L_T - L_{th}}{L_T}\right)h_{i_\pi}f(h_{i_T}\Sigma R) \quad (29)$$

The local individual heat-transfer coefficient may be represented more accurately by using several line segments with appropriate slopes. In this case, the lower limits of the integrals must be introduced, and both h_i as well as $f(h_{i_T}\Sigma R)$ must be evaluated at the lower and upper limits. The difference between U_{i_m} at the upper limit and U_{i_m} at the lower limit is multiplied by the fraction of the tube length to yield the contribution of each segment.

Application of Theory

The procedure is simple in that first a solution is obtained by assuming that it lies on the sloping branch of the curve in Fig. 1. If the thermal-entrance length is exceeded, the heat transfer for that length is calculated; subsequently, the remaining length is calculated by using the appropriate constant heat-transfer coefficient.

Example 1—A counterflow oil cooler is to be designed to cool a light hydrocarbon from 135°F to 115°F. The oil flows at a velocity of 2 ft/s inside a tube that is cooled with water entering at 70°F and leaving at 90°F. The exchanger will be built from a ¾-in O.D. (0.620-in I.D.), 16 BWG, copper tube surrounded by a 1½-in stainless-steel tube. The sum of the resistances through the tubewall and at the outside of the tube are assumed to be 1/200 (h)(ft²)(°F)/Btu. The fluid properties of the oil are:

Temperature, °F	125	100	80
c_p, Btu/(lb$_m$)(°F)	0.47	0.46	0.44
k, Btu/(h)(ft)(°F)	0.075	0.076	0.077
μ, lb$_m$/(ft)(s)	0.0090	0.0153	0.0278
ρ, lb$_m$/ft³	55	56	57

How long must the heat exchanger be to perform satisfactorily?

Let us assume that the solution will be completely on the Sieder-Tate line. The equation for the rate of heat transfer is then:

$$qT = (\pi D_i L)(h_{i_m})\Delta t_{lm} f(h_{i_T}\Sigma R)$$
$$= 1.86\pi k(N_{Re}N_{Pr}D_i^{1/3}(\mu/\mu_w)^{0.14}L^{2/3}\Delta t_{lm}f(h_{i_T}\Sigma R) \quad (30)$$

Mean fluid properties of the oil will be taken at 125°F, and the wall temperature as 80°F.

Solutions for the Function Comprising the Bracketed Terms in Eq. (26)—Table IV

Function, $h_{i_T}\Sigma R$	Integer, n			
	2	3	4	∞
0.001	0.9931	0.9980	0.9985	0.9995
0.002	0.9876	0.9961	0.9970	0.9980
0.005	0.9735	0.9903	0.9926	0.9950
0.01	0.9539	0.9809	0.9853	0.9910
0.02	0.9214	0.9632	0.9711	0.9804
0.05	0.8478	0.9152	0.9304	0.9524
0.1	0.7602	0.8480	0.8728	0.9091
0.2	0.6417	0.7433	0.7770	0.8333
0.5	0.4507	0.5493	0.5880	0.6667
1	0.3069	0.3863	0.4206	0.5000
2	0.1891	0.2437	0.2688	0.3333
5	0.0884	0.1161	0.1294	0.1667
10	0.0469	0.0620	0.0695	0.0909
20	0.0242	0.0321	0.0361	0.0476
50	0.0099	0.0131	0.0148	0.0196
100	0.0050	0.0066	0.0074	0.0099

We will now proceed to calculate the numerical values for the several quantities for Eq. (30):

$$w = \rho VS = 55(2 \times 3,600)(\pi/4)(0.620/12)^2 = 830.2 \text{ lb}_m/h$$

$$q_T = wc_p(t_2 - t_1) = 830.2(0.47)(135 - 115) = 7,804 \text{ Btu/h}$$

$$N_{Re} = \frac{DV\rho}{\mu} = \frac{0.620(2)55}{12(0.0090)} = 631 \text{ (laminar flow)}$$

$$N_{Pr} = \frac{c_p\mu}{k} = \frac{0.47(0.0090)(3,600)}{0.075} = 203$$

$$\frac{\mu}{\mu_w} = \frac{0.0090}{0.0278} = 0.324, \left(\frac{\mu}{\mu_w}\right)^{0.14} = 0.854, \left(\frac{\mu}{\mu_w}\right)^{0.42} = 0.623$$

$\Delta t_{lm} = 45°F$ because $\Delta t_1 = \Delta t_2 = 45$

Solving Eq. (3) for h_i, yields:

$$h_i = 1.24\left(\frac{k}{D_i}\right)(N_{Re}N_{Pr}D_i)^{1/3}\left(\frac{\mu}{\mu_w}\right)^{0.14}\left(\frac{1}{L}\right)^{1/3}$$

$$h_i = 1.24\left[\frac{0.075(12)}{0.620}\right]\left[(631)(203)\left(\frac{0.62}{12}\right)\right]^{1/3}(0.854)\left(\frac{1}{L}\right)^{1/3}$$

$$h_i = 1.24(1.45)(18.77)(0.854)/L^{1/3} = 28.8/L^{1/3}$$

$$h_{i_T}\Sigma R = 28.8/200L^{1/3} = 0.144/L^{1/3}$$

Substituting the appropriate numerical values into Eq. (30) and simplifying, yields:

$$7,804 = 316L^{2/3}f\left(\frac{0.144}{L^{1/3}}\right)$$

$$L^{2/3} = \frac{24.7}{f\left(\frac{0.144}{L^{1/3}}\right)}$$

$$L = 123/[f(0.144/L^{1/3})]^{2/3}$$

This last equation is easily solved by successive approximations. As a first approximation, let the value of the function equal 1, then $L = 123$ ft. Using this value, we now find that the function equals 0.029. From the graph of Fig. 2 or from Table IV, when the function is 0.029, its value becomes 0.926. Hence the second approximation produces a value for L of $123/0.926 = 132.5$. After a third approximation, we find that the value for L is about 132 ft.

We now use Eq. (5) to determine the thermal-entrance length, as follows:

$$L_{th} = 0.039(631)(203)(0.623)(0.62/12) = 161 \text{ ft}$$

In this example, the length of the heat exchanger ($L = 132$ ft) does not exceed the thermal-entrance length.

Example 2—All conditions remain the same as in Example 1 except that the oil is now to be cooled from 140°F to 110°F. Find the length of the exchanger for this case.

Since the log mean-temperature difference is nearly the same, 44.8°F instead of 45°F, and q_T is:

$$q_T = 830.2(0.47)(140 - 110) = 11,706 \text{ Btu/h}$$

the needed length will exceed the thermal-entrance length.

Heat transfer in the thermal-entrance section becomes:

$$q_{th} = 316\left(\frac{44.8}{45}\right)(L_{th})^{2/3}f\left[\frac{0.144}{(L_{th})^{1/3}}\right] = 315(161)^{2/3}f\left[\frac{0.144}{(161)^{1/3}}\right]$$

$$q_{th} = 315(29.6)(0.95) = 8,860 \text{ Btu/h}$$

The balance of the heat transfer must take place in the developed flow regime where:

$$h_i = 3.66(k/D_i) = 3.66(0.075)(12/0.620) = 5.31$$

Therefore, the local overall heat-transfer coefficient, U_{II}, is calculated as:

$$U_{II} = \left(\frac{1}{h_i} + \Sigma R\right)^{-1} = \left(\frac{1}{5.31} + \frac{1}{200}\right)^{-1} = 5.17 \text{ Btu/(h)(ft}^2)(°F)$$

The amount of heat to be removed in the developed flow regime is the difference between the total heat and that removed in the thermal-entrance section, or $11,706 - 8,860 = 2,846$ Btu. We can now compute the length of this section from:

$$q = \pi D_i L(\Delta t_{lm})U_{II}$$

$$2,846 = \pi(0.620/12)L(44.8)(5.17)$$

$$L = 2,846/37.6 \approx 76 \text{ ft}$$

Hence, for this example, the total length of the heat exchanger will be the sum of L_{th} and L, or $161 + 76 = 237$ ft.

These examples illustrate the application of the derived equations. In practice, it would frequently be advantageous to introduce mixing chambers, and to have the heat exchanger operate under initial entrance conditions where the heat-transfer coefficients are high. #

References

1a. Graetz, L., *Ann. Physik*, **18**, 79 (1883).
1b. Graetz, L., *Ann. Physik*, **25**, 337 (1885).
2. Nusselt, W., *Z. Ver. Deut. Ingr.*, **54**, 1154 (1910).
3. Pohlhausen, K., *Z. Angew. Math. Mech.*, **1**, 252 (1921).
4. Lévêque, A., *Ann. Mines*, **12**, No. 13, 201, 305 and 381 (1928).
5. Drew, T. B., *Trans. AIChE*, **26**, 26 (1931).
6. Sellars, J. R., Tribus, M. and Klein, J. S., *Trans. ASME*, **78**, 441 (1956).
7. Kays, W. M., *Trans. ASME*, **77**, 1265 (1955).
8. Sieder, E. N. and Tate, G. E., *Ind. Eng. Chem.*, **28**, 1429 (1936).
9. Hausen, H., *Z. VDI Beih. Verfarenstech.*, **4**, 91 (1943).
10. Schiller, L., *Z. Angew. Math. Mech.*, **2**, 96 (1922).
11. Prandtl, L. and Tietjens, O., "Hydro- und Aeromechanik," Vol. 2, p. 28, Springer, Berlin, 1931.
12. Langhaar, H. L., *Trans. ASME*, **64**, p. A-55 (1942).
13. Gröber, H., Erk, S. and Grigull, U., "Fundamentals of Heat Transfer," 3rd ed., p. 234, McGraw-Hill, New York, 1961.
14. Kreith, F., "Principles of Heat Transfer, 3rd ed., pp. 573-575, Intext Educational Publishers, New York, 1973.

Meet the Author

Niels Madsen is professor of chemical engineering at the University of Rhode Island, Kingston, RI 02881. Since coming there, he has spent two years as visiting professor at the Technological University of Eindhoven, The Netherlands. Previously, he was a chemical engineer with Foster Wheeler Corp. His principal interest is in heat transfer, particularly boiling heat transfer and two-phase flow. He has a B.Ch.E. from Cooper Union, an M.S. from Stevens Institute of Technology and a Ph.D. from Columbia University. He is a Fellow of AIChE, a member of the American Soc. for Engineering Education and of Sigma Xi, and a licensed professional engineer in Rhode Island.

Design Parameters for Condensers and Reboilers

Operating characteristics, equipment internals, liquid velocity and distribution, vertical or horizontal position, and filmwise or nucleate-boiling heat transfer are among the many factors that influence the design of condensers and reboilers.

R. C. LORD, P. E. MINTON and R. P. SLUSSER, Union Carbide Corp.

Condensers and reboilers are two of the most important types of heat-transfer equipment used by the chemical process industries. Some of the factors that influence their design and performance will be reviewed here.

A method for the thermal design of heat exchangers was discussed in an earlier article.[1] This method may also be used for the following types of condensers and reboilers:
- Total condensers (condensate may contain components that are either miscible or immiscible).
- Total condensers with subcooling of the condensate (miscible or immiscible).
- Partial condensers.
- Condensing (noncondensable gases present).
- Natural-circulation calandrias (thermosyphon reboilers).
- Forced-circulation calandrias.
- Horizontal kettle-type reboilers.

Condensers

Vertical vapor-in-tube downdraft condensers have several advantages over horizontal condensers. Many of the same advantages are offered by vertical vapor-in-shell condensers that use baffles designed to permit condensate to remain on the tube. The advantages of and differences between vertical and horizontal condensers are:

1. Condensate subcooling is more efficiently accomplished due to falling-film heat transfer. Horizontal tubeside cooling uses only a small portion of the available area. Appreciable horizontal shellside subcooling can be done only by flooding part of the shell, unless the exchanger is designed so that the vapor passes vertically across the tube bundle only once (horizontal one-pass crossflow), with subcooling occurring in the falling-film mode. With a flooded shell, vent losses are higher.

2. Vapors enter the tube at relatively high velocities affording good distribution, producing high film coefficients by reducing condensate thickness. Noncondensable gases are forced to the outlet, and subsequent gas binding is prevented.

3. Better heat transfer is obtained at high condensate loading. The Dukler correlation (to be discussed later) applies for vertical tubes. High heat-transfer coefficients are obtained with high loadings and Prandtl numbers. Vertical exchangers are generally more economical than horizontal condensers employing low-finned tubes.

4. Better heat transfer is achieved in knockout condensers because the condensate and uncondensed vapors and gases always coexist. For horizontal vapor-in-shell condensers, this is not true because the condensate drops from the tubes as it is formed.

5. Countercurrent flow is accomplished, resulting in maximum possible subcooling. Noncondensable gases, which always exist in a condenser, are contacted with the lowest available temperature before removal. Vent losses are, therefore, at a minimum.

6. Fluted tubes can be used. These improve heat transfer via a mechanism to be described later.

7. Pressure drop is generally lower.

8. More-accurate control is afforded.

9. Shellside baffling can provide higher cooling-medium coefficients.

Vapors Containing Noncondensable Gases

The presence of noncondensable gases in the vapors decreases the film coefficient. This reduction depends

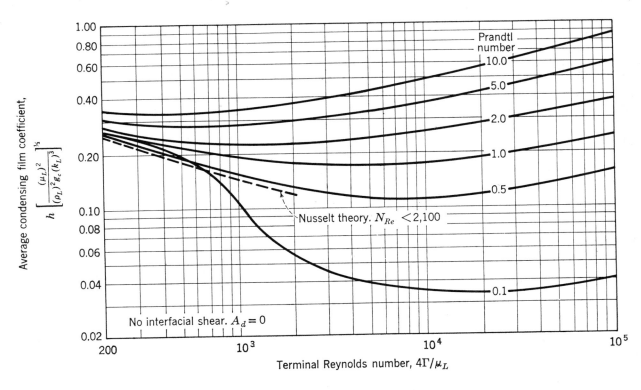

DUKLER PLOT gives average condensing-film coefficients at various Prandtl numbers—Fig. 1

upon the relative sizes of the gas-cooling load, and the total cooling and condensing duty.

One mechanism for the analysis of condenser-coolers assumes that the latent heat of condensation is transferred only through the condensate film, whereas the gas-cooling duty is transferred through both the gas and condensate films. This mechanism yields good results for cases in which the condensate and cooling vapor-gas mixture are in substantial equilibrium, such as in vertical, vapor-in-tube or horizontal, one-pass, crossflow condenser-coolers. This mechanism applies only when the condensing surface is colder than the dewpoint of the mixture.

Therefore, the following equations may be written:

$$\frac{(Q/\theta)_g}{h_g} + \frac{(Q/\theta)_T}{h_c} = A \Delta t_o$$

$$\frac{(Q/\theta)_T}{A \Delta t_o} = h_{cg}$$

From these equations, we obtain:

$$h_{cg} = \frac{1}{\frac{(Q/\theta)_g}{(Q/\theta)_T}\left(\frac{1}{h_g}\right) + \frac{1}{h_c}}$$

Some comments on these equations must be made:

1. Sensible heat $(Q/\theta)_g$ includes the sensible heat for cooling the noncondensable gases, plus the vapors that remain in the vent stream and are subsequently cooled.

2. Total heat $(Q/\theta)_T$ includes the gas-phase sensible heat $(Q/\theta)_g$, plus the latent heat of condensation of the vapor that condenses, plus the sensible heat of the liquid-phase subcooling.

3. The gas-film coefficient, h_g, is calculated by using a mean mass velocity based on the entering and exit gas-vapor rates.

4. The condensing-film coefficient, h_c, is calculated for the loading rate represented by the condensate film.

5. Value of the net-effective coefficient, h_{cg}, will be between the values for the gas and condensing coefficients.

6. The greater the ratio of sensible heat to total heat, the more closely the net-effective coefficient approaches the gas-film coefficient.

7. The gas-film coefficient is usually controlling. Therefore, higher gas velocities result in higher net-effective coefficients.

When the condensing-cooling curves are other than straight lines, the mean temperature difference will vary from the logarithmic-mean temperature difference. For such cases, the condensing curve must be known, and the variance between the actual and log-mean temperature differences calculated. In most cases, the actual condensing curve will yield temperature differences that are greater than the log mean obtained with straight-line condensing.

Provided the condensing surface is cooler than the saturation temperature of the steam, superheat is removed by a mechanism of condensing and re-

Nomenclature

A	Heat-transfer area, sq. ft.
A_o	Crest area, sq. ft.
D	Dia., ft.
d_i	Inside tube dia., in.
G	Mass velocity, lb./(sq. ft.)(hr.)
g_c	Gravitational constant, 4.18×10^8 ft./hr.^{2}
h	Film coefficient of heat transfer, Btu./(hr.)(sq. ft.)(°F.)
k_L	Liquid thermal conductivity, Btu.(hr.)(sq. ft.)(°F./ft.)
K_1	Dimensionless number in Fig. 3
K_2	Dimensionless number in Fig. 3
L	Length, ft.
l	Height of liquid in tube, in.
n	Number of tubes per pass, or in parallel
N_{Nu}	Nusselt number
Q	Heat transferred, Btu.
Q_F	Flowrate, gpm.
q	Flowrate, cu. ft./sec.
r_o	Crest radius, in.
R_o	Trough radius, in.
s	Specific gravity
Δt_o	Temperature difference across sensible and condensing film.
Δt_c	Temperature difference across condensing film
v	Superficial vapor velocity, ft./sec.
W_1	Flowrate, lb/(hr.)(1,000)
a	Contact angle, deg.
Γ	Tube loading, lb./(hr.)(ft.)
θ	Time, hr.
ρ_L	Density of liquid, lb./cu. ft.
ρ_V	Density of vapor, lb./cu. ft.
ρ_G	Density of gas, lb./cu. ft.
μ_L	Viscosity, lb./(hr)(ft.)
μ	Viscosity, cp.
λ	Latent heat of vaporization, Btu./lb.
σ_o	Surface tension, lb./ft.
σ	Surface tension, dynes/cm.

Subcripts

c	Condensing
g	Gas
cg	Net effective
T	Total

evaporation. This mechanism maintains the condensing-surface temperature at essentially the same level as that obtained with saturated steam. For superheated steam (within limits at constant heat level), condensate loading is lower and, therefore, the condensing coefficient is slightly higher. Desuperheating occurs at the expense of the temperature difference available between the superheat temperature and the saturation temperature.

Updraft vs. Downdraft Condensers

With few exceptions, downdraft condensers provide better heat transfer. Occasionally, updraft condensers are used when absolutely no subcooling is acceptable or when the remaining vapors will not condense in the presence of a previously obtained condensate.

The primary disadvantages of updraft condensers are associated with flooding. In an updraft condenser, the liquid must fight with incoming vapor in order to escape from the tubes. If sufficient area is not available for both liquid and vapor, surging will occur, with slugs of liquid building up and escaping. This results in a serious control problem.

Several methods for predicting flooding velocities are available. The best one involves using the data of Holmes.[2] For the same vapor rate, the liquid-handling capacity of a tube can be increased 40% by tapering the inlet end 60° or 75°.

Dukler Correlation

The Dukler theory[3] predicts heat tranfer for vertical condensers. Coefficients are much higher than predicted by the Nusselt theory or by the Colburn equation. According to the Dukler theory, three fixed factors must be known to establish the value of the average film coefficient: (1) terminal Reynolds number, (2) Prandtl number of the condensate, and (3) a dimensionless group designated A_d and defined as:

$$A_d = \frac{0.250(\mu_L)^{1.173} G^{0.16}}{g^{2/3} D^2 (\rho_L)^{0.553} (\rho_G)^{0.78}}$$

For the case of no interfacial shear ($A_d = 0$), the graph of Fig. 1 applies.

Fluted Tubes

The fluted condensing surface is described by Gregorig,[4] and has a profile similar to that shown in Fig. 2. Surface tension of the curved liquid-vapor

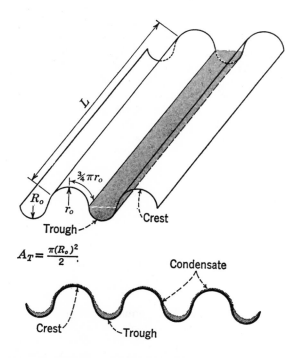

$$A_T = \frac{\pi (R_o)^2}{2}$$

FLUTED condensing surface of Gregorig's[4]—Fig. 2

interface produces a large excess pressure in the condensate film adjacent to the crests of the flutes. This causes a thinning of the film in that region, resulting in very high local heat transfer. The surface-tension mechanism causes the condensate to accumulate in the troughs. Condensate is removed by flowing vertically downward into the troughs.

Enough condensate accumulates in the troughs within a short distance from the top of the tube to make heat transfer there negligible. Thus, heat transfer, averaged around the circumference of the tube, is essentially independent of tube length as long as the troughs are not flooded.

Although only a portion of the tube surface is available for condensation, the mean heat-transfer coefficient over the total surface is 5 to 10 times that of a smooth tube of equivalent length. A similar enhancement may be obtained by loosely attaching vertical wires to a vertical tube.[5]

Average coefficients, based on superficial area, were calculated by Gregorig by finite-difference calculus for the condensation of steam for two flute geometries, and the results were confirmed by tests.

Of the two geometries, one was milled and the other drawn. Since tolerances obtained by drawing are not as close as in milling, a less than ideal profile was accepted for the drawn tube. Experimental performance of both tubes was in good agreement with that predicted, indicating that close tolerances for the fluted profile are not mandatory. Experimental data for other fluids, and different flute geometries, have further confirmed Gregorig's original calculations.

Gregorig correlated his results in terms of dimensionless numbers. These are given in Fig. 3, along with a general solution to condensing on fluted surfaces.

Assuming laminar flow, the length of a tube at which the troughs flood can be calculated from the Bernoulli equation. The crest may be assumed to be an ellipse with a perimeter as shown in Fig. 2. From these assumptions, we obtain:

$$L = \frac{(R_o)^4 \lambda \rho^2 g_c}{24 \mu_L (Q/A_c \theta) r_o}$$

Flooding in Horizontal Tubes

Condensate partially fills a horizontal tube. When subcooling, the depth of liquid in the tube can be calculated in order to determine the heat-transfer coefficients. The following diagram illustrates the relationship of the brink depth l to the inside diameter of the tube d_i:

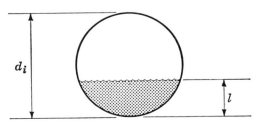

An equation[6] for calculating the height of liquid at the outlet of a horizontal tube is:

$$q = 9.43 \left(\frac{d_i}{12}\right)^{2.56} H^{1.84}$$

$$H = l/d_i$$

where q is flowrate, cu.ft./sec.; d_i is tube I.D., in.; and l is brink depth, in. The height is greater at other points along the tube. Therefore, the value of H should not exceed 0.5, and preferably should be less than 0.3. When H is 0.25, the value of $W_1/ns(d_i)^{2.56}$ is 0.3.

This criterion provides a guide to determine whether proper venting will occur, and permits calculation of subcooling heat transfer for horizontal tubes. If a more rigorous approach is desired, two-phase flow correlations may be used to predict the proper flow regime.

Condensate Connections

Condensate must often flow by gravity from the condenser. Hence, nozzles must be properly sized to ensure that the condensate line will be self-venting. Nozzles should also be sized to prevent excessive liquid levels, especially in horizontal, vapor-in-tube condensers. An equation for finding proper nozzle size is:

$$d_i = 0.92(Q_F)^{0.4}$$

where d_i is nozzle I.D., in.; and Q_F is flowrate, gpm. This equation predicts nozzles that act as circular weirs. No whirlpool will exist; the line will be self-venting; and ratio of liquid height above the nozzle to its diameter will be 0.25.

The amount of condensate held back by shellside baffles in a horizontal condenser can become sufficient to flood a good percentage of the tubes in the bundle. This can easily happen in small-diameter shells carrying a large condensate flow. The height of liquid in the shell can be estimated from weir formulas. Sometimes, it will be necessary to alter the baffle design or to provide several condensate outlets for proper operation of the condenser.

Constant-Pressure Vent Systems

Inert gases introduced in the condenser to maintain a constant pressure in the condenser and distillation column provide good control. The gases are either injected or vented to maintain the desired pressure. In effect, they blanket part of the heat-transfer surface when it is not needed for condensing or subcooling. Without such a system, the pressure in the condenser would reflect the vapor pressure of the subcooled condensate, and the column pressure would not remain constant. With a proper vent system, the pressure drop is only that due to friction and does not reflect subcooling of condensate.

Other means of control can be used but the response is inferior to that of a constant-pressure vent system. Sometimes, the process inerts are allowed to accumulate to provide a sort of constant-pressure vent system. Control is sluggish, especially when the condenser load is decreased.

Control can also be achieved by flooding part of the condenser with condensate. This procedure is also sluggish when loads are decreased. Increasing loads can also cause sluggish control, if the driving force for fluid flow is small. Venting is inadequate, and only the inerts soluble in the condensate will be removed. This is especially undesirable when the inerts are corrosive.

Column-Mounted Condensers

Column-mounted (integral) condensers are often used when condensing under vacuum. Integral condensers eliminate large vapor lines and reflux pumps. Shell-and-tube, spiral, air-cooled, and bayonet exchangers can be used as integral condensers. Shell-and-tube exchangers are frequently constructed to provide a single vapor pass across a horizontal tube bundle. This construction has the advantages of low pressure-drop, good vapor distribution, and possibility of subcooling the condensate.

Integral condensers are not limited to vacuum service. If no steel superstructure exists, they can provide the necessary hydraulic head for gravity-flow systems. They are also used to eliminate structural steel, and when the refluxed condensate is difficult or hazardous to pump.

Vaporizers and Reboilers

Natural-circulation calandrias (thermosyphon reboilers) depend upon density differences to produce required flowrates. Vaporization creates an aerated liquid with a density less than that of the liquid in the system. The hydraulic head resulting from this density difference circulates the fluid in the system. Circulation rates are high, with liquid-to-vapor ratios ranging from 1 to 1 to 50 to 1.

Boiling on the tubeside has many advantages over boiling on the shellside. Since calandrias are often used in fouling services, the tubeside can be more readily cleaned. Tubeside distribution is more uniform, and pressure drop is lower. Consequently, heat transfer is higher because of the resulting higher circulation rates.

Vertical units provide more hydraulic head and higher circulation rates. However, such units have higher boiling-point elevations. To offset this, they can be inclined but not less than 15° from horizontal. Horizontal units frequently have poor distribution and vapor binding, especially when boiling is on the shellside. The vertical boiling-in-tube unit is generally the most economical choice, both in first and operating costs. The greater boiling-point elevation is seldom a serious penalty when all the advantages are weighed.

Liquid levels are generally maintained at the top tubesheet. Greater capacity can be obtained by lower-

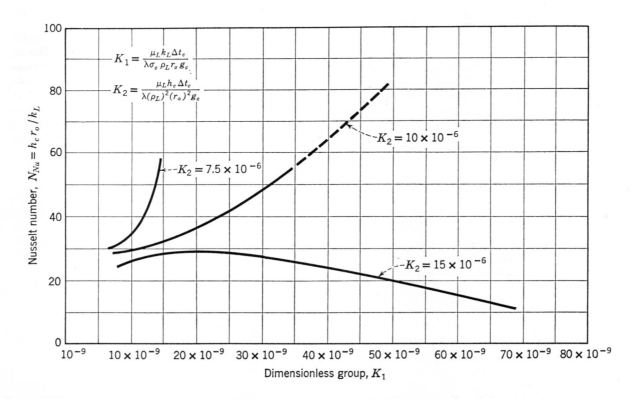

NUSSELT numbers for condensing on fluted tubes, from the data of Gregorig[4]—Fig. 3

DESIGN PARAMETERS FOR CONDENSERS AND REBOILERS

No re-entrant angles in either channel. Smooth transition from shell diameter to nozzle diameter.

$D_V = d_i \sqrt{N}$

$D_L = \tfrac{1}{2} D_V$

D_V = Vapor outlet nozzle I.D., in.
D_L = Liquid inlet nozzle I.D., in.
d_i = Tube I.D., in.
N = Number of tubes in bundle

CHANNEL and nozzle criteria for the design of natural-circulation calandrias produce high capacities—Fig. 4

ing the liquid level—to a point. Beyond this, the top portion of the tubes operates free of liquid, and the heat transferred is used in superheating the vapor. The increase in capacity obtained by lowering the liquid level to this point is modest, but the decrease at levels beyond this point is rather drastic. Lower liquid levels may require surge capacity in the base of the column. Such levels also reduce circulation rates, which in turn increase fouling and maintenance, especially if the top portion of the tubes runs dry.

The circulation rate in calandrias may not be sufficient to provide all the volatile component required for the boilup. When this is the case, surging occurs. The calandria will operate until all the circulated volatile component is vaporized. At this point, the boiling point of the remaining liquid increases, and the vapor rate is greatly decreased. Consequently, not enough vapor is produced to satisfy the column requirements, and some of the liquid on the trays is dumped to the base. This replenishes the volatile component; the vapor rate increases; and column requirements are satisfied. Eventually the concentration of the volatile component will be decreased to the point where dumping will reoccur. Under this condition, the column cannot be properly controlled.

The mechanism of operation of natural-circulation calandrias is quite complicated. The final design is based on the net effect of several interdependent variables. Because of the nature of these variables, natural-circulation calandrias are not advantageous for vacuum service. Because of the pronounced rise in boiling point due to hydrostatic head, the temperature difference is low, the sensible-heat area (low coefficient) is large, the boiling area is small, and the circulation rate is low. If an attempt is made to reduce the effect of hydrostatic head by reducing tube length, the result is an uneconomical design that consists of many tubes in a large-diameter shell.

Design of the top channel and nozzle (Fig. 4) is one of the most important variables affecting the performance of natural-circulation calandrias. An improperly designed top head can reduce the unit's capacity by 40%. The top channel should be constructed to give a smooth transition from shell diameter to nozzle diameter. Reentrant angles should be eliminated to reduce eddies and internal recirculation. Area of the vapor-outlet nozzle should be approximately equal to the flow area of the tubes. Diameter of the liquid-inlet nozzle should be one-half that of the vapor-outlet nozzle.

When the shell diameter is large, the required vapor nozzle is large, and is quite some distance above the tubesheet. A high percentage of the available driving force may be required to lift the two-phase mixture into the column. When shell diameters exceed 72 in., it is often better to design the calandria to be integral with the base of the column. This will give a more economical design because circulation rates will be higher, and no inlet or outlet heads will be required. Economies often result if one large unit is fabricated instead of several smaller ones.

In general, for steam-heated, natural-circulation calandrias, the saturated-steam temperature should not be more than 50° (C.) above the process boiling temperature. Large temperature differences between the heating medium and the boiling liquid may result in a vapor film being formed on the boiling surface. When boiling inside a tube, the fluid velocity caused by the two-phase condition helps to remove the vapor film from the tube wall. However, there will be cases in which the fluid velocity may be low, and the highest flux for vertical tubes must be predicted. The maximum velocity at which a vapor escapes from a vertical surface is three-fourths of the value for a horizontal surface.[9]

Natural-circulation calandrias are generally heated by condensing a vapor. The best control is achieved by controlling the condensing pressure and, therefore, the temperature difference. Control can also be done by flooding part of the surface with condensate. The problems associated with this procedure have been discussed previously. When steam is condensed in a steel exchanger, the problems are even greater. Carbon dioxide, usually present in steam, will cause extensive corrosion if not properly vented.

Forced-Circulation Calandrias

No vaporization occurs in forced-circulation calandrias. The process fluid is recirculated by pumping, heated under pressure to prevent boiling, and subsequently flashed to obtain the required vaporization. Forced-circulation calandrias are used in vacuum service, or when a specific velocity is needed to reduce fouling.

Circulating pumps should be selected so that the developed head is dissipated as pressure drops through the system. Circulation rates are determined by the allowable temperature rise of the pumped fluid, or the back-pressure needed to prevent boiling at the outlet temperature of the fluid. The desired velocity may at times determine the circulation rate.

It is important that the pump and system match. The fluid being pumped is at or near its boiling point, and the required NPSH (net positive suction head) may be critical. The pump should operate at its design level. If it develops excessive head, it will handle more volume at a lower head. At the new operating point, the required NPSH may be more than is available, and cavitation will occur in the pump. If insufficient head is provided, the velocities may not be sufficiently high to prevent fouling, or the fluid may boil in the exchanger, with subsequent fouling or decomposition.

Forced-circulation calandrias should be designed with velocities ranging from 10 to 15 ft./sec. Velocities above 15 ft./sec. may cause erosion and are of diminishing value in reducing fouling. These calandrias may be installed horizontally or vertically. Horizontal units often make cleaning easier. The longer piping that they require can be used to create the necessary back-pressure. For vertical multiple-tube-pass installations, velocities must be sufficient to prevent vapor blanketing. Control is the same as for natural-circulation calandrias.

Horizontal Kettle-Type Reboilers

Kettle-type reboilers are horizontal units boiling on the shellside. A vapor space above the tube bundle provides for vapor disengaging. Kettle-type reboilers have low liquid velocities and are more readily fouled. Retention time is higher, and may cause some product degradation. Proper blowdown must be maintained to prevent residue buildup or changes in the liquid concentration, which result in higher boiling points.

Vapor bubbles can escape at a velocity determined by the pressure of the system. When vapor is generated faster than it can escape, all or a portion of the bundle will become blanketed by a film of vapor. Heat transfer will suffer because nucleate boiling will give way to film boiling. The maximum vapor velocity can be determined from curves in Ref. 7, or from the following equation:[8, 9]

$$v = 0.0825 \frac{(\rho_L - \rho_V)^{1/4}}{(\rho_V)^{1/2}} \sigma^{1/4}$$

where v is velocity, ft./sec.; ρ_L is liquid density, lb./cu.ft.; ρ_V is vapor density, lb./cu.ft.; and σ is surface tension, dynes/cm.

For a tube bundle, this is the superficial vapor velocity and must be based on the area projected above the bundle. The bundle diameter should be used, not the shell diameter.

For small bundles, a minimum of 12 in. should be provided above the bundle to reduce entrainment. The shells for larger units should be 1.3 to 1.6 times the bundle diameter. In long units, more than one vapor outlet should be provided.

Falling-Film Vaporizers

Falling-film vaporizers are designed so that vaporization occurs on the surface of a fluid as it flows down the tube filmwise. They may be operated with or without recirculation. The primary advantage is higher heat tranfer at low temperature differences. Fouling is lower in falling-film vaporizers because the film is highly turbulent and vaporization occurs on the surface of the liquid film and not on the tube wall.

Distribution is usually done with an orifice. In some applications, this may not be acceptable because of solids in the fluid. If the fluid flashes as it enters the exchanger, the flashed vapors must be removed to achieve good liquid distribution.

Downdraft operation is best. With updraft operation, the rising vapors interfere with liquid distribution. Slugs of liquid can be blown upward if the tubes are not properly sized for the vapor load. Downdraft operation is less susceptible to fouling and provides better heat transfer.

Temperature difference across the liquid film must be kept relatively low [less than 15° (C.)]. For this reason, large temperature differences between the utility and process fluids should be avoided. High temperature differences will result in film boiling and poor heat transfer. In some cases, the fluid is vaporized on the tube surface and increases fouling.

Inert gases are sometimes injected into a falling-film evaporator in order to reduce the partial pressure required to vaporize the volatile component. This technique will often eliminate the need for vacuum operation, and is frequently used in updraft operation. Enough inert gas must be injected to achieve the desired results, but too much can produce flooding and entrainment.

For falling-film exchangers, a minimum flowrate is required to induce a film. This minimum rate can be determined by the method of Hartley and Murgatroyd:[10]

$$\Gamma_{min} = 19.5(\mu s \sigma^3)^{1/5}$$
or: $$\Gamma_{min} = 41.1(\mu s \sigma^3)^{1/5}(1 - \cos \alpha)^{3/5}$$

where Γ_{min} is minimum loading, lb./(hr.)(ft.); μ is viscosity of liquid, cp.; s is specific gravity of liquid; σ is surface tension of liquid, dynes/cm.; and α is contact angle, deg.

Once the film is induced, a lower terminal flowrate can be achieved without destroying the film. However, little work has been done to determine the magnitude of this reduction. If less than the minimum rate is achieved, the film will break and form rivulets. Part of the tube surface will not be wetted, and the result is reduced heat transfer and increased fouling.

Long-Tube Vertical Vaporizers

Long-tube vaporizers (LTV) may be operated with or without recirculation. With recirculation, heat transfer is generally less than that obtained in natural- or forced-circulation calandrias, or in falling-film vaporizers. Once-through operation usually does not provide for return of entrained liquid, and heat-transfer area must be supplied to assure that splashed or entrained liquid is vaporized. Consequently, the vapor is slightly superheated.

Entrance velocities for the liquid must be high (above 5 ft./sec.) to assure uniform distribution to a multiple-tube unit. LTV's should not be used for viscous fluids or for fouling applications.

Improved Boiling Surfaces

It has long been known that nucleate-boiling heat transfer can be increased by artificially roughening with techniques such as scratching, sandblasting and etching. More recently, fluted boiling surfaces,[11] and surface enchancement with fluorocarbon (Teflon)[12] have been presented. Union Carbide Corp.'s Linde Div. has developed improved boiling surfaces[13] that increase boiling heat transfer by a factor of 10 to 50 over that obtained with smooth surfaces.

Boiling on fluted tubes is a thin-film process that takes place on each crest. Fluted boiling surfaces must operate as a falling-film exchanger. Boiling heat transfer is increased by a factor of 5 to 10 with fluted surfaces. The performance of fluted boiling surfaces is relatively poor when compared to the results achieved when condensing. The reason for this is not clear but is probably due to the inability of maintaining thin films on the crests of the flutes. The crests may run dry; they may be flooded by liquid displaced from the troughs by axial flow of the accumulated vapor; or nucleate boiling may occur in the troughs.

Teflon is poorly wetted by most liquids. A multiplicity of poorly wetted spots can be used to achieve better boiling heat transfer. This method gives results comparable to those of the fluted tube. Cost of application and durability of Teflon spots has not been demonstrated.

The Linde surface consists of a thin layer of porous metal bonded to the heat-transfer substrate. Bubble nucleation and growth are promoted within a porous layer that provides a large number of stable nucleation sites of a predesigned shape and size.

Microscopic vapor nuclei in the form of bubbles entrapped on the heat-transfer surface must exist in order for nucleate boiling to occur. Surface tension at the vapor-liquid interface of the bubbles exerts a pressure above that of the liquid. This excess pressure requires that the liquid be superheated in order for the bubble to exist and grow. The porous surface substantially reduces the superheat required to generate vapor. The entrances to the many nucleation sites are restricted in order to contain part of the vapor in the form of a bubble and to prevent flooding. Many individual sites are also interconnected so that fresh liquid is continuously supplied.

Boiling heat-transfer coefficients, which are functions of pore size, fluid properties and heat flux, are 10 to 50 times greater than smooth-surface values at a given temperature difference. Boiling coefficients are stable and relatively independent of convection effects. Nucleate boiling exists at much lower temperature difference than required for smooth surfaces. ■

References

1. Lord, R. C., Minton, P. E. and Slusser, R. P., Heat Exchanger Design, *Chem. Eng.*, Jan. 26, 1970, pp. 96-118.
2. Perry, J. H., "Chemical Engineers' Handbook," 3rd ed., p. 686, McGraw-Hill, New York, 1950.
3. Dukler, A. E., Dynamics of Vertical Falling Film Systems, *Chem. Eng. Progr.*, Oct. 1959 p. 62
4. Gregorig, R., An Analysis of Film Condensation on Wavy Surfaces Including Surface Tension Effects, *Z. Agnew. Math. Phys.*, 5, 36 (1954).
5. Thomas, D. G., Enhancement of Film Condensation Heat-Transfer Rates on Vertical Tubes by Vertical Wires, *Ind. Eng. Chem. Fundamentals*, Feb. 1967, p. 97.
6. Greve, F. W., Measurement of Pipe Flow by the Coordinate Method, Bulletin No. 32, Engineering Experiment Station, Purdue University, Aug. 1928.
7. Perry, J. H., "Chemical Engineers' Handbook," 4th ed., p. 10-18, McGraw-Hill, New York, 1963.
8. Kutateladze, S. S., "Heat Transfer in Condensation and Boiling," (1952), Translated as U.S. Atomic Energy Commission Report, A.E.C.-TR-3770 (1953).
9. Sims, G. E., Akturk, U. and Evans-Lutterodt, K. O., Simulation of Pool Boiling by Gas Injection at the Interface, *Intern. J. Heat Mass Transfer*, 6, 531 (1963).
10. Hartley, D. E. and Murgatroyd, W., Criteria for the Break-up of Thin Liquid Layers Flowing Isothermally Over Solid Surfaces, *Intern. J. Heat Mass Transfer*, 7, 1003 (1964).
11. Carnavos, T. C., Thin Film Distillation, First International Symposium on Water Desalination, Oct. 5, 1965, Washington, D. C.
12. Young, R. K. and Hummel, R. L., Improved Nucleate Boiling Heat Transfer, *Chem. Eng. Progr.*, July 1964, p. 53.
13. Czikk, A. M. and O'Neill, P. S., Application of Enhanced Heat Transfer Surface to VTE Process Plant, Paper No. 10, Presented at the Symposium on Enhanced Tubes for Desalination Plants, Office of Saline Water, Mar. 11-12, 1969, Washington, D. C.

Meet the Authors

Richard C. Lord   **Paul E. Minton**   **Robert P. Slusser**

Richard C. Lord is a project engineer in the engineering department at Union Carbide Corp.'s Technical Center, P.O. Box 8361, South Charleston, W. Va., where he is part of the heat-transfer technology group. Formerly he was involved in activities related to process design and economics. He has a B.S. degree in chemical engineering from the University of Maine and is a member of AIChE.

Paul E. Minton is a project engineer in the engineering department at Union Carbide Corp.'s Technical Center, P.O. Box 8361, South Charleston, W. Va., where he is a part of the heat-transfer technology group. A B.S. graduate in chemical engineering from the Missouri School of Mines and Metallurgy, he is a member of AIChE.

Robert P. Slusser is a project engineer in the engineering department at Union Carbide Corp.'s Technical Center, P.O. Box 8361, South Charleston, W. Va., where for the last 10 years he has been designing equipment as a member of the heat-transfer technology group. His past work includes 12 years as production supervisor in the vinyl-resin facilities. He holds B.S. and M.S. degrees in chemical engineering from Massachusetts Institute of Technology and is a member of AIChE.

Designing Vertical Thermosyphon Reboilers

These empirical correlations give direct results accurate enough to design reboilers in standard commercial installations.

O. FRANK and R. D. PRICKETT, Allied Chemical Corp.

Natural-circulation, or thermosyphon, reboilers are generally the most economical vaporizers for distillation and evaporation equipment.

In the chemical process industries, vertical tubeside boiling is frequently preferred over shellside boiling because of (1) excellent heat-transfer factors possible inside vertical tubes, and (2) the smaller tubeside surface exposed to corrosive fluids. In addition, tubular heat exchangers and standard smooth tubes are preferred because of their low cost, relative ease of cleaning, and predictability of performance.

However, vertical thermosyphon reboilers are difficult to calculate strictly theoretically. The phenomenon of boiling inside their tubes is controlled by a combination of tube-wall film resistance, nucleation, and the extent of two-phase flow, in addition to the heat flow. A rigorous analysis of these reboilers requires not only the knowledge of heat-transfer relationships but also hydrodynamic effects, since the rate of vaporization and fluid circulation are closely interrelated.

Experiments by Simulation

A simplified solution to this problem has been developed by comparing the results of a computer model against empirical performance data.

Fair[1,2,3] and Hughmark[4,5] have published practical design procedures that incorporate both a pressure and a thermal balance. A method similar to that used by Fair was applied to the development of a computer program suitable for designing and rating tubeside reboilers. The procedure incorporated calculations for a sensible heat zone, the determination of an incremental temperature profile, a nucleate boiling correlation, a pressure-balance convergence scheme, and the option of using either an isothermal or nonisothermal heating medium.

Finally, the computer program was tested by simulating the measured performances of a number of commercial reboilers.

These tests showed that the calculated performances were generally conservative, falling within an accuracy range of -10% to $+25\%$. This was considered to be equally as reliable as all known proprietary thermosyphon programs. Therefore, we have assumed that commercial thermosyphon reboiler performance can be adequately represented by the computer program. We have generated a large number of simulated operating data, treating these as if they represented experimental values, in order to establish those critical variables that have the greatest effect on the performance of vertical thermosyphon reboilers.

It soon became clear, both from an analysis of the generated points and from actual performance data, that the thermal rating falls within a relatively small range. Given the limits of normal heat-exchanger configurations, the influence of tube dimensions was found to be minimal. It also appeared that, within a wide range, sump level has only a minor effect on heat flux. Two items influence performance most strongly: the nature of the process fluid, and the degree of tubeside fouling assumed by the designer.

Fluid property: A vast number of vertical thermosyphon reboilers handling heavy organic chemicals will demonstrate overall heat-transfer coefficients in the range of 100 to 160 Btu./(hr.)(sq.ft.)(°F.). Those reboilers vaporizing light hydrocarbons might operate between 160 and 220 Btu./(hr.)(sq.ft.)(°F.). And water or aqueous solutions would raise this range to 220–350. It thus appears that the fluid's approach to its critical temperature most directly expresses the effect of process fluid on thermal performance.

Fouling factors: A comparison of calculated and measured performances showed that a 0.001 fouling factor yielded the best reproduction of the plant data. Most of the commercial reboilers that were studied were found to be reasonably clean after at least six months of normal operation. Only polymerizing compounds caused early degradation of the heat-transfer surface.

Therefore, the 0.001 fouling factor was incorporated into the computer-generated test cases and subsequently

Originally published September 3, 1973

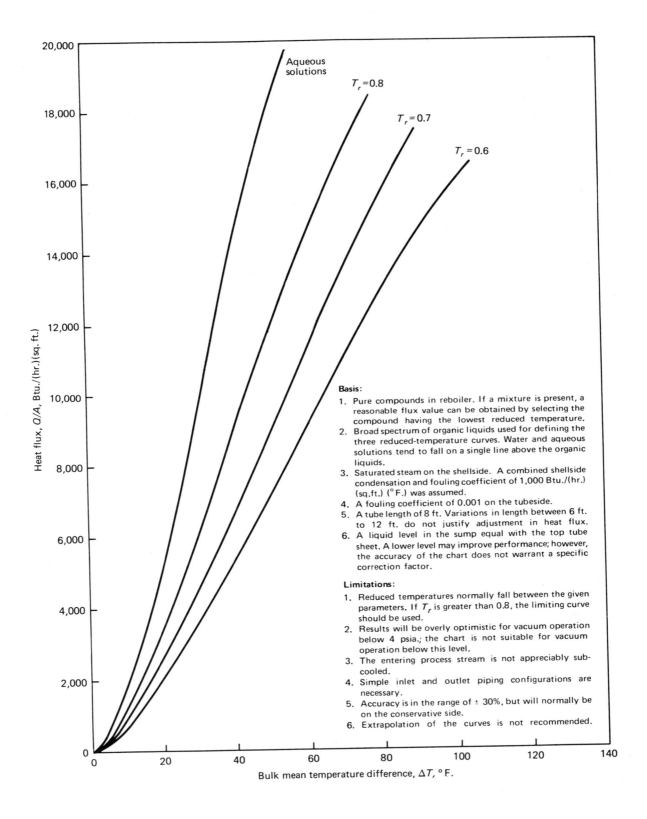

CRITICAL VARIABLE CORRELATION for vertical thermosyphon reboilers

into the summarized results. When dealing with a clean, nondegrading fluid, this fouling factor can be reduced to 0.0005, while it should be raised to 0.003–0.004 for polymerizing material.

It was thus determined that, given a standard exchanger configuration, performance could be best represented by plots of heat flux, Btu./(hr.)(sq.ft.), versus bulk mean-temperature difference, °F., at given parameters of reduced temperature for the tubeside fluid. The shellside coefficient was assumed to be condensing steam at 1,000 Btu./(hr.)(sq.ft.)(°F.). Metal resistance was based on 14-gage stainless steel tubes. These plots are in the figure.

Should the user want to determine a reboiler heat flux with a heating medium other than steam, or a process-side fouling coefficient other than 0.001, he may adjust the graph-derived heat flux through the following relations:

$$U_g = (Q/A)_g (1/\Delta T)_g$$
$$U_c = \frac{1}{(1/U_g) - (1/1,000) + (1/h_s) - 0.001 + f_p}$$

where: g = a subscript indicating graph-derived
c = a subscript indicating corrected
U = overall transfer coefficient, Btu./(hr.)(sq.ft.)(°F.)
A = the reboiler heat-transfer surface, sq.ft.
ΔT = the bulk log-mean-temperature difference between tubeside and shellside, °F.
h_s = the film coefficient for the shellside fluid, Btu./(hr.)(sq.ft.)(°F.)
f_p = the fouling factor for the process fluid, 1/Btu./[(hr.)(sq.ft.)(°F.)]

When the shellside heating medium is hot oil, this correction reduces to:

$$U_c = U_g/(1 + 0.0014)$$

In most cases, the accuracy of the procedure does not justify a correction for tubewall thickness or thermal conductivity.

Design Procedure

The following procedure is accurate enough to design thermosyphon reboilers for standard commercial installations. The mechanical recommendations were developed from personal experience or abstracted from the references.[6,7,8,9]

1. Establish the tubeside temperatures and the shellside heating-medium temperature. From critical-temperature data, obtain the reduced temperature of the process fluid. Read heat flux directly from the figure. Depending on the system, the maximum mean temperature difference should be 60–100 F. This limit should avoid excessive fouling from polymerizable fluids; it also avoids excess per-pass vaporization, when heat transfer in the upper parts of the tubes is reduced by replacing the normal slug and annular flow with a liquid-deficient mist flow.

2. From the total duty and heat flux, calculate the required number of tubes, based on selected diameter and length. Given the limits of standard heat-exchanger configurations, the influence of tube dimensions is found to be nominal, with 1-in. by 8-ft. tubes frequently used for commercial reboilers.

The 8-ft. length seems to strike a balance between complete vaporization in long tubes and predominantly sensible-heat transfer in short tubes. Thus, tubes are seldom longer than 12 ft., although short tubes, 4–6 ft. long, are sometimes used in vacuum service.

Cleaning considerations usually limit tube I.D. to 1 in. minimum, which is suitable for most applications, although 1¼-in., 1½-in., and even 2-in. tubes have been installed.

3. Size the reboiler inlet and outlet piping to achieve the proper pressure-drop relations. The cross-sectional I.D. area of the outlet piping should be approximately equal to that of the total tubes, whereas the cross-sectional I.D. area of the inlet piping should be no greater than half that of the total tubes. Also, tubeside fluids should leave the reboiler axially, in order to minimize pressure drop in the return line.

4. Assure that the reboiler is capable of stable operation over the widest possible range of conditions. This requires careful attention to the design of the tower's bottom section and the mode of instrument control over the entire distillation system. Flow fluctuations through the reboiler can be stabilized by incorporating a properly sized restricting orifice in the liquid inlet pipe. Taking between 25% and 40% of the total loop pressure-drop across an inlet restriction will provide a dampening effect.

5. Set the reboiler elevation according to the operating liquid level in the tower. Maximum heat flux may be achieved if most of the liquid is vaporized by the time it reaches the top of the tubes; and this occurs when the liquid level in the tower is about one-third the tube length above the bottom tubesheet. Unfortunately, operation at such a level also produces maximum fouling, and it is therefore the usual practice to start operation with the liquid level at the same elevation as the top tubesheet. It will be found that the optimum liquid level is frequently a function of the system pressure, being higher at higher pressures and lower at subatmospheric pressures.

Thermosyphon vs. Forced Circulation

Tubeside boiling may occur either in conjunction with naturally induced flow or via forced circulation. In either case, heat transfer along the tubes can be assumed to take place both in a sensible-heat zone, where the temperature of a subcooled liquid is raised to its boiling point, and in a boiling zone, where vaporization takes place.

At moderate or high pressures, the sensible-heat zone is small and does not affect the overall performance of the reboiler. However, under vacuum operation or with high-viscosity liquids, the hydrostatic head may suppress boiling along an appreciable distance of the tubes. In the case of a thermosyphon reboiler, this reduces the circulation rate and thereby the overall heat-transfer coefficient.

It has thus become frequent practice to use forced-circulation reboilers in low-pressure (under 4 psia.) or highly-fouling service, and to use natural-circulation

Comparison of Forced-Circulation and Vertical-Thermosyphon Reboilers

Forced Circulation	Vertical Thermosyphon
Advantages	
☐ Preferred in fouling service because circulation rate and degree of vaporization are controlled externally.	☐ Usually the cheapest reboiler installation in terms of capital and operating cost.
☐ Suitable for viscous and solids-bearing fluids.	☐ Permits simple, compact piping arrangement.
☐ Can be designed for high sensible-heat transfer at low vaporization rates and is therefore suitable for low-pressure-drop systems.	☐ Provides excellent thermal performance.
Disadvantages	
☐ Power costs are appreciable in large installations.	☐ More heat-transfer area is required for vacuum operation because suppressed boiling causes a large, inefficient, sensible-heat zone.
☐ Requires a costly pump installation.	☐ Not suitable for viscous or solids-bearing fluids.
☐ May present a leakage problem at pump seals.	☐ Because of its vertical arrangement, requires additional column skirt-elevation.*

*This restriction is eliminated if a horizontal, shellside thermosyphon reboiler can be used.

(thermosyphon) reboilers for most standard distillation and evaporator services. The table lists advantages and disadvantages of these two types of vaporizer and may be used as a guide for selecting the proper equipment.

Reboiler Programs

In recent years, industry-sponsored research organizations, such as Heat Transfer Research, Inc. in the U.S., and Heat Transfer and Fluid Flow Service in the U.K., have developed proprietary thermosyphon reboiler programs for the use of their members. Also, a number of companies offer similar programs for sale.

Many of these design or modeling procedures have been based, at least in part, on Fair's correlations, which, because they can be rationalized on a mechanistic basis, readily lend themselves to practical applications. Frequently, such programs have been "standardized" by comparing calculated results with the published performance data of Shellene and others,[10] Lee and others,[11] and A. E. Johnson.[12]

In conclusion, it has been shown that the thermal performance of commercial thermosyphon reboilers can be realistically predicted by a correlation of system bulk-temperature difference and fluid reduced temperature. The curves representing this relationship have been found useful for estimating new reboiler installations or evaluating the performance of existing systems, frequently within the range of accuracy normally expected of computer programs. In combination with the commercially practiced design procedures listed above, the chart is suitable for specifying vertical, intube, natural-circulation reboilers. ■

References

1. Fair, J. R., What You Need to Design Thermosiphon Reboilers, *Pet. Ref.*, **39**, No. 2, p. 105 (1960).
2. Fair, J. R., Vaporizer & Reboiler Design, Part I, *Chem. Eng.*, **70**, No. 14, p. 119 (1963).
3. Fair, J. R., Vaporizer & Reboiler Design, Part II, *ibid.* **70**, No. 16, p. 101 (1963).
4. Hughmark, G. A., Design of Thermosiphon Reboilers, *Chem. Eng. Prog.*, **57**, No. 7, p. 43 (1961).
5. Hughmark, G. A., Designing Thermosiphon Reboilers, *ibid.* **60**, No. 7, p. 59 (1964).
6. Kern, Robert, Thermosyphon Reboiler Piping Simplified, *Hyd. Proc.*, **47**, No. 12, p. 118 (1968).
7. Jacobs, J. K., Reboiler Selection Simplified, *Hyd. Proc.*, **40**, No. 7, p. 189 (1961).
8. Chexal and Bergles, "Two Phase Instabilities in a Low Pressure Natural Circulation Loop," 13th Natl. Heat Transf. Conf. 13, Denver (1972).
9. McKee, H. R., Thermosiphon Reboilers—A Review, *Ind. Eng. Chem.*, **62**, No. 12, p. 76 (1970).
10. Shellene, Sternling, Church & Snyder, Experimental Study of a Vertical Thermosiphon Reboiler, *Chem. Eng. Prog.* Symp. Ser. 82, **64**, 102 (1968).
11. Lee, Dorsey, Moore & Mayfield, "Design Data for Thermosiphon Reboilers," *Chem. Eng. Prog.*, **52**, No. 4, p. 161 (1956).
12. Johnson, A. E., "Circulation Rates and Overall Temperature Driving Forces in a Vertical Thermosyphon Reboiler", *Chem. Eng. Prog.*, Symp. Ser. 18, **52**, p. 37 (1956).

Meet the Authors

◀ **Otto Frank** is Group Leader of Process Design and Methods at Allied Chemical Corp., where he is responsible for developing and testing design procedures for the Corporate Engineering Dept. With Allied Chemical since 1960, he was formerly employed by Air Products, Inc., and by Du Pont. A registered professional engineer in New York, he holds a B.S.ch.E. from Clarkson College of Technology and an M.S. in Ch.E. from Princeton University.

Richard D. Prickett is (since April 1973) a process engineer with the M. W. ▶ Kellogg Co. in Houston. Prior to joining Kellogg, he worked for four years with Allied Chemical's Corporate Engineering Dept., during which time he collaborated with Mr. Frank on this article. A member of AIChE, he holds a B.S.ch.E. from Drexel University and an M.S. in Ch.E. from Newark College of Engineering.

Double-Tubesheet Heat-Exchanger Design Stops Shell-Tube Leakage

When leakage of shellside fluid into the tubeside fluid would present a hazard, consider the double tubesheet. Two types are available: the conventional double tubesheet and the integral design.

STANLEY YOKELL, Process Engineering and Machine Co.

No known method of joining tubes to tubesheets in heat exchangers entirely eliminates the possibility of leakage.

In many situations, some leakage of shell- and tubeside fluids can be tolerated. But for those situations (Table I) where mixing of these fluids cannot be allowed, double-tubesheet construction should be considered.

The usual methods for joining tubes to tubesheets are: roller expanding; a combination of roller expanding and welding; welding only; roller expanding and beading or flaring the tube ends, followed by welding; using a ferrule or similar device to interpose packing or seals between the outside of the tube and the inside of the tube hole. If substantial axial forces are anticipated on the tubes, tube holes prepared for roller expanding are grooved with one or more grooves, ⅛ in. wide by 1/64 in. deep. During expansion, some tube metal flows into the groove, providing an anchor against the tube's being pulled or pushed through the tube hole.

Regardless of the method, there is always the possibility of leakage at the joints of the tubes to the tubesheets. Joints that are tight under shop-testing conditions may leak after a startup, operating upset or shutdown.

The double-tubesheet design does *not* eliminate leak-

Process Conditions That Might Require Double-Tubesheet Construction — Table I

Condition	Process Factors	Examples
Corrosion	Fluid streams are not corrosive separately, but the mixture of streams is highly corrosive.	HCl gas coolers, oleum coolers, chlorine vaporizers.
Health	One fluid stream is lethal. If it mixes with the second stream, lethal material is carried to areas of the plant not designed for lethal service.	Phosgene reboilers, phosgene condensers, phosgene coolers, cyanogen coolers.
Safety	The streams ignite or explode when they mix.	Alkali-metal coolers.
Equipment fouling	Resinous substances or polymers are generated when the streams mix.	Methacrylic-acid coolers, latex heaters.
Catalyst poisoning	Catalyst form may be altered or its chemical composition may be changed if the catalyst contacts a second stream.	Fluid-bed catalyst regeneration coolers.
Yield reduction	A chemical reaction may be stopped or reversed if the fluid streams mix.	Esterification over-head condensers.
Impure product	Product may become contaminated with fluid from the second stream.	Purification-column condensers, distillation-column economizers.

Originally published May 14, 1973

DOUBLE-TUBESHEET HEAT-EXCHANGER DESIGN STOPS SHELL-TUBE LEAKAGE

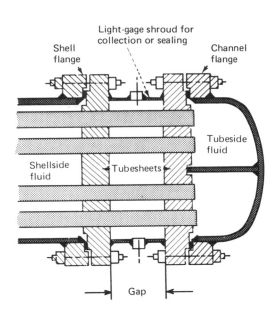

CONVENTIONAL double-tubesheet construction in U-tube exchanger has two flanged connections—Fig. 1

age; but it does eliminate leakage of shellside fluid into the tubeside fluid.

Two designs are available: the "conventional" double tubesheet, and the integral tubesheet.

These double tubesheets are usually found in fixed-tubesheet and U-tube heat exchangers, but use in floating-head equipment is possible.

Conventional Double Tubesheet

The conventional double-tubesheet heat exchanger requires two tubesheets at each end of the tubes. Adjacent tubesheets are joined to each other only by the tubes.

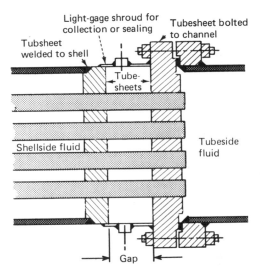

DOUBLE-TUBESHEET design usually has one sheet welded to the heat-exchanger shell—Fig. 2

Shrouding may be used to seal the gap between adjacent tubesheets for collecting leakage. If the gap is sealed, it can also be filled with an inert substance at a pressure higher than either shell or tubeside pressures. This permits leakage to be detected without losing process fluids (after a leak is discovered, operation may be continued until it is convenient to make a repair).

Conventional U-tube double-tubesheet units may use each tubesheet as a separate bolting flange. In such a situation, the shellside tubesheet is bolted to the shell flange and the channelside tubesheet is bolted to the channel flange (Fig. 1). Ordinarily, conventional fixed double-tubesheet units are built with the shellside tubesheet welded to the shell and the channelside tubesheet bolted to the channel (Fig. 2). For either type, the channelside tubesheet design is similar to the design of a floating tubesheet.

Problems With the Design

There are some problems in the manufacture of double tubesheets. The sheets are often stacked for drilling. If the drill drifts, holes occupying the same position in adjacent tubesheets will be displaced. Then, when the tube bundle is assembled with adjacent tubesheets close to each other, lateral shearing and bending forces are imposed on the tubes. It may then be impossible to get a tight seal of the tubes to the tubesheets.

Design of double-tubesheet heat-transfer equipment must also provide for the effects of temperature on the tubesheets. The tubesheets expand or contract when the metal temperature varies from ambient. If adjacent tubesheets reach different temperatures, the radial growth of the tubesheets will be different. This generates forces on the tubes producing shearing and bending that may cause leakage. If temperature changes are cyclical and

58 HEAT EXCHANGER DESIGN AND SPECIFICATION

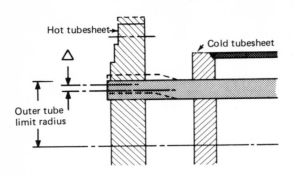

TEMPERATURE differences on tubesheets place stresses on the tubes that can cause failure—Fig. 3

frequent, tubes may fail because of metal fatigue (Fig. 3).

For conventional double-tubesheet construction, the effects of drill drift and temperature are handled by including a suitable gap between adjacent tubesheets large enough to permit the tubes to deflect harmlessly.

If the tubes are considered to be partially built-in beams, a useful relationship for determining the gap can be derived:

$$G = [(1.5\ E d_o\ \Delta)/(S)]^{1/2}$$

Assuming allowable stress in the tubes is limited to 40% of the yield, this equation becomes

$$G = [(E d_o\ \Delta)/(0.27\ Y)]^{1/2}$$

Where d_o = Tube O.D., in.
        E = Modulus of elasticity in tension, lb./sq.in.
        G = Required gap (beam length), in.
        S = Stress in tension, lb./sq.in.
        Y = Yield stress in tension, lb./sq.in.
        Δ = Deflection of tube (beam), in.

$$\Delta = (OTL/2)\ [\alpha_t\ (t-70) - \alpha_s\ (T-70)]$$

Here OTL = Outer tube limit circle, in.
        t = Channelside tubesheet metal temperature, °F.
        T = Shellside tubesheet metal temperature, °F.
        α_t = Thermal coefficient of expansion of the channelside tubesheet between 70 F. and T°F., in./(in.) (°F.)
        α_s = Thermal coefficient of expansion of the shellside tubesheet between 70 F. and T°F., in./(in.) (°F.)

These equations assume the heat exchanger is assembled at an ambient temperature of 70 F.

The differential displacement between the tubesheets should be selected at the most severe condition of process startup, steady state, upset or shutdown situations to be encountered.

Rolling the tubes into the tubesheets separated by a gap is achieved by using specially designed rolling equipment, although the rolling technique is basically the same as that for single-tubesheet heat-exchanger construction.

Of course, the gaps between tubesheets must be considered in the thermal design of the equipment. The effective heat-transfer surface is the surface between the inner faces of the innermost tubesheets. Tubing in the gap and in the four tubesheets is lost to heat transfer, but contributes to frictional resistance to flow in the exchanger tubes.

An Alternate Design

An alternate double-tubesheet design restrains the differential radial expansion by joining adjacent tubesheets beyond the outer-tube-limit circle with a thick-walled cylinder welded to the tubesheets, or with hubs integrally forged to the tubesheets and welded together. This design has a limitation: the tubesheets behave like a heated

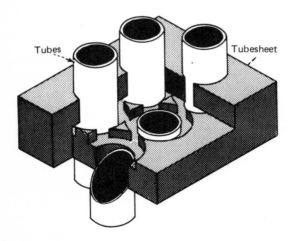

INTEGRAL double tubesheet has no gap between tubesheets; but this design is more expensive—Fig. 4

Economics of Single- and Double-Tubesheet Construction — Table II

Capital Costs of Heat Exchanger, $/Sq. Ft. Effective Surface

Nominal T.E.M.A. Shell Size, In.	Carbon Steel*			Stainless Steel 304*			Stainless Steel 316*		
	Single	Double Conv.	Double Integral	Single	Double Conv.	Double Integral	Single	Double Conv.	Double Integral
6	19.62	22.91	23.54	24.19	27.92	30.73	26.13	29.92	31.99
12	8.30	10.20	10.58	11.25	13.61	15.98	12.28	14.79	17.13
17	5.73	7.31	7.87	7.89	9.73	10.15	9.05	11.02	11.36
23	4.77	6.37	6.05	6.75	9.10	7.75	7.90	10.32	9.26
29	4.64	6.33	5.07	6.69	9.00	7.33	7.80	10.17	8.37
35	4.19	6.27	4.87	6.58	8.13	7.00	7.68	9.75	8.17
39	4.02	6.25	4.65	6.32	8.10	6.75	7.50	9.71	7.95

*Costs are for tubular heat exchangers fabricated of carbon-steel shellside and material of construction indicated on the tubeside.

bimetallic strip. If there is enough metal at the joint of the tubesheets to restrain the differential radial deflection, the tubesheets will bow. Forces are then imposed on the tubes between the innermost and outermost faces of the adjacent tubesheets.

Stress-corrosion cracking and fatigue failures have occurred in tubes rolled into tubesheets constructed in this way. Such failures took place within the rolled region and in the gap between the tubesheets.

Integral Double Tubesheets

There is a different type of tubesheet, shown in Fig. 4, known as the integral double tubesheet.*

In this design, a single tubesheet is drilled to the desired pattern. Then, a ¼-in.-wide groove is machined into the tube hole surface, halfway between the faces, until adjacent grooves break through the tube hole walls into each other. Thus, two separate tubesheets are created. They are integrally joined to each other between adjacent tubes, at pass-partition lanes and beyond the outer tube limit.

The grooves straddle the neutral plane of the tubesheet so there is little reduction in the bending strength of the tubesheet. And since there is only one tubesheet for each tube end to enter, the drill-drift problem is eliminated.

Heat is conducted through the metal in the tube walls, in the ligaments joining the halves of the tubesheet and in the solid peripheral section of the tubesheet. There is a temperature gradient between the hot and the cold faces of the tubesheet. This is not a sharp interface, and forces resulting from the hotface-coldface effect are restrained by the integral ligaments between the tubesheets.

Effective heat-transfer surface is slightly reduced from that of single-tubesheet construction. But frictional resistance to fluid flow is the same as for single tubesheets.

Separate vents and drains in the gap in each pass of multipass units are required. This is advantageous in leak testing because when a leak is found, the tubefield to be rerolled is limited to the pass that contains the leaking joint.

Integral double-tubesheet construction requires fairly expensive special tooling. The technique calls for careful attention to tool speeds and coolants. If the tubesheet material is easily work hardened, extreme care is required in cutting the gap; thus the method is usually limited to relatively free-machining materials.

Economics of Special Tubesheets

The economics of single tubesheet vs. conventional and integral tubesheet construction are shown in Table II. In this evaluation, it was assumed that the inner-tubesheet faces contact a coolant at an average temperature of 80 F. and the outer-tubesheet faces contact a vapor condensing isothermally at 360 F. The shells of the exchangers were designed with expansion joints. All units were single pass, containing ¾ in. O.D. x 16 BWG welded tubes, 16 ft. long. A T.E.M.A. type BEM configuration was chosen.

According to this study, for T.E.M.A. shell sizes 17 in. and below, the conventional double tubesheet is more economical; for sizes 23 in. and above, the integral type is the economical choice.

Of course, in all cases, single-tubesheet construction is the lower-cost construction. The double-tubesheet design should be chosen only when no other alternative is acceptable. ∎

Meet the Author

Stanley Yokell is the President of Process Engineering and Machine Co., York Street and Dowd Ave., Elizabeth, NJ 07201. He is a graduate of New York University, College of Engineering, with a B.S. in chemical engineering. A member of AIChE and NSPE, Mr. Yokell is a licensed Professional Engineer in New Jersey.

* U.S. Patent No. 1,987,891.

GUIDE TO TROUBLE-FREE
Heat Exchangers

The designer of heat-transfer equipment can do much to minimize maintenance costs and production losses by anticipating the many problems associated with the cleaning, leakage, startup and shutdown, and repairs to heat exchangers.

R. C. LORD, P. E. MINTON and R. P. SLUSSER, Union Carbide Corp.

Heat-transfer equipment provides the economic and process viability for many plant operations. The basis for successful application of such equipment depends in large measure on the designer.

By anticipating problems, he can avoid high maintenance or cleaning expenses and costly production shutdowns. When corrosion is likely, he can assist by selecting corrosion-resistant materials, and by specifying types of heat exchangers that are easy to repair. He can minimize leakage between the shellside and tubeside by anticipating situations in which thermal cycling, vibration, and differential thermal expansion will overstress or loosen tube-to-tubesheet joints, or cause tubes to rupture. He can design to minimize fouling when suspended solids or fouling chemicals are present.

To anticipate maintenance problems, the designer needs to be familiar with the plant location, process flowsheet, and anticipated plant operation. Some of the questions that he must consider are:

1. Will the heat exchanger need to be cleaned? How often? What cleaning method will be used?

2. What penalty will the plant pay for leakage between the tubeside and shellside?

3. What kinds of production upsets can occur that could affect the heat exchanger? Will cycling occur?

4. How will the heat exchanger be started up and shut down?

5. Will the heat exchanger be likely to require repairs? If so, will the repairs present any special problems?

Good Design Reduces Fouling

Anticipated fouling and the need for cleaning have a major effect on the choice and cost of equipment. For example, a forced-circulation calandria using a tubeside velocity of 10 to 15 ft./sec. would be selected in place of a natural-circulation calandria if the chemicals in the base were likely to foul the tubes frequently. Addition of the pump, and the cost of power for pumping, add considerably to the costs of the equipment. These extras would be compared to the cost of production losses and cost for cleaning in order to arrive at an economical design for a particular process application.

Normally, the fouling fluid would be placed in the tubeside so that the unit could be cleaned without the high cost of removing the tube bundle. Easily removable, flat, cover plates would be installed on the channels to facilitate cleaning if frequent physical cleaning were necessary. A horizontal installation would probably be chosen to avoid the cost of a scaffold usually required for physically cleaning a vertical exchanger.

The time between cleanings can often be lengthened

Originally published June 1, 1970

by increasing tubeside velocity to 10 to 15 ft./sec. An aromatics preheater having a velocity of about 1 ft./sec. in an existing benzene plant was being cleaned about once a month. A revised design for a new plant used a velocitiy of 10 ft./sec. in 12 tubeside passes, with a tubeside pressure drop of 30 psi. The redesigned preheater was cleaned once during the first year of operation.

Fouling due to sedimentation will often be reduced by using heat-transfer equipment with a single-flow channel. For example, a spiral-plate exchanger may be selected in place of a multipass shell-and-tube unit to avoid settling of suspended solids in the shellside and at the bottom of the tubeside channels. In one instance, two multipass shell-and-tube units that fouled in the shellside and tubeside after two to three weeks of operation were replaced with a spiral exchanger. The spiral exchanger performed successfully, and was cleaned about once every three months by flushing with hot water and sparging with air.

Designs that allow cooler tubewall temperatures in heaters and reboilers also make for reduced fouling and cleaning costs. But, the size and cost of the equipment is increased, because the available temperature difference for heat transfer is reduced. The following is an example of the use of a cooler wall-temperature together with a single-flow channel and a somewhat higher velocity to reduce fouling.

An organic liquid was preheated with 200-psig. steam (198 C.) in the shell; tubeside velocity was 4 ft./sec. in 1-in. O.D. tubes. This preheater was cleaned about twice a week. The design was changed to obtain 7 ft./sec. in a pipe coil (3 in. I.D.) installed in the top of a distillation column with a head temperature of 95 C. Cleaning was unnecessary during the first year of operation.

Proved Design Practices Are Most Useful

Ideally, the designer would have data available concerning the fouling rates with various tubewall temperatures and liquid velocities. He could then optimize the cost of the heat-transfer area and the pumping costs versus the costs of cleaning and production losses. In practice, these data are seldom available. Consequently, the designer must follow proved design practices, including the use of tubeside velocities of 10 to 15 ft./sec., reduced tubewall temperatures, and single-flow channels when convenient.

Other techniques for avoiding fouling by using good design practices have been presented by Gilmour.[2] Good shellside design avoids eddies and dead zones where solids can accumulate. Inlet and outlet connections should be located at the bottom and top of the shellside and tubeside to avoid creating dead zones and unvented areas. The use of metals that will not foul due to accumulation of corrosion products is important, especially with cooling waters. Copper, copper alloy and stainless steels are satisfactory for most cooling waters.

An outstanding example of the use of a revised design to reduce cleaning costs is a horizontal ethylene reboiler that was redesigned by our plant heat-transfer specialist.

Low-finned tubes were used to replace bare tubes. Because of the additional area available with fins and because of the reduced fouling of the sharp-edged fins during operation, fewer tubes were required in the reboiler. The 5/8-in. finned tubes were placed on a 1 1/2-in.-square pitch. Labor for cleaning with high-pressure water was reduced from 250 to 16 hr.

Fortunately, fouling is not generally a serious design problem with most heat exchangers for our chemical plants. Low fouling factors of 1,000 Btu./(hr.)(sq.ft.)(°F.) are used for condensers, heaters and coolers. For reboilers and vaporizers, a fouling factor of 500 Btu./(hr.)(sq.ft.)(°F.) is usually adequate. Most properly designed and operated heat exchangers never require cleaning.

Leakage Problems in Heat Exchangers

Sometimes a small leakage from the tubeside to the shellside, or vice versa, can cause a large production loss or maintenance expense. When the designer is aware of this situation from his knowledge of the process, he can recommend equipment that will eliminate or minimize the chances of leakage. For example, air-cooled condensers have been selected for some services to avoid the possibility of seawater entering the process equipment. If steam leaking into the process side of a heater would contaminate a 100,000-gal. tank of chemical production, the designer might specify steam-heated patterned plates to be installed on the outside of the tank. Or, he might decide that a U-bend exchanger with the tubes welded into the tubesheets would provide enough extra protection against leakage.

Leaks may develop at the tube-to-tubesheet joints of fixed-tubesheet exchangers because differential thermal expansion between the tubes and the shell causes overstressing of the rolled joints. Or, thermal cycling caused by frequent shutdowns or batch operation of the process may cause the tubes to loosen in the tubeholes. Floating heads or U-bend exchangers would be considered first for this type of service. If a fixed-tubesheet unit is required, an expansion joint will be specified. An exchanger that will be thermally cycled two or three times a day will require superior mechanical construction such as the strength-welding of tubes to the tubesheet, complete inspection of the shell and channel welds during fabrication, specific limitations on the types of acceptable ASME Code welds for the tubesheet-to-shell joint, and a bellows-expansion joint for fixed-tubesheet units.

The designer needs to be especially alert to the unusual situations that can cause unexpected differential thermal expansions in fixed-tubesheet units. For example, the tubeside of a fixed-tubesheet condenser may be subjected to steam temperatures, with no coolant in the shellside whenever a distillation column is "steamed out" in preparation for maintenance. Or, an upset in the chemical process may subject the tubes to high temperatures.

In a specific case, we observed copper tubes that had pushed themselves about ½ in. through the tubesheet when the fixed-tubesheet unit was started up with 125 C. water in the tubes, and no coolant in the shell. The designer should change the design to floating head or U-bend construction or place the hot fluid in the shellside to avoid damaging the tube-to-tubesheet joints.

Welding the tubes to the tubesheets does not guarantee that a leak will not occur. We have had numerous instances of weld failure because of porosity in the welds, or just one poorly welded tube out of the hundreds of welds that are required in some exchangers.

The designer can attempt to obtain the best welds possible by:

1. Reviewing the fabricator's proposed cleaning and welding procedures.

2. Specifying that samples of the proposed welds be submitted to the purchaser for testing.

3. Specifying air tests and dye-penetrant tests to detect welds that need to be repaired during fabrication.

For severe services, such as 600-psig. steam service, the designer may specify strength welds made with two welding passes.

The use of double tubesheets to minimize the chances of leakage between the tubeside and shellside appears to be a good solution to the problem. Nevertheless, double tubesheets are an exotic solution that has caused considerable maintenance problems, and should not be specified unless the mixing of the two fluids must absolutely be avoided.

Because the outboard and inboard tubesheets may be subjected to considerably different process temperatures, there will be a differential expansion between the tubesheets that will bend the tubes.

Double tubesheets may need to be spaced 6 to 10 in. apart to avoid overstressing the tubes at the outer rows. If leaks develop at the inboard tubesheets, they cannot be stopped by plugging the tubes. If rerolling the tubes does not stop the leaks, the bundle would have to be retubed. An integral type of double tubesheet (described later) can be used to minimize tube failures caused by differential expansion between the inboard and outboard tubesheets. But again, leaks at the inboard side may require retubing the bundle if rerolling fails to stop the leaks.

The designer needs to be alert to process conditions that can lead to tube vibration. If a two-phase (gas-liquid) mixture is likely to be entering or leaving the shellside of a heat exchanger, the nozzles or vapor belts must be large enough to give velocities well below the critical. We have recently experienced costly failures in the shellside of two 5,400-sq. ft. reactor-cycle exchangers. Critical-velocity calculations, based on an average two-phase density, were in agreement with the actual location of the tube failures in front of the inlet and outlet connections.

Repair Considerations

If a heat exchanger will be used in corrosive service and if the components will have a short life expectancy, the designer should select units that are easy to repair. Removable-bundle shell-and-tube units would be the first choice. Fixed-tubesheet units are sometimes easily repaired if the plant shop has the necessary saws to cut through the shell and tubes at the same time. The shell is rewelded after the corroded tubes have been replaced.

Standard heat exchangers are not generally used for corrosive services. Shell-and-tube standard heat exchangers often have thin (18 to 22 BWG) tubes. They also have sensitized welds in stainless-steel construction (low-carbon stainless steels are not usually available in standard units). A spiral plate-heat exchanger would have to be returned to the factory for repairs if corrosion caused leaks to develop at inaccessible locations inside the spiral.

The designer can often reduce maintenance costs and production losses by specifying equipment with standard components. For example, the same type of fans and controllers can be specified for all of the air coolers in a particular plant. Or, the size and material for tubing is standardized for a particular process. In one plant, located some distance from large repair shops, we designed six interchangeable 5,600-sq.ft. natural-circulation calandrias to minimize possible production losses.

The designer of heat exchangers can contribute to the selection of materials of construction because of his experience with various kinds of corrosion and erosion failures. But he must also rely on the materials specialists for information concerning metal properties and corrosion rates with the process chemicals. Table I lists a few examples of heat-exchanger failures and the resulting recommendations.

Design Standards

The design engineer needs to establish standard practices to achieve an economical balance between the cost of the equipment and the cost of maintenance and production losses. These practices may vary in different locations of the same plant as the penalty for heat-exchanger failures, and the severity of the service, vary. In general, we use the requirements of TEMA-B for all heat exchangers. Some of our standards are:

1. Tubing thickness—Thickness for tubes is usually specified to be at least 16 BWG (average wall). This thickness is preferred because the tubes can be re-

Causes and Cures of Heat-Exchanger Failures—Table I

Failure	Recommendation
Steel baffles and tie rods corroded away in river-water service. Tubes were stainless steel.	Make baffles and tie rods with same materials as tubes. Rebuilding of baffles is expensive and causes costly production losses.
A corrosive process-chemical leaked from the tubeside to the shellside. The steel baffles and tie rods were completely corroded.	Same as above.
An all-aluminum air cooler was severely corroded by production upsets that permitted traces of caustic to enter the cooler.	Aluminum exchangers should be used only where trace quantities of destructive impurities cannot enter the equipment.
Brass shells cracked because of ammonia vapors from a neighboring processing unit.	Use copper shells in chemical-plant environment.
Type 316 stainless-steel ferrules at inlet end of waste-heat boiler tubes deteriorated at the point where temperature was estimated to be 1,000 F.	Use Inconel 600 or Incoloy 800 ferrules for waste-heat boilers.
Cast-aluminum fan blade in an air-cooled condenser failed by intergranular corrosion caused by mercury contamination (perhaps from a broken manometer).	Aluminum fan-blades and hubs should be coated with pigmented epoxy for extra protection in chemical-plant atmosphere. Plastic fan-blades are preferred for chemical plants.
High process-temperature at inlet end of tubeside caused stress cracking of Type 304 stainless-steel tubes from chlorides in shellside cooling water.	Lengths of Incoloy 800 tubing should be butt-welded to the top of the Type 304 stainless tubes (16 BWG tubes have adequate thickness for recently developed welding techniques).
Erosion occurred at inlet end of 90:10 copper-nickel tubes, inlet-water channels, and tubesheets, in seawater service.	Use 4-in.-long nylon inserts with tapered downstream edge in inlet of all tubes in seawater service. Limit nozzle velocity to 2 ft./sec. higher than tube velocity.
Tubes in seawater service became coated with calcium salts.	Limit maximum tubewall temperatures to 55 C. for seawater. Use 6 to 8 ft./sec. water velocity for 90:10 copper nickel tubes.
Duplex tubes in a fixed-tubesheet exchanger failed by stress cracking when copper on outside cracked, exposing stainless-steel liner to cooling water.	Use metallurgically bonded duplex tubes to avoid differential expansion and failure of the copper.

rolled once or twice without excessive loss of the strength of the metal. Thus, tube-to-tubesheet leaks can be repaired in the field. However, we occasionally use 18 BWG alloy tubes in large exchangers in which small amounts of leakage can be tolerated, such as a cycle-gas exchanger. Also, we use standard exchangers with 18 to 22-BWG tubes in noncritical services where the thermal stresses and corrosion rates are at low levels.

2. Grooving of tubeholes—Two grooves (1/8 in. wide by 1/64 in. deep) are specified for all tubesheet holes. The use of grooves provides inexpensive insurance against the development of leaks by slippage of the tubes in the tubesheets. The maintenance shops have omitted grooves in some floating-head units.

3. Tubesheet thickness—A minimum thickness of 1 in. is used for all new designs. Some of our maintenance shops use 1¼-in. tubesheets in rebuilt units. The thicker tubesheet provides additional assurance of leakproof tube-to-tubesheet joints. The tubes can also be rerolled without developing out-of-roundness in the holes. The added contact area in the tubeholes gives a joint that is not easily loosened by vibration or the unexpected stresses occurring in the tube-to-tubesheet joints during startups or process upsets. If necessary, the shop can machine an eighth of an inch or more, from the face of a corroded tubesheet during retubing operations. This saves fabrication of a new tubesheet.

4. Minimizing crevices—Corrosion and the buildup of salts and other chemicals in the crevices of a heat exchanger will cause premature failure. Normally, we specify that all tubes shall be rolled over the entire thickness of the tubesheet. Rolling beyond backface of the tubesheet could cause stress cracking in stainless steels because the metal will be under tensile stress. Therefore, it is important not to roll beyond the backface. As a cost reduction, partial rolling of tubes in tubeholes has been specified for some exchangers containing clean, noncorrosive fluids and requiring thick (over 2¼ in.) tubesheets.

The crevice between the tubesheet and the shell should also be eliminated by welding from the inside, or specifying the desired type of weld as shown in Fig. UW-13.2 of the ASME Code, Section VIII.

5. Tubesheet venting—Three or four tubesheet vents are specified for all vertical units for the purpose of venting gases trapped under the top tubesheet. The top shellside nozzle in nearly all heat exchangers is located several inches below the top tubesheet. The dead area between the top of the outlet nozzle and the bottom of the tubesheet will fill with air or other inert gases unless vented.

Corrosion and stress cracking occur at the interface between the liquid and the trapped gas because solids generally build up on the tubes at this point. This is the most common cause of failure of vertical condensers and heat exchangers.

To be certain that there is flow through the vents, the piping connecting the vents to a drain should include an open funnel in which the flow can be regularly observed. The outlet-nozzle piping should also include a loop that will fill the shell with liquid above the tubesheet vent. Both items are shown in the following diagram:

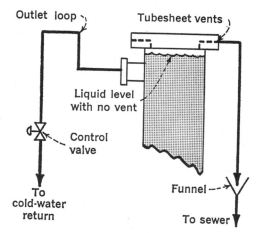

In vertical, external-packed floating-head exchangers, the floating end may be located at the top. Venting of the floating tubesheet is inconvenient. We specify that the highest point of the top shellside nozzle shall be flush with the inboard face of the floating tubesheet so that the gases trapped under the top tubesheet will be swept out through the top nozzle. In addition, a 1-in. coupling is welded to the shell opposite the outlet nozzle and as close to the tubesheet as possible for venting gases.

Tubesheet vents are also used in vertical thermosyphon reboilers to vent inerts and carbon dioxide that can accumulate under the top tubesheet at the corners. We have had corrosion of shells opposite the top steam-inlet nozzle where carbon dioxide accumulated, as shown in the following sketch:

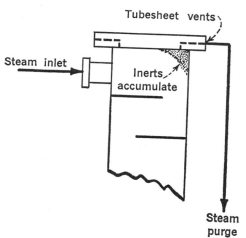

A small flow of steam is vented to the atmosphere.

6. Shells—Standard pipe sizes are specified for all heat-exchanger shells up to 24-in. O.D. This allows the maintenance shops to buy standard piping for replacement of heat-exchanger shells. The expense of rolling shells is saved. Steel is generally used for shells, with cooling water in the shell.

7. Shellside fouling—Shellside fouling with once-through cooling water in the shell is minimized by:

a. Using screens in the cooling-water supply headers to remove particles above 1/16 in.

b. Providing the valves and piping that will permit periodic reversal of the coolant-flow direction during operation if screens in the water-supply headers are not provided. This flow reversal (backwashing) removes silt and also solids that may block the inlet nozzle.

c. Injecting air periodically into the bottom-inlet shellside nozzle—especially during periods when the water contains excessive silt. The shellsides of fouled exchangers have been cleaned without shutting down by injecting air every 1 to 2 min.. through a timing mechanism over a period of several days.

d. Designing for a minimum velocity of 3 ft./sec. to keep silt suspended when using dirty cooling water. Tubes have also been omitted from the baffle windows to minimize the effect of silting.

e. Using 20% baffle cuts, dummy tubes in the pass lanes, and sealing devices between the tube bundle and the shell. Deposits of silt occur where the flow is low, or where there are larger eddies. Improved flow distribution minimizes fouling and silting.

f. Restricting cooling-water outlet temperatures to a maximum of 45 to 50 C. in order to avoid buildup of solids on the tubes.

g. Selecting tube metals that resist fouling. For example, steel tubing is not used when the cooling-water flow is in the shellside. The tubes are normally placed on triangular pitches for cooling-water services. If removal of the bundle and cleaning by high-pressure water are anticipated, a square pitch will be specified.

8. Impingement protection—Impingement plates are used for all steam-inlet nozzles on the shellside of heat exchangers. The plates are normally installed in enlarged sections of the inlet line, which provide twice the inlet-pipe area in the annular space around the plates. Impingement plates are also used for high velocity, two-phase, and abrasive fluids (TEMA-B requirements).

Vapor belts with the shell extended to serve as an impingement plate are used to reduce shellside inlet velocities and to avoid flow-induced vibration. The following sketch shows a typical vapor belt used for introducing gases into the shellside:

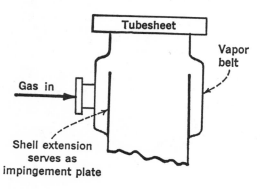

Flow-Induced Vibration

No heat-exchanger design is complete without considering the possibility of flow-induced vibration. This is especially true as larger exchangers, greater fluid-flow rates and higher shellside velocities become more prevalent.

In flow-induced vibration, the force causing the vibration is usually associated with the shedding of vortices from the downstream side of the tube. As a vortex is shed, the flow pattern (and hence, the pressure distribution) changes, causing oscillations in the magnitude and direction of the fluid-pressure forces acting on the tube. If the frequency of vortex shedding approaches the natural frequency of this tube section, the tube will vibrate with large amplitude. And eventually, it will fail, resulting in leakage of one fluid into another.

Several articles have been presented on flow-induced vibration.[1,4,5,7] Fig. 1 may be used to predict flow-induced vibration, with fluid flowing normal to a bank of tubes in a tube bundle. This curve is based on the modulus of elasticity at 70 C. For most materials of construction, the modulus of elasticity decreases as temperature increases, and the values from Fig. 1 will be lower by less than 10% at 600 F. (316 C.). Admiralty metal and aluminum are two exceptions, with reductions of 20% and 10%, respectively, at 600 F.

Axial Load Affects Fluid Velocity

The action of an axial load in a tube may have a significant influence on the acceptable fluid velocity around a tube. For an axial, compressive load:

$$\frac{P}{P_{CR}} + \left(\frac{V}{V_0}\right)^2 = 1$$

where P is axial load, lb.; P_{CR} is buckling axial load, lb.; V is fluid velocity, ft./sec.; and V_0 is critical velocity of fluid, ft./sec.

From this equation, it may be seen that if the compressive load is only half the critical value ($P/P_{CR} = 0.5$), the fluid velocity should not exceed 70% of the velocity that would be acceptable if no axial load were present.

Compressive loads frequently exist under operating conditions in certain tubes of fixed-tubesheet exchangers. The mechanical design of such exchangers often causes the outer peripheral tubes to be subjected to compressive loads. Also, these outermost tubes are usually subjected to higher velocities associated with the fluid inlet and outlet.

Vibration may be eliminated by reducing velocities, decreasing the unsupported span or, in some cases, by

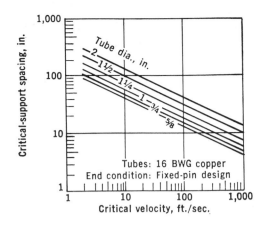

CRITICAL conditions for flow-induced vibration—Fig. 1 ▶

Correction Factors

Tube Arrangement and Tube Spacing				
	Tube Pitch			
TUBE SPACING S/D	Flow ↓ Triangular	Flow ↓ Square	Flow ↓ Rotated Square (45°)	Flow ↓ Rotated Triangular
Velocity Correction When Spacing is Known				
1.25	1.000	0.485	0.375	0.295
1.3	0.985	0.495	0.375	0.305
1.4	0.940	0.590	0.375	0.315
1.5	0.905	0.690	0.375	0.325
Spacing Correction When Velocity is Known				
1.25	1.000	0.696	0.612	0.543
1.3	0.992	0.704	0.612	0.552
1.4	0.970	0.768	0.612	0.561
1.5	0.951	0.831	0.612	0.570

Tube Material Correction		
Material	Velocity When Spacing is Known	Spacing When Velocity is Known
Admiralty	1.023	1.011
Aluminum	1.469	1.212
Aluminum bronze	1.125	1.061
Copper	1.000	1.000
Cupro-nickel	1.087	1.043
Inconel	1.453	1.205
Monel	1.281	1.132
Nickel	1.375	1.172
Stainless steel (austenitic)	1.399	1.183
Stainless steel (ferritic)	1.450	1.204
Steel (carbon)	1.438	1.199

Tube End Conditions	Velocity When Spacing is Known	Spacing When Velocity is Known
Fixed-pinned	1.000	1.000
Fixed-fixed	1.453	1.205
Pinned-pinned	0.641	0.801
Cantilevered	0.230	0.480

Tube Wall Thickness					
Tube O.D.	18 BWG.	16 BWG.	14 BWG.	12 BWG.	10 BWG.
Velocity Correction When Spacing is Known					
5/8"	1.025	1.000	0.973	0.935	0.903
3/4"	1.021	1.000	0.977	0.945	0.916
1"	1.016	1.000	0.982	0.958	0.935
1¼"	1.013	1.000	0.986	0.966	0.947
1½"	1.011	1.000	0.988	0.971	0.956
2"	1.008	1.000	0.991	0.978	0.966
Spacing Correction When Velocity is Known					
5/8"	1.012	1.000	0.986	0.967	0.950
3/4"	1.010	1.000	0.988	0.972	0.957
1"	1.008	1.000	0.991	0.979	0.967
1¼"	1.006	1.000	0.993	0.983	0.973
1½"	1.005	1.000	0.994	0.985	0.978
2"	1.004	1.000	0.995	0.989	0.983

Problems With Heat Exchangers and Their Cures—Table II

Symptom	Diagnosis	Cure
1. Pressure surging in horizontal, three-phase vaporizer.	Vacuum below control valve caused flashing and unsteady flow. Pipe size and flowrate fall within pulsating flow area in Fig. 12 of Simpson's article.[6]	An orifice was installed at vaporizer inlet. Pressure drop across orifice was 6 psi.
2. Karbate tubes broke in horizontal condenser. (Cooling water in shellside.)	Surging flow in cooling-water return line (vertical downflow) caused tube breakage.	Valve was installed at top of water-outlet line, and air admitted to relieve vacuum and give steady flow.
3. Shellside outlet temperature cannot be controlled within desired range (55 to 62 C.) by controlling flow of 125 C. water to tubes. Exchanger has 4-pass tubeside.	Heat exchanger is considerably oversized for the duty (because of an alternate service). Temperature-correction factors for mean-temperature difference (MTD) fluctuate widely with small changes in tubeside flow.	Tubeside tempered-water temperature reduced to 70 C., and control valve removed. Control valve installed in new shellside bypass line.
4. Vacuum jets overloaded with uncondensed vapor from vacuum condenser. (Cooling water in shellside, vapor in tubes.)	Large eddies behind 45% baffle-cuts provided inefficient condensing for many tubes.	Replaced condenser with new unit having 20 to 25% baffle cut.
5. Shell assumed banana-shape, and piping connections leaked. Leakage between tube and shellside (see photo).	Vertically cut baffles, and inlets and outlets on top of shellside, caused stratification of gases at top of shell. Poor distribution of hot gases led to unequal expansion of tubes.	Increased the number of baffles from two to three; welded baffles in shell; installed sealing strips at edges of bundle; installed three concentric cones in tubeside inlet; installed vapor belt for shellside inlet nozzle; changed baffles from vertical to horizontal cut.
6. Air-cooled steam condenser performing below design capacity.	Careful measurement of tube levels disclosed that tubes sloped ¼ in. in wrong direction (rising toward condensate end).	Raised inlet end to obtain 2-in. slope toward condensate outlet.
7. Horizontal vaporizer superheater (four tubeside passes) was operating well below calculated capacity.	Film boiling of ethyl acetate occurred because of the large temperature differences.	Installed pressure controller in outlet vapor line to raise vaporizer pressure from 120 mm. Hg to 27 psig., thus raising vaporization temperatures from 32 to 105 C.
8. Falling-film vaporizer performed little vaporization with 75-psig. steam in shell, and 340-mm. Hg tubeside pressure. Liquid was mixture of "heavies" boiling at 130 C., and "lights" boiling at 60 C.	High MTD caused falling film to move away from tubewall. Film boiling was occurring.	Warm water (80 to 85 C.) was used in shellside for heating in place of 75-psig. steam. Preheater was added to the process-inlet line to increase flashing in the inlet channel.
9. Tubes in front of shellside inlet nozzle failed one-by-one during first five days of operation.	Inlet nozzle velocity (17.2 ft./sec.) caused tubes to vibrate at natural frequency. Excitation velocity calculated to be 16.7 ft./sec.	Two rows of tubes in front of nozzle were plugged after inserting rods inside tubes. Rods prevented broken tubes from damaging other tubes.
10. Low heat-transfer rate (U = 11 Btu./(hr.) (sq. ft.) (°F.) in vertical, spiral-plate heat exchanger was cooling latex with water.	Channeling in vertical spiral permitted cold latex to accumulate at bottom, and hot thinner latex to flow in top half of spiral channel.	Changed spiral to horizontal position. Heat-transfer rate was almost doubled [U = 19 Btu./(hr.) (sq. ft.) (°F.)].

altering the method of fixing or pinning the ends of the unsupported span.

Integral Double Tubesheets

Special, integral double tubesheets can be constructed to overcome the problems associated with conventional designs. A single tubesheet can be machined with interconnecting passages between the tubeholes leaving metal lands (see Fig. 2) to provide strength and maintain uniform temperature profiles throughout the tubesheet. The tubesheet passages can be machined by conventional mechanical methods or by electrochemical machining techniques. The passages can be purged, vented and monitored, or pressurized to detect leaks.

Plant Problems

The testing of a heat exchanger does not necessarily have to be a costly project. If a quick answer is needed to permit corrective action for a poorly performing exchanger, quick readings of temperatures, pressures and flowrates can be made (flow measurements may be possible for one side only).

The troubleshooter may be able to locate the problem by observing the operation, checking the temperatures of the channels and shell at different points, studying the adequacy of the piping to and from the exchanger, and examining the exchanger drawings. From the data, he can make calculations that may pinpoint the problem.

Gilmour's article,[3] Trouble-Shooting Heat-Exchanger Design, contains a step-by-step procedure for testing a heat exchanger. He recommends preliminary testing with available instruments to calculate heat loads. The need for additional or more-accurate instruments may then be apparent. After the test is completed and the data sheet filled out, the final step at the test site is to be certain that there is a heat balance within 10% between the tubeside and shellside streams. An evaluation of the data is made to compare the observed overall coefficient and the design coefficient. Discrepancies may be caused by deviations in physical properties of fluids, flowrates, inlet temperatures, fouling, mechanical construction, or those caused during the installation of the heat exchanger.

In summary, Table II presents some recent plant problems with heat exchangers and provides answers.

Maintenance Information

For more information on maintenance and troubleshooting, see Design of Heat Exchangers, pp. 14-36, and especially Ref. 147 through 153 dealing with maintenance, and Ref. 154 through 156 dealing with troubleshooting.

References

1. Chen, Y. N., Flow-Induced Vibration and Noise in Tube-Bank Heat Exchangers due to von Karman Streets, ASME 67-VIBR-48 (1967).
2. Gilmour, C. H., No Fooling—No Fouling, *Chem. Eng. Progr.*, July 1965, p. 49.
3. Gilmour, C. H., Trouble-Shooting Heat-Exchanger Design, *Chem. Eng.*, June 19, 1967, p. 22.
4. Lentz, G. J., Flow-Induced Vibration of Heat-Exchanger Tubes Near the Shellside Inlet and Exit Nozzles, AIChE Preprint No. 20, National Heat Transfer Conference, Minneapolis, Aug. 1969.
5. Nelms, H. A. and Segaser, C. L., Survey of Nuclear Reactor System Primary Circuit Heat Exchangers, ORNL-4399, UC-38-Engineering and Equipment, Oak Ridge National Laboratory, Apr. 1969.
6. Simpson, L. L., Sizing Piping for Process Plants, *Chem. Eng.*, June 17, 1968, Fig. 12, p. 204.
7. Thompson, H. A., Fatigue Failures Induced in Heat-Exchanger Tubes by Vortex Shedding, ASME Paper 69-PET-6, Sept. 1969.

Meet the Authors

Richard C. Lord   Paul E. Minton   Robert P. Slusser

Richard C. Lord is a project engineer in the engineering department at Union Carbide Corp.'s Technical Center, P.O. Box 8361, South Charleston, W. Va., where he is part of the heat-transfer technology group. Formerly he was involved in activities related to process design and economics. He has a B.S. degree in chemical engineering from the University of Maine and is a member of AIChE.

Paul E. Minton is a project engineer in the engineering department at Union Carbide Corp.'s Technical Center, P.O. Box 8361, South Charleston, W. Va., where he is a part of the heat-transfer technology group. A B.S. graduate in chemical engineering from ical engineering from the University of Maine and is a member of AIChE.

Robert P. Slusser is a project engineer in the engineering department at Union Carbide Corp.'s Technical Center, P.O. Box 8361, South Charleston, W. Va., where for the last 10 years he has been designing equipment as a member of the heat-transfer technology group. His past work includes 12 years as production supervisor in the vinyl-resin facilities. He hold B.S. and M.S. degrees in chemical engineering from Massachusetts Institute of Technology and is a member of AIChE.

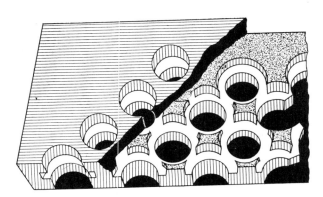

INTEGRAL double tubesheet has passages that are machined from single thickness of steel plate—Fig. 2

Horizontal-thermosiphon-reboiler design

This step-by-step procedure—proven by successfully operating installations—enables a process engineer to size and layout a horizontal thermosiphon reboiler system, including piping, elevations and exchanger shells.

Gerald K. Collins, El Paso Products Co.

☐ Any engineer occasionally responsible for designing fractionating equipment usually has on hand the data and methods for establishing the distillation heat requirements. When it comes to the reboiler design, however, he must either use a rough analogy or spend time and/or money obtaining an engineering proposal from a heat-exchanger manufacturer.

Usually, the choice of a reboiler is contingent on prior decisions that have been made with respect to the distillation system. If the tower pressure is high with respect to subsequent operations, for example, it may be possible to transfer the tower's bottoms under pressure without the aid of a pump. This in turn may permit the tower to be located a minimum distance above grade, so that thermosiphon reboilers would be impossible and the most economical solution would probably be a kettle-type reboiler.

If, on the other hand, the tower's pressure is such that the bottoms product must be pumped, the tower must be elevated to provide net positive suction head for that pump; and the same elevation might be used for thermosiphon reboiling. The choice is then that between thermosiphon reboiling and pump-around reboiling by means of an oversized bottoms pump. Usually, the thermosiphon type is the most economical of these two systems, because with proper drawoff arrangements it can be made to provide one theoretical step of distillation; and, depending on the tray efficiencies and spacing, this theoretical step can be equivalent to 4–9 ft of tower height and 2–3 trays.

Even when the decision has been made for a thermosiphon reboiler, there remains the choice between vertical thermosiphon (with the reboiling liquid in the tubes) and horizontal thermosiphon (with reboiling in the shell). One rule of thumb has been that, if the viscosity of the reboiler liquid is less than 0.5 cp, the vertical thermosiphon should be considered, but that if the viscosity is more than 0.5 cp, the horizontal reboiler is likely more economical. More importantly, horizontal thermosiphon reboilers have a number of advantages (see box) that, with the general characteristics of distillation systems, cause them to be the most common of all three types of reboiler.

Although a number of good articles have discussed thermosiphon reboilers, [1–6] and one article deals rigorously with the hydraulics of reboiler systems [7], specific design information for horizontal reboilers is limited, even though many installations of this type are used in the chemical and petroleum industries.

The following method, proven in actual practice, includes as an example the calculation of the reboiler for a C_3-splitter that has been installed, and presently operates at design specifications.

Guidelines

Once you have decided that a horizontal thermosiphon reboiler is the best type for your service, you should:

1. Establish the amount of vaporization and reboiler circulation.
2. Determine the allowable pressure drop through the system.
3. Prepare a layout based on the required difference of elevations and on the exchanger surface.
4. Size the piping.
5. Determine the locations of the outlet and return nozzles on the tower.
6. Specify the reboilers.

The normal range of vaporization is 10–20% of the liquid entering the reboiler, with a maximum of 30% to avoid pressure drop due to acceleration and flow surges. The allowable pressure drop across the exchanger alone should be 0.25–0.5 psi, while the total pressure drop must be less than the difference in hydrostatic head

Originally published July 19, 1976

between the liquid-filled outlet to the reboiler and the vapor-liquid-filled return line.

The normal range of liquid velocity in the reboiler inlet is 2–7 ft/s, while the maximum velocity of the vapor-liquid mixture in the return line may be taken as $(4,000/\rho_m)^{1/2}$ where ρ_m is the density of the mixture, and

$$\rho_m = \frac{100}{\% \text{ vap.}/\rho_v + \% \text{ liq}/\rho_l}$$

In preparing the layout, it is usually good practice to provide for two inlets to each reboiler shell (see figure), for distribution of the inflowing liquid along the exchanger, as well as a dome on the top of each shell, for flow transition into the return line. The number of these shells must be estimated, and this requires a step-by-step estimation of the exchanger design. The overall duty has been set by the tower heat balance, and the temperature of the reboiling liquid by a bubble-point calculation on the inlet liquid and a vapor-liquid equilibrium calculation on the allowable percent flash.

The mean temperature difference between these reboiling temperatures and the tubeside fluid should be limited to about 90°F, and in many cases should be less, to avoid operating problems and degradation of the bottoms. The film coefficient can be taken from specific data; or, when this is lacking, a rough estimate can be based on: 500 Btu/(h)(ft²)(°F) for 100 psig and above, 300 Btu/(h)(ft²)(°F) for atmospheric pressure, and 200 Btu/(h)(ft²)(°F) for vacuum. The overall transfer coefficient calculated from these film coefficients can be checked against rule-of-thumb maximum heat fluxes of: 15,000 Btu/(h)(ft²) for 3/4-in tubes and 18,000 Btu/(h)(ft²) for 1-in tubes. These fluxes may be increased 30–35% for water inside tubes and should be decreased 20–30% for vacuum service.

The overall transfer coefficient may then be used to calculate the total exchanger surface from the known heat duty and mean temperature difference. This total surface is then translated into one or more exchanger shells operating in parallel, taking into account the problems of symmetrical piping arrangements and the cost per shell.

With the number of shells and the layout established, it is possible to locate relative elevations of equipment and the tower's drawoff and return lines. The minimum elevation difference between the liquid level in the bottom of the column and the centerline of the reboiler bundles (see figure) may be calculated by:

$$H_1 = \frac{144 \Delta P + \rho_m H_2}{\rho_l - \rho_m}$$

where: ΔP = total pressure drop through the system, from column outlet to column return, psi
ρ_1 = density of liquid entering the reboiler, lb/ft³
ρ_2 = density of vapor-liquid mixture leaving the reboiler, lb/ft³
H_2 = difference in elevation between the centerline of the return nozzle on the column and the liquid level in the column, ft.

A factor of 1.3–2.0 is often introduced with the ΔP term in this equation, as a margin of safety in the first estimate.

In summary, thermosiphon reboiler design involves relating a predicted circulation rate to a predicted heat-transfer area. These relationships are further revealed through the following calculation:

Sample calculation

A reboiler system for a large C_3-splitter column will be designed, using some plant data and the guidelines given above:

Known:
- Reboiler duty = 80 million Btu/h.
- 170°F hot water available for heat, with 145°F return desired.
- Column bottoms is 99% propane at 260 psia and 126°F.
- The maximum difference in elevation between the centerlines of the reboiler bundle and of its return line to the column is 18 ft.

Data for propane at column bottom conditions:

$\rho_l = 28.08$ lb/ft³ liquid density
$\rho_v = 1.88$ lb/ft³ vapor density
$\lambda = 121$ Btu/lb heat of vaporization

Design basis for reboiler:
Amount vaporized = 15% (typical for horizontal unit reboiling C_3)
Maximum shellside pressure drop = 0.3 psi (This will be specified to the manufacturer.)
Density of the vapor-liquid mixture:

$$\rho_m = \frac{100}{\frac{85}{28.08} + \frac{15}{1.88}} = 9.08 \text{ lb/ft}^3$$

Amount of hydrocarbon vaporized = 80,000,000/121
= 660,000 lb/h

Reboiler circulation = 660,000/0.15
= 4,400,000 lb/h
= 1,222 lb/s

Estimated elevation, H_1, including safety factor:

$$H_1 = \frac{1.5 \times 144 \Delta P + \rho_m H_2}{\rho_l - \rho_m}$$

From the known limitations, we have $H_1 + H_2 = 18$ ft, maximum. Combining the two relations for H_1 and H_2, we have:

$$18 = \frac{216 \Delta P + \rho_l H_2}{\rho_l - \rho_m}$$

Substituting values of ρ_l and ρ_m, we have:

$$7.69 \Delta P + H_2 = 12.2$$

From this, we assume some reasonable values for H_2 and calculate a table of corresponding values for the estimated allowable system ΔP, as follows:

H_2, ft	ΔP, psi
6.0	0.81
5.0	0.94
4.0	1.07
3.0	1.20

70 HEAT EXCHANGER DESIGN AND SPECIFICATION

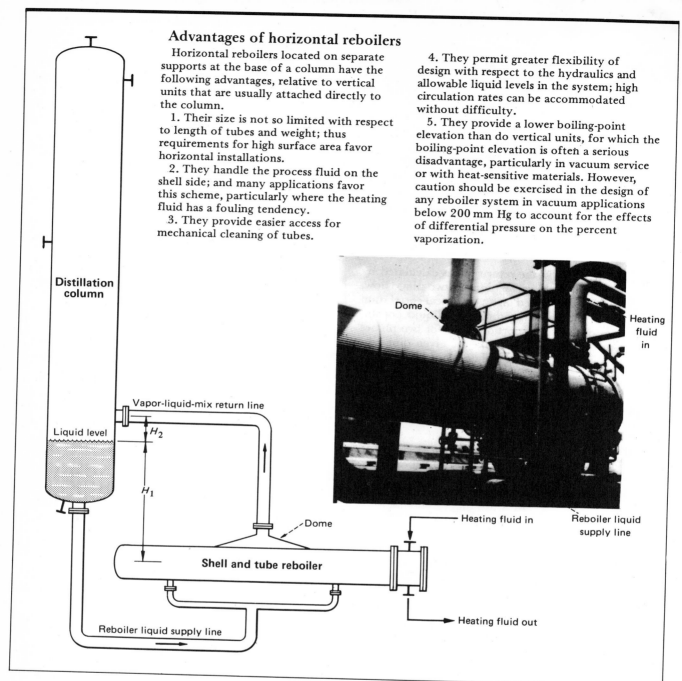

Advantages of horizontal reboilers

Horizontal reboilers located on separate supports at the base of a column have the following advantages, relative to vertical units that are usually attached directly to the column.

1. Their size is not so limited with respect to length of tubes and weight; thus requirements for high surface area favor horizontal installations.
2. They handle the process fluid on the shell side; and many applications favor this scheme, particularly where the heating fluid has a fouling tendency.
3. They provide easier access for mechanical cleaning of tubes.
4. They permit greater flexibility of design with respect to the hydraulics and allowable liquid levels in the system; high circulation rates can be accommodated without difficulty.
5. They provide a lower boiling-point elevation than do vertical units, for which the boiling-point elevation is often a serious disadvantage, particularly in vacuum service or with heat-sensitive materials. However, caution should be exercised in the design of any reboiler system in vacuum applications below 200 mm Hg to account for the effects of differential pressure on the percent vaporization.

The next step is to determine the probable number of reboiler exchangers required for transferring the design heat load, so that the layout can be made and the system ΔP calculated.

$$\Delta T = \frac{(170 - 126) - (145 - 126)}{\ln(44/19)} = 29.7°F$$

For the first estimate, assume an overall transfer coefficient, U, equal to 100 Btu/(h)(ft²)(°F). Then:

$$A = \frac{80 \times 10^6}{100 \times 29.7} = 27,000 \text{ ft}^2$$

Considering both capital cost and ease of installation (and since conventional smooth tubes are desirable for this service), a total of three reboiler shells will be required, each having tube bundles with about 10,000 ft².

Flow to each reboiler = 1,222/3 = 407 lb/s

Assume a velocity of 5 ft/s in each of the reboiler supply lines. Then:

Required cross-sectional area =
(407 lb/s)/(28.08 lb/ft³)(5 ft/s) = 2.9 ft²

This is equivalent to a 24-in-dia. schedule 20 pipe. Since these exchangers have very large shells, each supply line should branch into two 18-in legs for divided flow to each reboiler.

Using pipe-flow pressure-drop charts for this inlet

piping, the following ΔPs are obtained:

	ΔP, psi
About 20 ft of 24-in pipe with two 90-deg elbows	0.08
Tee plus split flow in the 18-in leg (6 ft plus elbow)	0.04
Entrance loss at exchanger inlet nozzle	0.02
Subtotal to reboiler	0.14

Taking the guideline maximum velocity for the mixture leaving the top of each shell as $(4,000/\rho_m)^{1/2}$, we have the maximum velocity as $(4,000/9.08)^{1/2} = 21$ ft/s. Since this is the maximum, the actual pipe will be chosen to give a lower velocity. Thus a 24-in-dia. schedule 20 pipe will give a velocity of:

$$(407 \text{ lb/s})/(9.08 \text{ lb/ft}^3)(2.9 \text{ ft}^2) = 15.5 \text{ ft/s}$$

The pressure drop in the return line is calculated by the Lockhart-Martinelli correlation for two-phase flow. [1] This calculation is based on an empirical correlation of the ratio of the pressure drop of the liquid portion (ΔP_L) to that of the vapor portion (ΔP_V) for two phases in turbulent-turbulent flow within the given pipe segment. A numerical value is calculated for $(\Delta P_L/\Delta P_V)^{1/2}$ and used as the X_{tt} value on a graph to obtain a factor, ϕ_L. Then the two-phase pressure drop equals $\Delta P_L(\phi)^2$. Thus a 24-in return line that is about 20 ft long and has one 90-deg elbow gives a pressure drop of 0.86 psi.

Allowing 0.3 psi through each shell, the total pressure drop through the system becomes:

$$\Delta P = 0.14 + 0.30 + 0.86 = 1.3 \text{ psi}$$

From the comparison of estimated ΔPs with H_2, we see that the nearest practical value for H_2 is 3 ft. Now, we can repeat the calculation of H_1, using this value of H_2 and eliminating the safety factor:

$$H_1 = \frac{144 \Delta P + \rho_m H_2}{\rho_l - \rho_m} = \frac{144(1.3) + 9.08(3)}{28.08 - 9.08}$$

$$H_1 = 11.3 \text{ ft}$$

Since the allowable maximum value of $H_1 + H_2$ is 18 ft, we can plan to use $H_1 = 15$ ft and $H_2 = 3$ ft. It is typical to have values of H_1 equal to 2–8 times those of H_2. However, the size of H_2 should never be so small that fluctuations in the liquid level would block the return of the vapor-liquid mixture.

With the reboiler-system elevations established, the design heat-exchange area for each of the three reboiler tube bundles needs to be calculated. Areas are sized for 115% of normal duty, in accordance with usual reboiler design policy.

Each reboiler duty = $(80 \times 10^6)(1.15/3) = 30.7 \times 10^6$ Btu/h

Design hot-water temperature drop = $(170 - 145) = 25°$F.

Boiling-propane temperature = $126°$F.

Published literature and plant data provide the following film coefficients for these conditions:

	Btu/(h)(ft^2)(°F)
propane boiling shellside	1,000
hot water inside tubes	500

	Resistance, (h)(ft^2)(°F)/Btu
tube wall (13 BWG steel)	0.0002
propane-side fouling	0.0018
hot-water-side fouling	0.0050

From the expression for the overall coefficient:

$$U = (1/h_s + 1/h_i + \Sigma R)^{-1}$$
$$U = (0.001 + 0.002 + 0.007)^{-1}$$
$$= 100 \text{ Btu/(h)(ft}^2)(°F)$$

This calculated value of U corresponds to that assumed earlier and is typical of the heat-transfer coefficient expected for smooth tubes with normal fouling in this service. From this, the required surface per tube bundle may be calculated as:

$$A = (30.7 \times 10^6)/(100 \times 29) = 10,600 \text{ ft}^2$$

The corresponding heat flux, which is 2,900 Btu/(h)(ft^2), is well within the guideline values of limiting flux.

To complete the reboiler design, a choice of tube length, pitch and diameter must be made. However, in this area the plant engineer can heavily rely on exchanger manufacturers or vendors. When submitting requests for bids on reboilers, the engineer should remember that standardized sizes, among other things, may cause some variation in actual heat-transfer area or number of shells proposed by various bidders. A company bidding on the exchangers will have made its own calculation using the data supplied, and is offering its reputation as a guarantee that its proposed equipment should perform as specified in the request. Most vendors will also recommend a configuration for the outlet dome design.

References

1. Fair, J. R., What you need to design thermosiphon reboilers, Pet. Ref., 39, No. 2, p. 105 (1960).
2. Jacobs, J. K., Reboiler selection simplified, Pet. Ref., 40, No. 7, p. 189 (1961).
3. Hughmark, G. A., Designing thermosiphon reboilers, Chem. Eng. Prog., 57, No. 7, p. 43 (1961).
4. Ibid, July, 1964, p. 59.
5. Ibid, July, 1969, p. 67.
6. Fair, J. R., Vaporizer and reboiler design—Part I, Chem. Eng., July 8, 1963, p. 119.
7. Kern, Robert, How to design piping for reboiler systems, Chem. Eng., Aug. 4, 1975, p. 107.

The author

Gerald Collins is a senior process design engineer with El Paso Products Co., P.O. Box 3986, Odessa, TX, 79760. Prior to joining El Paso Products, in 1969, he was a process engineer with Texaco, Inc. A registered engineer in the State of Texas and a member of the Permian Basin chapters of the AIChE and the Texas Society of Professional Engineers, he holds a B.S. in Chemical Engineering from Texas University (1962) and has done graduate study at the University of Texas of the Permian Basin.

How to design piping for reboiler systems

Interactions between hydraulic requirements and piping configurations require close attention to many fluid and mechanical details, in order to obtain the most efficient and economical distillation units.

Robert Kern, Hoffmann - La Roche Inc.

☐ Familiarity with graphic piping design is an essential requirement for the designer of hydraulic systems. The accuracy of his calculations, predictions of flowrate and pressure differential, reliability of operation, and the economy of capital, energy, maintenance and operating costs depend to a great extent on pipe configurations and pipe components.

In these articles, we have recognized the importance of graphic piping design, and to a limited degree have presented its fundamentals. We will now evaluate the flow systems and piping design for a distillation column, which is a more integrated unit than the individual systems discussed in our earlier articles.

Layout for distillation columns

A process flow diagram of a typical distillation column with bottom pump, thermosyphon reboiler, overhead condenser, reflux drum and reflux pump are shown in Fig. 1 (F/1). The equipment components are located adjacent to each other in the actual plant. Also in F/1, we find elevation and plan drawings for the column. These show how the principal elements of a distillation column are usually integrated into an overall plant arrangement. Manholes face access roads (or access aisles at housed installations). Each manhole has a platform for maintenance. Valves and instruments are located above these platforms for convenient access.

For economy and easy support, piping should drop immediately upon leaving the tower nozzle, and run parallel, and as close as possible, to the tower itself. A vertical line lends itself as a suitable location for the straight run of an orifice. The horizontal elevations, after the lines leave their vertical run, are governed by the elevations of the main pipe rack. Lines that run directly to equipment at grade (more or less in the direction of the main pipe rack) often have the same elevation as the pipe bank.

Lines from tower nozzles below the pipe rack should approach the pipe bank roughly 2 ft below the pipe-rack elevation. The same elevation is used for those lines that run to pumps located below the pipe rack.

Pump-suction lines can also be arranged on this elevation. They should be as short as possible and run without loops or pockets. Pipelines, dropping from above the pipe-rack elevation, will approach the pipe bank roughly 2 ft higher than the elevation of the main pipe bank. This elevation is also used for steam lines to reboilers. These steam lines usually connect to the top of the headers to avoid excessive condensate drainage toward process equipment.

The plan view (F/1) of the tower shows the segments of its circumference allotted to piping, nozzles, manholes, platform brackets and ladders. Such a pattern usually leads to a well-organized arrangement for the process equipment and auxiliary components.

From a layout standpoint, it is preferable to have equal platform-bracket spacing, and the orientation of brackets lined up along the entire length of the tower. This will minimize interferences between the piping and structural members. According to OSHA, ladders between platforms should not be longer than 30 ft.

Area segments for piping going to equipment at grade are available between the ladders and on both sides of the manholes. Lines approaching the pipe rack

Originally published August 4, 1975

HOW TO DESIGN PIPING FOR REBOILER SYSTEMS

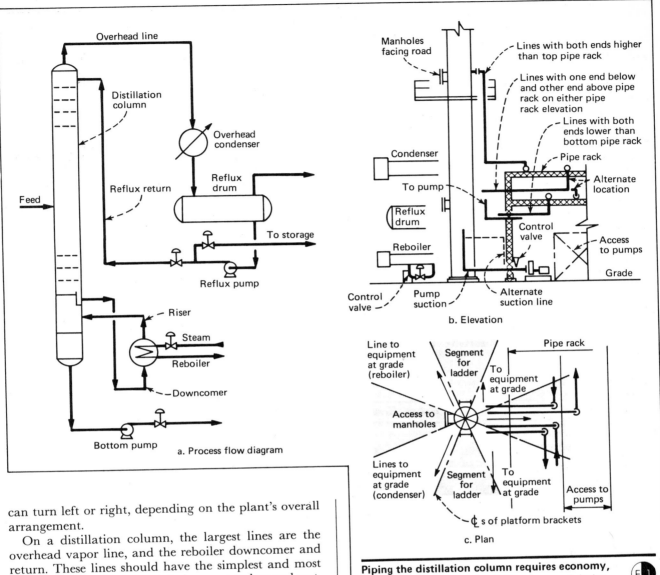

Piping the distillation column requires economy, ease of support and accessibility for good design.

can turn left or right, depending on the plant's overall arrangement.

On a distillation column, the largest lines are the overhead vapor line, and the reboiler downcomer and return. These lines should have the simplest and most direct configurations, to minimize pressure loss and cost.

During normal operation, the pump in Fig. 1 transports liquid at equilibrium. This means that the distillation column and reflux drum are elevated to satisfy NPSH (net positive suction head) requirements. The discharge lines often have two destinations. Total-head requirements should be designed and calculated so that operating points fall on the pump's head-capacity curve when pumping to an alternative destination. Alternative discharge lines can have equal capacity and alternative operation, or partial capacity with simultaneous operation. All alternatives should be investigated for the process pumps.

The design for a pumped reboiler circuit is similar to that of a reflux pump system. A bottom pump transports the liquid through an exchanger or fired heater and returns it to the distillation column. Close attention should be paid to possible two-phase flow in pipelines coming after the heater—especially when we want to locate the heater close to the column.

Inserted-type reboilers have no process piping. Larger-diameter towers can have one to four U-tube stub bundles inserted directly into the liquid space through tower nozzles, and extending across the tower diameter. Reboilers with small heat duties are usually designed as helical coils.

Reboiler arrangements

In horizontal thermosyphon reboilers, liquid flows from an elevated drum or tower bottom or tower-trapout boot through a downcomer pipe to the bottom of exchanger shell. The liquid is heated, leaves the reboiler in the return piping as a vapor-liquid mixture, and flows back to the tower or drum.

In vertical reboilers, heating usually occurs on the shellside. In horizontal reboilers, heating is on the tubeside. For a large evaporation rate (for example, 90% of total flow), a kettle-type reboiler is used.

Piping to horizontal reboilers is designed as simply and directly as possible within the limitations of thermal-expansion forces.

Symmetrical arrangements between the drawoff and

74 HEAT EXCHANGER DESIGN AND SPECIFICATION

reboiler-inlet nozzles, as well as between the reboiler outlet and return connection on the tower, are preferred in order to have equal flow in the reboiler circuit. A nonsymmetrical piping configuration may also be accepted for a more-economical or more-flexible piping design.

Reboilers often have two outlets and two parallel-pipe segments. When sizing and arranging nonsymmetrical piping, an attempt should be made to equalize the resistance through both legs of the reboiler piping. More resistance in one leg produces a smaller flow than in the other. Hence, uneven heat distribution will occur in the reboiler—one segment of the riser will be hotter than the other.

At startup in reboilers having high, liquid drawoff nozzles, a gravity-flow bypass is usually provided from the tower's liquid space to a low point of the downcomer.

Valves are rarely included in reboiler piping, except when a standby reboiler is provided, or when two or three reboilers are used and operated at an extremely wide heat-capacity range. Some companies require line blinds to blank off the tower nozzles during shutdown, turnaround and maintenance.

The heating media (steam or a hot process stream) connect to the tubeside of horizontal reboilers. The inlet piping usually has a temperature-regulated control valve (with block valves and bypass globe valve, if required). This is normally arranged at grade near the reboiler's tubeside inlet.

Reboiler elevations

Most reboilers are at grade next to the tower, with centerline elevations of about 3 to 5.5 ft above ground level for exchangers about 1 to 3 ft dia. Exchangers at grade provide economical arrangements—valves and instruments are accessible, tube-bundle handling is convenient, and maintenance is easy. In this arrangement, the static heads are well determined between the exchanger's centerline and the drawoff and return nozzles on the tower. Vertical reboilers are usually supported on the distillation column itself.

Some reboilers have a condensate or liquid-holding pot located after the tubeside outlet, as shown in F/2. In such cases, the centerline elevation of the reboiler is somewhat higher than units that do not have these control vessels.

The arrangement in F/2a is a high-capacity steam trap. The top of the condenser pot should not be higher than the bottom of the exchanger shell, to avoid flooding the tubes with condensate and adversely affecting the exchanger's heat-transfer duty.

The arrangement in F/2b maintains a required condensate level in the reboiler, to provide for a wide range of heat-transfer control. Process conditions determine the precise relationship between the exchanger and the vertical condensate-control pot.

In F/3, we show an example where a reboiler has been elevated to meet the NPSH requirement of the centrifugal pump. The elevated reboiler, in turn, raises the tower because the minimum liquid level in the bottom of the tower must be higher than the liquid level in the exchanger. The elevation difference (dimen-

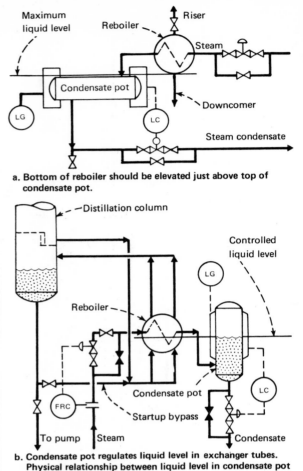

a. Bottom of reboiler should be elevated just above top of condensate pot.

b. Condensate pot regulates liquid level in exchanger tubes. Physical relationship between liquid level in condensate pot and required liquid level in exchanger tubes is important.

Process conditions determine equipment elevations. F2

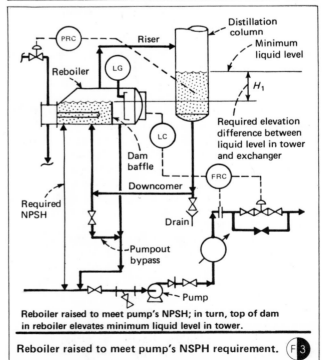

Reboiler raised to meet pump's NPSH; in turn, top of dam in reboiler elevates minimum liquid level in tower.

Reboiler raised to meet pump's NSPH requirement. F3

HOW TO DESIGN PIPING FOR REBOILER SYSTEMS

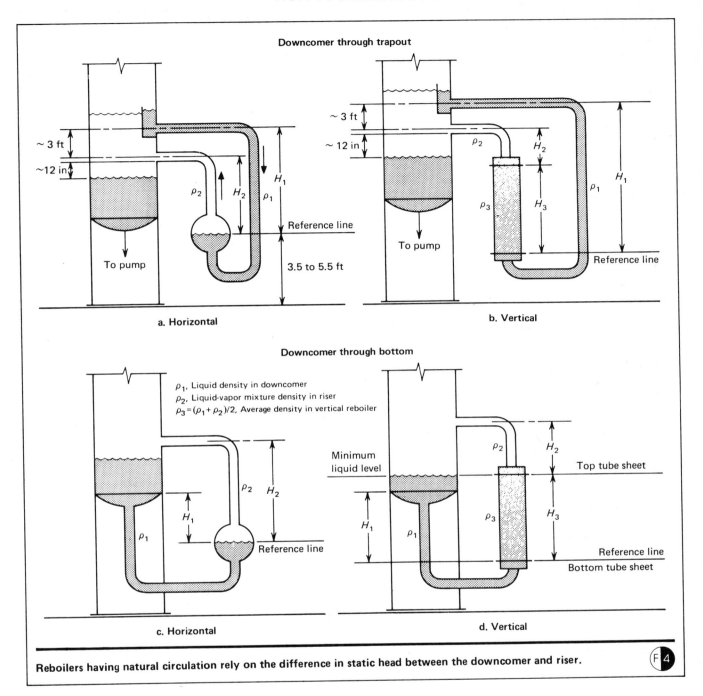

Reboilers having natural circulation rely on the difference in static head between the downcomer and riser.

sion H_1 in F/3) provides the positive static head for flow in the reboiler circuit, and overcomes friction losses in the exchanger, and downcomer and return lines.

Designing the reboiler system

Typical reboiler arrangements are shown schematically in F/4. In all cases, the vessel pressure is the same at the tower's outlet and return nozzles. Circulation is forced by the static-head difference between the liquid column in the downcomer and the vapor-liquid column in the riser. For convenience, reference lines are chosen at the exchanger's centerline for horizontal reboilers, and at the bottom tubesheet for vertical reboilers.

If P_1 is liquid pressure in the downcomer at the reference line, and P_2 is backpressure in the riser's vapor-liquid column, the pressure difference ($\Delta P = P_1 - P_2$) must overcome the exchanger and piping friction losses. Therefore, P_1 must be greater than P_2. If ρ_1 is the hot liquid density in the downcomer, then $\rho_1 H_1 / 144 = P_1$, psi. The backpressure, P_2, can have two alternative expressions:

1. For horizontal exchangers (see F/4a and F/4c):

$$P_2 = \rho_2 H_2 / 144, \text{ psi} \qquad (1)$$

where:

$$\rho_2 = \frac{W}{\dfrac{W_l}{\rho_l} + \dfrac{W_v}{\rho_v}} = \frac{100}{\dfrac{\% \text{ Liquid}}{\rho_l} + \dfrac{\% \text{ Vapor}}{\rho_v}} \qquad (2)$$

2. For vertical exchangers (see F/4b and F/4d):

$$P_2' = (\rho_2 H_2 + \rho_3 H_3)/144, \text{ psi} \quad (3)$$

where ρ_2 is again the mixture's density, as expressed by Eq. (2) for horizontal exchangers, and ρ_3 is the average density of liquid and liquid-vapor mixture in the reboiler:

$$\rho_3 = (\rho_1 + \rho_2)/2 \quad (4)$$

Eq. (4) provides a conservative estimate of the density gradient in vertical reboilers. Actual density will be less than that expressed by Eq. (4). In all equations, the units for ρ are lb/ft^3, and for H, ft. We also note that the vertical reboiler should be flooded. The maximum elevation of the top tubesheet should not be higher than the minimum liquid level in the tower.

Hydraulics in horizontal reboilers

In the following discussion, the hydraulic conditions only in horizontal exchangers will be developed. (The derivations are the same for vertical exchangers, except that P_2' will replace P_2.) For horizontal exchangers:

$$P_1 - P_2 = \Delta P = (1/144)(\rho_1 H_1 - \rho_2 H_2) \quad (5)$$

If a safety factor of 2 is introduced, then the available pressure difference for friction losses is halved, and:

$$\Delta P = (1/288)(\rho_1 H_1 - \rho_2 H_2) \quad (6)$$

The quantity $(H_1 - H_2)$ is usually 3 ft (see F/4a). Consequently, a minimum driving force of $\Delta P_{min} = (3/288)\rho_1 \approx 0.01\rho_1$ is always available at horizontal exchangers.

The maximum possible driving force depends on the elevation difference between the drawoff nozzle and exchanger centerline (dimension H_1) and on the total evaporation taking place in the reboiler. Neglecting the vapor-column backpressure in the return line, the maximum usable driving force is:

$$\Delta P_{max} = (H_1/288)\rho_1 \quad (7)$$

In most applications, the actual driving force is not much below this maximum. H_1 can range from 6 to 24 ft, depending on the size of the arrangement and on NPSH for a pump taking suction at the bottom of the tower. For these H_1 values:

$$\Delta P_{max} = (6/288)\rho_1 \text{ to } (24/288)\rho_1$$
$$\Delta P_{max} = 0.02\rho_1 \text{ to } 0.08\rho_1 \quad (8)$$

Thus, the driving force is reduced to a function of the downcomer liquid density at operating temperature. For example, if the piping geometry produces $H_1 = 12$ ft, and $\rho_1 = 50$ lb/ft^3 for kerosene:

$$\Delta P_{max} = (12/288)50 \approx 2.0 \text{ psi}$$

These simple relationships are useful when the evaporation rate is not known and line sizes have to be estimated. The available driving force will be near but less than ΔP_{max}.

ΔP_{max} as evaluated here is, of course, an extreme value taken at total evaporation. In reboilers, partial evaporation usually takes place, which will reduce ΔP_{max}. However, even if the driving force is assumed

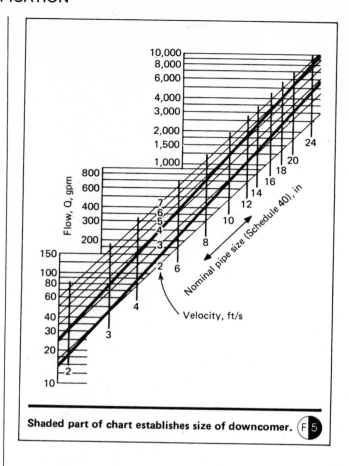

Shaded part of chart establishes size of downcomer. F5

at the maximum value, any inaccuracy is well compensated for by the safety factor of 2, and by the necessity to use commercially available pipe sizes that are normally larger than calculated pipe diameters.

Friction losses in reboilers

The total friction losses in a thermal-circulating reboiler system must be smaller than the available driving force. The pressure loss caused by friction takes place in two main locations: in the exchanger itself, Δp_e, and in the piping, Δp_p. Hence:

$$\Delta p_e + \Delta p_p < \Delta P$$

Friction losses in reboilers, Δp_e, are generally given as 0.25 to 0.5 psi. (A note should indicate whether entrance and exit losses are included.) Unit losses in downcomers and risers are in fractions of 1 psi/100 ft.

Calculation procedures for liquid lines have been outlined in Part 2 of this series, *Chem. Eng.*, Jan. 6, 1975, pp. 115–120 (Example 1); and for the two-phase flow risers in Part 8, June 23, 1975, pp. 145–151 (Example 1). We calculate reboiler returns as dispersed flow, regardless of the intersection of the Baker parameters (see Fig. 1 in Part 8).

To avoid trial-and-error calculations, a selection chart for reboiler pipe size is presented in F/5. This chart is based on limiting velocities for flow in downcomers of 2 to 7 ft/s. We enter the graph with known liquid-flow quantities. We obtain downcomer pipe sizes from the shaded portion of the graph, and also find the corresponding flow velocities for computing the

Reynolds numbers. The riser can be assumed as one or two sizes larger than the downcomer pipe.

In vertical reboiler circuits, reboiler losses are greater, and pipe losses smaller, than in horizontal circuits. In this case, a safety factor of 1.25 applied to the driving force can be used. In kettle-type reboilers, evaporation rates are high. For these reboilers, a large-diameter return line is usually necessary.

Elevation of the drawoff nozzle

The minimum elevation, H_1, for the downcomer tower nozzle above that for the centerline of the horizontal reboiler may be found from Eq. (6), where $H_2 = H_1 - 3$, ft:

$$H_1 = \frac{288\Delta p - 3\rho_2}{\rho_1 - \rho_2} \quad (9)$$

The downcomer nozzle cannot be lower than H_1. Δp replaces ΔP in Eq. (6), and is the sum of the downcomer, riser and exchanger friction-losses:

$$\Delta p = \Delta p_d + \Delta p_r + \Delta p_e$$

The value of H_1 is useful when elevation adjustments are made to vessel heights during graphic piping design, or when the vessel can be located at a minimum elevation. The coefficient for ρ_2 in Eq. (9) is the elevation difference between the downcomer and riser nozzles. If this is other than 3 ft, the correct dimension should be inserted.

Many towers have a bottom drawoff pump. NPSH requirements usually cause the process vessel and the reboiler drawoff nozzle to be raised higher than that of the reboiler's minimum elevation. This increases the static head in the vertical legs, and also the driving force in the circuit. With increased tower height, it is worthwhile to check the reboiler circuit for a possible reduction in size of the liquid and return lines—especially where large-diameter lines are necessary. Uneconomical reboiler lines are just carelessly oversized or poorly routed.

Example demonstrates calculations

Let us size the reboiler lines for the kerosene distillation unit, as sketched in F/6. Flow data are:

	Downcomer	Riser	
	Liquid	Liquid	Vapor
Flowrate, W, lb/h	85,000	59,500	25,500
Density (hot), ρ, lb/ft³	36.7	36	1.31
Viscosity, μ, cp	0.6	0.5	0.01
Molecular weight, M	—	—	53

The flow data for the riser reflect that 70% of total flow is liquid and 30% vapor. We obtain the vapor density,* ρ_v, in the riser from:

$$\rho_v = MP'/10.72Tz$$
$$\rho_v = 53(181.7)/(10.72)(682) = 1.31 \text{ lb/ft}^3$$

We find the mixture density,* ρ_2, in the riser by substituting in Eq. (2):

$$\rho_2 = \frac{100}{(70/36) + (30/1.31)} = 4.05 \text{ lb/ft}^3$$

And, we calculate volume flowrate,* Q, from:

$$Q = W/500S$$
$$Q = 85,000/500(36.7/62.37) = 289 \text{ gpm}$$

We will begin by sizing the downcomer and then calculate the overall pressure loss in it. We follow up with similar computations for the riser—remembering that the line has vapor/liquid flow. The simplest part of this analysis is finding the preliminary size of the reboiler lines. The computations for checking whether these line sizes are adequate require considerable detail. We should also note that the reboiler has two inlets and two outlets (F/6). Consequently, we find full flow (100%) in each line from the column to the "tee" connection on the reboiler's inlet and outlet piping; but only 50% of total flow in each line segment after the tee. In the following procedures, we will see how this flow arrangement affects our design calculations.

Downcomer—For a liquid flow of 289 gpm, we find the pipe size for the downcomer as 6 in from the selection chart of F/5. (For a 6-in, Schedule 40 pipe, I.D. = 5.761 in, $d^5 = 6,346$ in⁵.)

In order to calculate the unit loss for the 100% flow section, we must calculate the Reynolds number from:

$$N_{Re} = 50.6(Q/d)(\rho/\mu)$$
$$N_{Re} = 50.6\left(\frac{289}{5.761}\right)\left(\frac{36.7}{0.6}\right) = 155,300$$

For this Reynolds number, we obtain the friction factor, f, as 0.0182 from the chart in Part 8 of this series (*Chem. Eng.*, June 23, 1975, p. 147), and calculate the unit loss from:

$$\Delta p_{100} = 0.0216 f \rho_1 (Q^2/d^5)$$
$$\Delta p_{100} = 0.0216(0.0182)(36.7)[(289)^2/6,346]$$
$$\Delta p_{100} = 0.19 \text{ psi}/100 \text{ ft}$$

We can then find the unit loss at 50% flow:

$$N_{Re} = 0.5(155,300) = 77,650; \; f = 0.022$$
$$\Delta p_{100}(50\%) = 0.19\left(\frac{144.5}{289}\right)^2\left(\frac{0.022}{0.0182}\right)$$
$$\Delta p_{100}(50\%) = 0.0574 \text{ psi}/100 \text{ ft}$$

We now determine the equivalent length of pipe and fittings for each segment from tables in Part 2 of this series (*Chem. Eng.*, Jan. 6, 1975, pp. 115–120), as follows:

	Segment for 50% Flow, Ft	Segment for 100% Flow, Ft
Actual length	6	26
Entrance loss	—	18
Elbows*	10	20
Sharp tee	30	—
Exit loss	36	—
Total	82	64

*One elbow for 50% flow segment, 2 for 100% flow.

Overall pressure loss of the downcomer:

$$\Delta P = 0.19(64/100) + 0.057(82/100) = 0.167 \text{ psi}$$

*See Part 1 of this series, *Chem. Eng.*, Dec. 23, 1974, pp. 58–66, for complete details for these calculations.

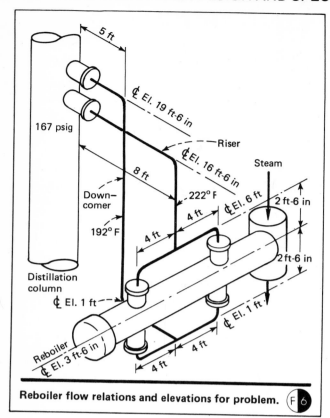

Reboiler flow relations and elevations for problem. (F/6)

Riser—Since the downcomer is a 6-in pipe, let us use a riser having a nominal size of 8 in. (For an 8-in Schedule 40 pipe, I.D. = 7.981 in, d^5 = 32,380 in,[5] A = 0.3474 ft².)

In order to calculate the unit loss for the 100% flow section, we must calculate the vapor-phase Reynolds number from:

$$N_{Re} = 6.31 W_v/d\mu_v$$
$$N_{Re} = 6.31(25,500)/(7.981)(0.01)$$
$$N_{Re} = 2 \times 10^6$$

For this Reynolds number, we obtain a friction factor, f, of 0.014 from the chart in Part 8 of series; and then calculate the vapor-phase unit loss from:

$$\Delta p_{100} = 0.000336(f/\rho_v)(W_v^2/d^5)$$
$$\Delta p_{100} = 0.000336(0.014/1.31)[(25,500)^2/32,380]$$
$$\Delta p_{100} = 0.072 \text{ psi}/100 \text{ ft}$$

Since the riser handles vapor/liquid flow, we must now determine the two-phase flow modulus* from:

$$X^2 = (W_l/W_v)^{1.8}(\rho_v/\rho_l)(\mu_l/\mu_v)^{0.2}$$
$$X^2 = \left(\frac{59,500}{25,500}\right)^{1.8}\left(\frac{1.31}{36}\right)\left(\frac{0.6}{0.01}\right)^{0.2} = 0.38$$

With this value of X^2, we find the two-phase flow modulus, ϕ^2, as 14 from a chart in Part 8 of this series.* We calculate the two-phase-flow unit loss* from:

$$\Delta p_{100} = 0.072 \times 14 = 1.01 \text{ psi}/100 \text{ ft}$$

*See Part 8 of this series, *Chem. Eng.*, June 23, 1975, pp. 145–151 for complete details of these calculations.

Let us estimate that the unit loss in the 50% flow section is one-third that in the 100% flow segment. Hence:

$$\Delta p_{100}(50\%) = 0.34 \text{ psi}/100 \text{ ft}$$

Again, we determine the equivalent length of pipe and fittings for each segment of the riser as follows:

	Segment for 50% Flow, Ft	Segment for 100% Flow, Ft
Actual length	6	18
1 Elbow	14	14
1 Sharp tee	24	—
1 Entrance loss	24	—
1 Exit loss	—	48
Total	68	80

Overall pressure loss of the riser:

$$\Delta P = 1.01(80/100) + 0.34(68/100) = 1.04 \text{ psi}$$

In summary, the total pressure loss is obtained by adding the loss in the downcomer, riser and reboiler:

6-in Downcomer	0.167 psi
8-in Riser	1.02 psi
Reboiler	0.35 psi
Total ΔP	1.557 psi

By substituting the appropriate values into Eq. (6) for this example, we determine the available pressure difference, as:

$$\Delta P = (1/288)[(36.7)(13.5) - (4.05)(10.5)] = 1.57 \text{ psi}$$

The available pressure difference of 1.57 psi is greater than the calculated pressure losses of 1.557 psi. Therefore, the design and sizes are acceptable.

Finally, we check the minimum elevation of the drawoff nozzle above the centerline of the reboiler by substituting into Eq. (9):

$$H_1 = \frac{288(1.535) - 3(4.05)}{31.7 - 4.05} = 15.55 \text{ ft}$$

Since the drawoff nozzle is actually 16.0 ft above the reboiler's centerline (F/6), the minimum value of 15.55 ft is acceptable.

The next article in this series will appear in the Sept. 15, 1975, issue. This article will review the design of pipelines for the hydraulic and thermal conditions occurring in overhead condensing systems.

The author
Robert Kern is a senior design engineer in the corporate engineering department of Hoffmann-La Roche Inc., Nutley, NJ 07110. He is a specialist in hydraulic-systems design, plant layout, piping design and economy. He is the author of a number of articles in these fields, and has taught several courses for the design of process piping, plant layout, graphic piping and flow systems, both in the U.S. and South America. Previously, he was associated with M. W. Kellogg Co. in England and the U.S. Mr. Kern has an M.S. in mechanical engineering from the Technical University of Budapest.

How to find the optimum layout for heat exchangers

Internal elements such as tubes, baffles and tubesheets, and external components including covers, channels and nozzles, determine the extent to which simplicity in piping can be achieved.

Robert Kern, Hoffmann - La Roche Inc.*

☐ Piping connected to heat exchangers is generally uncomplicated. Piping economy depends more on knowing what alterations can be made to exchangers than on piping design. Consequently, the layout designer can influence the mechanical design of heat exchangers in the interest of economical and functional piping. Alterations to exchangers should not affect their duty and cost.

The variety of exchangers in type, duty and application is very large; our design analyses will be limited to shell-and-tube exchangers in chemical plants. With modifications, the design principles are also applicable to other types of exchangers arranged in outdoor (i.e., open) or indoor (i.e., housed) chemical plants.

In order to evaluate alternative possibilities of exchanger piping, we must be familiar with the construction details of exchangers, the wide range of exchanger types, and the functions and duties of exchangers in chemical plants. We can then achieve the most eco-

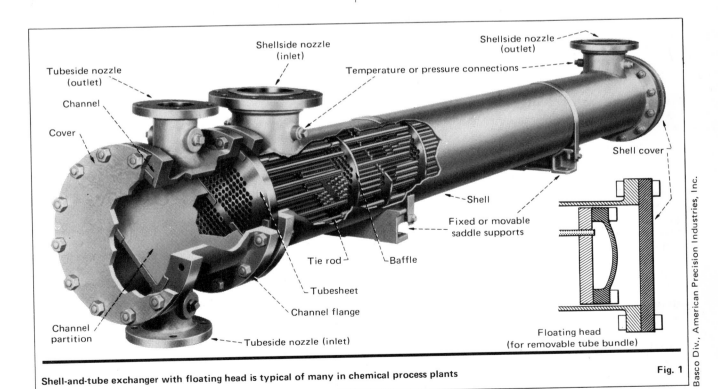

Shell-and-tube exchanger with floating head is typical of many in chemical process plants Fig. 1

Originally published September 12, 1977

80 HEAT EXCHANGER DESIGN AND SPECIFICATION

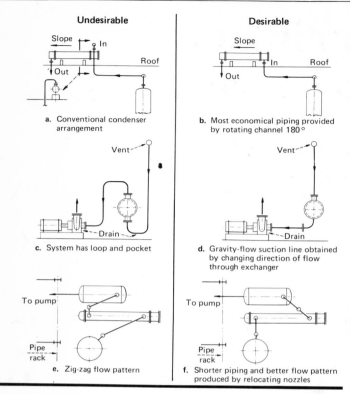

Simplifying the flow path improves piping design   Fig. 2

nomical piping arrangements to satisfy engineering, plant-operation, maintenance and safety requirements.

Let us start by reviewing the design principles and construction details for heat exchangers as these affect piping design.

Shell-and-tube exchangers

Most shell-and-tube exchangers used in chemical plants are of welded construction (Fig. 1). The shells are built as a piece of pipe with flanged ends and necessary branch connections. Up to 24 in. dia., the shells are of seamless pipe; above 24 in., they are made of bent and welded steel plates. Channel sections are usually of built-up construction, with welding-neck forged-steel flanges, rolled-steel barrels and welded-in pass-partitions. Shell covers are either welded directly to the shell, or are built-up constructions of flanged and dished heads and welding-neck forged-steel flanges.

Heat is exchanged between two fluids of differing temperatures. One fluid flows in the tubes; the other through the shell in a predetermined pattern around the tubes. Generally, cooled streams flow downward, and heated streams upward, through the tubeside or shellside of heat exchangers. When phase change takes place, this principle is essential. For liquid streams, it is a matter of preference. Vapor and gas streams can flow upward or downward, regardless of temperature.

Some typical examples illustrate these principles. In most cases, water inlets are at the bottom side of exchangers, and outlets at the top. Steam enters at the top channel nozzle of reboilers, and condensate exits at the bottom nozzle. Vapor inlet for condensers is at the top, and liquid outlet at the bottom.

When large quantities of vapor have to be condensed or released, the physical change usually takes place in the shell, where more volume can be provided than in the restricted space of tubes.

Counterflow provides better heat transfer than concurrent flow and is preferred. This can only be fully satisfied with single-pass exchangers on both the tubeside and shellside, or with double-pipe units. With multipass tubes and crossflow shells, this principle loses its importance.

Using these principles, a single-tube-pass condenser has been arranged after a distillation column, as shown in Fig. 2a. Here, inlet and outlet have been designed by the exchanger specialist and piped up accordingly. There is nothing wrong with it. But the arrangement of Fig. 2b works just as well, and saves welds and fittings in the large-size overhead line.

The essential parts of an exchanger, as shown in Fig. 1, are: (1) tubes or tube bundle with tubesheets, baffles and tie rods; (2) the shell, housing the tube bundle, with inlet and outlet nozzles; (3) channel (front head) inlet and in most cases outlet nozzles, partitions and occasionally channel cover; (4) shell cover (rear head) with or (mostly) without outlet nozzles; and (5) exchanger supports located on the shell.

Tubes and tube bundle—The most popular tube sizes range from $5/8$ in. to 1 in. Tubes are in square or triangular arrangement. After the baffles have been placed around the tubes, the tubes are roller-expanded into the tubesheets at each end. The tubesheets can be welded to the shell. If the tubesheets are separate (held between flanges), they form with tubes and baffles a removable bundle. The close spacing of tubes in triangular pitch makes mechanical cleaning difficult and increases shellside pressure drop. When low shellside pressure drop is required, square pitch is used. Because of wider tube spacing, with square pitch, the shell becomes larger and the exchanger will be more expensive. In plant design, access must be provided to the tubesheets because the exchanger covers are removed for inspecting the tube ends periodically.

Baffles—These are an integral part of tube bundles, and serve to (a) direct the flow horizontally and vertically in the shell for optimum heat transfer, and (b) support the tube between the tubesheets. Cross-baffles are $1/8$- to $1/4$-in.-thick plates suitably cut and arranged for directing flow through the shell. Exchangers usually have vertical segmental baffles; condensers have horizontal baffles. Baffles also determine the locations for inlet and outlet nozzles on the shell.

Typical baffle arrangements for various types of exchangers are shown in the sketches of Table I. Segmental baffles provide a single shell pass, with nozzles at each end of the shell (see Fig. 1 and Item 1 and 2 in Table I.). Single split-flow design has one inlet and two outlets (Item 4). Single split-flow design with horizontal baffles necessitates one inlet and one outlet (Item 7). The double-split flow shell, with horizontal baffles, has two inlets and two outlets (Item 5). The segmental two-or-more shell-pass design has nozzles at one end

How to find the optimum layout for heat exchangers

Alterations to typical heat exchangers for better piping — Table I

Tube bundle	Item No.	Typical heat-exchanger arrangements	Tube passes	Typical use	Possible alterations*					
					Interchange flow media	Change direction of flow on			Change nozzle location	
						Shell-side only	Tube-side only	Both sides same time	Turn tube-side nozzle radially 180°	Turn shell-side nozzle radially 180°
Floating-head	1	Single-pass shell	1	Exchanger Cooler Heater	●			●	●	●
	2	Single-pass shell	2 or 4		●	●	●	●		●
	3	Double-pass shell	2 or 4		●			●		
	4	Single split-flow shell	2 or 4	Condenser Evaporator Reboiler			●			
	5	Double split-flow shell	2 or 4				●			
U-tube	6	Single-pass shell	2 or 4	Exchanger Cooler Heater (in clean tubeside service)	●	●	●	●		●
	7	Double-pass shell	2 or 4		●			●		
	8	Kettle-type reboiler	—	Reboiler Steam generator Vaporizer						
Fixed-tube	9	Single-pass shell	1	Exchanger Cooler Heater (in low-temperature clean service)	●			●	●	●
	10	Single-pass shell	2 or 4		●	●	●	●		●
	11	Two-pass shell	2 or 4		●			●		

*These possible alterations to typical arrangements do not affect thermal design.

82 HEAT EXCHANGER DESIGN AND SPECIFICATION

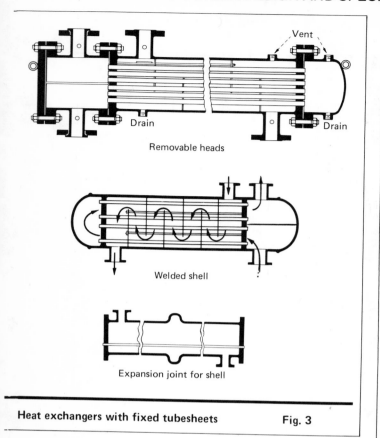

Heat exchangers with fixed tubesheets — Fig. 3

Basic types of exchangers

Many combinations of shells, shell covers and channel sections, tubes and baffles are possible for heat exchangers [1,2]. We will discuss four types:

Fixed-tubesheet—Exchangers with completely enclosed tubes (Fig. 3) can be used only in clean service. Cleaning can be done by flushing through the tubeside and shellside. Cleanout connections are provided in the piping as close as practical to the exchanger nozzles between the exchanger and block valve. Two bolted covers facilitate inspection and cleaning of tubes. Because no provision is made for tube expansion, these exchangers are built for low-temperature service. When differential expansion between tubes and shell exists, an expansion joint is built into the shell.

U-tube—In this type of exchanger, the tube bundle is hairpin-shaped and can freely expand. The bundle is removable from the shell. For inplace maintenance, space must be provided in the back and front of the exchanger. Space must also be allowed for mobile tube-removal facilities. The U-tube design is used when fouling inside the tubes is not expected.

Floating-head—This is the most-often used type in chemical plants (Fig. 1), and is more expensive than fixed-tube or U-tube exchangers. One end of the tube bundle has a stationary tubesheet held between shell and channel flanges. The floating head can freely expand and contract with temperature changes.

Kettle-type—For high evaporation rates, the kettle-type exchanger is chosen (Item 8 in Table I). This type can have a U-bundle or a floating-head bundle. The shell is larger to accommodate the generated vapor.

These examples of exchangers show details of interest to the piping designer. Exchanger type, size, construction details, nozzle arrangements, flow direction through shell and tubes are determined by the exchanger designer, usually without taking into account external piping. Most often, the piping designer has no influence in the design and selection of heat exchangers, but can request alternate flow and nozzle arrangements in the interest of economical piping.

The piping designer is responsible for the physical arrangement of piping, and for the efficient and trouble-free operation of process equipment, including exchangers. For example, if the piping designer blindly followed the nozzle locations and flow requirements provided by the exchanger specialist, he could end up with the piping arrangements shown in Fig. 2a, 2c and 2e. The suction line in Fig. 2c has a pocket and a loop in the line, which means longer piping, more fittings, more vents and drains, and unreliable pump operation. By reversing the flow through the exchanger, a loop, pocket, vent and drain are eliminated, as shown in Fig. 2d. Moreover, the suction line has been shortened and simplified.

Another comparison is illustrated in Fig. 2e and 2f. The long zig-zag flow in Fig. 2e has been simplified with the more functional arrangement of Fig. 2f.

Exchanger modifications for improved piping

Table I lists alterations to heat exchangers that can be made without a cost increase in order to achieve an

when the number of passes is even, and nozzles at each end if the number is odd (Item 5, 6, 9 and 10).

Heads—A channel head at the front and a cover head at the back enclose both ends of the exchanger tubesides. Inlet and outlet connections may be on one or both heads for the return flow. The heads can be cast or built-up constructions of carbon steel or alloys. The choice of head design depends on inspection and maintenance frequency for exchangers. Heads can be welded to the shell, or they can be flanged with additional head covers (Fig. 1) for tube inspection without disconnecting piping.

Piping has to be disconnected to remove the channel. The cover usually has no piping connections. Often a davit is provided, pivoted on the shell, for cover removal. (See Fig. 7.)

Inlet and outlet nozzles, and flow direction, on both tubeside and shellside are arranged to give the required flow through the unit for optimum heat-transfer. The shell and channel have vent and drain connections. Pressure, gage-glass, level-controller and relief-valve connections can also be arranged on the shell, if they are required.

Two supports are usually welded on the shell, one with elongated anchor-bolt holes to permit thermal expansion. When exchangers are stacked, additional supports are necessary on the top of the bottom shell. A light structure can be designed for spacing stacked exchangers between exchanger footings.

HOW TO FIND THE OPTIMUM LAYOUT FOR HEAT EXCHANGERS

optimum piping arrangement. These may range from changing the flow direction to relocating the nozzles. The possible alterations are:

☐ Interchange flowing media between the tubeside and shellside. This change is often possible, and more so when the flowing media are similar, such as liquid hydrocarbons. Preferably, the hotter liquid should flow in the tubes to minimize heat losses through the shell or to avoid the use of thicker shell insulation.

☐ Change direction of flow on tubeside, on shellside, or on both sides. These changes are frequently possible and are accepted by the exchanger designer if the tubes are in double-pass or multipass arrangement and the shell has a cross-flow baffle arrangement. In exchangers having counterflow conditions, changing the direction of flow should be made simultaneously on tubeside and shellside.

☐ Change the exchanger's nozzle location on tubeside or shellside. These changes are frequently possible on the shell and channel without affecting the required duty on most exchangers in chemical plants.

The following factors can also influence the decision of the piping designer when considering changes in exchanger construction:

Viscous liquids—Exchanger performance is usually improved when viscous material flows through the tubes, particularly when cooling. On the shellside, pockets can form, thereby reducing the effective heat-transfer surface.

Shell leakage—When gases, liquid hydrocarbons or chemicals are watercooled, the water usually passes through the shell. A tube leakage will result in contaminating the water; a shell leakage, on the other hand, can vent process materials to the atmosphere, with potential hazard.

High-pressure service—If high-pressure fluid flows on the tubeside, only the tubes, tubesheets, channels and cover have to be designed for high pressure. (A relief valve must be provided on the shell.) High pressure on the shellside requires a much heavier shell and covers, and considerably increases the exchanger cost.

Pressure drop—Where pressure drop must be minimized, the flow passes through the shellside. By opening up tube-and-baffle spacing, low mass velocities can be obtained and pressure drop reduced. The larger shell increases the cost of the exchanger.

Shellside flow—Shellside volume can be designed much larger than tubeside volume. Vaporization or condensation of freely flowing fluid is more effective than through the tubes.

Corrosion—Corrosive liquids should pass through the tubes so that the shell can be made of carbon steel. Only the tubes and channel have to be made of alloy steels.

Fouling—If one medium is dirty and the other clean, passing the clean one through the shell will result in easier tube-bundle removal for cleaning, or even simpler exchanger design.

Mechanical modifications to exchangers

Mechanical alterations to the basic exchanger do not affect the thermal design, and can save money, provide better access, and improve piping layout. Often, the slight increase in cost caused by a special nozzle ar-

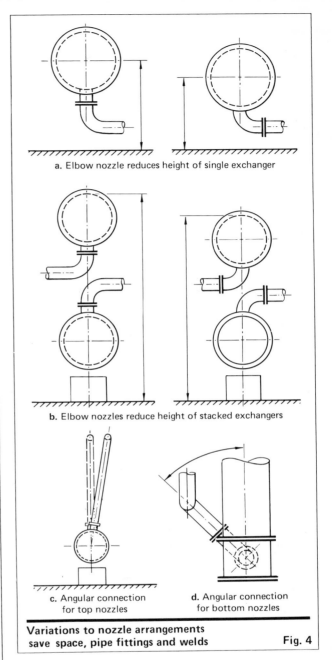

a. Elbow nozzle reduces height of single exchanger

b. Elbow nozzles reduce height of stacked exchangers

c. Angular connection for top nozzles

d. Angular connection for bottom nozzles

Variations to nozzle arrangements save space, pipe fittings and welds   Fig. 4

rangement is more than offset by more-economical piping.

A few variations in nozzle arrangements are shown in Fig. 4. Elbow nozzles permit the lowering of exchangers to bring them closer to grade. Elbow nozzles also enable stacked exchangers, in parallel or dissimilar service, to be arranged closer to each other. This facilitates better access and easier maintenance of exchanger valves and instruments. Angular connections can save one or two bends in the pipeline. These are more often applied to the top nozzle of the shellside or tubeside. Too many angular connections at the bottom can mean a separate drainage point on the shell or channel. The maximum angle from a vertical centerline can be about 30°. This

84 HEAT EXCHANGER DESIGN AND SPECIFICATION

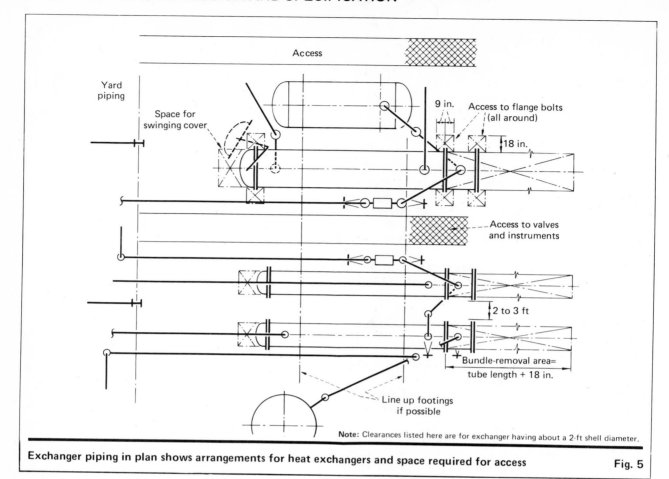

Exchanger piping in plan shows arrangements for heat exchangers and space required for access   Fig. 5

angle depends on nozzle and shell sizes, and internals of the exchanger, such as the baffle arrangement in the shell, and partitions in the channel. Tangential connections can save fittings, make piping arrangement simpler, and improve access to valves.

For restricted space at indoor installations, and for exchangers in structures or supported on vessels, the originally designed unit might be too long. It is possible to shorten exchangers to satisfy space limitations. However, the practical rule is that a more economical heat exchanger can be designed by using small-diameter shells and maximum shell-lengths.

Horizontal exchangers can be turned vertical for conserving floor space. Verticals can be changed to horizontal when installation height is restricted.

Exchanger footings relative to shell and channel nozzles can also be relocated to adjust to a more-economical overall (lined-up or combined) foundation design.

Heat-exchanger piping

The information required for piping design as applied to heat exchangers is the same as that required for vessel piping.*

After all information has been collected, with the exchangers located on the plot plan and their elevations established, the first step is to outline clearance and

*Part 3, *Chem. Eng.*, Aug. 15, 1977, p. 153

working space in front, and around both ends, of the exchangers. These working spaces should be kept clear of any piping and accessories to facilitate channel, shell-cover and tube-bundle removal, as well as maintenance and cleaning. (The clearances shown in Fig. 5 are for exchangers having about a 2-ft shell diameter. Smaller units will require somewhat smaller clearances, and larger ones slightly greater.)

Piping in plan

The overall plant layout influences the main arrangement of exchanger piping and access (see Fig. 5). The channel-ends of exchangers face the main plant road for convenience in tube removal. The shell cover faces the pipe rack.

If piping is arranged on one elevation only (between exchanger and yard piping), one pipeline will be located over the exchanger centerline. It is suggested that a top shellside nozzle be chosen for this location. The top tubeside connection can be placed on a slight angle (Fig. 4c) to miss the top shellside pipeline, in order to avoid an offset in this line.

Pipelines turning to the right in the yard should be run to the right of the exchanger centerline. Those turning left should approach the yard on the left-hand side of the exchanger centerline. Pipelines from bottom connections of exchangers should also turn up on the

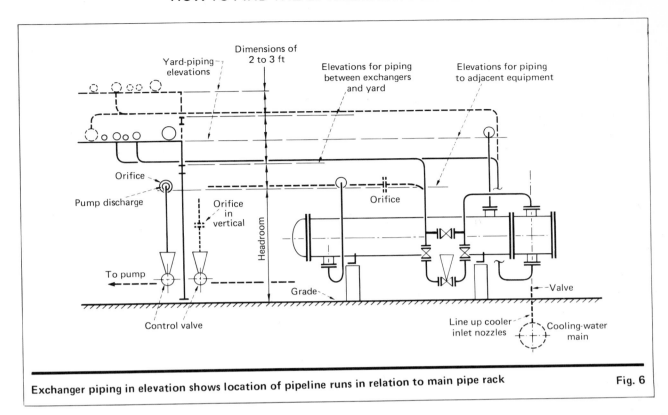

Exchanger piping in elevation shows location of pipeline runs in relation to main pipe rack Fig. 6

right or left side of the centerline, depending upon which way the pipeline turns in the yard. Pipelines with valves and control valves should turn toward the access aisle, which is arranged close to the exchanger.

Pipelines interconnecting exchangers with adjacent process equipment can run point-to-point, just above required headroom or about on the same level as the yard piping. Reboiler-line elevations are established by the drawoff nozzle and return nozzle on the tower.

Steam lines connecting to a header in the yard can be arranged on either side of the exchanger centerline without increasing the pipe length.

Cooling-water lines, in most cases, are below grade, and should run right under the lined-up channel nozzles of all coolers. The warm-water return header is usually adjacent to the cooling-water supply main.

Access to valve handwheels and instruments will influence the piping arrangement around heat exchangers. Valve handwheels should be accessible from grade and from a convenient accessway. These accessways should be used for arranging manifolds, control valves and instruments.

Piping in elevation

Fig. 6 shows an exchanger in elevation with adjacent process equipment and single-level yard piping. The main elevation for lines between the exchanger nozzles and yard piping is about 2 to 3 ft lower than the yard elevation. This elevation can be used for pump-discharge lines, if the pump is under the yard piping and near the exchanger, and for lines connecting to equipment arranged below the yard piping. To avoid condensate drainage toward the exchanger, the preferred connection for steam lines is to the top of the header. However, there is nothing wrong in having a steam connection to the bottom of the header if steam traps are placed at the low point.

Orifice flanges in exchanger piping are usually in horizontal pipe runs that should be just above headroom. The orifice should be accessible from a portable ladder. When convenient, pipes having an orifice and differential-pressure-cell measuring element can be located about $2\frac{1}{2}$ ft from grade to the centerline of the pipe. Orifices in a liquid line and using a mercury-type measuring element require more height. Long, vertical measuring U-tube gages must be just below the orifice. In gas lines, the U-tube can be above the pipeline containing the orifice. Lines containing orifice flanges should have the necessary straight runs before and after the orifice.

Locally mounted pressure and temperature indicators on exchanger nozzles, on the shell, or on process lines should be visible from the access aisles. Similarly, gage glasses and level controllers on exchangers should be visible from this aisle, and associated valves accessible. Instrument connections on exchangers should have sufficient clearances between flanges and exchanger supports, and between instruments and adjacent piping. Insulation of piping and exchangers should also be taken into account. Consideration should be given to internal details of heat exchangers when arranging instruments.

Excessive piping strains on exchanger nozzles from the actual weight of pipe and fittings and from forces of thermal expansion should be avoided.

The data in Fig. 7 emphasize often overlooked di-

86 HEAT EXCHANGER DESIGN AND SPECIFICATION

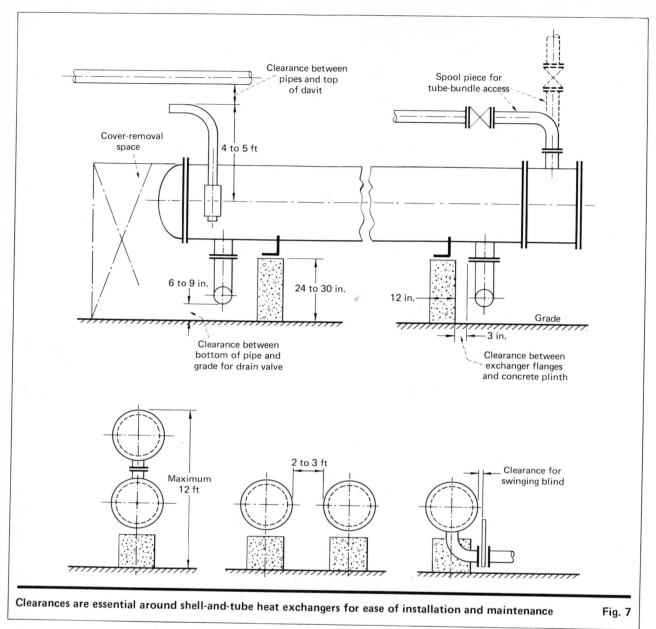

Clearances are essential around shell-and-tube heat exchangers for ease of installation and maintenance   Fig. 7

mensions and interferences related to exchanger-piping design.

Layout of exchangers affects piping design

Process equipment in most plants is arranged in the sequence of process flow. Whatever layout system is used, the general evaluation regarding exchanger locations is similar.

In plant layout, the fractionation towers should be arranged first, and the other equipment after the proper tower sequence has been established. The position of an exchanger in chemical and petrochemical plants usually depends on the location of distillation columns. The relative position of exchangers can be readily evaluated from flow diagrams. For exchanger positions, the following general concepts apply:

1. Exchangers should be immediately adjacent to other equipment—for example, reboilers should be located next to their respective towers; condensers should be next to their reflux drums close to the tower.

2. Exchangers should be close to other process equipment—for example, exchangers in closed pump-circuits (some reflux circuits). In the case of drawoff flow through an exchanger from a vessel bottom, the exchanger should be close to and under the tower or drum in order to have short pump-suction lines.

3. Exchangers between two distant pieces of process equipment, as shown in Fig. 8a—for example, exchangers with process lines connected to both shellside and tubeside—should be located where the two streams meet in the pipe rack and have a parallel run, and on that side of the yard where the majority of related

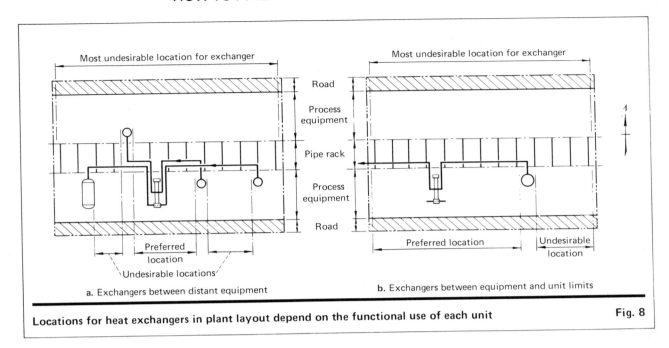

Locations for heat exchangers in plant layout depend on the functional use of each unit — Fig. 8

equipment is placed. Fig. 8 shows that northside locations will cost more in pipe runs.

4. Exchangers between process equipment and the unit limit, as shown in Fig. 8b—for example, product coolers—should be located near the unit limit in order to minimize pipe runs.

A further step in layout is to establish those exchangers that can be stacked in order to simplify piping and save plot space. Most units in the same service are grouped automatically. Two exchangers in dissimilar services can also be stacked. Sufficient clearances must be provided for shellside and channelside piping between the two exchangers. Reboilers and condensers usually stand by themselves, alongside their respective towers.

Design specifications usually limit the maximum height of exchangers to about 12 ft to the top of the shell, so that mobile equipment can conveniently handle the tube bundles.

Maintenance procedures

For shop maintenance, piping has to be disconnected in order to remove the exchanger from its location. Inplace maintenance can consist of changing gaskets, and cleaning, reaming or plugging of tubes. Removable bundles are pulled out of the shell for cleaning and repair; the shell is cleaned in place.

The piping designer can help maintenance operations by:

☐ Designing and supporting piping so that no temporary support will be required for removing the channel and tube bundle. On the other hand, temporary supports can easily be built.

☐ Providing easily removable spool pieces (Fig. 7), flanged elbows, break flanges, or short pipe runs to provide adequate clearances for the operation of tube-removal equipment.

☐ Leaving space and access around exchanger flanges and head, and bundle-pulling space in front and in line with the shell (Fig. 5).

Bundle pulling at grade is facilitated by hitching points. A chain or rope is fixed to the bundle and hitching point, and a pulley transmits the necessary force and motion for removal. The hitching point can be at grade. The distance between the front of the exchanger and the hitching point is about twice the bundle length.

A more positive pulling force can be exerted horizontally if the hitching point is in line with the exchanger centerline. Existing structures, in convenient positions, can be used for rigging the pulling beam for bundle removal. Trolley and pulling beams in a permanent structure over exchangers can also be arranged for single or stacked exchangers. Structural clearances are equal to one bundle length plus 12 to 18 inches.

For handling a row of single or stacked exchangers at grade, a travelling gantry can be provided. The gantry consists of trolley beam, trolley, and pulling beam on a structural frame that can be moved on rails along the front of a row of exchangers.

With a mobile bundle-puller, four tubular legs are placed in front of the exchanger. These support the bundle at grade and enable a removal cart to be raised into position. The reaction to the pulling forces is taken up by struts attached to the exchanger footing. Ropes and pulley attached to the bundle transfer the force and motion while the bundle slides out on the elevated cradle, which is then lowered to grade.

References

1. "Heat Exchangers," Manual No. 700-A, The Patterson-Kelley Co., East Stroudsburg, PA 18301.
2. "Standards of Tubular Exchanger Manufacturers Association," Tubular Exchanger Manufacturers Assn., New York, NY 10017.

Predicting heat-exchanger performance by successive summation

This programmable method lets the engineer calculate heat-transfer coefficients and operating temperatures without resorting to the correction factors associated with the classical, log-mean temperature-difference approach.

Robert A. Spencer, Jr., E. I. du Pont de Nemours & Co.

☐ Many plant heat-exchange operations are not accurately analyzed by the log-mean temperature-difference, $\Delta \bar{T}_{LM}$, method for calculating heat transfer. To improve this situation, the method presented here was developed. Today, it is being used to calculate heat-transfer coefficients on a time-shared terminal system available in one of Du Pont's plants. It has replaced the classical method in many applications because it is compatible with programmable calculators and requires less data input. This is significant in plant situations where complete sets of data are rare.

$\Delta \bar{T}_{LM}$ method

The classical method of calculating the overall heat-transfer coefficient for an exchanger is to integrate the basic differential equation describing heat transfer as shown in Eq. (1) and (2).

$$dq = UdA(T - t) \qquad (1)$$

$$q = UA(\Delta T - \Delta t)/\ln(\Delta T/\Delta t) = UA\Delta \bar{T}_{LM} \qquad (2)$$

$\Delta \bar{T}_{LM}$ results from this integration.

For single-pass, cocurrent and countercurrent flow, calculation of $\Delta \bar{T}_{LM}$ is straightforward. Complications arise when multipass tube-bundles, condenser or reboiler liquid-levels, or cooling medium crossflows (such as on air coolers) must be considered. In such cases, correction factors must be used to modify $\Delta \bar{T}_{LM}$ to fit the particular design configuration. Correction factors are empirical and lead the engineer away from Eq. (1), the mathematical base of heat transfer. In effect, the engineer is reacting to his equipment instead of predicting its performance.

Successive summation method

Correction factors are avoided by directly using the differential equation. By programmed calculation, Eq. (1) can be manipulated in differential elements, and

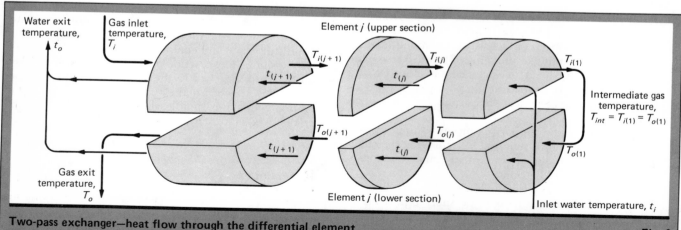

Two-pass exchanger—heat flow through the differential element

Fig. 1

then, through successive summation, the numerical value of the integral is obtained without using a $\Delta \overline{T}_{LM}$.

As an example of this method, a program will be constructed to determine (a) the overall heat-transfer coefficient, U, or (b) the exit gas temperature, T_o, for a two-pass heat exchanger.

The two-pass tube bundle shown in Fig. 1 permits a hot gas to pass in a U pattern through the exchanger. A baffled head controls the gas flow. The shell side contains one-pass cooling water, which enters at the right and leaves at the left. (For simplicity, water crossflow due to shell-side baffles has been neglected but could be incorporated into the program.) Water enters the shell at temperature t_i, and increases in temperature as it proceeds through each element by the amount:

$$\Delta t = \Delta q / m_w C p_w \quad (3)$$

Written in accordance with Fig. 1, the temperature of the water leaving the differential element, j, is equal to the entering temperature plus Δt:

$$t_{(j+1)} = t_j + \Delta t \quad (4)$$

The rate of heat transfer in the differential element, $dq_{(j)}$, is equal to the sum of the rates in the upper and lower sections:

$$dq_{(j)} = dq_{(j)} \text{ (upper)} + dq_{(j)} \text{ (lower)} \quad (5)$$

The sections must be analyzed separately because they contain gas at two different temperatures. Referring to Fig. 1, T_i is the inlet gas temperature, and $T_{i(j)}$ is the first-pass gas temperature through the j element. T_o is the exit gas temperature and $T_{o(j)}$ is the second-pass gas temperature through the j element. In the upper section of the element, the rate of heat transfer is:

$$dq_{(j)} \text{ (upper)} = U(A/2)(T_{i(j)} - t_{(j)})/N \quad (6)$$

and in the lower section it is:

$$dq_{(j)} \text{ (lower)} = U(A/2)(T_{o(j)} - t_{(j)})/N \quad (7)$$

The Δt of the cooling water through the differential element is calculated from Eq. (3), (4) and (5) to be:

$$\Delta t = t_{(j+1)} - t_{(j)} = \frac{dq_{(j)} \text{ (upper)} + dq_{(j)} \text{ (lower)}}{m_w C p_w} \quad (8)$$

Similarly, for the upper section of the element:

$$\Delta T_i = T_{i(j+1)} - T_{ij} = \frac{dq_{(j)} \text{(upper)}}{m_g C p_g}$$
$$= \frac{U(A/2)(T_{i(j)} - t_{(j)})}{N m_g C p_g} \quad (9)$$

For the lower section:

$$\Delta T_o = T_{o(j)} - T_{o(j+1)} = \frac{dq_{(j)} \text{ lower}}{m_g C p_g}$$
$$= \frac{U(A/2)(T_{o(j)} - t_{(j)})}{N m_g C p_g} \quad (10)$$

Eq. (4 to 10) are the components of the calculation, which will now be programmed.

Program structure

The structure of the program is shown in Fig. 2. After calling for general input, it sets up the information

Nomenclature

A	area of heat transfer, ft²
Cp	heat capacity, Btu/(h)(°F)
D	dia. of tube, ft
g	subscript denoting gas
i	inlet condition, or first tube-pass condition
j	denotes differential element
l	length of tube, ft
m	mass flowrate, lb/h
N	number of differential elements
o	exit condition, or second tube-pass condition
q	heat-transfer rate, Btu/h
t	temperature, shellside (water), °F
T	temperature, tubeside (gas), °F
U	heat-transfer coefficient, Btu/(h)(ft²)(°F)
w	subscript denoting water

needed for the type of calculation to be done, for either T_o or U. An initial guess of the intermediate temperature, T_{int} (see Fig. 1), is used to start the successive summation at the right (water inlet) and progress to the left (water exit). For the first element, $j = 1$, T_{int} is equal to both $T_{o(1)}$ and $T_{i(1)}$. The calculation then proceeds through N elements (15 to 20 elements are practical, based on computing time and accuracy considerations) to produce values for the gas inlet temperature, $T_{i(N)}$, and gas exit temperature, $T_{o(N)}$.

Next, the program compares the calculated $T_{i(N)}$ with the given input T_i. If they are not equal, a new T_{int} is calculated and the program is restarted. Convergence of T_{int} with the correct value will allow $T_{i(N)}$ to be equal to T_i. If they are equal (or within convergence criteria limits), then $T_{o(N)}$ is the predicted gas exit temperature and the calculation is finished.

If U is to be calculated, $T_{o(N)}$ is compared to the given (input) T_o. If U is correct, $T_{o(N)}$ will be equal to T_o. If they are not equal, a new value for U is calculated and the program is restarted. Once U converges with the correct value, $T_{o(N)}$ will be equal to T_o, $T_{i(N)}$ will be equal to T_i, and the calculation is finished.

See p. 123 for a worked-out sample problem.

The described calculation has been specifically structured for a two-pass heat exchanger, but its concept is applicable to many other heat-transfer problems. In general, if three of the four inlet and exit temperatures are known, the successive summation method will calculate the fourth. If two of the four are known, the third can be determined by trial and error.

Computing time depends on the number of differential elements and trial-and-error loops, and on the success of the convergence routines.

Accuracy depends on the number of differential elements and the temperature used to represent each element. Using an average element temperature rather than the inlet or exit temperature of the element increases accuracy, especially for cocurrent flow. Relative errors of 1% are not difficult to achieve.

Predicting performance

The predictive nature of the method is an important tool. Consider the cooling of a recycle gas stream by 9 exchangers arranged in 3 sets of 3 parallel exchangers

90 HEAT EXCHANGER DESIGN AND SPECIFICATION

An example problem demonstrating successive summation.

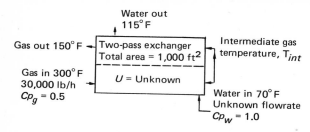

Based on the information given above, calculate (a) the heat-transfer coefficient, U; then predict the effect on exit gas temperature when (b) the gas flowrate is increased to 35,000 lb/h (water flow remains constant), and (c) the inlet water temperature rises to 80° F.

The structure of the program requires values for the water flowrate, intermediate gas temperature, and heat-transfer coefficient. The water flowrate calculation is based on a heat balance:

$$m_w C p_g \Delta T_w = m_g C p_g \Delta T_g$$

$$m_w (1.0)(115° - 70°) = 30,000 (0.5)(300° - 150°)$$

$$m_w = 50,000 \text{ lb/h (or 100 gal/min)}$$

An initial guess of 180° F is made for the intermediate gas temperature, T_{int}, and of 30 for the overall

PROGRAM LOCATION NUMBER	PROGRAM LISTING FOR TI-SR52 CALCULATOR	MEMORY LOCATION ASSIGNMENTS
000	LBL A	00: Decrement register
002	RCL 01 STO 00	01*: N, number of iterations
008	RCL 13 STO 14 STO 15	02*: A
017	RCL 09 STO 16	03*: Cp_g
023	RCL 11 × RCL 02 ÷ 2 ÷ RCL 01 = STO 99	04*: Cp_w
040	0 STO 17 STO 18 STO 19	05*: m_g
		06*: m_w
050	LBL A'	07*: T_i
052	RCL 14 − RCL 16 = × RCL 99 = STO 17 ÷ RCL 05 ÷ RCL 03 = SUM 14	08*: T_o
080	RCL 15 − RCL 16 = × RCL 99 = STO 18 ÷ RCL 05 ÷ RCL 03 = INV SUM 15	09*: t_i
109	RCL 17 + RCL 18 = SUM 19 ÷ RCL 06 ÷ RCL 04 = SUM 16	10:
132	dsz A'	11*: U
		12: Temporary storage
134	RCL 14 − RCL 07 = STO 12 χ^2 √ − 1 = if pos E'	13*: T_{int} (current)
152	if flg 1 D'	14: $T_{i(j)}$
155	RCL 15 HLT	15: $T_{o(j)}$
		16: $t_{(j)}$
159	LBL D'	17: $dq_{(j)(upper)}$
161	RCL 15 − RCL 08 = STO 12 χ^2 √ − 1 = if pos C'	18: $dq_{(j)(lower)}$
179	RCL 11 HLT	19: $q_{(j)}$
183	LBL C'	
185	1 + RCL 12 ÷ RCL 08 = PROD 11 A	99: $UA/N\,2$
199	LBL E'	FLAG 1*:
201	RCL 12 ÷ 3 = INV SUM 13 A	set = calculate U
212	HLT	reset = calculate T_o

*Note: These are inputs to the calculation.

heat-transfer coefficient. The appropriate values are U loaded into the assigned memory locations; flag 1 is set (to calculate U), and the calculation is started by pressing "A".

(a) After the calculation is complete, the value shown (19.45) is the heat-transfer coefficient, U, based on the given data. It is already stored in memory location 11 for future use. Flag 1 should be reset so that exit gas temperatures can be predicted based on this calculated value of U.

(b) Using 19.45 for U, the gas flowrate of 35,000 lb/h is entered into memory location 5, and the calculation is started again by pressing "A". The predicted gas-exit temperature is 162° F.

(c) Again using 19.45 for U, the gas flow is left at 35,000 lb/h in memory location 5, and the new water-inlet temperature (80° F) is stored in memory location 9. The calculation is started by pressing "A". The predicted gas-exit temperature is 168° F, for U = 19.45, gas flow = 35,000 lb/h, and water inlet temperature = 80° F.

Calculation times for the program as shown are: (a) 19 min, 6 s (b) 3 min, 39 s, and (c) 2 min, 56 s. The calculation time for part (a) can be reduced by optimizing the convergence routine for U contained in line 185 of the program. If the factor "RCL 12 ÷ RCL 08" is multiplied by 2 to accelerate convergence, the calculation time for (a) is 9 min, 27 s.

PREDICTING HEAT-EXCHANGER PERFORMANCE BY SUCCESSIVE SUMMATION

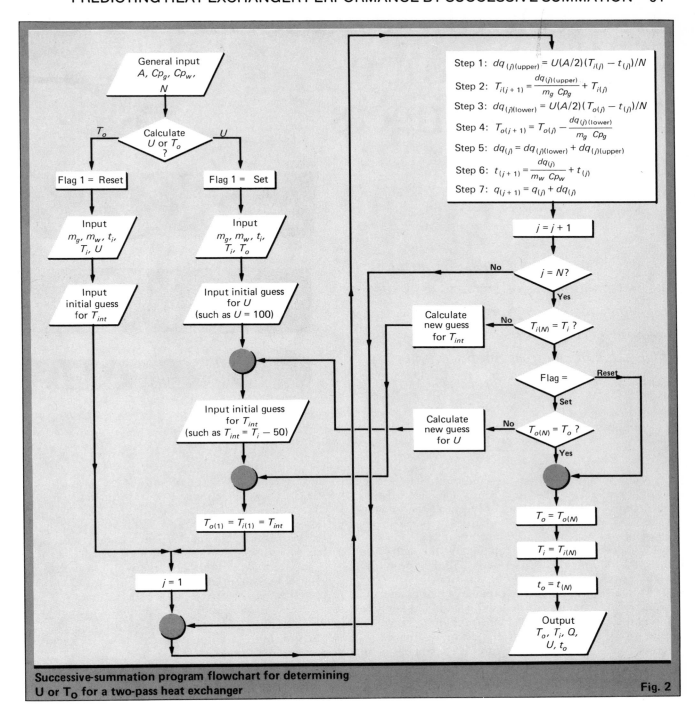

Successive-summation program flowchart for determining U or T_o for a two-pass heat exchanger

Fig. 2

connected in series by manifold for gas stream intermixing between successive sets. Gas recycle rate is the limiting factor in production. The exchangers must be periodically cleaned. Which ones and when?

The successive summation method can be used to make the decision. Heat-transfer coefficients are calculated for each of the heat exchangers, based on plant data. These coefficients are then used to predict the consequences of removing a particular exchanger from service. The optimum arrangement is the one that provides the maximum heat-transfer capacity.

Robert S. Glaubinger, Editor

The author

Robert A. Spencer, Jr., is a developmental engineer for Du Pont's Explosive Products Div., Repauno Plant, Gibbstown, NJ 08027. He is currently developing composite diamond coatings for various end-uses. He has been a process engineer and a manufacturing supervisor. His experience includes production of intermediate chemicals for nylon and polyethylene manufacture.

Mr. Spencer holds a B.S. in chemical engineering from Virginia Polytechnic Institute and State University.

Preventing vibration in shell-and-tube heat exchangers

Sound engineering principles can produce heat exchangers free of problems associated with tube vibrations. Since natural frequencies can be predicted accurately, flow-induced vibration can be prevented by establishing reasonable fluid velocity limits.

George W. Schwarz, Jr., Chemicals and Plastics Div., Union Carbide Corp.

☐ Vibrations of heat exchanger tubes can cause tube-wall thinning at baffles and consequent leaks. Leaks also occur when fatigue cracks develop at the tube-to-tubesheet joint, or when tubes flatten at midspan because of collisions with neighboring tubes.

Loud flow-induced noises can damage the shell foundations and piping [1,2]. Here, we shall deal primarily with flow-induced vibrations of the tubes.

Most fabricating shops are capable of manufacturing sound heat exchangers at reasonable cost. The key to a sound design is the shellside baffle configuration, which must give the tubes sufficient support to withstand flow forces. Much of the technology presented here has been, since 1970, part of the standard design practices of the Chemicals and Plastics Div. of Union Carbide Corp. Since that time, hundreds of heat exchangers have been designed without incident of tube-vibration failure. Careful attention given to potential tube vibration has undoubtedly contributed to this success.

Causes of flow-induced vibration

All heat exchanger tubes vibrate to some degree, but damaging flow-induced vibrations occur more often in large heat exchangers in which the unsupported span approaches or exceeds the TEMA (Tubular Exchanger Manufacturers Assn.) recommended maximum. Damage to a large number of tubes is more likely to occur with gases or vapors on the shellside than with

Originally published July 19, 1976

liquids, especially if the pressure is higher than about 100 psi. Except in extremely poor design, damage is not likely to produce leaks in more than about 30% of the tubes. Flow-induced vibrations also occur with liquids on the shellside, but the damage is usually limited to a few tubes in localized areas of high velocity.

In severe cases, tubes can leak within a few days or even a few hours after the heat exchanger has been placed in service. More often, leaking tubes will show up a year or so after startup. Additional tube leaks will develop after the initial leaking tubes have been plugged, but the number of new leaks will eventually decrease with time.

Currently available methods for predicting flow-induced vibration damage are inadequate for predicting heat-exchanger failures. At best, they identify the exchangers that are most susceptible to damage. The primary reason for this lack of precision is that flow-induced vibrations are extremely complicated. Although much has been learned over the past few years, the technology is not yet complete.

Some of the difficult areas are: (1) the complex pattern of the flow through the bundle; (2) the complicated fluid mechanics of a bank of vibrating tubes; (3) the role of damping; and (4) the rates of wear and fatigue. Nevertheless, it is possible to develop design criteria—especially when tempered with experience—to ensure that equipment will be reasonably safe from vibration damage.

PREVENTING VIBRATION IN SHELL-AND-TUBE HEAT EXCHANGERS

Design criteria should predict conditions under which large-amplitude vibrations are possible. Here, a limit is set on the frequency of potentially damaging forces. For flow-induced vibrations, this limit takes the form of a maximum velocity, above which damage-free operation cannot be assured.

Several mechanisms should be checked, including vortex shedding, fluid-elastic excitation, and turbulent buffeting. When a new exchanger design does not satisfy the criteria, the designer should consider the performance of exchangers in similar services, the consequences of leaking tubes, and the cost of the design changes necessary to make the exchanger satisfy the criteria.

There are many exchangers in service that have never developed leaks, even though they do not satisfy the vibration criteria. Past experience, however, can be misleading. A case in point could be that of two nearly identical heat exchangers, in the same service, and operated at the same rates. Although one exchanger has developed leaks in about 30% of the tubes, the other one has no leaks at all.

Unless amplified by resonant phenomena, the flow forces normally encountered in heat exchangers are not sufficient to cause damage to units built according to standard design practices, such as those suggested by TEMA. Resonance, which can increase the tube deflection by a factor of 100 or more, occurs when the frequency of a cyclic exciting force coincides with the natural frequency of the tube. The exciting force could be the result of: (1) a fluid dynamic mechanism, such as vortex shedding; (2) the pulsations of a compressor; or (3) a mechanical vibration that is passed through the structure.

Damping

Damping limits the amplitude of vibrations, although in heat exchangers it does not occur due to simple damping forms, such as viscous or structural. For convenience, one can distinguish forms of damping by their sources rather than their mechanisms.

Fluid damping, for instance, refers to damping resulting from the interaction of the tubes with the fluid, whereas mechanical damping refers to damping of a mechanical nature, such as the rubbing friction and rattling at baffle holes. Internal damping refers to dissipative forces that occur within the tube metal.

When in doubt, damping should be neglected. Internal damping is negligible for most flow-induced vibrations. Mechanical damping should also be neglected, because mechanical damping does not become significant until the amplitude of the vibration becomes large. Since the object is to keep the amplitude of the vibration small, mechanical damping becomes negligible.

Fluid damping could be important, but there is no ready source of information that relates fluid properties with damping in heat exchangers. There are indications, however, that for low-viscosity fluids, damping forces are small, relative to flow forces.

Safety design factor

A safety factor can be developed logically by drawing an analogy with the spring-and-mass system. If the frequency of an exciting force is less than 80% of the natural frequency, the amplitude of a forced vibration cannot exceed three times the displacement from a static force that is equal in magnitude to the amplitude of the exciting force [3].

The frequency of a cyclic exciting force should not exceed 80% of the natural frequency of the tube, even though a flow-induced vibration is not a forced vibration but a self-excited one. Since, however, the response to a forced vibration agrees reasonably well with the response of flow-induced vibrations on single cylinders, such a safety factor should be valid for design purposes. This 80% safety factor assumes that all quantities relating to the exciting and natural frequency are known with precision, which is unrealistic. Therefore, additional safety factors are applied, which are left to the discretion of the designer, who should be able to predict flow velocities accurately, and be fairly sure that the method he is using will permit operation at off-design conditions.

Natural frequency

Natural frequency is that at which the tube will vibrate—in the absence of external forces—when given an initial displacement and then allowed to vibrate freely. A heat exchanger tube has several natural frequencies, each of which corresponds to a different mode-shape. Each mode is capable of resonance if subjected to an exciting force at the proper frequency.

The recommended method to calculate the natural frequency of heat-exchanger tubes is a modified Holzer technique [4,5], whereby the tube is assumed to be a series of springs and masses. This method provides both the natural frequency of the tubes and the corresponding mode shape. It can also be used to calculate the dynamic response of the tube to local, or distributed, periodic forces, including the effects of damping. Since the calculation is complex and high precision is required, the method has been computerized.

Another equally valid but simpler method involves the calculation of the determinate of a characteristic matrix [6]. Although the natural frequency is easier to

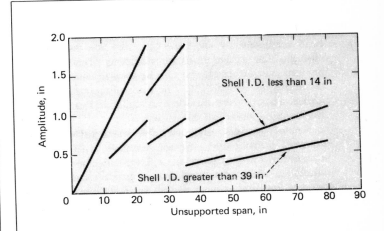

Vibration amplitude at which standard baffles provide greater stiffness than pinned support Fig. 1

calculate by this method, the mode shape and dynamic response are more difficult.

A common design error is to consider the individual spans of the tube independently of one another, and to assume that only the forces that act on a particular span are able to cause damage to that span. But questions arise as to the stiffness of the individual spans. Should the end supports be considered rigidly fixed, simply pinned, or something in between? The stiffness of the end supports makes a substantial difference in the calculated natural frequency.

Because the forces on one span are transmitted by the tube through the supports to neighboring spans, one particular span cannot vibrate independently of neighboring spans. All the spans must vibrate together, at the same frequency, for resonance to be attained.

A heat-exchanger tube should be treated as a continuous, multiple-supported elastic structure with rigidly fixed supports at the tubesheets and pinned supports at the baffles. It has been suggested that baffles provide more stiffness than simple, pinned supports, but it is fairly easy to show that the amplitude of a vibration exceeds the normal gap between tubes, and that collisions occur before any appreciable stiffness is provided by standard baffles. This conclusion is supported by tests; other authors agree [7]. The amplitude at which additional stiffness begins is shown in Fig. 1 for the TEMA standard.

Rolling the tubes into the baffles will increase the stiffness at the supports, and therefore increase the natural frequency. Rolling also isolates the spans to some extent. This procedure is costly and not practical except, perhaps, for a few tubes that are particularly susceptible to damage. In addition, rolling is limited to those baffles less than 6 ft from the tubesheet face.

Virtual mass, axial load

Two additional considerations must be taken into account when calculating the natural frequency of a tube. These are the masses of fluid inside and outside the tube, and the axial load on the tube. The mass of the fluid inside the tube should be added to the mass of the tube when calculating the natural frequency.

It is less obvious that the mass of some of the fluid outside the tube should also be added. A generally accepted procedure is to add the mass of the shellside fluid displaced by the tube [8], but this may not be enough. The added mass for vibrating objects in close proximity to other objects can be several times the mass of the displaced fluid [9]. The mass of the tube plus the added mass of the tubeside and shellside fluids is called the "virtual mass" of the tube.

Tensile stresses increase the natural frequency, whereas compressive stresses decrease it [10]. These stresses are most important in fixed-tubesheet heat exchangers, where standard design practices allow compressive loading as large as half the buckling load. Such a stress is important in floating-tubesheet units, if the operating pressure of one side is considerably higher than in the other. The effect of a compressive stress, at half the buckling load, is to lower the natural frequency of the tube to 70% of the neutral stress value, as calculated by the Holzer or any other common method.

Calculating the tube natural frequency by rigorous means has lead to some important observations. First, the natural frequencies of heat-exchanger tubes are very close to each other. Fig. 2 shows the first few modes for a 1-in, 16-BWG (Birmingham wire gage), stainless steel tube with thirteen 72-in spans. The shape of the fundamental mode, or first natural frequency is similar to a sine wave with the nodes (zero-amplitude points) at the tubesheets and baffles.

Unexpectedly, the second mode does not have nodes at the tubesheets, baffles, and midpoints of all the spans. The second mode has only one additional node located at the midpoint of the least-stiff span. If there are more than about five spans, the frequency of the second mode is only a few Hertz higher than the first mode. Each subsequently higher mode has an additional node, or perhaps a double node, where a single node previously existed. Equipment should be designed so that the frequency of exciting forces is below the fundamental frequency, because natural frequencies are too closely spaced to attempt to design between higher modes.

Another important observation is that a variable baffle spacing will lower the natural frequency of a tube, and increase the resonant amplitude. Analysis, test work and experience all indicate that variable baffle spacing makes a heat exchanger more susceptible to damage from vibrations.

Calculating the natural frequency

Because of the desirability of having a quick method for determining the natural frequency of heat-exchanger tubes, a very simple approximation has been developed. The first step is to determine the least-stiff span, which is usually the longest span supported on both ends by baffles. The approximate method calculates the fundamental natural frequency of that pinned-pinned span (Eq. 1). The added mass of the fluids inside and outside should be taken into account, and the calculated natural frequency should be decreased to 70% if compressive stresses are present.

The span adjacent to the tubesheet will be the least-stiff span provided it is longer than 150% of the longest baffle-to-baffle span. The approximate natural frequency for this instance should be calculated using the length of the span adjacent to the tubesheet, with fixed-pinned end conditions (Eq. 2).

This approximation is conservative but surprisingly accurate, if there are more than five spans and the least-stiff span is between baffles. It should be clear that this simplified method is an approximation to the natural frequency of the entire tube, but does not imply that the individual spans behave independently, although the natural frequencies of individual spans are used in the calculation.

For pinned-pinned supports:
$$f_N = 1.56 \sqrt{gEI/Wl^4} \qquad (1)$$

For fixed-pinned supports:
$$f_N = 2.45 \sqrt{gEI/Wl^4} \qquad (2)$$

where f_N = fundamental natural frequency with no axial load, Hz; g = gravitational force constant, in/s^2;

E = module of elasticity, psi; I = area moment of inertia, in^4; W = virtual weight per unit length, lb/in; and l = span length, in.

U-bend tubes

There is evidence that the bend portion of a U-bend tube is somewhat less stiff than a straight span of the same length. The span containing the bend should be assumed to be a straight span of a length 12% greater than the actual developed length of the span [11].

Reiterating, the natural frequency of a heat-exchanger tube can be calculated accurately by assuming it to be one, long, multiple-supported structure. The tube-to-tubesheet joints should be considered fixed-end supports, and the baffles should be considered as pinned supports. The effect of the added masses of the tubeside and shellside fluids should be included, in addition to the effect of axial stresses. Simply stated, the design criterion is to avoid a large-amplitude resonant condition, and not to allow the frequency of any cyclic exciting force to exceed 80% of the fundamental natural frequency of the tube.

Exciting mechanisms

There are a number of mechanisms that cause periodic exciting forces in heat exchangers, any one of which is a potential source of damaging vibrations. The pulsations in the flow caused by the passage of the blades of a compressor is one source of a periodic force, and certain vibrations transmitted through supporting structures is another. Since the frequency of these types of exciting forces is relatively simple to predict, it is fairly simple to check for susceptibility to damage from them. Flow-induced mechanisms are more difficult to predict.

Flow-induced mechanisms can be divided into two categories; those induced by flow parallel to the tube axis, and those induced by flow perpendicular to it. Parallel-flow mechanisms can be further divided into those resulting from flow inside the tubes [12], and those resulting from outside flow [13,14].

None of the parallel-flow mechanisms are normally troublesome in heat exchangers, because either the amplitudes are small, or the velocities are much higher than those normally encountered. Perpendicular-flow mechanisms, on the other hand, can cause significant amplitudes at common velocities. The forms of perpendicular flow that will be discussed are vortex shedding, fluid-elastic excitation, and turbulent buffeting.

Vortex shedding

The most common flow-induced cyclic force is that produced by vortex shedding, which is a flow pattern in which vortices shed alternately from the sides of a bluff body in crossflow, such as a heat-exchanger tube. In the Reynolds number range usually found in heat exchangers, the vortex-shedding frequency is proportional to the velocity.

Alternate shedding of the vortices produces cyclic forces on the tube, with components in two directions. The frequency of the perpendicular or transverse component coincides with that of the vortex shedding, whereas the frequency of the parallel, or streamwise

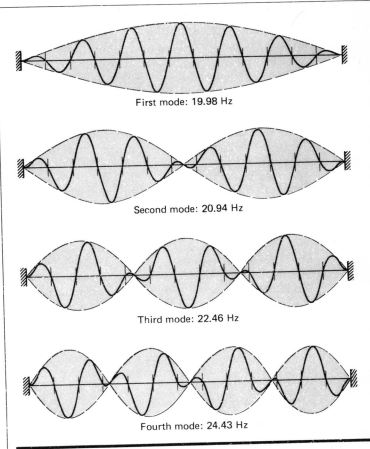

First modes of a 1-in, 16-BWG, stainless steel tube with 13 equal 72-in spans   Fig. 2

component is twice the vortex-shedding frequency. At low Reynolds numbers and at large pitch-to-diameter ratios, the transverse component is the stronger of the two. As the Reynolds number increases, or the pitch-to-diameter ratio decreases, the streamwise component becomes the larger force.

The magnitude of the resultant lift force is on the order of 0.1–1.0 velocity heads [15]. The larger lift coefficients can be expected with triangular pitch and with large pitch-to-diameter ratios. The lower lift coefficients correspond to pitch-to-diameter ratios fairly close to 1.0.

Although most design methods consider only the transverse force at the vortex-shedding frequency, the procedure is a dubious one for heat exchangers. A designer should be aware that the streamwise force has a significant magnitude for standard pitch-to-diameter ratios, and that its frequency is twice the vortex-shedding frequency. Recent tests with liquids have shown large-amplitude vibrations for a single cylinder from the streamwise force [19].

The most widely accepted method for predicting the vortex-shedding frequency in tube banks is that of Y. N. Chen [2]. The vortex-shedding frequency is described in terms of a dimensionless frequency number called the Strouhal number, which is the vortex-shedding frequency multiplied by the tube diameter and divided by the velocity. Chen plots the Strouhal number, in an ideal tube bank, as a function of geometrical parameters of the tube-layout pattern for both staggered and inline tube banks. A word of caution: in the region of the Chen plots, where most standard tube patterns lie, the data are less accurate and the Strouhal number is sensitive to the pitch-to-diameter ratio. It is advisable to make allowances for this when designing equipment.

Questions arise as to what velocity to use when calculating the vortex-shedding frequency from the Strouhal number. The flow may not be perpendicular to the tube axis, as it was in the ideal tube bank, and the velocity may vary along the length of the tube because of flow distribution. Therefore, the velocity that should be used when calculating the exciting force is the highest normal component of the velocity in the velocity distribution [*16,17*]. Because the vortex shedding may "lock in" at the natural frequency, such shedding may not actually occur at the frequency calculated from the Strouhal number on the plot. However, this procedure provides the highest potential vortex-shedding frequency; it is therefore the proper value to use when designing equipment.

There is an important deficiency in the data for vortex shedding in tube banks. The Chen graphs suggest that the Strouhal number is strongly dependent on the angle at which the flow moves across the bundle. For example, if the tubes are spaced on a triangular pitch, with a pitch-to-diameter ratio of 1.25, the Chen graphs predict a Strouhal number of 0.25. But if the direction of the flow is rotated 90 deg, the graphs predict a Strouhal number of 0.80. There is no way, however, to predict the Strouhal number for flow at an oblique angle. Fluid does flow at oblique angles in heat exchangers, particularly near the entrance and exit regions, and near the baffle cuts. Since these are areas in which experience shows that tube damage is likely to occur, the designer must exercise caution when selecting a Strouhal number for tubes in these areas.

In Chen's work, the tubes in his ideal tube banks were too rigid, and therefore did not vibrate at amplitudes large enough to substantially affect the flow. Although there are no published data on tube banks, there are for flexible, single cylinders, which show that the vortices will shed at roughly the same frequency they would from a rigid tube, if the amplitude of the vibration is small. However, when the vortex-shedding frequency reaches a value of about 0.8 the natural frequency, the vortex-shedding frequency locks-in on the natural frequency of the tube. As the velocity is increased, the vortices continue to be shed at the natural frequency, even though the frequency calculated from the Strouhal number is much higher than the natural frequency [*18,19*].

Since serious questions have been raised concerning the lock-in region in tube banks [*2*], further studies of this phenomenon are required. It is reasonable to assume, however, that lock-in occurs and that the highest potential vortex-shedding frequency should not be allowed to overlap this lock-in region. Again, the designer must exercise caution, because the onset of the lock-in region may be dependent on amplitude as well as frequency.

Designing for an exciting frequency 0.8 of the natural frequency leaves no margin of safety with vortex shedding, because of the lock-in region. When designing for transverse vibration, the vortex-shedding frequency should not exceed 65% of the natural frequency; and when designing for streamwise vibration, the vortex-shedding frequency should not exceed 35% of the natural frequency.

Fluid-elastic excitation

Another mechanism for flow-induced tube vibration, originally introduced by H. J. Connors [*8*] is fluid-elastic excitation. Here, the actual tube motion is an elliptical orbit, with the major axis of the ellipse perpendicular to the flow in one tube, and parallel to the flow in neighboring tubes. Connors developed a method for predicting when this mechanism can produce a large-amplitude vibration (Eq. 3), which includes the effects of damping. Test data upon which the method is based has been developed only for a row of tubes perpendicular to the flow and not for a bank of tubes. Nevertheless, the method can be applied to a bank of tubes, provided one has the data to describe the damping. Connors' equation is:

$$U/f_0 D = 9.9 \sqrt{m_0 \delta_0 / \rho D^2} \qquad (3)$$

where U = velocity between tubes, in/s; f_0 = fundamental natural frequency of the tube, Hz; D = tube diameter, in; m_0 = virtual mass of tube, [(lb-ft)(s^2)]/in^2; δ_0 = logarithmic decrement; ρ = fluid density, [(lb-force)(s^2)]/in^4.

Chen has suggested a modification to Connors' criterion to make it applicable for pitch-to-diameter ratios other than 1.41, which is the only ratio Connors investigated. Chen's modification includes a dependence on Reynolds numbers:

$$U/f_0 D = (0.35 \times 10^6)(m_0 \delta_0 / \rho D^2)^{0.6}(X_t / N_{Re}) \qquad (4)$$

where X_t = transverse-pitch-to-diameter ratio, dimensionless; and N_{Re} = Reynolds number, dimensionless.

The logarithmic decrement—a parameter indicative of the damping in the system—poses a problem. There are little data available, and the logarithmic decrement can vary over several orders of magnitude. In addition, Connors evaluated the effects of damping by introducing internal damping with other forms negligible. Fluid and mechanical damping are likely to predominate over internal damping in heat exchangers. In tubes, the value of the logarithmic decrement is on the order of 0.001 for internal damping, and 0.05 for fluid-mechanical damping in water. Whether it is proper to use a fluid-mechanical logarithmic decrement in Connors' criterion is open to question.

A very serious deficiency with the criterion for the fluid-elastic mechanism is the complete lack of definitive data for liquids. The value of the parameter $m_0 \delta_0 / \rho D^2$ is about 0.08 for a 1-in tube in water. This is a factor 100 below Connors' lowest test point. J. H. Kissel has suggested that Connors' criterion is appropriate for liquids [*7*], basing his recommendation on heat-exchanger test data. However, for the conditions of his tests, the predictions of the fluid-elastic mechanism and the vortex-shedding mechanism are essentially indistinguishable. More comprehensive studies of this mechanism are necessary before it can be applied with confidence to liquids.

With gases, the velocity required to excite a large-amplitude, fluid-elastic vibration is considerably higher than that required to produce resonance from vortex shedding. Since it has been established that such shedding can cause damaging tube vibrations, a design criterion should not be based solely on the fluid-elastic mechanism. There are cases, however, in which vortex shedding either does not occur or, when it does, it produces forces of very small magnitude. This could be the case for flows with very high Reynolds numbers. When condensing steam or low-density organics, experience indicates that the vortex-shedding criterion is too stringent. In such cases, the criterion based on the fluid-elastic mechanism may be more appropriate.

Turbulent buffeting

Turbulence is another mechanism sometimes considered in tube-vibration analysis. P. R. Owens [21] developed a criterion for predicting damaging vibrations by this mechanism (Eq. 5), which applies only to gases at high Reynolds numbers. He sets forth a theory that describes the frequency at which the most energetic eddies encounter the tubes. To produce large-amplitude vibration, the tube must be able to filter out and store energy from these eddies at its natural frequency. It has not been clearly established that this mechanism, as described by Owens, is possible in heat exchangers with standard pitch-to-diameter ratios. The theory upon which the mechanism is based ceases to be valid as the pitch-to-diameter ratio approaches 1.0. Nevertheless, the data that Owens collected from several sources agreed well with his theory. The data contained points at a pitch-to-diameter ratio of up to 1.25. The correlation of these data, if assumed applicable to shell-and-tube heat exchangers, provides results of the same order of magnitude as vortex shedding:

$$(fl/u)(T/d) = 3.05(1 - d/T)^2 + 0.28 \quad (5)$$

where f = frequency at which most energetic eddies encounter tubes, Hz; l = longitudinal spacing between tubes, ft; u = velocity between tubes, ft/s; T = transverse spacing between tubes, ft; and d = tube diameter, ft.

Comparison of vibration mechanisms

Fig. 3 shows a comparison of the three flow-induced vibration mechanisms presented. The basis is a bank of 1-in by 16-BWG admiralty tubes on a 1.25-in triangular pitch, where the natural frequency is approximated as the natural frequency of the least-still span with pinned-end conditions. The span length is plotted on the abscissa, and the velocity at which the excitation frequency is equal to the fundamental natural frequency is plotted on the ordinate. This velocity is sometimes called the "critical velocity." There are no safety factors built into this plot.

This graph demonstrates that there is need for careful thought on the part of the designer. To design for the mechanism with the lowest critical velocity may be very difficult, and the resulting design can be a costly one.

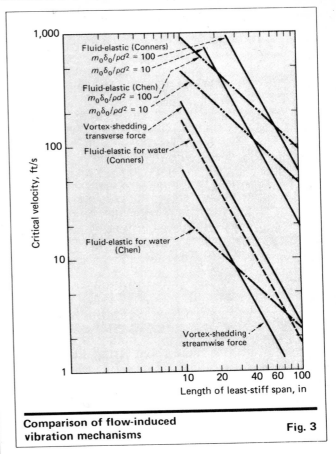

Comparison of flow-induced vibration mechanisms   Fig. 3

Designing one of the other mechanisms leaves the possibility that damaging vibrations may occur. In the end, the designer must rely heavily on experience, to judge whether the risk of vibration damage has been reduced to a reasonable level.

Design guidelines

The following guidelines are suggested:

For liquids, the Reynolds number should probably be in the 300–50,000 range. It is reasonable to assume that the vortex-shedding mechanism will be active. If the velocity head is high ($\rho V^2/9,266 > 0.5$ psi),* the force in the streamwise direction may be sufficient to cause damage. Under these conditions, the design should apply to the streamwise vortex-shedding mechanism. If the velocity head is low ($\rho V^2/9,266 < 0.5$ psi), the design should apply to the transverse vortex-shedding mechanism.

For gases or vapors, follow the same guidelines as for liquids, if the Reynolds number is below 300,000. But if this number is larger, the design should apply to the fluid-elastic mechanism.

If the pitch-to-diameter ratio is greater than 1.5, apply the design to the turbulent-buffeting mechanism. In addition, with square-pitch or widely spaced triangular pitch, check for acoustic vibrations.

These suggested guidelines are based on information in the literature and on experience, but no guarantees can be offered. There is no substitute for sound engi-

*ρ = fluid density, (pound-mass)/ft^3; V = velocity, ft/s.

neering judgement that is based on the particular environment the exchanger will experience.

Large heat exchangers with gases on the shellside and low allowable pressure drop are particularly difficult to design. When designing equipment in which tube vibrations are likely to be a problem, the following design practices are recommended:

1. Design the shellside so that the flow pattern is predictable. Avoid large (greater than 35%) and small (less than 15%) baffle cuts, because both conditions provide poor velocity distribution.

2. Block any bypass flowpaths between the bundle and the shell, as well as through the pass-rib lanes; high local velocities in these areas can cause local damage in an otherwise sound design. Sometimes it is possible to use parallel-flow baffles (also called window baffles and triple-segmental baffles), which provide a flow that is primarily parallel to the tubes. The velocities will be lower, and the low-crossflow component of velocity will produce a low vortex-shedding frequency.

3. In cases where the only troublesome velocity is that near the entrance or exit nozzles, questions arise as to whether this velocity can produce damage to the more limber spans downstream. It may be very costly, however, to alter the entire design to meet the vibration criteria for this localized velocity. In such instances, good results have been obtained by one of these three methods:

(a) Installing a nozzle-velocity-reducing device. A standard pipe reducer installed with an impingement plate is often sufficient. In more severe cases, a distribution belt may be required.

(b) Installing a tube-support baffle directly under the nozzle. This puts the exciting force at a node,* and significantly diminishes the amplitude of the vibration that can be produced in the other spans.

(c) Rolling the tubes into the first baffle in the vicinity of the nozzle. Such rolling partially isolates the exciting force from the rest of the tube, and lessens the amplitude of the vibration.

Of the three design practices just mentioned, the nozzle-velocity-reducing device is preferred.

Extra-thick baffles reduce the rate of wear. They do not, however, increase the natural frequency unless the holes are precision-machined to very close tolerances. Such machining is costly, and the close fit can result in assembly difficulties. The additional cost for precision-machined baffles is not usually justified, but the cost for extra-thick baffles is.

The heat-exchanger design most resistant to tube-vibration damage is the segmental-baffle type, without tubes in the baffle cuts. There are two advantages to such a design. First, the most troublesome tubes—those that are supported only at every other baffle—are eliminated. Second, the intermediate tube supports—baffles with the segments removed from both ends—can be installed between the segmental baffles. The designer should install as many of these intermediate supports as are necessary to increase the natural frequency beyond the exciting frequency. With this design, a nozzle-velocity-reducing device is recommended at the inlet.

If pressure drop is a problem with a segmental-baffle

*At a point relatively free of vibratory motion.

design, the tube pitch may be increased. Because of the larger shell and possibly greater surface, an exchanger of this construction will cost more than a conventional one. The additional expense is justified if leaking tubes can shut down the unit, contaminate a product, or cause pollution.

References

1. Barrington, E. A., Acoustic Vibrations in Tubular Exchangers, *Chem. Eng. Progr.*, Vol. 69, No. 7, July 1973, pp. 62–68.
2. Chen, Y. N., "Flow Induced Vibration and Noise in Tube-Bank Heat Exchangers Due to Von Karman Stress," American Soc. of Mechanical Engineers, Paper No. 67-VIBR-48, 1967.
3. Thomson, W. T., "Vibration Theory and Applications," pp. 51–57, Prentice-Hall, Englewood Cliffs, N.J. (1948).
4. Prohl, M. A., A General Method for Calculating Critical Speeds of Flexible Rotors, *J. Applied Mechanics*, Sept. 1945.
5. Myklestad, N. O., A New Method of Calculating Natural Modes of Uncoupled Bending Vibration of Airplane Wings and Other Types of Beams, *J. Aeronautic Sciences*, Apr. 1944.
6. Moretti, P. M., and Lowery, R. L., "Natural Frequencies and Damping of Tubes in Shell-and-Tube Heat Exchangers," Oklahoma State University, Stillwater, Okla.
7. Kissel, J. H., "Flow Induced Vibrations in Heat Exchangers—A Practical Look," AIChE Preprint No. 8, Thirteenth National Heat Transfer Conference, Aug. 1972.
8. Connors, H. J., Jr., Fluidelastic Vibration of Tube Arrays Excited by Cross Flow, in "Flow Induced Vibration in Heat Exchangers," p. 42, published by American Soc. of Mechanical Engineers (1970).
9. Kennard, E. H., "Irrotational Flow of Frictionless Fluids, Mostly of Invariable Density," Dept. of the Navy, Report 2299, Feb. 1967, pp. 380–396.
10. Lentz, G. L., "Flow Induced Vibrations in Heat Exchangers Near the Shell-Side Inlet and Exit Nozzles," AIChE Preprint No. 20 (1969).
11. Den Harton, J. P. "Mechanical Vibration," p. 168, 4th ed., McGraw-Hill, New York (1956).
12. Clinch, J. M., Prediction and Measurement of the Vibration Induced in Thin-Walled Pipes by the Passage of Internal Turbulent Water Flow, *J. Sound Vibration*, Vol. 12, No. 4, 1970, pp. 429–451.
13. Burgrun, D., Byrnes, J. J., and Benforado, D. M., "Vibration of Rods Induced by Water in Parallel Flow," Transactions of the American Soc. of Mechanical Engineers, July 1958, pp. 991–1003.
14. Paidoussis, M. P., Dynamics of Flexible Cylinders in Axial Flow, Part 1—Theory, and Part 2—Experiments, *J. Fluid Mechanics*, Vol. 26, Part 4, 1966, pp. 717–751.
15. Chen, Y. N., "Fluctuating Lift Forces of the Karman Vortex Stress on Single Circular Cylinders and in Tube Bundles, Part 3—Lift Forces in Tube Bundles," American Soc. of Mechanical Engineers, Paper No. 71-VIBR-13, 1971.
16. "Vortex Wakes From Flexible Circular Cylinders at Low Reynolds Numbers," Naval Air Development Center, Report No. NADC-AE-7011, July 1970.
17. Chiu, W. S., and Lienhard, J. H., On Real Fluid Flow Over Yawed Circular Cylinders, *J. Basic Eng.*, Transactions of the American Soc. of Mechanical Engineers, Vol. 89, pp. 815–857 (1967).
18. Hartlen, R. T., and Currie, I. G., Lift-Oscillator Mode of Vortex-Induced Vibration, *J. Eng. Mechanics Div.*, Proceedings of the American Soc. of Civil Engineers, Oct. 1970, pp. 577–591.
19. Halle, H., and Lawrence, W. P., "Crossflow-Induced Vibration of a Circular Cylinder," American Soc. of Mechanical Engineers, Paper No. 73-DET-68, 1973.
20. Chen, Y. N., "The Orbital Movement and the Damping of Fluidelastic Vibration of Tube Banks Due to Vortex Formation, Part 2—Criterion for the Fluidelastic Orbital Vibration of Tube Arrays," American Soc. of Mechanical Engineers, Paper No. 73-DET-146, 1973.
21. Owens, P. R., Buffeting Excitation of Boiler Tube Vibration, *J. Mechanical Eng. Science*, Vol. 7, No. 4, 1965, pp. 431–439.

The author

G. W. Schwarz, Jr., is a senior engineer in the engineering department at Union Carbide Corp.'s Technical Center, P.O. Box 8361, South Charleston, WV 25303, where he is a member of the heat-transfer and fluid-dynamics technology group. He holds B.S. and M.S. degrees in mechanical engineering from the University of Illinois, and is a member of the American Soc. of Mechanical Engineers.

Selecting tubes for CPI heat exchangers—I

Enhanced heat-transfer tubes, proved in water desalting applications, promise greater efficiencies in many chemical-process-industries applications.

D. H. Foxall and P. T. Gilbert
Yorkshire Imperial Metals Ltd., U.K.

☐ The past few years have seen new types of profiled tubes that give more-economical heat-transfer performance in certain applications than do plain tubes. Development and acceptance of such tubes has been stimulated by advances in the technology of seawater distillation, but many other uses can be envisaged.

The various types of heat-transfer tube now available can be made from a wide range of materials, and the purpose of these articles is to give some guidance on the choice of tubes from the viewpoint of both heat-transfer performance and material suitability, with particular reference to the chemical process industries. With the universal need to minimize cost, designers of new heat exchangers, and engineers who need to uprate the performance of existing plants, will wish to ensure that no suitable product is overlooked.

Heat-transfer performance

The use of finned tubes is widely accepted, and the mechanism that enables fins to be used economically is generally understood. The newer types of tubes, however, give an advantage in a different way, which does not depend on providing extra surface per unit length of tube; these are becoming known as enhanced-heat-transfer (EHT) tubes to distinguish them from extended-surface tubes. To show that the various products now available have distinct fields of application, it will be useful to consider briefly the manner in which heat is transferred between a metal surface and a fluid.

Fundamental heat-transfer considerations

Taking the case of a single-phase fluid in turbulent flow parallel to a plane surface from which it is receiving heat, the rate at which heat flows is:

$$q = hA\Delta T \qquad (1)$$

where h, the film or surface heat-transfer coefficient (which is a function both of the physical properties of the fluid and the geometry of the system) indicates how efficiently the surface transfers heat.

However high the main stream velocity may be, the fluid very close to the solid surface moves slowly—and it is the ease or difficulty with which heat flows across this boundary layer that determines the value of h. A velocity gradient exists across the boundary layer, part of the latter being in laminar flow. Movement of fluid particles perpendicular to the surface in the laminar sublayer is negligible, and heat transfer here is essentially by conduction. There is also a buffer layer in which, at different times and places, flow may be laminar or turbulent, with an average velocity and turbulence level lower than those in the main stream. Fig. 1a illustrates the situation.

Heat flows relatively easily across the buffer layer, aided by the movement of fluid particles into and out of the main stream, but this layer does contribute to the total heat-transfer resistance.

If the simplifying assumption could be made that there was a completely laminar boundary layer of thickness, t, with heat flow by conduction only, followed by a sudden transition to the fully turbulent conditions of the main stream (i.e., accepting Fig. 1b as a reasonable approximation to Fig. 1a), then the film coefficient would be given by:

$$h = K_f/t \qquad (2)$$

Heat transfer would then take place in much the same way as through a solid barrier. Knowing the value of h, determined by experiment or calculated from accepted data, enables us to estimate the corresponding value of t, by using Eq. (2). But it must be remembered that the true thickness of the laminar sublayer is somewhat less than t, whereas the total thickness, including the buffer layer, is greater than t. Eq. (2), therefore, presents an oversimplified but useful indication of the relative magnitude of boundary-layer thickness in different situations.

The range of values of h that may be encountered varies by a factor of about 10,000 and, as there are two surfaces to consider, the most economical tubular solution will vary greatly from one case to another. Fig. 2 shows numerous typical levels of h, and although it cannot be assumed that all situations are covered (the

Originally published March 15, 1976

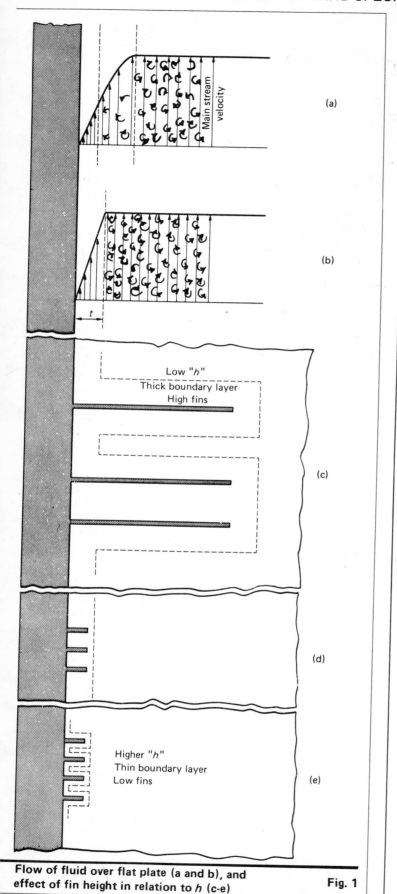

Flow of fluid over flat plate (a and b), and effect of fin height in relation to h (c-e) Fig. 1

variations with temperature, velocity and the geometry of the system make it difficult to specify absolute limits), this figure helps to determine which types of heat-exchanger tube will be appropriate in a given situation.

The lowest values of h are, of course, obtained when the fluid conductivity is low and the boundary layer thick, as is the case with gases at low velocity; the highest values occur with liquids of much higher conductivity, and thin boundary layers (resulting from low viscosity and high velocity). The thermal conductivity, K, can vary by a factor of about 100 from one situation to another (even excluding liquid metals) and t, the nominal boundary-layer thickness, can be assumed to vary by a comparable factor, the minimum values occurring with the highest values of h being about 1 or 2 mils (0.02 – 0.05 mm).

Transversely finned tubes (Fig. 3)

Where fins seem appropriate, the fin height and pitch that can be used effectively is fundamentally related to the boundary-layer thickness, and if the wrong type of profile is employed, performance may be disappointing.

The amount of heat transferred can be increased by adding to the initial surface area, A—see Eq. (1)—provided that the boundary-layer area is also increased. The advantage that can be obtained from fins, therefore, depends on their pitch being not too small (see Fig. 1c); if the fins are too close, the boundary-layer area will not be significantly increased because the main stream will not adequately penetrate the fin spaces.

The largest area-multiplication factors are justified when h is extremely low and there is superficially an incentive to consider high and closely pitched fins; it is just in this situation, however, that the thickest boundary layers occur, and so, if the fin pitch is too small, the result may not come up to expectations. High fins, therefore, should generally be relatively more widely spaced.

In large air-cooled heat exchangers, for example, the external film coefficient is often about 10, and ⅝-in-high fins are used with either 8 or 11 fins per inch; the advantage of 11 rather than 8 fins is often marginal and there seems little incentive to consider even-closer spacing. If the minimum tube pitch is used, of course, bypassing cannot take place and the air stream must penetrate the fin spaces, but an excessive pressure drop will then be obtained if the fin pitch is too small.

With gases at exceptionally low velocity, or in natural convection, very low values of h are obtained. Even higher fins can then be used but fin pitch should also be increased.

Although the metal's thermal conductivity, K, is not usually a significant factor in choosing a tube material, it is more important in fins because of the relatively long and narrow heat-flow path, and this factor sometimes limits the fin height that can economically be used. The effective material cost is then to some extent a function of thermal conductivity. In duties such as those mentioned above, aluminum, having a high K and relatively low cost, is attractive. An inner tube of different material may be required to provide strength

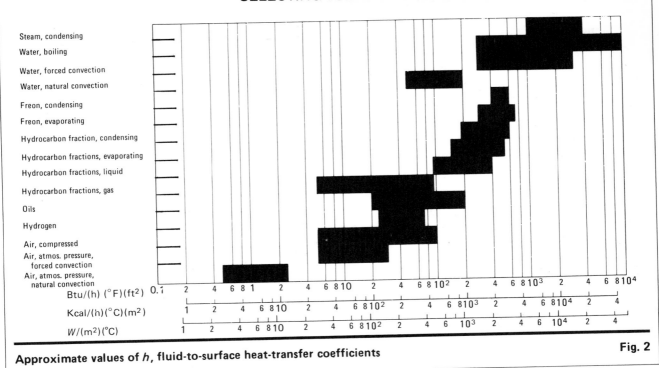

Approximate values of h, fluid-to-surface heat-transfer coefficients — Fig. 2

or corrosion resistance (we will cover this in a later article).

Further up the heat-transfer-coefficient range, there are many fluids that are used on the shell side of shell-and-tube heat exchangers for which the value of h is an order of magnitude higher than in the above examples. With hydrocarbon liquids in cross-baffled flow, for example, h will not usually be much below 100 and may be considerably higher; comparable values are obtained with many other liquids. Fins can be used with advantage in these circumstances, but resistance to heat flow through a fin is large enough to make the high-fin type of profile uneconomical. With these higher values of h, however, the boundary layer is usually relatively thin, and so the fins can be more closely spaced.

With the fins of nominal 1/16-in height that have become standard for the low-fin type of duty, it is usual to reduce the tube diameter during the finning process so that the fin diameter is confined within the limits of the initial stock-tube diameter; the tube can thus be fed through holes in tubesheets and baffles just as easily as can plain tube. Low-fin tube is therefore ideally suited for shell-and-tube use and can replace plain tube in existing equipment as well as being used in new plants.

There is relatively little overlap between the fields of application of high-fin and low-fin tubes, as low-fin types may be ineffective in some typical high-fin duties. It is desirable that the fin height should be significantly greater than the boundary layer thickness (Fig. 1c and 1e); otherwise, the fundamental objective of increasing the surface area of the boundary layer will not be achieved (Fig. 1d).

The most commonly used low-fin tube has approximately 19 fins per inch, the distance between the surfaces of adjacent fins being about 0.040 in, and long experience confirms that this is satisfactory at the intermediate values of h. Based on unit surface area, the external film coefficient, h_o, in single-phase flow is usually about equal to the plain-tube value at the same Reynolds number, confirming that the turbulent stream penetrates adequately between the fins. Experimental data suggest, however, that low-fin heat transfer begins to fall (relatively) as Reynolds number decreases below the range of about 800–1,000, possibly indicating that penetration is then less effective*.

Apart from heating or cooling liquids, the low-fin type of surface is effective with compressed gases, i.e. in intercoolers and aftercoolers, and is also used in condensing and evaporating duties, particularly with refrigerants and hydrocarbons. In condensing duties, the h_o for low-fin tubes can be significantly higher than the corresponding plain-tube value because the fins have excellent drainage characteristics that lead to a lower mean condensate-film thickness.

* Briggs, D. E., others, How to Design Finned-Tube Exchangers, *Chem. Eng. Progr.*, Vol. 59, Nov. 1963, pp 49–59.

Finned tubes come in many styles and fin heights — Fig. 3

Nomenclature

- A Area of heat-transfer surface, ft^2
- h Individual heat-transfer coefficient between fluid and surface, $\text{Btu}/(\text{h})\,(\text{ft}^2)\,(°\text{F})$
- K Thermal conductivity, $\text{Btu}/(\text{h})\,(\text{ft}^2)\,(°\text{F/ft})$
- K_f Thermal conductivity at mean temperature of boundary layer, $\text{Btu}/(\text{h})\,(\text{ft}^2)\,(°\text{F/ft})$
- q Overall rate of heat transfer, Btu/h
- t Nominal thickness of boundary layer, ft
- U Overall heat-transfer coefficient, normally based on tube external surface, $\text{Btu}/(\text{h})\,(\text{ft}^2)\,(°\text{F})$
- T Temperature difference across boundary layer, °F

Subscripts
- o External tube surface
- i Internal tube surface

In most low-fin duties, water flows through the tubes, and it is usually the water conditions that determine the choice of tube material. For instance, copper tubes are widely used in refrigeration and air-conditioning equipment.

In recent years, a 26-fin-per-inch tube, giving a further increase of about 25% in surface area, has gained some popularity, and there is an incentive to investigate its wider use. The reduction in the distance between the fins to about 0.027–0.028 in. may rule out some of the single-phase 19-fin duties at the lower end of the range, and it seems likely that a Reynolds number of well over 1,000 may be required for the newer profile to prove attractive. It may be significant that the main use of the 26-fin profile at present appears to be with Freon refrigerants, where the film coefficients are relatively high; and even here, its advantage over 19-fin does not appear to be substantial.

The acceptance of 26-fin appears to have persuaded some fabricators to investigate an even-closer pitch, e.g., 35 fins per inch, but there are probably few instances in which such a profile will have a commercial advantage over the low-fin tubes in general use.

Enhanced-heat-transfer (EHT) tubes

If a finned tube is considered for use with very high h, conduction between the tip and the root of the fin becomes an increasingly important factor, and the fin height that can be economically used is severely limited. Even lower and more-closely-pitched fins can be considered as h increases, but the benefit from following such a trend must progressively diminish. Furthermore, externally finned tubes are only economical when h_i is substantially higher than h_o and suitable applications obviously become less frequent as h_o increases. If h_o approaches or exceeds 1,000, therefore, plain tube is at present an almost universal choice, but with the development of EHT tubes there is now an interesting alternative available.

Referring again to Eq. (1) and (2), instead of increasing the surface area A, q can be increased by making the boundary layer thinner or more turbulent—i.e., h is increased because t is reduced. A valuable characteristic of such a tube is that its advantage can be obtained with essentially the same amount of metal per unit length as the alternative plain tube. The cost increment, therefore, is usually modest, and the case for using the tube does not necessarily require a very large improvement in performance.

Because there is no significant increase in resistance through the metal, the use of EHT tubes is not ruled out by high values of h—and even though the plain-surface h is 2,000 or more, an increase may be worth considering. Furthermore, the economical use of these tubes does not depend on a substantial ratio between h_i and h_o, and their fields of application are therefore generally distinct from those of finned tubes.

With a plain-tube design, when both h_o and h_i are very high the resistance to heat transfer through the wall will already be of some significance; and clearly if an EHT tube is used to promote even-higher film coefficients, the tube-wall resistance must become still more important. In extreme cases, one should consider questioning established views on choice of tube material and wall thickness, because a conservative attitude may unnecessarily restrict the advantage obtainable.

In the seawater-distillation field, where plain-tube film coefficients are high, the need to reduce capital cost has quickly led to the acceptance of EHT tubes, and similar benefits appear to be available elsewhere.

(The next article in this series will examine roped tubes and fluted tubes and a third article will cover materials of construction.)

Acknowledgement

The authors wish to thank the Directors of Yorkshire Imperial Metals Ltd. for their permission to publish this paper.

The authors

P. T. Gilbert is technical director of Yorkshire Imperial Metals Ltd., P. O. Box No. 166, Leeds LS1 1RD, England. He was educated at Elmhurst Grammar School, Street, Somerset, and at University College of the South West of England (now University of Exeter). He was then awarded a B.Sc. (Chemistry) degree of the University of London (1st Class Honours) and, later, a Ph.D. for work on corrosion of zinc and zinc coatings in supply waters. He is a Fellow of Institution of Metallurgists, Institute of Marine Engineers, Institution of Corrosion Science & Technology, and is a member of Institution of Water Engineers and Scientists, Soc. for Metals, and National Assn. of Corrosion Engineers.

D. H. Foxall is Manager of Heat Transfer Services for Yorkshire Imperial Metals Ltd., where he is concerned with product development, sales promotion, publicity and technical service on heat-transfer tubes. He was educated at King Edward VI School, Birmingham, and at Birmingham University, England, where he obtained a B.Sc. (1st Class Honours, Mathematics), followed by two years of graduate study in engineering production. He has worked for the British Motor Corp. as a development engineer, and with Imperial Chemical Industries on product development.

Selecting tubes for CPI heat exchangers—II

Here are some things you need to know in order to properly select and use fluted and roped tubes.

D. H. Foxall and *P. T. Gilbert**
Yorkshire Imperial Metals Ltd., U.K.

□ In the first article of this series, we considered the different mechanism of heat transfer through extended surface and EHT (enhanced heat transfer) tubes. Various types of finned tubes were discussed and we will now consider EHT tubes in more detail.

Roped tubes

The roped type of tube (Fig. 1) is now being used in multistage flash preheaters for vertical-tube seawater evaporators and could find numerous other applications including, for instance, power plant condensers. A useful feature is that the profile can be varied to give whatever groove depth, pitch and helix angle appear suitable for a particular requirement. The roped tube is used with horizontal axis, steam condensing on the external surface, and coolant flowing internally.

In a typical steam condenser, assuming a medium-sized tube bank using ¾-in. O.D. tubes, the plain tube condensing coefficient, h_o, may be in the region of 1,750, and h_i (assuming a water velocity of 6 ft/s) about 1,500—giving a design value of about 600 for the overall coefficient (U).

This is clearly a situation in which a finned tube is not appropriate; in fact, the performance of the usual low-fin profile is poorer than might be expected from the excellent results that it gives in other condensing duties. With steam, because of the high surface tension in the liquid film, the condensate tends to be held between the fins and this counteracts the tube's otherwise good drainage characteristics.

Roped tubes giving about three times the plain tube h_i can be provided, but the accompanying increase in friction factor is not likely to be acceptable. For economical design, it is preferable to use a more modest internal-heat-transfer factor of about 1.6 and, if circumstances permit, it seems good practice to increase the tube diameter, e.g., from ¾-in. O.D. plain to 1-in. O.D. roped tube. The water velocity is then reduced and a

Profiles of roped tubes adapt to special needs   Fig. 1

profile having a fairly high enhancement characteristic can be used without excessive pressure drop.

However, if only the internal conditions are considered, the maximum advantage in using roped tubes for steam condensers will not be obtained. A suitable choice of profile can also give a useful increase in the condensing coefficient, by improving the drainage, probably in much the same way as occurs when using a low-fin profile with refrigerants. Tubes are at present available, for example, that combine a condensing enhancement factor of about 1.4 with a water-side factor of about 1.6, giving an overall gain of about 40%, with only a modest increase over the plain-tube cost per unit length.

To obtain good drainage, the indentations need to be nearly vertical—i.e., with spiral grooves a low helix angle should be used, whereas if the profile were optimized solely on internal performance, a larger helix angle with widely pitched and deep grooves would be chosen, and this could give a condensing coefficient lower than the plain-tube value. For a steam-condenser tube, the ideal profiles for the two surfaces are therefore

*For information on the authors, see the first article of the series, p. 102.

Originally published April 12, 1976

Assortment of fluted tubes in which suitable profiling increases the condensing coefficient   Fig. 2

conflicting to some extent, and development work to establish the best compromise continues. Condensing-enhancement factors of up to 2 have been obtained on tubes tested recently, and the 40% overall advantage mentioned above may ultimately prove to be conservative.

This type of tube should also be of interest when both fluids are in a single-phase flow, particularly if h_o is much higher than h_i (e.g., oil inside, and water outside, the tube). In such a duty, internal fins can be considered, but such tubes tend to be expensive and of limited value. Inserts are a possible alternative, but they may prevent cleaning of the internal surface, whereas normal methods can be used to clean roped tubes. For duties in this category, it would be reasonable to disregard the external performance and to use a profile that gives maximum advantage in the bore, the resulting helix angle probably being somewhat larger than would be appropriate for a steam-condenser tube. The overall benefit in this case should not be much influenced by the direction of flow of the external fluid. The same type of profile should similarly be of value when a fluidized bed is used to improve external heat transfer and some compensating improvement is justified internally.

Given two single-phase fluids with comparable values of h, as may be the case when heat is transferred between two water streams or two gas streams, the use of roped tubes should be attractive if the external fluid is in essentially longitudinal flow, e.g., in double-pipe heat exchangers and also perhaps in units containing a small number of relatively long tubes. In such duties, the external effect of the profiles should be approximately predictable from data available on internal flow, but experiments are being performed to obtain confirmation.

Fluted tubes

The fluted tube, Fig. 2, which is an EHT tube for use in the vertical position, is also in commercial use in seawater distillation plants. The development of this type of tube followed the work of Gregorig [1], who pointed out that if the surface of a condensate film takes the form of a series of convex and concave curves, surface tension would cause a flow of condensate from the "peaks" to the "troughs"; at the peaks, therefore, condensate thickness would be reduced and the condensing coefficient increased.

This situation can be brought about by suitably profiling the surface on which condensation takes place and, according to theory, the smaller the radius of curvature, particularly at the crest of the flute, the greater the effect. It must be borne in mind that the trough dimensions must be adequate to carry a condensate flow that increases progressively as it descends the tube, because if the tube surface becomes inundated the mechanism can no longer work.

Practical experience suggests that the profile specification is not so critical as might be expected from theory, which is perhaps fortunate in view of the limitations of realistic production processes. A very sharply ridged profile could be manufactured, but would be expensive, whereas the tubes illustrated, which have a somewhat larger crest radius, provide extremely good performance at relatively low cost.

The mean steam-condensing coefficient for a vertical surface is a function of the height—a typical figure for a seawater distillation plant, assuming a tube length of 10 ft, being about 1,300. The corresponding figure for a suitably profiled fluted tube has been found to be at least six times as high, even when taken as an average over the whole surface (considering that some of the surface is relatively ineffective, the coefficient in the region of the crest must be substantially higher).

In the VTE (vertical-tube evaporator) type of distillation equipment, seawater feed evaporates inside the tubes, and fluted tubes can be used with both upward flow and downward flow. The initial use of such tubes was in downflow, with the feed flowing nominally in layer form, although the characteristics of seawater are such that foaming, which substantially increases the evaporating coefficient, tends to occur at the higher operating temperatures.

Overall heat-transfer coefficients that are at present accepted for design purposes are somewhat higher for downflow than upflow, but this advantage may to

Nomenclature

h	Individual heat-transfer coefficient between fluid and surface, Btu/(h) (ft^2) (°F)
K	Thermal conductivity, Btu/(h) (ft^2) (°F/ft)

Subscripts

i	Internal tube surface
o	External tube surface

some extent be offset by the cost of pumping the feed to the upper header. The upflow type of plant operates in a similar way to the traditional natural-circulation evaporator, the feed being 100% liquid at the lower end of the tube, and becoming two-phase with increasing steam quality as it flows up the tube.

A typical, downflow, plain tube, internal evaporating coefficient for water at 210°F is about 1,900, the corresponding U value being about 625. There is some evidence that fluting the surface increases h_i, but not substantially; in this duty, therefore, the fluted-tube advantage is essentially on the outside, leaving a relatively high resistance to heat flow internally. The overall effect is approximately to double the plain-tube U. In addition, because of the relatively deep flutes, the surface area is generally about 25% more than that of a plain tube of the same nominal diameter, yielding an overall multiplication (heat transfer per unit length of tube) of about 2.5. Similar but marginally lower benefits can be expected when using fluted tubes in upflow.

The lack of balance between h_o and h_i that results when using fluted tubes in these circumstances has stimulated research [2] in which h_i, the evaporating coefficient, is increased by adding foaming agents to the feed. This work, carried out mainly in the upflow mode, shows that substantial gains are possible; U values of considerably over 3,000 have been reported in experiments using small amounts of foaming additive. At such levels, the heat-transfer resistance of the tube materials becomes an important factor, the high K of copper giving a significant advantage even when compared with aluminum brass, which is the material of highest thermal conductivity currently accepted for seawater evaporating duties.

There is interest in the possibilities of vertically-tubed steam condensers, stimulated by the availability of fluted tubes. Again, the very large increase obtainable in the condensing coefficient should provide sufficient justification for using fluted tubes, even without a significant increase in the water-side coefficient. Investigations carried out [3] have been mainly on tubes with spiral flutes, which do not appear to have a significant adverse effect on the condensing gain and also help to promote an increase in h_i (typically 25–30% if the velocity is not reduced, but with some increase in pressure drop). The overall gain is likely to be up to about 90%, or a little lower if the water velocity is reduced to yield the same pressure drop as the plain tube.

The fluted tube, therefore, merits careful consideration in this situation and, indeed, in others in which steam condenses externally and h_i is also high. This will obviously rule out some vertical-tube evaporator duties in which the internal fluid becomes highly concentrated or viscous, but suitable conditions may be found in other cases.

Commercially available fluted tubes provide similar internal and external profiles, and they should therefore also be of interest if steam is condensed inside vertical tubes, provided that the external coefficient is fairly high.

From the brief outline of the "Gregorig effect," [1] it will be apparent that the very large increase in condensing coefficient obtainable with steam must be associated with its high surface tension, and it may be rather optimistic to hope for a comparable gain with, for example, condensing hydrocarbons. The plain-tube condensing coefficients for most fluids are much lower than for steam, however, and a much more modest enhancement factor should in many situations give a substantial overall gain. Thus, the performance of fluted tubes for condensing duties should be worth investigating more generally whenever a vertical-tube axis can be used.

Fouling

The effect of fouling-deposits on heat-exchange performance is usually unimportant when the "clean" U is low, but becomes progressively more significant with increasing U. It follows that with typical high-fin applications, the effect of fouling on heat transfer is rarely noticed, though periodical cleaning is sometimes desirable because a deposit can build up to the extent that air or gas flow between the fins is reduced. In low-fin designs, fouling must be allowed for (if it is likely to occur), though the proportional effect on overall heat transfer is usually no greater than with the corresponding plain-tube design.

The fairly lengthy intervals between cleaning operations that are generally acceptable in process heat exchangers can also be tolerated in steam condensers if the cooling water is clean. Under less-favorable conditions, however, use of a continuous cleaning technique may well prove economical. Similarly, in seawater distillation plants the use of fouling factors that are normal in other situations will lead to a prohibitive increase in heat-exchange surface area, and dosing of the feed to avoid scaling is essential. In general, with very high overall coefficients significant fouling is unacceptable whether using plain or EHT tubing. Therefore, the precautions that must be taken are not greatly influenced by the type of surface chosen.

References

1. Gregorig, R., Hautkondensation auf feingewellten Oberflächen bei Berücksichtigung der Oberflächenspannungen, *Zeitschrift für Angewandte Mathematik und Physik*, Vol. 5, 1954.
2. Sephton, H. H., Desalting by Upflow Vertical Tube Evaporation with Interface Enhancement, IDEA Conference, Puerto Rico, Apr. 1975.
3. Newson, I. H. and Hodgson, T. K., The development of enhanced heat transfer condenser tubing, 4th International Symposium on Fresh Water from the Sea, Heidelberg, Sept. 1973.

Selecting tubes for CPI heat exchangers—III

Here are some hints, and some tables, to help you to select materials for heat exchangers using extended-surface and enhanced-heat-transfer tubes.

D. H. Foxall and P. T. Gilbert
Yorkshire Imperial Metals Ltd., U.K.

☐ Having gone into the theory of heat transfer in heat-exchanger tubes and considered the applications of extended-surface and enhanced-heat-transfer tubes, let us now turn to the tube materials themselves.

It is impossible in a short article to give comprehensive advice on the choice of materials for tubular heat exchangers. The number of variables is so great that a textbook is needed to deal with the subject. Information is given, for example, in Ref. 1–5.

In addition to the choice of material for the tubes themselves, consideration must be given to the following matters:

(1) The tubesheet material.
(2) The method of fixing of the tubes into the tubesheets (e.g., roller expansion, explosive expansion, fusion welding, explosive welding). If bimetal tubes are to be used special fixing methods may have to be considered.
(3) Materials for waterboxes/headers and associated pipework.
(4) In the case of water-cooled units there may be a need for: protective measures such as coatings on waterboxes; cathodic protection (sacrificial anode or applied current) to protect waterboxes, tubesheets and tube ends; or the fitting of plastic or metal inserts at the tube inlets. Also water treatment may be necessary to prevent scaling, fouling, corrosion (e.g., with seawater, chlorination to combat marine growths, ferrous sulfate dosage to give protective films on tubes). Screening of cooling water to prevent ingress of debris is likely to be essential, and consideration may need to be given to the use of in-use tube-cleaning systems (e.g., the Taprogge sponge-rubber-ball system).
(5) Materials for the shell and for support/baffle plates.
(6) In addition to these various components being suitable for the environments concerned, they must be compatible with one another in any electrolytes involved so that excessive galvanic corrosion effects are avoided. They must also be suitable for the methods of construction involved (e.g., fusion welding) and be of adequate thickness and tensile strength to withstand the operating temperatures and pressures involved, as well as any other stress imposed during service.

In the chemical and petrochemical industries, the environments that may be encountered are very diverse. There may be heat exchange between two liquids, or two gases, or a liquid and a gas. In some cases, a liquid may be evaporating or a vapor condensing, or both. In choosing the material for the tubes in the heat exchanger, a primary factor is the chemical nature of the substances involved, together with the nature and content of any impurities that may occur. Temperatures, pressures and velocities are other factors that need to be taken into consideration. When cooling some gases the presence of water vapor may be important, since if the temperature falls below the dewpoint, corrosive solutions may be formed. In the case of aqueous solutions, the content of dissolved oxygen is frequently a major factor determining the rate of corrosion.

In the case of cooling waters, many factors—in addition to those already mentioned—can exert an influence, e.g., the presence of air bubbles entrained in the water, or the presence of suspended solid particles that may cause abrasion, or pollution of the water (for instance by hydrogen sulfide as a result of the activity of sulfate-reducing bacteria).

With plain tubes, the choice of materials is largely determined by compatibility with the environments—the thermal conductivity of the metal being generally a secondary consideration. Even with highly efficient steam condensers (e.g., in power stations) the difference in condenser size to give equal performance is only about 10% as between aluminum brass (K value approximately 60) and 70/30 copper-nickel (K value approximately 20) tubes. However, with profiled tubes operating at significantly higher overall heat-transfer rates, increased importance has to be attached (in selecting the material) to the thermal conductivity of the metal, the wall thickness, and the mechanical fasibility of fabricating the material into the required profiled tube.

An indication of the ease with which the various classes of materials used for heat exchangers can be fabricated into tubes of various types is given in Table I. An indication is also given of materials whose use would give notable benefits in heat-transfer perform-

Originally published May 10, 1976

Materials for heat-exchanger tubes — Table I

Material	High-fin*	Low-fin*	Roped	Fluted
Copper	++P	++P	++P	++P
Brasses		++P	++P	++P
Copper-nickel and nickel-copper alloys		++	++	+
Stainless steel		+	++	+
Titanium		+	+	+
Aluminum and Al alloys	++P	++P	++P	+P
Mild steel		+	++	++	+

Key:
++ Fabrication presents no particular problems.
+ Fabrication is possible, but with more-restricted range of dimensions attainable.
P Preferred on heat-transfer performance grounds (if suitable on other grounds.)
* These tubes can be provided with bimetal construction, having as a lining a plain tube of any of the other materials listed.

ance, though these materials can, of course, be used only if they are otherwise suitable for the operating conditions concerned.

Salt waters

Some brief comments may be given on the selection of materials for use in corrosive environments. Salt waters (brackish, seawater or brines) are frequently used as the coolant in tubular heat exchangers in many industrial applications. Usually the cooling water flows through the tubes, but occasionally (e.g., when the product being cooled is at very high pressures) the cooling water may be on the shell side with the product in the tubes. As already indicated, the performance of materials in salt water depends on many factors such as water speed, local turbulence effects, presence of air bubbles, temperature, composition (including dissolved gases and polluting substances), presence of solid suspended matter, use of chlorination or other water treatments, tube-cleaning procedures, efficiency of screening, etc. Various types of corrosion can occur, and a qualitative summary of the extent to which the various classes of material are susceptible to these is given in Table II. This gives a rough guide to circumstances in which caution must be exercised in selecting materials.

Other points that may be mentioned are:

(1) Small amounts of copper leached from copper-base materials tend to discourage marine growths and confer inherent antifouling characteristics. Marine organisms can grow unhindered on all other materials.

(2) For roped tubes, where the configuration promotes some degree of turbulence in the bore, it is desirable to use materials with relatively good resistance to impingement corrosion.

(3) Austenitic, ferritic and mixed-structure stainless steel can all suffer pitting, crevice corrosion and possibly intercrystalline corrosion and stress corrosion, in seawater and other chloride solutions. Certain compositions, notably those containing considerable amounts of molybdenum (e.g., iron with 23% nickel, 20% chromium, 6.5% molybdenum, 0.5% titanium) are, however, much more resistant to these forms of corrosion than other compositions.

Different considerations apply in other environments, and some brief comments are as follows:

Fresh waters

In most fresh waters corrosion is not a problem with many of the materials under consideration. With copper and copper alloys, the limiting velocities are considerably higher than in seawater. Aluminum brass is rather susceptible to pitting corrosion in hard fresh waters (although not in clean seawater) and in a few cases stress-corrosion cracks have been observed to propagate from the base of many of the corrosion pits. In view of the pitting corrosion problems that may arise with aluminum brass it is probably best to avoid the use of this material for fresh-water service. This limitation does not apply to pure water (e.g., condensate).

Aluminum and aluminum alloys can suffer severe pitting attack in fresh waters, particularly in hard waters containing minute traces of dissolved copper salts at normal temperatures. The use of these materials for fresh-water service must therefore be approached with considerable caution.

In general, mild steel rusts too rapidly in fresh waters for it to be of practical value as a heat-exchanger tube material.

Pure water (condensate)

All the materials listed (except mild steel) are resistant to pure condensate. However, condensates containing dissolved oxygen and other noncondensable gases (particularly carbon dioxide, ammonia and hydrogen sulfide) can be aggressive to copper and brasses and to a lesser extent to the copper-nickel alloys. This must be taken into account when selecting materials.

Alkaline aqueous solutions

In alkaline solutions aluminum and its alloys are unsuitable and there are certain limitations on the use of copper and brasses (depending on concentration, temperature and flow velocity). The other materials listed can all be used satisfactorily in most alkaline solutions. Copper and copper alloys are not suitable for ammoniacal environments because general attack is likely to be severe, and stress corrosion is a serious hazard.

Acid aqueous solutions

It is very difficult to make any generalizations about acid solutions because rates of corrosion are highly dependent on the nature of the acid, its concentration, the temperature, the flow velocity, and the degree of aeration of the solution. There are a few specific circumstances where copper alloys, mild steel, or aluminum can be used, but generally these materials are unsuitable. Stainless steels, titanium, and Monel and other high-nickel alloys can all be used over a wide range of operating conditions, but none is satisfactory under all circumstances. Published information or the advice of tube suppliers therefore needs to be sought in cases where no results of previous operating experience are available.

Behavior of materials in salt waters (brackish, seawater, brines) — Table II

Material	General corrosion	Pitting	Crevice corrosion, deposit attack	Stress corrosion	Impingement corrosion. Approximate limiting velocity in plain tubes, m/s
Copper	+	−¶	++	++	1
70/30* or admiralty brass*	+	−¶	++	−	1½
Aluminum brass*	++	−¶	++	−	2½
90/10 or 70/30 copper-nickel alloys	++	−¶	+	++	3-4
Monel	++	+	+	++	N
Stainless steels	++	−	−	−†	N
Titanium	++	++	−†	++	N
Aluminum and Al alloys	+	−	−	−‡	2
Mild steel	−	−¶	−	++	N

Key:
++ Immune, for all practical purposes.
+ Slightly susceptible, but usually not to an extent that is of practical significance.
− Susceptible to an extent that in adverse circumstances could give rise to problems in service.
N Velocity not a limiting factor in practice.
* Contains arsenic as a dezincification inhibitor.
† Only at elevated temperatures.
‡ Only in certain alloys.
¶ Only in certain polluted-water situations.

Summary

The range of application of externally finned tubes is well established, high fins being appropriate with low film-coefficients and low fins with an intermediate range; in both cases the internal heat-transfer coefficient must be substantially higher than the external for such tubes to be used economically.

With the availability of enhanced-heat-transfer (EHT) tubes, it should no longer be assumed that plain tubes must be used in all other situations. Plain tubes will remain the best choice in many cases, but EHT tubes will often deserve consideration and their use should increase. These newer products should be investigated when internal and external plain-tube coefficients are of similar magnitude (whether the actual values are high or low) and in numerous other situations in which finned tubes are unlikely to be attractive, e.g., when the internal coefficient is controlling, and in various steam-condensing duties.

Guidelines for the selection of tube materials are reasonably well established for plain tubes, although there remain numerous situations in which it is advisable to consult the manufacturer. The basic considerations are similar when choosing materials for finned or EHT tubes but the relative importance of some of the factors may be different, e.g., there is generally more incentive to use high-conductivity materials and thinner walls and, in the case of roped tube, the higher turbulence characteristic of the profile may in some circumstances suggest the use of a more resistant material.

References

1. Uhlig, H.H., "The Corrosion Handbook," Wiley, New York, 1948.
2. Shreir, L. L., ed., "Corrosion," George Newnes Ltd., London, 1963.
3. Cairnes, J.H., and Gilbert, P.T., "The Technology of Heavy Non-Ferrous Metals and Alloys," George Newnes Ltd., London, 1967.
4. Gilbert, P. T., Copper Alloys for Sea Water Service, Inst. of Marine Engineers Materials Section Symposium, London, 1968.
5. Gilbert, P. T., Corrosion Problems in Condensers and Heat Exchangers, Trans. Inst. Marine Eng. (London), vol. 82, no. 7, p. 6 (1970).

The authors

P. T. Gilbert is technical director of Yorkshire Imperial Metals Ltd., P. O. Box No. 166, Leeds LS1 1RD, England. He was educated at Elmhurst Grammar School, Street, Somerset, and at University College of the South West of England (now University of Exeter). He was then awarded a B.Sc. (Chemistry) degree of the University of London (1st Class Honours) and, later, a Ph.D. for work on corrosion of zinc and zinc coatings in supply waters. He is a Fellow of Institution of Metallurgists, Institute of Marine Engineers, Institution of Corrosion Science & Technology, and is a member of Institution of Water Engineers and Scientists, Soc. for Metals, and NACE.

D. H. Foxall is Manager of Heat Transfer Services for Yorkshire Imperial Metals Ltd., where he is concerned with product development, sales promotion, publicity and technical service on heat-transfer tubes. He was educated at King Edward VI School, Birmingham, and at Birmingham University, England, where he obtained a B.Sc. (1st Class Honours, Mathematics), followed by two years of graduate study in engineering production. He has worked for the British Motor Corp. as a development engineer, and with Imperial Chemical Industries on product development.

Use of computer programs in heat-exchanger design

Accurate computer techniques are now widely used by the chemical process industries for the design of all types of heat exchangers. Read about what such computer programs can accomplish, and about the considerations involved in selecting a program.

David Butterworth and *Lionel B. Cousins, Heat Transfer & Fluid Flow Service, Harwell*

☐ By means of computer programs, many more parameters that affect flow conditions and heat transfer in an exchanger can now be taken into account than was previously possible using manual techniques. In addition, many more configurations may be tried to find a design that satisfies all the imposed constraints in the cheapest possible way. However, for computer programs to be accepted by an engineer, they must be validated by experimental data taken from process equipment.*

Types of computer programs

Equipment manufacturers use computer programs to design exchangers that will meet a required duty within certain operating constraints. Equipment users, on the other hand, use programs to check what a given exchanger will achieve if operated in a specified way. Three general types of computer programs can be considered for dealing with most demands:

*The Heat Transfer and Fluid Flow Service (HTFS) has written several computer programs for the design and performance of industrial heat exchangers. These programs are verified by producing experimental data on the industrial-scale loops at Harwell and the National Engineering Laboratory at East Kilbride, and by obtaining feedback and data from HTFS member companies.

Design program—This type of program determines the geometry of the equipment that will meet the thermal duty within the imposed constraints.

Checking program—Here, both the geometry and the heat duty are specified. The program checks both the geometry and the duty to determine whether the two are compatible.

Simulation program—This type of program predicts the behavior of an exchanger under conditions in which the independent process variables are specified. Usually, this means that the outlet conditions of the flows are calculated for known inlet conditions.

As far as program logic is concerned, design programs—of which there are many—are undoubtedly the most complicated. The simplest type is that in which the program can find only a single design that meets all the constraints. Often, however, several answers are possible, so the program then requires criteria for selecting the best exchanger. Cost is the most frequently used criterion; reliability and safety are others.

At first sight, it would seem fairly simple to determine the amount of material and machining that goes into a given design, from which the cost could be determined.

Originally published July 5, 1976

But to include such calculations in the logic of a thermal-design program is usually very expensive. Therefore, the procedure often adopted is to use crude costing methods during the thermal-design stage. The program user can then assess the chosen design and, if satisfied, continue to the mechanical-design stage and costing. Crude costing methods for the more important geometric configurations usually involve curves of cost per unit area against surface area. Ordinarily, only manufacturing costs are involved, but it is possible to include operating costs as well.

During the design of shell-and-tube exchangers, hundreds of alternative designs are possible. It is the job of the program author to develop search procedures that do not use excessive computer time. The most costly type of search is the factorial method, in which the program systematically evaluates every single possible geometry. The efficiency of this method can, however, be greatly improved by a skilled programmer with heat-transfer design experience. Such a programmer will arrange the calculations for each geometry so that the simplest one is done first. Relevant checks on design limitations are then performed. In this way, some of the geometries can be discarded at an early stage.

Another method for cutting short the calculations is to redirect the search sequence as soon as a given line of investigation looks unpromising. It is not always necessary to start at the lowest position in the table of alternative geometries. The program may use a few simple preliminary calculations to show that it can jump in at a higher position. Time can also be saved by storing selected results from previous calculations when it is known that these results may be used again later.

An alternative and, in principle, the fastest design method, is to use standard optimizing techniques of the "hill-climbing" type to arrive at a design. This method, however, suffers from two serious drawbacks that often render it useless in exchanger design: (1) optimization techniques assume that the variables are continuous, rather than having certain discrete values; and (2) problems arise when the theoretical optimum is outside the allowable geometry range.

This does not mean that optimization has no place to play in exchanger design, because the different methods may be combined to achieve a solution. For example, in a shell-and-tube exchanger design, the number of tubeside passes could be incremented in a factorial manner, after which the tube length and the number of tubes could be estimated by conventional optimization techniques. Since the tube length and number of tubes are usually restricted to certain standard values, it would still be necessary to conduct a factorial search in the region close to the optimum.

Simulation and checking are examples of performance calculations. Checking calculations have a simple logic, because they can often be constructed in a way that avoids extensive iterations. Simulation programs are ordinarily more complicated, because they involve extra iterations; they are also more powerful, because they model the behavior of an actual exchanger. The main usefulness of checking calculations is that they provide a program core around which design or simulation calculations can be constructed. Programs built

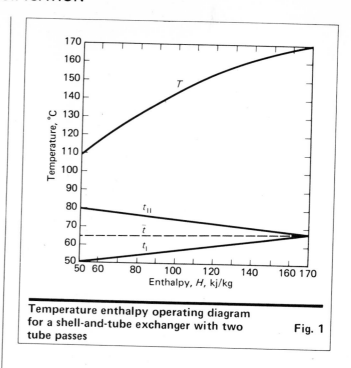

Temperature enthalpy operating diagram for a shell-and-tube exchanger with two tube passes Fig. 1

up in this manner will usually have options for design, simulation and checking.

Design programs often have to use simple, and therefore approximate, methods to estimate the performance; otherwise, computer times would be excessive. It is sometimes useful to produce programs solely for performance, because such programs can incorporate much more rigorous methods without producing excessive computer time. The reason for this is that the basic calculations are carried out less frequently.

The above general points are illustrated by two existing computer programs. The first is a shell-and-tube program, with design and performance options, while the second is a simulation program for air-cooled heat exchangers.

Shell-and-tube exchanger design

TASC is a shell-and-tube program that was originally written for condensers, but also has options for single-phase and boiling. The heat-transfer calculation methods used in the core of the program have been kept deliberately simple so that design calculations can be conducted without generating excessive computer time. Hence, the program is flexible because it has design, simulation and checking options.

The capacity for handling a wide range of geometries has also been achieved by using simple generalized methods, rather than more precise ones that are limited to specific geometries. Considerable experience has gone into the choice of the approximate methods, so that the user may have confidence that the results produced by the program are accurate enough for his purpose.

The shell types that can be handled by TASC are TEMA (Tubular Exchanger Manufacturers Assn.) E and J types, which have shells with both combined and divided flow, with all the rear-end head-types specified by TEMA [1]. Designs with up to 16 tubeside passes can be dealt with, and the baffles may be of single- or double-segmental type. Special situations can be handled, such as designs with no tubes in the baffle windows, or even designs with no baffles at all.

The program first calculates the temperature-enthalpy curves for the two fluid streams, assuming phase equilibrium. The curve for the shellside is shown in Fig. 1, where T is the temperature and H is the specific enthalpy of that stream.

It has been shown [2] that, if the overall heat-transfer coefficient varies only along the shell but not from pass to pass, the tubeside temperatures can be determined using these equations:

$$dh_{II}/dh_I = -(T - t_{II})/(T - t_I) \quad (1)$$
$$H = H_{in} - (w/W)(h_{II} - h_I) \quad (2)$$

where h_I and h_{II} are the tubeside enthalpies in Pass I and Pass II, respectively, and t_I and t_{II} are the corresponding temperatures. W and w are the shellside and tubeside mass flowrates, respectively, and H_{in} is the shellside inlet enthalpy.

Using Eq. (2) and the temperature-enthalpy curves for each stream, Eq. (1) can be readily integrated by starting with $h_{II} = h_{out}$, when $h_I = h_{in}$. Integration is stopped when $h_I = h_{II}$.

The tubeside temperatures for the two passes are shown plotted in Fig. 1. This plot shows the values of the tubeside temperatures at points along the shell, where the shellside enthalpy is H. The mean tubeside temperatures corresponding to each H value are calculated from:

$$\bar{t} = (t_I + t_{II})/2 \quad (3)$$

Up to this stage in the calculation, detailed information about the geometry is not required, except to say that the exchanger has two tube passes. Hence, when incorporated into the design logic, the above calculation need only be done once for all two-pass exchangers, thereby saving computer time.

A heat balance over area da of the exchanger shown in Fig. 2 yields:

$$U(T - \bar{t})da = WdH \quad (4)$$

where U = local overall heat-transfer coefficient. This equation may be rearranged and integrated as follows:

$$\int_0^A U da = W \int_{H_{out}}^{H_{in}} dH/(T - \bar{t}) \quad (5)$$

It is common practice in heat-exchanger design to work with an equation of the following form:

$$Q = W(H_{in} - H_{out}) = U_m A \theta_m \quad (6)$$

where Q = heat load (duty); A = heat-transfer area; U_m = mean overall coefficient; and θ_m = mean-temperature difference. Comparing Eq. (5) and (6) leads to the following definitions for θ_m and U_m:

$$1/\theta_m = [1/(H_{in} - H_{out})] \int_{H_{out}}^{H_{in}} (dH)/(T - \bar{t}) \quad (7)$$

$$U_m = (1/A) \int_0^A U da \quad (8)$$

Using the type of information shown in Fig. 1, the program can integrate Eq. (7) to determine the mean-temperature difference. Again, this calculation requires no detailed information about the geometry. Therefore, when used in design calculation, θ_m can be calculated once for all two-pass exchangers.

At this calculation stage, the program has to consider the geometry in detail. The stream coefficients are calculated at selected positions in the unit, and from these the overall coefficients are calculated. The mean coefficient is subsequently calculated from Eq. (8), and the required heat-transfer area, A_r, is calculated from a rearranged form of Eq. (6):

$$A_r = Q/(U_m \theta_m) \quad (9)$$

In a checking calculation, the program simply compares the required area calculation to the actual heat-transfer area. The final step in the checking procedure is to calculate the pressure drops on the two sides of the exchanger.

As previously stated, the heat-transfer calculation methods in this computer program have been kept deliberately simple. For example, one of the simplifications is that the shellside conditions are considered averaged across the shell. Ordinarily, this is a very reasonable approximation, but for partially flooded shells, such an approach can lead to errors. Simple methods are used in TASC to allow for such likely errors, but other programs exist that can deal with the problem of flooded shells more rigorously.

Air-cooled heat-exchanger simulation

An air-cooled heat-exchanger program that can cope with either single-phase or condensing streams on the process side is called ACOL. This program takes account of detailed local variations throughout the exchanger. The program has been limited to simulation calculations, because the many repetitive calculations involved in a complete design would require long computer times. Another advantage of the simulation option is that a wide range and variety of geometries can be allowed for.

In a design program, the number of alternative geometries have to be limited so that the number of

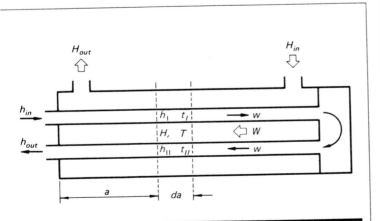

Shell-and-tube exchanger with two tube passes   Fig. 2

112 HEAT EXCHANGER DESIGN AND SPECIFICATION

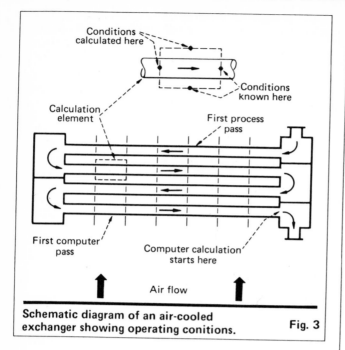

Fig. 3 Schematic diagram of an air-cooled exchanger showing operating conitions.

alternative designs tried are kept to manageable proportions. Flexibility has been built into this air-cooled exchanger program. For example, up to 16 transverse rows of tubes can be handled, and it is not necessary for all the tubes in a row to be in the same pass. The maximum number of passes allowed at present is 20. The finning on the outside of the tubes may be of the L, G, integral shoulder-grooved, or bimetallic types. A variety of common fin materials may be used. Examples of how the program can provide for detailed local variations are row-to-row variations in the air-side coefficients, as well as changes in air flow along the tube length.

ACOL uses a technique of stepping backward along the process stream to calculate the conditions throughout the exchanger. The reason for working backward, instead of forward, is that the air temperature is only known at the bottom passes, which correspond to the process outlet, as shown in Fig. 3, in which each row corresponds to a tubeside pass. The bottom pass is split into a larger number of hypothetical sections.

The program works from the last element, calculating the change in process conditions and air temperature for that element. This provides process conditions for the next element along the tube, as well as the air temperature approaching the element of tube that is directly above it. This process can be repeated along the bottom row of tubes to provide the air-approach temperature for the whole length of the row above. This procedure is repeated for all the rows in the exchanger, using the calculated air-approach temperatures.

The idealized calculations described here are often more complicated in practice, because other parameters must be considered, including the fact that successive rows may be in the same or in a different tubeside pass, or that a given row may contain more than one pass.

Programming standards

Before a computer program for design is coded, it is essential to consider who will want to use it. For a

Cost of the several designs corresponding to the air-cooled, heat-exchanger example

Type	Shells		Tube passes	Shell I.D., in	Tube length, in	Number of tubes per shell	Segmental baffle type	Number of baffles	Over surface, %	Cost*
	Number									
	Parrallel	Series								
E	1	1	4	39	144	1,060	single	4	8.0	6,690
E	2	1	4	25	144	408	single	5	1.8	7,627
E	1	1	4	37	192	942	double	8	14.9	6,673
E	1	1	4	39	168	1,060	double	7	5.6	6,827
E	1	1	4	42	144	1,268	double	4	0.5	7,834
J	1	1	2	33	240	750	double	21	5.1	6,175
J	1	1	4	37	192	963	double	15	2.0	6,793
J	1	1	4	41	168	1,225	double	13	2.6	7,651
J	1	1	2	31	240	652	single	9	5.9	5,528
J	1	1	2	33	216	750	single	9	10.0	5,899
J	1	1	2	35	192	855	single	9	1.2	6,181
J	1	1	4	37	192	963	single	9	12.7	6,789

*The cost figures provided roughly correspond to pounds sterling in January 1974, but the values can be made to correspond to dollars by including an appropriate cost factor.

program to be well used by industry, it is important for it to contain the most accurate data available. The program must also be easy to use with many options, be portable, contain data checking for user errors, and be validated by data contained in industrial process plants.

Computer programs of this type must have a clear and concise objective, which should be documented in the technical specifications, together with the models to be used, the mathematical routines, and the full details of the data input and output, as required. It is also necessary that the proposed users of the program be given the opportunity to comment on the specifications.

If programs are to be accepted by industry for use on a wide variety of problems, they must conform to rigorous standards.* Also, all of the individual methods used in the programs must be validated both separately and combined. The latter is achieved by comparing predictions from the programs with actual exchanger-performance data.

Example

The capability of the TASC program for condenser design is illustrated by an example based on an industrial process:

A mixture of 25% butane, 73% octane and 2% hydrogen (all percentages by weight) is to be cooled from 280°F at a mean pressure of 50 psia. The maximum pressure drop allowed is 2 psi. Cooling is to be provided by 650,000 lb/h of water at 80°F and 30 psia. The outlet temperature of the cooling water is 115°F, and the outlet pressure must not be less than 20 psia; water velocity is to be restricted to a 3–20 ft/s range.

The equipment is to consist of a baffled shell-and-tube heat exchanger built of carbon steel, with ¾-in, 16-BWG (Birmingham wire gage) tubes (i.e., I.D. = 0.62 in) arranged on a 1.0-in rotated-triangular pitch. The tubes are to be between 10 and 20 ft long. The exchanger may consist of up to four horizontal shell-and-tube units with removable bundles.

*HTFS programs are written in standard FORTRAN, as specified by the American National Standards Institute.

For protection of the tubes, the maximum allowable distance between the baffles is to be 30 in (with a maximum unsupported length of 60 in). Also, the top row of tubes under the inlet nozzle must be protected by an impingement baffle.

Objective: to find the most economical unit for the above specifications.

Data were prepared to run this example, assuming an E shell, with single-segmental baffles. The first two lines in the accompanying table summarize the alternative designs, which achieve the heat-duty within the pressure-drop and velocity constraints. The program was asked to consider only those configurations whose cost would be within 25% of the cheapest unit. This resulted in possible designs for two exchangers, as shown.

Since it was thought worthwhile to consider whether there might be an advantage to using double-segmental baffles, the program was asked to repeat the design taking this into consideration. This was achieved by including one card in the data, known as a change card, which told the program to rerun the previous data set with the specified item changed. The next three lines in the table show that there is no advantage in using double-segmental baffles.

Using the change card again, J shells were investigated with both types of baffles. The type of J shell involved was one with two inlet nozzles and one outlet (i.e., a combining-flow type). As can be seen, some cost advantage was gained by going to J shells with double-segmental baffles, and even more so with single-segmental baffles.

References

1. "Standards of the Tubular Exchangers Manufacturers Assn.," 5th ed., New York (1968).
2. Butterworth, D., "A calculation method for shell-and-tube heat exchangers in which the overall coefficient varies along the length," Conference on Advances in Thermal and Mechanical Design of Shell-and-Tube Heat Exchangers, Nov. 28, 1973, National Engineering Laboratory, East Kilbride, Glasgow, Scotland.
3. Roberts, K. V., The publication of scientific FORTRAN programs, *Computer Physics Communications*, No. 1, pp. 1–9, 1969.

The authors

David Butterworth is associated with the Heat Transfer & Fluid Flow Service, Thermodynamics Div., Harwell, Oxfordshire, OX11 ORA, England, where most of his work has been in the fields of heat transfer and fluid flow. Recently, he completed an introductory booklet on heat exchangers that is to be published by the Design Council in association with Oxford University Press. He has a degree in chemical engineering from University College, London.

Lionel B. Cousins is computer-services manager of the Heat Transfer & Fluid Flow Service, Harwell, Oxfordshire, OX11 ORA, England. He took a course at Battersea Polytechnic in London that led to associateship of the Royal Institute of Chemistry in 1957. He then joined the Chemical Engineering Div., AERE Harwell, and worked on a series of nuclear projects in separation, fallout, and two-phase gas-liquid flow and heat transfer.

(B) Heat Exchangers of Other Designs

Designing direct-contact coolers/condensers
Designing spiral-plate heat exchangers
Designing spiral-tube heat exchangers
Evaluating plate heat-exchangers
How heat pipes work
Uses of inflated-plate heat exchangers
Where and how to use plate heat exchangers

Designing Direct-Contact Coolers/Condensers

In cooling a hot gas, there are times when it is not practical to use a tubular heat exchanger. Such situations—involving partial or total condensation as well as cooling—can be accommodated by baffle-tray columns, spray chambers, packed columns, and pipeline contactors. Here are design equations that apply to these cases, and calculations for each equipment type.

JAMES R. FAIR, Monsanto Co.

Tubular heat exchangers play a major role in cooling and/or condensing gases and vapors. So much so that many a designer rarely considers that an alternative exists. But there are times when it is not desirable or economically feasible to use an intervening metal wall between hot gas and coolant liquid. Barometric condensers and pyrolysis-gas quenchers are examples where direct-contact cooling is preferable. There are also refinery applications where pressure drop and fouling considerations make tubular heat exchangers impractical.

I have recently reviewed the literature on direct-contact heat transfer,[10] and have extracted sufficient information to develop design procedures. We will consider in this article various types of gas coolers/condensers, as well as the types of contacting devices used. Performance characteristics of the equipment (e.g., flooding, entrainment, pressure drop) will also be covered to permit the engineer to completely design the heat exchanger.

The four general classifications of direct-contact gas-liquid heat transfer are:
- Simple gas cooling.
- Gas cooling with vaporization of coolant.
- Gas cooling with partial condensation.
- Gas cooling with total condensation.

A particular set of equations describes each of these situations.

Most of the direct-exchange applications listed above are accomplished with the following devices:
- Baffle-tray columns.
- Spray chambers.
- Packed columns.
- Crossflow-tray columns.
- Pipeline contactors.

Example applications of these devices are shown in

Originally published June 12, 1972

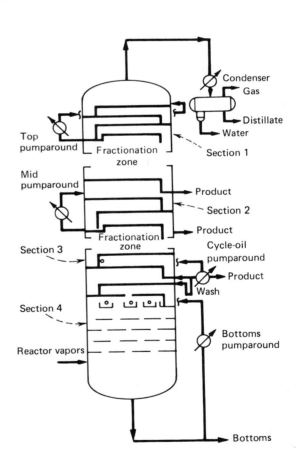

HEAT-TRANSFER sections of a heavy hydrocarbon fractionator use both baffle and crossflow trays—Fig. 1

HEAT EXCHANGER DESIGN AND SPECIFICATON

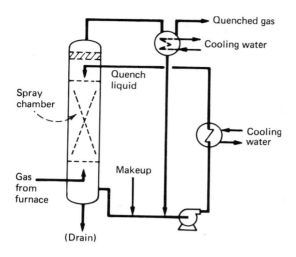

SPRAY QUENCHER for pyrolysis gases—Fig. 2

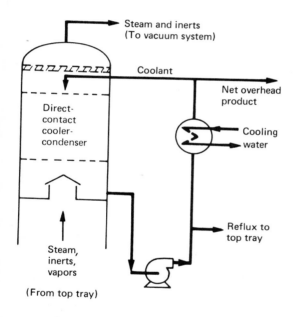

CONDENSER for vacuum steam fractionator—Fig. 3.

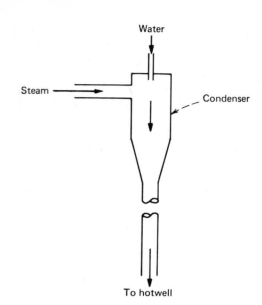

BAROMETRIC UNIT for condensing steam—Fig. 4

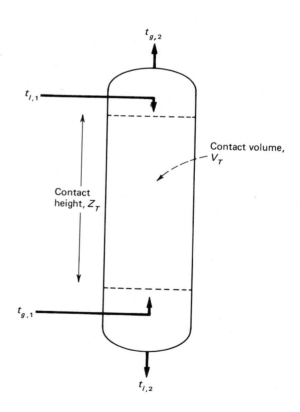

COUNTERCURRENT direct-contact cooler—Fig. 5

Figs. 1–4. In Fig. 1, both baffle and crossflow trays find use in a heavy hydrocarbon fractionator. Fig. 2 shows a spray chamber cooling pyrolysis gases. In Fig. 3, special packing condenses reflux in a vacuum steam fractionator. And Fig. 4 presents a barometric condenser for steam (a special case of a cocurrent-flow pipeline contactor).

Not shown is an important gas-liquid direct-contact heat-transfer application: the cooling tower. Since the technology for design has been well covered in the literature and by proprietary publications of cooling-tower manufacturers, it will not be discussed in this article.

TYPES OF COOLING/CONDENSATION
Simple Gas Cooling

For simple gas cooling, little or no mass transfers between gas and liquid. See Fig. 5 for a typical contactor.

The required volume can be calculated from:

$$V_T = \frac{Q_T}{(Ua)_{avg}\Delta t_M} \quad (1)$$

where the average volumetric coefficient, $(Ua)_{avg}$, and the mean temperature difference, Δt_M, are characteristic of

Nomenclature

a	Surface area, sq. ft./cu. ft.	Q_l	Sensible heat-transfer rate for liquid stream, Btu./hr.
A_c	Curtain area, sq. ft.	Q_T	Total heat-transfer rate, Btu./hr.
A_t	Tower cross-sectional area, sq. ft.	Sc	Schmidt number
A_w	Window area, sq. ft.	S_t	Tray spacing, ft.
c_g	Gas heat capacity, Btu./(lb.)(°F.)	t	Temperature, °F.
c_l	Liquid heat capacity, Btu./(lb.)(°F.)	U	Overall heat-transfer coefficient, Btu./(hr.)(sq. ft.)(°F.)
C_o	Factor defined by Eq. 9	Ua	Volumetric overall heat-transfer coefficient, Btu./(hr.)(cu. ft.)(°F.)
C_1–C_5	Constants	v	Gas velocity, ft./sec.
C_F	Flooding factor	V_T	Volume of contacting zone, cu. ft.
d_p	Particle diameter, ft.	Z_f	Height of froth on tray, ft.
D_T	Tower diameter, ft.	Z_T	Height (or length) of contacting zone, ft.
D_L	Liquid diffusion coefficient, sq. ft./hr.		
E_h	Heat-transfer efficiency		
E_{mv}	Murphree mass-transfer efficiency		
f_1, f_2, f_3	Factors defined by Eq. 27–29		
F_p	Packing factor		
g	Conversion constant, 32.17 (lb.-mass)(ft.)/(lb.-force)(sec.²)		
G	Gas mass velocity, lb./(hr.)(sq. ft. of superficial area)		
G_F	Maximum allowable gas velocity for packing, lb./(sec.)(sq. ft.)		
h	Surface heat-transfer coefficient, Btu./(hr.)(sq. ft.)(°F.)		
ha	Volumetric heat-transfer coefficient, Btu./(hr.)(cu. ft.)(°F.)		
H_g	Height of a gas-phase transfer unit, ft.		
H_l	Height of a liquid-phase transfer unit, ft.		
H_{og}	Height of an overall transfer unit, ft.		
k	Mass-transfer coefficient, ft./hr.		
k'_l	Thermal conductivity, Btu./(hr.)(ft.)(°F.)		
L	Liquid mass velocity, lb./(hr.)(sq. ft. of superficial area)		
m_1	Exponent in Eq. 25		
n_1	Exponent in Eq. 25		
N_A	Rate of diffusion of species A, lb./hr.		
N_B	Number of baffles		
N_g	Number of gas-phase transfer units		
N_l	Number of liquid-phase transfer units		
N_{og}	Number of overall transfer units		
$N_{s,h}$	Number of stages for heat transfer		
Pr	Prandtl number		
q	Volumetric gas flowrate, cu. ft./sec.		
Q_c	Heat-transfer rate for condensation, Btu./hr.		
Q_g	Sensible heat-transfer rate for gas stream, Btu./hr.		

Greek Letters

α	Correction factor for simultaneous heat and mass transfer (Eq. 8)
Δp	Pressure drop, in. flowing liquid
$\Delta p'$	Pressure drop, in. water
Δt_m	Mean-temperature driving force, °F.
θ	Exposure time, sec.
θ_q	Characteristic quench time, sec.
λ_h	Stripping factor for heat transfer
μ_l	Viscosity of liquid, lb./(ft.)(hr.)
ρ	Density, lb./cu. ft.
σ	Surface tension, dynes/cm.
ϕ	Parameter in Eq. 26 and Table IV
ψ	Parameter in Eq. 25 and Table IV

Subscripts

A	Species A
avg	Average
c	Baffle curtain
$corr$	Corrected
d	Mass transfer
$design$	Design case
g	Gas phase
h	Heat transfer
l	Liquid phase
max	Maximum allowable, at flooding
n	Equilibrium-stage number
t	Tower
w	Baffle window

the system and the contacting device. At a particular point in the contactor:

$$Ua = \frac{1}{\frac{1}{h_g a} + \frac{1}{h_l a}} \quad (2)$$

and $(Ua)_{avg}$ represents an average of Ua values.

The height of the contacting zone comes from:

$$Z_T = \frac{V_T}{A_t} = (N_{og,h})(H_{og,h}) \quad (3)$$

where the height of a heat-transfer unit, $H_{og,h}$, is characteristic of the system and device. The number of heat-transfer units, $N_{og,h}$, depends upon the temperature profiles in the contactor. By using an analogy to mass transfer, $H_{og,h}$ can be described as:

$$H_{og,h} = H_{g,h} + \lambda_h H_{l,h} \quad (4)$$

A graphical approach is a convenient way to evaluate $N_{og,h}$, as in Fig. 6a. The shapes of the profiles depend on axial mixing of the phases, amount of mass transfer, flowrates, and specific heats of the streams.

When tray devices are used, a stagewise concept is a more appropriate calculation tool:

$$Z_T = V_T/A_t = (N_{s,h}/E_h) S_t \quad (5)$$

The height, Z_T, includes contacting space above and below the tray section. Evaluation of a theoretical heat-transfer stage is shown in Fig. 6b; it is based on how heat-transfer efficiency, E_h, is defined:

$$E_h = \frac{t_{g_n} - t_{g_{n-1}}}{t_{l_n} - t_{g_{n-1}}} \quad (6)$$

When the temperature of the exit gas equals the temperature of the exit liquid ($t_{g_n} = t_{l_n}$), the efficiency is 100%.

Gas Cooling With Coolant Vaporization

Frequently, some of the coolant liquid vaporizes while

120   HEAT EXCHANGER DESIGN AND SPECIFICATION

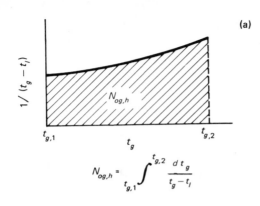

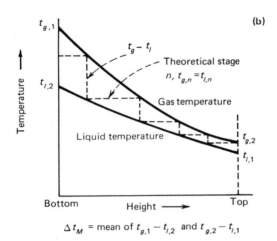

EVALUATION of transfer units and theoretical stages for direct-contact heat transfer via graphical techniques—Fig. 6

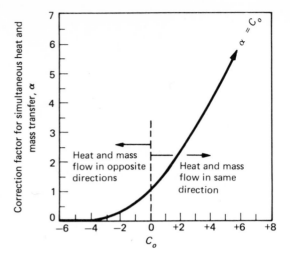

VARIATION of correction factor α with C_o—Fig. 7

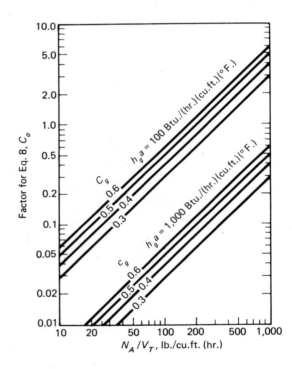

EVALUATION of C_o for use in Eq. 8—Fig. 8

the gas is being cooled, as in an adiabatic humidifier. For this case:

$$Ua = \frac{1}{\frac{1}{\alpha h_g a} + \left(\frac{1}{h_l a}\right)\left(\frac{Q_l}{Q_T}\right)} \quad (7)$$

where

$$\alpha = \frac{C_o}{1 - e^{-C_o}}, \quad (8)$$

and

$$C_o = \frac{N_A c_{g,A}}{h_g a \, V_T} \quad (9)$$

In humidification, heat and mass will flow in opposite directions and N_A is negative, hence $\alpha < 1.0$. Fig. 7 shows how α varies with C_o, and Fig. 8 permits evaluation of C_o over ranges of practical interest.

For a transfer-unit approach:

$$H_{og,h} = H_{g,h} + H_{l,h}\left(\frac{Gc_g}{Lc_l}\right)\left(\frac{Q_l}{Q_T}\right) \quad (10)$$

The stage concept, Eq. 5, may also be used.

Gas Cooling With Partial Condensation

When partial condensation of the gas stream accompanies cooling, the overall volumetric heat-transfer coefficient can be described by the equation:

$$Ua = \frac{1}{\frac{1}{h_l a} + \left(\frac{1}{\alpha h_g a}\right)\left(\frac{Q_g}{Q_T}\right)} \quad (11)$$

where α is as defined in Eq. 8 and 9. Note that N_A is now positive because heat and mass flow in the same direction (see Fig. 7).

When the transfer-unit approach is being used,

$$H_{og,h} = H_{g,h} + H_{l,h}\left(\frac{Gc_g}{Lc_l}\right)\left(\frac{Q_T}{Q_g}\right) \quad (12)$$

For theoretical-stage analysis, use Eq. 5.

Gas Cooling With Total Condensation

For this case, the total duty, Q_T, is the same as the condensation duty (no net sensible-heat duty, Q_g). Thus, Eq. 11 becomes:

$$Ua = h_l a \quad (13)$$

If the coolant liquid is the same as the condensate (e.g., in a barometric condenser), there is relatively little resistance in the liquid phase, and extremely high Ua values result.

The four types of gas cooling occur in specific types of equipment. The designer must consider the nature of this equipment to apply the general heat-transfer equations presented above.

EQUIPMENT PERFORMANCE

Baffle-Tray Columns

This common counterflow contactor contains segmental or disk/donut baffles. Gas-liquid contacting occurs in the curtain of liquid that cascades from plate to plate, as shown in Fig. 9.

A typical baffle column contains segmental baffles, 50% cut (i.e., with no overlap). The baffles should slope if the liquid contains solids. Experience shows that a well-designed baffle column can handle slurries and fouling-prone systems without difficulty.[15,16]

Entrainment flooding determines the vapor capacity of baffle columns. The maximum allowable vapor velocity is based either on window area or curtain area, whichever is limiting:

$$v_{c,max} = 1.15\left(\frac{\rho_l - \rho_g}{\rho_g}\right)^{1/2} = \frac{q}{A_c} \quad (14)$$

$$v_{w,max} = 0.58\left(\frac{\rho_l - \rho_g}{\rho_g}\right)^{1/2} = \frac{q}{A_w} \quad (15)$$

Souders-Brown coefficients of 1.15 and 0.58 are recommended.[7] For both v_c and v_w to be limiting (i.e., $v_{c,max} = v_{w,max}$), A_w can be twice A_c. In such a case, for a 50% baffle cut, vertical spacing between baffles would be approximately 20% of the tower diameter.

Generalized pressure-drop data for baffle columns are not available. However, handbooks provide velocity-head data on a dry basis (no liquid flow). For 50% cut baffles with $A_w = A_c$, 1.2–2.0 velocity heads/baffle is reasonable. This is equivalent to orifice coefficients of 0.9–0.7. For such cases,

$$\Delta p_{dry} = 0.186\left(\frac{\rho_g}{\rho_l}\right)\left(\frac{v_w}{0.7}\right)^2 N_B \quad (16)$$

However, this approach does not account for flow-reversal effects. For example, some investigations have found

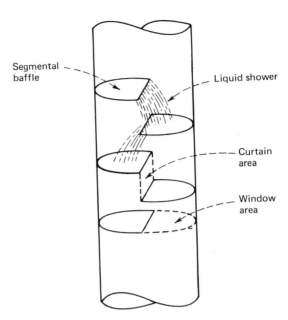

LIQUID FLOW in baffle-tray column—Fig. 9

an effective orifice coefficient of 0.27 for a baffle column with A_w and A_c equal to 32% and 20%, respectively, of total tower cross-section.[23] Allowance for liquid flow should also be considered, particularly under loading/flooding conditions. These effects have been studied.[16]

For heat-transfer performance, assume a column with 10 or more baffle sections operates in countercurrent-flow, and that Δt_M is equal to the log mean temperature difference. Values of Ua from data on air-oil systems[22] have been correlated by the author,[10] using the following equation:

$$Ua = C_1 G^{0.70} L^{0.40} \quad (17)$$

where C_1 is a constant fixed by the baffle spacing and oil type, as shown in Table I. The effect of liquid viscosity indicates that liquid-phase resistance was significant in this system. More details are available from the references.[10,22]

Eq. 17 has been applied to air/water studies[13] and hydrogen-light hydrocarbon/oil data.[21,15] These results are shown in Table II; the higher values of C_1 when water was the cooling liquid, and the lower values of C_1 for oil and for the larger plate spacings, are as expected. The range of C_1 values in Table II shows that Eq. 17 does not account for all the influential variables; but much more experimental work is required before there can be any improvement. Until such studies are completed, use the following recommended values of C_1:

Tray Spacing, In.	C_1
18	0.026
24	0.022
30	0.018

These values yield conservative results for systems not having liquid-phase resistance.

For designing baffle columns, the heat-transfer-unit approach is not suitable. The stagewise technique using

Constants for Eq. 17—Table I*

Plate Spacing, In.	Oil Viscosity, Cp.	C_1
18	27	0.026
18	97	0.057
24	27	0.022
24	97	0.038

Approximate L range = 1,300 to 16,000 lb./(hr.) (sq. ft.)
Approximate G range = 900 to 2,500 lb./(hr.) (sq. ft.)
*Constants are based on data from Ref. 22.

Cooling in Baffle-Tray Columns—Table II

Service	Air	Hydrogen/Hydrocarbon Systems	Catcracker Gas
Ref.	13	21	15
Coolant	water	oil	oil
L range	3,800–9,100	2,800–5,900	2,300–4,200
G range	1,100–3,800	600–1,000	1,500–2,700
Plate spacing, in.	18	30	30
C_1 range*	0.041–0.083	0.014–0.022	0.013–0.022

*C_1 is the coefficient for Eq. 17.

Coefficients for Eq. 19—Table III

Column Dia., In.	Orifice Dia., In.	C_2	C_3
8.0	0.17	5.15×10^{-5}	0.93
15.2	0.17	4.90×10^{-4}	0.70
15.2	0.22	8.97×10^{-4}	0.62
15.2	0.33	4.28×10^{-3}	0.47

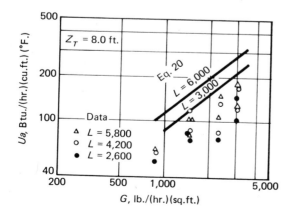

COMPARISON of Eq. 20 with spray-column data[12]—Fig. 10

Eq. 5 and 6 is much more appropriate. Values of E_h for slurry sections of catcracker fractionators usually run in the range of 0.2–0.3. And, as a general rule, E_h values of baffles columns are about 40–50% of the efficiency values obtained for crossflow trays in the same service.

Spray Columns

Spray columns are inexpensive, low-pressure-drop devices that are moderately efficient for heat and mass transfer. Since most experimental studies of spray devices have dealt with mass transfer, analogies must be used for estimating heat-transfer coefficients. Heat-transfer data for commercial spray columns (where temperature driving forces were available) were correlated by the author[9] by using the equation:

$$h_g a = \frac{0.043 \, G^{0.8} \, L^{0.4}}{Z_T^{0.5}} \qquad (18)$$

Test data were derived from hydrocarbon systems as well as from air and water systems. Solid-cone spray nozzles were used in towers of 2.5 to 6 ft. dia.

A later study[12] offers experimental data for air-water systems in a 3-ft.-dia. column, using both hollow- and solid-cone nozzles. Heat-transfer coefficients were calculated from measured mass-transfer data. These coefficients were found to be about 50% below these predicted by Eq. 18,[10] but observed rates in refinery vacuum-tower spray sections were twice those observed in the experimental air-water unit.[12]

Another correlation based upon the mass-transfer/heat-transfer analogy[18] is:

$$h_g a = \frac{1.62 \, C_2 \, G^{0.82} \, L^{C_3}}{0.205^{C_3} \, Z_T^{0.38}} \qquad (19)$$

Values for the constants are presented in Table III.

For a 15.2-in.-dia. column with a typical solid-cone nozzle used in commercial applications (0.33 in.), Eq. 19 reduces to:

$$h_g a = \frac{0.0146 \, G^{0.82} \, L^{0.47}}{Z_T^{0.38}} \qquad (20)$$

A comparison of Eq. 20 with Eq. 18 indicates that they are quite similar. Fig. 10 contains a plot of Eq. 20 with the air-water data from Ref. 12. Since these data were recorded for large nozzle orifices, the coefficients in Eq. 19 could be adjusted to give a better correlation.

Eq. 19, based on assumed log-mean-temperature differences, appears suitable for design. It is important to remember that the equation assumes countercurrent flow. Axial mixing in spray columns has not been studied; however, there are reports that indicate departure from counterflow is not severe.[5,18] Significant backmixing in larger columns can be expected, but the height term (Z_T) in the denominator of Eq. 19 offers partial compensation for this effect.

Other design parameters for spray columns should be noted. For gas pressure drop, the following correlation has been suggested:[12]

$$\Delta p' = 5.88 \times 10^{-5} \, L + 0.065 \qquad (21)$$

where $\Delta p'$ is in inches of water. For maximum allowable vapor capacity, the Souders-Brown coefficient may be applied:

$$v_{t,max} = C_4 \left(\frac{\rho_l - \rho_g}{\rho_g} \right)^{1/2} \quad (22)$$

The coefficient C_4 has not yet been determined by flood tests. Although values as high as 0.40 have been used successfully, a design range of 0.15–0.25 appears more appropriate. One warning: very fine atomization by the nozzles can cause entrainment at relatively low gas rates.

Packed Columns

Even though packed gas cooler-condensers find wide use in industry, their performance has been largely described on the basis of mass transfer. The large amount of absorption/desorption data for packings makes using the analogy particularly attractive. The heat-transfer coefficients for the gas and liquid phases are:

$$h_g a = \left(\frac{Sc_g}{Pr_g} \right)^{2/3} \frac{c_g G}{H_{g,d}} \quad (23)$$

$$h_l a = \left(\frac{Sc_l}{Pr_l} \right)^{2/3} \frac{c_l L}{H_{l,d}} \quad (24)$$

Values of $H_{g,d}$ and $H_{l,d}$ are obtained from mass-transfer data for the particular type and size of packing used. Data for Raschig rings and Berl saddles were summarized and generalized some years ago.[6] This work led to methods for predicting transfer units:

$$H_{g,d} = \frac{\Psi Sc_g^{0.5} D_t^{n_1}}{(L f_1 f_2 f_3)^{m_1}} \quad (25)$$

$$H_{l,d} = \phi C_F (Sc_l)^{0.5} \quad (26)$$

$$f_1 = (\mu_l/2.42)^{0.16} \quad (27)$$

$$f_2 = (62.4/\rho_l)^{1.25} \quad (28)$$

$$f_3 = (72.8/\sigma)^{0.8} \quad (29)$$

Values of the parameters are given in Table IV. It should be noted that the maximum value of the tower diameter (D_T) should be 2 ft., even though larger diameter towers are under consideration. Use of Eq. 23–29 yields conservative heat-transfer coefficients.[9] Unfortunately, more-recent general methods have not been published.

Eq. 23 and 24 are based on film theory. However, a penetration theory also successfully predicts liquid-phase heat-transfer coefficients:[14]

$$\frac{h_l}{k_l} = \frac{\sqrt{(2C_l \rho_l k'_l)/\pi \theta_l}}{\sqrt{2D_l/\pi \theta_l}} = \sqrt{\frac{C_l \rho_l k'_l}{D_l}} \quad (30)$$

For a small packed condenser, with liquid-phase heat transfer controlling, volumetric coefficients ($h_l a \approx Ua$) fell in the range of 50,000–150,000 Btu./(hr.)(cu. ft.)(°F.).[14]

Sizing of packed columns for pressure drop and flooding is covered in many references. The generalized correlations presented in Ref. 8 are the most popular.

Crossflow Tray Columns

Sieve, bubble-cap and valve trays find applications for heat transfer when their cost and fouling susceptibility can be tolerated. In these situations, it is usually adequate to use the heat-transfer efficiency analogy (Eq. 5 and 6),

Parameters for Packed-Column Heat/Mass Transfer—Table IV
Terms for Design Eq. 25–28

	Raschig Rings		Saddles	
Gas-Phase Transfer	1 In.	2 In.	1 In.	2 In.
Exponent m_1	0.6	0.6	0.5	0.5
Exponent n_1	1.24	1.24	1.11	1.11
Parameter Ψ:				
40% flood	110	210	60	(95)
60% flood	105	210	60	(95)
80% flood	80	(210)	(60)	(95)
Liquid-Phase Transfer				
Parameter ϕ				
L = 500	0.045	0.059	0.032	(0.044)
L = 1,000	0.048	0.065	0.040	(0.050)
L = 5,000	0.048	0.090	0.068	(0.075)
L = 10,000	0.082	0.110	0.090	(0.090)
Parameter C_F:				
< 50% flood	1.00	1.00	1.00	1.00
60% flood	0.90	0.90	0.90	0.90
80% flood	0.60	0.60	0.60	0.60

Note: Values in parentheses are extrapolated. See Ref. 6.

with mass-transfer performance obtained from experience or prediction methods.

The AIChE method is the one most frequently applied to predict mass-transfer performance. It is suggested that no advantage be taken of liquid crossflow effects—i.e., Murphee efficiency E_{mv}, should be taken as equal to point efficiency predicted by the model. Thus:

$$E_h = E_{mv} = 1 - e^{-N_{og,d}} = 1 - e^{-Z_l/H_{og,d}} \quad (31)$$

where $H_{og,d}$ results from phase-resistance calculations.

Flooding and pressure-drop calculations for crossflow trays have been widely presented. (A summary of the calculation methods appeared a few years ago in CHEMICAL ENGINEERING.[11]) Heat-transfer coefficients for crossflow trays tend to be high; characteristic data for heavy hydrocarbon service are reported in the literature.[19]

Pipeline Contactors

It is possible to cool a hot gas by injecting a cooler liquid into the gas line. Line desuperheaters for steam, and line quenchers for pyrolysis gas, are examples of this type of cooler. Sizing a pipeline contactor involves several considerations:
- Phases must be well dispersed.
- Flow pattern must be predicted.
- Volumetric mass/heat transfer rates must be determined.
- Pressure drop, if critical, must be calculated.

General principles covering design of gas-liquid pipeline contactors have been summarized.[1] A typical flow arrangement is shown in Fig. 11.

Pipeline heat-transfer devices usually involve high gas/liquid ratios, and either mist or annular flow results. The pattern that is likely to develop can be predicted (see

124 HEAT EXCHANGER DESIGN AND SPECIFICATION

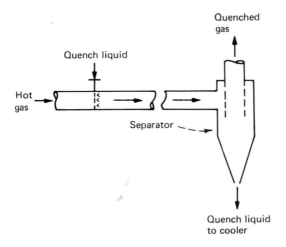

LINE QUENCHER for hot-gas cooling—Fig. 11

Spray Quenching—Typical Physical Property Data—Table V

Material	Temperature, °F.	c_g	ρ_l	k_l'	$\dfrac{c_g \rho_l}{k_l'}$
Water	100	1.00	62.0	0.364	170
	200	1.00	60.1	0.392	154
	300	1.03	57.3	0.395	150
n-octane	200	0.56	40.0	0.079	283
	300	0.60	36.6	0.075	292
Mineral oil A	200	0.54	51.2	0.075	368
	300	0.58	48.7	0.072	392
Mineral oil B	200	0.48	58.3	0.066	425
	300	0.53	56.2	0.064	465

Ref. 1). Initial dispersion is then controlled to produce the desired flow regime. For mist flow, spray nozzles in the line should be used. Assuming no liquid slipping at the pipe wall, droplets quickly approach the gas velocity. For this case, a Nusselt-type relationship applies:[10]

$$h_g = C_5 \left(\frac{k_g'}{d_p} \right) \qquad (32)$$

where the effective droplet diameter, d_p, represents a range of actual sizes. The breakup influence of the turbulent gas limits the size of the droplets.

Spray quenching of a hot gas at a linear velocity greater than 200 ft./sec. has been studied, and a characteristic quench time, θ_q, proposed:[17]

$$\theta_q = \left(\frac{d_p^2}{C_5} \right)\left(\frac{c_g \rho_l}{k_l'} \right)\left(\frac{600}{L/G} \right) \qquad (33)$$

where d_p is volume-diameter mean droplet size, and C_5 varies from 2 to 10. Typical values of $c_g \rho_l / k_l'$ are listed in Table V. Fig. 12 permits rapid estimation of $\theta_q (L/G)$. It should be noted that L/G can vary if coolant is vaporized and/or liquid is condensed from the gas.

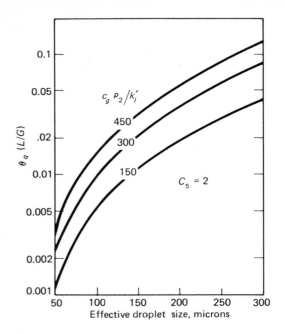

ESTIMATION of characteristic quench time—Fig. 12

The constant C_5 absorbs many uncertainties in the calculation, such as poorly defined volume-diameter mean droplet size, changes in the effective thermal conductivity of the liquid, and the possibility of liquid slipping. The most conservative approach is to use a C_5 value of 2.

Temperature profiles along the pipe can be calculated from:

$$\frac{t_g - t_l}{t_{g,in} - t_l} = e^{-\theta/\theta_q} \qquad (34)$$

When gas and liquid temperatures approach each other, θ is much greater than θ_q.

Droplet diameter, d_p, is a function of spray-nozzle configuration and gas turbulence. The Nukiyama-Tanasawa equation[20] has been found to give reasonable values when the Sauter mean diameter is used.[17] For general design purposes, assume a coarse spray (d_p in range of 200 microns or 6.6×10^{-4} ft.). However, for critical designs, employ nozzle-vendor droplet-size distribution data and a more accurate value of C_5.[17] Breakup effects of the flowing gas should also be considered.

Annular flow patterns for pipeline heat-transfer cases rarely occur. When they do exist, a disperser should be used to promote annular flow.[1] Use of the heat/mass transfer analogy is probably best for determining temperature profiles. For those interested, mass-transfer studies for annular gas-liquid flow have been reported.[3,4]

Example Calculation

Hot pyrolysis gases are to be cooled from 1,100 F. to 100 F. by direct contact with 85 F. quench water. Baffle trays, sprays, and packings are to be considered for a heat-transfer column; line quenching is also to be evaluated.

DESIGNING DIRECT-CONTACT COOLERS/CONDENSERS 125

The gases are composed of hydrogen, light hydrocarbons, and steam. A heat balance is made:

	Entering Gas, Lb./Hr.	Exit Gas, Lb./Hr.
H_2-hydrocarbons	6,000	6,000
Water	600	228
Total	6,600	6,228
ρ_g, lb./cu. ft.	0.027	0.078
c_g, Btu./(lb.) (°F.)	—0.60 avg.—	
Pr_g	—0.90 avg.—	

Q_g, sensible cooling duty = 3,960,000 Btu./hr.
Q_c, condensation duty = 387,000 Btu./hr.
Q_T, total duty = 4,347,000 Btu./hr.
Water condensed = 375 lb./hr.

Baffle-Tray Column

Assume countercurrent flow and permit the coolant temperature to increase to 115 F. (30 F. rise). Then, Δt_M = log mean (Δt) = 232 F. Coolant rate = (4,347,000)/(1.0 × 30) = 144,900 lb./hr.

For rating, choose a 3.5-ft.-I.D. column with 50%-cut segmental baffles at 18-in. vertical spacing. There are no weirs or perforations on the baffles. Key dimensions are:

A_t, tower cross-section = 9.62 sq. ft.
A_w, window area = 4.81 sq. ft.
A_c, curtain area = 5.25 sq. ft.

Column gas capacity:
Base rating on bottom of column, where gas and liquid rates are highest.
Gas flow = 6,600/(3,600 × 0.0271) = 67.65 cu. ft./sec.
Liquid flow = (144,900 + 372)/(0.99 × 500) = 293 gpm.

$$v_{w,max} = 0.58 \left(\frac{61.8 - 0.03}{0.0271}\right)^{1/2} = 27.7 \text{ ft./sec.}$$

$v_{w,design}$ = 67.65/4.81 = 14.1 ft./sec. Therefore, this velocity is within design limits.

$$v_{c,max} = 1.15 \left(\frac{61.8 - 0.03}{0.0271}\right)^{1/2} = 55.0 \text{ ft./sec. (Eq. 14).}$$

$v_{c,design}$ = 67.65/5.25 = 12.89 ft./sec., and this value is acceptable.

Heat transfer:
Using Eq. 17, with C_1 = 0.026 for 18-in. spacing:
G = 6,600/9.62 = 686 lb./(hr.)(sq. ft.).
Ua = 0.026 (686)$^{0.7}$ (15,100)$^{0.4}$ = 117.3 Btu./(hr.) (cu. ft.) (°F.).
L = 145,272/9.62 = 15,100 lb./(hr.) (sq. ft.).

For this case, it is assumed that essentially all heat-transfer resistance is in the gas phase (i.e., $Ua = h_g a$). For mass transfer of N_A = 375 lb./hr. water, $N_A/V_T \approx$ + 2 and $C_o \approx$ 0.01 (Fig. 8). Hence, $\alpha \approx$ 1 (Fig. 7). Then, by Eq. 11, with Q_g/Q_l = 0.91:

$$(Ua)_{corr} = \frac{1}{\frac{1}{117}(0.91)} = 129 \text{ Btu./(hr.)(cu. ft.)(°F.)}$$

Tower dimensions:

$$V_T = \frac{Q_T}{(Ua)\Delta t_M} = \frac{4,347,000}{(129)(232)} = 145 \text{ cu. ft.}$$

Included height = 145 (9.62) = 15.1 ft.
Baffle zones = 16.6/1.5 = 10 calculated; use 12.
Therefore, use 3.5-ft. column, 13 baffles on 18-in. spacing.

Spray Column

For a spray column, it may not be possible to achieve true countercurrent flow. Hence a coolant outlet temperature of 115 F. may not be possible with a gas outlet of 100 F. (For complete mixing, coolant outlet temperatures could not exceed 100 F.). However, since it is not uncommon to have at least a 15 F. difference between the exiting gas and liquid streams in spray columns, the 30 F. coolant rise will be retained, along with the assumption of logarithmic-mean-temperature driving force that is equal to 232 F.

For rating purposes, choose a 3.5-ft.-I.D. column with a bank of solid-cone spray nozzles at the top and gas distributor at the bottom. Column gas capacity, based on bottom conditions, is calculated from Eq. 22:

$$v_{t,max} = 0.20 \left(\frac{61.8 - 0.0271}{0.0271}\right)^{1/2} = 9.6 \text{ ft./sec.}$$

$v_{t,design}$ = 67.65/9.62 = 7.1 ft./sec. This velocity is less than $v_{T,max}$ and therefore is acceptable.

Heat transfer:
Use Eq. 1, 11 and 20, assuming negligible liquid-side resistance to transfer:

$(h_g a) (Z_T)^{0.38}$ = 0.0146 (686)$^{0.82}$ (15,100)$^{0.47}$ = 285 (from Eq. 20)

From Eq. 11, $(Ua)(Z_T)^{0.38}$ = 285/0.91 = 313
Now, using Eq. 1 and remembering that $V_T = A_t Z_T$

$$V_T = 9.62 Z_T = \frac{4,347,000}{(313)(232)} Z_T^{0.38} = 59.9 Z_T^{0.38}$$

$Z_T^{0.62}$ = 6.22
Z_T = 19.2 ft. (V_T = 185 cu. ft.)

Therefore, use a 3.5-ft.-dia. column with 20 ft. between gas inlet and spray nozzles. For safety, be prepared to install additional nozzles lower in the column and/or increase coolant rate.

Packed Column

For rating, choose a 3.5-ft.-I.D. column packed with 2-in. metal Raschig rings. Retain 30 F. coolant temperature rise.

Column gas capacity:
The flow parameter is first calculated:

$$\frac{L}{G}\left(\frac{\rho_g}{\rho_l}\right)^{1/2} = \frac{15,100}{686}\left(\frac{0.0271}{61.8}\right)^{1/2} = 0.46$$

From Ref. 8, for 0.46 flow parameter,

$$\frac{G_F^2 F_p (\mu_l/2.42)^{0.2}}{g\rho_g\rho_l} = 0.045$$

$$G_F^2 = \frac{(0.045)(32.2)(0.0271)(61.8)}{F_p (0.58)^{0.2}} = \frac{2.71}{F_p}$$

The packing factor for 2-in. metal Raschig rings is 57.
Then: $G_F = \sqrt{2.71/57} = 0.218$ lb./(sec.)(sq. ft.)
$= 785$ lbs./(hr.)(sq. ft.)

Approach to flooding $= 686/785 = 0.87$ or 87%. This is too close for the accuracy of the correlation; hence go to a 4-ft.-I.D. column (or use 2-in. Pall rings). For the larger column:
$G = 6,600/12.56 = 525$ lb./(hr.)(sq. ft.) This value corresponds to 67% of flooding.
$L = 145,272/12.56 = 11,600$ lb./(hr.)(sq. ft.)

Heat transfer:
From Table IV and Eq. 27 through 29,

$\psi = 210 \quad f_1 = 0.92$
$m_1 = 0.6 \quad f_2 = 1.01$
$n_1 = 1.24 \quad f_3 = 1.044$
$\quad\quad\quad\quad f_1 f_2 f_3 = 0.97$

Also $Sc_g = 0.60$, $Pr_g = 0.90$, $Sc_g/Pr_g = 0.667$
Then, by Eq. 25:

$$H_{g,d} = \frac{(210)(0.60)^{0.5}(2.0)^{1.24}}{[(11,600)(0.97)]^{0.6}} = 1.41 \text{ ft.}$$

And by Eq. 23,

$$h_g a = (0.667)^{2/3} \frac{(0.6)(525)}{1.41} = 171 \text{ Btu./(hr.)(cu. ft.)(°F.)}$$

From Eq. 11, $Ua = 171 (0.91) = 187$ Btu./(hr.)(cu. ft.)(°F.)

Tower dimensions:
$V_T = \dfrac{4,347,000}{187(232)} = 100$ cu. ft. (From Eq. 1)
$Z_T = \dfrac{V_T}{A_t} = 100/12.56 = 7.96$ ft. Use 10 ft.

Therefore, a 4-ft. column packed with 10 ft. of 2-in. metal Raschig rings will be sufficient.

Pipeline Contactor

It would not be appropriate to use a pipeline for complete cooling of these gases. However, the design principles may be demonstrated by assuming that the gases will be cooled to 200 F. by injecting and vaporizing water (effective latent heat: 1,100 Btu./lb.) in a transfer line.

Water required: $\dfrac{6,600(0.6)(1,100-200)}{1,100} = 3,240$ lb./hr.

Inlet $L/G = 3,240/6,600 = 0.49$
Average $L/G = 1,620/8,220 = 0.20$

For rating, use 6-in. Schedule-40 pipe. Gas velocity is $(67.65$ cu. ft./hr.$)/(0.201$ sq. ft.$) = 337$ ft./sec. at inlet and $50.90/0.201 = 253$ ft./sec. at exit, with the arithmetic average being 295 ft./sec.

Heat transfer:
From Eq. 33, the quench time is:

$$\theta_q = \left(\frac{d_p^2}{C_5}\right)\left(\frac{c_g \rho_l}{k'_l}\right)\frac{600}{(L/G)}$$

A physical property factor of 150 is obtained from Table V; the effective droplet diameter is 200 microns, and $C_5 = 2.0$. From Fig. 12:

$(L/G)\theta_q = 0.022$
$\theta_q = 0.022/0.20 = 0.11$ sec.

Then, for a 1 F. difference between gas and liquid exit temperatures, $\theta = 6.8\theta_q$ (from Eq. 34). The line length is then $(295$ ft./sec.$)(0.11$ sec.$)(6.8) = 221$ ft. Note that overall averages have been employed, and that a careful stepwise analysis should be used for design purposes.

References

1. Alves, G. E., *Chem. Eng. Progr.* **66**, No. 7, 60 (1970).
2. AIChE, "Bubble Tray Design Manual," New York (1958).
3. Anderson, J. D., others, *AIChE J.*, **10**, 640 (1964).
4. Bollinger, R. E., and Lamb, D. E., "Liquid-Phase Controlled Mass Transfer in Annular, Two-Phase Flow of Air and Water Through a Horizontal Pipe," ACS Div. of Ind. & Eng. Chem., Newark, Del., Meeting, Dec. 28-29, 1961.
5. Bonilla, C. F., others, *Ind. Eng. Chem.*, **42**, 2521 (1950).
6. Cornell, D., others, *Chem. Eng. Progr.* **56**, No. 7, 68 (1960).
7. Davies, J. A., and Gordon, K. F., *Petro./Chem. Eng.*, Oct. 1961, p. 54.
8. Eckert, J. S., *Chem. Eng. Progr.*, **66**, No. 3, 39 (1970).
9. Fair, J. R., *Petro./Chem. Eng.*, August 1961, p. 57.
10. Fair, J. R., "Process Heat Transfer by Direct Fluid-Phase Contact," *Chem. Eng. Progr. Symp. Ser.*, No. 118, **68**, 1 (1972).
11. Fair, J. R., and Bolles, W. L., Modern Design of Distillation Columns, *Chem. Eng.*, Apr. 22, 1968, p. 156.
12. Friend, L., others, "Proc. Seventh World Petroleum Congress," pp. 77-85, Spottiswoode, Ballantyne and Co., London (1967).
13. Gautreaux, M. F., Ph.D. Thesis, Louisiana State University (1958).
14. Harriott, P., and Wiegandt, H. F., *AIChE J.*, **10**, 755 (1964).
15. Houghland, G. S., others, *Oil Gas J.*, July 26, 1954, p. 198.
16. Katzen R., others, *Chem. Eng. Progr.*, **64**, No. 1, 79 (1968).
17. Laskowski, J. J., and Ranz, W. E., *AIChE J.*, **16**, 802 (1970).
18. Mehta, K. C., and Sharma, M. M., *Brit. Chem. Eng.*, **15**, No. 11, 1,440; No. 12, 1,556 (1970).
19. Neeld, R. K., and O'Bara, J. T., *Chem. Eng. Progr.*, **66**, No. 7, 53 (1970).
20. Nukiyama S., and Tanasawa, Y., *Trans. Soc. Mech. Engrs.* (Japan), 4, 14, 15, 86, 138 (1938); 5, 18, 63, 68 (1939); 5, 22, 23, II-7, II-8 (1940).
21. O'Donnell, R. J., others, *Chem. Eng. Progr.*, **47**, 309 (1951).
22. Winkler, D. A., Ph.D. Thesis, Louisiana State University (1961).
23. Zuiderweg, F. J., others, "Distillation-1960," Inst. Chem. Engrs., London (1960).

Summary of Example Calculations—Table VI

Contacting Device	Dia., Ft.	Calculated Height, Ft.	Calculated Volume, Cu. Ft.
Baffle column	3.5	15.1	145
Spray column	3.5	19.2	185
Packed column	4.0	8.0	100
Pipeline*	0.51	221	44.3

* For cooling to 200 F. An additional cooler-condenser is required to completely meet requirements of problem.

Meet the Author

James R. Fair is an Engineering Director in Monsanto Co.'s Central Engineering Dept., 800 N. Lindbergh Blvd., St. Louis, MO 63166. He holds B.S., M.S.E. and Ph.D. degrees from Georgia Institute of Technology, and the Universities of Michigan and Texas, respectively. Dr. Fair, a well-known authority on heat and mass transfer, has written numerous technical articles and is presently working on the upcoming fifth edition of "Perry's Chemical Engineers' Handbook." He is a Fellow of AIChE and a member of the executive committee of Fractionation Research, Inc.

Designing Spiral-Plate Heat Exchangers

Spiral-plate exchangers offer compactness, a variety of flow arrangements, efficient heat transfer, and low maintenance costs. These and other features are described, along with a shortcut design method.

P. E. MINTON, Union Carbide Corp.

Spiral heat exchangers have a number of advantages over conventional shell-and-tube exchangers: centrifugal forces increase heat transfer; the compact configuration results in a shorter undisturbed flow length; relatively easy cleaning; and resistance to fouling. These curved-flow units (spiral plate and spiral tube*) are particularly useful for handing viscous or solids-containing fluids.

Spiral-Plate-Exchanger Fabrication

A spiral-plate exchanger is fabricated from two relatively long strips of plate, which are spaced apart and wound around an open, split center to form a pair of concentric spiral passages. Spacing is maintained uniformly along the length of the spiral by spacer studs welded to the plates.

For most services, both fluid-flow channels are closed by alternate channels welded at both sides of the spiral plate (Fig. 1). In some applications, one of the channels is left completely open (Fig. 4), the other closed at both sides of the plate. These two types of construction prevent the fluids from mixing.

Spiral-plate exchangers are fabricated from any material that can be cold worked and welded, such as: carbon steel, stainless steels, Hastelloy B and C, nickel and nickel alloys, aluminum alloys, titanium, and copper alloys. Baked phenolic-resin coatings, among others, protect against corrosion from cooling water. Electrodes may also be wound into the assembly to anodically protect surfaces against corrosion.

Spiral-plate exchangers are normally designed for the full pressure of each passage. Because the turns of the spiral are of relatively large diameter, each turn must contain its design pressure, and plate thickness is somewhat restricted—for these three reasons, the maximum design pressure is 150 psi., although for smaller diameters the pressure may sometimes be higher. Limitations of materials of construction govern design temperature.

Flow Arrangements and Applications

The spiral assembly can be fitted with covers to provide three flow patterns: (1) both fluids in spiral flows; (2) one fluid in spiral flow and the other in axial flow across the spiral; (3) one fluid in spiral flow and the other in a combination of axial and spiral flow.

For spiral flow in both channels, the spiral assembly includes flat covers at both sides (Fig. 1). In this arrangement, the fluids usually flow countercurrently, with the cold fluid entering at the periphery and flowing toward the core, and the hot fluid entering at the core and flowing toward the periphery.

This type of exchanger can be mounted with the axis either vertical or horizontal. It finds wide application in liquid-to-liquid service, and for gases or condensing vapors if the volumes are not too large for the maximum flow area of 72 sq. in.

For spiral flow in one channel, and axial flow in the other, the spiral assembly contains conical covers, dished heads, or extensions with flat covers (Fig. 2). In this design, the passage for axial flow is open on both sides, and the spiral flow channel is welded on both sides.

This type of exchanger is suitable for services in

---

*Although the spiral-plate and spiral-tube exchangers are similar, their applications and methods of fabrication are quite different. This article is devoted wholly to the spiral-plate exchanger; pp. 137-144 will take up the spiral-tube exchanger.

For information on shell-and-tube exchangers, see Ref. 8, 9.

The design method presented is used by Union Carbide Corp. for the thermal and hydraulic design of spiral-plate exchangers, and is somewhat different from that used by the fabricator.

Originally published May 4, 1970

128 HEAT EXCHANGER DESIGN AND SPECIFICATION

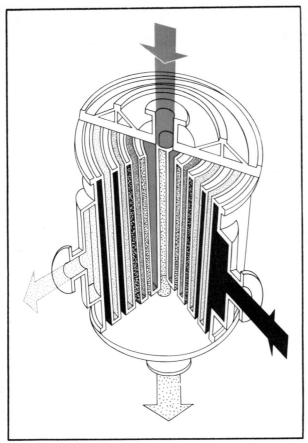

SPIRAL FLOW in both channels is widely used—Fig. 1

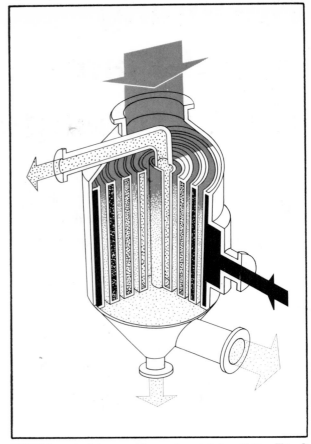

FLOW is spiral in one channel, axial in other—Fig. 2

which there is a large difference in the volumes of the two liquids. This includes liquid-liquid service, heating or cooling gases, condensing vapors, or as reboilers. It may be fabricated with one or more passes on the axial-flow side. And it can be mounted with the axis of the spiral either vertical or horizontal (usually vertically for condensing or boiling).

For combination flow, a conical cover distributes the fluid into its passage (Fig. 3). Part of the spiral is closed at the top, and the entering fluid flows only through the center part of the assembly. A flat cover at the bottom forces the fluid to flow spirally before leaving the exchanger.

This type is most often used for condensing vapors (mounted vertically). Vapors first flow axially until their volume is reduced sufficiently for final condensing and subcooling in spiral flow.

A modification of this type: the column-mounted condenser (Fig. 4). A bottom extension is flanged to mate with the column flange. Vapor flows upward through a large central tube and then axially across the spiral, where it is condensed. Subcooling may be by falling-film cooling or by controlling a level of condensate in the channel. In the latter case, the vent stream leaves in spiral flow. This type is also designed to allow condensate to drop into an accumulator without appreciable subcooling.

The spiral-plate exchanger offers many advantages over the shell-and-tube exchanger:

(1) Single-flow passage makes it ideal for cooling or heating sludges or slurries. Slurries can be processed in the spiral at velocities as low as 2 ft./sec. For some sizes and design pressures, eliminating the spacer studs enables the exchanger to handle liquids having a high content of fibers.

(2) Distribution is good because of the single-flow channel.

(3) The spiral-plate exchanger generally fouls at much lower rates than the shell-and-tube exchanger because of the single-flow passage and curved-flow path. If it fouls, it can be effectively cleaned chemically because of the single-flow path and reduced bypassing. Because the spiral can also be fabricated with identical passages, it is used for services in which the switching of fluids allows one fluid to remove the scale deposited by the other. Also, because the maximum plate width is 6 ft., it is easily cleaned with high-pressure water or steam.

(4) This exchanger is well suited for heating or cooling viscous fluids because its L/D ratio is lower than that of tubular exchangers. Consequently, laminar-flow heat transfer is much higher for spiral plates. When heating or cooling a viscous fluid, the spiral should be oriented with the axis horizontal. With

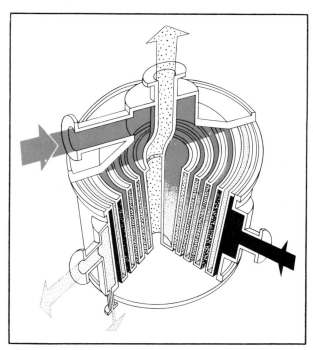

COMBINATION FLOW is used to condense vapors—Fig. 3

the axis vertical, the viscous fluid stratifies and this reduces heat transfer as much as 50%.

(5) With both fluids flowing spirally, flow can be countercurrent (although not truly so, because, throughout the unit, each channel is adjoined by an ascending and a descending turn of the other channel; and because heat-transfer areas are not equal for each side of the channel, the diameters being different). A correction factor may be applied;[1] however, it is so small it can generally be ignored. Countercurrent flow and long passages make possible close temperature approaches and precise temperature control.

(6) The spiral-plate exchanger avoids problems associated with differential thermal expansion in non-cyclic service.

(7) In axial flow, a large flow area affords a low pressure drop, which becomes especially important when condensing under vacuum.

(8) This exchanger is compact: 2,000 sq. ft. of heat-transfer surface in a 58-in.-dia. unit with a 72-in.-wide plate.

Limitations Besides Pressure

In addition to the pressure limitation noted earlier, the spiral-plate exchanger also has the following disadvantages:

(1) Repairing it in the field is difficult. A leak cannot be plugged as in a shell-and-tube exchanger (however, the possibility of leakage in a spiral is less because it is fabricated from plate generally much thicker than tube walls). Should a spiral need repairing, removing the covers exposes most of the welding

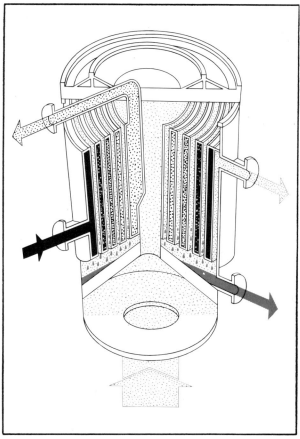

MODIFIED combination flow serves on column—Fig. 4

of the spiral assembly. However, repairs on the inner parts of the plates are complicated.

(2) The spiral-plate exchanger is sometimes precluded from service in which thermal cycling is frequent. When used in cycling services, its mechanical design sometimes must be altered to provide for much higher stresses. Full-faced gaskets of compressed asbestos are not generally acceptable for cycling services because the growth of the spiral plates cuts the gasket, which results in excessive bypassing and, in some cases, erosion of the cover. Metal-to-metal seals are generally necessary.

(3) This exchanger usually should not be used when a hard deposit forms during operation, because the spacer studs prevent such deposits from being easily removed by drilling. When, as for some pressures, such studs can be omitted, this limitation is not present.

(4) For spiral-axial flow, the temperature difference must be corrected. The conventional correction for cross flow applies. Fluids are not mixed; flows are generally single pass. Axial flow may be multipass.

SHORTCUT RATING METHOD

The shortcut rating method for spiral-plate exchangers depends on the same technique as that

HEAT EXCHANGER DESIGN AND SPECIFICATION

Empirical Heat-Transfer and Pressure-Drop Relationships

Eq. No.	Mechanism or Restriction	Empirical Equation—Heat Transfer
	Spiral Flow	
(1)	No phase change (liquid), $N_{Re} > N_{Rec}$	$h = (1 + 3.54\, D_e/D_H)\, 0.023\, cG\, (N_{Re})^{-0.2} (Pr)^{-2/3}$
(2)	No phase change (gas), $N_{Re} > N_{Rec}$	$h = (1 + 3.54\, D_e/D_H)\, 0.0144\, cG^{0.8} (D_e)^{-0.2}$
(3)	No phase change (liquid), $N_{Re} < N_{Rec}$	$h = 1.86\, c\, G\, (N_{Re})^{-2/3} (Pr)^{-2/3} (L/D_e)^{-1/3} (\mu_f/\mu_b)^{-0.14}$
	Spiral or Axial Flow	
(4)	Condensing vapor, vertical, $N_{Re} < 2{,}100$	$h = 0.925\, k\, [g_c \rho_L^2/\mu \Gamma]^{1/3}$
(5)	Condensate subcooling, vertical, $N_{Re} < 2{,}100$	$h = 1.225\, k/B\, [cB/kL_B]^{5/6}$
	Axial Flow	
(6)	No phase change (liquid), $N_{Re} > 10{,}000$	$h = 0.023\, c\, G\, (N_{Re})^{-0.2} (Pr)^{-2/3}$
(7)	No phase change (gas), $N_{Re} > 10{,}000$	$h = 0.0144\, c\, G^{0.8} (D_e)^{-0.2}$
(8)	Condensing vapor, horizontal, $N_{Re} < 2{,}100$	$h = 0.76\, k\, [g_c \rho_L^2/\mu \Gamma]^{1/3}$
(9)	Nucleate boiling, vertical	$h = 4.02\, c\, G\, (N_{Re})^{-0.3} (Pr)^{-0.6} (\rho_L \sigma/P^2)^{-0.425} \Sigma$
	Plate	
(10)	Plate, sensible heat transfer	$h = 12\, k_w/p$
(11)	Plate, latent heat transfer	$h = 12\, k_w/p$
	Fouling	
(12)	Fouling, sensible heat transfer	$h = assumed$
(13)	Fouling, latent heat transfer	$h = assumed$

Eq. No.	Mechanism or Restriction	Empirical Equation—Pressure Drop
	Spiral Flow	
(14)	No phase change, $N_{Re} > N_{Rec}$	$\Delta P = 0.001\, \dfrac{L}{s} \left[\dfrac{W}{d_s H}\right]^2 \left[\dfrac{1.3\, Z^{1/3}}{(d_s + 0.125)} \left(\dfrac{H}{W}\right)^{1/3} + 1.5 + \dfrac{16}{L}\right]$
(15)	No phase change, $100 < N_{Re} < N_{Rec}$	$\Delta P = 0.001\, \dfrac{L}{s} \left[\dfrac{W}{d_s H}\right]^2 \left[\dfrac{1.035\, Z^{1/2}}{(d_s + 0.125)} \left(\dfrac{Z_f}{Z_b}\right)^{0.17} \left(\dfrac{H}{W}\right)^{1/2} + 1.5 + \dfrac{16}{L}\right]$
(16)	No phase change, $N_{Re} < 100$	$\Delta P = \dfrac{LsZ}{3{,}385\,(d_s)^{2.75}} \left(\dfrac{Z_f}{Z_b}\right)^{0.17} \left(\dfrac{W}{H}\right)$
(17)	Condensing	$\Delta P = 0.0005\, \dfrac{L}{s} \left[\dfrac{W}{d_s H}\right]^2 \left[\dfrac{1.3\, Z^{1/3}}{(d_s + 0.125)} \left(\dfrac{H}{W}\right)^{1/3} + 1.5 + \dfrac{16}{L}\right]$
	Axial Flow	
(18)	No phase change, $N_{Re} > 10{,}000$	$\Delta P = \dfrac{4 \times 10^{-5}}{s\, d_s^2} \left(\dfrac{W}{L}\right)^{1.8}\, 0.0115\, Z^{0.2}\, \dfrac{H}{d_s} + 1 + 0.03\, H$
(19)	Condensing	$\Delta P = \dfrac{2 \times 10^{-5}}{s\, d_s^2} \left(\dfrac{W}{L}\right)^{1.8} \left[0.0115\, Z^{0.2}\, \dfrac{H}{d_s} + 1 + 0.03\, H\right]$

Notes:
1. $N_{Rec} = 20{,}000\, (D_e/D_H)^{0.32}$
2. $G = W_o \rho_L / (A \rho_v)$
3. Surface-condition factor (Σ') for copper and steel = 1.0; for stainless steel = 1.7; for polished surfaces = 2.5.

for Rating Spiral-Plate Heat Exchangers—Table I

Numerical Factor	Physical Property Factor	Work Factor	Mechanical Design Factor	
$\frac{\Delta T_f}{\Delta T_M} = 20.6$	$\times \quad \frac{Z^{0.467} M^{0.222}}{s^{0.889}}$	$\times \quad \frac{W^{0.2}(T_H - T_L)}{\Delta T_M}$	$\times \quad \frac{d_s}{LH^{0.2}}$	(See Note 1)
$\frac{\Delta T_f}{\Delta T_M} = 19.6$		$\times \quad \frac{W^{0.2}(T_H - T_L)}{\Delta T_M}$	$\times \quad \frac{d_s}{LH^{0.2}}$	(See Note 1)
$\frac{\Delta T_f}{\Delta T_M} = 32.6$	$\times \quad \frac{M^{2/9}(Z_f)^{0.14}}{s^{8/9}(Z_b)^{0.14}}$	$\times \quad \frac{W^{2/3}(T_H - T_L)}{\Delta T_M}$	$\times \quad \frac{d_s}{LH^{2/3}}$	(See Note 1)
$\frac{\Delta T_f}{\Delta T_M} = 3.8$	$\times \quad \frac{M^{1/3} Z^{1/3}}{cs^2}$	$\times \quad \frac{W^{4/3} \lambda}{\Delta T_M}$	$\times \quad \frac{1}{L^{4/3} H}$	
$\frac{\Delta T_f}{\Delta T_M} = 1.18$	$\times \quad \frac{M^{1/18} Z^{1/18}}{s^{1/3}}$	$\times \quad \frac{W^{2/9}(T_H - T_L)}{\Delta T_M}$	$\times \quad \frac{1}{H^{1/6} L^{2/9}}$	
$\frac{\Delta T_f}{\Delta T_M} = 167$	$\times \quad \frac{Z^{0.467} M^{0.222}}{s^{0.889}}$	$\times \quad \frac{W^{0.2}(T_H - T_L)}{\Delta T_M}$	$\times \quad \frac{d_s}{HL^{0.2}}$	
$\frac{\Delta T_f}{\Delta T_M} = 158$		$\times \quad \frac{W^{0.2}(T_H - T_L)}{\Delta T_M}$	$\times \quad \frac{d_s}{HL^{0.2}}$	
$\frac{\Delta T_f}{\Delta T_M} = 16.1$	$\times \quad \frac{Z^{1/3} M^{1/3}}{cs^2}$	$\times \quad \frac{W^{4/3} \lambda}{\Delta T_M}$	$\times \quad \frac{1}{L^{4/3} H^{4/3}}$	
$\frac{\Delta T_f}{\Delta T_M} = 0.619$	$\times \quad \frac{M^{0.2} Z^{0.3} \sigma^{0.425}}{cs^{1.075}} \frac{\rho_v{}^{0.7}}{P^{0.85}}$	$\times \quad \frac{W^{0.3} \lambda}{\Delta T_M}$	$\times \quad \frac{d_s{}^{0.3} \Sigma'}{L^{0.3} H^{0.3}}$	(See Notes 2 and 3)
$\frac{\Delta T_w}{\Delta T_M} = 500$	$\times \quad \frac{c}{k_w}$	$\times \quad \frac{W(T_H - T_L)}{\Delta T_M}$	$\times \quad \frac{p}{LH}$	
$\frac{\Delta T_w}{\Delta T_M} = 278$	$\times \quad \frac{1}{k_w}$	$\times \quad \frac{W \lambda}{\Delta T_M}$	$\times \quad \frac{p}{LH}$	
$\frac{\Delta T_s}{\Delta T_M} = 6{,}000$	$\times \quad \frac{c}{h}$	$\times \quad \frac{W(T_H - T_L)}{\Delta T_M}$	$\times \quad \frac{1}{LH}$	
$\frac{\Delta T_s}{\Delta T_M} = 3{,}333$	$\times \quad \frac{1}{h}$	$\times \quad \frac{W \lambda}{\Delta T_M}$	$\times \quad \frac{1}{LH}$	

(See Note 1)

(See Note 1)

for shell-and-tube heat exchangers (which were discussed by Lord, Minton and Slusser[9]).

Primarily, the method combines into one relationship the classical empirical equations for film heat-transfer coefficients with heat-balance equations and with correlations that describe the geometry of the heat exchanger. The resulting overall equation is recast into three separate groups that contain factors relating to the physical properties of the fluid, the performance or duty of the exchanger, and the mechanical design or arrangement of the heat-transfer surface. These groups are then multiplied together with a numerical factor to obtain a product that is equal to the fraction of the total driving force—or log mean temperature difference (ΔT_M or LMTD)—that is dissipated across each element of resistance in the heat-flow path.

When the sum of the products for the individual resistance equals 1, the trial design may be assumed to be satisfactory for heat transfer. The physical significance is that the sum of the temperature drops across each resistance is equal to the total available ΔT_M. The pressure drops for both fluid-flow paths must be checked to ensure that both are within acceptable limits. Usually, several trials are necessary to get a satisfactory balance between heat transfer and pressure drop.

Table I summarizes the equations used with the method for heat transfer and pressure drop. The columns on the left list the conditions to which each equation applies, and the second columns gives the standard forms of the correlations for film coefficients that are found in texts. The remaining columns in Table I tabulate the numerical, physical property, work and mechanical design factors—all of which together form the recast dimensional equation. The product of these factors gives the fraction of total temperature drop or driving force ($\Delta T_f/\Delta T_M$) across the resistance.

As stated, the sum of $\Delta T_h/\Delta T_M$ (the hot-fluid factor), $\Delta T_c/\Delta T_M$ (the cold-fluid factor), $\Delta T_s/\Delta T_M$ (the fouling factor), and $\Delta T_w/\Delta T_M$ (the plate factor) determines the adequacy of heat transfer. Any combinations of $\Delta T_f/\Delta T_M$ may be used, as long as the orientation specified by the equation matches that of the exchanger's flowpath.

The units in the pressure-drop equations are consistent with those used for heat transfer. Pressure drop is calculated directly in psi.

Approximations and Assumptions

For many organic liquids, thermal conductivity data are either not available or difficult to obtain. Because molecular weights (M) are known, the Weber equation, which follows, yields thermal conductivities whose accuracies are quite satisfactory for most design purposes:

$$k = 0.86 \, (cs^{4/3}/M^{1/3})$$

If, on the other hand, the thermal conductivity is known, a pseudomolecular weight may be used:

$$M = 0.636 \, (c/k)^3 s^4$$

In what follows, each of the equations in Table I is reviewed, and the conditions in which each equation applies, as well as its limitations, are given. In several cases, numerical factors are inserted, or approximations made, so as to adapt the empirical relationships to the design of spiral-plate exchangers. Such modifications have been made to increase the accuracy, to simplify, or to broaden the use of the method. Rather than by any simplifying approximations, the accuracy of the method is limited by that with which fouling factors, fluid properties and fabrication tolerances can be predicted.

Equations for Heat Transfer—Spiral Flow

Eq. (1)—No Phase Change (Liquid), $N_{Re} > N_{Rec}$—is for liquids with Reynolds numbers greater than the critical Reynolds number. Because the term $(1 + 3.54 \, D_e/D_H)$ is not constant for any given heat exchanger, a weighted average of 1.1 has been used for this method. If a design is selected with a different value, the numerical factor can be adjusted to reflect the new value.

Eq. (2)—No Phase Change (Gas), $N_{Re} > N_{Rec}$—is for gases with Reynolds numbers greater than the critical Reynolds number. Because the Prandtl number of common gases is approximately equal to 0.78 and the viscosity enters only as $\mu^{0.2}$, the relationship of physical properties for gases is essentially a constant. This constant, when combined with the numerical coefficient in Eq. (1) to eliminate the physical property factors for gases, results in Eq. (2). As in Eq. (1), the term $(1 + 3.54 \, D_e/D_H)$ has been taken as 1.1.

Eq. (3)—No Phase Change (Liquid), $N_{Re} < N_{Rec}$—is for liquids in laminar flow, at moderate ΔT and with large kinematic viscosity (μ_L/ρ). The accuracy of the correlation decreases as the operating conditions or the geometry of the heat-transfer surface are changed to increase the effect of natural convection. For a spiral plate:[11]

$$(D/L)^{1/3} = [12^{1/2} D_e/(D_H d_s)^{1/2}]^{1/3} = 2^{1/3} (d_s/d_H)^{1/6}$$

The value of $(d_s/d_H)^{1/6}$ varies from 0.4 to 0.6. A value of 0.5 for $(d_s/d_H)^{1/6}$ has been used for this method.

Heat Transfer Equations—Spiral or Axial Flow

Eq. (4)—Condensing Vapor, Vertical, $N_{Re} < 2,100$—is for film condensation of vapors on a vertical plate with a terminal Reynolds number $(4\Gamma/\mu)$ of less than 2,100. Condensate loading (Γ) for vertical plates is $\Gamma = W/2L$. For Reynolds numbers above 2,100, or for high Prandtl numbers, the equation should be adjusted by means of the Dukler plot, as discussed by Lord, Minton and Slusser.[8] To use Eq. (4) most conveniently, the constant in it should be multiplied by the ratio of the value obtained by the Nusselt equation to the Dukler plot.

The preceding only applies to the condensation

of condensable vapors. Noncondensable gases in the vapor decrease the film coefficient, the reduction depending on the relative sizes of the gas-cooling load and the total cooling and condensing duty. (A method for analyzing condensing in the presence of noncondensable gases is discussed by Lord, Minton and Slusser.[8])

Eq. (5)—Condensate Subcooling, Vertical, $N_{Re} < 2,100$—is for laminar films flowing in layer form down vertical plates. This equation is used when the condensate from a vertical condenser is to be cooled below the bubble point. In such cases, it is convenient to treat the condenser-subcooler as two separate heat exchangers—the first operating only as a condenser (no subcooling), and the second as a liquid cooler only. Fig. 5 shows the assumptions that must be made to determine the height of each section, so as to calculate intermediate temperatures that will permit in turn the calculation of the LMTD.

Eq. (4) is used in combination with appropriate expressions for other resistances to heat transfer, to calculate the height of the subcooling section. In the case of the subcooling section only (See Fig. 5), the arithmetic mean temperature difference, $[(T_{hm} - T_{cm}) + (T_{hL} - T_{cL})]/2$, of the two fluids should be used instead of the log mean temperature difference.

Equations for Heat Transfer—Axial Flow

Eq. (6)—No Phase Change (Liquid), $N_{Re} > 10,000$—is for liquids with Reynolds numbers greater than 10,000.

Eq. (7)—No Phase Change (Gas), $N_{Re} > 10,000$—is for gases with Reynolds numbers greater than 10,000. Again, because the physical property factor for common gases is essentially a constant, this constant is combined with the numerical factor in Eq. (6) to get Eq. (7).

Eq. (8)—Condensing Vapor, Horizontal, $N_{Re} < 2,100$—is for film condensation on spiral plates arranged for horizontal axial flow with a terminal Reynolds number of less than 2,100. For a spiral plate, condensate loading (Γ) depends on the length of the plate and spacing between adjacent plates. For any given plate length and channel spacing, the heat-transfer area for each 360-deg. winding of the spiral increases with the diameter of the spiral. The number of revolutions affects the condensate loading in two ways: (1) the heat-transfer area changes, resulting in more condensate being formed in the outer spirals; and (2) the effective length over which the condensate is formed is determined by the number of revolutions and the plate width. The equations presented depend on a value for the effective number of spirals of $L/7$. Therefore, the condensate loading is given by:

$$\Gamma = W (1,000) 7 (12)/4HL = 21,000 \, W/HL$$

This equation can be corrected if a design is obtained with a significantly different condensate loading. It does not include allowances for turbulence due to vapor-liquid shearing or splashing of the condensate. At high condensate loadings, the liquid condensate on the bottom of the spiral channels may blanket part of the exchanger's effective heat-transfer surface.

Eq. (9)—Nucleate Boiling, Vertical—is for nucleate boiling on vertical plates. In a rigorous analysis of a thermosyphon reboiler, the calculation of heat transfer is combined with the hydrodynamics of the system to determine the circulating rate through the reboiler. However, for most design purposes, this calculation is not necessary. For atmospheric pressure and higher, the assumption of nucleate boiling over the full height of the plate gives satisfactory results. The assumption of nucleate boiling over the entire height of the plate in vacuum service produces overly optimistic results. (The mechanism of thermo-

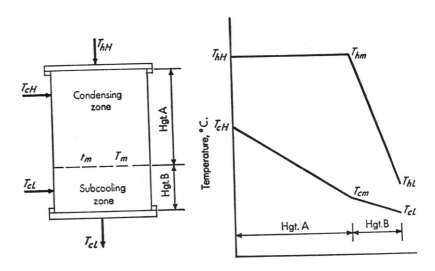

SUBCOOLING-ZONE calculations depend on arithmetic-mean temperature difference of the two fluids instead of log-mean temperature differences—Fig. 5

syphon reboilers has been already discussed by Lord, Minton and Slusser.[8, 9])

A surface condition factor, Σ, appears in the empirical correlations for boiling coefficients. This is a measure of the number of nucleation sites for bubble formation on the heated surface. The equations for $\Delta T_f/\Delta T_M$ contain Σ' (the reciprocal of Σ), which has values of 1.0 for copper and steel, 1.7 for stainless steel or chrome-nickel alloys, and 2.5 for polished surfaces.

Equations for Heat Transfer—Plate

Eq. (10) and (11)—Heat Transfer Through the Plate—are for calculating the plate factor. The integrated form of the Fourier equation is $Q/\theta = (k_w A \Delta T_w)/X$, with X the plate thickness. Expressed in the form of a heat-transfer coefficient, $h_w = 12k_w/p$. Eq. (10) is used whenever sensible heat transfer is involved for either fluid. Eq. (11) is used when there is latent heat transfer for each fluid.

Equations for Heat Transfer—Fouling

Eq. (12) and (13)—Fouling—is for conduction of heat through scale or solids deposits. Fouling coefficients are selected by the designer, based upon his experience. Fouling coefficients of 1,000 to 500 (fouling factors of 0.001 to 0.002) normally require exchangers 10 to 30% larger than for clean service.

The selection of a fouling factor is arbitrary because there is usually insufficient data for accurately assessing the degree of fouling that should be assumed for a given design. Generally, fouling for a spiral-plate exchanger is considerably less than for shell-and-tube exchangers. Because fouling varies with material, velocities and temperature, the extent to which this influences design depends on operating conditions and, to a great degree, the design itself.

Eq. (12) is used for sensible heat transfer for either fluid, and Eq. (13) when latent heat is transferred on both sides of the plate.

Equations for Pressure Drop—Spiral Flow

Eq. (14)—No Phase Change $N_{Re} > N_{Rec}$—is based on equations proposed by Sander.[4, 12] Term A in Sander's equation can be closely approximated by the value of $28/(d_s + 0.125)$. Term B in Sander's equation accounts for the spacer studs. The factor 1.5 assumes 18 studs/sq. ft. and a stud dia. of 5/16 in.

Eq. (15)—No Phase Change $100 < N_{Re} < N_{Rec}$—again is based upon the equation proposed by Sander. For this flow regime, the term A can be closely approximated by the value of $103.5/(d_s + 0.125)$. As in Eq. (14), the factor of 1.5 accounts for the spacer studs.

Eq. (16)—No Phase Change $N_{Re} < 100$—also is based on the Sander equations. For this flow regime, term A can be closely approximated by the value of $2,170\, d_s^{1.75}$. For this flow regime, the studs have little effect on the pressure drop, and any such effect is included in the Sander equation.

Eq. (17)—Condensing—is for calculating the pressure drop for condensing vapors and is identical to that for no phase change, except for a factor of 0.5 used with the condensing equation. For total condensers, the weight rate of flow used in the calculation should be the inlet flowrate. Because the average flow for partial condensers is greater than

Nomenclature

A	Heat-transfer area, sq. ft.
B	Film thickness $(0.00187\, Z\, \Gamma/g_c s^2)^{1/3}$, ft.
C	Core dia., in.
c	Specific heat, Btu./(lb.)(°F.)
D_e	Equivalent dia., ft.
D_H	Helix or spiral dia., ft.
D_s	Exchanger outside dia., in.
d_s	Channel spacing, in.
f	Fanning friction factor, dimensionless
G	Mass velocity, lb./(hr.)(sq. ft.)
g_c	Gravitational constant, ft./(hr.)² (4.18 x 10⁸)
H	Channel plate width, in.
h	Film coefficient of heat transfer, Btu./(hr.)(sq. ft.)(°F.)
k	Thermal conductivity, Btu./(hr.)(sq. ft.)(°F./ft.)
L	Plate length, ft.
M	Molecular weight, dimensionless
P	Pressure, psia.
p	Plate thickness, in.
ΔP	Pressure drop, psi.
Q	Heat transferred, Btu.
s	Specific gravity (referred to water at 20 C.)
ΔT_M	Logarithmic mean temperature difference (LMTD), °C.
U	Overall heat-transfer coefficient, Btu./(hr.)(sq. ft.)(°F.)
W	Flowrate, (lb./hr.)/1,000
Γ	Condensate loading, lb./(hr.)(ft.)
Z	Viscosity, cp.
θ	Time, hr.
λ	Heat of vaporization, Btu./lb.
μ	Viscosity, lb./(hr.)(ft.)
ρ_L	Liquid density, lb./cu.ft.
ρ_v	Vapor density, lb./cu.ft.
Σ, Σ'	Surface condition factor, dimensionless
σ	Surface tension, dynes/cm.

Subscripts

b	Bulk fluid properties
c	Cold stream
f	Film fluid properties
H	High temperature
h	Hot stream
L	Low temperature
m	Median temperature (see Fig. 5)
s	Scale or fouling material
w	Wall, plate material

Dimensionless Groups

N_{Re}	Reynolds number
N_{Rec}	Critical Reynolds number
N_{Pr}	Prandtl number

for total condensers, the multiplying factor should be 0.7 instead of 0.5. Because the estimation of the pressure drop for condensing vapors is not clear-cut, the equation should be used only to approximate the pressure drop, so as to prevent the design of exchangers with excessive pressure losses.

Equations for Pressure Drop—Axial Flow

Eq. (18)—No Phase Change $N_{Re} > 10,000$—is an expression of the Fanning equation for noncompressible fluids, in which the friction factor f in the Fanning equation $= 0.046/N_{Re}^{0.2}$. The equation has been revised to account for pressure losses in the inlet and outlet nozzles, and the inlet and outlet heads. The equation also includes the correction for the spacer studs in the flow channels.

Eq. (19)—Condensing—again is identical to that for no phase change, except for a factor of 0.5. Again, for partial condensers, a value of 0.7 should be used instead of 0.5. For condensing pressure drop, only approximate results should be expected, which themselves should be used only to prevent designs that would result in excessive pressure losses.

For overhead condensers, the pressure drop in the center tube must be added to the pressure drop calculated from Eq. (19).

SAMPLE CALCULATIONS

This example applies the rating method to the design of a liquid-liquid spiral-plate heat exchanger under the following conditions:

Conditions	Hot Side	Cold Side
Flowrate, lb./hr.	6,225	5,925
Inlet temperature, °C.	200	60
Outlet temperature, °C.	120	150.4
Viscosity, cp.	3.35	8
Specific heat, Btu./lb./°F.	0.71	0.66
Molecular weight	200.4	200.4
Specific gravity	0.843	0.843
Allowable pressure drop, psi.	1	1
Material of construction	stainless steel ($k = 10$)	
$(Z_f/Z_b)^{0.14}$	1	1

Preliminary Calculations

Heat transferred $= 6,225 \times (200-120) \times 1.8 \times 0.71 = 636,400$ Btu./hr.

ΔT_M (or $LMTD$) $= (60 - 49.4)/\ln(60/49.4) = 54.5$ C.

For a first trial, the approximate surface can be calculated using an assumed overall heat-transfer coefficient, U, of 50 Btu./(hr.)(sq. ft.)(°F.):

$$A = 636,400/(50 \times 1.8 \times 54.5) = 130 \text{ sq. ft.}$$

Because this is a small exchanger, a plate width of 24 in. is assumed. Therefore, $L = 130/(2 \times 2) = 32.5$ ft. A channel spacing of 3/8 in. for both fluids is also assumed. The Reynolds number for spiral flow can be calculated from the expression:

$$N_{Re} = 10,000 \, (W/HZ)$$

Therefore:

Hot side $N_{Re} = (10,000 \times 6.225)/(24 \times 3.35) = 774$
Cold side $N_{Re} = (10,000 \times 5.925)/(24 \times 8) = 309$

Because the fluids will be in laminar flow, spiral flow is selected for the heat exchanger design. From Table I, the appropriate expressions for rating are: Eq. (3) for both fluids, Eq. (10) for the plate, Eq. (12) for fouling and Eq. (15) for pressure drop.

Heat-Transfer Calculations

Now, substitute values:
Hot side, Eq. (3):

$$\frac{\Delta T_h}{\Delta T_M} = 32.6 \left[\frac{200.4^{0.222}}{0.843^{0.889}} \right] \times$$
$$\left[\frac{6.225^{2/3} \times 80}{54.5} \right] \left[\frac{0.375}{24^{2/3} \times 32.5} \right]$$
$$= 32.6 \times 3.775 \times 4.967 \times 0.001387 = 0.848$$

Cold side, Eq. (3):

$$\frac{\Delta T_c}{\Delta T_M} = 32.6 \left[\frac{200.4^{0.222}}{0.843^{0.889}} \right] \left[\frac{5.925^{2/3} \times 90.4}{54.5} \right] \times$$
$$\left[\frac{0.375}{24^{2/3} \times 32.5} \right]$$
$$= 32.6 \times 3.775 \times 5.431 \times 0.001387 = 0.927$$

Fouling, Eq. (12):

$$\frac{\Delta T_s}{\Delta T_M} = 6,000 \left[\frac{0.66}{1,000} \right] \left[\frac{5.925 \times 90.4}{54.5} \right] \left[\frac{1}{32.5 \times 24} \right]$$
$$= 6,000 \times 0.00066 \times 9.828 \times 0.001282 = 0.050$$

Plate, Eq. (10):

$$\frac{\Delta T_w}{\Delta T_M} = 500 \left[\frac{0.66}{10} \right] \left[\frac{5.925 \times 90.4}{54.5} \right] \left[\frac{0.125}{32.5 \times 24} \right]$$
$$= 500 \times 0.066 \times 9.828 \times 0.0001603 = 0.052$$

Some Spiral-Plate Exchanger Standards—Table II

Plate Widths, In.	Outside Dia., Maximum, In.	Core Dia., In.
4	32	8
6	32	8
12	32	8
12	58	12
18	32	8
18	58	12
24	32	8
24	58	12
30	58	12
36	58	12
48	58	12
60	58	12
72	58	12

Channel spacings, in.: 3/16 (12 in. maximum width), 1/4 (48 in. maximum width), 5/16, 3/8, 1/2, 5/8, 3/4 and 1.
Plate thicknesses: stainless steel, 14-3 U.S. gage; carbon steel, 1/8, 3/16, 1/4 and 5/16 in.

Sum of Products (SOP):

$$SOP = 0.848 + 0.927 + 0.050 + 0.052 = 1.877$$

Because SOP is greater than 1, the assumed heat exchanger is inadequate. The surface area must be enlarged by increasing the plate width or the plate length. Because, in all the equations, L applies directly, the following new length is adopted:

$$1.877 \times 32.5 = 61 \text{ ft.}$$

Pressure-Drop Calculations

Hot side, Eq. (15):

$$\Delta P = \left[\frac{0.001 \times 61}{0.843}\right]\left[\frac{6.225}{0.375 \times 24}\right] \times$$
$$\left[\frac{1.035 \times 3.35^{1/2} \times 1 \times 24^{1/2}}{(0.375 + 0.125) 6.225^{1/2}} + 1.5 + \frac{16}{61}\right]$$

$$\Delta P = 0.07236 \times 0.6917 \times 9.202 = 0.461 \text{ psi.}$$

Cold side, Eq. (15):

$$\Delta P = \left[\frac{0.001 \times 61}{0.843}\right]\left[\frac{5.925}{0.375 \times 24}\right] \times$$
$$\left[\frac{1.035 \times 8^{1/2} \times 1 \times 24^{1/2}}{(0.375 + 0.125) 5.925^{1/2}} + 1.5 + \frac{16}{61}\right]$$

$$\Delta P = 0.07236 \times 0.6583 \times 13.55 = 0.645 \text{ psi.}$$

Because the pressure drop is less than the allowable, the spacing can be decreased. For the second trial, ¼ in. spacing for both channels is adopted.

Because the heat-transfer equation for every factor except the plate varies directly with d_s, a new SOP can be calculated.

$$\Delta T_h/\Delta T_M = 0.848 (0.25/0.375) = 0.565$$
$$\Delta T_c/\Delta T_M = 0.927 (0.25/0.375) = 0.618$$
$$\Delta T_s/\Delta T_M = 0.052 (0.25/0.375) = 0.035$$
$$\Delta T_w/\Delta T_M = 0.050$$

$$SOP = 0.565 + 0.618 + 0.050 + 0.052 = 1.285$$
$$L = 1.285 \times 32.5 = 41.8 \text{ ft.}$$
$$A = 41.8 \times 2 \times 2 = 167 \text{ sq. ft.}$$

The new pressure drop becomes:

Hot side:

$$\Delta P = \left[\frac{0.001 \times 41.8}{0.843}\right]\left[\frac{6.225}{0.25 \times 24}\right] \times$$
$$\left[\frac{1.035 \times 3.35^{1/2} \times 1 \times 24^{1/2}}{0.375 \times 6.225^{1/2}} + 1.5 + \frac{16}{41.8}\right]$$

$$\Delta P = 0.04958 \times 1.037 \times 11.80 = 0.607 \text{ psi.}$$

Cold side:

$$\Delta P = \left[\frac{0.001 \times 41.8}{0.843}\right]\left[\frac{5.925}{0.25 \times 24}\right] \times$$
$$\left[\frac{1.035 \times 8^{1/2} \times 1 \times 24^{1/2}}{0.375 \times 5.925^{1/2}} + 1.5 + \frac{16}{41.8}\right]$$

$$\Delta P = 0.04958 \times 0.9875 \times 17.59 = 0.861$$

The pressure drops are less than the maximum allowable. The plate spacing cannot be less than ¼ in. for a 24-in. plate width; decreasing the width would result in a higher than allowable pressure drop. Therefore, the design is acceptable.

The diameter of the outside spiral can now be calculated with Table II and the following equation:

$$D_S = [15.36 \times L (d_{sc} + d_{sh} + 2p) + C^2]^{1/2}$$
$$D_S = \{15.36 (41.8) [0.25 + 0.25 + 2 (0.125)] + 8^2\}^{1/2}$$
$$D_S = 23.4 \text{ in.}$$

For a spiral-plate exchanger, the best design is often that in which the outside diameter approximately equals the plate width.

Design summary:
Plate width............... 24 in.
Plate length.............. 41.8 ft.
Channel spacing.......... ¼ in. (both sides)
Spiral diameter........... 23.4 in.
Heat-transfer area........ 167 sq. ft.
Hot-side pressure drop.... 0.607 psi.
Cold-side pressure drop... 0.861 psi.
U........................ 38.8 Btu./(hr.)(sq.ft.)(°F.)

■

Acknowledgements

The author thanks American Heat Reclaiming Corp. for providing figures and for permission to use certain design standards. He is also grateful to the Union Carbide Corp. for permission to publish this article.

References

1. Baird, M. H. I., McCrae, W., Rumford, F. and Slesser, C. G. M., Some Considerations on Heat Transfer in Spiral Plate Heat Exchangers, *Chem. Eng. Science*, **7**, 1 and 2, 1957, p. 112.
2. Blasius, H., Das ähnlichkeitsgesets bei Riebungsvorgängzen in Flussigkeiten, *Forschungsheft*, **131**, 1913.
3. Colburn, A. P., A Method of Correlating Forced Convection Heat Transfer Data and a Comparison With Fluid Friction, *AIChE Trans.*, **29**, 1933, p. 174.
4. Hargis, A. M., Beckmann, A. T. and Loiacona, J. J., Applications of Spiral Plate Heat Exchangers, *Chem. Eng. Progr.*, July 1967, p. 62.
5. "Heliflow Coolers and Heaters," *Bull. 58G*, Graham Mfg. Co., Great Neck, N.Y.
6. Ito, H., Friction Factors for Turbulent Flow in Curved Pipes, *Trans. ASME*, **81**, 2, 1959, p. 123.
7. Lamb, B. R., The Rosenblad Spiral Heat Exchanger, *Trans. Inst. Chem. Engrs.* (London), June 1962, p. A108.
8. Lord, R. C., Minton, P. E. and Slusser, R. P., Design Parameters for Condensers and Reboilers, *Chem. Eng.*, Mar. 23, 1970, p. 127.
9. Lord, R. C., Minton, P. E. and Slusser, R. P., Design of Heat Exchangers, *Chem. Eng.*, Jan. 26, 1970, p. 96.
10. Noble, M. A., Kamlani, J. S. and McKetta, J. J., Heat Transfer in Spiral Coils, *Petr. Eng.*, Apr. 1952, p. 723.
11. Perry, J. H., Ed., "Chemical Engineers' Handbook," 4th ed., McGraw-Hill, New York, 1963, **10**, p. 24.
12. Sander, J., (unpublished), A. B. Rosenblads, Patenter, Stockholm, Sweden, 1955.
13. "Spiral Heat Exchangers," *Bull. S.A. 1410 3-69 HR5M*, American Heat Reclaiming Corp., New York.
14. Tangri, N. N. and Jayaraman, R., Heat Transfer on a Spiral Plate Heat Exchanger, *Trans. Inst. Chem. Engrs.* (London), **40**, 3, 1962, p. 161.
15. "Thermal Handbook," Alfa-Laval/DeLaval Group, Sweden, 1969.
16. White, C. M., Streamline Flow Through Curved Pipes, *Proceedings Royal Soc.* (London), Series A, **123**, 1929, p. 645.

Meet the Author

Paul E. Minton is a project engineer in the engineering department at Union Carbide Corp.'s Technical Center (P.O. Box 8361, So. Charleston, W. Va. 25303), where he is a part of the heat-transfer technology group. A graduate in chemical engineering with a B.S. degree from the Missouri School of Mines and Metallurgy, he is a member of AIChE.

Designing Spiral-Tube Heat Exchangers

Compact and easily installed, the spiral-tube exchanger handles low flows and small heat loads, and heats or cools viscous fluids efficiently. Its features are presented, along with a shortcut design method.

P. E. MINTON, Union Carbide Corp.

Spiral-tube heat exchangers (like spiral-plate heat exchangers*) offers several advantages over conventional shell-and-tube heat exchangers: secondary flow caused by centrifugal force that increases heat transfer; compact, short, undisturbed flow lengths that make spiral-tube exchangers ideal for heating and cooling viscous fluids, sludges and slurries; less fouling; relatively easy cleaning.

Spiral-tube exchangers are generally more expensive than shell-and-tube exchangers having the same heat-transfer surface. However, the spiral-tube exchanger's better heat transfer and lower maintenance cost often make it the more profitable choice.

Fabrication Methods and Materials

Spiral-tube exchangers consist of one or more concentric, spirally wound coils clamped between a cover plate and a casing. Both ends of each coil are attached to a manifold fabricated from pipe or bar stock (Fig. 1, 2).

The coils, which are stacked on top of each other, are held together by the cover plate and casing. Spacing is maintained evenly between each turn of the coil to create a uniform, spiral-flow path for the shellside fluid.

Coils can be formed from almost any material of construction, with some of the more common ones being carbon steel, copper and copper alloys, stainless steels, and nickel and nickel alloys. Tubes may have extended surfaces. Casings are made of cast iron, cast bronzes, and carbon and stainless steel.

Tubes may be attached to the manifolds by soldering, brazing, welding or, in some cases, rolling. Draining or venting can be facilitated by various manifold arrangements and casing connections. Flow through both the coil and casing may be single- or multipass (the latter by means of baffling).

Spiral-tube exchangers are available in sizes up to 325 sq. ft., and pressures up to 600 psi. Tubeside pressures may be even higher.

Advantages and Limitations

The spiral-tube exchanger offers the following advantages over the shell-and-tube exchanger: (1) it is especially suited for low flows or small heat loads; (2) it is particularly effective for heating or cooling viscous fluids; since L/D ratios are much lower than those of straight-tube exchangers, laminar-flow heat transfer is much higher with spiral tubes;

---

*Although spiral-tube and spiral-tube heat exchangers are similar in many ways, their applications and methods of fabrication are different enough to warrant being discussed separately. This article deals only with the spiral-tube exchanger. An article devoted wholly to the spiral-plate exchanger appears on pp. 127-136. To meet the author, see p. 136.

The design method presented in this article is that used by Union Carbide Corp. for the thermal and hydraulic design of spiral-tube exchangers, and is somewhat different from that used by the fabricator.

Originally published May 18, 1970

CASING is removed without disconnecting pipes—Fig. 1

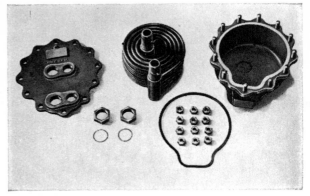

EXCHANGER'S design makes cleaning easy—Fig. 2

(3) its flows can be countercurrent (as with the spiral-plate exchanger, flows are not truly countercurrent; but again, the correction for this can be ignored); (4) it does not present the problems usually associated with differential thermal expansion; and (5) it is compact and easily installed.

The following are the chief limitations of the spiral-tube exchanger: (1) Its manifolds are usually small, making the repair of leaks at tube-to-manifold joints difficult (leaks, however, do not occur frequently); (2) it is limited to services that do not require mechanical cleaning of the inside of tubes (it can be cleaned mechanically on the shellside, and both sides can be cleaned chemically); (3) for some of its sizes, stainless steel coils must be provided with spacers to maintain a uniform shellside flow area—and these spacers increase pressure drop (this increase is not accounted for in the equations presented later).

SHORTCUT RATING METHOD

The shortcut rating method for spiral-tube exchangers depends on the same technique as used for shell-and-tube exchangers (which is discussed by Lord, Minton and Slusser[6]).

Primarily, the method combines into one relationship the classical empirical equations for film heat-transfer coefficients with heat-balance equations and with correlations that describe the geometry of the heat exchanger. The resulting overall equation is recast into three separate groups that contain factors relating to: the physical properties of the fluid, the performance or duty of the exchanger, and the mechanical design or arrangement of the heat-transfer surface. These groups are then multiplied together with a numerical factor to obtain a product that is equal to the fraction of the total driving force—or log mean temperature difference (ΔT_M or LMTD)—that is dissipated across each element of resistance in the heat-flow path.

When the sum of the products for the individual resistances equals 1, the trial design may be assumed satisfactory for heat transfer. The physical significance is that the sum of the temperature drops across each resistance is equal to the total available ΔT_M. The pressure drop for both fluid-flow paths must be checked to ensure that they are within acceptable limits. Usually, several trials are necessary to get a satisfactory balance between heat transfer and pressure drop.

Table I summarizes the equations used with spiral-tube exchangers. The column on the left presents the conditions to which each equation applies, and the second column gives the standard form of the film-coefficient correlation found in texts. The remaining columns tabulate the numerical, physical property, work, and mechanical design factors—all of which together form the recast dimensional equation. The product of these factors gives the fraction of the total temperature drop or driving force ($\Delta T_f/\Delta T_M$) across the resistance.

As stated, the sum of $\Delta T_i/\Delta T_M$ (the tubeside factor), $\Delta T_o/T_M$ (the shellside factor), $\Delta T_s/T_M$ (the fouling factor) and $\Delta T_w/\Delta T_M$ (the tube wall factor) determine the adequacy of heat transfer. Any combinations of $\Delta T_f/\Delta T_M$ may be used as long as the orientation specified by the equation matches that of the exchanger's flowpath.

The units in the pressure-drop equations are consistent with those used for heat transfer. Pressure drop is calculated directly in psi.

For many organic liquids, thermal conductivity data are either not available or difficult to obtain. Because molecular weights (M) are known, the Weber equation (which follows) yields thermal conductivities whose accuracies are satisfactory for most design purposes:

$$k = 0.86\ (cs^{4/3}/M^{1/3})$$

If, on the other hand, the thermal conductivity is

known, a pseudomolecular weight may be used:

$$M = 0.636 (c_f/k)^3 s^4$$

In what follows, each of the equations in Table I is reviewed, and the conditions in which each equation applies, along with its limitations, are given. In several cases, numerical factors are inserted, or approximations made, so as to adapt the empirical relationships to the design of the spiral-tube exchanger. Such modifications have been made to increase the accuracy, to simplify or to broaden the use of the method. Rather than by any simplifying approximations, however, the accuracy of the method is limited by the precision with which fouling factors, fluid properties, and fabrication tolerances can be predicted.

The details of the velocity profile in a curved path are not completely known. However, most investigators agree that there is a secondary flow in curved paths, which is caused by the action of a centrifugal force. The extra energy absorbed in this secondary flow, which causes a form of mechanical mixing, raises pressure drop and lowers heat-transfer resistance. Because the nature of the secondary flow and the distorted mean-velocity profile are not known, verified theoretical models for pressure drop and heat transfer in "curved" flow geometries have not been developed. However, the equations presented agree in general with the results of the majority of past investigators. Test data on operating spiral-tube exchangers attest to the method's applicability.

Equations for Heat Transfer—Tubeside

Eq. (1)—No Phase Change (Liquid), $N_{Re} > N_{Rec}$—is for liquids with Reynolds numbers larger than the critical Reynolds number. Because the term $(1 + 3.54 D_e/D_H)$ is not constant for any particular heat exchanger, a weighted average of 1.15 has been used for this method. If a design is selected with a different value, the numerical factor can be adjusted.

Eq. (2)—No Phase Change (Gas), $N_{Re} > N_{Rec}$—is for gases with Reynolds numbers larger than the critical Reynolds number. Because the Prandtl number of common gases is approximately equal to 0.78 and the viscosity enters only as $\mu^{0.2}$, the relationship of physical properties for gases is essentially a constant. The physical-property factor has been combined with the numerical factor. The factor $(1 + 3.54 D_e/D_H)$ has again been taken as 1.15.

Eq. (3)—No Phase Change (Liquid), $N_{Re} < N_{Rec}$—is for liquids in laminar flow at moderate ΔT and with large kinematic viscosity (μ_L/ρ). As in the case of the spiral-plate exchanger, this equation also neglects the effects of natural convection. For a spiral tube:[8]

$$(D/L)^{1/3} = (D_i D_H)^{1/2}$$

Because the value of $(D_i/D_H)^{\frac{1}{2}}$ varies from 0.15 to 0.25, a value of 0.2 has been used for this method.

Eq. (4)—Condensing Vapor, Horizontal, $N_{Re} < 2,100$—is for film condensation of vapors in horizontal tubes with a terminal Reynolds number $(4\Gamma/\mu)$ of less than 2,100. Tube loading (Γ) for condensing inside horizontal tubes is $W_i/2nL$. This equation does not include allowances for turbulence due to vapor-liquid shearing or splashing of condensate. At high loadings, the liquid condensate on the bottom of the tube blanks off part of the condensing surface. To correct for this, the heat-transfer factor $(\Delta T_i/\Delta T_M)$ should be multiplied by 1.3 if the term $W_i/ns_i d_i^{2.56}$

Nomenclature

A	Heat-transfer area, sq. ft.
a	Net free-flow area, sq. in.
B	Film thickness $(0.00187 Z \Gamma/g_c s^2)^{1/3}$, ft.
c	Specific heat, Btu./(lb.)(°F.)
D_e	Equivalent dia., ft.
D_H	Helix or spiral dia., ft.
D_i	Inside tube dia., ft.
d	Tube dia., in.
f	Fanning friction factor, dimensionless
G	Mass velocity, lb./(hr.)(sq. ft.)
g_c	Gravitational constant, ft./(hr.)² (4.18 x 10⁸)
h	Film coefficient of heat transfer, Btu./(hr.)(sq. ft.)(°F.)
k	Thermal conductivity, Btu./(hr.)(sq. ft.)(°F./ft.)
L	Tube length, ft.
M	Molecular weight, dimensionless
N_{PT}	No. tube passes/shell, dimensionless
n	No. tubes/pass, dimensionless
P	Pressure, psia.
ΔP	Pressure drop, psi.
Q	Heat transferred, Btu.
s	Specific gravity (referred to water at 20 C.)
T	Temperature shellside, °C.
t	Temperature tubeside, °C.
ΔT_M	Logarithmic mean temperature difference (LMTD), °C
U	Overall heat-transfer coefficient, Btu./(hr.)(sq. ft.)(°F.)
W	Flowrate, (lb./hr.)/1,000
Γ	Condensate loading, lb./(hr.)(ft.)
Z	Viscosity, cp.
θ	Time, hr.
λ	Heat of vaporization, Btu./lb.
μ	Viscosity, lb./(hr.)(ft.)
ρ_L	Liquid density, lb./cu.ft.
ρ_v	Vapor density, lb./cu.ft.
Σ, Σ'	Surface condition factor, dimensionless
σ	Surface tension, dynes/cm.

Subscripts

b	Bulk fluid properties
f	Film fluid properties
H	High temperature
i	Tubeside conditions
L	Low temperature
o	Shellside conditions
s	Scale or fouling material
w	Wall, tube material

Dimensionless Groups

N_{Re}	Reynolds number
N_{Rec}	Critical Reynolds number
N_{Pr}	Prandtl number

140 HEAT EXCHANGER DESIGN AND SPECIFICATION

Empirical heat-transfer and pressure-drop relationships

Eq. No.	Mechanism or Restriction	Empirical Equation — Heat Transfer
	Tube Side	
(1)	No phase change (liquid), $N_{Re} > N_{Rec}$	$h = (1 + 3.54 D_e/D_{II})\, 0.023\, cG\, (N_{Re})^{-0.2} (Pr)^{-2/3}$
(2)	No phase change (gas), $N_{Re} > N_{Rec}$	$h = (1 + 3.54 D_e/D_{II})\, 0.0144\, c\, G^{0.8} (D_i)^{-0.2}$
(3)	No phase change (liquid), $N_{Rec} < N_{Rec}$	$h = 1.86\, cG\, (N_{Re})^{-2/3} (Pr)^{-2/3} (L/D_e)^{-1/3} (\mu_f/\mu_b)^{-0.14}$
(4)	Condensing vapor, horizontal, $N_{Re} < 2{,}100$	$h = 0.76\, k\, [g_c \rho_L^2/\mu \Gamma]^{1/3}$
	Shell Side	
(5)	No phase change (liquid), $N_{Re} > N_{Rec}$	$h = (1 + 3.54 D_e/D_{II})\, 0.023\, cG\, (N_{Re})^{-0.2} (Pr)^{-2/3}$
(6)	No phase change (gas), $N_{Re} > N_{Rec}$	$h = (1 + 3.54 D_e/D_{II})\, 0.0144\, c\, G^{0.8} (D_e)^{-0.2}$
(7)	No phase change (liquid), $N_{Re} < N_{Rec}$	$h = 1.86\, cG\, (N_{Re})^{-2/3} (Pr)^{-2/3} (L/D_e)^{-1/3} (\mu_f/\mu_b)^{-0.14}$
(8)	Condensing vapor, horizontal, $N_{Re} < 2{,}100$	$h = 0.76\, k\, [g_c \rho_L^2/\mu \Gamma]^{1/3}$
(9)	Nucleate boiling, horizontal	$h = 4.02\, cG\, (N_{Re})^{-0.3} (Pr)^{-0.6} (\rho_L \sigma/P^2)^{-0.425}\, \Sigma$
	Tube Wall	
(10)	Tube wall, sensible heat	$h = (24 k_w)/(d_o - d_i)$
(11)	Tube wall, latent heat	$h = (24 k_w)/(d_o - d_i)$
	Fouling	
(12)	Fouling, sensible heat	$h =$ assumed
(13)	Fouling, latent heat	$h =$ assumed

Eq. No.	Mechanism or Restriction	Empirical Equation — Pressure Drop
	Tube Side	
(14)	No phase change, $N_{Re} > N_{Rec}$	$\Delta P = 0.00268 \left[\dfrac{Z_i^{0.15}}{s_i}\right] \left[\dfrac{W_i}{n}\right]^{1.85} \left[\dfrac{L/d_i + 16}{d_i^{3.75}}\right] N_{PT}$
(15)	No phase change, $100 < N_{Re} < N_{Rec}$	$\Delta P = 0.00195 \left[\dfrac{W_i^{4/3} Z_i^{2/3} Z_f^{0.14} L}{s_i Z_b^{0.14} n^{4/3} d_i^{13/3}}\right]$
(16)	No phase change, $N_{Re} < 100$	$\Delta P = 0.000544 \left[\dfrac{L W_i Z_i}{n s_i d_i^4}\right]$
(17)	Condensing	$\Delta P = 0.00134 \left[\dfrac{Z_i^{0.15}}{s_i}\right] \left[\dfrac{W_i}{n}\right]^{1.85} \left[\dfrac{L/d_i + 16}{d_i^{3.75}}\right] N_{PT}$
	Shell Side	
(18)	No phase change, $N_{Re} > N_{Rec}$	$\Delta P = 0.00152 \left[\dfrac{Z_o^{0.15}}{s_o}\right] [L(nd_o)^{1.15}] \left[\dfrac{W_o^{1.85}}{a^3}\right] N_{PT}^{1.15}$
(19)	No phase change, $100 < N_{Re} < N_{Rec}$	$\Delta P = 0.0011 \left[\dfrac{W_o^{4/3} Z_o^{2/3} (nd_o)^{5/3} L\, Z_f^{0.14}}{s_o a^3 Z_b^{0.14}}\right] N_{PT}^{5/3}$
(20)	No phase change, $N_{Re} < 100$	$\Delta P = 0.000308 \left[\dfrac{L W_o Z_o}{s_o}\right] \left[\dfrac{(nd_o)^2}{a^3}\right] N_{PT}^2$
(21)	Condensing (spiral flow)	$\Delta P = 0.000757 \left[\dfrac{Z_o^{0.15}}{s_o}\right] \left[\dfrac{W_o^{1.85}}{a^3}\right] [L(nd_o)^{1.15}] N_{PT}^{1.15}$
(22)	Condensing (axial flow)	$\Delta P = 4 \times 10^{-5} \times \dfrac{1}{s_o}\, [(W_o/L d_s)^2\, n\, N_{PT}]$

for rating spiral-tube heat exchangers—Table I

Numerical Factor		Physical Property Factor		Work Factor		Mechanical Design Factor	
$\frac{\Delta T_i}{\Delta T_M} = 9.07$	$\times$	$\frac{Z_i^{0.467} M_i^{0.222}}{s_i^{0.889}}$	$\times$	$\frac{W_i^{0.2}(t_H-t_L)}{\Delta T_M}$	$\times$	$\frac{d_i^{0.8}}{n^{0.2} L}$	(See Note 1)
$\frac{\Delta T_i}{\Delta T_M} = 8.58$			$\times$	$\frac{W_i^{0.2}(t_H-t_L)}{\Delta T_M}$	$\times$	$\frac{d_i^{0.8}}{n^{0.2} L}$	(See Note 1)
$\frac{\Delta T_i}{\Delta T_M} = 13.0$	$\times$	$\frac{M_i^{0.222} Z_f^{0.14}}{s_i^{0.889} Z_b^{0.14}}$	$\times$	$\frac{W_i^{2/3}(t_H-t_L)}{\Delta T_M}$	$\times$	$\frac{d_i^{1/3}}{n^{2/3} L}$	(See Note 1)
$\frac{\Delta T_i}{\Delta T_M} = 2.92$	$\times$	$\frac{Z_i^{1/3} M_i^{1/3}}{c_i s_i^2}$	$\times$	$\frac{W_i^{4/3} \lambda}{\Delta T_M}$	$\times$	$\frac{1}{(nL)^{4/3} d_i}$	
$\frac{\Delta T_o}{\Delta T_M} = 12.3$	$\times$	$\frac{Z_o^{0.467} M_o^{0.222}}{s_o^{0.889}}$	$\times$	$\frac{W_o^{0.2}(T_H-T_L)}{\Delta T_M}$	$\times$	$\frac{a}{(nd_o)^{1.2} L}$	(See Note 1)
$\frac{\Delta T_o}{\Delta T_M} = 11.5$			$\times$	$\frac{W_o^{0.2}(T_H-T_L)}{\Delta T_M}$	$\times$	$\frac{a}{(nd_o)^{1.2} L}$	(See Note 1)
$\frac{\Delta T_o}{\Delta T_M} = 15.4$	$\times$	$\frac{M_o^{0.222} Z_f^{0.14}}{s_o^{0.889} Z_b^{0.14}}$	$\times$	$\frac{W_o^{2/3}(T_H-T_L)}{\Delta T_M}$	$\times$	$\frac{a}{(nd_o)^{5/3} L}$	(See Note 1)
$\frac{\Delta T_o}{\Delta T_M} = 2.92$	$\times$	$\frac{Z_o^{1/3} M_o^{1/3}}{c_o s_o^2}$	$\times$	$\frac{W_o^{4/3} \lambda}{\Delta T_M}$	$\times$	$\frac{1}{nL^{4/3} d_o}$	
$\frac{\Delta T_o}{\Delta T_M} = 0.352$	$\times$	$\left[\frac{Z_o^{0.3} M_o^{0.2} \sigma_o^{0.425}}{c_o s_o^{1.075}}\right]\left[\frac{\rho_v^{0.7}}{P_o^{0.85}}\right]$	$\times$	$\left[\frac{W_o^{0.3} \lambda}{\Delta T_M}\right]$	$\times$	$\left[\frac{1}{n^{0.3} d_o^{0.3}}\right] \Sigma'$	(See Notes 2 and 3)
$\frac{\Delta T_w}{\Delta T_M} = 3{,}820$	$\times$	c/k_w	$\times$	$W(t_H-t_H)/\Delta T_M$	$\times$	$(d_o-d_i)/nd_o L$	
$\frac{\Delta T_w}{\Delta T_M} = 2{,}120$	$\times$	$1/k_w$	$\times$	$W\lambda/\Delta T_M$	$\times$	$(d_o-d_i)/nd_o L$	
$\frac{\Delta T_s}{\Delta T_M} = 159$	$\times$	c/k_w	$\times$	$W(t_H-t_L)/\Delta T_M$	$\times$	$(d_o-d_i)/nd_o L$	
$\frac{\Delta T_s}{\Delta T_M} = 88$	$\times$	$1 k_w$	$\times$	$W\lambda/\Delta T_M$	$\times$	$(d_o-d_i)/nd_o L$	

(See Note 1)

(See Note 1)

(See Note 1)

(See Note 1)

Notes:
1. $N_{Rec} = 20{,}000 (D_e/D_H)^{0.32}$
2. $G = W_{\rho L}/(A_{\rho v})$
3. Surface-condition factor (Σ') for copper and steel = 1.0; for stainless steel = 1.7; for polished surfaces = 2.5.

Spiral-Tube Exchanger Design Standards —Table II

No. Tubes	Tube Spacing, In.	Shell-side Flow Area, Sq. In.	Standard Lengths, Ft. Tube-side	Standard Lengths, Ft. Shell-side	Heat-Transfer Area, Sq. Ft.
Tube O.D., ¼ in.					
8	1/8	0.358	4.92	5.7	2.56
12	1/8	0.537	8.11	9.4	6.31
18	3/16	1.08	9.77	11.5	11.5
18	3/16	1.08	15.06	17.5	17.66
30	3/16	1.80	9.77	11.5	19.02
30	3/16	1.80	15.06	17.3	29.5
Tube O.D., ⅜ in.					
8	1/8	0.618	5.51	7.9	4.4
12	1/8	0.93	9.9	11.5	11.6
12	1/8	0.93	14.79	15.5	17.4
20	1/8	1.55	9.9	11.5	19.4
20	1/4	2.48	10.98	13.1	21.5
20	5/16	2.95	12.87	15.5	25.2
Tube O.D., ½ in.					
4	1/8	0.466	5.25	6.5	2.75
6	1/8	0.699	5.25	6.5	4.13
9	1/8	1.04	8.16	9.75	9.63
9	1/8	1.04	10.88	13.0	12.7
15	1/8	1.75	8.16	9.75	16.0
15	3/16	2.23	10.62	12.75	20.9
15	1/4	2.68	12.4	14.9	24.5
15	5/16	3.16	19.25	22.2	37.5
15	5/16	3.16	27.5	30.8	54.0
15	5/16	3.16	33.41	37.2	66.07
15	5/16	3.16	43.2	47.9	84.87
30	1/4	5.37	12.38	14.9	48.9
30	5/16	6.30	19.14	22.2	75.0
30	5/16	6.30	27.5	30.8	108.0
30	5/16	6.30	33.41	37.2	132.15
30	5/16	6.30	43.2	47.9	169.14
Tube O.D., ⅝ in.					
12	1/8	1.94	6.6	8.25	13.0
12	3/16	2.42	8.46	11.6	16.6
12	1/4	2.88	12.2	14.5	24.0
12	5/16	3.35	18.13	21.2	35.5
12	5/16	3.35	23.87	27.25	46.8
12	5/16	3.35	29.36	33.4	57.66
12	5/16	3.35	38.05	42.1	74.95
24	1/4	5.76	12.2	14.5	48.0
24	5/16	6.68	18.13	21.2	70.3
24	5/16	6.68	23.87	27.25	93.6
24	5/16	6.68	29.36	33.4	115.32
24	5/16	6.68	38.05	42.1	149.9
Tube O.D., ¾ in.					
10	3/16	2.62	8.12	9.7	15.9
10	1/4	3.09	9.92	12.3	19.4
10	5/16	3.59	15.2	18.25	29.8
10	5/16	3.59	20.61	24.0	40.3
10	5/16	3.59	25.81	29.5	50.76
10	5/16	3.59	33.75	38.0	66.29
10	5/16	3.59	40.3	45.3	79.0
10	5/16	3.59	47.72	53.4	94.0
10	5/16	3.59	55.58	62.2	109.0
20	1/4	6.16	9.87	12.3	39.2
20	5/16	7.10	15.2	18.25	59.6
20	5/16	7.10	20.61	24.0	80.6
20	5/16	7.10	25.81	29.5	101.52
20	5/16	7.10	33.75	38.0	132.58
20	5/16	7.10	40.3	45.3	158.0
20	5/16	7.10	47.72	53.4	188.0
20	5/16	7.10	55.58	62.2	218.0
30	5/16	10.62	33.75	38.0	198.9
30	5/16	10.62	40.3	45.3	237.0
30	5/16	10.62	47.72	53.4	282.0
30	5/16	10.62	55.58	62.2	327.0

equals or is larger than 0.3. The height of liquid in the tube can be calculated by the correlation discussed by Lord, Minton and Slusser.[5]

Equations for Heat Transfer—Shellside

Eq. (5)—No Phase Change (Liquid), $N_{Re} > N_{Rec}$—is for liquids with Reynolds numbers larger than the critical Reynolds number. A value of 1.1 has been used as a weighted average for the term $(1 + 3.54\, D_e/D_H)$.

Eq. (6)—No Phase Change (Gas), $N_{Re} > N_{Rec}$—is for gases with Reynolds numbers larger than the critical Reynolds number. Again, the physical property factor has been combined with the numerical factor, and a value of 1.1 has been used for the term $(1 + 3.54\, D_e/D_H)$.

Eq. (7)—No Phase Change (Liquid), $N_{Re} < N_{Rec}$—is for liquids in laminar flow at moderate ΔT and having large kinematic viscosity (μ_L/ρ). Again, the effects of natural convection are neglected. For the shellside of a spiral-tube exchanger:[8]

$$(D/L)^{1/3} = (D_e 12^{1/2}/D_H^{1/2} d_s^{1/2})^{1/3}$$

Values for this term range from 0.57 to 0.68. A value of 0.63 has been used for this method.

Eq. (8)—Condensing Vapor, Horizontal, $N_{Re} < 2{,}100$—is for film condensation of vapors with a Reynolds number of less than 2,100. Tube loading is obtained from $\Gamma = W_o/2L$. This equation does not take into account the effect of vapor shearing on the condensate layer.

Eq. (9)—Nucleate Boiling, Horizontal—is for rating kettle-type reboilers. Because the equation applies only to the nucleate-boiling mechanism, which in turn is a function of the temperature difference across the boiling film, criteria that establish maximum heat flux should be set. (Correlations for the maximum superficial vapor velocity and vapor disengaging are discussed by Lord, Minton and Slusser.[5, 6])

Equations for Heat Transfer—Tube Wall

Eq. (10) and (11)—Tube Wall—are for calculating the tube-wall factor. The integrated form of the Fourier equation—$Q/\theta = k_w A \Delta T_w / X$, with X the tube-wall thickness—expressed in the form of a heat-transfer coefficient is: $h_w = 24\ k_w/(d_o - d_i)$. Eq. (10) is used whenever sensible heat is involved on either the tube- or shellside; Eq. (11), when there is latent heat transfer on both sides of the tube wall.

Equations for Heat Transfer—Fouling

Eq. (12) and (13)—Fouling—is for conduction of heat through scale or solid deposits. Eq. (12) is used for sensible heat applications; Eq. (13), for latent-heat transfer.

Equations for Pressure Drop—Tubeside

Eq. (14)—No Phase Change, $N_{Re} > N_{Rec}$—is for fluids with Reynolds numbers larger than the critical Reynolds number. It combines the Fanning equation with the Colburn j factor,[2] the Ito[4] and the Blasius equations.[1] It has been revised to account for pressure losses in the inlet and outlet headers, which are approximately equal to 200 tube dia. for each pass.

Eq. (15)—No Phase Change, $100 < N_{Re} < N_{Rec}$—combines the Fanning equation with the White equation.[9] The friction factor ($f = 16/N_{Re}$) has been modified by the Colburn viscosity ratio.[2] For laminar flow, pressure losses in the inlet and outlet headers can be neglected.

Eq. (16)—No Phase Change, $N_{Re} < 100$—is for a flow regime in which the pressure drop for curved tubes is the same as for straight tubes. The Fanning equation applies with the friction factor $f = 16/N_{Re}$.

Eq. (17)—Condensing—is for calculating the pressure drop for condensing vapors, and is identical to that for no phase change, except for a factor of 0.5. Only approximate results can be expected, which should be used only to preclude designs whose pressure losses would be excessive.

Equations for Pressure Drop—Shellside

Eq. (18)—No Phase Change, $N_{Re} > N_{Rec}$—is for fluids with Reynolds numbers larger than the critical Reynolds number. It also combines the Fanning equation with the Colburn j factor,[2] the Ito[4] and the Blasius equations.[1]

Tube length is used for calculating pressure drop. Shellside flow length is approximately $1.166 \times$ tube length. This has been incorporated into the numerical factor. The equation accounts for pressure losses in the inlet and outlet nozzles.

Eq. (19)—No Phase Change, $100 < N_{Re} < N_{Rec}$—combines the Fanning with the White equation.[9] The friction factor ($f = 16/N_{Re}$) has been modified by Colburn's viscosity ratio.[2] As for Eq. (18), pressure drop is calculated from tube length; again the correction has been incorporated into the numerical factor. For laminar flow, the inlet- and outlet-connection pressure losses can be neglected.

Eq. (20)—No Phase Change, $N_{Re} < 100$—is for a flow regime whose pressure drop is the same for both curved and straight tubes. The Fanning equation applies, with $f = 16/N_{Re}$. As for Eq. (18), pressure drop is calculated from tube length, and the correction has again been incorporated into the numerical factor.

Eq. (21)—Condensing, Spiral Flow—is for calculating pressure drop for condensing vapors in spiral flow. It is identical to that for no phase change, except for the factor of 0.5. Again, only approximate results can be expected, which should only be used to preclude designs of exchangers whose pressures losses would be excessive.

Eq. (22)—Condensing, Axial Flow—is for calculating pressure drop for condensing vapors in axial flow. It is based on expansion and contraction losses as vapors flow across the tube bundle. For axial-flow exchangers, space is left between the tube bundle and the cover at the top and the casing at the bottom. This equation accounts for nozzle losses; however, only approximate results should be expected.

SAMPLE CALCULATIONS

This example applies the rating method to the design of a liquid-condensing spiral-tube exchanger under the following conditions:

Conditions	Tubeside	Shellside
Flowrate, lb./hr.	30,000	3,422
Inlet temperature, °C	20	121
Outlet temperature, °C	80	121
Viscosity, liquid, cp.	0.55	0.23
Viscosity, vapor, cp.	.....	0.013
Specific heat, (Btu./lb.)(°F.)	0.998	1.015
Thermal conductivity, Btu./(ft.²)(hr.)(°F./ft.)	0.368	0.398
Specific gravity	0.999	0.95
Heat of vaporization, Btu./lb.	.....	945
Vapor density, lb./cu. ft.	.....	0.0727
Material of construction	steel ($k = 26$)	
Allowable pressure drop, psi.	15	5

Preliminary Calculations

Heat transferred = $3,422 \times 945 = 3,234,000$ Btu./hr.

Tubeside:
$$M = 0.636 \times 0.998^3 \times 0.999^4 / 0.368^3 = 12.63$$

Shellside:
$$M = 0.636 \times 1.015^3 \times 0.95^4 / 0.398^3 = 8.59$$

ΔT_M (or LMTD) $= (101 - 41)/\ln(101/41) = 66.5$ C.

For a first trial, the approximate surface can be

calculated by using an assumed overall heat-transfer coefficient, U, of 300 Btu./(hr.)(sq.ft.)(°F.):

$$A = 3{,}234{,}000/300 \times 1.8 \times 66.5 = 90 \text{ sq. ft.}$$

From Table II, the following exchanger is chosen: 30 ½-in.-outside-dia. tubes, 27.5 ft. long; $a = 6.3$ sq. in.; $A = 108$ sq. ft.; and $d_s = 0.3125$ in.

The maximum number of tubes for turbulent flow can be approximated by the term $W_i/2\, d_i\, z_i$:

$$n_{max} = 30/2 \times 0.402 \times 0.55 = 68$$

Because flow will be turbulent for the exchanger selected, the proper heat-transfer equations from Table I are: Eq. (1), tubeside; Eq. (8), shellside; Eq. (11), tube wall; and Eq. (13), fouling.

Heat-Transfer Calculation

Tubeside, Eq. (1):

$$\frac{\Delta T_i}{\Delta T_M} = 9.07 \left[\frac{0.55^{0.467} \times 12.63^{0.222}}{0.999^{0.889}} \right] \times$$

$$\left[\frac{30^{0.2} \times 60}{66.5} \right] \left[\frac{0.402^{0.8}}{30^{0.2} \times 27.5} \right]$$

$$= 9.07 \times 1.329 \times 1.781 \times 0.008884 = 0.191$$

Shellside, Eq. (8):

$$\frac{\Delta T_o}{\Delta T_M} = 2.92 \left[\frac{0.23^{1/3} \times 8.59^{1/3}}{0.95^2 \times 1.015} \right] \left[\frac{3.422^{4/3} \times 945}{66.5} \right] \times$$

$$\left[\frac{1}{30 \times 27.5^{4/3} \times 0.5} \right]$$

$$= 2.92 \times 1.370 \times 73.28 \times 0.0008032 = 0.235$$

Tube wall, Eq. (11):

$$\frac{\Delta T_s}{\Delta T_M} = 88 \left[\frac{1}{26} \right] \left[\frac{3.422 \times 945}{66.5} \right] \left[\frac{0.098}{30 \times 27.5 \times 0.5} \right]$$

$$= 3.385 \times 48.63 \times 0.0002376 = 0.039$$

Fouling, Eq. (13):

$$\frac{\Delta T_w}{\Delta T_M} = 2{,}120 \left[\frac{1}{1{,}000} \right] \left[\frac{3.422 \times 945}{66.5} \right] \left[\frac{1}{30 \times 27.5 \times 0.5} \right]$$

$$= 2.120 \times 48.63 \times 0.002424 = 0.250$$

Sum of the Products (SOP):

$$\text{SOP} = 0.191 + 0.235 + 0.250 + 0.039 = 0.715$$

The assumed exchanger is adequate with regard to heat transfer. The next step is to check both sides for pressure drop.

Pressure-Drop Calculation

Tubeside, Eq. (14):

$$\Delta P = 0.00268 \left[\frac{0.55^{0.15}}{0.999} \right] \left[\frac{30^{1.85}}{30^{1.85}} \right] \left[\frac{(27.5/0.402) + 16}{0.402^{3.75}} \right]$$

$$= 0.00268 \times 0.9151 \times 1.0 \times 2{,}574 = 6.3$$

Shellside, Eq. (22):

$$\Delta P = 4 \times 10^{-4} \left[\frac{62.4}{0.00727} \right] \left[\frac{3.422}{27.5 \times 0.3125} \right]^2 \times 30$$

$$= 0.00004 \times 858.3 \times 0.1586 \times 30 = 0.163 \text{ psi.}$$

Both the shellside and tubeside pressure drops are much lower than the allowable. Therefore, the next step is to try the next-smaller exchanger, which has a tube length of 19.14 ft.

Correction of Heat-Transfer Factors

The four preceding heat-transfer factors must now be corrected as follows:
Tubeside: $0.191\,(27.5/19.14)^{0.2} = 0.205$
Shellside: $0.235\,(27.5/19.14)^{4/3} = 0.381$
Fouling: $0.250\,(27.5/19.14) = 0.359$
Tube wall: $0.039\,(27..5/19.4) = 0.056$
The new SOP becomes:
SOP $= 0.205 + 0.381 + 0.359 + 0.056 = 1.001$

This exchanger, which is adequate for heat transfer, will have the following pressure drops:

Tubeside:

$$\Delta P = 6.3 \left[\frac{(19.14/0.402) + 16}{(27.5/0.402) + 16} \right] = 4.8 \text{ psia.}$$

Shellside:

$$\Delta P = 0.163(27.5/19.14)^2 = 0.336 \text{ psi.}$$

The second exchanger selected is adequate with respect to heat transfer and pressure drop for both fluids. An exchanger with the same transfer area but with larger tubes would be more expensive. One with fewer or smaller tubes would have an excessive tubeside pressure drop.

Design summary:

No. of tubeside passes	1
Tube length, ft.	19.14
Casing length, ft.	22.2
Tube, outside dia., in.	0.50
Tube, inside dia., in.	0.402
Heat-transfer area, sq. ft.	75.0
Tubeside pressure drop, psi.	4.8
Shellside pressure drop, psi.	0.336
U, Btu./(hr.)(sq. ft.)(°F.)	359

■

Acknowledgments

The author thanks Graham Mfg. Co. for providing photographs, and for permission to use certain design standards. He is also grateful to the Union Carbide Corp. for permission to publish this article.

References

1. Blasius, H., Das Ähnlichkeitsgesetz bei Riebungsvorgängen in Flussigkeiten, *Forschungsheft*, **131**, 1913.
2. Colburn, A. P., A Method of Correlating Forced Convection Heat Transfer Data and a Comparison With Fluid Friction, *AIChE Trans.*, **29**, 1933, p. 174.
3. "Heliflow Coolers and Heaters," *Bull. 58G*, Graham Mfg. Co., Great Neck, N. Y.
4. Ito, H., Friction Factors for Turbulent Flow in Curved Pipes, *Trans. ASME*, **81**, 2, 1959, p. 123.
5. Lord, R. C., Minton, P. E. and Slusser, R. P., Design Parameters for Condensers and Reboilers, *Chem. Eng.*, Mar. 23, 1970, p. 127.
6. Lord, R. C., Minton, P. E. and Slusser, R. P., Design of Heat Exchangers, *Chem. Eng.*, Jan. 26, 1970, p. 96.
7. Noble, M. A., Kamlani, J. S. and McKetta, J. J., Heat Transfer in Spiral Coils, *Petr. Eng.*, Apr. 1952, p. 723.
8. Perry, J. H., Ed., "Chemical Engineers' Handbook," 4th ed., McGraw-Hill, New York, 1963, 10, p. 24.
9. White, C. M., Streamline Flow Through Curved Pipes, *Proceedings Royal Soc. (London)*, Series A, **123**, 1929, p. 645.

Evaluating Plate Heat-Exchangers

Design improvements in plate units are enhancing their use for heat-exchange and evaporation duties. This article describes the advantages and limitations, costs, and performance characteristics.

J. D. USHER, A.P.V. International Ltd.

Heat transfer between parallel corrugated plates clamped in a frame (each plate sealed by peripheral gaskets and with four corner ports, one pair for each of the two fluid media, as in Fig. 1) has been known for many years. But the advantages and limitations of this operation have only been fully recognized in the past few years and described in a number of papers presented at recent heat-transfer symposia.[1,2]

Briefly, by way of introduction, here is a rundown of the pros and cons:

The reason for the widening popularity of plate heat-exchangers is in their greater compactness and accessibility of heat-transfer surfaces as compared with those of normal shell-and-tube heat exchangers. Plate units can also be easily assembled or reassembled in any size or arrangement of passes; and by providing special connector plates that permit fluid flow at various points along the plate-pack, these exchangers can perform multiple duties such as heating, cooling and recuperation in a single frame.

However, since the entire plate area must withstand full hydraulic pressure, plate units are limited in their maximum operating pressure, and only now can some units withstand pressures up to 300 psi. Also, the dependence on rubber gaskets restricts both the operating temperature and the type of fluids that can be handled. This unsatisfactory condition has been improved somewhat by the development of gaskets that enable operations up to 300 F.; and new plates that have been designed with gaskets of compressed asbestos fiber are likely to increase the range of applications into higher temperatures and into the handling of organic solvents.

Operating conditions can be further extended in heat exchangers of the lamella type; here, the parallel plates clamped together in a frame are sealed by welding around their edges and communicating ports. Although the flexibility of the gasket plate-unit is lost, the elimination of the rubber seal enables operations up to 600 psi. and 1,470 F.

Cost-Cutting Is Possible

Modern heat-exchanger plates are highly complex pressings and, since they require a considerable expenditure in tool costs, they only come in a limited number of standard sizes. Nevertheless, plates are available with developed areas from 44 sq. in. to 15 sq. ft., and the largest fully plated units have a total surface area of 4,000 sq. ft. Port sizes are large enough to enable flows of at least 100,000 gal./hr. to be handled in a single unit. Available plate materials include stainless steel, copper-based alloys, Monel, Incoloy 825 and titanium.

Because of the high cost of some of these, a minimum thickness is necessary; it can be achieved by designing plates with a large number of support points to minimize deflection under the differential pressure that can exist between the two fluid media. This, coupled with high heat-transfer coefficients and the absence of welding, can result in marked cost savings over shell-and-tube units in the more expensive materials.

Determining the Number of Passes

The corrugations pressed into the plate have the effect of strengthening it; and, by inducing artificial turbulence, they increase the heat-transfer coefficients of the two liquid media in association with a greater

Originally published February 23, 1970

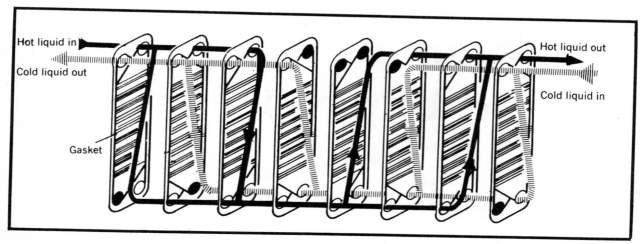

CORRUGATED PLATES can be assembled as shown, or rearranged to optimize heat transfer—Fig. 1

pressure drop. Since a plate is a standard production item, its performance can be conveniently assessed in terms of the temperature ratio (T.R.) that it can achieve. This ratio, defined as the temperature rise per unit of log-mean-temperature-difference (for water flowing in a pair of single passages having equal flowrates), can be expressed as:

$$T.R. = 2\, UA/(wC_p) \qquad (1)$$

where U is the overall heat-transfer coefficient; A, the developed area of each plate; w, the water flow in each passage; and C_p, the specific heat of water.

From Eq. (1), it is clear that the temperature ratio—unlike the overall coefficient, U—decreases with velocity and, for most plates, its value drops to 70 to 50% of its maximum over a practicable range of flowrates. Then the necessary number of passes is easily determined by dividing the overall $T.R.$ by the plate temperature-ratio.

Plate Performance-Characteristics

Since velocity conditions in the plate passage determine the overall heat-transfer coefficient, the temperature ratio of a plate depends both on the configuration into which it is pressed and on the plate length, each factor having its own limitations. If too short, a plate is likely to encounter distribution problems across its width because of its low aspect ratio; if excessively long, a plate will cause handling difficulties.

High overall coefficients will call for high velocities, so that a high temperature-ratio plate will involve a narrow plate-gap that imposes obvious limitations on certain types of duty. It is important to realize that a plate yielding the highest possible temperature ratio is not necessarily desirable in all cases because it entails a loss of flexibility in an arrangement on low T.R. duties; and the type of thermal duty to which a plate is most suited is determined, to a certain extent, by the plate T.R. range, which is from 0.3 to about 2.5 in current plate designs.

Fig. 2 and 3 show typical performance characteristics. Fig. 2 shows the relationship between overall coefficient, temperature ratio, and average velocity for a representative plate on water at unity flow ratio, i.e. the ratio obtained when the liquid flowrates on either side of the plate are equal. The relation between heat transfer and energy (Fig. 3) was obtained by plotting the film coefficient against the horsepower

---

Nomenclature

A	Developed area of each plate, sq. ft.
C_p	Specific heat of water, Btu./(lb.)(°F.)
c	Specific heat, Btu./(lb.)(°F.)
D	Diameter, ft.
f	Friction factor, dimensionless
G	Mass velocity, lb./(hr.)(sq. ft.)
h	Film heat-transfer coefficient, Btu./(hr.)(sq. ft.)(°F.)
k	Thermal conductivity, Btu./(hr.)(ft.)(°F.)
N_{Nu}	Nusselt number, dimensionless, hD/k
N_{Pr}	Prandtl number, dimensionless, $c\mu/k$
N_{Re}	Reynolds number, dimensionless, DG/μ
$T.R.$	Temperature ratio, dimensionless, $2UA/(wC_p)$
U	Overall heat-transfer coefficient, Btu./(hr.)(sq. ft.)(°F.)
w	Water flow in each passage, lb./hr.
μ	Viscosity, lb./(hr.)(ft.)

temperature ratio available.

Although plate heat-exchangers are not subject to the cross-flow factors of shell-and-tube units, other corrections must be made according to the pass arrangement. Briefly:

1. An unequal number of passes is usually adopted when the flowrates and permissible pressure drop for the two liquid media are different. Although the passes are in countercurrent arrangement, cocurrent conditions will exist in certain of the flow passages; hence, the plate T.R. has to be factored by a figure that depends on the flow ratio, number of passes, and end-temperature conditions.

2. Equal passes are adopted for recuperative duties; here, cocurrency exists across the plates at the end of each pass. This requires another correction factor, which causes a decrease in the actual T.R. achieved as the number of passes is increased but as their size is reduced.

For commercial purposes, both types of factors must be determined empirically, although studies of a number of plate arrangements have been made.[3,4]

Effects of Plate Configuration

Plate design has as its object the creation of induced turbulence by imposing continual changes in velocity and direction on the fluid.

This turbulence can be achieved in a number of generally accepted ways. One way is by pressing the plates into a series of parallel, relatively deep troughs that mate with the adjacent plate to form an undulating flow passage between the plates (Fig. 4); in this case, the gap between the plates is preserved by spacing depressions pressed into the troughs. Another method is by pressing a series of parallel, shallower troughs inclined to the vertical axis of the plate at opposite angles for adjacent plates; thus, when the plates are clamped together, the peaks of the troughs of one plate rest on the base of the troughs of the next. This type of plate is mutually self-supporting (Fig. 5).

The precise effect of trough geometry on heat transfer and pressure drop is a fruitful field for research and has already been the subject of recent investigations in Russia, Japan and the U.S.[5,6,7,8,9] It is well known by now that troughs of the type described bring about an appreciable reduction in the critical Reynolds Number to a value that can lie between 23 and 400, depending on the particular plate configuration.

Analysis of a number of plate characteristics has indicated that the relationship between friction factor and Reynolds Number in the turbulent regime can vary quite widely, according to the following expression:

$$f = (1)^x/N_{Re} \qquad (2)$$

where $x = 0.1$ to 0.4. The value of the friction factor is likely to be 10 to 60 times greater than that relating to a tube at the same Reynolds Number.

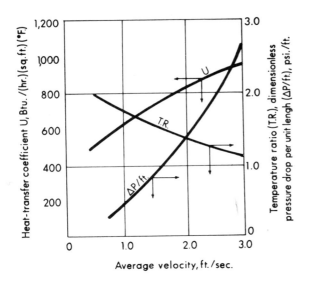

OVERALL COEFFICIENT, as well as temperature ratio and pressure drop, change sharply with fluid velocity—Fig. 2

absorbed per square foot of heat-transfer surface for the same plate and for a tube of equal cross-sectional area over an identical velocity range (the plate data were obtained experimentally, and the tube data from the Dittus-Boelter equation).

As indicated in Fig. 3, the plate shows that higher coefficients can be obtained for both fluid media, whereas a lower performance has to be accepted on the shell side of a tubular unit. Moreover, since a multipass plate heat-exchanger can operate under countercurrent flow, it is capable of very high temperature ratios and can thus attain high efficiency in recuperative duties. This is not possible with a shell-and-tube unit where the cross-flow factor limits the

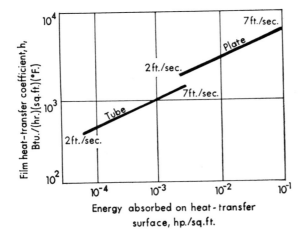

PLATE PERFORMANCE is higher than tube performance within the same fluid-velocity range—Fig. 3

CONTINUOUS VELOCITY CHANGES and turbulence are obtained by plate design and arrangement—Fig. 4

A typical expression for heat transfer under turbulent conditions is the following:[11]

$$N_{Nu} = 0.2536 \, (N_{Re})^{0.65} \, (N_{Pr})^{0.4} \quad (3)$$

Under laminar-flow conditions, the heat-transfer coefficients benefit from the narrow gap between the plates, and the resulting high shear-rates are of particular significance on non-Newtonian liquids with pseudoplastic properties. Liquid distribution along the ports becomes a problem at very high viscosities, particularly on cooling duties, when any stagnation brings about an increase in local viscosity.

General Applications

High coefficients can be achieved in plate heat-exchangers at relatively moderate velocities so that high temperature-ratio duties can be fulfilled in a comparatively small number of passes. Thus, the overall pressure drop is minimized. Where fouling is concerned, a word of warning: a pessimistic value of a reciprocal fouling factor will have a much greater effect on the overall coefficient of a plate exchanger than on that of a heat exchanger having lower coefficients.

In view of the foregoing, a plate unit finds its applications in pasteurization and sterilization that involve multiple duties and where accessibility and short retention time are important; in waste-heat recovery schemes where high recuperative efficiencies can be achieved; and in coping with corrosive conditions.

Plate Units for Evaporators

The very factors that make the plate heat-exchanger so effective for liquid heat-transfer at moderate viscosities make it less suitable for condensing duties. This applies particularly to vapors under vacuum because the narrow plate-gaps and induced turbulence result in appreciable pressure drops on the vapor side, which can cause a serious reduction in the log-mean temperature difference across the plates, while the vapor velocity through the ports themselves provides another limitation.

Early attempts to use the plate heat-exchanger as a flash evaporator (by circulatory heating of the liquid under pressure followed by discharge into a separator) failed because at that time plate heat-exchangers were incapable of handling the high flowrates involved.

The need was then recognized for a plate specially designed for evaporative duties, in which the ports were large enough to handle the vapor, and the plates were flat and relatively widely spaced in order to minimize the pressure drops both on the boiling and condensing sides. This led to the introduction of the plate evaporator,[12] which is a climbing- and falling-film unit of short retention time, in which the plates are assembled in units of four, as shown in Fig. 6.

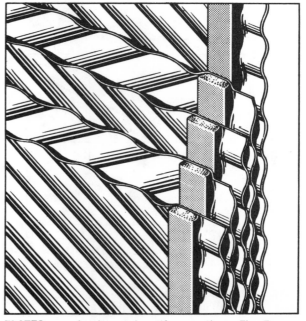

PLATES can also be made self-supporting—Fig. 5

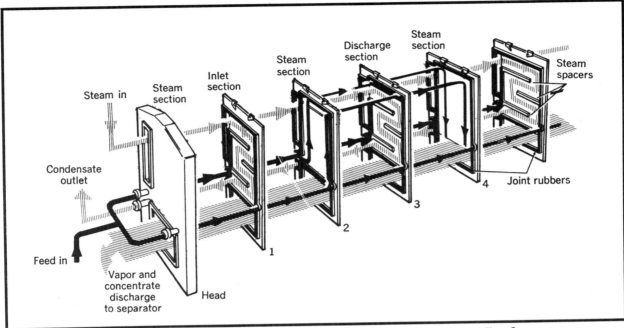

PLATE EVAPORATORS, such as the one above, are easy to assemble, clean and operate—Fig. 6

Feed enters through ports at the bottom two corners of the climbing-film plate; the partially concentrated liquid and vapor pass through a transfer port at the top to a falling-film plate and leave through a large rectangular vapor port at the base. Steam is supplied to the passages between the two boiling plates.

Such plate evaporators are used on a wide range of duties for concentrating products up to 50-poise viscosity, and combine accessibility with a low headroom of about 10 ft. They are suitable, however, only for vacuum operation and pressures slightly above atmospheric; dependence on rubber gaskets excludes the handling of organic solvents. Although the port size restricts the maximum rate of evaporation, multiple-effect units are capable of evaporation rates up to 50,000 lb./hr. Plate evaporators are largely used for milk, fruit juices, glucose solutions, distillery and other liquids.

Another method of using plates for evaporation fulfills a different class of duties than the orthodox plate evaporator. In this alternative system, standard heat-exchanger plates are arranged in a series of passes through which the liquid feed is pumped with heating steam on the other side of the plates. The pressure of the liquid through the plates will fall, whereas the temperature will rise and, as soon as it exceeds the saturated vapor temperature of the liquid, boiling takes place, so that very high vapor velocities will ultimately occur in the final passes and give rise to high shear rates of the product and high heat-transfer coefficients.

Since this system is not limited to vacuum operation and can handle high temperatures and pressures, it is particularly suitable for attaining high concentrations in a finishing stage and has been used for such applications as soap drying and for concentrating tomato and gelatin solutions. ∎

Acknowledgement

The author wishes to thank his collegues at A.P.V. for their assistance in preparing this paper.

References

1. Troupe, R. A., Morgan, J. C., Prifti, J., *Chem. Eng. Prog.*, 56, No. 1.
2. Lane, D. E., *Chem. Proc. Eng.*, Heat Transfer Survey, 1966.
3. Buonopane, R. A., Troupe, R. A., Morgan, J. C., *Chem. Eng. Prog.*, 59, No. 7.
4. Foote, M. R., *N.E.L. Report* No. 303.
5. Maslov, A., *Kholodil'naya Tekhnika*, (6), 1965.
6. Maslov, A., *Dairy Industry* (U.S.S.R.), No. 3, Mar. 1965.
7. Maslov, A., *Food Technology* (U.S.S.R.), No. 5, (42), 1964.
8. Konno, H., Okada, K., Ohtani, S., *Kagaka Kogaku, Chem. Eng. Japan*, 31, No. 9.
9. Smith, V. C., Troupe, R. A., *A.I.Ch.E. J.*, 11, No. 8
10. Emerson, W. H., *N.E.L. Reports*, Nos 284, 285, 286.
11. Hargis, A. M., Beckmann, A. T., Loiacoma, J., "The Plate Heat Exchanger", A.S.M.E. Publication.
12. Usher, J. D., *Chem. Proc. Eng.*, May 1965.

Meet the Author

J. D. Usher is design and development manager at A.P.V. International Ltd. (Manor Royal, Crawley, Sussex, U.K.), where his interests are related to heat exchangers, evaporators and other equipment for the chemical process industries. Educated at Sutton Coldfield and Cambridge University where he studied mechanical engineering, he holds an M.A. degree, and is a chartered engineer and member of the Institution of Mechanical Engineers.

How Heat Pipes Work

Heat pipes are compact, self-contained, self-maintaining and self-controlling devices that have no moving parts and transport heat from hot to cold zones.

DON W. NOREN, Noren Products, Inc.

The heat pipe presents a novel and versatile approach to the solution of heat-transfer problems. Yet, its acceptance in common industrial applications has been slow.

Designing a heat pipe is extremely simple and relies on elementary concepts. A heat pipe is essentially nothing more than a boiler for heat addition, connected to a condenser for heat removal with a vapor, with a condensate pump for returning condensed fluids to the boiler section. The entire mechanism is housed in a simple casing. Fig. 1 represents a cutaway view of a heat pipe. In its simplest form, there are only three parts: (1) a sealed enclosure of metal or glass, (2) a wick that acts as a condensate pump to move fluid by capillary action, and (3) a working fluid that actively transports the heat. A simple heat pipe is symmetrical throughout, and any portion may become the boiler or condenser.

When heat is absorbed in one part of the container, that part becomes the boiler. The working fluid in that area evaporates, absorbing the latent heat of vaporization. As vapor evolves in the input section, it flows, by seeking lower vapor pressure, to the cooler sections of the container where it surrenders its latent heat. Thus, a net transport of heat occurs from the boiler to the condenser. The wick then transports the condensed fluid back to the boiler section.

The overall heat-transfer resistance can be analyzed in three distinct stages: (1) that into the vapor, (2) that down the length of the pipe, and (3) that out of the pipe. The resistance into the pipe consists of wall conduction, wick conduction, and boiling. The resistance down the length is minute, and the only temperature drop results from the necessary pressure generation to force the vapors the length of the tube. The output resistance is the sum of the vapor condensation, conduction through the wick, and conduction through the wall. By capillary action, the wick controls the fluid and improves the boiling and condensing coefficient slightly over smooth-surface performance. But this improvement is offset by the low conduction of the wick, which must necessarily be porous for proper heat-pipe operation.

The net effect of the total resistances may be considered as equivalent to a rod having a superconductance many thousands of times the conductivity of copper and being overlaid with a much-less-conductive material. Such a model makes possible a clearer evaluation of thermal-resistance restrictions affecting possible uses. The outside sheath represents the total input and output resistances associated with a heat pipe. Many factors influence the level of resistance, including temperature of operation, design, working fluid, and orientation. It is difficult to give a general value for all designs, but resistance will normally fall between 0.1 and $0.5°C/(W)(in^2)$ for pipes running around 130°F (54°C)—a resistance equivalent to that of many inches of copper.

Hence, the input and output resistances are the primary reason why the heat pipe is not a pure superconductor, and why it presently does not represent a universal cure-all for heat-transfer problems. In the early high-temperature applications, the liquid metals (used as working fluids) saturated the wick sufficiently to cause a fully conductive smooth-surface interface for heat from the input wall to vaporize the fluid. In the lower-temperature applications of the chemical process industries, the working fluids are nonmetallic, and consequently the crossover resistance between heat-pipe wall and working fluid is greatly increased.

Applications and Limitations

Heat pipes can be used anywhere to transport, level out or control heat flow. However, a certain minimum transport distance is required to make their use worthwhile. Short-distance transport can generally be handled as efficiently at lower cost by solid metal. The minimum distance may actually be quite short—for example, ¾ in for a ⅛-in-dia. heat pipe when cooling electronic components, or 7 in when sinking a high-powered solar collector.

Frequently, the advantages or limitations of heat pipes are not fully comprehended, and some engineers will misapply or force heat pipes into a design that is unsatisfactory. Typically, such a misapplication occurs in a de-

Originally published August 19, 1974

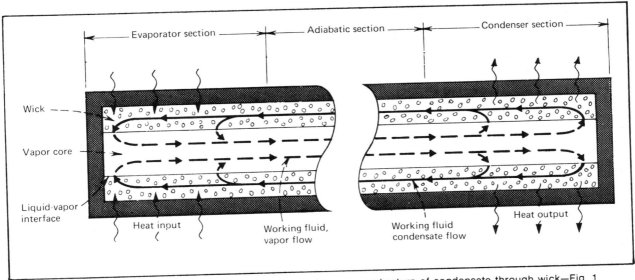

FLOW patterns in a simple heat pipe show transport of heat by vapor and return of condensate through wick—Fig. 1

sign where heat input is at the very tip of a rod-type heat pipe. Despite the extremely efficient transport down its length, the pipe has the same problems as any other sink when collecting or dumping heat. The device works more efficiently if the heat input is over a relatively large fraction of the pipe's surface area, thus reducing the input resistance.

In general, heat pipes have come into their own for handling previously insoluble problems at the prototype stage. But in some fields, notably electronic cooling, the heat pipe has proved an economical high-production device for improving heat-removal efficiency, spreading hot spots, leveling temperatures, and allowing for remote heat-sink operation.

With the advent of solid-state electronic components, overheating due to component density has become common. Surface-to-air heat exchangers called heat sinks have been developed that spread the heat from the semiconductors into either still or forced air. Each time the power of the semiconductors is boosted, the heat sinks have to undergo a corresponding increase in size. Eventually the growing sinks become conduction limited, and water cooling has to be used.

The light weight, compactness and efficiency of heat pipes render them ideal for heat-sinking transistors. Since heat pipes are not conduction limited, the option of adding more fins always improves the heat-transfer efficiency, and thus allows heat pipes to replace expensive and complex water-cooling systems. Where fan or blower noises are objectionable, a forced-air heat-sink system can be replaced with a natural-convection heat-pipe system. Besides its efficiency, the heat pipe allows separation of the heat-source and heat-sink components, and can even provide for electrical isolation between them.

Heat pipes have also been successfully adapted to solve difficult problems in a variety of other fields. Let us briefly review some of them:

Medicine—Heat pipes can operate at constant low temperatures. Hence, they have proven invaluable in cryosurgery, freezing skin tumors, as a constant-temperature probe in brain surgery, and as a nerve block in medical research. Higher-temperature units are also being used in mechanical hearts. Here the heat pipe transfers waste heat from a radioisotope hot-air engine to a blood pump. The engine and pump must be located in different areas of the body, and thus require a means of heat transport between them. Such a heat pipe needs to be flexible to allow implanting and sustained use. The heart application is a nearly ideal one because it takes advantage of all the refinements of heat pipes—namely, light weight, super efficiency, freedom from maintenance, self-con-

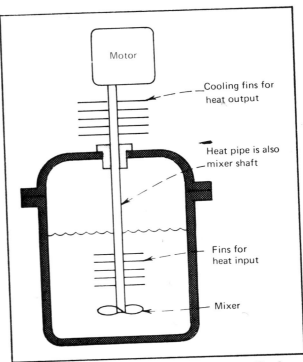

HEAT PIPE keeps reactor temperature constant—Fig. 2

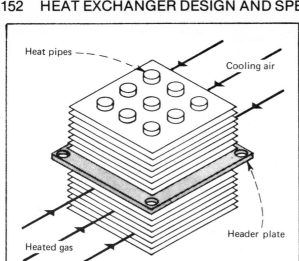

GAS-to-gas exchanger uses heat pipes—Fig. 3

tainment, structural flexibility, and the sterilization made possible by the fabrication process.

Space—Since heat pipes are lightweight and gravity-insensitive, they are well-adapted for use in outer space for controlling temperatures and for transporting heat. Several have been included in a number of missions.

Pilot plant—A problem facing the engineers in cooling a pilot-plant reactor was to remove heat from an exothermic reaction in order to maintain reactant temperature at a constant 80°C. In this case, the heat pipe was used as the mixer shaft, as shown in Fig. 2. The shaft was equipped with additional stirring blades to act as fins for the heat pipe. The output fins spun just below the motor and were shaped to scoop and funnel air. This heat pipe had been designed as a constant-temperature input device, factory-set to 80°C. As the reaction commenced and the temperature rose to that point, heat was then selectively removed to hold the contents at that level. The system is compact, self-contained, self-maintaining and self-controlling.

Exchangers—Air-to-air heat exchangers using heat pipes are currently available in a variety of sizes. These feature compactness, moderate pressure drops, and good performance when operated with a countercurrent flow. At a 1-to-1 flow ratio, such units recover more than 75% of the available heat. These units can be built to handle temperatures ranging from those accompanying liquid nitrogen to a high of over 1,000°F. The small size and flexibility easily justify the premium price when space limitations present a problem. The most frequent configuration is one that allows reactor gas to be cooled countercurrently by another air stream. Fig. 3 shows one arrangement. Here, a header plate is constructed with finned heat pipes projecting through to thermally connect the streams.

Heat sinks—An unusual application for special-function constant-temperature heat pipes is in remote, battery-powered repeater stations in Antarctica. When each station transmits, two large TWT's (travelling wave tubes) inside the station put out 2,000 W of waste heat—enough to overheat and damage the components. When standing by in receiving mode, the electronics emit a steady 100 W, an amount insufficient to keep the components from freezing. Batteries adequate to provide the necessary heat would need to be 20 times as massive, but a constant-temperature heat pipe provides a better solution. Each station is well-insulated to permit the standby output of 100 W to keep the components warm in extreme weather. The heat pipe is set to switch on at 68°F, at which point it carries any excess heat (especially the TWT emissions) through the insulation to a heat sink outside the station. Heat pipes, called Cabinet Coolers, are now available in several standard sizes. A worthwhile advantage of this system is the lack of fans or blowers that frequently suffer mechanical breakdowns.

When explosion-proofing solid-state devices, the protective container must act as the heat sink. Consequently, its size is dictated by the fin area required for cooling. The use of finned heat pipes for removing the heat enables the container to be much smaller, and also reduces the cost. Such a heat pipe fits through a packing gland or is fitted with a bolting flange, either one of which is heavy enough to maintain the explosion-proof integrity of the container.

Future Applications

In light of the energy crisis, new applications under investigation for heat pipes include solar-energy collection and waste-heat recovery. Many companies are delving into these areas, but much of the information is considered proprietary.

One major endeavor, representing the most ambitious heat-pipe project to date, concerns the trans-Alaska pipeline. This application will call for 110,000 heat pipes, up to 20 meters long, whose purpose will be to prevent thawing of the permafrost layer due to heat radiated from the pipeline.

Heat pipes are bound to find their way into more and more applications in the chemical process industries. Although they may appear now and again as primary heat exchangers, it is likely that their innate adaptability will make them useful on a far larger scale as built-in control components for automatic consoles, motors, generators, remote electronic packages, and other self-contained heat exchangers. Such applications will take advantage of the many specialized variations possible with heat pipes to a much greater degree than would simple process-heat exchanging. #

Meet the Author

Don W. Noren is president of Noren Products, Inc., 846 Blandford Blvd., Redwood City, CA 94062. His industrial experience includes five years as assistant plant manager for Pacific Vegetable Oil Corp., one year at Oliver Mfg. Co. and five years with Litton Industries Tube Div. In 1968, he founded Noren Products to manufacture standard and special heat pipes. He holds patents covering areas of seed processing, materials processing, tube components, heat-transfer systems, heat pipes, and bonding methods. He has a B.S. in chemical engineering from the University of California.

Uses of Inflated-Plate Heat Exchangers

Inflated, panel-type heat exchangers are prime candidates for operations involving single or two-phase flow. They are particularly adaptable to batch processes but can also be used in many continuous operations.

DAVID C. HARVEY and GERALD E. GLASS, Paul Mueller Co.

Normally, the most economical type of heat exchanger is the one that can do the job with the least amount of heat-transfer surface. Because of this, panel-type heat exchangers have a definite place in the chemical process industries. Applications include: heating or cooling the contents of a tank, effluent water, or conveyor beds; evaporation of cryogenic fluids; drying shelves for pharmaceuticals and textiles; freeze-drying shelves; falling-film cooling or heating; plating; etc.

Special kinds of panel exchangers are the inflated (single or double-embossed), and the preformed types. Here, we will concentrate on the inflated type, because it offers many more advantages than the preformed kind.

Panel-type heat exchangers are formed by three techniques. One of them is by mechanical pressure exerted by dies on a corrugated cross-sectional surface. One corrugated surface may be welded to another one, yielding passageways that are somewhat tubular in shape, or it may be joined to a flat surface, which results in semi-circular cross-sections. The edges of the plates are joined by electrical-resistance seam welds; resistance welds are also placed in the valleys of the corrugated surface.

The second, dimpled-surface technique, on the other hand, consists of swaging a portion of the material and concurrently punching a hole at the bottom of the formed area. The surface then stands away from the other sheet of metal (or vessel wall) a distance equal to the depth of the swaged portion. This provides a large cross-sectional area through which the heat-transfer medium may flow. A fillet weld is applied to the perimeter of the hole to secure the dimpled plate to the other one.

Both types of such preformed plate-type heat exchangers are widely used in industry. The term "preformed" applies to plates that are mechanically formed prior to welding. The obvious departure from this is to form the plate after welding. This is the third technique used in making heat-transfer plates, and the one that is applied in forming the improved inflated plates.*

*Known as Temp-Plate surfaces, these heat exchangers are manufactured by Paul Mueller Co., Springfield, Mo.

Originally published November 11, 1974

These are made by intermittently spot welding two sheets in a flat position, while the perimeter is continuously welded. The welded sheets are then held by a "stretcher-leveler" machine—which holds the plate ends in tension—while hydraulic pressure is applied to the space between the sheets. After inflation, the plate is sheared or drilled to weld the desired connections.

Inflated plates are either single or double embossed. The single kind has two different thicknesses of metal, so that the thinner one yields during inflation. The double-embossed type consists of two sheets of the same thickness, which results in the same amount of "pillowing" on each side. Materials of construction include carbon and stainless steels, Monel, Inconel, Hastelloy, Carpenter 20, and other chrome-nickel alloys. Thickness may vary from 24 to 7 gage, and overall sizes up to 6 ft × 16 ft without splicing. Customer needs dictate the shape (rectangular, triangular, curved).

How To Choose a Heat Exchanger

The merits of different types of heat exchangers are based on parameters such as: space or weight savings, economy, types of fluids, flowrates, allowable pressure drop, type of process, cleanability, ease with which system can be expanded, maintainability, etc.

Processes—either continuous or batch-type—may involve fluids in one phase throughout, or one of these (at least) may change phase. Batch-type processes virtually eliminate all heat exchangers except the panel type, pipe coils, or heat-transfer surfaces that are integral parts of vessel walls. In general, the heat-transfer coefficient, U, depends on the fluid in the vessel, because heat flow across the panels or coils is limited by the degree of agitation or of natural convection.

Values of U can range from 3-4 Btu/(h)(ft^2)(°F) for cooling viscous products, to 500 for heating low-viscosity liquids with steam. Preformed or inflated sections that are part of the vessel walls are easy to clean and save space, but both sides cannot be used for heat transfer.

154 HEAT EXCHANGER DESIGN AND SPECIFICATION

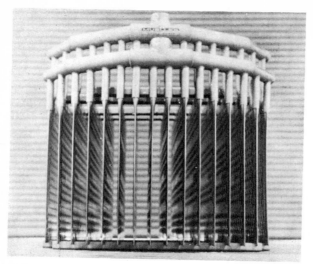

INFLATED-PLATE BANK made to fit inside a tank—Fig. 1

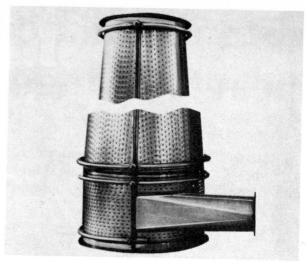

FLUE GASES in jacketed stack heat makeup water—Fig. 2

Almost any type of heat exchanger can be adapted to operate continuously. The shell-and-tube exchanger has good versatility, due to the ease with which fluid velocities can be changed by varying tube size and number, or varying the shell diameter. Liquid flow through panel-type exchangers can be controlled by circuiting passageways for series or parallel flow. This is done in preformed panels by headering tubes together or placing them in series. In the inflated plate, it is done by wide-flow passes, or by additional seam welds in the plate to provide a serpentine pattern.

The problem with the panel-type of heat exchanger is optimizing the flow of the fluid outside the panel. Unless a very high flowrate is encountered, it becomes difficult to place the panels close enough together to obtain high film coefficients on the outer side. Even with reasonable spacing between the panels, a means must be provided to contain the flow of fluid in the other direction. This can be done by clamping the panels between two plates, or placing the panels in a covered sump. Since the flat sides of the outside panels in such a fixture cannot withstand high pressures, velocities may be governed by this factor, if a high pressure-drop is allowable.

Advantages of Inflated Plates

The inflated double-embossed or single-embossed plate-type exchanger offers many advantages over the preformed plate-type coils:

1. The quilted pattern provides greater efficiencies because of more steam turbulence and less condensate buildup. Gas pockets are allowed to escape, thus minimizing back pressures and starving of passages.

2. All resistance seam welds are pressure tested during inflation up to as high as 5,000 psig. The preformed plate-type coils are pressure tested after welding at 1½ times working pressure.

3. The material thinning during inflation is so minute that it is difficult to measure. Thinning of the center of passes in preformed plate-type coils during the pressing operation can range to a high of 15-18%.

4. The inflated design reduces condensate buildup that is common with typical header and multiple-header, die-formed, embossed-plate heat exchangers. This significantly reduces fatigue failure associated with condensate buildup and water-hammer conditions. These problems are common in conventional preformed heat-transfer surfaces. The automotive and appliance industries in their spray phosphate and other metal-cleaning lines have recognized this fact and have replaced preformed plate-type coils with the inflated plates.

When To Use Inflated Plates

Depending upon the application and size of units required, the relative cost of inflated plates will range from par to 25% less than the preformed units. The more heat-transfer surface needed, the greater the percentage of savings on the initial cost. As already stated, the most economical heat exchanger is normally the unit that can do its task with the least amount of heat-transfer surface. This is especially true with stainless and alloy metals.

The heat-transfer coefficient, U, that can be achieved will normally determine the most economical type of heat exchanger. Values of U in excess of 500—when exchanging heat between two contained water streams—to as low as 3 for water-to-air have been experienced. Based on such values, it can be determined which type of heat exchanger would be most economical. For air-to-air, a fin-tube-type unit is recommended.

Pressure drops must also be taken into consideration in selecting embossed-plate exchangers. If large flowrates are needed—i.e., 300-500 gal/min with minimal pressure drops—the shell-and-tube exchanger would definitely be more desirable.

In actual tests with the double-embossed inflated plate (4 ft × 16 ft, with a slotted header pipe at each end at 100 gal/min), an 11-12 psig pressure-drop has been experienced. By installing, say, five plates in a bank, the 500 gal/min flow will distribute itself equally through the five plates at a rate of 100 gal/min or less, depending on the number of plates. Thus, pressure drop can be min-

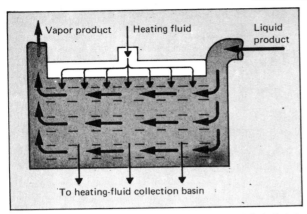

CASCADING FLUID over both outside faces of inflated plate vaporizes cryogenic liquid flowing inside—Fig. 3

imized also with the double-embossed, plate-type unit.

Inflated plates are furnished in various sizes and shapes.* The immersed kind is perhaps the most economical. By installing these units on the interior of a tank or vessel, you benefit from both sides of the plate (Fig. 1). Also, greater U values are normally achieved.

For rectangular batch-type tanks, for example, the plate is often suspended inside by hangers, usually about 1½ in. off the sidewall. Heating or cooling lines are then connected to the plate. If a union is used, the lines can be readily removed for cleaning. The quilted pattern is easier to clean than the preformed or pipe-coil type.

If immersed plate sections are installed perpendicular to the vessel's sidewall, or at 15-20 deg off the sidewall axis, they act both as heat-transfer surfaces and as baffles to ensure good agitation. By keeping the width of the plate less than the diameter of the manway in a tank, these units can be installed in the tank and connected to common inlet and outlet header pipes. This is done in the chemical, pharmaceutical and food industries.

Another application is the use of curved or dished sections of inflated plates as clamp-on units. Two half-sections rolled to the outside diameter of the vessel—or dished to fit the bottom head—can be clamped over the outside of the vessel and bolted together with spring-loaded bolts. It is recommended that heat-transfer mastic be applied about 1/16 to 1/8 in. thick between the interior of the clamp-on section and the exterior wall of the tank. This increases the heat-transfer coefficient.

Increasing costs of energy have caused the development of regenerative-type operations, which were formerly cost prohibitive. A common application in the textile industry is the heating of clean water with dirty effluent. The ease of cleaning the exterior surface, as well as the relatively low cost, makes the panel-type exchanger attractive for such a process. It is not uncommon to recover initial equipment costs in just a matter of months from such energy savings. Also, by using banks of inflated plates within exhaust stacks (Fig. 2), the heat from the flue gas can be recovered to heat the fluid circulating through the plates.

Panel Types for Cryogenic Systems

Panel-type heat exchangers are particularly suitable for cascade systems, where a liquid is allowed to flow by gravity over a vertical panel. With relatively low pumping costs, film coefficients in the order of 1,800 $Btu/(h)(ft^2)(°F)$ have been achieved.

A significant process involves the vaporization of liquefied natural gas (LNG), using water as the heating medium (Fig. 3). The panel's advantage here is that the possibility of destruction of the heat exchanger by ice formation does not exist, as it happens with systems that contain water in tubes, passages or shells.

The process that the LNG undergoes is much like that of a primary refrigerant in a cooling system. This two-phase flow presents problems due to the tremendous change in volume of the LNG as it is vaporized. Tests on preformed panels indicated a lack of stability, but it was later found that this problem was virtually nonexistent with inflated plates. The reason for this is that the multidirectional flow possibilities provide a means for the vapor to escape without the gas being carried completely through the plate with the remaining liquid. Because of the good overall heat-transfer coefficient that can be obtained—and the large temperature difference that exists— a heat flux as high as 75,000 $Btu/(h)(ft^2)$ is possible.

Inflated plates are also applicable to other cryogenic fluids such as nitrogen, ethylene, oxygen, ammonia, chlorine, etc., as well as for other types of gases.

In actual practice, the nature of a process may be particularly well suited for a shell-and-tube exchanger, if the surface-area requirements are significantly lower than for a panel-type system. However, this does not rule out the panel heat exchangers, because the total initial cost may still be comparative. In addition, maintenance and ease of cleaning and replacement may make the panel exchanger more attractive. #

*This refers specifically to the Mueller Temp-Plate kind.

Meet the Authors

▶ **David C. Harvey** is industrial processing equipment specialist with the Paul Mueller Co., P.O. Box 828, Springfield, MO 65801. Before, he worked for Panhandle Mfg. Inc., Ryan Industries, Tranter Mfg. Co., and some other concerns. He has experience in supervision, manufacturing engineering, structural and heavy-plate fabrication, machine design, welding, plant layout, production management, purchasing and sales promotion. He holds a B.S. degree in mechanical engineering from the University of Rochester, and is a member of the Industrial Management Club, American Welding Soc., Junior Achievement Companies and other societies.

Gerald Edward Glass is chief engineer with the Paul Mueller Co., P.O. Box 828, ▶ Springfield, MO 65801. Previously, he was graduate teaching assistant in the Mechanical Engineering Dept. of the University of Arkansas, from which he received an M.S.M.E. degree in 1968. He is currently pursuing a Ph.D. degree in mechanical engineering from the same university. His doctoral research is in transpiration cooling. He is associate member of the American Soc. of Mechanical Engineers.

Where and How To Use Plate Heat Exchangers

The special characteristics of plate heat exchangers make them ideally suited for specific applications in the chemical process industries. Here are performance, cost and fouling data, to help you in evaluating this equipment.

J. MARRIOTT, Alfa-Laval AB

The continuous search for greater economy and efficiency has led to the development of many different types of heat exchangers, other than the popular shell and tube. Some of these have been highly successful in particular fields of application.

One of the most successful is the gasketed plate heat exchanger (PHE), originally introduced in the 1930's to meet the hygienic demands of the dairy industry. Unfortunately, the plate unit is still regarded by some engineers as suitable only for the foodstuff industries. Actually, the PHE possesses many unique characteristics of significance to the chemical and allied industries.

Plate heat exchangers are fully described in previous articles;[1-19] some[1-14] are highly technical in nature. Here, I will attempt to summarize the most important practical features of PHE's. Some information which, to the best of my knowledge has not been published before, is also included.

Briefly, a plate heat exchanger consists of a number of corrugated metal sheets (Fig. 1) provided with gaskets and corner portals (to achieve the desired flow arrangement, each fluid passes through alternate channels). The fluids are at all times separated by two gaskets, each open to the atmosphere. Gasket failure cannot result in fluid intermixing but merely in leakage to atmosphere. Proper selection of gasket material and operating conditions will eliminate the leakage risk, but for applications where the remote possibility of such a leakage cannot be ignored, a protective cover can be fitted.

The plates are clamped together in a frame that includes connections for the fluids. All wetted parts are accessible for inspection by removing the clamping bolts and rolling back the movable cover. By including intermediate separating/connecting plates, three or more separate fluid streams can be accommodated. Frames are usually free standing (Fig. 2); or, for smaller units, they are attached to structural steel work.

Originally published April 5, 1971

Characteristics of Exchanger

Plates are spaced close together, with nominal gaps ranging from 2 to 5 mm. (0.08 to 0.2 in.), giving hydraulic mean diameters in the range 4 to 10 mm. (0.15 to 0.4 in.). The plates are corrugated so that very high degrees of turbulence are achieved—critical Reynolds numbers are in the range of 10 to 400, depending on geometry. These factors

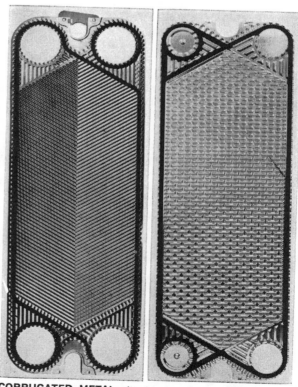

CORRUGATED METAL sheets are clamped together in plate heat exchangers to provide flow channels—Fig. 1

WHERE AND HOW TO USE PLATE HEAT EXCHANGERS

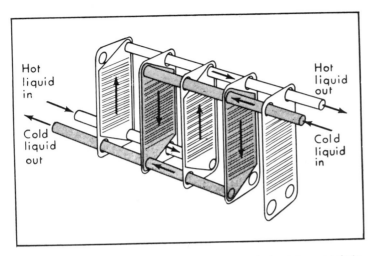

PLATE HEAT EXCHANGER includes corrugated plates, gaskets, frame, and corner portals to achieve desired flow—Fig. 2

combine to produce exceptionally high film coefficients of heat transfer.

In the general relationship for turbulent flow:

$$N_{Nu} = (\text{const.}) N_{Re}^n N_{Pr}^m (\mu/\mu_w)^x$$

Typical reported values[1 to 12] are:

const. = 0.15 to 0.40
n = 0.65 to 0.85
m = 0.30 to 0.45 (usually 0.333)
x = 0.05 to 0.20

One of the most widely used plates has the following relationship:

$$N_{Nu} = (0.374) N_{Re}^{0.668} N_{Pr}^{0.333} (\mu/\mu_w)^{0.15}$$

Attempts have been made to derive relationships which include plate geometry. Troupe, Morgan and Prifti[1] report the following:

$$N_{Nu} = (0.383 - 0.0505^{l/s}) N_{Re}^{0.65} N_{Pr}^{0.4}$$

where l is the straight length of channel before a directional change, and s is the spacing normal to local flow direction. For plates with single lateral corrugations, l/s will be in the range 1.5 to 10;
but for many types, such as the cross-corrugated herringbone pattern (Fig. 1, left) and the double corrugated, l/s is impossible to determine. Buonopane and Troupe[11] present generalized relationships for a number of geometries. For laminar flow, it seems that the following Sieder-Tate type relationship applies:

$$N_{Nu} = \text{const.} (N_{Re} N_{Pr} d_h/L)^{0.333} (\mu/\mu_w)^{0.14}$$

where const. = 1.86 to 4.50 depending on geometry.

Nominal velocities for "waterlike" liquids in turbulent flow are usually in the range 0.3 to 1.0 m./sec. (1 to 3 ft./sec.), but true velocities may be higher by a factor of up to about four due to the effect of the corrugations. All heat-transfer and pressure-drop relationships are, however, based on nominal velocity or on flow per channel. Erosion is seldom a governing factor, since pressure-drop limitations will generally determine the maximum feasible fluid velocities. Plate heat exchangers of relatively soft materials (e.g. aluminum brass) handling sand or grit suspensions may be subject to impingement attack, but experience will show the maximum flow per channel that can safely be handled for a given application.

If the characteristics of various plates are examined, it is found that the majority have similar film coefficients at a given pressure drop and for a given set of fluid property data. In the following, water at a mean temperature of 40C. (104F.) will be used as a basis.

In assessing the performance of any heat exchanger, the term "specific pressure drop" (J) can be used.[3] This is defined as the pressure drop per NTU, i.e.:

$$J = \Delta P/\theta$$

The term θ represents the number of transfer units and is itself defined as:

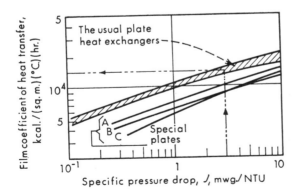

MOST PLATES have similar pressure drops—Fig. 3

$$\theta = (\partial t)_{max}/\Delta t_m = kA/(V\rho c_p)_{min}$$

While the values of θ achieved per pass for various plate types varies enormously (0.15 to about 4), Fig. 3 shows that most plates (shaded area) have a very similar J vs. h (film coefficient) relationship.

Jensen[3] reports that optimum values for J for all commercially available plates are close to 4.5 psi./NTU, so that for most plates the film coefficient for water at 40C. (104F.) under optimum conditions for pumping costs, depreciation, etc., will be about 13,500 kcal./(sq.m.)(°C.)(hr.) or 2,800 Btu./(sq.ft.)(°F.)(hr.).

Three lines in Fig. 3 are shown having lower h values at a given J, labelled A, B and C. Curve A refers to a special plate having greater than normal spacing, designed for duties having very low values of θ or for applications involving fluids of high solids content.

Curve B is for a plate designed for a similar purpose but having relatively few corrugations per unit area (and therefore fewer contact points).

Curve C represents a plate that for all intents and purposes is flat—plate separation is achieved with impressed dimples that have little effect on the flow pattern.

Even for these three types of "lower" efficiency, h values of 8,000 to 10,000 kcal./(sq.m.)(°C.)(hr.) or 1,600 to 2,000 Btu./(sq.ft.)(°F.)(hr.) at optimum J can be expected.

Since a PHE normally has identical channel geometries for the two fluids, similar film coefficients will be achieved for similar pressure drops per pass when the two fluids have similar physical properties. Under optimum conditions ($J = 3$ mwg./NTU = 4.5 psi./NTU), a water/water PHE can therefore be expected to achieve a clean overall heat-transfer coefficient of the order of $U = 5,000$ kcal./(sq.m.)(°C.)(hr.) or 1,000 Btu./(sq.ft.)(°F.)(hr.) when using stainless steel (or other material of similar conductivity) plates of thickness 0.5 to 1.0 mm. (0.02 to 0.04 in.).

Pressure Drop in Plate Exchangers

Any reasonable pressure-drop requirement can normally be closely approached in a plate heat exchanger, due to the extremely flexible nature of the design parameters—basic type, size, number, and arrangement of plates.

Channel geometries for the two fluids are normally identical, so that unlike a shell- and -tube heat exchanger, a PHE will produce equal pressure drops

Nomenclature

A	Heat transfer area, sq.m. or sq.ft.
c_p	Specific heat, kcal./(kg.)(°C.) or Btu./(lb.)(°F.).
d_h	Hydraulic mean diameter, m. or ft.
f_k	LMTD correction factor.
F_f	Friction factor (Fanning).
g	Acceleration due to gravity.
h	Film coefficient of heat transfer, kcal./(sq.m.)(°C.)(hr.) or Btu./(sq.ft.)(°F.)(hr.).
J	Specific pressure drop, mwg./NTU or psi./NTU.
j_H	Colburn heat transfer factor.
k	Thermal conductivity of plate.
L	Nominal plate length, m. or ft.
mwg.	Meters water gage.
ΔP	Pressure drop, mwg. or psig.
R_f	Fouling factor, (sq.m.)(°C.)(hr.)/kcal. or (sq.ft.)(°F.)(hr.)/Btu.
Δt_m	Mean temperature difference, $(f_k) \times$ (LMTD), °C. or °F.
U	Overall heat transfer coefficient excluding fouling, kcal./(sq.m.)(°C.)(hr.) or Btu./(sq.ft.)(°F.)(hr.).
U_k	Overall heat transfer coefficient including fouling.
u	Fluid velocity, ft./sec.
V	Total flowrate, cu.m./hr. or gal./min.
v	Flow per channel.
θ	Number of transfer units (NTU).
ρ	Density of fluid, kg./cu.m. or lb./cu.ft.
μ	Viscosity at bulk temp., cp.
μ_w	Viscosity at wall temp., cp.
Dimensionless groups	
N_{Re}	Reynolds number
N_{Pr}	Prandtl number
N_{Nu}	Nusselt number

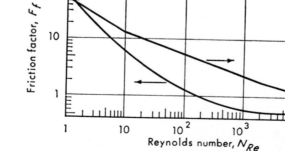

FRICTION FACTORS for plate heat exchangers are considerably higher than for similar flow in tubes. Curves are for one type of plate—Fig. 4

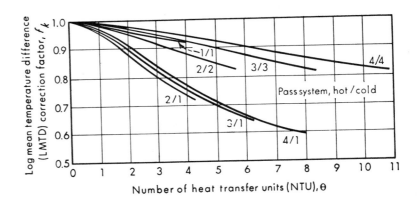

CORRECTION FACTORS must be applied to the log mean temperature difference (LMTD), particularly for multipass systems—Fig. 5

for equal flows of similar fluids. For certain well-defined applications, however, it may be possible to employ special channels for one fluid—the use of turbulence promoters for viscous low-conductivity fluids is one example.

Friction factors, calculated from nominal velocity, are reported by Usher[19] to be some 10 to 60 times higher for turbulent flow in a PHE channel than for flow inside a tube at the same Reynolds number. His work, however, covered a comparatively small range of plate types, and it seems that some types can have friction factors up to 400 times those inside a tube at the same N_{Re} number.

Fig. 4 shows friction factor F_f vs. N_{Re} for one plate type, and it can be seen that in fully developed turbulent flow, the friction factor is about 100 times that for a normal tube at the same N_{Re} value.

Nominal velocities are low, and nominal plate lengths do not exceed about 6 ft., so that the term $(u^2/2g)L$ in the general pressure drop equation is very much smaller than would be the case in a tubular unit. Furthermore, relatively few passes are required—single-pass operation will often achieve the required NTU value (up to 4 at low J values)—so that pressure drop is effectively utilized for heat transfer and losses due to "unfruitful" flow direction changes are minimized.

Correcting the Log Mean Temperature Difference

Some previously published work[19] appears to suggest that the PHE can only operate under equal flow conditions. This is not so—while the plate unit is at its most efficient when having equal pass systems (Fig. 5) for the two fluids, and when the flow ratio $V_{min.}/V_{max.}$* is not less than 0.7, upsymmetrical pass systems enable widely different flow ratios to be handled. This is, however, at the expense of a substantial log mean temperature difference (LMTD) correction factor to allow for the partially cocurrent conditions that then exist.

Buonopane, Troupe and Morgan[8] report that for both fluids in single-pass flow (countercurrent), a correction factor averaging 0.95 should be applied to the LMTD. This is indicated on Fig. 5, where it can be seen that at about 2.5 NTU, the factor for a 1/1 arrangement is indeed 0.95.

The correction factor that must be applied for multipass systems having unequal numbers of passes for the two fluids is substantial, though not as severe as would be required in tubular designs due to absence of cross-flow and baffle leakage. Fig. 5 gives approximate values for f_k for various pass systems at NTU's up to 11.

The mathematical derivation of these factors is extremely complex, and even for the simpler cases requires extensive computer facilities for the solution of the differential equations involved. Buonopane, et al[8] and Foote[13] present mathematical models for various plate arrangements.

Most theoretical derivations are, however, too limited in scope and complex in nature for practical use, so that a more empirical approach is normally adopted. This method (upon which Fig. 5 is based) relies on a number of assumptions and simplifications, such as that the film coefficient does not change appreciably for either fluid in its passage through the unit and that the channel flow ratio $V_{min.}/V_{max.}$ is between 0.7 and 1. Spot checks using the more accurate computer-based method indicate a surprisingly good agreement.

Consider Pumping Costs

Usher[19] shows that for a given energy loss (horsepower per unit area), the PHE produces higher film coefficients than does a tubular unit (considering only flow inside the tube). The diagram he gives is valid for one plate type, but this type appears to be representative of most commercially available types. Fig. 6 indicates that all plate types follow this general rule—even those plates labeled A, B and C on Fig. 3.

When assessing various heat exchanger types, the question of pumping costs should be considered, since

---
* Ratio of flows on either side of a representative plate.

HEAT EXCHANGER DESIGN AND SPECIFICATION

these will probably represent by far the greatest part of operating costs. Plate heat exchangers are clearly advantageous in this respect.

How To Handle Fouling

The overall heat-transfer coefficients so far given have excluded fouling, the almost inevitable presence of which will reduce overall heat-transfer coefficient values. At optimum J, the PHE will achieve a U_k value for water/water of about 3,000 to 4,000 kcal./(sq.m.)(°C.)(hr.) or 600 to 800 Btu./sq. ft.)(°F.)(hr.) when a reasonable fouling factor is included.

Fouling factors required in plate heat exchangers are small compared with those commonly used in shell and tube designs for six reasons:

1. High degrees of turbulence maintain solids in suspension.
2. Heat transfer surfaces are smooth. For some types, a mirror finish may be available.
3. No "dead spaces" where fluids can stagnate, as for example near the shell-side baffles in a tubular unit.
4. Since the plate is necessarily of a material not subject to massive corrosion (being relatively thin), deposits of corrosion products to which fouling can adhere are absent.
5. High film coefficients tend to lead to lower surface temperatures for the cold fluid (the cold fluid is usually the culprit as far as fouling is concerned).
6. Extreme simplicity of cleaning. The small holdup volume and very high turbulence in a PHE (plus the absence of dead spaces) mean that chemical cleaning methods are rapid and effective. If mechanical cleaning is required, all wetted parts within a PHE are extremely readily accessible.

Recommended fouling factors for plate heat exchangers (at least those of the corrugated plate) are as follows, assuming operation at economic pressure drop ($J = 3$ mwg./NTU = 4.5 psi./NTU):

Fluid	Fouling factor
	(sq.m.)(°C.)(h.)/kcal. x 4.88 = (sq.ft.)(°F.)(hr.)/Btu.
Water	
Demineralized or distilled	0.00001
Towns (soft)	0.00002
Towns (hard) heating	0.00005
Cooling tower (treated)	0.00004
Sea (coastal) or estuary	0.00005
Sea (ocean)	0.00003
River, canal, borehole, etc.	0.00005
Engine jacket	0.00006
Oils, lubricating	0.00002 to 0.00005
Oils, vegetable	0.00002 to 0.00006
Solvents, organic	0.00001 to 0.00003
Steam	0.00001
Process fluids, general	0.00001 to 0.00006

In no circumstances, even under low pressure-drop conditions, is a total fouling factor exceeding 0.00012 (sq.m.)(°C.)(hr.)/kcal. or 0.0006 (sq. ft.)(°F.)(hr.)/Btu. recommended, for the same reasons that current work suggests that tubular fouling factors should not be so high as are commonly specified.

Under high pressure-drop conditions, these fouling

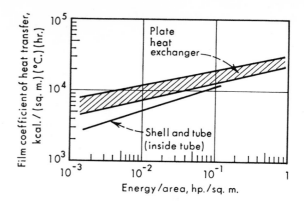

PUMPING COSTS: Plate heat exchangers have an advantage over flow in tubes—Fig. 6

factors can be reduced somewhat; for instance, coastal sea water at 10 mwg./NTU requires $R_f = 0.00003$.

Comparing With Tubular Heat Exchanger

The following typical water/water duty has been calculated for both a PHE and a tubular heat exchanger:

	Hot Side	Cold Side
Flow	50 cu.m./hr. = 210 gal./min.	50 cu.m./hr. = 210 gal./min.
Temperature in/out	80/40 C. =176/104 F.	20/60 C. =68/140 F.

The following are the results of the calculation for both types:

	Plate	Tubular
Heat-transfer area	25 sq.m. =269 sq. ft.	85 sq.m. =915 sq.ft.
Overall heat-transfer coefficient (clean)	5,200 kcal./(sq.m.)(°C.)(hr.) 1,067 Btu./(sq.ft.)(°F.)(hr.)	1,750 kcal./(sq.ft.)(°C.)(hr.) 359 Btu./(sq.ft.)(°F.)(hr.)
Fouling factor	0.00006 (sq.m.)(°C.)(hr.)/kcal. 0.0003 (sq.ft.)(°F.)(hr.)/Btu.	0.0001 (sq.m.)(°C.)(hr.)/kcal. 0.0005 (sq.ft.)(°F.)(hr.)/Btu.
Overall heat-transfer coefficient (service)	3,960 kcal./(sq.m.)(°C.)(hr.) 812 Btu./(sq.ft.)(°F.)(hr.)	1,500 kcal./(sq.m.)(°C.)(hr.) 307 Btu./(sq.ft.)(°F.)(hr.)
Pass system (hot/cold)	1/1	8/Baffled
Channels/tubes per pass	40	56
Plate/tube size	0.32 sq.m. =3.44 sq.ft.	17/20 mm. x 3 m.
Calculated pressure drop	4 mwg. =5.7 psi.	6 mwg. =8.5 psi.
J	2 mwg./NTU = 3 psi./NTU	3 mwg./NTU = 4.5 psi./NTU
Pumping power	1.1 hp.	1.65 hp.
Weight (empty)	615 kg. =1,350 lb.	2,400 kg. =5,280 lb.
Weight (full)	720 kg. =1,580 lb.	3,100 kg. =6,820 lb.
Overall size (including cleaning space)	1.5 x 0.7 x 1.4 m. = 4.9 x 2.3 x 4.6 ft.	7.0 x 0.7 x 0.7 m. = 23 x 2.3 x 2.3 ft.
Floor area required	1 sq.m. =11 sq.ft.	5 sq.m. =54 sq.ft.

Particularly worth noting are the relative figures

WHERE AND HOW TO USE PLATE HEAT EXCHANGERS

for weight, holdup (PHE: 105 kg. = 230 lb.; THE: 700 kg. = 1,540 lb.), floor area required, and pumping costs. Even though the tubular unit has been designed with a very low fouling factor, the PHE has obvious advantages on all these points.

The relative costs of the two units described are difficult to establish with any high degree of accuracy. Assuming, however, that for a design pressure of 142 psig., stainless steel (AISI 304) is required for all wetted parts, the PHE is likely to cost between $2,700 and $4,000 in 1970. The tubular unit would cost, according to various published pricing methods, about $5,500 to $6,000 (fixed tubesheet). In mild steel throughout, the tubular would cost about $3,500, or about the same as the stainless steel PHE. These data do not include installation costs.

Materials of Construction

One of the reasons for corrugating the plates is to provide a number of contact points between adjacent plates (and to impart a degree of stiffness). Plate thicknesses as low as 0.6 mm. (0.024 in.) can therefore be used for working pressures as high as 230 psig., particularly when using the cross-corrugated (herringbone) pattern.

Economic use is therefore made of relatively expensive materials—in fact, the more expensive the material, the more economic the PHE becomes, compared to other exchangers.

Any material which can be cold formed may be used, irrespective of the material's welding characteristics, since very little or no welding is involved in the construction of plate exchangers.

The most commonly used material is stainless steel, either 18/10 (AISI 304) or 18/12/2.5Mo (AISI 316), preferably with a carbon content of less than 0.07%, since this eliminates the need for titanium stabilization.

Titanium (99.8%) and palladium stabilized (0.2%) titanium are available for many plate types. These materials are used for duties involving chloride solutions (including brackish cooling water) because of their outstanding corrosion resistance.

Other materials available for many types include the high nickel alloys (Monel 400, Incoloy 825, Inconel 600 and 625, Hastelloy B and C) and copper based alloys (Cu/Ni 70/30 and 90/10, aluminum brass 76/22/2, etc.).

Pure metals such as copper, aluminum, nickel, silver, and tantalum are also used. Pure zirconium is too brittle to press, but zirconium alloys possess properties suitable for PHE manufacture.

Gaskets, Frames

A wide range of gasket materials is available, including natural rubber styrene, resin-cured butyl, nitrile, and silicone rubbers. Elastomers such as neoprene, hypalon, and Viton can also be classed as "standard" materials for some types. Compressed asbestos fibre gaskets can be used on many plate types.

Plastic materials such as PTFE (Teflon or Fluon) are unfortunately unsuitable as PHE gaskets due to their viscoelastic behavior.

Frames are normally carbon steel with a synthetic resin finish to provide some resistance to atmospheric corrosion. Stainless-steel clad frames may be available for some types, for heavily corrosive environments.

Connections are usually the same material as the plates to eliminate electrochemical effects. Rubber-lined connections can be used for some duties.

Pressure Limitations

Maximum allowable working pressure may be determined by frame strength, gasket retainment, or plate deformation resistance. Of these, it is often the frame that limits operating pressure, so that many manufacturers produce a low cost frame for low pressure duties (say up to 6 atm. g. = 85 psig.) and a more substantial frame for higher pressures, for a given plate size.

All PHE's used in the chemical and allied industries are capable of operating at 85 psig., most at 142 psig., many at 230 psig. and some at pressures as high as 300 psig.

Temperature Limitations

Normally, it is the gaskets that limit the maximum operating temperature for a plate heat exchanger. In the absence of chemical attack, the following may serve as a rough guide:

Natural, styrene, neoprene max.	70C. (160F.)
Nitrile, Viton	100C. (210F.)
Resin-cured butyl	120C. (250F.)
Ethylene/propylene, silicone	140C. (280F.)
Compressed asbestos fibre	200C. (390F.)

Operating temperatures may also be limited by plate corrosion effects. For instance, refrigerant quality brine can be handled by a PHE in 18/12/2.5Mo (AISI 316) stainless steel provided the surface temperature does not exceed 10C. (50F.). At higher temperatures, plate failure due to pitting and/or stress corrosion is inevitable.

Other Limitations

Individual plate sizes range from 0.03 sq.m. (0.3 sq. ft.) to about 1.5 sq.m. (16 sq. ft.). Complete plate heat exchangers may possess surface areas ranging from 0.03 sq.m. to over 600 sq.m. (6,500 sq. ft.).

The largest connection size currently available is 300 mm. (12 in.), permitting flows of waterlike liquids of up to 1,000 cu.m./hr. (4,500 gal./min.) to be handled in single units. When single pass operation is involved, double connections may be used.

As with any form of heat exchanger, very high viscosity fluids present some problems due to flow distribution effects, particularly when cooling is taking place. These problems are not, however, insoluble, and a plate heat exchanger can be used successfully for fluids with viscosities in the range 100 to 1,000 poise, provided that a reasonable pressure drop is allowed (say 30 to 45 psi.). With plastic and pseudoplastic fluids, the plate corrugations induce high shear rates, therefore reducing the effective viscosity and improving performance.

Where To Apply Plate Exchangers

The characteristics of plate heat exchangers are such that they are particularly well suited to liquid/liquid duties in turbulent flow (note that a fluid sufficiently viscous to produce laminar flow in a "smooth surface" heat exchanger may well be in turbulent flow in a PHE). It is for such applications that the majority of plate heat exchangers are supplied in the chemical and allied industries.

Condensation of vapor (including steam) at moderate pressure, say 6 to 60 psia., is also economically handled by PHE's, but duties involving large volumes of very low pressure gas or vapor are better suited (generally speaking) to other forms of heat exchanger.

One of the more widespread applications for plate heat exchangers is that of central cooling. This is the cooling of a closed circuit of fresh, noncorrosive and nonfouling water for use inside a plant, by means of readily available, cheap, but highly corrosive brackish water. Central coolers are normally in titanium, since this is the only material that will withstand indefinitely the action of brackish water of whatever source—coast, estuary, river, canal, lake or bore-hole. Plate heat exchangers are ideal for this application, since high-efficiency (with close temperature approaches thereby practicable) compact units capable of extension to meet future increased plant capacities are clearly required. The advantages of PHE's for this application have been recognized by the nuclear industry in the UK, where every nuclear power station under construction or projected uses plate heat exchangers in the secondary cooling circuits.

The list of successful plate heat-exchanger installations within the chemical industry is virtually endless, but it can be said that where high efficiencies are required and where corrosion is a problem, plate heat exchangers can be and are being used, subject of course to the pressure and temperature limits that the basic design imposes. In the biochemical and foodstuff industries, hygienic demands also lead to the selection of the plate type, with the other advantages as a bonus.

True flexibility is unique to the plate heat exchanger both in initial design and after installation. In the initial design, the basic size, geometry, total number and arrangement of standard plates can normally be selected to precisely fit the required duty. An existing plate heat exchanger can very easily be extended or modified to suit an increased or changed duty. ■

References

1. Troupe, R., Morgan, J., and Prifti, J., "The Plate Heat Exchanger—Versatile Chemical Engineering Tool," Northeastern Univ. publication, Boston, Mass., 1960.
2. McKillop, A., and Dunkley, W., "Plate Heat Exchangers—Heat Transfer," *Ind. & Eng. Chem.*, 52, 9, pp. 740-744 (1960).
3. Jensen, S., "Assessment of Heat Transfer Data," *Chem. Eng. Progr.* "Symposium Series," "Heat Transfer," 56, 30, pp. 195-201 (1960).
4. Jackson, B. and Troupe, R., "Laminar Flow in Plate Heat Exchangers," *Chem. Eng. Progr.*, 60, 7, pp. 62-65 (1964).
5. Jackson, B. and Troupe, R., "Plate Heat Exchanger Design by the NTU Method," AIChE-ASME 8th National Heat Transfer Conf., Los Angeles, Calif. Aug. 1965.
6. Crozier, R., Booth, J., and Stewart, J., "Heat Transfer in Plate and Frame Exchangers," *Chem. Eng. Progr.*, 60, 8, pp. 43-45 (1964).
7. Smith, J., and Troupe, R., "Pressure Drop Studies in a Plate Heat Exchanger," AIChE 55th National meeting, Houston, Tex. Feb. 1965.
8. Buonopane, R., Troupe, R., and Morgan, J., "Heat Transfer Design Method for Plate Heat Exchangers," *Chem. Eng. Progr.*, 59, 7, pp. 57-61 (1963).
9. Morgan, J., Troupe, R., Anderson, R., et al, "The Plate Heat Exchanger—A Continuous Flow-Type Reactor," *Ind. & Eng. Chem.*, 52, 10, pp. 821-824 (1960).
10. Alfa-Laval Thermal Handbook, 1969.
11. Buonopane, R., and Troupe, R., *AIChE J.*, July 1969, pp. 585-596.
12. Emmerson, W., Nat. Eng. Lab., Scotland, reports No. 283-286.
13. Foote, R., Nat. Eng. Lab., Scotland, report No. 303.
14. Watson, E., McKillop, A., Dunkley, W., and Perry, R., "Plant Heat Exchangers—Flow Characteristics," *Ind. & Eng. Chem.*, 52, 9, pp. 733-740 (1960).
15. Lawry, R., "Plate Type Heat Exchangers," *Chem. Eng.*, June 29, 1959, pp. 89-94.
16. Hargis, A., Beckmann, A., and Loiacono, R., "The Plate Heat Exchanger," ASME publication 66-PET-21 (1966).
17. Flack, P., "The Feasibility of Plate Heat Exchangers," *Chem. & Proc. Eng.*, Aug. 1964, pp. 468-472.
18. Jones, R., and Usher, J., "Plate Heat Exchangers in Chemical Engineering," 4th International Congress of the European Federation of Chemical Engineering.
19. Usher, J., "Evaluating Plate Heat Exchangers," *Chem. Eng.*, Feb. 23, 1970, pp. 90-94.

Acknowledgments

The author expresses his thanks for the many helpful suggestions and constructive criticisms made by his colleagues in the Alfa-Laval De Laval Group. In the U.S., American Heat Reclaiming Corp. (N.Y.C.) and the De Laval Separator Co. (Poughkeepsie, N.Y.) are members of this group. Thanks are also due to Dr. Ralph A. Buonopane of Northeastern University, Boston, Mass., for his suggestions and encouragement.

Meet the Author

Jan Marriott is a group leader in a specialist team at Alfa-Laval AB, Lund, Sweden, engaged in the development of applications for thermal equipment in the chemical and allied industries. A graduate chemical engineer, he was educated at the Battersea Institute of Technology, London, England. In 1958 he obtained graduate membership in the Institution of Chemical Engineers and he is a member of the Heat Transfer Society. His prior industrial experience was with a manufacturer of graphite heat exchangers.

(C) Materials of Construction for Heat Exchangers

Choosing materials of construction for plate heat exchangers—I
Choosing materials of construction for plate heat exchangers—II
Graphite heat exchangers—I
Graphite heat exchangers—II
Selecting steel tubing for high-temperature service
Stainless-steel heat exchangers—Part I
Stainless-steel heat exchangers—Part II

PLATE heat exchanger (boxed, lower right) replaces 14 tubular exchangers (some shown at left)—Fig. 1

Choosing Materials of Construction For Plate Heat Exchangers—I

Plate exchangers have many advantages, but they require special consideration in material selection. Here are the basics that you must know.

C. T. COWAN, The A.P.V. Co., U.K.

It is only recently that plate heat exchangers have gained significant acceptance in the chemical process industries, although they have held a prominent position in the dairy, food and brewing industries for the past 50 years.

The many advantages of plate heat exchangers include:
- Very high heat-transfer coefficients on both sides of the exchanger.
- Close approach temperatures, and fully countercurrent flow.
- Ease of inspection of both sides of the exchanger.
- Ease of cleaning.
- Ease of maintenance. Also, heat-transfer area can be added or subtracted without completely dismantling the equipment.
- Savings in required floor space (see Fig. 1).
- Low holdup volume.
- Low cost, especially when made of expensive metals.

These advantages are being increasingly exploited by chemical engineers. Because the chemical process industries have to deal with a wide range of corrosive environments, the engineer is particularly conscious of the need for detection and replacement of corroded equipment. The ease with which this is achieved with the plate heat exchanger accounts, in no small part, for its wide acceptance.

However, even the need for plate replacement can be obviated, by careful attention to detail at the design stage and by implementing correct operating and maintenance schedules.

Design Features of Plate Exchangers

The design principles, methods of construction, and operating characteristics of these exchangers differ fundamentally from the more traditional forms of heat-transfer equipment (such as the shell-and-tube exchanger). These differences have to be taken into account when specifying materials of construction.

The plate heat exchanger consists of a series of plates (having peripheral elastomeric gaskets) that are all sandwiched together in a frame (Fig. 2). The inlets and outlets for both the heating (or cooling) medium and the process stream are located in the four corners of the plates. The desired flow pattern through the plate pack is

Originally published June 9, 1975

166 HEAT EXCHANGER DESIGN AND SPECIFICATION

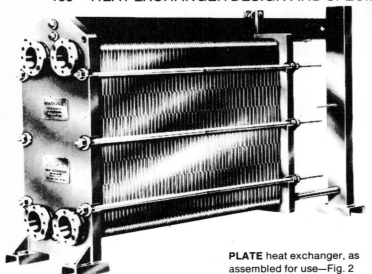

PLATE heat exchanger, as assembled for use—Fig. 2

achieved by the positioning of an elastomeric gasket to blank off flowpaths, and by blanking-plates at the ports of individual heat-exchange plates (Fig. 3).

The plates of the exchanger are made of thin-gage material (between 0.036 and 0.025 in), which is formed into a wide-pitched, horizontally troughed, or a diagonally troughed pattern. (Fig. 4). This pattern has two functions. First, it induces turbulence in the process stream and thus achieves a high heat-transfer coefficient. Second, it provides reinforcement and plate-support points that maintain interplate separation. The large number of plate-to-plate contact points that result when the plates are formed into a pack provide full plate support on a relatively fine matrix (especially with the diagonally troughed pattern) providing high operating pressures (up to 300 psig) even with very thin-gage metals.

The configuration and geometry of the plate heat-transfer surface differs clearly from that of, say, a shell-and-tube heat exchanger. Consequently, a new materials-specification philosophy often has to be adopted when considering materials of construction for a plate exchanger operating in a particular process stream.

Combating Erosion-Corrosion

As good-quality water supplies become scarcer and more expensive, engineers look to alternative water sources for cooling purposes. By far the largest reserve available to plants located on the coast and on tidal rivers is seawater, and it is becoming increasingly common to rely on this. But because of the general corrosivity of such seawater supplies, some of the larger processing plants are installing freshwater cooling circuits.

That is, a large heat exchanger with seawater on one side is used to cool a freshwater supply that is circulated around the installation for all cooling duties. An advantage of this system is that it requires only one piece of equipment to be constructed of seawater-resistant materials. The exchangers in the rest of the plant can be constructed either of carbon steel or of alloys suitable to the process streams, without having to bother about special corrosion resistance on the cooling-water side.

Copper-based alloys, such as aluminum brass or 90/10 and 70/30 copper-nickel, have conventionally been specified for shell-and-tube heat exchangers. These materials have given (and are continuing to give) excellent service life in such units. Successful as these alloys are, though, their susceptibility to erosion-corrosion is well known. The presence of a mussel or other marine growth in the tube of a heat exchanger has disastrous results because it induces turbulence. This causes a classical horseshoe-shaped erosion pattern to develop, which in a relatively short time results in tube perforation.

Much has been written about critical seawater velocities, and maximum design-velocities for the various copper-based alloys are well established. Despite this, erosion-corrosion is encountered in poorly designed shell-and-tube heat exchangers, especially at the entry to the tubes; sometimes it is necessary to fit plastic inserts to overcome this problem. These so-called "entry conditions," with their characteristically high level of turbu-

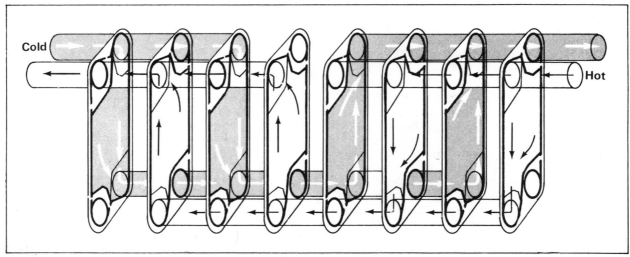

DIAGRAMMATIC REPRESENTATION of the flow pattern existing in a plate heat exchanger—Fig. 3

CHOOSING MATERIALS OF CONSTRUCTION FOR PLATE HEAT EXCHANGERS—II

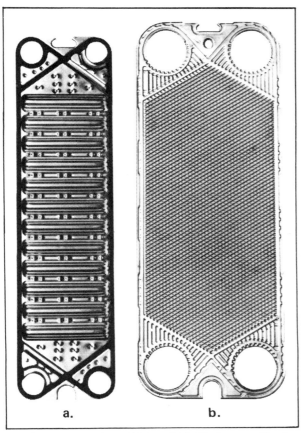

HEAT-exchanger plates: (a) troughed, (b) chevron—Fig. 4

lence, result from a change of direction of water flow from the water box into the tube.

With the plate heat exchanger, the geometric form of the plate is specifically designed to produce a high level of turbulence and, in effect, "entry conditions" prevail over the entire heat-transfer surface. Consequently, materials that have been successfully employed for the construction of shell-and-tube heat exchangers that handle seawater or brackish water, will not necessarily give erosion-free service in plate heat exchangers.

When specifying materials for plate-heat-exchanger duty, it is advisable to consider metals that will combine excellent corrosion resistance to chloride ion with immunity to erosion-corrosion. Materials such as Monel or titanium are the obvious answer. Although these are somewhat more expensive on a pound-for-pound basis, a plate heat exchanger made of these more-expensive alloys is frequently cheaper than a shell-and-tube exchanger in the conventional metals, when the cost is based on a particular thermal duty.

Designing for General Corrosion

As previously stated, the plates of a plate heat exchanger are thin as compared with the heat-transfer surface of other forms of heat-transfer equipment (25 to 36 mils, compared with 48 to 64 mils). Hence, corrosion allowances quoted for chemical equipment in the standard reference books tend to be meaningless. Such books will accept a corrosion rate of 5 mils/yr or less, which is realistic when considering material of ten times this thickness. However, with metal 25 mils thick, a corrosion rate of 5 mils/yr represents more than 50% reduction in three years. This would be unacceptable to most engineers.

As a general guideline, the maximum permissible corrosion rate for a plate heat exchanger is 2 mils/yr. At this level, the general corrosion of plate heat exchangers is brought into line with normally accepted levels for other pieces of process-plant equipment. Unfortunately, there is a dearth of published information on corrosion at these very low levels. Therefore, the design engineer must rely heavily on the equipment supplier for specifying material suitable for a specific duty.

The obvious implications of all this are that, in a particular corrosive environment, the changeover from, say, a shell-and-tube heat exchanger to a plate heat exchanger may require an upgrading of the alloy used for the heat-transfer surface. For example, where Type 316 stainless has been used in cooling sulfuric acid, use of a plate heat exchanger for similar duty may require specifying a 25 Ni, 20 Cr, 4 Mo, 2 Cu alloy, or even Incoloy 825 (40 Ni, 25 Cr, 3 Mo, 2 Cu).

Despite this occasional need for a more-expensive material of construction for plate heat exchangers, the advantages of using thin-gage material, together with the inherently high heat-transfer coefficient of the equipment, frequently means that the cost/unit-of-heat-transferred is less. There is also an added attraction in using a more highly alloyed metal. That is, some degree of added corrosion protection is built into the equipment. This permits excursions from design conditions into the more corrosive regimes frequently encountered during the startup and shutdown of a modern process plant.

Sometimes, in the chemical process industries, the choice of material for the construction of heat-transfer equipment is between a metal such as tantalum or silver, and a nonmetal such as graphite. Here, construction of conventional metal heat-transfer equipment (such as the shell-and-tube exchanger) is usually ruled out on the grounds of cost; graphite heat exchangers have other limitations. But by exploiting the plate-heat-exchanger principle, it is possible to design an exchanger with plates only 15 mils thick; indeed, because of the high heat-transfer coefficients obtained, the actual metal usage is low and the unit may be very competitively priced.

Meet the Author

Colin T. Cowan is Laboratory Manager of The A.P.V. Co., Manor Royal, Crawley, Sussex, England, where he is responsible for the R and D laboratory that covers the company's activities in the manufacture of heat-exchange, distillation and evaporation equipment. He received an Associateship of the Royal Institute of Chemistry at Leeds College of Technology, followed by a Ph.D. from Leeds University. His special interests are in materials of construction and materials science.

Choosing Materials of Construction For Plate Heat Exchangers—II

Here is further information on picking materials for plate heat exchangers; it covers crevice corrosion, "concentration" corrosion, and gasket selection.

C. T. COWAN,* The A.P.V. Co., U.K.

In the first article of this series, we considered the design features of plate heat exchangers as well as general corrosion and erosion corrosion. This article treats crevice corrosion, "concentration" corrosion and the selection of gasket materials.

Crevice Corrosion

In the shell-and-tube heat exchanger, the shellside invariably contains a large number of crevices, where the tubes project through the tubesheet. Elimination of such crevices is difficult if not impossible. On the other hand, the tubeside of such an exchanger can be regarded as crevice-free. Consequently, where the combination of the material of construction and one of the process fluids is likely to give rise to crevice corrosion, the engineer can design the flow pattern to minimize the risk.

In plate heat exchangers, pips or troughs are formed on the plate surface, and these abut similar profiles on both the adjacent plates (to maintain interplate separation). Each of these contact points represents a potential crevice, and one must be aware of this when specifying materials of construction.

The great majority of plate heat exchangers are supplied with Type 316 stainless-steel plates. As is well known, this material is susceptible to crevice corrosion by chloride-containing environments, especially at pH values below 7. Reasonable levels of chloride in water and other process streams can be tolerated without running a serious risk of crevice corrosion in stainless steel. However, there are certainly environments that would cause rapid initiation of crevice attack in a plate heat exchanger.

It is impossible to give specific figures for the maximum chloride content of a process steam that can be handled by such an exchanger because many other factors enter into consideration, e.g., temperature, pH, heating or cooling duty, dissolved-oxygen content, and inhibitory effects of other constituents. Consequently, every case must be reviewed in the light of past experience on similar materials.

Many of the more highly alloyed stainless steels have been successfully employed where Type 316 has failed. But for this class of environment, titanium is assuming increasing importance as the most economically attractive material. Even in the food industry, which grew up around tinned copper, aluminum, nickel and stainless steel, users are now specifying titanium for plate heat exchangers that process piquant sauces containing high sodium chloride levels at low pH levels (acetic acid content of the vinegar used).

"Concentration" Corrosion

"Concentration" corrosion is a form of attack that may assume any of the classical forms of corrosion (general, crevice, pitting, or stress-corrosion cracking) resulting from the development of an environment in discrete regions that differs from that of the main process stream the equipment was designed to handle. It may be an environment that is a concentrated solution of the process stream or, alternatively, it may be one in which the pH differs vastly from that of the process stream and which is produced, for example, by the loss of acidic or alkaline dissolved gases.

In any heat-transfer equipment, there is always a tendency for the transfer surfaces to become fouled. This is particularly the case in heating duties where there is a large ΔT across the heat-transfer surface. In such cases,

*To meet the author, see p. 167.

Originally published July 7, 1975

CHOOSING MATERIALS OF CONSTRUCTION FOR PLATE HEAT EXCHANGERS—II

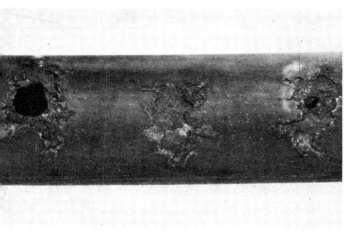

"CONCENTRATION" corrosion in a 3-in, Type 316 coil.

two factors come into play that can have serious effects on the corrosion of the surface. First, the scale deposits formed may contain disproportionately large amounts of one species present in the process stream—material that at its normal concentration may be noncorrosive, but in its concentrated condition will badly attack the heater surface. Second, because the scale will have a poor heat-transfer coefficient, the temperature of the metal under the scale will approach that of the heating medium. The overall effect is to produce an area on the heat-transfer surface where the combined effect of high temperature and high corrodent concentration will result in corrosion and eventual failure.

The extreme case in which this type of failure can occur is in a steam heating coil in the base of a storage vessel. Turbulence at the heating-coil surface is very low indeed and, in certain environments, scaling can occur quite rapidly. However, with the plate heat exchanger, because of both the induced turbulence and a thermal efficiency that will permit effective heating at a low ΔT, the fouling tendency is greatly reduced.

The net effect is that less highly alloyed—and hence less expensive—materials can be specified for the heating surface of a plate heat exchanger when it replaces a failed heating coil. The figure shows typical failure (after one year's service) in a Type 316 steam coil that would have normally required replacement with an alloy tube such as Carpenter 20 or Incoloy 825. However, replacement was made with a Type 316 plate heat exchanger on a recirculation loop; after many years of service, it shows no signs whatever of corrosion.

In another case, a Type 316 plate heat exchanger now in service for five years (again on a recirculation loop) replaced an Incoloy 825 heating coil in an electrolyte bath. The coil had failed after only six weeks of operation, due to concentration corrosion under the heavy scale that had rapidly formed.

These are only two cases where a plate heat exchanger permitted the use of a lower-cost alloy for replacement of conventional heating coils. However, the same principles apply wherever shell-and-tube heat exchangers are used, because these are inevitably more prone to fouling than their plate-heat-exchanger counterparts.

Gasket Materials

One obvious difference between the shell-and-tube heat exchanger and a plate heat exchanger is in the need for a gasket. Individual plates of the plate heat exchanger carry a gasket, the functions of which are to effect an overall seal as well as to close off the flowpath of one of the process streams. Although non-elastomeric materials (e.g., compressed asbestos fiber) have been used in plate heat exchangers, it is advantageous to specify an elastomer whenever possible.

The prime requirement for such a gasket is to provide a long-lasting seal of high integrity. Two properties are of paramount importance—thermal stability (good aging characteristics) and low compression set.

Of the many possible materials, five polymers will handle most applications. These are:

SBR (Styrene Butadiene Rubber)—General-purpose applications involving aqueous systems at up to 180°F.

Medium Nitrile (Acrylonitrile Butadiene Rubber)—This combines the general-purpose characteristics of SBR with an excellent resistance to fat and aliphatic hydrocarbons, and with an upper working temperature of 285°F. This, however, is being superseded by EPDM (a terpolymer elastomer made from ethylenepropylene diene monomer) which offers similar or, in some cases, superior chemical resistance and better thermal stability.

Butyl (Isobutylene and Isoprene Copolymer)—Excellent resistance to wide range of chemical environments such as acids, alkalis, some ketones, amines, etc., but has poor fat resistance.

Silicone—Limited applications, but especially suited to sodium hypochlorite and general low-temperature use.

Fluoroelastomer (e.g., Viton)—Use is restricted by high cost, but it has no equal for temperatures above 300°F or for use in oil.

As in selection of plate materials, gasket selection is based on a number of interrelated factors—temperature and the chemical composition of the process stream being most important. In a heating duty, one must consider the highest prevailing temperature, i.e., the inlet temperature of the steam or hot water. When only high-pressure (over 60 psig) or superheated steam is available, it may be necessary to reduce this to a somewhat lower temperature. (Of course, because of the high heat-transfer coefficients achieved, there is little need for high-temperature heating media.)

When high temperature is required, it may be necessary to specify compressed asbestos-fiber gaskets, which raise the working temperature of a plate heat exchanger to 400°F. Use of these less-resilient gaskets does, however, require a somewhat more substantial frame and tightening arrangement.

Conclusions

Specifying a plate heat exchanger in place of more-conventional types introduces certain corrosion phenomena while eliminating others. By carefully considering all of the operating conditions, and the compositions of process streams, it is possible to minimize (or eliminate) failure due to corrosion. #

Graphite Heat Exchangers—I

Graphite heat exchangers can take many forms. In this first of two articles, we discuss the shell-and-tube type and the cubic (or rectangular) block.

DENNIS E. G. HILLS, Graphite Equipment Ltd.

SHELL-AND-TUBE graphite heat exchanger—Fig. 1

Originally published December 23, 1974

Graphite is unique: it has a most valuable combination of properties. It may be regarded as a material standing midway between metal and non-metal. From metals, it borrows the properties of being a good conductor of both heat and electricity; from non-metals, such as rubber and refractories, the property of being resistant to chemical corrosion. Graphite's other valuable properties are:

- Ability to withstand rapid and uneven change in temperature without fracture.
- Low friction coefficient.
- Stability at high temperatures—at 1,500°C, steel is molten but graphite is improving in strength.

Carbon has, on the whole, the same properties as graphite except that it is not such a good heat conductor, hence graphite is preferred for heat-transfer equipment; but, since it costs approximately twice as much as carbon, carbon is used in those cases where good heat transfer is not essential.

Different grades of carbons and graphites having varying properties have been developed and the manufacturers can alter the combination of raw materials to suit various engineering requirements. Due to the fine grinding of the coal and the special processing developed for these grades of carbons and graphites, it is possible to manufacture materials with a finer grain structure and a strength three to four times that of other carbons and graphite produced by the old methods. The properties of the newer materials can be varied considerably to suit the application for which they are required—they can be made dense and compact or light and porous. It is possible to machine them into very intricate shapes that have good surface finishes.

Chemical Applications

It is in the field of chemical applications that the most outstanding development of carbons and graphite have been achieved. Owing to the dense structure and low porosity that can be obtained with certain grades, the

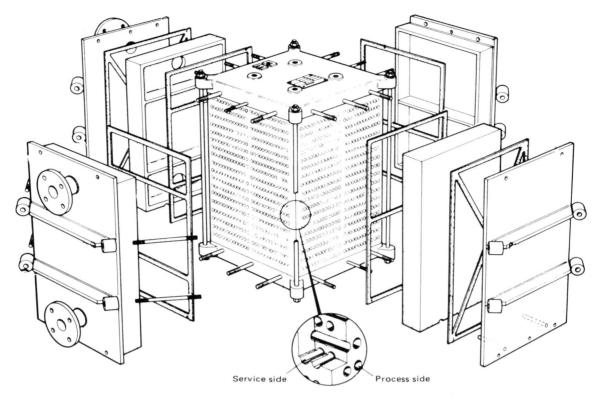

RECTANGULAR BLOCK graphite heat exchanger takes advantage of good compressive strength of graphite—Fig. 2

materials have resistance to all but the strongly oxidizing chemicals well beyond that of ordinary carbons. In addition, graphite's mechanical strength enables it to compete with precious and semiprecious metals, as a material for chemical equipment and plant construction where the brittleness of the ordinary carbons is a disadvantage, and where the ordinary materials do not sufficiently resist chemical attack.

Carbon and graphite find many applications in the construction of chemical equipment—since they are stable over a wide range of temperatures and are chemically resistant to most corrosive reagents. Graphite, as it has been stated, offers the additional advantages that it has a high thermal conductivity and good machining properties, which make it an ideal material for heat-transfer equipment. However, the construction of this equipment introduces a number of special problems, since all forms of graphite are brittle with a low tensile strength, and the manufactured varieties, as stated previously, are porous and permeable to gases and liquids. The porosity is overcome by impregnation of the graphite with synthetic resins, but the mechanical properties call for techniques in design and fabrication that differ from those employed with metals.

Construction of Heat Exchangers

There are numerous types of graphite exchangers; these articles will present an appraisal of their advantages and disadvantages, as well as a guide to process industries applications.

Shell-and-Tube Heat Exchangers

The usual approach is to make a graphite heat exchanger of shell-and-tube design, with exchanger tubes and tubesheets out of graphite. Although the resulting heat exchanger is the best solution to many problems, graphite used this way is not being used to its greatest potential—good in compression, poor in tension.

In general, the full corrosion resistance of graphite is needed only on the tube side; the shell can be either plain lined or metal. Provided the shell-side fluid is not too corrosive, this is not a problem. However, if it is corrosive, other types of block-construction graphite heat exchangers will be more effective. The shell-and-tube construction has a lower price structure than the block designs and also limitations. The graphite tubes are normally ⅞ in. I.D. and 1¼ in. O.D., although several diameters are available that have correspondingly thick walls. The usual tube length is 9 ft. Longer lengths are fabricated from these, using a lap-over or scarf joint that is cemented before impregnation. This joint is as strong as the tube itself; in fact, in many cases it may be stronger.

The shell-and-tube design offers a variety of graphite exchangers that are similar to the conventional metallic types, and can be made in many combinations of pass arrangements on both the shell and tube sides. The standard graphite heat exchangers are normally offered in 6, 9, 12, 14 and 16-ft lengths with ⅞ in. I.D. × 1¼ in. O.D. Heat-transfer areas vary from 10 to over 5,000 ft^2.

Shell-and-tube heat exchangers are commonly described in terms of their shell-side areas. Care must be taken in choosing a unit if the tube-side coefficient is the controlling factor because of the thickness of the tubes: the ratio of outside to inside areas is 1.43:1.

The graphite tubes are set in graphite headers and encased in a steel shell; the shell can also be supplied in other metals, such as alloy steel, copper, aluminum or lead, rubber- or glass-lined steel, and sometimes graphite. If the inlet and outlet ports are required radially on the side of the header, there is a considerable extra charge, since longer headers may be required.

The limitations on this design are the pressure limits, which are relatively low. The highest recommended operating temperature is approximately 180°C, and the operating pressure 75 psig with liquids, and up to 50 psig with steam. These designs take up a very large amount of valuable process space and are not recommended for use in the fine-chemical and pharmaceutical industries. They are mainly used where very large throughputs of chemicals are required—for example, in the heavy-chemical, fertilizer and synthetic-fiber industries.

If there is tube leakage, locating the leak is often difficult, especially if the flow is into the shell area (normally, the service side). Repairing the leakage is carried out by removing the headers, boring out the tube from the tubesheet by means of a special tool, and extracting the rest of the faulty tube. A new tube is then cemented in place (with sleeves), which is subsequently compressed by bulk loading to achieve the effective joint. The cement joint is subjected to heat by means of a fan heater for approximately 23 h before retesting. After a satisfactory test, the headers may then be replaced, complete with new gaskets, and the unit put back into service.

Block—Cubic and Rectangular

The cubic or rectangular block heat-exchanger consists of a graphite block perforated with rows of parallel holes for conveying the two fluids. The graphite blocks themselves consist of a series of accurately machined plates of graphite laminated together. The graphite plates are bonded with the latest thermosetting resins, and the block is permanently compressed between cast-iron clamping plates, with tie bolts passing through lugs at the corners. After lamination, the complete block is impregnated and oven-cured to ensure permanent bonding of the graphite. The particular feature of this construction is that the graphite is prestressed in such a way as to neutralize tensile stresses and to utilize the compressive strength of the material to the best advantage—good in compression—poor in tension.

The joints formed between the graphite plates are in no sense "cemented" joints, in that the bonding medium is less than 0.001-in, and the oven-curing process results in a complete and virtually indestructable union.

Mechanical and excess-pressure tests over a long period, some at elevated temperatures, have proved that joints between plates are, in fact, stronger than the parent material. Graphite blocks without the cast-iron clamping plates have been pressure tested by applying steam at 100 psig, and there has been no breakdown of the joints.

Gasketed joints are the weakest point of any type of equipment where corrosive liquids or gases are handled. In the graphite-block and -cartridge designs, gaskets are only employed between the graphite and the removable headers. These gaskets are inexpensive and easily replaced when necessary.

In good designs, if gasket failure should occur, there can be no resultant leakage between the service and process sides of the heat exchanger, but only leakage to the atmosphere, where it is immediately detectable.

Headers are bolted to the four faces of the graphite block for the supply and discharge of the fluids. The headers may be made of carbon, cast iron or steel, fitted if necessary with a rubber lining or Penton coating, according to the corrosive nature of the fluids. A variety of multipass arrangements can be effected by incorporating partitions in the inside of the headers. This feature is particularly useful for the heating or cooling of fluids or heat-recovery duties, since the velocity of each fluid can be adjusted to the most economic figure, and the multipass exchanger gives a close approximation to true countercurrent flow. Headers of split or graduated passes can also be supplied. Block heat exchangers are ideal for duties where corrosion is involved on both the process and service sides of the heat exchanger.

Experience has shown that a hole diameter of ⅜ or ½ in is the most economic when the fluids are reasonably clean and not inclined to deposit scale.

It is important to remember that the blockage and scaling up of metal heat exchangers is often due to the formation of a scale on the surface of the metal, due to the corrosion attack of the liquor. This scale then acts as a collector of all suspended matter in the liquor, quickly causing excessive blockage. This does not occur with carbon and graphite materials because the surface retains its normal smoothness and so resists the buildup of scale.

Larger holes, usually ¾ in. dia., can be provided and can also be supplied in exchangers used as boilers or evaporators. When one is handling process solutions that are free from suspended solids and do not tend to scale, the effective heat-transfer area can be doubled by employing slotted holes, instead of the normal round holes, on the process side. In this manner, it is possible to obtain twice the effective heat-transfer area for the same block unit-size and for approximately the same basic price. Slotted heat exchangers are ideally suited for the heating or cooling of gases and the condensing of vapors where the heat-transfer coefficient on one side is low.

The block design allows for a variety of graphite exchangers for different working pressures. #

Meet the Author

Dennis E. G. Hills is Managing Director of Graphite Equipment Ltd., Arundel Road, Industrial Estate, Uxbridge, Middlesex, UB8 2SA, England. He is a trained mechanical engineer, a member of the Institute of Plant Engineers, a member of the Institute of Marketing and a fellow of the Institute of Directors. For the past 16 years he has been actively engaged in the graphite equipment field.

Graphite Heat Exchangers—II

Here are descriptions of a number of specialized graphite heat exchangers, and a discussion of their applications in the chemical process industries.

DENNIS E. G. HILLS, Graphite Equipment, Ltd.*

The previous article of this two-part series discussed the properties of graphite and its applications. In addition, it described two kinds of graphite heat exchangers—the shell-and-tube and the cubic- (or rectangular) block exchangers.

The rectangular-block exchanger may be had in either inch or metric dimensions.

Modular-Block Rectangular Exchanger

Process and service holes are drilled in the solid block. Steel headers are used on the service side, and glass piping on the process side. Units are designed for 4, 6, 8, 9, 12 and 18-in glass piping. These heat exchangers are being sold to replace conventional glass, coiled heat-exchangers.

They are ideally suited for condensing, evaporation and reboiler work. The units occupy only one-fourth to one-third of the space of the glass heat-exchanger, and the heat-transfer area can be reduced by four to five times. For example, a 15-ft² glass unit can be replaced by a 3.5-ft² graphite unit.

Cylindrical Exchangers

The solid impervious-graphite blocks have holes drilled in them. These blocks can be multi-stacked in a cylindrical steel shell that has gland fittings. The process holes are axial and the service holes are transversal. Multipass header arrangements are available. The units have been designed for use as evaporators and reboilers. Working pressures are as high as 300 psig. Various process-hole-diameters are available in exchangers having areas ranging from 3m² to 130m².

The Cartridge-Block Exchanger

This design consists of a graphite core and process headers housed in a cylindrical steel shell, with gland fittings at each end.

The graphite is perforated with holes for the process

*To meet the author, see p. 172.

Originally published January 20, 1975

fluid opening into headers at each end, and having slots on the shell side of the core to convey the cooling or heating fluids.

Such design permits a normal working pressure of 100 psi on the process and service sides. Special units are available for a working pressure of 300 psig, which makes them attractive for use where the limits of the shell and tube would be exceeded. The standard units are normally offered in heat-transfer areas from 1¼ to 1,000 ft². The standard construction can easily be adapted and modified in design to act as absorbers and evaporators.

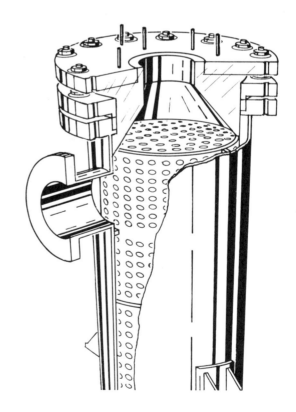

MODULAR-BLOCK exchanger has stackable blocks—Fig. 1

173

174 HEAT EXCHANGER DESIGN AND SPECIFICATION

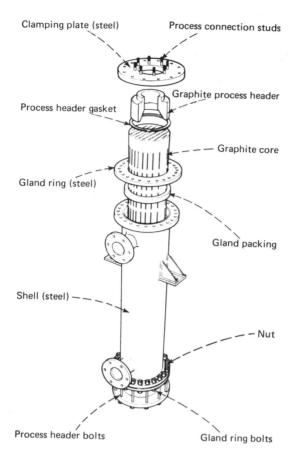

CARTRIDGE-BLOCK exchanger has steel shell—Fig. 2

CASCADE cooler is designed for dirty liquids—Fig. 3

The Cascade Cooler

This type of heat exchanger was designed for cooling corrosives using scaling or dirty water that cannot be passed through shell-and-tube or block-construction heat exchangers.

The cascade cooler consists of a vertical bank of graphite tubes having solid graphite blocks at each end that act as elbows, enabling the corrosive fluid to pass to the next tube. A distributor tray is located on the top of the bank of tubes. The dirty scaling water is allowed to cascade down over the graphite tubes that contain the corrosive fluid to be cooled.

The cascade coolers can be joined in parallel or series, to give more heat-transfer surface. The standard range is normally in 3 series of 9, 18 and 27 ft length, in a variety of tube diameters from 1 to 1½, 2, 3, and 4 in. Heat-transfer areas are from 4 to 40 ft².

Pressure and temperature limitations are generally similar to those of the shell-and-tube design. However, they can be used for dirty, scaling, service-water conditions.

Immersion Heaters

The immersion heater is generally described as a bayonet heater. The sheath is of graphite, with an inner concentric metal tube. Steam is sent through the tube, transferring the heat through the graphite. For some applications, a coolant is used in place of steam. Some manufacturers have produced more-elaborate plate-and-grid heaters, using the same principle.

Pressure and temperature limitations are similar to the shell-and-tube design. The bayonet heaters are only available in a limited number of sizes.

The plate-and-grid heaters vary from 4 to 110 ft².

A new graphite immersion-heater that is similar to an electric heater is now available. The sheath is of graphite (like the standard bayonet heater), but instead of an inner concentric metal tube, this is replaced by an electric element. Units are available for single- and three-phase supply of 1 and 3-kW rating.

Heat-Transfer Characteristics

The path of heat through the graphite plate between adjacent holes is rather complicated and not amenable to mathematical analysis. However, the conductivity of the graphite is high—40 CHU*/(hr) (°C) (ft²/ft), and the heat path is short, so under most practical conditions the temperature drop in the graphite is not a significant factor. When an allowance is necessary, it is usually based

*CHU = centigrade heat unit, the energy required to raise 1 lb of water through 1 deg. C. One CHU = 1.8 Btu.

on an approximate wall coefficient of 1,000 CHU/(hr) (°C) (ft²).

The heat-transfer coefficients for heating or cooling fluids are calculated using the well-known correlations for fluids in pipes. The allowance for scale and dirt deposition is considerably lower than that recommended for metal heat exchangers, since the graphite surfaces are less liable to foul. This allowance varies according to the conditions of the fluids, but under normal conditions, it amounts to 1,000 CHU/(hr) (°C) (ft²).

Applications of Graphite Heat Exchangers

Impervious graphite resists a wide variety of inorganic and organic chemicals. In general, only those substances that exhibit strong oxidizing characteristics, such as nitric acid, concentrated sulfuric acid and wet chlorine, cannot be handled. Heat exchangers with improved resistance to oxidizing agents are being developed.

Typical examples of applications with inorganic chemicals are the heating of acids in metal-pickling and -treatment plants (plating and anodizing baths), and heating and cooling of hydrochloric, sulfuric or phosphoric acids, and of corrosive salt solutions. Graphite heat exchangers are also employed as boilers and condensers in the distillation by evaporation of hydrochloric acid and in the concentration of weak sulfuric acid, spin-bath acid, and rare-earth-chloride solutions.

As stated earlier in this text, the principal limitations in the application of graphite equipment lie in the synthetic resins used both for impregnation and as a laminate. These undergo decomposition at temperatures in excess of 180°C, and this is the upper limit to which the graphite may generally be exposed. Special conditions and applications can be considered and applied; for example, graphite heat exchangers are working successfully with an inlet gas temperature in excess of 800° C. (This is a specially designed unit.) Lower maximum-temperature limits are necessary when the graphite is in contact with certain corrosive reagents or strong solvents.

The synthetic resins employed in the early stages of impregnated-graphite development were attacked by many organic solvents. However, now available are a range of furane resins that resist most organic solvents and reagents. Exchangers impregnated with these resins have proved satisfactory for many purposes in the organic chemical industry. Aromatics—benzene, toluene, xylene; chlorinated solvents—carbon tetrachloride, ethylene chloride and dichloride; methyl and butyl alcohol and phenol, as well as boiling and condensing chlorinated hydrocarbons, organic acids and acid chlorides. Graphite equipment is also used in the absorption of hydrogen chloride formed as a byproduct of organic chlorinated reactions.

Developments for use in chemical processes involving oxidizing fluids have moved very rapidly over the last five years. Impregnants now being used are polyesters and PTFE (polytetrafluoroethylene). Oxidizing fluids fall into two principal groups, those that provoke chemical oxidation and those that induce a physical-chemical reaction to form interstitial compounds.

Both effects may overlap. The true oxidation symp-

IMMERSION heater is of "bayonet" type—Fig. 4

toms on graphite are the softening and degradation of the graphite surface and can normally be clearly observed. The effects of a fluid are dependent upon its composition and temperature as well as its impurities.

Oxidizing fluids now being handled in graphite equipment include 10% nitric acid and pickling liquor (heating), the mixture of nitric and hydrofluoric acid pickling liquor (heating), 93-94% sulfuric acid (cooling), and hypochlorite solutions (cooling). The hypochlorite is produced by passing chlorine into concentrated sodium hydrochloric acid containing chlorine. It should be remembered that the handling of oxidizing fluids in graphite equipment is still relatively new. Great care should be taken in the process control of the plant so as to maintain conditions specified by the supplier of the equipment.

In addition to corrosion resistance, graphite has the advantage that there is no risk of metallic contamination, which is an important feature in the preparation of drugs and foodstuffs. Graphite exchangers have been employed for the evaporation of fruit juices, heating of liquors for the canning of fruit and vegetables (carrots, peas, etc.), and concentration of pickles and sauces. Other applications are distillation of hydrochloric acid from glutamic acids, and a heating of demineralized water. One unusual application is the heating and cooling of water in tropical, temperate and arctic aquariums. Here the advantage lies in preventing poisoning of the fish by metallic impurities.

Another interesting application of graphite exchangers is for the heating or cooling of viscose in the rayon-spinning industry. This liquid has a viscosity of 50 to 400 poises and it is therefore necessary to design the exchanger for streamline flow. Tubular exchangers of conventional design are unsuitable for such a material, owing to the low heat-transfer coefficient achieved. It can be shown theoretically that the best results are obtained in an exchanger having a large number of short-diameter tubes. This has been realized by a modification of the cubic block. #

Selecting steel tubing for high-temperature service

Four classes of steel are commonly used in high-temperature tubing: carbon, chromium-molybdenum, ferritic stainless and austenitic stainless. The author discusses four important material properties—strength, thermal conductivity, thermal expansion and oxidation resistance—that the engineer must consider when choosing a steel for high-temperature service.

William C. Mack, Babcock & Wilcox Tubular Products Div.

☐ Steel tubing remains a favorite for high-temperature applications in heat exchangers and pressure boilers. Because of its strength and resistance to high-temperature oxidation, it makes an ideal heat-transfer surface for steam, oils and many other chemical process fluids.

Four classes of steel are commonly used in high-temperature tubing: carbon, chromium-molybdenum, ferritic stainless, and austenitic stainless. Each has characteristics that suit it to particular applications.

Four major properties

Before selecting steel for a high-temperature application, one should consider four major properties.

Strength—This property is always crucial at high temperatures. When the steel is used at a temperature within its creep range, its strength is time-dependent. Within its creep range, a steel behaves somewhat like taffy: it will flow (creep) and deform when stressed. The greater the stress on the steel, the faster the flow, or creep.

Suppose, for example, that a steel rod—30 ft of ½-in type 304—is used to support 1-½ tons of bricks in a 1,200°F furnace. The rod will slowly creep, until after four months it will have reached a length of 30 ft 1 in.

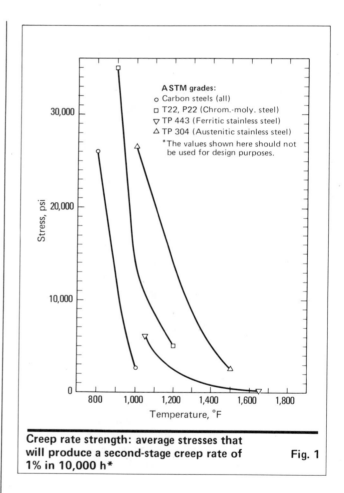

Creep rate strength: average stresses that will produce a second-stage creep rate of 1% in 10,000 h*   Fig. 1

After four years, it will break. This happens even though the rod, when new, was strong enough to support 5 tons of bricks at 1,200°F.

Double the brick load to 3 tons, or raise the furnace temperature to 1,350°F, and the 30-ft rod will creep to 30 ft 1 in. in less than a day. It will break within a week.

Originally published June 7, 1976

SELECTING STEEL TUBING FOR HIGH-TEMPERATURE SERVICE 177

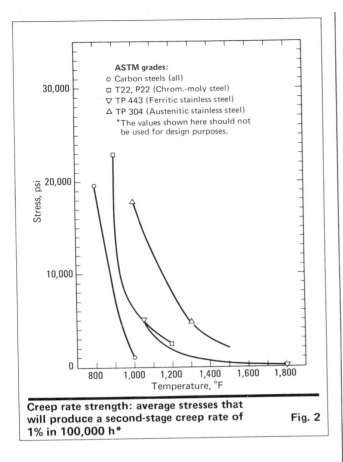

Creep rate strength: average stresses that will produce a second-stage creep rate of 1% in 100,000 h* Fig. 2

Typical hot tensile properties of several steels* Table I

ASTM grade	Temperature, °F	Tensile strength, psi	Yield strength, psi	Elongation, %
Carbon steels (all)	75	68,000	41,000	32
	200	60,000	29,000	32
	400	57,000	26,000	28
	600	54,000	25,000	28
	800	46,000	23,000	20
	1,000	28,000	18,000	15
	1,200	17,000	13,000	15
T22, P22	75	74,000	45,000	34
	200	67,000	37,000	31
	400	60,000	32,000	30
	600	61,000	31,000	24
	800	61,000	24,000	24
	1,000	50,000	21,000	27
	1,200	28,000	16,000	30
	1,400	14,000	7,500	30
TP 443	75	74,000	40,000	34
	200	70,000	35,000	34
	400	66,000	35,000	32
	600	60,000	34,000	30
	800	52,000	31,000	32
	1,000	38,000	24,000	36
	1,200	24,000	13,000	53
	1,400	10,000	6,000	72
TP 304	75	84,000	38,000	60
	200	75,000	32,000	52
	400	68,000	25,000	45
	600	65,000	24,000	45
	800	65,000	23,000	45
	1,000	60,000	19,000	42
	1,200	55,000	17,000	36
	1,400	35,000	16,000	50

Values listed here should not be used for design purposes.

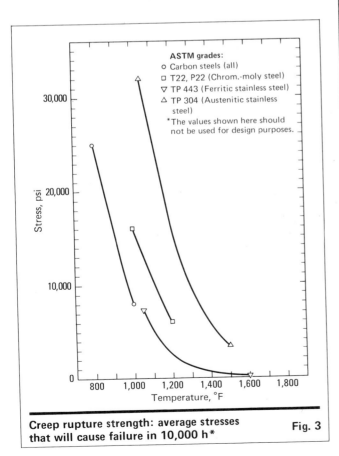

Creep rupture strength: average stresses that will cause failure in 10,000 h* Fig. 3

The two important creep properties—creep rate strength and rupture strength—cannot be ignored at high temperatures. Creep rate strength at different temperatures is shown in Fig. 1 and 2 as the stress (psi) that will cause a second-stage creep rate of 1% in 10,000 h, or in 100,000 h, respectively. Creep rupture strength is shown in Fig. 3 and 4 as the stress that will cause failure after 10,000 h, or after 100,000 h. (Values given in these figures are average ones and should not be used for design.)

As indicated in Fig. 1–4, creep rate strength and rupture strength come up rapidly as temperature falls off. The generalized curves in Fig. 5 show that strength at high temperatures is limited by creep properties. At lower temperatures, it is limited by hot tensile properties. (Table I lists hot tensile properties for several steels.)

Thermal conductivity—In equipment involving heat transfer, such as heat exchangers and boilers, thermal

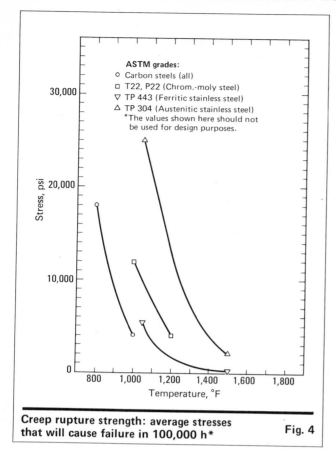

Creep rupture strength: average stresses that will cause failure in 100,000 h* Fig. 4

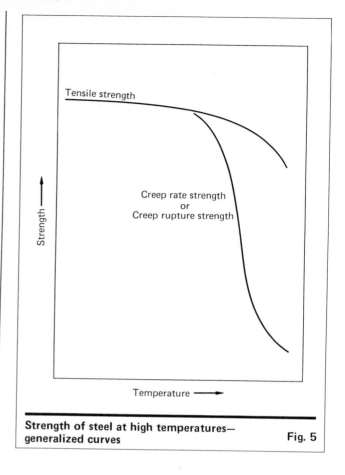

Strength of steel at high temperatures—generalized curves Fig. 5

conductivity becomes important. Fig. 6 shows thermal conductivities for various steels at different temperatures. At low temperatures, an increase in alloy content reduces thermal conductivity. However, at higher temperatures, the effect of alloy content on thermal conductivity is diminished.

Thermal expansion—Stresses caused by thermal expansion and contraction can shorten the life of metals. Whenever a system is repeatedly heated and cooled, the materials engineer must take steps to reduce the effects of these stresses. He does this either by using special design techniques, or by specifying metals with less thermal expansion. Table II lists thermal expansion for the four classes of steel.

Oxidation resistance—Corrosion due to oxidation also limits the temperatures at which a steel can be used. Adding chromium boosts a steel's oxidation temperature limit. Processes that demand nil corrosion, for example, often require stainless steels. Table III lists oxidation temperature limits in air and in steam for several steel alloys.

Both austenitic and ferritic stainless steels are used in high-temperature food processing in order to prevent corrosion that could alter either the color or taste of the food.

Although there are other properties that must be considered at high temperatures, these are the principal ones. Others will be dealt with in the following discussion of the four common classes of high-temperature steels.

Thermal expansion data for several steels Table II

ASTM grade	Mean coef. of thermal expansion from 32 to 1,200°F, in/(in)°F
Carbon steels (all)	8.3×10^{-6}
T22, P22	8.9×10^{-6}
TP 443	6.7×10^{-6}
TP 304	10.4×10^{-6}

The four classes of steels

In general, carbon steel should be specified for applications up to 950°F. Chromium-molybdenum steel should be used at temperatures between 800°F and 1,200°F. Above 1,000°F, austenitic and ferritic steels should be considered.

Carbon steels—Of the four classes, carbon steels are the cheapest and have the highest thermal conductivities. At low temperatures, carbon steel tubes can handle considerably greater heat fluxes than similar tubes fabricated from other steels.

Carbon steels also have good strength up to 750°F. But above this, their long-term strength plunges as temperature rises (Fig. 1, 2). Moreover, because of excessive oxidation, carbon steels are not recommended for service in air above 950°F, or in steam above 800°F.

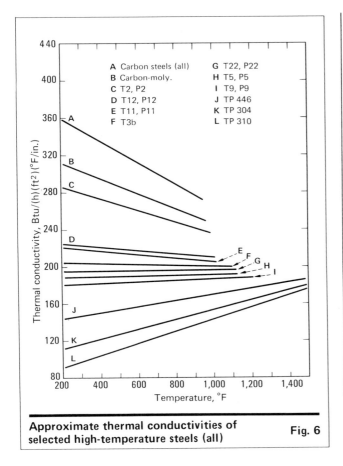

Approximate thermal conductivities of selected high-temperature steels (all)   Fig. 6

Oxidation temperatures limits of steels

ASTM grade	In air, °F	In steam, °F
Low-carbon steels (all)	950	800
Carbon-moly. steels (all)	975	850
T2, P2	1,025	950
T11, P11	1,050	1,050
T22, P22	1,125	1,100
T5, P5	1,150	†
T5b, P5b	1,200	†
T9, P9	1,200	†
TP 410	1,250*	1,200
TP 430	1,550*	†
TP 304	1,650*	1,600
TP 443	1,750*	†
TP 304 Si	1,750*	†
TP 310	2,000*	†
TP 446	2,000*	†

*It may be necessary to reduce these temperature limits if vanadium-containing ash or sulfur-containing gases are present.
† Not available

Heavy oxide scale on a carbon steel tube will insulate the tube and lessen its ability to transfer heat. For these reasons, the thermal conductivity advantage enjoyed by carbon steels drops off at high temperatures.

Chromium-molybdenum steels—These steels usually contain ½ to 1% molybdenum and ½ to 9% chromium. Their cost is higher than that of the carbon steels, but is still less than that of the ferritic and austenitic stainless steels.

Chromium-molybdenum steels are generally used at temperatures from 800 to 1,200°F. Adding chromium (or silicon) to a steel enhances its resistance to high-temperature oxidation. Unfortunately, increasing the chromium content can also reduce thermal conductivity and raise costs.

Ferritic stainless steels—After too long exposure to high temperatures, these alloys can weaken or embrittle. Despite these limitations, ASTM types 430, 443 and 446 have become popular in low-pressure, high-temperature heat exchangers. These alloys have higher thermal conductivity, lower thermal expansion, and lower cost than most other candidates for such applications.

Ferritic stainlesses have often proved to be the best for service in flue-gas conveyance systems. Up to 2,000°F, Type 446 resists attack from sulfur- and carbon-containing flue gases. Type 446 also will not scale excessively because of oxidation at the same high temperatures. For this application, the ferritic stainlesses are preferred over nickel-alloy austenitic stainlesses which have comparable oxidation resistance but are vulnerable to attack from hot, sulfur-containing gases.

Austenitic stainless steels—The strongest steels commonly used in high-temperature tubing are the austenitic stainlesses. They resist oxidation as well as ferritic stainlesses having the same chromium content. But unlike the ferritics, they do not embrittle after prolonged high-temperature exposures.

Austenitic steels cost more per pound than other high-temperature steels. Their higher cost, however, can sometimes be offset by materials savings realized from their greater strength. To contain a given pressure at temperatures above 1,000°F demands a far thinner tube wall with austenitic stainless than with carbon steel. Since less metal is needed for a thinner-wall tube, austenitic tubes can be cheaper than similar ones fabricated from other steels.

A thinner tube wall also offers less resistance to heat transfer, thus compensating for the lower thermal conductivity of austenitic stainlesses. And in systems that must transport hot fluids with minimal heat losses, the lower thermal conductivity of the austenitic steels may actually prove to be an advantage.

ASTM type 304 is commonly used in boilers, superheaters, condensers, and other high-temperature exchangers. The chromium-nickel alloys are specified in still tubes for refinery service.

The author

William C. Mack is Supervisor of the Technical Service Section, Metallurgical Dept., of Babcock & Wilcox Co.'s Tubular Products Div., P.O. Box 401, Beaver Falls, PA 15010. A registered professional engineer in Pennsylvania, he holds a B. S. in metallurgy from Pennsylvania State University and an M. S. in metallurgical engineering from Michigan Technological University. At Babcock & Wilcox, he has been concerned primarily with the applications and properties of steel. He has written several articles on the use of steels in high-temperature and corrosive atmospheres and is a member of the American Soc. for Metals, the National Assn. of Corrosion Engineers and the American Soc. for Testing and Materials.

Stainless-steel heat exchangers— Part I

The different types of heat exchangers. The different types of tubes. And the types, properties and characteristics of stainless steels available as construction materials.

*Ralph M. Davison and Kurt H. Miska,
Climax Molybdenum Co., div. of Amax Inc.*

Shell-and-tube heat exchangers are the standard design for many CPI applications. Fig. 1a

☐ Heat exchangers suffer a double-barreled attack— from the process stream and from the heat-transfer medium. As construction materials, stainless steels are highly desirable because of a wide range of corrosion-resistant properties and an ability to maintain excellent thermal transfer performance.

Paradoxically, it is usually somewhat easier to select a stainless steel for the process-stream side than for the heat-transfer-medium side. Despite the fact that the chemical content may be virtually anything, including organic and inorganic acids, bases, oils, food products and pharmaceuticals (all in a wide range of concentrations and temperatures), it is relatively easy to characterize the corrosivity of the process and hence to select an appropriate stainless steel.

By contrast, the heat-transfer medium is often water, whose corrosivity is very difficult to characterize. Selecting the appropriate alloy for the water side can be more difficult than choosing a material for the process stream.

Types of heat exchangers

The most common design employed in CPI (chemical process industries) applications is the shell-and-tube heat exchanger (Fig. 1a). The high strength and high modulus of stainless steel tubing, availability of the required product forms, and compatibility of stainless steel with the process stream and the rest of the process equipment are all factors that make stainless steel an excellent choice. Another design increasingly used is the flat-plate heat exchanger, formed by inflating two sheets resistance-welded together (Fig. 1b). Such a plate may be immersed in the product stream, or the heat transfer surface may be integral with the vessel itself, the vessel wall serving as one side of the plate (Fig. 2). When high contact area is required, an efficient exchanger is made by stacking flat-plate exchangers consisting of alternating layers of thin flat sheets and corrugated sheets. Fabrication is relatively simple and performance is excellent.

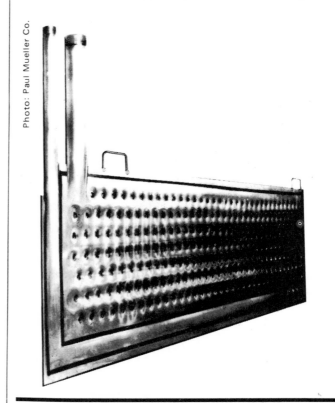

Flat heat-transfer surfaces are immersed in tanks for heating or cooling liquids. Fig. 1b

Originally published February 12, 1979

Characteristics and thermal properties of stainless steels used in CPI heat exchangers — Table I

Grade	Thermal conductivity (at 100°C, 212°F) W/mK, [(Btu)(ft)/(h)(ft²)(°F)]	Coefficient of thermal expansion[a] ×10⁻⁶ m/(m)(K), (in./(in.)(°F)	Characteristics
304	16.3 (9.4)	17.3 (9.6)	General-purpose grade
304L	" "	" "	Low-carbon version of 304
316	" "	16.0 (8.9)	More corrosion-resistant than Type 304
316L	" "	" "	Low-carbon version of 316
317	" "	" "	More corrosion-resistant than Type 316
317L	" "	" "	Low-carbon version of 317
321	16.1 (9.3)	16.7 (9.3)	Stabilized for welded applications and for service at 425-870°C (800-1,600°F)
329	—	14.4 (8.0)[b]	Duplex steel; more corrosion-resistant than Type 316
347	16.1 (9.3)	16.7 (9.3)	Similar to Type 321; better in highly oxidizing environments; some welding difficulties
348	" "	" "	Similar to Type 347
410	24.9 (14.4)	9.9 (5.5) / 11.7 (6.5)[c]	Martensitic grade; significantly inferior to Type 300 grades in corrosion resistance
430	26.1 (15.1)	10.4 (5.8)	Ferritic grade; slightly inferior to Type 300 grades in corr. res.
446	20.9 (12.1)	10.4 (5.8)	Similar to Type 442, but Cr increased for maximum scale resistance
Nitronic 50 (Armco)	15.6 (9.0)	16.2 (9.0)[d]	Corr. res. exceeds that of Type 316, plus it has twice the yield stress at room temperature
AL-6X (Allegheny Ludlum)	13.7 (7.9)	8.5 (15.3)[d] / 10.0 (18.0)[e]	Developed for seawater condenser service
20Cb-3 (Carpenter)	13.1 (7.6)	15.0 (8.3)[d]	Developed for H₂SO₄ service but extensive use in CPI
No. 20 Mod (Haynes-Cabot Corp.)	—	—	Corr. res. and res. to chloride pitting is midway between austenitic stainless steels and Ni-base superalloys
JS700 (Joslyn)	14.7 (8.5)	16.5 (9.15)	Resists H₃PO₄ in presence of chlorine and fluorine. Very good chloride res.
18Cr-2Mo[1]	26.3 (15.2)	11.0 (6.1)[f]	Excellent res. to chloride stress-corrosion cracking; ductile after welding. Comparable or superior to Type 304 in corr. res., especially chloride crevice attack
E-Brite™ 26Cr-1Mo[1,2,5]	19.5 (11.3)	10.6 (5.9)[g]	Made by vacuum induction melting; excellent res. to wide range of corrosives; excellent ductility and toughness after welding
26Cr-1Mo-Ti[1,3]	—	—	Similar in corr. res. to E-Brite 26-1 but lower in toughness. Available only as light-gauge sheet and tube
29Cr-4Mo[1,4] (Allegheny Ludlum)	—	9.4 (5.2)[d] / 13.0 (7.2)[e]	Particularly good res. to organic acids
29Cr-4Mo-2Ni[1,4] (Allegheny Ludlum)	—	9.4 (5.2)[d] / 13.5 (7.5)[e]	Developed for use in H₂SO₄ production; also good res. to organic acids
3RE60 (Sandvik)	20.0 (11.5)	12.2 (6.8)[e] / 14.9 (8.3)[h]	Two-phase alloy meets ASTM A669

A variety of exchangers are available for special applications. For example, a spiral heat exchanger, consisting of intertwined spirals, will provide similar flow characteristics for each medium (Fig. 3). Tube-in-tube exchangers are also quite commonly used.

Types of tubing

Seamless and welded tubing each have advantages. Some grades of stainless steel tubing are readily produced as seamless but are very difficult to weld. Although product uniformity and assurance against tube defects are likely to be superior for seamless tubing, welded tubing is readily available and, generally speaking, the most common grades are significantly less expensive. Welded tubing comes in several styles, including as-welded; welded and annealed; welded, bead-conditioned and annealed; and welded, drawn, and annealed. The choice is determined by process requirements and economics.

- Austenitic stainless steels are generally available in both seamless and welded tubing. Welded austenitic-stainless tubing is easily produced and widely available, especially AISI (American Iron and Steel Institute) Type 304; Type 316 is also commonly available.

- Standard ferritic-stainless steels are much more troublesome to weld, so welded AISI Type 430 is available only from a limited number of producers. Because of welding problems and inherent notch sensitivity, the standard ferritics are usually considered only in the seamless form. On the other hand, the low interstitial ferritics, including 18Cr-2Mo, 26Cr-1Mo (both grades), 26Cr-3Mo-2Ni, 29Cr-4Mo, and 29Cr-4Mo-2Ni, all have been produced satisfactorily as welded tubing.

- Martensitic stainless steels will air-harden, and therefore are seldom used to make welded tubing. However, they are available as seamless tubing in a wide range of sizes.

It is possible to increase heat-transfer performance by using tubing that has been "enhanced." Such tubing has had its surface area increased by a variety of processes, for example, spiral indentation or double fluting. Flow characteristics are modified significantly both inside and outside the tube. Another type gaining wider usage is "integrally finned tubing," on which fins are cold-formed

The heat-transfer surface is integral with the vessel wall.   Fig. 2

from the tube wall by a roll-threading operation. Outside contact area is thereby greatly increased. Further increases can be achieved by attaching a separate fin to the tube in various alignments, including a spiral winding.

Heat-transfer properties of stainless steel

If thermal conductivity were the only criterion for selecting an alloy for heat-exchange service, stainless steel would not be frequently selected. The thermal conductivity of stainless steels, shown in Table I, is lower than that of other metals and alloys. However, for long-term service, the overall heat-transfer performance of stainless steel may prove superior.

Fig. 4 shows the results of a two-year test of overall heat-transfer coefficients of AISI Type 304 stainless steel and arsenical admiralty brass in identical heat-exchanger service. Early in the test, the relative performance of the materials corresponded to the values established by the Heat Exchange Institute (Cleveland, Ohio). However, in only 240 days, the overall heat-transfer rate of the stainless steel was superior to that of the brass. Both materials showed some decrease in heat transfer due to fouling, but the important difference was in the nature of

Notes:
1. ASTM specifications A176, A240 and A268; all values maximum except as shown.
2. ASTM grade XM27; all values maximum except as shown.
3. ASTM grade XM33; all values maximum except as shown.
4. ASTM A268 standard specification for seamless and welded ferritic stainless-steel tubing for general service; all values maximum except as shown.
5. E-Brite 26-1 is the registered trademark of Allegheny Ludlum Industries Inc. (AL).

Temperature ranges
a. 0-100°C, 32-212°F   e. 20-1,000°C, 68-1,832°F
b. 20-816°C, 68-1,500°F   f. 20-200°C, 68-392°F
c. 0-650°C, 32-1,200°F   g. 32-106°C, 91-223°F
d. 20-100°C, 68-212°F   h. 20-700°C, 68-1,290°F

The same flow characteristics are obtained for each medium in spiral heat exchangers.   Fig. 3

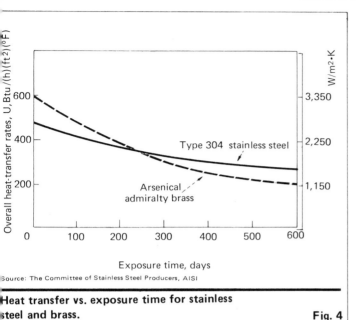

Heat transfer vs. exposure time for stainless steel and brass. Fig. 4

surface corrosion. Stainless steel forms only its atomically thin passive film, which has a negligible effect on the thermal conductivity. Brass forms an oxide layer of finite thickness for corrosion protection, and it is this film that interposes an additional thermal barrier.

The nature of the corrosion barrier is also an important factor in choosing the method used to clean tubes in service. To maintain efficiency, biofouling must be removed. For stainless steel, this can be done mechanically, without damaging the corrosion resistance of the tubing. However, for metals relying on a thick oxide layer, care must be taken that the layer is not broken or removed. Exposure of new metal will produce more oxidation, and cumulative damage from regular cleaning maintenance can lead to significant reduction of sound metal thickness.

Another comment that may be made on thermal conductivity as a measure of heat-exchanger performance is that the metal contributes little to the heat-exchange barrier. As shown in Fig. 5, films and scale contribute most of the barrier, while the metal typically accounts for only 2% of the resistance to heat flow.

The thermal conductivity of ferritic stainless steels is approximately 50% greater than that of the austenitic stainlesses.

The coefficient of thermal expansion of the ferritic stainless steels is low, roughly comparable to that of carbon steel. Thermal expansion of the austenitic stainless steels is about 50% higher, but is still less than that of some alloys considered for heat-exchange service. This lower thermal expansion can be very convenient in designing heat-exchange equipment for compatibility with the steels commonly used by the CPI.

Types and characteristics of stainless steels

Stainless steels are iron-base alloys that contain at least 10.5% chromium, the minimum level required to form the passive film that characterizes their corrosion resistance. The corrosion resistance, mechanical properties, weldability and other properties important to application are modified and improved by control of alloy elements— adding, for example, chromium, nickel, molybdenum, manganese, silicon, carbon, nitrogen, sulfur, titanium and niobium. The result is that there are 57 standard grades, as approved by AISI, and over 200 proprietary grades. As many as 15 of the standard grades have been used at one time or another for heat-exchange equipment. Many of the proprietary grades are specially formulated for heat-exchange service, but there is not enough space to review all of them. Omission is not meant as criticism.

Stainless steels may be grouped in five families, according to their metallurgical structure:
- Austenitic.
- Ferritic.
- Duplex (austenitic/ferritic).
- Martensitic.
- Precipitation-hardening.

It is useful to discuss stainless steels in this manner because of the similarities in properties within each family.

Austenitic stainless steels

The austenitic stainless steels are the most commonly used. Alloying elements—usually nickel, but sometimes also manganese and nitrogen—are used to stabilize the austenitic crystal structure. Austenite has a high solubility for carbon and nitrogen, which tends to minimize the damaging effects these elements have on toughness and weldability. Austenitic stainless steels are distinctive in having very great toughness, excellent ductility and a high rate of hardening by cold work. These stainless steels are generally nonmagnetic, although some of the grades become weakly magnetic after cold work. The austenitic structure provides for a somewhat lower thermal conductivity and higher thermal expansion than the other stainless families.

AISI Type 304 is far and away the most commonly produced and used stainless steel in all kinds of applications, including heat-exchange service. It has good corrosion resistance in a wide range of moderately aggressive chemical environments, and resists pitting and crevice attack in most fresh waters at moderate temperatures. It is readily weldable and provides excellent mechanical properties.

When higher corrosion resistance is required, the next alloy typically considered is AISI Type 316, in which 2-3% molybdenum has been added, along with additional nickel to assure the stability of the austenite. The molybdenum strongly increases resistance to chloride pitting and crevice corrosion. Type 316 is generally similar to Type 304 in its mechanical properties and fabricability. When greater corrosion resistance is needed, use is made of AISI Type 317, which features 3-4% molybdenum, a slight increase in chromium, and again more nickel to provide for the stability of the austenitic structure.

Ferritic stainless steels

As mentioned before, a stainless steel may be produced simply by alloying chromium with iron. However, the ferritic structure of such an alloy poses problems with toughness and weldability unless the carbon and nitrogen contents are minimized.

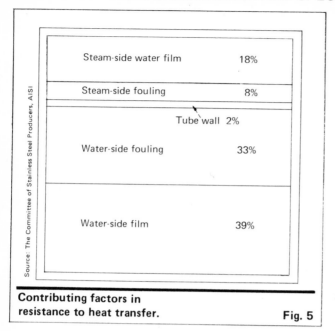

Contributing factors in resistance to heat transfer. Fig. 5

The rapid development of several new methods of lowering the carbon and nitrogen in the ferritic stainless steels, notably argon/oxygen decarburization and vacuum induction melting, have opened up a wide range of alloy levels not previously available. The ferritic stainless steels are all very resistant (possibly immune) to chloride stress-corrosion cracking, and now range in pitting resistance from being comparable to Type 304 to exceeding the resistance of the most highly alloyed austenitic stainless steels. As is the case with the austenitics, molybdenum is added for improved corrosion resistance, and there seems to be increased flexibility in the use of higher chromium contents in the ferritics.

At the top of the new ferritic-stainless-steel family are two steels unmatched in corrosion resistance by other stainlesses—29Cr-4Mo and 29Cr-4Mo-2Ni. They are produced by vacuum induction melting and rely on their low carbon and nitrogen contents to prevent sensitization by welding. The corrosion resistance of these grades is outstanding, permitting use in seawater and very severe chemical environments. These grades are readily weldable, but contamination by carbon and nitrogen must be prevented.

All ferritic stainlesses, including these new grades, are susceptible to the phenomenon called "885°F embrittlement." For long-time exposures between about 330 and 595°C (625 and 1,100°F), the steel is embrittled by the precipitation of an intermetallic phase. The steel generally retains its mechanical properties at higher temperatures but it is brittle when cycled to ambient temperatures. This embrittlement occurs most rapidly at 475°C (885°F), but more slowly as the temperature deviates more widely from this middle range. The phenomenon is unavoidable, and ferritic stainless steels should not be used under these conditions.

Duplex (austenitic/ferritic) stainless steels

When limited amounts of nickel, and sometimes manganese and nitrogen, are added to the ferritic stainless steels, it is possible to produce a stainless containing both austenite and ferrite. The resultant duplex steel has a yield strength significantly above that of the common austenitic grades. Best known of these grades is AISI Type 329, sometimes known as Carpenter 7-Mo, which provides good general corrosion resistance, and stress-corrosion-cracking resistance somewhat better than that of AISI Type 304. Sandvik 3RE60 is a similar alloy with a lower chromium content. In using the duplex steels, attention must be given the final high-temperature exposure because of the tendency for chromium and molybdenum to segregate between the phases.

Martensitic stainless steels

With lower chromium and higher carbon, it is possible to produce stainless steels that are hardenable by heat treatment. These steels are usually selected when their excellent strength is required. AISI Type 410 is the most widely used of the martensitic grades. It is occasionally used for heat-exchange service, but because of its lower chromium and higher carbon, its corrosion resistance is substantially less than that of AISI Type 304. The straight-chromium martensitic stainless steels are resistant to chloride stress-corrosion cracking but are susceptible to cracking when fully hardened.

Precipitation-hardening stainless steels

The precipitation-hardening steels use an aging reaction to produce very-high-strength stainlesses. These grades are seldom used for heat-exchange service.

Kenneth J. McNaughton, *Editor*

The authors

Kurt H. Miska is a Senior Technical Writer in the Technical Information Dept. at Climax Molybdenum Co. in Greenwich, Conn. Tel: (203) 622-3598. Before joining the company in June 1978, he was an Associate Editor of *Materials Engineering* for seven years. There he covered developments in metals, composites, adhesives and welding in news and feature articles. He has written hundreds of papers on a wide range of technical subjects.

Ralph M. Davison is Manager of Stainless Steel Development at the Climax Molybdenum Co., div. of Amax Inc., 3072 One Oliver Plaza, Pittsburgh, PA 15222. Tel: (412) 281-6238. Previously, he did research in the physical metallurgy of stainless steels at the Climax Molybdenum laboratory in Ann Arbor, Mich. He has a B.S. and Sc.D. in metallurgy from Massachusetts Institute of Technology, and an M.B.A. from the University of Michigan. He is a member of the American Inst. of Mining, Metallurgical and Petroleum Engineers; American Soc. for Metals; American Soc. for Testing and Materials; and National Assn. of Corrosion Engineers.

Stainless-steel heat exchangers—Part II

Stainless steel is not immune to corrosion. But if you select the right type for heat exchangers, you can reduce the risk of attack.

Ralph M. Davison and Kurt H. Miska,
Climax Molybdenum Co., div. of Amax Inc.

☐ The first part of this two-part article (pp. 188-192) examined the types of heat exchangers and tubing that are available. It also looked at the different kinds of stainless steels, their characteristics and heat-transfer properties. This part examines the corrosion of stainless steel—how it occurs and how it can be prevented. There is also a guide on how to select a specific grade of stainless steel for a particular chemical-process-industries (CPI) service.

Corrosion of stainless steel

There are many ways in which materials, including stainless steels, can fail. However, corrosion problems generally can be characterized in several fairly distinct groupings. Most corrosion failures fall within the following categories:

■ **General corrosion.** Most materials, including stainless steel, corrode due to uniform chemical or electrochemical attack. Uniform electrochemical corrosion occurs in environments that cause a constant shifting of anodic and cathodic areas. A prime example of general corrosion is rusting of metal surfaces exposed to the atmosphere.

■ **Galvanic corrosion.** When two dissimilar metals are in contact, or otherwise connected electrically, there is the opportunity for accelerated corrosion attack on the less-noble metal. This phenomenon is employed to great advantage when sacrificial anodes are used to protect equipment. However, there can be a problem when dissimilar metals are combined in construction. For example, the thin-walled tubes of a stainless-steel heat exchanger must be noble with respect to the shell, heads and supports of the unit, to prevent early failures of the tubing. Using thick carbon-steel tubesheets to protect the light-walled stainless is a positive application of galvanic corrosion. The size of the relative areas of the two metals can play a significant role. A very large cathodic area coupled with a very small anodic area can lead to highly accelerated corrosion of the anode.

■ **Pitting.** Pitting is a form of localized attack that can be highly destructive to CPI equipment because perforation can occur very quickly with almost negligible weight loss of the structure. Because of the nature of the passive film of stainless steel, pitting is one of the likely forms of attack. Imperfections in the film (associated with mechanical damage to the surface), and inhomogeneities in the metal surface (such as inclusions, or surface scale), can initiate localized corrosion. The presence of chloride ion in the environment greatly accelerates the pitting attack. Once a pit is initiated, the local chemical environment within the pit becomes much more aggressive than the bulk environment, accounting for the rapid propagation of pits. Flowrate of the corrosion medium over the surface plays an important role, with high flowrates preventing this concentration and acceleration effect. The stability of the stainless-steel passive film, and thereby its resistance to pitting, can be increased by using grades with higher chromium and molybdenum levels. The role of minor elements is also important—for example, the presence and composition of sulfide inclusions that could be pitting-initiation sites.

■ **Crevice corrosion.** Crevice attack may be thought of as a severe form of pitting. It can start at any joint, or at any deposit resulting from fouling or scaling. In these areas, flow is restricted, corrosion is initiated, and the corrosivity of the local environment is greatly increased, just as occurs in the bottom of a pit. It is difficult to totally avoid crevices in a practical operation, but an effort should be made to do so in construction and to clean fouling from

Originally published March 12, 1979

Stainless steels in heat-exchanger service

Environment	Grades	Environment	Grades
Acids			ferritic stainless steels, such as 29Cr-4Mo-2Ni, resist boiling H_2SO_4 to concentrations of at least 10%.
Hydrochloric acid	Stainless generally not recommended except when solutions very dilute and at room temperature.		
"Mixed acids"	Usually no appreciable attack on Type 304 or 316 as long as sufficient nitric acid present.	Sulfurous acid	Type 304 may be subject to pitting, particularly if some sulfuric acid present. Type 316 usable at moderate concentrations and temperatures. Fe-Cr-Mo ferritic steels can be used for modest concentrations.
Nitric acid	Type 304L or 430.		
Phosphoric acid	Type 304 satisfactory for storing cold phosphoric acid up to 85% and for concentrations up to 5% in some processes. Type 316 more resistant and generally used for storing and manufacture if fluorine content not too high. Type 317 somewhat more resistant than Type 316. At concentrations up to 85%, metal temperature should not exceed 100°C (212°F) with Type 316; slightly higher with Type 317. Oxidizing ions inhibit attack, and other inhibitors such as arsenic may be added. JS 700 and 904L recommended for more-severe service. Carpenter 20Cb3 also much more resistant than Type 316.	**Bases** Ammonium hydroxide, sodium hydroxide, caustic solutions	Steels in the 300 series generally have good corrosion resistance at virtually all concentrations and temperatures in weak bases, such as ammonium hydroxide. In stronger bases, such as sodium hydroxide, some attack, cracking or etching in more-concentrated solutions and at higher temperatures. Commercial-purity caustic solutions may contain chlorides, which will accentuate attack and may cause pitting of Type 316 as well as Type 304. 26Cr-1Mo resists hot caustic and does not crack.
		Organics Acetic acid	Acetic acid seldom pure in chemical plants but generally includes numerous minor constituents. Type 304 used for wide variety of equipment—including stills, base heaters, holding tanks, heat exchangers, pipelines, valves and pumps—for concentrations up to 99% at temperatures up to about 50°C (120°F). Type 304 also satisfactory for contact with 100% acetic acid vapors, and—if small amounts of turbidity or color pickup can be tolerated—for room-temperature storage of glacial acetic acid. Types 316 and 317 have broadest range of usefulness, especially if formic acid also present or with unaerated solutions. Type 316 used for fractionating equipment, for 30 to 99% concentrations where Type 304 cannot be used, for storage
Sulfuric acid	Type 304 can be used at room temperature for concentrations over 80%. Type 316 can be used in contact with sulfuric acid up to 10% at temperatures up to 50°C (120°F) if solutions are aerated; attack is greater in air-free solutions. Type 317 may be used at temperatures as high as 65°C (150°F) with up to 5% concentration. Presence of other materials may markedly change corrosion rate. As little as 500 to 2,000 ppm of cupric ions permits Type 304 in hot solutions of moderate concentration. Other additives may have opposite effect. Cr-Ni-Mo-Cu austenitic stainless steels such as Carpenter 20Cb3 resist sulfuric acid over broad range of concentrations and temperatures. Ni-bearing high-chromium Fe-Cr-Mo		

the steel. More-highly-alloyed stainless steels, particularly those with molybdenum additions, are used when crevices are unavoidable.

■ **Intergranular corrosion.** Several mechanisms lead to preferential attack at grain boundaries of a stainless steel. The most important of these is sensitization that results from welding. In this case, carbon at grain boundaries precipitates as chromium carbide. The resulting chromium-depleted zone adjacent to the carbide is subject to accelerated attack. Once material has been sensitized, it can be restored to a corrosion-resistant condition by annealing to permit (1) resolution of the carbides and (2) chromium diffusion that eliminates the chromium gradient. A better solution is to avoid sensitization by alloy selection. The low-carbon grades, or "L-grades," of austenitic stainless steels are not sensitized by welding, nor are the grades stabilized with titanium or niobium. The ferritic stainless steels meant for welding before service are generally stabilized with titanium or niobium. The stabilized grades also resist sensitization by very long-time high-temperature exposure.

■ **Stress-corrosion cracking** (SCC). Critical combinations of temperature, stress, chloride and oxygen can cause stress-corrosion cracking. Most ferritic stainless steels are highly resistant or immune to chloride stress-corrosion cracking. However, all austenitic stainless steels are at least partially susceptible, although the highly alloyed grades are extremely resistant in most sodium chloride solutions. Although it is generally true that localized corrosion occurs before the initiation of SCC, the amount of the corrosion may be so small as to be virtually undetectable even after a failure. This is a very insidious kind of corrosion because it is difficult to detect while in progress, yet can lead to catastrophic failures. Even water with very low chloride levels can lead to SCC when the right conditions of temperature and stress are present.

STAINLESS-STEEL HEAT EXCHANGERS—PART II

Table I

Environment	Grades	Environment	Grades
	vessels, pumps and process equipment handling glacial acetic acid, which would be discolored by Type 304. Type 316 likewise applicable for parts having temperatures above 50°C (120°F), for dilute vapors and high pressures. Type 317 has somewhat greater corrosion resistance than Type 316 under severely corrosive conditions. None of the stainless steels adequately resists corrosion by glacial acetic acid at the boiling temperature or at superheated vapor temperatures. 18Cr-2Mo, 26Cr-1Mo, 29Cr-4Mo all have good resistance to organic acids.	Phthalic anhydride	Type 316 usually used for reactors, fractionating columns, traps, baffles, caps and piping.
		Soaps	Type 304 used for parts such as spray towers, but Type 316 may be preferred for spray nozzles and flake-drying belts to minimize off-color product. Ferritic grades useful when chloride stress corrosion expected.
		Synthetic detergents	Type 316 used for preheat, piping, pumps and reactors in catalytic hydrogenation of fatty acids to give salts of sulfonated high-molecular alcohols. Ferritic grades useful when chloride stress corrosion expected.
Aldehydes	Type 304 generally satisfactory.		
Amines	Type 316 usually preferred to Type 304.	Tall oil (pulp and paper industry)	Type 304 has only limited usage in tall-oil distillation service. High-rosin-acid streams can be handled by Type 316L, with minimum molybdenum content of 2.75%. Type 316 can also be used in more-corrosive high-fatty-acid streams at temperatures up to 245°C (475°F), but Type 317 probably required at higher temperatures. Higher-molybdenum stainless steels, such as JS700 or Type 317LM, should be considered for more-severe environments.
Cellulose acetate	Type 304 satisfactory for low temperatures, but Type 316 or Type 317 needed for high ones.		
Citric, formic and tartaric acids	Type 304 generally acceptable at moderate temperatures, but Type 316 resistant to all concentrations at temperatures up to boiling. 18Cr-2Mo, 26Cr-1Mo and 29Cr-4Mo excellently resist these acids.		
Esters	From corrosion standpoint, esters comparable with organic acids.	Tar	Tar distillation equipment almost all Type 316 because coal tar has high chloride content; Type 304 does not adequately resist pitting.
Fatty acids	Up to about 150°C (300°F), Type 304 resists fats and fatty acids, but Type 316 needed at 150 to 260°C (300 to 500°F) and Type 317 at higher temperatures. Higher-molybdenum stainless steels, such as JS700 or Type 317LM, should be considered for more-severe environments.	Urea	Type 316 generally required. Higher-chromium alloys, such as 25Cr-24Ni-2.3Mo and 26Cr-1Mo, give better service in stripper columns.
		Pharmaceuticals	Type 316 usually selected for all parts in contact with product because of inherent corrosion resistance and greater assurance of product purity.
Paint vehicles	Type 316 may be needed if exact color and lack of contamination important.		

■ **Erosion-corrosion.** It is possible for a combination of erosion and corrosion to cause metal deterioration much more rapidly than either process acting separately. For a metal other than stainless steel (one that depends on an oxide layer for corrosion protection), this mode of attack is particularly dangerous. There can be a sequence of initial corrosion, abrasive removal of the protective layer, and more-rapid corrosion. With stainless steels, severe abrasion can cut through the passive film, leading to initiation of corrosion attack. Erosion-corrosion is especially troublesome in piping systems at fittings and other geometric restrictions to flow.

Avoiding corrosion

Careful materials selection and good design practice are two important factors in minimizing corrosion. Understanding the nature of corrosion and the general capabilities and limitations of various alloys is essential. Table I lists stainless steels used in heat-exchange service and comments on their characteristics.

The successful use of stainless steels depends on avoiding localized attack by: pitting or crevice attack, sensitization to intergranular attack, stress-corrosion cracking, and galvanic corrosion.

Pitting and crevice corrosion most frequently result from chlorides in the process stream or in the water side. However, the severity of a particular level of chlorides is strongly influenced by temperature, flowrate, oxygen, and various dissolved ions. The opportunity for localized concentration must be considered, as in a splash zone or an area of low flowrate over a hot surface. With the austenitic stainless steels, resistance to pitting is generally increased by molybdenum additions, with a progression of levels in AISI (American Iron and Steel Institute) Types 304, 316 and 317.

When further corrosion resistance is required, "Type

317LM," Jessop 700, and Alloy 6X continue the series, to name a few of the many special proprietary steels. Among the ferritic stainless steels, both chromium and molybdenum are increased with the series: AISI Type 430, 18Cr-2Mo, 26Cr-1Mo, 26Cr-3Mo-2Ni, 29Cr-4Mo and 29Cr-4Mo-2Ni spanning a range of corrosion resistance even greater than that of the austenitics. Good design practice that eliminates crevices and areas where deposits are likely to build up and restrict flow will contribute to equipment performance and lessen the likelihood of corrosion damage in normal operation.

It should also be remembered that corrosion damage can occur very easily during downtime or scheduled maintenance. Most chemical cleaning solutions are highly corrosive and should be flushed from equipment as soon as cleaning is completed.

Attention must be paid to temperature excursions during downtime. Equipment may be subjected to temperatures that deviate considerably from the normal operating conditions. High temperatures may suddenly accelerate chloride attack. However, cooling can also be dangerous, leading to condensation of sulfuric acid in a flue gas. Original alloy selection should consider the extremes of maintenance schedules and possible unscheduled excursions of the process conditions, not just the scheduled operation.

Sensitization to intergranular corrosion occurs when the metal is exposed to high temperatures, whether for short times, as in welding, or for long times in planned operation. There are several methods to avoid sensitization, and each should be considered when applicable to a particular system:

- Use annealed material. Properly annealed material is not susceptible to intergranular attack. Annealing can restore corrosion resistance to sensitized material. However, annealing fabricated equipment is frequently expensive or impractical.
- Use low-carbon alloys. These do not get sensitized during the short time exposures of welding. The standard austenitic grades with the "L" suffix are intended for welding, including AISI Types 304L, 316L and 317L. Many proprietary grades are made with low carbon levels. Of the ferritics, only the most highly alloyed proprietary grades, 29Cr-4Mo and 29Cr-4Mo-2Ni, contain low carbon. The low-carbon austenitic grades are sensitized by long exposures to high temperatures.
- Use stabilized grades. Addition of titanium or niobium will lead to precipitation of a stable carbide, preventing sensitization by either short or long exposure. AISI Types 321 and 347 use this approach. Most of the proprietary ferritic stainless steels are stabilized for weldability.

As already noted, **stress-corrosion cracking** can be a serious problem for any susceptible material when the right conditions of stress, temperature, chloride and oxygen are present. All austenitic stainless steels are susceptible to some degree, although high-molybdenum and high-nickel in the proprietary austenitic stainlesses produce a level of SCC resistance for most practical heat-exchange environments, including seawater. The ferritic stainless steels are all either immune or extremely resistant to SCC, and are frequently selected for heat-exchange service on this basis.

Tube-in-tube exchangers are used in food processing, sewage treatment and numerous other applications.

Galvanic interactions can be used to great advantage when selecting metals in a system. Although it is generally best to use as few different metals as possible within a single assembly, mixing of alloys is frequently unavoidable. Galvanic interaction can be made to help if the large, noncritical elements of construction can be made anodic to the small, thin or critical elements. For example, heavy tubesheets and tube supports can be made to protect the thin-walled tubing where any corrosion may lead to leakage. In cases where dissimilar metals must be used and contact is undesirable, an insulating lacquer or gasket can provide a nonconducting separation at the point of contact.

Stainless steel is often used in heat exchange equipment because of its resistance to both the water and the process environments in a wide variety of applications (see photo above). It also resists corrosion due to fouling and resists oxidation and scaling in high-temperature service. Finally, it has low "lifetime costs"—the combination of initial costs and maintenance requirements.

Kenneth J. McNaughton, Editor

The authors

Ralph M. Davison is Manager of Stainless Steel Development at Climax Molybdenum Co., div. of Amax Inc., 3072 One Oliver Plaza, Pittsburgh, PA 15222. Tel: (412) 281-6238. Previously, he did research in the physical metallurgy of stainless steels at the Climax Molybdenum laboratory in Ann Arbor, Mich. He has a B.S. and Sc.D. in metallurgy from Massachusetts Institute of Technology, and an M.B.A. from the University of Michigan. He is a member of the American Inst. of Mining, Metallurgical and Petroleum Engineers; American Soc. for Metals; American Soc. for Testing and Materials; and National Assn. of Corrosion Engineers.

Kurt H. Miska is a Senior Technical Writer in the Technical Information Dept. at Climax Molybdenum Co. in Greenwich, Conn. Tel: (203) 622-3598. Before joining the company in June 1978, he was an Associate Editor of *Materials Engineering* for seven years. There he covered developments in metals, composites, adhesives and welding in news and feature articles. He has written hundreds of papers on a wide range of technical subjects.

Section II
HEAT EXCHANGE EQUIPMENT—OPERATION AND MAINTENANCE

Design and specification of air-cooled steam condensers
Continuous tube cleaning improves performance of condensers and heat exchangers
Evaluate reboiler fouling
Finding the natural frequency of vibration of exchanger tubes
In-place annealing of high-temperature furnace tubes
Preventing fouling in plate heat exchangers
Operation and maintenance records for heating equipment—I
Operation and maintenance records for heating equipment—II
Operating performance of steam-heated reboilers
Performance of steam heat-exchangers

Design and specification of air-cooled steam condensers

Even the nonspecialist purchaser of an air-cooled steam condenser can apply these guidelines to ensure that the unit selected avoids common design deficiencies. The components of a steam condensing system, and the key considerations that should underlie a purchaser's inquiry specifications, are also reviewed.

M. W. Larinoff, W. E. Moles and R. Reichhelm, Hudson Products Corp., Houston, Tex.

☐ Steam turbines are finding increasing use in electric-utility powerplants, industrial plants, process plants and commercial installations. Such turbines drive not only electric generators but also all types of pumps, fans, compressors, shredders, mills, paper machines, and so on.

Steam condensers coupled to the exhaust of these turbines return condensate to the power cycle and boiler. Either surface-type or air-cooled condensers can be selected. The former have once-through or recirculating water as the cooling medium, while the latter are once-through systems employing the atmosphere as the heat sink. Among the advantages of air-cooled steam condensers, compared with wet systems, are elimination of: makeup water supply, blowdown disposal, water-freezing problems, water vapor plumes, and concerns over governmental water-pollution restrictions. Because of the dry nature of the equipment, lower system-maintenance costs also result.

Air-cooled steam condensers have been used since the 1930s. Some are as small as 1 million Btu/h, condensing at 20 psi, while others are as large as 2 billion Btu/h, condensing at 2 in. Hg absolute pressure. Units can be installed at grade, on pipe racks or on top of buildings; they can be mechanical draft or natural draft; their bundle arrangement adapts to mounting vertically, horizontally or on an incline.

A typical A-frame arrangement is shown in Fig. 1. Fig. 2 presents the basic flow diagram of a system.

Many evolutionary design changes have been effected during the past 40 years as a result of field experience. The main problem plaguing the industry has not been how to condense steam but rather how to prevent a unit from suffering a loss in thermal performance during the summer, and freezing during winter.

Originally published May 22, 1978

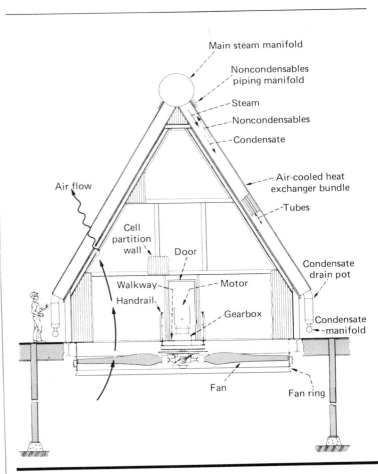

Air-cooled steam condenser of A-frame design   Fig. 1

192 HEAT EXCHANGE EQUIPMENT—OPERATION AND MAINTENANCE

Today, the mechanics of this steam-condenser problem are understood, fortunately, so that the deficiencies of past designs can be identified and corrected.

Identifying the steam-condenser problem

A successful air-cooled steam condenser must continuously and completely gather and discharge all of the noncondensables in the system. These are the gases that result from atmospheric air leaks into the vacuum portions of the steam-cycle equipment, and from the chemicals used for boiler feedwater treatment. The noncondensables are left behind inside the tubes and headers when the steam condenses. They accumulate if not removed from the system at the release rate.

Such trapping of noncondensables is responsible for the steam condenser problem. During the winter, the trapped noncondensables can cause freezing of condensate; during the summer, they blanket heat-exchange surfaces and reduce heat-transfer capability. In addition, the noncondensables are absorbed by condensate in the trapped pockets and promote metal corrosion.

How do pockets of trapped noncondensables form? Typically, they arise when steam enters the same area of the condenser from different directions—the most common location is the condenser tubes themselves. Turbine exhaust steam flows into the tubes from the inlet end, while "backflow" steam flows (from higher rows of tubes) into the same tubes from the rear end, via the rear header. With both ends blocked by the flow of steam, the noncondensables become trapped inside the tubes.

Fig. 3 illustrates in more detail the trapping of noncondensables in a simple steam condenser having just two rows of tubes and a conventional, nondivided, rear header. Since the first row is exposed to the lower, ambient-air temperature, while the second row is contacted by already-heated air, the second row condenses less steam than the first, and therefore has a lower steam-pressure drop. The pressure in the rear header equals the front-header pressure minus the pressure drop in the second row. The pressure in the rear header thus exceeds the pressure at the outlet end of the first row.

So, steam flows into both ends of the first row of tubes, and noncondensables become trapped inside. They cannot flow into the rear header until their pressure equals the rear-header steam pressure (point C in Fig. 3). By then, noncondensables extend for the tube length G-H. Since there is very little steam flowing with the noncondensables, that length of the metal tube

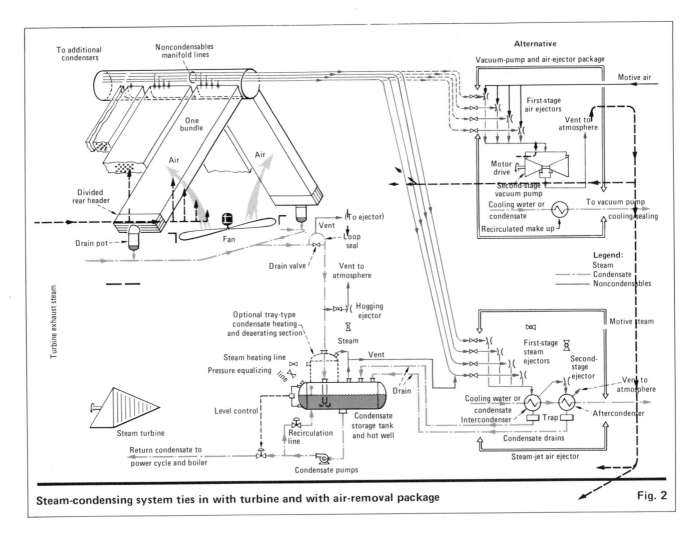

Steam-condensing system ties in with turbine and with air-removal package — Fig. 2

DESIGN AND SPECIFICATION FOR AIR-COOLED STEAM CONDENSERS

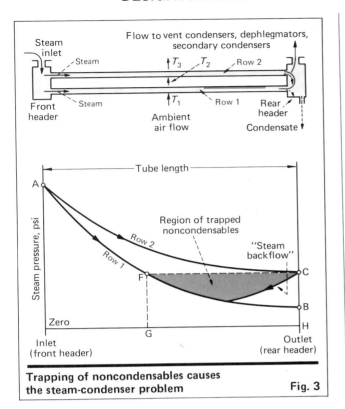

Trapping of noncondensables causes the steam-condenser problem   Fig. 3

becomes cold. Condensate freezes enroute as it flows downward by gravity toward the rear header through this cold section.

Design alternatives

Some condenser designs try to cope with this backflow problem by steam "blowthrough" to dephlegmators or secondary condensers. The objective is to equalize the steam pressure-drop across each tube row in the main condenser by using larger steam flows, with the remainder of the steam being condensed downstream in the secondary or vent condenser.

Fig. 4 shows a typical arrangement of main and vent condensers. This design has an open rear header on both condenser sections. If any one of the many variables of turbine exhaust-steam flow, ambient-air temperature or air flowrate is upset so that the blowthrough steam quantity is less than it should be, steam backflow to the first row of the main condenser can occur—which is the steam condenser problem. Similarly, steam backflows to the first row in the open rear headers of the vent condenser, thereby trapping noncondensables once more.

A variation to this A-frame configuration is the horizontal bundle arrangement, built with a slight inclination for condensate drain purposes. Here, the vent condenser has cocurrent flow of steam and condensate

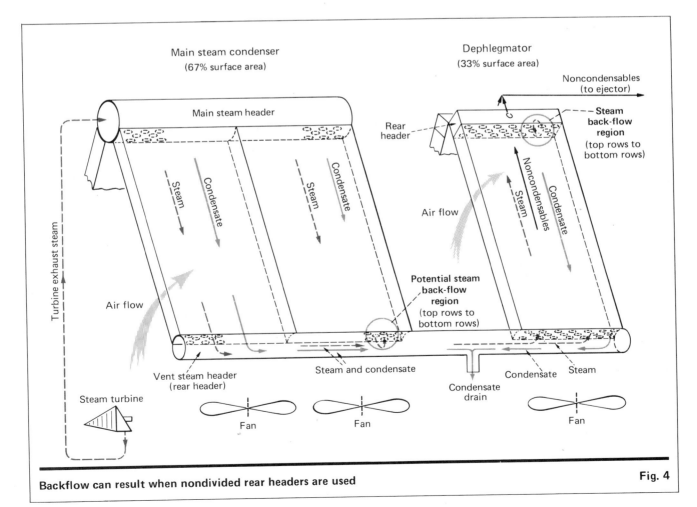

Backflow can result when nondivided rear headers are used   Fig. 4

rather than countercurrent, but the steam condenser problem remains a reality.

Fig. 5 shows a steam condenser in which each bundle has its own main and vent section. The horizontal tubes have a two-pass arrangement and are interposed so as to minimize steam pressure differences in the rear headers and at the connection to the steam-jet air ejector. The 1-4 mains connect to the 2-3 vents, and the 2-3 mains connect to the 1-4 vents. But since the face velocity of the air flow in the upper regions of the bundle exceeds that at the base, where the vent condenser is located, the steam pressures are not completely equal where the tubes are connected to the common exhaust manifold system (leading to the steam-jet air ejector). There can thus be a backflow between the vent condenser manifolds serving rows 3 and 4. Also, the first row of the vent condenser is here exposed to the ambient air (to try to achieve balanced steam pressures at the outlet of the vent condenser), rather than being protected from the cold ambient air, as dictated by the low heat content of the low-partial-pressure steam.

Other designs employ internal flow-control devices in the front header, such as fixed orifices or flapper valves. This equalizes steam pressure drop among rows, but only at the design operating point. A change in any operating variable changes the flow relationship among rows, and thus the steam pressure drops; the net result again is steam backflow in the undivided rear headers.

Still other designs may vary fin height, fin spacing or finned length from row to row in an attempt to achieve balanced steam-pressure drop between rows and avoid rear header backflow.

Summing up, all of the above methods are undesirable because they can degrade the fluid energy on the steam side, and/or are heat-transfer inefficient on the finned side. They function properly at the design point but run into difficulty when any one of several operating variables (ambient air temperature, air flow, steam flow) is changed.

A better steam condenser is the single-row design shown in Fig. 6. As the steam flows through the tubes, it condenses and pushes the noncondensables forward until they reach the rear header. The rear header is purged of noncondensables by means of vent tubes connected to the steam-jet air-ejector system. The vent tubes provide a more effective scavenging action by inducing additional mass flow through the rear header. As a further freeze protection feature, the vent tubes are installed in the warm-air portion of the bundle.

Note that there is no need to balance steam pressure drops because there is only one row, and each tube in the row experiences the same air temperature. The movement of noncondensable gases is always forward because steam backflow does not occur.

To make the single-row concept commercially practical, several such steam condensers must be stacked together, one on top of the other (Fig. 7). The internal fluid flows of this multirow condenser must be completely independent at all times, and the condensate and noncondensables must be withdrawn separately.

A new condenser from Hudson Products Corp. (the Stac-Flo) has such a design. Condensate from each row is withdrawn from the rear headers through hydraulic pressure-seals of a water-leg loop design, into a common, heated drain pot. The noncondensables are removed from each row by individual first-stage ejectors; these connect to a common header for flow to the intercondenser, second-stage ejector and finally the aftercondenser. There is positively no passage of steam, condensate or noncondensables among rows inside the condenser at any time.

Fig. 8 illustrates the operation of a typical divided rear-header vacuum steam condenser in turbine service. Note the wide spread of air temperatures, steam condensing rates and steam pressure drops between the rows.

Scope of the purchased package

The purchaser has many options to consider and many questions to answer in preparing the inquiry specification to be presented to manufacturers of air-cooled steam condensers. First, the scope of the system package to be purchased must be decided, and the more important specification details established.

An air-cooled steam condenser system starts at the turbine exhaust flange. It includes all of the equipment necessary to condense the steam and return the condensate to the boiler feedwater piping. These items are:

1. Air-cooled steam condenser tower.
2. Air-flow control equipment.
3. Wind and/or cell-partition walls.
4. Steam-bypass heating system.
5. Air removal equipment.
6. Condensate storage tank.
7. Condensate pumps.
8. Steam ducts and expansion joints.
9. Condensate drain and air-removal piping.
10. Instrumentation, controls and alarms.
11. Pressure-relief device for protection of steam-turbine exhaust casing.
12. Steam-duct condensate drain system.

The purchaser has the option of buying this complete system package, or requesting only a portion of it.

The basic air-cooled steam condenser (Item 1) includes the bundles, steam distribution manifold, fans, motors, gear boxes and supporting steel. In large installations, the cost of the tower structure supporting the condenser bundles can be a substantial portion of the total cost. The structure's design specifications for wind load, snow load, live load and seismic requirements should be carefully chosen. Generally, grade-mounted towers cost less than roof-mounted ones.

Limitations on plan dimensions must be made clear in the inquiry specification. Heat sources located close to the proposed tower and discharging into the atmosphere must be identified. The prevailing wind directions define the proper location and orientation of the tower with respect to other large structures and heat sources. Summer winds are important in the consideration of thermal performance, and winter winds in prescribing freeze-protection measures. Noise limitations should also be stated, since lower fan noise generally requires lower tip speed, more fan blades and possibly wider blades.

The purchaser should specify whether the thermal performance guarantees are to be based on steam pres-

DESIGN AND SPECIFICATION FOR AIR-COOLED STEAM CONDENSERS

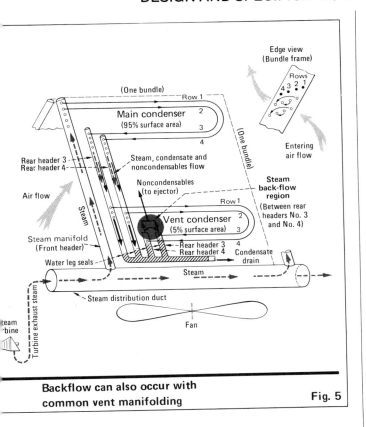

Backflow can also occur with common vent manifolding — Fig. 5

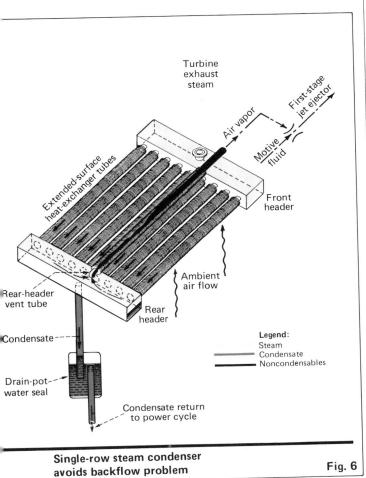

Single-row steam condenser avoids backflow problem — Fig. 6

sure measured at the turbine exhaust flange, or at the steam manifold inlet at the condenser. Other options are an all-welded system to reduce the potential for air leaks into the condenser, and the use of extruded aluminum fins (Fig. 9), which provide longer troublefree operation than embedded or wrap-on fins (these are prone to galvanic corrosion because of their bimetallic tube-to-fin interface).

Air-flow control equipment for freeze protection (Item 2), though an integral part of the engineered package supplied by the manufacturer, nevertheless reflects the purchaser's preferences and needs. Consideration should be given to variable-pitch fans, air-flow control louvers, steam isolating valves and two-speed motors. The extra price of electric starters needed for two-speed motors should be included.

Wind walls (Item 3) are sometimes necessary to protect the bundles from wind gusts that can upset equilibrium operating conditions and at times cause freezing in some remote parts of the tower. Partition walls between fan cells isolate operating cells from nonoperating ones. Without partition walls, a nonoperating fan would induce bypass of air intended for the bundles.

Depending upon the minimum design ambient-air temperature, the type of turbine, and the type of plant operation, it may be economic to provide a steam-bypass heating system for cold-weather startup (Item 4). This would operate directly off the boiler, requiring both a steam pressure-reducing station and a desuperheating station, with steam flow exhausting directly into the main steam duct. Part of the condenser heating steam during startup would be supplied by the turbine exhaust, and the remainder from this bypass system. Alternatively, large steam-isolating valves can be installed, to supply condenser sections sequentially, with steam flows only from the turbine exhaust.

The equipment extracting noncondensables from the system (Item 5) consists of the hogging ejector and the operating ejectors. During startup, the hogging ejector removes air from inside the turbine, steam ducts, steam manifolds and bundles. It reduces the air pressure within the system from atmospheric to about 10 in. Hg absolute in a time period specified by the purchaser.

For the usual full-vacuum steam condenser, a two-stage operating ejector system complete with condensers is normally provided, with or without standby. Its capacity is generally specified by the purchaser in accord with the Heat Exchange Institute Standards for steam surface condensers. Some purchasers add a safety allowance by doubling the venting capacity recommended in the standard. The costliest parts of the ejector package are the inter- and aftercondensers, which are shell-and-tube construction. These can be smaller and lower-cost if a separate, colder, cooling-water supply is used instead of the hot condensate.

Motor-operated vacuum pumps can also be chosen; these adapt readily to automated remote operations.

The purchaser's inquiry specification should establish, for the air removal package, these points: choice of steam-jet air ejector or motor-driven vacuum pump; motive steam pressure and temperature; hogging-ejector minimum operating time; evacuating capacity of operating ejector package (compared with Standards recommendation); standby requirements for condensers and ejectors; and condenser cooling-water supply source and temperature.

The condensate storage tank (Item 6) is generally sized for a 5- to 10-min operating storage capacity. Total tank size exceeds this operating storage capacity by an amount representing the total condensate held in the drain pots and drain piping.

The condensate pumps (Item 7) are generally either two 100%-size units or three 50%-size units, to provide standby capability for emergency situations. The system generally has a very low net positive suction head availability so the pumps should be installed close to the condensate storage tank. The pump's total dynamic head must be sufficient to deliver the condensate into the purchaser's boiler feedwater system.

The steam duct system (Item 8) connects the condenser inlet-steam manifold to the turbine exhaust flange. It includes expansion joints, anchor points, elbows, turning vanes and duct supports. The purchaser should specify the preferred corrosion allowance for the manifolds and steam ducts since this affects system cost.

Economics dictate the steam-duct diameter. The smaller the size, the greater the steam pressure drop and the greater the required heat-transfer-surface area in the condenser. The tradeoff lies between heat-transfer-surface cost and steam-duct cost. (The steam-turbine thermal performance and power output depend on condenser pressure at the turbine exhaust flange, not on the steam pressure at the inlet to the bundles.) Past evaluations for full-vacuum systems have generally indicated an optimum steam velocity of about 200 ft/s at 6 in. Hg absolute steam pressure.

The condensate drain piping and manifold system (Item 9) starts at the bottom of the bundles and ends at the condensate storage tank. The air-removal piping and manifold system starts at the top of the bundles and terminates at the steam-jet air ejector package.

The instrumentation package (Item 10) includes such devices as temperature indicators and thermocouples; pressure indicators and transducers; vibration-pickup transducers; liquid-level devices; status lights; annunciator panel; and recorders. The controls might include storage-tank condensate level; low-flow condensate pump bypass; fan pitch control; air louver control; steam-valve control; and fan-motor control. These controls can be computerized from startup to shutdown, to maximize the turbine's thermal efficiency and power output, minimize the auxiliary-fan power consumption, and protect the condenser from freezing.

In the event of complete electric-power failure to the steam-condenser fans, an atmospheric-relief diaphragm safety device (Item 11) should be installed in the turbine exhaust system, to protect the turbine exhaust hood from excessive steam pressure. This diaphragm generally ruptures and relieves at about 5 psi for turbines designed for full-vacuum service. Some turbine manufacturers provide such a device on the exhaust hood; if not, the purchaser can provide external protection by installing an atmospheric relief valve(s) in the exhaust steam duct close to the turbine.

The large steam duct connecting the turbine exhaust to the steam-condenser manifold condenses a consider-

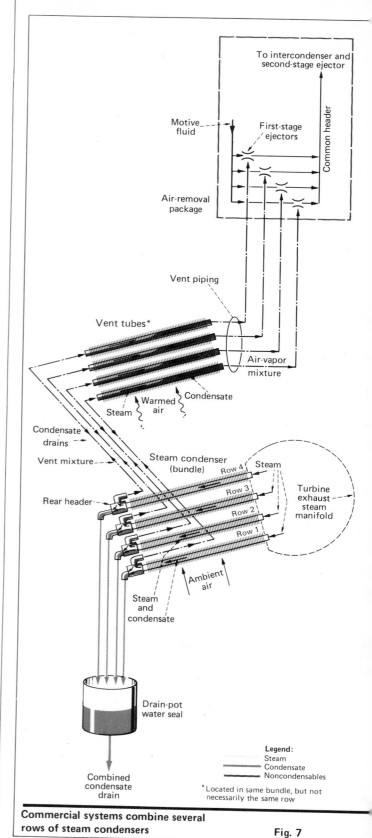

Commercial systems combine several rows of steam condensers

Fig. 7

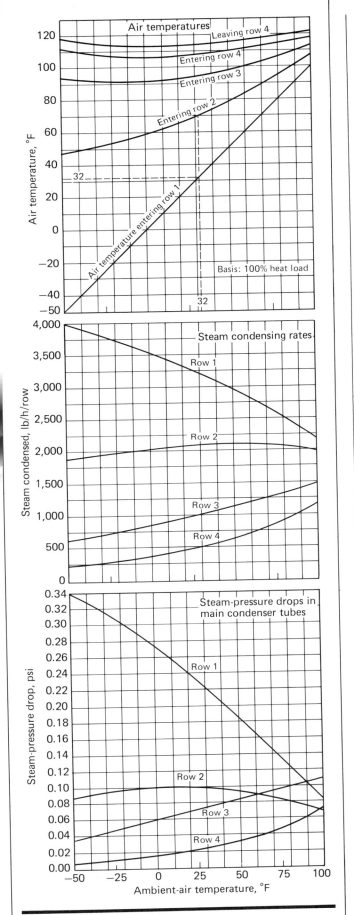

Operating characteristics of an air-cooled steam condenser   Fig. 8

able quantity of steam during a cold startup, while the metal temperature rises to some equilibrium level. This condensate must be drained to an appropriate low point in the duct system and then pumped or ejected (Item 12) into the condensate storage tank.

Thermal specifications

The more-important thermal data that the manufacturer requires from the purchaser for the design and optimization of the steam condenser are:

1. Exhaust steam flowrate.
2. Exhaust steam enthalpy.
3. Design exhaust pressure.
4. Design ambient-air temperature.
5. Maximum ambient-air temperature.
6. Minimum ambient-air temperature.
7. Lowest optimum turbine-exhaust pressure.
8. Highest permissible turbine-exhaust pressure.
9. Economic optimization of fan power usage.

The first three items define the full-load fluid properties entering the air-cooled steam condenser. If there are any condensate drains or other waste-heat streams entering the condenser, these must of course be detailed.

The design exhaust pressure (Item 3) is measured at the turbine exhaust flange if the manufacturer supplies all of the steam duct from turbine to condenser. When the purchaser supplies the steam duct, the pressure is generally measured at the connecting point of the purchaser's duct to the manufacturer's steam manifold.

The design exhaust pressure is the pressure that exists simultaneously with the design ambient-air temperature (Item 4). Since the heat-transfer driving force of an air-cooled steam condenser decreases with a rising ambient-air temperature, the unit's pressure and temperature design points should be established for the relatively adverse operating conditions. This should be the highest exhaust pressure that can be routinely tolerated by the turbine during a hot summer day. The higher the design exhaust pressure and the lower its companion ambient-air temperature specification, the smaller and less costly the steam condenser.

The upper limit for the design exhaust pressure is set by economic or turbine-mechanical considerations. The higher the value, the lower the horsepower available from the turbine to drive the electric generator, the compressor or the pump. There will be plant constraints or losses if horsepower falls below a prescribed minimum. Also, the steam-turbine manufacturer may have a turbine-exhaust-pressure limitation for mechanical and metallurgical reasons; typically, the limit may be 5 or 6 in. Hg absolute for a vacuum turbine, which must not be exceeded during normal full-load operation.

The companion design ambient-air temperature (Item 4) can range from 60 to 110°F. It should be selected on the basis of economics; the figure can, but does not always, turn out to be an annual average, a summer peak, or the temperature that is not exceeded more than 5% of the time.

The economic design ambient-air temperature is determined by selecting several potential values, sizing the steam condensers, estimating the capital cost of each, and then calculating their average annual performance. The higher-temperature cases will have larger steam condensers of higher capital cost, which can pay for themselves only by yielding a larger plant output. When the higher annual capital cost for a larger condenser just equals its annual savings, the corresponding ambient-air temperature becomes the economic optimum and establishes the condenser size.

When time restraints do not permit such a comprehensive economic study, the alternative is to select the lowest ambient-air temperature that experience prudently allows. The purchaser must be aware that performance will suffer when that temperature is exceeded (even if for only a few hours a year), and the turbine output may have to be cut back somewhat. The dollar penalty of such cutbacks must be balanced against the higher capital cost of a greater-capacity condensing system that could avoid cutbacks.

The maximum ambient-air temperature (Item 5) establishes the maximum turbine-exhaust pressure at full load for a given condenser. The minimum ambient-air temperature specification (Item 6) determines the type and degree of freeze protection.

The lowest optimum turbine-exhaust operating pressure (Item 7) is a characteristic of the turbine's particular design and construction. Below a given exhaust pressure, the turbine's last-stage leaving losses become so large as to reduce the turbine shaft output. This specification comes from the turbine manufacturer. Similarly, the highest permissible turbine-exhaust operating pressure (Item 8) is set by the turbine maker. This pressure cannot be exceeded during the maximum ambient-air temperature (Item 5) even if it requires reducing the throttle steam flow to the turbine.

The purchaser normally is not concerned with the internal steam-pressure drop of the air-cooled condenser. Such pressure drop is optimized by the manufacturer, taking into account the performance of the steam-jet air ejector and the specification for the lowest optimum turbine-exhaust pressure (Item 7). Purchasers who specify pressure drop can in fact limit the condenser designer's choice of tube diameter and length, and thus prevent the optimum capital-cost selection.

The steam condenser manufacturer optimizes designs by balancing the cost of fan power (Item 9) against the capital cost of heat-transfer surface. The cost of the purchaser's electric power, both demand and energy charges, must be known. This should reflect the actual increase in the annual utility power bill for each brake horsepower of fan power. It should be priced on the basis of the lowest-cost increment on the utility's rate schedule applicable to the purchaser.

This annual power cost must be converted into a lifetime cost figure, by capitalizing it to reflect the present value of all monies to be paid for power over the life of the plant. Adding this capitalized power cost (in $ per hp-lifetime) to the equipment capital cost gives the total owning and operating cost of the steam condensing system. Such power cost data submitted by the purchaser allows the manufacturer to trade off between heat-transfer surface and fan power, to provide the purchaser with the most economic condenser design.

If the purchaser does not provide the manufacturer with the dollar value of fan power, an indication should at least be given of whether the steam condenser should

Extended heat-exchanger surface consists of extruded aluminum fins over a carbon-steel tube Fig. 9

be designed for a) lower fan-power cost at the expense of higher capital cost, or b) lower capital cost at the expense of higher fan-power cost.

Cold-climate considerations

The factors involved in warmup, startup and freeze protection during cold weather are:
1. Minimum-available steam flow.
2. Bypass steam flow.
3. Air flow control.
4. Ambient air preheat.

In general, the lower the minimum ambient-air temperature, the more costly the system equipment. Similarly, the smaller the minimum-available steam flow from the turbine (for immediate warmup of the condenser surfaces), the more costly the system.

A steam turbine must be started with steam flow to the throttle not exceeding the maximum rate prescribed by the turbine manufacturer. Turbines require careful startup to protect the rotors and stators from thermal distortions, which can occur as a result of too-fast loading that produces large metal-temperature gradients.

While a slow startup is desirable for the turbine, it is, however, undesirable for the condenser. Metal surfaces in the steam condenser must be brought up to a temperature above freezing quickly, to prevent condensate from freezing in some remote part of the tower.

There are several remedies if the immediate, minimum-available steam flow to the air-cooled condenser is too low for safe startup (considering the minimum-controllable air flow and natural-draft effects). One is to isolate the condenser into several sections by means of large steam valves, for sequential startup. Another is to increase the steam available to the condenser by bringing in live bypass steam from the boiler. An occasionally used method is to heat the incoming ambient air with open-flame torches that burn natural gas or fuel oil.

Once the metal of the air-cooled steam condenser is heated, the next hurdle is to condense steam safely on a continuous and controlled basis. Two independent variables that can upset equilibrium conditions are ambient-air temperature and wind. Another upsetting factor is a decreasing exhaust-steam flowrate. Controlling air flow through the bundles is the only technique available to counterbalance these effects.

There are several means for achieving air flow control. The selection depends upon the severity of the cold weather and the minimum-available steam flow. Some of the more common means are listed below, in order of increasing control capability and cost:

Fan-blade pitch	Motor speed	Other	Air flow
Fixed	Single	—	S% or 100%
Fixed	Two	—	S% or 50% or 100%
Variable	Single	—	S% to 100%
Fixed	Single	Louvers	S% to 100%
Variable	Single	Valves	S% to 100%

The air flow quantity S% refers to a "small" amount as induced by natural draft, wind effects, blade eddies, or leakages through louvers. Even this small amount can be critical during extremely cold weather, when the heat-transfer driving force becomes very large and only a very small air flow is needed to maintain the desired thermal equilibrium. However small the air flow, it must be operator controlled at all times, for example by using variable-pitch fans that can be set into negative pitch to counteract natural draft when necessary.

Other questions

In the review of alternative steam-condenser proposals, the purchaser should question the manufacturers to ensure that their designs do not trap noncondensables under any operating conditions, and thus are not freeze-prone. Typical questions and concerns include:

■ Are there open rear headers or common rear manifolds that connect different rows of tubes together? Check both the main condenser bundles (or tubes) and the vent condenser bundles (or tubes).

■ If the answer is yes, ask the manufacturer to provide steam flowrates, steam condensing rates and steam pressure drops in each row of the main and vent condensers (for a typical bundle), for the full range of operating steam loads, ambient-air temperatures and air flow velocities. Find out how the manufacturer maintains identical steam pressures in the rear header (or rear manifold) for each row over the full operating range, to avoid steam backflow.

■ How does the total travel length of purged noncondensables compare among units? This is the longest distance noncondensables must travel through the rear header of the main condenser before reaching the entrance of the vent condenser tubes. The longer this travel length, the more difficult it is to purge the main condenser tubes that are farthest away from the vent condenser tubes.

■ Do the vent tubes contact cold ambient air, or are they installed in a heated section of the bundle where they cannot freeze? Tubes in the vent condenser carry some steam along with the noncondensables. Since the steam partial pressure is low, heat content is low.

■ How are the condensate-drain water-seal loops protected from freezing?

■ Can the condenser function with all fans off for indefinite periods, without steam backflow?

■ What is the degree of steam flow upset that occurs in operating cells when the fan of an adjacent cell is turned off? The nonoperating cell will have a higher steam pressure in its rear headers. How will this higher pressure affect the main and vent condensers of the adjacent operating cells, and what does the manufacturer recommend to relieve steam backflow?

■ Does the vent section of the condenser have a separate set of fans from the main condenser? If so, is it necessary to run the vent fans in some prescribed manner in relation to the main condenser fans? What happens when the vent condenser fans are operated differently from the prescribed speed regimen?

■ Where will the major components, such as the bundles and the fans, be manufactured, and, if job shops are used, how is quality control maintained?

The authors

M. W. Larinoff   W E. Moles   R. Reichhelm

M. W. Larinoff has been a vice-president of Hudson Products Corp. (P.O. Box 36100, Houston, TX 77036) since joining the firm in 1969. He is responsible for the engineering development and marketing of steam-condensing dry and wet/dry cooling-tower systems. For 23 years previously, he held various positions with Ebasco Services, Inc. (New York City), in the field of electric-utility power plants and systems. His B.S. and M.S. degrees in mechanical engineering were earned at Illinois Institute of Technology and Massachusetts Institute of Technology, respectively. He has published many technical papers, holds several patents, and has served as a member of several technical society committees including a U.S. cooling-tower delegation exchanging technical information with the U.S.S.R.

William E. Moles is east coast manager of Hudson Products (Ridgefield Park, N.J.). He has been responsible for application and sales of air-cooled heat exchangers in the process and power industries for over ten years, both domestically and internationally. Formerly, he held positions with Gilbert & Barker Mfg. Co. and Sier-Bath Gear & Pump Co. A member of ASME, he attended Rutgers U.

Robert Reichhelm is west coast manager of Hudson Products (Sherman Oaks, Calif.). His duties include application and sales of air-cooled heat exchangers in the process and power industries. He formerly worked for the turbine division of Intl. Silver Co., and in jet-engine testing for Pratt & Whitney Aircraft. He attended Massachusetts Institute of Technology.

Continuous tube cleaning improves performance of condensers and heat exchangers

Continuous mechanical tube cleaning of condensers and heat exchangers through the recirculation of sponge-rubber balls results in improved performance and greater efficiency

William I. Kern, Amertap Corp.

☐ Because of higher fuel costs and shortages, pressure from environmentalists, and shrinking availability of sites and cooling media, it is very important to obtain greater reliability and efficiency from surface condensers and heat exchangers. Condensers for fossil-fueled turbine units continue to increase in surface area, and nuclear units now online or planned for the future are a quantum jump in size. Moreover, many more special functions have been imposed to a condenser, such as reheating, regenerative preheating, multipressure units, etc. Thus, designers and operators can no longer rely on standards of performance prevalent a decade ago.

Turbine performance

The most effective way to improve heat transfer is to improve turbine performance through better maintenance of design back-pressure. Back pressure, however, is a function of condenser efficiency. When the circulating cooling water produces an absolute pressure at the turbine flange that induces choking flow, the optimum heat rate is obtained. At a pressure above or below that point, there is a loss in heat transfer because, at choking flow, the steam passing through the turbine exit annulus has reached sonic velocity. When this occurs, no additional energy can be imparted to the turbine by lowering the absolute pressure [1].

Improved heat-transfer rates can result in both increased generation capacity and reduced fuel requirements. For example, consider a 600-MW generating unit. As little as 0.2 in. Hg absolute-pressure improvement in condenser back-pressure can equal a 0.5% reduction in turbine heat rate. This is equal to about three additional megawatts of generating capacity. In addition, significant fuel savings result. At a fuel cost of 40¢ per million Btu, and a 60% loading factor for this unit, savings of over $60,000/yr can be realized.

A condenser or heat exchanger that is kept free from corrosion, leakage and blockage is a more reliable unit. In addition to the cost of unnecessary fuel consumption, plant outages, chemical treatment, and inefficient use of cooling media, there are pressures applied to industry to meet environmental regulations.

Effect of water velocity

Inlet water temperature affects the back pressure of a condenser. If once-through cooling water is used, inlet temperature is not a factor that can be controlled, but mechanical-draft cooling towers can regulate the temperature. Where once-through water is used, seasonal variations magnify the effects of increased back pressure. One study [2] points out the seasonal effects of an increase as small as 0.1 in. Hg absolute pressure, occasioned by waterside deposits in the condenser of a 265-MW generating unit. Increased operating costs that resulted from this fouling were as high as $1,000/month during periods of high inlet temperature.

Waterside resistance accounts for 72% of the total resistance to heat transfer of a tube. There are two components to waterside resistance: the laminar film of stagnant water next to the tube wall, and the building sedimentary deposits called "fouling."

The laminar film has insulating qualities that can account for 39% of the total heat-transfer resistance of a condenser. In addition, it isolates the turbulent water within the tube from the fouling buildup on the tube surface, which contributes to a fast rate of deposit. Tests [4] have shown that there is a rapid decrease in heat transfer after a thorough manual cleaning of condenser tubes—as much as 15–20%—after only 10 h of operation on a 90,000-ft²-surface condenser. If the

Originally published October 13, 1975

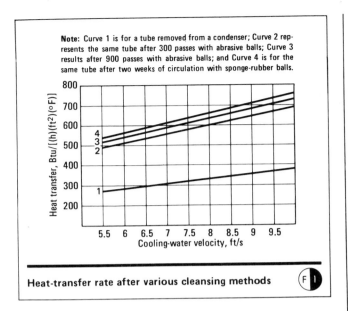

Heat-transfer rate after various cleansing methods

Note: Curve 1 is for a tube removed from a condenser; Curve 2 represents the same tube after 300 passes with abrasive balls; Curve 3 results after 900 passes with abrasive balls; and Curve 4 is for the same tube after two weeks of circulation with sponge-rubber balls.

laminar film is disrupted by continuous mechanical cleaning, the heat transfer through the tube can be increased (F/1), and the rate of waterside fouling significantly reduced.

Tube-material considerations

Among the many factors that affect the type of deposit are dissolved solids, pH, chlorination, bacteria and water velocity. Deposits contribute to corrosion, particularly of stainless steel tubing [5,6,7]. The most troublesome deposits [6] have been oxygen-excluding patches of calcium carbonate containing chlorides, and the soft, porous precipitate of a hydrous oxide of manganese found in seaboard locations.

Another fouling source is organic in nature. It varies from bacterial action that produces a heavy slime, to the growth of marine organisms that thrive in the cooling media. Non-copper-bearing metals are especially susceptible to slime formation. The presence of bacteria can also increase the manganese content of cooling water. It has been noted [6] that, though a river may contain no detectable concentration of manganese, once the water is impounded, its content of manganese may increase as a result of biological activity.

Power plants and industrial facilities must rely more than ever on cooling water that is increasingly subjected to pollution. A large portion is either sea water or brackish water from estuaries that contain marine life. The result is either outright plugging of the tube or localized erosion leaks. As cooling water flows past an obstruction, it must necessarily increase in velocity, which results in rapid erosion in the vicinity. Copper-bearing alloys are highly vulnerable to this; stainless steel is affected to some extent. Only titanium seems resistant to localized velocity increases [8].

Cleaning methods

A number of methods have been used to increase condenser and heat-exchanger efficiency. They range from cleaning techniques, to new tube materials and cooling towers.

Various chemicals (acids, chlorine) have been used to reduce fouling and restore tube cleanliness. Acids may either be strong (which damages the equipment) or weak (citric, formic, sulfamic); these are less effective and probably more expensive. Acid cleaning is limited to once a year or less often; between cleanings, fouling can severely affect condenser performance. The use of chlorine is being cut back or eliminated in many regions by government regulations. Thus, mechanical cleaning of condenser tubing—manual or automatic—seems to be assuming more of the burden of maintaining condenser back pressure.

Manual methods include periodic cleanings with rubber plugs, nylon brushes, metal scrapers, or turbining tools. Where clean water exists and the deposits are light, manual cleaning can be effective in reducing fouling deposits, which can account for 33% of the heat transfer resistance. However, manual cleaning is becoming more expensive as condensers and heat exchangers grow in size, with higher hourly rates and downtime playing important roles. Moreover, it is difficult for manual cleaning to keep up with fouling and blockage where the cooling water has high deposition due to its content of chlorides and solid debris. The greatest drawback to manual cleaning is that it is intermittent; between cleanings, fouling builds up rapidly.

Automatic cleaning by means of sponge-rubber balls is economical in areas where deposition, pollutants, chlorides and other corrodents exist. These balls distribute themselves at random through the condenser, passing through a tube at an average of one every five minutes. Slightly larger in diameter than the tube, they wipe the surface clean of fouling and deposits.

In an effort to reduce solid debris ingestion, particularly in seawater locations, most facilities have used "traveling" or "sliding" screens. These have not been very effective because the screens sometimes act as handy conveyors for flotsam and jetsam right into the water mains.

There are also several filters available, but until recently none could be flushed out online. A debris filter now is available that can be located adjacent to the waterbox and can be flushed without interrupting the flow of screened water to the condenser.

As good sites and once-through cooling water become less available, industry is turning more and more to cooling towers, even though they do not solve all condenser problems. Among the drawbacks are a buildup of dissolved solids and chloride concentrations. This is accentuated as plants strive for zero discharge.

Rubber-ball cleaning

The basic principle of cleaning with sponge-rubber balls is to frequently wipe clean the inside of the tube while the unit is in operation. Since the balls are slightly larger in diameter than the tube, they are compressed as they travel the length of the tube. This constant rubbing action keeps the walls clean and virtually free from deposits of all types. Thus, suspended solids are kept moving, and not allowed to settle, while bacterial fouling is wiped quickly away. Pits do not form because deposits are prevented.

The balls are selected in accordance with the installation, their specific gravity being nearly equal to that of the cooling media. Therefore, they distribute themselves in a homogeneous fashion. They travel the length of the tube forced by the pressure differential between the inlet and the outlet. The ball's surface allows a certain amount of water to flow through the area of contact with the wall, flushing away accumulated deposits ahead of the ball. They are available in various degrees of resiliency, depending on the requirements.

An abrasive-coated ball is also available for situations where the cooling water tubes have already been heavily fouled. Here, the effect is a gentle scouring that removes the scale slowly but steadily, until the tube is ready to be maintained by the normal sponge-rubber ball. Heat-transfer efficiency climbs steadily throughout this treatment.

The balls are circulated in a closed loop, including the condenser or heat exchanger (F/2). At the discharge end, they are caught in a screen installed directly in the line. They are then rerouted through a collector back to the condenser ball-injection nozzles to ensure that the balls are uniformly distributed.

At the collector unit, the balls can be counted or checked for size. The number required for a particular service is a function of the number of cooling tubes. Naturally, some wear occurs so that the balls must eventually be replaced. The labor needed to count and check a charge is usually about 1 h/wk.

This cleaning system can be retrofitted into most existing condensers or heat exchangers, although some modification of piping or unit design may be required. The slight increase in pumping resistance due to the

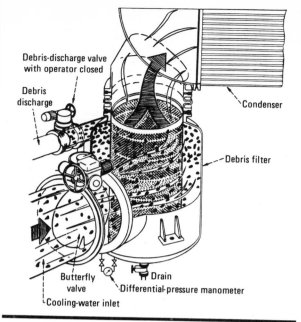

Debris filter is cleaned without stopping water flow F/2

pressure drop across the screening device is more than offset by the reduction in fouling resistance in the condenser or heat exchanger tubes. The most effective way to take advantage of this system is to provide for its installation at the design stage. A filter prevents solid debris from entering the waterbox of the condenser or heat exchanger. Located in the cooling-water inlet, it is flushed as needed without shutting down or bypassing the filter (F/3).

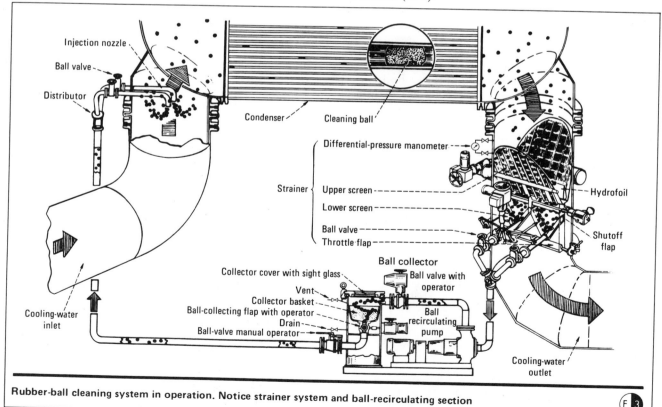

Rubber-ball cleaning system in operation. Notice strainer system and ball-recirculating section F/3

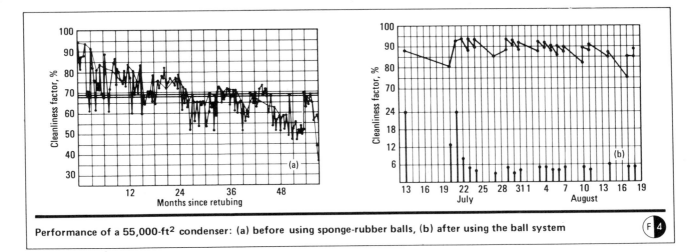

Performance of a 55,000-ft² condenser: (a) before using sponge-rubber balls, (b) after using the ball system

Examples of continuous tube cleaning

There are approximately 2,000 installations of sponge-rubber-ball cleaning systems in Europe, and about 200 in the U.S. All have achieved an outstanding record of success in maintaining condenser efficiency and reliability. A typical case is shown in the "before" and "after" graphs in F/4. Here, a 95-MW unit was retubed with stainless steel 304. Data kept for five years showed an average 27% deterioration in cleanliness from the original 94%, despite 69 manual cleanings. After a sponge-rubber system was installed in the same condenser, the cleanliness approached 95% within one week of operation.

One of the most extensive studies of continuous tube cleaning was conducted by United Illuminating at its English Station [11], which uses brackish, polluted water. It was found that the rubber-ball system can reduce the fouling rate of new brass tubes by as much as 50%, when compared with tubes operated without continuous cleaning.

Another instance involved stainless steel tubing, where the rubber-ball system maintained a cleanliness factor of 98% and a back pressure of 1.49 in. Hg. After 1,800 h of operation, the tube-cleaning system was taken out of service for testing purposes. During a month of operation without cleaning, the condenser back pressure climbed to 1.65 in. Hg, and the cleanliness factor dropped from 98 to 81%. When the cleaning was restarted, the original back pressure and cleanliness were recovered in 10 days.

Extensive tests at the Tennessee Valley Authority Widows Creek Plant [12] for over a year showed that the continuous system was highly economical and produced superior performance over manual cleaning. Results from two nearly identical condensers were compared; continuous cleaning by ball recirculation showed a 17% better performance than manual cleaning. This projected to an annual net savings of fuel (in 1969) of $20,000.

Summarizing, continuous cleaning and filtering systems maintain a high level of condenser and heat-exchanger efficiency. The ball-cleaning scheme results in fuel savings, fewer outages, and the reduction or elimination of cleansing chemicals.

References

1. Stoker, R. J., and Seavy, E. F., The Selection of Large Steam Surface Condensers, *Combustion*, Sept. 1967.
2. Heidrich, A., Jr., Roosen, J. J., and Kunkle, R. J., "Calorimetric Evaluation of Heat Transfer by Admiralty Condenser Tubes," ASME Publication 65-WA/CT-2, presented in Chicago, Nov. 7-11, 1965.
3. Lustenader, E. L., and Staub, F. W., "Development Contribution to Compact Condenser Design," INCO Power Conference, Wrightsville Beach, N.C., May 5, 1964.
4. Detwiler, D. S., "Improving Condenser Performance With Continuous In-Service Cleaning of Tubes," American Soc. for Testing Materials, Publication STP 538.
5. Maurer, J. R., The Use of Stainless Steel Tubing in Condenser and Related Power Plant Equipment—A Progress Report, *Combustion*, July 1967.
6. Long, N. A., "Recent Operating Experiences With Stainless Steel Tubes," paper presented at the 1966 American Power Conference.
7. "Experiences With Stainless Steel Tubes in Utility Condensers," *Nickel Topics*, Vol. 24, No. 5, 1971, International Nickel Co.
8. Feige, N. C., "Titanium Tubing for Surface Condenser Heat Exchanger Service," Bull. SC-1, Timet, Div. of Titanium Metals Corp. of America, West Caldwell, N.J.
9. Papamarcos, J., Condenser Tube Design Directions, *Power Eng.*, Vol. 77, No. 7, July 1963.
10. O'Keefe, W., Better But Costlier Tube Metals Tackle Present and Future Condenser Problems, *Power*, Vol. 117, No. 8, Aug. 1973.
11. Kuester, C. K., and Lynch, C. E., "Amertap at English Station," ASME Publication 66-WA/CT-1, contributed to the ASME by the Research Committee on Condenser Tubes, New York, Nov. 27 to Dec. 1, 1966.
12. Condenser Cleaning Improves Economics, *Electrical World*, Dec. 15, 1969.
13. Kuester, C. K., Here's How to Eliminate Debris from Heat Exchangers, *Electric Light and Power*, Energy/Generation ed., Aug. 1973.

The author

William I. Kern is General Manager of Amertap Corp. (Div. of Fa. Taprogge, West Germany), P.O. Box 151, Mineola, NY 11501, with responsibility for operations in the U.S. and Canada. Before, he was regional sales manager. He also worked as development and sales engineer for the Hamilton Standard Div., U.S. Aircraft Corp. He holds a B.S. degree in metallurgical engineering and a B.A., both from New York University, has presented papers on electron-beam welding of aerospace materials, and has conducted welding-development programs for the U.S. Air Force. He belongs to the Air Pollution Control Assn., National Assn. of Corrosion Engineers, and the American Soc. for Testing and Materials.

Evaluate reboiler fouling

Product quality drops and manufacturing costs rise when scale forms on reboiler tubes.

John A. Lowry, E. I. du Pont de Nemours & Co.

☐ Fouling of heat-transfer surfaces causes problems in the operation of thermosiphon reboilers, and can seriously affect the quality of a process stream.

These reboilers find wide use in the chemical process industries with distillation columns, reactor towers and evaporators. Their performance is the key to product quality and yield. In some cases, reboilers may process heat-sensitive materials, and provide the driving force for evaporation and chemical reaction.

It is common experience that reboilers foul. The effect of fouling on the process depends on the nature of the particular unit operation involved. In general, the two main effects are reduced product quality and reduced throughput.

Let us discuss the mechanisms involved in fouling and show how the process variables such as the driving force, ΔT, and the throughput are related to the amount of fouling in reboilers. This will enable us to understand how fouling occurs in many process plants.

Vertical reboiler

A vertical natural-circulation reboiler is shown schematically in Fig. 1. This reboiler is fed process liquid, L, and its function is to vaporize a certain fraction of the liquid. Let us assume that the feed, L, contains two chemicals, A and B, which react to form product C. The product is a material that deposits on the tubes of the reboiler and causes fouling.

Let us now discuss observations that we have made on some reboilers located at the bottom of reaction towers. The important variables are throughput and ΔT. $\Delta T = (T_D - T_p)$, where T_D is the temperature of the heating-medium condensate (in this case, Dowtherm vapor) and T_p is the temperature of the process liquid entering the reboiler. The practical observations are:

1. Reboilers foul with time. Fouling rate is proportional to throughput (as indicated by $d(\Delta T)/dt$).
2. The higher the ΔT, the greater the amount of degraded product in the column bottoms.
3. Many tubes in the reboiler are blocked with scale when the column is shut down for cleaning.

Let us derive a relationship between ΔT, throughput, and fouling. We will also explain the increase in product degradation with ΔT.

*The author is now employed by Celanese Polymer Specialties Co.

Originally published February 13, 1978

Heat transfer in the reboiler

The amount of heat transferred to the process is given by:

$$q = kA(\Delta T/X)$$

where k = thermal conductivity, Btu/(h)(ft²)(°F/ft); A = area of heat-transfer surface, ft²; ΔT = difference in temperature between Dowtherm condensate and process liquid, °F; X = distance at right angles to surface area over which heat is transferred, ft; and q = rate of heat flow, Btu/h.

Fig. 2 shows a longitudinal cross-section through the wall of a tube in the reboiler. If the throughput is

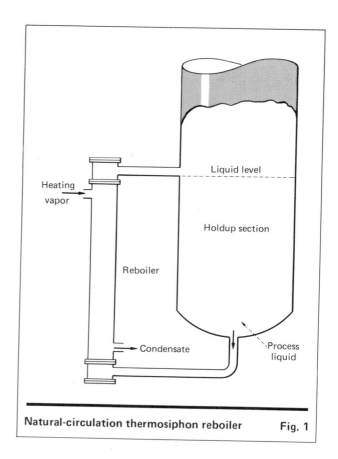

Natural-circulation thermosiphon reboiler   Fig. 1

EVALUATE REBOILER FOULING

constant, the total flow of heat from the Dowtherm to the process liquid can be derived.

Heat transfer from Dowtherm to tube wall:

$$q = k_D A_D (T_D - T_1)/X_F$$

where k_D = heat-transfer coefficient for Dowtherm film, X_F = film thickness, and A_D = total heat-transfer area for Dowtherm. Note that $A_D = n\pi D_o L$, where n = number of tubes in reboiler, D_o = tube O.D., and L = tube length.

The Dowtherm film coefficient is $h_o = k_D/X_F$. Therefore, heat transfer from Dowtherm to tube wall becomes:

$$q = h_o A_D (T_D - T_2) \qquad (1)$$

Heat transfer through the tube wall:

$$q = k_T A_D (T_2 - T_3)/X_T$$

where k_T = heat-transfer coefficient for tube material, and X_T = thickness of tube wall. If we define $h_T = k_T/X_T$ as the tube coefficient, heat transfer through the tube wall becomes:

$$q = h_T A_D (T_2 - T_3) \qquad (2)$$

Heat transfer through the scale on the process side:

$$q = k_s A_s (T_3 - T_4)/X_s \qquad (3)$$

where k_s = thermal conductivity of scale, X_s = scale thickness, and A_s = total area of scale over tube surface. Note that A_s can be defined as $A_s = n\pi D_i L$, where D_i is tube I.D. Heat transfer into the process:

$$q = k_L A_L (T_4 - T_L)/X_L$$

where k_L = thermal conductivity of process liquid, X_L = film thickness of liquid, and A_L = surface area available for heat transfer. As before, we define the film coefficient h_L as being equal to k_L/X_L. Now, A_L will depend on the thickness of the scale adhering to the inside of the tube wall. Hence, A_L is defined by:

$$A_L = 2\pi n L (r_T - X_s)$$

where r_T = inside radius of a reboiler tube, and X_s = scale thickness. Substituting this relation into the equation for heat transfer, we find that q becomes:

$$q = h_L [2\pi n L (r_T - X_s)](T_4 - T_L) \qquad (4)$$

We will now add Eq. (1), (2), (3) and (4), and simplify:

$$q \left[\frac{1}{h_o A_D} + \frac{1}{h_T A_D} + \frac{X_s}{k_s A_s} + \frac{1}{h_L [2\pi n L (r_T - X_s)]} \right] =$$
$$\Delta T = (T_D - T_L)$$

If we define $K_D = (1/h_o A_D) + (1/h_T A_D)$, and substitute K_D into the previous equation, we get:

$$\Delta T = q \left[K_D + \frac{X_s}{k_s A_s} + \frac{1}{2\pi h_L n L (r_T - X_s)} \right] \qquad (5)$$

For a metal-clean reboiler, $X_s = 0$, and Eq. (5) reduces to:

$$\Delta T = q \left[K_D + \frac{1}{2\pi h_L n L r_T} \right] \qquad (6)$$

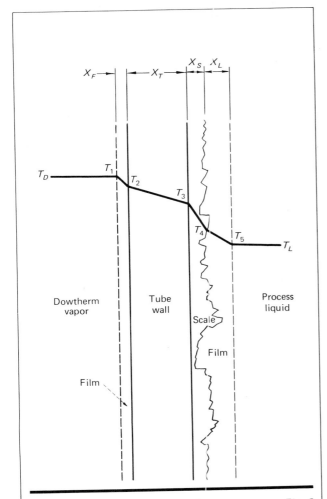

Section across tube wall of reboiler   Fig. 2

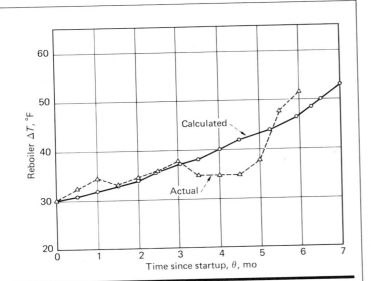

Predicted temperature rise in the reboiler   Fig. 3

How to obtain numerical data for the constants — Table I

Parameter	Assumptions/Method	Value
Heat input, q	Heat required is equal to amount needed to raise process liquid from ambient to boiling temperature.	$q = W(378 + 0.7\,\Delta T)$, Btu/h (Note 1) $q = W(\lambda + C_M\,\Delta T)$, Btu/h
Heat-transfer coefficient for Dowtherm, k_D	$k_D = X_F h_o \cdot\; h_o = 300$ Btu/(h)(ft²)(°F) (Note 2)	$k_D = 1.4 \times 10^{-5}$ (h)(°F)/Btu
Heat-transfer area, A_D	---	$A_D = 1{,}374$ ft²
Thermal conductivity of scale, k_s	Scale material is known.	$k_s = 3.41$ Btu/(h)(ft²)(°F/in.)
Heat of vaporization, λ	For process liquid.	$\lambda = 378$ Btu/lb (Note 1)
Density of scale, ρ_s	Known.	$\rho_s = 108.6$ lb/ft³
Heat-transfer coefficient for tube, h_T	Material is known.	$h_T = 450$ Btu/(h)(ft²)(°F) (Note 3)
Film coefficient for process liquid, h_L	Use eq (6). Solve for h_L and use known value of ΔT for metal-clean reboiler.	$h_L = 58$ Btu/(h)(ft²)(°F)
Fouling factor, f	Scale material forms from reaction, $A + B \rightarrow C$. Amount of A lost in reboiler is known from material balance; B is in excess, so amount of C can be calculated stoichiometrically.	$f = 4.6 \times 10^{-5}$ lb scale/lb process liquid (Note 4)

Note 1: Typical values for latent heat of vaporization and heat capacity for certain esters. W = throughput, lb/h; ΔT = temperature difference, °F; λ = heat of vaporization, Btu/lb; C_M = heat capacity of material, Btu/(lb)(°F).

Note 2: From Dowtherm Handbook, p. 70, 1970. Adapted from Kern, D.Q., "Process Heat Transfer," pp. 261-270, McGraw-Hill, New York, 1950.

Note 3: Perry, R.H. and Chilton, C.H., eds., "Chemical Engineers' Handbook," 5th ed., p. 23--39, McGraw-Hill, New York, 1973. Assume ¼-in. tubewall.

Note 4: f is a fundamental factor that can be expressed as $f = ca$, where c = concentration of scale in process liquid, and a = fraction of scale that adheres to wall. This factor is derived from a material balance.

Let us assume that scale is deposited by evaporation at the heat-transfer surface. The total scale deposited, m_s, will be proportional to the concentration, c, in the process liquid and the amount of liquid vaporized. Therefore, the mass of scale formed in the reboiler is:

$$m_s = qf\theta/\lambda$$

where q = heat transferred to process liquid, θ = time, λ = latent heat of liquid vaporization, and the term f is a factor that includes the concentration (c, weight %) of scale in the process liquid and the amount of it that adheres to the tube wall. We will designate f as the fouling factor.

Total volume of scale formed is $V_s = m_s/\rho_s = qf\theta/\rho_s$, where ρ_s = density of scale.

Let us further assume that the scale is spread evenly over the reboiler surface. Then, scale thickness, X_s, is given by:

$$X_s = qf\theta/\lambda\rho_s A_s$$

We will substitute this expression for X_s into Eq. (5). After noting that $2\pi n L r_T = A_s$, or $2\pi n L = (A_s/r_T)$, and simplifying, we get:

$$\Delta T = q\left[K_D + \frac{qf\theta}{k_s(A_s)^2\lambda\rho_s} + \frac{\lambda\rho_s r_T}{h_L A_s \lambda\rho_s r_T - qf\theta h_L}\right]$$

For reasons to be discussed later, it is desirable to express ΔT as a function of n. Since $A_s = 2\pi n L r_T$, we will designate $2\pi L r_T$ as the reboiler constant, G. Therefore, $A_s = Gn$, and:

$$\Delta T = q\left[K_D + \frac{qf\theta}{k_s n^2 G^2 \lambda\rho_s} + \frac{\lambda\rho_s r_T}{h_L(nG\lambda\rho_s r_T - qf\theta)}\right] \quad (7)$$

Numerical data for the parameters

Reasonable values for K_D, q, f, k_s, λ, ρ_s and h_L are required. Table I shows how these were obtained.

The values from Table I were substituted into Eq. (7) and a curve for ΔT vs. time was computed. The value of n used in these calculations was based on observations of reboilers when they were disassembled for mechanical cleaning. After about six months of continuous operation, the value of n was decreased linearly until it had reached the time at which the vessel was cleaned. (The actual plot of ΔT vs. time is shown in Fig. 3, and is compared with the computed values. The variation at

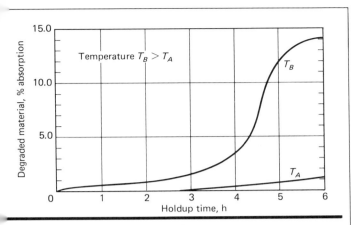

Degradation of process liquid   Fig. 4

(7) is the controlling factor in determining $d\Delta T/d\theta$. If this term is large enough, $d\Delta t/d\theta$ can rise almost exponentially. In addition, $n = F(f)$, so that a high value of the fouling factor, f, also implies a large decrease in $\Delta n/\Delta\theta$.

Throughput dependence—Assuming that the boilup ratio of the column is held constant, an increase or decrease in the reboiler throughput will call for an increase or decrease in heat flux. Because q appears in the $qf\theta$ product, small changes in throughput or boilup rate can drastically affect $\Delta T/\Delta\theta$.

A key observation of the effect of ΔT on product quality was that degraded material increased as ΔT increased. Likewise, swings in the concentration of degraded material increased as ΔT increased. (As ΔT increases, the tube-wall temperature increases proportionally, and the volume of liquid held up for long periods of time increases due to an increase in the number of plugged tubes.)

The effect of time and temperature on the process liquid is shown in Fig. 4, where the concentration of degradation products is expressed as a relative absorption of light of wavelength λ_A for two different temperatures of the process liquid. (The degraded material absorbs strongly at λ_A but pure material does not.) The concentration of degraded material can be expressed mathematically as:

$$D_M = V(n)b(T)e^{-[a(T)/t]^2} \qquad (8)$$

where D_M = degraded material, $V(n)$ = volume of process liquid held up in plugged tubes, $b(T)$ = final concentration of degradation products, $a(T)$ = coefficient related to temperature effect on degradation kinetics, and t = holdup time.

Fig. 5 is a plot of Eq. (8) for different values of t. The value of $V(n)$ for each calculation is derived from the theoretical volume of one tube, assuming an infinitely thin plug, times the number of tubes that must be plugged to produce the given ΔT. The temperature of the process liquid in the plugged tubes will then reach the value of the Dowtherm due to lack of circulation. The shape of the curves in Fig. 5 explains the deterioration and variability of product quality. The higher the ΔT, the greater the probability that liquid is being held up for long periods of time.

Reboiler performance can have important effects on product quality and manufacturing costs. Knowledge of how fouling affects them can aid in planning, operating and designing process plants.

three to five months is due to a significant decrease in throughput during this period of time.) The computer program for calculating ΔT assumed a constant value of q and, therefore, predicted too high a value for ΔT.

Reboiler performance

The importance of the reboiler equation derives from the qualitative insights we can draw from its form, and not the quantitative applications. Certain phenomena that occur as reboilers foul are easily understood in terms of the dependence of ΔT on n, q, f and θ:

Physical mode of fouling—The heat-transfer area is affected in two ways. First, scale buildup on the walls of the tubes increases resistance to heat transfer. If this material flakes off the tube wall, it can plug the tube and eliminate it as a heat-transfer surface and, thereby reduce the effective number of tubes in the reboiler. Second, plugged tubes become a reservoir for process liquid that is held up for a long period of time at high temperature. This can result in degradation of the liquid and unwanted side-reactions.

Product of $qf\theta$—This term in the denominator of Eq.

The author

John A. Lowry is a research chemist at Celanese Polymer Specialties Co., Technical Center, 9800 East Bluegrass Parkway, Louisville, KY 40299, where he is engaged in polymer process and product development. Previously, he was with E. I. du Pont de Nemours & Co. as a process chemist responsible for plant startup and development work on polymer processes. He has a B.S. in chemistry from the University of Florida, an M.S. in physical chemistry from the University of Arizona, and is a member of ACS.

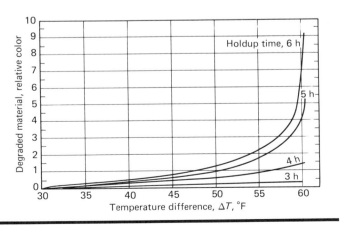

Fouling affects degradation rate of the process liquid   Fig. 5

Finding the natural frequency of vibration of exchanger tubes

V. Ganapathy, *Bharat Heavy Electricals*
Product Development (Boilers), Tiruchirapalli-620 014, India

☐ The vibration of heat exchanger tubes due to cross-flowing fluids can lead to extensive damage, if not avoided in the design. Consequently this is one aspect of exchanger design that should be checked by chemical engineers making bid comparisons. However, tube vibration is a complex calculation, which requires tabulated data for different mechanical properties of the tubes.

This nomograph may ease the situation. It eliminates the need to look up densities of tube materials, Young's modulus for the tube, and the radius of gyration. It assumes that heat exchanger tubes act as continuous beams with points of support at the baffles, and the ends fixed in the tubesheet. The following formula then applies:

$$f = (CR/L^2)(Ep_s/E_s p)^{0.5} \times 10^4$$

where: f = natural frequency of vibration, cycles/s
C = a constant depending on the mode of vibration
R = radius of gyration for the tubes, in
L = length of span between baffles or tube supports, in
E = Young's modulus for the tube material, lb/in^2
p = density of the tube material, lb/in^3
s = a subscript denoting steel

Both the value of Young's modulus and the density of the tube material are incorporated, on the nomograph, in the position of the lettered points, where:

A = brass, bronze and copper
B = magnesium

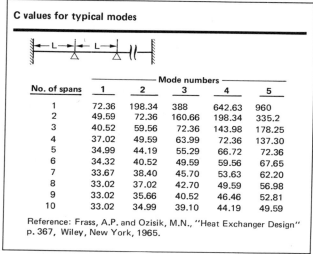

C values for typical modes

No. of spans	Mode numbers				
	1	2	3	4	5
1	72.36	198.34	388	642.63	960
2	49.59	72.36	160.66	198.34	335.2
3	40.52	59.56	72.36	143.98	178.25
4	37.02	49.59	63.99	72.36	137.30
5	34.99	44.19	55.29	66.72	72.36
6	34.32	40.52	49.59	59.56	67.65
7	33.67	38.40	45.70	53.63	62.20
8	33.02	37.02	42.70	49.59	56.98
9	33.02	35.66	40.52	46.46	52.81
10	33.02	34.99	39.10	44.19	49.59

Reference: Frass, A.P. and Ozisik, M.N., "Heat Exchanger Design" p. 367, Wiley, New York, 1965.

C = nickel
D = aluminum alloys 25,35,45,175,245,255,515 and 525
E = steel

Values of C are given in the table above.

Example: A heat exchanger is designed with 5/8-in. O.D. tubes of 20 BWG copper, and with five cross-flow baffles on 20-in. spacing between tubesheets 120 in. apart. Determine the natural frequency of vibration in the first mode for its tubes.

On the nomograph extend a vertical line from D = 5/8 to t = 20BWG, then horizontally to find the Radius of gyration, R. Connect this with C = 34.99 (from the table) and extend the line to its intersection with Reference line 1. From this intersection extend a line through L = 20 in to its intersection with Reference line 2, and from this intersection connect a line with point A for copper to read f = 120 cycles/s.

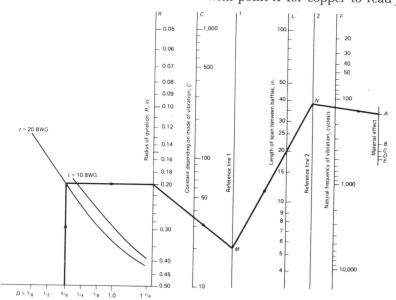

In-place annealing of high-temperature furnace tubes

Repair welds of high-temperature furnace tubes cannot be accomplished after the furnace has been in service for a time, without first annealing the tubes. Here is how this can be done in place with electrical heaters.

Darrell W. Maukonen, Exomet, Inc.
John Wagner, Air Products and Chemicals, Inc.

☐ How do you repair 5-in (125 mm) dia., 42-ft (13 m) long furnace tubes that have ruptured—when there's no way to get them out of the furnace? Consider the solution to the problem at Air Products' Texas hydrogen-generating plant. HK 40 furnace tubes that ruptured in service were solution-annealed in place, practically returning the metal to its original state of weldability. Moreover, the heaters did not burn up as if they were throwaway tools.

Maintenance engineers at the La Porte, Texas gas plant faced the problem common to many process plant operations using direct-fired furnaces: repair-welding the high-alloy HK 40 furnace tubes that had failed in service—without removing them from the furnace. The tubes, operating at 1,400 to 1,700°F (760 to 925°C), contain the catalytic reaction for generating hydrogen and carbon monoxide in a direct-fired hydrogen-reforming furnace.

Metallurgical problem

Although HK 40 is readily weldable before going into service, sustained high temperatures and combustion atmospheres cause much carbide precipitation, resulting in loss of mechanical strength and reduced weldability. Failures most often occur across carbide "islands" formed in carbide precipitation areas. The carbides inhibit "wetting" by filler metal when repair welds are made. The problem stems from the high-alloy composition and the cast structure of HK 40 tubes (0.35 to 0.45% C; 1.50% Mn; 1.75% Si; 0.04% S; 0.04% P; 23 to 27% Cr; and 19 to 22% Ni).

Furnace tubes in place to stay

Removing furnace tubes for solution-annealing in a conventional way is a near to impossible job, particularly for 10 specific tubes in each of three 25-tube reformer banks. They are buried in refractory material that would require removing first, before one could begin to remove the tubes.

Estimates of the cost of disassembly run into thousands of dollars. The job would require a crew of 6 men, a crane and other special tools. Even more costly, complete disassembly of the furnace would add two weeks of expensive downtime. The "repair-in-place" solution eliminated all these extra costs.

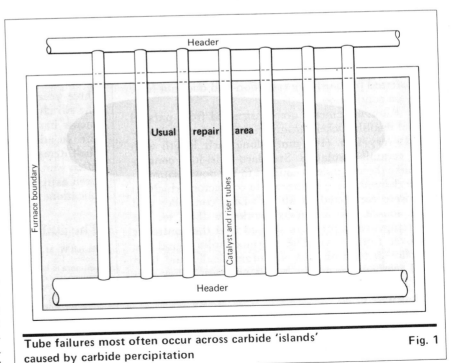

Tube failures most often occur across carbide 'islands' caused by carbide percipitation — Fig. 1

Originally published July 19, 1976

Ribbon heaters present more heating area to workpiece

Problem/solution

Metallurgical examinations indicated the need to solution-anneal the failed tubes before attempting repair welds, so the carbides would go back into solution and good grain structure would be restored. As previously mentioned, furnace construction ruled out removing several tubes from the furnace. If the tubes were to be repaired at all, they had to be repaired in place. (Removable tubes are repaired externally to the reformer.)

First attempts centered around special "finger" resistance heating elements. But they failed after only one to three heats, due to embrittlement arising from temperatures up to 2,200°F (1,200°C). Failures most often occurred during setup or knockdown, rather than during the cycle itself.

Then, maintenance engineers tried the new standard Ribbon* resistance heating elements. Because the heating element is flat and exposed in Ribbon heaters, heat is transferred primarily by radiation and does not have to rely on conduction.

The Ribbon elements are constructed from parallel strips of flexible nickel chromium resistance metal, supported every 4 in. (100 mm) along their length with heavy ceramic insulators. Standard twist-lock connectors make them simple to connect to the power source.

The elements were connected to contactors which, in turn, were connected to an Exo-Lec* controller and power source. Air Products' workmen then set the time/temperature program desired and the controller took over from there. The thermocouples signal the controller to turn on and off automatically and hold actual temperature to the preset program. They also feed a recorder, which generates a permanent record of the heat-treat cycle.

Savings, stemming from greater heater life, more mileage from the HK 40 tubes and simpler mainte-

---

* "Ribbon" and "Exo-Lec" are trademarks of Exomet, Inc., Conneaut, OH 44030, a part of the Metallurgical Systems Div. of Air Products and Chemicals, Inc.

nance, are estimated at better than $10,000 a year—not including downtime savings.

Step-by-step procedure

Here is a quick rundown of the procedure:

Step 1: Arc-gouge out the failed section, usually the top 4 to 13 ft (1¼ to 4 m) of the tubes, where the tubes are hottest.

Step 2: Grind bevels on the cut-off ends of the sound section of tubing.

Step 3: Dye-penetrant check the freshly ground section to be sure the tube has been cut back to sound metal.

Step 4: Cut, from the cool end of used HK 40 tubing, a section to replace the failed tube section.

Step 5: Tack-weld the fresh section in place with ⅛-in. (3 mm) gaps, using Inconel-type welding rod.

Step 6: Tack-weld thermocouples in the area of the joints.

Step 7: Wrap the Ribbon heaters in place, connect the wires and insulate with 6 in. (150 mm) of ceramic insulating blanket.

Step 8: Turn power on and permit weld area to be heated to a temperature of 2,250°F (1,230°C) ±25°F (15°C).

Step 9: Remove heating materials from area.

Step 10: TIG-weld the root pass.

Step 11: Stick-weld the rest.

Depending on the type of furnace and location of the damaged tube in the furnace, maintenance men may weld the tube right in place, lift a single tube partially out of the furnace to get at the damaged section, or pull the entire tube or tube bank. Several repairs are made with tubes "jogged out" just far enough to replace a damaged tube section while working from outside the furnace.

The solution-anneal step improves grain structure by encouraging dispersal and refinement of the carbide precipitates throughout the matrix. In effect, it restores the metal to nearly the same condition as found in a new furnace tube.

In-place solution annealing of repair areas in the tubes has extended life of the tubes by about two to three years—doubling their life, on the average. And the switch from finger heating elements to Ribbon elements has reduced the cost to accomplish this added life. In addition, Ribbon heaters are standard off-the-shelf items, whereas the special finger heating elements often were not immediately available. The practice has been extremely successful at La Porte and in other applications at Air Products' New Orleans facility.

The authors

Darrell W. Maukonen is product manager for heat treating systems for Exomet, Inc., Conneaut, OH 44030. He spends much time providing technical assistance in heat treating to major chemical-construction companies and maintenance organizations of process companies. He holds a degree in mechanical engineering from Tri-State College, Angola, Ind., and a correspondence degree in metallurgical engineering from International Correspondence School, Scranton, Pa. He is a member of the American Soc. for Metals and the American Welding Soc.

John Wagner is a welding engineer in the metallurgy department of Air Products and Chemicals, Inc., where he has had 20 years of experience. He attended the University of Minnesota, Minneapolis, Minn., and the University of Wisconsin, Madison, Wis. He is a member of the American Welding Soc.

Preventing fouling in plate heat exchangers

These units can foul as a result of several factors, including scaling, suspended solids, corrosion and heat degradation. Here the author discusses the more-common fouling mechanisms—and how, in each case, to ensure clean operation.

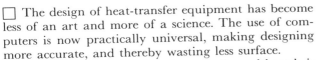

P. H. Cross, Alfa-Laval Co.

☐ The design of heat-transfer equipment has become less of an art and more of a science. The use of computers is now practically universal, making designing more accurate, and thereby wasting less surface.

However, little is known about fouling, although it strongly influences the design—and size—of the units. Manufacturers of plate heat exchangers claim that these units have far lower fouling tendencies than do tubular devices. Some purchasers use Tubular Exchanger Manufacturers Assn. (TEMA) fouling resistances that are for shell-and-tube exchangers, even though plate-type units have different fouling characteristics. The purchaser feels safe by using TEMA fouling resistances. However, using resistances that often are too large adds to cost and limits the merits of plate exchangers.

How does fouling take place in plate exchangers, how is heat transfer affected, and how can such fouling be prevented? Different fouling mechanisms are explained below, with examples given from industrial plants.

(The construction and use of plate heat exchangers are well described in the literature [1 and 2] and will not be repeated here.)

Scaling

In cooling water, scaling due to calcium carbonate and calcium sulfate is most common. Such scaling occurs because these salts exhibit inverse solubility effects.

Calcium carbonate scale-deposition increases with increasing temperature, concentration of the dissolved salt and pH. Scaling tendency is expressed by the Langelier index, which is equal to the difference between the actual pH and the saturated pH. Values are calculated from calcium hardness and alkalinity. A positive Langelier index indicates scaling.

Cooling towers often pick up SO_2, which reduces the pH and lessens the tendency toward fouling, even if the dissolved solids are concentrated. Experiments on scaling of heat-exchanger tubes by calcium carbonate are described by Watkinson and Martinez [3]. Fouling resistance increased with velocity until velocity reached 0.5 to 1.0 m/s, depending on tube diameter. The resistance then decreased as velocity increased. Fouling resistances were also plotted as a function of Reynolds numbers, R_e.

Above $R_e = 12,000$, fouling decreased with increasing Reynolds number. It appears that high turbulence and small hydraulic diameters help decrease this type of fouling. Plate heat exchangers induce high turbulence to fluids; in some units, turbulent flow occurs at Reynolds numbers as low as $R_e = 10$. In typical plate exchangers, hydraulic diameters are small, varying between only 0.0048 and 0.011 m.

The temperature rise of the cooling water and the metal wall are also important, as this type of scaling normally occurs at temperatures above 35°C.

With plate heat exchangers, heat-transfer coefficients for coolants are typically three times those for shell-and-tube units. Thus, hot-side metal temperatures are kept close to the cooling-water temperature.

Scale prevention

Calcium carbonate scale-formation may be prevented by reducing water hardness or by adding a mineral acid, typically sulfuric. Care must be taken to avoid creating low pHs that may promote corrosion.

Exchanger with fibrous matter clinging to plates. No filter used. Compare to Fig. 2 Fig. 1

There are also hazards of handling the acid during dosing. Adding overdoses may cause calcium sulfate scales to form.

Crystal growth can be inhibited by adding polyphosphates to the cooling system. Recently, organophosphorus compounds have been used for this purpose. These compounds are absorbed onto crystal surfaces, distorting their structure and thus slowing down growth. Polyacrylates and polymethacrylates can also be used.

Another common method is to first pass the cooling water through an ion-exchange column to replace the calcium and magnesium ions with sodium ones, which have a higher solubility. The method that is used is dependent upon the economics of the individual plant concerned.

Crystallization

Crystallization typically occurs on the process-stream side of the exchanger. As an example, consider a case where plate heat exchangers, used for crude oil cooling on North Sea platforms, did not give the required oil outlet-temperature. The correct temperature was obtained initially, but deteriorated over a 4-h period to a value 4°C above the required one. The exchangers were not operating at full load.

Cooling water, however, was at design conditions. Shutting it off for a few minutes, then turning it on dropped oil outlet temperatures by 4°C to give the required value. Evidently, wax had crystallized on the plate surface; shutting off the cooling water melted the wax.

Later on, the oil flowrate was increased to 25% above design conditions, yielding the required oil outlet-temperature.

Oil samples were tested on a small unit in the laboratory. The rate of wax deposition was found to be a function of plate metal temperature and oil velocity. In general, units gave the required oil outlet-temperature at full load but failed to achieve this at part load.

If the exchanger is not at full load, it can be treated as an oversized unit. In this case, the cooling-water control system does not reduce the flowrate. Further, at partial load, the oil velocity decreases, and this decreases the rate of removal of any fouling. The reduced velocity results in a lower heat-transfer coefficient, reduced metal-wall temperature, and further fouling.

There are various remedies for dealing with duties where crystallization occurs on the product side:

1. Do not include overly large fouling resistances in the design. This will result in an oversized unit, which presents the problems discussed above.
2. Use cocurrent flow.
3. Consider recirculating the cooling water from the outlet to the inlet, to raise the temperature of the transfer surfaces.
4. Specify systems controls, so that for flows of less than full load the cooling-water flowrate is reduced to maintain a high cooling-water outlet temperature (and, thus, high metal temperatures). The careful design of the heat exchanger and control system can often eliminate this type of fouling.

Suspended solids

Such solids may be present in the cooling water as mud, silt or sand, or may come from the plant itself, such as from corrosion of carbon-steel piping.

How the particles in the liquid adhere to the heat-exchanger surface depends upon surface roughness, liquid velocity and particle size.

Rough surfaces encourage particulate deposition. A surface may be rough to begin with, or may become so due to scale formation, formation of corrosion products, or erosion. High velocities (above 1 m/s) help prevent particulate fouling. However, there is a maximum velocity that can be used, due to erosion. Aluminum brass, employed commonly for heat exchangers, suffers from erosion at velocities above about 2 m/s.

Since particulate fouling is velocity dependent, prevention is assisted if stagnant areas are avoided. These often occur on the shell side of shell-and-tube heat exchangers. C. M. Gilmour [4] shows how an even flow distribution on tubular exchangers can prevent fouling. He also explains how a high surface finish (which plate units have) does the same.

Because exchangers produce high induced turbulence and have a high surface finish, they resist fouling by fine suspended solids. The ease of cleaning and effective backflushing of these devices are also important in keeping the units free of particles.

In one case, plate heat exchangers were installed in parallel with existing tubular units to test with water that had a high silt content. There were large pieces of debris in the water, and a backflushing system was fitted to plate units. Pressure drop for the plate units was set at 15 psi to keep the silt in suspension.

After one week's operation, the pressure drop exceeded 25 psi. Backflushing reduced this to 20 psi. The unit was opened for inspection, and the heat-transfer surface was found to be clean. High pressure drop was caused by large pieces of debris lodged in the inlet port.

A twin-basket, backflushing strainer was added, and the unit performed to specification for seven months without being opened, cleaned or backflushed. Upon opening the unit, the heat-transfer surface was found to be clean.

A high degree of induced turbulence kept the silt in suspension, and the lack of dead areas left nowhere for it to settle. Absence of organic fouling could be due to the scouring action of the silt. Since the plates were made of titanium, no corrosion occurred. Titanium also resists erosion by silt in seawater.

The tubular heat exchangers had to be opened for cleaning every three to four months.

Biological material

Fouling is caused by a variety of organisms—such as algae, weeds, fungi, bacteria, mussels, eels, barnacles and fish.

Suspensions of seaweed and other organic fibers often cause fouling, sometimes completely clogging heat exchangers. One example is cellulose remains of decayed sea grasses. These grow at depths of 6 to 90 ft in sand or mud.

These fibrous remains usually settle on the sea bed. Storms or rough seas may send a sudden deluge of them into the cooling-water system. *Large* amounts of fiber must not be allowed to get into plate heat exchangers. Recommended filter meshes depend on the type of plate heat exchanger—its pressing depth and type of corrugation. Generally, meshes are between 1 and 4 mm.

The usual treatment for biological growths is chlorination; a level of 1 or 2 ppm at the exchanger outlet is normally effective. However, other biocides can be used. Walko [6] gives a broad account of biocides used for controlling biological fouling in cooling systems.

Fig. 1 shows an exchanger with some fibrous matter clogging its plates. No filter was used. Compare this to the unit in Fig. 2, in which an automatic backflushing filter with 1-mm mesh was employed.

After six months, the plates were quite clean, with just odd fiber sticking to them.

The only answer to such fiber fouling is effective filtration.

Many types of bacteria will deposit slime on heat-transfer surfaces, and other types of foulants can adhere to these deposits. Larger growths restrict flow and often cause pitting of metal. (For a study of the effects of velocity and temperature on this type of fouling, see Bott and Pinheiro [5].)

Plate exchangers have a major advantage with this type of fouling: Good flow distribution makes chlorination very effective.

Filtration and chlorination are needed to keep plate

Automatic backflushing kept the filter plates relatively clean of fiber   Fig. 2

units clean when biological growths and fibrous solids are present. In one such case, fouling was severe and caused a considerable increase in pressure drop. Moreover, backflushing was only partially successful in reducing pressure drop and had no effect on fouling resistance.

Examination of the plates showed that fouling consisted of:

1. long, slender, coarse particles and grass at the inlet ports and
2. fine, slimy material adhering to the plates.

The coarse material increased the pressure drop, while the fine material decreased the heat-transfer coefficients.

Tests were made to establish the effects of backflushing, electro-polished plates and chlorination, as well as increased velocity.

Only chlorination had an effect. It was first applied for one hour every six hours. Fouling resistance decreased to zero; pressure drop increased by about 30%. Later on, chlorination was done in 15-min treatments, every 6 h.

Chlorination kept fouling minimal on clean plates, but did not clean up dirty plates.

These tests also showed that fouling by biological slimes was independent of velocity and could not be removed by backflushing.

Also, the type surface finish involved was shown to have no bearing on fouling.

Corrosion

The most common form of this type of fouling is iron oxide scale on carbon-steel tubes in water. Corrosion can be widespread if the pH is below 6, and minimal if it is above 10. Corrosion will only take place, of course, if there is dissolved oxygen in the water, except at very low pH values. Also, the presence of chlorides, sulfates and carbonates encourages corrosion. Prevention is achieved via:

1. Removing oxygen by thermal/mechanical means or by reducing agents.
2. Raising the Langelier index. This, however, encourages carbonate scaling, as the index varies with the pH of the water.
3. Adding corrosion inhibitors.
4. Using cathodic protection.

The best way to prevent corrosion is to use a material that is not corroded by the heat-transfer medium. Plates in plate heat exchangers are thin (0.6 to 1.2 mm) and thus are made of materials of construction selected for their corrosion resistance. Stainless steel, titanium, Hastelloy, Incoloy, Inconel and aluminum brass are standard. Plates are not made in carbon steel.

In fresh-water cooling service, stainless steel is not subject to the uniform type of corrosion that causes fouling problems. For seawater systems, titanium is completely immune to corrosion.

Heat degradation

Heat-sensitive liquids—particularly foods—may cause fouling due to thermal degradation. In general, relatively low temperatures and small temperature approaches are required in heating foods. Careful control of temperature is needed to ensure against heat spoilage.

Small temperature approaches are easily obtained in plate heat exchangers, as they give total countercurrent flow, and have excellent liquid distribution with no bypass streams. Therefore, "burning on" of the product is minimized.

For example, milk during pasteurization is normally heated to 165°F. Usually, 90% of this heat is regenerated from the pasteurized milk, so that only about 15°F is put in by some external heating medium, often vacuum steam. Large temperature differences would cause protein to precipitate on the plates, spoiling the product and causing fouling. Therefore, the vacuum steam is only 4°F above the milk outlet-temperature.

Conclusions

Using overly large fouling resistances in designing plate exchangers will result in equipment that is too large.

A high fouling resistance increases the required surface area and thus the number of plates. This, in turn, yields lower velocities. Moreover, since the unit contains excess surface, cooling water must be reduced further to obtain the specified process outlet temperature. If the cooling water temperature is below specification, a further reduction in cooling-water flowrate is required. Thus, we end up with low velocities, and, inevitably, high fouling results.

Fouling increases and so does the cooling-water flowrate. When the specified fouling resistance is finally reached, the specified cooling-water flowrate is also reached. At this point, further fouling is inhibited by the high velocities. Thus, the specified fouling resistance is obtained in practice, and the user feels that the original specification has proved accurate.

Sometimes, for a flow reduction of 3:1 below design, fouling resistances have been increased by ten times. Thus, whenever possible, units should be designed for low fouling factors—and fouling minimized by the methods we have discussed.

Individual film heat-transfer coefficients for plate-type units may be on the order of 12,000 to 17,000 $W/(m^2)(°C)$ for water.

Thus, clean overall heat-transfer coefficients may be as high as 5,200 to 7,000 $W/(m^2)(°C)$. With an overall clean coefficient of 7,000, a total fouling resistance of 0.00009 $(m^2)(°C)/W$ yields a fouling margin of 60%. A shell-and-tube unit with an overall heat transfer coefficient of 1,400 $W/(m^2)(°C)$ could well have a fouling resistance of 0.0003 $(m^2)(°C)/W$ and have a fouling margin of less than 50%.

For these reasons, on water/water duties for plate heat exchangers, fouling resistances should be between 0.00002 and 0.000075 $(m^2)(°C)/W$, depending on water quality. For other liquids, any fouling resistance that produces a margin of more than 25% is excessive.

In plate exchangers, heat-transfer coefficients are easier to predict than those for shell-and-tube units, as flows in all sections of a plate unit are identical. Thus, fouling resistances can be specified without safety margins that allow for uncertain correlations.

For heat duties with overall heat-transfer coefficients below 1,000 $W/(m^2)(°C)$, higher fouling resistances may be tolerated, as they do not represent so much in percentage terms.

Richard Greene, Editor

References

1. Marriott, J., Where and How to Use Plate Heat Exchangers, *Chem. Eng.*, Vol. 78, No. 8, Apr. 5, 1971, p. 127.
2. Clark, D. F., Plate Heat Exchanger Design and Recent Developments, *The Chemical Engineer*, May 1974, p. 275.
3. Watkinson, A. P. and Martinez, O., Scaling of Heat Exchanger Tubes by Calcium Carbonate, *Trans. of the Amer. Soc. of Mech. Eng.*, Vol. 1, No. 4, Nov. 1975, p. 504.
4. Gilmour, C. H., No Fooling—No Fouling, *Chem. Eng. Progr.*, Vol. 61, No. 1, July 1965, p. 49.
5. Bott, T. R. and Pinheiro, M. M. P. S., Biological Fouling in Cooling Systems, paper presented at 16th Natl. Heat Transfer Conf., St. Louis, Aug. 1976.
6. Walko, John F., Controlling Biological Fouling in Cooling Systems, *Chem. Eng.*, Vol. 79, No. 24, Oct. 30 (Part I of II parts), p. 128, No. 26, Nov. 27 (Part II), p. 104.

The author

P. H. Cross is thermal product manager at Alfa-Laval Co., Great West Road, Brentford, Middlesex, U.K. TW8 9BT. He has served a five-year apprenticeship with Foster Wheeler Ltd. and subsequently was made design engineer. Cross has also worked for Babcock & Wilcox Ltd. and Woodall-Duckham Ltd. Throughout his career, he has been involved with a wide variety of heat-transfer equipment, ranging from shell-and-tube units, to fired boilers, to plate exchangers. He is a chartered engineer and a member of the Institution of Mechanical Engineers.

Operating and maintenance records for heating equipment—I

Here is a description of forms used for recording the design, installation, inspection and operating data that are useful for operations and maintenance.

Edgar C. Sharp, Jr., King-Wilkinson, Inc.

☐ When making a detailed survey of the heaters in a plant, it is worthwhile to devise forms that list all the data required. Such forms save time and effort in the survey, and ensure that no data are overlooked.

My company uses a 12-sheet checklist for recording the design, installation and inspection details for each piece of heating equiment (Sheets 1—9), and for recording operating conditions (Sheets 10—12).

The first nine sheets are particularly useful. If filled in after each maintenance turnaround, fouling can be studied, and maintenance scheduling and techniques adjusted accordingly.

Sheet 1 is shown in Fig. 1. Sheets 2 through 9 provide space for recording:*

Steel structure and casing

Heater bottom: Plate and support oxidation? Plate cracks?

Cylinder- and cabin-type heater walls: Plate and support oxidation (NESW)?

Transition between radiant and convection sections; convection section (NESW); transition to stacks; stack: Plate and support oxidation?

Lining

Heater bottom: Material (concrete or refractory); repairs required? Burner blocks: Repairs required?

Cylinder or wall; transition; convection walls; transition to stack; stack: Material (concrete or refractory); repairs required?

Coils

Radiation tubes: Bent tubes?; oxidation?; dirty tubes?; black spots?

Bends: Inside or in boxes; oxidation?; weldings (normal or bad).

Shock tubes: Bent?; oxidation?; black spots?

Convection tubes: Extended surface?; fins or studs?;

*In these lists, a question mark indicates that a "yes" or "no" answer is called for. The letters "NESW" (north, east, south, west) mean that the location of the inspection point is indicated on the sheet.

Sheet 1

Client _____ KW Job No. _____
P.O. No. _____ Heater No. _____
Location _____ Date: _____ By: _____

1. General state
1.1 Type of heater _____
    _____
    _____
1.2 Years of service
1.3 Stack   On top   _____
                              Separated   _____
1.4 Free access for   From 4 sides   _____
    combustion air   From 3 sides   _____
                              From 2 sides   _____
                              From 1 side   _____
1.5 Twelve feet of space between heater and other installations or pipes
    North side — More   _____
               Less   _____
    East side — More   _____
               Less   _____
    South side — More   _____
               Less   _____
    West side — More   _____
               Less   _____
1.6 Estimate of pollution   Yes   _____
    near the heater   No   _____
                             Little   _____
1.7 Foundations   Pillars _____
                             Blocks   _____
                             Others   _____

---

dirty?; kind of deposits (carbon or other; less than 5 mm thick, less than 20 mm, more than 20 mm).

Bends: Inside or in boxes; weldings (normal or bad).

Castings, parts, etc.

Tube guides: Normal or with faults.

Intermediate guides or supports: Oxidation?; changeable?; individual or plates.

Convection supports: Oxidation?; number.

Peepholes, doors, etc.

Peepholes: Number; dimensions; well handled?; inner lining (good or bad); hotter than usual?; visibility (good, normal or bad).

Originally published April 25, 1977

Sheet 10

Client _____ KW Job No. _____
P.O. No. _____ Heater No. _____
Location _____ Date: _____ By: _____

Process conditions		Operating data	Basic design
1. Product			
2. Charge, lb/h			
3. Specific gravity, 15°C			
4. °API			
5. Molecular weight			
6. Temperature	Inlet		
	Bet. rad. & conv.		
	Outlet		
7. Pressure, psig	Inlet		
	Bet. rad. & conv.		
	Outlet		
8. Evaporation, wt. %	Inlet		
	Bet. rad. & conv.		
	Outlet		
9. Heat-absorbed total, million Btu/h			
10. Heat rate, Btu/(ft^2)(h)	Radiant		
	Conv. bare		
11. Liberation, million Btu/h			

Remarks:

Process operating conditions

Sheets 10–12 are for recording heater and process operating conditions. They are useful for correlating with inspection results and can also be used as a log of operations. Sheet 10 is shown in Fig. 2. Sheet 11 has spaces for the following information, with operating and basic design data shown.

Combustion (Sheet 11)

Fuel oil: Lower heating value (Btu/lb); consumption (lb/h); °API/sp. gr., 15°C; temperature (°F); viscosity at temperature; pressure at burner (psig).

Fuel gas: Lower heating value (Btu/std ft^3); consumption (std ft^3/h or lb/h); density; molecular weight.

Atomizing steam: Temperature (°F); pressure (psig); consumption (lb/lb oil).

Flue-gas temperature: Outlet radiation; between first and second row; convection section; outlet of convection section; stack base.

Draft (in. water gage): At heater bottom; between radiation and convection sections; outlet of convection section; before and after damper.

Excess air (%): Oil; gas.

Fuel and stack analyses

Fuel-oil analysis (%): C; H_2; O_2; N_2; S; H_2O.

Fuel-gas analysis (%): H_2; CH_4; C_2H_6; C_3H_8; n-C_4H_{10}; n-C_5H_{12}; H_2S.

Flue-gas analysis (%): O_2; CO_2 (half-high radiant section and inlet of convection section).

Stack-base gas analysis (%): O_2; CO_2; H_2; hydrocarbons.

The second part of this article is on page 216-217. It shows how heater design and operating parameters can be conveniently recorded on a four-sheet list that provides more-detailed information than is included on Sheets 10 and 11 of the checklist in this article.

Entrance doors: Number; dimensions; easy to handle?; lining (good or bad); entrance possibility (good or bad).

Explosion doors: Number; dimensions; work properly?; lining?

Convection cleaning doors: Number; dimensions; easy handling?; inspection possibility (good, normal or bad).

Damper: Heater bottom handling?; work?

Convection tubes: Cleaning possibility (by hand?; soot blowers?); number; horizontal or vertical.

Burners

Burners: Manufacturer; model number; number of burners; draft (natural or forced); fuel (oil, gas or oil/gas).

Duty per burner (from manufacturer): Oil Btu/h (minimum, normal, maximum); gas Btu/h (minimum, normal, maximum).

Atomizing design (steam or air).

Fuel: Temperature (°F); viscosity; pressure (psig); oil or gas; fuel quantity per hour.

Pilots: Present?; number; type; lighted?

Flame control: Present?; quantity; type.

Flame: Shape; color (dark, normal, very light); distribution inside the heater.

The Author

Edgar C. Sharp, Jr., is Manager, Energy Conservation Services, of King-Wilkinson, Inc., 3701 Kirby Drive, Houston, TX 77034. For more than 25 years, he has held managerial and engineering responsibilities with companies associated with the process industries, including Combustion Engineering, Inc.; Air Preheater Co.; Curtin Scientific, Inc.; Pritchard Products; and Southwestern Engineering and Equipment Co. He has a B.S. in mechanical engineering from Rice University.

Operation and maintenance records for heating equipment—II

Here is a description of forms used for recording detailed heater design and operating data for operations and maintenance.

Edgar C. Sharp, Jr., King-Wilkinson, Inc.*

☐ The previous article of this two-part series gave a detailed description of a checklist for heater design, installation, inspection and operating details, for use in making a survey of heating equipment in a plant or project.

This section presents the features of a four-sheet checklist used by my company for recording more-detailed heater design and operating data. This list can be used separately from the first one as a log of heater operating conditions, for comparison with design data. However, for an in-depth study to correlate fuel quality, operating conditions and heater fouling, it is recommended that the two checklists be used jointly.

Fig. 1 and 2 show Sheets 1 and 2 of the four-sheet checklist. Sheets 3 and 4 provide spaces for listing design data for individual furnaces and burners, respectively.

Furnace design data

Furnace Type:

Radiant section:
    Number of tubes.
    Number of passes.
    Diameter × thickness of tubes (in.).
    Exposed length of tube (in.).
    Exposed surface (total ft^2).
    Tube material.
    Center-to-center distance (in.).
    Center tube-wall (in.).
    Bend type (welded/rolled).
    Bends exposed (yes or no).
    Tube supports, material (top, bottom, intermediate).
    Tube configuration (horizontal or vertical).

Convection section:
    Number of tubes (bare, studded, finned).
    Number of tubes per row.
    Number of passes.
    Diameter × thickness of tubes (in.).

The first article of this two-part series appears on pp. 215-216. Information about the author appears on p. 216.

Originally published May 23, 1977

Sheet 1

Heater No. _____ KW Job No. _____ Date _____
Location _____ P.O. No. _____ Originator _____
Service _____ Client _____

Process conditions		Normal	Basic design	Operating data	Required operating
Charge, lb/h					
Barrels/d					
Sp. gr. (15°C)					
°API					
Mol. weight					
Enthalpy, Btu/lb	inlet				
	outlet				
Latent heat, Btu/lb					
Analysis of feed					
Temperature, °F	inlet				
	cross-over				
	outlet				
Pressure, psig	inlet				
	cross-over				
	outlet				
Pressure drop, psi					
Evaporation, wt.%	inlet				
	cross-over				
	outlet				
Heat absorbed, million Btu/h (process)	radiant				
	conv. bare				
	studded				
	finned				
Heat abs., million Btu/h aux. coil conv.					
Heat abs., million Btu/h					
Heat rate, Btu/(ft^2)(h)	radiant				
	conv. bare				
	studded				
	finned				
	aux. coil				
Liberation million Btu/h					
Thermal efficiency (LHV),%					

Sheet 2

Heater No. _____ KW Job No. _____ Date _____
Location _____ P.O. No. _____ Originator _____
Service _____ Client _____

Combustion		Normal	Basic design	Operating data	Required operating
Fuel oil	(LHV), Lower heating value Btu/lb				
	Consumption, lb/h				
	°API/Sp. gr. (15°C)				
	Viscosity at temp.				
	Temperature, °F				
	Pressure at burner, psig				
Fuel gas	LHV Btu/scf or Btu/lb				
	Consumption, (scfh), (lb/h)				
	Density (15 °C)				
	Molecular weight				
Atomizing Steam	Temperature, °F				
	Pressure, psig				
	Consumption, lb/lb oil				
Flue-gas temperature, °F	Outlet rad. sect.				
	Intermed. conv. sect.				
	Outlet conv. sect.				
Draft, water gage	At heater bottom				
	At outlet rad. sect.				
	At outlet conv. sect.				
	Before damper				
	After damper				
	Discharge forced-draft blower				
Excess air, weight %	Oil				
	Gas				
Flue-gas analysis (Orsat), vol. % Date of measurement					
Composition of fuel oil, weight %					
Composition of fuel gas, vol. %					

Exposed surface (ft² — bare, studded, finned).
Extended surface (description; extension ratio; material; stud/fin tip temperature, °F).
Center-to-center distance (horizontal or vertical).
Bend type (welded or rolled).
Bends exposed (yes or no).
Tube support material.

Burner design data

Burners:
  Manufacturer, type, number.
  Natural or forced draft.
  Number of burners.
  Liberation per burner in million Btu/h:
    Gas (minimum, normal, maximum),
    Oil (minimum, normal, maximum).
  Burner capacity curve (yes or no).
  Design excess air, wt. %:
    Gas (normal, maximum),
    Oil (normal, maximum).
  Atomizing design (steam, air).
  Quantity of material used for atomization (lb/lb oil).
  Number of pilots.
  Flame control (type).

Blower:
  Manufacturer, type (forced, induced).
  Capacity, std. ft³/h:
    Normal (forced, induced),
    Maximum (forced, induced);
    Med. inlet temperature, °F (forced, induced).
  Forced-draft discharge pressure (in. water gage).
  Induced-draft inlet pressure (in. water gage).

The checklists described in these two articles have been used by King-Wilkinson, Inc. in its heater-service operations during the past ten years. In some instances, the company's clients have adapted these checklists to fit their particular needs in recording heater design, installation and operating data.

Operating performance of steam-heated reboilers

Detailed computations show how fouling affects: the operation of reboilers; heat-transfer coefficients; steam-chest pressures and temperatures, control-valve pressure drops.

Albert E. Helzner, Badger America, Inc.

☐ Reboilers or heat exchangers are an integral part of most chemical process systems. Yet many engineers have only a hazy idea about how the steam side of an exchanger operates in conjunction with the steam-flow controller and the condensate removal system. In our discussion, we will explain how the reboiler system operates under various service conditions.

To put heat into a reboiler (Fig. 1), we set the steam-flow controller at the desired flow and wait for heating to occur. Heat input by the steam is determined by:

$$Q = w\Delta H \qquad (1)$$

For the exchanger, the relationship is:

$$Q = UA\Delta T_m \qquad (2)$$

where:
$$\frac{1}{U} = \frac{1}{h_i} + \frac{1}{h_o} + r_w + r_i + r_o \qquad (3a)$$

This form of Eq. (3a) helps us to understand the operation of the reboiler. The quantity $(r_i + r_o)$ is the total fouling resistance. Every heat exchanger has some fouling resistance built into the design to ensure long onstream time [3, 4].

Heat-exchanger specifications

Specifications for the typical heat exchanger in Table I show the rated design of an actual reboiler. For this exchanger, $Q = 12.8 \times 10^6$ Btu/h, $U = 83.6$ Btu/(h)(ft²)(°F), $A = 3,195$ ft² and $\Delta T_m = 48.0$ °F. We also find that design fouling factors are $r_i = 0.005$ and $r_o = 0.0005$. Substituting into Eq. (3a), we get:

$$\frac{1}{83.6} = \left(\frac{1}{h_i} + \frac{1}{h_o} + r_w\right) + (0.005 + 0.0005)$$

$$\left(\frac{1}{h_i} + \frac{1}{h_o} + r_w\right) = 0.01196 - 0.0055 = 0.00646$$

Thus, for this exchanger, the heat-transfer rate at design flows is:

$$(1/U) = 0.00646 + (r_i + r_o) \qquad (3b)$$

The exchanger has been designed to operate on the basis of a maximum $(r_i + r_o)$ equal to 0.0055. However, this is not how the exchanger will actually operate. When the exchanger is new and clean, fouling has not yet taken place, and the quantity $(r_i + r_o)$ is zero. Hence, the heat-transfer rate for zero fouling is:

$$(1/U_c) = 0.00646 + (0 + 0) \qquad (3c)$$

And, $U_c = 154.7$, as shown in Table I.

Temperature drop with clean exchanger

Since the reboiler is clean, U_c is now quite different from our design value of 83.6. How does this affect the reboiler operation? Examining Eq. (2), we find that Q is fixed at 12.8×10^6 Btu/h by the steam-flow controller, U_c is 154.7 as determined above, and $A = 3,195$ ft² from Table I. Therefore, ΔT_m must change to keep Eq. (2) in balance. Solving for ΔT_m, we get:

$$12.8 \times 10^6 = 154.7(3,195)(\Delta T_m)$$

$$(\Delta T_m)_c = 26°F$$

This value is compared to the design ΔT_m of 48.0, as shown in Table I. What has happened?

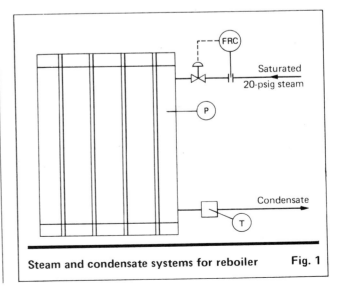

Steam and condensate systems for reboiler   Fig. 1

Since no fouling has taken place in the tubes, the resistance to heat transfer across them is less than design, and a temperature difference of only 26°F is required to transfer the desired heat to the 180°F process fluid. The condensing temperature in the steam chest is now 180 + 26 = 206°F. The pressure in the steam chest, P, is the saturation pressure corresponding to 206°F, which is 13.0 psia.

The steam flowrate is kept constant because our flow controller is set to hold a constant flow of steam, and thus heat input to the reboiler is kept constant. (The actual duty of the reboiler will change slightly at constant steam flow under the reduced-pressure condition because the enthalpy of the steam will vary slightly with the change in condensing temperature. But in the field a constant steam flow is normally maintained and is considered as constant duty to the reboiler.)

Because the pressure in the steam chest is less than the design operating pressure shown in Table I, the pressure drop across the flow-control valve has increased. To maintain constant steam flow at 13,250 lb/h with increased pressure drop across the valve, the flow controller will automatically change the signal to the control valve. In effect, the controller will set the port to a smaller opening for steam flow at the desired setpoint.

In our example, the design operating pressure in the steam chest is 5.3 psig. With an assumed steam-header pressure of 20.0 psig, the design pressure-drop across the control valve is 20.0 − 5.3 = 14.7 psi. But when the reboiler is clean, the steam-chest pressure is 13.0 psia, and the pressure drop across the control valve becomes 20 + 14.7 − 13 = 21.7 psi.

Steam temperature after control valve

When the exchanger is clean, the temperature of the steam immediately downstream of the control valve will not be the saturation temperature. Pressure loss across the control valve is an adiabatic operation (no heat added or removed). Hence, the steam immediately downstream of the control valve must have the same enthalpy as the 20-psig saturated steam, that is, 1,167.0 Btu/lb, and its temperature is 240°F, with 34°F superheat above the condensing temperature of 206°F at 13.0 psia.

Similarly, when the operating pressure in the steam chest is 5.3 psig (i.e., 20.0 psia), the enthalpy of the steam is also 1,167.0 Btu/lb, and the temperature downstream of the control valve is 249°F, with 21°F superheat above the condensing temperature of 228°F at 5.3 psig.

The temperature of the superheated steam at the tubes is not normally shown as the inlet temperature to the exchanger (on the specification) unless it is significant, because the film of condensate on the tubes will tend to desuperheat the steam. Consequently, the inlet and outlet conditions shown are the saturation temperatures at which the steam condenses.

Importance of condensate removal

The liquid condensate that results from the condensing steam will fall to the bottom of the steam chest and will flow out the bottom outlet of the shell. Under conditions where the steam chest operates at a high

Nomenclature

A	Heat-transfer surface, ft²
H	Enthalpy, Btu/lb
ΔH	Change in enthalpy, Btu/lb
h_i	Inside-tube heat-transfer coefficient, Btu/(h)(ft²)(°F)
h_o	Outside-tube heat-transfer coefficient, Btu/(h)(ft²)(°F)
P	Steam-chest pressure, psig or psia
ΔP	Pressure drop, psi
Q	Duty, Btu/h
r_i	Inside-tube fouling resistance to heat transfer, (h)(ft²)(°F)/Btu
r_o	Outside-tube fouling resistance to heat transfer, (h)(ft²)(°F)/Btu
r_w	Tube-wall resistance to heat transfer, (h)(ft²)(°F)/Btu
T	Temperature, °F
ΔT_m	Mean temperature difference, °F
U	Overall heat-transfer coefficient, Btu/(h)(ft²)(°F)
U_c	Overall heat-transfer coefficient (clean), Btu/(h)(ft²)(°F)
w	Steam flow, lb/h

condensing pressure, the condensate will flow to a steam trap, which will discharge condensate to a collecting header operating at lower pressure. The trap will also prevent any steam from passing through. Correct design of the steam-trap installation is important to ensure proper operation of the reboiler [1, 2].

In our example, the condensing pressure is only 13.0 psia, and condensate cannot flow to a header operating at atmospheric pressure or above. Under these conditions, other means must be used to recover the condensate. Mathur [1] states that "condensate removal from exchangers where steam pressures are . . . below return-header pressure . . . will require use of a 'pumping trap' or a vacuum pump with condensate pot and condensate pump." In the past, we have successfully used a condensate pot with a pressure-balanced line to the steam chest, and an external pump. Or, a pressure-balanced condensate pot, with an internal vertical pump. More recently, self-priming condensate pumps without condensate pots have been used successfully to remove condensate from the exchanger.

Failure to remove condensate as it forms can cause it to back up in the reboiler, and block off the heat-transfer surface. The relationship of Eq. (2) must always hold. Thus, as the available surface decreases, the temperature difference, ΔT_m, must increase in order to maintain constant heat transfer. Effectiveness of the reboiler is thus reduced. If condensate continues to build up and cover more tube area, the reboiler will not be able to effect the desired heat-transfer duty.

Fouling affects performance

As the reboiler continues to operate over a period of time, some fouling will begin to take place on the tubes. This may occur inside the tubes due to the nature of the process fluid, and is represented by r_i. Fouling may also take place outside the tubes due to the accumulation of noncondensables or other impurities in the steam, and is represented by r_o. Since the process fluid on the tubeside of the reboiler in this example has a high tendency to foul, we have assigned a fouling factor of 0.005 for this service. Steam is generally considered a clean heating medium, and a fouling factor of 0.0005 has been assigned for this service. The heat-exchanger specialist has used these fouling factors in the design of the exchanger.

What takes place as the reboiler starts to foul? The tubes will be coated by a deposit that will increase the resistance to heat transfer. After some time, fouling resistances, r_i and r_o, [Eq. (3b)] will slowly increase from their initial values of zero, as shown in Eq. (3c).

If we assume that the reboiler has been operating for some months and develops a total fouling resistance of 0.001, we substitute this value into Eq. (3b) to determine a new heat-transfer coefficient of:

$$1/U_I = 0.00646 + (0.001) = 0.00746$$

Now, the heat-transfer coefficient, $U_I = 134.0$, compared to an original $U_c = 154.7$ when the exchanger was clean. Fouling has added a barrier to heat transfer.

If the steam in the steam chest remains at 206°F, the temperature differential across the tubes will remain at 26°F. Assuming no change occurs in the operating conditions, the duty that would be transferred with a fouling resistance of 0.001 is equal to: $Q = 134.0 \times 3,195 \times 26 = 11.13 \times 10^6$ Btu/h. The steam flow would also drop, and its new flowrate would be:

$$w = \frac{11.13 \times 10^6 \text{ Btu/h}}{960 \text{ Btu/lb}} = 11,595 \text{ lb/h}$$

Actually, the above situation will not take place. The flow controller is set to maintain a constant flow of 13,250 lb/h of steam to the reboiler, and it will make adjustments to maintain the desired flow. The flow meter will sense any drop in steam flow, and will send a signal to the control valve to open and admit more steam to the reboiler, until setpoint is achieved.

As a result of opening the control valve, the pressure drop across the valve decreases, and the pressure in the steam chest increases. To calculate the temperature and pressure in the steam chest, we substitute the new value of the heat-transfer coefficient, $U_I = 134.0$, into Eq. (2):

$$12.8 \times 10^6 = 134.0(3,195)(\Delta T_m)$$

$$\Delta T_m = 30°\text{F}$$

The steam-chest temperature is now $180 + 30 = 210°F$, and steam-chest pressure is 14.1 psia (this is subatmospheric pressure). Pressure drop across the flow-control valve becomes: 20 psig + 14.7 − 14.1 psia = 20.6 psi.

In the field, the valve indicator will move toward the 100% open position as the reboiler fouls, because the control valve opens to maintain the steam flow constant. The local-pressure gage on the shellside of the reboiler will show higher steam pressure in the steam chest.

As time goes on, the reboiler will continue to foul, and the steam-chest temperature will continue to increase, with corresponding increase in steam-chest pressure. When the total fouling resistance reaches 0.0055 (the sum of the fouling resistances shown in Table I), the indicated steamside temperatures and pressures will be reached. (Note that the total fouling resistance could all be due to process-fluid fouling only.)

Operating in excess of design conditions

What will happen if we continue to operate after design conditions have been reached? Following our previous reasoning, we expect that fouling will continue to increase. Assuming total fouling resistance increases to 0.007, we calculate a new heat-transfer coefficient, U_{II}, as follows:

$$1/U_{II} = 0.00646 + 0.007 = 0.01346$$

$$U_{II} = 74.3$$

Again, we substitute into Eq. (2) to find ΔT_m:

$$12.8 \times 10^6 = 74.3 \times 3,195 \, \Delta T_m$$

$$\Delta T_m = 54°\text{F}$$

Steam-chest temperature is now $180 + 54 = 234°F$, and steam-chest pressure is 22.4 psia (7.7 psig). The temperature and pressure are higher than shown in Table I, but the reboiler will continue to deliver the

Tubular heat-exchanger specification — Table I

Description of system
Service: Column reboiler
Size: 67-60   Type: TEMA CEN
Number of units: 1   Surface per unit: 3,195 ft²
Shells per unit: 1   Surface per shell: 3,195 ft²

Required performance of one unit

	Shellside	Tubeside
Fluid circulated	Steam	Crude
Total fluid entering, lb/h	13,250	381,500
Liquid, lb/h	—	381,500
Steam, lb/h	13,250	—
Fluid vaporized, lb/h	—	76,300
Steam condensed, lb/h	13,250	—
Latent heat, Btu/lb	960	168.5
Temperature (in and out), °F	228|228	180|180
Operating pressure (outlet), psig	5.3	80 mm Hg abs.
Passes per shell	1	1
Fouling resistance, (ft²)(h)(°F)/Btu	0.0005	0.005
Heat exchanged, Btu/h	12.8 × 10⁶ Btu/h at ΔT_m = 48.0°F	
Heat-transfer rate, Btu/(h)(ft²)(°F)	Service: 83.6 Clean: 154	
Design pressure, psig	54	Full vacuum and 50
Design temperature, °F	650	250

Description of unit
Tubes, shell, tube sheets, tube supports, channel and cover are carbon steel. A total of 2,441 tubes are contained in the 67-in.-I.D. shell. Tube array is 1¼-in triangular pitch. Tubes are 1-in O.D., wall thickness=0.075 in., length=5 ft.

Summary of the operating characteristics for the reboiler — Table II

			Operating conditions		
	Design	Initial	Fouling, Case I	Fouling, Case II	Fouling, Case III
Total fouling, $(r_i + r_o)$	0.0055	0.0	0.001	0.007	0.015
Heat duty, Q, 10^6 Btu/h	12.8	12.8	12.8	12.8	11.0
Steam flow, w, lb/h	13,250	13,250	13,250	13,250	11,460
Steam-chest pressure, P, psia	20.0	13.0	14.1	22.4	31.7
Steam-chest temperature, T, °F	228	206	210	234	254
Process temperature, °F	180	180	180	180	180
Mean-temperature difference, ΔT_m, °F	48	26	30	54	74
Overall coefficient, U, Btu/(h)(ft^2)(°F)	83.6	154.7	134.0	74.3	46.6
Steam-main pressure, psia	34.7	34.7	34.7	34.7	34.7
Control-valve ΔP, psi	14.7	21.7	20.6	12.3	3.0

design duty as long as the control valve can open to maintain constant flow to the reboiler at a ΔT_m of 54°F.

Fouling becomes severe

When fouling becomes severe, the control valve will reach the wide-open position, and the maximum temperature and pressure in the steam chest will have been reached. If the maximum temperature does not give sufficient ΔT_m to transfer the heat, the flow of steam to the reboiler will decrease and the desired reboiler duty will not be able to be maintained. Under these conditions, the reboiler must be shut down and cleaned.

For example, let us assume that total fouling resistance reaches 0.015. For this fouling factor, we calculate the heat-transfer coefficient, U_{III}^*, as:

$$1/U_{III}^* = 0.00646 + 0.015 = 0.02146$$

$$U_{III}^* = 46.6$$

Substituting into Eq. (2) yields:

$$12.8 \times 10^6 = 46.6 \times 3{,}195\, \Delta T_m$$

$$\Delta T_m = 86.0°F$$

The required steam-chest temperature is $180 + 86 = 266°F$, and steam-chest pressure is 39.2 psia (24.5 psig). But, the available steam pressure equals 20.0 psig minus the control-valve pressure drop with the valve wide open. Assuming control-valve drop is 3.0 psi when full open, maximum pressure available in the steam chest is $20.0 - 3.0 = 17.0$ psig. Since the required steam pressure is greater than the available pressure, the required temperature difference across the tubes of the exchanger cannot be attained for transferring the desired heat.

We will now evaluate the limiting conditions under which the reboiler may possibly operate. In the wide-open position, control-valve pressure drop = 3.0 psi, and pressure in the steam header is 20.0 psig. Hence, the maximum steam-chest temperature is 254°F, corresponding to a steam pressure of 17.0 psig (31.7 psia). Maximum ΔT_m is $254 - 180 = 74°F$. Just before the reboiler must be shut down for cleaning, we can calculate the minimum overall heat-transfer coefficient, U_{III}, at which the reboiler can continue to operate, by substituting into Eq. (2) the fixed area 3,195 ft^2, and the required heat duty of 12.8×10^6 Btu/h when ΔT_m is 74°F.

$$U_{III} = \frac{12.8 \times 10^6}{3{,}195 \times 74} = 54.0 \text{ Btu/(h)(ft}^2\text{)(°F)}$$

When excessive fouling has taken place and the overall heat-transfer coefficient, U_{III}^*, reaches a value of 46.6 (i.e., when the total fouling resistance, $r_i + r_o$, is equal to the previously assumed value of 0.015), the coefficient is below the minimum value to transfer the desired heat duty. Under these conditions, the maximum heat that can be transferred is:

$$Q = 46.6(3{,}195)(74) = 11 \times 10^6 \text{ Btu/h}$$

Well before the exchanger reaches this condition, the control-valve indicator will show that the steam valve is wide open, and the flow instrument will indicate that steam flow has dropped below the 13,250 lb/h needed to maintain the desired reboiler duty. By observing the steam-control valve, the operator can anticipate when cleaning of the reboiler is required. Table II summarizes the various operating conditions.

References

1. Mathur, J., Performance of Steam-Heated Exchangers, *Chem. Eng.*, Sept. 3, 1973, pp. 101–106.
2. Monroe, E. S., Install steam traps correctly, *Chem. Eng.*, May 10, 1976, pp. 121–126.
3. Taborek, J., Aoki, T., Ritter, R. B. and Palen, J. W., Fouling: The Major Unresolved Problem in Heat Transfer, *Chem. Eng. Progr.*, Feb. 1972, pp. 59–67.
4. Frank, O. and Prickett, R. D., Designing Vertical Thermosyphon Reboilers, *Chem. Eng.*, Sept. 3, 1973, pp. 107–110.

The author

Albert E. Helzner is a process engineer with Badger America, Inc., One Broadway, Cambridge, MA 02142, and has been employed by the company for more than 14 years. He has a B.S. in chemical engineering from Tufts University and an M. Ed. with a speciality in mathematics from Northeastern University. Mr. Helzner is a member of AIChE and is licensed as a professional engineer in Massachusetts, New York and Florida.

Performance of Steam Heat-Exchangers

When you use steam as the heating medium in a heat exchanger, there are problems of condensate removal and control that you may not be fully aware of.

JIMMY MATHUR, Brown & Root, Inc.

Heat exchangers, with steam as the heating medium, are commonly used in the chemical process industries. Although such equipment is in universal use, many engineers do not understand the difficulties inherent in its design, or the obstacles to satisfactory performance.

This article explores the behavior of steam heat-exchangers and discusses certain features concerning process control and condensate-removal systems.

Steam-Inlet Control—System Analysis

Fig. 1 illustrates an exchanger in which process-fluid outlet-temperature is varied by a control valve located on the steam-inlet line (with condensate removal through a steam trap). In this system, it is assumed that total heat-transfer surface, (A), is always available, but this is true only if the steam trap removes condensate as it is formed—that is, no flooding.

Under these conditions, a reduction in heat duty (Q), caused, for example, by reduced process-fluid flowrate, is satisfied by a reduction in mean temperature difference (ΔT). Since process fluid-side temperatures are being held constant regardless of process flowrate, ΔT can be varied only by a change in steam-side temperature (T_2).

This change in T_2 is effected by throttling of the control valve, which closes to a level where it supplies enough steam to maintain the new value of T_2 at the reduced load.

An analysis is presented in the box on p. 103.

The temperature ($T_{2,M}$) required in the steam-side at a load factor M is given by the following equation for the system under consideration:

$$T_{2,M} = \left(\frac{T_3 + T_4}{2}\right)\left(1 - \frac{M}{100}\right) + \left(\frac{M}{100}\right)(T_1)$$

T_3, T_4 are, respectively, process-fluid inlet and outlet temperatures, T_1 is the steam-supply temperature upstream of the control valve, M is the load-factor percent.

Let us examine the application of this equation for an exchanger that is *designed* for the following conditions: Steam supply is at 250 psig. (P_1), 406 F. (T_1); there is a constant back-pressure of 50 psig. (P_4) at the steam-trap outlet; process-fluid inlet is at 105 F. (T_3); outlet at 325 F. (T_4) is desired.

Temperatures ($T_{2,M}$) have been calculated at various load factors and are shown in Fig. 3. Pressures ($P_{2,M}$) for saturated steam corresponding to temperatures ($T_{2,M}$) have been taken from the steam tables. Pressures ($P_{5,M}$)

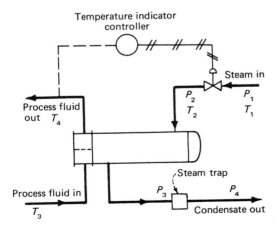

EXCHANGER with steam inlet control—Fig. 1

Originally published September 3, 1973

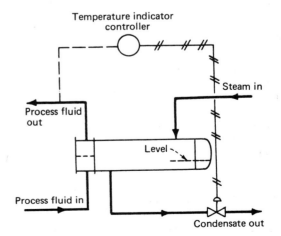

HEAT EXCHANGER with condensate outlet control—Fig. 2

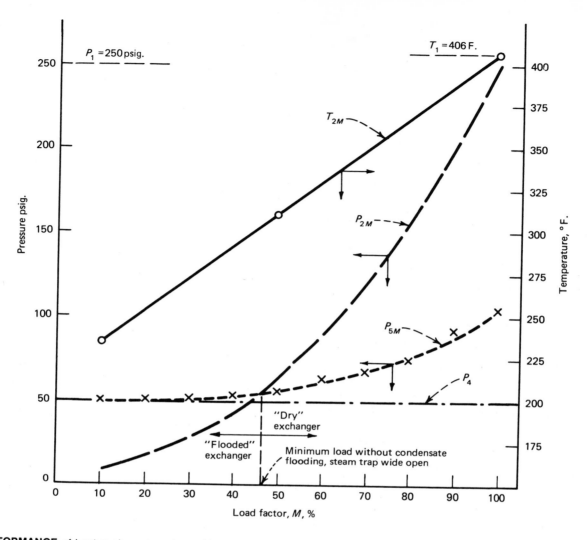

PERFORMANCE of heat exchanger system with steam inlet control—Fig. 3

are those required at the steam-trap inlet to provide the differential pressure across the trap that enables the trap to discharge the condensate produced at the various load factors. The exchanger in question has a 100% load of 2.7 million Btu./hr., requiring condensation of 3,280 lb./hr. of steam. The pressures ($P_{5,M}$) have been taken from the condensate capacity (lb./hr.) versus trap-differential-pressure data of a reputable manufacturer of steam traps.

A study of Fig. 3 reveals that operation of this exchanger system at a load factor less than about 47% will result in flooding. This is because steam-side pressure ($P_{2,M}$) at 47% load has dropped to a level where it is no greater than the minimum pressure ($P_{5,M}$) required at the trap inlet to discharge all condensate produced at this load. At loads higher than 47%, steam-side pressures ($P_{2,M}$) are higher than the minimum pressures ($P_{5,M}$) required at the trap inlet, and therefore the trap successfully discharges all condensate produced at those loads. At loads below 47%, the trap operates wide open but is unable to remove all condensate. This unremoved condensate backs into the exchanger, and covers exchange-surface area; following this, a new equilibrium is (possibly) established, with a portion of the total surface area flooded, and steam-side pressure ($P_{2,M}$) thus riding along the line for trap-inlet pressures ($P_{5,M}$). That is, trap performance at steam-side pressure will control exchanger performance by either flooding or exposing surface.

On the other hand, for example, if the load reduction has been sudden, a new equilibrium with a flooded exchanger may not arise. Instead, there could be cycling (with hunting of the steam-control valve) as exchanger surface is covered and uncovered. There may also be noise and hammering within the exchanger as steam bubbles collapse on contact with the colder condensate —which has been in prolonged contact with process-fluid inlet temperature (T_3); physical damage to equipment (collapsed float trap); and back-flow of condensate from the condensate-return header if there is no check valve in the trap-outlet line (most traps, including float traps, cannot prevent reverse flow).

It is worth noting that so long as there is a control

Analysis of Steam-Inlet-Control Exchanger for Conditions of Varying Load

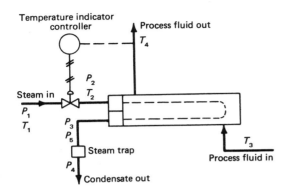

A. At Design 100% Load Conditions:

$Q_{100} = UA\Delta T_{100}$

Ignoring the normal minor ΔP across the control valve at full load conditions, $P_1 = P_2$ and $T_1 = T_2$.

Then, $\Delta T_{100} = \dfrac{(T_1 - T_3) - (T_1 - T_4)}{2.3 \log\left(\dfrac{T_1 - T_3}{T_1 - T_4}\right)}$

$= \dfrac{T_4 - T_3}{2.3 \log \dfrac{T_1 - T_3}{T_1 - T_4}}$

For simplicity, using the arithmetic mean ΔT instead of the log-mean ΔT,

$\Delta T_{100} = \dfrac{(T_1 - T_3) + (T_1 - T_4)}{2} = T_1 - \dfrac{(T_3 + T_4)}{2}$

Since $T_{B,100} = \dfrac{T_3 + T_4}{2}$

Therefore $\Delta T_{100} = T_1 - T_{B,100}$

Hence $Q_{100} = UA\,(T_1 - T_{B,100})$

B. At M% Load Conditions:

$Q_M = UA\,(T_{2,M} - T_{B,M})$

or $T_{2,M} = \dfrac{Q_M}{UA} + T_{B,M}$

Since $M = \dfrac{Q_M}{Q_{100}}$, or $Q_M = \left(\dfrac{M}{100}\right) Q_{100}$

Therefore $T_{2,M} = \left(\dfrac{M}{100}\right)\left(\dfrac{Q_{100}}{UA}\right) + T_{B,M}$

$= \left(\dfrac{M}{100}\right)(\Delta T_{100}) + T_{B,M}$

$= \dfrac{M}{100}(T_1 - T_{B,100}) + T_{B,M}$

For the system shown in the sketch above, T_3 and T_4 are held constant at all load conditions.

Hence $T_{B,M} = \dfrac{T_3 + T_4}{2}$

Therefore $T_{2,M} = \dfrac{M}{100}\left(T_1 - \dfrac{T_3 + T_4}{2}\right) + \dfrac{T_3 + T_4}{2}$

$= \left(\dfrac{T_3 + T_4}{2}\right)\left(1 - \dfrac{M}{100}\right) + \left(\dfrac{M}{100}\right)(T_1)$

The final equation given above enables calculation of the temperature (and corresponding steam pressure) existing in the steam side of the exchanger at various operating loads.

This analysis has been adapted from Ref. 1, to which the reader is referred for further detail, and for a thought-provoking consideration of flooded exchangers, condensate temperatures, and rapidity of response to load variation.

---

valve capable of throttling steam flow, for the purpose of meeting reduced heat duty, there is going to be a falloff in the pressure $(P_{2,M})$ at which steam condenses within the exchanger. It does not matter whether the steam-control valve is cascaded with a steam-flow controller, exchanger steam-side pressure controller, distillation-column-temperature controller or any other control system.

Steam-Inlet Control—Excessive Exchanger Surface

Besides reduced steam-side pressures (with their attendant problems of condensate disposal, resulting from reduced-load operation), the same situation occurs if excessive surface area has been provided because of the use of conservative heat-transfer coefficients, high fouling-factors or overdesign for emergency high-load conditions.

There can of course be sound reasons for conservative exchanger design. To quote from the AICHE "Equipment Testing Procedure for Heat Exchangers" manual,[2]—where an example is provided of a reboiler with an actual heat-transfer coefficient far greater than design—the procedure's writers have pointed out:

> This is not unusual practice in the design of reboilers, for many reasons; for example, (1) Fouling of an indeterminate amount is to be expected, (2) Excess capacity due to higher reflux ratios to attain purity is often desired, (3) Steam pressure and superheat may vary due to overloaded lines. Losses in pressure are to be expected in control valves, (4) The reboiler is the source of heat for fractionation in a distillation system. If the reboiler cannot supply sufficient heat, the product will fail to meet specifications if the normal feed rate is continued, or the capacity of the entire unit will be reduced if the product specification is maintained, (5) There are many expressions for obtaining film coefficients, which all give different answers, and so a designer is inclined to put an ample safety factor in his design.

Oversurfacing of exchangers for the reasons mentioned above would result in performance similar to that depicted in Fig. 3 even at 100% design load. The steam-control valve has to throttle, thereby resulting in P_2 being substantially lower than P_1, in order to provide the reduced ΔT value that fulfills Q_{100} in the presence of a high value of UA.

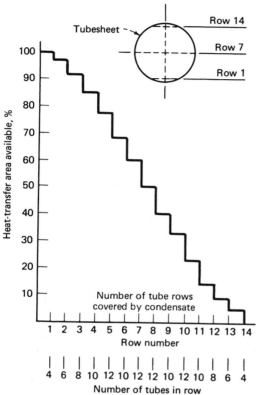

PERFORMANCE with condensate level control—Fig. 4

Nomenclature

A	Effective heat-transfer area, sq. ft.
M	Load factor = $(Q_M/Q_{100})(100)$, %
P_1	Pressure upstream of steam-control valve, psig.
P_2	Pressure downstream of steam-control valve, psig.
$P_{2,M}$	Pressure downstream of control valve, at M% load factor, psig.
P_3	Pressure at steam trap inlet, psig.
P_4	Pressure at steam-trap outlet (assumed constant back-pressure), psig.
$P_{5,M}$	Pressure required at steam-trap inlet for discharge of condensate at load M
Q_M	Heat exchanged at M% load factor, Btu./hr.
Q_{100}	Heat exchanged at 100% load factor, Btu./hr.
T_1	Temperature upstream of control valve, °F.
T_2	Temperature downstream of control valve, °F.
T_3	Process-fluid inlet temperature, °F.
T_4	Process-fluid outlet temperature, °F.
$T_{2,M}$	Temperature downstream of control valve, at M% load factor, °F.
$T_{B,M}$	Average bulk temperature, process-side, at M% load factor, °F.
U	Heat-transfer coefficient (assumed constant at all loads M), Btu./(hr.) (sq.ft.) (°F.)
ΔT_M	Mean temperature difference at M% load factor, °F.
ΔT_{100}	Mean temperature difference at 100% load factor, °F.

Steam-Inlet Control—Condensate Removal

The data depicted in Fig. 3 highlight the problem of condensate removal at low steam-side pressures. Proper specification and selection of the steam trap is important—and it is as difficult as it is important. The plot of minimum trap-inlet-pressures ($P_{5,M}$) shown in Fig. 3 is for a float trap claimed to pass 1.75 times 100% load condensate flow at 200-psi. trap-differential-pressure (i.e., the differential applicable at full load conditions).

If the safety factor of 1.75 used here is considered inadequate, we could settle for a trap with a larger capacity (larger orifice area) that may extend the possibility of low-load operation but could result in mechanical problems—larger diameter float (more susceptible to rupture) or operation at normal loads with the orifice valve running almost completely closed (and therefore liable to wire-drawing or sticking shut while closed).

The steam trap is eliminated with the use of a condensate-level-pot depicted in Fig. 5. The equal-percentage control valve has a wider range of flow capability than the steam trap, and is also more robust than the float-type trap generally used in heat-exchanger service. Where very high condensate rates are involved, making steam-trap selection difficult or trap-size cumbersome, the system of condensate pot with control valve has few if any competitors. The arrangement in Fig. 5 does not, however, solve the difficulty of removing condensate when steam-side pressure has fallen to a low level compared to return-header pressure.

In the example of Fig. 3, steam shell-side design-pressure-drop was only ½ psi., and P_2 was therefore taken to be identical with pressure P_3 at the trap inlet. There can be cases (with steam in the tubeside and lower steam-supply pressures), where steam-side pressure drop cannot be ignored. In such cases, a plot of P_3 pressures would be required and the P_3 plot compared with P_5 pressures to derive the intersection corresponding to minimum load for complete condensate removal.

The exchanger used as an example in this discussion and analyzed in Fig. 3 was located at a plantsite where 250-psig. condensate was flashed down to 50 psig. to meet 50-psig. steam-header demand. It is evident that the operating range without exchanger flooding could be extended if steam-trap outlet-pressure were less than 50 psig. The factors influencing choice of condensate-header pressure are integrated with a study of the steam-condensate balance at the overall plantsite, and are outside the scope of this article.

Condensate removal from exchangers where steam-side pressures are, in normal circumstances, below return header pressure or below atmospheric pressure will require use of a "pumping trap,"[3] or a vacuum pump to pull down pressure in the condensate pot and a condensate pump to remove condensate.[4]

Condensate-Outlet Control

Fig. 2 illustrates an exchanger where process-fluid outlet-temperature is controlled by throttling conden-

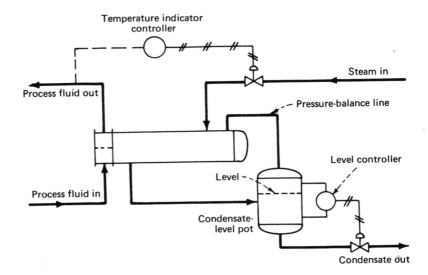

STEAM INLET control without steam trap—Fig. 5

sate-flow from the exchanger while steam continues to condense at essentially the supply-header pressure within the exchanger. At 100% design-load conditions, this system is no different from steam-inlet control shown in Fig. 1, in that both exchangers are dry, with all the heat-transfer surface being fully utilized, and with steam-side pressure corresponding to header pressure. At less than 100% design load, the condensate-outlet control system varies the available heat-transfer area (A), by flooding some fraction of the total surface to meet reduced heat-load demand, while mean-temperature difference (ΔT) remains essentially unchanged from design ΔT. Percent heat-transfer area available, at any load condition, is identical with percent load-factor; operation at 75% load-factor requires availability of only 75% of total surface area, and so on.

Fig. 4 depicts condensate-level control performance of the exchanger that was earlier analyzed in Fig. 3 for steam-inlet control. Consideration of the geometry of this exchanger shows that a load increase from 25% to 75% of design takes approximately 15 min., and a decrease from 75% to 25% takes approximately 45 min. before appropriate heat-transfer surface has been uncovered or covered, respectively. A ± 10% load change at half load, requiring covering/uncovering of a single row of tubes at midsection of the exchanger takes about 3 min. If steam were on the tube-side, these response times would be approximately halved.

Response times depend on geometry of the exchanger and rate of condensation. For another exchanger checked by the writer, response times were 1.33 to 4 minutes for 25% to 75% load changes, and only 1.33 sec. for a 5% load swing caused by covering/uncovering of a single row of tubes at the exchanger midsection. In the first example, we could say that response is sluggish but likely to be stable, while in the second we could expect unstable operation at around 50% load. (Response times referred to here have been calculated simply from time for displacement of condensate and do not involve consideration of instrumentation lag time.) If it is considered desirable to add surge volume to the condensate side and to prevent passage of steam into the condensate-return header, a condensate pot can be placed in the condensate line upstream of the control valve in Fig. 2. A level controller in this pot can override the temperature controller and close the control valve on low level.

Condensate-Outlet Control—Pros and Cons

The main advantages of this system are that (a) positive pressure is always available for removal of condensate regardless of load or excessive surfacing of exchangers for any of the reasons mentioned earlier; (b) steam-trap specification and selection with all their attendant difficulties for modulating control of exchangers are avoided; (c) sizing and specification of the condensate-outlet control valve is a comparatively simple matter; (d) sizing and specification of the steam-inlet control valve is not so simple[5]; (e) the control valve is a robust, low-maintenance item.

The disadvantages of this system tend to be emphasized in the literature.[5-8] They are (a) sluggish response, (b) unpredictable performance, (c) thermal shock, (d) corrosion due to the condensate level maintained in the exchanger. Each of these features should, however, be seen in the context of the particular exchanger. In cases where operating-load levels are not frequently varied; where exchanger tubesheet design has been checked for differential thermal stresses; and where the steam supply is free of carbon dioxide and oxygen—the disadvantages cited may be of no great significance. In fact, there are some operating companies that very definitely prefer condensate-outlet control. The disadvantages tend to be highlighted when it is assumed that a steam trap is always available for proper condensate removal in the steam-inlet control system. Our earlier analysis has shown that this assumption can be false, really through no fault of the trap. For a consideration of the limitations of trap

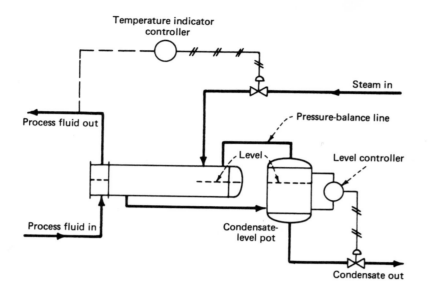

STEAM-INLET and condensate outlet control—Fig. 6

design, the reader is referred to my previous article.[9]

While it would seem that heavily oversurfaced exchangers are more readily handled by condensate-outlet control systems, there can be cases of hammering within an exchanger or in the outlet line, which may lead to equipment damage. The hammering may be caused by surging of condensate level within the exchanger. For an interesting example of this situation, see Ref. 10.

Steam-Inlet and Condensate-Outlet Control

Fig. 6 illustrates a system that combines the virtues of both types of control system. The virtue of the steam-inlet control system is that control is exercised on the primary variable (namely, steam flow), thus providing quick response. The virtue of condensate-outlet control is that exchanger surface can be varied, thereby providing positively high steam-side pressure to ensure condensate removal. Also, the steam trap, with its operating vagaries and maintenance problems, is eliminated.

Location of the condensate pot is such that condensate level can be varied within the exchanger by changing the setpoint on the level controller.

If we refer once again to the exchanger performance shown in Fig. 3, it will be seen that—at 40% load—steam-side pressure is not adequate for removal of condensate by the trap. At this load level, the mean-temperature-difference required is 76.4 F. and is met by steam temperature of 291.4 F., equivalent to a pressure of 44 psig., which is less than 50 psig. in the return header. If we cover 50% of exchanger surface by a 50% setting on the level controller in Fig. 6, the mean-temperature-difference required for 40% load operation will be 152.8 F., necessitating a steam temperature of 367.8 F., which is equivalent to a steam-side pressure of 153.5 psig. This pressure is well above return-header pressure, so the control valve can easily remove condensate.

The combination system of Fig. 6 provides the best solution for exchanger performance in conditions where there is operation over a wide range of load level, where exchangers are deliberately oversurfaced to meet a variety of contingencies, and quick response with fine control is required for fluctuating operating conditions. ∎

Acknowledgement

The writer thanks Mr. Eric Jenett of Brown & Root, Inc., for his encouragement and valuable suggestions during review of this article.

References

1. Ferwerda, G. G. J., Some Observations on What Happens Inside Steam Heated Condensers, *Heating, Piping Air Conditioning*, July 1965, pp. 129–137.
2. "Equipment Testing Procedure: Heat Exchangers," 2nd ed., American Institute of Chemical Engineers, New York, 1968, p. 35.
3. Sarco Co., "Sarco Pressure Powered Pump," Bulletin No. 165.
4. Kern, D. Q., "Process Heat Transfer," McGraw-Hill, New York, 1950, pp. 778–780.
5. Perry, J. H., others, eds., "Chemical Engineers' Handbook," 4th ed., McGraw-Hill, New York, 1963, pp. 22–105 to 22–107.
6. Bland, W. F., Davidson, R. L., "Petroleum Processing Handbook," McGraw-Hill, New York, 1967, pp. 10–11.
7. Sanders, C. W., Better Control of Heat Exchangers, *Chem. Eng.*, Sept. 21, 1959, pp. 145–148.
8. Forman, E. Ross, Control Dynamics in Heat Transfer, *Chem. Eng.*, Jan. 3, 1966, pp. 91–96.
9. Mathur, J., Steam Traps, *Chem. Eng.*, Feb. 26, 1973, pp. 47–52.
10. Wild, N. H., Noncondensable Gas Eliminates Hammering in Heat Exchangers, *Chem. Eng.*, Apr. 21, 1969, pp. 132–134. See also, *Letters Pro & Con*, they're Both Right, *Chem. Eng.*, Oct. 6, 1969, p. 7.

Meet the Author

Jimmy Mathur is a senior engineer in the Process Engineering Dept. of Brown & Root, Inc., P. O. Box 3, Houston, TX 77001, where he is concerned with the design of refinery and petrochemical plants and specializes in utilities systems, including environmental control. Previously, he worked 15 years in India, primarily in production management of polyethylene, synthetic rubber and petrochemical units. He has an M.S. in chemical engineering from the Indian Institute of Science, Bangalore, India, and is a member of AIChE.

Section III
HEAT TRANSFER IN REACTION UNITS

Controlling heat-transfer systems for glass-lined reactors
Heat more efficiently—with electric immersion heaters
Heat transfer in mechanically agitated units
Heating and cooling in batch processes
Picking the best vessel jacket

Controlling heat-transfer systems for glass-lined reactors

Generally, agitated glass-lined reactors have heat supplied and removed by a heat-transfer medium in the jacket. But the glass lining is sensitive to rapid temperature changes. Here are methods for circulating and controlling the thermal liquid in the jacket to avoid excessive temperature differentials.

Richard E. Hinkle and Joel Friedman, American Hydrotherm Corp.

☐ When selecting a method of heating and cooling a glass-lined reactor, the process engineer's objectives are rapid uniform heat-up and cool-down, and precise control. If these objectives are attained, the reaction yields and productivity are higher because less time is wasted at transition temperatures. Here, we will describe how thermal-liquid systems, and their associated reactor temperature controls, can help achieve this in glass-lined reactors, with their inherent limiting conditions.

Using glass-lined reactors

Glass-lined reactors are used to process very corrosive materials, and where exceptionally high product quality is required or the material is strongly adhesive. Generally, the reactors are jacketed and agitated with heat supplied and removed by a heat-transfer medium in the jacket, as long as there are not excessive temperature differentials. The differential temperature must be limited, while still maintaining responsive, uniform heating and cooling, without temperature overshoot.

Temperature overshoot is a serious problem in batch exothermic reactions. As the reaction reaches its initiating temperature, the heat of reaction tends to increase the batch temperature, which increases the reaction rate and therefore generates more heat. If cooling is inadequate at this point, the batch can be ruined. The glass lining in the reactor imposes a restriction because of cracking possibilities. How do you protect the lining?

Producers of glass-lined reactors are very careful to list the maximum allowable temperature difference between the product in the reactor and the heat-transfer liquid in the jacket. Exceeding this temperature difference during either heat-up or cool-down can cause sufficient uneven thermal expansion between the metal jacket and its glass lining to crack the lining. A check with two of the largest manufacturers of glass-lined reactors indicates that the cost of relining a reactor amounts to about 65% of the cost of a new reactor (without agitator or baffles), and takes about eight weeks for delivery (ten weeks of lost production). Extreme care, therefore, is advised as to the selection of proper instrumentation to prevent the temperature differential from being exceeded during both heat-up and cool-down.

The heat-transfer fluid

Organic thermal liquids are available that can operate between 0 and 600°F without pressurization or phase change. This makes them well suited for use in systems that use flow-control valves, circulating pumps, and coolers to achieve precise, responsive control of reactant temperature (Table I).

Pressurized hot water offers some of the same advantages, but it is usually limited to 300 to 320°F (52.3 to 75.0 psig equilibrium pressure) by the allowable design pressure of the jacket (90 to 100 psig). Alternating steam and cooling water in the same jacket is sometimes used on simple, manually controlled systems, but control of temperature overshoot and differential temperature is difficult when two phases are involved.

High-temperature molten salt can be used to heat and cool reactors to 1,000°F, with salt as the heat-transfer medium. This is well above the temperature limit for standard glass-lined reactors (450°F), and strict limitations would apply even to special units designed to withstand temperatures exceeding 500°F. The operation would have to be continuous, and a proprietary dilution technique must be used to heat the reactor and its contents slowly from ambient to operating temperature in the liquid phase. System design and controls for a salt system are considerably different from those that will be discussed here [2].

Typical heat-transfer system

Fig. 1 shows a thermal-liquid system serving multiple, independently-operated reactors and providing automatic heating and cooling. The heat source may be a fired, steam, or electric-resistance heater. A fired helical-coil heater is shown here. The main heater loop has

Originally published January 30, 1978

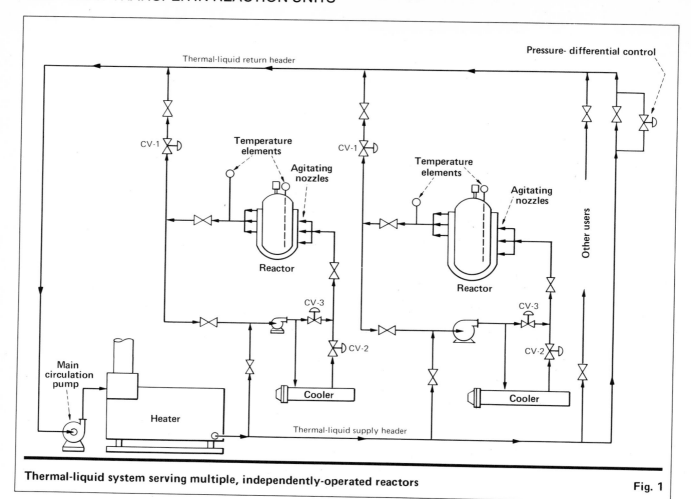

Thermal-liquid system serving multiple, independently-operated reactors   Fig. 1

Typical thermal liquids for glass-lined reactors			Table I
Thermal liquid	Recommended bulk temperature, °F		Equilibrium temperature,* °F
	Minimum	Maximum	
DuPont's Hitec	50	1,000	none
Dowtherm A	60	750	495
Therminol VP-1†	60	750	495
Dowtherm G	60	650	575
Therminol 66	50	650	>650
Therminol 60	−50	600	>600
Ucon HTF-X600**	−60	600	630
Mobiltherm 600	40	600	>600
Mobiltherm 603	70	600	>600
Gulf Security 53	70	600	740
Dowtherm J	−100	575	358
Therminol 55	70	575	>600
Shell Thermia 33	70	550	>600
Dowtherm HP	70	550	700
Ucon HTF-500	0	500	>500
Exxon Caloria HT43	70	500	>500
Texatherm	70	500	675

*Boiling point at atmospheric pressure, 14.7 psia.
†Monsanto Industrial Chemicals Co.
**Union Carbide Corp.

its own temperature controls, circulating pump and differential-pressure control, so that a source of constant-temperature thermal liquid is available at the supply header for use by each reactor independently of the others.

The secondary thermal-liquid loop, serving each reactor, also has its own circulating pump. The thermal liquid enters the reactor jacket through several agitating nozzles [4]. These nozzles impose a circulating turbulent flow through the jacket to obtain the highest practical heat-transfer coefficient. Overall coefficients, U, are in the range of 10 to 50 Btu/[(h)(ft^2)(°F)] with an organic thermal-liquid circulating through the jacket and agitated process-liquid in the reactor. Usually, the reactor-side film coefficient is the limiting factor in the overall coefficient.

The heat-transfer coefficient of the thermal liquid in the reactor jacket, h_o, will usually range from 200 to 400 Btu/[(h)(ft^2)(°F)], depending on its velocity. However, batch coefficients, h_i, are usually low. Their effect on the overall heat-transfer coefficient, U, is shown in Table II.

The thermal-liquid flowrates through the reactor loops are determined primarily by process heating and cooling requirements, but the reactor design and jacket-agitating nozzle selection also enter into the picture. If the process requires that the reaction take place within a narrow temperature band, the thermal-

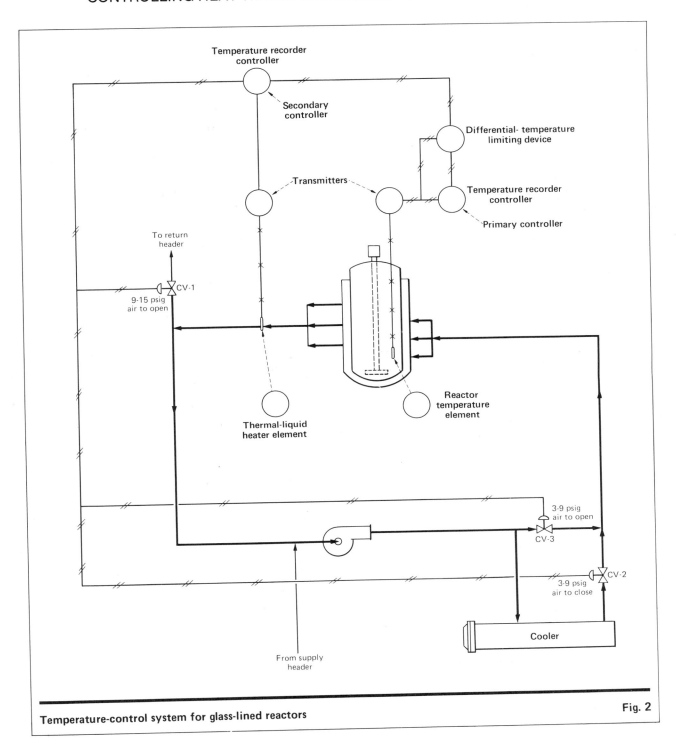

Temperature-control system for glass-lined reactors   Fig. 2

liquid inlet and outlet temperatures will be closer. This will require greater flowrates and surface area to transfer the required amount of heat.

On the other hand, if the allowable thermal-liquid differential temperature is greater, the flow (and piping costs) will be reduced, but the required pump head will be greater, to create an adequate amount of turbulence in the reactor jacket. In addition, as the temperature differential of the thermal liquid increases, the difference between the thermal-liquid and product temperature will increase. This differential must be limited to protect the glass lining.

Notice in Fig. 1 and 2 that the basic control concept for the reactors does not involve modulation of the primary heating and cooling media (in this case fuel flow to the fired heater and cooling-water flow to the cooler). This would still be the case, even if there were only one reactor in the system. Rather, the source and relative

Overall heat-transfer coefficient — Table II

Overall heat-transfer coefficient, U
Coefficient of thermal liquid in reactor jacket, h_o

h_i	200	250	300	350	400
10	9.5	9.6	9.7	9.8	9.8
15	14.1	14.3	14.5	14.7	14.7
20	18.2	18.5	18.7	19.2	19.2
25	22.2	22.7	23.7	23.8	23.8
30	26.3	27.0	27.8	28.6	28.6
35	30.3	31.3	32.3	33.3	33.3
40	33.3	34.5	35.7	37.0	37.0
45	37.0	38.5	40.0	41.7	41.7
50	40.0	41.7	43.5	45.5	45.5

amount of the thermal liquid, either from the hot-supply header or from the cooler, is selected by the reactor temperature controls. This minimizes the relatively long response times usually found in temperature-control loops in systems with a large heat content.

Heating, cooling and bypass valves, CV-1, CV-2, and CV-3, respectively, for each reactor are operated by the reactor's temperature controller.

When heating is required, CV-1 and CV-3 open and CV-2 closes. This discharges cooler liquid to the return header and allows hot liquid to circulate through the reactor jacket. During cooling, CV-1 and CV-3 are closed, and CV-2 opens so that cooled liquid is circulated through the jacket. Transition between heating and cooling is smooth, helping to avoid thermal shock to the glass lining.

The volume of thermal liquid in the reactor loop should be kept as small as possible, so that system response to a control demand is fast. This is done by assembling the control valves, pump and cooler into a compact package, which can be located close to the reactor. The MC (mass × heat-capacity) of the thermal liquid and metal—in which the temperature must be changed along with the reactants—is thus minimized.

Reactor temperature controls

A common control technique for jacketed reactors, especially batch reactors requiring heating and cooling, is a cascade control loop as shown in Fig. 2. The cascade arrangement provides a faster control-valve response than a single controller loop, and allows a narrower proportional band to be set on the primary controller. The output from the reactant temperature controller (primary) becomes the set point for the jacket temperature controller (secondary).

For glass-lined reactors the set point of the jacket temperature controller is limited to a preset maximum above or below the reactor temperature. Above 350°F, the temperature differential limit is usually 150 to 175°F, depending on reactor design. The limiting device comprises biasing, high-select, and low-select relays, so that it protects on both heating and cooling. This approach limits the temperature that the jacket controller will request, but does not control the actual differential temperature.

Another approach is to subtract the reactant temperature from the jacket temperature, the differential being the controlled variable of the secondary loop. The set point to the secondary controller will also be restricted between the upper and lower limits.

A batch-reactor-control technique that can be used to optimize operations is a dual-mode control [1]. On-off control is used for heat-up, applying full heat until the temperature is just below the set point. Heat is then shut off, and cooling applied so that the set point is reached with zero-approach velocity. Control is then switched to the preloaded output of the cascade-control loop. This technique heats the reactor to the desired temperature in minimum time, without overshoot, and allows the control loop to be tuned for the best steady-state control.

In general, additional controls for limiting differential temperature in glass-lined reactors are inexpensive since measurement and control components are already included in the system for reactant temperature control. It is important, however, that the relatively sophisticated instrumentation be incorporated into a carefully designed heating-cooling system. This system should be sufficiently versatile and responsive to operator or preprogrammed demands to yield optimum results with maximum economy, and with complete protection for the reactor and its glass lining.

References

1. Shinskey, F. G., The Foxboro Co., Foxboro, Mass., and Weinstein, J. L., Foster Grant Co., Leominster, Mass., "A Dual-Mode Control System for a Batch Exothermic Reactor."
2. American Hydrotherm Corp. salt-system literature, "Data Sheet 16-R."
3. Culotta, J. M., and Chang, B. O., Pfaudler Co., "Measurement and Control of Process Variables in Glasteel Reactors, Pfaudler Co. Technical Bull."
4. "The Pfaudler Agitating Nozzle," Pfaudler Bull. No. 950.

The authors

Richard E. Hinkle is assistant vice-president and project manager with American Hydrotherm Corp., 470 Park Ave. South, New York, NY 10016, where he specializes in design of industrial high-temperature water systems. His professional background includes being a sales engineer with Alco Products, and a heat-exchanger rating and process design engineer with Foster Wheeler Corp. He holds a B.S. degree in chemical engineering from Case Institute of Technology, and is a licensed professional engineer in the state of New York.

Joel Friedman is sales manager and assistant to the director of marketing with American Hydrotherm Corp., 470 Park Ave. South, New York, NY 10016, where he has been for ten years. Prior to this, he was general manager of W. R. Barry Pump Co., and a high-pressure design consultant for Cardwell Mfg. Co. He attended Cornell and New York Universities and is a member of the American Soc. of Mechanical Engineers and the American Petroleum Institute. He is also a frequent contributor of articles dealing with high-temperature applications.

Heat more efficiently— with electric immersion heaters

Electric immersion devices are recommended for heating a wide variety of liquids. Different types of units are described, and their assets and liabilities listed. Several factors that must be considered before selecting an immersion heater are reviewed and explained.

David R. Martignon, Watlow Electric Manufacturing Co., Rochester, Mich.

☐ When heating water, oil, chemicals, molten metals, molasses, or any of a wide variety of liquids, the most efficient approach is the use of an electric immersion heater. That is because, compared with other sources, electric heat is usually more efficient. One kilowatthour can be converted to heat with an efficiency of 100%, whereas when such fuels as gas, oil and coal are converted into heat, the efficiency may range from 5% to 80% because of losses in burning and applying the heat.

This article describes a variety of electric immersion heaters for all types of application parameters. Advice is given on how to choose the right type of heater, how to select the proper sheath material, and how to calculate the watt density of the heating elements.

Key advantages

Electric immersion units can be controlled by automatic devices—there are several types of temperature controllers that are used with electric heating. Among these, the solid-state controls provide accurate performance and long life. With these, there are no mechanical parts that wear out, or valves or gates, as in gas and oil systems.

The heater's elements are simple, of rugged construction, and easily replaceable. They cost less and have a longer life than elements for fossil-fuel heaters. They also need less floor space, since they are placed directly in the processing medium.

Oil- and gas-fired heaters require large burners and exhaust systems. Often, large exhaust fans are needed to clear the fumes from the working area. In some cases, storage tanks for fuel must also be purchased. In addition, the device's air filter must be cleaned and replaced on a regular basis.

With combustion systems, insulation must be kept away from the flames, to prevent charring. Electric systems, on the other hand, are well insulated; they lose less heat, so the surrounding areas will be cooler. Because there are no products of combustion, no ashes or dirt can soil the working area. And because there are also no fumes, fire and explosion hazards are greatly reduced.

Immersion heaters can be manufactured in accordance with ASME boiler codes, military guidelines, and other specifications. All of these illustrate how electric heating can save money and energy.

Available units

Immersion heaters are available in a variety of forms to fit all types of applications. The screw-plug immersion heater, for example, provides a permanent installation, which can nevertheless be removed without too much difficulty for cleaning or replacement. This unit simply screws into a threaded hole in the tank wall; if the wall is thick enough it can be drilled and tapped to accept the heater. If it is too thin, a pipe coupling can be welded or brazed in. Screw-plug units are normally available in 1–2.5-in. NPT sizes, and a variety of voltage and wattage values. Screw-plug elements are normally used when 20 kW or less are required to do the job.

Flanged immersion heaters are installed through the wall of the tank by bolting them to a matching pipe flange welded to the tank wall. The connection can be specially gasketed to withstand high pressures or contain penetrating liquids such as salts and ethylene glycol. Flanged immersion units are normally used where large kW packages are required. Flanged units are bulkier and somewhat more difficult to install than screw-plug heaters.

Over-the-side immersion units are ideal for applications in which easy installation and quick removal are desired. The lightweight portable immersion unit can be used for heating a wide variety of liquids in containers of different shapes and sizes. Hairpin-shaped tubular heaters are brazed or welded into a liquid-tight junction box. A riser takes the electrical connection up

Allowable watt densities for various liquids

Material being heated		Approximate temperature, °F	Allowable watt density, W/in²
Water and solutions containing at least 80% water, including alkaline and acid cleaning solutions		210	40 to 300
Metals in liquid state, such as solder, lead, sodium and potassium		500	600
		700	500
		900	400
		1,100	300
		1,300	200
Cooking oil	Based on velocity of 1 ft/s	400	30
Dowtherm A		500	20
Machine oil SAE-30		250	18
Therminol FR-2		500	12
Bunker C fuel oil		160	10
Asphalt		300	8
Molasses		100	5
Glue		100	No direct heating

to a liquid-tight terminal housing. The elements can be supplied straight or curved to fit a circular vessel.

Circulation heaters are available for oil, water and gas heating. They are used when indirect heating is required, or when processing will not permit placement of the immersion heater directly in the tank.

Indirect heating uses a heat-transfer medium that carries heat from the heater to the process material. Indirect heating units utilize a jacketed container for heating heavy, viscous materials that require extremely low watt-densities in order not to alter or ruin the process material. The immersion heaters are placed in the heat-transfer media, and can operate at watt densities that allow a more economical approach to heating those heavy, viscous materials.

Corrosion and watt density

These are factors that should be considered before buying an electric immersion heater.

The corrosion characteristics of the liquid must be weighed before selecting the proper sheath material for the heating elements. Copper is normally used in water applications in which high temperatures and pressures are not involved. Steel elements are employed for most oil applications. Incoloy is used for chemicals, some acid solutions, and also in steam generators and other applications in which high temperature of pressurized-water vessels is required. Other sheath materials are Monel, stainless steels, aluminum and titanium. These last materials are specially ordered, which may require long lead times. Copper, steel and Incoloy sheaths are standard, and can be obtained from most heater manufacturers in a reasonable time. Another factor that affects the life and performance of an immersion element is the watt density—the heat flux or total amount of heat being generated for each square inch of heated surface area of the element. The table gives some typical liquids and the recommended watt density for heating elements. Water allows for relatively high watt-density elements, whereas in oil and asphalts, the element must operate at a much lower watt density. If an immersion element is too high in watt density, the liquid will carbonize and insulate the sheath of the element, burning it out.

In the case of water heating, a normal expansion and contraction of the immersion element, when cycled on

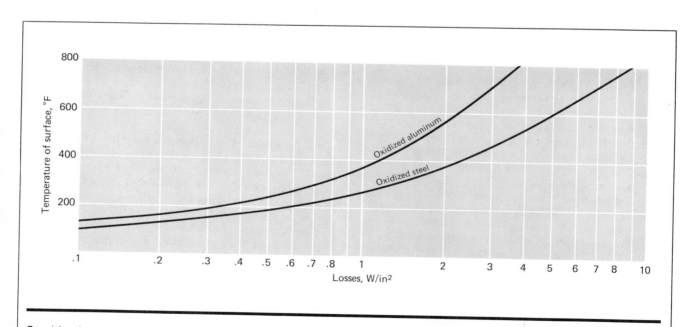

Combined radiation and convection losses for two uninsulated surfaces

Fig. 1

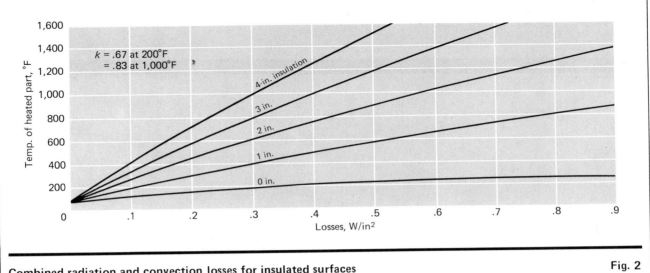

Combined radiation and convection losses for insulated surfaces — Fig. 2

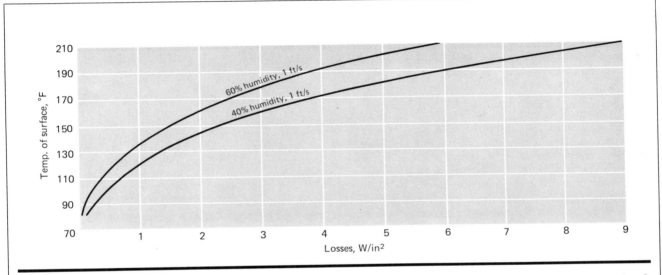

Combined radiation and convection losses for water surfaces — Fig. 3

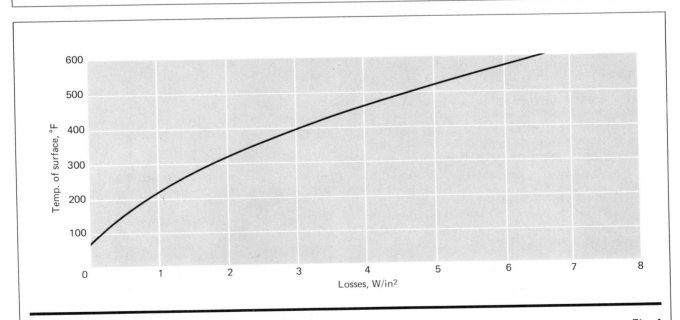

Combined radiation and convection losses for oil or paraffin surfaces — Fig. 4

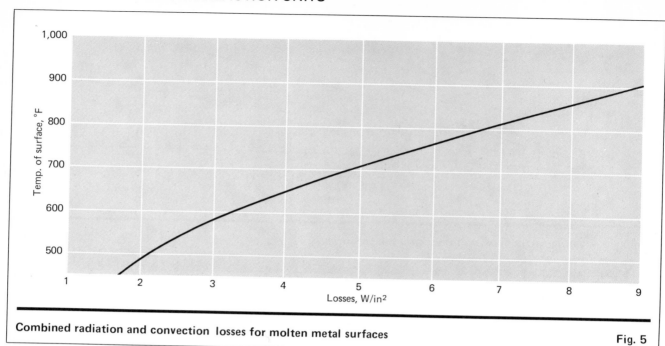

Combined radiation and convection losses for molten metal surfaces   Fig. 5

and off by controllers, tends to remove any lime deposits, so no maintenance is required. Where extreme hard-water conditions prevail, it is advisable to inspect the element after three months' use. If there is a lime buildup on the sheath, it should be cleaned off, and a periodic inspection of the elements should be made thereafter.

Check with the heater manufacturer and be sure to describe the material being heated. If it is a mixture of various chemicals, let the manufacturer know. Most suppliers have information allowing them to pick the correct sheath material and watt density, so that long life can be expected from the immersion element.

It is easy to determine the required wattage for an immersion-heater application. The power needed for any given use is either the watts required for startup, or the watts necessary for operation, whichever of these two is greater.

$$\text{Startup watts} = A + C + 3/4\, D \qquad (1)$$

In Eq. 1, A = watts absorbed in raising the temperature of the machine platen, tank, liquid, etc., in the required time; C = watts absorbed in melting or vaporizing material in startup or operation; and D = watts lost from surfaces by radiation, convection and conduction.

$$A = \text{lb/h} \times (c)(\Delta T)/3.412 \qquad (2)$$

In Eq. 2, lb/h = total weight divided by total time allowed for startup, or pounds of material processed per hour; c = specific heat of the material; and ΔT = temperature rise of the material, °F.

$$C = \text{lb/h} \times r/3.412 \qquad (3)$$

In Eq. 3, r = heat of fusion or vaporization.

D is the wattage lost from the surface of the liquid and/or the container by radiation, convection, and conduction. Several heater manufacturers have engineering application guides containing curves that give the watts lost through radiation and convection for such surfaces as water, oil and steels (Fig. 1-5) at an ambient temperature of 70°F. In the case of watts lost by conduction:

$$D = (k)(a)(\Delta t)/491d \qquad (4)$$

In Eq. 4, k = thermal conductivity, (Btu)(in.)/(h)(ft²)(°F); a = area of conduction, in²; Δt = difference in temperature across the insulator (usually an estimate), °F; and d = thickness of insulator, in.

$$\text{Operating watts} = B + C + D \qquad (5)$$

In Eq. 5, B = watts absorbed in raising the temperature of parts or material during the working cycle. Eq. 2 is used to estimate B.

If additional assistance is needed, most heater manufacturers maintain a staff of application and product engineers to help with problems.

As the availability of fossil fuels diminishes, the use of electric immersion heating as an easily controlled, reliable, clean, and overall economical means of heating liquids and gases will be increasing steadily.

The author

David R. Martignon is sales manager for Watlow Electric Manufacturing Co., at which he has also held the positions of chief applications engineer and sales representative. He has had previous experience in research and development, and in mold design. He holds an associate degree in Electrical Technology from Washington University (St. Louis).

Heat Transfer in Mechanically Agitated Units

Jacketed, mechanically agitated units can be used for heating and cooling viscous liquid, slurries and solids. Here is a discussion of common mixer types, together with information on ways to solve typical problems associated with their use.

WILLIAM L. ROOT, 3rd, and R.A. NICHOLS, Bethlehem Corp.

This article will discuss the types of mechanically agitated, indirectly heated units used in cooling, drying, heating, melting and reacting. A brief review of the theory applicable to the transfer of heat in this type of equipment will be presented for the processing of solids, slurries and viscous materials. Typical application problems and their solution will be suggested.

*This article covers the kinds of mixers most commonly used in these operations, and explains their general characteristics. No attempt has been made to include all of the numerous special designs available.

Originally published March 19, 1973

Classification of Equipment

The types of equipment treated in this article are limited to those in which heat is transferred to (or from) a heat-transfer medium through a metal barrier (vessel and/or agitator wall) to the mass being processed.* Thus, a distinct separation of heat-transfer medium from the process mass is provided. This medium may be in the liquid or vapor phase.

A second factor in the classification is that the material being processed is given motion by means of an agitator driven by a mechanical device. Mechanical agitation is

HEAT TRANSFER IN REACTION UNITS

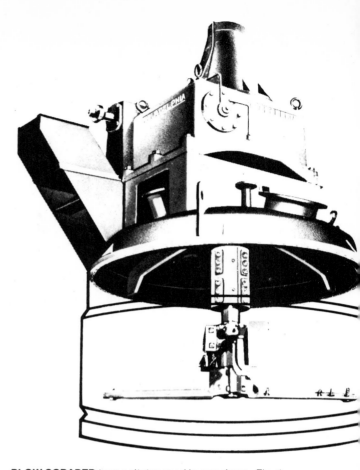

PLOW SCRAPER type agitator used in pan dryer—Fig. 1

employed to promote contact between the particles and the heat-transfer surface, thus reducing the dependence on the thermal conductivity of the bed and its individual particles for heat transfer. As a secondary function, blending of the mass can be accomplished, provided a proper selection of agitator design schemes is made.

A third factor in the classification of this type of equipment is that the containing shell is stationary. Sealing of the protrusions from the shell is thus simplified because of the smaller diameters involved.

A fourth factor; residence time associated with the equipment is relatively long, with high volume/heat-transfer-surface ratios being evident in most cases.

Within the scope outlined, the shell configuration can be used to further differentiate the equipment, as in Table I. The range of operating conditions will cover the complete spectrum. As a general rule, this type of equipment is not competitive in capital-equipment cost with direct-fired equipment or equipment using air as the heat-transfer medium, unless labor becomes a big factor in the cost of operation, such as with tray dryers.

Shell Configurations Used—Table I

Shell Configuration	Axis of Agitator	Common Name	Fig. No.	Heat-Transfer Surface	Normal Operating Conditions				
					Vacuum*	Atmospheric	Pressure†	Continuous	Batch
Cylindrical	Vertical	Pan	4	Jacketed bottom and/or side-walls	Yes	Yes	Yes	No	Yes
		Cone		Jacketed tiers					
		Multiple-tiered	5		Yes	Yes	Yes	Yes	No
	Horizontal	Horizontal (rotary vacuum)	6		Yes	Yes	Yes	Yes	Yes
				Jacketed shell					
Trough	Horizontal Sloping Vertical	Trough	7	Jacketed shell	No	Yes	No	Yes	Yes

* Normally up to 29 in. Hg.
† Normally up to 250 psig.

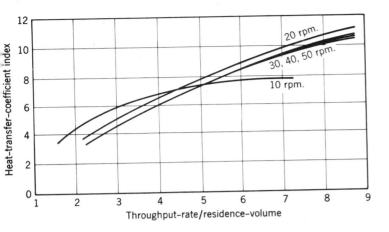

AGITATOR-speed effect on heat transfer—Fig. 2

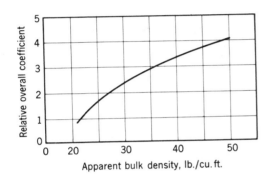

BULK DENSITY effect on overall heat transfer—Fig. 3

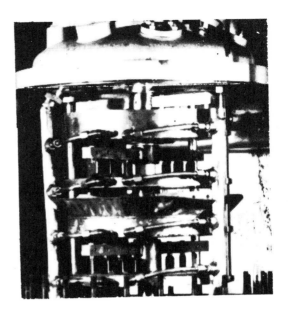

MULTIPLE TIER pan-type vessel—Fig. 5

The variety of agitation components is broad, with horizontal equipment as shown in Table II. Vertical cylindrical shells lend themselves to rabbling-type agitators or vertical screws. These rabbling-type agitators can be likened to the T-bar agitator shown in Fig. 1.

Area of Application

Another method of distinguishing the equipment would be its application in the field of heat transfer. The addition or removal of sensible heat; the removal of heat of fusion; the addtion of heat of vaporization; or the addition or removal of the heat of reaction, typify the function of this equipment group.

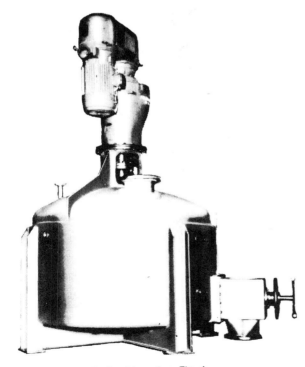

PAN type vessel for heat transfer—Fig. 4

HORIZONTAL jacketed-shell type vessel—Fig. 6

Agitators Used With Horizontal-Shell Configurations—Table II

Type Agitator	Fig. No.	Extended Heat-Transfer Surface	Normal Length of Heat-Transfer Paths, In.	Ratio of Heat-Transfer Surface to Process Volume	Heat Transfer Rates	Folding Action of Agitator
Hollow interrupted flight	8	Yes	2-6	Good	Excellent	Good
Hollow, disk	9	Yes	2-6	Excellent	Good	Fair
Hollow, screw	10	Yes	2-6	Excellent	Good	Fair
Tube bundle	11	Yes	2-6	Fair	Good	Fair
Paddle	12	No*	6-36	Poor	Poor	Poor
T-bar	13	No*	6-36	Poor	Poor	Poor
Ribbon	14	No*	6-36	Poor	Poor	Poor
Sigma blade	15	No	6-36	Poor	Poor	Excellent
Vertical screw	16	No	6-36	Poor	Poor	Excellent

*Shaft may be hollow and used for heat transfer.

A classification by industrial application could be as follows:
- Drying for recovery of solids, solvent or both.
- Reacting.
- Cooling.
- Heating.
- Sterilizing.
- Melting.
- Crystallizing.

Theory of Heat Transfer

The basic formula for heat transfer applies, i.e.:

$$Q = U A \Delta t$$

where Q = amount of heat to be transferred, Btu./hr.; U = overall heat-transfer coefficient, Btu./(hr.)(sq.ft.)(°F.); A = heat-transfer surface, sq.ft.; Δt = log mean temperature difference between the process mass and the heat-transfer medium, °F.

The overall heat-transfer rate, U, is a function of:
- Characteristics of the mass being processed, such as its thermal conductivity, specific heat, latent heat of vaporization (for drying), latent heat of fusion (for melting), heat of reaction, and bulk density.
- Effectiveness of the mixing action of the agitator.
- Flowability of the process mass.
- Fouling characteristics of the process mass on the heat-transfer surface.
- Resistance to heat transfer of the metal wall (a function of metal thickness and its conductivity).
- Fouling characteristics of the heat-transfer medium on the metal surface.
- Heat-transfer characteristics of the heat-transfer medium, i.e., specific heat, specific gravity, thermal conductivity, and viscosity.

Available data indicate that with each specific material, agitator speed will improve the overall heat-transfer coefficient up to a certain point, after which the mixing action of the agitator loses control, and the characteristics of the processing bed, such as thermal conductivity, apparent bulk density, etc., take over. Fig. 2 shows the effect of agitator speed and throughput-rate/residence-volume. Fig. 3 illustrates the effect of apparent bulk density on the relative overall coefficient.

Badger and Banchero[1] state, and we definitely concur, "as one goes into the subject further, he sooner or later realizes that in practice all such dryers must be designed by making a test run of the material in question on a

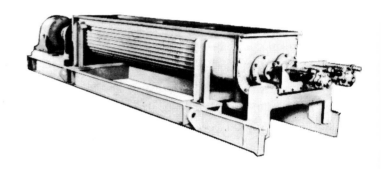

JACKETED TROUGH vessel for heat transfer—Fig. 7

HOLLOW interrupted-flight agitator—Fig. 8

small scale model of the type of dryer under consideration."

This idea can also be carried further to cover heating and cooling problems. Bailey[2] states (referring to dryers but still applicable to other forms of heat-transfer devices, as discussed in this article) that rates in small apparatus run about 40 to 60% of those in larger equipment; thus careful analysis of test data must be made.

When a piece of test equipment is used to determine U, certain conditions are fixed, such as the amount of heat to be transferred, the heat-transfer area of the unit, and the agitator configuration and, often, its speed. Practical field experiences and laboratory experiences indicate that in most instances the calculated U is erroneous and equipment improperly selected because of this.

An example might be cited of a process requiring low quantities of heat to be transferred, as in the case of dry-

Problems and Possible Cures—Table III

Problem	Cause	Effect	Possible Cure
Fouling	Phase change during processing.	Lower apparent heat-transfer rate.	Provide better mixing (to essentially eliminate local phase-changes that promote fouling) by use of breakers,
	Case-hardening of static films, as between agitator and jacket shell.		Recycle of portion of the dried product to mix with feed to produce an apparent feed that provides a more mobile nonsticking mass.
			Provide positive scraping of heating surface.
Heat sensitivity	An inherent characteristic of mass being processed.	Degradation of material being processed.	Better mixing to move mass off the heating surface quicker, allowing for higher temperature-driving-forces.
			Higher heat-transfer surface to process volume ratios, allowing for use of lower-temperature driving-forces.
			Lower boiling point, if drying is a problem, by lowering the vapor pressure either by vacuum, or by sparging with an inert gas to reduce partial pressure.
Residence time	Kinetic characteristic of process, related to a time-pressure-temperature relationship.	Need to hold for fixed period of time.	Having good data on process kinetics allows one to select correct heat-transfer-surface/volume ratio, temperature driving force, and operating pressure.
Varying bulk density	Speed, kinetic characteristic of process mass.	Lower U values.	Lower speed.
			Proper selection of equipment type to ensure maximum loading of heat-transfer surface.
Dusting	High vapor-velocities with small particle size.	Need for more-elaborate dust-collection equipment.	Lower agitator-speed.
			Higher vapor-release-surface to process-volume ratio, giving lower escape-velocities.
			Use of high-trough-type vessel (providing good setting volumes at minimum cost), large ports to reduce exit velocities.
Maintenance	Improper consideration given to following during design: Abrasiveness of process mass. Corrosion. Bearings. Drive. Packing and gaskets.	High maintenance costs.	Selection of proper materials of construction.
			Selection of mechanical-design parameters compatible with use, such as bearing loads, overhung loads, deflection, etc.
			Ability to hard-surface and/or to rehabilitate equipment locally (as with abrasive materials.)
Space	Limited space for equipment. High cost of real estate & building construction.	High capital costs.	Use of extended-surface units with maximized heat-transfer flux.
Cleaning	Required to keep batch identity. Sanitary design required.	High cleaning costs.	Use a trough-type unit with hinge cover or bottom, or both.
			Use a pan-type dryer.

244 HEAT TRANSFER IN REACTION UNITS

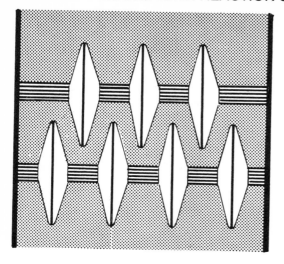

HOLLOW DISK agitator gives good heat transfer—Fig. 9

PADDLE agitator has no heat-transfer surface—Fig. 12

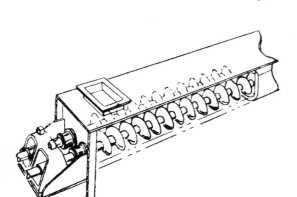

HOLLOW SCREW design of agitator—Fig. 10

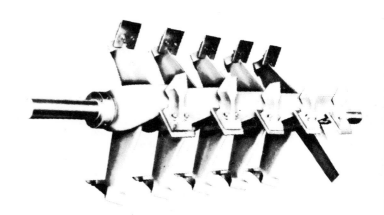

T-BAR agitator as used in horizontal shells—Fig. 13

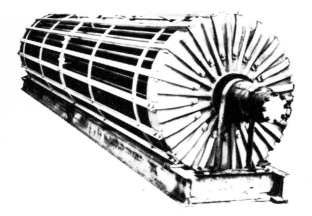

TUBE BUNDLE type agitator for heat transfer—Fig. 11

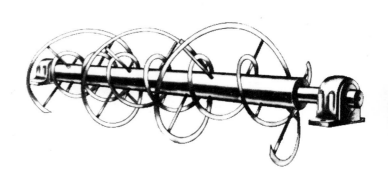

RIBBON agitator gives excellent folding action—Fig. 14

ing or reacting where a time-temperature relation may be involved in the process, and where extended heat-transfer equipment is used for piloting. Since Q is fixed by the charge to the pilot unit, and A is likewise fixed, only Δt can be varied. If by varying Δt, U also varies, then the pilot equipment has been misapplied. This could be remedied by continuing the variation of Δt until U became constant. This would give a more reasonable U for design if further piloting in more-suitable equipment were not contemplated. If rate, U, is high and the thickness of the heat-transfer surfaces increases, corrections must be made for this higher resistance due to the metal barrier. We cite these points just to show some of the problems associated with piloting.

Typical Data Needed From Pilot Test Runs

1. Change of condition or rate against time (and composition data if required for plot):
    Change in temperature of heat-transfer medium vs. time.
    Change of temperature of process mass vs. time.
    Change of composition of process mass vs. time.
    Inlet and outlet operating pressures of heat-transfer medium and process mass vs. time.
    Input rates of heat-transfer medium and process mass (by component).
    Output rates of process mass (by component).
2. Properties of process mass (by component).
    Particle size.
    Bulk density.
    Specific heat, heat of fusion, heat of vaporization.
    Vapor pressure.
3. Heat-transfer area.
4. Speed, and power consumption.
5. Visual observations.
    Stickness.
    Mixing characteristics.
    Scaling (and other pertinent observations).

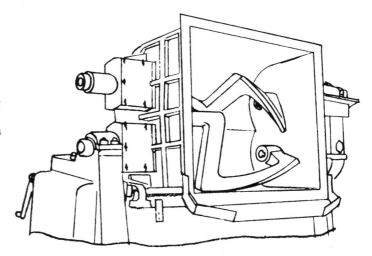

SIGMA BLADE agitator for jacketed shells—Fig. 15

Conclusions

There is no substitute for the collection of sufficient data to allow for sound judgment in the selection of heat-transfer equipment of the mechanically agitated, indirectly heated type, as with all other types of equipment. For economic reasons, this equipment should be used only where required, and then proper selection should be made from the wide array of available vessel and agitator designs. Each supplier has his strong points that must be weighed in proper perspective to obtain the most for one's dollar. ∎

References

1. Badger, W. L., Banchero, J. T., "Introduction to Chemical Engineering," McGraw-Hill, New York, 1955, p. 511.
2. Bailey, L. H., *Ind. Eng. Chem.,*, **30,** p. 1,008 (1955).
3. Uhl, V. W., Root, W. L., Heat Transfer to Granular Solids in Agitated Units, *Chem. Eng. Progr.,* **63,** No. 7, pp. 81-92 (1967).
4. Uhl, V. W., Root, W. L., Heat Transfer in Hollow Cut-Flight Jacketed Units to Viscous Fluids, *Chem. Eng. Progr. Symposium Series,* **66,** No. 102, pp. 199-208.
5. Uhl, V. W., Root, W. L., Indirect Drying in Agitated Units, *Chem. Eng. Progr.,* **58,** No. 6, pp. 27-44 (1962).
6. Horzella, T., Practical Indirect Drying of Solids, *Chem. Eng. Progr.,* **59,** No. 3, pp. 90-92 (1963).
7. Holt, A. D., Heating and Cooling of Solids, *Chem. Eng.,* Oct. 23, 1967, pp. 145-166.

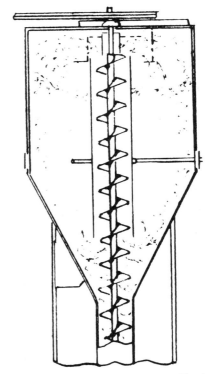

VERTICAL SCREW agitator for vertical shells—Fig. 16

Meet the Authors

◀ **William L. Root, III** is Manager of the Process Equipment Div. at The Bethlehem Corp., 225 West Second St., Bethlehem, PA 18106. He received his B.S. in chemical engineering from Drexel Institute of Technology, and is a member of AIChE, ACS and the Institute of Food Technologists.

R. A. Nichols is a District Sales Manager and Laboratory ▶ Director at the Bethlehem Corp. He received his B.S. in chemical engineering from Lehigh University and is a member of AIChE as well as being a registered professional engineer in the state of Pennsylvania.

Heating and Cooling In Batch Processes

Understanding the relationships between time, batch temperature, rate of heat transfer and flowrate or outlet temperature of the heat-transfer medium will reduce design errors and poor startups and will lower engineering costs.

T. R. BROWN, Procter & Gamble Co.

Heating and cooling operations for batch processes are some of the commonest stages in chemical process plants. Inadequate understanding of these operations often results in a higher than average number of design and startup problems. We can minimize these problems if we have a good appreciation of how the different batch heat-transfer variables relate to each other.

To obtain this, several easy-to-use equations have been developed for two general cases: change of phase in the heat-transfer medium, and no change of phase in the medium. These relations will allow the following quantities to be calculated as a function of time: (1) batch temperature, (2) heat load or rate of heat transfer, (3) flowrate for a heat-transfer medium that changes phase (i.e., condensing steam or vaporizing Freon), and (4) return or final temperature for a heat-transfer medium that does not change phase (i.e., cooling-tower water or cold-brine solutions).

Design and Startup Problems

Without a proper knowledge of the above variables, design and startup problems can occur. Some typical difficulties are:

1. *Insufficient process capacity*—Unless there is a basic plant-utility shortage (i.e., steam, refrigeration, cooling water, etc.), this problem is caused by having insufficient heat-transfer area to complete the heating or cooling within the desired cycle time.

2. *Insufficient plant-utility capacity*—Two such difficulties can occur. Either the utility cannot supply the average requirements of the process or, having sufficient average capacity, the utility cannot meet the instantaneous demands. The latter is the least understood, and thus more difficult to handle.

By knowing the magnitude and duration of the instantaneous demands, we can design or adapt the utility system to handle varying loads. For example, steam boilers have both normal and peak-demand capacity ratings. With cooling water, hot water, chilled brine and similar systems, surge tanks or basins can be installed on both sides of the utility, which allow it to run at some average rate while the tanks supply and absorb the high demand.

3. *High-cost designs*—When there is uncertainty about the technical factors in a design, we must oversize the system to ensure that it works. Although the oversized system usually works, it costs more to build.

4. *Excessive flowrate of the heat-transfer medium*—This problem results primarily from a lack of understanding of the instantaneous requirements of the heating or cooling process. Without a proper knowledge of these, greater-than-expected flows of the heat-transfer medium may result. With media such as steam or Freon, these greater flows can cause pressure drops that are high enough to reduce their effectiveness in heating or cooling the material in the batch.

5. *Excessive return temperature of the heat-transfer liquid*—A very common difficulty is that excessively hot water is being returned to a cooling tower. Ignoring capacity problems, high temperatures can shorten the life of the tower if they are not taken into account in the original design. For example, most wooden towers should not be subjected to water temperatures over 130 F. Hotter water will dissolve the lignin in the wood, causing the baffles to deteriorate.

6. *Poor startups*—Any equipment or process startup is fraught with difficulty. If the details of the design are not fully understood, the situation can become worse because the data necessary to successfully trouble-shoot the system will not be available. If the problems cannot be promptly resolved, extra construction and engineering costs can occur. In extreme cases, there can be a loss of business because of the plant's inability to meet sales commitments.

Let us now examine the derivation of several equations, and the solving of sample problems—manually from the equations and by means of a computer program.

Heat-Transfer Medium Changes Phase

To develop the relationship between time, temperature, heat load, and flowrate of the heat-transfer medium, let us consider the following system:

Originally published May 28, 1973

HEATING AND COOLING IN BATCH PROCESSES

Nomenclature

A	Area of the heat-transfer surface upon which U is based, sq.ft.
B	Weight of the batch, lb.
C_b	Specific heat of the batch, Btu./(lb.)(°F.)
C_w	Specific heat of heat-transfer medium, Btu./(lb.)(°F.)
h_{fg}	Heat of condensation or vaporization, Btu./lb.
T_b	Temperature of the batch, deg. F.
T_m	Temperature of heat-transfer medium, deg. F.
ΔT_{lm}	Log mean temperature difference, deg. F.
U	Overall heat transfer coefficient, Btu./(hr.)(sq.ft.)(°F.)
W	Flowrate of heat-transfer medium, lb./hr.
θ	Time, hr.

A tank, having a heating coil, is filled with a batch of liquid that is to be heated by a condensing vapor. Although this system involves heating, the same equations apply when a batch is cooled with a vaporizing fluid such as Freon. This system is diagrammed as:

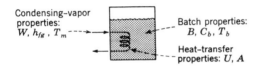

Making a heat balance across the coil gives:

$$q = UA\Delta T_{lm} = Wh_{fg} = BC_b(dT_b/d\theta) \tag{1}$$

Diagramming the temperature profiles of the batch and of the heating medium gives:

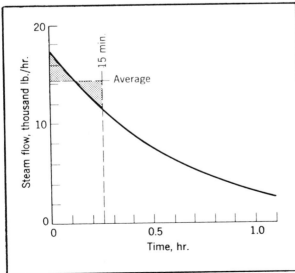

STEAM flow plot provides average rate—Fig. 1

Since dT_b is very small, ΔT_{lm} is approximated by ΔT, which equals $T_m - T_b$. Thus, Eq. (1) can be rewritten as:

$$UA\Delta T_{lm} = UA\Delta T = UA(T_m - T_b) = BC_b(dT_b/d\theta)$$

By rearranging and setting up limits of integration, the relationship between batch temperature and time can be found:

$$\int_0^\theta d\theta = \int_{T_{b1}}^{T_b} \frac{BC_b}{UA}\left(\frac{dT_b}{T_m - T_b}\right) = -\frac{BC_b}{UA}\ln(T_m - T_b)\Big|_{T_{b1}}^{T_b}$$

$$\theta = \frac{BC_b}{UA}\ln\left(\frac{T_m - T_{b1}}{T_m - T_b}\right) \tag{2}$$

Eq. (2) can also be written as:

$$e^{K\theta} = (T_m - T_{b1})/(T_m - T_b)$$

where $K = UA/BC_b$. Rearranging this equation gives:

$$T_b = T_m - (T_m - T_{b1})e^{-K\theta} \tag{2a}$$

Differentiating Eq. (2a) with respect to time:

$$dT_b/d\theta = K(T_m - T_{b1})e^{-K\theta} \tag{3}$$

Substituting Eq. (3) into Eq. (1) gives a solution for the heat load:

$$q = BC_bK(T_m - T_{b1})e^{-K\theta} \tag{4a}$$

If the batch is being cooled rather than heated, the term $dT_b/d\theta$ in Eq. (1) is given a negative sign. Thus, for the cooling situation, Eq. (4a) becomes:

$$q = -BC_bK(T_m - T_{b1})e^{-K\theta} \tag{4b}$$

The solution for the flowrate of the heat-transfer medium can be found by rearranging Eq. (1) to give:

$$W = q/h_{fg} \tag{5}$$

Examples Illustrate Calculation Method

Example 1—It is desired to increase the capacity of a batch process by decreasing the time required to heat the batch. By scaling-up existing conditions, it has been determined that a coil having an area of 1,000 sq.ft. will produce the required decrease in cycle time.

A check must now be made to ensure that the average steam demand is not greater than 15,000 lb./hr. for a 15-min. period, or the pressure in the boiler will drop below acceptable limits. The following data apply:

Overall heat-transfer coefficient	40 Btu./(hr.)(sq.ft.)(°F.)
Batch size	44,000 lb.
Average specific heat of batch	0.56 Btu./(lb.)(°F.)
Initial batch temperature	120 F.
Final batch temperature	400 F.
Heating medium, saturated steam	435 psig.
Temperature	456 F.
Heat of condensation	768 Btu./lb.
Total heating time	1.1 hr.

By using Eq. (4a) and (5), we can calculate the relationship between time and steam flow. Combining the two equations gives:

$$W = BC_bK(T_m - T_{b1})/h_{fg}e^{K\theta}$$

248 HEAT TRANSFER IN REACTION UNITS

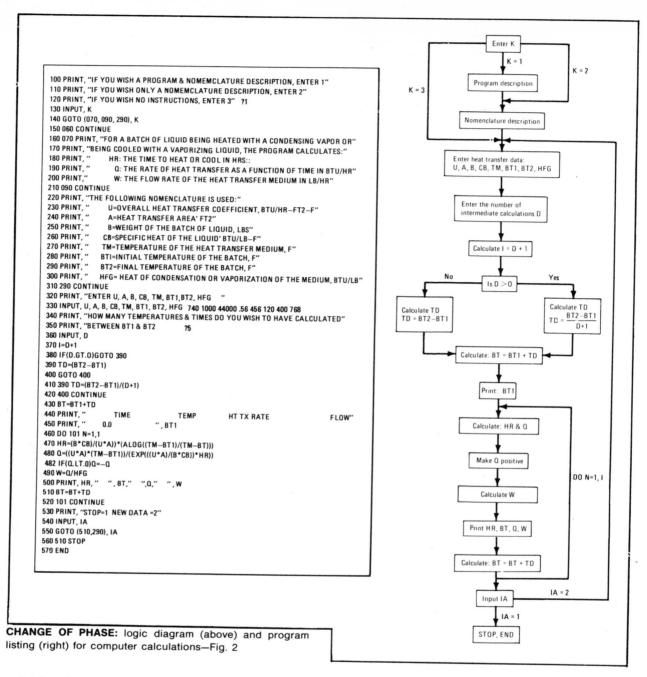

CHANGE OF PHASE: logic diagram (above) and program listing (right) for computer calculations—Fig. 2

Solving for K in $K = UA/BC_b$ gives the following:

$$K = \frac{40(1,000)}{44,000(0.56)} = 1.62$$

At $\theta = 0.20$ hr., or 12 min.:

$$W = \frac{44,000(0.56)(1.62)(456 - 120)}{768 e^{(1.62)(0.2)}} = 12,600$$

After substituting other values of θ in the equation, a plot is made of θ vs. W, as shown in Fig. 1. By visually inspecting this plot, we can approximate the average steam flow at 14,300 lb./hr. during the first 15 min. of the heating cycle. This shows that the boiler could supply the required steam flow.

The calculations could also have been done by using a computer program written in Fortran IV for a General Electric Mark II Time Sharing System. The logic diagram and program listing are shown in Fig. 2, along with the user input, which is shown in color. The results are:

TIME	TEMP	HT TX RATE	FLOW
0.0	1.2000000E+02		
9.2111547E−02	1.6666667E+02	1.1573333E+07	1.5069444E+04
2.0046019E−01	2.1333333E+02	9.7066668E+06	1.2638889E+04
3.3202184E−01	2.6000000E+02	7.8400001E+06	1.0208333E+04
4.8953302E−01	3.0666667E+02	5.9733333E+06	7.7777776E+03
7.3034418E−01	3.5333334E+02	4.1066667E+06	5.3472222E+03
1.1037239E+00	4.0000000E+02	2.2399999E+06	2.9166665E+03

Example 2—It is desired to determine the size of the cooling coil for a process in which a batch of liquid will

be cooled to -2 F. by vaporizing Freon 22. In anticipation of startup problems and questions, we should also calculate the relationships between batch temperature, pounds of Freon vaporized, and time. The following data apply:

Desired cooling time	0.5 hr.
Overall heat-transfer coefficient	30 Btu./(hr.)(sq.ft.)(°F.)
Batch size	25,000 lb.
Average specific heat of batch	0.7 Btu./(lb.)(°F.)
Initial batch temperature	80 F.
Final batch temperature	-2 F.
Refrigerant	Freon 22
Temperature	-10 F.
Heat of vaporization	96 Btu./lb.

To calculate the coil area, rearrange Eq. (2):

$$A = \frac{BC_b}{U\theta} \ln\left(\frac{T_m - T_{b1}}{T_m - T_b}\right)$$

$$A = \frac{25{,}000(0.7)}{30(0.5)} \ln\left[\frac{-10 - 80}{-10 - (-2)}\right] = 2{,}820 \text{ sq.ft.}$$

We then use the computer program mentioned in Example 1 and detailed in Fig. 2 to calculate the other values.

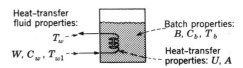

The same results could have been calculated by using Eq. (2) to find the time-temperature relationship, and by using Eq. (4b) and (5) to solve for the relationship between vapor flow and time.

No Change of Phase in Heat-Transfer Medium

To find the relationship between time, batch temperature, heat load and outlet temperature of the heat-transfer medium, let us consider the following system. A tank, equipped with a coil, contains a batch of liquid that is to be heated or cooled by pumping a heat-transfer liquid through the coil at a constant flowrate.

The resulting equations apply to both cooling and heating, but will be developed for the situation where the batch is cooled. A diagram of the system is:

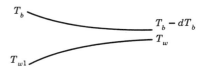

A heat balance across the coil gives:

$$q = UA\Delta T_{lm} = WC_w(T_w - T_{w1}) = BC_b\left(-\frac{dT_b}{d\theta}\right) \quad (6)$$

Diagramming the temperature profiles of the batch at any given instant gives:

The general expression for the logarithmic mean temperature difference is:

$$\Delta T_{lm} = \frac{\Delta T_1 - \Delta T_2}{\ln(\Delta T_1/\Delta T_2)}$$

Hence, for this system, ΔT_{lm} can be written as:

$$\Delta T_{lm} = \frac{(T_b - T_{w1}) - [(T_b - dT_b) - T_w]}{\ln\left[\dfrac{T_b - T_{w1}}{(T_b - dT_b) - T_w}\right]}$$

$$T_{lm} = \frac{T_w - T_{w1} + dT_b}{\ln\left(\dfrac{T_b - T_{w1}}{T_b - T_w - dT_b}\right)} \quad (7)$$

Since dT_b is very small compared to $(T_w - T_{w1})$ and $(T_b - T_w)$, Eq. (7) is approximated by:

$$\Delta T_{lm} \approx \frac{T_w - T_{w1}}{\ln\left(\dfrac{T_b - T_{w1}}{T_b - T_w}\right)} \quad (8)$$

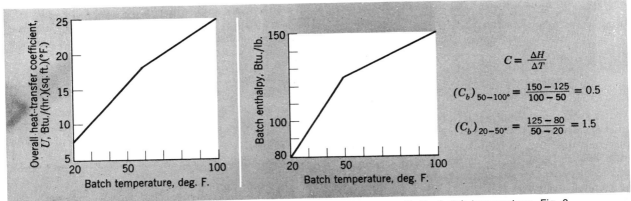

DATA for Example 3 show overall heat-transfer coefficient and enthalpy vary with the batch temperature—Fig. 3

HEAT TRANSFER IN REACTION UNITS

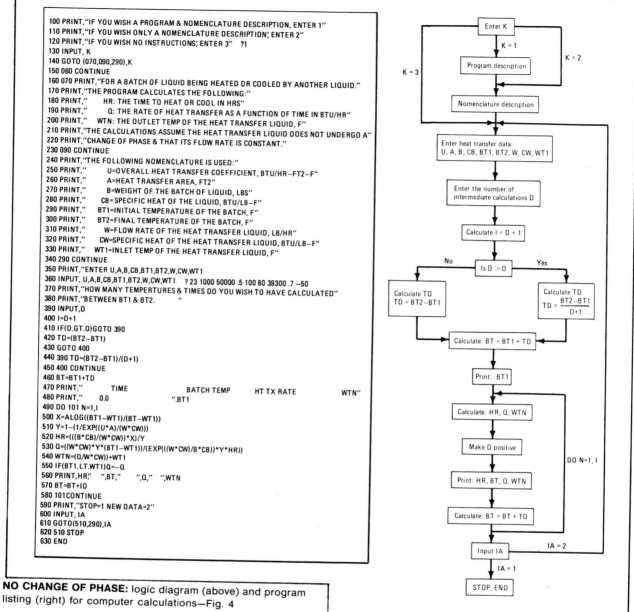

```
100 PRINT,"IF YOU WISH A PROGRAM & NOMENCLATURE DESCRIPTION, ENTER 1"
110 PRINT,"IF YOU WISH ONLY A NOMENCLATURE DESCRIPTION; ENTER 2"
120 PRINT,"IF YOU WISH NO INSTRUCTIONS; ENTER 3"   ?1
130 INPUT, K
140 GOTO (070,090,290),K
150 060 CONTINUE
160 070 PRINT,"FOR A BATCH OF LIQUID BEING HEATED OR COOLED BY ANOTHER LIQUID."
170 PRINT,"THE PROGRAM CALCULATES THE FOLLOWING:"
180 PRINT,"     HR: THE TIME TO HEAT OR COOL IN HRS"
190 PRINT,"     Q: THE RATE OF HEAT TRANSFER AS A FUNCTION OF TIME IN BTU/HR"
200 PRINT,"     WTN: THE OUTLET TEMP OF THE HEAT TRANSFER LIQUID, F"
210 PRINT,"THE CALCULATIONS ASSUME THE HEAT TRANSFER LIQUID DOES NOT UNDERGO A"
220 PRINT,"CHANGE OF PHASE & THAT ITS FLOW RATE IS CONSTANT."
230 090 CONTINUE
240 PRINT,"THE FOLLOWING NOMENCLATURE IS USED:"
250 PRINT,"     U=OVERALL HEAT TRANSFER COEFFICIENT, BTU/HR-FT2-F"
260 PRINT,"     A=HEAT TRANSFER AREA, FT2"
270 PRINT,"     B=WEIGHT OF THE BATCH OF LIQUID, LBS"
280 PRINT,"     CB=SPECIFIC HEAT OF THE LIQUID, BTU/LB-F"
290 PRINT,"     BT1=INITIAL TEMPERATURE OF THE BATCH, F"
300 PRINT,"     BT2=FINAL TEMPERATURE OF THE BATCH, F"
310 PRINT,"     W=FLOW RATE OF THE HEAT TRANSFER LIQUID, LB/HR"
320 PRINT,"     CW=SPECIFIC HEAT OF THE HEAT TRANSFER LIQUID, BTU/LB-F"
330 PRINT,"     WT1=INLET TEMP OF THE HEAT TRANSFER LIQUID, F"
340 290 CONTINUE
350 PRINT,"ENTER U,A,B,CB,BT1,BT2,W,CW,WT1"
360 INPUT, U,A,B,CB,BT1,BT2,W,CW,WT1   ? 23 1000 50000 .5 100 80 39300 .7 -50
370 PRINT,"HOW MANY TEMPERTURES & TIMES DO YOU WISH TO HAVE CALCULATED"
380 PRINT,"BETWEEN BT1 & BT2.      "
390 INPUT,D
400 I=D+1
410 IF(D.GT.0)GOTO 390
420 TD=(BT2-BT1)
430 GOTO 400
440 390 TD=(BT2-BT1)/(D+1)
450 400 CONTINUE
460 BT=BT1+TD
470 PRINT,"     TIME          BATCH TEMP      HT TX RATE       WTN"
480 PRINT,"     0.0                  ",BT1
490 DO 101 N=1,I
500 X=ALOG((BT1-WT1)/(BT-WT1))
510 Y=1-(1/EXP((U*A)/(W*CW)))
520 HR=(((B*CB)/(W*CW))*X)/Y
530 Q=((W*CW)*Y*(BT1-WT1))/(EXP(((W*CW)/B*CB))*Y*HR))
540 WTN=(Q/W*CW))+WT1
550 IF(BT1.LT.WT1)Q=-Q
560 PRINT,HR," ",BT," ",Q," ",WTN
570 BT=BT+TD
580 101 CONTINUE
590 PRINT,"STOP=1 NEW DATA=2"
600 INPUT, IA
610 GOTO(510,290),IA
620 510 STOP
630 END
```

NO CHANGE OF PHASE: logic diagram (above) and program listing (right) for computer calculations—Fig. 4

Rearranging Eq. (6) gives:

$$T_w = -\frac{BC_b}{WC_w}\left(\frac{dT_b}{d\theta}\right) + T_{w1} \qquad (9)$$

Substituting Eq. (9) into Eq. (8) gives:

$$\Delta T_{lm} = \frac{-\frac{BC_b}{WC_w}\left(\frac{dT_b}{d\theta}\right)}{\ln\left(\frac{T_b - T_{w1}}{T_b - T_{w1} + \frac{BC_b}{WC_w}\left(\frac{dT_b}{d\theta}\right)}\right)} \qquad (10)$$

Substituting Eq. (10) into Eq. (6) gives:

$$q = UA\left[\frac{-\frac{BC_b}{WC_w}\left(\frac{dT_b}{d\theta}\right)}{\ln\left(\frac{T_b - T_{w1}}{T_b - T_{w1} + \frac{BC_b}{WC_w}\left(\frac{dT_b}{d\theta}\right)}\right)}\right] = BC_b\left(-\frac{dT_b}{d\theta}\right)$$

Rearranging and converting to the exponential form to eliminate the logarithm gives:

$$e^{UA/WC_w} = \frac{T_b - T_{w1}}{T_b - T_{w1} + \frac{BC_b}{WC_w}\left(\frac{dT_b}{d\theta}\right)}$$

Rearranging gives:

$$\frac{BC_b}{WC_w}\left(\frac{dT_b}{d\theta}\right) = -(T_b - T_{w1})(1 - e^{-UA/WC_w})$$

Rearranging again, and setting up limits of integration gives:

$$\int_0^\theta d\theta = \frac{BC_b}{WC_w}\left(\frac{1}{1 - e^{-UA/WC_w}}\right)\int_{T_{b1}}^{T_b} -\frac{dT_b}{T_b - T_{w1}}$$

$$\theta = \frac{BC_b}{WC_w}\left(\frac{1}{1 - e^{-UA/WC_w}}\right)\ln\left(\frac{T_{b1} - T_{w1}}{T_b - T_{w1}}\right) \qquad (11)$$

Rearranging and converting to the exponential form to eliminate the logarithm gives:

$$\frac{T_{b1} - T_{w1}}{T_b - T_{w1}} = e^{K\theta}; \text{ where } K = \frac{WC_w}{BC_b}(1 - e^{-UA/WC_w})$$

Rearranging:

$$T_b = (T_{b1} - T_{w1})e^{-K\theta} + T_{w1} \quad (11a)$$

Differentiating, Eq. (11a) yields:

$$\frac{dT_b}{d\theta} = -K(T_{b1} - T_{w1})e^{-K\theta} \quad (12)$$

Substituting Eq. (12) into Eq. (6) gives a solution for the heat load:

$$q = BC_bK(T_{b1} - T_{w1})e^{-K\theta} \quad (13a)$$

If the batch is being heated rather than cooled, $dT_b/d\theta$ in Eq. (6) is given a positive sign. Thus, for heating, Eq. (13a) becomes:

$$q = -BC_bK(T_{b1} - T_{w1})e^{-K\theta} \quad (13b)$$

To solve for the outlet or return temperature of the heat-transfer medium, Eq. (6) can be rearranged as:

$$T_w = \frac{q}{WC_w} + T_{w1} \quad (14)$$

Example Illustrates Calculation Method

Example 3—How much can the capacity of the cooling step in a crystallization process be increased when the coolant temperature is reduced to -5 F. The coolant is a $CaCl_2$ brine solution. The existing cooler is a mechanically agitated tank having a 1,000-sq.ft. coil. Batch size is 50,000 lb.; the batch is presently cooled from 100 to 20 F. in 11 hr., using 40,000 lb./hr. of a $+5$ F. brine. When the brine temperature is dropped to -5 F., the flowrate must be reduced to 30,000 lb./hr. Other data that pertain to the process are: average brine specific heat is 0.7 Btu./(lb.)(°F.), and the overall heat-transfer coefficient and batch specific heat are obtained from the charts of Fig. 3.

Since the overall heat-transfer coefficient and the batch specific heat vary with the batch temperature, we should break the calculation for cooling time into several steps. Cooling time can be calculated either by using Eq. (11) or by using a computer program. The logic diagram and program listing are shown in Fig. 4, along with user input data, which are shown in color.

The computer printout that is obtained for the first two calculations is:

```
HOW MANY TEMPERATURES & TIMES DO YOU WISH TO HAVE CALCULATED
BETWEEN BT1 & BT2.       ?0
          TIME           BATCH TEMP      HT TX RATE        WTN
          0.0            1.0000000E+02
          3.3892415E-01  8.0000000E+01   1.3248741E+06     4.3159726E+01
STOP=1 NEW DATA=2  ?2
ENTER U,A,CB,BT1,BT2,W,CW,WT1    ?20 1000 50000 .5 80 60 39300 .7 -5
HOW MANY TEMPERATURES & TIMES DO YOU WISH TO HAVE CALCULATED
BETWEEN BT1 & BT2.       ?0
          TIME           BATCH TEMP      HT TX RATE        WTN
          0.0            8.0000000E+01
          4.7186497E-01  6.0000000E+01   9.2384261E+05     2.8582065E+01
```

The calculations for the several steps are summarized as follows:

Temperature, Deg. F.	Time, Hr.	Heat-Transfer Rate, Btu./Hr.	T_{w2}, Deg. F.
100	0	—	—
80	0.34	1.32×10^6	43.2
60	0.81	0.92×10^6	28.6
50	1.14	0.70×10^6	20.4
40	2.51	0.49×10^6	12.9
30	4.52	0.33×10^6	7.0
20	7.97	0.18×10^6	1.6

The heat-transfer rate and brine outlet temperature can also be calculated by using Eq. (13a) and (14).

Using the new calculated cooling time, and adding 2 hr. for filling and emptying the cooler, the percentage increase in capacity is:

$$\left[\left(\frac{11 + 2}{8 + 2}\right) - 1\right]100 = 30\%$$

Summary

The equations and the computer programs in this article provide the means for determining the relationships between time, batch temperature, rate of heat transfer, and flowrate or outlet temperature of the heat-transfer medium in batch heating or cooling. By understanding how these variables relate to each other, we can reduce the probability of design errors and oversights, overly expensive designs, and poor startups. The end-result of this improvement will be lower engineering and construction costs.

The important relations are: Eq. (2), (2a), (11) and (11a) for temperature and time; Eq. (4a), (4b), (13a) and (13b) for heat load and time; Eq. (5) for the heat-transfer medium's flowrate and time; and Eq. (14) for the medium's outlet temperature and time. The computer programs provide the ability to make many calculations in a short period of time, which is ideal for the study of alternates and for system optimization. ∎

Meet the Author

Thane R. Brown, 7947 Burgundy Lane, Cincinnati, OH 45224, is an engineer with Procter & Gamble Co., where heat-transfer design is one of his main areas of interest. He has a B.S. in chemical engineering from Oregon State University, is a member of the local section of AIChE, and is a registered chemical engineer in Ohio.

JACKETED VESSELS will normally employ one of the varieties of jackets shown above—Fig. 1

Picking the Best Vessel Jacket

Conventional, dimple and half-pipe coil jackets have advantages and faults that depend on such things as vessel size, heat-transfer medium and allowable pressure drop. Here is how to evaluate the factors.

RICHARD E. MARKOVITZ, Brighton Corp.

Jacketing provides the optimum method of heating and cooling process vessels in terms of control, efficiency and product quality.

Using a jacket as a means of heat transfer offers many advantages:

- All liquids can be used, as well as steam and other high-temperature vapors.
- Circulation, temperature and velocity of heat-transfer media can be accurately controlled.
- Jackets may often be fabricated from a much less expensive metal than the vessel itself.
- Contamination, cleaning and maintenance problems are virtually eliminated.
- Maximum efficiency, economy and flexibility is achieved.

Alternatives to jacketing include electrical-resistance immersion heaters, external electrical heaters, and induction heating.

TYPES OF JACKETS

Jacketing of process vessels is usually accomplished by using one of the three main available types—conventional jackets, dimple jackets, and half-pipe coil jackets. In designing reactors for specific processes, this variety gives the chemical engineer a great deal of flexibility in the choice of heat-transfer medium.

Dimple Jacket

The design of the dimple jacket (Fig. 1a, 3) permits construction from light-gage metals without sacrificing the strength required to withstand specified pressures. This results in considerable cost saving over conventional jackets. Savings increase when higher jacket pressures and larger reactors are produced.

The dimple-jacket design is approved by the National Board of Boiler and Pressure Vessel Inspectors, and can be stamped in accordance with the ASME Unfired Pressure Vessel Code up to pressures of 300 psi. Metal thicknesses are calculated to withstand the required pressure.

Where the dimple-jacket design is used for cooling, stainless steel baffles can be supplied, spaced as required to ensure positive circulation of the cooling medium. Dimple jackets can be manufactured from

Originally published November 15, 1971

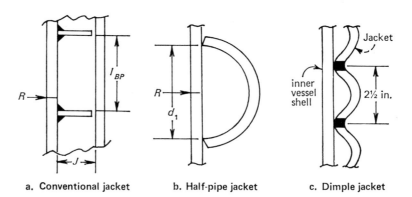

JACKETS of the three types discussed showing dimensions used in heat-transfer and pressure-drop calculations—Fig. 2

a. Conventional jacket   b. Half-pipe jacket   c. Dimple jacket

a number of materials, including Type 304, 304-L, 316 and 316-L stainless steel, as well as Incoloy and Inconel alloys.

In a cost comparison of the three types of jacket design, the dimple jacket will be:

1. More economical than the conventional jacket if the jacket pressure is the controlling factor in determining the vessel-wall thickness. As a rule of thumb, if the internal pressure is greater than 1.67 times the external pressure, the external pressure will not be governing.

2. More economical than the half-pipe jacket if the vessel's internal pressure is very low and the vessel size so small that clad steel is not economical.

(However, on small vessels, 500 gal. or less, it is usually not practical to apply the dimple jacket.)

Half-Pipe Coil Jacket

The half-pipe design (Fig. 1b, 4) provides high velocity and turbulence, and circulation can be closely controlled to effect an unusually high film coefficient. This, plus the built-in structural rigidity, makes the half-pipe design particularly well-suited to a wide range of processing services.

The half-pipe coil jacket is especially recommended for high-temperature services and for use with all liquid heat-transfer services. The design is ideal for hot-oil applications because of the equipment's high structural strength. It can be used effectively with water, and in many cases is better than either conventional or dimple jackets because of positive channeling and because pressure drop can be calculated simply. For steam, additional connections should be provided to carry away condensate and avoid two-phase flow. In addition to the other heat-transfer media, glycols can be used for cooling.

Because there are no limitations to the number and location of inlet and outlet connections, the half-pipe coil jacket can be divided into multiple zones (as shown in Fig. 4) for maximum flexibility and efficiency. Thus, either the entire jacket may be used, or only as much as needed, so that various-size batches can be processed economically in the same vessel. (This is not feasible with either conventional or dimple jackets, since the cost of zoning is considerably higher for these designs.) Multiple zoning reduces the pressure drop of the heat-transfer medium in the jacket.

In addition to zoning, other economies can be realized through the use of a half-pipe jacket. Most important, it usually allows reductions in the thickness of the inner wall of the vessel. In the fabrication of high-strength-alloy or clad-steel vessels, significant savings can be effected.

For maximum heat transfer, the space between the half-pipe coil is ¾-in. However, if the inside heat-transfer film coefficient is extremely low, the space between the half-pipe coils can be increased without any loss in heat-transfer surface, because of the fin efficiency of the half-pipe.

Half-pipe coil jackets are normally fabricated from carbon steel; but for temperatures above 300 F. and below −20 F., jackets are fabricated from the same material as the vessel, to avoid problems of differential thermal expansion. The various stainless steels, as well as Inconel, Monel and nickel, are among the materials that can be used.

Standard sizes of the half-pipe jacket are 2⅜, 3½ and 4½-in. O.D. When fabricated from carbon steel, thickness is 3/16-in. for the 2⅜-in. O.D. size and ¼-in. for both the 3½ and 4½-in. O.D. sizes. In alloy fabrication all sizes are ⅛ in. thick.

In a cost comparison with the other types of construction, the following guidelines should apply:

1. As in the case of the dimple jacket, if the jacket pressure is the controlling factor in determining the vessel wall thickness, the half-pipe jacket will be more economical than the conventional jacket.

2. If the vessel internal pressure is sufficiently high or the vessel size is large enough that the use of clad

Nomenclature

c_p	Specific heat, Btu./(hr.) (°F.)
D_H	Equivalent hydraulic radius, ft.
d_i	Half-pipe coil dia., in.
f	Friction factor
G	Mass velocity, lb./(hr.) (sq. ft.)
G_1	Mass velocity, lb./(sec.) (sq. ft.)
g	Mass acceleration, ft./sec.²
h_i	Heat-transfer film coefficient, Btu./(hr.) (sq. ft.) (°F.)
J	Jacket space, in.
K	Thermal conductivity, Btu./(hr.) (sq. ft.) (°F./ft.)
L	Length of path of travel, ft.
l_{BP}	Baffle pitch, in.
N_{Pr}	Prandtl number ($c_p\mu/K$), dimensionless
N_{Re}	Reynolds number ($VD_H\rho/\mu$), dimensionless
R	Radius of inner shell, in.
V	Velocity of jacket fluid, ft./hr.
μ	Viscosity (at average temp.), lb./(hr.) (ft.)
μ_w	Viscosity of fluid at wall temp., lb./(hr.) (ft.)
ρ	Density, lb./cu. ft.
ΔP	Pressure drop, lb./sq. in.

steels is economical, then the half-pipe will be lower in cost than the dimple jacket, and of course, if condition No. 1 is met, less costly than the conventional jacket.

3. With vessels of less than 500 gal., it is usually not practical to apply the half-pipe jacket.

Conventional Jacket

Conventional jackets (Fig. 1c, 5) are best applied on small-volume vessels (less than 500 gal.) and in high-pressure applications where the internal pressure is more than twice the jacket pressure.

The conventional jacket, simply described, is an extra covering around all or part of a vessel, with an annular space (generally concentric) between the outer vessel wall and the inner wall of the jacket. Baffles are provided as required to control the heat-transfer medium. Various configurations are shown in Fig. 5.

While the conventional jacket is still the most popular type, it is not as widely used as in the past. This is especially true in the jacketing of vessels made from the various alloys. The normal configuration of the conventional jacket is that shown in Fig. 6a. This assures the most efficient heat transfer to the maximum surface area of the vessel. A cross-section is also shown in this figure. Note that the heat-transfer surface available inside the jacket extends up to the top of the upper jacket-closure bar.

An often-used variation of this configuration is made (Fig. 6b) by dividing the straight side into two or more separate jackets.

If desired, the vessel can be jacketed on the straight side only (Fig. 6c), varying from complete to partial vertical coverage. A jacket can also be fabricated to cover the bottom head only (Fig. 6d).

CALCULATING HEAT TRANSFER

The basic equation for calculating heat transfer in an annulus is the Stanton Equation. This is applicable to conventional, dimple and half-pipe coil jackets:

$$h_i = 0.023 \, N_{Re}^{-0.2} \, N_{Pr}^{-2/3} \, c_p G (\mu_w/\mu)^{0.14}$$

The formula varies for each type of jacket in that each has a different equivalent hydraulic radius, D_H. The D_H equals 4 times the flow area divided by the wetted perimeter.

For a conventional jacket, the D_H is computed as follows (see Fig. 4):

Flow area = $J \times l_{BP}$

Wetted perimeter for heat transfer = l_{BP}

$$D_H = \frac{4 \times J \times l_{BP}}{l_{BP} \, (12 \text{ in./ft.})} = J/3 \text{ ft.}$$

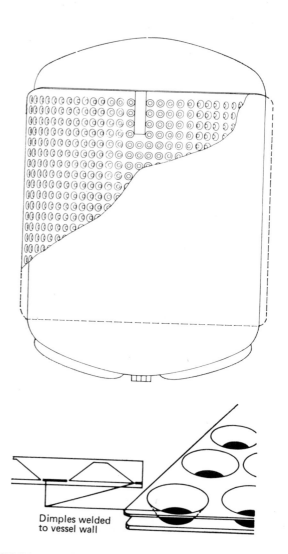

Dimples welded to vessel wall

DIMPLE jacket is made this way—Fig. 3

For the half-pipe coil jacket (Fig. 5):

$$D_H = \frac{4 \times \tfrac{1}{2}(\pi/4)(d_1)^2}{d_1 \,(12 \text{ in./ft.})} = 0.1308\, d_1$$

For the dimple jacket (Fig. 6):

$$D_H = \frac{4 \times 0.412}{2.5\,(12 \text{ in./ft.})} = 0.05493 \text{ ft.}$$

Note: There are 0.412 (average) sq. in. flow area between the center lines of the dimple welds, which are on 2½-in. centers (square pitch).

In checking the film coefficient in the dimple jacket, the calculated coefficient will usually work out rather low because the velocity in the dimple jacket has to be limited to around 2 ft./sec., due to the high pressure drop per foot (the dimples create a great deal of turbulence). However, in comparing the pressure drop in the dimple jacket to that of an annulus having the same flow cross-section with no dimples, the pressure drop in the dimple jacket is approximately 10 to 12 times higher than that of an open channel. Using the relationship that the pressure drop is proportional to the square of the velocity, and the film coefficient is proportional to the ratio of the velocities to the 0.8 power, the film coefficient could be increased 2 to 2.5 times to account for the added turbulence caused by the dimples. (In some cases, the Reynolds number for the dimple jacket works out to be in the laminar flow region; however, the added turbulence caused by the dimples would, in my opinion, disqualify the flow as laminar.)

Computing the Pressure Drop

In many cases where a high velocity is required in order to obtain a high heat-transfer coefficient on the jacket side, calculation of the pressure drop through the jacket becomes important. If the required pressure drop is higher than that available from a positive-displacement pump, the pressure in the jacket may possibly exceed the jacket design pressure. In the case of a centrifugal pump, the velocity of the fluid will adjust to the pump pressure available—with a lower velocity and a resultant lower heat-transfer film coefficient. With a dimple jacket or half-pipe jacket, the pressure drops are higher than the conventional jacket for a given velocity, due to this added turbulence.

In cases where the heat-transfer film coefficient inside the vessel is controlling, the overall heat-transfer coefficient can be optimized with a relatively low velocity of the jacket fluid.

The basic formula for calculating the pressure for fluids inside an annulus is the Fanning equation, which is stated as follows:

$$\Delta P = \frac{4f\, G^2 L}{2g\, P\, D_H\, (144)} = \text{lb./sq. in.}$$

In flowrates where N_{Re} is greater than 10,000:

$$f/2 = \frac{0.023}{\left(\dfrac{D_H G_1}{\mu}\right)^{0.2}}$$

With a conventional jacket, the equivalent hydraulic radius is as follows:

$$D_H = \frac{4 \times \text{cross-sectional flow area}}{\text{wetted perimeter}}$$

$$D_H = \frac{4 \times l_{BP} \times J}{(2\, l_{BP} + 2J)(12 \text{ in./ft.})} = \frac{l_{BP} \times J}{6(l_{BP} + J)}$$

For a half-pipe jacket:

$$D_H = \frac{4 \times \tfrac{1}{2}(\pi d_1^2/4)}{(d_1 + \tfrac{1}{2}\pi d_1)(12 \text{ in./ft.})} = 0.0509\, d_1$$

The friction factor, $f/2 = \dfrac{0.023}{(D_H G_1/\mu)^{0.2}}$, is basically for smooth tubes.

On page 156 of "*Heat Transmission*" by McAdams,*

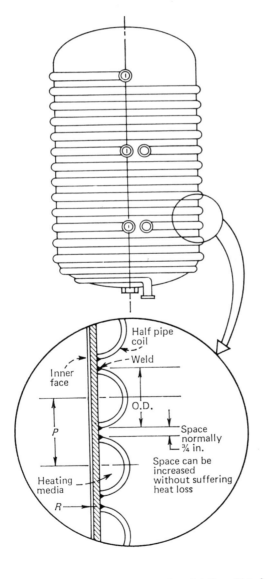

HALF-PIPE coil jacket with fabrication details—Fig. 4

---
* McAdams, William H., "Heat Transmission," 3rd ed., McGraw-Hill, New York, 1954.

the chart shows that for the same Reynolds numbers commercial pipe has a friction factor approximately 15% greater than equivalent smooth tube. Hence, for carbon-steel-jacketed vessels, the writer recommends that 15% be added to the pressure drop per foot.

For determining the length (travel path) in the Fanning equation, the conventional jacket and half-pipe jacket can be compared to a pipe coil. In Crane Co.'s technical paper No. 410, pages 2-12, pressure drop in a pipe coil is described. Each 90-deg. arc is basically a 90-deg. elbow. On page A-27, Crane has a chart, "Resistance of Bends". For "R," the mean radius of the coil, use the outside radius of the vessel diameter; for "d," the coil diameter, the equivalent hydraulic radius should be used. Based on a given R/D_H ratio, the bend resistance, length resistance and total resistance can be obtained from the chart "Resistance of Bends." As explained on pages 2-12, the total equivalent in pipe diameters is obtained by multiplying the number of turns on the coil* by four, which gives the number of 90-deg. bends. Subtract one from the number of 90-deg. bends and multiply by the resistance due to length plus one half of the bend resistance, and then add the total resistance of a 90-deg. bend. After determining the total equivalent length in pipe diameters, multiply by D_H, which then gives the total length. The pressure drop per foot multiplied by the total length will give the total pressure drop except for inlet and outlet losses.

For dimple jackets, due to the turbulence created by the dimples, calculations of pressure drop become very complex, and there are very large reservations as to their accuracy. Because of this, information should be obtained from the manufacturer.

MATCHING THE JACKET WITH THE MEDIUM

The design engineer must analyze several possibilities before deciding which type of jacket to specify for a given process vessel. Not the least of these is determining the most efficient employment of the heat-transfer medium.

Water

In order to avoid the possibility of stress-corrosion cracking due to chlorides in the water, the dimple jacket requires the use of a high nickel alloy, which is a very expensive material of construction.

Steel companies are currently developing stainless steels for this application. However, the material will still be a rather expensive high-alloy steel. The half-pipe jacket can use ¼-inch-thick carbon steel for the jacketing, as has been used on conventional jackets for water service for numerous years. The thickness of the inner shell can be reduced considerably, as in the case of dimple jackets, effecting an economy over the conventional jacket.

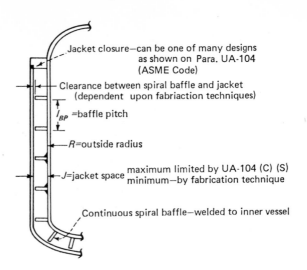

CONVENTIONAL jacket in cross-section—Fig. 5

The use of a carbon-steel half-pipe jacket will be applicable for larger vessels made of clad steel. If the application is for water cooling only, or for steam service below 45 psig., the carbon-steel half-pipe jacket can be applied to an alloy vessel.

With service involving large volumes of water to maintain a high LMTD (log mean temperature difference), the conventional jacket usually offers the best answers.

The small space in the dimple jacket often creates too high a pressure drop. The half-pipe coil jacket may have to be divided into multiple sections, or zones, to overcome the problem of a high volume of water per flow section (which results in a high pressure drop). An outside header would then have to be supplied to connect to the jacket connections. However, savings in the vessel cost might justify addition of the header.

Dowtherm Vapor

The conventional jacket, which may have any desired space between the outer vessel wall and the inner jacket wall (Fig. 5), normally handles Dowtherm* vapor most effectively. Also, since Dowtherm vapor has a low enthalpy (approximately one-tenth that of steam), it requires a larger jacket space for a given heat flux then does steam. Jacket space (Fig. 6a) must be designed in accordance with ASME Code specifications. Maximum allowable space is limited by Section UA-104, Paragraph (c) and (s). Minimum jacket space is determined by the production techniques and skills of the fabricator.

In the case of the dimple jacket, the space between the shell and jacket is sufficient to handle the volume of Dowtherm vapor without a high pressure drop. However, many Dowtherm-vapor installations require

---

* In the case of the half-pipe jacket, the number of turns is the number of turns of the half-pipe. In the conventional jacket, the number of turns is based on the baffle pitch of the spiral baffle.

* Dowtherm (a product of Dow Chemical Co.) is a mixture of 26.5% diphenyl and 73.5% diphenyl oxide. At atmospheric pressure, it boils at 496 F.; at 80 psig., it boils at 670 F. Equivalent steam pressures would be 635 and 2,500 psig., but steam would require uneconomically heavy-walled vessels.

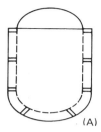 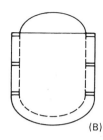 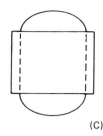 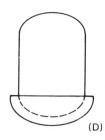

TYPICAL configurations of conventional jackets—Fig. 6

a vapor feedline as large as 6-in. pipe. Normally, the largest connection that can be installed in a manifold is 3 in. This requires that the dimple jacket be fitted with several smaller inlets. Also, the manifolds must be enlarged to handle the large Dowtherm-vapor flow.

If a half-pipe jacket is used for Dowtherm, several inlets and outlets may be required to avoid having the jacket partially flooded with condensate. Only if the process requires a multiple-zone jacket could this be an advantage.

Steam Service

Both the dimple jacket and the half-pipe coil jacket offer great economical advantages over the conventional jacket in the use of steam. The dimple jacket is limited to 300 psi. by Section VIII of the ASME Code, while half-pipe jacket is approved for pressures of up to 750 psi. The higher the steam pressure, the greater the economies effected by the dimple jacket and (especially) the half-pipe jacket. Also, since steam has a higher specific weight at the higher pressure, the narrow jacket space creates no problems.

For half-pipe jackets, the higher flux rates may require multiple sections of jacket to avoid condensate covering too much of the heat-transfer area. If the inside film coefficient is rather low (in the 10-25 range), economic advantages can be realized by reducing the number of turns of the half-pipe jacket on the shell and head, and taking advantage of the "fin" efficiency of the half-pipe jacket.

Liquids: Hot Oil, Dowtherm, Therminol*

Although the pressures are low when using the above liquids, the temperatures are high, with resulting low allowable-stress values for the inner-vessel material. Therefore, both the half-pipe jacket and the dimple jacket provide a more economical solution than does a conventional jacket.

Besides requiring a greater shell thickness than the other two designs, the conventional jacket requires expansion joints to eliminate the stresses induced by the difference in metal temperature between the jacket and the shell, or by the difference in thermal expansion when the jacket is not the same material as the shell (e.g., a stainless steel shell and a carbon steel jacket). Also, the conventional jacket would require a rather expensive, jacket closure bar.

The half-pipe coil jacket is superior to the dimple jacket because of its greater structural strength. In high-temperature applications, temperature growth of both the jacket pipelines and the vessel itself develop a severe bending moment on the jacket connections. The dimple jacket manifolds are a light-gage material and have neither the structural capacity to absorb the stresses developed nor the ability to transmit the forces to the heavier inner shell. The half-pipe jacket is of a heavier-gage metal and is also attached directly to the inner shell by two substantial welds. As a result, it is capable of withstanding piping loads. It should be emphasized however, that these forces are not to be ignored when designing a half-pipe jacketed vessel.

Dimple jacketed vessels can be, and in fact are being, used successfully with hot oil and Dowtherm service. However, the following special considerations must be made and included in the cost of such vessels:

- Use the proper jacket material with respect to thermal coefficient of expansion in relation to the inner shell of the vessel.
- Employ proper testing for pinholes in all of the small fillet welds.
- Design manifolds to avoid stress concentration due to discontinuity stresses; use flexible hoses to eliminate all external forces on the jacket connections and their manifolds.

In both the conventional and dimple jackets, the flowpath and the jacket fluid velocity are controlled by circulation baffles that have positive attachment to the inner shell only. In cooling applications where the media have high viscosities (such as oils, Dowtherm, etc.), the pressure drop through the jacket will be appreciably higher. With this high pressure drop, there will be a greater tendency for the fluid to bypass these circulation baffles. This bypassing can be minimized by proper fit-up and fabrication; how-

---
* Therminol is a registered trademark of Monsanto Chemical Co.

ever, this will add somewhat to the cost of the jacket.

The half-pipe coil jacket is so constructed that any bypassing is automatically precluded. This helps in maintaining a certain desired jacket-side heat-transfer coefficient under all conditions. Velocities of 15 ft./sec. have been possible in some applications of the half-pipe coil jacket. If the film coefficient inside the vessel is the controlling factor, then any bypassing of the jacket fluid can be ignored.

If the combination of the jacket film coefficient and the coefficient inside the vessel is less than a value of 50, the metal wall thickness will not have any appreciable effect of the overall "U" value. If, however, the coefficient is above 50, the half-pipe coil and dimple jackets, which require a greatly reduced wall thickness over that needed for a conventionally jacketed vessel, will have a higher overall "U" coefficient.

For vessels where the internal pressure is more than twice the jacket pressure, the conventional jacket will be the most economical. The vessel should be checked for both internal and external pressure. If the thickness required for the external pressure is 1/8 in. greater than the thickness required for internal pressure on a high-alloy vessel, then the half-pipe coil and the dimple jacket will both usually offer significant economic advantages over the conventional type.

Pressure and Temperature Limitations

Conventional Jackets—These have no limitations as far as design is concerned. However, if the jacket pressure exceeds 100 lb., vessel-wall thickness becomes large, and heat transfer is greatly reduced. In the case of an alloy reactor, a very costly vessel results. For high-temperature applications, thermal-expansion differentials must be considered between the metals used in the vessel and the jacket, and the difference in thickness between the vessel and the jacket walls. Design and construction details are contained in Div. I of the ASME Code, Section VIII, Appendix IX, Jacketed Vessels.

Dimple Jackets—These are limited to a maximum pressure of 300 psi. by Par. UW-19 (b) (1) of Section VIII, Div. I of the ASME Code. As to temperature limitations, dimple jackets are presently in service at temperatures of 700 F. At such high temperatures, it is mandatory that the jacket be fabricated from a metal having the same thermal coefficient of expansion as that used for the inner vessel. Additional special precautions should include the use of flexible hoses to prevent external loads on the relatively light-gage manifolds.

Half-Pipe Coil Jackets—These are not covered at present in Section VIII, Div. I of the ASME Code. Our company has designed and built half-pipe jackets up to 600 psig. at 720 F. under Par. UG-101. As for temperature limitations, a carbon steel half-pipe jacket can be applied to a stainless steel vessel

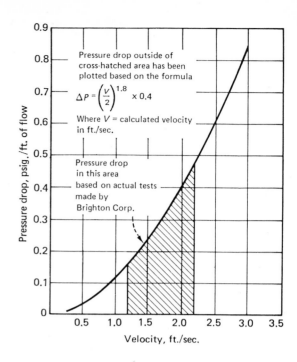

PRESSURE-DROP chart for dimple jackets—Fig. 7

Pressure drop outside of cross-hatched area has been plotted based on the formula

$$\Delta P = \left(\frac{V}{2}\right)^{1.8} \times 0.4$$

Where V = calculated velocity in ft./sec.

Pressure drop in this area based on actual tests made by Brighton Corp.

up to 300 F. Over 300 F., the jacket should also be stainless steel. This type of jacket is best for high-temperature and high-pressure applications.

Surface Area Available for Heat Transfer

With conventional jackets, there are no questions regarding surface that is available. It is always that area covered by the jacket.

On half-pipe coil jackets, some engineers believe that the space between the half-pipes is not useful for heat transfer. This is not accurate because the space between the coils acts like the fin on a finned tube.

With the dimple jacket, some engineers are also concerned that the dimple holes are not available for heat transfer. The same basic analogy for heat conduction is applicable to dimple jackets as for the half-pipe coil design. The temperature of the jacket fluid is transferred by conduction to the inner shell. ■

Meet the Author

Richard E. Markovitz is manager of engineering at Brighton Corp., 11861 Mosteller Rd., Cincinnati, OH 45241, where he is responsible for the design and performance of pressure vessels, heat exchangers and related chemical process equipment, including tank heads. He has been with Brighton Corp. for the past 8 yr., after working as a mechanical engineer for other firms since his graduation from the State University College at Buffalo, N. Y., with a B.S. in mechanical engineering in 1950.

Section IV
HEAT TRANSFER IN PIPING SYSTEMS

Calculating heat transfer from a buried pipeline
Designing steam tracing
Electric pipe tracing
Steam tracing of pipelines
Heating pipelines with electrical skin current

Calculating heat transfer from a buried pipeline

S. J. Amir, *Ebasco Services Inc.*

☐ The heat transferred from buried pipelines is becoming more and more important, whether the concern is for a utility pipeline, a heated oil pipeline, or underground steam pipes for keeping ice from sidewalks and driveways. Although solutions to this heat transfer problem are found in the literature [3,5], they are partial in nature and do not account for all the variables.

The most important of these variables are:

1. The type of fluid flow, which affects the film coefficient of heat transfer on the inside of the pipe.
2. The pipe material, which affects conductance through the pipe wall.
3. The type of soil, which affects dissipation of heat away from the pipe line.
4. Moisture content of the soil, which affects dissipation of heat through the soil.
5. Wind velocity and ground surface characteristics, which affect removal of heat from the soil around the pipe.

All of these effects are combined into one expression for overall thermal resistance to heat transfer out of the pipeline, as:

$$R = \frac{1}{2\pi L}\left[\frac{1}{h_f} + \frac{1}{K'_g}\ln\frac{2}{r}\left(Z + \frac{K'_g}{h_a}\right)\right]$$

Where: R = overall thermal resistance
 L = length of pipe
 h_f = heat transfer coefficient of contained fluid to the pipe wall
 r = radius of the pipe
 Z = distance from the ground surface to the center of the pipe
 K'_g = effective thermal conductivity of the ground
 h_a = effective heat transfer coefficient from the ground to the air

All of these expressions can be determined through familiar measurements or correlations, except K'_g and h_a. K'_g, the effective thermal conductivity of the ground, allows for the migration of moisture under the effects of a temperature gradient. Values of K'_g are shown in the figure; also, it can be expressed in terms of the conductivity of ground, K_g, and pipe-burial dimensions, as [2]:

$$K'_g = K_g \frac{\left(\frac{1}{d} - \frac{1}{2z}\right)}{\left(\frac{1}{d} - \frac{1}{2z}\right) + \frac{B}{12}\left(\frac{1}{d^3} - \frac{1}{(2z)^3}\right) + \frac{B^2}{40}\left(\frac{1}{d^5} - \frac{1}{(2z)^5}\right)}$$

Where: K_g = thermal conductivity of the ground
 d = diameter of the pipe
 B = a constant = 140 for English units [2]

The thermal resistance to heat transfer from ground to air, h_a, has been developed by the author from point-source analysis, by assuming that the heat transfer area is a point on a horizontal surface where natural convection is taking place; and this is corrected for wind speed, as:

$$h_a = (0.137)\left(\frac{K_a}{l}\right)(N_{Gr} \cdot N_{Pr})^{1/3} + 0.667S$$

Where: K_a = thermal conductivity of air
 l = characteristic length of pipeline
 N_{Pr} = Prandtl number for air flow
 N_{Gr} = Grashof number for air flow
 S = wind speed

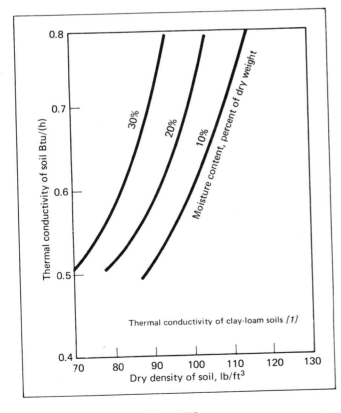

Thermal conductivity of clay-loam soils [1]

Originally published August 4, 1975

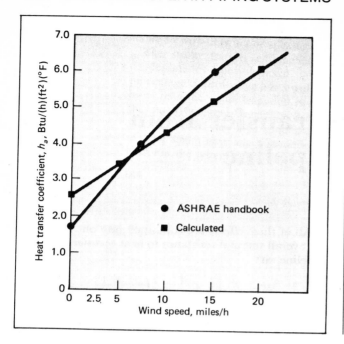

This equation is then simplified to:

$$h_a = (0.137)Ka\left[\frac{N_{Pr} \cdot \beta \cdot g(T_g - T_a)}{\nu^2}\right]^{1/3} + 0.667S$$

Where: β = the coefficient of thermal expansion for air
 g = the gravity constant = 32.16 for English units
 T_g = temperature of the ground surface
 T_a = temperature of air
 ν = kinematic viscosity of air

The value of h_a has been calculated for various wind speeds with this formula, and plotted in the figure, where the calculated values are compared with experimental values from the ASHRAE handbook [6]. Note that the equation for h_a is good for wind speeds up to 20 mph.

With R thus determined, the amount of heat transferred from a buried pipe can be determined from the temperature of the flowing fluid, T_f, by:

$$q = (T_f - T_a)/R$$

References

1. Mickley, A. S., Thermal Conductivity of Moist Soil, *AIEE Trans.*, Vol. 70, pp. 1789–1797.
2. Neher, J. H., The Temperature Rise of Buried Cables and Pipes, *AIEE Trans.*, Vol. 68, Part I 1949, pp. 9–22.
3. Kutateladze, S. S., others, "A Concise Encyclopedia of Heat Transfer," Pergamon Press, New York, 1966.
4. Boersma, L., Warm Water Utilization, paper presented before the conference on beneficial uses of thermal discharges, sponsored by the New York State Dept. of Environmental Conservation, New York City, Sept. 1970.
5. Brober, H., others, "Fundamentals of Heat Transfer," McGraw-Hill Book Co., New York, 1961.
6. Handbook of the American Soc. of Heating, Ventilation and Air Conditioning.

Designing Steam Tracing

Use this analytical approach to determine the effect of all variables, in order to obtain the optimum steam-tracing design.

CARL G. BERTRAM, VIKRAM J. DESAI and EDWARD INTERESS, The Badger Co.

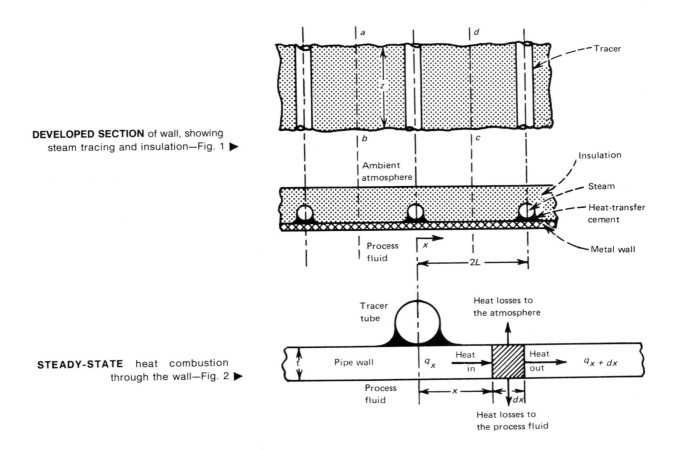

DEVELOPED SECTION of wall, showing steam tracing and insulation—Fig. 1 ▶

STEADY-STATE heat combustion through the wall—Fig. 2 ▶

In the chemical process industries, pipes and equipment often need to be heated. There are many reasons for this. Many materials become viscous or even solidify at ambient temperatures—phthalic anhydride solidifies at 268 F.

Some materials are relatively noncorrosive in the vapor phase but become highly corrosive when they condense—hydrogen sulfide and water vapor condense to form corrosive hydrosulfuric acid.

Although there are other ways of heating pipes and equipment, such as by electrical heating cables or by jacketing, steam tracing is becoming more widely used. An important reason for increased use has been the development of plastic heat-transfer cements.

The high thermal conductivity and good bonding characteristics of these cements, when used properly in well-designed installations, have virtually eliminated most of the disadvantages associated with steam tracing, e.g.,

- Unpredictable heat-transfer rate.
- Slow heatup.
- Uneven temperature distribution.
- Need for high temperature-differentials between tracer surface-temperature and process-fluid temperature.

To date, most installations of steam tracing that use heat-transfer cement have been designed by using either an empirical approach or one based on somewhat simplified equations that often do not accurately represent the system. Consequently, the general policy for steam-tracing design has been conservative. The cost of overdesign has been justified by considering the potential for increased operating and maintenance costs, and plant shutdowns, created by insufficient tracing (underdesign). However, steam-tracing costs may be upwards of 10% of the total piping cost of the plant. The money spent on overdesign is well worth saving.

The analytical approach to steam-tracing design presented here eliminates the over- and underdesign problems, and gives an optimum solution. A general solution can be obtained that determines the following:

1. The optimum pitch of steam-tracing. To be optimum, the tracing tubes must be spaced so that no part of the pipe wall can fall below the required temperature.
2. The temperature profile of the wall. The general solution gives the temperature profile on the wall at various distances from the tracer tube.
3. The heat transferred to the process fluid.
4. The total heat transferred by each tracer.

The parameters used in the general solution are wall thickness, wall material (thermal conductivity), insulation thickness and material, heat-transfer coefficient between wall and process fluid, process temperature, steam temperature, and ambient temperature.

Theory and Derivation

The heat balance at steady state around the differential element $t\,dx$ of the developed section of the wall in Fig. 1 and 2 is described by:

$$-Ktz\frac{dT}{dx} - \left(-Ktz\frac{dT}{dx} - Ktz\frac{d^2T}{dx^2}dx\right) - [h_o(T - T_{amb}) + h_i(T - T_p)]z\,dx = 0 \quad (1)$$

Letting:

$$A = \frac{h_o + h_i}{Kt} \text{ and } B = \frac{h_i T_p + h_o T_{amb}}{Kt},$$

Eq. (1) reduces to:

$$\frac{d^2T}{dx^2} - AT + B = 0$$

This linear differential equation with constant coefficients gives the following general solution for the wall-temperature distribution:

$$T = C_1 e^{\sqrt{A}x} + C_2 e^{-\sqrt{A}x} + (B/A) \quad (2)$$

C_1 and C_2, and hence the actual temperature distribution, are dependent on the following boundary conditions:

$$T = T_o \text{ at } x = 0$$
$$\frac{dT}{dx} = 0 \text{ at } x = L$$

The following solution results from Eq. (2):

$$\frac{T - (B/A)}{T_o - (B/A)} = \frac{e^{\sqrt{A}L\alpha}(1 + e^{2\sqrt{A}L(1-\alpha)})}{1 + e^{2\sqrt{A}(L)}}$$
$$= \frac{\cosh(\sqrt{A}L(1-\alpha))}{\cosh(\sqrt{A}L)}, \quad (3)$$

where $\alpha = x/L$.

Eq. (3) gives the temperature of the wall between the two parallel tracers, and is plotted in Fig. 3 as

$$\frac{T - (B/A)}{T_o - (B/A)} \text{ versus } \sqrt{A}(L)$$

for various values of α.

For $\alpha = 1$, the point described is midway between the two tracers. The minimum wall temperature occurs at this point. When the minimum allowable wall temperature is specified, the maximum tracer pitch, $2L$, can be calculated.

The heat transferred to the process fluid through the surface $abcd$ in Fig. 1 is:

$$\frac{Q}{z} = 2 \int_o^L (T - T_p) h_i \, dx \text{ Btu./(hr.)(ft.) of tracer}$$

Substitution of T from Eq. 3 gives:

$$\frac{Q}{z} = 2 \left(\frac{[T_o - (B/A)] h_i}{\sqrt{A}} \right) \times$$
$$\tanh (\sqrt{A} L) + 2 h_i [(B/A) - T_p] L \quad (4)$$
Btu./(hr.)(ft.) of tracer

The heat transferred to the process fluid, and the heat losses to the surrounding atmosphere (i.e., the total heat transferred by the tracer) are:

$$\frac{Q_T}{z} = 2 \int_o^L (T - T_p) h_i \, dx + 2 \int_o^L (T - T_{amb}) h_o \, dx$$
$$= \frac{2(h_i + h_o)}{\sqrt{A}} [T_o - (B/A)] \tanh (\sqrt{A} L)$$
Btu./(hr.)(ft.) of tracer (5)

Application

Following are stepwise procedures for solving the two most common tracing-design problems.

Problem Type I—Maintain stagnant-process-fluid bulk temperature at a specified minimum.

Step 1—Determine h_i and h_o from physical parameters and empirical equations.

Step 2—Calculate the values of constants A and B and calculate the ratio of B/A.

Step 3—Set T_o equal to the expected lowest temperature in the tracer, specifically the saturated steam temperature at the tracer outlet. Normally, trapping distances are based on a 10% or 10-psi. pressure drop, whichever is greater.

Step 4—From Eq. (4) calculate L by using a value of Q/z of 100 Btu./(hr.)(ft.) of tracer. For certain problems, a value other than 100 may be preferable, depending on the reheat time required if the steam supply were lost. The designer must, therefore, verify that the heatup time possible meets this requirement while not resulting in an uneconomical tracing design.

Step 5—The maximum pitch is $2L$ for both spiral-wound tracing (shown below) and parallel tracing. For

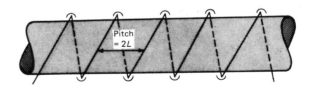

the case of multiple parallel tracers, the number of tracers required is:

$$\text{Number of tracers} = \frac{\text{Pipe circumference}}{2L} = \frac{\pi D}{2L}$$

where a fractional number of tracers is rounded off to the next integral number. These tracers are equally spaced around the circumference of the pipe.

Step 6—Determine the total heat transferred by the tracer by substituting the final value of L from Step 5 in Eq. (5).

Step 7—Calculate the steam consumption per foot of tracer from the following:

Nomenclature

A	A parameter = $(h_o + h_i)/Kt$, dimensionless
B	A parameter = $[(h_i T_p) + (h_o T_{amb})]/Kt$, dimensionless
C_1, C_2	Constants
c_p	Heat capacity of process fluid, Btu./(lb.)(°F.)
D	Pipe diameter, ft.
f	Darcy friction factor, dimensionless
G	Mass velocity of the process fluid, lb./(hr.)(sq. ft.)
h_{air}	Heat-transfer coefficient between ambient air and insulation, Btu./(hr.)(sq. ft.)(°F.)
h_i	Process-fluid heat-transfer coefficient, Btu./(hr.)(sq. ft.)(°F.)
h_o	Overall heat-transfer coefficient between the wall and ambient air, Btu./(hr.)(sq. ft.)(°F.)
k	Thermal conductivity of the process fluid, Btu./(hr.)(sq. ft.)(°F./ft.)
K	Thermal conductivity of pipe or vessel wall, Btu./(hr.)(sq. ft.)(°F./ft.)
K_{ins}	Thermal conductivity of insulation, Btu./(hr.)(sq. ft.)(°F./in.)
L	One-half the steam tracing pitch, ft.
N_{Nu}	Nusselt number, $(h_i D)/K$, dimensionless
N_{Pr}	Prandtl number, $(C_p \mu)/k$, dimensionless
N_{Re}	Reynolds number, $(DG)/\mu$, dimensionless
Q	Heat transferred to the process fluid, Btu./hr.
Q_T	Total heat transferred by the tracer, Btu./hr.
t	Pipe or vessel wall-thickness, ft.
T	Pipe or vessel wall-temperature, °F.
T_{amb}	Ambient air temperature, °F.
T_{mid}	Midpoint temperature at $x = L$, °F.
T_o	Temperature of pipe or vessel wall at tracer, °F.
T_p	Process-fluid temperature, °F.
T_s	Steam temperature, °F.
V_v	Specific volume of vapor, cu. ft./lb.
V_L	Specific volume of liquid, cu. ft./lb.
w'	Steam consumption per unit length of tracer, Btu./(hr.)(ft.)
x	Distance along the pipe or vessel wall, ft.
X_{ins}	Insulation thickness, in.
z	Length of the tracer, equivalent ft.
ΔH	Enthalpy change of steam through tracer, Btu./lb.
ΔP	Total steam-pressure loss through tracer, psi.
α	A parameter = x/L
μ	Viscosity, lb./(hr.)(ft.)
ρ	Density of fluid, lb./cu. ft.

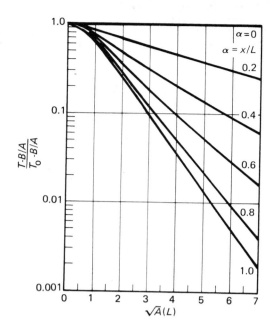

GRAPHICAL solution for steam-tracing design—Fig. 3

$$w' = \text{steam consumption (lb./(hr.)(ft.) of tracer)} = \frac{Q_T}{\Delta H}$$

where ΔH is the change in enthalpy of the steam from the tracer inlet to the outlet.

Problem Type II—Maintain wall temperature of process equipment or piping at a specific minimum.

Steps 1 to 3—Same as Problem Type I.
Step 4—Calculate:

$$\frac{T_{mid} - (B/A)}{T_o - (B/A)}$$

Step 5—From Fig. 3 (for $\alpha = 1$), determine $\sqrt{A}(L)$ and solve for L. The required tracing pitch is $2L$. The number of tracers is:

$$\frac{\text{Pipe circumference}}{2L} = \frac{\pi D}{2L}$$

A fractional number of tracers is rounded off to the next integral number.

Step 6—Substitute the value of L from Step 5 into Eq. (4) and (5) and calculate the heat transferred to the process fluid and the total heat transferred by the tracer.

Step 7—Same as Step 7 in Problem Type I.

Example—Maintaining Bulk Temperature

Problem 1—A 10-in. stainless-steel pipe is carrying phthalic anhydride at 300 F. The line is to be steam-traced using heat-transfer cement and is insulated with 1½ in. of calcium silicate insulation to maintain the bulk temperature at 300 F. The design case is for the process fluid being stagnant with an average, inside, natural-convection coefficient of 20 Btu./(hr.)(sq. ft.)(°F.). Determine the number of parallel tracers required, the heat transferred to the process fluid, and the steam consumption, using 150-psig. saturated steam.

Data:
Ambient temperature, $T_{amb} = -10$ F.
Supply steam temperature, $T_s = 366$ F.
Process fluid temperature, $T_p = 300$ F.
Thermal conductivity of stainless-steel pipe wall, $K = 9.8$ Btu./(hr.)(sq. ft.)(°F./ft.)
Thermal conductivity of insulation $K_{ins} = 0.3$ Btu./(hr.)(sq. ft.)(°F./in.).
Pipe wall thickness = 0.165 in.
Copper tracer, O.D. = 0.50 in., I.D. = 0.43 in.

Assume the heat-transfer coefficient between the insulation and the ambient atmosphere, $h_{air} = 2$ Btu./(hr.)(sq. ft.)(°F).

Solution—use Problem Type I procedure.

Step 1—Calculate h_o, the overall heat-transfer coefficient between the wall and the ambient atmosphere:

$$\frac{1}{h_o} = \frac{1}{h_{air}} + \frac{X_{ins}}{K_{ins}}$$

$$\frac{1}{h_o} = \frac{1}{2.0} + \frac{1.5}{0.3}$$

Therefore, $h_o = 0.182$ Btu./(hr.)(sq. ft.)(°F.) and, given, $h_i = 20$ Btu./(hr.)(sq. ft.)(°F.)

Step 2—Determine the constants A and B.

$$A = \frac{h_o + h_i}{Kt} = \frac{0.182 + 20}{(9.8)(0.165/12)} = 150$$

$$\sqrt{A} = 12.25$$

$$B = \frac{h_i T_p + h_o T_{amb}}{Kt}$$

$$= \frac{(20)(300) + (0.182)(-10)}{(9.8)(0.165/12)} = 44{,}510$$

$$B/A = \frac{44{,}510}{150} = 296.7$$

Note: $B/A = 296.7$ F corresponds to the equilibrium temperature of the pipe wall in the absence of steam-tracing.

Step 3—Assume a 10% pressure drop for the tracer circuit. Outlet pressure = 135 psig., corresponding to a saturated steam temperature of 358 F.

Set $T_o = 358$ F.

Step 4—Design for a net heat input of 100 Btu./hr.-ft. of tracer. Using Eq. (4), solve for L assuming tanh $[\sqrt{A}(L)] = 1$.

$$L = \frac{Q/z - 2(T_o - B/A)h_i/\sqrt{A}}{2h_i(B/A - T_p)}$$

$$L = \frac{100 - 2(358 - 296.7)(20)/12.25}{2(20)(296.7 - 300)} = 0.76 \text{ ft.}$$

Checking assumption,

$$\tanh(\sqrt{A}L) = \tanh(12.25)(0.76) = \tanh 9.3 \approx 1.0.$$

Therefore assumption was valid and $L = 0.76$ is correct.

Step 5—The minimum required pitch = $2L = 2 \times 0.76 = 1.52$ ft.

Number of parallel tracers $= \dfrac{\pi D}{2L} = \dfrac{\pi\left(\dfrac{10.75}{12}\right)}{(1.52)} = 1.85$

With 2 parallel tracers, the resultant distance between the tracers is:

$$\dfrac{\pi D}{2} = 1.4 \text{ ft or } 16.8 \text{ in.}$$

Step 6—The total heat transferred by the tracer is:

$$\dfrac{Q_T}{z} = \dfrac{2(20.182)}{12.25}(358 - 296.7)\tanh(12.25)(0.7)$$

$$= 202 \text{ Btu./(hr.)(ft.)}$$

Step 7—The steam consumption is:

$$w' = \dfrac{202}{865} = 0.234 \text{ lb./(hr.)(ft.)}$$

Note: To determine the maximum equivalent feet of tracing run per trap, corresponding to the tracer circuit pressure-drop assumed in Step 3, use the following equation:*

$$z^3 = \dfrac{1.48 \times 10^{11} D^5 (\Delta P)}{f(w')^2 (V_V - V_L)}$$

Substituting w' from Step 7,

$$z^3 = \dfrac{1.48 \times 10^{11}(0.43/12)^5(15)}{(0.012)(0.234)^2(3.02 - 0.018)}$$

$z = 405$ maximum equivalent feet

*This equation is derived by integrating the Darcy equation for a fluid with a changing specific volume. No account is taken for the fact that the fluid has two phases, since the largest portion of the total pressure drop is taken where the fluid is substantially all vapor. Furthermore, the steam pressure is taken to be that at the outlet of the tracer circuit, thereby rendering a somewhat-conservative steam specific volume for the circuit.

Problem—Prevent Condensation

Problem 2—An air stream saturated with hydrogen sulfide and water vapor is flowing at 150 F. through a 28-in.-dia. carbon steel duct insulated with $1\frac{1}{2}$ in. of calcium silicate insulation. It is necessary to maintain the duct wall above 150 F. to prevent condensation of hydrogen sulfide and water.

Determine the required pitch of steam-tracing and number of tracers to be applied with heat-transfer cement, using 30-psig. saturated steam at 274 F. to maintain the duct wall above 150 F. Also calculate the heat transferred to the process fluid, and the steam consumption.

Data:
Air stream flowrate = 53,500 lb./hr.
Ambient temperature, T_{amb} = 10 F.
Steam temperature, T_s = 274 F.
Process fluid temperature, T_p = 150 F.
Thermal conductivity of carbon-steel duct wall,
  K = 28 Btu./(hr.)(sq. ft.)(°F./ft.)
Thermal conductivity of insulation,
  K_{ins} = 0.3 Btu./(hr.)(sq. ft.)(°F./in.)
Thermal conductivity of air =
  0.0168 Btu./hr. sq. ft.(°F./ft.)
Viscosity of air = 0.02 cp.
Specific heat of air = 0.25 Btu./(lb.)(°F.)
Duct wall thickness = 0.25 in.
Copper tracer, O.D. = 0.50 in., I.D. = 0.43 in.
Assume the heat-transfer coefficient between the insulation and the ambient atmosphere, h_{air} = 2 Btu./(hr.)(sq. ft.)(°F.)

Solution—Use problem Type II procedure.

Step 1—Calculate h_i, the heat-transfer coefficient between the wall and the process air stream.

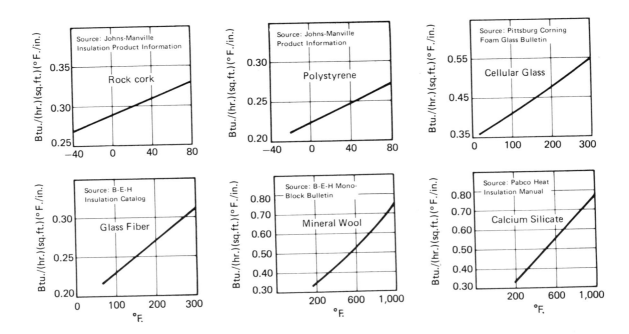

THERMAL conductivities of insulation materials—Fig. 4

268 HEAT TRANSFER IN PIPING SYSTEMS

Mass velocity, $G = \dfrac{53{,}500}{\frac{\pi}{4}\left(\frac{28}{12}\right)^2} = 12{,}500$ lb./(hr.)(sq. ft.)

$$N_{Re} = \frac{DG}{\mu} = \frac{(28/12)(12{,}500)}{(0.02)(2.42)} = 603{,}000.$$

$$N_{Pr} = \frac{C_p \mu}{k} = \frac{(0.25)(0.02 \times 2.42)}{0.0168} = 0.72$$

From the Dittus-Boelter equation,

$$N_{Nu} = 0.023(603{,}000)^{0.8}(0.72)^{0.4}$$
$$= 850 = \frac{h_i D}{k}$$

and

$$h_i = \frac{(850)(0.0168)}{(28/12)} = 6.12 \text{ Btu./(hr.)(sq. ft.)(°F.)}$$

Calculate h_o, the overall heat-transfer coefficient between the wall and the ambient atmosphere:

$$\frac{1}{h_o} = \frac{X_{ins}}{K_{ins}} + \frac{1}{h_{air}}$$

$$\frac{1}{h_o} = \frac{1.5}{0.3} + \frac{1}{2.0}$$

$$h_o = 0.18 \text{ Btu./(hr.)(sq. ft.)(°F.)}$$

Step 2—Determine the constants A and B:

$$A = \frac{h_o + h_i}{Kt} = \frac{0.18 + 6.12}{(28)(0.25/12)} = 10.8$$

$$\sqrt{A} = 3.29$$

$$B = \frac{h_i T_p + h_o T_{amb}}{Kt}$$

$$= \frac{(6.12)(150) + (0.180)(10)}{(28)(0.25/12)} = 1{,}575$$

and

$$B/A = 146$$

Step 3—Assume a 10-psi. pressure drop for the tracer circuit. Outlet pressure = 20 psig., corresponding to a saturated steam temperature of 260 F.

Set $T_o = 260$ F.

Step 4:

$$\frac{T_{mid} - (B/A)}{T_o - (B/A)} = \frac{150 - 146}{260 - 146} = 0.035$$

Step 5:

From Fig. 3, for $\alpha = 1$:

$$\sqrt{A}(L) = 4.05$$
$$L = 4.05/3.29 = 1.215 \text{ ft.} = 14.6 \text{ in.}$$

The maximum required pitch $2L = 2.43$ ft = 29.2 in.
For parallel tracers:

$$\text{number of tracers} = \frac{\pi(28/12)}{2.43} = 3.02.$$

Since this solution is dependent on the calculated parameter, h_i, which can vary over about a 20% range, verification shows that a larger value of h_i requires four tracers located 1.82 ft. apart for a sound design.

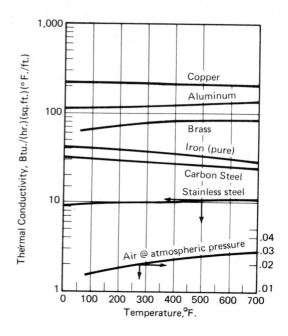

THERMAL conductivities of metals and air—Fig. 5

Step 6—The heat transferred to the process fluid is:

$$Q/z = \frac{2(260 - 146)(6.12)}{3.29} \tanh(3.29)(0.91) +$$
$$\frac{2(6.12)(146 - 150)(0.91)}{}$$
$$= 377 \text{ Btu./(hr.)(ft.) of tracer.}$$

The total heat transferred by the tracer is:

$$\frac{Q_T}{z} = \frac{2(6.12 + 0.18)(260 - 146)}{3.29} \tanh(3.29)(0.91)$$
$$= 434 \text{ Btu./(hr.)(ft.) of tracer.}$$

Step 7—The steam consumption is:

$$w' = \frac{434}{976} = 0.445 \text{ lb./(hr.)(ft.) of tracer}$$

Note: The maximum equivalent length (feet) of tracing run per trap is:

$$z = \left[\frac{(1.48 \times 10^{11})(43/12)^5(10)}{(0.012)(0.444)^2(11.9 - 0.017)}\right]^{1/3} = 146 \text{ ft.}$$

(See Problem 1.)

Problem—Heating a Vessel Bottom

Problem 3—The bottom of a 4-ft.-dia., stainless-steel, solvent-recovery column holds a liquid that freezes at 320 F. and polymerizes at 400 F. The bottom head must be traced to keep the material fluid after a shutdown. Determine the required pitch of the tracing using 150-psig. saturated steam.

Data:

Ambient temperature = −20 F.
Supply steam temperature, $T_s = 366$ F.

Thermal conductivity of stainless steel =
9.8 Btu./(hr.)(sq. ft.)(°F./ft.)
Insulation thickness = 2 in.
Thermal conductivity of insulation =
0.3 Btu./(hr.)(sq. ft.)(°F./ft.)
Wall thickness = 0.375 in.
Assume the heat-transfer coefficient between the insulation and air is 2.0 Btu./(hr.)(sq. ft.)(°F.), and assume an inside convection coefficient of 20.

Solution—Use Problem Type II.
Step 1:

$$\frac{1}{h_o} = \frac{1}{h_{air}} + \frac{X_{ins}}{K_{ins}} = \frac{1}{2} + \frac{2}{0.3}$$

$$h_o = 0.14 \text{ Btu./(hr.)(sq. ft.)(°F.)}$$

and

$$h_i = 20 \text{ Btu./(hr.)(sq. ft.)(°F.) (assumed)}$$

Step 2:

$$A = \frac{h_o + h_i}{Kt} = \frac{0.14 + 20}{(9.8)(0.375/12)} = 65.6$$

$$\sqrt{A} = 8.05$$

$$B = \frac{h_i T_p + h_o T_{amb}}{Kt} =$$

$$\frac{(20)(320) + (0.14)(-20)}{(9.8)(0.375/12)} = 20{,}900$$

$$B/A = \frac{20{,}900}{65.6} = 318$$

Step 3—Using a 15-psi. pressure drop for tracer circuit, outlet pressure = 135 psig., corresponding to a saturated-steam temperature of 358 F.

Set $T_o = 358$ F.

Step 4:

$$\frac{T_{mid} - (B/A)}{T_o - (B/A)} = \frac{320 - 318}{358 - 318} = 0.05$$

Step 5—From Fig. 3, for $\alpha = 1$,

$$\sqrt{A}(L) = 3.7,$$
$$L = 3.7/8.05 = 0.46 \text{ ft.} = 5.5 \text{ in.}$$

Therefore, the maximum allowable pitch for tracing the bottom head of the column = $2L$ or 11 in. A typical tracing layout for this problem is illustrated in Fig. 6.

References

1. Desmon, L. G., Sams, E. W., Natl. Advisory Comm. Aeronaut Res. Mem. E50H23 (1950).
2. House, F. F., Pipe Tracing and Insulation, *Chem. Eng.*, June 17, 1968.
3. McAdams, W. N., "*Heat Transmission,*" 3rd ed., McGraw-Hill, New York, 1954.
4. Sieder, E. N., Tate, G. E., *Ind. Eng. Chem.*, **28,** pp. 1429-36 (1936).

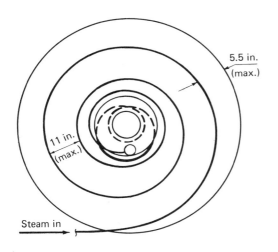

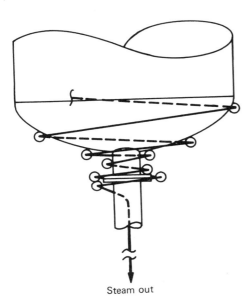

STEAM-TRACED vessel bottom (Problem 3)—Fig. 6

Meet the Authors

Carl G. Bertram

Vikram J. Desai

Edward Interess

Carl G. Bertram is a process engineer with The Badger Co., One Broadway, Cambridge, MA 02142. He has had experience in many aspects of chemical-plant design and construction, including process design, pollution control, economic evaluations, utility systems and startup. He holds a B.S. from the University of Pennsylvania and an M.S. from Villanova University, both in chemical engineering.

Vikram J. Desai is on leave from The Badger Co. and is studying for his doctorate at Massachusetts Institute of Technology—from which he also received his S.M. He obtained his B.S. from Birla Institute of Science and Technology, India. He has previously worked for American Cyanamid Co. and Oak Ridge National Laboratories as well as several companies in India, and is a member of AIChE.

Edward Interess is a process engineer with The Badger Co. His experience has included all aspects of process design of several processes, including acrylonitrile, vinyl chloride and phthalic anhydride. He holds a B.S. Ch.E. and an M. Eng. from Rensselaer Polytechnic Institute and an M.B.A. from Boston University, and is a member of AIChE and ACS.

Electric Pipe Tracing

These guidelines should acquaint a project engineer with aspects of electric pipe tracing that are necessary to achieve desired heating results on time and on budget.

CHARLES W. BROWN, Nelson Electric, Div. of Sola Basic Industries

In the past, steam tracing of pipes maintained temperatures in process lines that would otherwise have dropped due to losses through the insulation. However, for the last several years, electric tracing has been gaining importance and capturing many of the tracing applications formerly left to steam. One reason is that today's high interest rates penalize severely for downtime.

Although steam tracing is generally more economical for large, high-temperature pipes, or when a product needs to be heated, electric heating to maintain temperatures in pipelines has these advantages over steam:
- Heat is simple to provide with automatic controls.
- Heating is more uniform.
- Operating and maintenance costs are lower.
- Heating can be applied equally to short and long process lines.

Other applications for electric pipe tracing may involve: heating of certain storage tanks and vessels; heating of subsoil under liquefied natural and synthetic natural-gas storage tanks; heating of hoppers for electrostatic precipitators; and melting of snow and ice at loading docks and ramps.

Organizing an Electric-Tracing Project

Downtime or missed startup dates caused by frozen valves or broken pipes create severe economic penalties to a plant due to high interest rates. Therefore, a major electric heat-tracing system must be well planned because it may involve as many as 300 heating circuits. With heaters, controls, junction boxes, pipe straps, tools and accessories, such circuits may contain 800 to 900 items. These must be designed, fabricated, shipped, installed, tested and calibrated.

Pipe erection may complicate matters. Pipes 2½-in. O.D. and larger are usually purchased on a spool schedule, and installed to agree with piping drawings. Pipes 2-in or smaller are frequently rerouted in the field. Consequently, a major portion of a piping system must be verified by field measurements; instrument piping is especially subject to field changes.

Even though there is a shortage of experienced heat-tracing engineers, at least a design engineer and a field supervisor must be involved in any electrical tracing system. These people must have daily access to and full cooperation from the piping group. Most cable manufacturers will train a client's personnel at no cost to ensure proper installation of equipment. Planning meetings should be held, in which process, piping, instrument and electrical personnel should all be involved. Responsibilities and authorities must then be clearly defined. The project manager will be responsible for construction per-

This is an adaptation of a paper presented by the author at the Petroleum Mechanical Engineering Conference, Dallas, Tex., Sept. 15-18, 1974.

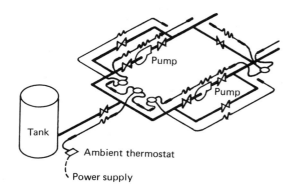

FREEZE PREVENTION for piping systems—Fig. 1

Originally published in June 23, 1975

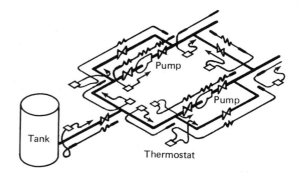

PIPE SEGMENTS with individual tracer, controller—Fig. 2

formance, while the engineer will be judged by the heating results obtained.

Determining Circuits

A task to be completed early in the project design is the determination of the heating circuits, which includes identifying the piping systems to be traced and specifying the schemes that are to be used.

Fig. 1 shows a sample piping system with the heating circuit and control added. This tracing circuit includes ten cables, connected in series, and controlled from a single, ambient-sensing thermostat. This scheme would be applicable for freeze protection of a water system, or of a noncritical system with infrequent flow.

Note that the valves and pumps are traced by the same cables used to heat the pipes. The heating cables must have additional length to replace the higher losses at valves, pumps, filters, pipe supports and wall penetrations. Cable manufacturers generally tabulate cable-length additions required for standard-size piping components. The engineer should therefore send the vendor outline drawings for motor-operated valves, pumps, filters, and all instruments not covered by standard-instrument drawings.

Separate heaters with removable insulated enclosures are recommended for: valves, pumps, or filters that require frequent access for maintenance; and for circuits requiring multiple tracers for high wattage.

Fig. 2 shows the same system with a different tracing scheme. Each pipe segment has an individual tracer and controller. This scheme would be used for processes having close temperature tolerances, changing-flow patterns, or those that are critical to plant operation or safety. Redundant heat tracings from separate power sources are sometimes used for critical circuits.

Isometric drawings similar to Fig. 1 and 2 are usually prepared by cable manufacturers. The complete drawing includes a bill of material and all pertinent reference data, including the location coordinates. These drawings, along with a manual of installation instructions, are used by field personnel to install the tracing system.

The engineer, on the other hand, must supply the manufacturer with specifications for determining the tracing circuits for various piping systems. The preparation of such specifications will require input from the process, piping, and instrumentation engineers. Construction and operating personnel should understand why tracing circuits and control requirements vary from system to system.

In addition to the specifications, the cable manufacturer should be supplied with: (1) plot plan; (2) plan and elevation drawings; (3) process-flow diagrams; (4) piping and instrument diagrams; (5) isometric piping-spool pieces; (6) company standard-instrument drawings; (7) vendor outlines for special instruments, valves, pumps, filters, etc.

Special Conditions and Situations

Normally, project managers are not involved in the calculation of heat losses, wattage applied, safety margins, etc., because the electrical engineer covers the methods to be used in the specifications.

The electrical engineer should also specify special conditions that may exist at the jobsite. Since most of this information must be gathered from operating, construction and process people, the project manager should be aware of these needs.

For instance, a thermostat that performs well in Chicago may fail in Alaska at -60°F. Also, the atmosphere in Borger, Tex., is nothing like that of Linden, N.J. A cable sheath temperature of 900°F may be acceptable in a coal gasification application, while 330°F is not acceptable for coal dust. Pipes that are periodically steam purged may require special techniques.

There may also be special situations of which the project manager should be aware. Such situations may best be solved by a redesign of the piping system or an alternative method of heating. Some such situations are the following:

Dike Penetration—Here, the preferred system is an overhead rack that provides access to the pipe and heating cable. If the pipe must go underground through a dike or roadbed, consult the cable manufacturer before making your field decision on pipe design. Access to cables may require conduit welded to the pipe, and 45-deg instead of 90-deg elbows.

Gut Tracing—Placing heating cables inside a buried pipe is recommended only as a last resort, because of special problems such as turbulence, chemical properties of the product, and cable-entry method.

Submerged Pipes—Pipes buried in a location with a high water table need special treatment. To assure that the water does not saturate the pipe insulation, a pipe-within-a-pipe construction may be used.

Temperature Controls

Another task to determine is the controls that are to be used for the various piping systems. Typically, such controls represent about one-half the cost of the heat-tracing system, and may represent 90% of the operation and maintenance problems.

Enclosure types must be specified for the particular location, such as NEMA* 1 for indoors, NEMA 4 for outdoors, and NEMA 7 for hazardous locations.

Water may enter outdoor enclosures through overhead conduit systems, if these are inadequately sealed, or moisture may be breathed into the enclosure with heat cycling. In humid climates and sea-coast locations, this moisture may corrode the thermostats or contactors to the extent that operation is impeded. Breathers, drains, and space heaters are recommended for these areas.

Thermostats—the workhorses of electric heat tracings—are used in the vast majority of control applications, because they are fairly rugged devices and are also relatively inexpensive. One drawback is that the maximum length of the capillary tube is 10 ft, or 25 ft with compensation.

If it is desirable to have the temperature probe remote from the controller, then it is necessary to use a thermocouple or resistance temperature-detector-type of con-

---
*National Electrical Manufacturers Assn.

HEAT TRANSFER IN PIPING SYSTEMS

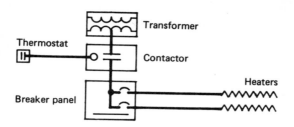

SINGLE THERMOSTAT for freeze-protection system—Fig. 3

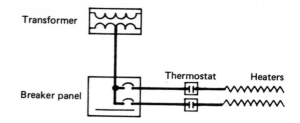

PROCESS CONTROL: one thermostat per heater—Fig. 4

trol. Since the thermostat also can only stand a limited amount of excess temperature, special analysis should be made for thermostats on a freeze-protected system, when steam purge or higher temperature is involved. The thermostat bulb is sometimes sandwiched in the insulation at a lower gradient temperature and setpoint. A thermocouple or resistance temperature-detector type of static control can be remote from a sensor probe, and can meet excess-temperature needs.

Control Schemes

In a freeze-protection system, a single thermostat may be used to control a contactor, and both in turn supply power to as many as 300 heaters (Fig. 3). This scheme offers a lower initial cost and fewer controls to maintain, although more energy is used because the heaters are on when the ambient temperature drops below the setpoint.

Fig. 4 shows a process control with an individual thermostat for each heater. This provides individual control that may be needed by changes in process flow and product requirements. Although this scheme has a higher initial cost and more controllers to maintain, less energy is required, and closer product temperatures maintained.

Sometimes it is desirable to have some indication of the status of the heat-tracing circuits. With the two previous methods shown, the first indication that something has gone wrong with the heating system is when an operator notices that a pipe is frozen or plugged.

Fig. 5 shows a thermostat control with individual controls, but with ammeters and a high-low temperature alarm added. This alarm would consist of two additional thermostats, one set higher and one lower than the control thermostats, as desired. It is common to place a red mark on the ammeter scale when the heater circuit is checked out during startup. Thus, the operator can detect any change in the heating circuit, if the ammeter indicating needle is not close to the red mark.

There are two shortcomings to this system. One is that the thermostat and heating circuits are energized only about one-third of the time. An operator seeing the ammeter's dial on zero would not be able to tell if the circuits are normal. Therefore, a bypass switch may be placed around the thermostat, so that the operator can observe the ammeter at any time. This switch should be in the anmeter panel. Since the thermostat may be remote, it will require additional wires to each heater-control circuit. The other shortcoming is that the ammeter system monitors the circuit only when the operator chooses to observe the ammeters. In a plant with 300 circuits, the ammeters are likely to be observed infrequently.

Fig. 6 shows a critical process-control system with a current monitor alarm. An auxiliary contact on the thermostat is placed in series with an electrically operated relay. If this contact does not open, a light is turned on in the annunciator panel and an alarm sounded.

The advantages of the current transformer and monitoring alarm are: (1) it is in operation 24 h/day; (2) it will detect if the heater has shorted out and tripped the supply circuit breakers; and (3) it provides a remote alarm that can be located in the operator's control room. Neither of these schemes, however, will detect a thermostat failure, such as a broken capillary tube.

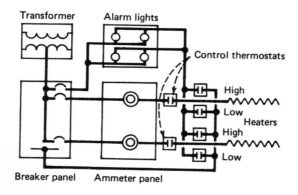

CONTROL with an ammeter and a high-low alarm—Fig. 5

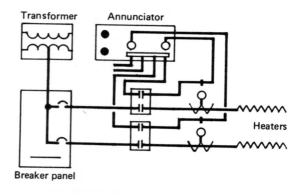

MONITOR ALARM for critical process control—Fig. 6

A minimum of 20% spare controls should be allowed for future additions and maintenance purposes. Also, a scheme should be selected that is flexible to accommodate last-minute changes, as well as load balancing in the field. The alternative is to receive the control package several weeks after the heating cables are installed.

Field Installation

Even an experienced supervisor in electric tracing needs the cooperation of other departments to accomplish his job successfully. He must be informed of all changes and of the magnitude of the project. He must also understand the specifications, and have copies of installation instructions; piping, instrument, and plant and elevation drawings; and heat-tracing isometrics. His responsibilities will include receiving, storing and issuing the hundreds of material items involved, as well as preinstallation and postinstallation testing, checkout, and record-keeping.

One common problem is the shortage of experienced electric-heat tracing electricians. A sufficient number—maybe only six to eight—should be assigned to heat tracing exclusively, and given special training.

Final cable design and fabrication should not begin until all piping is installed and verified by field measurement. This precaution is necessary because of field routing of 2-in and smaller pipes, and instrument piping. Most cable manufacturers make preliminary isometric drawings to facilitate such measurements, and provide service engineers to supervise the measurement and recording. On major projects, when the pipe installation may take several months, it is economical to train field people to make pipe measurements.

If changes are made in the piping and other piping systems are added, material requirements change. The supervisor should be authorized to requisition additional materials as required, and should make periodic status reports to the project manager to ensure that the project is on schedule and budget.

Applicable Codes and Standards

The subject of codes and standards for electric-heat tracing is indeed complex and beyond the scope of this article. The following comments should assist the project manager in determining a course of action:

Local Codes—In the majority of heat-tracing projects, local codes incorporate national codes and standards. A couple of noticeable exceptions are California and New York City. In general, cable manufacturers do not profess to meet all local codes. They will submit drawings and instructions so that the user can gain acceptance. But they reserve the right to decline the order if modifications or testing needed to gain acceptance are not economically feasible. The engineer should specify the local codes to be met and the name of the local enforcing authority.

Occupational Safety and Health Act—OSHA places the burden for compliance on the purchaser. OSHA inspectors may issue special requirements for particular installations. The engineer for the employer-purchaser should seek maximum acceptability by: (1) selecting a system that has been accepted by Underwriters' Laboratories or Factory Mutual; (2) having the system inspected by another federal agency, if no nationally recognized testing laboratory has certified the system; a state, municipal or other local authority responsible for enforcing occupational safety of the National Electric Code may also be contacted; and (3) receiving a certificate from the manufacturer that the equipment was designed and fabricated for the particular customer, where it is stated that the equipment is safe for the intended use, on the basis of test data that the employer keeps and makes available for inspection by the Assistant Secretary of Labor or his authorized representatives.

National Electrical Code—NEC is an installation guide, not a design specification. A manufacturer will supply equipment that—when installed in accordance with NEC—will be acceptable to the authority enforcing the code. As with OSHA, the burden of compliance remains with the employer-purchaser.

Underwriters' Laboratories—UL is a nationally recognized testing laboratory that certifies devices, systems and materials to have acceptable limits for life, fire and casualty hazards. The certificate is a report available to the user from the manufacturer upon request, and by application of the UL label to the equipment.

The reports include design, factory fabrication, field installation and testing. Separate test reports are available for pipe tracing, for general industrial, and for Class 1, Group D, Div. 2, hazardous locations.

A project manager should know that material and installation costs will be substantially higher for a hazardous location due to two factors: (1) all controllers must be in enclosures appropriate for the location; and (2) additional heating cable may be needed for large pipes having high wattage and high process temperatures.

National Electrical Manufacturers Assn.—NEMA defines the composition, construction, dimensions, tolerances, safety operating characteristics, performance, quality, rating, testing, and intended service of a product. A manufacturer supplies controllers that have been designed, built, and tested by NEMA standards. #

References

1. Holstein, W. H., Jr., What It Costs To Steam and Electrically Trace Pipelines, *Chem. Eng. Progr.*, Mar. 1966.
2. Butz, C. H., When Is Electricity Cheaper Than Steam for Pipe Tracing?, *Chem. Eng.*, Oct. 10, 1966, p. 230.
3. Tucker, R. T., "Steam Heating Versus Electric Heating," paper presented at the Petroleum Industry Electrical Assn. and Petroleum Equipment Suppliers Assn. conference held in 1969.

Meet the Author

Charles W. Brown is Manager of Heating Products at Nelson Electric, Div. of Sola Basic Industries, P.O. Box 726, Tulsa, OK 74101. Previously, he was substation and relay engineer with the Public Service Co. of Oklahoma. Holder of a B.S.E.E. degree in electrical engineering from the University of Oklahoma, he is also a registered professional engineer in Oklahoma, and a senior member of the Institute of Electrical and Electronic Engineers.

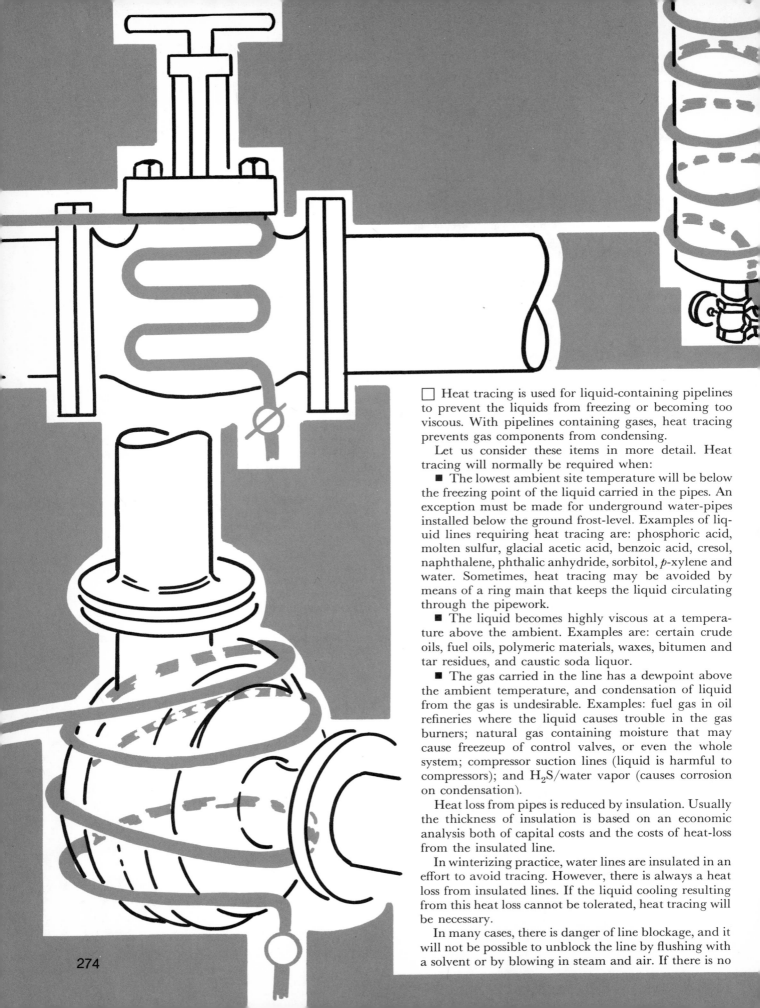

☐ Heat tracing is used for liquid-containing pipelines to prevent the liquids from freezing or becoming too viscous. With pipelines containing gases, heat tracing prevents gas components from condensing.

Let us consider these items in more detail. Heat tracing will normally be required when:

■ The lowest ambient site temperature will be below the freezing point of the liquid carried in the pipes. An exception must be made for underground water-pipes installed below the ground frost-level. Examples of liquid lines requiring heat tracing are: phosphoric acid, molten sulfur, glacial acetic acid, benzoic acid, cresol, naphthalene, phthalic anhydride, sorbitol, p-xylene and water. Sometimes, heat tracing may be avoided by means of a ring main that keeps the liquid circulating through the pipework.

■ The liquid becomes highly viscous at a temperature above the ambient. Examples are: certain crude oils, fuel oils, polymeric materials, waxes, bitumen and tar residues, and caustic soda liquor.

■ The gas carried in the line has a dewpoint above the ambient temperature, and condensation of liquid from the gas is undesirable. Examples: fuel gas in oil refineries where the liquid causes trouble in the gas burners; natural gas containing moisture that may cause freezeup of control valves, or even the whole system; compressor suction lines (liquid is harmful to compressors); and H_2S/water vapor (causes corrosion on condensation).

Heat loss from pipes is reduced by insulation. Usually the thickness of insulation is based on an economic analysis both of capital costs and the costs of heat-loss from the insulated line.

In winterizing practice, water lines are insulated in an effort to avoid tracing. However, there is always a heat loss from insulated lines. If the liquid cooling resulting from this heat loss cannot be tolerated, heat tracing will be necessary.

In many cases, there is danger of line blockage, and it will not be possible to unblock the line by flushing with a solvent or by blowing in steam and air. If there is no

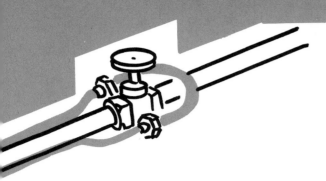

Steam tracing of pipelines

Here is a rundown on the uses of heat tracing, and the procedures for designing both internal and external steam-tracers.

*I. P. Kohli, Consultant**

heat tracing, the line may have to be disassembled, with associated high costs and long shutdown times. The reassembled pipe system may have to be pressure tested, and in some cases dried, before startup.

It is often said that tracing should provide for heat loss from the line—i.e., tracing should not be used for process heating. But in actual practice, facilities may have to be provided for preheating the line before startup, or for quickly thawing a frozen-up line.

The length of a traced line is quite variable. It may be a few feet in a process area, a few thousand feet between offsites and the process area, or over a hundred miles as in the case of underground lines carrying crude oil or fuel oil.

Heat for pipe tracing

Piping may be heated by using fluid heating media (steam, hot water, hot oil, or Dowtherm), or by electricity. The fluid heating systems are simple in operation; any part of the pipework may be isolated by shutting valves. However, there is a possibility of fluid leakage, leading to insulation damage and product contamination. Also, temperature control is rather poor due to the large heat capacity of the fluid-heated system. Still, steam tracing is commonly used because there is surplus low-pressure steam available in most plants.

Since steam has a high latent heat, only a small quantity is required for a given heating load. Also, steam has a high film heat-transfer-coefficient, condenses at constant temperature, and flows to the point of use without pumps.

Liquid heating media (hot water or oil) have to be pumped at a high rate to satisfy a given heating load, ensure uniform heating and provide a reasonable heat-transfer coefficient. The use of liquid and vapor media requires heated tracing pipes either inside or outside the pipeline. Vapor systems need two additional lines (vapor header and condensate return line—see Fig. 1). Liquid systems also require a return header.

Heating by electricity is a clean operation; there is no chance of fluid leakage. An electrical hazard appears only if the system is improperly installed. Temperature is readily controlled by using on/off thermostats. Thus, energy is used only when needed. Usually, electrical heating is employed when surplus steam is not available, or the temperature required would call for high-pressure steam. It is ideal for long-distance lines, because a supply of electricity is usually available along the line. (For details on electrical pipe tracing, see *Chem. Eng.,* June 23, 1975, pp. 172-178.)

Pipeline heating using steam

Heat to pipelines may be provided by steam pipes either inside or outside the line. Steam jacketing (Fig. 2) may also be used, but owing to its high cost it is employed only for special situations involving high heating loads. Jackets (which require oversized insulation) may also be considered for short process runs.

Internal steam tracing

When internal-trace pipes are installed, all the available heat-transfer surface is utilized. The disadvantages are: (a) reduction in the equivalent internal diameter of the pipeline, (b) loss of ability to clean the pipeline by pigging or by using rotary brushes, (c) difficulty in cleaning fouled heat-transfer surfaces.

The trace pipe can only be installed in straight lengths of pipeline that are free of valves. Trace-pipe lengths have to be short, to prevent problems in supporting the trace pipe. It must enter and leave the pipeline frequently, increasing the possibility of leaks. Stresses arising due to differential expansion of the trace and the pipeline should be considered.

One possible application of the internal type of trace is for a fuel-oil unloading line, where the operating temperature is such that it is uneconomical to insulate the line, and where the line is used only intermittently. (Insulation is essential for externally traced lines.) In-

*Now working with Davy International, 8 Baker St., London, WI, U.K.

276 HEAT TRANSFER IN PIPING SYSTEMS

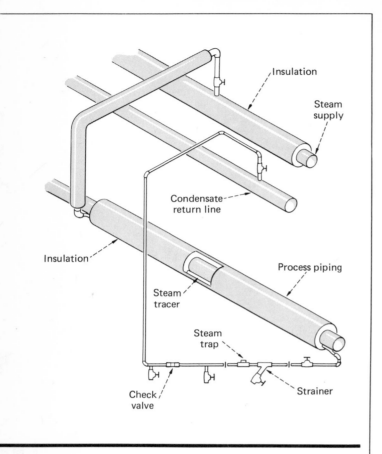

Typical steam-tracing system using external tracer Fig. 1

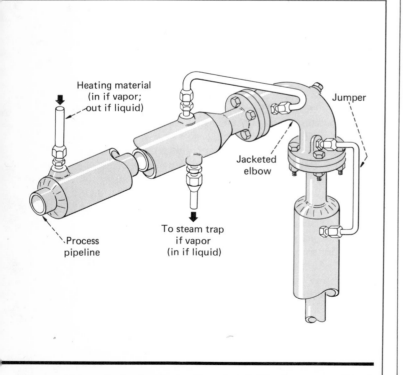

Steam jacketing is used in special situations Fig. 2

Nomenclature

A	Area of pipe, ft^2/ft of length
A_p	Outside area of tracer pipe, ft^2/ft length
A_1	Outside area of pipe lagging, ft^2/ft length
C_p	Specific heat, Btu/(lb)(°F)
D	Dia., in.
D_e	Equivalent pipe dia., ft
D_i	Inside dia., in.
D_{ip}	Pipe I.D., in.
D_o	Outside dia., in
D_o'	Outside dia. of tracer pipe, in.
d_i	I.D. of lagging, in.
d_o	O.D. of lagging, in.
G	Mass velocity, lb/(h)(ft^2)
h_a	Film coefficient to air, Btu/(h)(ft^2)
h_i	Inside film coefficient, Btu/(h)(ft^2)
h_o	Outside film coefficient, Btu/(h)(ft^2)
j_H	Colburn factor, dimensionless
k_a	Thermal conductivity, Btu/(h)(ft^2)(°F/ft)
k_1	Thermal conductivity of lagging, Btu/(h)(ft^2)(°F/ft)
L	Pipe length, ft
N_{Re}	Reynolds number, dimensionless
Q_a	Heat flux to air, Btu/(h)(ft^2)
Q_{conv}	Heat flux by convection, Btu/(h)(ft^2)
Q_p	Heat loss from pipe without tracer, Btu/(h)(ft^2)
q_2	Heat loss, Btu/(h (ft of length)
t_a	Air temperature, °F
t_p	Temperature inside pipe, °F
t_s	Temperature, surface of lagging, °F
t_{st}	Average steam temperature, °F
t_w	Pipewall temperature, °F
T_p	Pipe temperature, °F
U_o	Heat transfer coefficient based on outside surface, Btu/(h)(ft^2)(°F)
W_1	Mass flowrate, lb/h

Greek letters

μ	Viscosity, lb/(ft)(h)
μ_w	Viscosity at wall temperature, lb/(ft)(h)
ρ	Fluid density, lb/ft^3

ternal tracing is acceptable only if leakage of steam into the product can be tolerated.

External tracing

External tracing involves the placing of steam pipes or tubes outside the pipeline. The traced line is then insulated by using preformed sectional insulation.

Heat transfer from trace to pipeline is by conduction, convection in the air space, and radiation. The contact area between the trace and the pipeline is quite small. However, when heat-sensitive liquids are being heated, or when the pipe is plastic-lined, this contact may give rise to undesirable hot spots. In such cases, asbestos packing rings are often used to eliminate any direct contact between the trace and the pipeline.

The simplest method of external tracing is to wrap copper tube around the pipeline, and then to cover the traced pipe with insulation. This is the only way to trace around valves, pipe fittings and instruments

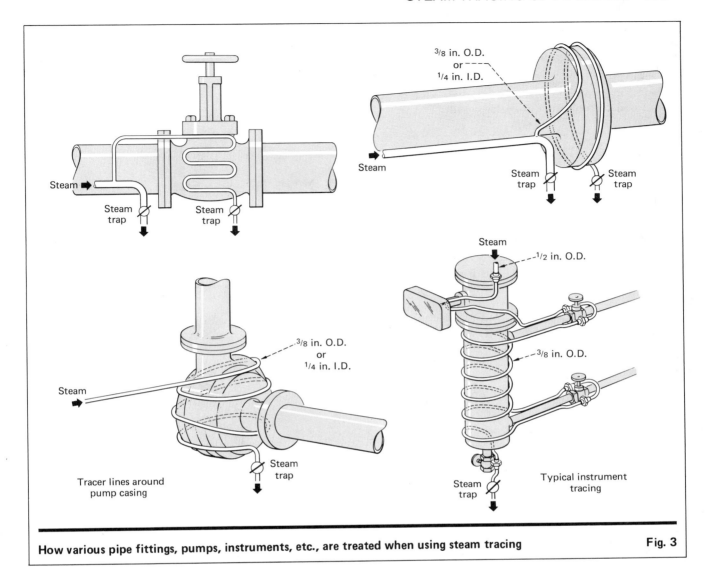

How various pipe fittings, pumps, instruments, etc., are treated when using steam tracing   Fig. 3

(Fig. 3). This procedure is unsuitable for horizontal runs because steam condensate collects at low points and may freeze during a shutdown.

It is essential to ensure that the trace lines are self-draining. Copper tubing of 0.5-in. O.D. and 0.035-in. wall thickness is usually used with straight piping. If the tubing has to be bent into a small radius (as when tracing valves), 0.375-in.-O.D. tubing may be used.

The length of a single trace tube (from steam supply valve to steam trap) is limited by pressure drop in the trace. The trap should have a condensate drainage capacity to match the heating load. At a steam pressure of 100 psig or higher, the length of a single trace should not exceed 200 ft. If the steam pressure is lower, a tracer length of 100 ft is recommended.

To improve contact, the trace tube may be wired to the pipeline. Even then, the conductive heat-transfer rate is quite low. It may be increased by putting a layer of heat-conducting cement (graphite mixed with sodium silicate or other binders) between the trace and the pipeline. This provides much more surface for conductive heat transfer.

When higher heat loads are desirable, straight lengths of ½-in. carbon-steel pipe are clipped along the pipeline. The number of tracer pipes depends upon the heat load—large-diameter pipes carrying liquids that have a high melting point may require up to ten tracers. Steel bands, fitted by using a packing-case banding machine, help to minimize air gaps between the pipeline and tracers. At pipeline bends, the tracers are also bent.

At valves and flanges, the tracers are formed into loops that also function as expansion joints. The loops are formed in a nearly horizontal plane, to ensure self-drainage. The traced pipeline is covered with larger-sized preformed lagging.

In some cases, the traced line is wrapped with aluminum foil and then covered with shaped lagging (Fig. 4). The foil increases the radiation heat transfer. It is essential that the space between the pipeline and the tracer be kept free of particles of lagging material.

Conductive heat transfer may be enhanced by welding the trace on the pipeline. However, welding causes problems due to differential expansion of the pipeline and trace. Because horizontal pipelines are traced on the lower half, welding is a difficult operation. It is more convenient to use heat-transfer cement, as noted earlier.

Before applying the cement, it is essential that both the pipeline and tracer pipes be wire-brushed.

Internal steam tracing—design example

Basis

Size a steam-tracing system to maintain an intermittently used fuel-oil unloading line at 140°F.

Data

Average temperature	20°F
Wind speed	20 mi/h
Line length	500 ft
Size	12-in. nominal bore
Max. unloading rate	500 tons/h
Steam available at 65 lb/in² gage	
Specific gravity of oil	0.985
Viscosity (see Fig. 5)	
Note: Viscosity in cP × 2.42 = viscosity in lb/(ft)(h)	
Thermal conductivity	0.0778 Btu/(h)(ft²)(°F/ft)
Specific heat	0.47 Btu/(lb)(°F)

Calculation

As the line will only be used intermittently, use internal steam tracing and an unlagged pipe.

Heat loss—Neglecting the effect of the internal tracer at this point:

$$\text{Reynolds No., } N_{Re} = \frac{6.31W}{\mu D_{ip}}$$

where W_1 = mass flowrate
= 500 × 2,240 lb/h
μ = viscosity
= 200 cP at 140°F
D_{ip} = pipe dia.
= 12 in.

$$\therefore N_{Re} = \frac{6.31 \times 2,240 \times 500}{200 \times 12}$$
$$= 2,940$$
$$L/D = 500$$

Hence, from Fig. 6:

$$\left(\frac{h_i D_{ip}}{k_a \times 12}\right)\left(\frac{C_p \cdot \mu}{k_a}\right)^{-1/3}\left(\frac{\mu}{\mu_w}\right)^{-0.14} = 8.0$$

whence, $h_i = \left(\frac{12 k_a}{D_{ip}}\right)\left(\frac{C_p \mu}{k_a}\right)^{1/3}\left(\frac{\mu}{\mu_w}\right)^{0.14} \times 8.0$

Assuming for the first trial that $t_s = 90°F$:

D_{ip} = 12 in.
μ = 200 × 2.42 lb/(ft)(h) [at 140°F]
μ_w = 520 × 2.42 lb/(ft)(h) [at 90°F]
C_p = 0.47 Btu/(lb)(°F)
ρ = 0.985 × 62.4 lb/ft³
k_a = 0.0778 Btu/(h)(ft³)(°F/ft)

$$\therefore h_i = \left(\frac{8 \times 12 \times 0.0778}{12}\right)\left(\frac{0.47 \times 200 \times 2.42}{0.0778}\right)^{1/3}$$
$$\left(\frac{200 \times 2.42}{520 \times 2.42}\right)^{0.14}$$
$$= 0.621 \, (2,920)^{1/3} \, (0.385)^{0.14}$$
$$= 0.621 \, (14.3) \, (0.875)$$
$$= 7.78 \text{ Btu/(h)(ft}^2\text{)(°F)}$$

take $h = 8.0$ Btu/(h)(ft²)(°F)

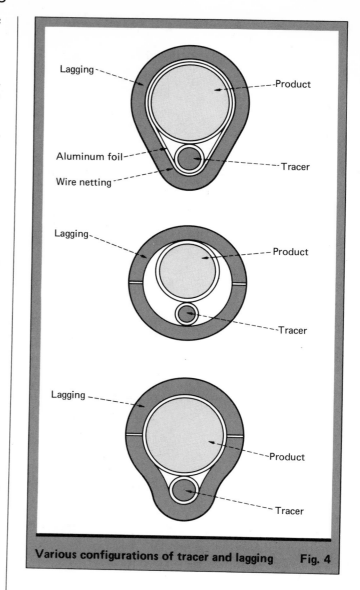

Various configurations of tracer and lagging   Fig. 4

Correcting to O.D. of pipe:
$h_i = 8.0 \times (12.7/12) = 8.5$ Btu/(h)(ft²)(°F)
Hence, $Q_{conv} = h_i A (t_p - t_s)$
$= 8.5A (140 - t_s)$ Btu/(h)(ft²)
and $Q_a = h_a A (t_s - t_a)$

From Tables I and II,
$h_a = 1.83 \times 2.76$
$= 5.04$ Btu/(h)(ft²)(°F)
[corrected for wind velocity]

$\therefore Q_a = 5.04A (t_s - 20)$ Btu/(h)(ft²)
Since $Q_{conv} = Q_a$
$8.5A (140 - t_s) = 5.04A (t_s - 20)$
i.e., $1,190 - 8.5 t_s = 5.04 t_s - 100.8$
$\therefore t_s = 1,290/13.54 = 95°F$

There is an insignificant difference in μ_w between 90° and 95°F; similarly, h_a does not change.

$\therefore$ heat loss = $(8.5) \pi (12.7/12)(140 - 95)$
$= 1,290$ Btu/(h)(ft of pipe)

Due to the low temperature of the pipe, radiation heat loss has been neglected.

Determine tracer size—For first trial, assume tracer is 1.5 in. I.D. (1.9 in. O.D.)

Equivalent diameter, $D_e = \dfrac{144 - (1.9)^2}{12} = 11.7$ in.

$$N_{Re} = D_e \frac{G}{\mu}$$

where: G = mass velocity, lb/(h)(ft²)

$$= \frac{500 \times 2{,}240}{\dfrac{\pi}{4}\left(1 - \dfrac{(1.9)^2}{144}\right)}$$

$$= 1.45 \times 10^6 \text{ lb/(h)(ft}^2)$$

$D_e = 0.975$ ft

$\mu = 200$ cP at 140°F

$$\therefore N_{Re} = \frac{11.7 \times 1.45 \times 10^6}{12 \times 200 \times 2.42} = 2.8 \times 10^3$$

From Fig. 6, $j_H = 9.0$; $N_{Re} = 2{,}800$; $L/D_e = 150/0.975 = 154$

Heat-transfer coefficient for heat transferred to the fuel oil based on I.D. of the 12-in. nominal-bore pipe at the steam-tracer surface where the fuel-oil viscosity is:

$$\mu_w = \frac{h_i D_e}{k_a}\left(\frac{C_p \mu}{k_a}\right)^{-1/3}\left(\frac{\mu}{\mu_w}\right)^{-0.14} = p$$

Assuming a 20-lb/in.² pressure drop along the steam tracer, the average steam pressure = 55 lb/in² gage (steam available at 65 lb/in² gage).

Average steam temperature = 303°F

$\therefore \mu_w = 1.0 \times 2.42$ lb/(ft)(h)[for fuel oil]

and heat-transfer coefficient at steam-tracer outside surface,

$$h_o = 9 \times \left(\frac{0.0778}{0.975}\right)\left(\frac{0.47 \times 200 \times 2.42}{0.0778}\right)^{1/3}\left(\frac{200}{1.0}\right)^{0.14}$$

$$= 21.6 \text{ Btu/(h)(ft}^2)(°F)$$

$$\frac{1}{U_o} = \frac{1}{h_o} + 0.005 = \frac{1}{21.6} + 0.005$$

$$= 0.051 \text{ Btu/(h)(ft}^2)(°F), \text{ approx.}$$

$\therefore U_o = 19.6$ Btu/(h)(ft²)(°F)

Heat transfer from steam tracer = heat lost from pipe without steam tracer:

$Q_p = U_o A_p (t_{st} - t_p)$

$\therefore 1{,}290 = 19.6 A_p (303 - 140)$

Hence, $A_p = \dfrac{1{,}290}{19.6 \times 163}$

$= 0.40$ ft²/ft run of pipe

Surface area of 1-in. nominal-bore pipe = 0.344 ft²/ft
Surface area of 1½-in. nominal-bore pipe = 0.498 ft²/ft
Therefore, 1½-in. nominal-bore tracing will be acceptable.

Check effect of increased fluid velocity on outside film coefficient at pipe wall (internal tracer increases the fluid velocity).

$$N_{Re} = \frac{D_e G}{\mu} = 1 \times \frac{1.45 \times 10^6}{2.42 \times 200} = 3{,}010$$

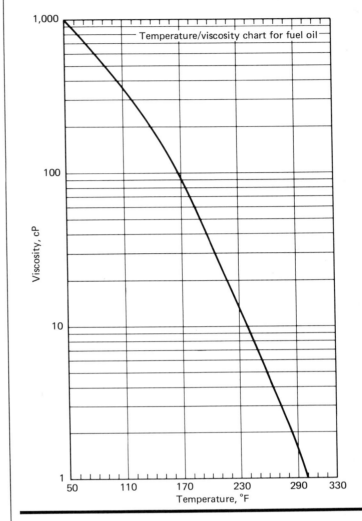

Typical temperature/viscosity chart for fuel oil Fig. 5

where G is based on the equivalent diameter of the pipe, and μ is at 140°F, the desired fuel-oil temperature.

$j_H = 8.05$ [from Fig. 6]

There will be no significant increase in the calculated heat loss.

Check pressure-drop increase for process line

Correction factor

$$= \frac{(D_i)^{2.4}(D_i + 0.5 D_o')^{1.2}}{(D_i^2 - D_o'^2)^{1.8}}\left(\frac{\mu_w}{\mu}\right)^{0.14}$$

$$= \frac{(12)^{2.4}(12 + 0.95)^{1.2}}{(144 - 3.6)^{1.8}}\left(\frac{520}{200}\right)^{0.14} = 1.33$$

Check steam tracer pressure-drop

Maximum tracer length = 150 ft
Heat load = 150 × 1,290 Btu/h
Latent heat of steam = 901 Btu/lb

$\therefore$ Steam load $= \dfrac{150 \times 1{,}290}{901}$

$= 215$ lb/h per tracer

280 HEAT TRANSFER IN PIPING SYSTEMS

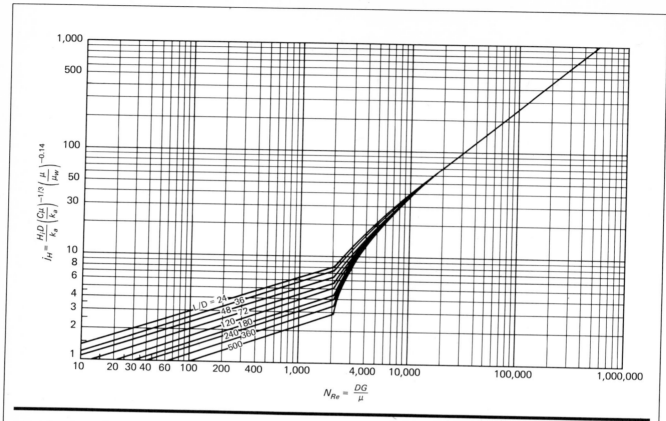

Chart for determining tube-side heat transfer, based on Reynolds number and Colburn factor Fig. 6

From Fig. 7, steam pressure drop at inlet conditions:

$$0.55 \text{ lb/(in}^2)(100\text{-ft run})$$

$\therefore$ pressure-drop/tracer = 1.5×0.275
$= 0.412$ lb/in^2

$\therefore$ the average steam temperature =
$312°F$ (65 lb/in^2 gage)

$U_o = 19.6$ Btu/(h)(ft^2)(°F)
$\Delta T = 312 - 140 = 172°F$

$\therefore$ area of tracer required = $\dfrac{1,290}{(19.6 \times 172)}$
$= 0.383$ ft^2/ft

The surface area of 1½-in. nominal-bore pipe is 0.498 ft^2/ft

Designing external steam tracing

Basis

Size an external tracing system to maintain a 3-in. line at 320°F. The line will be lagged with 1½ in. of insulation.

k_1 = 0.033 Btu/(h)(ft)(°F/ft)
Wind speed = 20 mi/h
Minimum air
 temperature, T_a = 20°F
Line length = 100 ft
Steam available at 150 lb/in^2 gage
Pipe temperature (t_p) = 320°F

Calculation

Determine heat load—Assuming 5 lb/in.^{2} pressure drop along the tracer, then the average steam temperature is:

$$t_{st} = \frac{366 + 363°F}{2} = 364.5°F, \text{ say, } 364°F$$

$\therefore$ average temperature = $0.5(t_{st} + t_p)$ [°F]
$= 0.5(364 + 320)$ [°F]
$= 342°F$

Allowing for tracer, I.D. of lagging = 4 in.

$$\therefore q_2 = \frac{2\pi k_1(t_p - t_s)}{\log_e \dfrac{d_o}{d_i}} = h_a A_1(t_s - t_a)$$

$d_o = 7$ in., $d_i = 4$ in.

Assuming for the first trial that $h_a = 4.0$ Btu/(h)(ft^2)(°F):

$$q_2 = \frac{2 \times \pi \times 0.033(342 - t_s)}{2.303 \log_{10}(7.0/4.0)}$$
$= 4.0\pi \times (7.0/12) \times 1(t_s - 20)$

or $\dfrac{0.0895(342 - t_s)}{0.245} = 7.35(t_s - 20)$

or $0.368(342 - t_s) = 7.35 t_s - 147$
or $126 - 0.368 t_s = 7.35 t_s - 147$

$\therefore t_s = \dfrac{273}{7.718} = 35°F$

STEAM TRACING OF PIPELINES 281

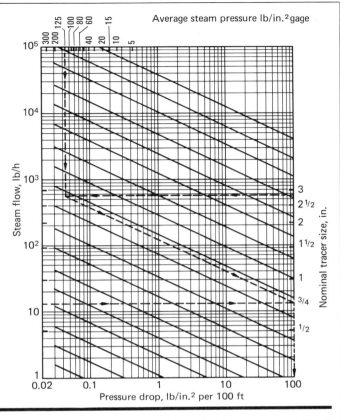

Determination of pressure drop in pipes   Fig. 7

Values of h_a for pipes in still air   Table I

Nominal pipe dia., in.	$(t_s - t_a)$, °F [For an unlagged pipe $t_s = t_w$]							
	50	100	150	200	250	300	400	500
½	2.12	2.48	2.76	3.10	3.41	3.75	4.47	5.30
1	2.03	2.38	2.65	2.98	3.29	3.62	4.33	5.16
2	1.93	2.27	2.52	2.85	3.14	3.47	4.18	4.99
4	1.84	2.16	2.41	2.75	3.01	3.33	4.02	4.83
8	1.76	2.06	2.29	2.60	2.89	3.20	3.83	4.68
12	1.71	2.01	2.24	2.54	2.82	3.12	3.83	4.61
24	1.64	1.93	2.15	2.45	2.72	3.03	3.70	4.48

h_a = Btu/(h)(ft²)(°F,) based on still air.
From McAdams, W. H., "Heat Transmission," 3rd ed., McGraw-Hill, New York, 1954, p. 179.

Correction factor for h_a at different wind velocities   Table II

Wind velocity, mi/h	$(t_s - t_a)$, °F [For an unlagged pipe, $t_s = t_w$]				
	100	200	300	400	500
2.5	1.46	1.43	1.40	1.36	1.32
5.0	1.74	1.69	1.64	1.59	1.53
10.0	2.16	2.10	2.02	1.93	1.84
15.0	2.50	2.42	2.33	2.27	2.08
20.0	2.76	2.69	2.58	2.45	2.30
25.0	2.98	2.89	2.78	2.64	2.49
30.0	3.15	3.06	2.94	2.81	2.66
35.0	3.30	3.21	3.10	2.97	2.81

From Thermon Data Book, Premaberg (GB) Ltd.

C_t, Thermal conductance, tracer to pipe   Table III

Tube size	C_t with no heat-transfer cement, Btu/(h)(°F)(ft of pipe)	C_t with heat-transfer cement Btu/(h)(°F)(ft of pipe)
3/8	0.295	3.44
1/2	0.393	4.58
5/8	0.490	5.73

Tables I and II are not sufficiently accurate in this region, but the value of 4.0 taken for h_a is on the safe side, allowing for wind speed.

$$\therefore q_2 = 7.35(t_s - 20) \text{ Btu/(h)(ft run of pipe)}$$
$$= 7.35(35 - 20)$$
$$= 113 \text{ Btu/(h)(ft run of pipe)}$$

Determine the tracer size—This is done by use of thermal-conductance data on tracer to pipe: Btu/(h)(°F)(ft of pipe). This takes into account the heat-transfer coefficient and pipe surface area. These have been found by experiment.

If one ½-in. tracer without cement is used, heat transfer

$$= 0.393(t_{st} - t_p)$$
$$= 0.393(364 - 320) = 17.3 \text{ Btu/(h)(ft run)}$$

This does not meet the heat load requirement. Two ½-in. tracers without cement do not overcome the problem. If one ½-in. tracer with heat-transfer cement is installed, the heat-transfer rate is:

$$4.58(364 - 320) = 201 \text{ Btu/(h)(ft of run)}$$

A ⅜-in. tracer with cement will yield a heat-transfer rate of:

$$3.44(364 - 320) = 151 \text{ Btu/(h)(ft of run)}$$

Standard practice requires the installation of ½-in. tracer pipes. There is always some overdesign with steam tracers.

R. V. Hughson, Editor

The author

I. P. Kohli, 118 Chadwell Heath Lane, Romford, Essex, U.K., is a consultant process engineer specializing in the design of offshore oil/gas production facilities. He has had experience in several of the chemical process industries, including pharmaceuticals, fibers, oil, gas and petrochemicals. He was graduated from Bombay University, and holds an M.Sc. in hydrocarbon chemistry and petrochemicals from Manchester University. He is a chartered engineer and a committee member of the London/South East branch of the Institution of Chemical Engineers.

Heating Pipelines With Electrical Skin Current

Skin effect of conductance—a long-known principle of electrical circuitry—makes pipeline heating easier to control, less costly and more dependable.

MASAO ANDO, Chisso Engineering Co., and DONALD F. OTHMER, Polytechnic Institute

Less power is required to pump fluids through pipelines when the viscosity of the fluid is lower. Even slightly lower temperatures than the ambient increase the viscosity of many liquids enough to raise pressure drops excessively, freeze liquids or condense vapors.

Jacketed pipes heated with a fluid are expensive to install, maintain and operate. Tracing pipes with tubing that carries the heating fluid in close contact for good heat transfer is less expensive to install but only slightly less expensive to maintain and operate. Temperatures far below steam's condensing temperature are difficult to control. Heated liquids (such as oils, water with antifreeze, etc.) are more flexible as to temperature range, but require pumps, heat exchangers and controls.

Although heat from electricity is much more expensive than from steam, advantages in control may reduce the amount of heat actually used so that the cost of heat from electricity may be less than that from steam, particularly if the pipeline temperatures are to be maintained below about 150 F.

Using the pipe itself as an electrical resistor presents insulation and safety problems. Resistor wires or cables, insulated from the pipe, are the usual quick and easy solution for short, complicated pipelines, but there are dangers from resistors shorting and burning out. For long-distance pipelines, many

Originally published March 9, 1970

HEATING PIPELINES WITH ELECTRICAL SKIN CURRENT

power substations and paralleling transmission lines are necessary.

Heating With Skin Current

Electrical skin-current tracing—developed in Japan recently[*]—depends on the fact that alternating current passing through a conductor does not flow uniformly through its cross section but concentrates near its surface. For mild steel, this "skin effect" is ten times as important as for copper. Thus, the theory of electromagnetism generated by alternating currents shows that, in a long steel conductor of large cross section, such as a pipe, practically all of the current is carried within the 1/25 in. of the inner surface, and virtually none on the outer surface.

The skin metal is the effective resistor, even though it may be only a small fraction of the total metal of the cross-section. The large mass of metal is practically neutral electrically, yet it is at substantially the same temperature as the "skin" because of the excellent thermal conduction. The depth of the skin carrying the current depends on the frequency of the alternating current, but only standard 60 or 50 cycle/sec. frequencies can be considered.

Because transport pipe may be too large for its skin to be used as the resistor, a smaller heating, or tracing, tube is welded continuously (or at appropriate points) to it, or attached by cement of high thermal conductivity. The inner skin of the heating tube is then the resistor. The heating tube and the transport pipe are thus in excellent electrical contact throughout their lengths. Because both are also grounded throughout their length, there is no possibility of electrical shock, or of a sparking or flammable hazard. To improve the uniformity of heating a large pipeline, several separately heated small tubes may be distributed uniformly around its circumference.

In the simplest circuit (see fig.), each end of the heating tube, the inner skin of which forms the resistor, is connected to the terminals of a 1,000 v. a.c. source, an alternator or transformer.

Grounding is accomplished, as far as the electrical circuit and the heating tube is concerned, by close contact with the transport pipe (which is grounded) throughout its length. Voltages measured between the transport line and the ground approach zero (usually only 0.05 v.), because the line is uniformly grounded throughout its length. Where nearby high-tension power lines cause stray currents, a maximum of 0.15 v. has been noted (with the heating circuit power either on or off). Current intensities from pipe to ground usually total no more than 0.1 amp., even in 100 kva. heating systems.

Heating or tracing tubes are usually ½-to-1½ in. standard steel pipe. The steel tube should be welded to the transport pipe at fixed intervals. If welding is impractical because of the material of the transport pipe, or because it has been installed in a hazardous area, heat conduction can be improved by a heat-transfer cement.

The tracing tube and transport pipe are completely connected electrically throughout their common length. Furthermore, the transport pipe is grounded every several hundred feet, and even in the case of breakdown of the insulation of the wire carrying the current through the heating pipe, danger from a short circuit to the tracing pipe is effectively prevented by the grounding.

Copper wire, insulated with polyvinyl chloride resistant to 220 F., may be used with 600 v. as the conductor inside the heating pipe for pipelines up to 3,000-4,000 ft. long carrying heavy fuel oil at 120-140 F. For longer lengths, copper wire insulated with heat-resistant polyethylene may be used with 1,500 v. or more. Silicon rubber, Teflon, asbestos, or similar standard insulations may be used for higher temperatures.

The heat output from one tracing tube may ordinarily be based on 10-50 w./ft., and voltage drop per foot as 0.1-0.2 v. These values are decided by considering the size of the tracing tube, wire size, heat resistant quality of insulation, fluid characteristics and temperature, voltage available, and other factors.

The power factor of the heating circuit is 90-95%.

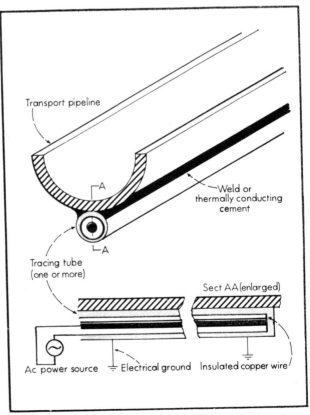

HEAT from tracing tube carrying electrical current along inner surface is conducted to transport pipe.

---

[*] Called SECT (for Skin Electric Current Tracing), the system was invented by M. Ando, and is covered by patents and patent applications throughout the world.

Comparative Costs of Heat Tracing System*

Installation Costs, $	Electric System	Steam Tracing
Pipeline	137,200	137,200
Thermal insulation	66,700	66,700
Steam header, insulation, etc.		79,450
Electrical system (including royalty)	47,400	
Overall field expenses	14,150	14,150
General expenses	47,050	52,500
Total	$312,500	$350,000

Annual Energy Costs

Heating time	1,860 hr./yr.	8,760 hr./yr.
Energy consumption	890,000 kwh./yr.	20,000 tons/yr.
Annual overall expense	$9,750	$28,330

* The cost comparison is based on a 14-in.-dia. pipe, 2.3 mi. long, transporting heavy fuel oil (Bunker C) at 140 F. for 4 hr./day and an ambient temperature of 32 F. Energy costs are based on those in Japan: 0.63¢/1,000 lb. steam and 1.11¢/kwh. electricity.

Advantages of Skin-Current Heating

In addition to convenience, flexibility and ease of control, skin-current heating also offers the following advantages:

(1) *Low-Cost Installation*—The arrangement is simple; aside from the insulated conductor wire inside the tube, the materials and methods of construction are ordinary. A conductor is necessary for only one side of the circuit, and it is protected by being inside the heating tube. Even with long-distance pipelines, there is no need for many feeder cables, and additional external wires of high voltage are usually not required. When the insulated inner wire is designed for 10,000 v., the length of one circuit may be over 30 mi. Additional high-tension lines will not be necessary because power will be available from the a.c. sources for pumping stations.

(2) *Uniform Heating*—Heat conduction between the heating tube and transport pipe is extremely good because of the close thermal contact. The temperature difference between tube and pipe, with as much as 30 w./ft. of heat flowing, is only about 5-9 F. When the transport pipe is of large diameter, heat is distributed circumferentially by a number of tubes. One tracing tube is usually required for a 12-in.-dia. transport pipe. So far, six has been the largest number of heating tubes required.

(3) *High Thermal Efficiency*—Because the insulated wire in the tracing tube feeds the electricity, the heat loss due to the resistance of the copper wire making up one side of the circuit is immediately and efficiently used for heating. In other systems, there is a loss of about 10-30% of the electric power carried in the two external feeder wires, depending on the distance from the generator.

(4) *Safety*—Because the pipeline is always entirely grounded, there can be no static electricity buildup caused by the flowing of the fluid, or by induction from nearby power transmission lines. Therefore, the maximum voltage difference between the pipe and the ground is only 0.1-0.2 v.; and the total current leaking to the ground from the pipeline is not over about 0.01-0.1 amps. Even if the grounding conductor is disconnected, there is no sparking, and no danger of flashing, even in the presence of an explosive gas.

(5) *Ease of Repair*—If the insulation, and hence the feeder wire in the tracing tube, should fail, there is no need to dismantle the covering insulation of large sections of the transport pipe to repair this electric wire, as with heating cables.

(6) *Versatility*—The optimum thickness of thermal insulation is determined as usual. However, there is absolutely no need to insulate electrically the outside of either the transport or the heating tube. The transport pipe may be exposed, may rest on normal steel pads or structures, and may even be buried in wet or swampy soil, or under water, because it is always grounded throughout its length.

(7) *Economics*—The table, which compares the cost of a skin-current tracing and steam tracing for a particular installation, shows that the total installation cost of a skin-current system (including royalty) is only about 60% of that for steam tracing, and the longer the pipeline, the greater the advantage.

The annual maintenance cost for steam tracing (which is not indicated) is estimated at 5-10% of the installation cost, but that of the electric system is almost nothing.

The table also shows that the cost of the energy for heating with the electric system is about one-third that of the steam tracing system for typical costs in Japan, where both electricity and fuel are relatively expensive compared to many parts of the U.S. To maintain the temperature of the pipeline at 140 F. (for transporting the oil only 4 hr./day), the steam must be supplied continuously throughout the year in Japan, even though this heats the oil to a much higher temperature than necessary, particularly that remaining in the pipelines when there is no flow. On the other hand, electric heating requires only about 1,900 hr./yr. of electrical energy. ■

Meet the Authors

Masao Ando is the Managing Director of Chisso Engineering Co. (No. 3-6, 1-Chome, Uchisaiwai-cho, Chiyoda-ku, Tokyo, Japan). He is a mechanical engineering graduate of the University of Tokyo.

Donald F. Othmer, Distinguished Professor of Chemical Engineering at Polytechnic Institute of Brooklyn (333 Jay St., Brooklyn, N.Y. 11201), was head of the department for many years. A co-editor of the well-known "Kirk-Othmer Encyclopedia of Chemical Technology," he received his Ph.D. degree in chemical engineering from the University of Michigan.

Section V
FIRED HEATERS

Fired Heaters—I Finding the basic design for your application
Fired heaters—II Construction materials, mechanicals features, performance monitoring
Fired heaters—III How to reduce your fuel bill operation
Fired heaters—IV How combustion conditions influence design and
Generalized method predicts fired-heater performance
Guide to economics of fired heater design

FIRED HEATERS—1

Finding the basic design for your application

Herbert L. Berman, Caltex Petroleum Corp.

☐ Energy conservation is not new. The economic success of any competitive process requires the efficient use of energy. Of the energy consumed in typical chemical processing or petroleum refining plants, approximately 75% is burned in the form of hydrocarbon fuel in fired heaters and steam boilers. Conservation, therefore, affords a strong and timely incentive for scrutinizing the design criteria and construction features commonly used in fired heaters.

What is a fired heater?

A fired heater, for our purposes, will include a number of devices in which heat liberated by the combustion of fuel within an internally insulated enclosure is transferred to fluid contained in tubular coils. Typically, the tubular heating elements are installed along the walls and roof of the combustion chamber, where heat transfer occurs primarily by radiation, and if economically justifiable, in a separate tube bank, where heat transfer is accomplished mainly by convection.

Industry identifies these heaters with such names as process heater, furnace, process furnace and direct-fired heater, all of which are interchangeable.

The fundamental function of a fired heater is to supply a specified quantity of heat at elevated temperature levels to the fluid being heated. It must be able to do so without localized overheating of the fluid or of the structural components.

Fired-heater size is defined in terms of its design heat-absorption capability, or duty. Duties range from about a half-million Btu/h for small, specialty units to about one billion Btu/h for superproject facilities such as the mammoth steam hydrocarbon-reformer heaters. By and large, the vast majority of fired-heater installations fall within the 10- to 350-million-Btu/h range.

Process industry requirements for fired heaters are divided, in the main, into a half-dozen general service categories. These categories can be designated and described as follows:

Column reboilers. This is normally considered one of the mildest and least critical of fired-heater applications. The charge stock taken from a distillation column is a recirculating liquid that is partially vaporized in the fired heater. The mixed vapor-liquid stream reenters the column, where the vapor condenses and releases the heat of vaporization. Reboiler applications are characterized by relatively small differentials between the inlet and outlet fluid temperatures across the fired heater, and by substantial vaporization (typically, 50% or more of the charge stock is vaporized). Depending on the particular application, reboiler heater outlet temperatures generally fall in the range of 400°F to 550°F.

Fractionating-column feed preheaters. Fired heaters in this service tend to be the workhorses of many process operations. The charge stock (usually all liquid, although some feeds may contain a nominal amount of vapor at the inlet) is sent to the fired heater following upstream preheating in unfired equipment. In the fired heater, the fluid temperature is usually raised high enough to achieve partial vaporization of the charge stock.

A typical example of this service is the feed heater for an atmospheric distillation column in the crude-oil unit of a petroleum refinery. Here, crude oil entering the fired heater as a 450°F liquid might exit near 700°F with about 60% of the charge stock vaporized.

Reactor-feed preheaters. Fired heaters in this application raise the charge-stock temperature to a level necessary for controlling a chemical reaction taking place in an adjoining reactor vessel. The nature of the charge stock and the heater operating temperatures and pressures can vary considerably, depending on the process. The following examples illustrate the diversity of the applications performed by reactor-feed preheaters:

■ Single-phase/single-component heating such as steam superheating in the reaction sections of styrene manufacturing processes. In this service, the fluid temperature across the fired heater increases from an inlet temperature of about 700°F to an exit temperature of approximately 1,500°F.

■ Single-phase/multicomponent heating, such as the heating of mixtures of vaporized hydrocarbons and recycle hydrogen gas prior to catalytic reforming in a refinery. In this service, the charge stock enters the fired heater at about 800°F and exits at approximately 1,000°F.

In reformers, the fluid pressure may range from about 250 to 600 psig. Severe restrictions on fluid pressure drop are normally associated with this service.

Originally published June 19, 1978

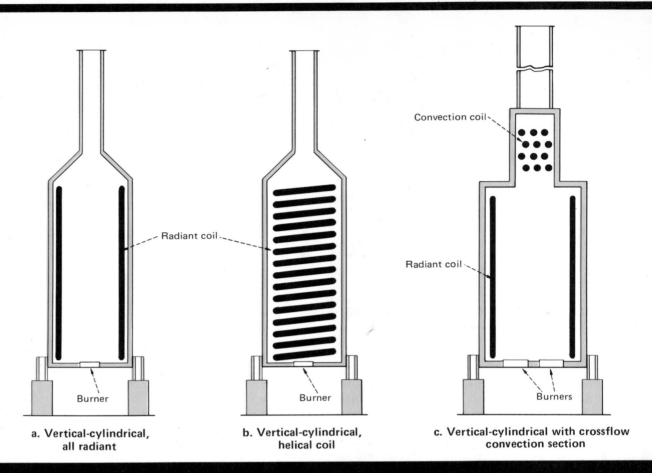

a. Vertical-cylindrical, all radiant

b. Vertical-cylindrical, helical coil

c. Vertical-cylindrical with crossflow convection section

Vertical-tube fired heaters can be identified by the vertical arrangement of radiant-section coil   Fig. 1

■ Mixed-phase/multicomponent heating, such as the heating of mixtures of liquid hydrocarbons and recycle hydrogen gas for reaction in a refinery hydrocracker. Fluid temperatures typically run from 700°F at the inlet to 850°F at the outlet. Operating pressures may reach 3000 psig, depending on the process.

Heat supplied to heat-transfer media. Many plants furnish heat to individual users via an intermediate heat-transfer medium. A fired heater is generally employed to elevate the temperature of the recirculating medium, which is typically a heating oil, Dowtherm, Therminol, molten salt, etc. Fluids flowing through the fired heater in these systems almost always remain in the liquid phase from inlet to outlet.

Heat supplied to viscous fluids. Oftentimes heavy oil must be pumped from one location to another for processing. At low temperatures, where the oil may have so high a viscosity as to render pumping infeasible, a fired heater is employed to warm the oil to a temperature that will facilitate pumping.

Fired reactors. In this category are heaters in which a chemical reaction occurs within the tube coil. As a class, these units represent the fired-heater industry's most sophisticated technology. The following two applications typify the majority of installations:

■ Steam hydrocarbon-reformer heaters, in which the tubes of the combustion chamber function individually as vertical reaction vessels filled with nickel-bearing catalyst. In reformers that yield hydrogen, fluid outlet temperatures range from 1,450 to 1,650°F.

■ Pyrolysis heaters, used to produce olefins from gaseous feedstocks such as ethane and propane and from liquid feedstocks such as naphtha and gasoil. In cracking heaters, where chemical reactions occur in the coil, the tubes and burners are arranged so as to assure pinpoint firing control. Fluid outlet temperatures in heaters designed for liquid feedstocks are in the 1,500°F to 1,650°F range.

Many variations in design and layout

There are many variations in the layout, design, and detailed construction of fired heaters. A consequence of this flexibility is that virtually every fired heater is custom-engineered for its particular application.

The simplest type of fired heater follows the so-called "all radiant" design, wherein the entire tube coil is arranged along the walls of the combustion chamber or radiant section. This design is characterized by low thermal efficiency and normally represents the lowest capital investment for a specified duty. The terminology "all radiant" is somewhat of a misnomer. Convection currents do exist, due to the flow of flue gases through the combustion chamber, and these currents

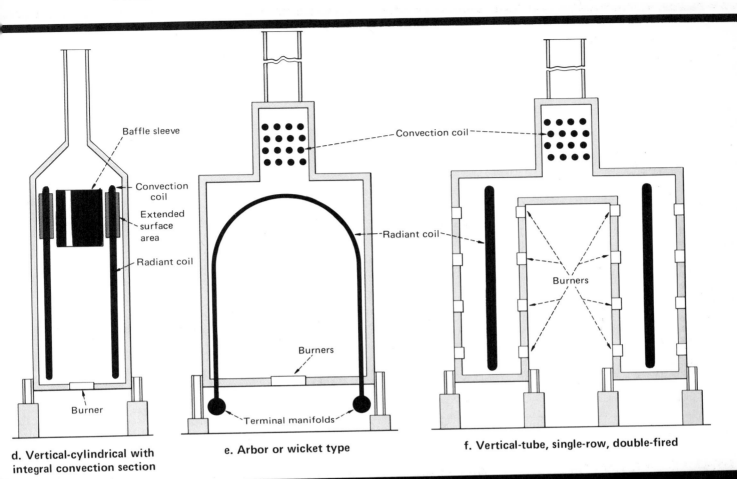

d. Vertical-cylindrical with integral convection section

e. Arbor or wicket type

f. Vertical-tube, single-row, double-fired

(continued) Fig. 1

account for a portion of the total heat absorbed in the radiant section.

In addition to the radiant section, most modern fired heaters include a separate convection section. The residual heat of the flue gases leaving the radiant section is recovered in this section, primarily by convection. Using this heat for preheating the charge stock, or for other supplementary heating services, increases the thermal efficiency of the fired heater.

The first few rows of tubes in the convection section are subject to radiant heat transfer, in addition to convective transfer from the hot flue gases as they flow across the tubes. Because these tube rows are usually being subjected to the highest heat-transfer rates in the fired heater, they are aptly termed "shield" or "shock" tubes.

Horizontal vs. vertical

The principal classification of fired heaters, however, relates to the orientation of the heating coil in the radiant section; i.e., whether the tubes are vertical or horizontal. Vertical arrangements are shown in Fig. 1; horizontal arrangements in Fig. 2. Salient features of each configuration are noted here:

Vertical-cylindrical, all radiant. Here the tube coil is placed vertically along the walls of the combustion chamber. Firing is also vertical, from the floor of the heater.

Heaters of this type represent a low-cost, low-efficiency design, which requires a minimum of plot area. Typical duties are 0.5 to 20 million Btu/h.

Vertical-cylindrical, helical coil. In these units, the coil is arranged helically along the walls of the combustion chamber, and firing is vertical from the floor. Although these heaters are grouped with others having vertical tube designs, their in-tube characteristics resemble those of horizontal-tube fired heaters.

This design also represents low cost, low efficiency, and requires a minimum of plot area. The tube coil is inherently drainable. One limitation on these units is that generally only one flow path is followed by the process fluid. Heating duties run from 0.5 to 20 million Btu/h.

Vertical-cylindrical, with crossflow convection. These heaters, also fired vertically from the floor, feature both radiant and convection sections. The radiant-section tube coil is disposed in a vertical arrangement along the walls of the combustion chamber. The convection section tube coil is arranged as a horizontal bank of tubes positioned above the combustion chamber.

This configuration provides an economical, high-efficiency design that requires a minimum of plot area. The majority of new, vertical-tube fired-heater installa-

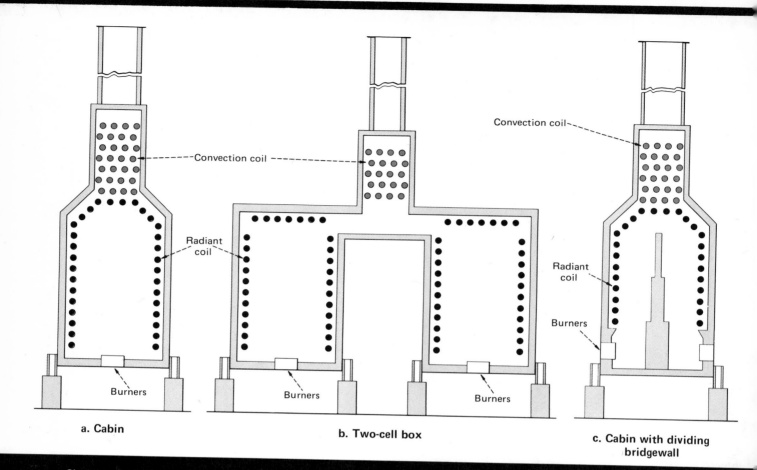

Six basic designs used in horizontal-tube fired heaters. Radiant-section coil is horizontal   Fig. 2

tions fall into this category. Typical duty range is 10 to 200 million Btu/h.

Vertical-cylindrical, with integral convection. Although this design is rarely chosen for new installations, the vast number of existing units of this type warrants its mention in any review of fired heaters.

As with the previous types, this design is likewise vertically fired from the floor, with its tube coil installed in a vertical arrangement along the walls. The distinguishing feature of this type is the use of added surface area on the upper reaches of each tube to promote convection heating. This surface area extends into the annular space formed between the convection coil and a central baffle sleeve. Medium efficiency can be achieved with a minimum of plot area. Typical duty for this design is 10 to 100 million Btu/h.

Arbor or wicket. This is a specialty design in which the radiant heating surface is provided by U-tubes connecting the inlet and outlet terminal manifolds. This type is especially suited for heating large flows of gas under conditions of low pressure drop. Typical applications are found in petroleum refining, where this design is often employed in the catalytic-reformer charge heater, and in various reheat services. Firing modes are usually vertical from the floor, or horizontal between the riser portions of the U-tubes.

This design type can be expanded to accommodate several arbor coils within one structure. Each coil can be separated by dividing walls so that individual firing control can be attained. In addition, a crossflow convection section is normally installed to provide supplementary heating capacity for chores such as steam generation. Typical duties for each arbor coil of this design are about 50 to 100 million Btu/h.

Vertical-tube, double fired. In these units, vertical radiant tubes are arranged in a single row in each combustion cell (there are often two cells) and are fired from both sides of the row. Such an arrangement yields a highly uniform distribution of heat-transfer rates (heat flux) about the tube circumference.

Another variation of these heaters uses multilevel side-wall firing, which gives maximum control of the heat-flux profile along the length of the tubes. Multilevel side-wall firing units are often employed in fired-reactor services and in critical reactor-feed heating services. In addition to the twin-cell furnaces already mentioned, single-cell models are available for smaller duties. As a group, these represent the most-expensive fired heater configuration. The typical duty range for each cell runs from about 20 to 125 million Btu/h.

Horizontal tube cabin. The radiant-section tube coils of these heaters are arranged horizontally so as to line the sidewalls of the combustion chamber and the sloping roof or "hip." The convection-section tube coil is posi-

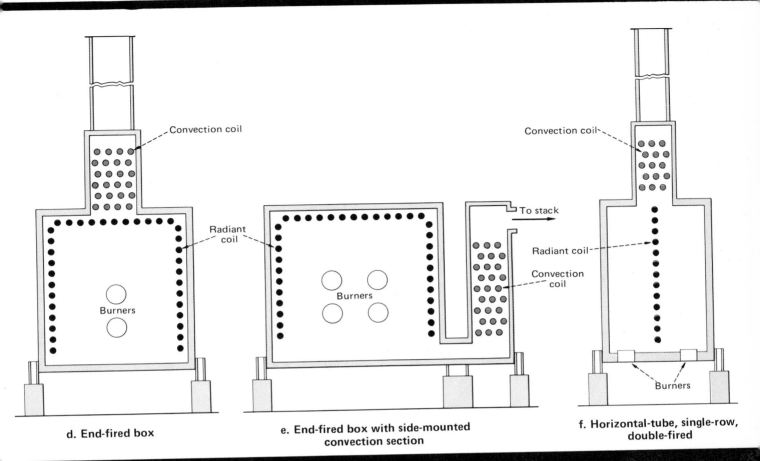

d. End-fired box

e. End-fired box with side-mounted convection section

f. Horizontal-tube, single-row, double-fired

(continued) Fig. 2

tioned as a horizontal bank of tubes above the combustion chamber. Normally the tubes are fired vertically from the floor, but they can also be horizontally fired by sidewall-mounted burners located below the tube coil. This economical, high-efficiency design currently represents the majority of new, horizontal-tube, fired-heater installations. Duties run 10 to 100 million Btu/h.

Two-cell horizontal tube box. Here the radiant-section tube coil is deployed in a horizontal arrangement along the sidewalls and roof of the two combustion chambers. The convection-section tube coil is arranged as a horizontal bank of tubes positioned between the combustion chambers. Vertically fired from the floor, this is again an economical, high-efficiency design. Typical duties range from 100 to 250 million Btu/h. For increased capacity, the basic concept can be expanded to include three or four radiant chambers.

Horizontal tube-cabin with dividing bridgewall. Again the radiant-section tube coil is arranged horizontally along the sidewalls of the combustion chamber, and along the hip. The convection-section tube coil takes the form of a horizontal bank of tubes positioned above the combustion chamber. A dividing bridgewall between the cells allows for individual firing control over each cell in the combustion chamber. Available options permit horizontal firing with sidewall-mounted burners (as shown), or vertical firing from the floor along both sides of the bridgewall. A typical duty range for this design is 20 to 100 million Btu/h.

End-fired horizontal tube box. The radiant-section tube coil is disposed in a horizontal arrangement along the sidewalls and roof of the combustion chamber. The convection-section tube coil is arranged as a horizontal bank of tubes positioned above the combustion chamber. These furnaces are horizontally fired by burners mounted in the end walls. Typical duty range for this design is 5 to 50 million Btu/h.

End-fired horizontal tube box, with side-mounted convection section. Here the radiant section tube coil is disposed in a horizontal arrangement along the sidewalls and roof of the combustion chamber. The convection-section coil is arranged as a horizontal bank of tubes positioned alongside the chamber. The unit is horizontally fired from burners mounted on the end wall.

These furnaces are found in many older installations, and occasionally in new facilities that burn particularly poor grades of fuel oil containing a high ash concentration. This relatively expensive design provides duties ranging from 50 to 200 million Btu/h.

Horizontal-tube, double-fired. Horizontal radiant tubes are arranged in a single row and are fired from both sides to achieve a uniform distribution of heat-transfer rates around the tube circumference. Such heaters are normally fired vertically from the floor. They are often

Evolution of the modern fired heater

The development of the design concepts used in the modern fired heater was spurred by numerous adverse operating experiences.

The pot still of the late nineteenth century was the first example of industrial direct-fired heating. A major drawback with the pot still was its unsuitability for heating viscous or flammable liquids. With such fluids, intense localized overheating, product degradation, and coke formation were frequently followed by disastrous failures and fires.

The horizontal shell still was the next development in direct-fired heating. The chief advantages of the shell still over the fired pot were improved equipment reliability and safety. However, heat transfer rates and thermal efficiency remained poor.

At the beginning of the twentieth century, industry moved to direct-fired heating of fluids contained in tubular elements. The occurrence of tube ruptures and fires remained commonplace, largely due to a lack of engineering knowhow in preventing the formation of undesirable coke and salt deposits, which resulted from overheating of the tubes.

The need for more-even heating led to the concept of separating the heat-absorbing tubes from the combustion zone by means of a partial wall, or "bridgewall." Consequently, these early tubular fired heaters were primarily convection types.

Despite this more-conservative approach, which removed the tubes from the flame zone, overheating, coke deposition, and tube failures persisted. Because combustion was carried out with about 40% excess air, flue-gas temperatures reached 3000°F as the gases left the combustion zone and entered the heat-absorbing section. Intense convection at this temperature level, coupled with a substantial radiative component, resulted in heat transfer to the tubes facing the combustion zone at better than 100,000 Btu/(h)(ft^2).

To reduce the heat-transfer rate in the combustion zone to tolerable levels, vast amounts of excess air (several hundred percent) were required. Not only was thermal efficiency low at these high excess-air rates, but also the quenching effect of large volumes of cold air resulted in poor combustion characteristics.

As the technology of radiant heat transfer has improved, fired-heater designs have reverted more and more to the inclusion of radiant-heat-absorbing tubes, resulting in the modern, primarily radiant, fired heater of today.

selected for critical reactor-feed heating services. For increased capacity, the concept can be expanded to provide for a dual combustion chamber. A typical duty range for each cell of this design is about 20 to 50 million Btu/h.

Air supply and flue-gas removal

Aside from the major classifications according to service and configuration, fired heaters can also be grouped according to their methods of combustion-air supply and flue-gas removal.

The capability for inducing the flow of combustion air into a fired heater exists when hot flue gas of relatively low density is confined in a structure and isolated from higher-density air at ambient temperature. The buoyancy of the hot flue gas contained in the fired heater creates "draft" (less than atmospheric pressure), which induces flow of air into the combustion chamber. Since this draft results from a natural stack effect, it is termed natural draft. The majority of fired-heater installations are such natural-draft types, where a stack effect introduces the combustion air and removes the flue gas.

Obstruction to the flow of flue gas through a fired heater can result in a condition of pressure greater than atmospheric (positive pressure) in the structure. It is the function of the stack in a natural-draft heater to generate draft sufficient to overcome such obstructions and to maintain a negative pressure throughout.

An induced-draft fired heater incorporates an induced-draft fan, in lieu of a stack, to maintain a negative pressure and to induce the flow of combustion air and the removal of flue gas.

A forced-draft fired heater is one wherein the combustion air is supplied under positive pressure by means of a forced-draft fan. It is to be noted that even with air supplied under positive pressure, the combustion chamber and all other parts of the fired heater are maintained under negative pressure, and the flue gas is removed by the stack effect.

A forced-draft/induced-draft heater uses a forced-draft fan to supply combustion air under positive pressure; an induced-draft fan maintains the combustion chamber and all other parts of the fired heater under negative pressure and removes the flue gas.

Most fired heaters equipped with air preheaters are of the forced-draft/induced-draft type.

The author

Herbert L. Berman is a staff engineer with Caltex Petroleum Corp., 380 Madison Avenue, New York, NY 10017, specializing in fired-heater equipment. He has over 25 years of engineering and management experience, primarily in the fired-heater industry. He received a B.Ch.E. degree from the Polytechnic Institute of Brooklyn and did graduate work at New York University. A member of the AIChE, ASME and AACE, he holds several patents and is a licensed professional engineer in New York.

FIRED HEATERS—II
Construction materials, mechanical features, performance monitoring

This second of four parts looks closely at some of the design options and mechanical features that are available in today's fired heaters. Items covered include refractories, tube coils, extended-surface devices, tube supports and burners.

Herbert L. Berman, Caltex Petroleum Corp.*

☐ A variety of process, structural and environmental factors influence the choice of materials and mechanical design features used in fired heaters. For example, high operating temperatures, or poor fuel quality (excessive trace metals and ash), may force the selection of costly, highly alloyed materials. Environmental considerations may necessitate extremely tall stack heights, and available plot-plan area may restrict dimensions.

An economic factor that has assumed greater importance in recent years is the shift toward shop assembly and fabrication in order to reduce field construction time and costs. A consequence of this shift is that provision for shipping clearance along the route from fabricator to jobsite has assumed the importance of a primary design criterion.

Fig. 1 shows a schematic for a typical fired heater, depicting the major structural components and their relationship to each other. These components will be taken up here.

Casing and structural framework

Typically, the outer wall or casing of the heater is fabricated from $3/16$-in. steel plate, reinforced against warping. However, for vertical cylindrical heaters (see Part I), in which the shell itself serves as a load-carrying structural member, the normal plate thickness can be as low as $1/4$ in. Floor plates, too, are normally designed for a thickness of $1/4$ in. Prevailing design guidelines call for sealing the heater casing plate by welding, in order to prevent air and water infiltration.

The heater's structural-steel framework provides load-carrying members, which permit lateral and vertical expansion of all parts of the fired heater. The framework also supports the tube coil, independently of the refractory. Common design practice calls for fireproofing the main structural columns to a specified height above grade, as well as the main floor beams.

When header boxes are provided to receive return-bend fittings, the minimum plate thickness is normally $3/16$ in., reinforced against warping.

Where appurtenances such as ladders and platforms are provided or anticipated, the structural design must be adjusted to carry such loads.

Refractories

The casing described above is lined internally with insulating materials. Aside from the basic function of preventing the steel structure from overheating, the insulation also serves to contain the firebox heat at high temperature by reradiating it to the tube coil. In addition, the internal insulation serves to minimize casing heat loss, and also functions as a barrier to prevent flue-gas particle migration to the steel casing. Such migration, in the case of sulfur-bearing fuels, may lead to acid corrosion of the steel plate.

In order to properly select and design a refractory lining for a fired heater, concern must be given to several important factors:

Extreme temperature. Exposure to temperatures beyond the design limitation of a refractory material can cause melting or fusion, and failure under load.

Thermal shock. Extreme or frequent temperature fluctuations can cause disintegration and spalling of refractory linings.

Mechanical stress. Abnormal vibration can contribute to the deterioration of some materials. Stresses due to the expansion and contraction of the structure can

---

*To meet the author, see Part I, which appears on p. 292. Part III appears on pp. 303-314.

Originally published in July 31, 1978

cause the loss of lining integrity unless proper allowance is made in the mechanical design.

Erosion. Extremely fine particles such as flyash or catalyst being carried at high velocity in a flue-gas stream can cause erosion of the refractory material.

Chemical attack. Some fuels contain impurities that can react with various refractory constituents, causing slagging and failure of the refractory lining. Alkalis and acids, depending on the temperature and dewpoint of the flue gases, can attack the components of a refractory lining, causing corrosion and deterioration.

Cost. The economic evaluation of refractory materials and construction types is complicated because materials having the best insulating properties often fall short on mechanical strength. Thus, most refractory choices represent a compromise between insulating value and mechanical serviceability.

Insulation

Insulating systems for modern fired heaters fall into three basic categories:

Insulating firebrick (IFB). This is a porous brick with good insulation characteristics, manufactured by firing mixtures of sawdust, coke and high-alumina fireclays. Design temperature ratings of IFB range from 1,600°F to 2,800°F.

Typical IFB walls are suspended from the heater casing, and supported by horizontal steel angles. With this arrangement, heavy loadings on the lower bricks of a wall are avoided. Several concepts are employed to retain the wall in place. One method anchors individual bricks, at frequent intervals, to the steel casing with steel hooks or rods. Another method utilizes long steel rods placed through holes in the bricks, with the rods then anchored at intervals to the steel casing. Fig. 2 illustrates a typical wall construction.

When IFB walls are designed for vertical cylindrical casings, the "keying" effect of the bricks on the curved wall normally holds the wall in place without the use of tieback hardware such as hooks or rods.

Improved insulating effectiveness of IFB walls can be achieved by incorporating a backup layer of mineral-wool block insulation. This "brick and block" construction serves in countless operating installations.

With the current emphasis on shop preassembly of heater sections in modular form, the problems associated with shipping brick-lined modules have discouraged their use in new installations. By far, the more popular setting employed in conventional heaters today is the castable refractory-wall construction. Nevertheless, brick and block settings remain the standard for high-temperature specialty units such as steam-hydrocarbon reformer heaters and pyrolysis heaters.

Castable refractory. Castable refractory used in heater settings is normally an insulating castable applied by pouring or gunning. For shop-assembled modules, gunning under controlled conditions has been shown to be a very economical method of application. However, pneumatic placement of castable is a skilled craft, and the techniques of the applicator can mean the difference between success or failure of the installation.

Lumnite-Haydite-Vermiculite insulating castable (1:2:4 mix by volume) is an inexpensive material having excellent insulation characteristics. Because of its low cost, LHV is used extensively in heater applications. In addition, it has a very low expansion coefficient and is therefore used on large wall areas, without expansion joints.

A maximum temperature limit of 1,800° to 1,900°F precludes the use of LHV on exposed walls (unprotected by tubes) in close proximity to the flame burst. For such applications, proprietary castable refractory mixes are available.

It is to be noted that as the service temperature and density of the castable increase, the insulating effectiveness decreases, resulting in a need for additional thickness to achieve the same cold-face temperature. In many cases, dual-layer constructions are employed in which the high-temperature, high-density material is exposed to the flame, and the lower-grade, better-insulating material is provided as a backup layer.

Thicknesses are typically 5 in. for convection-section walls and radiant section walls protected by tubes, and 6 to 8 in. for exposed radiant walls, arches and floors. It is now fairly common to provide a facing of first-quality firebrick over castable areas on the floor, particularly when the heater is designed for liquid fuels.

As important as the choice of insulating material is, the key to a properly applied castable refractory wall is the anchoring system. The most popular system employs "V" clips or modifications thereof, welded to the steel casing. These clips are normally $1/8$ to $3/16$ in. dia., with anchor heights usually not less than 70% of the castable thickness. The preferred anchor material is austenitic stainless steel. Typical anchor spacings are a maximum of twice the lining thickness but not exceeding 12 in. on a square pattern for walls, and 9 in. on a square pattern for arches.

Ceramic fiber. Ceramic-fiber construction is the most current development in the field of fired-heater insulation. These linings normally consist of a hot-face layer followed by one or more layers of backup material. For the hot-face layer, ceramic fiber or blanket with a minimum density of 8 lb/ft^3 is recommended, with a minimum thickness of 1 in. Backup layers should also have a minimum thickness of 1 in. and a minimum density of 4 lb/ft^3. Mineral-wool block insulation can also be employed as backup material, provided that the fuel does not contain more than 1% (wt.) sulfur if it is fuel oil, or $1\frac{1}{2}$% (vol.) H$_2$S if it is gas.

Advantages of ceramic-fiber installations derive primarily from their light weight, which permits a reduction of structural steel, and from their immediate availability for operation without special startup procedures such as curing, dryout or cold-weather precautions.

Because ceramic fiber is more porous than insulating castable, it is desirable to provide an internal protective coating on the casing plate to prevent corrosion.

Ceramic-fiber construction should not be employed in convection sections when soot blowing or steam lancing is contemplated.

The tube coil

The tube coil of a fired heater is the most important component of a heater installation. It is also a major cost contributor to the overall heater investment.

Normally, the tube coil consists of a number of tubes connected in series by 180-deg return bends. In the event of internal coke deposition in the tubes, all-welded coils can usually be cleaned by a steam-air decoking procedure, or by a relatively new process using an abrasive propelled by a high-velocity gas stream. In many older installations—as well as in current high-temperature designs, wherein heavy oils having the potential for substantial coke deposition are heated—plug-type headers are employed to connect the tubes. In this construction, the tubes are rolled into, or welded to, the headers. The tubes can be cleaned by reaming or turbining. Occasionally, all-welded coils are provided with plug-type headers at certain key locations, simply to permit internal tube inspection.

Tube design

Principal factors affecting the selection of tubing material for elevated temperatures are service life, environment, and cost.

Specified service life varies substantially for individual heating applications, and even for the same application within different operating companies. For example, one company might use Type 304 stainless steel as a coil material with an expected life of, say, 8 to 11 years for a particular service. Another company might select a chromium-molybdenum steel, expecting to make tube replacements after about five years and perhaps replace all the tubes within seven years.

In the case of the Type 304 steel, the initial investment is substantially greater, as is the cost of the replacement of any tubes lost through faulty operation. More-continuous operation, however, with less downtime for tube replacement, may make Type 304 a more-economical long-term choice.

The temperature and stresses to which the tubing is subjected are as crucial as the media to which it is being exposed. In fired heaters, the tube metal temperature is always higher than the bulk fluid temperature at a given location. In addition, the temperature differential between tube wall and bulk fluid may increase as coke or scale is deposited on the inner wall. Therefore, consideration must be given not only to the initial tube-wall temperature but also to the maximum metal temperature that may result at the end of a run.

Also important is whether the stress is constant or cyclic. Thermal stresses that are created when a unit is starting up or shutting down can result in failure of steel components. Under certain conditions, such stresses can be of greater magnitude than the steady-state operating stresses.

The medium to which the tubing is exposed affects the oxidation or corrosion behavior of the steel. If anticipated oxidation levels are severe, tube material selection must provide a high resistance to scaling. As for corrosion activity, the selection should reflect, where possible, data obtained from actual operating conditions at commercial units or pilot plants.

Probably the most important factor in the selection of tubing material is cost. A steel with excellent elevated-temperature properties will have limited application if its cost is prohibitive. One of the reasons for the

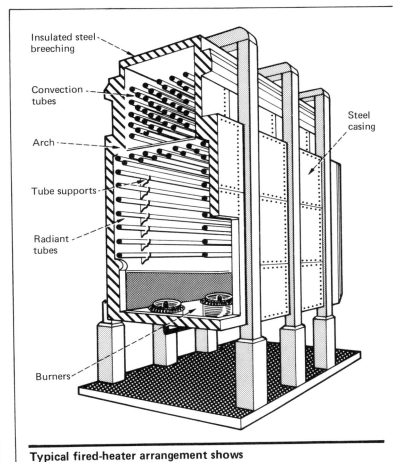

Typical fired-heater arrangement shows tube sections and structural components — Fig. 1

relatively wide selection of fired-heater tube-coil materials is that a particular material is often the most economical for a given application. Further, within an individual fired heater, it is not unusual to include two or more different tube materials in the most economical tube-coil design.

The use of a steel at high temperature levels will result in creep or permanent deformation even at stress levels well below the yield strength of the material. The tube will eventually fail by creep-rupture, even when a corrosion or oxidation mechanism is not active. For steels at lower temperature levels, creep effects are negligible, indicating that under such circumstances the

Sectionally supported construction avoids heavy loadings on the lower bricks in fired-heater wall — Fig. 2

a. Serrated fins

b. Solid fins

c. Studs

Three types of extended-surface devices used in the convection section   Fig. 3

tube will last indefinitely unless corrosion or oxidation effects manifest themselves.

There are, therefore, two different design considerations for heater tubes. At lower temperatures, in the "elastic range," the design stress is based on the yield strength. At higher temperatures, in the "creep-rupture range," the design stress is based on the rupture strength. For the temperature range where elastic and rupture stresses cross, the design tube-wall thickness requirement must satisfy both conditions.

Tubing materials

Carbon steel, the most widely used material for heater tubing, is suitable where corrosion or oxidation is relatively mild. The widespread usage of this material reflects its relatively low cost, generally good service performance and good weldability.

The alloy steels (Table I) used for elevated temperature service generally contain either molybdenum, chromium or silicon. The molybdenum is added principally to give higher strength; chromium is added to suppress graphitization and to yield improved oxidation resistance; and silicon is added to provide a further improvement in oxidation resistance.

The austenitic stainless steels are essentially alloys of iron, chromium and nickel and, as a group, are used for handling many corrosive materials or for resisting severe oxidation. Type 304, the most popular of the austenitic stainless steels, has excellent resistance to corrosion and oxidation and has high creep-strength. Types 321 and 347 are similar to Type 304 except that titanium and columbium, respectively, have been added. These additives combine with carbon and minimize intergranular corrosion that may occur in certain media after welding.

Type 316, which contains molybdenum, is used for high-strength service up to about 1,500°F and will resist oxidation up to about 1,650°F. For service above 1,600°F, Types 309 and 310, which contain about 25% chromium and 12 and 20% nickel, respectively, are used. These steels have excellent strength at these temperatures and, because of their chromium content, can be used in applications where extreme corrosion or oxidation is encountered.

Alloy 800 (20% chromium and 32% nickel) has excellent strength at up to 1,800°F and resists oxidation and carburization. It is used for tubes in pyrolysis heaters and steam superheaters, and for outlet pigtails and manifolds on steam hydrocarbon-reformer heaters.

Centrifugally cast materials, such as HK-40 (25% chromium and 20% nickel), are widely used for tubing in steam hydrocarbon-reformer heaters and in pyrolysis heaters.

Return bends. The least expensive methods of connecting the tubes is to join them by means of 180-deg return bends. The return bends are welded onto the tube ends in this arrangement, which typifies the majority of modern heaters. As noted already, internal cleaning of all-welded coils is performed by steam-air decoking procedures, or by recently developed high-velocity abrasive techniques. All-welded design permits return bends to be positioned either in the path of the flue gases, where they function as heat absorbing surfaces, or in header boxes external to the firebox and the flue-gas flow. Return bends can be of wrought or cast material.

Plug-type headers. Many types of plug-type headers using several closure designs have been developed. Compared to 180-deg return bends, these headers are somewhat more expensive and their use in new fired-heater equipment is relatively rare. As noted above, plug-type headers are used where mechanical cleaning of the tubes by turbining is anticipated and, occasionally, where tube internal inspection is planned. Plug-type headers cannot be placed in the firebox or in the path of the flue gases; they must be installed in header boxes external to the firebox and the flue-gas flow. Plug-type headers consist of cast material and, since they are external to the heat-transfer zone, can be designed for a lower temperature than the heat-absorbing tubes.

Extended surface improves convection

The surface area required in the convection section is controlled by film resistance on the flue-gas side. As a means of increasing the convection transfer rate per lineal foot of tubing, extended-surface devices have found almost universal acceptance. In today's designs, bare-tube convection sections are generally reserved for those relatively rare applications where the combustion of extremely poor grades of fuel oil presents the risk of heavy ash deposition on the convection-section tubes.

Popular devices employed in heater convection sections are reviewed here. Note that extended surface is never installed in the radiant section.

Serrated fins. This arrangement uses a V-notched or serrated fin that is helically wrapped around and continuously welded to the tube (Fig. 3a). Fins can be supplied in many combinations of thickness, height and density (number of fins per unit length of tube). Typically, thickness ranges from 0.035 to $3/16$ in., height from $1/4$ to $1 1/2$ in., and density from 2 to 7 fins per in.

Solid fins. This type is a noninterrupted fin that is helically wrapped around and continuously welded to the tube (Fig. 3b). These fins are available in the same ranges of thickness, height and density as serrated fins. Solid fins are mechanically stronger than serrated fins, but generally display a slightly lower heat-transfer rate for the same fin configuration and flue-gas mass flow.

Studs. Here, nominally cylindrical studs are flash-welded to the tube circumference (Fig. 3c). A stud diameter of $1/2$ in. is fairly standard for the industry, although $3/8$-in. studs are sometimes specified. Stud height ranges from $1/2$ to 2 in.

Cylindrical studs are the only extended surface that can be effectively employed on tubes positioned normal to, as well as parallel to, the flow of flue gas. Studs ordinarily cost more than finned tubing.

Dimensions of extended-surface devices used in gas-fired heaters ($1/2$-in. dia. studs or 0.05-in. thick [minimum] fins) are somewhat smaller than those used in oil-fired units ($1/2$-in. dia. studs or 0.10-in. thick [minimum] fins). Fin dimensions should preferably be limited to a maximum height of $3/4$ in. and a maximum density of 3 fins per in.

Table II presents calculated maximum tip temperatures for various extended-surface devices.

Choice of heater-tube materials is restricted by limiting design metal temperature — Table I

Material	Type or grade	Limiting design metal temperature, °F
Carbon steel	B	1,000
Carbon-½ Mo	T1 or P1	1,100
1¼ Cr-½ Mo	T11 or P11	1,100
2¼ Cr-1 Mo	T22 or P22	1,200
5 Cr-½ Mo	T5 or P5	1,200
7 Cr-½ Mo	T7 or P7	1,300
9 Cr-1 Mo	T9 or P9	1,300
18 Cr-8 Ni	304 or 304H	1,500
16 Cr-12 Ni-2 Mo	316 or 316H	1,500
18 Cr-10 Ni-Ti	321 or 321H	1,500
18 Cr-10 Ni-Cb	347 or 347H	1,500
Ni-Fe-Cr	Alloy 800H	1,800
25 Cr-20 Ni	HK-40	1,850

Maximum tip temperatures for some materials used in extended-surface devices — Table II

Extended-surface material	Tip temperature, °F
Fins	
Carbon steel	850
5 Cr	1,100
11-13 Cr	1,200
18 Cr-8 Ni	1,500
Studs	
Carbon steel	950
5 Cr	1,100
11-13 Cr	1,200
18 Cr-8 Ni	1,500

Limiting metal temperatures for some materials used to fabricate tube supports — Table III

Material	Type or grade	Limiting design metal temperature, °F
Carbon steel	A-283 Grc	800
5 Cr-½ Mo	GrC5	1,150
Alloy cast iron	A319 Class III Type C	1,200
18 Cr-8 Ni	Gr CF8	1,400
25 Cr-12 Ni	Type II	1,800
50 Cr-50 Ni		1,800
50 Cr-50 Ni-Cb	IN657	1,800
60 Cr-40 Ni		1,900
25 Cr-20 Ni	Gr HK40	2,000

Tube supports and guides

Proper mechanical design of a fired heater requires that the tube coil be adequately supported by tube hangers and tubesheets that are connected to the structural framework of the heater and not to the refractory.

Horizontal tubes. Horizontal coils having internal return bends are supported by intermediate tube hangers in the radiant section, and by tubesheets in the convection section. When the return bends (or plug-type headers) are located external to the heater environment in header boxes, end tubesheets and intermediate tube supports are provided. A common arrangement features header boxes having external return bends for the convection section, and internal return bends for the radiant section.

The design of intermediate tube supports used in a horizontal radiant section should ensure that the supports can be removed without tube removal and with a minimum of refractory replacement. It is also desirable for intermediate radiant tube supports to have restraints, or "keepers," to prevent the tubes from lifting off the supports during operation. Intermediate tubesheet castings in the convection section should be sectionalized to minimize the amount of tube removal needed when a casting is replaced. Typically, the maximum unsupported length of horizontal tubes should not exceed 35 times the tube outside diameter, or 20 feet, whichever is less.

Vertical tubes. Vertical coils can be supported either from the top or from the bottom. Top-supported tubes are provided with bottom guides, bottom-supported tubes with top guides. No intermediate supports are used. If necessary, intermediate guides are provided to restrain the vertical tubes from bowing inward toward the flame, or laterally toward adjacent tubes.

Materials used for tube supports

End tubesheets for tubes with external return bends or plug headers are normally of carbon-steel plate, $\frac{1}{2}$ in. thick. If the tubesheet temperature exceeds 800°F, alloy material should be used. End tubesheets are insulated on the hot side with castable refractory—generally 3 in. minimum in the convection section and 5 in. minimum in the radiant section.

In the radiant section, intermediate supports for horizontal tubes—as well as top supports for vertical tubes—are usually of cast 25% chromium-12% nickel, although many companies prefer the higher-grade 25% chromium-20% nickel. The same materials are used for intermediate guides and bottom guides of vertical tubes, although the use of 18% chromium-8% nickel for bottom guides is fairly common. Bottom supports for vertical tubes normally consist of alloy cast iron and are shielded from flame radiation by the floor refractory.

When the firing of fuel oil containing more than 100 ppm of vanadium is contemplated, radiant-section supports and guides should be made of higher alloys such as 50% chromium-50% nickel, 60% chromium-40% nickel, or IN-657, which is a proprietary 50% chromium-50% nickel alloy stabilized with columbium for additional high-temperature strength. These materials, however, are substantially more expensive than the conventional, lower-grade alloys.

Intermediate tubesheet castings exposed to hot gases in the convection section are normally fabricated from the same materials used for radiant-section supports exposed to flame radiation. In the colder flue-gases, alloy cast iron is commonly employed. Table III lists a variety of tube-support materials, and their usual temperature limits.

Burners

The fundamental criteria for selecting a burner include (1) the ability to handle fuels having a reasonable variation in calorific value, (2) provision for safe ignition and easy maintenance, (3) a reasonable turndown ratio between maximum and minimum firing rates, and (4) predictable flame patterns for all fuels and firing rates.

Gas-fired burners

Burners designed for gaseous fuel only are classified into two basic categories: premix inspirating and raw-gas burning.

Premix inspirating. The premix burner relies on the kinetic energy made available by the expansion of the fuel gas through an orifice to inspirate and mix combustion air prior to ignition at the burner tip. Approximately 50 to 60% of the combustion air is inspirated as primary air into the burner ahead of the ignition point.

Some of the advantages of this type of burner are:

1. Operating flexibility is good over a range of conditions. The amount of air inspirated varies with the fuel-gas pressure, and consequently requires only limited adjustment of secondary (noninspirated) combustion air. Premix burners can operate at low excess-air rates and are not significantly affected by changes in wind velocity and direction.

2. Flame length is short, and flame pattern sharply defined at high heat-release rates.

3. Burner orifices or spuds are fairly large, and, since they are located in a cold zone, are less subject to plugging than the smaller openings on noninspirating gas burners.

Some of the disadvantages of inspirating burners are:

1. Relatively high gas pressures must be available. Below a gas pressure of 10 psig at the burner, the percentage of inspirated air falls rapidly and flexibility is greatly reduced.

2. Flashback of the flame from the burner tip to the mixing orifice may occur at low gas pressures, or when the fraction of gases having high flame-propagation velocities, such as hydrogen, becomes too high.

3. The noise level of premix inspirating burners is higher than that of noninspirating types.

Raw-gas burning. The nozzle-mixing, raw-gas burner receives fuel gas from the gas manifold without any premixing of combustion air. The gas is then burned at a tip equipped with a series of small ports (Fig. 4a).

Some of the advantages of this type of burner are:

1. It has the greatest available turndown ratio for any given combustion condition.

2. It can operate at very low gas pressures on a wide variety of fuels and without flashback.

3. Noise level is reasonably low.

FIRED HEATERS—II CONSTRUCTION MATERIALS, MECHANICAL FEATURES

Some of the disadvantages of raw-gas burners are:

1. Flexibility is limited over its wide turndown range. Because no primary air is inspirated, combustion-air adjustments must be made over the full operating range of the burner.

2. The drilling of the burner ports is very sensitive, and any enlargement of the port opening will generally result in unsatisfactory flame conditions.

3. Flames tend to lengthen, and flame conditions become unsatisfactory as the burner is pushed beyond its design level.

4. The gas orifices or burner ports are exposed to the hot zone and are subject to plugging at low velocities and high temperatures.

Oil-fired burners

Special measures must be provided for burning fuel oil, since mixing of fuel and combustion air occurs in the gaseous phase. To accomplish this, all liquid-fuel burners use atomizing devices to break up the liquid mass into micron-size droplets. This increases the surface-to-mass ratio, thereby allowing extremely rapid heating and vaporization of the oil mass.

Oil burners in fired heaters almost always utilize steam as the atomizing medium. Such burners are designed with a double pipe in the feed tube to inject the steam and oil separately into a mixing chamber or atomizer immediately ahead of the burner tip. Steam pressure is slightly higher than oil pressure upon entering the atomizer, where the steam mixes with the oil due to the shearing action. Furthermore, oil in contact with hot water vapor tends to foam or emulsify, thus contributing to the atomization process. The steam and the finely dispersed oil then issue through a series of orifices into the turbulent air stream.

For proper combustion of oil, the following important requirements should be met:

1. The oil must be heated high enough that its viscosity is not greater than 150–200 Ssu.

2. Oil pressure must be held constant, typically at about 75 psig.

3. The steam delivered at the burner must be absolutely dry. If available, a moderate superheat of approximately 50°F is preferred. Typical steam pressure is about 100 psig.

Fig. 4b depicts a typical oil burner. To maintain flame stability during the firing of high-viscosity fuels, natural-draft oil burners should be of the double-block design, as illustrated. Atomizing steam needs fall in the range of about 0.15 to 0.35 lb steam per lb of oil.

When volatile fuels such as naphtha or gasolines are burned, a safety interlock should be provided on each burner. The interlock sequentially shuts off the fuel, purges the oil gun with steam, and shuts off the steam purge before the oil gun can be removed.

For those rare instances when oil must be burned and steam is not available, air atomization or mechanical atomization can be employed. The operating requirements of air-atomized oil burners are similar to those of steam-atomized ones, although a slightly higher oil temperature may be needed, to compensate for the cooling effect of the atomizing air. In addition, air (supplied by either a blower or a compressor) must be

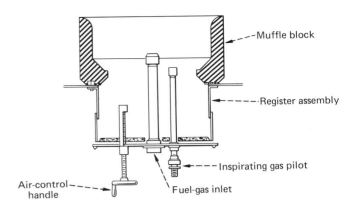

a. Raw-gas burner

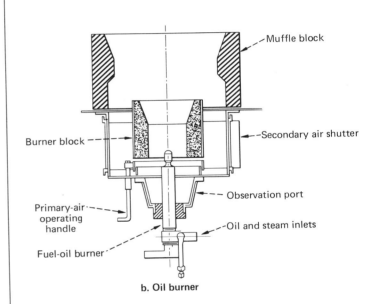

b. Oil burner

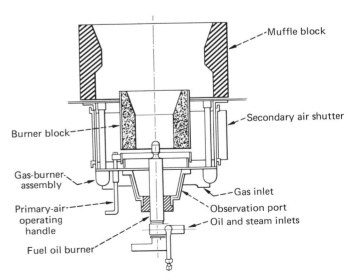

c. Combination oil and gas burner

Burner designs must offer safe operation and easy maintenance, and be able to handle variations in fuel quality Fig. 4

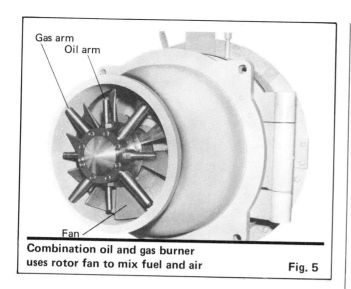

Combination oil and gas burner uses rotor fan to mix fuel and air Fig. 5

automatically modulated to adjust for changes in the firing rate. Uniformity of air supply is essential.

Mechanically atomized units take advantage of the oil's kinetic energy to atomize the fuel stream in the tip itself. Oil-temperature and viscosity ranges are similar to those needed for steam atomization. Oil must be available at a supply pressure in the area of 300 psig. When wide turndown ratios are required, oil pressures may go as high as 1000 psig.

Combination oil and gas burners

Combination burners (Fig. 4c) are designed to burn all-oil, all-gas, or any combination of oil and gas simultaneously. Typically, these burners feature a double-block design in which a single oil gun is arranged in the center of an array of gas nozzles.

Gas firing in the newer combination burners is almost always accomplished with raw-gas nozzles. Separate register adjustments allow independent control of primary air for oil combustion, and secondary air for fuel-gas combustion.

Radiant wall burners

The use of radiant-wall burners is generally confined to high-temperature specialty units such as steam hydrocarbon-reformer heaters and pyrolysis heaters. Because the capacity of an individual burner is generally low, these heaters require many more burners than a conventional fired heater having the same total heat release. However, this feature affords the advantages of improved heat distribution and pinpoint firing control. Radiant-wall burners generally burn only fuel gas, and are available as either inspirating or as raw-gas types.

The former type is designed to inspire 100% of the combustion air requirement. The flame from inspirating burners is flat and radial, rather than projected into the firebox. Impingement of the flame on a refractory burner block heats the block to incandescence, thus achieving the radiant-wall effect.

The raw-gas design is linear in shape, with the burners installed along two or three firing levels on the heater wall. The flames are thrown upward along the vertical or inclined firebox wall to create a radiant-wall effect. This burner type has been adapted to burn light oils in a radiant-wall mode of firing.

Rotor-fan burners

These burners (Fig. 5) exploit the kinetic energy of fuel under pressure to generate, by the reaction principle, mechanical work sufficient to rotate a built-in fan. The design of the fan assures proper proportioning of air in relation to fuel flow over a wide operating range.

Fuel-air mixing is produced by the discharge of fuel from the fan's rotating driver arms—through multiple orifices—into the air stream created by the fan's action. Thorough and instantaneous mixing of fuel and air is achieved. Relatively low amounts of excess air are needed to obtain complete, short-flame combustion.

The rotor-fan burner is used primarily to debottleneck existing heaters by reducing flame impingement in overloaded fireboxes, or by increasing heat input into draft-limiting heaters.

High-intensity burners

Several high-intensity burners are available for fired-heater applications. In general, they feature a large, cylindrical-shaped, refractory-lined combustion chamber. Combustion is fully established in this chamber, but not completed. By means of the circulation patterns developed in the chamber, flames of controlled shape and size can be produced at relatively low excess air. High-intensity combustion expells the flue gases at high velocity and temperature, producing very uniform firebox-temperature profiles.

High-intensity burners require a forced-draft air supply at relatively high pressures. Depending on the specific design, air pressure ranges from 6 to 20 in. H_2O.

Pilot burners

These units are supplied as an integral part of the main burner. They are most often installed where it is desired to simplify burner lightoff procedures, particularly with oil firing; where an extreme turndown to a fixed, minimum load is required; where intermittent on-off operation is required; and where extreme modulation of firing rate is needed.

The primary disadvantage of pilot burners is that they constitute a potential source of gas leakage into the firebox. The possibility always exists of a pilot being accidentally extinguished, permitting gas to be admitted to the heater during a shutdown. Also, because of their small port drillings, pilot burners clog easily and should be routinely inspected and cleaned.

Pilot burners are almost always gas-fired and are usually fueled from an outside source such as a propane or LPG drum which is not part of the process system. If the pilots are fueled from the main burner supply line, the gas offtake to the pilots must be upstream of the control and block valves.

Burner noise—what to do about it

Burner noise results from the flow of fuel and air through the burner and from the combustion process itself. Although at present little can be done to reduce combustion noise, it is possible to minimize propagation of sound to the outside of the heater.

Noise emitted by premix-inspirating gas burners originates primarily at the premix-venturi inspirator, and secondarily at the outlet-burner orifices. A method of reducing the inspirator noise is to replace the single-orifice inspirator with a multi-orifice inspirator. Further noise suppression can be achieved by fitting a silencer, or mute, on the primary air intake.

In raw-gas burners, the noise is emitted by the fuel gas as it passes through the burner nozzles. Noise from liquid-fuel burners is generated in the same way as in raw-gas burners, except that the noise is lower in frequency. Combustion noise from these burners can be effectively suppressed by acoustically absorbent air-intake plenums. These guidelines for plenum-chamber design will ensure reasonable noise reduction:

1. As much of the plenum interior as possible, including the heater wall, should be lined with acoustic material.
2. Absorbent surfaces should be arranged so that prior to escaping, sound waves from the burner air intake will undergo several reflections. For greatest effectiveness, the acoustic material should have a density of about 6 lb/ft^3 and its thickness should exceed 4 in.
3. Steel plate with a minimum thickness of $\frac{1}{8}$ in. should be used for plenum fabrication. Plenums should enclose the burner registers and be undercoated with sound-deadening material.
4. Provision should be made for inspection and for draining oil leaks. Acoustic lining should be omitted where drips collect.
5. All of the combustion air should be taken from the plenum chamber. There should be no line of sight from the burner or the plenum interior to the outside. Hand holes, and openings for external air register control, oil gun, etc., should be well sealed.

Dampers

The function of the stack damper is to control the heater's draft by maintaining a negative pressure of approximately 0.05 in. H$_2$O in the region directly below the convection section. The damper adjusts as necessary to maintain this negative pressure.

Stack dampers of single-leaf construction are used for small-diameter stacks, and multi-leaf construction for large-diameter stacks. In rectangular ductwork, which carries the flue gases between a heater and a separate stack, louvre-type dampers are normally used.

Stack dampers are usually manually operated from grade by means of cables. In the case of large dampers, it is now fairly common to maneuver them with pneumatic operators.

Table IV gives maximum flue-gas temperatures for various damper materials.

Cleaning the convection section

In order to maintain maximum thermal efficiency in a fired heater, it is necessary to keep convection heat-transfer surfaces clean. Although the extended surface of the convection section improves heat transfer, its physical arrangement renders it susceptible to the accumulation of fuel-ash deposits when oil fuels are fired. The major fouling constituents of an oil fuel are sulfur,

Flue-gas temperature restricts use of damper materials	Table IV
Material	Gas temperature, °F
Carbon steel	900
Cast iron	900
Alloy cast iron	1,000
11-13 Cr	1,200
18 Cr-8 Ni TP304	1,500
25 Cr-12 Ni TP309	1,800
25 Cr-20 Ni TP310	1,800

vanadium, sodium and ash. The latter is a most important component, since a high ash content will significantly increase the deposition rate.

The viscosity range of the fuel also influences ash deposition. The larger ash particles that accompany the burning of high-viscosity fuels carry over into the convection section to increase the fouling rate. Current methods used for onstream cleaning include manual lancing, soot blowing and water washing.

Manual lancing. This method requires a minimum of capital investment. Lance doors provided on the convection section sidewall permit workers to insert a steam or air lance and blow deposits off convection surfaces in the vicinity.

The effectiveness of lancing is poor by comparison with that of soot blowing. In addition, lancing requires the participation of more than one operator to transport the equipment from grade to the convection section, which is normally located at a relatively high elevation. Consequently, the success of a lancing program is dependent on operator initiative.

At the present time, manual lancing facilities are generally confined to retrofit installations on existing heaters—especially where a soot-blower retrofit would involve substantial modifications to convection coils.

Some operators have had reasonably good success in removing convection-section deposits by sandblasting using a nozzle inserted manually through the lance doors. However, the erosive potential of sandblasting on the convection-section refractory and on the heating surfaces must not be overlooked.

Rotating-element soot blowers. Also known as a fixed rotary type soot blower, this device is actually a multi-nozzle steam-lance tube installed in the convection section. A mechanical valve and drive assembly external to the convection section rotates the element, and automatically opens and closes the steam-supply valve.

The effectiveness of this type of soot blower is limited by the physical strength of the ash. Some combinations of ash chemistry and flue-gas temperature produce deposits that cannot be removed with such soot blowers. Further, highly corrosive fuels will materially reduce the life of the element or the lance tube.

Retractable-lance soot blowers. This apparatus differs from the rotating-element type in that the lance tube remains retracted from the convection section when not in use. During soot-blower operation, the lance is extended into the heater by means of an external drive.

The lance is equipped with two cleaning nozzles, of larger diameter than those installed along the length of the multi-nozzle rotating-element design. The retractable-lance blower cleans more effectively than does the fixed-rotary type, because its two nozzles permit a greater concentration of the cleaning-fluid stream. In addition, the retractable lance is exposed to high-temperature flue gases only during the cleaning cycle.

Notes on soot blowers

The effective cleaning range of a soot blower is dependent on its orientation, the flue-gas temperature, the arrangement of heating surface, and the blowing pressure. As a general guide, the vertical blowing capability of fixed-rotary soot blowers will not exceed 3 rows up and 3 rows down, with a maximum horizontal blowing radius of 3.5 feet. Corresponding limits for retractable lances are 4 rows up and 4 rows down, with a maximum horizontal blowing radius of 4 feet.

The rate of steam consumption of a retractable-lance soot blower will vary from 8,000 to 12,000 lb/h, depending upon the pressure at the nozzle. The cycle time required to operate a retractable lance is approximately $2\frac{1}{2}$ min. The fixed-rotary soot blower consumes steam at a rate of about 10,000 to 14,000 lb/h with a blowing time of approximately 40 s. Soot blowers are normally operated once per 8-h shift.

Steam for soot blowing is normally taken from supply lines at the relatively low pressures of 150 to 200 psig. Below 150 psig, soot blowing is virtually ineffective. More effective cleaning can be achieved with pressures higher than 200 psig. When available, pressures of 400 to 600 psig should be considered.

A completely automatic sequencing control system is recommended for multi-blower installations. Upon activation by one pushbutton, each of the soot blowers will operate in a predetermined sequence until the entire blowing cycle is completed.

Water washing

Onstream water washing of convection sections represents a very effective method of removing deposits from the tubes. Permanently installed networks of high-alloy headers and nozzles are positioned at selected elevations in the convection bank. Copious quantities of condensate or fresh water flow across the tube surfaces. Contact of the water with the hot tubes instantly vaporizes the water, thereby dislodging soot deposits.

Frequency of water washing depends on the nature and extent of the deposits. Typically, water washing is peformed about once per week. Understandably, care must be exercised to avoid getting water on the refractory or on intermediate tube-support castings. Water washing of stainless-steel tube coils is not recommended in view of the danger of chloride stress corrosion.

Performance monitoring

Getting the optimum performance from a heater requires the close monitoring of key variables on both the process and combustion sides. Data can be taken that will reveal excess air, thermal efficiency, and heat absorption. In addition, certain data provide an indication as to how well the heater is being fired.

Process stream flowrate. In most applications, where provision is made for individual-pass flow control, the flowrate to each parallel pass should be monitored. This is recommended for multi-pass heaters processing liquid hydrocarbons, where low flowrate in an individual pass can lead to excessive vaporization, increased pressure drop and further flow reduction—culminating in overheating and possible rupture of the tube.

Fuel firing rate. The rate of fuel input is normally controlled by the process-fluid outlet temperature. Measuring the fuel firing rate permits direct determination of firebox heat release from the fuel's heating value.

Process-stream temperatures. If individual-pass flow control is planned, it is recommended that indicators be installed to show the fluid outlet temperature after each parallel pass. These temperatures can be used as a guide for adjusting the flowrates of each pass, as well as for determining the process-side heat absorption. Measuring the outlet fluid temperatures after each parallel pass in the radiant and convection sections enables one to determine the duty split on the process side between the radiant and convection sections.

Flue-gas temperatures. Measurement of the flue-gas temperature leaving the radiant section serves as an index to the firing balance in the firebox, and also as an indicator of overfiring conditions. This temperature measurement should be made at 50-ft intervals along the length of the firebox. Such measurements are often useful for establishing the maximum firing rate.

Flue-gas temperatures should also be monitored at the inlet to each convection coil, and at the outlet from the convection section. These temperatures provide an indication of heater efficiency and of convection tube fouling.

Flue-gas draft profile. Draft measurements should be taken at the firebox near the burner level, at the inlet to the convection section, at the outlet from the convection section, and at a point downstream of the stack damper. Draft readings provide information on the pressure drop of combustion air and flue gas. The information is helpful for adjusting the burner registers and the stack-damper. Draft readings also indicate how close the heater is to its limiting operating conditions.

Flue-gas sampling. Provisions for flue-gas sampling are recommended at the exit of the radiant section and at the outlet of the convection section. Sampling to measure the oxygen in the flue gas at the first site provides an indication of the operator's firing technique. Measurement of combustibles is also recommended for this location.

A determination of the oxygen content in the flue gas leaving the convection section is also needed for calculating the heater's combustion efficiency. If the oxygen content of the gases leaving the radiant section is known, the extent of air leakage into the convection section can be estimated.

Tubeskin temperatures. Tubeskin thermocouples are recommended, as a minimum, for the outlet tube of each pass and for one shock tube of each pass. Skin temperatures help set maximum firing rates. They also serve as indicators of local overheating.

FIRED HEATERS—III
How combustion conditions influence design and operation

This third of four articles highlights some of the major process considerations that affect the design of fired heaters. Basic combustion principles are reviewed, as well as methods for determining coil arrangements, surface area, fluid pressure drops, and stack height.

Herbert L. Berman, Caltex Petroleum Corp. *

☐ The process design for a fired heater must reconcile complex relationships between a number of variables. Among these are the physical properties and phase behavior of the process fluid, the heating value and combustion behavior of the fuel(s), the economic proportioning of heating duty between the convection and radiant sections, the pressure drops of the process fluid and the flue gas, and the economic stack height. The relative importance of these variables will be examined here.

Combustion basics

For those fuels containing hydrogen, two sets of heating values are reported. The gross or higher heating value, HHV, is determined by assuming that all of the water vapor produced in the combustion process is condensed and cooled to 60°F. The net or lower heating value, LHV, assumes that the water vapor formed by combustion remains in the vapor phase and is numerically equal to the gross heating value less the latent heat of vaporization of the water.

In concept, the gross heating value can be considered as the *actual* heat release, whereas the net heating value can be considered as the *useful* portion of the actual heat release. For those fuels that do not contain hydrogen, e.g., CO, only one heating value is reported.

Unlike the boiler industry, which works on the basis of the gross heating value, the fired-heater industry almost always uses a net value. In this discussion, any reference to heat of combustion, heat input, or thermal efficiency will imply a net basis.

*To meet the author, see Part I, which appears on p. 292. Part II appears on pp. 293-302. Part IV, which concludes the series, appears on pp. 315-319.

Originally published August 14, 1978

Although the chemistry of the combustion process is exceedingly complex, it can be readily simplified in terms of final reaction products. As with all chemical reactions, in order for the combustion process to go to completion within a reasonable time, an excess of one of the reactants must be present. Under no circumstances can an excess of fuel be tolerated. Combustion in a fired heater normally uses air as the source of oxygen. Consequently, there must be an excess of air above stoichiometric requirements to ensure complete combustion of the fuel.

Reasonable excess-air requirements for natural-draft fired heaters are 20% for gas firing and 25% for oil firing. For forced-draft heaters, in which a greater degree of air control is possible, reasonable design excess-air values are 15% and 20% for gas and oil firing, respectively.

Fuel gas

Table I presents basic combustion constants for components of most industrial gaseous fuels. The following example illustrates the determination of heating value and stoichiometric air requirement for a multicomponent gaseous fuel. (Basis: 100 lb-moles of fuel.)

Component	Molecular weight	Mole %	Lb
CH_4	16.042	86.0	1,379.6
C_2H_6	30.068	8.6	258.6
C_3H_8	44.094	1.3	57.3
H_2	2.016	1.5	3.0
CO_2	44.010	2.6	114.4
			1,812.9

Properties used to predict heating values and air requirements for gaseous fuels — Table I

Gas	Formula	Molecular weight	Heat of combustion, Btu per lb		Combustion air, Lb per lb of combustible
			Gross	Net	
Carbon monoxide	CO	28.01	4,347	4,347	2.462
Hydrogen	H_2	2.016	61,095	51,623	34.267
Methane	CH_4	16.042	23,875	21,495	17.195
Ethane	C_2H_6	30.068	22,323	20,418	15.899
Propane	C_3H_8	44.094	21,669	19,937	15.246
n-Butane	C_4H_{10}	58.12	21,321	19,678	14.984
n-Pentane	C_5H_{12}	72.146	21,095	19,507	15.323
n-Hexane	C_6H_{14}	86.172	20,966	19,415	15.238
Ethylene	C_2H_4	28.052	21,636	20,275	14.807
Propylene	C_3H_6	42.078	21,048	19,687	14.807
Butylene	C_4H_8	56.104	20,854	19,493	14.807
Benzene	C_6H_6	78.108	18,184	17,451	13.297
Toluene	C_7H_8	92.134	18,501	17,672	13.503
p-Xylene	C_8H_{10}	106.16	18,633	17,734	13.663
Acetylene	C_2H_2	26.036	21,502	20,769	13.297
Naphthalene	$C_{10}H_8$	128.164	17,303	16,708	12.932
Ammonia	NH_3	17.032	9,667	7,985	5.998
Hydrogen sulfide	H_2S	34.076	7,097	6,537	6.005

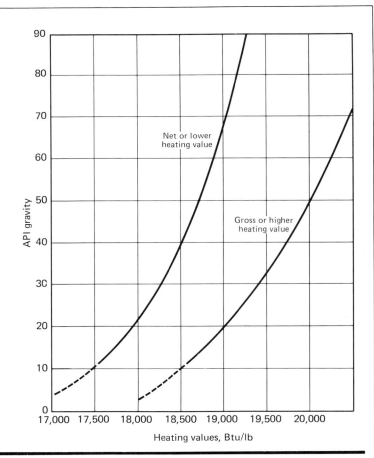

Heating values of liquid fuels have been correlated as a function of API gravity — Fig. 1

Component (contd.)	LHV Btu/lb	Btu	Lb air per lb fuel	Lb air
CH_4	21,500	29,654,500	17.195	23,720
C_2H_6	20,420	5,280,100	15.899	4,110
C_3H_8	19,940	1,142,400	15.246	874
H_2	51,620	154,900	34.267	103
CO_2	—	—	—	—
		36,231,900		28,807

Average molecular weight $= \dfrac{1,812.9}{100} = 18.13$

Net heating value $= \dfrac{36,231,900}{1,812.9} = 19,990$ Btu/lb

Air requirement $= \dfrac{28,807}{1,812.9} = 15.9$ lb air/lb fuel

Fuel oil

Unlike gaseous fuels, the heating value of a fuel oil is normally not developed from a component analysis. Instead, the heating value of a liquid fuel can be expressed as a function of the specific gravity of the oil, with an accuracy sufficient for most engineering computations. Fig. 1 depicts the relationship between the gross and net heating values of fuel oil as a function of API gravity. The API gravity, in turn, is related to the specific gravity, as follows:

$$°\text{API} = \frac{141.5}{\text{Sp. gr.}} - 131.5 \qquad (1)$$

If a fuel oil analysis is available, the stoichiometric air requirement can be approximated from the weight

FIRED HEATERS—III COMBUSTION CONDITIONS INFLUENCE DESIGN & OPERATION

Nomenclature

A_c	Convection surface area, ft²
A_f	Fin surface of extended-surface tube, ft²/ft
A_i	Internal tube surface, ft²/ft
A_o	External tube surface, ft²/ft
A_R	Radiant surface area, ft²
A_t	Total surface of extended-surface tube, ft²/ft
c_p	Specific heat, Btu/lb/°F
d_i	Tube inside dia., in.
d_o	Tube outside dia., in.
D	Tube outside dia., ft
D'	Stack dia., ft
e	Thermal efficiency
E	Fin efficiency
f	Fanning friction factor
F_i	In-tube fouling resistance, [Btu/h/(ft²)(°F)]⁻¹
F_o	External fouling resistance, [Btu/h/(ft²)(°F)]⁻¹
g	Flue-gas mass velocity, lb/s/ft²
g'	Fluid mass velocity, lb/s/ft²
G	Flue-gas mass velocity, lb/h/ft²
h_c	Convection film coefficient, Btu/h/(ft²)(°F)
h_i	In-tube fluid film coefficient, Btu/h/(ft²)(°F)
h_o	Total convection heat-transfer coefficient, Btu/h/(ft²)(°F)
h_{rg}	Gas radiation coefficient, Btu/h/(ft²)(°F)
h_w	Tube-wall coefficient, Btu/h/(ft²)(°F)
H_A	Heat available in flue gas, Btu/lb of fuel
H_F	Net heating value of fuel, Btu/lb
k	Thermal conductivity, Btu/h/(ft²)(°F)/ft
K_c	Thermal conductivity of coke/scale layer, Btu/h/(ft²)(°F)/in.
K_m	Thermal conductivity of tube wall, Btu/h/(ft²)(°F)/in.
L_e	Hydraulic length, ft
LMTD	Log-mean temperature difference, °F
L_s	Stack height, ft
Δp	Pressure drop, psi
p'	Ambient pressure, psia
P	Design pressure, psig
q	Transfer rate, Btu/h/ft² of outside tube surface
q_i	Transfer rate, Btu/h/ft² of inside tube surface
Q	Heat absorbed, Btu/h
Q_c	Convection-section heat absorption, Btu/h
Q_F	Heat fired, Btu/h
Q_R	Radiant-section heat absorption, Btu/h
R_i	In-tube film resistance [Btu/h/(ft²)(°F)]⁻¹
R_o	External film resistance [Btu/h/(ft²)(°F)]⁻¹
R_w	Tube-wall resistance [Btu/h/(ft²)(°F)]⁻¹
R_t	Total resistance [Btu/h/(ft²)(°F)]⁻¹
S	Design stress, psi
t_c	Thickness of coke/scale layer, in.
t_m	Thickness of tube wall, in.
T_a	Ambient temperature, °R
T_B	Bulk fluid temperature, °F
T_g	Flue-gas temperature, °F
T_{ga}	Flue-gas temperature, °R
T_m	Tube metal temperature, °F
u	Bulk viscosity, lb/ft/h
u_w	Viscosity at wall temperature, lb/ft/h
U	Overall coefficient, Btu/(h)(ft²)/°F
V	Specific volume, ft³/lb
W	Flowrate, lb/h

percentages of carbon, hydrogen, oxygen and sulfur, according to the following expression:

lb air/lb fuel = (0.1159)%C + (0.3475)%H + (0.0435)%S − (0.0435)%O

For a fuel oil consisting by weight of carbon 84.6%, hydrogen 10.9%, sulfur 1.6%, and oxygen 2.9%, the stoichiometric air requirement is:

(0.1159)(84.6) + (0.3475)(10.9) + (0.0435)(1.6) − (0.0435)(2.9) = 13.54 lb air/lb fuel

Combustion products

For the two types of fuel, the determination of flue-gas quantity per unit quantity of fuel is performed simply by assuming 20% excess air for the fuel gas and 25% excess air for the fuel oil.

For the gas:
  Total air = (1.20)(15.9) = 19.07 lb air/lb fuel
  Flue gas = 19.07 + 1 = 20.07 lb flue gas/lb fuel

For the oil:
  Total air = (1.25)(13.54) = 16.93 lb air/lb fuel
  Flue gas = 16.93 + 1 = 17.93 lb flue gas/lb fuel

A generalized relationship that gives the weight ratio of flue gas to fuel is presented in Fig. 2 for several representative fuel gases and fuel oils.

Heating-coil arrangement

The normal flow of process fluid through a fired heater starts at the inlet to the convection section. The stream moves through the convection section counter-current to the flow of flue gases, then into the shock bank. After leaving the shock bank, it passes into the radiant section, where the major portion of the heat is absorbed.

As described in Part II, adjacent tubes are connected by means of 180-deg return bends or plug-type headers. Each bank of consecutive tubes in which the fluid travels from entry until exit is known as a "pass," or parallel stream. (This definition differs from that applied to unfired heat-transfer equipment.) In a two-pass heater, the fluid is distributed into two streams at the inlet, with each stream flowing separately through its respective tube coil and then recombining after exiting from the heater. A single-pass or "series-flow" heater with 40 tubes, for example, would be equivalent in area to a two-pass heater having two coils of 20 tubes each.

The primary constraint affecting the selection of the number of passes and tube size is the allowable pressure

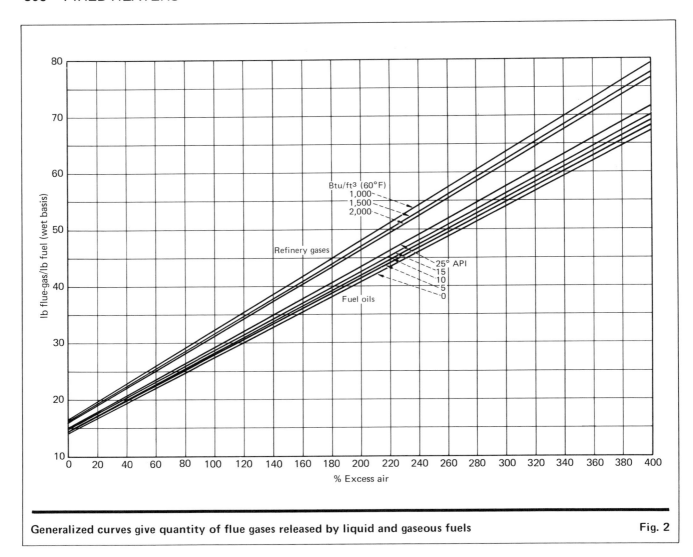

Generalized curves give quantity of flue gases released by liquid and gaseous fuels   Fig. 2

drop. The fewer the number of passes, the less the potential for flow maldistribution.

Each heater application must be evaluated individually with regard to the number of passes and tube size. Such evaluation must examine not only the cost of the heater but also the cost of external distribution and collection manifolds, pass control valves, and other ancillaries. For normal heating applications, a tube coil consisting of 4-in. ips (iron pipe size) tubes often represents the lowest heater investment (excluding externals). As tube diameter increases or decreases from this size, the heater cost tends to become progressively more expensive.

Thermal efficiency

For each fuel, profiles can be developed at constant values of excess air to express the heat extracted from the products of combustion as a function of flue-gas temperature. The percentage of the heat extracted from the flue gas will range from 0% at the flame temperature, to 100% at the datum flue-gas temperature of 60°F. This heat includes the heat absorbed by the charge stock plus the heat lost from the heater casing. At any flue-gas temperature, the differential between the percent heat extracted and 100% constitutes the percent heat loss up the stack.

Fig. 3 and 4 illustrate for two typical fuels (a 19,700 Btu/lb refinery gas and a 15° API fuel oil), profiles of heat available in the flue gas as a function of temperature and excess air. Data for other fuels, identified by their net heating values (Btu/lb), are given in Ref. [9].

Assuming that the fuel and combustion air are supplied at the datum temperature of 60°F, for a given combination of flue-gas temperature and excess air, the percent heat extracted from the flue gas is:

% Heat extracted =

$$\frac{(100)(\text{Heat available, Btu/lb fuel at flue gas temp.})}{(\text{Net heating value of fuel, Btu/lb})}$$

$$= (100) H_A / H_F$$

The heat transmitted from the heater casing to the atmosphere via radiation and convection (the so-called "radiation loss") is generally assessed at $1\frac{1}{2}$ to 2% of the heat fired for conventional installations, and at 2 to $2\frac{1}{2}$% for heaters with extensive hot-duct runs and/or air preheaters.

FIRED HEATERS—III COMBUSTION CONDITIONS INFLUENCE DESIGN & OPERATION

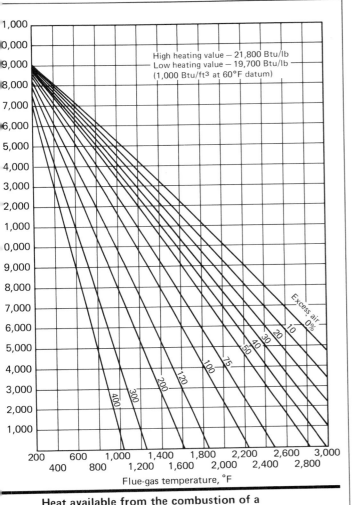

Heat available from the combustion of a 19,700 Btu/lb (LHV) refinery gas — Fig. 3

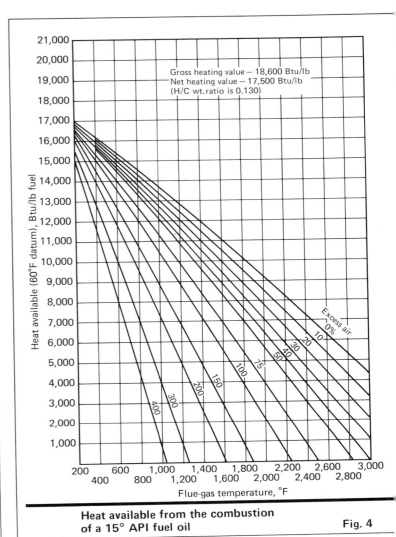

Heat available from the combustion of a 15° API fuel oil — Fig. 4

The actual or calculated thermal efficiency of the heater is computed as the percent heat extracted less the percent radiation loss. The heat fired is then determined from the relationship:

Thermal efficiency = e =
$$\frac{\text{Heat absorbed, Btu/h}}{\text{Heat fired, Btu/h}}(100) = \frac{Q}{Q_F}(100)$$

The quantity of fuel consumed is obtained as follows:

Fuel consumed, lb/h =
$$\frac{\text{Heat fired, Btu/h}}{\text{Fuel heating value, Btu/lb}} = \frac{Q_F}{H_F}$$

The flue-gas mass flowrate is approximated by multiplying the fuel consumption rate by the appropriate weight ratio of flue gas to fuel, obtained from the curves in Fig. 2.

The foregoing relationships define the interdependence of excess air, flue-gas temperature and thermal efficiency. In the design of heaters having convection sections, thermal efficiency is determined by selecting the flue-gas temperature. Specification of this temperature is normally contingent upon that of the process fluid at the inlet. A typical design basis would assume a temperature approach of about 150°F between flue-gas temperature and inlet fluid temperature.

A temperature differential of this magnitude represents a reasonable balance between thermal efficiency and capital investment. However, many criteria affect the temperature approach for a particular application. These include fuel value, project payback time, minimum stack-height requirement, type and material of extended surface, auxiliaries such as sootblowers, etc. In the case of all-radiant heaters without convection sections, the designer does not have the degree of freedom to select the design flue-gas temperature. For these heaters, the temperature is equal to the residual radiant-section gas temperature.

Radiant section

In order to evaluate the split in heat absorption between radiant and convection sections, it is necessary to determine the radiant efficiency—the fraction of heat liberated that is absorbed by the heat-transfer surface in the combustion chamber. For a given fuel at a given value of excess air, the radiant efficiency can be shown to be a function of the residual radiant-section gas

Typical heat-transfer rates across the radiant sections of heaters in various services	Table II
Service	Average radiant rate* Btu/h/ft² (based on O.D.)
Atmospheric crude heaters	10,000-14,000
Reduced-crude vacuum heaters	8,000-10,000
Reboilers	10,000-12,000
Circulating-oil heaters	8,000-11,000
Catalytic-reformer charge and reheat heaters	7,500-12,000
Delayed coker heaters	10,000-11,000
Visbreaker heaters—heating section	9,000-10,000
Visbreaker heaters—soaking section	6,000-7,000
Propane deasphalting heaters	8,000-9,000
Lube vacuum heaters	7,500-8,500
Hydrotreater and hydrocracker charge heaters	10,000
Catalytic-cracker feed heaters	10,000-11,000
Steam superheaters	9,000-13,000
Natural-gasoline plant heaters	10,000-12,000

*For tubes spaced at twice the nominal diameter, fired on one side and backed by refractory.

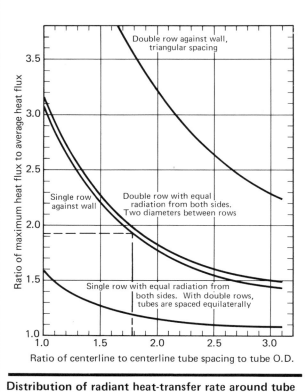

Distribution of radiant heat-transfer rate around tube depends upon coil arrangement and firing mode   Fig. 5

temperature [3]. In turn, for a particular heater configuration, this temperature can be expressed graphically as a function of excess air, tube metal temperature, and radiant transfer rate [4].

The selection of the average radiant transfer rate is, essentially, the first step in the process design of a fired heater. By definition, the average radiant rate represents the heat transferred to the charge stock in the radiant section, divided by the total radiant-section heat-transfer surface, based on the tube O.D. The higher the design radiant rate, the less the amount of heat-transfer surface, the smaller the heater, and the lower the investment cost.

Unduly high radiant rates, however, result in higher maintenance costs. Because the refractories and tube supports are exposed to higher temperatures, they have shorter service lives. Furthermore, high tube-wall temperatures reduce tube life and raise the potential for coke deposition and product degradation. Actual radiant transfer rates will reflect the experience of both the user and the designer. A tabulation of typical rates for various commercial heating services is shown in Table II.

Distribution of the radiant heat-transfer rate around a tube is not uniform. In fact, the rate varies substantially around the tube circumference, depending upon the ratio of tube spacing to tube diameter, and upon the firing mode—i.e., whether the tubes are fired from one side or from both sides. If the maximum radiant rate is assumed to occur at the front 60 deg of the fired tube face, the ratio of maximum radiant rate to average radiant rate can be obtained from Fig. 5 for various tube coil configurations and firing modes [5].

As noted in Part II, the tube metal temperature is always hotter than the bulk fluid temperature at any given location. The magnitude of the metal temperature depends on the fluid temperature, the radiant transfer rate, the tubeside film coefficient, the thermal conductivity of the tube material, and the thermal resistance of coke or scale deposits. The metal temperature can be expressed as the sum of the bulk fluid temperature plus the temperature differentials across the fluid film, the coke/scale layer, and the tube wall:

$$T_m = T_B + \frac{q_i}{h_i} + \frac{q_i t_c}{K_c} + \frac{q_i t_m}{K_m} \quad (2)$$

For vertical-cylindrical heaters and horizontal-tube heaters, Figs. 6a and b estimate the residual radiant-section gas temperature—the so-called bridgewall temperature (BWT)—as a function of the radiant transfer rate and the average metal temperature. The charts assume that the tube coil is fired from one side, and that the tubes are spaced at twice their nominal diameter. As an expedient, the metal temperature can be reasonably approximated for most conventional heating applications by adding 75°F to the average bulk-fluid temperature in the radiant section.

Once the bridgewall temperature and the flue-gas temperature (FGT) are established, the split in duty between radiant and convection sections can be obtained as a function of the heat available in the products of combustion at these temperatures. The split is governed by the relationship that the ratio of the heat available at the bridgewall temperature to that at the flue-gas temperature is equal to the ratio of the radiant duty to the total duty:

$$\frac{H_{A,BWT}}{H_{A,FGT}} = \frac{Q_R}{Q} \quad (3)$$

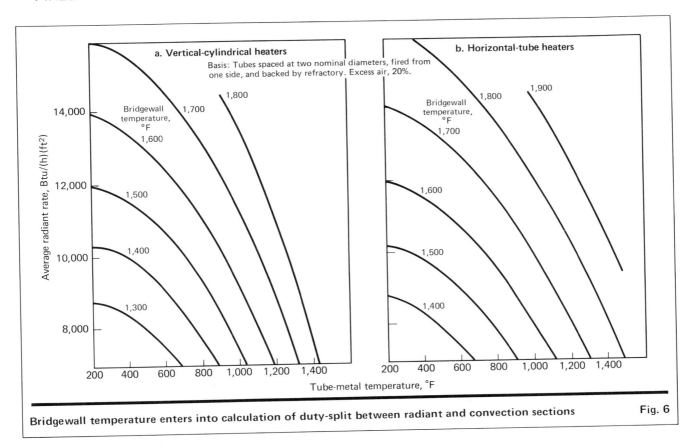

Bridgewall temperature enters into calculation of duty-split between radiant and convection sections — Fig. 6

The radiant heat-transfer surface area is obtained as the quotient of the radiant duty divided by the average radiant rate:

$$A_R = \frac{Q_R}{q} \qquad (4)$$

Convection section

All combustion should be completed before the flue gas reaches the convection section. Although convective heat transfer is the primary mode of heat transmission in this section, radiative effects also contribute.

The first two or three rows of tubes adjacent to the point at which the flue gas enters the convection section are known as the shield bank. Because these tubes are generally oriented at a staggered triangular pitch normal to the flue-gas flow, they will usually screen and absorb most of the residual radiative component. When the shield bank can "see" the firebox, as in most modern installations, industry practice is to consider the surface area of one row of shield tubes to be radiant transfer area. Consequently, this surface is usually included in the radiant-section surface requirement.

To estimate a film coefficient based on pure convection for flue gas flowing normal to a bank of bare tubes, a method was developed by Monrad [6] and was subsequently revised to the following form [7]:

$$h_c = \frac{2.14\, g^{0.6} T_{ga}^{0.28}}{d_o^{0.4}} \qquad (5)$$

where g = flue-gas mass velocity, lb/s/ft² at minimum cross-section; T_{ga} = average flue-gas temperature, °R; and d_o = tube outside diameter, in.

This equation does not take into account radiation from the hot gases flowing across the tubes, or re-radiation from the walls of the convection section. As an approximation, the radiation coefficient of the hot gas may be obtained from the following equation [8]:

$$h_{rg} = 0.0025\, T_g - 0.5 \qquad (6)$$

where T_g = average flue-gas temperature, °F.

Re-radiation from the walls of the convection section usually ranges from 6 to 15% of the sum of the pure-convection and the hot-gas-radiation coefficients. A value of 10% represents a typical average. Based on this value, the total heat-transfer coefficient for the bare-tube convection section can be computed as:

$$h_o = (1.1)(h_c + h_{rg}) \qquad (7)$$

Convection-section surface-area requirements are controlled by the film resistance on the flue-gas side. In order to increase transfer rates per unit length of tubing, convection sections in modern heater designs are almost always equipped with extended surface. Various types of extended surface employed in heater convection sections have been already described. With all of these extended-surface types, radiant transfer to the convection tubes is so small that it can be neglected. For example, the heat-transfer coefficient of a serrated-fin surface, h_o, can be obtained from Fig. 7 as a function of the Reynolds number:

$$h_o = J c_p G \bigg/ \left(\frac{u c_p}{k}\right)^{2/3} \qquad (8)$$

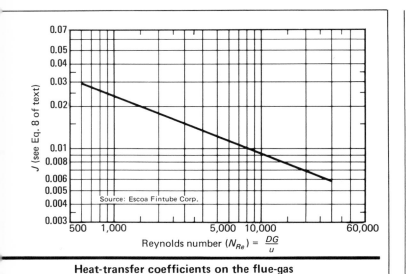

Heat-transfer coefficients on the flue-gas side of serrated fins — Fig. 7

Fluid mass velocities govern choice of tube size and number of passes	Table III
Service	**Mass velocity, lb/s/ft²**
Atmospheric crude heaters	175-250
Reduced-crude vacuum heaters (outlet tube)	60-100
Reboilers	150-250
Circulating-oil heaters	350-450
Catalytic-reformer charge and reheat heaters	45-70
Delayed coker heaters	350-450
Hydrotreater and hydrocracker charge heaters	150-200
Steam superheaters	30-75
Steam generators (forced circulation)	100-150
Catalytic-cracker feed heaters	300-400

where G = flue-gas mass velocity, lb/h/ft²; c_p = flue-gas specific heat, Btu/lb/°F; u = flue-gas viscosity, lb/h/ft; and k = flue-gas thermal conductivity, Btu/h/(ft²)(°F)/ft.

The effective outside heat-transfer coefficient is calculated from:

$$\text{Effective } h_o = h_o(EA_f + A_o)/A_t \tag{9}$$

The fin efficiency, E, takes into account the variability of fin effectiveness as a function of fin configuration, the thermal conductivity of the extended surface, and the convection film coefficient. For serrated fins, values of E may be approximated from Fig. 8.

The overall coefficient, U, is obtained from several relationships:

$$R_o = \frac{1}{\text{Effective } h_o} + F_o \tag{10}$$

$$R_w = \frac{A_t}{h_w A_i} \tag{11}$$

$$R_i = \frac{A_t}{A_i}\left(\frac{1}{h_i} + F_i\right) \tag{12}$$

$$R_t = R_o + R_w + R_i \tag{13}$$

$$U = \frac{1}{R_t} \tag{14}$$

The convection surface requirement is calculated as:

$$A_c = \frac{Q_c}{\text{LMTD}(U)} \tag{15}$$

The highest convection-transfer rate normally occurs at the lowest extended-surface row. In many designs, the density of the extended surface is reduced for the lowest row or two in order to keep the convection rate from becoming excessive. A reasonable estimated value for the maximum convection heat-transfer rate (expressed on an equivalent bare-tube basis) for conventional heating applications is twice the average radiant-transfer rate.

The maximum tube-metal temperature in the convection section is estimated in the same manner as for the radiant section, but with bulk fluid temperature, in-tube film coefficient, and transfer rate appropriate for the convection zone.

Fluid pressure drop

The optimum combination of tube size and number of passes is contingent on an accurate evaluation of the fluid pressure drop. Historically, for heaters processing heavy oils, an important criterion for choosing the tube size and the passing arrangement has been the cold-oil velocity. However, with the advent of numerous all-vapor applications, a more meaningful basis has proved to be the fluid mass velocity. Table III presents a listing of typical fluid mass velocities for the more frequently encountered heating applications.

The pressure drop for a single-phase fluid, either all vapor or all liquid, can be predicted with reasonable confidence using well-established hydraulic principles. The following relation allows pressure-drop computations to be made with an accuracy sufficient for most engineering purposes:

$$\Delta p = \frac{(0.00517)f(g')^2(V)(L_e)}{d_i} \tag{16}$$

where Δp = pressure drop, lb/in²; f = Fanning friction factor; g' = fluid mass velocity, lb/s/ft²; V = average specific volume, ft³/lb; d_i = tube inside diameter, in.; and L_e = hydraulic length, ft.

The total hydraulic length, L_e, is based on the sum of the actual tube lengths plus the equivalent lengths of return fittings and elbows. These equivalent lengths can be approximated for various fittings as a multiple of the tube inside diameter. Typical equivalent lengths are 50 diameters for 180-deg return bends, 30 diameters for 90-deg elbows, and 100 diameters for plug-type headers.

The pressure drop for a mixed-phase flow regime can be approximated using the foregoing relation—albeit with a lesser degree of confidence than for single-phase computations. In the case of mixed-phase flow, the specific volume is very sensitive to the quantity of vapor present at a given location. Consequently, a more accurate estimate of overall pressure drop can be obtained by raising the number of incremental pressure-drop determinations over the length of the coil.

In most vaporizing hydrocarbon services, vapor formation is not linear along the tube length; instead, it increases near the heater outlet. If an arithmetic average of the specific volumes at the inlet and outlet of a mixed-phase zone is used, the resultant pressure-drop value will be overly conservative. To compensate for this, it is suggested that the log mean average of the inlet and outlet specific volumes be used for each mixed-phase zone under consideration.

In view of the vast quantity of literature, both theoretical and empirical, concerning mixed-phase pressure drop, it would seem simplistic to apply a single friction factor to a mixed-phase flow system. Yet for many years, a single friction factor has been used with great success to design countless mixed-phase fired heaters. For example, in conventional hydrocarbon vapor-liquid systems, a factor of 0.0045 has proven to be of sufficient engineering accuracy for pressure-drop determinations.

Because it is unusual for a charge stock to enter a fired heater at its bubble point, vaporization normally begins part way through the tube coil. Under this condition, individual pressure-drop determinations should be made for an all-liquid, single-phase regime, and for a separate mixed-phase regime. The point of initial vaporization is arrived at via trial-and-error, mixed-phase pressure-drop calculations.

In heaters having high percentages of vaporization, it is not unusual for a temperature peaking condition to occur. As the mixed-phase fluid flows through the coil, it undergoes a substantial drop in pressure per unit length of flow. This causes the rate of vaporization to outstrip the rate at which latent and sensible heat can be supplied by the products of combustion. Thus the bulk fluid temperature drops even though the enthalpy increases. It may well be that at some intermediate location along the coil, the bulk fluid temperature will rise above the outlet temperature.

Although this temperature peaking effect is a perfectly normal condition, it is incumbent upon the designer to properly assess its magnitude and the consequences, particularly in high-temperature services.

Tube-wall thickness

As mentioned previously, there are two design criteria that have been applied to heater tubes. At lower temperatures in the elastic range, the design stress is based on the yield strength. At higher temperatures in the creep-rupture range, the design stress is based on the rupture strength. For both conditions, the recommended tube-wall thickness is determined from the mean diameter formula:

$$t_m = \frac{Pd_o}{2S + P} \quad (17)$$

In the elastic range, the corrosion allowance is added to the wall thickness as determined above. The design pressure, P, corresponds to the maximum pressure excursion during an upset condition. The design stress, S,

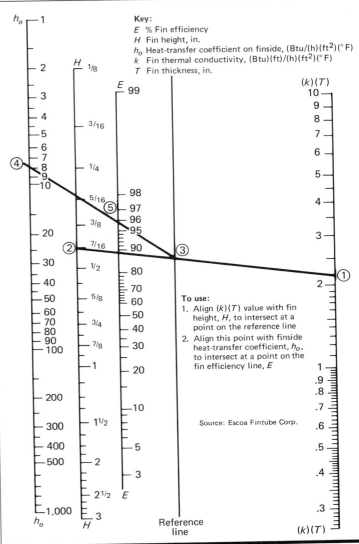

Fin efficiency varies with design and thermal conductivity of fin, and with convection film coefficient   Fig. 8

is based on two thirds of the yield strength for ferritic steels or 90% of the yield strength for austenitic steels.

In the creep-rupture range, a fraction of the corrosion allowance is added to the wall thickness, as determined by Eq. (17). The design pressure, P, corresponds to the operating pressure, and the design stress, S, is based on 100% of the minimum rupture strength for a specified service life. Typical design service lives range from 60,000 to 100,000 h.

A more detailed presentation of the foregoing, as well as design stress values and a procedure for evaluating the corrosion fraction, is to be found in the 2nd ed. of the API's Recommended Practice 530, "Calculation of Heater Tube Thickness in Petroleum Refineries" [10].

When the fluid being heated is water or steam, the requirements of the ASME Boiler and Pressure Vessel Code normally prevail. The tube wall thickness should be calculated in accordance with Section I, using Code formulas and stresses.

Design stresses are very sensitive to temperature, par-

Sample calculation

The following sequence of calculations illustrates the principal steps in the process design of a fired heater.

Fluid	Dowtherm A
Heat absorbed, Btu/h	27,390,000
Flowrate, lb/h	550,000
Inlet temperature, °F	490
Outlet temperature, °F	580
Inlet vaporization	Nil
Outlet vaporization	Nil
Outlet pressure, psig	150
Allowable pressure drop, psi	25
Design pressure, psig	250
Fuel gas, Btu/lb (LHV)	19,700

Design basis

Vertical-cylindrical heater with horizontal-tube convection section
Excess air, 20%
Average radiant rate, 10,000 Btu/h/ft²

Efficiency

Take flue-gas temp. (FGT) = 490 + 150 = 640°F
Heat available at FGT = 16,600 Btu/lb (Fig. 3)

$$\% \text{ Heat extracted} = \frac{(100)(16,600)}{(19,700)} = 84.3\%$$

Calculated efficiency = 84.3 − 2.0 (radiative loss) = 82.3% LHV

$$\text{Heat fired} = \frac{27,390,000}{0.823} = 33,280,000 \text{ Btu/h}$$

$$\text{Fuel consumed} = \frac{33,280,000}{19,700} = 1,689.3 \text{ lb/h}$$

Flue-gas flow = (1,689.3)(19.9) = 33,620 lb/h (Fig. 2)

Radiant/convection duty split

Select tube coil having 4 passes of 4-in. IPS, Sched. 40 (4.5-in. O.D. × 0.237-in. avg. wall)

$$\text{Fluid mass vel.} = \frac{550,000}{(3,600)(4)(0.0884)} = 432 \text{ lb/s/ft}^2$$

Assume fluid temp. at radiant inlet = 520°F

Radiant-section ave. fluid temp. = $\frac{520 + 580}{2}$ = 550°F

Take radiant-section average tube-metal temperature = 550 + 75 = 625°F
Bridgewall temp. (BWT) = 1,470°F (Fig. 6a)

$$\frac{H_{A,BWT}}{H_{A,FGT}} = \frac{11,700}{16,600} = \frac{Q_R}{27,390,000}$$

Q_R = radiant duty = 19,310,000 Btu/h
Q_c = convection duty = 8,080,000 Btu/h

$$A_R = \text{radiant surface} = \frac{19,310,000}{10,000} = 1,931 \text{ ft}^2$$

Assume 52 radiant tubes on 8-in. centers

Tube-circle dia. (TCD) = $\frac{(52)(8)}{(12)(\pi)}$ = 11.03 ft

Take effective tube length (ETL) of shield bank = 10 ft
With 8 tubes per row on 8-in. equilateral centers:

Shield-bank free area =
$$\frac{(8)(10.0)(8.0 - 4.5)}{(12)} = 23.33 \text{ ft}^2$$

g = flue-gas mass velocity = $\frac{33,620}{(3,600)(23.33)}$ = 0.400 lb/s/ft²

Consider surface equivalent to one shield row as radiant surface
Surface of one shield row =
$$(8)(10.0)(1.178) = 94 \text{ ft}^2$$

Vertical-tube radiant surface = 1,931 − 94 = 1,837 ft²

Vertical-tube ETL = $\frac{1,837}{(52)(1.178)}$ = 30 ft

Ratio: $\frac{\text{ETL}}{\text{TCD}} = \frac{30.0}{11.03} = 2.72$

Shield bank

Shield-bank heat absorption is approximated via a trial-and-error process
Use three-row shield bank
Take average fluid temp. in shield bank = 515°F
Fluid temperature increase in shield bank is small and can be neglected
Assume flue-gas temp. drop across shield = 210°F

LMTD calculation:
1,470	1,260
515	515
955	745

LMTD = 846°F

$$h_c = \frac{(2.14)(0.400)^{0.6}(1,825)^{0.28}}{(4.5)^{0.4}} = 5.54$$

$h_{rg} = (0.0025)(1,325) - 0.5 = 2.91$
$h_o = (1.1)(5.54 + 2.91) = 9.29$

Assume clean tube design and no internal or external fouling resistances

$$h_i = (0.027)\frac{k}{D}\left[\frac{(DG)}{(u)}\right]^{0.8}\left[\frac{(c_p u)}{(k)}\right]^{0.333}\left[\frac{(u)}{(u_w)}\right]^{0.14}$$

In view of the small variation in viscosity between bulk and wall temperatures, the value of $(u/u_w)^{0.14}$ may be taken to be 1.0.

$$h_i = (0.027)\frac{(0.064)}{(0.375)}\left[\frac{(0.375)(1,555,200)}{(0.605)}\right]^{0.8} \times \left[\frac{(0.543)(0.605)}{(0.064)}\right]^{0.333}(1.0) = 487$$

$R_i = \frac{A_o}{h_i A_i} = \frac{1.178}{(487)(1.054)} = 0.002295$

$h_w = \frac{K_m}{t_m} = \frac{324}{0.237} = 1,367$

$R_w = \frac{A_o}{h_w A_i} = \frac{1.178}{(1,367)(1.054)} = 0.000818$

$R_o = \frac{1}{h_o} = \frac{1}{9.29} = 0.10764$

$R_t = 0.11076$; $U = 9.03$

Shield-bank transfer rate = $(U)(\text{LMTD}) = (9.03)(846) = 7,639$ Btu/h/ft²
Three-row shield-bank surface = (3)(8)(10.0)(1.178) = 283 ft²
Shield-bank heat absorption = (283)(7,639) = 2,162,000 Btu/h

Check of flue-gas temperature above shield bank:
$$\frac{H_{A,BWT}}{H_{A,shield}} = \frac{11,700}{H_{A,shield}} = \frac{19,310,000}{21,472,000}$$
$H_{A,shield} = 13,010$ Btu/lb of fuel
Flue-gas temperature above shield bank = 1,260°F (as assumed)

Convection bank

Fin-bank heat absorption = 8.08 MM − 2.162 MM = 5,918,000 Btu/h
Fluid temperature leaving fin bank = 510°F
LMTD calculation:

```
1,260    640
  510    490
  ───    ───
  750    150
```

LMTD = 373°F

Try 3 fins/in., with each fin ¾ in. high × 0.05 in. thick ($A_t = 7.33$ ft²/ft)
Fin-bank free area = $(8)(10.0) \times \left[\frac{(8.0-4.5)}{(12)} - \frac{(2)(0.05)(0.75)(3)}{(12)}\right] = 21.83$ ft²

$G = \frac{33,620}{21.83} = 1,540$ lb/h/ft²

Reynolds number = $\frac{(4.5)(1,540)}{(12)(0.084)} = 6,875$

$J = 0.011$ (from Fig. 7)

$h_o = \frac{(0.011)(0.28)(1,540)}{\left[\frac{(0.084)(0.28)}{(0.030)}\right]^{2/3}} = 5.58$

Fin efficiency, E, = 84% (Fig. 8)

Effective $h_o = \frac{(5.58)}{(7.33)}[(0.84)(6.152) + 1.178] = 4.83$

$R_i = \frac{(7.33)}{(487)(1.054)} = 0.01428$

$R_w = \frac{(7.33)}{(1,367)(1.054)} = 0.00509$

$R_o = \frac{1}{4.83} = 0.20701$

$R_t = 0.22638$

$U = 4.42$

A_c = convection surface = $\frac{5,918,000}{(373)(4.42)} = 3,590$ ft²

Surface area per convection row = $(8)(10.0)(7.33) = 586$ ft²

Number of finned rows = $\frac{3,590}{586} = 6.1$ (use 6)

Convection surface = $(6)(586) = 3,516$ ft²

Pressure drop

Reynolds number = $\frac{(4.026)(1,555,200)}{(12)(0.581)} = 898,000$

Fanning friction factor = 0.0038

$L_{e, convection + shield} = (2)(13.0) + (16)(11.5) + \frac{(17)(50)(4.026)}{(12)} = 495$ ft

$L_{e, crossover} = 20 + \frac{(3)(30)(4.026)}{(12)} = 50$ ft

$L_{e, radiant} = (2)(33.0) + (11)(29.0) + \frac{(12)(50)(4.026)}{(12)} = 586$ ft

Total $L_e = 1,131$ ft
Average specific volume = 0.0194 ft³/lb
$$\Delta p = \frac{(0.00517)(0.0038)(432)^2(0.0194)(1,131)}{(4.026)} = 20.0 \quad \text{psi}$$
(vs. 25 psi allowed)

Tube-wall thickness

Average radiant rate = 10,000 Btu/h/ft²
Factor (Fig. 5) max. to ave. flux ratio = 1.93
Factor for local flux variation = 1.25 (assumed)
Factor for conductive/convective effects = 0.85 (assumed)
Maximum local radiant rate = $(10,000)(1.93)(1.25)(0.85) = 20,500$ Btu/h/ft²
At outlet fluid temperature, $h_i = 522$

$T_m = 580 + \frac{(20,500)(4.5)}{(522)(4.026)} + \frac{(20,500)(4.5)(0.237)}{(315)(4.263)} = 640°F$

To allow for fouling effects and to provide safety margin, use design temperature = 750°F
With carbon steel tubes, assume ⅟₁₆-in. corrosion allowance

$t_m = \frac{(250)(4.5)}{(2)(15,500) + 250} + 0.063 = 0.099$-in min. wall

As the least wall thickness, use Schedule 40 (0.237 in. average wall)

Stack design

Size stack for mass velocity of 0.8 lb/s/ft² at 125% of design gas flow

Cross-sectional area = $\frac{(1.25)(33,620)}{(3,600)(0.8)} = 14.59$ ft²

Stack diameter, $D' = 4$ ft 4 in.
Assume average gas temperature in stack = $640 - 75 = 565°F$.
Assume stack-exit gas temperature = 490°F.

Draft under arch = 0.050 in.
Shield-bank loss = $(3)(0.2)(0.0030)(1.25 \times 0.400)^2(46.3) = 0.021$ in.
Fin-bank loss = $(6)(1.0)(0.0030)(1.25 \times 0.428)^2(35.6) = 0.183$ in.
Stack-entrance loss = $(0.5)(0.0030)(0.8)^2(27.8) = 0.027$ in.
Damper loss = $(1.5)(0.0030)(0.8)^2(27.8) = 0.080$ in.
Stack-exit loss = $(1.0)(0.0030)(0.8)^2(24.0) = 0.046$ in.
Subtotal = 0.407 in.

Convection-section draft gain = $(0.52)(8.5)(14.69)\left[\frac{1}{540} - \frac{1}{1,515}\right] = 0.077$

Required stack draft = 0.330 in.

Stack draft gain/ft = $(0.52)(1.0)(14.69)\left[\frac{1}{540} - \frac{1}{1,025}\right] = 0.00669$ in.

Stack frictional loss/ft = $\frac{(0.8)^2(1,025)}{(211,000)(4.33)} = 0.00072$ in.

Net stack effect/ft = 0.00597 in.

Stack height required = $\frac{0.330}{0.00597} = 55$ ft 3 in.

ticularly in the higher regions. Consequently, the thickness calculation requires an accurate determination of the tube-wall temperature. It should be noted that the curves in Fig. 5, which show the ratio of the maximum radiant-transfer rate to the average rate do *not* consider convection heat transfer to the radiant tubes, circumferential heat transfer by conduction along the tube wall, or variations in the transfer rate at different zones of the combustion chamber.

The highest transfer rate in horizontal-tube cabin heaters generally occurs at the wall, 5 to 10 ft above the floor, or at the shield bank, where convection effects are most pronounced. The transfer rate in the lower third of cylindrical heaters may approach 1.1 to 1.5 times the average radiant-transfer rate, depending on the ratio of effective tube length to tube-circle diameter. The magnitude of this local variation is compensated somewhat by the combined conductive/convective effect, which tends to reduce the disparity between maximum and average transfer rates.

Stack design

The main functions of a stack are to induce the flow of combustion air into the heater, and to produce draft sufficient to overcome all obstructions to the flow of flue gas while maintaining a negative pressure system throughout. Never should a heater be operated with greater than atmospheric pressure at any point within the structure. Positive pressure within a heater setting creates a driving force for the outward movement of hot gases, which can lead to serious overheating and corrosion of the steel structure.

The draft produced by a column of hot flue gas depends on the density difference between the hot gas and the ambient air. The draft, in inches of water, can be expressed as:

$$\text{Draft} = (0.52)(L_s)(p')\left(\frac{1}{T_a} - \frac{1}{T_{ga}}\right) \quad (18)$$

where L_s = stack height, ft; p' = atmospheric pressure, psia; T_a = ambient temperature, °R; and T_{ga} = flue-gas temperature, °R.

Because of heat loss through the stack casing, the flue-gas temperature at the top of the stack is substantially lower than at the inlet at the stack base. The magnitude of the differential depends on several factors, including the stack dimensions and the effectiveness of the stack insulation.

For most applications, the average flue-gas temperature in the stack may be conservatively estimated at 75°F less than the inlet gas temperature.

The relationship between stack diameter and stack height is also affected by cost optimization. Assuming that there are no requirements dictating a minimum stack height, a reasonable basis for selecting the diameter would result in a flue-gas mass velocity in the range of 0.75 to 1.0 lb/s/ft².

As noted above, a major objective in stack design is to ensure negative pressure throughout the heater. The most critical location in this respect occurs where the flue gas enters the convection section. Positive pressure manifests itself first in this zone.

It is recommended that the stack design be based on a negative pressure of 0.05 in. H_2O at this entry point. The use of a design pressure less negative is not realistic considering the confidence level of the design computations. On the other hand, the use of still lower pressure (i.e., more negative) would tend to create too great a driving force for the leakage of air through various apertures and seams in the structures.

The frictional loss of the flue gas flowing through the stack, in in. H_2O is:

$$\text{Loss per foot of stack height} = \frac{(g)^2(T_{ga})}{(211,000)(D')} \quad (19)$$

where g = mass velocity in stack, lb/s/ft²; and D' = stack diameter, ft.

The remainder of the flue-gas losses can be expressed in terms of the velocity head, based on the flue-gas mass velocity at the location under consideration:

$$\text{Velocity head (in. } H_2O) = (0.0030)(g^2)(V_g) \quad (20)$$

The losses can be estimated as the product of the velocity head at each location and a factor. For bare convection tubes, this factor is 0.2 (per row); for finned tubes it is 1.0 (per row); for the stack entrance it is 0.5; for the damper it is 1.5; and for the stack exit it is 1.0.

It should be noted that the convection section exerts a stack effect due to its physical height and serves to reduce the overall draft requirement.

In order to assure good flue-gas distribution throughout the convection section, it is usual for a flue-gas withdrawal opening to be provided at every 40 ft of convection section length.

At best, the determination of flue-gas flow is but an approximation. Operation at excess air levels greater than design produces increased quantities of flue gas. Furthermore, efficiency falloff due to solids deposition on heat-transfer surfaces also results in greater flue-gas quantities. Consequently, a stack design based only on the design flue-gas flow affords virtually no overcapacity in terms of the firing rate, and it severely restricts increases in throughput. To ensure flexibility for nominal capacity increments, the designer will often add an overcapacity of 25% to the flue-gas design flow when estimating the various flue-gas frictional losses.

References

1. Gillis, R. D., Mullert, R. G., Kubus, J. M., and Popan, V. A., "New Sandjet Cleaning Process Improves Pipeline Efficiency and Cuts Downtime," presented at 1977 annual meeting of National Petroleum Refiners Assn.
2. Seebold, J. G., "Furnace Roar Control," *Chem. Eng. Prog.*, Vol. 71, No. 8, 1975, p. 53.
3. von Wiesenthal, P., Fired Tube Heaters, "Advances in Petroleum Chemistry and Refining," Vol. 3, Interscience Publishers, Inc., New York, 1960, p. 86.
4. Ibid, p. 90.
5. Derived from data by Hottel, H. C., Radiant Heat Transmission Between Surfaces Separated by Non-absorbing Media, *Trans. Amer. Soc. of Mech. Engrs.*, Vol. 53, 1931, p. 265.
6. Monrad, C. C., Heat Transmission in Convection Sections of Pipe Stills, *Ind. Eng. Chem.*, Vol. 24, 1932, p. 505.
7. Schweppe, J. L., Torrijos, C. Q., How to Rate Finned-Tube Convection Section in Fired Heaters, *Hydrocarbon Processing*, Vol. 43, No. 6, 1964, p. 159.
8. Nelson, W. L., Tubestill Heaters, "Petroleum Refinery Engineering," 4th ed., McGraw-Hill, New York, 1958.
9. Maxwell, J. B., "Data Book on Hydrocarbons," Krieger Publishing Co., Huntington, N.Y., 1968.
10. "Recommended Practice for Calculation of Heater Tube Thickness in Petroleum Refineries," API Recommended Practice 530, 2nd ed., American Petroleum Institute, Wash., D.C., May 1968.

FIRED HEATERS—IV
How to reduce your fuel bill

This final article in a four-part series takes a look at how energy conservation has affected the design and operation of fired heaters. Items of concern include excess-air reduction, enhanced heat recovery, combustion-air preheating, and the conversion of gas-fired heaters to liquid firing.

Herbert L. Berman, Caltex Petroleum Corp. *

☐ Demands on the CPI for greater energy conservation have had direct impact on the design and operation of fired heaters. Fuel savings have been realized with a variety of measures, ranging from the inexpensive—such as the fine-tuning of operating procedures and the upgrading of maintenance techniques—to the capital-intensive—such as the installation of complex heat-recovery facilities at substantial initial outlay. The following discussion will focus on methods that have proved successful in raising the thermal efficiency of fired heaters.

Reduction of excess air

In existing installations, excess air is the most important combustion variable affecting the thermal efficiency of a fired heater. Although heater operation is easier to control at high excess-air levels, it is very costly. The higher the excess air, the greater the fuel consumption for a given heat absorption. The extra fuel is consumed in heating the excess air volume from ambient temperature to the temperature of the exiting flue gases.

In order to exercise greater control over excess air, the oxygen content of the flue gas should be monitored above the combustion zone. Often, excess air in the combustion section may run as low as 10 to 15%, but stack-gas analysis reveals an oxygen content equivalent to as much as 100% excess air. This differential results from air leakage into the heater that occurs between the combustion zone and the stack. Such leakage cannot be corrected by burner adjustments.

Air leakage into a heater can occur at many locations. One route of entry is through the seams of the steel casing, between adjacent plates and stiffening members. Air can also enter through distorted or poorly gasketed header boxes. The terminal tubes in the tube-coil, where they enter and leave the heater casing, can likewise be a source of air leakage. Only through a rigorous, continuing maintenance effort can air leakage into heater settings be controlled effectively.

There are various schemes for controlling excess air via the monitoring of flue-gas oxygen content. Four such monitoring schemes are listed here, in order of increasing sophistication and cost [1]:

1. The least costly scheme requires only periodic checking of oxygen content using a portable oxygen analyzer and a portable draft gage. With these readings as a guide, the operator can make the adjustments necessary to operate at minimum excess-air levels.

2. One of the most common monitoring systems employs a continuous oxygen analyzer equipped with a local readout device and a permanently mounted draft gage. Continuous readings enable the operator to adjust heater operation whenever necessary.

3. Bringing the oxygen and draft readouts into the control room and adding a remotely controlled, pneumatic damper positioner is the next step toward improving excess-air control.

4. Although generally justifiable only for large heaters, further sophistication in excess-air control can be achieved with automatic stack-damper control. In this control scheme, the damper is positioned automatically according to a draft signal received from a probe located below the convection section. The controller changes the damper position so as to hold a set-point

---
*To meet the author, see Part I, which appears on p. 292.

Originally published September 11, 1978

Recuperative preheater warms combustion air. Unit is equipped with finned, cast-iron tubes. Fig. 1

draft corresponding to the targeted excess air. The draft set point is normally changed only to achieve variations in heater duty.

Further recovery of convection heat

The potential for additional heat recovery in the convection section exists when the heater operates at a relatively high flue-gas temperature. First consideration should be given to augmenting the convection-section surface area by adding several rows of convection tubes in the same heating service. Many current installations have anticipated such expanded service, and have been designed to accommodate the future installation of two rows of convection tubes.

The installation of supplementary heat-exchange units to reclaim additional convection heat often provides a quick economic return. Falling into this category are a number of units that recover process heat. Examples are feed preheaters for pyrolysis and steam hydrocarbon-reformer heaters, reboilers operated in conjunction with catalytic reformers, and superheaters providing process steam for petroleum refinery distillation units.

Very often the supplementary unit recovers convection heat for use in a steam-generating facility rather than in a process-stream heating application. Steam-generating units driven by convection heat are routinely coupled with catalytic-reformer heaters and steam hydrocarbon-reformer heaters. Fired heaters in such installations have convection sections equipped with several independent coils so that the same convection section furnishes heat for boiler-feedwater heating, steam generation, and steam superheating. Coils installed in the convection sections of steam-generating fired heaters are almost always of the forced-circulation type.

Occasionally, flue gases from several fired heaters are routed to a central waste-heat recovery facility. Typically, the recovered heat is utilized for steam generation. Waste-heat boilers of this type are normally designed with a flue-gas bypass around the boiler, which allows the heaters to operate even when the boiler is taken out of service. Where several heaters are involved, provisions for positive isolation at the flue-gas ducts from each heater will enable the user to take an individual heater out of service while the remaining heaters and the waste-heat boiler stay on line.

Before any add-on heat recovery device is retrofitted to a heater, the structural integrity of the existing steel-work and foundation must be examined to assess the effects of the increased loads.

Furthermore, it should be noted that the additional convection-section heat recovery will result in a lower stack-gas temperature, thereby reducing the stack draft. In addition, the installation of more surface area will increase the flue-gas pressure drop. Therefore, it is imperative that the effect of any alterations on the stack draft be analyzed beforehand in order to determine whether additional stack height will be required.

Replace bare tubes with extended surface

Many older generation fired heaters are equipped with bare-tube convection sections and operate with high flue-gas temperatures at 65 to 70% thermal efficiency. By replacing the bare tubes with extended surface tubes, efficiency improvement in the neighborhood of 10% may be realized in some installations. This corresponds to a stack-gas temperature reduction of about 300°F.

The most economical conversion from bare to extended-surface tubes can be made when the reduction in tube size is such that the tip-to-tip diameter across the extended surface tube is the same as the outside diameter of the original bare tube. On this basis, it is very likely that the existing convection section tube sheets can be retained. However, the reduction in tube I.D. will result in higher fluid pressure drops unless an increase in the number of convection-section parallel passes can be tolerated. Conversely, if the same tube I.D. is maintained, the conversion from bare to extended surface will necessitate the replacement of the convection-section tube sheets—with accompanying down-time and expense.

When conversion is contemplated for a heater firing liquid fuel, it is recommended that soot blowers be installed, in view of the greater fouling tendency of extended-surface devices. The cavities required to accommodate the soot-blower lances can usually be created simply by omitting the installation of a row or two of tubes at selected locations in the convection section.

Again, it is mandatory that the stack draft be assessed to ascertain whether the existing stack is adequate.

Preheat the combustion air

Fuel consumption in a fired heater can be reduced markedly by preheating the combustion air. In the preheater, heat is transferred from the flue gas to the combustion air, reducing the exit temperature of the flue gas and raising the thermal efficiency. With air-preheat systems, exit flue-gas temperatures often range in the 300 to 350°F range and efficiency levels commonly reach 90 to 92% (LHV). With such systems, the attainable thermal efficiency is no longer controlled by the approach between the flue-gas and inlet-fluid temperatures.

The salient features of the more popular air-preheaters are noted here:

■ *Recuperative-type air preheaters.* These devices, typically of tubular construction, transfer heat by convection from flue gas to combustion air. Customarily, the air flows inside the tubes, whereas the flue gas flows across the tube bundle. The preheater can be installed in the fired heater above the process convection section or, as is more usually the case, at grade alongside the heater.

Materials selected for heat-transfer surfaces vary from designer to designer. If the flue gas is well above the acid dewpoint, a manufacturer may select cast iron tubes having internal and external fins. In the zone approaching the dewpoint, fins are provided on the gas side only, in order to keep tube-metal temperatures as high as possible. Below the dewpoint, plain tubes of borosilicate glass are used in order to minimize acid corrosion.

Tubular-type air preheaters are essentially tight between air and gas, with no leakage of air into the flue-gas side. Fig. 1 shows a bank of tubular cast-iron heating elements equipped with internal and external fins.

A schematic arrangement for a typical recuperative air preheater installation is shown in Fig. 2. (The arrangement shown is equally applicable to regenerative preheater systems, described below.) The system employs a forced-draft fan to supply combustion air, and an induced-draft fan to maintain a negative pressure and draw the flue gas through the system to the stack. A cold-air bypass enables the operator to route a portion of the incoming combustion air around the preheater when ambient temperature is very low. The preheater surface area can thus be maintained above the acid dewpoint, at a nominal sacrifice in thermal efficiency.

■ *Regenerative-type air preheaters.* This apparatus consists of heat-transfer elements housed in a subdivided cylinder, which rotates inside a casing. Hot flue gases pass through one side of the cylinder, cold air through the other side. As the cylinder slowly rotates, the elements continuously absorb heat from the flue gas and release it to the incoming air stream.

The cylinder is subdivided by baffles which, like the seals between the cylinder and the casing, help to mini-

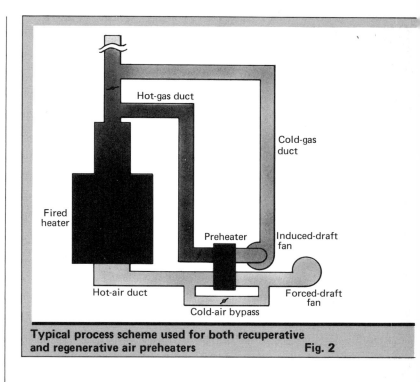

Typical process scheme used for both recuperative and regenerative air preheaters   **Fig. 2**

mize leakage of air into the flue-gas stream. This leakage, which results from the pressure differential that exists between the air side and the flue-gas side, is normally 10 to 15% of the total air flow. The preheater system, particularly the forced-draft and induced-draft

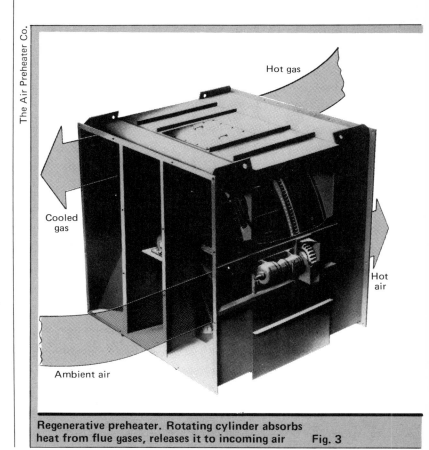

Regenerative preheater. Rotating cylinder absorbs heat from flue gases, releases it to incoming air   **Fig. 3**

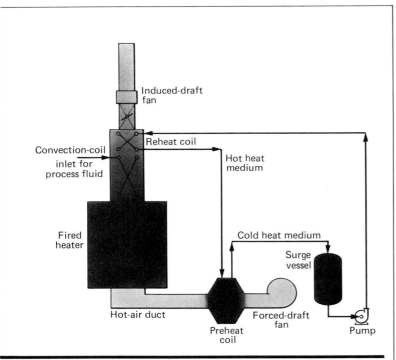

Heat-medium preheater transfers heat from flue gases to air via intermediate, circulating fluid   **Fig. 4**

fans, must be designed to accommodate this leakage.

The heating elements of the regenerative air preheater are installed in two or three layers. Corrosion of the elements is usually confined to the final portion of the cold-end layer, where cold air enters and cooled flue gas leaves. If such corrosion does occur, the elements of the cold-end layer can be removed and reversed to extend their service life. For those applications where the elements are exposed to very corrosive atmospheres, porcelain-enameled heating surfaces are available. Fig. 3 illustrates the basic construction of a regenerative air preheater. The main components of the overall process arrangement are shown in Fig. 2.

■ *Heat-medium air preheaters.* Instead of direct heat exchange between air and flue gas, these units employ an intermediate fluid to transfer heat from the flue gas to the incoming combustion air. The heat medium is contained in a closed loop that includes a reheat coil located in the flue-gas flow downstream of the process convection coil, and a preheat coil positioned in the air stream. The circulating fluid extracts heat from the flue gas, lowering the gas temperature and raising the fluid temperature. In turn, the hot fluid releases its heat to the incoming air.

The process scheme for a heat-medium preheat system (Fig. 4) includes, in addition to the preheat and reheat coils, a fluid-surge vessel and a circulating pump. Depending on the design of the heater and the available draft, the system may be provided with forced draft/induced draft, forced draft only, or induced draft only.

■ *Process-fluid air preheaters.* These systems [2] take off a portion of the process stream entering the convection section and send it to a preheat coil mounted in the air stream, which warms the air. The subcooled process fluid is then returned to a reheat coil located in the flue-gas flow downstream of the main-process convection coil, which reheats the fluid to its original temperature level and returns it to the main process stream ahead of the convection coil.

Only a portion of the process fluid stream is drawn off for preheat service. The mass flows and specific heats of the combustion air and process fluid must be carefully balanced in order to achieve the desired temperature changes in each stream.

A schematic arrangement for a process-fluid air preheater is shown in Fig. 5. As with heat-medium preheaters, these systems can be supplied with forced draft/induced draft, forced draft only, or induced draft only.

Process effects of air preheating

Fired-heater operation with preheated combustion air results in higher adiabatic flame temperatures than with ambient air. From the combustion kinetics, then, one might expect heaters using preheated air to generate higher amounts of NO_x than those using ambient air.

However, nitrogen fixation from the atmosphere is usually a relatively minor factor in NO_x production. The nitrogen which most readily converts to NO_x is that which is chemically bound up in the fuel. The quantity of oxides of nitrogen formed from this source is not appreciably affected by elevated combustion temperatures. In fact, with an air preheater, the reduction in the amount of fuel consumed corresponds directly to a reduction in the amount of NO_x formed from fuel-bound nitrogen, and is quite likely to offset the additional atmospheric fixation due to the higher flame temperature.

Another process effect that manifests itself when an air-preheat system is retrofitted to an existing heater is the shift in duty split between the radiant and convection sections. The higher combustion temperature due to preheating, coupled with the lower convection transfer rate at reduced flue-gas flowrates, causes a greater proportion of the total heat absorption to take place in the radiant section.

Consequently, operation with preheated combustion air results in higher radiant duties than with ambient air. The magnitude of the shift in duty split is, of course, contingent upon the degree of preheat applied to the combustion air.

Gas-turbine exhaust used as combustion air

The exhaust gas from gas turbines, widely used as drivers for compressors, pumps, and electrical generators, usually ranges from 800 to 900°F and contains from 17 to 18% oxygen. These gases are well-suited as a source of preheated combustion air for fired heaters, and can be used to cut heater fuel consumption.

In most systems utilizing gas-turbine exhausts, an auxiliary source of combustion air permits independent operation of the heater and the gas turbine. Typically, the overall system is designed so that the gas-turbine exhaust can be vented to atmosphere whenever the heater is operated on ambient air.

Conversion from gas to liquid firing

In the face of curtailments of natural gas deliveries, many industries are seeking alternative fuel sources, primarily liquid fuels. The following areas of concern should be examined closely before any conversion from gas to liquid firing is undertaken.

Impact on radiant section. The major factor affecting the combustion chamber is the size of the oil flame compared to the size of the gas flame. At the same level of heat release, an oil flame is generally longer and wider than a gas flame. For this reason, the distance from burner to tube should be examined to assess the potential for flame impingement on the tube coil. Similarly, vertical clearance dimensions should be reviewed, since the longer oil flame may impinge on the shield tubes, the reradiating cone, or the baffle provided at the top of some older vertical-cylindrical heaters.

If the contemplated liquid fuel contains substantial vanadium and sodium concentrations, consideration should be given to protecting exposed tube supports from the corrosive combustion environment. Also, if the heater contains a reradiating cone or baffle, the heater manufacturer should be consulted regarding the possibility of removing such vulnerable equipment.

Impact on convection section. Convection-section extended surface in gas-fired heaters is very often of the high-density, finned-tube type. Under heavy liquid-fuel firing, such tubing is difficult to keep clean using conventional onstream cleaning techniques. Before conversion, therefore, the designer should consider replacing high-density finned tubes with either heavy low-density tubes, or studded tubes. Since replacement will reduce the total effective surface area, additional rows of convection tubes will be required to maintain thermal efficiency.

Facilities for onstream convection cleaning, such as soot blowers, should be installed as part of the conversion. The addition of such equipment, as well as the necessary ladder and platform access, requires that the structural integrity of the steelwork and foundation be assessed.

Adjustments to draft. Many heaters designed for gas firing operate virtually at their draft limit, usually because they are pushed well beyond their original design capacity. Because oil firing requires higher excess-air levels to achieve acceptable combustion, draft-limited gas-fired heaters will suffer a capacity decrease when converted to liquid firing, unless additional stack height is provided.

Alterations to burners. Burner replacement in a gas-to-liquid fuel conversion can, in most instances, be made on a one-for-one basis. However, for those conventional gas-fired heaters that operate with relatively small burner-heat releases in the range of 2 to 3 million Btu/h, substitution of oil burners on a one-for-one basis may result in unstable combustion. In these cases, a totally new firing arrangement should be specified as part of the conversion. The final burner arrangement should take into account such parameters as the combustion-chamber configuration, burner turndown requirements and fuel-oil combustion behavior.

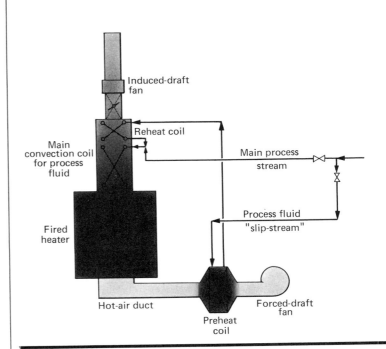

Typical arrangement for air preheater that uses process fluid as heat-exchange medium   Fig. 5

From a practical standpoint, an important consideration that must not be overlooked in any conversion study is the heater's mechanical condition and its remaining useful life. Investment costs for gas-to-oil conversions are expensive, particularly when convection sections must be revamped. Furthermore, the downtime necessary to effect the conversion is costly, due to lost revenues. Consequently, if conversion is contemplated for a heater that is in poor mechanical condition, an evaluation of maintenance costs, the conversion investment and revenue losses may well show replacement to be economically more attractive.

A good part of conversion cost lies in the fuel handling and distribution system. Conversion of only a few large units, particularly if they are located in the same general plot area, will reduce this cost.

References

1. Woodard, A. M., "Upgrading Process Heater Efficiency," *Chem. Eng. Progr.*, Vol. 71, No. 10, 1975, p. 53.
2. von Wiesenthal, P., "Furnace and Related Process Involving Combustion Air Preheating," *U.S. Patent 3,426,733*, 1969.

Generalized method predicts fired-heater performance

Direct-fired process heaters make up a large part of many plants' initial cost. Their fuel also represents a major operating expense. Both builders and operators of process plants, therefore, need a reliable rating method.

Norman Wimpress, C. F. Braun & Co.

☐ Purchasers and users frequently must make their own ratings of fired heaters in order to:
- Estimate heater sizes, fuel consumption, and waste heat recovery, during the conceptual design-stage.
- Evaluate vendors' offerings with regard to reasonableness of designs and consistency between competing proposals.
- Predict the effects of changes in feedrates, feed properties and similar operating variables.
- Anticipate the effects of proposed modifications to an existing heater.

Final design is usually done by the heater manufacturer, following the purchaser's duty specification. In general, each manufacturer has its own rating method for its particular design. Such a method can be expected to be more accurate for its range of application than any general procedure.

Suggested method validated by experience

The method proposed for computing the performance of both the radiant and convection sections of fired heaters is based on fundamental correlations for heat transfer by radiation and convection, adjusted on the basis of operating experience with industrial heaters. Experience has shown the method to be well suited for the type of calculations outlined.

Fig. 1 shows a cross-section of a typical process-plant fired heater. It consists of a firebox or radiant section, a convection section, and a stack system to dispose of the combustion products and provide draft.

The radiant section, where the fuel is burned, contains heat-absorbing tubes, which remove a portion of the heat from the combustion products before they flow to the convection section. In this type of furnace, the tubes are located around the outside of the firebox (in front of refractory walls), combustion taking place in the open central space.

There are also designs in which the tubes are suspended in the center of the furnace, with combustion space on both sides. In these furnaces, the burners may be arranged to fire against refractory walls so as to enhance radiant heat transfer to the tubes. In any case, the tubes must be arranged for uniform and efficient radiant heat absorption, and there must be adequate volume for complete combustion without local overheating or flame impingement on the tubes.

The convection section recovers additional heat from the flue gas at a lower temperature than in the radiant section. Here, since the primary heat-transfer mechanism is convection, the tubes are arranged to create high mass velocities and turbulence in the gas. Tubes having fins and other types of extended surface are frequently installed to improve convective heat transfer.

The first rows of tubes in the convection section, which are exposed to radiation from the hot gas and refractory in the radiant section, are generally called shield or shock tubes. They are usually a structural part of the convection section and are rated with it. However, special allowance must be made for the radiant heat added to these tubes.

The stack system collects and disposes of the flue gas. In natural-draft furnaces, the stack height must provide adequate draft to draw the gas through the firebox and convection section. In induced-draft furnaces, stack height is usually set by gas-dispersion requirements.

Because the governing heat-transfer mechanism is different in the radiant and convection sections, the two sections are rated by different methods.

Radiant-section design

Radiant heat transfer between solid surfaces in various arrangements, and between hot gases and solids, has been extensively investigated [1,2]. Applying basic radiation concepts to process-type heater design, Lobo and Evans developed a generally applicable rating method [3]. Almost all methods published since have followed their basic approach, as does the method described here (with simplifications made by eliminating minor variables and by including general correlations).

This article is based on a paper presented at the American Soc. of Mechanical Engineers' Energy Technology Conference and Exhibit, Houston, Tex., Sept. 18-22, 1977.

Originally published May 22, 1978

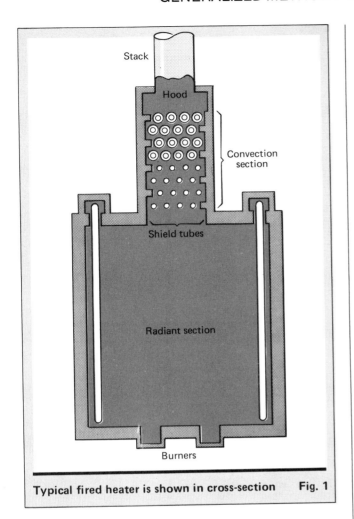

Typical fired heater is shown in cross-section   Fig. 1

Nomenclature

A	Area, m²
A_{cp}	Cold-plane area, m²
A_e	Total furnace envelope-area, m²
A_t	Tube surface area, m²
F	Radiant exchange factor
G	Mass velocity at minimum cross-section, kg/(m²)(s)
h	Film heat-transfer coefficient, kcal/(h)(m²)(°C)
L	Mean length of radiant beam, m
P	Partial pressure of radiating components, atm
p_v	Velocity head, mm H₂O
q	Heat rate, kcal/h
T	Temperature, °K
V	Volume, m³
W_B	Radiant emission from a black body, kcal/(h)(m²)
α	Factor of comparison between a tube bank and a plane
ρ	Density, kg/m³
σ	Stefan-Boltzmann constant, 4.92×10^{-8} kcal/(h)(m²)(°K⁴)

Subscripts

a	Combustion air
c	Convective heat transfer
f	Fuel
g	Flue gas
L	Losses
n	Net heat of combustion
R	Radiant section
r	Radiant heat transfer
S	Shield- and convection-tube radiant transfer
t	Tube surface

Radiant heat transfer is basically described by the Stefan-Boltzmann equation. A black body at absolute temperature T radiates energy at a rate W_B:

$$W_B = \sigma T^4 \qquad (1)$$

For radiant heat transfer between two real surfaces at temperatures T_a and T_b, the relation becomes:

$$q_r = \sigma A F (T_a^4 - T_b^4) \qquad (2)$$

Here, A is the area of one of the surfaces, and F is an exchange factor that depends on the relative area and arrangement of the surfaces, and on the emissivity and absorptivity of each. For transfer inside a furnace, it is generally best to use the heat-receiving, or "cold," surface as the basis for calculation.

Simplify with equivalent cold-plane surface

The usual heat-absorbing system consists of a number of parallel cylindrical tubes. Part of the radiation from the hot gas strikes the tubes directly and is absorbed, and the remainder passes through. If the tubes are in front of a refractory wall, the energy that passes through is radiated back into the furnace, where part of it is absorbed by the tubes, and the remainder passes through.

This complicated situation is handled by expressing the tube area as an equivalent plane surface, A_{cp}. It equals the number of tubes, times their exposed length, times the center-to-center spacing. Because the tube bank does not absorb all the energy radiated to the cold-plane area, an absorption efficiency factor, α, must be applied. Values for α as a function of tube arrangement and spacing have been developed and published by Hottel (in "Chemical Engineers' Handbook"), whose curves for the most-common tube arrangements in process furnaces are shown in Fig. 2.

The product αA_{cp}, called the equivalent cold-plane area, is the area of a plane having the same absorbing capacity as the actual tube bank.

In the type of heater shown in Fig. 1, the convection section forms part of the firebox enclosure. Because the convection section is a number of rows deep, it eventually absorbs all the radiation coming to it from the firebox. Therefore, α for the convection section equals unity. The working equation for calculating radiant heat transfer to the tubes thus becomes:

$$q_{Rr} = \sigma \alpha A_{cp} F (T_g^4 - T_t^4) \qquad (3)$$

Determining the exchange factor

The remaining term to be evaluated in Eq. (3) is the exchange factor, F. The gas in the firebox is a poor radiator, because the only constituents normally in flue gas that contribute significantly to the radiant emission are carbon dioxide and water. The amount of radiating components can be expressed by a single term, the partial pressure of carbon dioxide plus water multiplied by the mean beam length [3]. Fig. 3 shows the partial pressure, P, as a function of excess air for the usual hydrocarbon fuels. For the less-usual fuels, such as hydrogen, P can readily be calculated by simple stoichiometry. The mean beam length, L, is calculated by the equation:

$$L = 3.6\, V/A_e \qquad (4)$$

Here, V is the total firebox volume inside the centerline of tubes, and A_e is the total firebox envelope area.

Gas emissivity is also a function of the temperatures of the gas and the absorbing surface. However, because the tubewall temperature effect has been found to be minor, gas emissivity can be correlated as a function of the product of PL, and of the gas temperature (Fig. 4).

The exchange factor also depends on the amount of radiation reflected from exposed refractory. (Energy striking refractory is reflected back toward the tubes, where it has a second chance to be absorbed.) Thus, a furnace having a large amount of exposed refractory will transfer more heat per unit of tube surface than one whose walls are covered by tubes.

Lobo and Evans correlated this effect on the basis of the ratio of exposed refractory area to total equivalent cold-plane area. Fig. 5 is based on the same correlation, except that the ratio is of cold-plane area to total firebox envelope-area, which simplifies the calculations. Fig. 5 also takes into account that the tubes themselves are not perfect absorbers. The curves are based on a tube-surface absorptivity of 0.9, which is typical for oxidized metal surfaces.

Estimating convection transfer

Although radiation accounts for most of the heat transfer in the radiant section, convection cannot be neglected. The relationship for convective heat transfer is:

$$q_c = h_c A_t (T_g - T_t) \qquad (5)$$

Because convective heat transfer is not the major contributor and cannot be calculated precisely, some simplifying approximations are made. For the usual furnace, h_c in the radiant section is about 10 kcal/(h)(m^2)(°K), A_t is about two times αA_{cp}, and F is about 0.57. This allows putting Eq. (5) into a form similar to Eq. (3):

$$q_{Rc} = (10)(2\alpha A_{cp})(F/0.57)(T_g - T_t)$$
$$= 35\, \alpha A_{cp} F \qquad (6)$$

The total heat-transfer rate in the radiant section is the sum of the radiant and convective heat transfer:

$$q_R = \sigma \alpha A_{cp} F (T_g^4 - T_t^4) + 35\, \alpha A_{cp} F (T_g - T_t) \qquad (7)$$

$$\frac{q_R}{\alpha A_{cp} F} = \sigma (T_g^4 - T_t^4) + 35 (T_g - T_t) \qquad (8)$$

Thus, the ratio $q_R / \alpha A_{cp} F$ is a function of gas and tubewall temperatures only (Fig. 6).

Eq. (7) actually applies only to the tubes in the radiant coil and not to radiation to the convection section. Convective transfer to the convection section tubes, which is calculated separately, does not occur until after the flue gas leaves the firebox, so radiation to the convection section should be calculated by Eq. (3). Computer calculations can be easily programmed to maintain this distinction. For hand calculations, however, it is simpler to use Fig. 6 for all tubes, the error introduced usually being negligible.

Tubes mounted centrally

For the foregoing discussion, the furnace is assumed to be of the type shown in Fig. 1, with refractory-backed tubes around the periphery of the firebox. Certain ad-

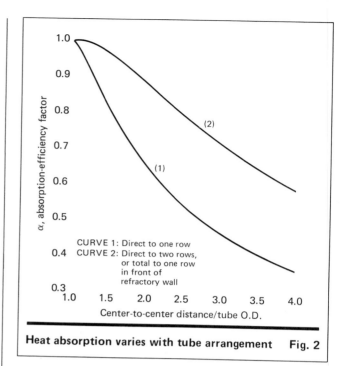

Heat absorption varies with tube arrangement   Fig. 2

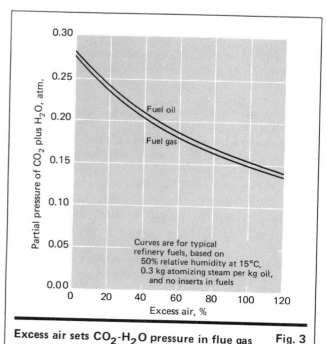

Excess air sets CO_2-H_2O pressure in flue gas   Fig. 3

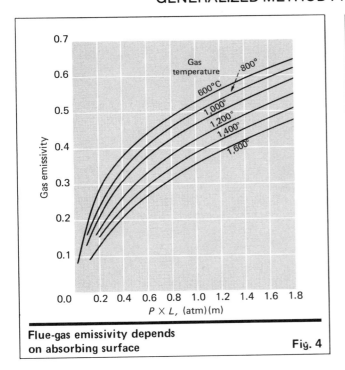

Flue-gas emissivity depends on absorbing surface — Fig. 4

justments are necessary when the tubes are arranged as in Fig. 7, with a single or double row of centrally mounted tubes. In such cases, advantage is taken of the fact that a plane of symmetry in the firebox can be replaced, for computational purposes, by a refractory wall (indicated by dashed lines in Fig. 7). Therefore, the cold-plane area, A_{cp}, for the tube bank is twice the projected cold-plane area, because each side is figured separately. On this basis, α for a single row is obtained

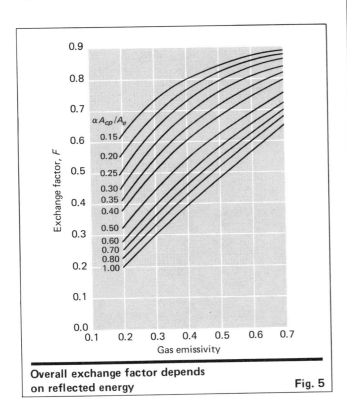

Overall exchange factor depends on reflected energy — Fig. 5

from Curve 1, and for a double row from Curve 2, in Fig. 2. Similarly, the mean beam length, L, must be calculated on the basis of half a furnace.

Temperature complexities

Fig. 6 gives the rate of heat transfer between a mass of gas at one uniform temperature and a tube surface at another uniform temperature. In most actual furnaces, however, such is not the case. Average effective temperatures must be selected in order to use Fig. 6.

Tube-wall temperature depends on the temperature of the fluid and its transfer coefficient inside the tube, the thermal resistance of the tube wall, and the total heat flux. At tube-wall temperatures below 500°C, when the radiant absorption rate is fairly insensitive to the receiving temperature, it is usually acceptable to use the average fluid temperature plus 50°C. At higher temperatures, more-detailed calculations are often necessary. In installations, such as pyrolysis furnaces, having extremely high tube-wall temperatures, it is advisable to divide the receiving area into zones of different average temperature, and to calculate each zone's heat-absorption rate separately.

Because about 70% of the radiation to the convection section is received by the front row of tubes, the average tube-wall temperature for these tubes should be used to calculate convection-section radiation.

Depending on the type of furnace, there may also be considerable variation in flue-gas temperature within the firebox. However, because of the large amount of turbulent mixing that occurs, the transparency of the radiating gas, and the effect of secondary radiation from exposed refractory, it is very difficult to quantitatively allow for these temperature variations when making the heat-flux calculations. In the absence of actual physical barriers between zones, it appears to be better to base all heat-transfer calculations on a single effective radiating temperature throughout the firebox.

Heat balance provides firing rate

The aforementioned procedure enables calculating the firebox temperature necessary to transfer a specific amount of heat into a specified radiant-section coil. The next step is to determine the firing rate necessary to maintain that temperature. This is done through a heat balance around the firebox.

Heat is put into the radiant section from three primary sources: (1) the net heat of combustion, q_n; (2) the sensible heat of the combustion air, q_a; and (3) the sensible heat of the fuel and any atomizing steam, q_f. Heat is removed via absorption by the radiant-section tubes (q_R), radiation to the shield and convection tubes (q_S), casing losses (q_L), and sensible heat of the exiting flue gas (q_{g2}):

$$q_n + q_a + q_f = q_R + q_S + q_L + q_{g2} \qquad (9)$$

The terms q_a and q_f, being generally proportional to the amount of fuel burned, can be expressed as ratios to q_n. Similarly, the loss q_L is usually taken as 1 to 3% of the net heat release, depending on the furnace design and experience. Finally, the fraction of net heat release remaining in the flue gas is a function of fuel composition, flue-gas temperature, and excess air. For the common hydrocarbon fuels, the ratio q_g/q_n can be correlated on a single set of curves, as in Fig. 8. On this basis, Eq. (9) can be rearranged to allow direct calculation of q_n:

$$q_n = \frac{q_R + q_S}{1 + \dfrac{q_a}{q_n} + \dfrac{q_f}{q_n} - \dfrac{q_L}{q_n} - \dfrac{q_{g2}}{q_n}} \quad (10)$$

The last function required to close the computational loop around the radiant section is the relationship between the effective gas radiating-temperature, T_g, and the exit flue-gas temperature, T_{g2}. For box-type heaters having an approximately square cross-section, and no areas of refractory with direct flame impingement, the two temperatures may be assumed equal.

In the opposite extreme of a high-temperature furnace having a tall, narrow firebox with wall-mounted radiant burners, T_g may be 100 to 150°C higher than T_{g2}. Other types, such as narrow, bottom-fired vertical cylindrical heaters, will fall somewhere in between. The magnitude of the difference must be determined empirically from experience with similar designs.

Convection-section design

The relative importance of the convection section in fired-heater design has increased markedly in recent years. One reason is the higher cost of fuel, which has resulted in setting furnace-efficiency targets far higher than could formerly be economically justified. Another major factor is the development of chemical conversion processes that demand extremely high heat-input fluxes, and correspondingly high firebox temperatures.

The result is a smaller fraction of the total heat released being removed in the radiant section, and a correspondingly greater convection-section loading. Whereas typical heat-distribution formerly was roughly 50% radiant section, 20% convection section, and 30% losses, it now may be, respectively, 40%, 50% and 10%.

Thus, there has been a trend toward larger convection sections and more use of extended surface. Also, there are more multiservice convection sections that—via steam generation, feedwater heating and similar extraneous services—take up heat that cannot be absorbed in the primary process. The result has been to make convection-section design more difficult and more critical.

Transfer coefficients depend on tube type

As in the radiant section, heat is transferred in the convection section by both radiation and convection. The classical basis for calculating convection-section heat transfer was developed by Monrad [4], who took into account direct convection, radiation from the gas, and radiation from the refractory walls. Monrad's basic method has been modified in light of later experimental results and adapted to extended-surface tubes by Schweppe and Torrijos [5].

Because of the previously mentioned trend toward increased convection-section heat recovery, most present-day heater designs incorporate extended-surface tubes. Because of the many types and sizes of extended

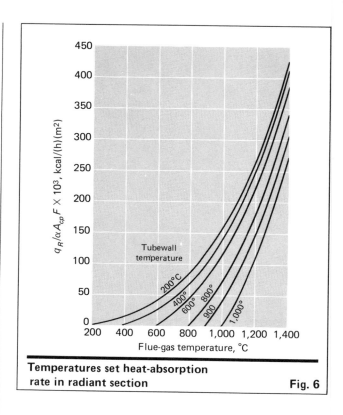

Temperatures set heat-absorption rate in radiant section   Fig. 6

surface available, it is not practical to present in this article a specific procedure for calculating coefficients. (For such information, refer to the Schweppe and Torrijos article, as the present article is limited to general principles of design.)

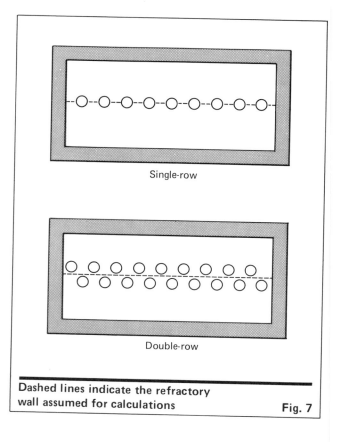

Dashed lines indicate the refractory wall assumed for calculations   Fig. 7

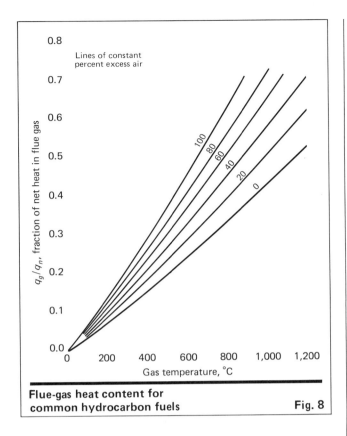

Flue-gas heat content for common hydrocarbon fuels — Fig. 8

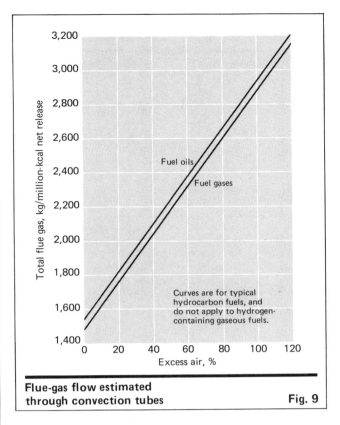

Flue-gas flow estimated through convection tubes — Fig. 9

In any case, the amount of flue gas flowing through the convection tube-bank must be known. As with the flue-gas heat content, the flue-gas quantity can be calculated from stoichiometric relationships involving the fuel consumption, the fuel heating value, and the amount of excess air. For normal hydrocarbon fuels, the quantity can be estimated closely from the net heat release and the excess air, as in Fig. 9.

For bare tubes, the inside-tube heat-transfer resistance is usually low enough that a reasonable estimate is satisfactory, at least for preliminary design. Fig. 10 presents calculated overall heat-transfer coefficients for typical sizes of bare tubes. These curves were calculated from the data of Schweppe and Torrijos, assuming usual staggered-tube arrangements, fluid temperatures, and inside-tube film coefficients. The average gas temperature used in this figure is the arithmetic average inside-tube fluid temperature plus the log-mean-temperature difference from flue gas to fluid. The mass velocity is that through the minimum cross-section of the tube bank.

Having established the duty and overall heat-transfer coefficients for each service in the convection section, one calculates the required surface via conventional heat-balance and heat-exchange procedures. For normal fuels, Fig. 8 can be used to establish flue-gas temperatures throughout the section.

Radiation from the firebox

In rating the radiant section, a quantity, q_s, representing the heat radiated directly from the firebox to the convection section, was calculated. This quantity of heat must be added to that derived by the convection-section calculations, to determine the total heat put into the lower tubes in the convection section. Because almost all the radiant heat is absorbed in the first two rows of the convection section, it is usually not necessary to make any detailed breakdown of the distribution.

Actual furnace performance

Some caution must be exercised when using published heat-transfer coefficients for industrial heater design. Experimental coefficients are usually measured under ideal conditions, with uniform flow distribution and no bypassing. Actual furnaces deviate from the ideal in some respects, as follows:

In tall convection sections, there can be considerable bypassing of hot flue gas through the tube header-boxes. Particularly with extended-surface tubes, there is often significant leakage area through the tubesheet holes. Under the driving force of the pressure drop through the convection section, hot flue gas flows out from the lower part into the header box, up through the box, and back into the upper convection section. This bypassing seriously undermines the overall performance of the convection section. Careful sealing of tubesheet holes, and in some cases partitioning of the header boxes, is necessary to avoid this difficulty.

Flow distribution may be far from uniform, particularly with long convection-section tubes. Flow is most commonly excessive in the center, near the stack opening, and is nearly stagnant in areas near the ends. Ample hood area above the convection tubes, and careful attention to flow patterns entering the section, are helpful in avoiding this problem.

Because of these and similar effects, some safety fac-

| Altitude corrections for gas density and stack draft ||
Height above sea level, m	Correction factor
0	1.000
500	0.942
1,000	0.887
1,500	0.835
2,000	0.785
2,500	0.738
3,000	0.694

Multiply sea-level density, Eq. (12), and stack draft, Fig. 11, by the correction factor.

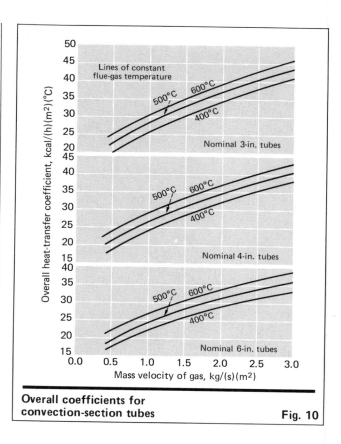

Overall coefficients for convection-section tubes — Fig. 10

tor should be applied when heat-transfer coefficients are calculated from laboratory data. This is particularly true for extended-surface designs. More reliance can usually be placed on good test data taken from actual furnaces.

Stack design

The stack system must create sufficient draft to draw the flue gases through the radiant and convection sections, and must discharge the gases at a suitable height. This is usually done in process furnaces by natural or induced draft.

Radiant-section draft

A pressure drop across the burners is usually required to draw in the combustion air. Its magnitude depends on the type of burner and the fuel, and is typically about 6 mm of water.

All sections of the furnace generally must also be below atmospheric pressure, so that flows through peepholes, tube openings and the like will be outside air moving in, rather than hot gas going out. A normal design point is about 2 mm negative pressure at the inlet to the convection section. In most modern heaters that have tall radiant sections, this provides more than enough draft for the burners.

Velocity head in the convection section

Pressure-drop calculations through the convection section and stack system are conveniently made in terms of velocity head. The velocity head in millimeters of water is given by:

$$p_v = 0.051\, G^2/\rho_g \quad (11)$$

Here, G is the mass velocity in kg/(m²)(s), and ρ_g is the gas density in kg/m³. If the flue-gas composition is known, its density can be readily calculated from the usual gas laws. Flue-gas density is relatively insensitive to fuel composition and amount of excess air. For the usual situation at sea level, the following equation is applicable:

$$\rho_g = 342/T_g \quad (12)$$

Here, T_g is the gas temperature in °K.

For banks of bare tubes, the frictional pressure drop is about one half of the velocity head per row. For ex-

tended-surface tubes, it is best to rely on the manufacturer's data, or correlations developed specifically for the particular configuration. In the absence of such information, the generalized relation of Schweppe and Torrijos is serviceable.

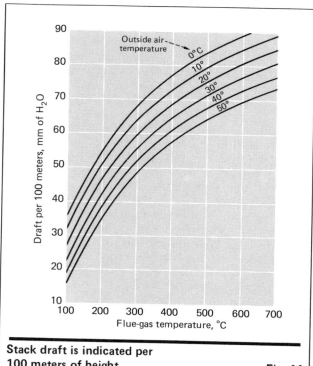

Stack draft is indicated per 100 meters of height — Fig. 11

Stack and damper head-losses

Remaining losses may be estimated as follows:

Source	Velocity head-loss
Stack entrance	0.5
Damper	1.5
Stack and ducts	1.0/50 dia.
Stack exit	1.0

Other losses may be similarly calculated from the usual published loss-coefficients.

Stack draft

Draft depends on the difference in density between the hot flue gas and the surrounding air. The molecular weight of the flue gas is quite insensitive to fuel composition, and is about 28.5 for the usual hydrocarbon fuels. On this basis, draft per 100 m of height is shown in Fig. 11. Note that tall convection sections have their own draft, for which allowance should be taken in the computations.

When calculating available draft, some allowance should be made for lower stack temperatures due to heat loss, air leaking in, and the like. The amount depends on the flue-gas temperature and the degree of insulation of the stack and ductwork. For an unlined stack, and flue gas at about 400°C, a reasonable allowance would be 50°C.

Most process plants are close to sea level. However, for installations above 300 m elevation, decreased atmospheric pressure must be taken into account. Lower pressure increases the volume of gas, and therefore the pressure drop, through the furnace and stack system. At the same time, it decreases the draft because of the smaller density differences. Corrections for gas density and stack draft are given in the table.

Overall furnace rating

Design of an actual furnace requires combining the various heat-exchange and flue-gas heat-content relationships with the specified flows, temperatures and heat duties for the process fluids. This must be done for each stream and for each section of the heater. In almost all cases, too many factors are involved to permit vigorous simultaneous solution of all the equations. Instead, an iterative procedure with successive approximations must be followed.

The approach to any specific problem depends on the type of heater under consideration, and on the purpose of the calculation. For example, the procedure for rating a vendor's offering will not necessarily be the same as that for checking an initial design. Still another approach would be necessary to estimate the effect of modifications to an existing heater.

The preceding correlations and procedures have been presented on the basis of performing the calculations by hand. This is to permit the occasional user to obtain valid results with a reasonable expenditure of time and effort. The many repetitive calculations, however, suggest the advantages of computerization, particularly for anyone who handles a significant number of heater design problems.

In addition to speeding up the iterative calculations, computerization allows substituting exact calculations fitted to the specific situation for some of the generalized correlations. For example, Fig. 3, 8 and 9 are based on the most common liquid and gaseous fuels. In a computerized operation, it is quite practical to calculate the flue-gas composition, amount, and heat content for the specific fuel and the specific combustion conditions. In this way, abnormal fuels or other unusual conditions can be accommodated.

When developing such a program, procedures should be included so that a heater can be designed to meet specified performance criteria, or so that the performance of a fixed design can be estimated for various operating conditions. Beyond that, there are a number of options that might be included, depending on individual needs. Typical requirements are the ability to handle multiple fuels, gas-turbine exhaust, combustion-air preheat, and extraneous heat-recovery streams in the convection section. The degree to which these variations can be accommodated is a major factor in determining the complexity of the program.

Computer programs of this type are valuable for handling the numerous repetitive calculations involved in optimizing a new plant design, for confirming the validity of vendors' proposals, and for predicting the effect of different feedstocks or operating conditions on the performance of existing heaters.

There are, of course, many factors in addition to the rating calculation that must enter into the basic design of a heater. Typical ones to consider are allowable heat flux rates, peak-to-average flux ratios, coil arrangement, combustion volume, and burner size and placement. These factors are all important but their discussion is beyond the scope of this article.

What has been presented here is a relatively simple procedure that can be used to calculate heat-transfer rates and efficiency for a wide variety of fired heaters. Experience has shown that the results are accurate enough to meet the needs of most engineers.

References

1. McAdams, W. H., "Heat Transmission," 3rd ed., McGraw-Hill, New York, 1954.
2. Perry, J. H., "Chemical Engineers' Handbook," 4th ed., McGraw-Hill, New York, 1963.
3. Lobo, W. E., and Evans, J. E., Heat Transfer in the Radiant Section of Petroleum Heaters, *Trans. of AIChE*, Vol. 35, 1939, pp. 743–778.
4. Monrad, L. L., Heat Transmission in Convection Section of Pipe Stills, *Ind. & Eng. Chem.*, Vol. 24, 1932, p. 505.
5. Schweppe, J. L., and Torrijos, C. Q., How to Rate Finned-Tube Convection Section in Fired Heaters, *Hydrocarbon Proc.*, Vol. 43, No. 6, 1964, p. 159.

The author

Norman Wimpress, assistant chief engineer for C. F. Braun & Co. (Alhambra, CA 91802), has had more than 30 years of experience in the design and operation of process-type fired heaters. He holds a B.S. from California Institute of Technology and an M.S. in chemical engineering from Massachusetts Institute of Technology.

Guide to Economics of Fired Heater Design

There are a lot of factors that have to be juggled in order to design a heater that is reliable and that provides an overall minmum cost per unit of throughput.

PETER VON WIESENTHAL and HERBERT W. COOPER,* Alcorn Combustion Co.

Economic design of a fired heater requires that three criteria be met: The heater must meet the required duty at the lowest capital investment; it should perform its function at the lowest possible level of fuel consumption; and, it should assure the lowest maintenance cost at a maximum of reliable onstream performance. In the following pages, we will show this can be done.

A designer must consider the factors listed above, and relate them to the user's specific process requirements. For example, an economical design for a preheater on a catalytic cracking unit targeted for a three-year run could be quite different from that of a low-temperature, carbon-steel reboiler with a minimal process liability.

In turn, an economical design for a 400-million-Btu./hr. crude-heater must place a substantially greater emphasis on fuel efficiency than a similar design on a 5-million-Btu./hr. hot-belt heater.

These two examples are obviously extremes, but they do emphasize the point that materials of construction and layout should by no means be the only consideration for which a good designer assumes responsibility. Any approach to a sound and economical engineering selection must consider the following:
- Heater configuration.
- Optimization of design.
- Materials of construction.
- Installation cost.

Heater Configuration

There has been a tendency in the chemical process industries to categorize generic heater design-types as either "economical" or "expensive" by configuration or layout.

*H. W. Cooper is now with Dynalytics Corp. See "Meet the Authors" on p. 336.

This article was presented as a paper to the 24th Annual Petroleum Mechanical Engineering Conference, Tulsa, Okla, Sept. 22, 1969.

Originally published April 6, 1970

A typical example is the long-held belief that vertical cylindrical heaters are always less expensive than horizontal-tube types. This conviction originated at a time when the vertical integral radiant-convection-tube heater, with extended surface (Fig. 1), came into use in competition with conventional horizontal-tube heaters. Plug-type cleanout fittings were in general use at this time, and very few heaters were equipped with extended-surface tubes. The great competitive advantage of the new vertical-tube unit was largely the result of the great reduction in the amount of bare convection surface required, and the attendant reduction in the number of expensive plug fittings needed, rather than the position of the tubes.

Within a few years, however, most horizontal-tube heaters were also being equipped with extended-surface tubes; further, with the advent of steam-air decoking resulting in the widespread elimination of plug fittings, the economic balance between vertical- and horizontal-tube heaters was largely restored.

Today, no categorical statements should be made regarding lower cost of one design approach vs. another purely on the basis of heater configuration. Each application must be studied, and a decision made on its own merits.

Basic Design Criteria

Specific process considerations and design requirements will definitely favor particular heater configurations. In full realization of the risk inherent in general statements, the following criteria are nevertheless useful in making basic design decisions:

A requirement for tall stacks 150 ft. or more high tends to favor the use of vertical-tube heaters when single units are involved. In the case of multiple-heater installations, a single common stack with collecting ductwork generally is the most economical approach and also leaves the designer greater free-

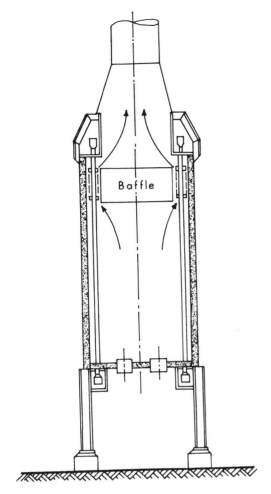

VERTICAL-TUBE cylindrical fired heater—Fig. 1

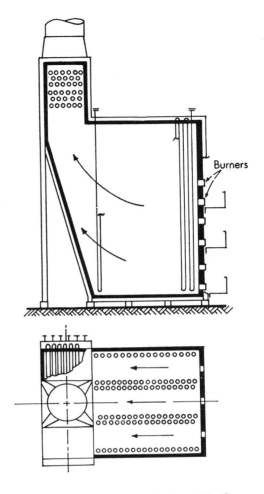

MULTI-LANE horizontally fired heater—Fig. 2

dom in selecting the optimum layout for each individual service.

Heavy-oil firing has a substantial effect on heater-design economics. This is particularly true with fuels that contain significant amounts of sulfur, metallic salts, and vanadium.

When no basic operating prejudice exists against upfiring of heavy fuels, the floor-fired cylindrical or box heater with vertically hung tubes presents the most economical approach. Both capital investment and maintenance cost are reduced since, in most instances, high-alloy tube supports can be completely eliminated in the radiant section. By limiting the convection tube length, convection supports can often be eliminated also.

Where horizontal fuel-oil firing is required by customer specifications, a multi-lane, horizontally fired vertical-tube unit can satisfy both the requirement for horizontal firing and the elimination of internal tube supports in the firebox by the use of hung tubes, with the return bends placed in header boxes (Fig. 2).

Cost savings are most substantial in those instances where fuel composition requires 50-50 or 60-40 chrome-nickel tube supports to resist vanadium attack. These special alloys cost approximately two to three times as much per pound as the conventional 25-20 or 25-12 supports. In addition, however, the lower allowable stresses of the vanadium-resistant materials require the use of heavier supports. In many cases, the added weight can drive the overall cost of intermediate tube supports up by a factor of five or six over a conventional design.

In the case of horizontal-tube heaters, similar heavy-fuel firing requirements generally justify the use of larger tube diameters, because of the greater permissible span between intermediate supports and the attendant reduction in the number, and consequently the cost, of these tube supports. Since, however, the use of larger tube sizes results in an increase in the cost of the heater coils, particularly in the case of alloy materials, the designer must check the economic balance between cost of tubes and cost of supports.

Apart from affecting heater configuration, the requirement for heavy fuel-oil firing also generally involves a modification of the convection section extended-surface. Many refiners consider the use of cylindrical stud surfaces mandatory when firing such fuels. This in turn has a profound effect on heater cost. The approximate relative investment in pressure parts for carbon steel studs vs. fins for a typical convection section in both carbon steel and 5% chrome tubing for a 100-million-Btu. heater is illustrated below.

	Fins, $	Studs, $	Increases, %
Carbon steel	12,750	32,700	156
5% Chrome	30,700	57,400	87

Combustion Air Preheat

Combustion-air preheat as a method for achieving maximum fuel efficiency is receiving ever-increasing interest for new heater installations. Based on present-day fuel costs and specified payout periods, air preheat can generally be justified on a majority of the new heater installations absorbing in the range of 100 million Btu./hr., or more. Unfortunately, the theoretical aspects of return on investment utilizing combustion-air preheat frequently do not conform to stringent-budget limitations that are often imposed on new projects.

For combustion-air preheat to receive serious consideration, the designer must consider all possible methods for reducing the needed capital investment.

The vertical-tube, floor-fired heater presents the most attractive layout for use of air preheat. This type of design generally requires fewer burners than a horizontal-tube heater. In addition, the entire flue-gas and combustion-air ductwork and burner plenum systems tend to be compact, simple, and as a result more economical.

It is also important to determine the optimum breakpoint between the use of heater convection surface and combustion-air preheat surface. In the case of a heater with carbon-steel tube coils, substantial use of convection surface in conjunction with a small air-preheater will probably show the greatest return. On the other hand, should the heater convection surface consist of alloy materials (or expensive stud surfaces) a greater emphasis on air preheater surfaces will in all probability result in a more economical approach.

Shop Fabrication and Field Erection

The continuing and rapid increase in field erection costs has resulted in an ever greater emphasis being placed on shop fabrication of heaters and heater components. This trend tends to favor horizontal-tube heater designs, since they can be supplied in substantially greater capacities on a shop-fabricated basis

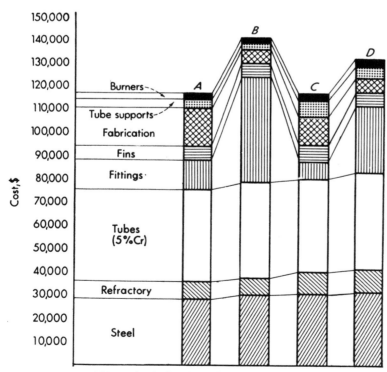

A. Vertical cylindrical; welding "U" bends.
B. Vertical cylindrical; rolled plug fittings.
C. Horizontal box; welding "U" bends.
D. Horizontal box; rolled plug fittings.

MATERIAL COSTS distribution for typical 100-million Btu. heaters having differing configurations and methods of fabrication—Fig. 3

than is possible with vertical-tube designs. This is primarily because the overall width of a horizontal-tube heater can generally be held within shipping limitations—a restriction compensated for by increasing the heater length and height.

Some limitation on this approach must be observed with regard to the heater height-to-width relationship. Excessive height will result in a substantial maldistribution of heat input along the vertical dimension of the heater, and a consequent increase in the radiant-heat transfer rates to the tubes in the lower portion of the unit.

This in turn can result in overheating of the lower tubes and potentially require a limitation on throughput to assure adequate run-lengths. Under such circumstances, the economics of maximum shop assembly could be completely negated by a reduction in furnace capability.

The type of return bend specified has a pronounced effect on the selection of an economical heater design. If welding "U" bends are called for, the cost of the fittings generally does not represent a sufficiently high percentage of the total materials cost to affect the selection of heater layout, either horizontal or vertical. Other considerations will tend to dominate the choice, particularly if the coil material is carbon steel.

If, on the other hand, cleanout-type plug-fittings are being considered, it becomes essential from an economics standpoint to maximize tube lengths in order to minimize the number of fittings required. The vertical cylindrical heater, with the usual cross-flow convection section, is at a definite disadvantage because of the relatively short length of the convection tubes and the resultant large number of fittings required. In contrast, the horizontal-tube heater, using convection tubes equal in length to the radiant tubes, requires far fewer fittings and generally presents a clear economic advantage. This relationship is illustrated in Fig. 3, which provides the material cost distribution for a 100-million-Btu. heater in a vertical as well as a horizontal layout for both welding "U" bends and plug fittings.

The same applies to any situation where the return bends are a high-cost item, as in the case of alloy-tube heaters operating at high temperatures and pressures. Consequently, reactor heaters in hydrocracking service, for example, are seldom designed as vertical cylindrical units if hydrocarbon convection surface is to be part of the design.

The cost of fittings relative to the overall cost of the heater increases on the smaller units. Economical design, therefore, requires a particularly careful review of alternative heater layouts when high-cost fittings are combined with small heaters. For such services, an attempt to achieve high efficiency by use of convection surfaces can be prohibitively expensive.

For many applications of this type, helical coil heaters have considerable merit, since return bends are eliminated. Efficiency must be sacrificed for economy, however, because the helical design does not lend itself readily to the use of convection surfaces.

Auxiliary Equipment

Auxiliary equipment, particularly soot blowers, can also have a marked effect on the economics of layout and design. The cost of soot blowers is largely a function of the number used, rather than the length of the soot-blower lance. The horizontal-tube heater, with its long, narrow convection section, requires far more soot blowers than the vertical heater with its relatively short (but wide) convection section. In practically all cases, the cost of the extra length of soot-blower lance needed for the vertical unit is negligible when compared to the cost of the extra blowers and attendant control equipment needed on the horizontal unit. Foreshortening of the convection section on the horizontal heater can partially overcome this situation, provided that process objections do not preclude such an approach.

It must be pointed out, however, that the very steps that reduce the cost of a soot-blower installation are exactly those that increase the cost of the convection surface by increasing the number of fittings and the cost of fabrication.

It is quite possible that a heater constructed with carbon steel or low-alloy return bends, and employing soot blowers, would be most economical as a vertical cylindrical unit: but if built with stainless steel pressure parts and fittings, it would be cheaper as a horizontal-tube heater (despite an increase in the number of blowers).

The foregoing covers some of the factors having a substantial effect on the selection of a basic heater design. It is apparent, however, that the several major considerations are not at all complementary and that, depending on the importance attached to each, the heater configuration can take a variety of forms.

The difficulty of determining the most-economical basic configuration for any given set of conditions should become quite apparent. A correct decision leading to selection of an inherently economical heater layout is only possible by a careful review of all the factors and their relationship to each other.

Optimization of Design

Once a basic heater layout has been selected, the next step is to optimize the design.

Primary consideration must be given to the pressure parts, since they provide substantial design options and thereby present the best opportunity for optimization and lowering of heater cost.

One of the first determinations the designer must make is to select the number of parallel flowpaths, or passes, to be used. He may choose a design based on using a minimum number of passes and larger tubes or, alternatively, more passes with smaller tubes. As a general rule, the cost of the pressure parts is directly related to the tube size. The larger sizes result in higher material costs, based on the need for greater wall thicknesses as the tube O.D. increases, and a higher per-pound cost for the tube metal in the case of alloys. The cost per pound of tube material in relation to tube size is given in Fig. 4, which clearly emphasizes the importance of this variable. For example, changing from 4 in. IPS (iron pipe size) to 6 in. IPS in 2¼% Cr material results in a 20% increase in the cost per pound of the tube metal (making no allowance for any increased wall-thickness requirements).

The cost relationship for return bends with changes in material and tube size is provided by Fig. 5. A move to the larger tube sizes, although increasing the unit cost of the fittings, is compensated for (in part) by a reduction in the number of fittings required for any given layout.

The composite effect on the overall cost of a tube coil with changes in tube size or material is shown in Fig. 6 for a typical, horizontal-tube heater. Variations in unit costs, the total weight of the tube coil, and the number of fittings needed have all been taken into consideration.

Practical use can be made of this information in conjunction with Fig. 7 and 8. The data on Fig. 7 show the cost of the major heater components for a 4-in. IPS horizontal-tube heater as a percentage of total heater cost over a wide range of heater sizes. The curves of Fig. 8 provide the cost in dollars per million-Btu.-absorbed for carbon-steel, high-efficiency, horizontal-tube heaters over the same sizes as Fig. 7.

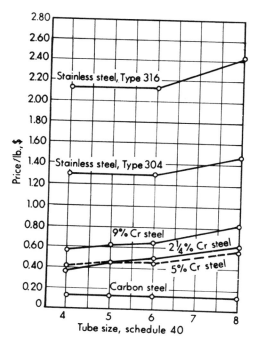

COMPARISON of tube price against tube size—Fig. 4

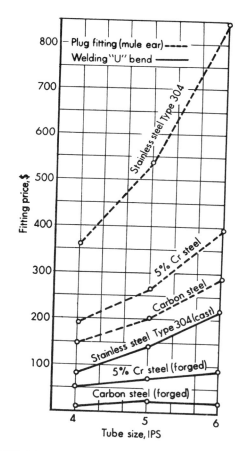

FITTINGS COST compared with tube size—Fig. 5

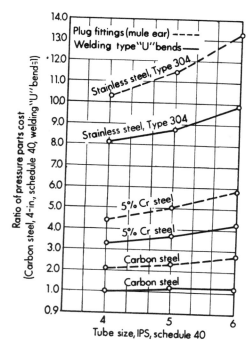

TUBE COIL cost for horizontal box heater—Fig. 6

Conversion of the information on Fig. 8 to other coil materials can be accomplished readily. First obtain from Fig. 8 the cost in dollars of the heater size being considered and multiply this number by the combined percentage of tubes, bends and fabrication as shown on Fig. 7. This will provide the cost of the carbon-steel pressure parts. Then apply against this the appropriate ratio factor for the desired alloy or tube size from Fig. 6, to obtain a new cost for the revised pressure parts. Adding this total back again to the balance of the heater cost obtained from Fig. 8 will give the approximate overall cost of the heater for the new conditions.

The foregoing indicates that increasing tube diameters will generally raise heater costs. For a wide range of refinery applications, the amount of this increase will be materially influenced by the criteria used for establishing tube-wall thicknesses. For most refinery applications other than catalytic reforming and hydrocracking, the methods in general use for determining tube-wall thickness do not result in a controlling criterion. In most instances, calculated values for wall thickness based on temperature and pressure are so low that the tubes would not be acceptable from the structural standpoint, i.e. their ability to withstand handling, thermal stresses, and buckling at points of supports. As a result, the tube walls are increased by adding a retirement thickness or a corrosion allowance, or possibly some other arbitrary method is used for establishing minimum allowable wall-thicknesses.

If IPS schedule thicknesses are called for, any increase in tube diameter will result in a substantial increase in coil metal weight. If, on the other hand, the design criteria are based on API RP-530, it is quite probable that tube-wall thickness for a 6-in. IPS coil would be no greater than that for a 4-in. IPS coil. This would tend to favor the use of larger pipe sizes and a reduction in the number of passes if the considerable savings in external manifolding, flow controls, and instrumentation were properly taken into account.

Heater Radiant Surfaces

The heater radiant surface is usually well defined, since most heater specifications generally are quite clear in stating maximum-allowable, average, radiant transfer rates. Considerable flexibility is nevertheless possible, based on varying tube center-to-center spacing, or possibly a decision to subject the tubes to direct radiation from both sides. Maximum-allowable transfer rates normally refer to tubes on a nominal two-diameter spacing backed by a refractory wall and which receives direct radiant heating from one side only.

An increase in specified, average, radiant transfer rates is generally permissible if this can be accomplished without exceeding the maximum peak front-face rate corresponding to the overall average rate. Fig. 9 provides a convenient method for determining

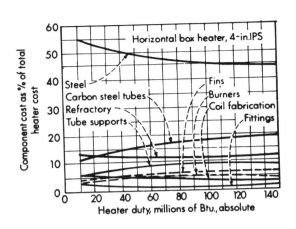

COMPONENT costs making up total heater cost—Fig. 7

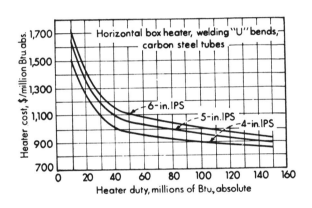

HEATER cost vs. duty for horizontal box heater—Fig. 8

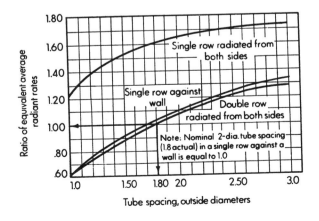

HEAT TRANSFER rates vs. tube arrangement—Fig. 9

how much the average heat-transfer rate can be increased by varying the tube center-to-center spacing or by firing on both sides of the radiant tubes (or both). Based on a single row of tubes on two nominal diameters, and fired from one side only, being equated to 1, the ratios given provide a direct multiplier for establishing the equivalent average rate for any other desired arrangement.

At first glance, modification of tube spacing or firing arrangement would appear to be an obvious method for reducing the amount and cost of radiant surface, and thereby provide a more economical heater design. However, any increase in tube spacing, or use of double firing, results in an attendant increase in the size, and hence cost, of the radiant enclosure of the heater. It again becomes a matter of establishing the economic balance, this time between the reduction in cost of the tubes, fittings, and supports versus the increased cost of the refractory and steel enclosure. Unfortunately, the breakpoint is not a clearly defined one, and from a practical standpoint its determination generally requires a relatively detailed layout and estimate for each alternative approach.

As an approximate guide, however, the use of double-firing will generally not show any savings in alloy materials up to, and including, 5% Cr in Schedule 40. The Schedule-80 wall thicknesses will probably present a different picture, particularly in the larger tube sizes. In most of the stainless alloys, the savings in tube material will substantially outweigh the extra cost of the heater enclosure for either wide tube spacing or double firing. The only probable exception is in the case of smaller heaters where the size range is such that a conventional single-fired approach permits complete shop fabrication, and the alternative double-firing results in a field-erected design.

Double firing will also result in a lower in-tube pressure drop as a consequence of the reduction in radiant surface and attendant flow length.

Convection-Section Optimization

In the case of high-efficiency heaters, the single most important area for design optimization and the associated material savings rests in the convection-section layout. The preponderant percentage of fired heaters built today utilize some form of extended surface to increase the heat transfer on the external surface of the convection tubes. The wide range of variations possible in the use of extended surface with respect to type, material, spacing, height and thickness presents a large number of variables to be considered. As a result, true optimization becomes a laborious and uncertain task. In addition, the position of the tubes with respect to horizontal and diagonal spacing has a significant effect on convection-section effectiveness.

The optimization of convection sections is an ideal application for the computer. It can be of tremendous assistance in permitting a rapid run-through of the involved iterative calculations essential to finding an economical design. A typical printout is illustrated in Fig. 10, for one row of a row-by-row calculation. This program checks a variety of arrangements, printing out complete data for each row, and at the end of its run gives the cost of all the convection-section materials, including tubes, fittings, extended surface, supports and the convection enclosure.

Experimental investigation of design and performance relationships for heater convection-sections would undoubtedly show substantial economic returns particularly when related to the computer programs. The generally accepted published data on convection heat transfer leave a great deal to be desired in this application to the size and configurations usually found in large fired heaters.

The very considerable maldistribution of heat input around the circumference of a convection tube, and its effect on both tube metal and extended surface tip-temperature, has received little consideration to date. The manner in which flue gases are evacuated from the convection section has a substantial influence on the flow of gases in the convection section and the effectiveness of the heat-transfer surface. Flue-gas baffles, although helpful in providing better flue-gas distribution along the length of a convection section, also have a substantial and undefined shielding effect on the surface

```
ROW NUMBER  5
0.562 INCH FINS,        4.00 FINS/INCH
88.7 SQUARE FEET (TOTAL) ON THIS ROW
FLUE GAS MASS VELOCITY = 0.409

TEMPERATURES                    IN          OUT
FLUE GAS                      946.8        813.2
PROCESS FLUID                 436.1        448.2
FILM                          460.1        472.3
TUBE METAL                    468.6        480.8
FIN TIP                       533.3        543.8

GENERAL INFORMATION
CUMULATIVE SURFACE (SQ FT)                 295.9
OUTSIDE FLUX (MAX)                        2814.2
BARE TUBE FLUX (MAX)                     16728.8
FIN EFFICIENCY                             0.893
H(OUTSIDE)                                 7.619
FLUE GAS DELTA P (IN H2O/ROW)             0.0076
CUMULATIVE FLUE GAS DELTA P (IN H2O)      0.0378
DUTY THIS ROW (MM BTU/HR)                  0.249
CUMULATIVE DUTY (MM BTU/HR)                1.331


PARTIAL MATERIAL AND COST ESTIMATE
(FROM BOTTOM OF FIRST ROW TO TOP OF TOP TUBE ROW)

                        TOTAL    TOTAL
                        AREA     WEIGHT       COST
ITEM            NO      SQ FT    POUNDS       DOLLARS
TUBES           20               531.2        69.06
FINS                             258.6        71.28
RETURN BENDS    18                             72.00
WELDS           40                            227.99
END TUBE SHEETS  2      8.5                   123.89
INTERMEDIATE T.S. 1     2.9      73.3         100.36
WALLS                   23.2                  169.91
HEADER BOXES     2      8.5                    64.08
                                             --------
                                    TOTAL     898.59
```

CONVECTION SECTION optimization printout—Fig. 10

ALCORN COMBUSTION COMPANY

INSULATION STUDY

INSULATION	THICKNESS (INCH)	DENSITY LB/CU.FT.	MAX WORKING HOT FACE T.	MAX. SERVICE TEMP DEG.F
HOT LAYER LHV 1,2,4 MIX	4.00	55.00	1700.00	2000.00
COLD LAYER MONOBLOCK	2.00	15.00	800.00	1900.00

VERTICAL FLAT SURFACE (SIDE WALL)

AMBIENT AIR TEMP		60 F		80 F		90 F	
WIND VELOCITY		0 MPH	5 MPH	0 MPH	5 MPH	0 MPH	5 MPH
HOT FACE TEMP (F)	1500						
HEAT LOSS		246.5	250.3	245.1	248.7	244.4	247.9
CASING TEMP (F)		184.7	145.2	198.6	162.2	205.6	170.6
HOT FACE TEMP (F)	1600						
HEAT LOSS		272.1	276.3	270.8	274.7	270.1	273.9
CASING TEMP (F)		194.6	152.2	208.1	169.1	214.9	177.4
HOT FACE TEMP (F)	1700						
HEAT LOSS		298.7	303.2	297.4	301.6	296.7	300.8
CASING TEMP (F)		204.4	159.3	217.6	176.1	224.1	184.4
HOT FACE TEMP (F)	1800						
HEAT LOSS		326.3	331.1	324.9	329.4	324.2	328.6
CASING TEMP (F)		214.2	166.7	227.1	183.2	233.6	191.3
HOT FACE TEMP (F)	1900						
HEAT LOSS		354.7	359.9	353.3	358.3	352.6	357.4
CASING TEMP (F)		224.0	174.0	236.6	190.2	243.0	198.3
HOT FACE TEMP (F)	2000						
HEAT LOSS		384.1	389.7	382.7	388.0	382.0	387.1
CASING TEMP (F)		233.8	181.3	246.1	197.5	252.3	205.4

INSULATION STUDY by computer can show design data for a wide range of conditions—Fig. 11

immediately below them. More definite data are badly needed in all these areas to permit development of correct design criteria that could be used for real optimization of convection-section heat-transfer surfaces.

An increasing percentage of new heaters use various grades of alloy tubes, either because of the elevated temperature and pressure levels of the service or because of the corrosive nature of the feedstock. In either instance, an accurate determination of the temperature profile within the coil will permit correlating the expected service conditions at any point with a suitable alloy to meet these conditions. With this knowledge at hand, a combination of several alloys (or possibly carbon steel and alloy) is a most effective method for decreasing the cost of heater pressure-parts while still providing adequate reliability and service life. It is essential, however, to carefully check the alloy breakpoints for reduced load conditions, since under these circumstances the temperature profile of the pressure-parts is often shifted substantially toward the inlet of the heater. Again the computer can be a most useful tool in aiding these determinations with respect to time and accuracy.

Materials of Construction

The effective and economical use of materials as related to their physical properties and performance is equally applicable to all types of design; heater layout generally is not a major factor.

A primary consideration is the effective use of refractory materials and associated refractory supports. It is most important to properly relate the temperature capability and insulating value of refractory materials both to service requirements and cost.

Without question, the use of lumnite, haydite and vermiculite (or perlite) in the well-known 1:2:4 mixture provides a maximum of service capability at minimum cost. It should, however, be limited to reasonable temperature levels, since linear shrinkage can become a real problem at levels above 1,800 F. Under more-severe service conditions, a good commercial premixed castable should be considered. In this connection, it must be noted that on a field-installed basis the use of either premixed castables or insulating firebrick generally balances off in cost. On a shop-installed basis, the castables will generally show a slight cost advantage.

Alternative approaches to insulation for heater

settings consist of combinations of castable and block insulation, or combinations of castables of different densities and insulating properties. The wide range of variables, including hot and cold face-temperatures, insulating properties, and cost make the evaluation difficult and time-consuming. Here once more is an ideal application for the computer and its ability to select quickly and accurately the optimum design from a wide range of alternative possibilities. Fig. 11 illustrates a typical computer printout for such an investigation.

Specifying Tube Supports

Tube supports should also receive a fair measure of attention, less from the standpoint of minimizing material cost than from that of heater maintenance. Tube-support elements are probably the single largest replacement item for fired heaters. The design of an economical heater in its broader sense should take into account the ease of replacing castings in all the critical areas of the heater. The use of individual castings for each support point in the radiant section, and of readily replaceable shock-tube-supports, deserves special considerations. For ease of maintenance, it is essential that the tube supports be replaceable in both the radiant and shock sections of the heater without disturbing the tubes.

Installation Cost

The total installed cost of any heater must be the final measure of whether or not it is economical design. All the design skill expended to achieve an optimum heater may be fruitless if erection costs do not receive adequate consideration. Much attention has been focused in the recent past on increasing the extent of shop fabrication and reducing field-erection problems.

This effort has centered primarily on modifying the heater layout to permit even very large heaters to be shop-assembled and shipped in a minimum number of pieces. Although well intentioned, this effort does not always achieve the desired results. In a number of instances, individual heater-components have reached weights that have made it difficult to move and position them on arrival in the field.

Field construction forces, accustomed to structurally rigid systems such as vessels and towers, have not always understood that similar weights in the form of heater components are much more flexible and can be very difficult to handle. They require far more careful rigging and additional pick-points, so that more and heavier crane equipment is needed at the construction site.

Modular Components

An effective and possibly more economical approach to the problem of the heater designer (who requires greater flexibility in his layout) and the field erector (who needs more manageable pieces) is the use of a modular-component approach. Instead of heater radiant sections being split into two (or at times three) sections, the heater is broken up into large flat-sided components that incorporate both refractory and tubes, and is arranged for ready field bolting.

Actual erection experience with this approach indicates that it may actually result in lower field costs and a more economical heater (on an installed basis) than the use of a smaller number of extremely large sections with their attendant clearance and handling problems.

Whatever route is chosen, there is no question that a major emphasis on shop fabrication is absolutely essential for any heater to be competitive on an installed basis.

Summary

The wide range of variables and their interrelationship makes it difficult, if not impossible, to establish simple, all-encompassing rules that will inevitably lead to the most economical designs. The aim of this article is not to cover all aspects of design optimization but to enumerate and discuss a number of the major factors affecting the economics of heater design, and thereby indicate basic approaches for coping with a variety of situations.

The true measure of an economical heater should properly be: that design that optimizes a combination of capital investment, good fuel efficiency, operating reliability and low maintenance, to provide an overall minimum cost per unit of throughput. ■

Meet the Authors

◀ **Peter von Wiesenthal** is president of Alcorn Combustion Co., 850 Third Ave, New York, N. Y. 10022. He previously worked for Petro-Chem Development Co., Petro-Chem Construction Co., and the Fired-Heater Div. of Foster Wheeler Corp. He received his degree from Massachusetts Institute of Technology, and is a member of American Soc. of Mechanical Engineers, American Petroleum Inst., American Soc. for Testing and Materials, and the New York Soc. of Professional Engineers.

Herbert W. Cooper is now president of Dynalytics Corp., 775 ▶ Brooklyn Ave., Baldwin, N. Y. 11510, specializing as a consultant in fired-heater design. At the time of writing this article, he was Director of Research for Alcorn Combustion Co. He has also worked for Scientific Design Co. and Bechtel Associates. He holds bachelor's and master's degrees in chemical engineering from City College of New York, and a doctorate from Columbia University. He is a member of AIChE and Sigma Xi.

Section VI
STEAM GENERATION AND TRANSMISSION

Balancing boilers against plant loads
Basic data for steam generators—at a glance
Converting boiler horsepower to steam
Converting gas boilers to oil and coal
Designing steam transmission lines without steam traps
Estimating the costs of steam leaks
How to select package boilers
How to size and rate steam traps
How to test steam traps
Improving boiler efficiency
Install steam traps correctly
Select the right steam trap
Short-cut calculation for steam heaters and boilers
Steam traps

Balancing Boilers Against Plant Loads

Although the units of a process plant can rarely all operate at their peak efficiencies, the overall plant can be maintained at an economical optimum. Here's how.

R. D. Smith and *R. B. Scollon, Allied Chemical Corp.*

☐ Process plants generally contain multiple-boiler units served by common feedwater and condensate-return facilities. The efficiency of each boiler and of the entire system varies inherently with the operating load.

Although it would be desirable to operate each boiler at its highest efficiency, the constraints imposed by the demands from plant users, plus the need for excess on-line capacity required for reliability, rarely permits each of the boilers to be operated at its optimum. Consequently, opportunities for energy conservation lie in establishing an overall optimum load schedule for the steam generators that also satisfies the plant's steam demands.

Requirements for maximum economy

Boilers generally operate most efficiently at 65–85% of full load, compared to 80–90% of design for centrifugal fans. The efficiencies tend to fall off either above or below the optimum, with the loss in efficiency most pronounced at low loads. Consequently, it is usually more efficient, overall, to operate a lesser number of boilers at higher loads, than to operate a larger number at low loads.

When choosing those boilers to operate at higher loads, consideration should be given to the comparative performance curves of the boilers. Generally, newer units with higher capacity are more efficient than older units with smaller capacity. The smallest and least-efficient units should be reserved for plant load swings.

Also, those boilers that operate at highest pressure are usually most efficient, and should supply as much of the plant demand as possible. However, the high-pressure steam from these units should be used efficiently, being let down through backpressure turbines rather than through control valves. Degrading high-pressure steam through a pressure-reducing and de-superheating station is the least efficient route, and direct generation at the required low pressure is usually more economical.

Although it is more efficient to operate fewer boilers at a high rating, plant engineers must be ready to maintain steam supply in the event of a forced shutdown of one of the operating boilers.

This can be accomplished by establishing a load-shedding schedule for nonessential equipment and/or maintaining a standby boiler in a "live bank" mode. The standby boiler is isolated from the steam system at no load, but kept at the system operating pressure by intermittent firing of either the ignitors or a main burner, in order to replace ambient heat losses. Guidelines for live-banking boilers are:

- Shut all dampers and registers to minimize heat losses from the unit.
- Establish and follow strict safety procedures for ignitor/burner light-off.
- For units supplying turbines, take precautions against any condensate (which may be formed during banking) being carried through to the turbines. (Units with pendant-type superheaters will generally form condensate in these elements).
- Operators should familiarize themselves with emergency startup procedures for the boiler, and it should be determined that the fall-off in system pressure that may be caused by bringing the boiler on-stream can be tolerated.

In addition, since a changing plant-load will affect the auxiliaries common to all boilers, a schedule should be established for cutting these auxiliaries in and out of service.

Determining the potential

A survey is required to determine the savings that can result from applying these principles; it should include establishing the magnitude and duration of the total steam demand, as well as the profiles of efficiency at various operating loads for all steam-producing and steam-consuming equipment.

Accurately calibrated chart recorders are the best source for the demand information. Readings of individual boiler-steam flow meters can be added for total plant output. Peaks and valleys in this output should be identified with plant demands, and their frequency estimated.

Manufacturers' performance curves should be consulted for the efficiencies of the boiler plant auxiliary equipment, while the efficiencies of boilers should be

Originally published March 29, 1976

Unit efficiencies as a function of flue-gas temperature and percent oxygen

Fig. 1

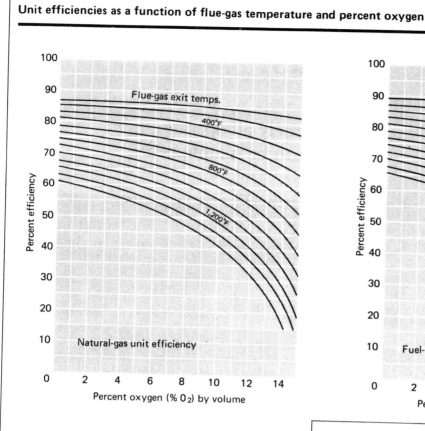
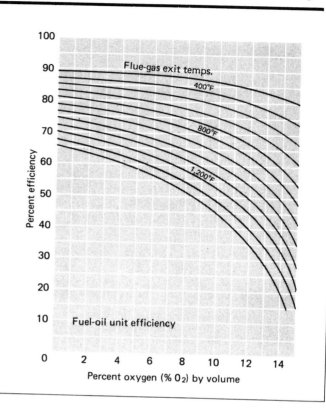
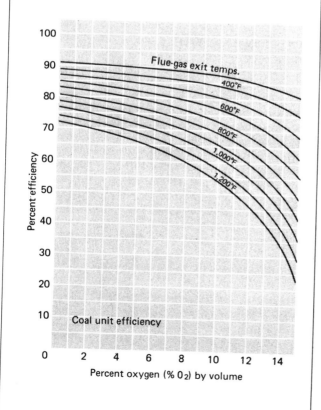

determined from operating test data from a minimum of four load points between ½ and maximum load.

A good estimate of boiler efficiencies can be made from measurements of flue gas temperatures and percent O_2 in the flue gas for a particular fuel classification. Since the percent O_2 in the flue gas correlates to the percent excess air, this can be used with the flue-gas temperature to read efficiencies from a chart like Fig. 1.

Step-by-step analysis

A typical analysis for an operating plant would proceed through three steps:

1. Determine the operating characteristics of each boiler in terms of both efficiency and fuel consumed per quantity of steam produced.
2. Determine the demand load and frequency for the plant.
3. Following the rules for maximum economy, preferrentially load the more efficient boilers; and by totaling heat inputs, determine the minimum overall plant energy usage by trial and error.

A further sophistication would involve computer control of boiler loading based on actual steam demand. This would permit continuous operating of the overall steam system at optimum efficiency.

Example: A plant has a total installed steam-generating capacity of 500,000 lb/yr, and is served by three boilers having a maximum continuous rating of 200,000 lb/h, 200,000 lb/h, and 100,000 lb/h respectively. Each unit can deliver superheated steam at 620 psig and 700°F with feedwater supplied at 250°F. Fuel

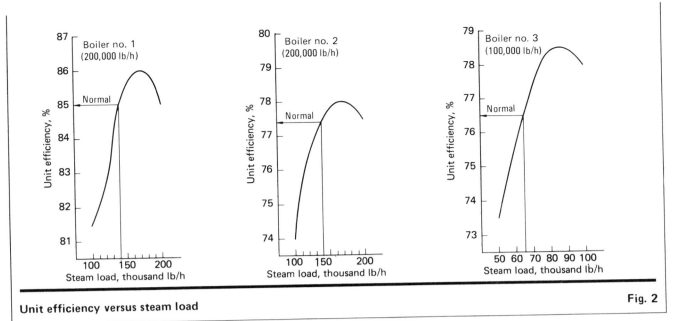

Unit efficiency versus steam load Fig. 2

fired is natural gas priced into the operation at $1.00/million Btu.

Total plant steam demand averages 345,000 lb/h and is relatively constant.

The boilers have been normally operated according to the following loadings.

Boiler design	Size boiler, thousand lb/h	Normal boiler load, thousand lb/h	Meas. exit temp., °F	Meas. % O_2	Unit eff.
1	200	140	290	5	85.0
2	200	140	540	6	77.4
3	100	65	540	7	76.5
Plant steam demand . . 345					

Step 1: Either through a consultant, or through plant measurements and Fig. 1, determine boiler efficiencies for each unit at three load points (Table I) and plot the results as in Fig. 2.

Step 2: The total heat input at the normal operating load is tabulated as follows:

Boiler design	Steam load, thousand lb/h	Heat input million Btu/h
1	140	186
2	140	204
3	65	96
Plant totals	345	486

Step 3: With the aid of Table I, the heat input versus load is plotted for each boiler (Fig. 3). Optimum steam-plant load-balancing conditions are satisfied when total plant steam demand is met according to:

$$\begin{bmatrix} \text{Boiler no. 1} \\ \text{heat input} \end{bmatrix} + \begin{bmatrix} \text{Boiler no. 2} \\ \text{heat input} \end{bmatrix} + \begin{bmatrix} \text{Boiler no. 3} \\ \text{heat input} \end{bmatrix} + \ldots = \text{Minimum}$$

By a trial-and-error exploration of Fig. 3, the minimum

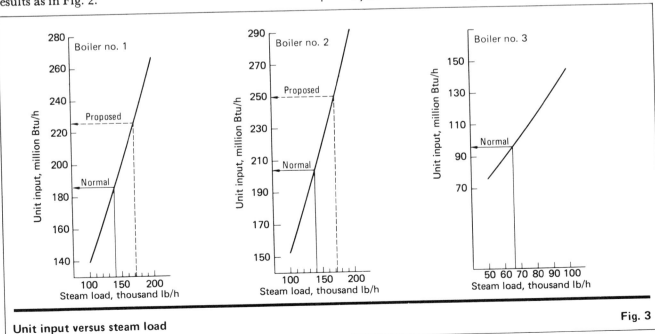

Unit input versus steam load Fig. 3

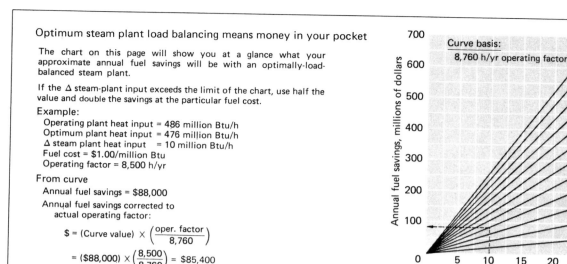

Annual fuel savings from optimum load balancing

Optimum steam plant load balancing means money in your pocket

The chart on this page will show you at a glance what your approximate annual fuel savings will be with an optimally-load-balanced steam plant.

If the Δ steam-plant input exceeds the limit of the chart, use half the value and double the savings at the particular fuel cost.

Example:
Operating plant heat input = 486 million Btu/h
Optimum plant heat input = 476 million Btu/h
Δ steam plant heat input = 10 million Btu/h
Fuel cost = $1.00/million Btu
Operating factor = 8,500 h/yr

From curve
Annual fuel savings = $88,000
Annual fuel savings corrected to actual operating factor:

$$\$ = (\text{Curve value}) \times \left(\frac{\text{oper. factor}}{8,760}\right)$$

$$= (\$88,000) \times \left(\frac{8,500}{8,760}\right) = \$85,400$$

Fig. 4

plant-heat input and corresponding optimum boiler-loading arrangement is:

Boiler design	Steam load, thousand lb/h	Heat input, million Btu/h
1	173	226
2	172	250
3	(Banked standby)	
Plant totals	345	476

Savings: The savings, which is 10 million Btu/h of fuel, can be related to dollars by the price of fuel, as in Fig. 4. Those savings, which amount to $85,400/yr, do not include either the additional savings from more-efficient fan operations or the additional costs of maintaining the third boiler in banked standby.

Part of the energy savings in this example was possible because the normal operation maintained a high ratio of total-capacity to total-demand, which permitted the banking of a boiler. However, it should be emphasized that even if a boiler cannot be banked, significant savings can be achieved through proper balancing of individual boiler loads to achieve a minimum energy consumption for the total system.

Other applications

While the example discussed above has been specific to steam generating systems, the principles of load balancing may be applied to various other multiple-equipment areas to achieve the same end-result of minimum energy consumption. These areas include the loading of turbine-generators, pumps, compressors, etc., or entire integrated systems composed of these components.

The Authors

R. Barry Scollon is a senior engineer in the Allied Energy Group of Allied Chemical Corp., where he has been involved in all phases of Allied's energy program, including the performance of plant audits and coauthoring several guides on conservation techniques. Before joining Allied Chemical, he worked in design and operation for Exxon Research and Engineering, where he specialized in utilities and offsites. He holds B.S. and M.S. degrees from Virginia Polytechnic Institute.

R. D. Smith is a senior engineer in the Allied Energy Group of Allied Chemical Corp., where he has been responsible for conducting in-plant energy surveys, performance evaluations and follow-up implementation programs. An author of several papers and guides on energy systems, he has also conducted in-house energy seminars. Before joining Allied, he worked for Exxon Research and Engineering in the design analysis of power system networks. He holds a B.S.M.E. degree from Newark College of Engineering.

Unit efficiency and input tabulation — Table I

Boiler design	Steam load, thousand lb/h	Meas. exit Temp., °F	Meas.[1] %O₂	unit eff., %	Output, million Btu/h	Input, million Btu/h
Boiler #1	200	305	2	85.0	226.2	266.1
	170	280	2	86.0	192.3	223.6
	130	300	7	84.0	147.0	175.0
	100	280	12	81.5	113.1	138.8
Boiler #2	200	625	2	77.5	226.2	291.9
	170	570	4	78.0	192.3	246.5
	130	520	7	77.0	147.0	190.9
	100	490	11	74.0	113.1	152.8
Boiler #3	100	600	2	78.0	113.1	145.0*
	85	570	2	78.5	96.1	122.5
	65	540	7	76.5	73.5	96.1
	50	500	11	73.5	56.6	76.9

[1] Percentage O₂ or equivalent excess air should be adjusted to minimum safe levels consistent with output requirements.

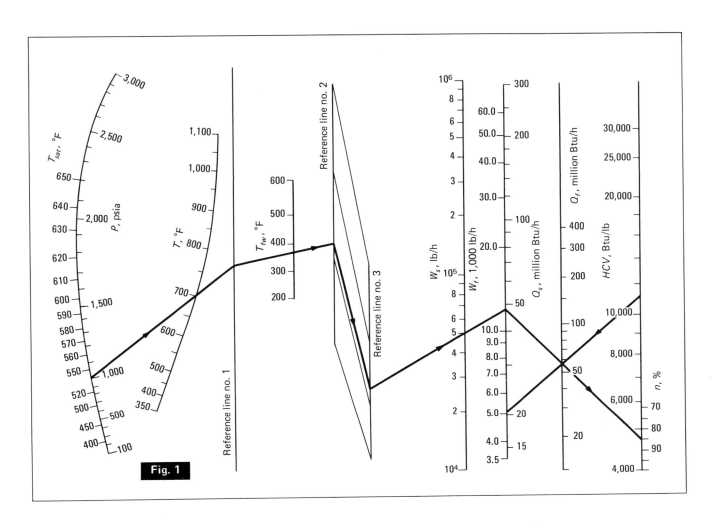

Basic data for steam generators—at a glance

V. Ganapathy, *Bharat Heavy Electricals Ltd.* *

☐ Chemical engineers frequently become involved with the steam generators supplying power and process steam for the operation of chemical industry plants. When they do, they need to have an idea of the following parameters for any given system:

- heat absorbed by the water and steam
- heat liberated by the combustion
- amount of fuel fired
- amount of combustion air
- combustion-air volume at operating temperature

These parameters can all be calculated by means of stoichiometry and heat balances, using steam tables and combustion data. However, the calculations are so involved that the furnace may be shifted to a different regime before the calculations are complete, so that the engineer's analysis is following rather than directing the performance.

These two charts may help with such situations. They incorporate steam-table data, so that reference to the tables is not necessary. Also, data for various fuels is included to allow comparison of the effects of the fuel on combustion air. The charts are based on the following equations:

$$Q_s = W_s(H_s - H_{fw})$$
$$Q_f = Q_s/n$$
$$W_f = Q_f/HCV$$
$$W_a = CaQ_f/10^6$$
$$V_a = W_a(460 + T)/2{,}400$$

*High Pressure Boiler Plant, Tiruchirapalli 620 014, India

Originally published June 6, 1977

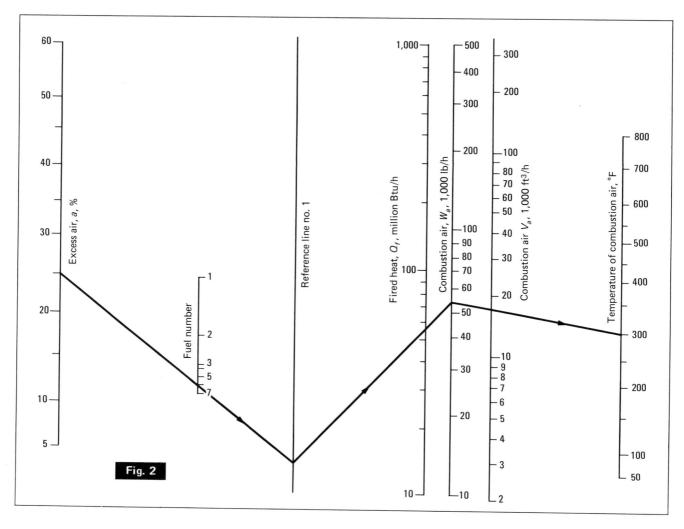

where: Q_s = heat absorbed by the water and steam, Btu/h
Q_f = heat fired, Btu/h
W_s = steam generated, lb/h
W_f = fuel fired, lb/h
W_a = combustion air, lb/h
V_a = combustion air, ft³/min
H_s = enthalpy of generated steam, Btu/lb
H_{fw} = enthalpy of boiler feedwater, Btu/lb
HCV = high heat of combustion of fuel, Btu/lb
n = efficiency of boiler, %
a = excess air factor (at 25% excess air, $a = 1.25$)
C = a constant depending on the fuel
T = temperature, °F

Values of C for various fuels are as follows:

Type of fuel	Constant	Fuel no.
blast-furnace gas	570	1
coke-oven gas	670	2
refinery gas and oil	720	3
natural gas	730	4
oil	745	5
bituminous coal	760	6
anthracite	780	7

Example: A boiler produces 50,000 lb/h of steam at 1,000 psia and 700°F from feedwater at 360°F by firing with 25% excess air a bituminous coal containing 10% ash and having a high calorific value of 11,000 Btu/lb. The boiler efficiency has been shown to be about 86%. What is the heat absorbed? the combustion heat liberated? the amount of fuel fired? the amount of combustion air?

On Fig. 1, connect $P = 1,000$ ($T_{sat} = 545\,°F$) with $T = 700$ and extend the line to intersect reference line no. 1. Connect this intersection with $T_{fw} = 360$, and extend the line to its intersection with reference line no. 2. Transfer this intersection point to reference line no. 3 along the parallel lines, connect that point with $W_s = 50,000$ and extend the line to the Q_s scale to read 49 million Btu/h heat absorbed. Connect this value with 86% on the n-scale, and read 58 million Btu/h of heat liberated on the Q_f scale. Connect this value with $HCV = 11,000$, and read 5,000 lb/h of fuel on the W_f scale.

On Fig. 2, connect $a = 25$ with fuel no. 6 (for bituminous coal) and extend the line to reference line no. 1. Connect this intersection with $Q_s = 58$ million Btu/h and extend the line to read 53,000 lb/h of combustion air on the W_a scale. To determine the air volume at, say, 300°F, connect this W_a value with $T = 300$ and read 17,000 ft³/min on the V_a scale.

Converting boiler horsepower to steam

F. Caplan, P. E., Oakland, CA

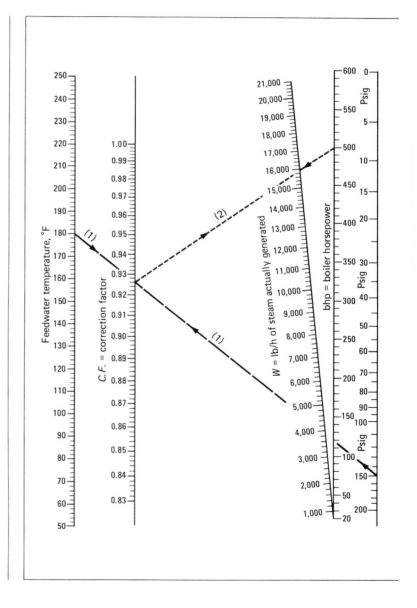

☐ Designs of package boilers have been described.* In addition to appreciating the usefulness of each design feature, the engineer who specifies package boilers must relate them to his plant requirements, and the first requirement is capacity.

Package boiler catalogue steam ratings are frequently expressed as "from and at 212°F" to signify from feedwater at 212°F to steam at 212°F. In order to determine the amount of steam generated at operating conditions, this rating must be corrected for the operating feedwater temperature and the saturated steam pressure. Such a conversion can be made by the formula:

$$W = \frac{(34.5)(\text{bhp})(970.3)}{(h_g - h_f)} = (34.5)(\text{bhp})(\text{C.F.})$$

Where: W = operating steam flow, lb/h
  bhp = boiler horsepower = 33,475 Btu/h = 34.5 lb/h steam from and at 212°F
  h_g = enthalpy of the saturated steam at the generation absolute pressure, Btu/lb
  h_f = enthalpy of the feedwater at operating conditions, Btu/lb
  C.F. = $970.3/(h_g - h_f)$

The nomograph permits a rapid calculation of this formula. If the desired values fall above or below the ranges indicated on the W and bhp scales, the nomograph can be converted by multiplying both scales by a convenient factor.

Example: How much steam does a 500-hp boiler generate when the feedwater temperature is 180°F and the steam pressure is 150 psig (164.7 psia)? On the nomograph align 180°F with 150 psig and read C.F. = 0.926; align this value with 500 bhp and read W = 15,970.

*See Buffington, Milton A., "How to select package boilers," pp. 359-367.

Originally published March 15, 1976

Converting gas boilers to oil and coal

A transition from gas-fired to oil- or coal-fired boilers requires extensive engineering to obtain satisfactory performance. Many factors must be considered, and new equipment is needed for the change.

Arlen W. Bell and *Bernard P. Breen,* KVB Engineering, Inc.

☐ Today, there is public pressure toward operating more boilers on coal—which is an abundant resource—rather than on oil or natural gas, both of which are scarce. Regulatory agencies will therefore ask plant managers and engineers to convert from gas to oil, or preferably to coal-burning boilers.

Natural gas and light distillate oils are in increasingly short supply, as proved by extensive shutdowns for lack of fuel, particularly in the winter of 1973. Even No. 6 fuel oil is considered a premium fuel due to increases in electric-utility demand.

Because of the complexities involved in each conversion, only elements to be evaluated and possible alternatives are presented here. Some relative cost estimates are also presented, but final decisions will undoubtedly be made for other reasons than the lowest steam-generating cost.

Considerations involved in boiler-fuel conversion are fuel availability, fuel purchase cost, environmental requirements, and socio-economic pressures.

For the boiler owner, the most important subject is cost. There may be overall increased product costs due to fuel-price increases, or production losses due to lack of fuel. Fuel costs to competitors may also prove to be significant.

Environmental requirements for low-sulfur oils and coals have further tightened the short- and long-term supply and have shifted geographical production plans. Environmental considerations have actually been responsible for the original shift to high natural-gas consumption, which is now being recognized as a waste of this valuable resource.

Socio-economic pressure on a local, regional or national level can also cause a change in fuel usage. There are strong justifications for maintaining local fuel sources in operation, or shifting to either more-economic or lower-sulfur fuels from a great distance (i.e., substitution of low-sulfur western coal for some of the higher-sulfur eastern coals.).

Technical problem areas

Many engineering problems must be solved in conversion of a gas- or oil-fired boiler to coal, but such problems can be solved by several alternate methods. Such problems can be grouped into three major classifications: coal handling and ash removal, boiler and related equipment, and emission control.

Coal handling and ash removal—Coal presents some unique requirements in contrast to gas and oil. Coal-storage areas for 30–90 days' supply are normally required, and mechanical conveying equipment is needed for transport from the storage area to the hopper, which is usually sized for a day's supply. Unlike oil, coal combustion generates large volumes of ash that must be temporarily stored in hoppers and shipped out for removal on a daily or weekly basis.

Boiler requirements—Each of the three fuels—gas, oil and coal—imposes unique requirements on the design of the boiler: the burner(s); fuel-supply system; accessory support-equipment such as soot blowers, forced-draft and induced-draft fans; and control systems. Particularly important is boiler-furnace sizing, as well as convective-tube spacing. A boiler designed originally only for gas firing may be almost impossible to convert economically to coal firing.

Emission-control requirements—Emission-control equipment is required to some degree on all boilers. For coal firing, the most troublesome area is the removal of ash and carbon carryover from the flue gas. SO_x control is usually done by limiting the sulfur content of the fuel. In the future, more-elaborate flue-gas cleaning devices, such as wet scrubbers, may be needed. NO_x control is ordinarily performed by burner design and minimum excess O_2 operation. CO can be controlled by proper air distribution to the burner(s), or coal grate.

Comparative efficiency—In general, conversion from gas or oil to coal will involve some loss in efficiency. For example, on boilers without combustion-air preheating, steam-generating efficiencies based on gross heating value are in the range of 81% for oil, 78% for gas, and 76% for stoker-fired coal. Efficiencies should be established by tests on operating boilers in normal service. Claimed efficiencies based on heat-transfer area, or tests on new boilers from factories, can be considerably higher than long-term, actual operating efficiency.

Fuel costs are really a larger economic factor than

Originally published April 26, 1976

CONVERTING GAS BOILERS TO OIL AND COAL

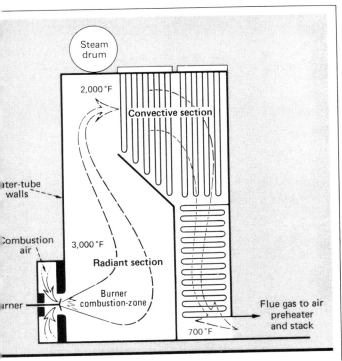

Sectional side-view of a vertical boiler   Fig. 1

just thermal efficiency alone. For instance, coal costs range from $0.50–$1.00/million (MM) Btu; oil from $1.50–$2.00/MM Btu; and gas from $0.34/MM Btu, for regulated interstate gas, to $2.00/MM Btu for unregulated intrastate gas.

Fuel supply and handling

Even though there may be sound economic and conservation reasons for boiler conversions, the logistics of fuel delivery and inplant handling can be severely misunderstood in the initial evaluation.

Outside plant boundaries—Inasmuch as gas-fired boilers receive their fuel through pipelines, there is very little interference with local developments. Oil and coal are more commonly delivered by tank truck, rail or barge. Therefore, switching from gas to oil, and particularly coal, may precipitate local traffic problems objectionable to nearby residents, or overextend existing rail or barge facilities. With the current financial collapse of several railroads, many local and spur rail lines will not be able to manage the load. And costs of a fuel-supply plan based on rail can drastically increase if switching to trucks is required.

Inside plant boundaries—Provisions will have to be made for oil tankage or coal-storage area within the plant's boundaries. The number of days' supply to be maintained on site as ready inventory has to be decided by management, as a hedge against strikes, severe weather, transportation-system breakdowns, or rapid changes in fuel availability. A 30–90-day supply of oil or coal is about normal.

Oil tanks usually require protective dikes to contain spills, as well as separation from buildings and equipment. Coal-storage areas should either be well paved or well drained, and should be located near the boiler to minimize conveyor length. Drainage from coal piles may require a holding and settling basin as well, to avoid water pollution. Dust from coal storage or reclaiming operations may also create in-plant and neighborhood problems.

If gravity does not suffice, a heater and pump (plus a backup pump) will be needed to move oil from storage to a tank adjacent to the boiler.

Belts and bucket conveyors are ordinarily used to move coal into a hopper adjacent to the boiler. Some installations use small tractors to move coal into the feed chute and to keep the coal pile in shape. Coal is ordered in the needed size, but classification and regrinding equipment are occasionally used.

How a boiler works

The boiler (Fig. 1) consists of a furnace (radiant section) and tube banks (convective section). For gas, oil and pulverized-coal firing, burners are usually inserted in the front face. For stoker-fired coal, the combustion takes place on a grate across the radiant-section floor. An integral part of boiler design is provision for soot blowers in the convective section. Provision must also be made for removal of the ash that falls to the boiler floor or ash hopper. Ash in the flue gas is removed by precipitators or cyclones.

Furnace—Designed to take advantage of the high radiant-heat flux near the burner, or coal stoker, the furnace is basically a watertube-lined box, with either a tube-lined or refractory floor. Normally, gas requires the smallest furnace, due to its lower emissivity, which results in a relatively even heat-flux to the tube walls. Oil, having a higher emissivity, requires moving the walls of the boiler further from the burner, to even out the peak-heat flux.

Coal firing presents a different problem, namely, that the flue-gas temperature at the entrance to the convection section must be at least 100°F below the ash-softening point. Such a reduction in temperature can result in about a 15% increase in radiant surface area. Coal-fired furnaces are therefore larger than oil or gas units. Comparative furnace sizes for gas, oil and coal are, respectively: 1, 1.05 and 1.10 for width; and 1, 1.2 and 1.5 for length.

To meet the coal-firing requirements on a unit originally designed for gas-firing, would require either an addition to the radiant-heat-transfer surfaces, or a reduced load to the boiler. The reason for this is the additional burnout time needed for coal particles, as well as the needed heat absorption in the radiant (waterwall) section, so that flyash temperatures will be below their softening point, prior to entering the convective section. Such a requirement can be quantified in terms of furnace volume versus fuel type. For gas, the furnace should be sized approximately 60,000 Btu/(ft^3)(h), while for overfed stokers the volume should be around 35,000 Btu/(ft^3)(h). For coal, in particular, the furnace volume is strongly dependent upon coal type and ash properties.

Convective-section design—The convective boiler section is designed to extract the maximum amount of heat from the partially cooled (2,000°F) flue gas. The convective section should be as compact as possible to obtain the most efficient use of the construction metal and the plant space. Design factors are flue-gas velocities between tubes, tube spacing, and the use of fins. Gas, oil and coal need different designs for optimum operation.

For gas, flue-gas velocities of 120 ft/s, and extensive use of tube fins would be representative, whereas for oil, fuel-gas velocities of 100 ft/s and fewer fins (due to fouling) would be common. For coal, flue-gas velocities should not exceed 60 ft/s, due to the highly erosive characteristics of coal ash; and the tube fins should be spaced much further apart than for gas.

These design restraints indicate that gas-to-oil conversions, or alternate day-to-day use of these fuels on a boiler, is usually satisfactory. Oil or gas to coal conversion, however, requires either extensive modifications to the convective section, or a drastically reduced load (up to 50%) on the boiler.

Soot blowers—Due to the particulate matter carried in the flue gas through the convective section of the boiler, soot blowers are generally required for oil-fired boilers and are almost universally needed for coal boilers. An alternative for oil-fired boilers is the use of frequent water washes of the tubes, with the boiler out of service.

Tube deposits from fuel ash have several adverse effects, including reduced flue-gas to tube-heat transfer coefficients; reduced space between tubes (which results in high horsepower requirements for both forced-draft or induced-draft fans); and increased corrosion of tubes by selective chemical attack from ash constituents.

Soot blowers are either retractable from high-temperature zones (greater than 1,900°F); or rotary (non-retracting, with a 15-ft-limit length for a 30-ft-wide boiler) for temperature zones less than 1,900°F. Both types have high-pressure lances that use either air or steam to blast tube deposits off. Operator attendance is reduced if blowers are automated to operate individually on a preselected schedule. The boiler normally remains in operation during soot blowing. When using high-pressure water jets for water washing, the boiler must be taken out of service.

Because gas-fired boilers do not require soot blowers, conversion to oil or coal can present serious problems in clearances between tubes for soot blowers, and in external clearances (equal to one-half the boiler width on each side) for the placement of the blower retraction mechanism. For a gas-to-oil conversion, soot blowers could be omitted with some oils, such as light distillates. For coal firing, soot blowers would be required.

Burner requirements

The burner is intended to evenly mix the fuel and combustion air, to be capable of load changes, to provide stable ignition or "flame holding," and to require minimal operator attention. The burner design plays a large role in overall efficiency, safety and reliability of the boiler. Gas, oil and coal impose somewhat unique requirements for the burner.

Gas—Industrial gas burners are usually of the ring type, with flame-holding being controlled by adjustments of air registers for swirl, and a ceramic quarl or

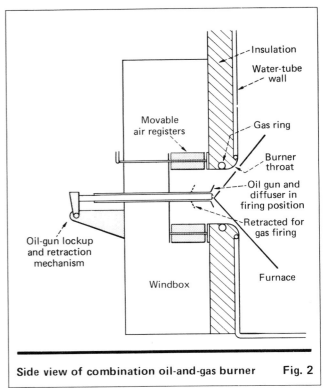

Side view of combination oil-and-gas burner   Fig. 2

burner "throat." A gas-fired igniter, flame detector(s), and a flame-failure shutdown system generally complete the burner package. Operation is relatively simple, and almost maintenance-free, which is one of the reasons for the popularity of gas firing.

Oil—Oil burners are geometrically similar to gas burners, with the oil gun being centered in the burner throat, and flame-holding generally established in the wake of a diffuser. Oil tips can be of different forms: straight mechanical (once through); constant differential (spill type); and steam atomizing (external and internal mixing). The type of atomizer selected depends upon boiler duty-cycle, fuel characteristics, availability of gas or diesel fuel for lightoff, and plant preference. A combination oil-and-gas burner is shown in Fig. 2.

Oil guns require more attention than gas ring-burners because, after shutdown, a steam purge is needed to prevent oil from freezing or coking in the tip and in the supply lines. The furnace should be visually inspected several times during each shift to verify flame quality and the absence of coking or of clinkers in the vicinity of the burner. Periodically, oil guns must be pulled for cleaning, and worn oil-tip components replaced.

Pulverized coal—Coal is delivered to coal burners (Fig. 3) from coal mills, in a finely pulverized form and suspended in a portion of the combustion air. The burner generally consists of a ceramic quarl, air registers, flame-shaping vanes, and a coal-supply tube centered on the burner throat. Pulverized-coal firing is ordinarily not used on boilers that produce less than about 200,000 lb/h of steam.

Operators must be concerned about flame quality,

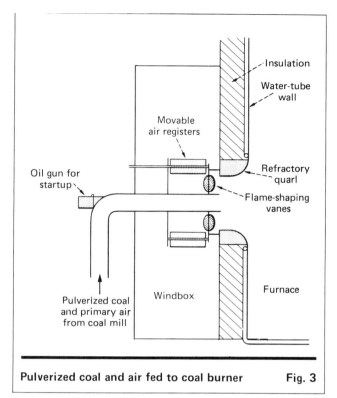

Pulverized coal and air fed to coal burner   Fig. 3

and the supply of fuel to and from the coal mill. The fuel-supply lines can erode rapidly from hard coal and contaminants, and often require annual patching or replacement. Lightoff is a more difficult and involved procedure than for gas or oil, often requiring the use of auxiliary oil or gas burners to preheat the furnace.

Stoker-fired coal—For coal firing in smaller industrial boilers, economics dictate one of the many variations of stoker designs, which require uniform distribution of the coal on the grate, and subsequent removal of the ash. Stokers can be broadly classified as underfeeding or overfeeding. The underfeeding stokers (Fig. 4) push coal up from a trough—in the center of a doubly inclined grate—so that the ash falls off the sides into ash pits. Frequently, the bed is mechanically agitated to maintain uniform coal flow. Typical capacity of an underfeeding stoker is 25,000 lb/h of steam flow. This limits its use to small plants; for large plants, multiple stokers must be used.

The overfeeding stokers are either spreader stokers (Fig. 5), or crossfed stokers, in which the fuel is mechanically thrown onto the grate, where burning takes place. Stokers are further classified according to their grate design: chain grate, traveling grate, or water-cooled vibrating grate. Heat-release rates are expressed in $Btu/(h)(ft^2)$, and vary with the coal used. For example, a spreader stoker with stationary or dumping grates is rated at 450,000 $Btu/(h)(ft^2)$; a stoker with vibrating, oscillating or reciprocating grate at 400,000–600,000; and one with a traveling grate at 750,000. The heat rate for crossfed stokers using chain or traveling grates is 450,000 $Btu/(h)(ft^2)$, and for those with water-cooled, vibrating grates it is 400,000.

Spreader stokers are the most versatile, being available in sizes from 5,000–400,000 lb/h of steam, and capable of burning virtually all coals, from lignite to high-rank bituminous. As the coal is mechanically thrown into the furnace, large lumps fall onto the grate in a thin layer and burn, while fines are usually burned in suspension.

The disadvantage of spreader stokers is that there is an appreciable carryover of combustible material in the flyash. However, carbon can be partially segregated from the flyash and can then be either reinjected into the furnace above the fuel bed or deposited on the fuel grate. This can raise boiler efficiency up to 3%.

Mass-burning stokers distribute coal onto a moving or vibrating grate through a hopper; ash falls into the ash pit at the far end. Almost any solid fuel except highly coking coals smaller than 1¼ in. can be burned, and sizes exist in the 6,000–200,000 lb/h of steam range. Typically, furnaces for mass-burning stokers include a long, rear furnace arch to direct the air-rich rear-end combustion gases forward. This enhances mixing with the highly fuel-rich gases from the front end.

Combustion air for both the spreader and mass stokers is introduced through the coal bed and also through overfire air-ports above the grate. This overfire air is injected at high velocity to obtain good mixing of combustion air with volatiles distilled off the coal.

Installation of a stoker in an existing oil- or gas-fired boiler entails, as a minimum, removal of the existing boiler front to accommodate the coal-feed system; mounting of the mechanical grate and drive system on the boiler floor; installation of an air-supply system under the grate, and overfire air above the grate; and installation of an ash pit. The placement of the ash pit for a spreader stoker makes it more advantageous for retrofitting purposes.

Fuel systems

Fuel systems increase in complexity from gas to oil to coal. For natural gas, a typical gas-main pressure of 50 psig is normal, but inasmuch as the burner requires around 10 psig, a stepdown regulator is needed. The boiler-control system opens or closes the control valve as load (or steam pressure) increases or decreases. With the exception of occasionally draining the condensate trap, the gas-fuel system is nearly maintenance-free.

For oil fuel, the system can become more complex, particularly for residual fuel. As a minimum, a filter set, booster pump, and control valve are required. For residual oil, insulated and steam-traced lines may be needed, as well as an oil header, a steam-purge system, and a recirculation system. If high-turndown, spill-type oil guns (wide-range mechanical atomizing) are applied, a return system is required, so another pump is ordinarily used. Oil systems generally require more maintenance than gas systems—because of filters, pumps, valves, and line cleaning—but a planned preventive-maintenance program makes an oil system nearly as flexible as one for gas.

Coal is more difficult to handle than gas or oil, because of the number of mechanical devices needed. Coal and associated contaminants are very abrasive, and all the handling equipment requires scheduled maintenance and planned component-replacement. A coal-handling system varies in complexity, depending upon whether the boiler is stoker-fed or whether it utilizes pulverized coal. Stokers require coal feeds within certain size limits provided by the supplier. For larger installations, a classifier and crusher for oversize lumps may be installed to improve fuel economy and minimize stoker jams.

For pulverized-coal firing, the fuel-supply system becomes mechanically much more complex. The coal is supplied to the pulverizer from the coal hopper by a variable-rate feeder. The coal is then finely pulverized and pneumatically conveyed by a portion of the combustion air to the burner.

Coal mills require heated air, which can either be supplied by the boiler forced-draft fan or through a separate primary air fan. If an exhauster is used to draw air through the pulverizer, it must resist the abrasion of coal particles. The heated air performs three tasks: it dries the coal, provides inertial energy for classification after the coal is initially ground, and transports finely ground coal to the burners.

Coal mills are sized on expected coal properties, i.e., heating value, grindability, moisture content, and desired ground-coal size. Departures from design conditions result in either oversized equipment or reduced capacity. Engineering judgment dictates some excess capacity. Typical pulverized-coal fineness is 70% through 200 mesh, with less than 2% larger than 50 mesh.

Coal mills can be classified in three types: ball or rod mills, roller/race pulverizers, and hammer mills. For abrasive coals, the ball or rod mill offers distinct advantages in lower overall cost. The coal is broken up by the impact from the tumbling balls, as well as from a grinding action within the mass of coal and balls. Speed is in the 2–25-rpm range.

Roller and race pulverizers are classified as either low (up to 75 rpm) or medium speed (75–225 rpm) units. The feed to these machines, which crush coal lumps between two rolling surfaces, is fed into the mill and circulated into the crushing zone by centrifugal force. When the coal is broken up, the finer particles are picked up by the air, carried to a classifier, and then conveyed to the burner. Oversized coal particles are recirculated back to the rolls. Races and rolls require replacement on an average of every two years, even though some units need annual replacement, and others last five years.

Hammer mills are high-speed machines (direct drive) that use hinged hammers to smash coal particles directly. Crushing is also performed between the hammers and fixed side-pieces. For small installations, these mills have the advantage of low capital cost per ton/h of output. They are also quiet, need only low space and direct drive, and are easy to maintain (hammer replacement). Because of the high speed, high-maintenance costs would occur with abrasive coals.

Auxiliary equipment

Requirements for auxiliary equipment are substantially different for gas and oil firing than they are for coal. Major items for coal burning include the furnace

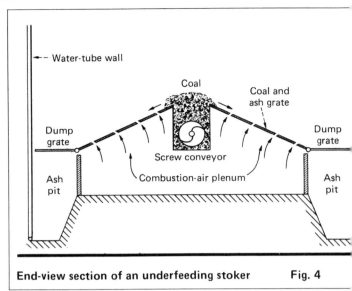

End-view section of an underfeeding stoker   Fig. 4

combustion-air-supply system, the air preheater, and the slag- and ash-handling methods.

Air fans—Gas- and oil-fired units are usually designed for pressurized firing. Combustion air is introduced into the furnace by a forced-draft fan, which supplies the total pressure required to overcome duct losses, burner-to-furnace pressure drop, convective-section pressure drop, and air-preheater pressure drop. As a result, the furnace operates under a positive pressure of 10–20 in H_2O.

For stoker firing, a pressurized furnace is impractical because it is difficult to maintain a sealed system. The coal-feed system and ash hopper present formidable sealing problems, which can be solved by using an induced-draft fan, following the air preheater. The furnace is then operated under a slight negative pressure of less than 0.5 in H_2O, with the I.D. of the fan carrying the pressure drop through the furnace, the convective section, and the air preheater. This additional fan would be mandatory on any gas- or oil-to-coal conversion. Also, a differential-pressure controller, to couple the forced-draft and the induced-draft fan operation, is required.

Air preheater—Although optional for stoker firing, air preheating is a must for pulverized-coal operation. Moisture content and mill air-flow determine the temperature necessary for pulverized-coal firing. If an existing preheater cannot supply the required temperature, a supplementary, direct-fired air heater may be necessary. Preheat temperature with stoker firing is normally limited to 350°F to avoid excessive maintenance of stoker parts. Air preheaters can improve overall efficiency of a boiler by approximately 2% per 100°F air preheat.

Air preheaters for gas firing use closely spaced, metal-mesh baffles (regenerative type), or closely spaced tubes (tube-and-shell type), constructed usually of mild steel. For oil firing, spacing is increased slightly, low-alloy material may be used, and provisions are ordinarily made for water-washing the oil-ash and sulfur

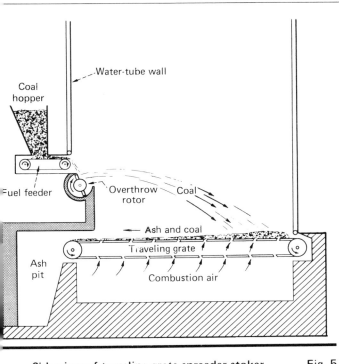

Side-view of traveling-grate spreader stoker Fig. 5

compounds off the surface. Because erosion can be a major problem with coal firing, reduced flue-gas velocities are required, as well as the use of low-alloy steel for longer life of the elements.

Emission-control considerations

Poorly maintained and operated boilers emitting large amounts of CO and carbon are wasteful of fuel, and require more frequent maintenance due to fouling and corrosion. Once a specific fuel is chosen, emissions must be controlled by equipment selection and operating techniques.

Recent changes in laws—and interpretation of laws—have severely narrowed the extent to which an industry can deteriorate the air. As a result, it is necessary to consider all the emissions from a boiler as constituting a potential hazard. Such emissions include carbon monoxide, carbon, flyash, sulfur dioxide, sulfur trioxide, nitric oxide, nitrogen dioxide—and even water from combustion processes and cooling towers, if objectionable artificial clouds or fog are formed.

SO_2 emissions are directly related to the sulfur content of the fuel. Natural gas contains insignificant amounts of sulfur compounds; oil can range from 0.1–3%; and most residual oil sold in the U.S. is at least partially desulfurized, so that 0.5%-sulfur oil can be obtained at some premium price. Washed coal has some reduction in pyritic sulfur but, in general, desulfurized coal is nonexistent. Coal from other sources must be used if local coals cannot meet sulfur requirements.

Nitric oxide emissions occur from the combustion of atmospheric nitrogen in the high-temperature combustion zone, and with the partial fixation of organic nitrogen within the oil or coal. Although natural gas does not contain organic nitrogen, oil may contain up to 0.5%, and coals are in the range of 1% or above. Also, oils that have been partially desulfurized by either hydrodesulfurization, deep cracking, or chemical means, are reduced in nitrogen content. As a rough estimate, NO_x emissions on an industrial boiler would average 100 ppm for gas, 150 ppm for light distillate, 250 ppm for residual oil, and 500 ppm for coal. Pulverized-coal firing has higher NO_x emissions than stoker-fired coal.

Gas, which minimizes objectionable emissions, usually is limited to nitric oxides (NO and NO_2). And because process boilers have relatively low NO_x concentrations and do not represent major concentrated sources, they have fared well from the legislative standpoint. Gas-fired boilers have generally been well accepted by residential neighbors.

Oil-fired boilers present increased emission problems. During lightoff, load changes, and poor maintenance, very visible carbon plumes occur. Visible dark plumes are what the public associates with pollution, even though other pollutants are more hazardous.

With oil firing, SO_2 emissions depend upon the sulfur content of the fuel. Sulfur oxides are odoriferous, and SO_3—which constitutes 2–5% of the sulfur oxides—combines with moisture to form sulfuric acid mist. Of particular concern is the combination of this mist with the very hygroscopic oil ash. Such an agglomeration accumulates on the outlet flue-gas duct-surfaces and on the air preheater; soot-blowing then forces these deposits out the stack. Sulfuric-acid-laden particles tend to settle in the vicinity of the plant, destroying most plants and causing severe corrosion.

Coal firing presents an even more serious emission-control problem. With spreader-stoker firing, flue-gas ash could be as high as 10% of the weight of the coal fired. SO_2, SO_3 and NO_x must also be considered.

Flue-gas cleanup

Coal ash can be separated from the flue gas by cyclone collectors, electrostatic precipitators, fabric filters, or wet scrubbers.

With cyclone collectors—the simplest and most reliable equipment during operation—the flue gas is given a whirling motion, so that the particles are thrown onto the walls by centrifugal force. Cyclones are characterized by a relatively high pressure drop (about 5 in H_2O) and are effective on particles 3 microns and larger. They are also widely used on stoker-fired boilers, because these units generally have flyash of 10 microns or larger.

Electrostatic precipitators are used for finer particles—1 micron or larger—or where very high collection efficiencies (up to 99%) are needed. These devices first charge the particles by passing them across high-voltage wires, and then collect the particles on oppositely-charged plates. The plates are then periodically vibrated or rapped, so that the flyash will fall into a collection hopper.

Precipitators must be carefully engineered to a specific application, with design variables being flowrates (gas and flyash), particle-size range, gas temperature, and sulfur content of the fuel. The sulfur content is im-

Properties	Low-volatile bituminous	High-volatile bituminous			Subbituminous		Lignite
	Pocahontas no. 3, W. Virginia	No. 9, Ohio	Pittsburgh, W. Virginia	No. 6 Illinois	Utah	Wyoming	Texas
Ash, dry basis, %	12.3	14.10	10.87	17.36	6.6	6.6	12.8
Sulfur, dry basis, %	0.7	3.30	3.53	4.17	0.5	1.0	1.1
Ash analysis, weight %							
SiO_2	60.0	47.27	37.64	47.52	48.0	24.0	41.8
Al_2O_3	30.0	22.96	20.11	17.87	11.5	20.0	13.6
TiO_2	1.6	1.00	0.81	0.78	0.6	0.7	1.5
Fe_2O_3	4.0	22.81	29.28	20.13	7.0	11.0	6.6
CaO	0.6	1.30	4.25	5.75	25.0	26.0	17.6
MgO	0.6	0.85	1.25	1.02	4.0	4.0	2.5
Na_2O	0.5	0.28	0.80	0.36	1.2	0.2	0.6
K_2O	1.5	1.97	1.60	1.77	0.2	0.5	0.1
Total	98.8	98.44	95.74	95.20	97.5	86.4	84.3
Initial deformation temperature at which first rounding of apex cone occurs*							
Reducing, °F	2,900+	2,030	2,030	2,000	2,060	1,990	2,130
Oxidizing, °F	2,900+	2,420	2,265	2,300	2,120	2,190	2,070
Softening temperature at which cone has fused to spherical lump where height = width at base*							
Reducing, °F	—	2,450	2,175	2,160	—	2,180	2,130
Oxidizing, °F	—	2,605	2,385	2,430	—	2,220	2,190
Hemispherical temperature at which cone has fused to hemispherical lump where height = 1/2 width of base*							
Reducing, °F	—	2,480	2,225	2,180	2,140	2,250	2,150
Oxidizing, °F	—	2,620	2,450	2,450	2,220	2,240	2,210
Fluid temperature at which fused mass is nearly flat layer with maximum height of 1/16 in*							
Reducing, °F	—	2,620	2,370	2,320	2,250	2,290	2,240
Oxidizing, °F	—	2,670	2,540	2,610	2,460	2,300	2,290

*According to standard of American Soc. for Testing and Materials (ASTM) D1857.

portant, because the SO_3 that is emitted forms sulfuric acid, which condenses on the particles. This acid lowers the resistivity of the particles sufficiently to allow them to be electrostatically charged.

Also very popular for smaller installations are fabric filters. These are generally arranged as a series of cylindrical bags, on the outside of which flyash collects; this is periodically removed by pulsing or shaking the bags, one at a time. Removal efficiency can be very high on submicron particles, once a thin layer of ash has built up on the bags.

Wet scrubbers are also effective in particle removal (95–98%), but they have high-pressure drops (up to 20 in H_2O), and create more of a disposal problem, because the ash becomes a saturated sludge. In the future—if ground-limestone scrubbing systems become more widely used for SO_2 removal—the combined SO_2 and ash removal functions should increase the use of wet scrubbers.

Stack requirements, burner adjustments

Stack-height requirements may be modified by switching boilers from gas to oil or from gas to coal. The most common reason for requiring taller stacks is high SO_2 ground-level concentrations near the plant. This problem can be evaluated by existing techniques (for calculating plume rise and dispersion), modified by local atmospheric conditions. An outgrowth of installing taller stacks is greater boiler visibility, which perhaps make necessary more-stringent operating techniques than legally required, because of excess smoke during lightoff and rapid-load conditions.

Carbon carryover, carbon monoxide, nitric oxides (NO and NO_2)—and to a lesser extent, SO_3—emissions can be affected or controlled by burner adjustments. Gas, oil and coal present different opportunities for emission reduction.

With gas, sufficient air must be mixed with the fuel to permit complete burnup to eliminate CO. Too much air results in reduced efficiency, and high nitric oxide emission. Overly intense mixing also increases nitric oxide formation. Typically, burner adjustment, or tuning, involves adjusting overall air-to-fuel flow and changing the degree of air swirl (mixing energy).

For oil, the major emissions to be controlled are smoke and coke. Because more adjustments can be made to an oil burner than to a gas burner—including overall air-to-fuel flow, air swirl, position of the oil gun

CONVERTING GAS BOILERS TO OIL AND COAL

and diffuser with respect to the burner throat—changes can be made in the burner tip geometry (fuel-spray angle), and the fuel atomization technique used (straight mechanical, steam atomization, or air-blast atomization). Much current package-boiler research has been performed to seek minimum excess air levels, while avoiding coke. Such minimum levels result in reduced NO_x emissions and in the amount of SO_2 converted to SO_3.

Pulverized-coal burners may have almost as many adjustments similar to those of oil burners. Often, in addition to air-swirl registers, flame-shaping vanes are installed. Also, both the size-distribution of pulverized coal and the amount of carrier or transport-air can be altered. Again, best operation and lower emissions are obtained when adjustments are made to minimize excess O_2.

For stoker-fired coal boilers, adjustments are generally made to obtain higher carbon burnout of the fuel. The effects of overall air flow, overfire air flow, and ash recycle to the stoker have not been examined in detail to show how NO_x and SO_x emissions are changed. At the present time, it can be assumed that gaseous emissions cannot be substantially modified.

Fuel storage and handling

Oil can impose requirements on storage-tank vapor recovery systems, varying with fuel volatility and ambient or storage temperatures. Crude-oil storage, in particular, poses difficulties because of the high volume of volatiles and H_2S outgassing.

Coal presents a dust-control problem, which can generally be controlled to a reasonable level by proper design of handling equipment, such as enclosing conveyors and good housekeeping (for example, maintaining high volume-to-surface coal piles). In extreme cases, all storage might have to be in silos. Ash storage and removal present analogous problems.

Gas-to-coal conversion methods

The starting point of a gas-to-coal conversion is to obtain analyses of the candidate coals. It is impossible to design a boiler that can handle all coals. As a minimum, data should be obtained on the amounts of volatile matter, fixed carbon, sulfur, ash, ash-fusibility and softening temperature, heating value, caking or coking, grindability, and sizing available. Such characteristics determine furnace-resizing needs, tube spacing, soot-blowing needs, type of combustor (pulverized or stoker), emission-control hardware, and added site requirements.

Coal-ash characteristics are extremely important to avoid serious convective-section fouling, because the flue-gas temperature leaving the radiant section must be lower than the ash-softening temperature. The table presents a summary of fusibility temperatures and ash analyses for several representative coals.

Costs vary so widely, particularly on custom-retrofit work, that only a rough order of the cost of conversions can be obtained. After detailed drawings are prepared, estimates of ±20% can probably be made. Some rules of thumb for new, factory-assembled equipment are costs of $3/lb of steam generation for oil- and gas-fired boilers. Field erection or modifications can easily cost twice as much as similar work performed in a shop.

Fuel savings on coal can be significant. For coal at $25/ton, with a heating value of 12,000 Btu/lb ($1.04/10^6$ Btu), as compared to oil at $10/bbl ($1.65/10^6$ Btu), or intrastate gas at $2.00/10^6$ Btu, fuel-cost savings of $140–230/h can result for a boiler of 200,000 lb/h steam output (gross firing rate of 240×10^6 Btu/h).

Three cases of gas-to-coal conversion or oil-to-coal conversion will now be discussed. These are in increasing order of difficulty: gas-fired boilers that originally operated as coal-fired units; older, larger boilers that have strong conversion potential; modern, shop-assembled package units designed optimally for gas or oil.

Reconverting back to coal firing

There are many older boilers, which originally operated as coal-fired units—utilizing multiple, under-feeding stokers, mass stokers, or spreader stokers. With the advent of cheap, plentiful natural gas, these were converted from coal to gas. Stokers were removed, ash pits were blocked or filled in, and gas burners installed. The use of natural gas practically eliminated soot blowing, as well as stack-gas cleanup.

The main items for conversion back to coal burning include: (1) rehabilitation or replacement of stoker; (2) verification that the system works as originally intended (this may require reinstallation of ductwork); (3) rehabilitation of ash-handling facility; (4) rehabilitation or replacement of soot blowers; (5) installation of equipment for flue-gas cleanup (particulates) to meet current emission requirements; and (6) consideration of the characteristics of the coal currently available versus the coal originally burned.

Older units with conversion potential

There is another class of boilers, which—although considered primitive by today's designs—should be able to be converted. These are relatively large-volume boilers designed for oil and gas firing, and induced or balanced draft, and probably incorporating soot blowers. For some coals, the boilers' convective-tube spacing would be adequate. Coal characteristics must be very carefully evaluated. Inasmuch as these boilers' furnaces are large enough, spreader-stoker equipment could be installed, although at some possible sacrifice in load. For units of 150,000-lb steam capacity or greater, pulverized-coal firing should be considered.

Major items required for conversion include: (1) installation of spreader stoker; (2) revision of ductwork to provide air through grates, and overfire-air through side-ports; (3) construction of ash pit, ash-removal system, and ash-recycle system; (4) installation of flue-gas cleanup equipment; (5) possible installation of added soot blowers; and (6) addition of air-supply capacity because of the additional ductwork and consequent increase of pressure drop, due to particulate removal.

Modern, shop-assembled package boilers

Package boilers are forced-draft, air-supply systems with high-heat-release furnaces, high velocities through the convective pass, all tubewall construction, combi-

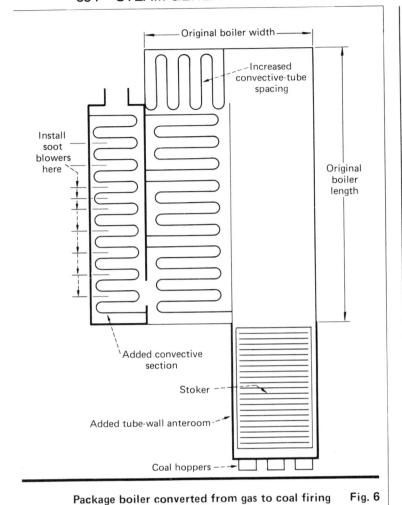

Package boiler converted from gas to coal firing   Fig. 6

tion of an entirely new coal-fired boiler, incorporating all the requirements for the intended coal.

Two techniques hold promise for gas-to-coal conversion: installation of pulverized-coal burner(s) in the position of the existing burners; and the addition of a refractory chamber, which should be large enough to accommodate a stoker.

For pulverized-coal firing, the following is required: (1) installation of coal mills; (2) installation of a pulverized-coal burner in the existing burner throat; (3) recontouring the floor of the furnace to facilitate partial bottom-ash removal; (4) installation of a stack-gas cleanup system; (5) installation of an induced-draft fan; (6) removal of a portion of tubes from the convective pass to reduce flue-gas velocity; (7) adding an economizer, or air-heater section, to replace the lost heating surface from the previous step and to provide sufficient preheated-air temperature; and (8) installation of soot blowers. A load reduction (up to about 25%) can be expected with this type of conversion.

For stoker firing, these would be necessary (Fig. 6): (1) installation of a combustion chamber to accommodate the stoker at the boiler front, including addition of sufficient heat-transfer surface; (2) installation of stoker and ash-removal systems; (3) removal of tubes from the convective section; (4) addition of tubes to the economizer section; (5) recontouring furnace floor; (6) installation of a special, soot-blowing system to clean tubes and floor surface; (7) installation of a stack-cleanup system; and (8) installation of an induced-draft fan. If all of these steps are properly performed, the rating (load) of the unit can be maintained.

From the foregoing discussion, it is apparent that extensive engineering is involved in changing from gas to oil or coal burning, particularly for the first few conversions of each boiler type. Prior to a final decision on conversion, similar, already-converted, boilers should be examined, and operating problems discussed with plant personnel.

nation oil or gas firing, and control systems designed for minimum operator attendance. Units intended for No. 6 oil normally have soot blowers, and horizontal convective passes that are adjacent to the furnace. The design is intended to obtain the most steam-generating capacity at the lowest cost, which results in a relatively lightweight, transportable boiler, with minimum onsite work required.

Prior to gas-to-coal conversion of a shop-assembled unit, serious consideration should be given to construc-

References

1. Babcock, G. H., and Wilcox, S., "Steam—Its Generation and Use," The Babcock & Wilcox Co., New York (1972).
2. Power from Coal, *Power*, Feb. 1974.
3. Leonard, R. R., "Influence of Fuels on Boiler Design," Riley Stoker Corp., Worcester, Mass., June 15, 1971.
4. "Boiler Rating Criteria for Nonresidential Boilers," Federal Construction Council, National Academy of Sciences, Washington, D.C. (1962).

The authors

Arlen W. Bell is associated with KVB Engineering, Inc., 17332 Irvine Blvd., Tustin, CA 92680. Before, he worked in aerospace, on projects such as aircraft and missile analysis, nuclear-weapon effects, structural dynamics, and research-facility design. He also has been program manager of nitric oxide reduction programs for various companies, and papers on this work have been presented at various professional-society meetings. He has also been program manager for burner development projects, and chemical desulfurization methods for oil and coal. He is a licensed mechanical, civil and structural engineer in California, and holds B.S. and M.S. degrees in engineering from the California Institute of Technology.

Bernard P. Breen is Vice-President of KVB Engineering, Inc., 17332 Irvine Blvd., Tustin, CA 92680. He cofounded this company with J. R. Kliegel, and has promoted KVB's growth in pollution reduction from utility boilers, gas turbines, and industrial combustors. This work has led to research and engineering in burner design, flame stability, and modified boiler operation. He has presented papers at meetings of the American Power Conference, Combustion Institute, and the Air Pollution Control Assn. He has a Ph.D. degree in chemical engineering from Iowa State University and is a licensed chemical engineer in California.

Designing steam transmission lines without steam traps

Here is proof that steam traps are not always needed, provided that certain conditions are met. The authors show how to design trapless transmission lines, with savings in equipment and energy.

Mileta Mikasinovic and *David R. Dautovich*, Ontario Hydro, Toronto, Canada.

☐ A steam transmission line carries energy from Point 1 to Point 2. This energy is a function of temperature, pressure and flowrate. Along the line, energy is lost through the pipe insulation and through steam traps. A design that would reduce the energy loss and the amount of required equipment would be highly desirable.

This article explains how to design steam transmission lines without steam traps, thereby saving the equipment cost of steam-trap stations and the energy normally lost through steam traps.

Design theory

The designer's primary concern is to ensure that steam conditions stay as close to the saturated line as possible. The steam state in the line changes according to the change in pressure due to a pressure drop, and the change in enthalpy due to heat loss through insulation. These changes of condition are plotted in Fig. 1. Point 1 is defined by P_1 and T_1 steam conditions. Due to the variability of such parameters as flowrate and ambient temperature, the designer should consider extreme conditions. Thus, P_2 would be defined by the minimum pressure drop produced by the minimum flowrate. Similarly, h_2 would be defined by the maximum heat loss produced by the lowest ambient temperature. Point 2 on the h-s diagram is defined by the above P_2 and h_2. If Point 2 is above or on the saturated steam line, no condensate is generated and steam traps are not required.

In some cases (small pressure drop, large heat loss), Point 2' is below the saturated line, and some condensate is generated. The usual practice has been to provide trap stations to collect this condensate, and steam traps to remove it. However, current research [3,4,6] in two-phase flow demonstrates that the turbulent flow, produced by normal steam velocities and reasonable steam qualities, disperses any condensate into a fine mist equally distributed along the flow profile. The trap stations do not collect the condensate and, once again, the steam traps are not required. Fig. 2 shows that for velocities greater than 110 ft/s, and a steam fraction more than 98% by volume, the condensate normally generated in a transmission line will exist as a fine mist that cannot be collected by steam trap stations.

Operating the line

A few basic points should be followed when operating a steam line without steam traps. All lines must be sloped. If a line is long, several low points may be required. Globe valves are used on drains for each low point and for a drain at the end of the line. Since trap stations are not required, drain valves should be located as close to the line as possible to avoid freezing. Vents are placed at all high points. All vents and drains are opened up prior to warming up the line. Once steam is flowing from all vent valves, they are closed. As each drain valve begins to drain steam only, it is partially closed so that it may still bleed condensate if necessary. When full flow is established, all drain valves are shut. If the flow is shut down, all valves are opened until the pipe cools, and are then closed to isolate the line from the environment. The above procedure would be the same if steam traps were on the line.

Conclusion

It has been demonstrated that for steam transmission lines, steam traps are not required, provided that one of the following parameters is met:
1. Steam is saturated or superheated.
2. Steam velocity is greater than 110 ft/s, and the steam fraction more than 98% by volume.

Using the above design, the steam energy normally lost through traps is saved, along with the construction, maintenance and equipment costs for the traps, drip leg, strainers, etc., associated with each trap station.

Example

Design a steam line for transporting a minimum of 6.0×10^5 lb/h and a maximum of 8.0×10^5 lb/h saturated steam at 205 psig and 390°F. The line is 3,000 ft long, with eight 90-deg elbows and one gate valve. Ambient temperatures range from −40°F to 90°F.

Using a steam velocity for maximum flowrate of

Originally published March 14, 1977

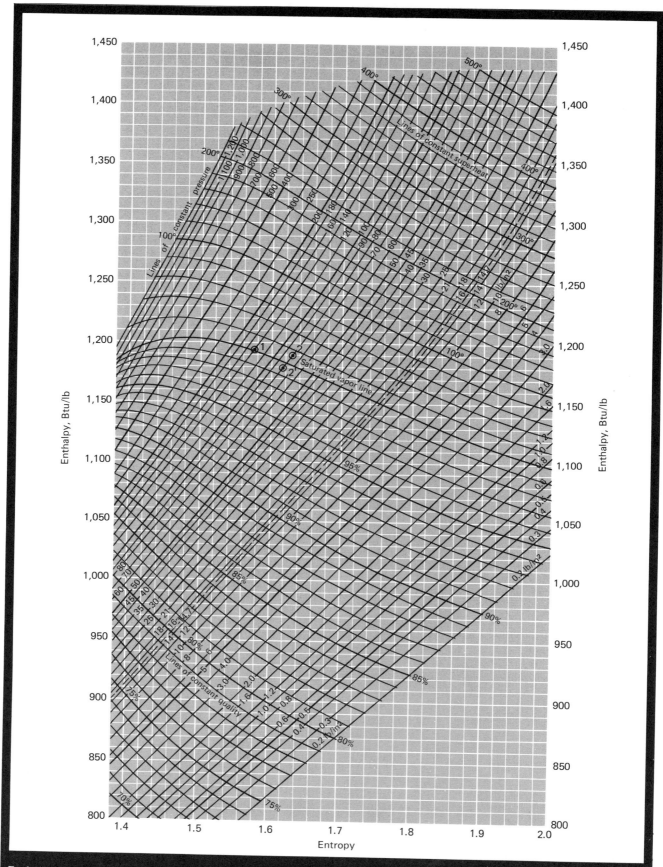

Enthalpy-entropy or Mollier diagram for steam — Fig. 1

Flow-pattern map   **Fig. 2**

160 ft/s to size the pipe:

$D = 24$ in., std. wt. (area = 2.94 ft²)

The velocities are:

    For maximum flow, 157.6 ft/s
    For minimum flow, 118.0 ft/s

and the pressure drops and final pressures at Point 2 are:

    For maximum flow, 23.3 lb/in² and 196.7 lb/in.²
    For minimum flow, 14.0 lb/in² and 206.0 lb/in.²

Therefore, the velocities at Point 2 are:

    For maximum flow, 173 ft/s
    For minimum flow, 124 ft/s

The line will be covered with a 3-in.-thick insulation with a thermal conductivity of 0.48 (Btu-in)/(h) (ft²)(°F).

The maximum heat loss at minimum ambient temperatures is:

$$Q_{max} = 1,452,931 \text{ Btu/h}$$

Therefore, the change in enthalpy becomes:

$$\Delta h_{max} = \frac{1,452,931}{6 \times 10^5} = 2.42 \text{ Btu/lb}$$

The minimum enthalpy at Point 2 is:

$$h_{2\,min} = h_1 - \Delta h_{max}$$
$$h_{2\,min} = 1,197.18 \text{ Btu/lb}$$

The minimum heat loss at maximum ambient temperatures is:

$$Q_{min} = 1,013,673 \text{ Btu/h}$$

$$\Delta h_{min} = \frac{1,013,673}{8 \times 10^5} = 1.26 \text{ Btu/lb}$$

$$h_{2\,max} = h_1 - \Delta h_{min}$$
$$h_{2max} = 1,198.34 \text{ Btu/lb}$$

Now we have two extreme states at Point 2:

(a) Minimum flow and lowest ambient temperature, where $P_{2\,max} = 206$ lb/in.²; $h_{2\,min} = 1197.18$ Btu/lb.
(b) Maximum flow and highest ambient temperature, where $P_{2\,min} = 196.68$ lb/in²; $h_{2\,max} = 1198.34$ Btu/lb.

The second condition is superheated and no condensate is generated. The first condition is below the saturated line and the amount of condensate and steam is:

             $2.22 \dfrac{\text{ft}^3 \text{ of dry steam}}{\text{lb of total mass}}$

and $.0000342 \dfrac{\text{ft}^3 \text{ of condensate}}{\text{lb of total mass}}$

or   .00154% condensate per volume
and   99.99846% dry steam per volume.

The velocity, 124 ft/s, >110 ft/s.

According to Fig. 2, we can see that the condensate will exist in the steam flow as a fine mist, and cannot be collected by steam traps.

References

1. Schwaigerer, S., "Rohrleitungen—Theorie und Praxis"; Springer Verlag (1967).
2. Simonson, J. R., "Engineering Heat Transfer," McGraw-Hill Co. Ltd., (1967).
3. Gorier, G. W., and Aziz, K., "The Flow of Complex Mixtures in Pipes," Van Nostrand Reinhold Co. (1972).
4. Tong, L. S., "Boiling Heat Transfer and Two-Phase Flow," John Wiley and Sons, Inc. (1967).
5. Shearer, C. J., and Hedderman, R. M., *Chem. Eng. Sci.,* Vol. 20, pp. 671–683, 1965.
6. Goldmann, K., Firstenberg, H., and Lombardi, C., Trans. ASME, *J. of Heat Transfer,* Ser. C, Vol. 83, pp. 158–162, 1961.

Fig. 2 is from "The Flow of Complex Mixtures in Pipes" by G. W. Govier and K. Aziz; © 1972 Litton Educational Publishing, Inc. Reprinted by permission of Van Nostrand Reinhold Co.

The authors

Mileta Mikasinovic is a design engineer at Ontario Hydro, 700 University Ave., Toronto, Canada. He holds a B.S. degree in mechanical engineering from the University of Belgrade. He has experience in the design and construction of mechanical equipment for the chemical and petrochemical industries, and is the author of a handbook of fluid flow published in Yugoslavia.

David R. Dautovich is a design engineer at Ontario Hydro, 700 University Ave., Toronto, Canada. Formerly employed by C-E Lummus, he has had experience in the design of mechanical equipment, pressure vessels and process piping in the petrochemical and thermal-generating fields. He holds a B.S. Sc. degree in mechanical engineering from the University of Waterloo, Ontario.

Estimating the costs of steam leaks

Jack Goyette, Tenneco Chemicals*

☐ In many processing plants, the repair of steam leaks can produce a larger saving and return on investment than any other energy-conservation effort. We found this out by developing a technique for placing a reasonably accurate dollar-estimate on the continuing costs of a steam leak. With such a technique, it becomes very easy to justify additional manpower, overtime, or outside services to make necessary repairs to even the smallest leaks.

We developed this method by devising a piece of equipment that has been fondly termed a "steam piccolo." It is a piece of 1-in. diameter pipe, placed vertically with three horizontal branches attached to it. The branches have 1-in. stainless-steel caps screwed on them. The top cap has a $\frac{1}{8}$-in. hole drilled in it, the middle cap a $\frac{1}{16}$-in. hole, and the bottom a $\frac{1}{32}$-in. hole.

An orifice calculation shows that, in 150 psig saturated steam service, these openings will leak out approximately 100 lb/h, 25 lb/h, and 6 lb/h, respectively, which relates to $3,000/yr, $750/yr and $190/yr. in incremental steam costs, based on steam at $3.42/1,000 lb (fuel value only). By placing this unit on an exterior wall with high visibility and painting the various criteria related to each orifice beside the orifices, we were able to produce a very effective demonstration of the cost of steam leaks.

We found that the most effective method for us to accomplish steam leak repairs was to perform most of the work on Saturdays, when a concentrated effort could be put forth. Of course, large leaks would receive top priority during the week. In preparing for weekend work, a survey is run on various departments during the week to determine the magnitude and number of the leaks. In order to educate plant personnel and keep them aware of the importance of steam leaks, the piccolo is turned on at each shift change, so that individuals can tune-in their eyes. This is necessary to show the changes in steam plume caused by temperature and relative humidity changes.

The individual goes out into the field and makes

*Organics & Polymers Div., Meadow Road, Fords, NJ 08861

Originally published August 29, 1977

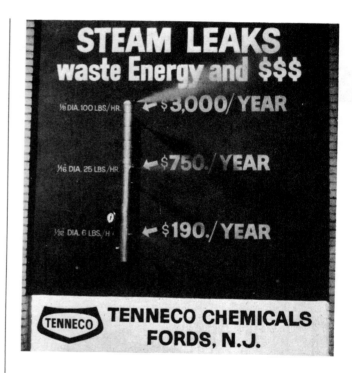

fairly accurate educated judgments on the costs of various leaks, and then totals them up for that particular department. An accurate account, which can be used for future justification, is kept on the savings accomplished.

The return on investment can be astounding. For instance, a 1-in. union was found to be leaking approximately 100 lb/h, representing a loss of $3,000/yr. The cost to repair the leak was $50, yielding a six-day payback and a 6,000% return on investment. It's obvious that this work is very easy to justify.

Total performance in energy conservation at our plant has shown a reduction from 4.51 million gal of No. 6 fuel oil burned in 1972 to 2.99 million in 1976.

Although many items contributed toward these savings, steam-leak repairs proved to be the single largest contributor, yielding well over $150,000/yr in savings.

How to select package boilers

Package boilers for steam generation arrive completely assembled at the plant site and require minimum fieldwork for installation. To specify such boilers, the engineer must compile and evaluate the steam requirements, fuels and firing, combustion efficiency, energy-saving auxiliaries, and operating and maintenance features

Milton A. Buffington, Bumstead-Woolford Co.

☐ You have just been handed an assignment to put a new boiler into your company's plant. Since most companies buy a new boiler at very infrequent intervals, it is likely you have never been through this experience—although you have probably been around a steam plant enough to have some general ideas about the equipment.

To get started, it is important to develop the basic information for making a preliminary selection of a boiler. It may also be necessary to obtain some field data to complete the compilation. Most boiler manufacturers furnish data sheets for pricing. These are useful for determining the technical, mechanical, thermal and construction requirements for a particular boiler. For our purposes, the principal design data are listed in Table I (T/I). In addition, it is important to know what type of fuel will be burned, and its characteristics.

First, let us define the two general types of boilers in relation to the method of assembly. For practical purposes, a package boiler is one assembled at a manufacturer's plant. The package boiler has only to be set on a foundation, be piped up, and have a few appurtenances connected before being ready for operation. In general, a field-erected boiler comes in pieces or partial assemblies and must be built in the field from the ground up. Almost any steam capacity is available. There are variations, but the foregoing is a general description.

Second, let us now examine some of the details of package boilers:

■ Package boilers are available in capacities ranging up to about 500,000 lb/h. A more practical upper limit is in the range of 300,000 lb/h. A unit the size of the upper limit would probably be shipped in two pieces and require some field assembly. In one such unit, the

Principal design data for boiler selection

Required Data	Source
Type of load	Process conditions
Steam flow, lb/h	Plant heat balance
Steam pressure, psi	Process conditions
Steam temperature, °F	Process conditions
Feedwater temperature, °F	Feedwater supply
Feedwater treatment	Plant equipment

Originally published October 27, 1975

Union Iron Works

drum section ships as one piece, and the furnace section ships as the other.

■ Package boilers are built in a wide range of sizes, and can be used for many service requirements. Units ranging up to about 150,000 lb/h steam-generating capacity can be obtained on a rental/purchase agreement for emergency use. Package boilers are often the only answer for increased steam capacity in an existing plant where space is paramount. Such boilers are used as the primary source of steam, as startup boilers in utility plants, and as emergency units, and are even mounted on trucks—in certain instances with their own feedwater systems. In this last case, capacities are limited.

The fuels to be burned, the quantities of steam required, and the amount of capital to be expended for needed steam are the three basic criteria for selecting between the two general types of boilers. If the fuel is other than gas or oil, or if more than 350,000 lb/h of steam are needed, the selected boiler most likely will be a field-erected unit.

Package boilers will generally have an installed cost of about \$5/lb of steam, compared with \$15/lb of steam for a field-erected unit.

Let us assume that the fuels are oil and gas, and that steam requirements are under 350,000 lb/h. We then proceed to investigate a package boiler for plant use.

Designs of package boilers

Package boilers have two basic designs: fire-tube or water-tube. As the terms imply, the hot combustion gases pass through the tubes of a fire-tube boiler, while water passes through the tubes of a water-tube boiler.

Fire-tube boilers are relatively simple and have many applications in the chemical process industries where pressures in excess of 150 psi are not required and saturated steam is satisfactory. Because of very simple steam capacity, such boilers readily accommodate swing loads.

Combustion usually takes place in a furnace under the vessel, and the hot gases traverse the tubes in two to four passes (heating the water to raise steam) before being discharged to the stack. The modern fire-tube boiler typically fires in a large circular chamber in the center of the boiler, as shown in Fig. 1 (F/1), and the combustion gases travel through four passes before being discharged. Units of this design are very competitively priced for capacities under 28,000 lb/hr (about

HOW TO SELECT PACKAGE BOILERS 361

800 boiler hp). A typical package fire-tube boiler is shown in F/2.

Water-tube boilers circulate water within the tubes. Heat transfer occurs from the hot combustion gases on the outside of the tubes to water on the inside. Usually, the tubes are arranged to provide natural circulation of the water within the unit. Water is heated in the tubes nearest the flame or heat source; and the heated water (being lighter from a volumetric standpoint) rises to the steam drum. Pockets of steam also rise to the steam drum and are there separated from the liquid.

The liquid within the drum travels down the connecting tubes to a lower drum, usually named the "mud" drum. Here, any dissolved solids in the feedwater (concentrated by the steam-generating process) settle out, and are removed, in part, from the system by the blowdown procedure, which can be either manual or automatic.

Package boilers have two basic forms: the "A" and "D" designs, as sketched in F/3. There are modifications of these basic types, and these will be described later. In the A-type package boiler, there is one steam drum at the top of the unit and two mud drums at the bottom. The configuration is in the shape of the letter "A", hence the designation. These units can be furnished in capacities ranging up to 300,000 lb/h, and for pressures up to 1,500 psig.

Superheaters, both radiant and convection types, are available. When the convection design is used, the outlet temperature is limited to around 810°F. Radiant superheaters allow the use of higher outlet temperatures.

In the "A" design, the generator or radiant tubes form the furnace sidewalls. Some builders also use water cooling for the floor and rear wall (see F/4).

Package water-tube boilers can be obtained with forced draft or induced draft, as shown in F/5. Recently, the trend has been to forced-draft units. The type of draft influences the casing design. Currently available package boilers generally have internal and external casings, except for the ends. The insulation (usually a blanket type) is installed between the inner and outer casings. As a result, the only refractory lining is at the ends, and often only at the firewall.

Advantages of the package boiler

Whether of fire-tube or water-tube design, the package boiler has these advantages:

1. Units meet or exceed the standards of the ASME Boiler Code for design, materials of construction, fabrication procedures, and workmanship.
2. Performance is predictable because of standard designs, and the user is assured of an integrated design that may include the boiler, firing equipment, and controls.
3. Compactness is a useful feature. These units require less floor space because the firing controls are mounted on the burner wall, and the stack is mounted directly on the boiler.
4. Installation costs are reduced. In the smaller package units, most of the trim is often factory installed. To place a unit into operation, fieldwork is confined to connecting fuel, water, power and steam lines to the boiler. Even in the larger units, much field labor is

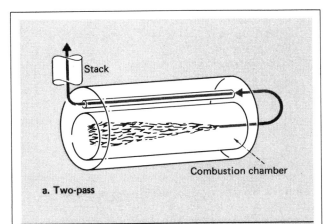

a. Two-pass

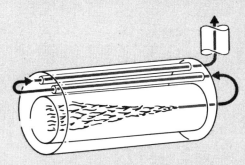

Another pass to get longer gas travel means adding a baffled chamber to stop gas from short-circuiting to the stack.

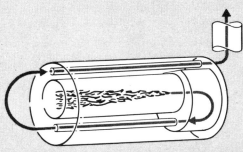

Wetback unit can be two or three pass, has submerged combustion chamber in which gas first changes direction.

b. Three-pass

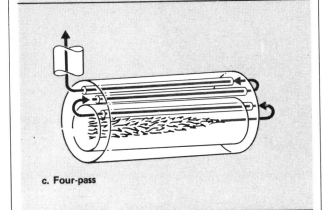

c. Four-pass

Basic gas-flow patterns for fire-tube boilers

362 STEAM GENERATION AND TRANSMISSION

saved. The auxiliaries have to be installed, but the pipe spools are ready, the firing equipment is packaged, and no refractory work is needed.

5. Efficiency is predictable and proven.

Other package water-tube designs are available for specialty applications. Such units are available for heating heat-transfer oil, for vaporizing Dowtherm, and for high-pressure and high-temperature water heating. For these services, the units are usually rated in Btu/h.

Once-through designs (having forced circulation) are available for specialized operations. These units generate steam in a very short time, and are available in capacities up to 300 hp and pressures up to 1,000 psi. One design of such a boiler is shown in F/6.

Fuel-burning equipment

Package boilers are generally equipped to burn two kinds of fuel, gas and oil. In the past, the oil has usually been a light one such as No. 2 diesel. Today, the oil-burning capability will probably include No. 6 oil or a fuel similar to Bunker C.

Completely integrated fuel-burning packages (F/7) are common for use with the package boiler. Standard burner packages are available from firms specializing in this equipment. In addition, many manufacturers of package boilers offer burner assemblies of their own design.

In selecting a burner for the package boiler, it is necessary to first select the fuel or fuels to be used. These fuels must be well specified and include the following information:

For gaseous fuels:
  a. Heating value, Btu/ft^3.
  b. Gas pressure.
  c. Gas temperature.
  d. Chemical analysis.

For liquid fuels:
  a. Heating value, Btu/lb.
  b. Fuel pressure.
  c. Fuel temperature.
  d. Viscosity curve.
  e. Chemical analysis.
  f. Atomization required for combustion.

Solid-fuel-burning package boilers, while not so common, are available. Fuel-burning equipment for solid fuels is special and not covered in this article.

Source of fuels

For gaseous fuels, we normally mean natural gas. The source of this fuel is the natural-gas distribution system. However, since the energy crisis, plants in relatively remote localities are turning to butane/propane gas systems as fuel. In this case, special tanks and in-plant distribution systems are necessary.

Oil for fuel requires special storage considerations. When burning light oils, it is necessary to have storage tanks and day tanks, together with pumping and recirculation systems. Due to the low viscosity of the light oils, heating equipment for the oil is not necessary.

A number of factors govern the size of storage tanks. Distance of the plant from the source of fuel is a primary consideration. In planning to burn heavy No. 6 oil, it is necessary to take into consideration the dis-

Package boiler of fire-tube design accomodates swing loads easily

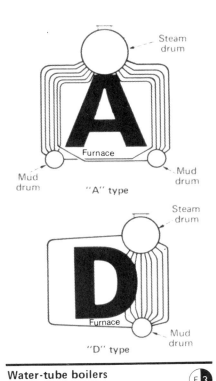

Water-tube boilers have two forms F3

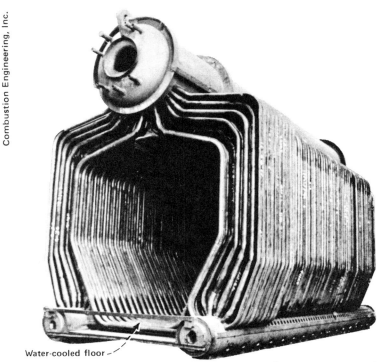

Combustion Engineering, Inc.

Tube arrangements in typical water-tube furnace feature a water-cooled rear wall and floor F4

tance of the plant from the source of fuel. In addition, special consideration must be given to the handling system within the plant because heavy oil must be heated to permit pumping. Also, filters and strainers must be provided. A typical arrangement for handling such a fuel oil is shown in F/8.

It is necessary to be specific in regard to the characteristics of the heavy oil to be burned in your fuel system. During the energy shortage, some plants found that while the oil had a satisfactory average pour point when the fuel was delivered to storage, solid paraffins were found to have settled out in the storage tanks. The higher pour-point components were the problem.

Safety equipment

All boilers require basic safety equipment. This will include high-pressure cutoff, low-water alarm, draft-failure alarm, and high-pressure relief valves.

With modern control equipment, it is possible to automate the package boiler so that it can operate without any attendant. Generally, the amount of supervision is dictated by applicable state laws governing the operation of a boiler rather than its mechanical requirements.

The boiler will probably be supplied with flame-monitoring equipment, i.e., a flame rod or a quartz element that senses ultraviolet rays of the flame. With these controls, the boiler startup sequence is programmed (automatic predetermined starting sequence). Before the boiler can be started, it must be run through a purge cycle. Then, the pilot light is lit. When the pilot light is proven, the main gas valve can be opened; and then, the main burner can be put online.

In the event of flame failure, low air pressure, low gas pressure, low water level, high steam pressure, or any combination of the foregoing, the boiler will automatically shut down.

The main difference between the foregoing system and that for oil-burning units is the type of pressure control and the line sizing due to the volumetric differences of the fuels. The basic equipment and function remain the same.

Energy-saving accessories

Energy-saving devices are available for package boilers just as they are for the larger field-erected ones. Here, we will describe these accessories and show what they accomplish when properly installed.

Flue gases leaving the typical boiler are in the range of 600°F. This temperature is an indication of the heat being discharged to the atmosphere. By using economizers or air heaters, the temperature of the discharged flue gas can be reduced to between 300 and 400°F. The overall boiler efficiency will increase about 2½% for each 100°F drop in the exit-gas temperature when the common fuels such as coal, oil and gas are burned under equal conditions.

Economizers

Economizers, as applied to package boilers, are generally heat-exchange units. Depending upon the fuel being fired and the heat recovery desired, the economizer may have either bare tubes or finned tubes.

In applying an economizer to the package boiler, the type of fuel being used is important. When firing natural gas or light oil, there is little to be concerned about

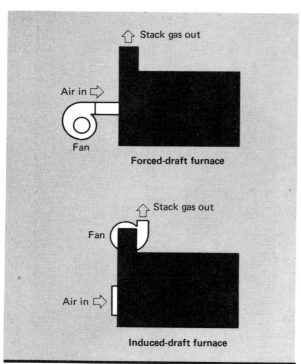

Draft arrangement influences casing design of boiler F/5

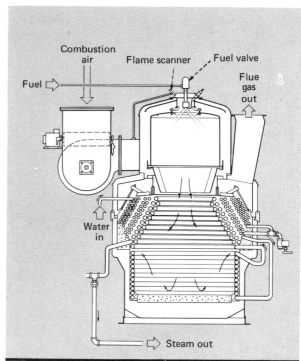

Once-through boiler has forced circulation F/6

in regard to sulfur dioxide content of the flue gases. Usually, finned-tube economizers will be used when firing these fuels because of higher heat-transfer rates to be realized with this design.

The most important consideration regarding the fuel is the amount of sulfur present. If sulfur is present in the fuel, it is most important to maintain the flue-gas temperature above the dew point (approximately 450°F) to prevent corrosion by the condensation of acidic liquids. A typical economizer and its arrangement in the boiler is shown in F/9.

There are many package boilers in service that have finned-tube economizers, are fired with a No. 6 oil or its equivalent, and operate satisfactorily. If you are confronted with these conditions, the use of soot blowers is important, and low skin temperatures should be avoided.

Air heaters

Air heaters operate either on the recuperative or regenerative principle. The recuperative type is generally a tubular design, with the heat being transferred from the flue gases. In one recuperative design, the flue gases pass through the inside of the tubes, and the incoming air circulates on the outside. Movement of the air and flue gas through the heater is generally in a counterflow manner. Air heaters are usually separately designed heat-exchange units with the gases being ducted to and from the heater. The incoming-air ducts are arranged so that air passes through the heater and into the burner box.

The most commonly used regenerative-type air heater is the Lungstrom unit, which is a rotary-plate

Integrated burner unit installs on package boiler F/7

HOW TO SELECT PACKAGE BOILERS 365

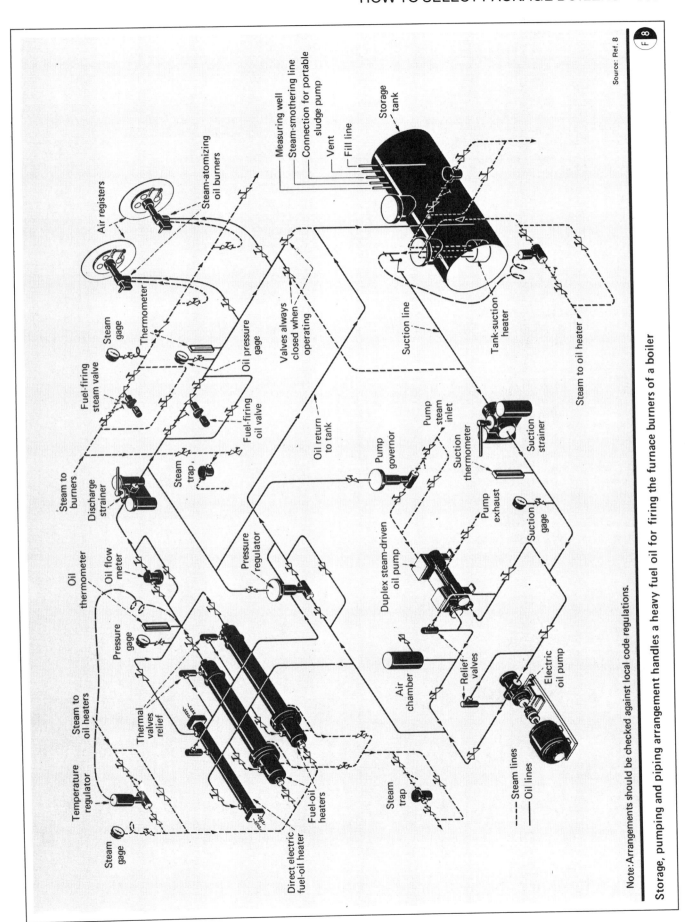

Storage, pumping and piping arrangement handles a heavy fuel oil for firing the furnace burners of a boiler

366 STEAM GENERATION AND TRANSMISSION

device. In this design, the plates pass through the exit-gas stream and are heated. The unit then revolves the hot plates into the incoming air, which absorbs the heat. Heated air is then passed to the burner box for combustion.

With air heaters, an increase in boiler efficiency of approximately 2% for each 100°F rise in the incoming-air temperature can be realized.

Soot blowers and automatic blowdown

While soot blowers and automatic blowdown controls might not be considered directly related to energy conservation, their use can result in fuel savings. Soot blowers assist in maintaining clean tube surfaces for better heat transfer. Such blowers are imperative when firing heavy oil or solid fuel.

The blowdown procedure removes the concentration of dissolved solids in the boiler water. Of course, with the discharge of this water, a certain amount of heat is lost. With manual operation, energy can be wasted if the boiler is blown down more frequently than necessary. With automatic control, the amount of heat lost can be held to a minimum while maintaining the proper level of solids in the boiler water.

Superheaters

Superheaters can increase the heat efficiency of the overall plant when considered in the plant's energy requirements. The amount of heat needed to raise the steam temperature to the superheat range is low. When superheaters are used in conjunction with steam turbines and generators, the plant's heat rate can be improved. Electric power can be supplied while process steam is obtained from the turbine exhaust.

Package-boiler manufacturers furnish two types of superheaters. Some builders supply convection superheaters, while others furnish radiant units. A few manufacturers make both types. Convection superheaters depend on the furnace gases for their heat source, and heat transfer is by convection only. These superheaters are installed in the convection section of the boiler. For these designs, the outlet temperature is limited to around 800°F. In the radiant design, heat transfer is primarily by radiation. This unit is mounted within the furnace and is subject to radiation from the burner flame. Higher steam temperatures can be realized with a radiant design—900°F is not unusual.

Specifying the boiler

By this time, you have probably established certain desirable features in your mind, such as the general type of boiler, and the fuel or fuels to be burned. Other basic factors have also been determined, such as feedwater source, and the requirements to be met for the air-pollution-control permit.

Now is the time for a word of caution. Do not impose arbitrary design features on boiler manufacturers un-

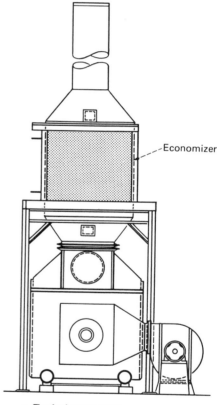

a. Typical economizer arrangement

b. Extended-surface economizer coil

Economizers on boilers recover heat that would otherwise be lost

less it is imperative to do so. All boiler manufacturers, bidding on your specifications, will have proven, efficient designs to offer. Furthermore, the boiler will be built to comply with the latest ASME Boiler Code. If you impose special features on a boiler supplier, you could be imposing an unwarranted increase in boiler cost. If you ask for the manufacturer's standard design to meet your requirements, you will be obtaining the least costly unit.

The general items of information for specifying and installing a boiler include: (1) type of boiler; (2) fuel or fuels; (3) components for the boiler such as burner and windbox, fan and ductwork, feedwater-control systems, safety and combustion controls; (4) design data: steam capacity, operating pressure and temperature, peak steam requirement; (5) utilities data for electricity and compressed air to enable selection of electric motors, and instruments and boiler controls; (6) physical data for plant site such as ambient temperature, site elevation, indoor or outdoor installation, wind loading, and where applicable seismic design zone; and (7) erection information: starting and completion dates, railroad spur track in relation to boiler site, tools and equipment to be furnished by the plant, site conditions, insurance requirements, and necessary building permits and licenses.

It is a good idea to list any special codes applicable to your project. In addition to the ASME Boiler Code, these include (a) OSHA regulations, (b) the Air Pollution Control District (APCD) having jurisdiction, (c) building codes, (d) state and/or local boiler codes, and (e) any special codes pertaining to your locality.

OSHA requirements should be checked, especially in the areas of ladder, platform and catwalk design; noise limits; permissable skin temperatures for casing and lagging; and valve and valve-operator locations.

Bid comparison

In obtaining the optimum unit for your plant, you will be faced with the task of selecting a boiler from the proposals presented. Preparing a good bid comparison is the best way to evaluate the engineering aspects of the various boilers.

A bid comparison should summarize on a single sheet all the data you have asked for. In addition, special items offered by the different boiler manufacturers should be included. You will then be able to evaluate each proposal in relation to the others on the basis of performance, engineering features, and control components—and, of course, price, delivery and terms. If you are receiving proposals for equipment only, it is helpful to have each bidder give you an evaluation of the number of man-hours required for installation of that unit. In addition, weights of the boiler and its components are helpful in completing the analysis.

Still other points to consider in evaluating the proposals are: (1) steam purity, (2) service, (3) hardware, (4) schedules, and (5) availability of spare parts.

Hardware should be reviewed with consideration for quality, manufacturer, service, and spare-parts availability. Hardware built by some manufacturers may offer better service and parts coverage in your particular locale than that made by others.

Operating considerations

Before the final selection is made, a review of the boilers with your operating people is advisable. Some features contribute to lower operating and maintenance costs than others. Here are some points to be checked:
- Burner-flame pattern.
- Location of operating valves.
- Platform arrangement.
- Instrumentation.
- Controls and their locations.
- Tools and accessories.

Accessories for good operation should include kits for flue-gas sampling and water-condition analysis.

Instrumentation should include the normal pressure and temperature information. Furthermore, an oxygen analyzer to enable the operator to accurately check his fuel/air ratio can be helpful. Reading the temperature of the exit flue gas is important. It gives a good indication of boiler operation and condition over a period of time. Higher-than-usual temperatures indicated wasted fuel. The reason may be scale on the tube interiors, soot-covered tube exteriors, improper burner operation, or other malfunctions in the steam-generator system.

After the boiler is placed in operation, the actual boiler efficiency should be determined. Boiler efficiency is determined at one load point, and is the heat absorbed by the steam, divided by the heat input.

When the boiler is working, its best performance can be assured by maintaining daily log sheets and by establishing a systematic preventive-maintenance program. With these management tools, your boiler installation will have the best chance for long life and low-cost operation.

The following firms have supplied information and/or illustrative material for this article: CEA Combustion, Inc.; Cleaver Brooks Div., Aqua-Chem, Inc.; Coen Co.; Combustion Engineering, Inc.; Riley Stoker Corp.; Struthers Wells Corp.; and Vapor Corp.

References

1. "Steam," 38th ed., p. 13-4, Babcock & Wilcox, New York 1972.
2. Todd Projects, Bull. 1000, Todd Shipyards Corp., New York, NY 10004.
3. Steam Generators, Bull. MH, Union Iron Works Div., Riley Stoker Corp., Worcester, MA 01613.
4. Clayton Steam Generators, Clayton Mfg. Co., El Monte, CA 91731.
5. Form AD 178-R-4, Cleaver-Brooks Div., Aqua-Chem, Inc., Milwaukee, WI 53201.
6. Va-Power Steam Generators, Vapor Corp., Chicago, IL 60648.
7. VP Boilers, Combustion Engineering, Inc., Windsor, CT 06095.
8. Bull. FC-68, Coen Co., Burlingame, CA 94010.

The author

Milton A. Buffington is district manager for Bumstead-Woolford Co., P. O. Box 1186, Lancaster, CA 93534. He has had considerable experience in boiler installation and operation in industrial plants. He is a registered engineer in California, and has a B.S. in mechanical engineering from Texas A&M University.

How to size and rate steam traps

Useful correlations and graphs provide rapid methods for sizing orifice traps and calculating flowrates of saturated water or steam-water mixtures through steam traps.

E. S. Monroe, Jr., E. I. du Pont de Nemours & Co.

☐ Practical methods of predicting flows through steam traps can help the user to prevent unnecessary steam wastage when using orifices, and to size other types of traps more accurately for better performance.

Flow of saturated water and steam-water mixtures (i.e., two-phase flow) can be easily defined by using discharge coefficients and formulas based on experimental observations. Exact formulas and graphs will be presented for determining flow through orifices.

One of the pioneer papers on this subject was that of Stuart and Yarnall [1]. Closely paralleling this was the work of Bottomley [2], in which he showed that actual flows were much higher than those theoretically calculated because of metastable conditions during the expansion. Benjamin and Miller [3] then showed that flows of saturated water could be calculated through single orifices by using a conventional flow equation with a discharge coefficient that varied with the equivalent head of water on the orifice. This author [4] showed that saturated-water flow through one or more orifices could be expressed by:

$$G = \frac{25{,}300(P_1 - P_2)^{0.8}(\mu)^{1.4}}{N^{0.75}(T/1{,}000)^{2.4}(\rho)^{1.2}} \quad (1)$$

where N = number of orifices of equal size in series; G = mass flow, lb/(s)(ft²); μ = initial absolute viscosity, lb/(h)(ft); $(P_1 - P_2)$ = total pressure drop, psi; T = initial temperature, °R; and ρ = initial density, lb/ft³.

Marriott [5] has derived an expression for the flow of two-phase, steam-water mixtures through a single orifice based on the experimental work of others (Benjamin and Miller; James, Monroe, Watson, et al.; Bottomley, Fitzsimons, Hoopes and Murdock) in which:

$$G = f(\beta)(k)(P_1)(P_2)(\rho_L)(\rho_V)(x) \quad (2)$$

where G = mass flow, lb/(s)(ft²); β = orifice ratio; k = isentropic expansion coefficient of steam; P_1 = initial pressure, lb/ft² abs.; P_2 = final pressure, lb/ft² abs.; ρ_L = density of initial water, lb/ft³; ρ_V = density of initial steam, lb/ft³; and x = initial quality of mixture flowing through orifice.

In Marriott's own words, "The problem has never been solved in closed form." In this article, we will show that by empirically reducing Eq. (1) to simple form,

Originally published April 12, 1976

and combining its use with a graphical plot for many solutions of Eq. (2), we can readily predict the amount of steam that will pass through an orifice not fully loaded with condensate. We will also show that a discharge coefficient can be used to accurately predict the maximum flow of condensate through a trap with varying upstream and downstream pressures.

Simplification and extension of Eq. (1)

Eq. (1) is only applicable to cases where the final pressure is atmospheric. Since the initial viscosity, temperature and density are all functions of the initial pressure, some curve-fitting techniques were used to reduce Eq. (1) to two equations that are a function of the initial pressure. These are:

$$N^{0.75}G = 900(P_1)^{1/3} \quad (3)$$

where P_1 = initial pressure, psig; and $P_1 \geq 64$ psig.

$$N^{0.75}G = (900/2)(P_1)^{1/2} \quad (4)$$

where $P_1 \leq 64$ psig.

Fig. 1 shows a plot of Eq. (1) that represents the

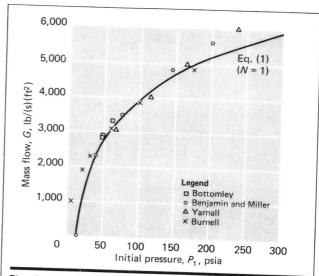

Plot for Eq. (1) of author's test data compared to published data of other investigators — Fig. 1

HOW TO SIZE AND RATE STEAM TRAPS 369

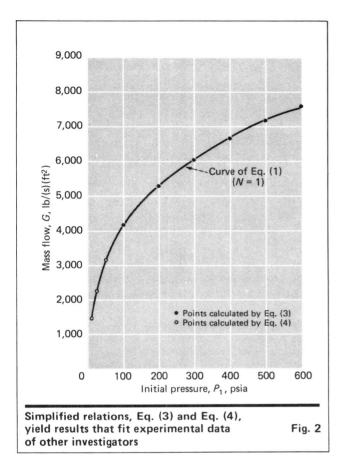

Simplified relations, Eq. (3) and Eq. (4), yield results that fit experimental data of other investigators   Fig. 2

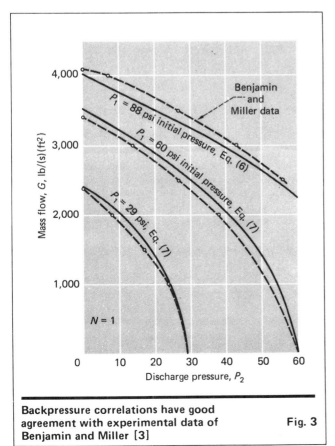

Backpressure correlations have good agreement with experimental data of Benjamin and Miller [3]   Fig. 3

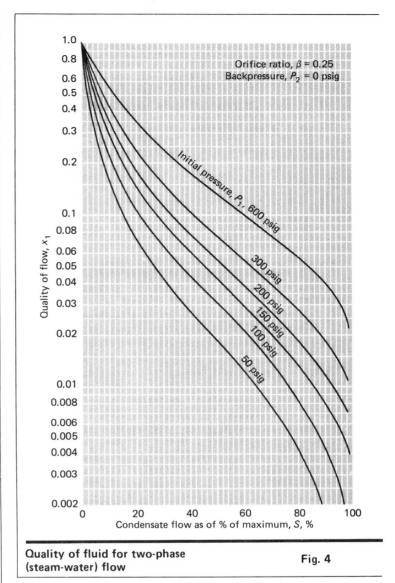

Quality of fluid for two-phase (steam-water) flow   Fig. 4

author's test data with data of Bottomley [2], Benjamin and Miller [3], Stuart and Yarnall [1], and Burnell [6].

Fig. 2 shows how closely Eq. (3) and (4) fit Eq. (1) over the usual range of steam-trap application.

Backpressures, other than atmospheric, are a common occurrence with steam traps; and the data of Benjamin and Miller [3] were examined for a possible correlation. It was found that the reduced mass flow, G^1, that occurs with backpressure could be expressed by:

$$G^1 = G\left(\frac{P_1 - P_2}{P_1}\right)^{1/2} \quad (5)$$

where P_2 is the backpressure, psig.

Fig. 3 shows such a correlation plotted against Benjamin and Miller's data. Since their data were read from a small curve, it is felt that the agreement is good. Furthermore, similar calculations for the author's own data for the flow through the first orifice in a series of orifices gave good agreement.

Flow of saturated water through one or more equal-sized orifices in series can, therefore, be given by the following two simple expressions:

$$W = \frac{17{,}700 D^2 (P_1)^{1/3}}{N^{0.75}} \left(\frac{P_1 - P_2}{P_1}\right)^{1/2} \quad (6)$$

where D = orifice dia., in; W = flow, lb/h; and $P_1 \geq 64$ psig.

$$W = \frac{8{,}850 D^2}{N^{0.75}} (P_1 - P_2)^{1/2} \quad (7)$$

where $P_1 \leq 64$ psig.

Graphical presentation of Eq. (2)

Because of their complexity, no attempt will be made here to repeat Marriott's expressions for correlating two-phase flow data through orifices. In Eq. (2), mass flow is a function of seven variables. If we assume a β ratio for the orifice small enough to have minimal effect in a $(1 - \beta^4)$ expression and a final pressure, P_2, of atmospheric, we can assume an initial pressure, P_1, and a quality of x_1, in order to solve for the mass flow.

Since the quality is assumed, the mass flow can then be separated into each of the two phases. If the liquid-phase flow is expressed as a fraction of liquid flow at zero quality, a relationship exists for the ratio of the vapor phase to liquid phase for each initial pressure. This relationship is shown in Fig. 4. While these curves are drawn for atmospheric discharges, it can be shown that two-phase flow will follow the form of Eq. (6) and (7) even if orifice sizes are not uniform. Fig. 5 shows how a curve of the form of Eq. (8) or Eq. (9) will adequately represent flow through four orifices, in series, of alternating sizes.

$$W = K_T (P_1)^{1/3} \left(\frac{P_1 - P_2}{P_1}\right)^{1/2} \quad (8)$$

where K_T = a discharge coefficient, and $P_1 \geq 64$ psig.

$$W = \frac{K_T}{2} (P_1 - P_2)^{1/2} \quad (9)$$

where $P_1 \leq 64$ psig.

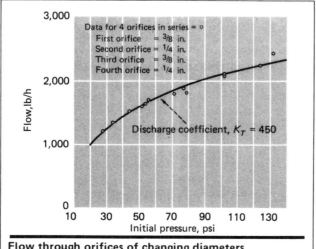

Flow through orifices of changing diameters and in series accounts for backpressure other than atmospheric Fig. 5

Since three of the four orifices must have two-phase flow, it is a logical conclusion that the curves of Fig. 4 may be applied to either multiple orifices or for flows other than to atmospheric.

How to calculate two-phase flow

Example 1—A $\frac{1}{4}$-in-dia. orifice is installed on a heat exchanger as a trap. Upstream pressure is 100 psig, and the condensate load is 500 lb/h. Is the orifice large enough, and how much steam (if any) will be passed by the orifice?

Step 1—Determine the maximum liquid flow by substituting into Eq. (6):

$$W = \frac{17{,}700 (0.25)^2 (100)^{1/3}}{1^{0.75}} \left(\frac{100 - 0}{100}\right)^{1/2}$$

$$W = 5{,}135 \text{ lb/h}$$

Step 2—Determine maximum liquid flow, S, % by substituting in the following relation:

$$S = \left(\frac{\text{Actual Flow}}{\text{Maximum Flow}}\right) 100$$

$$S = \left(\frac{500}{5{,}135}\right) 100 = 9.74\%$$

Step 3—Determine quality, x_1, from Fig. 4. At $S = 9.74\%$ and 100-psig inlet, the quality, x_1, is found to equal 0.23. Steam flow, y, is calculated from:

$$0.23 = y/(500 + y)$$

$$y = 149 \text{ lb/h}$$

Thus, the two-phase flow through the orifice will be 149 lb/h of steam and 500 lb/h of condensate for a total flow of 649 lb/h. The orifice is large enough because a maximum flow of 5,135 lb/h can be accommodated. This is ten times the required flow of liquid, as calculated in Step 1.

Example 2—Find the size of orifice required to limit steam flow to 50 lb/h when the condensate flow is 500 lb/h. Upstream pressure is 100 psig. Determine also the maximum condensate flow for the orifice.

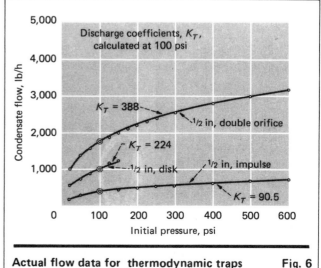

Actual flow data for thermodynamic traps Fig. 6

HOW TO SIZE AND RATE STEAM TRAPS 371

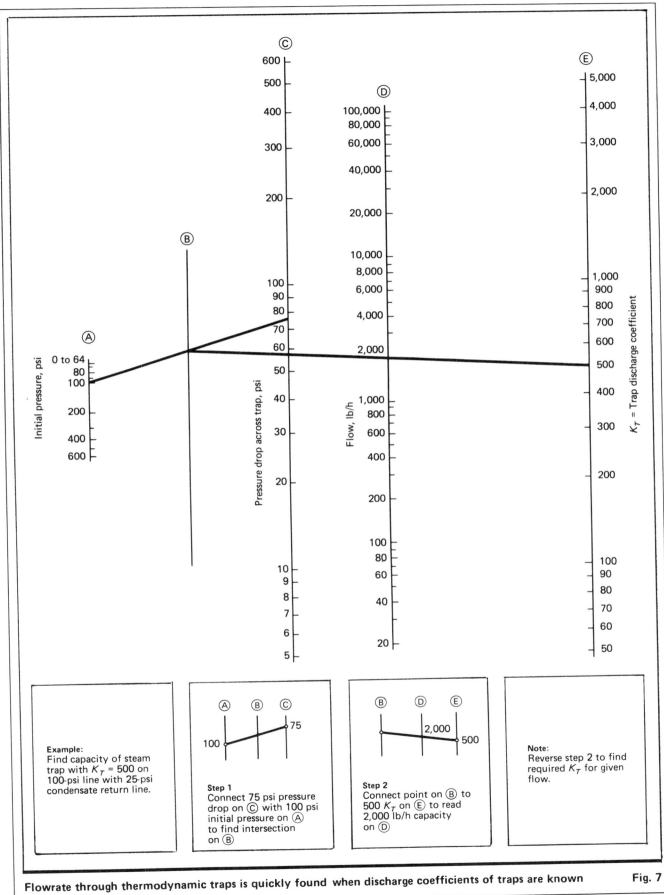

Flowrate through thermodynamic traps is quickly found when discharge coefficients of traps are known   Fig. 7

Step 1—Determine the quality, x_1, and then the value of S from Fig. 4.

$$x_1 = 50/(500 + 50) = 0.091$$
$$S = 23\% \text{ from Fig. 4}$$

Step 2—Determine orifice diameter from Eq. (6). First, we compute the maximum liquid flow, W, from:

$$W = 500/0.23 = 2{,}174 \text{ lb/h}$$

We now substitute the several values into Eq. (6):

$$2{,}174 = \frac{17{,}700 D^2 (100)^{1/3}}{1^{0.75}} \left(\frac{100-0}{100}\right)^{1/2}$$

$$D^2 = \frac{2{,}174(1)}{17{,}700(4.642)(1)} = 0.0264$$

$$D = 0.163 \text{ in}$$

Thus, to pass 500 lb/h of condensate with a flow of 50 lb/h of steam, orifice size should be 0.163-in dia. This orifice will handle up to 2,174 lb/h of condensate without flooding.

Example 3—Find the size of orifice required to handle a condensate flow ranging from 500 lb/h minimum to 1,000 lb/h maximum if the upstream pressure is 200 psig and the discharge is atmospheric. Determine the steam loss at minimum flow.

Step 1—Determine orifice diameter at maximum flow by substituting into Eq. (6).

$$1{,}000 = \frac{17{,}700 D^2 (200)^{1/3}}{1^{0.75}} \left(\frac{200-0}{200}\right)^{1/2}$$

$$D^2 = 0.00966$$
$$D = 0.098 \text{ in}$$

Step 2—Calculate the condensate flow at the minimum condition as a percent of maximum flow from:

$$S = (500/1{,}000)100 = 50\%$$

From Fig. 4 for $S = 50\%$, we find quality, x_1, = 0.055. Hence, steam loss, y, at minimum flow is:

$$y = \frac{500(0.055)}{1 - 0.055} = 27 \text{ lb/h}$$

Thus, a 0.098-in-dia. orifice will have a capacity range of up to 1,000 lb/h of 200-psig condensate. At 500-lb/h condensate capacity, it will also pass 27 lb/h of steam for a total flow of 527 lb/h.

Example 4—Two equally-sized orifices in series are to handle 1,000 lb/h of 200-psig condensate. What is the size of the orifices?

Again, Eq. (6) applies (note that $N = 2$).

$$1{,}000 = \frac{17{,}700 D^2 (200)^{1/3}}{2^{0.75}} \left(\frac{200-0}{200}\right)^{1/2}$$

$$D = 0.1275 \text{ in}$$

Application to other types of traps

Eq. (8) and (9) are also applicable to steam traps other than orifices if the discharge coefficient, K_T, is determined experimentally. Fig. 6 shows how closely the measured flows can match the equations.

The nomograph of Fig. 7 solves these equations for trap flow when the discharge coefficient of a trap is known. While discharge coefficients are not yet published by trap vendors, the author's company has used this method for years. Any engineer can readily determine the discharge capacity of any trap from good flow data by using either Eq. (8) and (9) or the nomograph (Fig. 7).

An example will show the advantage of this approach over that currently used by manufacturers.

Let us assume that a vendor rates his trap at 165 lb/h with a 100-psi drop and with 30° F subcooling. A test shows that the trap will pass only 100 lb/h with 0° F subcooling.

The discharge coefficient of this trap is:

$$100 = K_T (100)^{1/3}$$
$$K_T = 21.54$$

If the trap is applied to a system with an initial pressure of 150 psi and a backpressure of 50 psi, the usual practice is to rate the capacity of the trap at the same value as a 100-psi drop to zero backpressure, or 165 lb/h. Actually, by substituting into Eq. (8), we get:

$$W = 21.54(150)^{1/3} \left(\frac{150-50}{150}\right)^{1/2} = 93 \text{ lb/h}$$

Thus, the actual trap capacity will be only 56% of the published value. If installed as having the vendor's capacity, the trap may flood a heat exchanger, causing poor performance; and any carbon dioxide present will cause corrosion and high maintenance.

Future prospects

The American Soc. of Mechanical Engineers has formed a Performance Test Code Committee (PTC 31.1) to formulate test procedures for determining the capacity of steam traps. This committee has currently adopted the discharge coefficient method of reporting capacity. Rapid adoption of this method by the various manufacturers will facilitate trap application by users.

References

1. Stuart, M. C. and Yarnall, D. R., Fluid Flow Through Two Orifices in Series, *ASME Trans.*, Vol. 58, 479–484 (1936).
2. Bottomley, W. T., Flow of Boiling Water Through Orifices and Pipes, *Trans. Northeast Coast Institute of Engineers & Shipbuilders*, Vol. 53, Part 3 (Jan. 1937).
3. Benjamin, M. W. and Miller, J. G., Flow of Saturated Water Through Throttling Orifices, *ASME Trans.*, Vol. 63, 419–426 (1941).
4. Monroe, E. S., Flow of Saturated Boiling Water Through Knife-Edged Orifices in Series, *ASME Trans.*, Vol. 78, 373–377 (1956).
5. Marriott, P. W., Two-Phase Steam/Water Flow Through Sharp-Edged Orifices, General Electric Co. Report NEDO-10210, Aug. 1970.
6. Burnell, J. G., Flow of Boiling Water Through Nozzles, Orifices and Pipes, *J. Inst. Engrs., Australia*, Vol. 18, 41–49 (1946).

The author

E. S. Monroe has been a power consultant in the Engineering Dept. of E. I. du Pont de Nemours & Co., Louviers Building, Wilmington, DE 19898. Steam traps are one of his specialized areas and he has been issued five patents in this field. He is a graduate of Virginia Polytechnic Institute and Cornell University, is a member of the ASME Performance Test Code 39.1 committee (which is preparing test procedures for steam traps) and is a registered professional engineer in Delaware and New York.

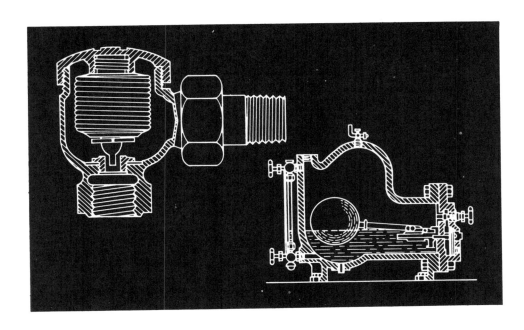

How to test steam traps

Defective steam traps can waste large amounts of energy. But how can you tell which ones are defective? Here are the testing procedures to follow.

E. S. Monroe, Jr., E. I. du Pont de Nemours & Co.

☐ Steam traps are well-known contributors to energy waste; they also need considerable maintenance.

The secret of improved trap operation is improved trap testing. The purpose of this article is to:
- Review the objectives of a trap test.
- Consider those methods of trap testing that have already been used.
- Select the trap-testing methods that are most meaningful.
- Develop a useful test procedure.

Meeting these objectives will make possible the achievement of improved trap installations, with resulting energy savings, lower maintenance costs and improved equipment performance.

Objectives of a steam-trap test

A steam-trap test should determine whether a trap is performing, properly and efficiently, the purpose for which it was installed.

It is possible for a trap to carry out its function properly without doing it efficiently. For example, an oversized orifice will drain a tracer completely, but the excess steam blown through the oversized orifice represents an inefficiency. Similarly, an undersized orifice may be efficient in not blowing steam, but the poorly drained tracer may not do its job properly.

The definition of "proper operation" varies with the trap application. The acceptable amount of subcooling of condensate is a good criterion, and this must be determined by the use to which the trap is put.

In general, steam tracing for freeze protection will function properly if the subcooling does not exceed 30°F. On the other hand, traps that drain well and do not exceed 5°F subcooling are required for applications that:
- Must achieve maximum equipment output.
- Must provide uniform heating.
- Contain noncondensable gases in the steam (such as CO_2).
- Start up frequently.

Originally published September 1, 1975

Nomenclature

- A Cross-sectional area of pipe, in^2
- P Pressure, lb/in^2
- V Volume, ft^3
- W Weight of fluid, lb

Subscripts

- 1   Before steam trap
- 2   After steam trap

The above degrees of subcooling are selected on the basis of the author's experience; each plant may choose to establish different values, based on its own requirements. The values of 5°F and 30°F subcooling will be used in this article as criteria for critical and noncritical trap applications, respectively.

Defining efficiency is more difficult. A trap that is working properly and closes completely without leakage can easily be judged to be efficient. Conversely, a trap that is designed for an on-and-off mode of operation, but that is blowing through continuously, can be judged inefficient. However, some of the traps that give the best performance on heat exchangers modulate, rather than going through on-and-off modes. The float-and-thermostatic trap is an example of this type.

For the purpose of this article, an efficient trap is one in which the steam in the condensate discharge does not exceed that which is theoretically formed by saturated water flashing to the lower pressure.

Trap tests

Four tests are commonly used to evaluate steam traps.

- Checking the inlet temperature.
- Checking the trap flow.
- Checking the outlet temperature.
- Checking the discharge pressure.

The last two procedures were discarded for the following reasons:

Outlet temperature—The outlet temperature is a function of either the saturation pressure of the condensate line, or the degree of superheat in the condensate line if sufficient steam blows through. If the condensate line is large enough not to build up backpressure, more than 20 lb of steam must blow through the trap for every pound of condensate that is passed for there to be any superheat.

A temperature rise under such conditions is unlikely, and even if it should occur, it would show up only for traps that are deteriorated far worse than those we wish to detect. Since discharge-temperature increases will not readily occur when traps are faulty, this technique was discarded as a meaningful steam-trap test.

Discharge pressure—Examination of discharge pressure was also eliminated for individual trap tests. It is true that when many traps are faulty, the pressure drop in the condensate line may increase. However, it will increase on the good traps as well as on the faulty ones. It indicates a system that may be faulty, but not necessarily an individual faulty trap.

Furthermore, if a trap *without condensate load* is blowing steam, the pressure drop in the condensate line may be less than when the trap is functioning properly. This is because the volume of flashed steam from condensate passing through an orifice will exceed the amount of steam that will pass through the same orifice.

The remaining two tests (inlet temperature and trap flow) were judged to be very meaningful, and to meet the objectives of a trap test as previously stated. These will be examined in detail.

Checking the trap flow

There are four common methods of checking the flow through a steam trap:

- Listening with a mechanics' stethoscope.
- Listening with an ultrasonic device.
- Observing the temperature fluctuations on the inlet of a trap by using a pyrometer.
- Observing the open discharge of a trap.

The first two are considered the best general methods to use, although each has its limitations.

Stethoscope—The mechanics' stethoscope is the cheapest and easiest technique to use. It suffers the drawback that when used on multiple banks of traps it requires discretion on the part of the listener to distinguish between background noise and the trap being checked. While this is a problem, it is possible to overcome it with practice.

Ultrasonic devices—Ultrasonic leak detectors are similar to mechanics' stethoscopes, except that they have electronic amplification. If they also have filters that limit their outputs to only true ultrasonic ranges, they will not transmit audible noises.

Since ultrasonic waves are quickly absorbed by solids, filtered units will also eliminate the sounds from other traps if a contact type of probe is used. One available filtered unit (Reglomat "Sonitor") has earphone, speaker and visual output. It has been proven to be superior by an order of magnitude over the mechanics' stethoscope.

It is not known whether ultrasonic devices with special filters can be used to distinguish between (a) no flow, (b) condensate flow, or (c) steam blowing through a trap. Consultants on acoustics indicate that this should be possible. It is one of the recommendations of this article that efforts be made to determine if it is indeed possible. Such a device in the hands of a trained operator would eliminate all guesswork and judgment from testing traps for flow.

Temperature fluctuations—The observation of temperature fluctuations with a pyrometer is useful in some trap applications, but not all. The best-performing traps that drain condensate completely and continuously will not respond to this test, whereas a trap that leaks because of wire drawing (but whose mechanism functions properly) may indicate satisfactory performance.

If a control valve is cycling, and thereby varying the steam pressure to the unit, it is possible that temperature variations may occur that are unrelated to the trap's operation.

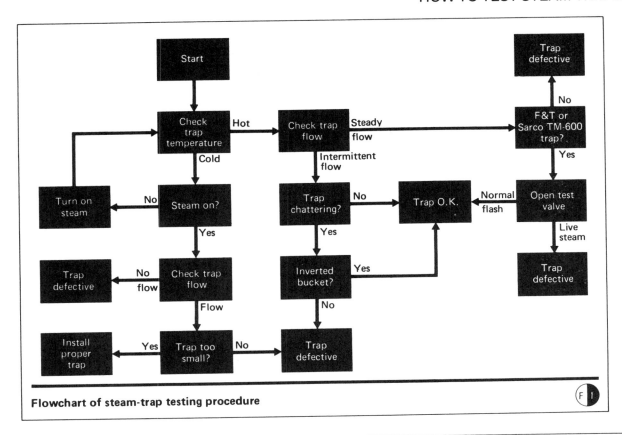

Flowchart of steam-trap testing procedure

For these reasons, the observation of temperature variations at the trap inlet is not considered a good method of checking trap flow.

Checking open discharge—Observing the open discharge of a trap is the fourth method of checking its flow. If a trap discharges to the atmosphere, it is an obvious method that is visual and—*for some types of traps*—foolproof. If a trap operates in an off-on mode, it is obvious when it is shutting off tightly and when it is open. Hence, for such traps as inverted buckets, thermodynamic disks and conventional thermostatic types, such observation is ideal—when open discharges exist.

For some other types, open discharges may be difficult to evaluate because of the problem of flash steam. The following example will serve to show the nature of the problem:

Assume a 100-psi trap with a ¼-in.-dia. orifice, discharging to atmosphere. If 100-psi saturated steam flows through the orifice:

$$W_1 = \frac{AP_1}{70} \text{ (Napier's formula)}$$

$$= \frac{114.7\,(\pi/4)(0.25)^2(3{,}600)}{70}$$

$$= 289 \text{ lb/h}$$

If there is adiabatic flow, the steam will be superheated 80°F and have a specific volume of 30 ft³/lb. The volume of steam issuing from the pipe will be:

$$V_1 = \left(\frac{30 \text{ ft}^3}{\text{lb}}\right)\left(\frac{289 \text{ lb}}{\text{h}}\right) = 8{,}670 \text{ ft}^3/\text{h}$$

Trap inspection procedure

Equipment: Surface pyrometer
  Mechanics' or ultrasonic stethoscope
  Gloves

Test No. 1: Check trap temperature with a pyrometer.
  Measure inlet temperature to trap.
  Measure steam temperature after control valve.
  Subtract one from the other.

Suggested allowable ΔT

Service	Example	Maximum ΔT, °F
Noncritical	Steam tracing	30
Critical	Process heater	5

Test No. 2: Check flow with stethoscope and determine whether flow is:
  ■ Zero
  ■ Intermittent
  ■ Steady

Test No. 3: For float-and-thermostatic and Sarco TM-600 types only. Open test valve and check for live (invisible) steam. If live steam is suspected, close trap inlet-valve for 5 min. Reopen and observe whether condensate flow is significantly larger than before. If so, the trap is O.K.

Instructions for testing of steam traps

If saturated condensate flows through the trap:

$$W_2 = 17{,}700 \, D^2 (P_1)^{1/3} \text{ (Author's formula derived from Ref. 1.)}$$
$$= 17{,}700 \, (0.25)^2 (100)^{1/3}$$
$$= 5{,}135 \text{ lb/h}$$

The percent of flash steam is calculable:
For saturated steam at 100 psig:
Enthalpy, sat. liquid, $h_f = 308.9$ Btu/lb
For saturated steam at atmospheric pressure:
Enthalpy, sat. liquid, $h_f = 180.7$ Btu/lb
Enthalpy, sat. vapor, $h_g = 1{,}150.4$ Btu/lb
Specific volume, $v_g = 26.8$ ft³/lb

$$1.150.4x + 180.07(1-x) = 308.9$$
$$970.4x = 128.83$$
$$x = 13.3\%$$

The volume of steam from the open pipe will be:
$$V_2 = 5{,}135(0.132)(26.8)$$
$$= 18{,}275 \text{ ft}^3/\text{h}$$

Thus, for the example chosen, there would be over twice as much steam volume escaping from a properly working trap as from a faulty trap blowing only live steam.

Further, the steam plume from the condensate flow is at saturation temperature and will condense to a white visible cloud much faster than the superheated steam discharge. It is this flash steam that makes open discharges so difficult to evaluate when traps modulate and discharge continuously.

It can be stated that when the steam issuing from the open discharge pipe is invisible for the first few inches, it is probably blowby or leakage steam, and not flash steam. Differentiating between normal flash steam and blowby steam is difficult, and the reaching of conclusions about trap conditions by this method should be avoided if possible.

Test valves should be avoided for the same reasons. An additional reason exists for the thermodynamic types of traps. Such traps are subject to misoperation when sufficient backpressure is applied to them. They will stay open and blow steam. The opening of a test valve may release the pressure in the condensate line and, during a test, the trap will function satisfactorily.

The best ways to test traps

For conducting tests to check the flow in a steam trap:

- Listening devices are superior to other methods for closed systems—*except* when the trap modulates.
- Open-drain observation is effective for off-and-on types of traps.
- For modulating traps, open-drain or test-valve observation is necessary, but difficult.

The float-and-thermostatic type of trap must be tested with an open drain or test valve (until better ultrasonic listening devices are available). Usually, the quantity of water issuing is a good indication of the proper operation of the trap. If the water is reduced in quantity from that obtained when the trap inlet-valve is closed for 5 min, it verifies that the float valve is modulating and functioning properly.

The Sarco TM-600 thermostatic valve (developed by the writer) will normally open and shut. Under certain loads it will modulate and, when this occurs, it should be treated like the float-and-thermostatic trap.

A third instance of modulating occurs under certain loads with inverted-bucket traps. This is distinguishable by a chattering sound in the trap as the trap opens and shuts rapidly, which indicates proper functioning. However, one must be sure that the chattering is not the metallic clanging of a broken linkage that allows the bucket to rattle against the side of the trap. If in doubt, treat the trap like the float-and-thermostatic type—the chattering should disappear momentarily during the higher flowrate that occurs when the trap inlet valve is reopened.

Although listening devices are normally used to determine whether traps are opening or closing, they can also be used to distinguish between condensate flow and steam blowby. However, this requires careful training in the use of the mechanics' stethoscope.

If an ultrasonic-type instrument can be developed that will distinguish between condensate flow and steam blowby, plants will be able to justify the training of an operator to properly use and maintain such an instrument. The time-consuming practice of closing inlet valves, judging the ambiguities of flash steam versus blowby steam, or interpreting the audible sounds of a mechanics' stethoscope, will all be avoided.

A generalized trap-testing procedure

The flowchart of Fig. 1 (F/1) illustrates a steamtrap inspection procedure that steers the user through the logic of a steam-trap test. The written trap-inspection procedure (F/2) was developed as a one-page guide to assist in using F/1.

Conclusions

- Steam traps should be tested frequently, and records kept, so that users can compile both efficiency and maintenance information on which steam traps perform best in their plants.
- Inlet-temperature subcooling and flow checks are the best tests for steam traps.
- Ultrasonic leak detectors with filters to respond only above 35 kHz are best for checking trap flow.
- Pyrometers are the best method of checking inlet temperatures.

References

1. Monroe, E. S., Jr., Flow of Saturated Boiler Water Through Knife-Edge Orifices in Series, *Trans. ASME*, Vol. 78, No. 2, pp. 373–377, Feb. 1956.

The author

E. S. Monroe is a power consultant in the Engineering Dept. of E. I. du Pont de Nemours & Co., Louviers Building, Wilmington, DE 19898. Steam traps are one of his specialized areas and he has been issued five patents in this field. He is a graduate of Virginia Polytechnic Institute and Cornell University, is a member of the ASME Performance Test Code 39.1 committee (which is preparing test procedures for steam traps) and is a registered professional engineer in Delaware and New York.

Improving boiler efficiency

Boilers can produce more steam from less fuel if heat losses are minimized and combustion is optimum. Downtime will be reduced with proper water quality.

J. C. Wilcox, Jr., Babcock & Wilcox

☐ Improving efficiency in the chemical process industries' boilers—boilers that produce steam for both process and power—can reduce consumption of expensive fuel by as much as 6%. More than 90% of the approximately 400 boilers sold to the U.S. petroleum industry and 95% of the approximately 850 boilers sold to the U.S. chemical industry in the past decade are burning gas or oil—fuels that are becoming more expensive and less available.

Controlling heat losses is one way to improve boilers. These losses, especially in older units, reduce the overall efficiency of many steam generating systems to below 70%. Another effective measure that is recommended for limiting fuel consumption is to install modern, effective combustion controls.

Minimizing heat losses

In any boiler system, heat lost up the stack represents a considerable waste of energy. Combustion products having temperatures above ambient, and moisture that leaves with the flue gas, constitute major heat losses.

Efficiency can be increased by reducing stack-gas temperatures through the addition of heat traps—either economizers or air heaters—that use the heat from the flue gas in other areas of the boiler.

Economizers heat feedwater with the captured energy, increasing boiler efficiency by 1% for every 10 to 11°F increase in feedwater temperature.

Air heaters transfer flue-gas heat to the furnace's incoming combustion air. For each 100°F increase in air temperature, efficiency is improved about 1.7%.

Generally, boiler efficiency increases about 2.5% for each 100°F drop in exit-gas temperature. And economizers and air heaters can improve combustion process efficiency by as much as 6%.

In deciding whether to install heat traps, a company must balance capital costs, maintenance expenses and possible added fan-power requirements against the substantial fuel savings. Most boilers in the chemical and petroleum industries are large enough to justify such equipment.

The typical boiler in these industries has a steam

Operator monitoring, even on automated boilers, can anticipate problems and increase efficiency

capacity of 100,000 lb/h. Current economic tradeoffs favor adding heat traps to boilers with steam capacities greater than about 30,000 lb/h.

Another factor in heat losses is radiation. Proper boiler insulation will minimize these losses. Burn spots and corrosion marks on outer casings indicate flue-gas leakage, which, as well as adversely affecting efficiency, can be irritating to operators.

Small leaks can occur, too, in welded inner casings.

Originally published October 9, 1978

These are often difficult to locate. One method of finding them is to discharge a smoke bomb in the furnace when the unit is down for maintenance. Leaking smoke streams will pinpoint locations where repair is needed. In these cases, it is strongly advised that the equipment manufacturer be contacted for recommendations.

Optimum combustion

In the furnace itself, combustion occurs most efficiently when the correct fuel-to-air ratio exists.

To maximize the efficiency in the boiler's furnace, three factors are important:

- Excess air in the combustion chamber should be held to the minimum consistent with complete combustion.
- The ash should contain no unburned combustible matter.
- The exit flue gas should contain no unburned combustible gases, such as carbon monoxide.

Efficient combustion-control systems can regulate fuel and air firing-rates in response to load changes, and they are extremely important when boilers are regularly subjected to wide load-swings. Oxygen monitoring equipment, specifically, can control the fuel-to-air ratio most precisely because it continuously fine-tunes the combustion control system.

This equipment analyzes the flue gas and provides continuous readings of oxygen, carbon dioxide and carbon monoxide. Oxygen monitors of this kind can be incorporated into the automatic control system or used as a constant visual indicator for manual adjustment.

Almost all chemical and petroleum plant boilers are equipped with fully automatic control systems. To make sure the controls work effectively operators must watch for changes that might affect reliability.

Burner flame scanners must be kept clean; they will fail to "see" the flame if soot builds up on the instruments' tubes. This could result in false trips.

Scanners are placed in the burner wall "behind" the flame or located in the rear wall of the furnace. This rear-wall location has proved effective for oil-fired package boilers. In such units, front-wall positions may give a false, bright reading when, in fact, the furnace may be smoky and dangerous.

Pneumatic combustion-control equipment should be watched for changes in relative control pressures. Deviations from constant pressure, relative to load, can mean potential problems.

Because a boiler normally burns four times its original cost in fuel every year, the importance of combustion control in maintaining high efficiency becomes very clear. Control changes should be noted and brought to the attention of the manufacturer's service engineer.

It is wise to keep spare control-system components on hand. For example, an extra program-controller module for the startup and shutdown interlock systems, along with spare scanner cells, will eliminate lead time and unnecessary boiler downtime.

The potential of furnace explosion is reduced by safety interlock flame-failure systems and startup programming. Operators should never bypass or jumper them out, and should be able to recognize danger signals that spell potential problems.

Fuel should never be allowed to enter the furnace without proper ignition, and the fuel- and air-flow should always be controlled within the proper specifications. Failure in either area could produce explosive conditions. Charts on the control panel often warn of impending danger.

Operators should be wary if changes occur in the normal relative air- and steam-flow pen position on the 24-hour chart. Abnormal changes in steam and flue-gas temperatures could also indicate trouble.

Wide fluctuations in furnace pressure also show unstable ignition at the burner. Pulsating or erratic burner operation, abnormally dark fire, unusual impingement on furnace walls or an unusually dark stack-gas should all alert the operator that the boiler is not operating most efficiently. Operators can also check and replace electron tubes in the program controller. If further work is required, however, the manufacturer's service engineer should be called.

Burner and furnace conditions

Burners should be inspected regularly to make sure they are free of fouling, which would detract from operating dependability. "Dual atomizer" burners are available. These allow fuel inputs for full boiler load while the main gun is being cleaned.

Improper burner assembly after cleaning also can cause poor operation, loss of efficiency and, in extreme cases, failure to start.

Normally, operators should not attempt to adjust burners and registers. If adjustment is required because of flame impingement, for example, it is advisable to request help from the manufacturer's service engineers. Changes in burner setting can often mean that the combustion-control equipment also must be reset. Because of the extensive and critical interrelationships between burners and controls, trial-and-error methods can prove costly.

A thorough purge before every lightoff is the only effective means of preventing puffs. The program controller's interlock system maintains this purge cycle and prevents the introduction of any fuel until the required number of furnace air changes are achieved and verified. However, to further ensure safe startups, operators should periodically clock the purge cycle independently. This will make sure the controller is operating reliably.

Operators also should make certain that the burner

Plant lab technician checks boiler water

immediately lights and continues to burn. Properly operating flame-scanners will shut off fuel feed immediately upon improper combustion. However, operators also should check the flame visually on a scheduled basis.

The more attention the equipment receives, the less likelihood that serious problems will occur. If any danger signal appears and the reason is obvious and immediately correctable, operators can take action.

Water quality increases reliability

Equipment availability and dependability are critical to economical operation. Both are enhanced by good boiler-water quality. In maintaining tight boiler-water standards there are three major concerns: boiler-feedwater makeup, condensate flowing into the boiler feedwater, and the boiler water itself.

Water leaving raw-water treatment for makeup should have essentially no hardness. This can be assured by any of several types of commercially available softeners. Condensate returned to the boiler can range from a very small to very high percentage of the boiler's total water requirements, depending on plant operations and design. Makeup water is added to the recovered condensate to provide the remainder of the boiler's feedwater.

Taking advantage of the heat contained in the condensate can reduce total fuel consumption. If contamination in the condensate can be treated economically, the condensate should be returned to the boiler rather than dumped. This has the additional benefit of reducing makeup requirements.

The most common of the many contaminants that can be picked up by the condensate stream are suspended iron, hardness salts, and organics. The most common organic substance is oil, which can incite foaming in the boiler water.

Operators should watch the condensate returns to be sure that contaminants are removed before they can enter the feedwater system.

Most of the common contaminants can be removed by filtration. In some plants, however, a sodium zeolite filter will be needed to further purify the condensate. Oxygen should always be removed from the feedwater stream by a deaerating heater, and neither condensate nor makeup should bypass it. The deaerating heater should be kept under pressure at all times. To further control oxygen, sulfite may be added as an oxygen scavenger.

To control hardness in the system's water, either phosphate treatment of the boiler water or chelation of the feedwater may be employed. Hardness, if left untreated, can cause scaling inside of boiler tubes and lead to overheating and tube rupture. Higher-pressure boilers require much more stringent water quality than low-pressure units; what is acceptable at 150 psi is well beyond the operating limits of a 900-psi system. Power-boiler water requirements, therefore, are much tighter than those for process steam units.

Plants that cannot maintain a completely equipped and staffed water laboratory should retain a competent water consultant to set up treatment procedures and perform periodic inspections to make sure quality remains at acceptable levels. In comparison to the costs of maintenance and loss of steam production caused by tube failure, the cost of the service is worthwhile.

Analyze alternative fuels

Most boilers equipped to burn either oil or gas can be retrofitted to fire both fuels. However, many of these units cannot be modified to burn waste fuels. Some alternative fuels require specially designed boilers.

After being ignored for many years, alternative energy sources have been pushed to the forefront of public attention by the recent shortages and escalating costs of oil and gas. Many forms of solid, liquid and gaseous hydrocarbon waste fuels are now being burned in boilers.

A hydrocarbon byproduct must be thoroughly analyzed to determine its suitability as a boiler fuel. Fuel-burning equipment and combustion-air intake methods should also be examined to make sure that the unit will operate with high efficiency and low carryover of solids in the stack gas. Special design considerations are required for liquid waste streams such as pitch, acid sludge or residual tars, and for waste gases such as carbon monoxide.

It is important to identify the waste types, their quantities and characteristics before a waste processing system is designed. The following chemical and physical properties of all byproduct fuels should be identified:
- Heating value.
- Ignition temperature.
- Volatile matter.
- Fixed carbon and hydrogen contents.
- Elemental analysis.
- Ash analysis.
- Ash and waste content.
- Density.
- Ash fusion temperature and fouling characteristics.
- Corrosiveness and erosiveness of fuel and combustion gases.
- Toxicity and odor of gases.
- Explosiveness.
- Viscosity characteristics (for liquid fuels).
- Variability of quantities and characteristics.
- Special handling problems (such as polymerization of liquid fuels).

The supplier of the boiler and burner should be called in to examine the potential waste fuel, and to determine whether the existing boiler can be modified to burn it. In many cases, these byproducts can readily supplement conventional fuel and cut energy costs.

The energy situation that the chemical and petroleum industries face today poses serious economic problems. More-efficient boiler operation is one way for these industries to meet these challenges.

The author

John C. Wilcox, Jr., is General Sales Manager, Industrial and Marine Div., Power Generation Group, The Babcox & Wilcox Co., P. O. Box 2423, North Canton, OH 44720. He received his degree in mechanical engineering from Stevens Institute of Technology, and is a member of the American Iron and Steel Institute, the American Institute of Mining, Metallurgical and Petroleum Engineers, the American Soc. of Mechanical Engineers, and currently represents Babcock & Wilcox as cochairman of the American Boiler Manufacturers Assn.'s Industrial Boiler Committee.

Install steam traps correctly

Improperly installed steam traps will operate inefficiently or not at all. Here are the many pitfalls you must avoid if your traps are to work.

E. S. Monroe, Jr., E. I. du Pont de Nemours & Co.

☐ The best steam trap will not function correctly if the piping to and from it is not properly designed and installed. Discussion of this important subject of installation completes the series of articles that have covered the characteristics [1]; testing [2], and flow [3] of steam traps.

This article will be concerned with:
- Double trapping.
- Multiple trapping of heat exchangers.
- Part-load operation.
- Steam locking.
- Inlet piping.
- Discharge piping.
- Piping details.

Double trapping

Double trapping is the poor practice of placing two traps in series with one another (Fig. 1). Unless the second trap is an orifice of sufficient size, flash steam from the saturated condensate entering the first trap will cause the second trap to close. The net effect is poorer drainage than with one trap.

Fig. 1a shows the most obvious arrangement of double trapping. Fig. 1b shows the first trap replaced with an orifice. This, too, is obviously bad. Fig. 1c, which shows an undersized inlet line connected to a single trap, is seldom recognized as the equivalent of a double-trapping situation.

Most engineers recognize the disadvantage of double trapping where two traps are placed in series. More engineers need to recognize that undersize inlet lines preceding a single trap are the equivalent of this. The recommendations given later for inlet piping must be met if the difficulties of double trapping are to be avoided.

Multiple trapping of heat exchangers

It is often tempting to save both space and investment by connecting two or more heat exchangers to a single trap (Fig. 2a). If the heat exchangers are physically identical, the load on each is identical, the exact same steam pressure is applied to each, and the condensate piping is identical, a single trap might work satisfactorily.

What is more likely to happen is that there will be both physical and load differences that cause the pres-

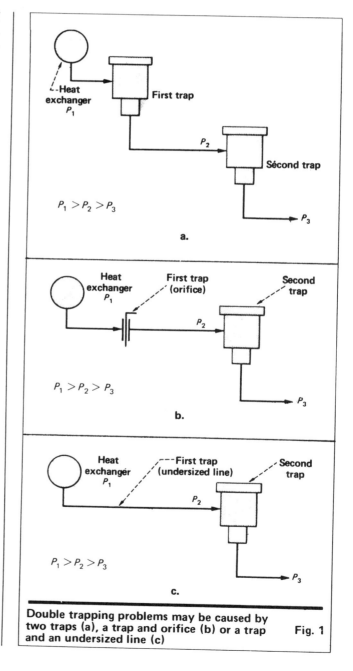

Double trapping problems may be caused by two traps (a), a trap and orifice (b) or a trap and an undersized line (c) Fig. 1

Originally published May 10, 1976

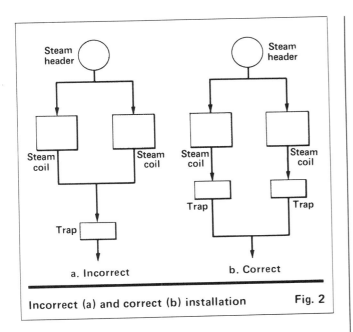

Incorrect (a) and correct (b) installation Fig. 2

sure differential across the heat exchangers to vary. When physical identicalness exists but loads are uneven, the more lightly loaded exchanger will drain completely, and the more heavily loaded exchanger will become waterlogged. Similarly, the heat exchanger with excessive pressure drop will be poorly drained even if the loads are equal. Hence, heat exchangers should be separately trapped (Fig. 2b) in all installations.

Part-load operations

When the output of a heat exchanger at light loads is controlled by throttling the steam supply, a trap that is adequate for full load operation may have a pressure drop inadequate to discharge the condensate without flooding the heat exchanger. This results in uneven heating, since the flooded part of such a heat exchanger is less effective. If the application is for heating outside air, frozen coils may result. If products are heat sensitive, throttling the supply may yield under- and overheated products (to give an average temperature), which causes off-quality production.

Two cases may exist. In the first, the steam pressure is always above atmospheric pressure. Good drainage can be assured in this case by making certain the trap is large enough to drain the reduced load at the reduced pressure differential.

In the second case, the coil goes to a vacuum on the steam side. This is typical when using low-pressure steam heating coils for outside air, and may occur before the air temperature reaches or drops below the freezing point. One answer is to put coils in series to the air flow. If this is done (Fig. 3a), the first coil is energized with an automatic valve to full steam pressure when the outside air is just above the freezing temperature. The second coil is then used to trim. Although this second coil may not freeze, it will not drain properly and will heat the air unevenly. For this reason, it is more common to provide (1) vacuum breakers and gravity drainage (Fig. 3b), or (2) vacuum return systems (Fig. 3c), or (3) both vacuum breakers and vacuum return systems (Fig. 3d). It should be noted that gravity drainage requires a head of water (usually about 2 ft) to provide the trap drainage, and that this will nullify any air venting capability of the trap. A separate thermostatic air-vent is thus shown in Fig. 3(b) to provide venting.

Steam locking

Steam trapping basically consists of separating condensate from steam vapor. This occurs naturally, by gravity. Whenever possible, drainage lines from heat exchangers to traps should slope downward. When they must run upward, a condition can occur called steam locking. This often comes about when rotary drums are steam heated (Fig. 4a). An internal pipe (siphon) is used to pick up the condensate. When the trap discharges, steam will enter the siphon. Since the siphon is surrounded by steam, there will be little or no condensation, and steam locking will occur. Condensate will accumulate in the roll and will not be discharged because the trap senses only steam.

Two solutions are possible. The first (Fig. 4b) bypasses the trap with a fixed orifice, while the second (Fig. 4c) provides an air- or water-cooled condenser ahead of the trap. Some traps have either a fixed or adjustable orifice built into them and are useful for such applications. Both methods waste steam but are necessary for proper operation. The siphon should be as short as possible to minimize the volume of trapped steam. Its pickup point should be as close to the inner drum surface as possible; this is particularly important when the roll is rotating fast enough to centrifugally hold the condensate around the entire inner circumference.

Another example of steam locking can occur with tank coil-heaters where the condensate discharge must be piped over the rim of the tank (Fig. 4d). To ensure that the condensate is lifted to the trap, a seal (to prevent steam entering the vertical pipe) must be provided, and the size and length of vertical piping must be controlled carefully. The same provisions for correction are applicable (Fig. 4b, 4c) as for rolls but are not as stringent since the vertical pipe is more of a condenser than it is in the case of the siphon surrounded by steam in a roll.

Inlet piping

Trap connection sizes as supplied by vendors frequently have little relation to the capacity of the trap. More often than not, they are undersized. Difficulty can occur with undersized inlet condensate pipes, as explained under double trapping. This is especially true of long lines that are undersized. If gravity drainage that can separate steam and condensate is not provided, severe water-hammer conditions can occur that will damage some traps. Examples are those traps using floats and bellows.

Excessive pressure drops in inlet piping to steam traps can, in the case of thermostatic traps, cause a phenomenon called "swell." The bellows will have an internal pressure equal to the steam pressure before flow. When the pressure on the outside of the trap bellows

382 STEAM GENERATION AND TRANSMISSION

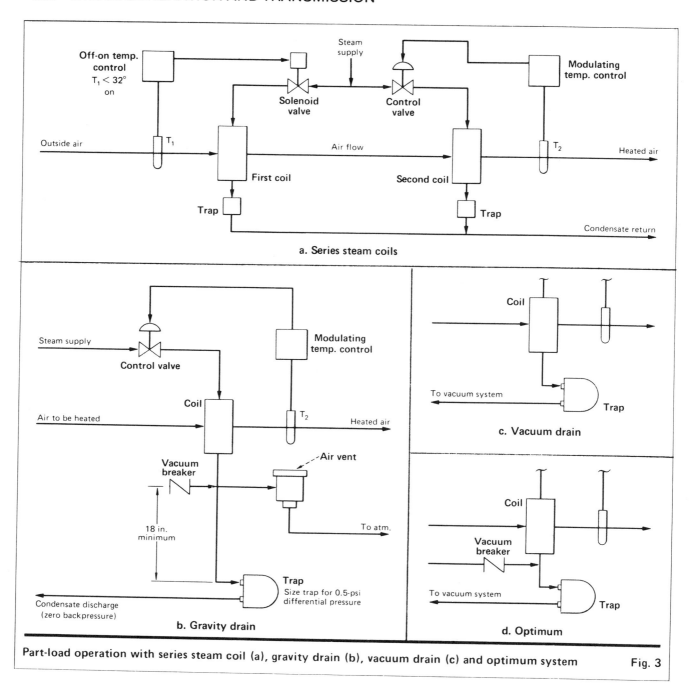

Part-load operation with series steam coil (a), gravity drain (b), vacuum drain (c) and optimum system   Fig. 3

drops suddenly, because the trap opens (and there are inlet-pipe flow restrictions), the trap bellows may elongate from its higher internal pressure. To avoid the sudden flashing of condensate, with its attendant problems in condensate inlet piping to traps, the lines should be:
- Adequate in size, to minimize pressure drop.
- Sloped downward to the trap, to ensure gravity drainage where possible.
- Individually trapped to avoid bypassing.
- Provided with anti-steam-locking features where gravity drainage is not possible.
- Sized for full trap capacity.

To assist in sizing inlet pipes, the nomogram shown in Fig. 6 may be used. This nomogram assumes that at least 24 in of vertical submergence exists, as shown in Fig. 5. In calculating inlet piping, the full trap capacity should be used and not the calculated average load. When the trap opens, this is the flow that occurs, and the piping must accommodate it if flashing problems are to be avoided. The only possible exceptions to this rule are traps that modulate, such as the float-and-thermostatic trap. If full flow can occur, it must be provided for.

Discharge piping

If traps are to function properly and erosion is to be avoided, the discharge piping is as important as the inlet piping. It is impossible to keep saturated condensate from partially flashing to steam when pressure is re-

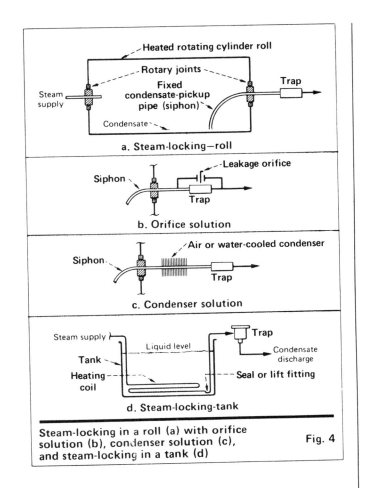

Fig. 4 Steam-locking in a roll (a) with orifice solution (b), condenser solution (c), and steam-locking in a tank (d)

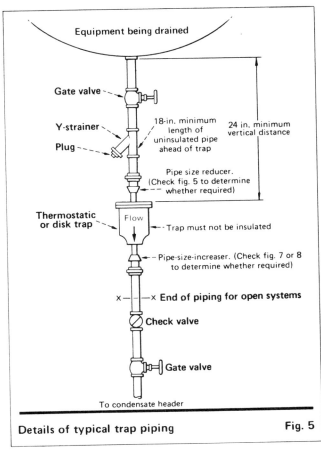

Fig. 5 Details of typical trap piping

duced by its flowing through a trap. A recent article by R. P. Ruskin, "Calculating Line Sizes For Flashing Steam-Condensate" (*Chem. Eng.* Aug. 18, 1975) [4] presents a nomogram for performing this function. A simplified form of this nomogram is shown in Fig. 7. Ruskin used conservative calculating methods, and the nomogram gives relatively low pressure drops. In his illustrated example, 1,000 lb/h of condensate flowing in a 1-in line from 600-psi steam pressure to 200 psi has less than 1-psi pressure drop per 100 ft. The nomogram in Fig. 8 is less conservative and would call for a ½-in. line for the same flow. It uses up to 20% of the initial pressure for the condensate flow and is cheaper to install. The trap capacity will suffer, however, because of the higher backpressure. For the case given, the loss will be 5%. This may or may not be negligible.

Either nomogram may be used, based on the engineer's evaluation of his needs, with the aim being to prevent such high backpressures that the trap either becomes inoperative or loses needed capacity.

The problem of flash steam cannot be avoided and so must be dealt with. Velocities resulting from using either Fig. 7 or 8 will be low enough to avoid excessive erosion. Undersized condensate piping will erode, often at impingement points such as at piping elbows.

Water hammer will generally not occur in flashing mixtures flowing in well-designed pipe or when flashing mixtures join at a common point. Admitting water that is not at saturation temperature into a flashing mixture will invariably result in water hammer, with potentially destructive effects. If possible, it should be avoided by repiping or by improved trapping. If not, specially designed mixers should be employed to reduce the water-hammer effect.

There is much debate about the merits of elevating condensate on the discharge side of a steam trap. With proper design, I believe it does not matter whether resistance to flow is static or dynamic as long as it is accounted for in the design stages. Proper design includes using check valves to prevent the condensate from flowing backward into the heat exchanger when steam pressure is turned off, and allowing for the static head of condensate when sizing the trap.

Piping details

A typical piping arrangement for a steam trap is shown in Fig. 5. A gate valve on the inlet is provided (for maintenance purposes), followed by a strainer. Strainers should have openings no greater than 3/64-in dia. Some orifices and impulse traps may require strainers with finer mesh. The blowoff connection is shown plugged rather than provided with a valve, for both cost and energy-conservation reasons. A maintenance joint follows the strainer, and a reducer to the trap inlet is installed if required. Following the trap is an enlarger (if required); a maintenance joint and, if the trap is connected to a common condensate line, a check valve; and a gate valve for maintenance isolation.

Fig. 9 shows another low-cost arrangement, consid-

384 STEAM GENERATION AND TRANSMISSION

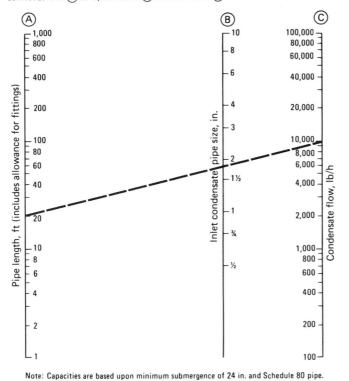

Nomograph for flow to trap (condenser inlet piping) — Fig. 6

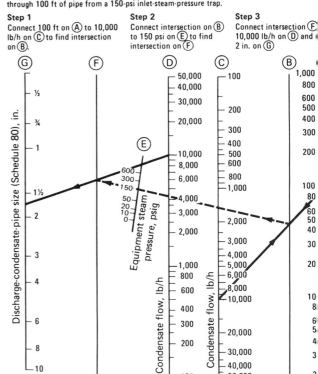

Nomogram (Du Pont) for condensate discharge piping — Fig.

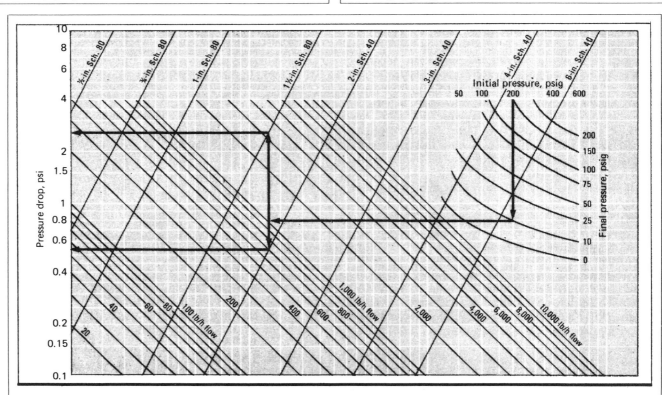

Nomograph for condensate discharge piping—Adapted from R. P. Ruskin (*Chem. Eng.*, Aug. 18, 1975) — Fig. 7

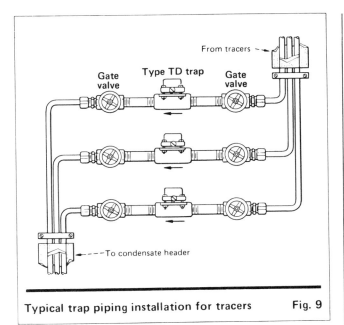

Typical trap piping installation for tracers — Fig. 9

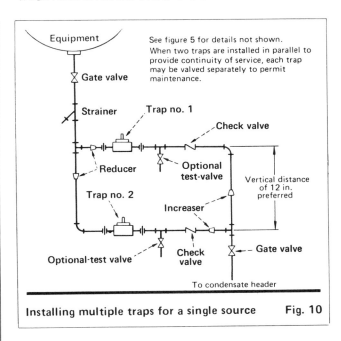

Installing multiple traps for a single source — Fig. 10

ered suitable only for tracing or for drip legs. It is not recommended for heat-exchanger service. A repairable-in-place trap is purchased as an assembly with two gate valves that have tubing connectors. A pressure bleedoff is added to the trap to permit safe maintenance. Integral strainers may be built into the trap or, if the tubing in the tracers is noncorrosive, a single strainer may be installed at the steam main, where it branches into various tracer lines. To economize, there is no check valve. Traps may be insulated or should not be insulated as shown in the table.

Traps that can be insulated	Traps that should be uninsulated
Float-and-thermostatic	Open bucket
Impulse	Inverted bucket
Orifice	Disk
Thermal expansion	Thermostatic
	Bimetallic

In addition, thermostatic and bimetallic traps should be installed with at least 18 in of bare condensate-inlet-piping immediately ahead of the trap.

Where traps are installed uninsulated and outdoors, simple sheet-metal covers to shield them from rain can reduce thermal losses. Insulation on traps and piping should be waterproof. It is not unusual to see the boiler load on an outdoor plant rise 10% during a rainstorm if such precautions are not taken.

Bypasses should never be installed around steam traps. They waste steam and, by lowering the steam pressure in the heat exchanger, may actually provide longer heat-up periods. If continuity of operation is essential, dual trapping as shown in Fig. 10 should be used; here, either trap is capable of handling the entire load.

Large loads may be handled in several ways. One way is to install multiple traps in parallel, as shown in Fig. 10. Another way is to use a single large trap if it is commercially available. A third way is to use a level pot with a level controller operating a conventional automatic control-valve. Economic evaluations should be made to determine which is the cheapest to install. If level pots are installed, one must not forget to use a thermostatic trap above the water level in the pot to act as a noncondensable-gas purge. Multiple banks of traps to a single load have the advantage that the loss of a single unit does not shut down the equipment.

Conclusion

Proper installation of the piping to and from a steam trap is as important as proper trap selection, testing and maintenance. Until attention is paid to all these features, the best possible trap performance will not be achieved.

Acknowledgement

The nomograms presented as Fig. 5 and 8 were prepared by Louis Bertrand and are based on the well-known work of Benjamin and Miller [5].

References

1. Monroe, E. S., Select the right steam traps, *Chem. Eng.*, Jan. 5, 1976, p. 129.
2. Monroe, E.S., How to test steam traps, *Chem. Eng.*, Sept. 1, 1975, p. 99.
3. Monroe, E. S., How to size and rate steam traps, *Chem. Eng.*, Apr. 12, 1976, p. 119.
4. Ruskin, R. P., Calculating line sizes for flashing steam-condensate, *Chem. Eng.*, Aug. 18, 1975, p. 101.
5. Benjamin, M. W., and Miller, J. G., *Trans. ASME*, vol. 64, p. 657 (1942).

Select the right steam trap

Experts say misapplication of steam traps squanders about 10% of the average plant's steam consumption. If you understand the characteristics of the various types of traps, you can select the right one for a specific service.

E. S. Monroe, Jr., E. I. du Pont de Nemours & Co.

☐ If the user of steam traps is aware of how the various types work, he can avoid misapplications that result in poor equipment performance, wasted steam and high maintenance.

There are three basic classes of steam traps, which may be distinguished from one another by the mode of operation:
- Liquid-level sensitive.
- Temperature sensitive.
- Fluid-property sensitive.

Each of these classes operates on a different physical principle—it is important that the user understand these principles fully if he is to get good trap performance. The various types of traps covered in this article are listed in Table I.

The float trap

Liquid-level traps respond to the liquid interface between the steam vapor and the condensate. The simplest version is the basic float trap (Fig. 1), which consists of a round, sealed, hollow ball that operates a lever, which in turn opens and closes a valve as the liquid level in the float chamber rises and falls. This trap has the clean-cut advantage that it readily recognizes the difference between condensate and steam. However, steam invariably contains noncondensable gases such as air and carbon dioxide. If these are not vented, the float trap becomes air-bound. Therefore, provisions must be made to vent such gases. For this reason, the float trap is combined with a thermostatic element that serves to release the noncondensables.

Even when equipped with such an air-relief device, this type of trap is limited to relatively low pressures because the sealed floats tend to collapse with higher pressures and because the closing forces on the condensate valves increase as the steam pressure rises. When properly fitted with an air-relief device, the float type of trap will provide good drainage of condensate.

Modifications to the float trap have included venting the float to equalize pressures, and elimination of the linkages. The former would seem to be susceptible to the same danger of water logging that exists for any expansion tank; while the latter is criticized by Northcroft [1] as having poor sealing characteristics. Insulation will not affect operating characteristics of these traps.

The open-bucket trap

This type (see Fig. 2) was developed to overcome the tendency of early cylindrical floats to collapse. When condensate initially enters, it causes the bucket to float and close the exit port. As condensate continues to enter the trap, it spills over into the float and sinks it. Steam forces the condensate from inside the bucket until the bucket again floats and seals the outlet. Open-bucket traps are almost never used today because they must be initially primed in order to work; they are extremely reduced in capacity; and they have no air or noncondensable gas handling capacity and in effect require a separate device for this.

It should be noted that the open-bucket trap will not open and drain condensate as long as steam fills the bucket and it floats. For this reason, the trap is an intermittent drainer and must be installed uninsulated. These traps are sealed by an initial prime of water.

Types of steam traps		Table I
Level sensitive	Temperature sensitive	Fluid-property-sensitive
Float	Thermostatic	Orifice
Open bucket	Bimetallic	Impulse
Inverted bucket	Liquid expansion	Disk

Originally published January 5, 1976

The inverted-bucket trap

The inverted-bucket trap (Fig. 3) operates with a floating bucket just as does the open-bucket trap, but because the bucket is turned over, it can operate the valve via linkage and therefore can have higher capacity. Whenever the inverted bucket is full of steam, the trap closes, and condensate does not flow from the system until the steam in the trap is condensed. Unfortunately, any noncondensable gases that are trapped under the bucket would stay there and keep the trap closed indefinitely. To eliminate this undesirable feature a small hole is put in the top of the bucket, which allows air to slowly escape.

Sometimes a bimetallic element is added to increase this rate when the trap is cold. Unfortunately, any bypass will leak steam when no air is present, so the air leakage rate is usually restricted. This means that the bucket trap will not start up very quickly on a cold system that is full of air. The small hole is also prone to plug with dirt, so sometimes a wire is used to mechanically clean the hole as the trap rises and falls. The inverted-bucket trap is an intermittent drainer in most cases (under some conditions, it will cycle rapidly and drain continuously) and must not be insulated. It is commonly sold with different outlet seat sizes in the same body size, and care must be taken not to use too large a seat at too high a steam pressure. If this is done, the weight of the bucket may be unable to open the outlet valve, and the trap will lock closed. These traps also require an initial prime of water to prevent steam blowby.

The balanced thermostatic traps

Temperature-sensitive traps are those that respond to either a temperature difference between the steam and the condensate or directly to a temperature of either steam or condensate. Their original form consisted of disks filled with volatile liquids such as alcohol, but these have generally been superseded by the balanced-bellows design of Clifford [2], which uses water. The balanced thermostatic trap (Fig. 4) consists of a flexible bellows containing water. The variation of the vapor pressure with condensate temperature within the closed bellows opens and closes the outlet seat.

In this trap, temperature differences required to open the trap may be very close to the saturation steam pressure despite variations in this pressure. Since the steam pressure helps open the trap, there is no need for the size of outlet orifice to decrease with increasing steam pressure; hence these traps have very high capacities at relatively low cost. Because they depend on subcooling, thermostatic traps should be installed uninsulated and it is best practice to leave 24 inches of uninsulated inlet piping ahead of the trap.

Modified units

Modifications to improve the thermostatic steam trap include calibration, solid filling of bellows, and installation of heat sinks. Calibration can be used both to narrow the subcooling of condensate required to open the trap and to give consistent performance. Calibrated units can provide continuous drainage equal to the

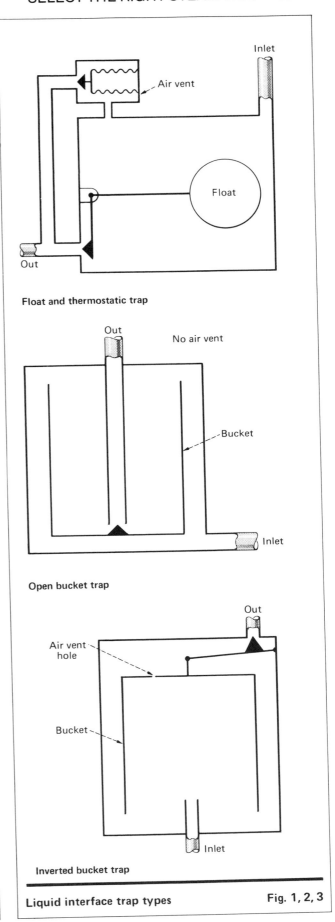

Float and thermostatic trap

Open bucket trap

Inverted bucket trap

Liquid interface trap types — Fig. 1, 2, 3

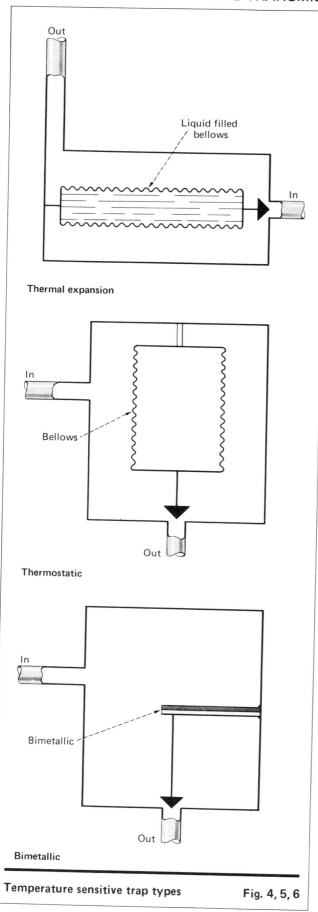

Temperature sensitive trap types   Fig. 4, 5, 6

float and thermostatic traps, while uncalibrated units with high subcooling requirements will give intermittent drainage. The solid filling of bellows with water will provide hydrostatic protection of the fragile bellows' convolutions against pressure increases on the steam side.

Equally important is to protect the bellows from excess pressures from inside when the pressure decreases on the steam side. This is done by installing an open cup in the center of the bellows to trap condensate. On a sudden decompression, this condensate will flash to a lower temperature and serve as a heat sink that removes potentially damaging energy from within the bellows. When so modified, the thermostatic trap can be used on higher pressures and will provide improved life at lower pressures.

Bimetallic traps

Bimetallic traps (Fig. 5) use the variations in thermal expansion of different metals to open and close steam valves. They can be simple levers like those in home thermostats or, more commonly, stacks of specially shaped plates that curve and force a valve to seat when they are heated. Since they are temperature responsive and do not recognize steam pressure, they must be calibrated for each operating pressure if they are to close at near saturation temperatures. In addition, despite many ingenious schemes to circumvent the hysteresis of bimetallic elements, they do suffer from this fault. Subcooled temperatures of over 100°F at the time of opening in order for the trap to close at saturation temperature are common.

These traps have good mechanical life. They are useful for very small loads where subcooling is not important, discharge is to the atmosphere, and the traps are carefully calibrated. They should not be insulated, and it is again best to leave 24 inches of uninsulated inlet piping ahead of them.

Thermal-expansion types

A thermal-expansion type of thermostatic trap (Fig. 6) usually consists of a liquid-filled bellows that expands against a spring to close a valve when it is heated. Since they are temperature sensitive and not pressure sensitive, they usually operate at a fixed temperature below 212°F. They are useful where considerable subcooling of the condensate in the steam system is desired; however, since condensate has a much lower coefficient of heat transfer than condensing steam, this is usually very undesirable.

One valuable application is in the tracing of steam-heated safety showers. Here they can be used to prevent water from freezing but not from getting so hot as to cause scalding when a shower is used. Almost no steam-heating applications other than tracing can economically use the high subcooling that results with thermal-expansion types of steam traps. If installed to prevent overheating of safety showers, they should be completely insulated.

Orifice types

Fluid-property traps respond to the differences in thermodynamic properties between steam and condensate. The orifice type (Fig. 7) is undoubtedly one of the older forms. Its origins are obscure in a welter of pinched pipe and gate valves with holes drilled in the valve plug. Bottomley gave them an impetus in 1937 in his well-known paper [3]. The orifice has limited self regulation, based on the fact that more pounds of water will pass through it than steam.

The two-orifice type of trap (Fig. 7) is often called a drainator. The flashing characteristic of hot condensate as it flows to a lower pressure can be used to choke the flow of a second downstream orifice. Such a device will automatically have a certain amount of self-regulation. Where loads are large, continuous, and steady in magnitude, these traps have a very long life. They have been used on large evaporators and in critical service such as on fuel-oil tank suction-heaters, but their characteristic of passing steam at light loads precludes widespread usage.

A recent entry into the commercial trap market is one having a single orifice. It may be used for constant pressure and constant loads with success but must be sized carefully to avoid costly steam losses. The next article in this series will give details on how to calculate two-phase flow of steam and water mixtures through such orifices. Orifice traps may be insulated completely without affecting their operation.

Impulse-type traps

The varying pressure between a pair of orifices when either steam or condensate flows can be used to act as a pilot for a main valve. This is the principle of the impulse trap (Fig. 8). It has the advantage that it samples continuously and therefore will keep equipment continuously drained. Since the sampling orifice represents a continuous steam bleed when there is no condensate, it is deliberately kept small. This means that the trap will have poor air-handling capability and is also susceptible to dirt plugging the orifice.

Two types of impulse traps are available commercially. Both are made by the same company—one uses an inner valve that serves both as the main valve and dual orifices, and the other employs a rocker arm that opens and closes two valves. The latter has large capacity but should be used with discretion when light loads occur, because of potential steam wastage. Impulse traps may be installed insulated.

Disk traps

The disk trap (Fig. 9) is the most popular one sold today. It uses the higher kinetic energy of steam over water to close a valve with static pressure. Its main advantages are its small size and low first cost. When condensate enters the trap, it simply lifts the disk and flows through the trap outlet. If steam reaches the trap, it flows with a higher velocity, which builds up sufficient pressure above the trap disk to press it closed. When the steam above the disk condenses or leaks off to the outlet, the cycle is repeated.

Air or noncondensable gases cause the trap to react similarly to steam. Since they cannot condense, it is necessary to provide a small controlled air vent. This is usually a scratch on the surface of the disk, or a controlled surface roughness between the space on top of

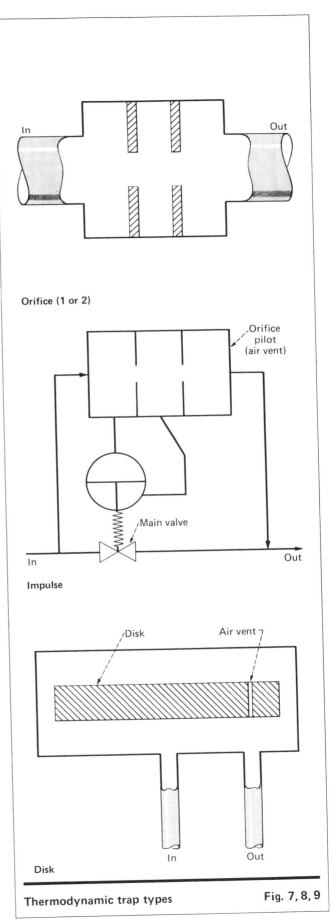

Thermodynamic trap types — Fig. 7, 8, 9

the disk and the outlet. The trap does not drain continuously and the degree of subcooling is a function of its geometry.

As the trap ages, the leak-off rate increases, cycling is more frequent, and wear is rapid until the trap begins cycling rapidly, which wastes steam. Causes of wear are high steam pressure, high back pressure, light loads, and oversizing of traps.

The air handling capacity of disk traps is poor. The kinetic energy available to close the trap is a function of the pressure drop, and the pressure drop required is a function of geometry. In general, larger sizes are more susceptible to back pressures, which can cause the trap to stay open. A good rule is to prevent the back pressure from exceeding 40% of the initial pressure. It is considered good practice to limit the initial steam pressure (say to 175 psi) and to use only the smallest sizes. If this is done and combined with a good testing program, disk traps may give acceptable service on steam tracing and drip legs.

Modifications available include capsulated seats and disks for easy maintenance, and steam jacketing of the chamber over the disk. The latter reduces the effect of ambient conditions on the time that the trap cycles. Disk traps should not be installed insulated.

Common trap problems

Some of the more common problems with traps are listed below:
- Improper sizing.
- Poor heat-exchanger performance.
- High maintenance costs.
- Steam wastage.

There are, of course, other special problems but the above are so frequent that they require extra mention. To cope with these problems, it is important to understand the characteristics of the nine types of traps. The four problems will be discussed individually.

Improper sizing

Traps sometimes may be either oversized or undersized. The effects of undersizing are obvious. Whether or not oversizing is serious or not depends on the type of trap and the magnitude of the oversizing. The capacity data published by vendors have not always been accurate and are often published with the recommendation that the trap capacity be specified several times larger than required for "safety factor" reasons. The problem has also been recognized by a Performance Test Code Committee of the American Soc. of Mechanical Engineers and the committee is working toward a solution. In the interim, care should be taken to avoid significant oversizing of disk, bucket, impulse, and orifice-type traps, as these are subject to poor performance if oversized.

Poor heat-exchanger performance

Steam when condensing has a much higher heat-transfer coefficient than water (condensate) or noncondensable gases. For this reason, if portions of a heat exchanger are flooded with condensate or noncondensable gases, the output of the exchanger will be very poor. If fluids or gases are flowing across the heat exchanger, uneven heating will occur. For example, in a preheat coil for a vapor catalyst bed, alternate flooding and draining of the steam coil can cause unvaporized oil to accumulate in the bed, which can cause a fire when volatized oil reaches the bed. Similarly, jacketed vessels for which uniform temperatures are desired may develop cold spots that result in poor yield of product.

There are only four types of traps that continually drain condensate when properly sized:
- Float and thermostatic.
- Modified thermostatic.
- Impulse.
- Orifice.

All other types drain condensate intermittently when properly sized. The disk type is closed when steam is on top of its disk and remains closed until the steam condenses or leaks off. The bucket types are closed until the steam in the bucket condenses or leaks off. The conventional thermostatic, liquid-expansion and bimetallic types are closed until sufficient subcooling takes place to open them. During these periods of closure, the condensate that is forming in the heat exchanger may back up, flooding the exchanger and impairing its performance.

It is often suggested that these types will give equal performance if sufficient surge capacity is provided between the heat exchanger and the trap, such as using posts or oversized lines. This is only true if there are no noncondensable gases in the steam. Usually there are. Air is common on startup, and carbon dioxide is common with many steam generating plants.

When condensate is intermittently discharged, it will subcool, allowing some of the noncondensable gases to be absorbed. If the remainder do not pass freely through the trap, they will accumulate, increasing their partial pressure. This causes a further deterioration in heat transfer, and absorption of gases into the condensate. The methods prescribed in the first article of this steam-trap series [3] will allow the reader to evaluate for himself how well his traps are performing. If the condensate entering the traps is subcooled more than 5°F, he can be assured that he does not have optimum heat-transfer conditions in his heat exchanger or other equipment.

Another reason for poor heat transfer is multiple assignment of units to a single trap. There are many arguments presented for doing this, such as cost and space requirements. These are poor arguments when the equipment doesn't perform because of bypassing and poor drainage. Every reputable trap vendor warns against this in his literature, yet it continues to happen.

Poor drainage at part load is another difficulty encountered with heat exchangers. An air-heating coil may work perfectly at −10°F, but at 30°F it freezes up, or the air is heated so nonuniformly that an area is either over- or underheated. For some unexplained reason, this is always the plant manager's office. The trap must be investigated to ensure that it will drain the heat exchanger adequately at the part-load condition. There are many resolutions to this problem, involving multiple coils controlled in sequence, air-mixing boxes, etc. One that is often overlooked is the vacuum condensate system.

High maintenance costs

Trap surveys in my company have disclosed that more than 25% of steam traps were in need of maintenance, primarily because seats and valve plugs were worn and wasting steam. In the purchase of traps, it is not customary for the buyer to specify anything other than brand, size and model. (The same buyer will rigorously insist that his steam valves have hardened seats.) I suggest that the specifying of steam-trap seats and valve plugs with a Rockwell C hardness greater than 50 and of noncorrosive materials might reduce overall costs.

If we assume that mechanically the designs are adequate (in general they are, the only major exception will be discussed shortly), the maintenance problems are created by erosion and corrosion. Flashing-condensate flow is very erosive. This is proportional to the initial trap pressure. Much work needs to be done in this field but, in my experience, high-hardness metals definitely last longer.

Corrosion is a second maintenance problem that is not confined to the working parts of the trap. Iron oxides formed anywhere in the system can cause malfunction of the trap mechanism. I have observed that iron oxides do not form significantly in condensate systems where the boilers use demineralized water. Corrosion is heaviest in those systems where the boiler waters generate carbon dioxide *and* the steam traps either do not vent noncondensable gases well *or* subcool the condensate significantly. This causes formation of carbonic acid ahead of the trap. This is well documented for condensate lines after the trap but is not appreciated that it happens upstream nearly so well. If the characteristics of the various steam traps are better appreciated when CO_2 is present in the steam, maintenance of traps can only improve. So can that of heat exchangers and piping.

It was noted above that most traps are adequately designed with one exception. This is the disk trap. Its mechanical life is directly a function of its size, operating pressure, and load. Large, high-pressure, and lightly loaded disk traps simply do not last long. If disk traps are to be used without frequent maintenance, they must be loaded nearly to their capacity, and the application pressure held below that specified as possible by the vendors. Small-capacity disk traps with inlet pressures limited to 175 psi can yield significantly better maintenance performance than large higher-pressure units.

Steam wastage

Steam waste can be a result of wear on seats and valve plugs, as noted above. There are other causes. An orifice-type trap may never wear, but if oversized it may waste significant quantities of steam. A bucket trap that loses its water seal may waste steam although it is in perfect working order. An oversized impulse trap may waste steam although it too is functionally perfect. A bimetallic trap that is calibrated to be fully open with minimum subcooling may blow steam if its load is reduced.

Even the disk trap that is working properly may fail to close if the condensate system pressure rises to a point where the kinetic energy developed across the trap is inadequate to build up enough static pressure above the disk to close it. Reducing its inlet pressure with an automatic steam-throttling valve has the same effect.

As the price of fuel advances, the cost of wasted steam cannot be ignored. It is sometimes argued that wasted steam from traps can be used in the powerhouse to preheat feedwater. But even if this is done, it is at the expense of using purchased and expensive electrically driven auxiliary drives instead of more-economical power from steam-driven auxiliaries. As noted in the introduction, a rule of thumb for many years has been that poor steam trapping will waste 10% of the fuel used to generate steam. A number of recent studies have confirmed the order of magnitude of that rule. Ten percent of your fuel bill can be a significant number.

Conclusion

Steam traps come in many types, sizes, and materials of construction. Understanding their characteristics better may help you select and maintain traps that will:

- Discriminate between steam and condensate and pass only the condensate.
- Drain heat exchangers better for uniform maximum output.
- Not subcool condensate excessively where it will affect corrosion or heat-exchanger performance.
- Discriminate between steam and noncondensables and pass only the latter.
- Adjust to varying loads when they occur.
- Adjust to varying steam pressures where necessary for proper performance.
- Adjust to varying condensate pressures if necessary.
- Be reliable with lower maintenance costs.
- Be freezeproof if steam is turned off.

References

1. Northcroft, L. G., "Steam Trapping and Air Venting," Chemical Pub. Co., New York, 1946.
2. Clifford, W. B., "Steam Trap," U.S. Patent 1816142, 1931.
3. Bottomley, W. T., Flow of Boiling Water through Orifices and Pipes, *Northeast Coast Institute of Engineers and Shipbuilders Transactions*, Vol. LIII, Part 3, Jan. 1937.
4. Monroe, E. S., Jr. "Steam Trap Testing," *Chem. Eng.*, Vol. 82, No. 18, Sept. 1, 1975, pp. 99–102.

The author

E. S. Monroe has been a power consultant in the Engineering Dept. of E. I. du Pont de Nemours & Co., Louviers Building, Wilmington, DE 19898. Steam traps are one of his specialized areas and he has been issued five patents in this field. He is a graduate of Virginia Polytechnic Institute and Cornell University, is a member of the ASME Performance Test Code 39.1 committee (which is preparing test procedures for steam traps) and is a registered professional engineer in Delaware and New York.

Short-cut calculation for steam heaters and boilers

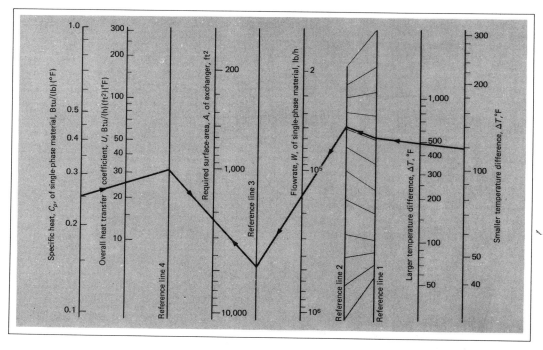

V. Ganapathy, *Bharat Heavy Electricals Ltd.*[*]

☐ One special type of heat transfer is quite common to the power and process industries, and is becoming still more common with increased emphasis on energy conversion and the co-generation of power. This type consists of heat transfer from or to a single component going through a change of phase. Examples include steam heaters, surface condensers, and reboilers and condensers on high-purity distillation operations.

Because the fluid undergoing phase change remains at constant temperature, this special type of heat transfer allows for the general equations to be simplified:

$$Q = UA(\text{LMTD}) + UA(\Delta T_1 - \Delta T_2)/\ln(\Delta T_1/\Delta T_2)$$
$$(\Delta T_1 - \Delta T_2) = (T_1 - t_2) - (T_2 - t_1)$$
$$t_2 = t_1$$
$$(\Delta T_1 - \Delta T_2) = T_1 - T_2$$

Also,
$$Q = WC_p(T_1 - T_2)$$

so these values can be substituted in the general equation:

$$WC_p(T_1 - T_2) = UA(T_1 - T_2)/\ln(\Delta T_1/\Delta T_2)$$
$$\ln(\Delta T_1/\Delta T_2) = UA/WC_p$$
$$(\Delta T_1/\Delta T_2) = e^{UA/WC_p}$$

Where

- Q = heat transfer rate or duty, Btu/h

[*]Research & Product Development, High Pressure Boiler Plant, Tiruchirapalli 620 014, India

- U = overall heat transfer coefficient, Btu/(h)(ft^2)(°F)
- A = heat transfer surface, ft^2
- (LMTD) = log mean temperature difference, °F
- T = temperature of fluid not undergoing phase change, °F
- t = temperature of fluid undergoing phase change, °F
- W = flow of fluid not undergoing phase change, lb/h
- C_p = specific heat of fluid not undergoing phase change, Btu/lb

This simplified equation has been incorporated into the accompanying nomogram, which can be quite useful in the difficult problem of arranging several heat-recovery exchangers on one hot medium, such as a furnace flue gas, or a hot stream of effluent leaving the secondary reformer in an ammonia plant.

Example: 150,000 lb/h of flue gases of 0.25 mean specific heat flow across a bank of tubes where water is converted into steam. The overall heat transfer coefficient is known to be about 25 Btu/(h)(ft^2)(°F). How much surface area is required? Connect $\Delta T_2 = 120$ with $\Delta T_1 = 500$, and extend the line to reference line 1. Proceed between the guidelines to the intersection with reference line 2. Connect this intersection with 150,000 lb/h and extend the line to its intersection with reference line 3. Join $C_p = 0.25$ with $U = 25$ and extend the line to its intersection with reference line 4. Connect this intersection to the intersection with reference line 3, and read 2,100 ft^2 of surface.

Originally published March 13, 1976

Steam Traps

Steam traps are critically important in maintaining efficient heat-transfer systems. Here is a rundown on the operating principles of the major types of mechanical, thermostatic and thermodynamic traps.

JIMMY MATHUR, Brown & Root, Inc.

Steam traps are used in process plants to obtain maximum steam utilization. But like other small items in the plant piping system, they often take a back seat to larger and more expensive process equipment in terms of time spent in engineering and maintenance efforts.

Besides lost heat-transfer efficiency, poorly functioning traps frequently leak steam. Even small leaks can add up to considerable sums of money in the course of a year.

The economy-minded plant engineer will make every effort to ensure that his unit is equipped with the best type of trap for his process requirements and that his traps are properly installed and maintained. To accomplish this, he must first understand how they function, what types are available, and what factors affect their performance.

Basic Function

Steam traps are devices that permit condensate and noncondensable gases to leave the system with minimum steam loss. Removal of these fluids is important because they form thin films on heat-transfer surfaces. Reference to the thermal conductivity values in Table I indicates that a film of condensate 0.01-in. thick offers the same resistance to heat transfer as a steel plate $\frac{5}{8}$-in. thick or a copper surface $6\frac{1}{4}$-in. thick. A film of air is even worse: a 0.01-in. thickness has a resistance equivalent to a 0.20-in. thick layer of water, steel 16-in. thick or copper 10-ft. thick. It is evident that water and noncondensables, like air, must be removed as rapidly and completely as possible to ensure maximum heat-transfer efficiency.

A steam trap handles only material that comes to it. Condensate can, in general, be expected to flow to the trap, but there is no guarantee that noncondensable gases will. When a heat exchanger is first placed onstream, the steam-side is full of ambient air. Colder, denser air will collect at the bottom of the exchanger surface remote from the steam inlet. At this stage, air is easily removed by the trap.

During operation, turbulence and diffusion mix air and steam together. Condensate forming on the heating surface is saturated with air present as a film on this surface. This assists in driving the air to the condensate outlet leading to the trap. But there can be an accumulation of air at locations remote from the condensate outlet nozzle.[2]

Besides being present throughout the steam system at startup, air enters whenever there is a shutdown that allows steam to condense and pull a vacuum. Also, it is constantly present if the deaerator (for deaerating boiler feed water including condensate) is malfunctioning and feedwater conditioning-chemical dosage is inadequate. Carbon dioxide is also present in steam produced from water that is not completely demineralized. Besides the adverse effect of these gases on heat transfer, their presence leads to corrosion in the steam-condensate system—another reason why steam traps are important for efficient plant operation.

Thermal Conductivity Values—Table I

Substance	Thermal Conductivity, Btu.(hr.)(sq. ft.)(°F.)(in.)
Carbon dioxide	0.12
Air	0.22
Mineral wool	0.28
85% magnesia	0.41
Water	4.36
Carbon steel	360
Copper	2,700

Originally published February 26, 1973

STEAM GENERATION AND TRANSMISSION

Operating Principles

The trap is required to first distinguish steam from condensate and noncondensable gases, and then to open a valve (or valves) that allows passage of condensate and noncondensable gases but closes to steam. The distinction is made on three principles: density difference, temperature difference and phase changes. These principles are used in the three major trap types: mechanical, thermostatic and thermodynamic.

There are very considerable differences in density between condensate (57.3 lb./cu. ft.), steam (0.15 lb./cu. ft.), and air (0.23 lb./cu. ft.) at a given operating condition (here 50 psig. and 297 F.). This principle is used by *mechanical-type* traps. This density difference always exists, except at the critical pressure when both water and steam have equal density. However, the densities of steam and air are too close for distinction. It is also important to note that any flash steam produced in the line from the heat exchanger is regarded as steam. The trap will close even though condensate is following the pocket of flash steam.

The difference in temperature between steam and cooled condensate is the principle used by *thermostatic type* traps. Note that condensate and the steam from which it is produced have the same temperature when in equilibrium. Therefore, to enable distinction by this type of trap, condensate must be cooled below the steam temperature. This may be achieved by locating the trap some distance from the exchanger and/or leaving the inlet line uninsulated; or, by deliberately imposing a delay in condensate removal rate, allowing it to back up into the exchanger or in the pipeline to the trap. Since temperature of condensate is sensed in this type of trap, it offers a method for controlling condensate discharge temperatures (e.g., in systems where it is worthwhile to extract some sensible heat from condensate).

Thermostatic traps can discharge line-flash steam, provided its temperature is measurably below that of live steam. And assuming that air is cooler than steam (which is true at startup when air removal is of the greatest significance), a thermostatic trap will discharge it.

The third principle used in trap design is the change of phase that occurs when condensate at any particular temperature experiences a drop in pressure, for example, when it flows through an orifice. This pressure drop results in formation of flash steam. The higher the condensate temperature upstream of the restriction, the greater the quantity of steam produced, for a fixed downstream pressure.

The flash steam is then used for closing a valve and/or as a means of choking off condensate flow, since steam occupies a much larger volume than water. Alternatively, the drop in pressure that occurs with increased steam velocity through a restriction is used to motivate a disk that closes off condensate flow. Traps that use a phase change are classed as *thermodynamic* steam traps.

The quantity of flash steam produced is a function of inlet condensate temperature (and corresponding steam pressure) and downstream pressure. The pressure in the condensate return line (trap back pressure) is important. If it is too high, there will not be enough flash steam and the trap will not function. After closure of the valve (or disk) there must be a lapse of time for bleed-off or condensation of steam before the valve can open. This causes the trap to operate on a time cycle and thus it does not remove condensate continuously.

Air is probably removed in the thermodynamic trap that operates by choking off flow, but heated air is likely to close off the disk trap in the same manner as steam.

Mechanical Traps

Float traps—Condensate enters the trap and raises a float that uses a lever mechanism to open the valve (see Fig. 1a). Line pressure then blows the condensate out. When there is no condensate flow, or when steam enters the trap, the float drops and closes the valve. While there is condensate flow, the float settles at an equilibrium level dictated by flowrate, upstream pressure, and pressure in the condensate return line. Discharge is continuous and there is no condensate backup as long as the available differential pressure (line pressure minus back pressure) is adequate for the particular valve orifice at that flowrate.

The buoyancy force of the float is used to open the discharge valve. For example, suppose the valve has an area of 1 sq. in. and there is a pressure of 50 psig. within

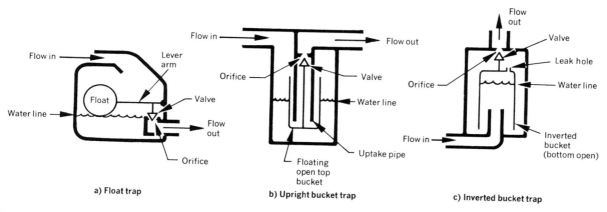

a) Float trap   b) Upright bucket trap   c) Inverted bucket trap

MECHANICAL-TYPE traps rely on density difference to distinguish between steam and water—Fig. 1

the trap. A buoyancy force of 50 lb. is needed to open the trap. Assuming the lever mechanism has a mechanical advantage of 25:1, a force of 2 lb. would be necessary. The buoyancy force available from the float is the weight of condensate displaced by it minus the float (and lever-arm) weight. For a 6-in. diameter float, weighing 1.5 lb., and condensate (at 297 F.) with a density 57.5 lb./cu. ft., the submergence would be 93.0%.

This means that the fully submerged float would be unable to open the valve if, for any reason, line pressure rose above 56.25 psig. Alternatively, an 8-in. float weighing 3.5 lb. would run 62% submerged at 50 psig., and would be capable of opening the valve even if the trap pressure were 135 psig. However, the 6 in.-float could be used for a higher operating pressure, provided the orifice and valve were reduced or the length of the lever-arm increased. Note that a smaller orifice and valve would reduce trap discharge capacity.

Floats cannot be increased in size with impunity, and this sets a limit on float-trap capacity and/or operating pressure. Note also that the density of condensate decreases with temperature; thus the higher the operating pressure (temperature), the less buoyancy force available for a particular size float.

A trip-action float trap uses a weight-operated mechanical device to fully open and close the discharge valve when the float has risen or dropped to certain levels. This trap operates fully open or completely shut, without flow modulation. However, the problem of opening the valve with the buoyancy force of the float, remains.

The float trap has no automatic means in itself for air removal. It is generally fitted with a thermostatic device for air removal from the top of the casing; this combination trap is called a float and thermostatic trap.

Location of the outlet valve below the lowest point of float travel results in a water seal at the trap bottom, thereby preventing steam leakage.

Upright Bucket Traps—This type of trap is pictured in Fig. 1(b). At startup, the floating bucket rests near the bottom and the valve is open. As condensate flows into the trap, the bucket floats up. When it has risen to a certain level, the valve is closed. Condensate flows in at a very slow rate (with the valve closed, the only driving force for flow is elevation difference). Eventually it overflows into the bucket. When sufficient water has entered the bucket to overcome its buoyancy and line pressure holding the valve shut, the bucket drops and opens the valve. Line pressure then forces condensate out of the bucket, up and out of the discharge pipe until the bucket is again sufficiently buoyant to float again. The buoyancy of the bucket is so arranged that the bucket never empties completely, thus maintaining a water seal between the uptake pipe and the bucket, so that steam cannot blow through.

In contrast to the float trap, the valve opening force is the weight of the bucket while the closing force is its buoyancy. This means that at lower pressures, the trap will operate more frequently since the force holding the valve shut is smaller and can consequently be overcome by less load in the bucket. For example, with a valve area of 0.5 sq. in. and a pressure of 50 psig., a weight of 25 lb. is required to open the valve. If the mechanical advantage is 2:1 (large leverage is not easily arranged), then a bucket weighting 12.5 lb. will open the valve. If the pressure is further raised, a greater bucket weight is required. Valve closing is not a problem since the buoyancy force is assisted by line pressure.

The trap acts intermittently. It has no means of discharging air and, as with the float trap, would require a thermostatic device fitted at the top casing for this purpose.

Inverted-Bucket Trap—The inverted bucket trap is shown in Fig. 1(c). At startup, the floating bucket rests near the bottom and the valve is open. As condensate enters, it accumulates inside the trap body and the bucket. Provided the condensate enters quickly, the small leak hole through which air escapes will be inadequate to equalize pressure inside and outside the bucket. The water level will climb faster outside the bucket than inside. This gives the bucket buoyancy and it rises to close the valve.

When sufficient condensate has risen inside the bucket its buoyancy disappears and the weight of the bucket opens the valve. Line pressure then blows the water out from inside the bucket and from around it. As soon as sufficient quantity of water has left the bucket, its buoyancy is restored. It then rises and closes the valve.

The bucket would retain its buoyancy if the steam and/or air could not escape, and the trap would lock shut, for example, if the bucket leak hole is blocked. Normally, air escapes through the leak hole and is discharged with the condensate. Steam also escapes through the leak and is condensed in the body of the trap, provided the top of the trap has not been totally insulated. During the discharge cycle, some steam is blown straight through the leak hole and out the discharge.

As in the upright bucket trap, the valve opening force is the weight of the bucket, and the valve closing force is the buoyancy of the bucket. The weight of the bucket depends on the amount of water displaced from within it. The greater the displacement (i.e., the lower the level within the bucket), the greater the upward buoyancy force and the less it weighs. At lower operating pressures, for a certain valve area, a lower weight is adequate to open the valve (as is the case for the upright bucket trap). Therefore, a lower level of condensate exists within the bucket. Valve closing is not a problem since the buoyancy force is assisted by line pressure. A water seal at the bottom prevents steam loss.

Operation of this trap is intermittent. It removes as much air as is allowed by the leak hole. To remove large quantities of air, it becomes necessary to fit some sort of small thermostatic device to open an additional vent hole at the top of the bucket. The device opens to vent cold air and shuts at steam temperature to prevent steam loss.

Thermostatic Traps

Metal-Expansion Traps—The tube and body of this trap, shown in Fig. 2(a), are made of metals with markedly different coefficients of thermal expansion. For example, with a carbon steel body and yellow brass tube, the respective coefficients are 6.7 and 10.5×10^{-6} in./

(in.) (°F.), for the range 32–212 F. This range is too low for our consideration, but the coefficient values are indicative of the difference in magnitude. Therefore, as the entering condensate increases in temperature, the tube expands towards the valve and at a preset temperature (set by the adjustment screw) the valve closes fully. The setting corresponds to a specific steam pressure and therefore if pressure falls, the trap will not be closed, resulting in steam loss. If pressure rises, the valve will close prematurely and condensate will not be discharged. At startup, the valve is fully open and air is eliminated. The temperature at which it opens must definitely be lower than the steam-saturation temperature.

Liquid expansion traps operate similarly. Since the coefficient of expansion for liquids is greater than for solids, they allow greater valve movement and are more compact traps. Liquid traps, however, are usually set to discharge below 200 F.

Bimetallic Traps—The thermostatic element consists of one or more elements of bimetal strips. They are fixed rigidly to the trap body at one end and free to deflect at the other. (See Fig. 2(b).) Each strip is made of layers containing two metals having different coefficients of thermal expansion so that the strip bends when the temperature increases. The force developed by this thermal deflection closes the valve. Each element exerts its deflecting force on the valve at a particular temperature, so that with several segments comprising the total thermostatic element, it is possible to cover a fairly wide range of saturation temperatures. The valve opening force is condensate line pressure. The valve opens when the opposing (bending) force exerted by the bimetallic strip relaxes at a temperature below the steam saturation temperature. There should be no steam loss unless bimetal action is sluggish. Condensate accumulates until the valve finally opens, resulting in intermittent action.

As long as air is at a temperature below that of the steam, it will be discharged readily. This type of trap is often used as the auxiliary air vent for float traps.

Balanced Pressure Traps—This type of trap, pictured in Fig. 2(c), uses a flexible metal structure in the form of bellows or a diaphragm that is partially filled with water, or a volatile fluid (i.e., having a boiling point lower than water), or with a mixture of water and volatile fluid. At a particular condensate temperature, part of the thermostatic fluid is vaporized. The bellows is fixed at one end and is free to expand or contract at the other, thereby closing or opening the valve. For the water-filled bellows at any particular temperature in the trap, the total pressure within the bellows is identical to the pressure outside. For the bellows with liquid having a lower boiling point than water, the pressure within is greater by a constant margin provided the liquid mixture has a vapor pressure temperature curve that is parallel to that of water. Therefore, trap operation is independent of pressure. At the steam saturation temperature, the valve location is such that a spring forces the bellows to close the valve. When the condensate cools to below the saturation temperature, the bellows fluid condenses and there is a drop in pressure within the bellows. Line pressure then compresses the bellows and the valve opens to permit discharge of cooled condensate. Operation is intermittent and air cooler than the steam temperature is readily discharged.

Piston Valve Impulse Traps—These devices, made by the Yarway Corp., are depicted in Fig. 3(a). Entering condensate lifts the piston valve so that flow is discharged through the main orifice. A small flow passes through the first orifice (the annular space between the piston disk and the tapered control cylinder) into the control chamber and then through the second orifice contained within the stem of the piston valve. This small control flow then rejoins the main condensate stream and is discharged from the trap.

Depending on its temperature, some of the control flow flashes after passage through the first orifice, and more flashes on passage through the second one. When inlet condensate temperature approaches steam temperature, the quantity of flash in the control chamber increases. The flash steam in the control chamber then chokes the second orifice and pressure in the control chamber increases. Since the effective area above the piston disk is greater, the downward force exerted by control-chamber pressure closes the piston valve at the main orifice. The valve is opened by line pressure when the inlet condensate temperature has fallen sufficiently to reduce the degree of flashing in the control chamber, thus decreasing the pressure there. A small, constant bleed of condensate (or steam) is maintained through the control orifices at all times.

The operation of this trap is essentially intermittent when condensate inlet temperatures are close to the

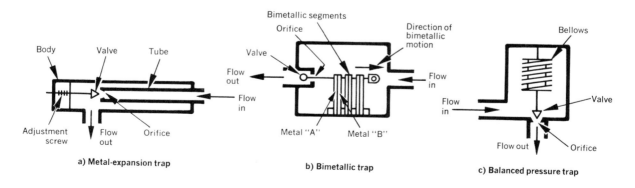

a) Metal-expansion trap b) Bimetallic trap c) Balanced pressure trap

THERMOSTATIC TYPE traps use the temperature difference between steam and condensate as their operating principle—Fig. 2

steam-saturation temperature. Air is readily handled, and there is no possibility of steam-binding since there is a constant control flow.

A modification called a Lever Valve Impulse Trap handles higher flowrates. The working principles, based on a control flow through two orifices in series, are similar to those of the Piston Valve Impulse Trap.

Velocity Disk Traps—As shown in Fig. 3(b) inlet condensate flow lifts the disk off the inner and outer seats, and discharges through the outer annular chamber. When condensate temperature approaches the saturation temperature there is increased flash steam. This steam passes at high velocity into the outer chamber, lowering the pressure below the disk. This reduction along with the pressure created by flash steam above the disk, causes the disk to seat simultaneously on both inner and outer seats, thus closing the inlet line and isolating the chamber. This closure occurs because the chamber steam pressure acts over the entire upper disk area while inlet line pressure acts over the smaller inlet seat area. Steam in the isolated chamber condenses by radiant heat loss, or slowly bleeds off through irregularities in the disk surface. In some models bleed-off is through a machined groove in the disk. When chamber pressure drops below line pressure, the cycle is repeated.

This trap operates essentially on an intermittent basis. Cycle time depends on steam pressure, temperature and the rate of heat loss or bleed-off from the control chamber. Condensate arriving in the middle of a cycle must wait until chamber pressure has dropped sufficiently to permit the disk to open. Air-binding may occur unless it is deliberately bled away. Back-pressure must be limited to maintain an adequate supply of flash steam and also to prevent an excessively high opposing line pressure from acting on the underside of the disk (thus allowing it to close properly).

Ref. 10 provides interesting test data on this class of trap.

Piston Traps

This class of traps, sometimes called compound traps or relay operated traps, is used for handling very high condensate flows, for example, in the order of 25,000 lb./hr. at a differential pressure across the trap of only 5 psi. Any of the mechanical or thermostatic type of trap could be used as a "relay" trap to activate the main valve. Fig. 4 shows a bimetallic trap used in conjunction with a piston valve.

When discharging condensate, the bimetallic trap is open and the piston rests on the piston seat, being held down by its own weight plus the line pressure force exerted on its upper area. (Effective area on piston upper face is greater than the effective area on its lower face.) When steam is detected by the bimetallic element, the relay valve closes and pressure above the piston falls off. Line pressure acting on its lower surface raises the piston and the main valve closes the main orifice. There must be a bleed-off of steam and/or condensate from the chamber above the piston to allow the pressure falloff mentioned. After condensate has cooled, the bimetallic element relaxes, the relay trap opens and pressure on its upper surface brings the piston back to its seat. The main valve then opens and discharges condensate.

The bimetallic trap is open on air and therefore air is readily handled—assuming that it is below the steam temperature.

Steam Trap Specification

Steam traps are integral and important units in a process. Their design calls for the same analysis that other process equipment receives. Although there is less literature available on steam traps than in other equipment areas, Ref. 2, 4, 5, 6 and 7 contain much useful information for the design engineer.

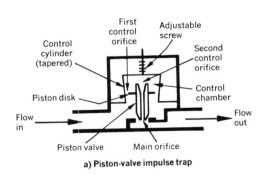

a) Piston-valve impulse trap

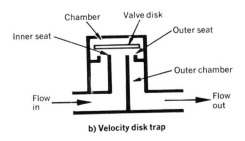

b) Velocity disk trap

THERMODYNAMIC traps use flash steam to close a valve and/or choke off condensate flow—Fig. 3

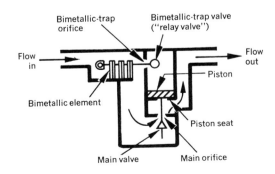

PISTON traps handle high condensate flows—Fig. 4

Effect of Inlet Temperature on Flowrate—Table II[1]

Temperature, °F.	Deg. F. Below Saturation	Flowrate, Lb./(hr.)(sq. in.)
60	278	196,000
308	30	107,000
318	20	90,000
328	10	70,000
333	5	47,000

Effect of Inlet Pressure on Fluid Properties—Table III

Inlet pressure, psig.	250	150	75
Inlet temperature, °F.	406	366	320
Outlet pressure, psig.	200	100	25
Outlet temperature, °F.	387	338	267
Differential pressure, psi.	50	50	50
Weight-percent flashed	2.35	3.36	5.84
Outlet volume, cu. ft./lb.	0.068	0.148	0.629

A trap that is undersized will cause condensate to interfere with heat-transfer efficiency; while traps having too much excess capacity not only waste money, but act sluggishly and produce a high back-pressure that may significantly reduce trap life.

The discharge capacity of the trap depends on the free-flow area of the valve orifice, the pressure drop across it and the inlet temperature of the condensate. Flow through a circular orifice is governed by the equation:[3]

$$W = 1,891 \, d_o^2 \, C \sqrt{\Delta P \rho}$$

where:
W Flowrate, lb./hr.
d_o Orifice diameter, in.
C Orifice flow coefficient
ΔP Differential pressure across orifice, psi.
ρ Fluid density, lb./cu. ft.

There is a considerable problem in measuring the pressure drop in the above equation because hot condensate flashes as soon as it passes through the valve orifice.

Trap capacity is not truly defined by orifice size and pressure differential. Pressures upstream and downstream of the trap are also subject to variation, depending on heat-exchanger performance, flowrates, temperatures and system back-pressure. And the orifice may never be fully open to flow because of the valve design. Nor can flow coefficients be measured with the same precision as for control valves, since the valve stem in the trap is often not definitely located with reference to the orifice.

The data in Table II[1] shows how flowrate varies with inlet temperature at constant ΔP (100 psi.). And Table III indicates variations in fluid properties as the inlet pressure is reduced ($\Delta P = 50$ psi.).

In order to compensate for these variations in operating conditions, the designer uses a capacity safety factor to increase the calculated condensate load. Although all manufacturers recommend safety factors, overdesign might be as high as 10 to 15 times the actual steam flowrate.[4]

One short-cut method that should be avoided: Sizing steam traps to equal line size. This practice is never a substitute for analyzing process conditions. It invariably leads to specification of the wrong size steam trap. ■

References

1. "Voluntary Standards for Determining Industrial Steam Trap Capacity Rating," Fluid Controls Institute, Inc., 1965.
2. Oliver Lyle, "The Efficient Use of Steam," Chapters 9 and 10, Her Majesty's Stationary Office, London.
3. Flow of Fluids, Technical Paper No. 410, Crane Company, pp. 3-5.
4. Arteca, N., Dictionary of Steam Traps, *Mech. Contractor*, pp. 17-21, 1970.
5. Pollard, R. S., "Industrial Steam Trapping Course," Vol. 1, 4th ed. Yarway Corp., Blue Bell, Pa., 1970.
6. Welker, J. W., What You Should Know About Steam Traps, *Heating, Piping & Air Conditioning*, Aug. 1965, pp. 129-135; Sept. 1965, pp. 125-128; Oct. 1965, pp. 105-109.
7. Northcroft, L. G., Steam Trapping and Air Venting, Hutchinson, 2nd ed., 1945.
8. Eland, K. G., New Guide to Steam Tracing Design, "Piping Handbook", pp. 139-141, Gulf Publishing Co., 1968.

Meet the Author

Jimmy Mathur is a Senior Engineer in the Process Engineering Dept. of Brown & Root, Inc., P. O. Box 3, Houston, TX 77001. He specializes in utility-systems design, including environmental control, for refineries and petrochemical plants. Before joining Brown & Root, Mr. Mathur worked for 15 yr. in India, primarily in production/management of polyethylene, synthetic-rubber and petrochemical units. He holds a M.S. degree in chemical engineering from the Indian Institute of Science in Bangalore and is a member of AIChE.

Section VII
COOLING SYSTEMS

Air cooler or water tower—which for heat disposal?
Cooling-tower basin design
Design of air-cooled exchangers
Operation and maintenance of cooling towers
Proper startup protects cooling-tower systems

Air Cooler or Water Tower—Which for Heat Disposal?

Climate, flowrate changes, space limitations, maintenance—these are some of the factors to weigh when choosing between air-cooler and water-cooling-tower systems. Yet a third alternative—a hybrid system—may be the best selection.

ROY W. MAZE, The Marley Co.

There is no hard and fast rule as to when an air cooler or water-cooling tower should be used. Both are excellent for dissipating waste heat.

What is the economic breakeven point for selecting between them? One is suggested: a 50°F approach to the summer dry-bulb temperature, with "average" conditions prevailing.

However, many factors can change this suggested point, such as: the availability of auxiliary power for fans and pumps, and its cost; the availability of water, and its quality and cost; whether there is an excess or shortage of cooling-tower and water-treatment capacity; blowdown disposal; fog problems; real estate size or cost; the relative cost of process and cooling-water piping; noise-level requirements; winter problems controlling the fluid being cooled (which depend on water or water vapor in the process stream; the stream's viscosity and pourpoint; the pressure drop allowed in the system; and the turn-down flowrates); the stream's toxicity; the fire hazard; summer and winter wet-bulb and dry-bulb temperatures; and basin or substructure costs and soil quality.

The 50°F rule is not an ultimate answer, only a starting point in an evaluation.

Rating Air Coolers and Water Towers

A shell-and-tube exchanger cooled with water will be considerably smaller and more compact than an air cooler in similar service, because a pound of water can contain many more Btus than a pound of air. Of equal importance is that an air cooler removes heat by sensible heat exchange, a water-cooling tower mainly by evaporation.

Because an air cooler is a sensible-heat exchanger, it is rated by reference to dry-bulb temperature, which is highest in summer and lowest in winter. Typically, in the U.S., this ΔT from summer to winter varies from 70°F to 125°F.

An air cooler selected for a 50°F approach to a 100°F dry-bulb temperature will cool the process fluid to 150°F. At –20°F, it can cool this same fluid to 58°F, which is a 78°F approach to the dry-bulb temperature, when the inlet process fluid is held at constant temperature (Table I).

A water-cooling tower is rated by reference to wet-bulb temperature, which also is highest in summer and lowest in winter. This ΔT is not too different than the dry-bulb ΔT, with winter wet-bulb temperatures about equal to winter dry-bulb temperatures, although the summer wet-bulb maximum is typically 20 to 30 degrees lower than the summer dry-bulb maximum.

A tower that cools water from 118°F to 88°F at an 80°F wet-bulb temperature in summer will cool water to about 45°F in zero weather. The approach will be greater than 45°F at subzero wet-bulb temperatures (Table II).*

At low wet-bulb temperatures, a tower cannot evaporate as much water per pound of air as at high wet-bulb temperatures. For example, during high wet-bulb periods, 80°F wet-bulb air entering a tower has an enthalpy (or total heat content) of 43.69 Btu/lb of air referred to zero enthalpy at zero dry-bulb, bone-dry air. Assuming each pound of air leaves a tower at a 100°F wet-bulb

---

This article is based on a paper presented at the American Soc. of Mechanical Engineers' Petroleum Mechanical Engineering Conference, Dallas, Texas, Sept. 15-18, 1974.

Originally published January 6, 1975

*Dickey and Cates describe the wet tower seasonal performance relationship for large mechanical- and natural-draft towers [2]. Their curves illustrate relationships for all application variables. The examples cited here for the chemical process industries will also vary for other applications in accordance with the curves.

PLUME from parallel-path wet-dry cooling tower (indicated by circle) is dramatically smaller than those from regular towers.

temperature, it would contain 71.73 Btus. This is a gain of 28.04 Btu/lb of air, which is picked up from the water circulating through the tower.

Any change in wet-bulb temperature alters the total heat content of air. Each pound of air entering a tower at 70°F contains 34.09 Btu/lb of dry air. Assuming it leaves the tower at 90°F, it would have 55.93 Btus—a gain of 21.84 Btu/lb of air.

(This is about 78% as much heat change per 20-deg temperature differential as in the foregoing example. The lower the wet-bulb temperature, the fewer Btus each pound of air can contain. At low winter wet-bulb temperatures, each pound of air cannot evaporate as much water because of the lower value of enthalpy differential per degree of temperature change.)

An air cooler can give much lower process fluid temperatures in cold weather. This can be an advantage or a disadvantage, depending on the fluid. Normally, a tower can operate at subzero temperatures; as long as there is a reasonable heat load, the water will not freeze. By cycling off tower fans, it is possible to hold any desired minimum water temperature. Because the air cooler responds to nearly constant enthalpy changes at all ambient temperatures, winter operation can present serious problems unless the air cooler is properly instrumented.

Water-cooling-tower systems have two distinct advantages over air coolers: fluids can be cooled to a lower temperature in warm weather, and process-fluid temperatures can be maintained to a closer tolerance without sophisticated controls.

On a given day, the dry-bulb temperature will vary over a considerably wider range than the wet-bulb temperature. Rain produces rapid changes in the dry-bulb

Air Cooler Designed To Cool From 321°F
Table I*

	Temperatures, °F		
Dry-Bulb	Fluid-Inlet	Fluid-Outlet	Approach
110	321	157	47
100	321	150	50
90	321	142	52
80	321	135	55
70	321	127	57
60	321	119	59
50	321	112	62
40	321	104	64
30	321	96	66
20	321	88	68
10	321	81	71
0	321	73	73
−10	321	65	75
−20	321	58	78

*All fans are operating, and throughout the year flow is at 100% and hot-fluid temperature is constant, i.e., the heat load increases as the dry-bulb temperature drops. Applications in which the heat load is constant will have a constant range (difference between entering and leaving fluid) at constant gal/min and a falling or reduced approach during colder weather because of higher ambient-air density. Normally, for either type of operation, a water-tower system would depend on staging off fans in cold weather to maintain a minimum process temperature or to keep the water from freezing. Cates and Nelson describe freezing tests with large power installations [1].

AIR COOLER OR WATER TOWER—WHICH FOR HEAT DISPOSAL?

Cooling Tower Designed To Cool From 118°F
Table II*

	Temperatures, °F		
Wet-Bulb	Hot-Water	Cold-Water	Approach
80	118.0	88.0	8.0
70	111.4	81.4	11.4
60	105.2	75.2	15.2
50	99.6	69.6	19.6
40	94.3	64.4	24.4
30	89.6	59.6	29.6
20	85.3	55.3	35.3
10	81.4	51.4	41.4
0	77.8	47.8	47.8

*All fans are operating; flow is 100% year-round; heat load is constant.

temperature and can flood the air cooler's finned exchanger, creating large, rapid changes in the process-fluid outlet temperature. Dry-bulb temperature changes, whether caused by natural or other means, can produce considerable variations in process-fluid outlet temperatures.

On the other hand, dry-bulb temperature variations have negligible effect on tower-system temperatures because of the relatively large volume of water suspended in piping and basins, which prevents rapid temperature fluctuations in the system.

Specifications for an Air Cooler — Table III

Heat exchanged	141,000,000 Btu/hr
Fluid	Crude tower overhead
Total fluid	471,218 lb/hr
Vapor, hydrocarbon	427,948 lb/hr
Steam	43,270 lb/hr
Vapor condensed	423,520 lb/hr
Steam condensed	42,840 lb/hr
Viscosity	0.22 cps at 270°F in, 0.31 cps at 210°F out
Molecular weight	104
Temperature	321°F in, 150°F out
Available pressure drop	5.0 psi
Pressure, inlet	30.0 psig
Pressure, design	75 psi
Temperature, design	371°F
Altitude	50 ft
Header	Fabricated box, carbon steel
Tube material	A214 welded carbon steel
Tube, O.D.	1-in 0.110 min. wall
Fins	5/8-in aluminum tension-wound footed
Number of passes	Even
Alternate bids	Cool to 140°F, 130°F, 120°F

Air-Cooler Prices

Table III presents a typical outline specification sheet for an air cooler as a crude tower overhead condenser. A tabulation of design features and price for the cooler is given in Table IV.

The base price with a 150°F outlet process temperature in the air cooler is given as X dollars. (Let us assume that a water-cooling-tower system—including pumps, piping, shell-and-tube exchangers, treating facilities, etc.—would also cost X dollars.)

From Table IV, it is apparent that the economics of an air cooler changes as does the approach to the dry-bulb temperature. Price is based on standard tension-wound footed fins on A214 welded carbon-steel tubes. Other services might require different fins. Some process fluids require alloy tubes to ensure reasonable life.

In Table IV, the price of Model 1M6-15.5 x 36 air cooler (with the outlet temperature of 150°F) is X dollars. The following options would increase this price by the indicated percentages of X: embedded fins on A214 carbon-steel tubes, 6%; extruded fins on A214 carbon-steel tubes, 30%; 18 gauge Admiralty tubes, tension wound, footed fins, 15%; copper-nickel tubes, tension wound, footed fins, 24%.

From a cost viewpoint, the air cooler becomes less attractive as more-expensive fins or tubes are required. Winter accessories and controls can also add to the price.

The Problems of Winter Operation

Winter operation cannot be defined as that at below 32°F. Dry-bulb ambient temperatures lower than the summer design dry-bulb temperature could cause problems with certain fluids.

Typical applications that require some lower dry-bulb-temperature controls include: water or water vapor in the process fluid, viscous fluid in a critical pressure-drop system, and a fluid stream having a high pourpoint or containing a corrosive condensate.

A properly selected cooling tower for a subzero wet-bulb temperature will have cold water leaving the tower well above 32°F (Table II). This is with full heat load and all fans operating at full speed. A cooling tower will not freeze up as long as there is a full heat load and the water is circulated.

Usually, individual fans of a multicell unit are staged-off to hold system temperature above 65°F. Some ice might form near the louvers, but not to a serious degree. On multifan towers, fans can be cycled off to maintain the desired minimum cold-water temperature.*

Any nuisance ice that forms near the louvers can be melted quickly by shutting off fans or by reversing fan motors. Cold-weather operation of a cooling-tower system is also not a problem, because water temperatures to the exchangers can be easily held to a narrow range from summer to winter.

Believing that one can resort to air cooling rather than water cooling to escape freezing problems is a common delusion. For services similar to those mentioned, air

*Hansen describes multifan staging in larger, common-plenum towers for power plants [3].

Design and Prices of an Air Cooler — Table IV

	Base	Alternates		
	150°F out	140°F out	130°F out	120°F out
Btu/hr, MM	141.0	144.0	146.5	149.0
Temperature, in, °F	321	321	321	321
Temperature, out, °F	150	140	130	120
Model number	1M6-15.5x36	1M7-13.5x36	1M7-14.5x36	1M8-14.5x36
Number of bays	6	7	7	8
Length and width of bay, ft	15.5 x 37	13.5 x 37	14.5 x 37	14.5 x 37
Overall length x width, ft	94 x 37	95.5 x 37	102.5 x 37	117 x 37
Number of 36 ft tubes/bay	363	328	353	338
Number of 36 ft tubes	2,178	2,296	2,471	2,704
Tubes, ft	78,408	82,656	88,956	97,344
Total number of fans	12	14	14	16
Bhp/motor	22.4	21.3	22	21
Total bhp	268.8	298.2	300	336
Prices	X	(1.055) $$X$	(1.14) $$X$	(1.25) $$X$

coolers can become a winter nightmare without the proper accessories and controls. Many more problems have occurred while operating air coolers at the lowest winter dry-bulb temperature than at the highest summer dry-bulb temperature.

There is no problem in selecting an air cooler to cool a process fluid to a 20°F approach to a dry-bulb temperature in summer. It is just a matter of spending enough to make the unit big enough. The closer the approach, the greater the amount of extended surface required. The more extended the surface on the air cooler, the greater the winter problem.

Several accessories are available to facilitate lower dry-bulb operation of air coolers: manual and automatic louvers, automatically-variable-pitch fans, steam coils, and warm-air-recirculation housing.

In addition to fluid characteristics, each air-cooler service must also be analyzed for: transient starting and stopping heat-transfer and fluid-flow conditions; lowest winter dry-bulb temperature; lowest turndown rate expected; lowest dry-bulb temperature and turndown rate at which the system, including the air cooler, will be brought onstream.

Manufacturers of air coolers can provide tube-wall temperature data for the lowest dry-bulb temperature. These can be calculated for 100% flow and 25% turndown. With these data, accessories can be added for operation during low dry-bulb temperatures.

For the Model 1M6-15.5 x 36 air cooler (a crude tower overhead condenser), Table IV gives: six bays, two fans per bay, 94 ft long x 37 ft wide, and a price of X dollars.

Accessories will increase the cost by the following percentages of X: manual louvers—6%; automatic operators for the louvers—3%; six 50% automatically-variable-pitch fans—4%; steam coils—14%; complete recirculation housing (including automatic louvers)—40%.

Many problems of winter operation have occurred because freeze protection was provided on the basis of 100% flow only. Most processes have a 25% turndown at times, especially at startups. Table V outlines recommended accessories and procedures for winter operation at 100% flow and 25% turndown.

For the base-case air cooler with 150°F outlet fluid and assuming an occasional 25% turndown, the price for 20°F operation would be: base unit—$$X$; for manual louvers—add 6% of $$X$; for 50% automatically-variable fans—add 3% of $$X$. This yields a total price of 109% of $$X$.

If the design winter dry-bulb temperature is 0°F or lower, and the turndown 25%, it would be necessary to purchase a complete recirculation system (Fig. 1) at a cost of 158% of $$X$ (which includes: base unit—$$X$; 50% automatically-variable fans—4% of X; steam coils—14% of X; and recirculation housing—40% of X).

Table IV alternates give designs and prices of air coolers for 140°F, 130°F and 120°F fluid outlet temperatures.

The closer the approach to the summer dry-bulb temperature, the greater the need for more cooling surface and horsepower. Also, it is practical to expect more control problems during winter. Air coolers sized for a 50°F approach to the summer dry-bulb temperature might require only louvers and 50% automatically-variable fans for mild winter operation.

If the air cooler were sized to give a 20°F approach at that same mild winter temperature, a complete recirculation system might be required, although cycling off fans or variable pitching is frequently effective. Recirculation systems would increase the air-cooler base price by 25%. Adding winter accessories would raise the base price 9% for a 50°F approach and double it for a 20°F approach.

It is not economical to use steam coils continuously as a heat source for air-cooler operation. However, if tube-wall temperatures indicate that heat is required during startup in cold weather, a steam coil should be installed. Coils are usually required in complete recirculation systems (obviously, during startup, heat cannot be recirculated if none is available). Steam is necessary to warm the tubes as well as the air in the plenum. In a complete recirculation system, the automatically-variable fan controls the fluid's outlet temperature; the plenum temperature controls the louvers; and the ambient temperature

Accessories for Crude Tower Overhead Condenser — Table V*

Temperatures, °F			
Ambient	Tube Wall Startup	Tube Wall Operating	Controls
For 100% flow:			
20	58	69	Control accessories are 50% automatically-variable fans, which sense the temperature of the outlet fluid. Manual fans are cycled off when the temperature drops below 55°F to 60°F. All fans are off at startup.
0	49	60	Control accessories are same as at 20°F.
−20	40	51	Control accessories are louvers and 50% automatically-variable fans. At startup, louvers are closed and fans are off. Operations are same as for +20°F.
For 25% turndown:			
20	35	75	Control accessories are louvers and 50% automatically-variable fans. The latter sense the exiting fluid temperature during normal operations. At startup, the louvers are closed and the fans shut off.
0	15	49	Control accessories are the complete recirculation system (50% automatically-variable fans with manual override and steam coil). At startup, the outside louvers on the recirculation housing are closed and the steam coil heated for about 20 min before fluid flow is begun in the air cooler. After a few minutes of fluid circulation, steam is turned off and the fan turned on. By the manual override on the fan, pitch blades move about one-half the air. A temperature of 80°F over that outside is maintained in the plenum. After 10 min, the fan is switched to fully automatic. The plenum temperature near the fan is sensed to maintain it at 80°F above that outside during severe winter weather.
−20	−2	38	Control accessories are same as for 0°F.

*The winter problem is water in the process fluid.

controls the fixed-pitch fan (assuming all bays have each type of fan).

Trim Cooling Saves Energy

Trim cooling has become standard practice in almost all process plants. Not long ago, all waste heat was dissipated through water cooling. But in the early 1960s, because of economic pressures, process engineers sought to cut down fuel and energy requirements, not so much because of an energy crisis but a profit one. It became common to exchange high-level heat from one process stream to another.

After all the "profitable" Btus were exchanged, the remaining, lower-level, waste heat had to be dissipated. However, scaling and metallurgical problems in water-cooled exchangers increased as process fluid temperatures went up. Process fluids over 200°F created many more exchanger problems than fluids under 200°F.

Air coolers were a logical answer to this "in-between" heat-exchange problem. At temperatures too low or too uneconomical to exchange heat, air cooling was the solution. When a process required low-level cooling (frequently lower than the dry-bulb temperature), a water-cooling-tower system was used as a trim cooler.

To illustrate, a fluid leaving a process unit at 500°F would be reduced to 350°F by exchanging heat with another process stream, then dropped to 150°F by an air cooler, and finally lowered to a 20°F to 30°F approach to the wet-bulb temperature by another shell-and-tube exchanger. The wet tower would cool the water in a shell-and-tube exchanger to a 7°F to 15°F approach to the wet-bulb temperature.

In process plants, economics dictates that some process fluids be cooled to low levels, while other streams simply have Btus removed at some higher level. For that reason, tower water is usually piped in series through two or more exchangers. Typically, water leaving a tower at 87°F would first be piped to critical exchangers, such as a condenser. After a temperature rise of 16°F, it would continue through another exchanger that would dissipate its heat at a higher level, then return to the cooling tower at 119°F to be cooled.

Exchangers in series reduce the price of towers be-

Economics of Series vs. Parallel Flow — Table VI

Exchanger Piping	Flow, Gal/Min	Water, °F Hot	Water, °F Cold	Cost, $	Plan Area, Ft²	Bhp
Series	15,000	119	87	(1.00)X	Y	Z
Parallel	30,000	103	87	(1.17)X	(1.28)Y	(1.25)Z

406 COOLING SYSTEMS

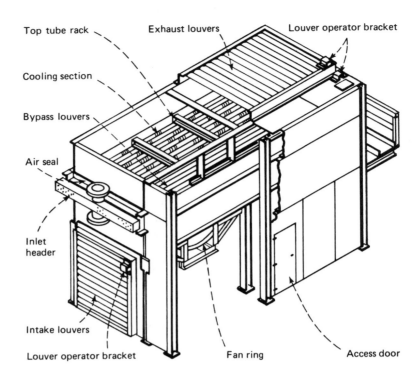

AIR COOLER shown is a forced-draft type having two fan bays and an over-the-side recirculation system. Complete recirculation is necessary if an air cooler is to operate at a design winter dry-bulb temperature of 0°F, or lower, and a 25% turndown rate—Fig. 1

cause the critical exchanger requires 87°F cold water, for which the tower would have to be sized. The hot-water temperature returning to the tower would depend on whether exchangers could be piped in parallel or series. Table VI—which is of a water-cooling tower that dissipates 240 MM Btu/hr at a 80°F wet-bulb temperature—indicates that the economics strongly favor series cooling-water flow through exchangers.

Energy Savings

No ironclad rule can be given regarding relative installed horsepower for water-cooling-tower and air-cooler systems. Again, as a starting point, the 50°F approach to the dry bulb should be about the breakeven point.

The installed horsepower of air coolers in the 40°F to 60°F range of approaches to the dry-bulb temperature is usually greater than for tower fan and pump combinations. However, during milder weather, the air-cooler fans may be staged off more frequently, or pitches varied, to conserve energy. Therefore, annual power consumption can be less for air coolers. Each case must be studied year round to determine the relative merits of wet and dry systems.

In a large air-cooler system having many bays of air coolers (each with two fans), low dry-bulb temperatures permit many hours of operation at lower horsepower. However, the savings may not be realized unless money is spent for instruments and controls to take advantage of mild dry-bulb temperatures. Operators can save energy by manually turning fans on and off, but the designer must decide whether the operators will or should make the effort.

Locating Air Coolers and Water Towers

In process plants, air coolers are usually located atop pipe racks so that ground space is not taken up, process piping can be shorter, and no fog is created.

Water-cooling towers are usually located centrally to serve numerous duties, but can also be remote from specific exchangers, requiring long runs of pipe. Although process piping to air coolers is usually much shorter and smaller in diameter, it is usually costly because of the fluids, temperatures and pressures involved.

Noise is usually more of a problem with air coolers than with cooling towers. Because fans are closely coupled to drivers on air coolers, inspection and maintenance walkways are close to the source of noise.

Tower fan-decks serve as inspection walkways. Fan noise is shielded from operating personnel by the 10-to-20-ft-tall fan cylinders. Fans are not easily heard from the ground near the tower because of the background of falling water.

What Causes Plumes

Air coolers have two distinct advantages over towers: they do not create vapor plumes and do not require makeup water or blowing down.

Air enters a tower at the ambient wet- and dry-bulb temperatures, and leaves, winter and summer, for all practical purposes fully saturated.

Fig. 2 illustrates what causes plumes. Point 1 indicates a typical winter condition of a 42°F dry-bulb and a 36°F wet-bulb temperature. Air leaves the fan cylinder fully saturated at 80°F, which is indicated by Point 2. A plume is caused by the 80°F saturated air returning in a straight

AIR COOLER OR WATER TOWER—WHICH FOR HEAT DISPOSAL?

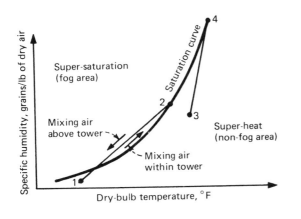

PSYCHROMETRIC CHART exhibits cause of plumes—Fig. 2

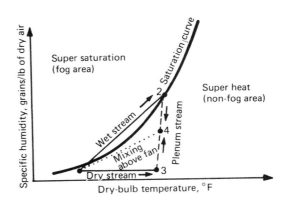

PSYCHROMETRY is of parallel-path wet-dry tower—Fig. 4

line to the original ambient condition of a 42°F dry-bulb and a 36°F wet-bulb temperature.

When the air plume is on the upper side of the saturation curve, it is in a supersaturated condition and is visible as fog. In this straight-line return to the original ambient-air condition, the plume crosses the saturation curve at about 50°F. From that point on, the plume is in the superheat area and is invisible.

From the same figure, it can be seen why towers do not produce plumes in summer. Point 3 represents a summer ambient condition of a 88°F dry-bulb and a 70°F wet-bulb temperature. The plume leaves the tower fully saturated at 100°F, shown as Point 4. As the plume leaving the tower at 100°F returns in a straight line to Point 3, it stays on the superheat side of the saturation curve, which is the nonfog area.

Because an air cooler does not add moisture to the air, it cannot create a visible plume. Using the psychrometric chart (Fig. 2) at the winter condition of a 42°F dry-bulb and a 36°F wet-bulb temperature, the line from Point 1 would move horizontally to the right to some higher dry-bulb temperature and at a much lower relative humidity.

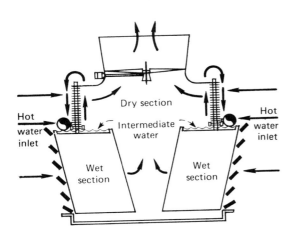

CROSS-SECTION shows parallel-path wet-dry tower—Fig. 3

Unitized Tower and Air Cooler

Under certain conditions, neither an air-cooler nor a water-cooling-tower system will be a completely satisfactory answer. One example of such a condition would be a process section that requires cold process temperatures but in which the only available tower site would create a fog problem on a public road. Another example might be where the scarcity of makeup water would be a problem for a water-cooling tower, but a high summer dry-bulb temperature would limit the outlet process temperature of an air cooler.

To solve the problem of excessive plume, the parallel-path wet-dry tower—in which the air cooler and the water-cooling tower are unitized—was designed (Fig. 3) [4]. Water is piped through the air-cooler sections, then flows by gravity through the water-cooling section. Air flows in parallel through the dry and wet sections. The psychrometrics involved.

Fig. 3 and 4 illustrate the psychrometrics involved. Point 1 indicates a winter air temperature. The air passing through the wet section of the tower leaves that section fully saturated, as shown by Point 2. Air passing through the dry section moves horizontally from Point 1 to Point 3. These two streams of air, one saturated and one at a low relative humidity, are mixed by the fan, and the theoretical plume leaves the fan cylinder at a point below the saturation curve.

Some plume is actually visible because the air mixing is not perfect. A typical parallel-path wet-dry tower of minimum air-cooler size has visible plumes that seem to disappear at 20 to 25 ft when standard towers in the area have dense plumes that extend several hundred feet.

A typical parallel-path wet-dry plume abatement tower has air-cooler sections that are smaller than its wetted fill section. This type increases the cost of a conventional tower by about 75%, depending on the air-cooler tube material. Towers that have larger proportions of dry to wet surface can further reduce the visible plume length. To get the same cold-water temperatures, the overall size of the parallel-path wet-dry tower must be increased, which raises the price accordingly.

The parallel-path wet-dry tower can also be used if

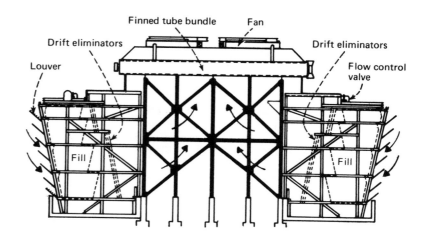

UNITIZED TOWER shown represents the second type of combination of the water-cooling tower and the air cooler: the series-path wet-dry air cooler. It is the logical choice when water is scarce and expensive, or of dubious quality, and the process requires that the fluid be cooled to temperatures that are too near, or below, the design dry-bulb temperature—Fig. 5

makeup water is a problem, because the Btus dissipated by the air-cooler sections do not evaporate water. Typically, the parallel-path wet-dry tower uses about 85% of the water of a standard wet tower, although coil size choice can reduce water consumption of the standard wet tower to only 15%.

The second example of a unitized system is the series-path wet-dry air cooler (Fig. 5). In arid climates, the summer design dry-bulb temperature is high and the design wet-bulb temperature is low. Where water is scarce, expensive, and of dubious quality, air coolers are a logical answer, but usually the process requires the fluid to be cooled to a temperature that is too near, or even below, the design dry-bulb temperature.

A tower will cool water to the wet-bulb temperature if heat load is not added to the system. The dry-bulb temperature of air leaving the tower under these conditions can be cooled by changing sensible heat into latent heat.

If the summer design dry-bulb temperature is 100°F and the wet-bulb temperature is 70°F, a cooling tower without a heat load will cool the water to 70°F. At the same time, the air leaving the tower is typically cooled to about a 78°F dry-bulb temperature. This air can then be directed "in series" to the inlet of an air-cooled exchanger. Adiabatic saturators of this type are used in arid climates for comfort "air conditioning," and the same principle can be applied to the process industry.

Cooling process fluids from 250°F to 100°F with air coolers is impossible if the design dry-bulb temperature is 100°F, but can be done with a series-path wet-dry air cooler. A standard air cooler could be installed to cool that same process fluid from 250°F to 100°F at a 78°F dry-bulb temperature. With a series-path wet-dry air cooler, anytime the dry-bulb temperature exceeds 78°F, water can be circulated over the wet portion of the system. As long as the dry-bulb temperature does not exceed 100°F and the wet-bulb temperature does not exceed 70°F, the coil sections of the series-path wet-dry air cooler will receive air at 78°F or less.

When water is circulating over the tower, it will be evaporated. The amount evaporated will be less than half that evaporated in the conventional tower. For the largest part of a year, when the dry bulb does not exceed 78°F, no water will be required. Annual water consumption will be typically less than 10% of the amount required for the conventional tower.

A Comparison of Maintenance

Evaporation rates, of course, vary over the operating year, primarily because of the psychrometric variable. For the average case, however, the rule-of-thumb frequently used is that for every 9°F that the water is cooled in the tower, 1% of the water circulated is evaporated.

The air leaving the fan cylinder for practical purposes is saturated. This effluent also contains free water droplets, called drift. Traditionally, tower manufacturers' drift loss guarantee was 0.2% of the circulating water, a figure that was accepted because there was no accurate method for measuring drift. During the past two years, however, techniques have been developed that measure drift loss to 0.001%.

Towers with standard drift eliminators (carefully installed) control drift to about 0.05% of the total water circulated. More costly eliminators can lower the drift loss to a guaranteed 0.0025%.*

Regardless of the amount of drift, humid conditions caused by the saturated plume are troublesome from a maintenance and operation standpoint. Obviously, chemicals in the drift can accelerate corrosion on nearby structures.

Shell-and-tube exchangers in the water system are subject to scale and corrosion on the water side. Of course, maintenance problems can be reduced if process temperatures are kept below 200°F in exchangers. Of paramount importance is the water treatment and the number of concentrations allowed to build up in the cooling-water system.

A typical tower system cooling the hot water from 118°F to 88°F will evaporate 2.7% of the water circulated. This loss must be continuously replaced, as must be the 0.05% drift—otherwise, the concentration of solids

---

*Holmberg and Kinney describe drift technology and physical requirements to meet low-drift objectives [5].

in the system would build toward equilibrium. The buildup would be 55 times:

$$\frac{2.7\% \text{ (evaporation)} + 0.05\% \text{ (drift)}}{0.05\% \text{ (drift)}}$$

Obviously, not many sources of makeup water could be used with solids building up to this concentration. More common are concentrations of 3 to 6 times.

If 4 times is the limit, for every 4 gallons of makeup water, 1 part of the circulating water should be removed as blowdown and drift. A small percentage of blowdown can drop the concentration of solids drastically. A blowdown of 1% of the water in the system would decrease the concentration of solids from 55 times to 3.6 times:

$$\frac{2.7\% + 0.05\% + 1.0\%}{0.05\% + 1.0\%} = 3.57$$

When salty and brackish water, which is being used more often, is the makeup water, the solids concentrations can be held at less than 2 times by increasing the blowdown rate. If salt water (30,000 ppm) is the makeup, however, drift can be a serious nuisance. It might be judicious to pay extra to reduce drift loss to 0.0025%.

A blowdown of 2.70%—equal to the evaporation ratio—would drop the concentration to two times.*

Towers are usually designed to produce cold water at a wet-bulb temperature that occurs 2% to 5% of a year. One fan out of service on a typical five-fan tower, for example, would not raise the cold-water temperature over 1.5°F during summer peaks. If the wet-bulb temperature were 2°F lower than design and a fan failed, the cold-water temperature would still be within specification.

A single mechanical failure in a multicell wet or dry tower at the design wet-bulb temprature would not shut down a plant. It could be operated at slightly reduced load. Water could be shifted from the cell out of service to other cells without a hydraulic problem. This would, in fact, improve the performance of the active cells.

Individual air-cooler bays lack the flexibility of a water-cooling system. Each bay typically has two fans. The failure of one fan in a typical "range condenser" air cooler, which has partitions between fans and is designed to cool a fluid from 321°F to 150°F at a 100°F dry-bulb temperature, will raise the outlet temperature to 190°F.

In the case of a six-bay air cooler, this would not be too severe because 150°F fluid from the other five bays would be blended with the 190°F fluid, so the temperature of the total stream would be 157°F. (The heat removed by the one-fan bay would be about 55% to 60% of that normal for a two-fan bay. Half the exchanger would be operating by natural convection.)

Gear-Box Vs. V-Belt Drives

Mechanical maintenance for air coolers can be more demanding than on water systems. With two fans per bay, a large plant can have more than 100 fan drives to maintain. This compares to five or six large fan drives for a water-cooling tower, plus two to four water pumps.

Specifications usually called for V-belt drives with 25-hp motors, or smaller. However, gear boxes are always used with power and process industry water-cooling towers because of problems with V-belts in a wet atmosphere. Typical specifications call for gear-box drives on 30-hp motors and larger. The higher horsepower gear-box drive is more economical in terms of initial cost.

A typical V-belt drive has three-groove cast-iron sheaves and three solid-back belts that, on pitted sheaves, wear very rapidly. Instructions usually call for belts to be tightened after several days of operation, which can be done by means of a belt tensioner outside the fan cylinder. The belts must be adjusted evenly to equalize the load on all of them without creating the too high tension that can cause bearings to fail. These grease-lubricated bearings should have extended grease lines and fittings outside the fan cylinder for easy access.

A loose belt will turn a fan at a slower speed, causing the process fluid temperature to rise. Slippage on sheaves also accelerates belt wear. V-belts are usually replaced every 12 to 18 months, depending on the preventative maintenance program, amount of fan cycling, and the environment. A replacement set of V-belts may cost about $75 or more.

Gear-box oil must be changed every 6 to 12 months.

Maintenance personnel generally prefer forced-draft air coolers because the mechanical equipment is accessible. Such maintenance can be performed below the hot finned tubes. Walkways and platforms under the finned tubes make inspections and maintenance easier than do those on top of induced-draft air coolers.

Mechanical failures on induced-draft coolers require access to the plenum and fan cylinder area for repair. If the process stream is left flowing in the bay when its fan is out, the repair area will become uncomfortably hot, often too much so for occupancy. Some specifications call for the drivers on induced-draft coolers to be under the finned tubes. Specifications may require the removal of the fan and fan shaft without personnel working in, or over, the hot fan-cylinder area. #

References

1. Cates, R. E. and Nelson, J. A., "Dry Cooling Towers for Large Power Installations," The Marley Co., March 1973.
2. Dickey, J. B., Jr. and Cates, R. E., "Managing Waste Heat with the Water Cooling Tower," 2nd ed., The Marley Co., 1970, 1973.
3. Hansen, E. P., "Dry Towers and Wet-Dry Towers for the Indirect Power Cycle," The Marley Co., 1973.
4. Hansen, E. P. and Cates, R. E., "The Parallel Path Wet-Dry Cooling Tower," The Marley Co., 1972.
5. Holmberg, J. D. and Kinney, O. L., "Drift Technology for Cooling Towers," The Marley Co., 1973.
6. Kadel, J. O., "Cooling Towers—A Technological Tool to Increase Plant Site Potentials," The Marley Co., April 23, 1970.

Meet the Author

Roy W. Maze is Southwest Regional Manager for The Marley Co. (2801 S. Post Oak Rd., Suite 482, Houston, TX 77027). He is responsible for the sales and field construction of cooling towers, air coolers and deaerators. After joining Marley as Director of Public Relations, he was Sales Manager of Air Conditioning Products.

He received a B.S. degree from College of Emporia and an M.S. degree from Kansas State University. A member of ASME, Process Heat Exchange Soc. and American Soc. of Refrigerations and Air Conditioning Engineers, he has published several articles on water-cooling towers.

*Curves described by Kadel illustrate the relationship of concentrations to blowdown, drift and makeup for large nuclear and fossil power stations. Proportionally to size, these same relationships apply to the process industries [6].

Cooling-Tower Basin Design

Designing the basin of a cooling-tower system is an important aspect—
often neglected—that depends on the needs of the individual plant.
There are, however, general rules that apply to all systems.

FREDERIC FRIAR, Shamrock Chemical Co.

Often, the concrete basin of a cooling tower receives little attention with respect to good design. As long as the tower fits on top and the basin is strong enough to support the tower, no other factors are considered.

Nevertheless, the basin is a very important part of the total system, because a well-designed basin results in a cleaner cooling system and in lower maintenance costs for the life of the system.

Five basic functions are performed by a tower basin:
- It acts as a foundation for the tower.
- It is a surge tank or reservoir for cooling water.
- It is a place where the recirculating water can slow down to allow suspended matter to settle out.
- It is a mixing zone where raw water, recirculating water, and treatment chemicals become uniform cooling water.
- It is a container for collecting and holding settled-out mud, silt and other debris.

Basic Considerations

The basin system should consist of three parts: the basin itself, the pump pit, and the connecting flume or trough (Fig. 1).

The pump pit requires little attention to design insofar as cleaning is concerned, because the turbulence created

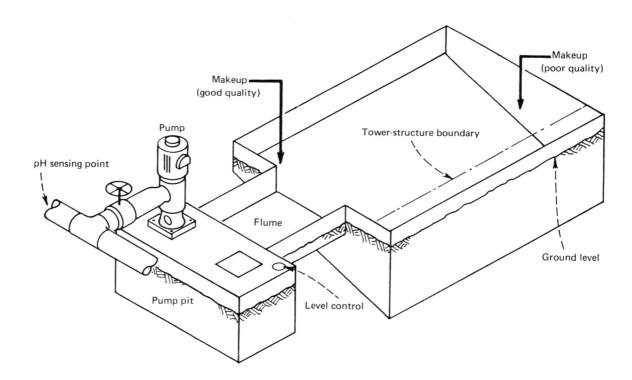

SIDE-SLOPE FLOOR BASIN SYSTEM shows relative positions of basin, flume (or trough) and pump pit—Fig. 1

Originally published July 22, 1974

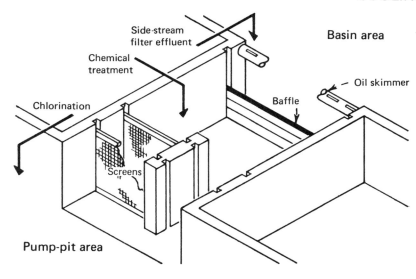

FLUME DETAIL shows location of various streams—Fig. 2

by the pump suction will not allow settling; and if the basin and flume are designed correctly, there should be but little heavy, suspended solids going into the pump pit. This pit should not form part of the basin, but should be a separate enclosure, so that cooling water from the basin flows into the pit through the flume.

To be considered when designing the pump pit, should be the location of makeup, filter-effluent and chemical-treatment lines (Fig. 2). If the makeup water is from a potable supply or has been clarified, it should enter the system at or near the pump pit, but if it is of poor quality (raw river water, reclaimed water, or highly turbid water), it should be introduced into the basin as far away from the flume and pit as possible. This arrangement permits any heavy matter in the makeup water to settle before it enters the system.

The effluent from an inline or sidestream filter should reenter the system at or near the pump pit. Chemical-treatment lines generally should be introduced into the flume or pump pit. This applies especially to chlorination, microbiocide and antifoulant feeds.

For the addition of chemicals, definite rules cannot be established because of the varied nature of cooling systems, as well as of the treatment used. Final consideration for the location of these lines should be made with the advice of a competent water-treatment specialist.

The basin should be constructed so that it can be quickly and easily cleaned. Because most industrial cooling systems operate continuously, a shutdown for cleaning is impossible; therefore, online cleaning should be the objective. Several design possibilities make online cleaning practical:

- Sloping the floor of the basin into a center blowdown pipe.
- Sloping the floor to one side, and extending this side about 2 ft beyond the tower structure; this allows for a suction hose to be dropped into the low side of the basin.
- Sloping the floor away from the pump pit, and extending the back wall of the basin to allow for the hose method of cleaning.

Designing the Flume

The connecting trough or flume deserves careful planning. Ideally, it should be divided into several passageways, each approximately 2½ by 2½ ft, but no more than 3 by 3 ft. Each passageway should have a set of fiberglass or steel screens. (Larger passageways would make such screens unmanageable by one man.)

Screen-mesh size should be such as to strain out any floating debris that might otherwise enter the system and plug heat-exchanger tubes or other small passageways in water-cooled equipment. Two screens per passage allow periodic cleaning of one while the other, installed in the system, remains to provide continuous screening of the cooling water.

On the basin side of the flume, there should be a baffle or weir to keep settled solids from entering the system. It is a good idea to design this baffle so that its height can be changed by adding or removing sections of the baffle plate. Since cooling water should flow over this baffle, care should be taken to prevent its being so high as to restrict cooling-water flow into the pump pit (Fig. 2).

The flume is a good place to introduce treatment chemicals, because in addition to good mixing occurring here, further mixing is achieved as the water passes through the set of screens.

Other features can be built into the flume area. A lightweight catwalk over the flume can be provided. And if the system is subject to oil contamination, a simple American Petroleum Institute skimmer can be built in. Although the possibilities are many and varied, the needs of the individual plant should be of prime concern when designing its cooling-tower basin. #

Meet the Author

Frederic Friar is President of Shamrock Chemical Co. (P.O. Box 1132, Huntington, WV 25714), a concern specializing in industrial water treatment. Prior to forming his own company, he had several years' experience in the water-treatment field, especially cooling-water systems. From time to time, he presents seminars and training sessions concerning some phase of water treatment. He attended West Virginia Institute of Technology, Concord College (Athens, W. Va.), and Ohio State University, majoring in chemistry, art and zoology.

Design of air-cooled exchangers

A procedure for preliminary estimates
Robert Brown, Happy Div. of Therma Technology, Inc.

Process design criteria
V. Ganapathy, Bharat Heavy Electricals Ltd.

Specifying and rating fans
John Glass, Happy Div. of Therma Technology, Inc.

Originally published March 27, 1978

Chemical engineers should find that these three articles will enable them to estimate the design parameters of an air cooler much more reliably than was thought possible for nonspecialists a few years ago. Thoughtful engineers will be able to use these procedures for estimating an optimum with almost the same degree of confidence as might be achieved through the computer programs now used by large-scale purchasers of air coolers.

Experience accumulated from many applications has produced standard designs suited to the calculated optimum for almost any condensing or cooling application. Air cooling is also expanding into new applications. Emergency core-cooling in liquid-metal fast breeder reactors offers an extreme but important example. Here air holds two dominant advantages over other methods of cooling: It is always available and no hazardous reaction is involved in the event of a leak of molten salt or metal from the process side of the tubes.

This application is illustrated by some "dump heat exchangers," which we were commissioned to design and fabricate for the Fast Flux Test Facility currently under construction at Richland, Wash. All of the heat generated in the reactor is to be dissipated to the atmosphere through twelve sodium-to-air exchangers, each with a rated heat-removal capacity near 114 million Btu/h.

In order for these exchangers to be built, it has been necessary to develop:

■ An extended-surface tube design that permits full nondestructive examination of the fin-to-tube attachment.

■ A means to preheat the tube bundle, to prevent solidification of sodium in the tubes.

■ A tube support system that does not impose severe stresses during thermal expansion from ambient to near 800°F during operation.

■ Resistance to vibration induced by flow of molten sodium through the tubes.

■ A louvred stack design to impose minimum back-pressure on the fans while ensuring protection against windblown rain.

■ A liner design to protect the casing insulation from erosion by the high-velocity air.

While most air coolers work at much lower temperatures, and in relatively routine applications, this design shows that difficulty of service is no longer a restriction to the use of air coolers.

J. P. Fanaritis, Struthers Wells Corp.

A procedure for preliminary estimates

Advance planning is needed to allocate space and power for air coolers. Here is a simple method by which these requirements can be estimated.

Robert Brown, *Happy Div. of Therma Technology, Inc.*

☐ The current practice, when planning to install air coolers in the U.S., is to solicit bids from one or more of about seven leading designer-manufacturers of this equipment. Usually, the purchaser accepts the design recommendations of the successful bidder, who not only manufactures proprietary finned tubes but also rates the coolers and specifies the fans, the air-flow requirements and the spatial requirements.

Unfortunately, the successful bid often fails to produce an optimum of operating costs (for fan horsepower requirements) versus initial investment that is suited to the user's requirements for heat-transfer duty and temperature. Also, with this current industry practice, suppliers sometimes find that they have wasted thousands of dollars of design effort because certain critical information has not been included in a requisition. Meanwhile, users often find that not enough advance planning has been done, so that attractive proposals have to be rejected because of restrictions due to ambient-temperature variations, space limitations, and so forth.

The industry in general—both suppliers and users—can benefit from a procedure whereby prospective purchasers could estimate and plan air coolers well in advance, so as to be able to write more-definitive purchase requisitions.

Parameters of air cooling

The primary variables affecting design of an air cooler can be arrived at only through trial and error. Consider the basic equation:

$$Q = U A \Delta T$$

where: Q = heat transferred
U = overall heat-transfer coefficient
A = heat-transfer surface
ΔT = the effective temperature difference, depending on the temperatures of the hot process fluid (T_2 and T_1) and of the air (t_2 and t_1).

Of these variables, the prospective user knows only Q, T_2, T_1 and t_1. The exchanger designer must assume an air flow, which for the given duty establishes t_2 and the mean temperature difference across the exchanger. Since the heat-transfer coefficient for the air film is almost directly proportional to the rate of air flow, an increase in the assumed rate of air flow increases both the overall transfer coefficient, U, and the mean temperature difference.

However, pressure drop across the coolers increases almost as the square of the rate of air flow, so that the required horsepower for the fans is increased as the transfer coefficient and mean temperature difference are increased. Finally, the type of fins used to extend the outside surface of the tubes affects both the transfer coefficient and the horsepower, as well as the cost.

Without some guidance, the novice in air-cooler design can waste weeks of calculation pursuing an optimum combination.

Estimating an optimum

The following procedure may help. The key is three assumptions: First, an overall heat-transfer coefficient is assumed, depending on the process fluid and its temperature range. Second, the air-temperature rise ($t_2 - t_1$) is calculated via an empirical formula. Third, the estimate is based on bare tubes, with a layout and fan horsepower estimated from that, so as to avoid the peculiarities of any one fin type.

Overall heat-transfer coefficients to be used are shown in Table I. An analysis of these numbers with values experienced for the inside film coefficients for the process fluids, and the equation for overall heat-transfer coefficient, $(1/U) = (1/h_o) + (1/h_i)$, will indicate that the effective transfer coefficient for the air film varies around 75, indicating that some sort of fin is required but that the actual fin design is left open.

Once the overall transfer coefficient is assumed, the exit-air temperature may be estimated as:

$$(t_2 - t_1) = 0.005\, U \left[\frac{(T_2 + T_1)}{2} - t_1 \right]$$

DESIGN OF AIR-COOLED EXCHANGERS

Approximate overall heat-transfer coefficient for air-cooled heat exchangers — Table I

Liquid coolers

Material	Heat-transfer coefficient, Btu/(h)(ft²)(°F)	Material	Heat-transfer coefficient, Btu/(h)(ft²)(°F)
Oils, 20° API:		Heavy oils, 8–14° API:	
200°F avg. temp.	10–16	300°F avg. temp.	6–10
300°F avg. temp.	13–22	400°F avg. temp.	10–16
400°F avg. temp.	30–40	Diesel oil	45–55
		Kerosene	55–60
Oils, 30° API:		Heavy naphtha	60–65
150°F avg. temp.	12–23	Light naphtha	65–70
200°F avg. temp.	25–35	Gasoline	70–75
300°F avg. temp.	45–55	Light hydrocarbons	75–80
400°F avg. temp.	50–60	Alcohols & most organic solvents	70–75
Oils, 40° API:		Ammonia	100–120
150°F avg. temp.	25–35	Brine, 75% water	90–110
200°F avg. temp.	50–60	Water	120–140
300°F avg. temp.	55–65	50% ethylene glycol & water	100–120
400°F avg. temp.	60–70		

Condensers

Material	Heat-transfer coefficient, Btu/(h)(ft²)(°F)
Steam	140–150
Steam-10% noncondensibles	100–110
20% noncondensibles	95–100
40% noncondensibles	70–75
Pure light hydrocarbons	80–85
Mixed light hydrocarbons	65–75
Gasoline	60–75
Gasoline-steam mixtures	70–75
Medium hydrocarbons	45–50
Medium hydrocarbons w/steam	55–60
Pure organic solvents	75–80
Ammonia	100–110

Vapor coolers

Material	Heat-transfer coefficient, Btu/(h)(ft²)(°F)				
	10 psig	50 psig	100 psig	300 psig	500 psig
Light hydrocarbons	15–20	30–35	45–50	65–70	70–75
Medium hydrocarbons & organic solvents	15–20	35–40	45–50	65–70	70–75
Light inorganic vapors	10–15	15–20	30–35	45–50	50–55
Air	8–10	15–20	25–30	40–45	45–50
Ammonia	10–15	15–20	30–35	45–50	50–55
Steam	10–15	15–20	25–30	45–50	55–60
Hydrogen — 100%	20–30	45–50	65–70	85–95	95–100
— 75% vol.	17–28	40–45	60–65	80–85	85–90
— 50% vol.	15–25	35–40	55–60	75–80	85–90
— 25% vol.	12–23	30–35	45–50	65–70	80–85

The air-temperature rise $(t_2 - t_1)$ calculated in this manner can usually be relied upon to establish a size within 25% of optimum. It may be adjusted for some increase in accuracy through use of a correction factor taken from Fig. 1.

Once an estimated surface is calculated from the assumed U and an effective temperature difference, the unit size of the air cooler may be estimated from Table II. Note that this table assumes 1-in. bare tubes on $2\frac{1}{2}$ in. triangular pitch, thus providing space for fins up to $2\frac{1}{4}$ in. O.D. (i.e., fins $\frac{5}{8}$ in. high). The fan horsepower, predicted from the estimated unit size and surface by Fig. 3, p. 121, also allows for finned tubes.

Since no existing computer program is capable of considering all variables in optimizing air coolers, this procedure is also useful as a first trial in calculating an optimum design.

Calculating temperature difference

The accuracy of this estimating procedure justifies a correction for effective temperature difference. Once the air-exit temperature, t_2, is calculated, it is a simple matter to calculate the log-mean temperature difference (LMTD) for counter current flow by means of one of the

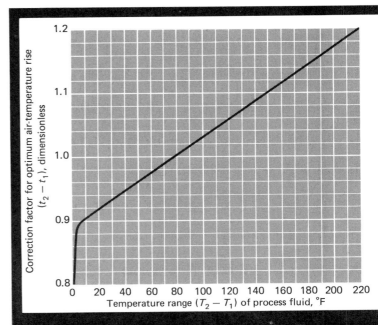

Correction factors for estimated temperature rise Fig. 1

Approximate bare-tube surface versus unit size

Table II

1-in. O.D. bare tube on 2 3/8-in. △pitch

Approximate unit width, ft	Tube length, ft	Fans per unit	No. of tube rows in depth			
			3	4	5	6
4	4	1	49	64	81	97
	6	1	73	97	122	146
	8	2	98	129	163	194
	10	2	123	162	204	243
6	6	1	121	160	201	240
	8	1	161	213	268	320
	12	2	242	320	402	481
	14	2	282	374	469	561
8	8	1	224	297	373	446
	10	1	280	372	466	558
	12	1	336	446	559	669
	14	1	392	520	652	781
	16	2	448	595	746	892
	20	2	560	744	932	1,116
	24	2	672	892	1,119	1,339
10	10	1	351	466	584	699
	12	1	421	559	701	839
	14	1	491	652	817	979
	16	1	561	746	934	1,119
	20	2	702	932	1,168	1,399
	24	2	842	1,119	1,402	1,678
	30	2	1,053	1,399	1,752	2,098
	32	2	1,123	1,492	1,869	2,238
12	12	1	515	685	858	1,028
	14	1	601	799	1,001	1,199
	16	1	687	913	1,144	1,370
	20	1	859	1,142	1,430	1,713
	24	2	1,031	1,370	1,716	2,056
	30	2	1,289	1,713	2,145	2,570
	32	2	1,374	1,827	2,288	2,741
	36	2	1,546	2,056	2,574	3,084
	40	2	1,718	2,284	2,861	3,426
14	14	1	700	931	1,166	1,397
	16	1	800	1,064	1,333	1,597
	20	1	1,000	1,330	1,666	1,996
	24	2	1,201	1,597	1,999	2,395
	30	2	1,501	1,996	2,499	2,994
	32	2	1,601	2,129	2,666	3,194
	36	2	1,801	2,395	2,999	3,593
	40	2	2,001	2,661	3,332	3,992
16	16	1	897	1,190	1,492	1,785
	20	1	1,121	1,488	1,865	2,232
	24	1	1,345	1,785	2,238	2,678
	30	2	1,682	2,232	2,798	3,348
	32	2	1,794	2,381	2,984	3,571
	36	2	2,018	2,678	3,357	4,018
	40	2	2,242	2,976	3,730	4,464
18	20	1	1,247	1,655	2,075	2,483
	24	1	1,496	1,987	2,490	2,980
	30	2	1,870	2,483	3,112	3,725
	32	2	1,995	2,649	3,320	3,974
	36	2	2,244	2,980	3,735	4,470
	40	2	2,494	3,311	4,150	4,967
20	20	1	1,404	1,865	2,337	2,798
	24	1	1,685	2,238	2,804	3,357
	30	2	2,106	2,798	3,505	4,197
	32	2	2,246	2,984	3,739	4,477
	36	2	2,527	3,357	4,206	5,036
	40	2	2,808	3,730	4,674	5,596

Notes:

1. Assume 4 rows of tubes in depth except for the following conditions:
   a. If the temperature range on the process side is 10°F or less, assume 3 rows.
   b. If the temperature range of the process fluid falls between 10°F and 20°F, and special materials of construction are required, assume 3 rows.
   c. If the temperature range of the process fluid is between 100°F and 200°F and/or the assumed overall heat-transfer rate is less than 60, assume 5 rows.
   d. If the temperature range of the process fluid is between 200°F and 300°F and/or the overall heat-transfer rate is less than 40, assume 6 rows.
   e. If the temperature range of the process fluid is greater than 300°F and/or overall heat-transfer rate is less than 30, assume 8 rows.

2. Relative to 14 BWG, the effect of tube-wall thickness on cost is:

Average gage	Cost factor
12 Bwg	1.025
14 Bwg	1.0
16 Bwg	0.99

3. Relative to 6 rows of tubes, the effect of the number of tube rows on cost is:

Rows	Cost factor
4	1.10
5	1.05
6	1.00
8	0.95

4. Relative to a length of 24 ft, the effect of tube length on cost is:

Tube length, ft	Cost factor
10	1.15
12	1.13
14	1.11
16	1.08
18	1.06
20	1.05
24	1.00
30	0.95
32	0.93
36	0.89
40	0.85

5. Because of shipping limitations the widest tube bundle that can be shop fabricated and shipped to a plantsite is 12 ft. Wider bundles must be field fabricated.

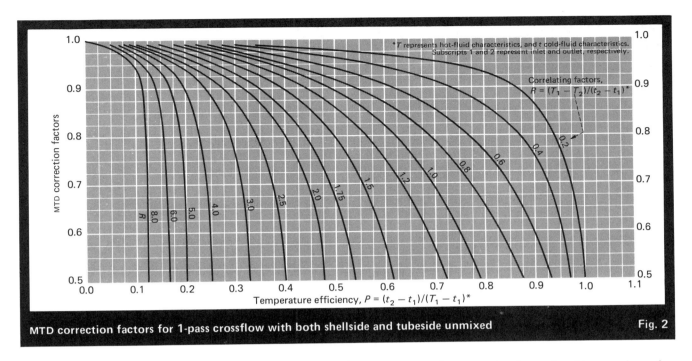

MTD correction factors for 1-pass crossflow with both shellside and tubeside unmixed — Fig. 2

many charts, or by employing the well-known formula:

$$\text{LMTD} = (\Delta t_2 - \Delta t_1)/\ln(\Delta t_2/\Delta t_1)$$

where Δt_2 is the greatest temperature difference and Δt_1 the least temperature difference taken at the inlets and outlets of the exchanger.

As for conventional shell-and-tube exchangers, the LMTD does not apply for air coolers and must be corrected according to the chosen flow pattern. A number of such patterns are available to air-cooler designers; and correction factors for these various flow patterns are given in the Standards of the Tubular Exchanger Manufacturers Assn. (TEMA) and the many handbooks on heat exchange. The flow pattern and correction factors assumed for this estimating procedure are those for one-pass crossflow with both tube-fluid and air unmixed as they flow through the exchanger (Fig. 2).

Sample estimation

Kerosene flowing at a rate of 250,000 lb/h is to be cooled from 160°F to 125°F, for a total duty of 4.55 million Btu/h. How large an air cooler would be required for this service, if the design dry-bulb temperature of the air were 95°F?

From Table I, estimate the overall transfer coefficient for a kerosene air-cooler at 55 Btu/(h)(ft²)(°F). Then the air-temperature rise is:

$$(t_2 - t_1) = 0.005\,(55)\,((160 + 125)/2 - 95)$$
$$= 13.06$$

From Fig. 1, the correction factor for a process-fluid temperature rise of $(160 - 125) = 35°F$ is 0.92, so:

$$(t_2 - t_1) = (0.92)(13.06) = 12.02 \text{ and } t_2 = 107.02$$

The LMTD is:

$$\text{LMTD} = ((160 - 107.02) - (125 - 95))/\ln((160 - 107.02)/(125 - 95)) = 40.41*$$

*Decimal places are maintained to help identify the terms.

From Fig. 2, the correction factor for this LMTD is read for a temperature efficiency of:

$$P = (107.02 - 95)/(160 - 95) = 0.185$$

and a correlating factor of:

$$R = (160 - 125)/(107.02 - 95) = 2.91$$

as (from Fig. 3) 0.95, so that the corrected LMTD is:

$$(0.95)(38.39) = 40.41$$

From this effective temperature difference, the assumed overall transfer coefficient, and the given duty, it is now possible to calculate the hypothetical bare-tube surface, as:

$$A = Q/U\,\Delta T$$
$$= (4,550,000)/(55)(38.39)$$
$$= 2,154 \text{ ft}^2$$

From Table II, this hypothetical bare-tube surface would indicate an air cooler 12 ft wide, with either four rows of 40-ft-long tubes with two fans, for a total bare surface of 2,284 ft² or five rows of 32-ft-long tubes with two fans for 2,288 ft² of surface. From Fig. 3, p. 121, the fan horsepower would be $(22.84)(1.56) = 35.63$.

The author

Robert Brown is general manager of Happy Div., Therma Technology, Inc., a manufacturer of air-cooled heat exchangers. Prior to joining Therma Technology, he worked for Yuba Heat Transfer Corp. and Struthers Wells Corp. He holds a B.S. in chemical engineering from Pennsylvania State University, and an M.S. in Industrial Administration from Carnegie-Mellon University.

Process-design criteria

New economic and environmental restrictions require chemical engineers to compare water cooling and air cooling at the outset of today's projects. Here are air-cooling design methods to use in such comparisons.

V. Ganapathy, *Bharat Heavy Electricals, Ltd.*

☐ Although water has been the standard cooling medium for condensers and coolers for many years, the fear of thermal pollution, shortages of cooling water, and economics of the newer air-cooler designs have brought about a recent switch to air coolers. Today, if cooling water is available, the choice between water coolers and air coolers is based mainly on economics. If cooling water presents problems, there is no choice; air coolers are required.

A brief review of typical transfer coefficients makes it obvious that the overall heat-transfer coefficient of air coolers is governed by the air-film heat transfer, which is low—on the order of 10 $Btu/(h)(ft^2)(°F)$—whereas a similar review of water coolers shows that the overall coefficient is governed by the films both inside and outside the tubes, and that the overall transfer coefficient can be from 10 to 30 times that of an air cooler.

In addition, the specific heat of air is about 0.245 $Btu/(lb)(°F)$, or only one fourth that of water. Hence, the same heat-transfer duty and cooling-medium temperature rise will require four times as much air as water. As a result, air coolers are very large, relative to water coolers; and there is a need to estimate their size early in the life of a design proposal, in order to avoid space problems later on.

Further, it is often difficult to get as low an exit temperature with an air cooler as with a water cooler, since air cooling is based on the dry-bulb temperature, whereas cooling-tower performance is based on the wet-bulb temperature.

On the other hand, air coolers can make use of finned tubes having an external surface about 20 times that of plain tubes; and this partly nullifies the effect of the low air-film heat transfer. Designers of air coolers have developed modules that can be easily assembled and erected at the plant site, thereby reducing labor costs. And finned exchangers have been improved.

Also, since water-cooling systems usually require a cooling tower, the space and cost of the tower must be added to the water-cooling system. When seawater is used, it causes corrosion, so that costly cupronickel tubes are required. Air is almost always clean, with only low fouling factors called for, whereas cooling-tower water must allow more for biological fouling and scale. Finally, a point that engineers generally forget: Air is free, whereas cooling water is not.

Additional features of air cooling versus water cooling are listed in Table I. These show that almost any new design proposal should at least consider air cooling and make a comparison against the complete cost of a water-cooling system, including: the exchangers; a cooling tower; associated water lines, pumps, valves and instrumentation; makeup water; and a water-treatment and tube-cleaning system.

The air-cooling system

Typical air-cooling systems are illustrated for forced-draft fans (Fig. 1) and induced-draft fans (Fig. 2). In these systems, the air cooler consists of one or more rectangular bundles of finned tubes arranged in staggered rows and suitably supported on a steel structure. Both ends of the tubes are fixed in tubesheets in channels (Fig. 3) that have holes opposite the tubes, or removable covers, for tube-rolling and cleaning. These channels are baffled to provide the desired number of tube passes. In some instances of high pressure, the channels may consist of drilled steel billets into which the tubes are rolled.

A fan below the tubes (Fig. 1) forces air up through the bundle, or a fan located above (Fig. 2) draws the air through. These are axial propeller fans varying from 4 to 12 ft dia. and having four to six blades, which may be of aluminum, plastic or, in the case of corrosive atmospheres, stainless steel. The drive can be an electric motor with gears or V-belts, or may be a steam turbine.

The tubes are usually 1 in. dia., with wrapped-on aluminum fins (Fig. 4) spaced from 8 to 16 per in. and varying from $3/8$ in. to $5/8$ in. high, and from 0.012 to 0.02 in. thick. These tubes are arranged in standard bundles ranging from 4 to 40 ft long and from 4 to 20 ft wide (see Table II, p. 110).

The temperature of a process fluid passing through

DESIGN OF AIR-COOLED EXCHANGERS

Air-cooling versus water-cooling systems — Table I

In favor of water cooling

Air cooling	Water cooling
Because of air's low specific heat, and a dependence on the dry-bulb temperature, air cannot usually cool a process fluid to low temperatures.	Water can usually cool a process fluid from 10°F to 5°F lower than air, and recycled water can be cooled to near the wet-bulb temperature of the site in a cooling tower.
Air coolers require large surfaces because of their low air-film heat-transfer coefficients and the low specific heat of air.	Water coolers require much less heat-transfer surface in compact, well-proven exchanger designs.
The seasonal variation in air temperatures can affect performance, while rain and sun can cause appreciable variations in the daily temperature. If winter temperatures are low, they may cause process fluids to freeze.	Water is less susceptible to temperature variations.
Air coolers should not be located near large obstructions such as buildings, trees, etc., since air recirculation can set in.	Water coolers can be located among the other equipment.
Air coolers require finned tubes—a specialized technology.	Well-established designs of shell-and-tube exchangers are satisfactory.

In favor of air cooling

Air cooling	Water cooling
Air is available free, with no preparation costs.	Water for cooling is generally scarce; and when it is available, it must be brought to the site by pump, pipeline, or from a well, etc., at an appropriate cost.
Plant location is not restricted by air cooling.	Sites for large plants in particular are dependent on suitable sources for water cooling.
Air is seldom corrosive, so that less provision need be made for fouling and cleaning.	Water is corrosive and requires treatment to control both scaling and deposition of dirt.
Operating costs for air coolers are lower, since the draft losses are on the order of 0.5-1.0 in. of water.	Operating costs for water coolers are higher, because the cooling-water circulation pump can have a head running to tens of feet of water, depending on the location of the cooler and the cooling towers.
There is less danger of contaminating the cooling medium with air cooling.	In many processes where a toxic fluid needs cooling, there is danger of contaminating the cooling water.
Maintenance costs for air-cooling systems are generally 20 to 30% those of water cooling systems.	Maintenance of cooling water is costly, because: there is more equipment; water contains living organisms that grow in the warm conditions and thus foul exchangers; minerals such as iron can deposit on tubes as oxides or hydroxides; the inside of the tubes (water side) requires shutdown and dismantling of the exchanger for cleaning.

these coolers may be controlled by: (1) louvers or shutters, either manual or automatic; (2) variable-speed, steam-turbine drives; (3) variable-pitch fans (for accurate control); or (4) bypass control of the process fluid.

Three important design criteria affecting these systems are: air data, type of draft, and type of fin.

Air data

Design ambient temperature is the most important single variable involved in rating an air cooler. If a higher-than-actual value is used, overdesign is built into the exchanger, so that it will have capacity to cool the process fluid to well below its desired outlet temperature; if a lower-than-actual temperature is specified, the cooler will not be able to perform its required duty.

Unfortunately, the dry-bulb temperature varies considerably throughout the year, so that current practice

Values of the correction factor for altitude and air temperature — Table II

Altitude, ft	Air temperature			
	0°F	70°F	100°F	200°F
0	1.151	1.000	0.916	0.803
1,000	1.110	0.964	0.913	0.774
2,000	1.070	0.930	0.880	0.747
3,000	1.032	0.896	0.848	0.720
4,000	0.995	0.864	0.817	0.694
5,000	0.958	0.832	0.787	0.668
6,000	0.923	0.801	0.759	0.643
7,000	0.889	0.772	0.731	0.619
8,000	0.855	0.743	0.703	0.596

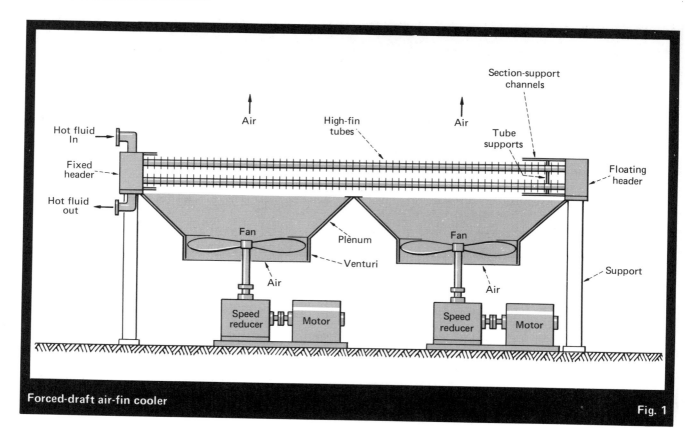

Forced-draft air-fin cooler — Fig. 1

is to assume a design temperature that is exceeded during only 2 to 5% of the annual time period. It is advisable to calculate the performance of an air cooler designed for one temperature against a few other dry-bulb readings known to occur frequently at the plant site.

The following data are needed for realistic estimates of the design air temperature:
- An annual temperature-probability curve.
- Typical daily temperature curves.
- Duration-frequency curves for the occurrence of the maximum dry-bulb temperature.

With these data, the designer can get a good idea of the most probable and most frequently occurring temperatures and thus err as little as possible.

Air density affects the required air flow and, thus, the fan capacity, the head to be developed by the fans, and the power consumption. Table II gives an idea of the effects that altitude and temperature have on the density of air [1].

Also, the nature of the atmosphere is important. If it is corrosive, as in a marine environment or an environment containing sulfur dioxide, the fans, fins, tubes and structures have to be designed accordingly. If the atmosphere is dusty, fouling is possible on the fins. A fouling factor has to be used; and tube pitches have to be increased, thus reducing air mass-velocities and heat-transfer coefficients.

The predominant winds must also be known. If wind velocities are high, the air pattern around the air cooler is likely to be affected. Recirculation patterns close to obstructions can cause difficulty in control. Also, if rains are frequent, the air temperature will fluctuate.

Finned tubes

The fins on air-cooler tubes provide an external heat-transfer area 15 to 20 times as large as the area of the bare tube. So, fins are used to compensate for the low air-film coefficient. While these fins can be solid or serrated, tension-wound or welded, and of a variety of metals, the fin most commonly used for air coolers is aluminum, tension-wound and footed (Fig. 4).

Aluminum is preferred because of its lightness and high thermal conductivity. However, either SO_2-containing or marine atmospheres may corrode aluminum fins, so carbon steel might be preferred at such times. Tension-wound carbon steel fins on carbon steel tubes can be hot-dip galvanized, so that the galvanic layer acts both to resist corrosion and to improve the heat transfer between fin and tube.

Embedded fins have less resistance to heat transfer than do tension-wound fins, but corrosion is likely to be induced at the groove between fin and tube.

The effect of fin design and the air-film transfer coefficient on fin efficiency are shown in Fig. 5. It can be seen that: (1) as the air-film heat-transfer coefficient increases, the efficiency drops, so that the benefits of increased air velocities are lost, particularly for carbon steel; (2) aluminum fins have much higher fin efficiencies than do carbon steel fins, due primarily to the higher conductivity of aluminum; (3) fin efficiency increases with both a reduction in fin height and an increase in fin thickness. Also, the efficiency is reduced as the number of fins per inch increases, so that fin design depends on a balance between many compensating factors.

DESIGN OF AIR-COOLED EXCHANGERS 421

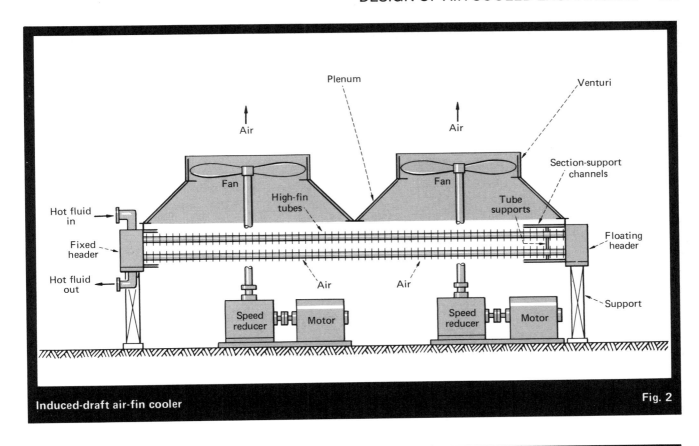

Induced-draft air-fin cooler — Fig. 2

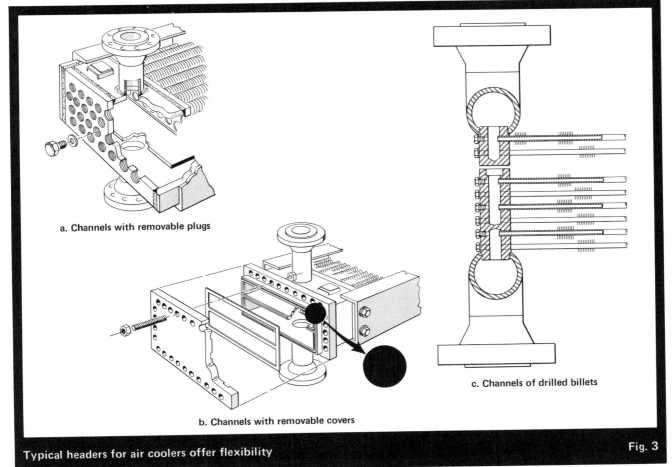

a. Channels with removable plugs

b. Channels with removable covers

c. Channels of drilled billets

Typical headers for air coolers offer flexibility — Fig. 3

COOLING SYSTEMS

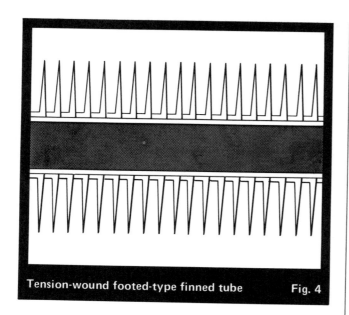

Tension-wound footed-type finned tube — Fig. 4

With 1-in.-O.D. tubes wrapped in fins 0.625 in. (⅝ in.) high, the tube pitch may vary from 2⅜ in. to 2.75 in., so that the clearance between fins varies from 0.125 to 0.50 in. This clearance should be considered in long tube-bundles since the tubes can sag, and the fins should not be allowed to touch.

Forced draft versus induced draft

Forced-draft fans have the advantage of handling cold air entering the exchanger, requiring smaller volumes of air and less horsepower. From the standpoint of maintenance and accessibility, forced-draft units generally offer better arrangements and are easier to place in multiple units. Also, forced-draft fans afford a higher heat-transfer coefficient (Fig. 6) relative to induced draft, because forced-draft fans cause turbulence across the rows.

On the other hand, the upper row of finned tubes of forced-draft units is exposed to the atmosphere with potential damage by hail, rain, etc. In some climates, these exposed fins are heated by the sun, adding to the required duty of the cooler.

Induced-draft fans mounted over the tube bundles nullify these effects. Also, induced-draft fans are less noisy than forced-draft ones. And, induced-draft fans are less susceptible to recirculation of air due to winds and obstructions.

Air-cooler calculations

As with all types of design, calculations for air coolers should begin by identifying the necessary assumptions and segregating the important from the unimportant variables.

At the top of the list is function. The role of "cooler" may vary from essential (e.g., cooling regenerated solvent recycled to an absorber) to incidental (e.g., cooling a refinery product to some arbitrary temperature for storage). Next come requirements for the plant site. Finally, at the bottom of the list, come the variables affecting design. For air coolers, these data are:

Process requirements: Tubeside temperatures (T_1 and T_2), tubeside flowrate (W_T), properties of the tubeside fluid, allowable pressure drop through the tubes (ΔP_T), and calculated duty (Q).

Site data: Dry-bulb temperature of air (t_1), density of air (ρ_a) and specific heat of the air (s_a), width limitations for shipping air coolers to the plant, space limitations, preferred driver and cost of horsepower delivered by that driver, and the payback time for balancing capital investment for adding surface area against operating costs of fan horsepower.

Design variables: Air outlet temperature (t_2), or mass rate of airflow (G_a) through the cooler, tube geometry, heat-transfer coefficients, air-flow pressure drop, required surface, and tube arrangement.

The difficulty of relating the design variables often obscures the relatively greater importance of variables relating to the process and the site. Accurate design calculations for finned-tube coolers have presented an almost insurmountable obstacle for the nonspecialized chemical engineer, because fin construction varies among different manufacturers, who bid on their own proprietary data.

The procedure to follow in air-cooler design is:
1. Identify all process and site data.
2. Assume the layout of the tube bundle, air-temperature rise or mass flowrate, and fin geometry.
3. For the assumed values, calculate film coefficients and overall heat-transfer coefficient, effective temperature difference, and surface; check this surface against the assumed layout.
4. When the required surface fits the assumed layout, calculate the tubeside pressure drop and check this against the allowable pressure drop.
5. When surface and tubeside pressure drop are veri-

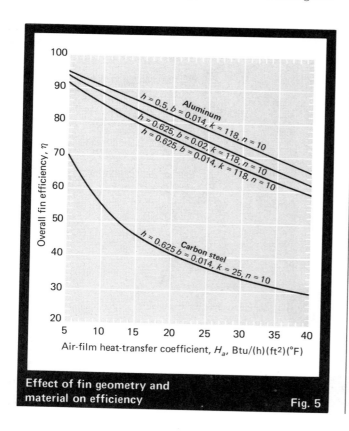

Effect of fin geometry and material on efficiency — Fig. 5

Nomenclature

A	surface area (of fins, obstruction, tubewall, etc.), ft² per ft of length	T	tubeside temperature, °F
b	fin thickness, in.	V	volume rate of flow, ft³/min, or velocity, ft/s
C_p	specific heat at constant pressure, Btu/(lb)(°F)	W	weight rate of flow, lb/h, or width, ft
d	diameter of tube, in.	Z	no. of tube rows per pass
E	efficiency, dimensionless	ρ	density, lb/ft³
f	friction factor, ft²/in.²	μ	viscosity, lb/(ft)(h)
F	correction factor, dimensionless	ΔP	pressure drop, in. of water
FF	fouling factor, (°F)(ft²)(h)/Btu	η	fin efficiency, dimensionless
g	acceleration due to gravity, ft/s²	ϵ	temperature efficiency (for corrected LMTD), $(t_2 - t_1)/(T_1 - t_1)$, dimensionless
G	mass velocity, lb/(ft²)(h)		
h	height (of a fin), in.		
H	heat-transfer coefficient of film, Btu/(h)(ft²)(°F)		
I	Bessel function, dimensionless		**Subscripts**
K	modified Bessel function, dimensionless	a	air
k	thermal conductivity, Btu/(h)(ft²)(°F/ft)	c	corrected
L	length, ft	e	entering
LMTD	log-mean temperature difference, °F	f	fin; process fluid
M	molecular weight, lb	g	gas or air
N	number (of tubes), dimensionless	i	inside
n	number of fins per inch	l	longitudinal; leaving
Nu	Nusselt no., dimensionless	m	tube; metal
P	pitch (of tubes), in.	o	outside
Pr	Prandtl no., dimensionless	ob	obstruction
Q	heat-transfer duty, Btu/h	P	provided; passes
R	correlation factor for correcting LMTD, $(T_2 - T_1)/(t_2 - t_1)$, dimensionless	R	required
		t	transverse
Re	Reynolds no., dimensionless	T	tubeside
S	spacing (of fins), in.	w	wall; at wall temperature; width
s	specific gravity, dimensionless	0, 1	order (for a Bessel function)
t	shellside temperature, °F	1	inlet (as for temperature)
		2	outlet (as for temperature)

fied, calculate the airside pressure drop and fan horsepower.

Air-cooler correlations

The most widely accepted correlations for relating air-cooler design variables are those resulting from experiments done on banks of tubes with copper and aluminum fins, at the University of Michigan, Ann Arbor, Mich., by Edwin H. Young, Dale E. Briggs, and Ken K. Robinson. These correlations, which were presented in A.I.Ch.E. symposia in the late 1960s, have been analyzed and summarized [3].

The general correlation for heat-transfer across fins is:

$$Nu = 0.134\, Re^{0.681}\, Pr^{1/3}\, (S/h)^{0.200}\, (S/b)^{0.1134} \quad (1)$$

This may be reduced to:

$$H_a = 0.295\, G^{0.681}\, d^{-0.319}\, k_a^{0.67}\, Cp_a^{0.33}\, \rho_a^{-0.351}\, S^{0.313}\, h^{-0.2}\, b^{-0.113} \quad (2)$$

The general correlation for pressure drop across finned tubes is:

$$f = 18.93\, Re^{-0.316}\, (R/d)^{-0.927}\, (P_t/P_l)^{0.515} \quad (3)$$

This may be reduced to:

$$\Delta P_a = 1.58 \times 10^{-8}\, G^{1.684}\, d^{0.611}\, \rho_a^{0.316}\, P_t^{-0.412}\, P_l^{-0.515}\, ((T+460)/M)N \quad (4)$$

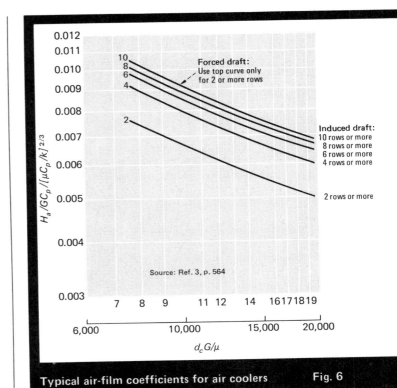

Typical air-film coefficients for air coolers   Fig. 6

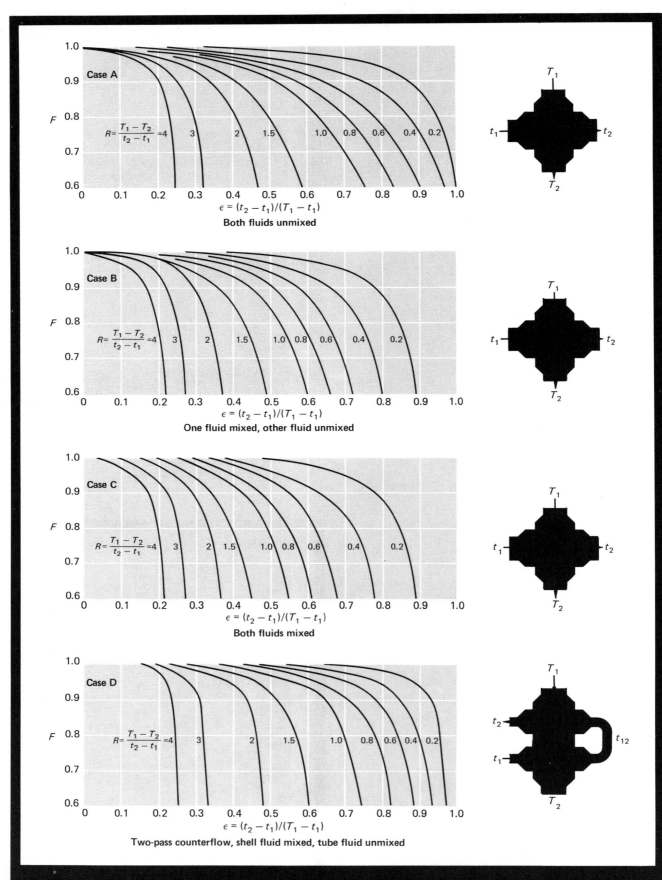

MTD correction factors for different flow arrangements Fig. 7

Heat-transfer coefficients calculated by Eq. (1) and (2) are the main heat-transfer coefficients for the air film in contact with the external surface. This coefficient should be corrected for fouling to obtain an overall heat-transfer coefficient for the air film:

$$H_{ac} = 1/((1/H_a) + FF_a)$$

Then the corrected air-film coefficient is again corected for the efficiency of the fin in conducting the heat to the external surface of the tube:

$$H_g = H_{ac}(EA_f + A_{mo})/A_o$$

where H_g is the apparent heat-transfer coefficient, E is the fin efficiency, A_f is the area of the fins, A_{mo} is the exposed external surface of the tubes, and A_o is the total outside surface. These data must be obtained for the specific finned tubes proposed for the exchanger. They should be available from manufacturers of finned tubes or of air coolers [6].

The heat transfer for the process fluid inside the tubes can be estimated from:

$$Nu_f = 0.023\, Re_f^{0.8}\, Pr_f^{0.4}$$

which can be reduced to:

$$H_f = 0.276\, k_f^{0.6}\, d_i^{-0.2}\, G_f^{0.8}\, \mu_f^{-0.4}\, Cp_f^{0.4}$$

To be consistent with the calculated apparent outside-coefficient, H_g, (1) the internal film coefficient, (2) the internal fouling factor and (3) the resistance of the tube wall must all be based on outside surface area. This correction is made through the ratio of inside to outside area:

$$H_{fc} = H_f(A_i/A_o)$$
$$FF_{fc} = FF_f(A_o/A_i)$$
$$R_{mc} = R_m(A_o/A_i)$$

so that the overall heat-transfer coefficient, U, is:

$$1/U = (1/H_g) + (A_o/A_i)(R_m + FF_f + (1/H_f))$$

The log-mean temperature difference is calculated for air coolers as for other types of exchangers, and is corrected for type of flow according to a temperature efficiency, ϵ, and correlation factor, R, analogously to shell-and-tube exchangers (Fig. 7).

Given the corrected LMTD, the overall transfer coefficient for the cooler, and heat-transfer duty, it is possible to calculate the surface from the basic heat-transfer equation:

$$\text{Surface} = AL = Q/U\,\Delta T$$

This surface determines a width and corresponding length, which can in turn be used to calculate the mass flow of air, G_a, from the weight of air necessary to provide the assumed temperature rise $(t_2 - t_1)$. The calculated value of G_a is compared to the G_a assumed for the heat-transfer calculations. Cooler geometry should be adjusted, or new values of $(t_2 - t_1)$ and/or G_a assumed for a new calculation, until the required surface is only slightly less than that corresponding to the assumed values.

When this trial-and-error procedure has yielded an acceptable arrangement, the air-pressure drop across the cooler, ΔP_a, can be calculated from Eq. (4) given above, and from this the fan horsepower.

One must still check the required pressure drop through the cooler's tubes against that allowed in the process design. Such tubeside pressure drops are calculated in the same manner as for conventional exchangers, except for the special dimensions of air coolers. TEMA correlations are widely accepted for these calculations (Fig. 8).

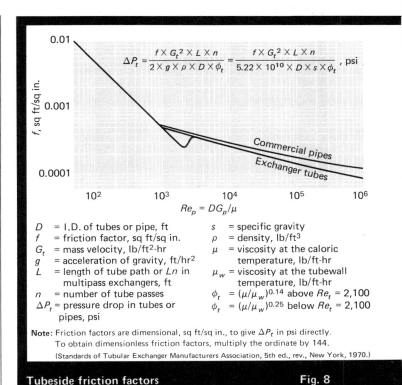

Tubeside friction factors — Fig. 8

References

1. Cook, E. M., Air-Cooler Heat Exchangers, *Chem. Eng.*, Aug. 3, 1964, p. 97.
2. Ganapathy, V., Charts simplify spiral finned-tube calculations, *Chem. Eng.*, Apr. 25, 1977, p. 117.
3. Kern, D. Q., and Kraus, A. D., "Extended Surface Heat Transfer," McGraw-Hill, 1972.
4. Ganapathy, V., Determining inside heat transfer of common gases and liquids, *Plant Engrg.*, May 12, 1977, p. 121.
5. Ganapathy, V., To get heat transfer coefficients, *Hyd. Proc.*, October 1977.
6. Weierman, C., Correlations ease the selection of finned tubes, *Oil & Gas J.*, Sept. 6, 1976, p. 95.

The author

V. Ganapathy is senior development engineer with Bharat Heavy Electricals Ltd. (High Pressure Boiler Plant, Tiruchiarapalli-620 014, India). Currently engaged in the design of boilers for utility and process services, he is particularly concerned with applications in the chemical and fertilizer industries. He holds a bachelor's degree in technology from the Indian Institute of Technology in Madras, India, and an M.Sc. (engineering) in boiler technology from Madras University. He has written over 25 articles for various U.S. and British publications.

Specifying and rating fans

Horsepower and noise have become important to the specification of air-cooler fans. Here are the relationships needed to incorporate these criteria into the specification.

John Glass, *Happy Div. of Therma Technology, Inc.*

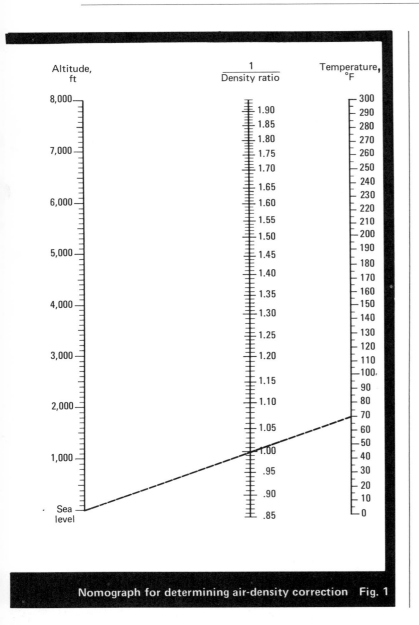

Nomograph for determining air-density correction   Fig. 1

☐ Although fan-rating procedures differ from one manufacturer to another, they all serve the same purpose: to specify the best fan for an application. In some cases, the best will be the least expensive fan that can deliver the required amount of air at the specified operating conditions; but in many instances, particularly when low energy and noise requirements are important, the least expensive will not be the best.

When horsepower and noise must be included in fan evaluation, specification becomes complicated. Nevertheless, both power and noise reduction are becoming more important for air-cooler fans; and these two factors can virtually control the design of the entire air cooler.

When high power costs and long payout periods emphasize costs of the required horsepower, it should be reduced so as to keep the total of operating costs plus investment costs to a minimum. Horsepower can be usually cut in four ways:

1. Design the air cooler with fairly low air velocity across the bundles, to hold down the pressure loss and thus the static pressure at the fan outlet.
2. Increase the tube pitch of the air-cooler bundles to reduce the pressure loss.
3. Increase, up to a point of optimum efficiency, the diameter of the fan, so as to reduce the velocity-head loss, and thus the horsepower.
4. Increase the fan efficiency by such a design method as the use of more blades.

As for noise, the easiest way to reduce it is to reduce blade-tip speed. The effect is indicated by the general formulas included in the new proposed revision of the API (American Petroleum Inst.) "Guideline on Noise" for predicting noise levels:

■ For induced-draft fans, the sound-pressure level, dBA, at 3 ft below the air-cooler bundles is:

$$dBA = 63 + 30 \log V + 10 \log hp + 20 \log D$$

■ For forced-draft fans, this formula is:

$$dBA = 66 + 30 \log V + 10 \log hp + 20 \log D$$

DESIGN OF AIR-COOLED EXCHANGERS

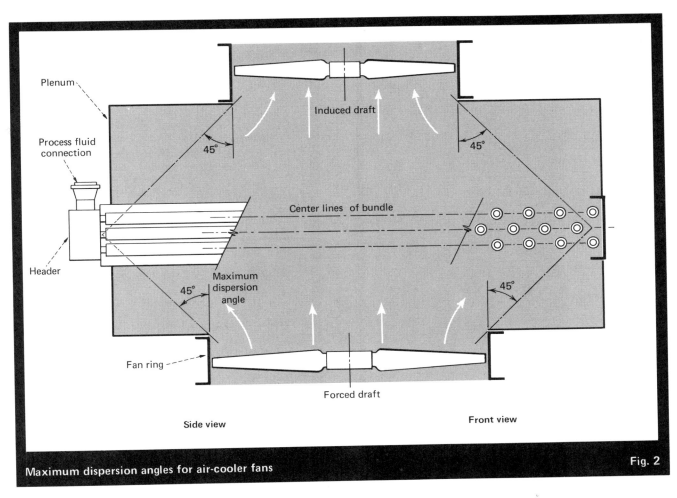

Maximum dispersion angles for air-cooler fans   Fig. 2

where: V = fan tip speed, 0.001 ft/min
hp = fan horsepower
D = fan diameter, ft

These formulas show that when noise requirements are strict, it is best to use fairly large fans with low horsepower, running at low rpm. This will usually require more fan blades than normal, and a larger chord-width.

Basic variables

With so many restrictions affecting fan specification and design, it becomes important to identify the basic variables in rating a fan. There are six: one or two of them independent, three or four related to the air-cooler design, and one dependent only on the design criteria used for the fan.

The one variable that is always independent of the design is elevation, which affects the air density (Fig. 1). Also, air temperature can be an independent variable, depending only on plant location, for a forced-draft fan; but it becomes dependent on the air cooler design if an induced-draft system is selected.

The three other variables that depend on the air cooler are: (1) actual ft³/min that the fan should deliver, (2) pressure drop across the bundle, and (3) fan diameter.

Finally, the variable depending solely on fan design is

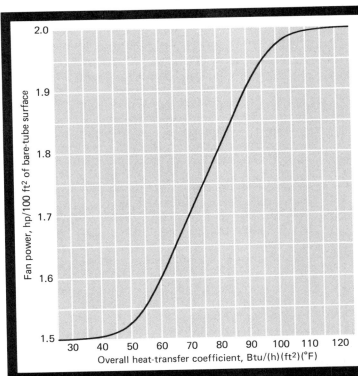

Approximate fan power-requirements for air coolers   Fig. 3

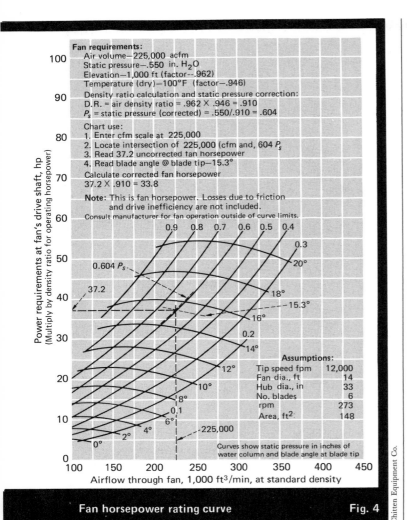

Fan horsepower rating curve   Fig. 4

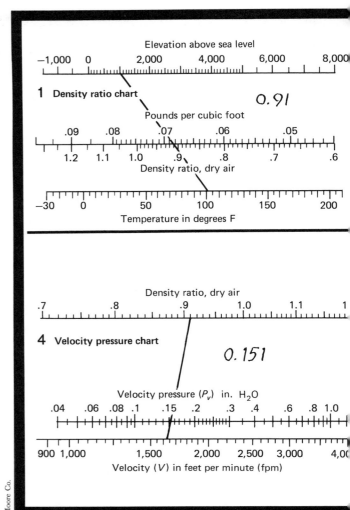

speed. A maximum tip-speed of 12,000 ft/min is usually maintained for fans over 5 or 6 ft dia.; but this tip speed may be reduced to decrease noise.

Induced vs. forced draft (air temperature): An accurate comparison of induced- vs. forced-draft systems requires a completely optimized design for each type, so that a decision between the two is usually made early in the project, based on factors related to the plant site (see p. 116). In general, the induced-draft design offers better air distribution over the bundles, and closer control of the process-fluid outlet temperature; but induced-draft fans are more difficult to service (since they are not accessible from below), their brake horsepower is higher, and the design is slightly more expensive. Also, induced-draft fans should not be used where the outlet air temperature is much above 200°F.

Pressure drop across the bundle: This depends on the bundle design (see p. 117, and Table II, p. 110).

Fan diameter: An air cooler must be sized so that a fan or fans occupy at least 40% of the area served, in order for the air to reach all portions of the bundle. Thus, the bay size should be selected so that a fan or fans of sufficient diameter may be used to get the required coverage.

The latest edition of API 661, "Air-Cooled Heat exchangers for General Refinery Services," recommends a maximum dispersion angle between fan and bundle of 45 deg. (see Fig. 2). It may be necessary to use larger fans to achieve this angle than would be required for 40% coverage.

The fan diameter is also related to the number per bay. Using longer and narrower bays, with two or more fans per bay (see Table II, p. 110), permits operation with one fan while the other is shut down for maintenance. Also, a two-fan bay provides a greater degree of temperature control, assuming that no other means (i.e., louvres, etc.) is used.

Fan rating procedures

An approximate fan horsepower can be obtained from an estimated bare-tube surface (Table II, p. 110) and Fig. 3 above. More precisely, the horsepower is:

$$\text{bhp} = (\text{ACFM})(\text{TP})/(6{,}356)(\text{eff.})$$
$$\text{TP} = \text{VSP} + \text{VP}$$
$$\text{VP} = 1.27(\text{ACFM})/((\text{fan dia.})^2 - (\text{hub dia.})^2)$$

where: TP = discharge pressure of fan, in. of water

          ASP = air static pressure, or pressure loss across the bundle, in. of water

DESIGN OF AIR-COOLED EXCHANGERS 429

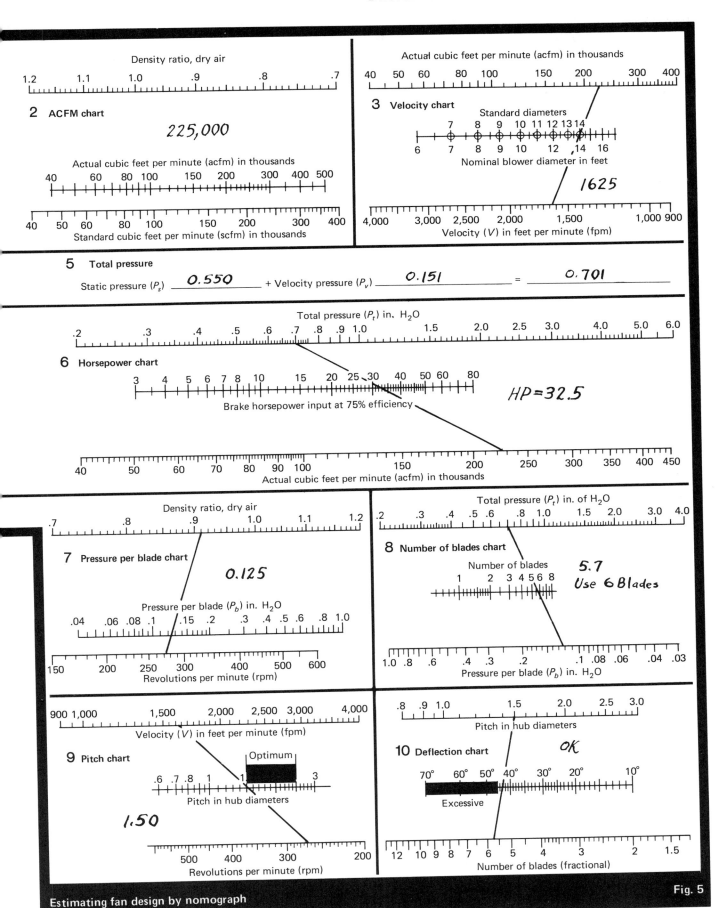

Fig. 5 Estimating fan design by nomograph

vp = the velocity head required to accelerate the air from standstill to the fan's exit velocity, in. of water = $6.2 \times 10^{-8}\, \rho v^2$

ρ = air density, factor, (lb/ft³)/0.075

v = axial air velocity through the fan, ft/min

eff. = efficiency of the fan, fraction

This equation shows that, with other things constant, the fan horsepower decreases as the fan diameter increases, and increases as the pressure drop increases. Otherwise, the horsepower is primarily affected by the efficiency, which can vary widely, depending on loading, speed, etc. Under normal conditions of correct specification and rating, efficiency should range from 0.70 to 0.80, so that a close estimation can be made by assuming 0.75 efficiency, and thus using the formula:

$$\text{bhp} = (\text{ACFM})(\text{TP})/4{,}767$$

When rating fans, when estimating the effects of weather on horsepower, or when modifying a drive, it is helpful to remember that, when other things are constant:

- ACFM is directly proportional to fan speed; pressure varies as the square of the speed; and horsepower varies as the cube of the speed.
- Horsepower is directly proportional to air density.
- Horsepower will vary roughly as the cube of the air velocity across the coils.

Rating procedures

We generally use fans manufactured by either one of two firms. Although we use a computer program for rating both types in actual practice, a comparison of the rating procedures offered by the companies is instructive. Both firms provide approximate methods for rating fans for air coolers. Both offer fans in varying hub diameters and blade-chord widths, so that it is possible to select just the right fan to do the job. However, the procedures for rating these fans can appear to be quite different.

In the first (Fig. 4), a set of curves is used to determine the fan performance for an assumed set of conditions for: diameter, rpm, number of blades, and chord width. If the number of assumed blades is not enough, or if the fan speed is too low, the operating point will be off the curve, and it will be necessary to assume higher values for one or the other, or both.

In the second (Fig. 5), a set of nomograms is used to determine the horsepower and the number of blades necessary for a certain rpm and a given chord width.

The velocity head and fan efficiency are built into the Fig. 4 rating curves, whereas the procedure in Fig. 5 actually calculates the velocity head and assumes a 75% efficiency, which can be corrected later.

To demonstrate these procedures, let us assume that a 14-ft fan rotating at 273 rpm (tip speed equals 12,007 ft/min) is required to deliver 225,000 ACFM of air against a static pressure of 0.550 in. of water at an elevation of 1,000 ft above sea level, with an air temperature of 100°F at the fan.

Using Fig. 4, it is first necessary to correct the actual static pressure by dividing it by the air-density ratio (Fig. 1), thus converting it to "standard conditions" of 70°F and 1.0 atm. and 50% relative humidity, which is the basis for the curves. The horsepower thus determined is then corrected to refer it back to the actual operating conditions at the fan.

Using Fig. 5, the actual horsepower and dispersion are determined directly. Using the nomograms numbered 1 through 10, one first makes the air-density correction (no. 1), then calculates the ACFM (no. 2). The ACFM with the assumed diameter is used to calculate the air velocity (no. 3) and this with the corrected density is used to calculate the velocity head (no. 4), which is added to the given pressure drop to get the total pressure (no. 5); and this, in turn, is used to calculate the horsepower at 75% efficiency (no. 6).

The air-density ratio with the assumed speed is used to calculate the pressure per blade (no. 7); the pressure per blade, with the total pressure, is used to calculate the number of blades (no. 8); the velocity is used with the speed to calculate the pitch (no. 9); and the number of blades, with the pitch, is used to calculate the deflection (no. 10), which can then be compared to the bundle geometry to see if the total tube area is covered (as in Fig. 2).

Either of these procedures can be repeated to obtain an optimum through trial and error, as the air-cooler design is also optimized through a trial and error. It is easy to see that a good set of starting assumptions are important to successful air-cooler design. Such assumptions for the many possible applications can only be derived through much experience.

The author

John Glass is chief rating engineer for the Happy Div. of Therma Technology, Inc. His department is responsible for the thermal design of air-cooled exchangers, air preheating coils and atmospheric sections. Before joining Therma Technology, he was employed by Yuba Heat Transfer Corp., where he was involved in rating air coolers and shell and tube exchangers, and also in developing computer programs for such calculations. A member of the AIChE, he holds a B.S. in chemical engineering from the University of Tulsa.

Operation and Maintenance Of Cooling Towers

Clearly written operating procedures and scheduled inspections keep up the operating efficiency of cooling towers and keep down the cost of maintenance.

A. M. KUEHMSTED, Ecodyne Corp./Fluor Cooling Products Co.

Preserving the operating efficiency of a cooling tower and protecting the capital investment it represents depends on establishing, and following through with, proper operating and maintenance practices.

Before Startup

Before a cooling tower is put onstream, the following items should be thoroughly checked:
- Distribution system clean, and nozzles or orifices properly installed.
- Basins clean and free of debris.
- Suction screen clean and in its proper position over the sump.
- Float valve for basin makeup water operating freely.
- All bolts properly torqued, especially those of the fan assembly and other mechanical equipment; and coupling alignments within manufacturer's tolerances.
- Speed reducer filled to the proper level with clean oil of the correct grade; and the breather vent unobstructed.
- Proper motor lubrication.
- Fan turns freely in both directions and, under normal power, rotates clockwise (looking down on it).
- Electrical connections safe under tower operating conditions.
- Motor and drive units run without excessive noise or heating (first, run the motor and drive units for at least half an hour, then carefully inspect all mechanical equipment).

After each of the above steps has been checked off, the collecting basin can be filled with water and the tower put into operation. To eliminate the possibility of the wood drying out and shrinking, water should be circulated daily from the time the tower is completed.

Originally published May 3, 1971

Cold-Weather Operation

Proper tower maintenance during winter months is the best preparation for summer operation—when temperatures get close to the wet bulb temperature for which the tower was designed.

The best method of operating a cooling tower under cold-weather ambient conditions depends on a number of factors besides the ambient temperature: the value of the cold water; the number of circulating pumps; the cost of fan and pump horsepower; whether motors are single or two-speed; and the effect that changes in ambient conditions may have on heat load or the need for cold water where stream pollution may be involved.

In freezing weather, a cooling tower should never be operated without a heat load. As the air temperature drops, and the need for colder water becomes marginal, fans should cut back to half speed to allow water temperature to be maintained. To prevent damage to the speed reducer, the motor should be always deenergized, before speed direction is changed, by holding down the "STOP" button for at least one minute. A time-delay relay can perform this function automatically.

A fan at half speed delivers half the air at one-eighth the horsepower. It is likely that a tower equipped with two-speed motors can be operated most of the winter with all of its fans at half speed. If the tower has only single-speed motors, fans should first be shut off.

The circulating-water system can often be used to combat icing. If there is more than one circulating pump, a pump can be shut off to increase the temperature rise (range) of the water, as well as to reduce power consumption.

Cells should not be operated at reduced flow-rate. If total throughput is reduced, some cells should

Cooling Tower Troubleshooting Guide

Trouble	Check	Remedy
Water Distribution		
Uneven water distribution	Broken or plugged distributors; plugged distribution piping; broken fill or packing; broken header or lateral piping; gravity distributing pan out of level; excessive or uneven water flow.	Replace or repair defective parts; clean distribution system and pump suction screen; adjust water flow to design conditions.
Louver splashout	Excessive water flow over tower.	**Counterflow:** Reduce water flow; install water-diverter shelf inside tower at top of louver intakes, where required; check longitudinal wind baffle. **Crossflow:** Reduce water flow; check for uneven distribution caused by control-valve adjustment, for plugged nozzles causing heavy water loading adjacent to the louvers, or for overflow of the longitudinal hot-water-basin walls over the louver section.
Excessive drift loss	Water hitting drift eliminators because of excessive water flow; broken or plugged distribution system; broken or missing drift eliminators; fan pitched above design; restricted sections of header or lateral piping; broken header or lateral piping.	Restore water flow to design conditions; replace or clean nozzles; pitch fan to design conditions.
Ladder, handrails, structure and panels loose	Tower dried out; excessive vibration.	Tighten all bolts; keep tower wet; circulate water over tower by using distributing system.
Tower panel leaks	Water stains.	Tighten loose panel bolts; seal leaks with water-resistant caulking.
Speed Reducers		
Noisy gears	Worn bearings or gear set; warped gearing; low oil level; contaminated oil; dust-and-moisture shield rubbing the gear case.	Check oil for level and contamination; adjust oil shield; replace worn bearings, oil seal or gear sets; check tooth contact.
Noisy bearings	Low oil level; contaminated oil; bearing fatigue.	Replace worn bearings and oil shield; check tooth contact; add oil or replace if contaminated.
Excessive movement in pinion shaft	Worn high-speed bearings (apply pressure on pinion and look carefully at oil seals for movement).	Replace worn bearings and oil seals; check tooth contact after replacing gears.
Excessive movement in low-speed shaft	Worn low-speed bearings (lift tip of fan blade up and down, and look for movement at low-speed shaft).	Replace worn bearings and oil seals; check tooth contact of gears after replacing bearings.
Water in oil	Broken or plugged vent line; loose inspection plug, or gear inspection plate; loose or missing cap screws; worn oil seals; gear units out of service for longer than one week; water in oil-storage container.	Inspect gear vent line and gear inspection holes; replace loose or missing cap screws, and all bearings and oil seals if noisy or throwing oil; replace oil; run gear units for 10 min. once a week during a lengthy shutdown; keep oil in air tight container and store in a dry place.
Oil throwing at oil seals	Worn oil seals and bearings; plugged oil vent; high oil level or oil foaming.	Replace oil seals if bearings are quiet and tight; replace worn bearings and oil seals; check oil vent line by blowing through it (oil should rise in oil fill line); do not fill gear unit while unit is running.
Oil leak from fittings, cap screws, gear case, inspection plug or drain plug	Loose or misaligned bearing containers; dirty or distorted adjustment shims; loose or cracked elbow plug or plate; sand hole in casting.	Tighten all cap screws; clean adjustment shims; apply sealing compound on metal-to-metal seals if necessary; renew inspection-plate gasket; tighten cap screws; replace cracked or distorted fittings.
Couplings and Drive Shaft		
Vibration	Misalignment of coupling; foreign matter encrusted on coupling; shaft out of balance, bent or off-center; worn bearings or bent shaft in motor or gear unit.	Realign coupling, and recheck alignment after 30 days; tighten motor and speed-reducer hold-down bolts.
Electrical Drivers		
Noisy motor	Worn bearings.	Replace bearings and grease seals.
Motor shuts down after short runs	Faulty or incorrect motor-starting equipment; overloaded or incorrect line voltage.	Check voltage available to all phases at motor terminal, overload protective devices, and fan pitch angle.
Vibration	Bent motor shaft; unbalanced rotor; worn bearings; loose hold-down bolts.	Tighten hold-down bolts and check coupling arrangement; disconnect motor from coupling if vibration persists, and check for bent motor shaft or unbalanced motor.

be shut off to keep those cells operating well flooded. Cells should be shut down in order from first to last so that water circulates past the inactive cells before discharging over the tower.

When multiple towers are involved, and it is desirable to remove a whole tower from service, the flowrate through the remaining towers may be substantially increased by removing some nozzles in the distribution system. Bypassing the hot-water supply from the hot-water riser to the cold-water basin can add still further flexibility. The bypass should be located as far as possible from the pump pit to allow maximum circulation of hot water in the basin.

If ice has formed around the tower louvers, the

Operation and Maintenance of Cooling Towers

Periodic Maintenance Checklist

	Weekly	Monthly	Semi-annually	Yearly
Water-makeup valve	Inspect			
Cold-water basin	Inspect			
Drift eliminators		Inspect		Clean
Drive shaft	Inspect	Check alignment		
Fan	Inspect		Clean	
Shaft bearings	Inspect	Lubricate		
Fill			Inspect	
Speed reducer	Check oil level	Check oil for contamination, and mounting bolts	Change oil	Check fill line for level piping
Hardware			Inspect	Clean and tighten bolts
Header		Inspect		
Laterals		Inspect		
Motor	Inspect operating temperature			Lubricate
Distribution system				Clean
Casing			Inspect	Clean
Structure				Inspect
Sump screen	Clean			
Distribution basin				Clean

first step in removing it is to shut the fans down one at a time for a brief period. This serves a twofold purpose: With the fan off, the water temperature will rise; and without the flow of air, the warm water will move toward the louver face, flooding the area.

Under more-severe icing conditions, the fans may be cycled in reverse to further force warm water and air on the louvers. The duration and frequency of this practice can be established through experimentation. A typical cycle is 10 to 15 min. every 8 hr.

If the motors are two speed, fan reversal normally should be done at half speed. Under very extreme weather conditions (−20 F. and colder), it may be necessary to run the fans in reverse at full speed to control icing.

Fans should be started up carefully in freezing weather, because ice and snow may collect on them and thus unbalance them. Ice may be removed with steam or hot water.

With counterflow towers, intake openings can also be reduced by adjusting louvers. With crossflow towers, the removal of the water distributors on the louver side of the hot-water basin will increase hot-water flow over the louver area.

Operating procedures should be written to limit the decisions that must be made by the operators. These procedures should be upgraded each year to take into account the tendency for a tower to ice up, or to compensate for changes in operating conditions.

The written procedures should cover the following: the number of circulating pumps to be operated during winter months; the number of cells to be operated with this number of pumps to assure a full distribution pan; the number of fans to be operated for various wet-bulb increments (for example: 0 to 10 F., four fans at half speed, one fan off); a deicing procedure for both mild and heavy icing (for example: 0 to 20 F., each fan off in sequence for 15 min. once each shift; below 0 F., reverse each fan in sequence for 10 min. once each shift.

Shutdown Procedure

During a shutdown, regardless of its length, the fan, gear and motor assembly should be checked visually. The oil in the speed reducer should be inspected for proper level and to ensure that it is free of water. If water is present the oil should be changed immediately to avoid rust.

Each cell that is to be out of operation longer than one week should be started and run once a week for at least 10 min. to reduce the danger of rust forming in the bearings and gears. The distribution system should also be inspected and cleaned, and the tower basin drained and cleaned.

Exposed piping and water basins should naturally be drained before a prolonged shutdown in freezing weather.

Water-Treatment Procedures

There are three areas of concern in water treatment: pH control, inhibitors to prevent corrosion of piping and exchangers, and agents to prevent the growth of microorganisms and to lessen fouling.

The pH, which is controlled by the addition of sulfuric acid or caustic, should be maintained at 6 to 7. A higher pH will result in the removal of the lignin that holds the wood fibers together and so cause the wood to deteriorate.

Corrosion inhibitors are selected on the basis of the chemicals found in the makeup water and the materials of heat-exchange equipment.

Generally, algaecides and fungicides need to be shock injected to maintain the level of chemical agents. The first injection should be at 5 ppm. for a period of 15 to 20 min., followed by a lower rate—to maintain the concentration at ½ to 1 ppm.—for another 2 hr. Usually, this procedure should be repeated two to three times a week. ■

Meet the Author

A. M. Kuehmsted is Manager, Service Dept., of Ecodyne Corp./Fluor Cooling Products Co. (P. O. Box 1267, Santa Rosa, Calif. 95403). Previously, he served in Fluor Cooling Products' marketing and engineering departments. He had also been chief engineer and manager of the Cooling Tower Div. of George Windele Co. and sales manager of the Fan Div. of Moore Fan Co. A licensed mechanical engineer, he received his B.S. in engineering from the University of California at Berkeley.

Proper startup protects cooling-tower systems

These techniques for testing and starting up a cooling-tower system will minimize corrosion and deposits that can cause severe damage to components in the system

Paul R. Puckorius, P. R. Puckorius & Associates

☐ The first weeks of startup of a new recirculating cooling-tower system either can get it off to a long trouble-free life, or doom it to a short life in which heat-transfer problems may crop up shortly after operation has started. The difference depends largely on how you prepare for and carry out the initial startup.

A typical cooling-tower system is tied together by the water circulating through all the different components (Fig. 1). Thus, whatever the water contacts, removes, carries, and deposits, may materially affect one, a few, or many components. Troubles in an individual component may cause severe damage to other components because of the transfer capability of the water. However, the circulating water can also move treatment chemicals to specific areas to prevent corrosion, deposits, algae, etc.

Let us review what frequently happens during the initial startup of a cooling-tower system. After construction of the cooling tower and the associated equipment, the exchangers are pressure-tested and flushed. The system is then filled with water from whatever supply is available at the time—often not the water to be used during operation. Frequently the water remains stagnant in the system until completion of construction, and it may not be treated with chemicals.

Operation is started—usually with little or no load—while the system is checked for leaks, etc., by circulating the water over the cooling tower and through the equipment. The water contacts the cooling-tower lumber, extracts the solubles in the wood, and carries them throughout the entire cooling system. These soluble extracts can deposit in the distribution lines or exchangers, or react with other materials in the water (again dropping out in the equipment). This is also true of any materials present in the lines, exchangers or tower basin, which can be picked up by the water and either dissolved or moved through the entire cooling system. Frequently pieces of wood, paper, plastic—and occasionally objects like flashlights, boots, and gloves, inadvertently left in the system during construction—can be found in the heat exchangers.

There are several critical areas to consider separately during this startup period. They are: (1) cooling-tower wood, (2) protection of heat exchangers, and (3) water-treatment controls.

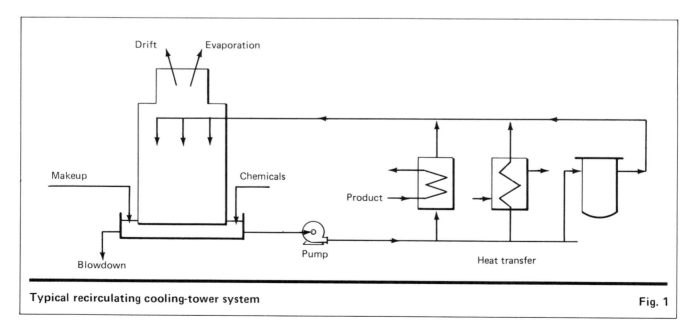

Typical recirculating cooling-tower system   Fig. 1

Originally published January 2, 1978

PROPER STARTUP PROTECTS COOLING-TOWER SYSTEMS 435

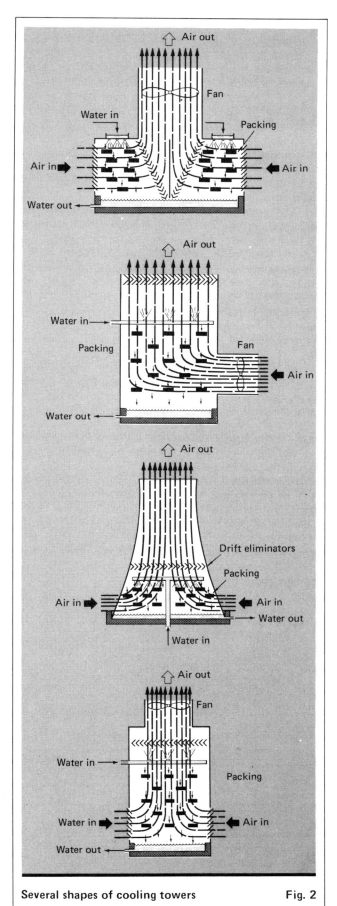

Several shapes of cooling towers   Fig. 2

Woods used for cooling towers	
Untreated	Treated
Redwood, California	Redwood, California
Canadian Red Cedar	Douglas Fir
Ipe (South America)	Canadian Red Cedar
	Baltic Redwood (Europe)
	Ipe & Peroba (South America)

Cooling-tower wood

Often the cooling tower receives less attention during startup than the heat exchangers, perhaps because it does not directly contact the manufactured product. If the tower becomes scaled or covered with algae, it does not affect cooling of the manufactured product nearly as quickly as do problems in the exchangers. Likewise, less attention is given to the tower's materials of construction than to those of other units. Yet the wood and its treatment can effect the critical heat exchangers.

The wooden parts of a cooling tower that are constantly washed by the cooling water include the fill, the structural members and the distribution decks or headers (Fig. 2). Over half of industrial sized towers still use wood fill.

In the U.S., the most commonly used wood, both treated and untreated, is California redwood. But Douglas fir or other locally available wood is often used in the U.S., and in other countries (Table).

With untreated redwood, the initial water circulating over the tower will leach out considerable quantities of organic materials. The liquid is then often termed "black water" since the water actually becomes very dark brown or even black. This condition lasts for several weeks and can cause corrosion to start on metal components either by interfering with water-treatment chemicals or by producing gummy deposits. In addition, the leached organics can support microbiological growth, resulting in slime in the system.

Treated lumber can cause even more-dramatic effects. Treatment chemicals, usually pressure-impregnated in the wood, improve resistance to decay from wood-rot microorganisms. The Cooling Tower Institute's Standard, WMS 112, describes the treating procedures. The most commonly used treatment is referred to as CCA (chromated copper arsenate), though other waterborne salts may be used. In the past, creosote was often employed.

The water initially circulating over the cooling tower will leach these preservatives from the wood. When using wood treated with copper salts, concentrations have been measured (after one week of circulation) as high as 200 ppm of copper. This level, or even much less, can result in extensive copper plating on metal surfaces, resulting in rapid and severe corrosion. Both mild steel and aluminum components are particularly susceptible to the effects of copper plating.

Even creosote-treated lumber will be leached, and the chemical carried by the water to redeposit on metal surfaces. If allowed to remain, these acidic tar-like deposits cause metal corrosion and interfere with corrosion inhibitors.

Special procedures must be employed to minimize the

adverse effects on metal components from the materials leached from cooling-tower lumber. These techniques start with the washing of the tower lumber with water that is not circulated through the rest of the system.

The water is simply circulated over the tower, preferably with the system pumps, by-passing the equipment. Or, auxiliary pumps can be used to circulate from the tower basin to the top of the tower. This should be continued for at least two weeks with several changes of water. The used water, when discarded, should go through proper waste-treatment procedures to prevent pollution.

Certain chemicals (mercaptobenzotriazoles and thiazoles) can be used with the initial water and should be continued during the first month of regular operation, to prevent copper plating. Selection of the proper chemicals depends upon the regular water-treatment program, the system metallurgy, the wood treatment, and levels of leached salts in the water. Control of pH (between 6 and 7) is also very beneficial during the initial leaching stage.

Heat-exchanger protection

In addition to leaching problems, heat-exchanger equipment in a recirculating cooling-tower system is most sensitive to corrosion and buildup of deposits, especially if fabricated using mild steel.

Often, such units as shell-and-tube exchangers, jacketed vessels, and barometric condensers do not receive sufficient protection initially, resulting in rapid corrosion and fouling and early need for cleaning or replacement. Corrosion protection depends not only on starting with a clean surface but on the use of sufficient inhibitor to rapidly and effectively protect the metal surfaces.

In a new cooling-tower system, hydrotesting is often done with untreated or poorly treated water. Often the exchangers then remain stagnant long enough to allow corrosion to start. Heat-exchanger fabrication and assembly results in an accumulation of lubricants, rust, and mill scale. This must be removed, and the clean metal protected before corrosion starts. (This protection must also be compatible with the inhibitor program to be used after startup.)

The common method of starting up cooling-water systems is to water-flush the exchangers. This removes only some of the foreign material. Then a corrosion inhibitor is applied to the system water at the normal level (or several times that level) to establish protection. Even on clean surfaces these levels may not be sufficient to establish good protection; on dirty surfaces, severe localized and pitting attack can result. Leaking tubes and restricted heat transfer may make maintenance necessary only a few weeks after startup [1].

To illustrate the corrosion that occurs, laboratory tests have been performed both on mild-steel heat-transfer tubes and corrosion coupons [2] using high levels of both chromate-based and polyphosphate-based inhibitors that are considered very effective. Corrosion and tuberculation were essentially controlled at high levels but not at the maintenance levels.

To properly start up a recirculating cooling-tower system and to prevent this type of deposit and attack, a chemical conditioning technique should be used that involves two steps: (1) the thorough cleaning of metal surfaces, and (2) the development of a passive film on the metal surfaces. A proposed NACE (National Assn. of Corrosion Engineers) Recommended Practice has been prepared and is in the final stages of approval. In essence, it recommends the use of an alkaline polyphosphate cleaner followed by either a chromate- or polyphosphate-based inhibitor to establish protection [3]. Treatments, dosages, time required, and temperatures depend on the metallurgy of the system and inhibitor to be used later. The results of laboratory testing shows that protection is excellent when this procedure is followed on mild-steel specimens. The initial conditioning product is a polyphosphate-based inhibitor.

Plants have shown similar protection with mild-steel heat exchangers. Some have changed from admiralty metal to mild-steel tubes because of the excellent protection and lower costs possible through initial conditioning. This technique will provide substantially longer life and better heat transfer not only when starting up new cooling-tower systems but also to replacement bundles, new sections to existing systems, and to freshly cleaned equipment.

Water-treatment controls

If the following controls are applied to the startup of a new cooling-tower system, they will go a long way toward maintaining the best protection possible with the treatment program utilized:

Excellent pH control.
Maintaining inhibitors at 2 to 3 times normal levels.
Adding biocides even if not apparently needed.
Feeding chemicals continuously.
Maintaining cycles of concentration.
Maintaining good flows through equipment.
Monitoring corrosion rates.
Preventing copper plating from tower chemicals.
Initially conditioning the entire system.
Taking immediate action with product contamination.

References

1. Comeaux, R. C., Operating Problems in Petroleum Plant Cooling Systems, Presented at National Assn. of Corrosion Engineers Annual Meeting, Chicago, Ill., 1974.

2. Puckorius, P. R. and Ryzner, W. J., Cooling Water Inhibitor Performance: Film Formation vs. Film Maintenance, Presented at National Assn. of Corrosion Engineers Annual Meeting, Dallas, Tex., 1960.

3. Initial Conditioning of Cooling Water Equipment, Proposed Standard Recommended Practice of National Assn. of Corrosion Engineers, Task Group T-7A5, P. R. Puckorius, Chairman.

The author

Paul R. Puckorius is president and owner of P. R. Puckorius & Associates, Box 4846, Cleveland, OH 44126, which is a consulting firm specializing in cooling water, wastewater and potable water for all kinds of industries. He has 25 years of experience with cooling-water corrosion, scale and microbiological problems, and all types of cooling towers. He holds a B.A. degree, with a major in chemistry, from North Central College, Naperville, Ill., has done graduate work in chemistry at Northwestern University, and has taken numerous management courses. He is the principal or co-inventor of five patents on products or processes, and has made presentations at many technical societies.

Section VIII
HEAT TRANSFER CALCULATIONS

Calculate enthalpy with a pocket calculator
Calculating radiant heat transfer
Charts simplify spiral finned-tube calculations
Estimating liquid heat capacities—Part I
Estimating liquid heat capacities—Part II
New correlation for thermal conductivity
Quick estimation for gas heat-transfer coefficients
Relating heat-exchanger fouling factors to coefficients of conductivity
Relating heat emission to surface temperature
Steam from flashing condensate

Calculate enthalpy with a pocket calculator

Programs presented for two calculators develop a data-base library on component cards. With the library, a polynomial expression is synthesized for calculating multicomponent enthalpies from temperature, or vice versa.

Raymond T. Schneider, Pridgen Engineering Co.

☐ These programs will develop a data-base library of component cards from polynomial heat-capacity data in the general form $C_p = a + bT + cT^2 + dT^3 + e/T^2$, and the standard heat of formation of the compound. Programs are presented for the Hewlett-Packard HP 67/97 and HP 65.

The data-base library is then used, under simple program control, to synthesize a polynomial expression of enthalpy vs. temperature for any multicomponent mixture of any proportions in which the physical heat of mixing is negligible.

Additional programs for each calculator solve the resulting polynomial expressions for either temperature or enthalpy, given one or the other. The programs that are presented apply to gases, liquids, or solids, or combinations of any of these, as long as the heat of mixing and the effect of pressure are negligible in the specific application.

Preparation of component cards, Program HP 67/97-1 Table I

Step no.	Key entry	HP-67 key code	HP 97 key code	Comments	Step no.	Key entry	HP-67 key code	HP 97 key code	Comments
001	*LBL a	32 25 11	21 16 11		033	.	83	-62	
002	STO 0	33 00	35 00	MW_i	034	1	01	01	
003	P⇄S	31 42	16-51		035	5	05	05	
004	CLRG	31 43	16-53		036	STO 7	33 07	33 07	
005	P⇄S	31 42	16-51		037	RCL 2	34 02	36 02	
006	RTN	35 22	24		038	×	71	-35	
007	*LBL A	31 25 11	21 11		039	RCL 7	34 07	36 07	
008	STO 2	33 02	35 02		040	x^2	32 54	53	
009	X⇄Y	35 52	-41		041	RCL 3	34 03	36 03	
010	STO 1	33 01	35 01		042	×	71	-35	
011	RTN	35 22	24		043	+	61	-55	I_i, Eq. (5)
012	*LBL B	31 25 12	21 12		044	RCL 7	34 07	36 07	
013	2	02	02		045	x^2	32 54	53	
014	÷	81	-24	$B_i = b_i/2$	046	RCL 7	34 07	36 07	
015	STO 3	33 03	35 03		047	×	71	-35	
016	RTN	35 22	24		048	RCL 4	34 04	36 04	
017	*LBL C	31 25 13	21 13		049	×	71	-35	
018	3	03	03		050	+	61	-55	
019	÷	81	-24	$C_i = c_i/3$	051	RCL 7	34 07	36 07	
020	STO 4	33 04	35 04		052	x^2	32 54	53	
021	RTN	35 22	24		053	x^2	32 54	53	
022	*LBL D	31 25 14	21 14		054	RCL 5	34 05	36 05	
023	4	04	04		055	×	71	-35	
024	÷	81	-24	$D_i = d_i/4$	056	+	61	-55	
025	STO 5	33 05	35 05		057	RCL 6	34 06	36 06	
026	RTN	35 22	24		058	RCL 7	34 07	36 07	
027	*LBL E	31 25 15	21 15		059	÷	81	-24	
028	CHS	42	-22	$E_i = -e_i$	060	+	61	-55	
029	STO 6	33 06	35 06		061	ST-1	33 51 01	35-45 01	
030	2	02	02		062	WDTA	31 41	16-61	
031	9	09	09		063	RTN	35 22	24	
032	8	08	08						

Originally published May 23, 1977

440 HEAT TRANSFER CALCULATIONS

Enthalpy computation program, Program HP 67/97-2 Table II

Step no.	Key entry	HP-67 key code	HP 97 key code	Comments	Step no.	Key entry	HP-67 key code	HP 97 key code	Comments
001	*LBL 9	31 25 09	21 09		066	X^2	32 54	53	
002	CLRG	31 43	16-53		067	RCL 3	34 03	36 03	
003	2	02	02		068	×	71	-35	
004	9	09	09		069	+	61	-55	$H = f(T)$
005	8	08	08		070	RCL i	34 24	36 45	
006	.	83	-62		071	X^2	32 54	53	
007	1	01	01		072	RCL i	34 24	36 45	
008	5	05	05		073	×	71	-35	
009	STO 7	33 07	35 07		074	RCL 4	34 04	36 04	
010	2	02	02		075	×	71	-35	
011	7	07	07		076	+	61	-55	
012	3	03	03		077	RCL i	34 24	36 45	
013	.	83	-62	Initialization routine	078	X^2	32 54	53	
014	1	01	01		079	X^2	32 54	53	
015	5	05	05		080	RCL 5	34 05	36 05	
016	STO B	33 12	33 12		081	×	71	-35	
017	4	04	04		082	+	61	-55	
018	5	05	05		083	RCL 6	34 06	36 06	
019	9	09	09		084	RCL i	34 24	36 45	
020	.	83	-62		085	÷	81	-24	
021	6	06	06		086	+	61	-55	
022	7	07	07		087	RTN	35 22	24	
023	STO C	33 13	35 13		088	*LBL 2	31 25 02	21 02	
024	1	01	01		089	RCL B	34 12	36 12	
025	.	83	-62		090	+	61	-55	
026	8	08	08		091	GTO 3	22 03	22 03	
027	STO D	33 14	35 14		092	*LBL B	31 25 12	21 12	
028	P⇄S	31 42	16-51		093	P⇄S	31 42	16-51	
029	RTN	35 22	24		094	F1?	35 71 01	16 23 01	
030	*LBL e	32 25 15	21 16 15	Error routine	095	GTO 2	22 02	22 02	
031	0	00	00		096	RCL C	34 13	36 13	
032	1/X	35 62	52		097	+	61	-55	
033	RTN	35 22	24		098	RCL D	34 14	36 14	
034	*LBL A	31 25 11	21 11		099	÷	81	-24	
035	STO A	33 11	35 11		100	*LBL 3	31 25 03	21 03	Calculate H from T
036	CF 3	35 61 03	16 22 03		101	STO 0	33 00	35 00	
037	6	06	06		102	0	00	00	
038	STO I	35 33	35 46		103	STO I	35 33	35 46	
039	RCL A	34 11	36 11		104	GSB 4	31 22 04	23 04	
040	MRG	32 41	16-62		105	STO 8	33 08	35 08	
041	PSE	35 72	16 51		106	F1?	35 71 01	16 23 01	
042	F3?	35 71 03	16 23 03		107	GTO 3	22 03	22 03	
043	GTO 0	22 00	22 00		108	RCL D	34 14	36 14	
044	GTO e	22 31 15	22 16 15		109	×	71	-35	
045	*LBL 0	31 25 00	21 00	Input components, molar quantities	110	*LBL 3	31 25 03	21 03	
046	ST × (i)	33 71 24	35-35 45		111	P⇄S	31 42	16-51	
047	DSZI	31 33	16 25 46		112	RTN	35 22	24	
048	GTO 0	22 00	22 00		113	*LBL C	31 25 13	21 13	
049	6	06	06		114	P⇄S	31 42	16-51	
050	STO I	35 33	35 46		115	F1?	35 71 01	16 23 01	
051	*LBL 1	31 25 01	21 01		116	GTO 3	22 03	22 03	
052	RCL i	34 24	36 45		117	RCL D	34 14	36 14	
053	P⇄S	31 42	16-51		118	÷	81	-24	
054	ST + i	33 61 24	35-55 45		119	*LBL 3	31 25 03	21 03	
055	P⇄S	31 42	16-51		120	STO 8	33 08	35 08	
056	DSZI	31 33	16 25 46		121	7	07	07	
057	GTO 1	22 01	22 01		122	STO I	35 33	35 46	
058	RTN	33 22	24		123	GSB 4	31 22 04	23 04	
059	*LBL 4	31 25 04	21 04		124	STO E	33 15	35 15	
060	RCL 1	34 01	36 01		125	RCL C	34 13	36 13	
061	RCL 2	34 02	36 02		126	*LBL 5	31 25 05	21 05	
062	RCL i	34 24	36 45		127	STO 0	33 00	35 00	
063	×	71	-35		128	0	00	00	
064	+	61	-55		129	STO I	35 33	35 46	
065	RCL i	34 24	36 45		130	GSB 4	31 22 04	23 04	

Table II (continued)

Step no.	Key entry	HP 67 key code	HP 97 key code	Comments
131	STO 9	33 09	35 09	
132	RCL 8	34 08	36 08	
133	−	51	-45	
134	ABS	35 64	16 31	
135	RCL 8	34 08	36 08	
136	EEX	43	-23	
137	4	04	04	
138	CHS	42	-22	Calculate T from H
139	×	71	-35	
140	ABS	35 64	16 31	
141	X>Y?	32 81	16-34	
142	GTO 3	22 03	22 03	
143	RCL 8	34 08	36 08	
144	RCL E	34 15	36 15	
145	−	51	-45	
146	RCL 9	34 09	36 09	
147	RCL E	34 15	36 15	
148	−	51	-45	
149	÷	81	-24	
150	RCL 0	34 00	36 00	
151	RCL 7	34 07	36 07	
152	−	51	-45	
153	×	71	-35	
154	RCL 7	34 07	36 07	
155	+	61	-55	
156	GTO 5	22 05	22 05	
157	*LBL 3	31 25 03	21 03	
158	F1?	35 71 01	16 23 01	
159	GTO 6	22 06	22 06	
160	RCL 0	34 00	36 00	
161	RCL D	34 14	36 14	
162	×	71	-35	
163	RCL C	34 13	36 13	
164	−	51	-45	
165	GTO 3	22 03	22 03	
166	*LBL 6	31 25 06	21 06	
167	RCL 0	34 00	36 00	
168	RCL B	34 12	36 12	
169	−	51	-45	
170	*LBL 3	31 25 03	21 03	
171	P⇄S	31 42	16-51	
172	RTN	35 22	24	
173	*LBL b	32 25 12	21 16 12	Select metric and initialize
174	SF1	35 51 01	16 21 01	
175	GTO 9	22 09	22 09	
176	*LBL D	31 25 14	21 14	
177	P⇄S	31 42	16-51	
178	RCL 8	34 08	36 08	
179	F1?	35 71 01	16 23 01	Recall last H and convert to proper units
180	GTO 3	22 03	22 03	
181	RCL D	34 14	36 14	
182	×	71	-35	
183	*LBL 3	31 25 03	21 03	
184	P⇄S	31 42	16-51	
185	RTN	35 22	24	
186	*LBL c	32 25 13	21 16 13	Select English and initialize
187	CF1	35 61 01	16 22 01	
188	GTO 9	22 09	22 09	
189	*LBL a	32 25 11	21 16 11	
190	STO A	33 11	35 11	
191	CF 3	35 61 03	16 22 03	
192	6	06	06	
193	STO I	35 33	35 46	
194	RCL A	34 11	36 11	
195	MRG	32 41	16-62	Input components, weight quantities
196	PSE	35 72	16 51	
197	F3?	35 71 03	16 23 03	
198	GTO 7	22 07	22 07	
199	GTO e	22 31 15	22 16 15	
200	*LBL 7	31 25 07	21 07	
201	RCL 0	34 00	36 00	
202	÷	81	-24	
203	GTO 0	22 00	22 00	

Basis of the programs

The enthalpy of any multicomponent mixture for which the physical heat of mixing is negligible can be expressed as:

$$H = \Sigma(x_i H_i^o) \quad (1)$$

Or, for total enthalpy:

$$nH = \Sigma(n_i H_i^o) \quad (2)$$

If the elements in their standard states at temperature $t_o = 298.15$ K are chosen as the datum:

$$H_i^o = \Delta H_{f_i}^o + \int_{t_o}^{T} C_{p_i} dT \quad (3)$$

$$C_{p_i} + a_i + b_i T + c_i T^2 + d_i T^3 + e_i/T^2 \quad (4)$$

Combining Eq. (3) and (4), and rearranging, yields:

$$H_i^o = \left(\Delta H_{f_i}^o - a_i t_o - \frac{b_i}{2} t_o^2 - \frac{c_i}{3} t_o^3 - \frac{d_i}{4} t_o^4 + \frac{e_i}{t_o}\right) +$$
$$a_i T + \frac{b_i}{2} T^2 + \frac{c_i}{3} T^3 + \frac{d_i}{4} T^4 - \frac{e_i}{T}$$

Or:

$$H_i^o = I_i + A_i T + B_i T^2 + C_i T^3 + D_i T^4 + E_i/T \quad (5)$$

In Eq. (5),

$$I_i = \Delta H_{f_i}^o - a_i t_o - \frac{b_i}{2} t_o^2 - \frac{c_i}{3} t_o^3 -$$
$$\frac{d_i}{4} t_o^4 + \frac{e_i}{t_o}; A_i = a_i; B_i = b_i/2;$$
$$C_i = c_i/3; D_i = d_i/4; \text{ and } E_i = -e_i.$$

Combining Eq. (5) and (1) yields:

$$H = I + AT + BT^2 + CT^3 + DT^4 + E/T \quad (6)$$

In Eq. (6),

$$I = \Sigma(x_i I_i); A = \Sigma(x_i A_i); B = \Sigma(x_i B_i); \text{ etc.}$$

Eq. (6) is solved for T by means of an iterative convergence routine that makes use of the fact that mean molar heat capacity does not vary a great deal over large temperature intervals.

$$\frac{H_{T_k} - H_{298.15}}{T_k - 298.15} \approx \frac{H - H_{298.15}}{T - 298.15} \quad (7)$$

Calculation of constants, Program HP 65-1 — Table III

Key entry	Key code	Comments	Key entry	Key code	Comments
LBL	23		RCL 3	34 03	
A	11		×	71	
STO 2	33 02		+	61	
X ≤ Y	35 07		RCL 7	34 07	
STO 1	33 01		ENTER	41	
RTN	24		ENTER	41	I_j, Eq. (5)
LBL	23		×	71	
B	12		×	71	
2	02	$B_i = b_i/2$	RCL 4	34 04	
÷	81		×	71	
STO 3	33 03		+	61	
RTN	24		RCL 7	34 07	
LBL	23		ENTER	41	
C	13		×	71	
3	03	$C_i = c_i/3$	ENTER	41	
÷	81		×	71	
STO 4	33 04		RCL 5	34 05	
RTN	24		×	71	
LBL	23		+	61	
D	14		RCL 6	34 06	
4	04	$D_i = d_i/4$	RCL 7	34 07	
÷	81		÷	81	
STO 5	33 05		+	61	
RTN	24		STO	33	
LBL	23		−	51	
E	15	$E_i = -e_i$	1	01	
CHS	42		RTN	24	
STO 6	33 06				
2	02				
9	09				
8	08				
.	83				
1	01				
5	05				
STO 7	33 07				
RCL 2	34 02				
×	71				
RCL 7	34 07				
ENTER	41				
×	71				

Master data card, Program HP 65-2 — Table IV

Key entry	Key code	Comments	Key entry	Key code	Comments
LBL	23		+	61	
A	11		5	05	
ENTER	41		CLX	44	
ENTER	41		NOP	35 01	Space for E_i
ENTER	41		×	71	
NOP	35 01	Space for I_i	STO	33	
×	71		+	61	
STO	33		6	06	
+	61		RTN	24	
1	01		LBL	23	
CLX	44		B	12	
NOP	35 01	Space for A_i	NOP	35 01	Space for MW_i
×	71		÷	81	
STO	33		A	11	
+	61				
2	02				
CLX	44				
NOP	35 01	Space for B_i			
×	71				
STO	33				
+	61				
3	03				
CLX	44				
NOP	35 01	Space for C_i			
×	71				
STO	33				
+	61				
4	04				
CLX	44				
NOP	35 01	Space for D_i			
×	71				
STO	33				

Nomenclature

- C_p Molar heat capacity
- H^o Enthalpy of the standard state
- ΔH_i^o Standard heat of formation
- n Number of moles
- T Absolute temperature, °K
- x Mole fraction
- x_w Weight fraction
- MW Molecular weight
- t_o 298.15 K

Subscripts

- i Component i of a mixture
- k Trial k in an iterative routine

$$T_{k+1} = \left(\frac{H - H_{298.15}}{H_{T_k} - H_{298.15}}\right)(T_k - 298.15) + 298.15 \quad (8)$$

Eq. (8) is the basis of the iterative routine for the program of each calculator.

When reaction heats are computed with the programs, the heats of formation must be consistent. Data in the Bureau of Standards Technical Notes 270-3 through 270-7 can be used for computations involving inorganic compounds and the simple organic compounds listed. For hydrocarbon calculations, the data of API Project 44 can be used. Heats of formation must be expressed in cal/g-mol.

For heat capacity data, it is only necessary that the equations be valid over similar temperature ranges, and be expressed in molar heat capacities. An excellent source of these data is the 24-part series "Physical and Thermodynamic Properties," by. C. L. Yaws [1]. Other sources of heat capacity data are K. A. Kobe et al. [2], T. P. Thinh et al. [3], and the U.S. Bureau of Mines [7].

If reliable heat of formation datum does not exist for a particular compound, a zero may be substituted for its heat of formation, with the restriction that only sensible

Enthalpy computation program, Program HP 65-3 — Table V

Key entry	Key code	Comments	Key entry	Key code	Comments
LBL	23		LSTX	35 00	
A	11	Enter T or	−	51	
STO 7	33 07	T (guess)	X≤Y	35 07	
RTN	24		÷	81	
LBL	23		RCL	34	
B	12	Enter	9	09	
STO 8	33 08	known H	×	71	
RTN	24		RCL 7	34 07	
LBL	23		+	61	
C	13		STO 7	34 07	
E	15	Calculate H from T	GTO	22	
STO 8	33 08		D	14	
RTN	24		LBL	23	
LBL	23		E	15	
D	14		RCL 7	34 07	
E	15		RCL 2	34 02	
ENTER	41		×	71	
ENTER	41		RCL 7	34 07	
RCL 8	34 08		ENTER	41	
−	51		×	71	
g	35		RCL 3	34 03	
ABS	06		×	71	
RCL 8	34 08		+	61	
EEX	43		RCL 7	34 07	
CHS	42		3	03	
4	04		g	35	
×	71		y^x	05	
g	35		RCL 4	34 04	
ABS	06		×	71	
X>Y	35 24		+	61	$H = f(T)$
GTO	22		RCL 7	34 07	
2	02		4	04	
R↓	35 08		g	35	
R↓	35 08		y^x	05	
RCL 7	34 07		RCL 5	34 05	
2	02	Calculate T	×	71	
9	09	from H	+	61	
8	08		RCL 6	34 06	
.	83		RCL 7	34 07	
1	01		÷	81	
5	05		+	61	
STO 7	33 07		RCL 1	34 01	
−	51		+	61	
STO	33		RTN	24	
9	09		LBL	23	
CLX	44		2	02	
E	15		RCL 7	34 07	
−	51		RTN	24	
RCL 8	34 08				

Gas composition of an ammonia oxidation converter — Table VI

Component	Inlet gas, lb-mol/h	Outlet gas, lb-mol/h
NH_3	482.15	
N_2	3,254.19	3,260.21
O_2	862.43	265.77
H_2O	82.33	805.56
NO		470.10

heat computations may be made with mixtures involving it.

As was noted, the method requires that the heat of mixing (e.g., heat of solution) be negligible for any calculations made with the programs. Latent heats for phase changes, and heats of solution that are not negligible, can be handled by incorporating the latent heat for the phase change, or the heat of solution, with the heat of formation, and treating each phase, or the solution, as a separate component. This, of course, requires polynomial, molar heat-capacity data for the phase or solution in question.

Programs for the HP 67/97

The first of the two programs for the HP 67/97 generates the data-base library of component cards (Table I). The constants I_i and A_i through E_i of Eq. (5) are calculated from heat of formation data and the coefficients of the polynomial heat capacity data. These constants, along with the molecular weight of the compound, are then stored on magnetic cards for later use by the second programs. The ability of these two calculators to magnetically record the data in storage registers is convenient for this purpose.

The second program generates the constants of Eq. (6) from the constants previously stored on the component cards, and an appropriate x_i or n_i for the mixture in question (Table II). For any given mixture, the input data may be in either weight fractions of each component, mole fractions of each component, total weight of each component, or total moles of each component. The enthalpy will then be consistent with the input, that is, enthalpy per unit mass of mixture, enthalpy per mole of

Data for first example data cards [1, 4] — Table VII

Component	MW	ΔH_f° (25°C)	a	$b \times 10^3$	$c \times 10^6$	$d \times 10^9$	e
NH_3 (gas)	17.03	−11,020	6.07	8.23	−0.16	−0.66	0
N_2 (gas)	28.01	0	7.07	−1.32	3.31	−1.26	0
O_2 (gas)	32.00	0	6.22	2.71	−0.37	−0.22	0
H_2 (gas)	18.02	−57,796	8.10	−0.72	3.63	−1.16	0
NO (gas)	30.01	21,570	6.93	−0.065	2.23	−0.98	0
H_2O (liquid)	18.02	−68,315	12.147	50.907	−150.845	154.990	0

444 HEAT TRANSFER CALCULATIONS

User instructions — calculation of constants, Program HP 65-1   Table VIII

Step	Instructions	Input data units	Keys	Output data units
1	Enter standard heat of formation, at 25 °C, cal/g-mol, $\Delta H_f°$		Enter	
2	Enter coefficient a of heat-capacity equation	*	A	
3	Enter coefficient b of heat-capacity equation	*	B	
4	Enter coefficient c of heat-capacity equation	*	C	
5	Enter coefficient d of heat-capacity equation	*	D	
6	Enter coefficient e of heat-capacity equation	*	E	
	*Note: These coefficients must be for the molar heat capacity in cal/(g-mol)(°K). If any coefficient is zero, a zero must be entered for it.			
7	Manually record the constants of Eq. (5) for use in preparing the Component card.		RCL 1	I_i
			RCL 2	A_i
			RCL 3	B_i
			RCL 4	C_i
			RCL 5	D_i
			RCL 6	E_i

User instructions — preparation of data cards, Program HP 65-2   Table IX

Step	Instructions	Input data units	Keys	Output data units
1	Enter program HP 65-2 and switch to W/PRGM mode			
2	Single-step until no operation code is displayed		SST	35 01
3	Delete NOP code and ENTER code will be displayed		g DEL	41
4	Key in constant I_i			
5	Single-step until next NOP code is displayed		SST	35 01
6	Delete NOP code and CLX code will be displayed		g DEL	44
7	Key in constant A_i			
8	Repeat Steps 5, 6 and 7 with constants B_i through E_i respectively			
9	Single-step until next NOP code is displayed		SST	35 01
10	Delete NOP code and B code is displayed		g DEL	12
11	Key in molecular weight for component			
12	Record program on blank card for future permanent use			

mixture, or total enthalpy. In addition, the program may operate in either the English or metric units for any given mixture. The program operates internally with calories, gram-moles, and °K, with the appropriate conversions being made by the program.

Programs for the HP 65

Three programs are required for the HP 65 because it cannot record data from the data registers directly onto the magnetic cards, and the program memory is limited to 100 steps. The first program for the HP 65 performs the same function as the first one for the HP 67/97; that is, the constants I_i and A_i through E_i of Eq. (5) are calculated from the heat of formation data and the coefficients of the polynomial expression for molar heat capacity (Table III). As before, the heat of formation must be expressed in calories per mole.

The second program of the set is the master component card, which contains the coding to generate the constants of Eq. (6) (Table IV). Component cards are prepared by reading this master component card into the program memory, manually keying the constants (I_i, A_i through E_i, MW_i) for a specific component into the program memory at the appropriate locations, and recording the resultant program onto a blank card. Each component card, therefore, contains the specific constants applicable to the compound, as well as the program for generating the consants of Eq. (6).

The third program solves Eq. (6) for either enthalpy or temperature, given one or the other (Table V).

The component cards require as input, for any given mixture, either the mole fraction, weight fraction, total moles, or total weight of the component in the mixture. This choice must, of course, be consistent for all the components of the mixture. The enthalpy will, as with the HP 67/97, be consistent with the input. Because of its limited program memory, the HP 65's programs are restricted to the metric system, that is, to calories, gram-moles, and °K.

Problems illustrate the programs

Data cards for the examples were prepared from the data in Table VII, according to the user instructions in Tables VIII and IX for the HP 65, and Table X for the HP67/97. Note that these component cards then become a part of the permanent library, and need not be prepared again for other problems involving these components.

Example 1—The composition of the gas at the inlet and outlet of an ammonia oxidation converter is given in Table VI. If the inlet gas is at 500°F, what is the temperature at the outlet, assuming conditions are adiabatic, and the effect of pressure is negligible? How much heat must be removed to cool the outlet gas to 500°F?

The composition in Table VI indicates that some of the ammonia is lost to side reactions and forms nitrogen and water vapor, with the main product of reaction being nitric oxide and water vapor. At the pressure that nitric acid plants are normally operated, the effect of pressure on enthalpy is approximately the same on the inlet gas as on the outlet gas, and for design purposes, the effect of pressure on the temperature rise can be neglected.

With these programs, the datum for enthalpy of the inlet gas is the same as that for the outlet gas: namely, the elements in the standard state at 25°C; therefore, the ΔH of the several reactions are included, and need not be computed separately. With the assumption of adiabatic conditions, the enthalpy of the outlet gas is the same as that of the inlet gas.

The procedure to solve this problem with the HP 67/97 is as follows:

1. Read program HP 67/97-2 into the calculator.

2. Press keys f and c to select English dimensions and initialize the routine.

3. Enter the lb-mol/h of each component of the inlet gas, press A, and read the component card during the pause. After this has been done for all the components, the constants of Eq. (6)—which relate the enthalpy flowrate of the inlet gas stream to temperature—are now stored in the calculator.

4. Enter the temperature 500, and press B. The calculator now displays the enthalpy flowrate of the inlet-gas stream: -3.619×10^6 Btu/h.

Table X. User instructions—preparation of component cards for HP 67/97, Program HP 67/97-1

Step	Instructions	Input data units	Keys	Output data units
1	Enter molecular weight		f / a	
2	Enter standard heat of formation, ΔH_f°, at 25°C, cal/g-mol		Enter	
3	Enter coefficient a of heat-capacity equation	*	A	
4	Enter coefficient b of heat-capacity equation	*	B	
5	Enter coefficient c of heat-capacity equation	*	C	
6	Enter coefficient d of heat-capacity equation	*	D	
7	Enter coefficient e of heat-capacity equation	*	E	Crd
8	Record data on data card			

*Note: These coefficients must be for the molar heat capacity in cal/(g-mol)(°K). If any coefficient is zero, a zero must be entered for it.

Table XI. User instructions—enthalpy computation program for HP 67/97, Program HP 67/97-2

Step	Instructions	Input data units	Keys	Output data units
1	Initialize and select metric dimensions (g or g-mol, cal, °C).		f / b	
	OR			
	Initialize and select English dimensions (lb or lb-mol, Btu, °F).		f / c	
2	Enter weight or weight-fraction data for component i of the mixture.	W_i or x_{w_i}	f / a	
	OR			
	Enter mole or mole-fraction quantity for component i.	n_i or x_i	A	
3	Read data card for component i during pause. (If the card is not read properly, or is not read during the pause, "error" will be displayed. This condition does not affect the data previously entered for other components. It is only necessary to clear the error, reenter the data for that component, and read the data card during the pause.)			
4	Repeat Steps 2 and 3 for all components of the mixture.			
5	Enter temperature and calculate enthalpy. (Temperature selection must be consistent with Step 1 above.)	°C or °F	B	H
	OR			
	Enter enthalpy and calculate temperature. (The input data for enthalpy must be consistent with the selections made under Steps 1 and 2 above.)		C	t
6	Optional: Recall the last H in appropriate dimensions.		D	Last H

*Note: Choice of W_i, x_{w_i}, n_i, or x_i must be consistent for all components of a mixture.

User instructions—enthalpy computation program for HP 65, Program HP 65-3   Table XII

Step	Instructions	Input data units	Keys	Output data units
1	Initialize by clearing registers.		f REG	
2	Read card for component i.			
3	Enter gram-mole or mole-fraction quantity for component i.	N_i or X_i	A	
	OR			
	Enter weight (grams) or weight-fraction quantity for component i.	W_i or X_{W_i}	B	
4	Repeat Steps 2 and 3 for all components of the mixture.			
	Note: Choice of N_i, X_i, W_i, or X_{W_i} must be consistent for all components.			
5	Read program HP 65-3.			
6	Enter T or first guess of T.	°K		
7	Calculate H from known T		C	H (cal)
	OR			
	Calculate T from known H.	H (cal)	B D	T (°K)

5. At this point, two methods of setting up the equation for the outlet-gas stream are possible. One is to initialize the routine and enter the components of the outlet stream (as was done with the inlet stream in Steps 1 and 2). The second method, followed here, is to adjust the constants of the equation already stored in the calculator by quantities representing changes in those components. The second method is especially useful for mixtures containing a large number of components when only one or two of the components change. When applying the second method, the flowrate of the component in the outlet stream is entered, and the flowrate of that same component in the inlet stream is subtracted.

Then press key A and read the appropriate component card during the pause. After this has been done for all the components of the mixture, the constants of Eq. (6)—which represent the enthalpy flowrate of the outlet stream vs. temperature—are now stored in the calculator.

6. Press key D and the previously computed enthalpy, -3.619×10^6 Btu/h, will be displayed.

7. Press key C and the temperature 1,731°F will be displayed.

8. Enter 500 from the keyboard, press key B, and the enthalpy flowrate for the outlet-gas stream at 500°F, −50,784,000 Btu/h, will be displayed.

9. Subtract from this quantity the previous enthalpy (−3,619,000), and the quantity −47,166,000 will be displayed, indicating that 47,166,000 Btu/h must be removed from the gas stream to cool it to 500°F.

Solution on the HP 65

Even though the programs for the HP 65 require operating in metric, note that it is not necessary to convert all of the flow quantities into metric dimensions in order to solve the problem. The temperature rise for adiabatic conditions would be the same if the flow quantities were gram-moles per hour, rather than pound-moles per hour. Similarly, for the second part of the problem, if no conversions are made, the answer obtained will be calories for the total gram-moles per hour of gas flow. To convert this quantity into Btus, it is merely necessary to multiply the final answer by 1.8. It is necessary, however, to convert temperatures into °K.

The procedure followed will be obvious from a comparison of the keystrokes for the HP 67/97, and the user instructions given in Tables XI and XII.

Example 2—How much heat must be supplied to vaporize and superheat 1 lb of water at low pressure from 100°F to 1,000°F?

In the solution of this problem, the liquid-phase water and the gas-phase water are treated as separate components. Again, because the datum for both of these components is the same, the heat required is the algebraic difference between the enthalpy of gas-phase water at a 1,000°F, and liquid-phase water at 100°F, calculated by means of these enthalpy programs.

The value obtained by the calculation with these programs is 1,471 Btu/lb. The corresponding values obtained from the steam tables are 1,463 Btu/lb at 1 and 1,461 Btu/lb at 100 psia.

References

1. Yaws, Carl L., *et al.*, Physical and Thermodynamic Properties, *Chem. Eng.*, 24-part series, June 10, 1974 to Nov. 22, 1976; especially parts 22 and 23, Aug. 16, 1976, and Oct. 25, 1976.
2. Kobe, K. A. *et al.*, Thermochemistry for the Petrochemical Industry, *Petr. Ref.*, Jan. 1949 to Nov. 1954.
3. Thinh, T. P., *et al.*, Equations Improve C_p Prediction, *Hydr. Proc.*, 50, Jan. 1971, pp. 98–1040.
4. Wagman, D. D., *et al.*, "Selected Values of Chemical Thermodynamic Properties," National Bureau of Standards, Technical Notes 270-3 through 270-7, Jan. 1968 through Apr. 1973.
5. Wagman, D. D., *et al.*, "Chemical Thermodynamic Properties of Compounds of Sodium, Potassium and Rubidium: An Interim Tabulation of Selected Values," National Bureau of Standards Report NBSIR 76-1034, NTIS Report PB 254–460, Apr. 1976.
6. American Petroleum Institute Research Project 44, "Selected Values of Properties of Hydrocarbons and Related Compounds," Apr. 30, 1969; from TRC Data Distribution Office, Texas A&M Research Foundation.
7. Wicks, C. E. and Block, F. E., "Thermodynamic Properties of 65 Elements, Their Oxides, Halides, Carbides and Nitrides," *Bull.* 605, U.S. Bureau of Mines, 1963.

The author

Raymond T. Schneider is chief process engineer for Pridgen Engineering Co., a div. of Jacobs Engineering Co. (P.O. Box 2008, Lakeland FL 33803). He has spent more than 20 years with several major engineering firms in a variety of process and project engineering assignments. He holds a degree in chemical engineering from the University of Cincinnati. A registered engineer in the states of Florida and Ohio, he is a member of AIChE, the Soc. of Mining Engineers of AIME, and ACS.

Calculating radiant heat transfer

F. Caplan, P.E.
409 Merritt Ave., Oakland, CA 94610

☐ In most practical problems of heat transfer, transmission occurs by radiation as well as conduction and convection. Numerous charts and nomographs have been published for calculating heat transfer by conduction and convection; but radiation is relatively neglected. These nomographs may thus be useful.

In many cases, radiant heat transfer can be described by the equation:

$$Q_1 = \frac{1.713 \times 10^{-9} A_1 (T_1^4 - T_2^4)}{\frac{1}{\varepsilon_1} + \frac{A_1}{A_2}\left[\frac{1}{\varepsilon_2} - 1\right]}$$

Where Q_1 = radiant heat transferred, Btu/h
A_1 = area of hot body, ft^2
ε_1 = emissivity of hot body, dimensionless
T_1 = hot-body temperature, °R
$_2$ = subscript signifying cold body

When a surface is exposed to open space, as a steam pipe in a room, A_2 of this equation approaches infinity, and A_1/A_2 approaches zero, so that the equation becomes:

$$(Q_1/A_1) = 1.73 \times 10^{-9} \varepsilon_1 (T_1^4 - T_2^4)$$

This equation is solved by nomograph Fig. 1, for any one of the four variables, when the other three are given. The ε and Q/A scale markings can be simultaneously divided by 10. Emissivities for use with this nomograph have been published in many tables.[†]

Example: The outside surface of an uninsulated steam pipe at 760°F is exposed outdoors to −40°F. What is the radiant heat loss, if the pipe wall emissivity is 0.50? On the chart, extend a line from $T_2 = -40$ to $T_1 = 760$, and note the intersection with the pivot line, then extend a line from $\varepsilon = 0.50$ through this intersection and read $Q/A = 1,870$ Btu/(h)(ft^2).

In other cases, as for example with parallel walls, A_1 is approximately equal to A_2, and $A_1/A_2 = 1.0$. In this case, the above equation becomes:

$$(Q_1/A_1) = 1.73 \times 10^{-9} F (T_1^4 - T_2^4)$$

where: $F = \dfrac{1}{\dfrac{1}{\varepsilon_1} + \dfrac{1}{\varepsilon_2} - 1}$

Fig. 2 permits a rapid solution of this equation for F, which can then be substituted for ε in Fig. 1 and used to solve for Q_1/A_1. *Example:* What is the effective emissivity between two parallel walls, if the emissivity of one wall is 0.80 and that of the other is 0.50? On Fig. 2 extend a line from $\varepsilon_1 = 0.80$ to $\varepsilon_2 = 0.50$, and read $F = 0.444$.

[†] See, for example, Perry, R. H., and Chilton, C. H., "Chemical Engineer's Handbook," 5th ed., Table 10-17, pp. 10-46.

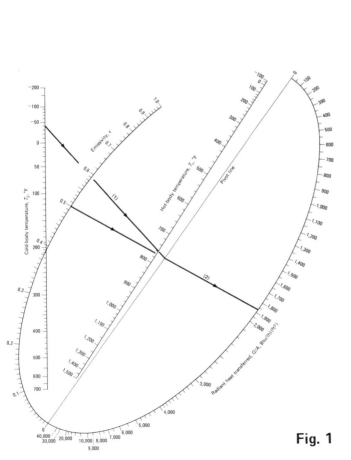

Fig. 1   Fig. 2

Originally published September 27, 1976

Charts simplify spiral finned-tube calculations

These nomographs can save you a lot of time figuring heat-transfer coefficients and pressure drops when a computer isn't handy and all you need are quick estimates.

V. Ganapathy, Bharat Heavy Electricals Ltd.

☐ Whether a finned tube represents a suitable choice for an application depends on many factors, such as cost, reliability of the bond between fin and tube, temperature and material limitations, the extent of corrosion and fouling, as well as heat-transfer and pressure-drop requirements.

For calculating heat transfer, the correlation of Robinson and Briggs, and for calculating pressure drop, the correlation of Briggs and Young are often recommended [8] (these equations are for staggered tubes having radial or spiral fins):

$$\frac{H_g d}{12k} = 0.134 \left(\frac{Gd}{12\mu}\right)^{0.681} \left(\frac{\mu C_p}{k}\right)^{0.33} \left(\frac{S}{h}\right)^{0.2} \left(\frac{S}{b}\right)^{0.113} \quad (1)$$

$$\Delta p = 18.93 \left(\frac{Gd}{12\mu}\right)^{-0.316} \left(\frac{P_t}{d}\right)^{-0.927} \left(\frac{P_t}{P_l}\right)^{0.515} \left(\frac{G^2 N}{g_c \rho}\right) \quad (2)$$

(If performance data on specific finned tubes or bundles are available from a manufacturer, these should be used in preference to the two correlations, which are only suitable for approximate design.)

Eq. (1) and (2) are complicated because of the many variables raised to difficult powers, and because thermal and transport properties must be known to solve them. A simplified procedure, such as was used to determine heat transfer over bare tubes, is preferable [3].

Simplifications lead to nomographs

Eq. (1) may be rewritten:

$$H_g = 0.295 \left(\frac{G^{0.681}}{d^{0.319}}\right)\left(\frac{k^{0.67} C_p^{0.33}}{\mu^{0.351}}\right)\left(\frac{S^{0.313}}{h^{0.2} b^{0.113}}\right) \quad (3)$$

$$H_g = H_g' F_g \quad (4)$$

$$H_g' = 0.295 \, (G^{0.681}/d^{0.319})(k^{0.67} C_p^{0.33}/\mu^{0.351}) \quad (5)$$

$$F_g = S^{0.313}/h^{0.2} b^{0.113} \quad (6)$$

Fig. 1 solves Eq. (5), and Fig. 2, Eq. (6).

With finned tubes, the calculation of gas heat-transfer coefficients must take into account the temperature drop along the length of the tube, i.e., fin efficiency. Curves for fin efficiency as a function of fin geometry and outside heat-transfer coefficient have been published [6,9]. The equation is:

$$\phi = f[h(H_g/b\lambda)^{1/2}, Z] \quad (7)$$

Fig. 3 solves Eq. (7).

Overall heat-transfer coefficients may be expressed on the basis of total extended surface area (or on bare tube area), and then corrected by an effectiveness factor, η:

$$H_{go} = [1 - (1 - \phi)A_f/A_t]H_g = \eta H_g \quad (8)$$

Fig. 4 solves for η, if A_f, A_t and ϕ are known. The outside resistance to heat transfer, R_o, becomes $1/H_{go}$. Values for A_f and A_t are available in tables, and for A_o, the area occupied by tubes, which is needed to calculate the gas flow area. (Such a table typically lists, for a particular tube diameter, values for n, number of fins/in.; h, fin height, in.; and b, fin thickness, in.; along with A_f, fin surface area, ft²/ft; A_t, total external surface area, ft²/ft; and A_o, area occupied by the tubes, ft²/ft.) Values for A_f, A_t and A_o may also be calculated with the following equations:

$$A_f = (\pi n/24)(4dh + 4h^2 + 2bd + 4bh) \quad (9)$$

$$A_t = (\pi h/24)[n(4dh + 4h^2) + 4nbh + 2d] \quad (10)$$

$$A_o = d/12 + nbh/6 \quad (11)$$

Eq. (2) for pressure drop may be simplified as follows:

$$\Delta p = 1.58 \times 10^{-8} \left(\frac{G^{1.684} d^{0.611} \mu^{0.316}}{P_t^{0.412} P_l^{0.515}}\right) \times \frac{(T+460)N}{M} \quad (12)$$

Fig. 5 solves Eq. (12) when d, G, T, P_t and P_l are known. The Δp obtained is per row of tubes. Also considered in Fig. 5 are viscosity and molecular weight as functions of temperature.

Originally published April 25, 1977

CALCULATING RADIANT HEAT TRANSFER

Nomenclature

- A_f Surface area of fins, ft²/ft
- A_i Inside surface area, ft²/ft
- A_o Area occupied by tubes, ft²/ft
- A_t Total external surface area, ft²/ft
- A_W Wall area, ft²/ft
- b Fin thickness, in.
- C_p Specific heat of gas, Btu/(lb)(°F)
- d Tube outside dia., in.
- d_i Tube inside dia., in.
- F_g Geometric factor
- F_T Temperature factor for gases, $\mu^{0.316}(460+T)/M$
- G Gas mass velocity, lb/(ft²)(h)
- h Fin height, in.
- H_g Outside gas heat-transfer coefficient, Btu/(ft²)(h)(°F)
- H_{go} Outside gas heat-transfer coefficient corrected for total outside area, Btu/(ft²)(h)(°F)
- h_i Inside heat-transfer coefficient, Btu/(ft²)(h)(°F)
- k Thermal conductivity of gas, Btu/(ft²)(h)(°F/ft)
- M Molecular weight
- N Number of tube rows
- n Number of fins/in.
- Δp Gas pressure drop, in. w.c.
- P_l Longitudinal pitch, in.
- P_t Transverse pitch, in.
- R_i Inside resistance to heat transfer, (ft²)(h)(°F)/Btu
- R_o Outside resistance to heat transfer, (ft²)(h)(°F)/Btu
- R_m Metal resistance to heat transfer, (ft²)(h)(°F)/Btu
- S Fin clearance, in.
- T Gas temperature, °F
- U Overall heat-transfer coefficient, Btu/(ft²)(h)(°F)
- Z $(2h+d)/d$
- η Fin effectiveness
- λ Thermal conductivity of fins, Btu/(ft)(h)(°F)
- μ Viscosity of gas, lb/(ft)(h)
- ρ Gas density, lb/ft³
- ϕ Fin efficiency

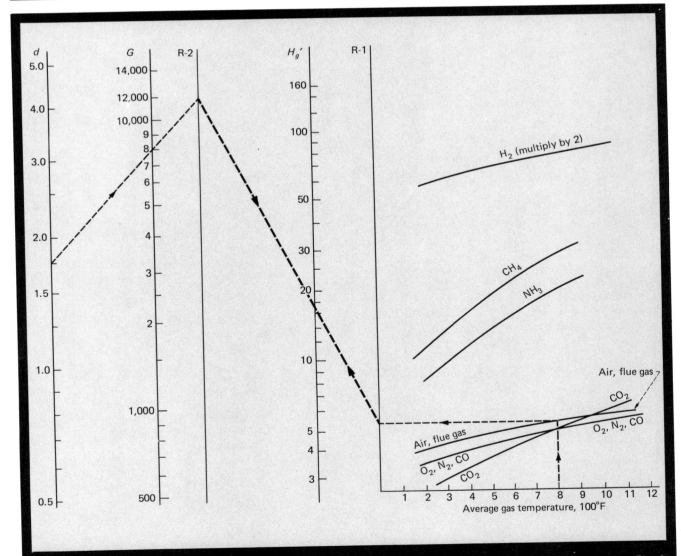

Outside heat-transfer coefficient, H_g, is calculated after H_g', defined by Eq. (5), is determined   Fig. 1

450 HEAT TRANSFER CALCULATIONS

Problem illustrates procedure

Hot air flows at 2.4 million lb/h over a bank of spiral-finned tubes inside a 20-ft × 25-ft duct. The air heats up water flowing inside the tubes.

Finned-tube details:

Tube and fin material	carbon steel
Tube outside dia., d, in.	1.75
Tube inside dia., d_i, in.	1.35
Fin height, h, in.	0.75
Fin thickness, b, in.	0.06
No. of fins/in.	2
Fin tube length. ft	24
Transverse pitch, P_t, in.	4.5
Longitudinal pitch, P_l, in.	4.5
Clearance, S, in.	0.44
No. of tubes across	52

Thermal data:

Conductivity of tube and fin, Btu/(ft)(h)(°F)	30
Inlet gas temperature, °F	900
Outlet gas temperature, °F	700
Heat transferred, Btu/h	120 × 10^6
Water flow, lb/h	500,000
Inlet water temperature, °F	100
Outlet water temperature, °F	340
Log mean temperature difference, °F	575
Waterside heat-transfer coefficient, Btu/(ft)(h)(°F)	1,700

(Because the objective is to present the calculation of the gas heat-transfer coefficient with a finned surface, the calculation of a heat balance and the inside coefficient, h_i, are not detailed; the latter is generally high for

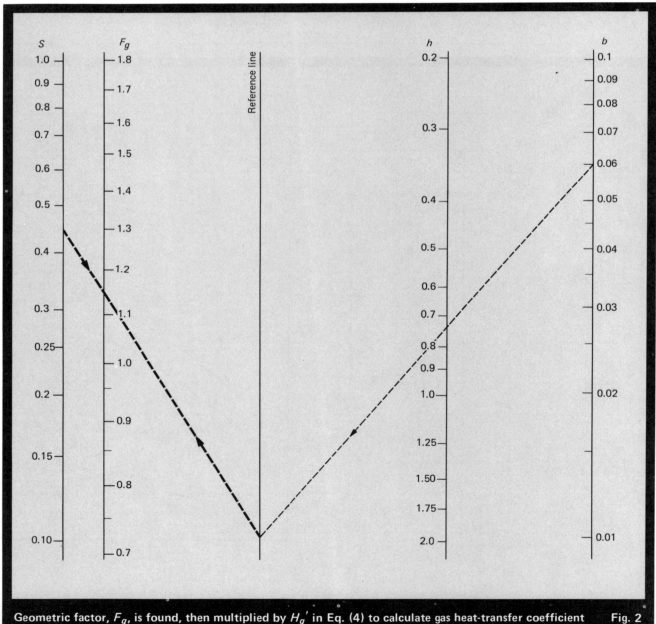

Geometric factor, F_g, is found, then multiplied by H_g' in Eq. (4) to calculate gas heat-transfer coefficient   Fig. 2

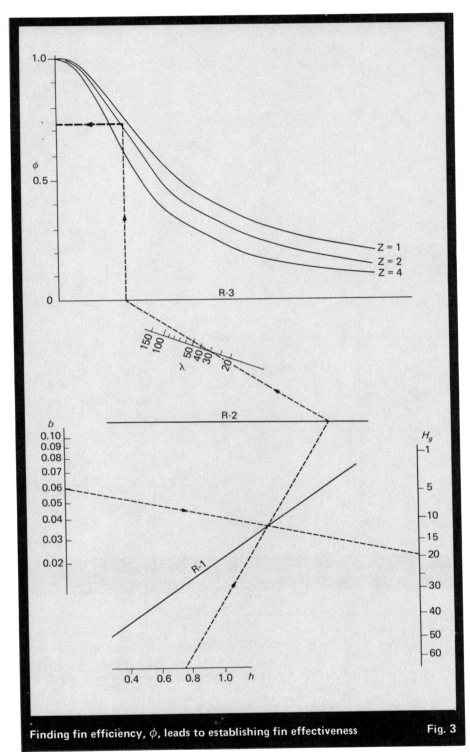

Finding fin efficiency, ϕ, leads to establishing fin effectiveness   Fig. 3

corrected for the total outside area (tube plus fin surface), using Eq. (4) and (8), and Fig. 1, 2, 3 and 4, successively. Resistances are subsequently calculated to finally determine the overall heat-transfer coefficient, U.

In Fig. 1, start from the average gas temperature of 800°F, go vertically to the *air, flue gas* curve, then move left to reference line R-1. Next, connect $d = 1.75$ with $G = 8,000$, and extend this line to reference line R-2. Connect the point determined on line R-1 with the point determined on line R-2 to establish a line crossing the vertical line H_g', which gives an H_g' value of 17 Btu/(ft²)(h)(°F).

In Fig. 2, connect $b = 0.06$ with $h = 0.75$ and extend the line to the vertical reference line. Connecting the established point on the reference line with $S = 0.44$ gives $F_g = 1.15$. Using Eq. (4), H_g is then calculated as 19.5 Btu/(ft²)(h)(°F).

In Fig. 3, connect $b = 0.06$ with $H_g = 19.5$ to intersect reference line R-1. Then draw a line from $h = 0.75$ through the intersection and extend the line to reference line R-2. Next, draw a line from this point on R-2 through $\lambda = 30$, and extend the line to reference line R-3. From this point on R-3, draw a vertical line to $Z = 1.85$—interpolating between $Z = 1$ and $Z = 2$; $Z = (2h + d)/d$. From $Z = 1.85$, draw a horizontal line to $\phi = 0.71$.

In Fig. 4, connect $A_f = 2.07$ with $A_t = 2.47$, and extend the line to the reference line. From this point on the reference line, draw a line through $\phi = 0.71$ and extend it to get $\eta = 0.76$.

Using Eq. (8), $H_{go} = 0.76 \times 19.5 = 14.8$ Btu/(ft²)(h)(°F). Calculate the outside resistance: $R_o = 1/H_{go} = 0.0676$ (ft²)(h)(°F)/Btu.

3. The inside coefficient, h_i, = 1,700; it may be used to calculate the total outside-area resistance:

$$R_o = \frac{A_t}{h_i A_i} = \frac{12 \times 2.469}{1,700 \times \pi \times 1.35}$$
$$= 0.0041 \text{ (ft²)(h)(°F)/Btu}.$$

4. Metal resistance is calculated:

$$R_m = \left(\frac{A_t}{A_w}\right)\left(\frac{d}{24\lambda}\right) \ln(d/d_i) = 0.00384 \text{ (ft²)(h)(°F)/Btu}.$$

liquids, so its effect on the overall heat transfer would be less than that of the gas.)

Calculation procedure:

1. Area obstructed by tubes, A_o = length of tubes × number of tubes × area obstructed by 1 ft of tube and fins = 24 × 52 × 0.161 = 200 ft².

Gas free-area = 20 × 25 − 200 = 300 ft². Gas mass-velocity = 2.4 × 10⁶/300 = 8,000 lb/(ft²)(h).

2. Find H_{go}, the outside (gas) heat-transfer coefficient

452 HEAT TRANSFER CALCULATIONS

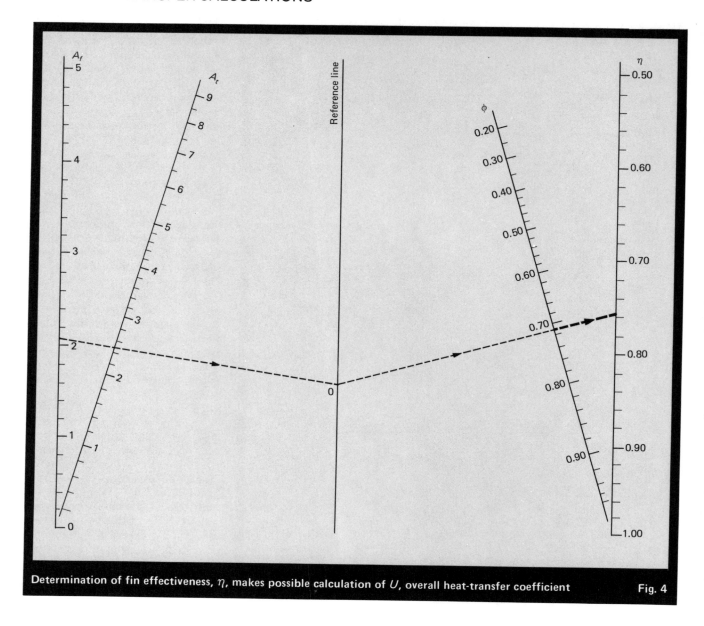

Determination of fin effectiveness, η, makes possible calculation of U, overall heat-transfer coefficient   Fig. 4

5. The overall heat-transfer coefficient, U, is:
$$U = 1/(R_o + R_i + R_m) = 13.2 \text{ Btu}/(\text{ft}^2)(\text{h})(°\text{F}).$$
(Fouling may be taken into account in the foregoing calculation.)

6. The surface area required $= (120 \times 10^6)/(575)(13.2) = 15{,}810 \text{ ft}^2$. The number of tubes deep in the bundle $= 15{,}810/(2.469)(24)(52) = 5.1$ (use 6).

7. To find the gas pressure drop, use Fig. 5. Starting from $T = 800°\text{F}$, draw a vertical line up to the *air, flue gas* curve, then move horizontally to reference line R-1. Next, draw a line from $P_t = 4.5$ to $P_l = 4.5$ through R-3, then connect this point with a line through $G = 8{,}000$ to the d line. Draw a connecting line between this point on the d line and the previously established point on line R-1 so as to intersect line R-2. Then connect this point on R-2 with $d = 1.75$ and extend it to obtain $\Delta p = 0.38$ in. water column.

The total gas $\Delta p = 6 \times 0.38 = 2.2$ in. w.c.

On the basis of experience, installation requirements, economy desired, etc., one can choose different finned-tube sizes and mass velocities, check the Δps, and figure the optimum exchanger without waiting for a detailed computer analysis, which may not be necessary in the proposal stage.

Factors in selecting finned tubes

An approximate rule-of-thumb for choosing finned-tube surface area is to take the inverse ratios of heat-transfer coefficients on the gas and air sides. Other guidelines based on experience should not be ignored [1,6].

The larger the fin area, the lower the fin efficiency. The greater the number of fins/in., the likelier the chance of dust and other particles settling on the tubes. The smaller the fin height and the thicker the fin, the lower the fin-tip temperature, which is important when the gas temperature is high.

For applications in which the gas heat-transfer coefficient is low, the thermal conductivity of steel is good

CHARTS SIMPLIFY SPIRAL FINNED-TUBE CALCULATIONS

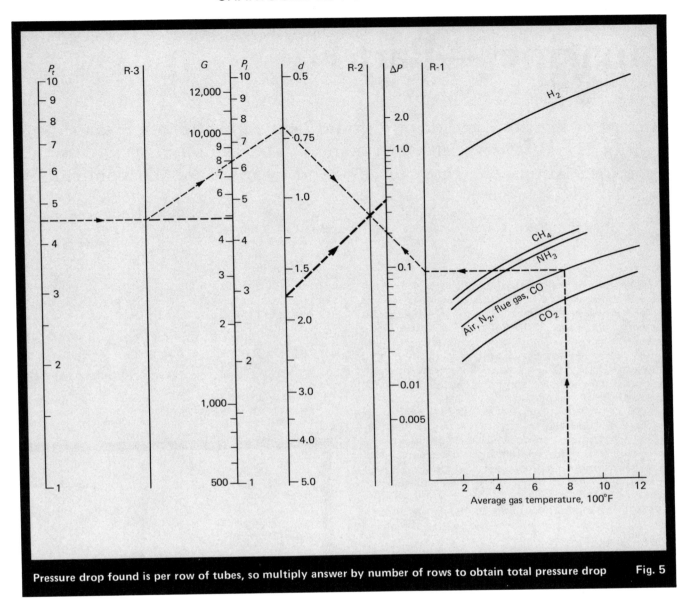

Pressure drop found is per row of tubes, so multiply answer by number of rows to obtain total pressure drop   Fig. 5

enough to allow adequate fin efficiency to be obtained with a reasonable fin thickness [2]. With higher fin-side heat-transfer coefficients and large fin heights, the thickness of steel becomes excessive, and it is worthwhile to use copper or aluminum.

Fin efficiency may be lowered if the fin is not bonded adequately to the tube. As the contact pressure between the fin and tube lessens, the gas pressure increases [4]. Tapering the fins so that they are thicker at the base reduces the fin weight and increases the gas flow area, but raises the manufacturing cost, which is why the tapered design is rarely used for welded fins.

References

1. Boyen, J. L., "Practical Heat Recovery," Wiley-Interscience, New York.
2. Fraas, A. P. and Ozisik, M. N., "Heat Exchanger Design," Wiley-Interscience, New York.
3. Ganapathy, V., Quick estimation of gas heat-transfer coefficients, *Chem. Eng.*, Sept. 13, 1976, p. 199.
4. Gardner, K. A., and Carnavos, T. C., Thermal Contact Resistance in Finned Tubing, *Trans. ASME J. of Heat Transfer*, Nov. 1960.
5. Kays, W. M. and London, A. L., "Compact Heat Exchangers," National Press, Palo Alto, Calif., 1955.
6. Kern, D. Q., "Process Heat Transfer," McGraw-Hill, New York.
7. Perry, J. H., "Chemical Engineers' Handbook," McGraw-Hill, New York.
8. Rohsenow, W. M. and Hartnett, J. P., "Handbook of Heat Transfer," McGraw-Hill, New York.
9. Kern, D. Q. and Kraus, A. D., "Extended Surface Heat Transfer," McGraw-Hill, New York, 1972, pp. 101–104.

The author

V. Ganapathy is Senior Development Engineer with Bharat Heavy Electricals Ltd. (High Pressure Boiler Plant, Tiruchiarapalli-620 014, India). Currently engaged in the design of boilers for utility and process services, he is particularly concerned with applications in the chemical and fertilizer industries. He holds a bachelor's degree in technology from the Indian Institute of Technology in Madras, India, and an M.Sc. (engineering) in boiler technology from Madras University. He has written over 25 articles for various U.S. and British publications.

Estimating liquid heat capacities—Part I

Current estimation methods for liquid heat capacities are classified, described, and tested with experimental data. Definitive recommendations are given for both polar and nonpolar compounds.

Robert C. Reid and *Juan L. San Jose,* Massachusetts Institute of Technology*

☐ In heating or cooling liquids, the energy requirements are proportional to the heat capacity. Values of this property are, therefore, necessary to size heat exchangers, and to specify the flowrates of process fluids. Experimental heat-capacity data for liquids are quite limited, and even when available cover mostly small temperature ranges. This situation has spawned numerous estimation methods; an evaluation of these is the objective of this two-part article.

Part I defines heat capacity, explaining and clarifying the differences that exist between the various versions of heat capacity in current use. It also delineates the differences and relationships between several group-contribution methods used to estimate the heat capacity of liquids at temperatures below the normal boiling point.

Part II will explain various techniques used to estimate liquid heat capacity over a wide range of temperatures and pressures.

Definitions

There are several heat capacities. All of them can be defined as a partial derivative of a thermodynamic property with respect to temperature—with certain restraints. For example, the constant volume heat capacity C_v is

$$C_v = (\partial U/\partial T)_v \quad (1)$$

where U = specific internal energy and $(\partial U/\partial T)_v$ = change of U with T, measured at constant specific volume.

Estimation methods for liquid heat capacity that are based upon molecular theory yield C_v. None are, however, particularly accurate. Even if experimental C_v data were available, these are not readily converted to heat capacities useful to the chemical engineer, as extensive and accurate liquid volumetric data as a function of pressure and temperature are also required.

Enthalpy changes in liquids are of more interest to engineers, and three liquid heat capacities are in common use. At constant pressure

$$C_p = (\partial H/\partial T)_p \quad (2)$$

where H = specific enthalpy, and $(\partial H/\partial T)_p$ is measured at constant pressure. When dealing with saturated liquids, the heat capacities of interest are

$$C_\sigma = (dH/dT)_\sigma \quad (3)$$
$$C_{s\sigma} = T(dS/dT)_\sigma \quad (4)$$

In Eq. (3) and (4), the derivatives of enthalpy and

*A biographical sketch of both authors appears on p. 462.

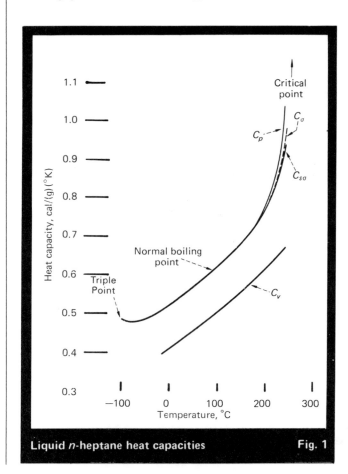

Liquid n-heptane heat capacities   Fig. 1

Originally published December 6, 1976

ESTIMATING LIQUID HEAT CAPACITIES—PART I

Nomenclature

a,b,c,A,B,C,D	General constants
C_p	Heat capacity at constant pressure
C_v	Heat capacity at constant volume
C_σ	Saturation liquid heat capacity, Eq. (3)
$C_{s\sigma}$	Saturation liquid heat capacity, Eq. (4)
ΔC_σ	Yuan and Stiel parameter
H	Enthalpy
ΔH_v	Enthalpy of vaporization
M	Molecular weight
P	Pressure; P_c, critical pressure; P_{vp}, vapor pressure
R	Gas constant
$\mathcal{R}$	Molecular radius of gyration
S	Entropy
T	Temperature; T_b at normal boiling point; T_c at critical point; T_f at freezing point.
U	Internal energy
V	Volume
β	Constant
κ	Association factor
χ	Stiel polar factor
ω	Pitzer acentric factor

Subscripts

σ	Saturated phase
p	Pressure
v	Volume

Superscripts

o	Ideal-gas state
v	Vapor

All units are equivalent.

Estimation methods for low-temperature liquid heat capacities — Table I

Method	Reference	Temperature range, °C	Approximate error, %
Johnson and Huang	9	20	4-10
Shaw	22	25	3-7
Chueh and Swanson	4	20	3-7
Missenard	14	T_b	3-7
Luria and Benson	12	T_b	±1 cal/(g-mol)(°K)

entropy are measured with a liquid that is in equilibrium with a vapor phase. (The subscript σ indicates a saturated liquid.)

$C_{s\sigma}$ is the liquid heat capacity most often reported experimentally. It reflects the energy required to cause a temperature change while maintaining the liquid in a saturated state. C_σ, on the other hand, shows the variation of the saturated liquid enthalpy with temperature. C_p is of value to calculate enthalpy changes in heat exchangers, as the pressure changes are small.

The three heat capacities defined in Eq. (2) through (4) are related from thermodynamics. For example,

$$C_\sigma = C_{s\sigma} + V_\sigma (dP/dT)_\sigma \qquad (5)$$

$$C_{s\sigma} = C_p - T(\partial V/\partial T)_p (dP/dT)_\sigma \qquad (6)$$

In Eq. (5) and (6), $(dP/dT)_\sigma$ = change in vapor pressure with temperature, and V_σ = specific volume of the saturated liquid.

For liquids, at temperatures below about $0.8 \times T_c$ (i.e., $T_r = T/T_c < 0.8$), C_p, C_σ and $C_{s\sigma}$ differ but slightly. At higher temperatures, approximate correlations that are generally accurate within 3–5% are given in Eq. (7) and (8).

$$1.0 > T_r \geq 0.8$$

$$\left[\frac{C_p - C_\sigma}{R}\right] = (1 + \omega)^{0.85} \exp(-0.7074 - 31.014 T_r + 34.361 T_r^2) \qquad (7)$$

$$\left[\frac{C_\sigma - C_{s\sigma}}{R}\right] = \exp(8.655 T_r - 8.385) \qquad (8)$$

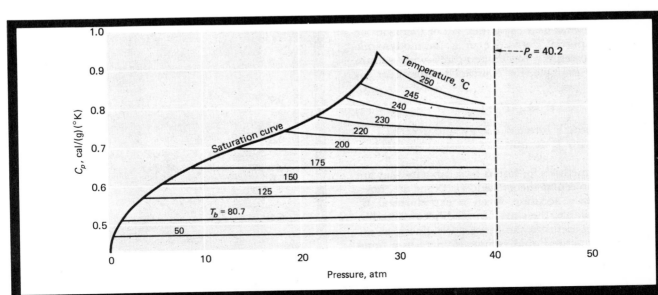

Liquid isobaric heat capacity of cyclohexane — Fig. 2

Polynomial coefficients for the Luria and Benson liquid heat-capacity equation — Table II

$$C_p = A + BT + CT^2 + DT^3, \text{ cal/(g-mol)(°K); } T \text{ in kelvins}$$

Group	A	B	C	D
C–(C)(H)$_3$	8.459	2.113 E-3*	–5.605 E-5	1.723 E-7
C–(C)$_2$(H)$_2$	–1.383	7.049 E-2	–2.063 E-4	2.269 E-7
C–(C)$_3$(H)	2.489	–4.617 E-2	3.181 E-4	–4.565 E-7
C–(C)$_4$	9.116	–2.354 E-1	1.287 E-3	–1.906 E-6
C$_d$–(H)$_2$	8.153	1.776 E-2	–1.526 E-4	2.542 E-7
C$_d$–(C)(H)	5.792	–1.228 E-2	6.036 E-5	–1.926 E-8
C$_d$–(C)$_2$	8.005	–9.456 E-2	4.620 E-4	–6.547 E-7
C$_d$–(C$_d$)(H)	8.127	–7.171 E-2	3.894 E-4	–5.462 E-7
C$_d$–(C$_B$)(C)	5.745	–1.085 E-1	5.898 E-4	–6.983 E-7
C–(C$_d$)$_2$(H)$_2$	9.733	–1.100 E-1	5.522 E-4	–6.852 E-7
C–(C$_d$)(C)(H)$_2$	3.497	–1.568 E-1	1.808 E-4	–3.277 E-7
C–(C$_d$)(C)$_2$(H)	–2.232	–1.773 E-2	2.812 E-4	–4.199 E-7
C–(C$_B$)(C)(H)$_2$	30.192	–2.812 E-1	1.002 E-3	–1.115 E-6
C$_t$–(H)	30.122	–2.081 E-1	5.945 E-4	–3.430 E-7
C$_t$–(C)	–10.407	1.662 E-1	–5.679 E-4	6.667 E-7
C$_B$–(H)	–1.842	5.778 E-2	–1.716 E-4	1.995 E-7
C$_B$–(C)	28.807	–2.824 E-1	9.779 E-4	–1.103 E-6
C$_B$–(C$_B$)	–3.780	2.563 E-2	1.190 E-5	–9.774 E-8
C$_a$	13.756	–1.338 E-1	6.553 E-4	–9.447 E-7

Corrections

Group	A	B	C	D
Cis with double bond	14.299	–1.646 E-1	6.069 E-4	–7.716 E-7
Cyclopropane ring	28.469	–2.696 E-1	6.534 E-4	–1.636 E-7
Cyclobutane ring	6.060	–3.114 E-2	–2.461 E-4	8.349 E-7
Cyclopentane ring	34.261	–3.803 E-1	1.161 E-3	–1.118 E-6
Spiropentane ring	32.469	–1.991 E-1	1.820 E-4	4.090 E-7
Cyclohexane ring	13.021	–1.468 E-1	2.802 E-4	–3.185 E-8
Cycloheptane ring	210.72	–2.344 E 0	8.235 E-3	–9.500 E-6
Cyclooctane ring	1691.9	–1.680 E+1	5.523 E-2	–6.033 E-5
Cyclopentene ring	13.650	–1.126 E-1	7.257 E-5	3.400 E-7
Cyclohexene ring	5.360	–3.456 E-2	–2.232 E-4	7.324 E-7
Cycloheptatriene ring	–22.158	3.985 E-1	–2.059 E-3	2.863 E-6
Cyclooctatetraene ring	–1.060	2.739 E-1	–1.972 E-3	3.167 E-6
Decahydronaphthalene ring (cis and trans)	141.85	–1.510 E 0	4.773 E-3	–4.872 E-6
1,2,3,4-tetrahydronaphthalene	–212.65	2.203 E 0	–7.571 E-3	8.565 E-6

* Where E-3, E-2, E-1, etc. correspond to 10^{-3}, 10^{-2}, 10^{-1}, etc.

Polynomial coefficients for the Missenard liquid heat-capacity equation — Table III

$$C_p = A + BT + CT^2, \text{ cal/(g-mol)(°K); } T \text{ in kelvins}$$

Group	A	B × 10^2	C × 10^4
–H	4.12	–1.51	0.43
–CH$_3$	9.45	–1.41	0.53
–CH$_2$–	7.21	–0.96	0.27
–CH	–3.1	4.68	–0.56
–C–	2.0	0	0
–C≡C–	11.0	0	0
–O–	5.91	0.40	0
–CO–	8.016	0.80	0
–OH–	–7.55	3.66	0.79
–COO– (ester)	9.32	1.65	0
–COOH	5.35	4.58	0
–NH$_2$	39.05	–20.04	4.00
–NH–	12.2	0	0
–N–	2.0	0	0
–CN	12.4	0.40	0
–NO$_2$	13.04	0.92	0
–NH–NH–	19.0	0	0
C$_6$H$_5$– (phenyl)	12.68	5.26	0
C$_{10}$H$_7$– (naphthyl)	50.6	–9.45	2.57
–F	8.31	–2.23	0.49
–Cl	5.70	0.47	0
–Br	7.00	0.55	0
–I	8.05	0.54	0
–S–	7.18	0.68	0

where ω = the Pitzer acentric factor (see nomenclature), used in many property correlations.

These three liquid heat capacities are shown in Fig. 1 for n-heptane. For comparison, C_v is also presented over a smaller temperature range.

The shallow minimum near the melting point is not found for many liquids. Note also that the liquid heat capacity is undefined at the critical point.

In Fig. 2, C_p for liquid cyclohexane is graphed as a function of pressure at several different temperatures [21]. Below 175°C ($T_r = 0.81$) there is essentially no effect of pressure, at least up to the critical pressure (40.7 bar). Above reduced temperatures of about 0.8, an increase in pressure results in a decrease in C_p, although there appears to be an asymptotic value at high pressures. In cyclohexane, data were available only to 250°C ($T_r = 0.95$).

For rough estimates of C_p: When expressed on a unit mass basis, most organic compounds have values of C_p between 0.5 and 0.8 cal/(g)(°K) at their normal boiling temperature (T_b). Within homologous series, the range is even less—e.g., hydrocarbons have C_p values between 0.5 and 0.65 cal/(g)(°K) at T_b. First members of a series are often anomolous and have a higher-than-expected C_p.

ESTIMATING LIQUID HEAT CAPACITIES—PART I

Sources of experimental data

Most experimental measurements yield $C_{s\sigma}$. At the low temperatures reported for most compounds, values of $C_{s\sigma} \sim C_\sigma \sim C_p$. Few studies have been made at high temperatures or at pressures greater than the vapor pressure. For polar liquids, above the normal boiling point, only water, ammonia, and isopropyl alcohol have been studied in any detail. The scarcity of high-temperature liquid heat capacity data restricts the evaluation of estimation methods in this region.

Low-temperature liquid heat capacities of hydrocarbons have been correlated in a nomograph [7] and with equations [8,24]. A tabulation of most experimental data is available in a thesis [20].

Group-contribution estimation methods

For temperatures below the normal boiling point, it is often assumed that various structural groups in a molecule contribute to the total liquid heat capacity in an additive manner. In Table I, the accuracy of the five available group-contribution estimation methods is summarized. Only the Missenard and Luria-Benson methods estimate C_p at different temperatures. Neither should be used above the normal boiling point.

For hydrocarbons, the Luria-Benson method is the most accurate. To delineate the appropriate structural groups for this method, for each carbon atom, one must specify the other atoms to which it is covalently bonded. A carbon with four single bonds is designated as C. A carbon double-bonded to another carbon is noted as C_d, and it is necessary to show only the atoms which are attached to the two single bonds. Similarly, a carbon triple-bonded to another is C_t, an aromatic carbon is C_B, and an allene group ($>C=C=C<$) is C_a. Contributions for various groups are shown in Table II to calculate polynomial constants for estimating C_p as a function of temperature. The method is illustrated in Table IV.

For nonhydrocarbons, the group-contribution values for Missenard are shown in Table III again as polynomial constants in temperature. An example showing the method is given in Table IV.

Examples of group-contribution estimation methods for low-temperature liquid heat capacities — Table IV

Luria-Benson

1,1-dimethylcyclopentane at 300 °K

Number	Type
4	C–(C)$_2$(H)$_2$
1	C–(C)$_4$
2	C–(C)(H)$_3$

With table 2 and the correction of the cyclopentane ring,

$$C_p = 4[-1.383 + (7.049)(10^{-2})T - (2.063)(10^{-4})T^2$$
$$+ (2.269)(10^{-7})T^3] + [9.116 - (2.354)(10^{-1})T$$
$$+ (1.287)(10^{-3})T^2 - (1.906)(10^{-6})T^3] + 2[8.459$$
$$+ (2.113)(10^{-3})T - (5.605)(10^{-5})T^2 + (1.723)$$
$$(10^{-7})T^3]$$
$$+ [34.261 - (3.803)(10^{-1})T + (1.161)(10^{-3})T^2$$
$$- (1.118)(10^{-6})T^3]$$
$$= 54.763 - 0.330T + (1.511)(10^{-3})T^2 - (1.772)$$
$$(10^{-6})T^3$$

At 300 °K, $C_p = 43.9$ cal/(g-mol)(°K). Luria and Benson [12] indicate the experimental value is 44.8 cal/(g-mol)(°K).

Missenard

Ethanethiol at 315 K

With Table III

Type	A	B × 10^2	C × 10^4
(CH$_3$–)	9.45	–1.41	0.53
(–CH$_2$–)	7.21	–0.96	0.27
(–S–)	7.18	0.68	0.0
(–H)	4.12	–1.51	0.43
	27.96	–3.20	1.23

$C_p = 27.96 + 315(-3.20 \times 10^{-2}) + (315)^2(1.23 \times 10^{-4})$
$= 30.10$ cal/(g-mol)(°K)

The experimental value is 28.7 cal/(g-mol)(°K).

References

1. Ambrose, D., and Towsend, R., "Vapor-Liquid Critical Properties," National Physical Laboratory, Teddington, Middlesex, England.
2. Aston, J. G., Kennedy, R. M., and Schuman, S. C., *J. Am. Chem. Soc.*, Vol. 62, p. 2,052 (1940).
3. Bondi, A., *Ind. Eng. Chem. Fund.*, Vol. 5, p. 442 (1966); also "Physical Properties of Molecular Crystals, Liquids, and Glasses," Wiley, N.Y. (1966).
4. Chueh, C. F., and Swanson, A. C., *Chem. Eng. Prog.*, Vol. 69, No. 7, p. 83 (1973); *Can. J. Chem. Eng.*, Vol. 51, p. 596 (1973).
5. Chueh, P. L., and Deal, C. H., Thermophysical Properties of Pure Chemical Compounds, Paper presented at 65th Annual AIChE Meeting, New York, N.Y., Nov. 1972.
6. Fishtine, S. H., *Ind. Eng. Chem.*, Vol. 55, No. 4, p. 20; ibid., No. 5, p. 49; ibid., Vol. 55, No. 6, p. 47 (1963); *Hydro. Proc. Pet. Ref.*, Vol. 42, No. 10, p. 143 (1963).
7. Gambill, W. R., *Chem. Eng.*, Vol. 64, No. 5, p. 263; ibid., No. 6, p. 243; ibid., No. 7, p. 263; ibid., No. 8, p. 257 (1957).
8. Hadden, S. T., *J. Chem. Eng. Data*, Vol. 15, p. 92 (1970).
9. Johnson, A. I., and Huang, C. J., *Can. J. Technol.*, Vol. 33, p. 421 (1955).
10. Jones, J. L., et al., *Chem. Eng. Progr. Symp. Ser.*, Vol. 44, No. 59, p. 52 (1963).
11. Lee, B. I., and Kesler, M. G., *AIChE J.*, Vol. 21, p. 510 (1975). Correction: ibid., p. 1,237 (1975).
12. Luria, M., and Benson, S. W., *J. Chem. Eng. Data*, Vol. 21, (1976).
13. Lyman, T. J., and Danner, R. P., *AIChE J.*, Vol. 22, p. 759 (1976).
14. Missenard, F. A., *Comp. Rend.*, Vol. 260, p. 5,521 (1965).
15. Passut, C. A., and Danner, R. P., *Chem. Eng. Prog. Symp. Ser.*, No. 140, Vol. 70, p. 30 (1974).
16. Peng, D., and Stiel, L. I., *AIChE Symp. Ser.*, Vol. 70, No. 140, p. 63 (1975).
17. Reid, R. C., and Sobel, J. E., *Ind. Eng. Chem. Fund.*, Vol. 4, p. 328 (1965).
18. Reid, R. C., Sherwood, T. K., and Prausnitz, J. M., "The Properties of Gases and Liquids," 3rd ed., McGraw-Hill, New York (in press).
19. Sage, B. H., and Lacey, W. N., *Ind. Eng. Chem.*, Vol. 30, No. 6, p. 673 (1938).
20. San Jose, J. L., Sc.D. thesis, Dept. of Chem. Eng., Mass. Inst. of Tech., Cambridge, Mass. (1975).
21. San Jose, J. L., Mellinger, G., and Reid, R. C., submitted to *J. Chem. Eng. Data*, Vol. 21, p. 414 (1976).
22. Shaw, R., *J. Chem. Eng. Data*, Vol. 14, p. 461 (1969).
23. Stull, D. R., Westrum, E. F., and Sinke, G. C., "The Chemical Thermodynamics of Organic Compounds," Wiley, New York (1969).
24. Tamplin, W. S., and Zuzic, D. A., *Hydro. Proc.*, Vol. 46, No. 8, p. 145 (1967).
25. Tyagi, K. P., *Ind. Eng. Chem. Proc. Des. Dev.*, Vol. 14, No. 4, p. 484 (1975).
26. Watson, K. M., *Ind. Eng. Chem.*, Vol. 35, p. 398 (1943).
27. Yesavage, V. F., Katz, D. L., and Powers, J. E., "Experimental Determinations of Some Thermal Properties of Propane: Heat Capacity, Joule Thompson Coefficient, Isothermal Throttling Coefficient and Latent Heat of Vaporization," Proc. of 4th Symposium on Thermophysical Properties, ASME, New York, N.Y. (1968).
28. Yuan, T.-F., and Stiel, L. I., *Ind. Eng. Chem. Fund.*, Vol. 9, p. 393 (1970).

Estimating liquid heat capacities—Part II

The conclusion of a two-part series that started in the Dec. 6 issue, this article describes techniques used to estimate liquid heat capacity over a wide range of temperatures and pressures.

Robert C. Reid and *Juan L. San Jose, Massachusetts Institute of Technology*

☐ The difference between the heat capacity of a liquid and that of an ideal gas at the same temperature depends upon intermolecular forces. As such, this difference is often correlated empirically as a function of reduced temperature and reduced pressure, as suggested by corresponding-states theory. The low-pressure (ideal-gas) heat capacity C_p^o must then be determined separately. Estimation methods for C_p^o are reviewed by Reid et al. [18], and tabulations of C_p^o from theory are available for many compounds [23].

Corresponding-states methods

Five corresponding-states methods are reviewed and summarized in Table V. In Table VI, suggestions are given for obtaining the necessary input parameters.

The Sternling-Brown and Rowlinson-Bondi methods require only that the critical temperature and acentric factor be available. Normally, for nonpolar liquids, they provide estimates of C_p within 4 or 5% (see Table VIII).

The Yuan-Stiel correlation also requires that the critical temperature and acentric factors be known. In addition, for polar liquids, the Stiel polar factor is required. This parameter is defined in Table VI. Generally, the method is quite accurate, although the results obtained are sensitive to the values of the acentric factor and Stiel polar factor selected. A table is necessary to determine ΔC_σ values (Table VII).

The Lyman-Danner equation requires that the critical temperature, the molecular radius of gyration, and the association factor be known. The latter two parameters are tabulated for about 250 compounds in [15] and [18]. Errors are low when predicting $C_{s\sigma}$.

The final method shown in Table V, that of Lee and Kesler, is based on a modified Benedict-Webb-Rubin equation of state. It is the only one which can be used to predict heat capacities of both saturated and subcooled liquids. In the testing of this method, the complex equations as described in [11] were programmed. However, for simplicity, the two deviation functions have been plotted in Fig. 3 and 4. In the former, the termination of any reduced temperature curve is on the saturated liquid curve, which itself varies with the acentric factor. It is clearly seen that at low reduced

Corresponding-states methods to estimate liquid heat capacities			Table V
Method	**Reference**	**Equations and comments**	
Sternling and Brown	3	Nonpolar liquids, $0.4 < T_r < 1.0$ $(C_p - C_p^o)/R = (0.5 + 2.2\omega)[3.67 + 11.64(1 - T_r)^4 + 0.634/(1 - T_r)]$	
Rowlinson and Bondi	3	Nonpolar liquids, $0.4 < T_r < 1.0$ $(C_p - C_p^o)/R = 2.56 + 0.436/(1 - T_r) + 0.17\omega[17.11 + (25.2/T_r)(1 - T_r)^{1/3} + 1.742/(1 - T_r)]$	
Lyman and Danner	13	Polar and nonpolar liquids, $0.35 < T_r < 0.96$ $(C_{s\sigma} - C_p^o)/R = (5.0968 + 0.15826\kappa - 0.03551\kappa^2) + T_r(-7.7275 + 1.6109\mathcal{R}) + T_r^2(1.2756\kappa + 0.03656\kappa^2) + T_r^5(9.9296 - 0.4504\mathcal{R} - 1.0187\kappa) - 0.00749\mathcal{R}^2/T_r^2 + 0.1128\mathcal{R}/T_r^3 - 0.02185/T_r^5$	
Yuan and Stiel	28	Nonpolar liquids, $0.4 < T_r < 0.96$ $(C_{s\sigma} - C_p^o)/R = (\Delta C_\sigma)^{(0)} + \omega(\Delta C_\sigma)^{(1)}$ Polar liquids, $0.44 < T_r < 0.94$ $(C_{s\sigma} - C_p^o)/R = (\Delta C_\sigma)^{(0p)} + \omega(\Delta C_\sigma)^{(1p)} + \chi(\Delta C_\sigma)^{(2p)} + \chi^2(\Delta C_\sigma)^{(3p)} + \omega^2(\Delta C_\sigma)^{(4p)} + \omega\chi(\Delta C_\sigma)^{(5p)}$	
Lee and Kesler	11	$(C_p - C_p^o)/R = [(C_p - C_p^o)/R]^{(0)} + \omega[(C_p - C_p^o)/R]^{(1)}$	

Originally published December 20, 1976

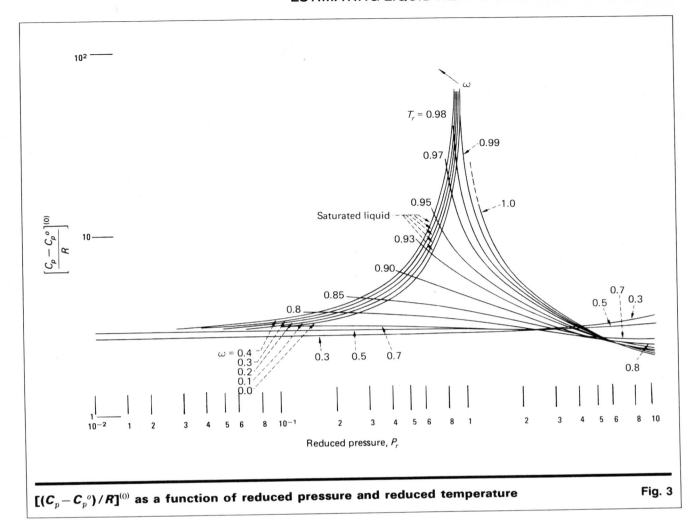

$[(C_p - C_p^o)/R]^{(0)}$ as a function of reduced pressure and reduced temperature Fig. 3

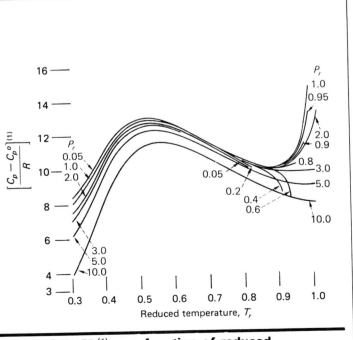

$[(C_p - C_p^o)/R]^{(1)}$ as a function of reduced pressure and reduced temperature Fig. 4

temperatures there is essentially no effect of pressure on the heat capacity of a liquid.

In Fig. 4, the value of $[(C_p - C_p^o)/R]^{(1)}$ is the same for all pressures up to the critical for reduced temperatures below about 0.9. Near the critical temperature, this function becomes a strong function of pressure. For those who desire to use the Lee-Kesler correlation near the critical temperature, it is recommended that the original equations be used, rather than extrapolating curves in Fig. 3 and 4.

The Lee-Kesler method is slightly more accurate for subcooled liquids than for the saturated phase. Results for the latter one are shown in Table VIII, while the average error was only 2.2% when tested with subcooled liquids and liquid mixture data [10,16,21,27].

The expected errors for all corresponding-states methods are shown in Table VIII. A note of caution is advisable relative to the applicability of any of the methods at high temperatures—especially for polar compounds. Too few reliable data are available to make a definitive test.

Fig. 5 shows the results of using all five corresponding-states methods to predict $C_{s\sigma}$ for liquid isobutane. Clearly, most are quite accurate above about 200°K ($T_r = 0.49$); below this temperature larger differences are found.

Parameters needed for Table V		Table VI
Symbol	Property	Sources
T_r	Reduced temperature	$T_r = T/T_c$
T_c	Critical temperature	Tabulations
P_c	Critical pressure	given in [1, 18]
R	Gas constant	8.314 J/(g-mol)(°K) 1.986 cal/(g-mol)(°K) 1.986 Btu/(lb-mol)(°R)
ω	Pitzer acentric factor $\omega = -\log(P_{vp}/P_c) - 1.000$ P_{vp} at $T_r = T/T_c = 0.7$	Calculate from vapor pressure at $T_r = 0.7$; tabulation given in [1, 18]
χ	Stiel polar factor $\chi = \log(P_{vp}/P_c) + 1.70\omega + 1.552$ P_{vp} at $T_r = T/T_c = 0.6$	obtain from ω and vapor pressure at $T_r = 0.6$; tabulation given in [18]
C_p^o	Ideal gas-heat capacity	See [18, 23]
$\mathcal{R}, \kappa$	Molecular radius of gyration and association factor	They are tabulated for 250 compounds in [15, 18]
$(\Delta C_\sigma)^{(i)}$	Yuan and Stiel deviation parameters	Table VII
$[(C_p - C_p^o)/R]^{(i)}$	Lee and Kesler deviation parameters	Fig. 3 and 4

Thermodynamic cycle calculations

Watson [26] first proposed a rigorous thermodynamic method that yields saturated liquid heat capacities. In essence, the difference in saturated liquid enthalpies at two temperatures is calculated by using a cycle that involves vaporizing the liquid at one temperature, expanding the vapor isothermally to a low-pressure, ideal-gas state, changing the temperature to a new value, compressing the vapor isothermally and, finally, liquefying. Choosing the two temperature levels to differ by only a small amount, one obtains

$$(C_\sigma - C_p^o)/R = -d/dT_r[(H^o - H_\sigma^v)/RT_c] - d/dT_r(\Delta H_v/RT_c) \quad (9)$$

H^o is the ideal-gas enthalpy, while H_σ^v refers to the enthalpy of the saturated vapors at the same temperature. ΔH_v is the enthalpy of vaporization.

Several ways have been proposed [4,5,17,25] to evaluate each of the two derivatives; all were tested in this study. Only one technique is described here and, in fact, combines ideas from several proposed methods.

The first derivative in Eq. (9) may be estimated from any equation of state applicable to the vapor phase. It is only a function of temperature for any specific material. A generalized correlation is shown in Fig. 6. At low reduced temperatures, this term is quite small and may

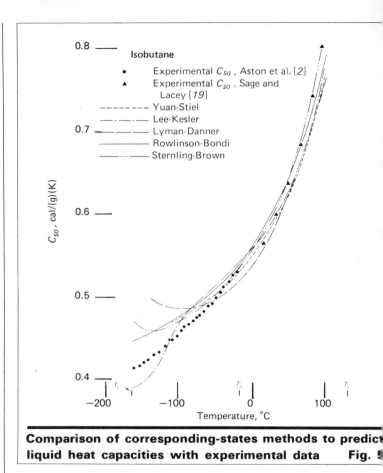

Comparison of corresponding-states methods to predict liquid heat capacities with experimental data Fig. 5

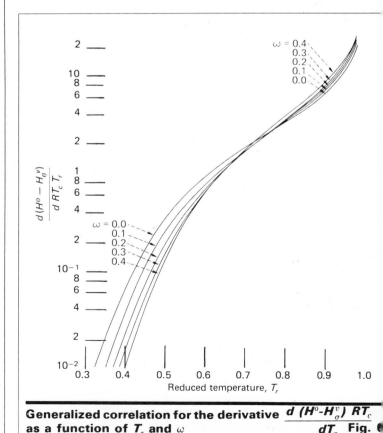

Generalized correlation for the derivative $\dfrac{d(H^o - H_\sigma^v)}{dT_r} \dfrac{RT_c}{}$ as a function of T_r and ω Fig. 6

Yuan and Stiel deviation functions for saturated liquid heat capacity — Table VII

Reduced temperature	$(\Delta C_\sigma)^{(0)}$	$(\Delta C_\sigma)^{(1)}$	$(\Delta C_\sigma)^{(0p)}$	$(\Delta C_\sigma)^{(1p)}$	$(\Delta C_\sigma)^{(2p)}$	$(\Delta C_\sigma)^{(3p)} \times 10^{-2}$	$(\Delta C_\sigma)^{(4p)}$	$(\Delta C_\sigma)^{(5p)}$
0.96	7.484	18.6						
0.94	6.175	14.7	6.19	14.7	−63.4			
0.92	5.335	13.7	5.37	13.8	−61.9			
0.90	4.76	13.1	4.80	13.0	−60.9			
0.88	4.33	12.8	4.36	12.5	−59.1			
0.86	3.99	12.5	4.03	12.2	−57.9			
0.84	3.75	12.2	3.82	11.8	−56.6			
0.82	3.57	11.9	3.65	11.6	−55.4			
0.80	3.43	11.7	3.56	11.4	−54.35			
0.78	3.31	11.5	3.42	11.2	−53.85			
0.76	3.21	11.3	3.33	11.0	−53.3			
0.74	3.14	11.2	3.23	11.3	−52.8	−0.35	−2.12	−14.8
0.72	3.07	11.0	3.06	11.9	−53.85	0.075	−3.62	−15.1
0.70	3.02	10.9	3.02	12.3	−55.4	0.66	−5.49	−14.6
0.68	2.97	10.9	2.99	12.9	−56.9	1.19	−7.65	−11.5
0.66	2.93	11.0	2.91	13.7	−59.4	1.54	−10.1	−4.0
0.64	2.89	11.2	2.80	14.7	−62.4	1.63	−12.6	7.45
0.62	2.84	11.5	2.68	16.0	−66.4	1.44	−15.35	21.6
0.60	2.79	11.8	2.58	17.4	−71.0	0.976	−18.3	36.8
0.58	2.73	12.3	2.48	18.9	−76.0	0.254	−21.4	51.3
0.56	2.67	12.9	2.36	20.7	−81.0	−0.69	−24.8	64.4
0.54	2.60	13.5	2.18	22.9	−86.6	−1.80	−28.3	75.0
0.52	2.53	14.3	1.88	25.6	−92.6	−3.03	−32.2	83.0
0.50	2.46	15.1	1.44	28.9	−99.6	−4.31	−36.3	90.1
0.48	2.38	16.0	0.886	32.7	−107	−5.59	−40.6	96.6
0.46	2.30	16.9	0.34	36.5	−115	−6.69	−45.0	104
0.44	2.22	17.8	0.096	39.5	−123	−7.55	−49.4	111
0.42	2.14	18.8						
0.40	2.05	19.8						

Data not available for $(\Delta C_\sigma)^{(3p)}$ to $(\Delta C_\sigma)^{(5p)}$ above $T_r = 0.74$; assume zero

be neglected. The correlation becomes less accurate at high reduced temperatures. It is important to note that the first derivative in Eq. (9) is always positive and, since $C_\sigma > C_p^o$, the second term in Eq. (9) must be negative. At high reduced temperatures, one is, therefore, subtracting two numbers, both of which are large.

To evaluate the second term in Eq. (9), an analytical expression for $\Delta H_v/RT_c$ as a function of T_r is usually assumed. Differentiation is then carried out as indicated. In the most general form,

$$\Delta H_v/RT_c = a(1 - T_r)^{0.38 - \beta(1-T_r)} + b(1 - T_r)^c \quad (10)$$

Eq. (10) has four undetermined parameters, a, b, c and β. Chueh and Swanson (4) assume b to be zero. Then, to determine a and β, they employ one experimental value of ΔH_v at some reference T_r and, in addition, a value of C_σ, either experimental or estimated from group contributions. To use the C_σ value, Eq. (10) is differentiated with respect to T_r and put into Eq. (9).

Reid and Sobel [17] assume b and β to be zero, and determine a from a known enthalpy of vaporization.

Chueh and Deal [5,18] choose c to be 6, and assume that $0.38 - \beta(1 - T_r) = n = f(T_{b_r})$ as given by Fishtine [6]

$$n = \begin{cases} 0.740\, T_{b_r} - 0.116 & 0.57 \leq T_{b_r} \leq 0.71 \\ 0.30 & 0.57 > T_{b_r} \\ 0.41 & T_{b_r} > 0.71 \end{cases}$$

To obtain a and b, either of two routes is suggested. If experimental values of C_σ and ΔH_v are known at any temperature, these may be used with Eq. (9) and (10). If a value of either C_σ or ΔH_v is available, b may be approximated as $19.6M/T_c$, and then a is determined.

In the testing of all these alternates, it was found that the Chueh and Deal method employing experimental values of C_σ and ΔH_v is the most accurate; this technique is tested in Table VIII.

Evaluation of the methods

Results from the five corresponding-states methods and the Chueh-Deal thermodynamic cycle method are compared with experimental liquid heat-capacity data in Table VIII. A low overall error was found in the Chueh-Deal method but, as noted before, in calculating heat capacities with this technique, one experimental liquid heat capacity was required as an input.

No liquid heat capacity correlations have been developed specifically for mixtures, and there are few experi-

Table VIII — Comparison between estimated and experimental liquid heat capacity data shows average absolute errors. Number of compounds tested are in parentheses.

	Sternling-Brown	Rowlinson-Bondi	Lyman-Danner	Yuan-Stiel	Lee-Kesler	Chueh-Deal Thermodynamic Cycle
Paraffinic hydrocarbons	3.64 (26)	4.24 (26)	1.20 (26)	1.99 (26)	4.79 (26)	1.67 (27)
Naphthenic hydrocarbons	2.89 (3)	2.06 (3)	2.51 (3)	1.97 (3)	2.72 (3)	1.22 (3)
Olefinic hydrocarbons	4.53 (5)	5.07 (5)	5.73 (4)	2.08 (5)	4.61 (5)	16.22 (4)
Aromatic hydrocarbons	2.81 (4)	1.74 (4)	3.01 (4)	2.49 (4)	2.12 (4)	0.81 (4)
All hydrocarbons and other nonpolar compounds	3.63 (39)	4.22 (39)	2.36 (39)	2.20 (39)	4.41 (29)	3.08 (38)
Alcohols	26.5 (12)	21.1 (12)	7.7 (6)	6.58 (12)	18.2 (12)	7.18 (12)
Ketones	2.86 (3)	3.09 (3)	1.9 (5)	7.82 (3)	5.44 (3)	1.80 (3)
Ethers	5.27 (4)	6.19 (4)	4.91 (3)	4.52 (4)	5.52 (4)	2.74 (4)
All polar compounds	14.5 (33)	12.4 (33)	5.3 (22)	8.20 (33)	12.1 (33)	4.56 (32)
All compounds	8.61 (72)	7.98 (72)	3.42 (61)	4.95 (72)	7.95 (72)	3.76 (70)

mentally measured values. A good estimation procedure employs a mole-fraction average of the pure component heat capacities. This technique neglects only the effect of temperature on the heat of mixing. When tested with available experimental data, errors were found to be less than 5%.

Recommendations

■ If $\Re$, κ, T_c and C_p^o values are available, use the Lyman-Danner correlation (Table V). Alternately, if $\Re$ and κ are not known, but the acentric factor (ω) and the Stiel polar factor (χ) can be determined, the Yuan-Stiel method (Table V) will yield substantially the same error.

■ For reasonably accurate estimation of liquid heat capacities below the normal boiling point, the group-contribution methods of Luria-Benson (Table II) or Missenard (Table III) are recommended.

■ For subcooled liquids above the boiling point, the Lee-Kesler method may be employed using either the original equation [11] or the simplified form shown in Fig. 3 and 4 and Table V.

In conclusion, it is noted that not every liquid heat capacity estimation method is included in this review. Those that are inaccurate or misleading were not considered. The recent paper by Tyagi [25] is included in the latter categories.

Acknowledgements

The research program dealing with liquid heat capacities was supported by the National Science Foundation. Its support is gratefully acknowledged.

The authors

Robert C. Reid is a professor of chemical engineering at the Massachusetts Institute of Technology. A co-author of several books, he has been a member of the AIChE Council, and editor of the *AIChE Journal*.

Juan L. San José has a B.S. in chemical engineering from Universidad Iberoamericana, and an Sc.D. from Massachusetts Institute of Technology. A post-doctoral researcher at the MIT Liquid Natural Gas Research Center, he recently joined the Research and Development Div. of the Hylsa Steel Group, Monterrey, Mexico.

New Correlation for Thermal Conductivity

The authors say their new correlation for predicting the thermal conductivity of organic liquids is simpler and more accurate than other relationships.

K. S. NARASIMHAN, K. M. SWAMY and K.L. NARAYANA, Regional Research Laboratory (India)

Chemical engineering design calculations frequently require thermal conductivities of organic liquids. Because of the difficulties involved in experimental determination, several attempts have been made to predict thermal conductivity, resulting in empirical or semiempirical correlations involving known parameters.

Thermal conductivity expressions proposed by several workers are summarized in Table I. As can be seen, these complex relationships require knowledge of properties such as heat capacity, density, sonic velocity, intermolecular free length, molar volume, critical temperature, etc.

Besides being tedious to handle, these expressions are only accurate to about 10%. The most accurate [7] is not a generalized correlation, and can be applied only in the case of hydrocarbons in a homologous series for which group-constants are known.

To overcome some of the drawbacks of previous methods, we offer another correlation that we believe is simpler to operate and that gives more-accurate results.

Combining the relationship proposed by Sakiadis and Coates [2]:

$$k = C_p \rho U_s L \quad (1)$$

with that derived by Jacobson [8]:

$$U_s L \rho^{1/2} = \text{constant} \quad (2)$$

one obtains the relationship:

$$k = y C_p \rho^{1/2} \quad (3)$$

However, Weber [1] has suggested that k is proportional to $C_p \rho^{4/3}$. Therefore, we considered a generalized correlation between k, C_p and ρ:

$$k = y C_p \rho^x \quad (4)$$

We evaluated x and y from known experimental values of k for 50 organic liquids at 293°K. From this, we

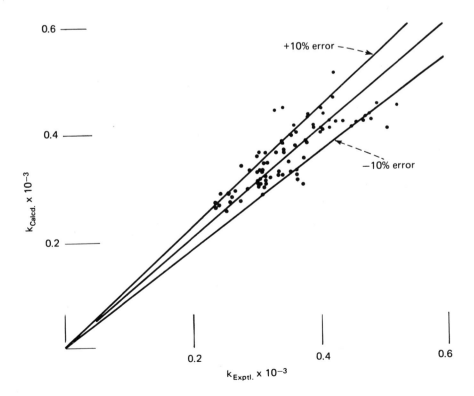

CALCULATED versus experimental values of k—Fig. 1

Originally published April 14, 1975

Nomenclature

k	Thermal conductivity, cal/(s)(cm)(°C)
C_p	Heat capacity, cal/(gm)(°C)
ρ	Density, gm/cm³
U_s	Sonic velocity, cm/s
L	Intermolecular free length, Å
σ	Molecular dia. at boiling point, Å
H_v	Latent heat of vaporization at b.p., cal/gm
V_m	Molar volume at normal b.p., gm/cm³
M	Molecular weight
c	Packing factor
C, n	Constants for a particular series
T_r	Reduced temperature, T/T_c
T_{rb}	Reduced temperature at normal b.p., T_b/T_c
T	Required temperature, °K
T_c	Critical temperature, °K
x, y	Emperical constants

Expressions for Thermal Conductivity—Table I

Expression	Average Error, %	Total No. of Points	Ref.
$0.869 \dfrac{C_p \rho^{4/3}}{M^{1/3}}$	14.8	46	1
$C_p \rho U_s L$	13.8	7	2
$41.2 C_p \left(\dfrac{\rho}{M}\right)^{4/3} \dfrac{T_b}{H_v}$	8.8	47	3
$\dfrac{3.6 \times 10^{-4} \Delta H_v}{c V_m^{2/3} M^{1/2} T^{1/2}} \left(\dfrac{1-T_r}{1-T_{rb}}\right)^{0.38}$	9.5	50	4
$\dfrac{5.6 \times 10^{-3} M^{1.26}}{V_m} C_p \Delta H_v^{1/2} \sigma$	10.7	51	5
$1.65 \times 10^{-3} C_p \rho \Delta H_v^{1/2} \left(\dfrac{T_c}{T}\right)^{1/2}$	11.0	81	6
$\dfrac{C(M)^n}{V_m}$	3.0	84	7

Comparison of $k \times 10^{-3}$ at 293°K—Table II

Liquid	Experi-mental	Present Study	Ref. 1	Ref. 2	Ref. 4	Ref. 5	Ref. 6
CCl₄	0.247	0.263	0.255	0.177	0.237	0.202	0.328
Ethyl alcohol	0.421	0.410	0.452	0.433	0.433	0.486	0.445
Bromobenzene	0.266	0.283	0.269	0.269	0.281	0.251	0.339
Ethyl benzene	0.316	0.313	0.261	0.335	0.324	0.317	0.328
O-xylene	0.321	0.331	0.268	0.251	0.342	0.339	0.334
Iodobenzene	0.244	0.268	0.253	0.219	0.269	0.230	0.323
Ethyl iodide	0.259	0.257	0.264	0.188	0.226	0.188	0.336
Avg. error, %		4.3	7.2	13.8	6.3	8.4	19.7

Accuracy of k Values by Different Methods—Table III

Authors	Total No. of Points	Points Within % 2	5	10	Above 10%
At 293°K					
Pachaippan, et al.	51	2	11	33	18
Pachaippan, et al.	48	4	16	30	18
Present study	51	9	19	39	12
Between 273°K and 343°K					
Pachaippan, et al.	81	6	26	53	28
Present study	84	12	34	68	16

obtained the following equation:

$$k_{293} = 0.877 \times 10^{-3} C_p \rho^{0.83} \qquad (5)$$

Further, we found it possible to incorporate a temperature factor into Eq. (5) to give the general relation:

$$k_T = 0.877 \times 10^{-3} C_p \rho^{0.83} \left(\frac{293}{T}\right)^{0.38} \qquad (6)$$

We tested the validity of Eq. (6) for 52 organic liquids at temperatures varying from 273°K to 343°K. Fig. 1 is a plot of the calculated vs. experimental values of k. Only 16 of 84 calculations fall outside of ±10%. The average error is ±8%.

Tables II and III compare the accuracy of Eq. (6) with relationships proposed earlier. These data show that the average errors of our correlation are lower at room and other temperatures. Thus we conclude that it is possible to predict the thermal conductivity of organic liquids with better accuracy than other methods, knowing only the liquids' heat capacity and density at 293°K. #

Acknowledgement

We wish to acknowledge the encouragement of Prof. P. K. Jena, director of the Regional Research Laboratory, and the financial assistance by CSIR, India.

References

1. Weber, H. F., *Ann. Phys. Chem.*, **10** (1880), p. 103.
2. Sakiadis, B. C., and Coates, J., *AIChE Journal*, **1** (1955), p. 275. [Also *ibid*, 3 (1957) 121.]
3. Palmer, G., *Ind. & Eng. Chem.*, **40** (1948), p. 89.
4. Viswanath, D. S., *AIChE Journal*, **13** (1967), p. 850.
5. Pachaiyappan, V., Ibrahim, S. H., and Kuloor, N. R., *J. Chem. Eng. Data*, **11** (1966), p. 73.
6. Pachaiyappan, V., *Brit. Chem. Eng.*, **16** (1971), p. 382.
7. Pachaiyappan, V., Ibrahim, S. H., and Kuloor, N. R., *Chem. Eng.*, **74** (1967), p. 140.
8. Jacobson, B., *J. Chem. Phys.*, **20** (1952), p. 927.

Meet the Authors

K. S. Narasimhan is assistant director of the Regional Research Laboratory, Council of Scientific & Industrial Research, Bhubaneswar 751004, India. He coordinates the process and equipment development activities of the process engineering group. Dr. Narasimhan is a chemical engineering graduate of the Indian Institute of Science, and obtained a Ph.D. in 1964 from the University of Sheffield, U.K.

K. M. Swamy is a scientist at the Regional Research Laboratory, working in the field of applied ultrasonics. He holds an M.S. degree (1967) from Andhra University, India.

K. L. Narayana is a senior research fellow at the Regional Research Laboratory, concerned with research and development in the field of ultrasonics. He has an M.S. degree (1965) in applied physics from Andhra University.

Quick estimation of gas heat-transfer coefficients

In many heat-exchange operations, gas-side coefficients control the overall coefficient. Here are several time-saving nomograms that will shorten your calculations for estimating these coefficients.

V. Ganapathy, Product Development, Bharat Heavy Electricals Ltd., India

☐ Preliminary sizing and layout of heat-recovery equipment such as waste-heat boilers, fired heaters, and exchangers can be a tedious chore that often involves an iterative procedure of assuming velocities, estimating gas-side coefficients, and then checking pressure drop and tube size.

To lighten the burden on the process engineer who does not have access to a computer routine for sizing exchangers, the author offers some time-saving data for predicting gas heat-transfer coefficients in two applications. The first is commonly encountered waste-gas streams flowing inside and outside heat-exchanger tubes. (These streams include off-gases from sulfuric and nitric acid plants, ammonia plants, and fertilizer plants.) The second is cross-flow of common process gases over tube bundles.

Gas-coefficient correlations

Data for calculating gas heat-transfer coefficients have been presented in nomograms (Fig. 1–3). The nomograms draw on two Nusselt relationships that describe heat transfer for fluids flowing in the turbulent region.

$$Nu = 0.023\, Re^{0.8}\, Pr^{0.4} \qquad (1)$$

$$Nu = 0.33\, Re^{0.6}\, Pr^{0.33} \qquad (2)$$

For heating and cooling waste-gases flowing inside exchanger tubes, Eq. (1) applies. Streams considered here include off-gases from ammonia combustion in nitric acid plants; sulfur combustion in sulfuric acid plants; steam-reforming in naphtha- and natural-gas-based fertilizer plants; synthesis reactions in ammonia plants; Shell gasifying in oil-based fertilizer plants; and coal gasifying in Koppers-Totzek fertilizer plants. The ranges of compositions, temperatures and pressures encountered in these various applications are summarized by Table I.

For calculating the heat-transfer coefficients of the same streams flowing across a staggered tube arrangement, Eq. (2) has been used. In addition, the coefficients

Compositions and conditions for various waste-gas streams — Table I

Gas stream	N_2	NO	H_2O	O_2	A	SO_2	CO_2	CO	CH_4	H_2S	H_2	NH_3	Pressure, atm	Temperature, °C
					Composition, volume %									
1	18–20				1–5						56–60	18–20	200–400	200–500
2	0.2–0.5						4–6	46–48	0.2–0.5	0–0.8	45–49		40–60	300–1,000
3	12–13		40–41				6–8	7–9	0.1–0.3		30–32		25–50	300–1,100
4	0.5–1.0		40–45		0.1–0.3		8–10	30–32	0–0.1	Traces	15–18		1	200–1,000
5	65–67	8–10	17–19	4–6									3–8	200–900
6	70–72		16–18	2–3			9–10						1	300–1,100
7	78–82			8–10		8–11							1	300–1,100

Gas stream key:
1 Synthesis gas   2 Shell gasifier exit-gases   3 Secondary reformer off-gases   4 Gases from Koppers-Totzek gasifier
5 Nitrous gases   6 Primary reformer flue-gases   7 Sulfur gases

Originally published September 13, 1976

466 HEAT TRANSFER CALCULATIONS

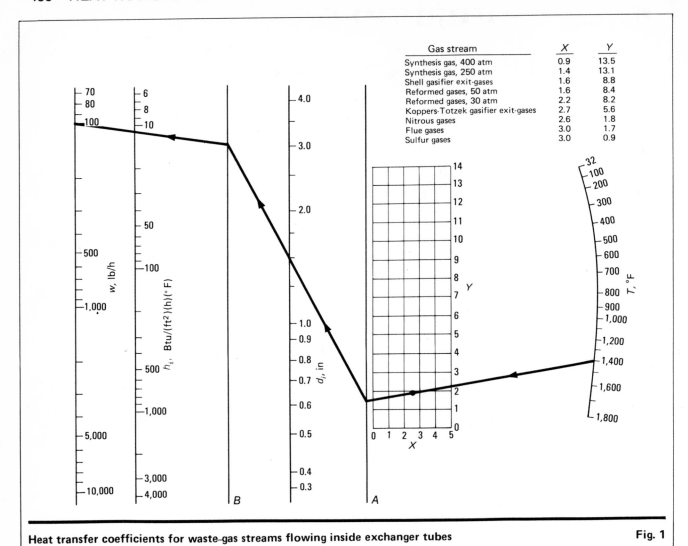

Heat transfer coefficients for waste-gas streams flowing inside exchanger tubes   Fig. 1

of a variety of common process gases (Fig. 3) have also been evaluated using Eq. (2).

In preparing the nomograms, the Nusselt relationships were rearranged, and thermal and transport properties substituted, to give the inside and outside gas coefficients as a function of stream properties.

Eq. (1) and (2) become, respectively:

$$h_i = 2.43(w^{0.8}/d_i^{1.8})f_i(P,T) \quad (3)$$
$$h_o = 0.90(G^{0.6}/d_o^{0.4})f_o(P,T) \quad (4)$$

where
$$f_i = k^{0.6}\left(\frac{C_P}{\mu}\right)^{0.4} \quad (5)$$

and
$$f_o = k^{0.67}C_P^{0.33}/\mu^{0.27} \quad (6)$$

To evaluate the functions f_i and f_o by computer, the effect of pressure on C_P was accounted for with Gambill's techniques [1]. The method of Stiel and Thodos [2] was used to adjust k and μ for pressure.

Using the nomograms

To estimate the inside gas coefficients from Fig. 1, the engineer should pick the I.D. of the tube (d_i, in), the flow per tube (w, lb/h), and the gas temperature (T, °F). Using the grid coordinates for the appropriate mixture, he then follows the diagrammed procedure to get h_i.

Example: What is the gas-side coefficient for 100 lb/h of nitrous gas flowing through a 1.5-in I.D. tube in a waste-heat boiler at 1,400°F?

Nomenclature

C_P Heat capacity at constant pressure, Btu/(lb)(°F)
d_i Tube I.D., in
d_o Tube O.D., in
G Gas mass flux, lb/(ft²)(h)
h_i Inside heat transfer coefficient, Btu/(ft²)(h)(°F)
h_o Outside heat transfer coefficient, Btu/(ft²)(h)(°F)
k Thermal conductivity, Btu/(ft)(h)(°F)
Nu   Nusselt number
P Pressure, atm
Pr   Prandtl number
Re   Reynolds number
T Temperature, °F
w Flow per tube, lb/h
μ Viscosity, lb/(ft)(h)

QUICK ESTIMATION OF GAS HEAT-TRANSFER COEFFICIENTS

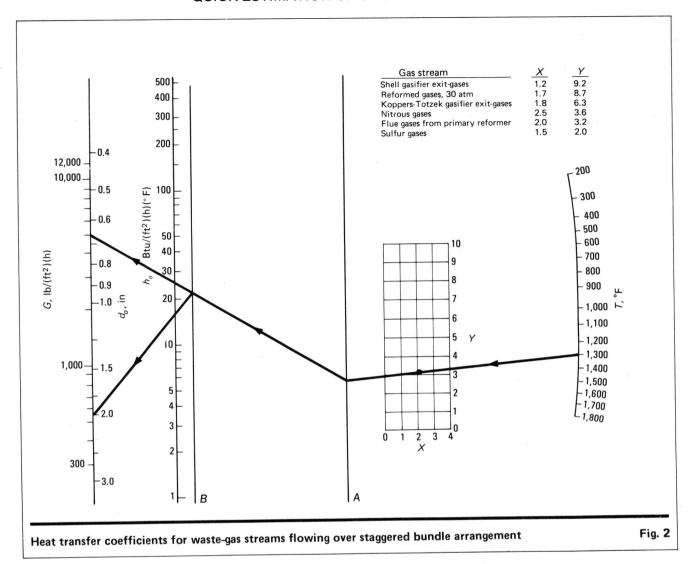

Heat transfer coefficients for waste-gas streams flowing over staggered bundle arrangement   Fig. 2

Solution: Locate the grid coordinates ($X = 2.6$, $Y = 1.8$) for nitrous gases on Fig. 1. Connect $T = 1,400°F$ with this point and extend to cut vertical line A. To this intersection point, connect a line through $d_i = 1.5$ and cut the next vertical line, B. From this second intersection, draw a line to $G = 100$. Where this line cuts h_i, the gas coefficient is given as 12 Btu/(ft²)(h)(°F).

Outside gas heat-transfer coefficients can be estimated from Fig. 2 if the mass flux of the gas (G, lb/(ft²)(h)), the outer tube diameter (d_o, in), and the gas temperature (T, °F) have been chosen.

Example: Flue gases from a primary reformer flow over a staggered bundle in the superheater of a flue-gas boiler. Gas mass flux is 5,000 lb/(ft²)(h) and tube O.D. equals 2.0 in. If the temperature is 1,300°F, what is h_o?

Solution: Find the coordinates ($X = 2.0$, $Y = 3.2$) for primary reformer gas. Connect $T = 1,300°F$ with this locus and cut vertical line A. From this intercept, extend a line to $G = 5,000$ that cuts line B. When the line-B intersection and $d_o = 2.0$ are joined, h_o is revealed to be 15 Btu/(ft²)(h)(°F).

For an inline tube arrangement, the values of h_o obtained from Fig. 2 should be multiplied by the factor 0.8. For baffled exchangers, multiply h_o by 0.6.

Also, if G falls outside the range of values in Fig. 2, the value of h_o can be arrived at by choosing some arbitrary value of G, say G_1, and determining h_o at that point. The actual value of h_o at G is obtained by multiplying the value at G_1 by $(G/G_1)^{0.6}$.

Cross-flow coefficients for other gases

Gas-side coefficients for a number of common process gases flowing across tube bundles have been provided in Fig. 3. Again, Eq. (2) was used to estimate h_o, and the calculation procedure outlined above applies for these gases as well. For the problem illustrated in Fig. 3, an h_o of 17 Btu/(ft²)(h)(°F) was calculated for 600°F air flowing at 5000 lb/(ft²)(h) over a staggered bundle of 1.0-in O.D. tubes.

General considerations for tube design

Once the gas heat-transfer coefficients have been established, the engineer must consider a variety of other factors that influence tube design. Pressure drop and mass velocity should be checked to ascertain that

468 HEAT TRANSFER CALCULATIONS

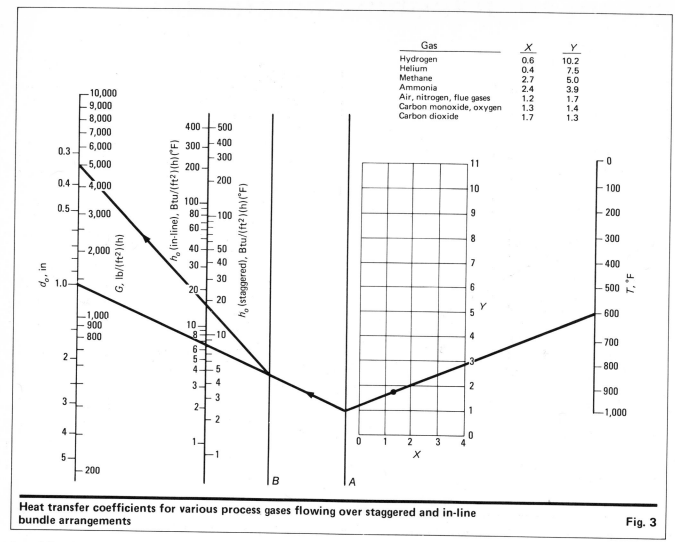

Heat transfer coefficients for various process gases flowing over staggered and in-line bundle arrangements

Fig. 3

they fall within allowable ranges; in fired heaters and waste-heat boilers, these variables are governed mostly by the available gas or steam pressure, and by the need to minimize vibration, fouling and corrosion. In coal-fired boilers, the gas mass flux, G, could be as low as 5,000 lb/(ft^2)(h), whereas in oil- or gas-fired heaters, it may range as high as 8,000–10,000. When gas density is high, G will also run high.

In fired heaters and waste-heat boilers, where available gas pressures are high, small tube diameters—usually less than 1.0 in—are desirable, in order to reduce the wall thickness and cost of the tubes; smaller tubes will also transfer greater heat fluxes. In fired heaters that use tube banks, tube sizes usually run from 1.0 to 3.0 in.

In shell-and-tube exchangers, tube sizes of 0.5 to 1.0 in are often encountered. Tube size will affect bundle configuration and pitch, as well as shell size and thickness. The final design will also be influenced by the availability of standard tube sizes and lengths, and tube sheet layouts. Often, the specifications may be altered because of factors unrelated to performance calculations; these factors include commercial availability and delivery times, the past practice of the organization, and convenience of maintenance and operation.

References

1. Gambill, W. R., *Chemical Engineering*, May 1957, p. 263.
2. Reid, R. C., and Sherwood, T. K., "Properties of Gases and Liquids," 2nd ed., McGraw-Hill, New York, 1966.
3. McAdams, W. H., "Heat Transmission," 3rd ed., McGraw-Hill, New York, 1954.
4. Rohsenow, W., and Hartnett, J. P., "Handbook of Heat Transfer," McGraw-Hill, New York, 1972.
5. Kern, D. Q., "Process Heat Transfer," McGraw-Hill, New York, 1950.
6. "Chemical Engineers' Handbook," ed. Robert Perry and C. H. Chilton, 5th ed., McGraw-Hill, New York, 1973.
7. "Steam: Its Generation and Use," Babcock & Wilcox, New York, 1972.
8. Lord, R. C., others, *Chemical Engineering*, Jan. 26, 1970, p. 96.

The author

V. Ganapathy is Senior Development Engineer with Bharat Heavy Electricals Ltd., High Pressure Boiler Plant, Tiruchirapalli-620014, India. Currently engaged in the design of boilers for utility and process services, he is particularly concerned with applications in the chemical and fertilizer industries. He holds a bachelor's in technology from the Indian Institute of Technology in Madras, India, and an M.Sc. (Engineering) in boiler technology from Madras University. He has authored over 25 articles in various U.S. and British publications.

Steam from flashing condensate

Bill Sisson, Nipak Inc.*

☐ When hot condensate under pressure is released to a lower pressure, some of it flashes to steam. The percentage of the condensate that will flash to steam can be computed as follows: Divide the difference between the heat contents of the high-pressure and low-pressure condensates by the latent heat of the low-pressure condensate and multiply by 100. Since the heat content of saturated condensate, as well as the latent heat of saturated steam, are dependent on the pressure, it is possible to make this calculation from pressures alone, without reference to the data in steam tables.

This nomograph makes that calculation. Example: If 8,000 lb/h of condensate is discharged from a 125-psig system into a flash tank operating at 10 psig, how much steam is formed? On the nomograph, connect 10 psig on the flash-steam pressure scale with 125 psig on the steam pressure scale, extend the line to the percent flashed scale and read 12.25%. Connect 12.25% with 8,000 lb/h on the condensate discharged scale, extend the line to the flash steam scale and read 980 lb/h. If the quantity of discharged condensate were 80,000 (i.e., beyond the limits of the scale) simply multiply by 10, for 9,800 lb/h of flashed steam.

*P.O. Box 338, Pryor, OK 74361.

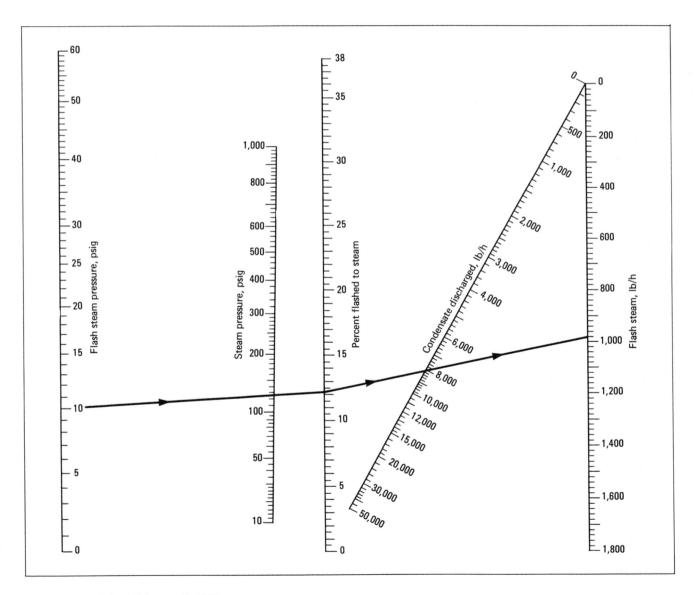

Originally published February 14, 1977

Relating heat emission to surface temperature

V. Ganapathy, Bharat Heavy Electricals Ltd.*

☐ In many problems involving room heaters and the selection of insulation and refractory, one needs to calculate the surface temperature of a pipe that will give a certain rate of heat emission, or vice versa. The relation, including both convective and radiant heat, is

$$Q = 0.296(T_s - T_a)^{1.25} + 0.174e[(T_s/100)^4 - (T_a/100)^4]$$

Where: Q = heat loss, Btu/(hr)(ft^2)
T_s = surface temperature, °R
T_a = ambient temperature, °R
e = surface emittance, ratio to a black-body

This equation is not too difficult to solve for Q, when the surface temperature is known, but it requires trial and error when Q is known and the surface temperature is sought. Trial and error is time-consuming, even with a pocket calculator. The accompanying nomograph solves the equation in either direction without the need for trial and error.

Example: Given a total heat-release rate of 210 Btu/(hr)(ft^2), and ambient temperature of 80°F, and an emissivity of 0.8, what is the surface temperature of the pipe? On the chart, connect $Q = 210$ with $e = 0.8$; note the intersection of this straight line with the curved line for $T_a = 80$, and read $T_s = 180$°F.

*Product Development (Boilers), High Pressure Boiler Plant, Tiruchirapalli-620 014, India.

Originally published December 19, 1977

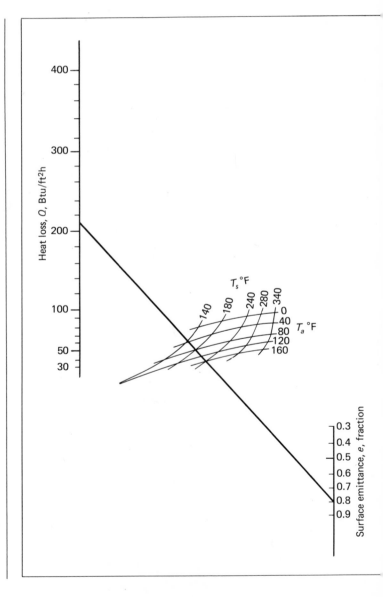

Relating heat-exchanger fouling factors to coefficients of conductivity

Andre Bestcherevnykh,
Ste. Chimique de la Grande Paroisse SA*

☐ It is convenient to think in terms of overall coefficients of heat transfer, U, or individual film coefficients, h, as heat transferred, i.e. Btu/(h)(ft^2)(°F). However, the individual and overall coefficients are related as the sum of reciprocals

$$(1/U) = (1/h_1) + (1/h_2) + (1/h_3) + \cdots (1/h_n)$$

whereas heat exchanger fouling factors, as well as insulating factors in general, are treated as the reciprocals of heat transfer—overall and individual coefficients of resistance, (h)(ft^2)(°F)/Btu, that are added directly

$$R = r_1 + r_2 + r_3 + \cdots r_n$$

Consequently, it is often difficult to get a quick sense of proportion for the contributions of individual components when evaluating the design of a heat exchanger or of built-up insulation.

This simple trick might help: Multiply resistance by 10,000 to convert the small fractions into whole numbers for the analysis; then multiply the overall heat transfer coefficient by 10,000 to recover the same units.

Example: the coefficients of heat transfer for an exchanger are 238, 3,200 and 52 Btu/(h)(ft^2)(°F) for the films inside the tubes, the tube walls and the films outside the tubes, respectively, and the fouling factors are specified as 0.002 and 0.001 (h)(ft^2)(°F)/Btu inside and outside the tubes respectively. What is the overall coefficient of heat transfer? and how much of the total surface is allowed for fouling?

Multiply resistances by 10,000:

$$10,000\ r_i = 10,000/238 = \quad 42$$
$$10,000\ r_w = 10,000/3,200 = \quad 3$$
$$10,000\ r_o = 10,000/52 = \quad 192$$
$$10,000\ r_{if} = 10,000(0.002) = \quad 20$$
$$10,000\ r_{of} = 10,000(0.001) = \quad \underline{10}$$
$$R = r_i + r_w + r_o + r_{if} + r_{of} = \quad 267$$
$$U = 10,000/267 = 37.5\ \text{Btu/(h)(ft}^2\text{)(°F)}$$

The resistance to this heat transfer is due $100(192/267) = 72\%$ to the outside film coefficient and $100(20 + 10)/ = 11\%$ to the fouling factors. Thus the best way to increase the efficiency of this exchanger would be to allow more pressure drop on the shell side for more baffling and a higher outside film coefficient.

*Division Engineering, 9, Ave. Robert Schuman, 75007 Paris, France

Originally published February 13, 1978

Section IX
HEAT TRANSFER MEDIA

Cooling-water calculations
Controlling corrosive microorganisms in cooling-water systems
Cycle control cuts cooling-tower costs
Heat-transfer agents for high-temperature systems
Low-toxicity cooling-water inhibitors—how they stack up
Organic fluids for high-temperature heat-transfer systems
Understanding vapor-phase heat-transfer media
Using wastewater as cooling-system makeup water
Water that cools but does not pollute

Cooling-water calculations

Here is a method to calculate evaporation, makeup and blowdown in an open recirculating cooling system, as well as the chemical composition of the water. Recommended limits are provided to determine allowable cycles of concentration, along with procedures to estimate the pH and conductivity of the concentrated water.

R. G. Kunz, A. F. Yen and T. C. Hess, Air Products and Chemicals, Inc.

☐ Cooling towers remove heat by evaporative cooling of a stream of water that continuously passes through heat exchangers (Fig. 1). The evaporation can be estimated by making a heat balance around the cooling tower:

$$E = 0.001(C_r)(\Delta T) \qquad (1)$$

where E and C_r are the evaporation and circulation rates in gpm, respectively, and ΔT is the temperature differential between the hot and cold water in °F. The evaporation amounts to 1% of the circulation for every 10°F ΔT. This equation assumes that the cooling effect arises solely from the latent heat of evaporation at a constant 1,000 Btu/lb, and that the heat capacity of the recirculating water is also constant at 1 Btu/(lb)(°F).

Since in the evaporation process the nonvolatile impurities in the makeup water are concentrated, to prevent excessive concentrations some of the recirculating water must be removed from the system. In addition to this deliberate "blowdown," a relatively small amount of entrained water is lost as fine droplets in the air stream. This is called "windage" or "drift loss." Unlike evaporation, windage carries dissolved impurities with it, and reduces dissolved solids in the recirculating water.

For typical mechanical draft towers that use fans to aid air circulation, 0.1% to 0.3% of the circulating rate is considered typical [1]. With patented mist-eliminator designs, some cooling-tower manufacturers warrant as low as 0.008% for windage loss [2].

Makeup water, M_u, must be added to replace the evaporation, E; blowdown, B; and windage, W, losses. By material balance:

$$M_u = E + B + W \qquad (2)$$

Windage is often included in the blowdown term. In this case, B would represent the upper limit for the amount of water purged from the system via the blowdown line.

Water in its natural form is never pure. It is a universal solvent and will dissolve a little of everything; it will also carry suspended material. The type and amount of foreign matter depends upon the water source.

The major dissolved impurities in cooling-tower makeup water are silica, sodium, calcium, magnesium, iron, bicarbonates, chlorides and sulfates. Other mineral impurities occur in smaller amounts. Concentrations are expressed as milligrams/liter (mg/L) or parts per million by weight (ppm). Analyses of several cooling-tower makeup waters are shown in Table I. Accounting of makeup-water impurities is necessary if one is to predict the composition of the recirculating water.

A cooling tower is also an efficient scrubber of suspended particulates and gases such as CO_2, SO_2 and NH_3 from the atmosphere. Other uncontrolled contamination can occur from leakage of process fluids into the recirculating water.

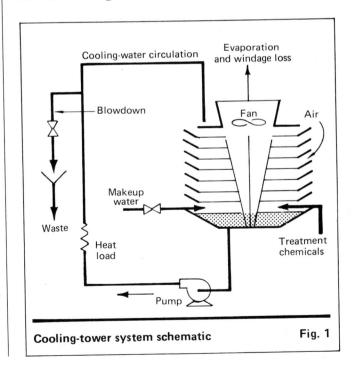

Cooling-tower system schematic   Fig. 1

Originally published August 1, 1977

Composition of cooling-tower-makeup waters					Table I
Constituents*	Water A	Water B	Water C	Water D	
Aluminum	0	0	0.3	0	
Calcium	15	147	31	15	
Magnesium	6	99	8	0	
Sodium†	279	602	21	39	
Potassium	0	0	0	2	
Iron	0	0	0	0	
Bicarbonate	334	179	31	31	
Carbonate	0	0	15	0	
Sulfate	5	137	51	80	
Chloride	60	1,272	30	9	
Fluoride	0	0	1	1	
Phosphate	1	0	0	0	
Hydroxide	**	**	1.7	**	
Carbon dioxide	0	0	0	0	
Silica	27	15	7	0	
pH	8.0	7.6	10	7.2	
Phenolphthalein alkalinity	0	0	25	0	
Methyl orange alkalinity††	274	147	50	25	

*All units are in mg/L of the constituent.
†Calculated values based on electroneutrality.
**Negligible.
††Rounded-off numbers.

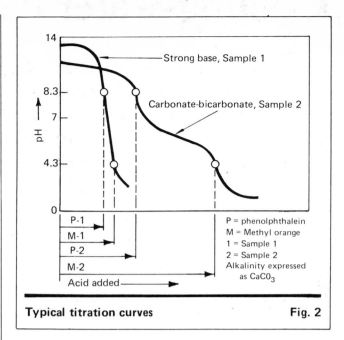

Typical titration curves   Fig. 2

pH, alkalinity and hardness

Among the most important measurements to characterize the impurities in cooling-tower water are pH, alkalinity and hardness. pH is defined as the negative logarithm to the base 10 of the hydrogen-ion activity, expressed in gram-moles/L; in a dilute solution the activity is equal to the concentration.

A knowledge of pH is important because pure water dissociates into hydrogen and hydroxyl ions as follows:

$$K_w = (H^+)(OH^-) = 10^{-14} \text{ at } 25°C \quad (3)$$

but $\quad pH = \log\left(\dfrac{1}{H^+}\right) = -\log(H^+)$

and $\quad pOH = \log\left(\dfrac{1}{OH^-}\right) = -\log(OH^-)$

then $\quad \log\left(\dfrac{1}{H^+}\right) + \log\left(\dfrac{1}{OH^-}\right) = -\log(K_w)$

and $\quad pH + pOH = pK_w = 14 \text{ at } 25°C \quad (4)$

Although the hydrated proton or hydronium ion (H_3O^+) is thought to be the actual species existing in solution, it is written as the hydrogen ion (H^+) for simplicity. The pH of a neutral solution at 25°C is 7.0, and decreases slightly with increasing temperature [3].

This equilibrium is shifted by acids that donate H^+ or by bases that provide OH^-. The acidic pH range is indicated by numbers ranging from 0 to 7, and the basic range from 7 to 14. A solution's pH—the measure of free hydrogen-ion activity—should not be confused with alkalinity, which indicates the resistance to change of pH upon addition of an acid.

Alkalinity, commonly expressed as mg/L of $CaCO_3$, is defined as the capacity to neutralize an acid; it is measured by titration of a sample with a strong acid. Typical titration curves are shown in Fig. 2. Phenolphthalein (or P-) alkalinity refers to the amount of acid necessary to reach the phenolphthalein color-change end point (about pH 8.3), and methyl orange (M-), or total alkalinity, is that necessary to cause a color change in the methyl-orange indicator (about pH 4.3). Alkalinity is a measure of the buffer capacity of the original sample and is not synonymous with the pH. Note that in Fig. 2 the sample with the higher initial pH has the lower alkalinity.

Since the principal contributors to the alkalinity of a natural water are the hydroxide, carbonate and bicarbonate ions, alkalinity measurements are often interpreted exclusively in terms of these ions, and the contribution of other species such as phosphates, silicates, borates, sulfides, sulfites, ammonia and salts of organic acids—which might also be present—is ignored. On this basis, simplified mathematical relationships [4] among the various contributors to the measured alkalinity are shown in Table II. This table assumes the incompatibility of hydroxide and bicarbonate alkalinities [5], an assumption that is not valid above pH 9.

The various alkalinity relationships may also be computed from methyl-orange alkalinity and an accurately determined pH value by solving Eq. 3, along with the following simultaneous equations:

$$(OH^-) + 2(CO_3^=) + (HCO_3^-) - (H^+) = \text{M-alkalinity}/50{,}000 \quad (5)$$

$$K_2 = (H^+)(CO_3^=)/(HCO_3^-) = 4.7 \times 10^{-11} \text{ at } 25°C \quad (6)$$

Dissolved CO_2 can then be calculated from:

$$K_1 = [(H^+)(HCO_3^-)]/(CO_2) = 4.5 \times 10^{-7} \text{ at } 25°C \quad (7)$$

The P-alkalinity can be used as a check of the sum of hydroxide and carbonate alkalinities.

The original definition of hardness included all polyvalent metal ions capable of precipitating soap. The principal contributors to hardness are calcium, magnesium, strontium, ferrous, manganous, and aluminum ions. Of these, only calcium and magnesium are of practical importance in natural waters; hardness is commonly understood to include only these two ions because they can deposit scale in a cooling tower by reacting with certain of the anions present.

Calcium, magnesium and total hardness are also expressed as mg/L of calcium carbonate. Total hardness can be determined directly by titration with ethylenediamine tetraacetic acid [4], or by separate measurements of calcium and magnesium. Magnesium hardness is either measured or computed by difference of total and calcium hardness. Calcium hardness is always determined by measurement. Waters containing up to about 50–75 mg/L hardness are considered soft; above 200–300 they are very hard. Waters in the intermediate range are classified as hard.

The term mg/L $CaCO_3$ is encountered in water-treating calculations. Alkalinity and hardness are invariably expressed in terms of this unit, but concentrations of other constituents such as sodium, chloride and sulfate ions are also frequently expressed as calcium carbonate. Care must be exercised in interpreting water analyses.

Expressed mg/L of $CaCO_3$ means the amount of material having the same number of chemical equivalents as the indicated quantity of $CaCO_3$, but has nothing to do with the composition. By definition:

mg/L $CaCO_3$ = (mg/L as X)(equivalent weight of $CaCO_3$)/(equivalent weight of X) (8)

where X represents any species under consideration.

Since $CaCO_3$ has a molecular weight of 100, and an equivalent weight of 50, it serves as a convenient common denominator. This enables one to perform calculations using whole numbers of the same magnitude as the original mg/L of each constituent, rather than with gram-equivalents of the order of 10^{-4}. Conversion factors for several important ions are listed in Table III.

Electroneutrality, cycles of concentration

The first step in water-treatment calculations is to examine the composition of the makeup water for electroneutrality. According to the principle of electroneutrality, the ionic positive and negative charges must be in balance. For example:

$$(H^+) + (Na^+) + 2(Ca^{++}) + 2(Mg^{++}) + \cdots = (OH^-) + (HCO_3^-) + 2(CO_3^=) + (Cl^-) + 2(SO_4^=) + 3(PO_4^\equiv) + \cdots \quad (9)$$

In the above equation, the bracketed terms are in moles/L. These terms, along with their numerical coefficients, are the chemical equivalents of each ion and can be expressed as milliequivalents/L or, more conveniently, as mg/L $CaCO_3$. Hence:

Total cations (as mg/L $CaCO_3$) = total anions (as mg/L $CaCO_3$) (10)

Theoretically, for a complete water analysis, the above equation will be satisfied. In practice, however, typical water analysis data are seldom complete, and certain deviations from electroneutrality as calculated can be expected. For example, the sodium ion concentration is often not measured but is instead computed by difference from the other constituents. Moreover, in the pH range of 6–9, the hydrogen and hydroxide ion concentrations tend to cancel each other and are small compared to the other ions. Therefore, they are normally neglected in ion-balance calculations. Results of an ion-balance computation are shown in Table IV.

As water evaporates, the impurities left behind will concentrate to an excessive level unless a portion of the recirculating water is bled off. By definition, cycles of concentration is the term used to indicate the degree of concentration of the recirculating water compared to that of the makeup water. Therefore:

X in recirculating water = (X in makeup)(cycles) (11)

where X represents any nonvolatile species dissolved in

Alkalinity relationships [4] — Table II

Condition*	Hydroxide alkalinity† as $CaCO_3$	Carbonate alkalinity† as $CaCO_3$	Bicarbonate alkalinity† as $CaCO_3$
P = 0	0	0	M
P < ½M	0	2P	M−2P
P = ½M	0	2P	0
P > ½M	2P−M	2(M−P)	0
P = M	M	0	0

*P = phenolphthalein alkalinity
M = methyl orange alkalinity
†Hydroxide alkalinity as $CaCO_3$ = $[OH^-] \times 50{,}000$
Carbonate alkalinity as $CaCO_3$ = $[CO_3^=] \times 100{,}000$
Bicarbonate alkalinity as $CaCO_3$ = $[HCO_3^-] \times 50{,}000$

Conversion factors — Table III

Constituents	To convert ion to $CaCO_3$ equivalent, multiply by	To convert $CaCO_3$ equivalent to ion, multiply by
Aluminum	5.56	0.18
Calcium	2.50	0.40
Iron, Fe++	1.79	0.56
Iron, Fe+++	2.69	0.37
Magnesium	4.10	0.24
Potassium	1.28	0.78
Sodium	2.18	0.46
Bicarbonate	0.82	1.22
Carbonate	1.67	0.60
Chloride	1.41	0.71
Fluoride	2.63	0.38
Phosphate	1.58	0.63
Sulfate	1.04	0.96
Sulfite	1.25	0.80
Sulfide	3.13	0.32

Checking electroneutrality for Water D			Table IV
	Concentration, mg/L as CaCO$_3$		
Constituents	Ions	Cations	Anions
Aluminum	0	0	—
Calcium	15	37.5	—
Magnesium	0	0	—
Sodium	—	—	—
Potassium	2	2.56	—
Iron	0	0	—
Bicarbonate	31*	—	25.4
Carbonate	0*	—	0
Sulfate	80	—	83.2
Chloride	9	—	12.7
Fluoride	1	—	2.63
Silica	0	—	—
Total	138	40.06	123.93

*Phenolphthalein alkalinity = 0; methyl orange alkalinity = 25.4 mg/L as CaCO$_3$.
Note: (total anions)−(total cations) = 83.87 mg/L as CaCO$_3$, assuming that the difference is entirely attributable to sodium. Sodium concentration is about 39 mg/L as Na.

the makeup water, unless significant precipitation or shift in ionic equilibrium occurs.

Cycles of concentration are controlled by comparing the concentration of some parameter in the recirculating water with its value in the makeup water. In practice, chloride, hardness, or conductivity (which is roughly proportional to the total dissolved solids, TDS) is used.

Chloride has been chosen because it remains soluble even at a high degree of concentration. However, use of the chloride concentration is not valid with less than 5 mg/L in the makeup water [7], and is questionable when the recirculating water has a high chlorine demand and chlorine is being used as a biocide. Chlorine demand is the difference between the amount of chlorine added and the amount of chlorine left in solution at the end of a specified contact period [8].

Hardness is likewise questionable if precipitation is at all likely, whereas conductivity is amenable to automatic control. Prediction of recirculating water conductivity will be discussed in a subsequent section.

The relationship between the amount of makeup, evaporation and cycles of concentration, C, is made according to the equation:

$$M_u = EC/(C - 1) \quad (12)$$

which is plotted in Fig. 3 for a ΔT of 10°F [1].

Based on Eq. 12, it is evident that maximizing the cycles of concentration can minimize water consumption. As shown in Fig. 3, the curve is nearly flat beyond about 3 cycles of concentration, reaching a point of diminishing returns at about 6 to 7 cycles. There is ordinarily little incentive to operate at a higher number of cycles because of the danger of scale formation or enhanced corrosion, when the concentrations of certain species are exceeded.

Acid is often added for pH control to allow operation at higher cycles. The added acid will react with the water's alkalinity to replace carbonate and bicarbonate ions with sulfate ions and depress the pH. The bulk of the CO_2 formed is stripped from the tower, and the added water is negligible:

$$2HCO_3^- + H_2SO_4 \rightarrow SO_4^= + 2CO_2 + 2H_2O \quad (13)$$
$$CO_3^= + H_2SO_4 \rightarrow SO_4^= + CO_2 + H_2O \quad (14)$$

The reason for acid addition in some cases is that the pH of the recirculating water tends to increase with increasing cycles of concentration, and a higher pH intensifies the tendency of scale deposition. As the recirculating water becomes more concentrated, its bicarbonate alkalinity increases, but dissolved CO_2 in equilibrium (Eq. 7) remains more or less constant. Left alone, the pH will naturally equilibrate between 8 and 8.5 at the operating cycles of concentration [9]. When high-alkalinity water is used as makeup, the equilibrium pH may fall above 8.5 [10]. With certain Gulf Coast well waters, which contain very high alkalinity, the pH may exceed 9.

These generalizations are borne out by the curves of Fig. 4. This figure portrays the expected pH change with cycles of concentration for the four different makeup waters of Table I. Without adding acid, the pH of Water A would reach 9.5 at 6 cycles. The pH for Water B, at 6 cycles, would be lower because this water contains a lower alkalinity. However, in practice, Water B would be limited to less than 2 cycles without acid addition because of hardness and potential scaling. The corresponding pH is 8.3.

Water C is a lime-softened, alum-coagulated water whose alkalinity is low despite its high pH, because of incomplete recarbonation. The pH would first decrease as the water is recirculated and CO_2 in solution achieves a more reasonable level. After this initial drop, pH would rise for this water also. Similarly, the initial pH dip for Water D indicates an unsaturated condition with respect to CO_2, typical of some well waters. The low alkalinity would permit the pH to increase only to about 8, even at 6 cycles. These curves demonstrate that the makeup water's alkalinity, and not its pH, is responsible for the pH of the recirculating water.

Predicting recirculating water pH

The pH of the recirculating water is affected primarily by the carbonate-bicarbonate-CO_2 system. The stripping or scrubbing of CO_2 in the cooling tower, the equilibrium distribution of bicarbonate and carbonate, and the pH of the recirculating water are interrelated. Even with adequate input information, to solve the simultaneous equations that describe the relationship for these species is, at best, cumbersome.

Instead, the empirical correlation of total alkalinity and pH of the recirculating water (Fig. 5), based on routine monitoring data gathered from over 40 cooling towers, provides an excellent, simple tool for estimating purposes. These towers represent different types of mechanical construction as well as different makeup-water compositions and water-treatment programs. Water temperatures are typically in the range of 90°F, with a 10 to 20°F ΔT.

Approximately 400 pH-alkalinity data points were subjected to least-squares regression, in which pH was

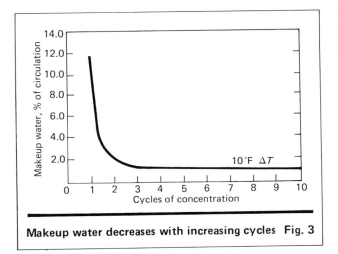

Makeup water decreases with increasing cycles Fig. 3

considered as the dependent variable. The 0.8 correlation coefficient obtained reflects the scatter, but indicates a definite correlation between pH and alkalinity [12]. At the 90% confidence level, pH can be predicted within ±0.75 units when the alkalinity is known. These data approximate the rule-of-thumb that 20 to 50 mg/L of methyl orange alkalinity will correspond to a pH of 6.0 to 7.0 [7]. Water quality, treatment, and aeration in the tower may affect both alkalinity and pH.

Scatter is caused primarily by differences in the concentration of dissolved CO_2. A 1 mg/L increase in CO_2 in the range of 1–10 mg/L produces a decrease of 0.1–0.2 pH units at constant alkalinity. The CO_2 concentration maintained in solution would be expected to vary both from tower to tower and with liquid and gas rates in a single tower. Analytical results depend on sample handling.

Variability in temperature and dissolved solids also contributes slightly to the scatter. An increase in temperature from 50 to 140°F would result in a decrease of 0.2 pH units; for an increase in TDS of 3,000 mg/L, the pH would decrease by only 0.1.

Changes in the ambient atmosphere from time to time and from place to place may also alter the pH characteristics. For example, Sussman [9] points out the difference in pH between two cooling towers in the same immediate vicinity caused by the pickup of SO_2 from a plant stack by one of them but not the other.

Scatter in the pH-alkalinity relationship is also caused by the presence of various water-treatment chemicals. Detailed consideration of additives to control corrosion, scaling and biological fouling is beyond the scope of this article; an excellent discussion is contained elsewhere [13].

Fig. 6 shows the line of best fit through the cooling-tower pH data from Fig. 5, plus several lines computed from theory with dissolved CO_2 as a parameter. Since the majority of the alkalinity between 4.3 and 8.3 is caused by the bicarbonate ion in equilibrium with CO_2, then from Eq. 7, a semilogarithmic plot should yield a straight line with a slope of unity, provided that the CO_2 is constant:

$$\text{pH} = \log(HCO_3^-) + pK_1 - \log(CO_2) \quad (15)$$

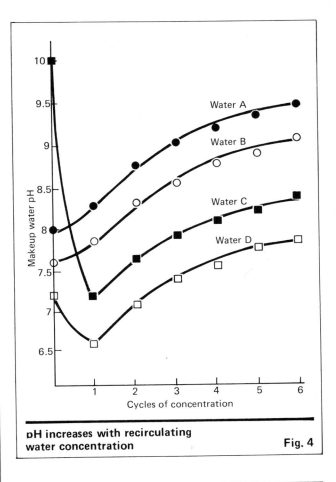

pH increases with recirculating water concentration Fig. 4

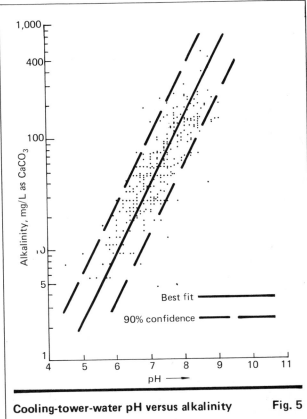

Cooling-tower-water pH versus alkalinity Fig. 5

However, the observed slope of the best-fit line implies that the dissolved CO_2 is not exactly constant, but is indeed a function of pH. To attain low pH values, much acid must be added and much CO_2 is formed in solution (Eqs. 13 and 14). The empirical fit in the middle pH range of 6.8 to 7.6 corresponds to a dissolved-CO_2 concentration between 5 and 10 mg/L. Other researchers indicate [14] concentrations of 5–10 mg/L CO_2 attainable with typical wooden degasifiers that have geometries similar to cooling towers. This range of CO_2 concentrations is considerably in excess of the 0.4 mg/L in equilibrium with the CO_2 in the atmosphere [3]. Dissolved-CO_2 values higher than anticipated imply a mass-transfer limitation in stripping this gas from the recirculating water.

Lower CO_2 values at the upper end of the experimental line are to be expected since little or no acid would be added for control in this region, with minimal formation of dissolved CO_2; the only sources of CO_2 are the makeup water and the thermal breakdown of the bicarbonate ion:

$$2HCO_3^- \rightarrow CO_3^= + CO_2 + H_2O \qquad (16)$$

and pickup from the atmosphere. Finally, the presence of carbonate above pH 8.3 should cause the line to curve upward, when the *total* methyl-orange alkalinity is plotted against pH, but the anticipated curvature in the empirical relationship is obscured by the scatter of the data in Fig. 5.

Quantity of acid required

Towers using chromate ion as the corrosion inhibitor normally operate slightly below a pH of 7 (6.2–6.8 is a common control range), and nonchromate inhibitor programs are usually accompanied by a somewhat higher pH. To control the pH at the desired level, sulfuric acid is added to the recirculating water (Eqs. 13 and 14). Since one molecular weight of sulfuric acid will destroy two equivalents of alkalinity, the quantity of sulfuric acid required can be formulated as indicated in this equation:

$$W = [(\Delta Alk)/50{,}000] \times 0.5 \times 98 \times B \times$$
$$(1{,}440/10^3)(1/0.9319) \times 8.34 =$$
$$0.0126 \times \Delta Alk \times B \qquad (17)$$

where $W = 66°$ Baumé sulfuric acid required, lb/d; ΔAlk = alkalinity to be reduced in the recirculating water, mg/L as $CaCO_3$; B = blowdown water flowrate (including windage loss), gpm.

Total dissolved solids and conductivity

Conductivity and TDS are two of the most critical parameters in cooling-tower control. The ability to estimate these two parameters permits consideration of design limits and allows the selection of a conductivity meter with an appropriate range.

TDS can be calculated by summation of all the dissolved species in the water, including silica [expressed as $Si(OH)_4$], but not including dissolved gases. A calculated TDS value should be higher than a measurement of filterable residue. In such an analysis, a sample is usually dried at 103–105°C, resulting in loss of CO_2 and conversion of bicarbonate ion to carbonate (Eq. 16).

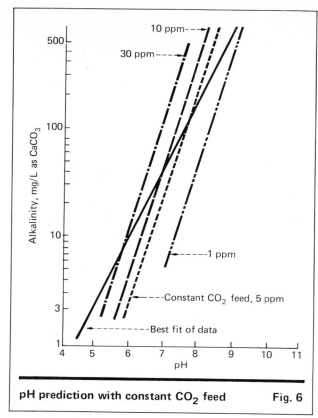

pH prediction with constant CO_2 feed Fig. 6

According to the stoichiometry, up to about 51% of the bicarbonate originally present will not be measured as TDS [4].

For a rough estimate, conductivity can be calculated by dividing the filterable residue by a factor within 0.55–0.7 [4]. A more rational estimate is attained from the conductivity-factor technique. Conductivity of the recirculating water can be predicted to within ±25% of measured values from the concentrations of its major ions, their activity coefficients, and a factor specific for each ion [4,6], as shown in Table V, followed by summation of the products according to the equations:

$$\text{Ionic strength} = I = 0.5 \sum_i C_i Z_i^2 \qquad (18)$$

Activity coefficient:

$$\log f_i = -A Z_i^2 \left(\frac{I^{1/2}}{1 + I^{1/2}} - 0.2 I \right) \qquad (19)$$
$$(I < 0.5)$$

$$\text{Activity} = a_i = f_i c_i \qquad (20)$$

where I = ionic strength, moles/L; C_i = concentration of the i ion, moles/L; Z_i = charge of the i ion; A = constant $\cong 0.5$, tabulated as a function of temperature [6]; f_i = activity coefficient for the i ion; a_i = activity of the i ion, mg/L; c_i = concentration of the i ion, mg/L.

Activity coefficients are necessary to account for the effect of concentration of the recirculating water. The Davies equation [15], which is Eq. 19, has been found to provide activity coefficients adequate for conductivity predictions. The method is valid in the pH range 6–9, where the concentrations of the highly conductive

H⁺ or OH⁻ ions are not significant, and for conductivities above 90 micromhos/cm [4].

Fig. 7 shows predicted versus measured conductivities at 18 or 25°C for the available 136 recirculating-water data points, and 116 makeup-water points. Ionic strength varied between 0.01 and 0.12 for the cooling waters and was as low as 0.001 for fresh makeup water. The 1:1 parity line appears to correlate the data quite well, even for low values of conductivity. Conductivity increases (or decreases) 2% for every 1°C above (or below) 25°C [4]. This relationship permits adjustment of conductivity from ambient to tower temperatures.

Several criteria for determining the maximum allowable cycles of concentration are summarized in Table VI. These guidelines that appear in the literature should be tempered by one's own experience and/or the recommendations of a reputable water-treatment service company. Unless otherwise stated, quoted limits refer to the recirculating water.

Langelier and Ryznar indexes

The theoretically based Langelier Index is defined as the difference between the actual pH and the pH at which a given water would be saturated with $CaCO_3$:

$$pH - pH_s = pH - (pK_2 - pK_s + pCa + pAlk) \quad (21)$$

The terms in the above equation are negative logarithms of the second ionization constant for carbonic acid, the solubility product [6] for $CaCO_3$ ($K_s = 4.5 \times 10^{-9}$ at 25°C), and the molar and equivalent concentrations of the calcium ion and the methylorange alkalinity, respectively. The constants are a function of temperature and total dissolved solids [3]. Nomographs are available to simplify this calculation [1,16].

When the $pH - pH_s$ is positive, the system has a tendency to deposit scale, but when it is negative, the system tends to dissolve $CaCO_3$ scale and, by inference, is corrosive. When the $pH - pH_s$ is zero, the system is theoretically in balance. However, an index of -0.5 to $+0.5$ is said to be unreliable for predicting scaling tendency [9].

Chromate treatments operate at negative values of the Langelier Index, and treat the water to prevent corrosion. Nonchromate treatments operate at positive values to take advantage of the corrosion protection offered by a thin film of calcium salts. Gross precipitation of $CaCO_3$ is inhibited by the addition of deposit-control chemicals, including chelating agents and crystal-growth modifiers.

Nonchromate treatments are operated at a $pH - pH_s$ between 0.5 and 1.0 [17,18], at 1.0 [9,18], or at a maximum of 2.5 for calcium carbonate precipitation, and 1.5 for calcium phosphate precipitation [19]. These values will vary according to the maximum temperature in the system [18]. A range of about 1.0–1.5 appears to be in order.

The Ryznar Index is an empirical index, defined as $2 pH_s - pH$, and is based on a study of operating data with water having various saturation indexes. For this particular index, 6.5 is the nominal neutral point. Values of 6 or less indicate scaling [7,9,17], and 7 or greater are corrosive [9]. The 6 to 7 range is too close to define.

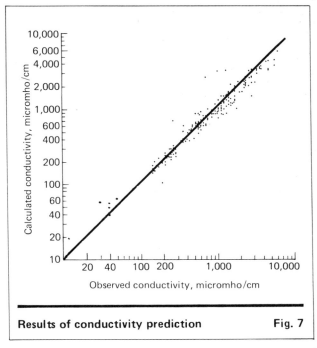

Results of conductivity prediction Fig. 7

The range 5.5–6 corresponds to a Langelier Index of about +1.0.

The pH should be kept within the control range for the corrosion inhibitor program in use. In general, chromate programs should be kept above a minimum of about pH 6 [13], with somewhat higher values recommended for other types of treatment.

At a high pH [7,18,20], especially with a high-alkalinity, low-hardness water [21], chlorination for biological control oxidizes the lignin in cooling-tower wood to aldehydes and acids [13]. Delignification reportedly commences at a pH of 7.8 [13]. Moreover, chlorine's biocidal properties are weak at a pH 8 and negligible at pH 9 [13]. The effectiveness of certain other biocides is also diminished in this same pH range.

Dissolved materials

Solubility of $CaCO_3$, 20–50 mg/L, at cooling-tower temperatures [9,18] is accounted for in the Langelier

Conductivity factors for ions commonly found in water Table V

Ion	Conductivity @ 25°C, micromho/cm per mg/L of ion
Aluminum	3.44
Bicarbonate	0.715
Calcium	2.6
Carbonate	2.82
Chloride	2.14
Fluoride	2.86
Iron, Fe++	1.93
Iron, Fe+++	3.65
Magnesium	3.82
Potassium	1.84
Phosphate	2.18
Sodium	2.13
Sulfate	1.54

Guidelines for cooling-tower operation — Table VI

Parameters	Limits Minimum	Limits Maximum	Remarks
Langelier Saturation Index	+0.5	+1.5	Nonchromate programs
Ryznar Stability Index	+6.5	+7.5	Nonchromate programs
pH	6.0	8.0	
Calcium, mg/L as $CaCO_3$	20–50	300	Nonchromate program
		400	Chromate program
Total iron, mg/L		0.5	
Manganese, mg/L		0.5	
Copper, mg/L		0.08	
Aluminum, mg/L		1	
Sulfide, mg/L		5	
Silica, mg/L		150	For pH < 7.5
		100	For pH > 7.5
$(Ca) \cdot (SO_4)$, product		500,000	Both calcium and sulfate expressed as mg/L $CaCO_3$
Total dissolved solids, mg/L		2,500	
Conductivity, micromhos/cm		4,000	
Suspended solids, mg/L		100–150	

Index. $MgCO_3$ requires a pH of 9 or greater for precipitation, and is usually not found in cooling systems.

With a nonchromate corrosion inhibitor, calcium hardness should be at least 20–50 mg/L as $CaCO_3$ [19,22] but less than 300 mg/L as $CaCO_3$ [13]; for a chromate program the upper limit can be extended to 400 [13]. The minimum concentration is to enable formation of a protective film [22], and the maximum is to ensure that calcium can be kept from precipitating, using normal concentrations of treatment chemicals.

Phosphates, whether originating in the makeup water or from added treatment chemicals, form sticky phosphate sludges with calcium and other metals. The tendency to precipitate is accentuated with increasing calcium concentrations and pH. The interested reader is referred elsewhere [13,23] for further details.

Calcium also forms a precipitate with sulfates $CaSO_4 \cdot 2H_2O$ (gypsum) at much higher concentrations, 1,200–2,000 mg/L [8,18]. To avoid scaling, the sum of calcium and sulfate (both expressed as mg/L $CaCO_3$) should be below 1,500, and their product should be less than 400,000–500,000 [9]. This corresponds to a gypsum concentration of about 1,200 mg/L and a solubility product of the same order of magnitude as $K_{sp} = 2.4 \times 10^{-5}$, as quoted by McCoy [13].

Silica in natural waters results from the degradation of silica-containing rocks in the earth's crust. This silica occurs as suspended particles, in a colloidal or polymeric state, and as dissolved material [4]. Dissolved silica in natural water exists in the form of $Si(OH)_4$, orthosilicic acid; silicate anions are formed only above a pH of about 9 [15]. Silica concentrations are not used in computing the cation-anion balance.

In a cooling system, silica may deposit as a scale on heat-transfer surfaces and may therefore be the limiting parameter in a cooling-water treatment program. The form of silica is important [19]. The scaling potential of magnesium silicate increases as the pH rises above 8 [19], but magnesium silicate is generally not actually encountered in cooling systems unless the pH is above 9 [7]. Dissolved silica can coprecipitate with iron and manganese hydroxides, and may be adsorbed on various hydroxides, including aluminum, manganese and magnesium. Examples of silica scale precipitated with calcium salts are pictured in a report [24]. Deposit-control agents, containing acrylate polymers, reduce the tendency of silica scale to form.

Levels in the makeup water of 40–50 mg/L, expressed as SiO_2, are a source of silica scale and may limit the cycles of concentration [18]. Recommended maximum values of silica in the recirculating water, expressed as mg/L of SiO_2, are 150–175 [13], 150–200 [7,18], and as a maximum 200 [19]. If the silica level is above 100, a pH of 7.5 or less is preferred [19]. Based on this information, a maximum silica concentration of 100–150 in the circulating water appears to be conservative, with values at the lower end of the range being indicated for a pH of 7.5 and above.

Heat-insulating deposits can also form by oxidation of soluble ferrous ions in the water supply, followed by precipitation of a hydrated ferric oxide sludge [17], which is converted by heat to iron oxides [7]. Iron deposits can also originate from corrosion products, and precipitation of iron compounds can be aggravated by iron bacteria [7].

Iron should be limited to 0.5 mg/L [18]; above this amount, a phosphonate/polymer treatment is required to control iron deposition [19]. Concentrations greater than 3 mg/L require higher phosphonate levels because of adsorption of the phosphonate on iron [19]. With levels approaching 10 mg/L, iron removal should be considered [19]. Manganese, also, should be limited to 0.5 mg/L [18].

Acceptable concentrations of copper range from 0.04–0.08 mg/L; concentrations greater than 0.1 mg/L cause significant galvanic corrosion [13]. High values of total dissolved solids will accelerate copper corrosion with nonchromate treatments [19].

		Table VII			
\multicolumn{6}{l}{Calculation procedure for Water D at 5 cycles}					

Constituents	Concentration, mg/L	$\tfrac{1}{2} C_i Z_i^2$ (10^3 mole/L)	f_i	A_i, mg/L	Conductivity, micromho/cm
Aluminum	0	0	0.300	0	0
Calcium	75	3.74	0.586	44.0	114
Magnesium	0	0	0.586	0	0
Sodium	195	4.24	0.875	171	364
Potassium	10	0.128	0.875	8.75	16
Iron	0	0	0.586	0	0
Bicarbonate	155	1.28	0.875	137	98
Carbonate	0.45	0.015	0.586	0.264	1
Sulfate	400	8.33	0.586	234	360
Chloride	45	0.636	0.875	39.4	84
Fluoride	5	0.13	0.875	4.38	13
Carbon dioxide	4	0	0	0	0
Silica	0	0	0	0	0
Sum	890	18.5	—	—	1,050

Notes: Calculated TDS = 890 mg/L, theoretical filterable residue = 810 mg/L, pH = 7.8.

Aluminum causes deposition problems. The recommended level is less than 1 mg/L [19]. Fluoride, added as sodium fluoride or sodium fluorosilicate, will complex the aluminum ion, but not suspended alumina-floc carryover from a clarifier [7]. Sulfides should be limited to less than 5 mg/L because of their corrosive effects [18].

TDS, suspended solids, conductivity

High values of total dissolved solids increase conductivities, cause galvanic corrosion, interfere with inhibitor-film formation, accelerate the corrosion of copper in nonchromate programs, and destroy the effectiveness of certain biocides. An upper limit of 2,000 mg/L has been recommended to minimize galvanic corrosion [18]. Conductivities above 4,000 micromhos/cm indicate extremely corrosive conditions and can cause marginal to poor results in a nonchromate program [19]. A conductivity of 4,000 would correspond to a TDS between 2,000 and 3,000. A good cutoff point for TDS appears to be approximately 2,500 mg/L.

Suspended solids enter the cooling tower with the makeup water, with the air passing through the cooling tower, and from debris within the cooling-tower circuit. They can be inorganic or biological in origin.

Suspended solids in a cooling tower are abrasive [17] and can lead to underdeposit corrosion [17,20]. Suspended solids also consume certain biocides [21,25], reducing the effectiveness of the microbiological control program. Recommended limits are 100 mg/L [9,18] up to a maximum of 150 mg/L [18]. These solids can be controlled by chemicals, which tend to disperse or agglomerate the particles, or they can be removed to the desired level by installing a sidestream filter.

Sidestream filters are frequently used in controlling suspended solids in the recirculating water [7,9,17,18,20,26], by filtering from 1–5% of the total circulation. This technique [25] has obtained as much as 80% reduction in suspended solids (10 microns and greater). The smaller, colloidal particles can be dispersed with a suitable silt dispersant [25]. Additional design information and operating experience on sidestream filters can be found elsewhere [27].

Example problem

Perform the following step-by-step calculation procedures for five cycles, using Water D (Table I) as makeup water:

1. *Compute evaporation, makeup and blowdown*—The circulation rate is 50,000 gpm and has a 15°F cooling range. Evaporation is therefore 750 gpm (Eq. 1); makeup water is 937.5 gpm (Eq. 12); and the sum of blowdown and windage losses is 187.5 gpm (Eq. 2). The windage contribution will amount to 100 gpm at 0.2% of the circulation rate.

2. *Check electroneutrality of makeup water*—A comparison of total cation and anion concentrations, all expressed as mg/L $CaCO_3$, shows anions to be in excess (Table IV). To satisfy electroneutrality, 39 mg/L (as Na) must be added to account for the difference.

3. *Cycle system up*—For all constituents except carbonate, bicarbonate and carbon dioxide, the concentration at 5 cycles is simply 5 times that of the makeup water (Tables I and VII). Total alkalinity expressed as $CaCO_3$ will be 5 × 25.4 mg/L = 127 mg/L. The corresponding pH is 7.8 (Fig. 5). The concentrations of carbonate and bicarbonate can be calculated from Eq. 5 and 6, and the CO_2 concentration from Eq. 7.

For five cycles and above, the pH is near the upper limit (Table VI), and provisions for acid addition may be justified. Concentrations of bicarbonate, carbonate, carbon dioxide and sulfate would change by adding acid. To achieve an arbitrarily selected pH of 7, the alkalinity must be reduced to 42 mg/L (Fig. 5). At this pH, the bicarbonate and carbonate concentrations are 51 and 0 mg/L of the ion, respectively. Carbon dioxide concentration is about 8 mg/L.

To reduce the alkalinity from 127 to 42 mg/L would require: (127 − 42) × (1/50) × (0.5 × 98) = 83.3 mg/L of H_2SO_4, or 82 mg/L as $SO_4^=$, which need 201 lb/d of 66° Baumé acid (Eq. 17). Total sulfate in the recirculating water at 5 cycles (482 mg/L) is the sum of the sulfate added (82 mg/L), plus that concentrated from the makeup water (5 × 80 = 400 mg/L).

4. *Calculate* TDS—Summing up the concentrations of all constituents yields a TDS of 890 mg/L, before adding acid (Table VII). Similarly, TDS after acid addition is 870 mg/L (Table VIII).

5. *Estimate conductivity*—Compute ionic strength (Eq. 18) and activity coefficients (Eq. 19). The product of concentration and activity coefficient is activity (Eq. 20). Multiply the activity of each ion by the ap-

Detailed calculation results for Water D — Table VIII

Constituents*	Adjusted M_u	2X BA†	2X AA†	3X BA	3X AA	4X BA	4X AA	5X BA	5X AA	6X BA	6X AA
Calcium	15	30	30	45	45	60	60	75	75	90	90
Magnesium	0	0	0	0	0	0	0	0	0	0	0
Sodium	39	78	78	117	117	156	156	195	195	234	234
Potassium	2	4	4	6	6	8	8	10	10	12	12
Bicarbonate	31	62	51	93	51	124	51	155	51	185	51
Carbonate	0	0	0	0	0	0	0	0.5	0	1	0
Sulfate	80	160	178	240	272	320	368	400	482	480	586
Chloride	9	18	18	27	27	36	36	45	45	54	54
Fluoride	1	2	2	3	3	4	4	5	5	6	6
Carbon dioxide	0	7	8	6	8	5	8	4	8	4	8
pH	7.2	7.1	7.0	7.4	7.0	7.6	7.0	7.8	7.0	7.9	7.0
Phenolphthalein alkalinity as $CaCO_3$	0	0	0	0	0	0	0	0	0	2	0
Methyl orange alkalinity as $CaCO_3$	25.4	51	42	76	42	102	42	127	42	152	42
Langelier Saturation Index at 100°F	−1.92	−0.84	−1.00	−0.20	−0.85	+0.24	−0.74	+0.63	−0.65	+0.89	−0.57
Ryznar Stability Index at 100°F	+10.44	+8.78	+9.00	+7.80	+8.70	+7.12	+8.48	+6.54	+8.3	+6.12	+8.14
TDS, calculated	180	350	360	530	520	710	700	890	870	1,060	1,040
Theoretical filterable residue	160	320	340	480	500	650	660	810	840	970	1,010
Condition: 25°C, micromho/cm	240	460	470	650	660	840	840	1,050	1,050	1,200	1,200

*All units are in mg/L of the constituent, unless otherwise indicated.
†BA = before adding acid, AA = after adding acid, X = cycles of concentration.

Values of critical parameters for various waters — Table IX

	Water A 2X* BA*	Water A 2X* AA*	Water A 4X BA	Water A 4X AA	Water B 2X BA	Water B 2X AA	Water B 3X AA	Water C 3X BA	Water C 3X AA	Water C 5X BA	Water C 5X AA
Langelier saturation index at 100°F	+1.85	−1.05	+2.92	−0.68	+2.04	−0.14	+0.02	+1.09	−0.45	+1.86	−0.2
Ryznar stability index at 100°F	+5.1	+9.1	+34.7	+8.36	+4.27	+7.28	+6.96	+5.77	+7.90	+4.58	+7.5
pH	8.8	7.0	9.31	7.0	8.35	7.0	7.0	7.95	7.0	8.3	7
Calcium, mg/L as $CaCO_3$	75	75	150	187.5	735	735	1,102.5	232.5	232.5	387.5	387
Total iron, mg/L	0	0	0	0	0	0	0	0	0	0	
Manganese, mg/L	0	0	0	0	0	0	0	0	0	0	
Copper, mg/L	0	0	0	0	0	0	0	0	0	0	
Aluminum, mg/L	0	0	0	0	0	0	0	0.9	0.9	1.5	1
Sulfide, mg/L	0	0	0	0	0	0	0	0	0	0	
Silica, mg/L	54	54	108	135	30	30	45	21	21	35	3
$(Ca) \cdot (SO_4)$, product = (mg/L as $CaCO_3)^2$	780	38,690	3,120	280,860	209,445	389,080	919,580	36,995	65,770	102,770	189,81
TDS, mg/L	1,200	1,070	2,400	2,670	4,740	4,820	7,350	670	640	1,120	1,07
Conductivity, micromho/cm	1,270	1,260	2,450	2,810	6,900	6,830	9,650	775	795	1,200	1,21

*BA = before adding acid, AA = after adding acid, X = cycles of concentration.

propriate conductivity factor (from Table V), to yield the conductivity contributed by that ion; then sum all the contributions to obtain 1,050 micromhos/cm, which is the conductivity of the water (Table VII).

6. *Compute parameters for other cycles of concentration*—These results are listed in Table VIII.

7. *Determine the approximate cycles allowable for Water D*—A comparison of the entries of Table VIII with the guidelines in Table VI reflects the good quality of Water D; 6 to perhaps 7 cycles appears to be reasonable.

8. *Determine allowable cycles for other waters*—Similarly, for waters A, B and C, the calculated values of limiting water-quality parameters at the critical cycles of concentration are presented in Table IX.

Water A—Because of high alkalinity, the natural pH at 2 cycles already far exceeds 8 with a Langelier Index of +1.85. Acid should be added to lower the pH and saturation index. At 4 cycles, silica barely exceeds 100 mg/L. TDS is marginally above 2,000 mg/L, and the natural pH is too high. At 5 cycles or above, TDS is too high. The maximum cycles of concentration for this poor-quality water would be limited to about 4 for a nonchromate program. Field experience has confirmed this prediction.

Water B—This water contains a high level of Ca hardness, exceeding the allowable Ca level even at 2 cycles. Its natural pH is also too high. The Langelier Index is exceeded at 2 cycles without acid, and $CaSO_4$ controls at 3 cycles with acid. Because TDS and conductivity are high, galvanic corrosion is likely to occur unless proper treatment is provided. In practice, this water is limited to about 1.5 cycles with acid addition.

Water C—Without adding acid, the maximum allowable cycles would probably be 3. The pH and Langeliers and Ryznar indexes are approaching their limits. Scaling is likely to occur at higher cycles. Decreasing the pH to 7, or slightly lower, and using a chromate program would be able to increase the allowable cycles to about 5. At this pH, the corrosive tendency can be controlled by chromate. Calcium exceeds the guideline for a nonchromate program. However, the negative Langelier saturation index suggests this would be academic. Aluminum is also a potential problem. The current control range for this water is 4 to 6 cycles.

References

1. "Betz Handbook of Industrial Water Conditioning," 6th ed., Chap. 31-33, Betz Laboratories, Inc., Trevose, Pa., 1962.
2. Furlong, D., "The Cooling Tower Business Today," *Env. Sci. Technol.,* Vol. 8(8), pp. 712-716, 1974.
3. Fair, G. M., Geyer, J. C., and Okun, D. A., "Water and Wastewater Engineering," Vol. 2, Chap. 28-29, Wiley, New York, 1968.
4. "Standard Methods for the Examination of Water and Wastewater," 14 ed., Amer. Public Health Assn., 1193 pp., Washington, D.C., 1976.
5. Sawyer, C. N. and McCarty, P. L., "Chemistry for Sanitary Engineers," 2nd ed., p. 331, McGraw-Hill, New York, 1967.
6. Dean, J. A., ed., "Lange's Handbook of Chemistry," 11 ed., Section 5, McGraw-Hill, New York, 1973.
7. "Cooling Water Treatment Manual," Technical Practices Committees Publication No. 1, 35 pp., National Association of Corrosion Engineers, Houston, Tex., 1971.
8. White, G. C., "Handbook of Chlorination," p. 295, Van Nostrand Reinhold Co., New York, 1972.
9. Sussman, S., "Facts on Water Use in Cooling Towers," *Hydrocarbon Processing,* Vol. 54(7), pp. 147-153, July, 1975.
10. Freedman, A. J., and Shannon, J. E., "Alkaline Cooling Water Treatments," Paper presented at the International Water Conference sponsored by the Engineers' Society of Western Pennsylvania, Pittsburgh, Pa., Oct., 1972.
11. Beychok, M. R., "Aqueous Wastes from Petroleum and Petrochemical Plants," p. 153, Wiley, London, 1967.
12. Lipson, C., and Sheth, N. J., "Statistical Design and Analysis of Engineering Experiments," pp. 387-390, McGraw-Hill, New York, 1973.
13. McCoy, J. W., "The Chemical Treatment of Cooling Water," 237 pp., Chemical Publishing Co., New York, 1974.
14. Appelbaum, S. B., "Demineralization by Ion Exchange," pp. 110-113, Academic Press, New York, 1968.
15. Stumm, W., and Morgan, J. J., "Aquatic Chemistry," pp. 83, 395-396, Wiley-Interscience, New York, 1970.
16. Caplan, F., "Is your water scaling or corrosive?" *Chem. Eng.,* Sept. 1, 1975, p. 129.
17. Brower, E. W., "Corrosion and Water Treatment," Chapter 36 in ASHRAE Handbook & Product Directory 1973 Systems, pp. 36.1 to 36.20, American Society of Heating, Refrigerating, and Air Conditioning Engineers, New York (1973).
18. Sussman, S., "Treatment of Water for Cooling, Heating, and Steam Generation," Chapter 16 in "Water Quality and Treatment," 3rd ed., pp. 499-525, American Water Works Assn., Handbook of Public Water Supplies, McGraw-Hill, New York (1971).
19. Rue, J. R., "Non-Chromate Treatment in Cooling Water Systems," Paper No. 47e presented at the AIChE 82nd National Meeting, Atlantic City, N.J., Sept. 1, 1976.
20. Capper, C. B., "The Protection of Open Recirculating Cooling Systems," *Effluent and Water Treatment J.,* Vol. 14(10), pp. 577-583, Oct. 1974.
21. Grier, J. C., and Christensen, R. J., "Biocides Give Flexibility in Water Treatment," *Hydrocarbon Processing,* pp. 283-286, Nov. 1975.
22. Harpel, W. L., and Donahue, J. M., "Effective Phosphate/Phosphonate Treatments Replace Chromate-Based Programs," Paper No. TP 117A presented at the Cooling Tower Institute Annual Meeting, Houston, Tex., Jan. 29-31, 1973.
23. Green, J., and Holmes, J. A., "Calculation of the pH of Saturation of Tricalcium Phosphate," *J. Am. Water Works Assn.,* Vol. 39, pp. 1090-1096, 1947.
24. Midkiff, W. S., and Foyt, H. P., "Amorphous Silica Scale in Cooling Towers," Paper No. TP 148A presented at the Cooling Tower Institute Annual Meeting, Houston, Tex., Jan. 19-21, 1976.
25. Ward, W. J., "Cooling Water Treatment Chemicals," in "Volume 2 Cooling Towers," a CEP Technical Manual, pp. 60-63, American Institute of Chemical Engineers, New York, 1975.
26. Donahue, J. M., and Hales, W. W., "Improve Cooling Water Treatment," *Hydrocarbon Processing,* pp. 101-106, June, 1968.
27. Hayes, J. W., Jr., "Current Practice in Sidestream Filtration for Cooling Towers," Paper No. TP 25A presented at the Cooling Tower Institute Annual Meeting, Houston, Tex., Jan. 23, 1967.

The authors

R. G. KUNZ   A. F. YEN   T. C. HESS

Robert G. Kunz is Manager, Environmental Engineering, Cryogenics Systems Div., Air Products and Chemicals, Inc., Allentown, PA 18105, where he supervises a group making recommendations for water-treatment and setting specifications for environmental control equipment. He holds a B.Ch.E from Manhattan College and a Ph.D. from Rensselaer Polytechnic Institute, both in chemical engineering, and an M.S. degree in environmental engineering from Newark College of Engineering. He is a member of AIChE, ACS, Air Pollution Control Assn., Sigma Xi and Tau Beta Pi. He is a licensed professional engineer in Louisiana, New Jersey, Pennsylvania and Texas.

Alan F. Yen is Senior Environmental Engineer in the Cryogenics Systems Div., Air Products and Chemicals, Inc., Allentown, PA 18105. His experience includes research and development, and consulting in municipal and industrial pollution control abatement. He has a B.S. in civil engineering from Cheng Kung University, Taiwan, an M.S. from Pennsylvania State University in sanitary engineering and a Ph.D. in environmental engineering from Cornell University. He is a member of the Water Pollution Control Federation, and is a registered professional engineer in Pennsylvania.

Thomas C. Hess is an Environmental Engineer, Cryogenic Systems Div., Air Products and Chemicals, Inc., Allentown, PA 18105. Formerly, he was an instructor at the U.S. Naval Nuclear Power School. He holds a B.S. in chemical engineering from Carnegie-Mellon University and an M.S. in environmental engineering from Johns Hopkins University. His main scientific interest is in the application of statistics and computer modeling in the water-treatment field.

Controlling corrosive microorganisms in cooling-water systems

Fungi, algae and bacteria in cooling water systems can cause rapid and severe wood rot and metal corrosion. This article tells how to control and identify the most important types of these harmful organisms.

*Paul R. Puckorius, P.R. Puckorius & Associates, Inc.**

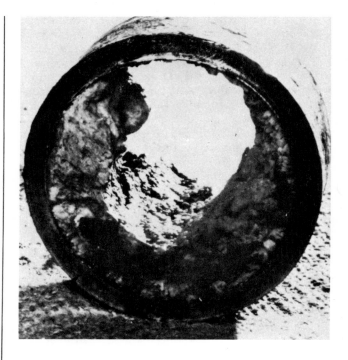

☐ Corrosion problems in cooling-water systems are usually due to dissolved solids, dissolved oxygen, contaminants, ineffective corrosion inhibitors or poor pH control. However, there is often rapid and severe corrosion that cannot be stopped, even when inhibitors are used. Such corrosion may be caused by microbiological organisms. Here we will discuss these organisms and provide guidelines for their identification and control.

Organisms typically found in cooling systems fall into three groups: fungi, algae, and bacteria.

Fungi

Fungi are yeasts and molds. They are microscopic plants, larger than bacteria. Fungal growths are often seen by the naked eye in the form of mold growing on bread, cheese and meat, and as large, shelf-type formations on stumps and limbs of dead trees. In cooling-water systems, fungi result in slime and wood decay. Fungal growths are found on cooling-tower wood and basin walls, and in heat exchangers.

One species of penicillin (a mold) causes wood decay in cooling towers. In wood decay, the cellulose in wood is consumed by organisms until it loses its strength.

Fungi are not directly corrosive to metals in a cooling-water system. However, the deposits these organisms produce establish differential corrosion cells that cause corrosion, and interfere with the action of a corrosion inhibitor by shielding the metal surfaces from it.

Wood deterioration may not be seen easily. It may be deep rot (Fig. 1), in which the surfaces of the wood appear unaffected while the inside has little or no strength. However, surface decay is easy to see, and soft wood can be broken apart with a screwdriver, or even a fingernail.

In the laboratory, samples are tested by microscopic examination and incubation of test-specimens to determine the presence of decay organisms (Fig. 2). Wood scrapings suspected of containing decay-causing organisms are placed in a petri dish containing only pure cellulose as a carbon source. Clear areas show where cellulose has been consumed, just as wood would be attacked in a cooling system.

Wood can be protected by impregnation with toxic salts that inhibit fungal growth. Excess salts must be removed, otherwise the cooling water will carry them. This may result in plating on metal surfaces, which causes corrosion. For example, copper salts are sometimes used, and they can cause corrosion of mild steel and aluminum.

Fungi can also be controlled by the periodic application of fungicides such as pentachlorophenol salts or tributyl tin compounds. Chlorine is not effective against these organisms.

Algae

Algae, like fungi, are relatively large organisms. They are generally colored blue, blue-green or green by the presence of chlorophyll; sunlight is necessary for growth. Algae commonly cause slimy deposits in cooling towers where sunlight and water are present. Deposits of dead algae provide food for bacteria and fungi, since they act as a filter to catch other organisms. Algae are not known to cause corrosion directly, except for occasional occurrences under their deposits.

Control can be effected by covering cooling-tower decks to prevent sunlight from reaching the tower water, or with

---

*Portions of this article were delivered at the March 1978 meeting of the Natl. Assn. of Corrosion Engineers, CORROSION/78, in Houston, Tex.

Originally published October 23, 1978

CONTROLLING CORROSIVE MICROORGANISMS IN COOLING-WATER SYSTEMS

Surface of wood affected by fungus-caused deep rot is relatively intact   Fig. 1

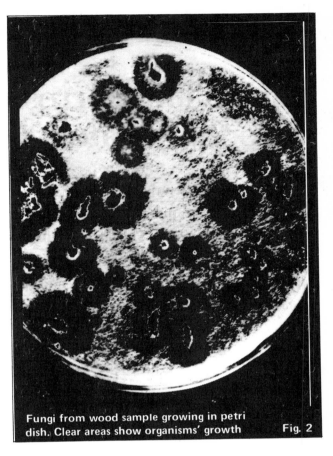

Fungi from wood sample growing in petri dish. Clear areas show organisms' growth   Fig. 2

chemicals such as chlorine, quaternary ammonium compounds and copper salts.

Bacteria

Bacteria can be seen only when large colonies are present. Species are identified with a microscope at a magnification of 800 times or more. Each has a specific action and often is referred to by its effect on materials. Some classes of bacteria cause slime, corrosion, gas-production and decay of wood. Bacteria common to cooling-water systems are the slime-formers, both the non-spore-forming and spore-forming types. They produce a slimy, gelatinous deposit that can clog heat-exhanger tubes and shield metal surfaces from inhibitors, promoting corrosion. However, they do not directly cause corrosion.

Iron-depositing bacteria cause corrosion directly by producing iron oxide as byproducts of their metabolism. (See table.) They convert soluble ferrous iron salts into ferric oxide. The lead photograph shows a cooling-water line partially plugged by the growth of iron-depositing bacteria. These deposits also shield metal surfaces from corrosion inhibitors, promoting corrosion via a concentration-cell mechanism. Because the organism removes the ferrous iron from the area of corrosion, the reaction is accelerated.

Iron-depositing bacteria can be controlled easily with chlorine and many non-oxidizing biocides such as quaternary ammonium compounds.

Sulfide-producing (or sulfate-reducing) bacteria produce chemicals that result directly in corrosion of metals. These convert water-soluble sulfur compounds to hydrogen sulfide. This conversion usually starts with sulfates that either occur naturally or come from the addition of sulfuric acid. The bacteria metabolize the sulfur and discharge hydrogen sulfide, creating and living in an anaerobic, reducing environment.

Hydrogen sulfide is acidic and aggressively attacks metals, principally mild steel, but also stainless steel and copper alloys. However, most metals are attacked by a combination of low pH, sulfides and reducing conditions. Nickel and nickel-based alloys are severely pitted under such conditions. This corrosion is often readily identified by concentric ridges formed in the pits (Fig. 3). These ridges can be seen without magnification, but severe attack can obliterate them.

In a recirculating cooling-water system, corrosion due to these organisms can occur at a rapid rate. Perforation of a 16-mil mild-steel corrosion coupon occurred within 60 days, a rate of approximately 100 mils/yr. Stainless steel, nickel and other alloys subject to this attack failed in 60 to 90 days in heat exchangers and vessels. Pitting rates vary from 50 to 200 mils/yr; penetration is dependent on the degree of contamination and the rate of growth.

Even with chromate-zinc-based corrosion inhibitors and good control of pH in a recirculating cooling system, these organisms have the ability to penetrate metals rapidly. The hydrogen sulfide reacts with chromate and zinc to remove the inhibitors from solution and from contact with the metals, resulting in reduced chromates and precipitates of zinc salts, thus causing fouling and lack of metal protection.

Using chlorine to control these organisms is not effective

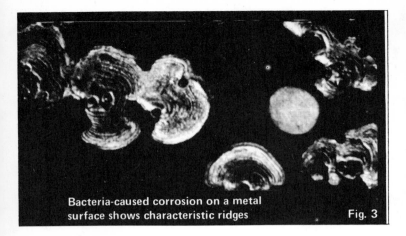

Bacteria-caused corrosion on a metal surface shows characteristic ridges — Fig. 3

Some types of bacteria that directly cause corrosion in metals

Bacterium	Action	Problem
Gallionella Crenothrix Spaerotilus	Convert soluble ferrous ions to insoluble ferric ions	Produces iron oxide deposits Increases corrosion
Desulfovibrio Clostridium Thiobacillus	Hydrogen-sulfide producers (sulfate-reducers)	Corrodes metals Reduces chromates Destroys chlorine Precipitates zinc
Thiobacillus	Sulfuric-acid producer	Corrodes metals
Nitrobacter Nitrosomonas	Nitric-acid producers	Corrodes metals

because: (1) the organisms are usually covered by slime masses that prevent the chlorine from reaching the sulfide producers, and (2) the hydrogen sulfide surrounding these organisms reacts with chlorine to form chloride salts that negate the effect of chlorine.

One part of hydrogen sulfide theoretically requires about 8.5 parts of chlorine for complete reaction. It has been shown by laboratory tests that a high concentration of hydrogen sulfide and low pH occur around the organism colony. Thus, even large amounts of chlorine can be consumed without killing the bacteria.

One of the organisms in this group is *Clostridium*. In addition to producing hydrogen sulfide gas, *Clostridium* yields methane, providing a nutrient source for slime-producing bacteria that grow in the immediate vicinity.

All of these bacteria are difficult to control. Chlorine can't get to them owing to the gases they produce. Thus special toxicants are necessary. A long-chain (fatty-acid) amine salt is effective. Other nonoxidizing biocides such as organo-sulfur compounds (methylene bis-thiocyanate) are toxic to all organisms in this group.

Nitrifying (or acid-producing) bacteria, often found in cooling water systems, produce nitric acid from ammonia. These bacteria are found in soil where they convert (or fix) ammonia to nitrates—which plants use.

When present in cooling systems, these organisms use ammonia either from the atmosphere or from equipment leaks. Normally, ammonia causes a rise in pH; however, if these bacteria are present the pH drops, because of the production of nitric acid. This results in corrosion of metals that are attacked by low-pH conditions—mainly mild steel, but also copper and aluminum.

These organisms are not adversely affected by oxygen and do not neutralize corrosion inhibitors such as chromate or zinc. Sometimes, pH control is effected in a cooling-water system by the controlled injection of ammonia in the presence of these organisms. Their presence can be determined by special microbiological tests not normally employed for routine testing of cooling systems. Fortunately, chlorine is extremely effective, as are many nonoxidizing biocides, in controlling these organisms. With any appreciable amount of ammonia, however, chlorine is neutralized and is unavailable for control; thus chlorine may not appear effective.

The last group (see table) of these corrosive organisms, *Thiobacillus*, converts soluble sulfur compounds into sulfuric acid. Hydrogen sulfide may be the source of sulfur used to produce the acid conditions.

Fig. 4 shows a typical attack on mild steel by sulfide-producing bacteria in a well-treated (from the standpoint of corrosion inhibitors) cooling tower system. Severe attack occurred in spite of the use of a high level of chromate-based corrosion inhibitor (20 to 25 ppm chromate) and good pH control (6.5 to 7.5). Periodic chlorination was used with no other toxicants. Sulfide-producing bacteria caused localized pitting, and resulted in equipment failures in 60 to 70 days. It took over 90 days of concentrated attention to correct the situation. Chlorine, a non-oxidizing biocide and a dispersant were used.

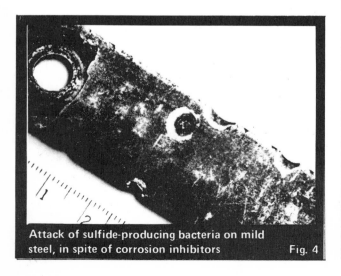

Attack of sulfide-producing bacteria on mild steel, in spite of corrosion inhibitors — Fig. 4

The Author

Paul R. Puckorius is president of P.R. Puckorius & Associates, Inc., Box 4846, Cleveland, OH 44126, a consulting firm specializing in cooling water, wastewater and potable water. He has 25 years of experience, primarily in cooling water corrosion, scale and microbiological problem-solving, and he is the author of over 40 articles dealing with such problems. He holds a B.A. in chemistry from North Central College and has done graduate work at Northwestern. His affiliations include the Natl. Assn. of Corrosion Eng. and the Cooling Tower Inst.

Cycle control cuts cooling-tower costs

The blowdown requirement of a cooling-tower system will be minimized if the concentration ratio of the system is maintained at a constant level.

Thomas Haupt, Logic Control Systems

☐ The effectiveness of pH or corrosion control is critically dependent on blowdown control. Scale inhibition is most commonly achieved by controlling pH or using threshold techniques (or both); corrosion is controlled by maintaining an inhibitor at a given concentration.

Control of pH maintains calcium bicarbonate in equilibrium with some other more-soluble and -stable calcium compound, depending on the type of acid used. The purpose of the blowdown, of course, is to ensure that the saturation point of the resulting compound is not reached. Needless to say, systems should always be run well below the saturation point of any compound, and these levels are guaranteed by analyzing the makeup water to determine the maximum concentration ratio allowable.

Scale control using threshold-type treatments is achieved by keeping the chemical additive at a prescribed level in the system, below that required to complex the calcium. The mechanisms by which this technique retards calcium-carbonate scale growth are believed to be adsorption with the crystal during the early stages of formation, and nuclei replacement within the crystalline structure. An amorphous material is then formed, whose structure and adhesive properties depend on the concentration of scale-forming ingredients in the circulating water.

Owing to this softening process, the residual calcium ion is maintained at the saturation point in equilibrium with calcium carbonate, permitting the system to operate at a relatively high concentration ratio. Since the

Originally published September 11, 1978

inhibitor is maintained at a prescribed concentration, its effectiveness is maximized with minimum blowdown.

Corrosion inhibitors, like threshold-type scale inhibitors, are maintained at prescribed concentrations. Hence, the consumption rate of corrosion additives depends directly on the blowdown rate. Using these water-conditioning techniques, there are no methods of water treatment known to the author that are inconsistent with cycle control.

In general, cycle control is performed by: analyzing the makeup water to determine the levels of the problem minerals that will be introduced into the system; deciding at what levels these minerals can be tolerated in the treated system; increasing the mineral concentrations to the level determined, by reducing the blowdown rate; and, as will soon be apparent, minimizing the blowdown requirement by holding the cycles constant.

Cycles of concentration

Cycles of concentration, henceforth called cycles, are defined precisely as the ratio of the total dissolved solids (TDS) of the circulating water to the TDS of the makeup water.

$$\text{Cycles }(C) = \frac{\text{TDS circulating water }(b)}{\text{TDS makeup water }(m)} \quad (1)$$

Cycles are seen to vary directly with the TDS of the circulating water and inversely with the TDS of the makeup water. To exactly determine the cycle variation due to changes in these variables, the total differential is used:

$$dC = (1/m)\left|\frac{db}{m=\text{constant}} - (b/m^2)\right|\frac{dm}{b=\text{constant}} \quad (2)$$

The first term on the right side of Eq. (2) describes how the cycles change in response to changing circulation-water TDS while the makeup-water TDS remains constant. Rewriting this:

$$dC = (1/m)db \quad [\text{for constant } m] \quad (3)$$

The actual change in C for a given change in b is calculated with the integral

$$\int_{C_1}^{C_2} dC = \int_{b_1}^{b_2} (1/m)db$$

resulting in

$$C_2 - C_1 = \frac{b_2 - b_1}{m} \quad (4)$$

where C_2 and C_1 are the upper and lower cycle levels, and b_2 and b_1 are the upper and lower circulating-water TDS levels.

The difference, $C_2 - C_1$ (i.e., the change in cycles) is directly proportional to the change in circulating-water TDS, and is attenuated by the factor $1/m$. If the TDS change is of the same order as m—for example, $(b_2 - b_1)/m > 0.1$—it becomes a significant source of cycle variation and, as will be seen shortly, a source of extraneous blowdown.

This source of changing cycles is directly related to blowdown control error (overshoot, span, etc.), and

unless cycles are controlled directly, such that $C_2 = C_1$, the error cannot be eliminated.

The second term on the right side of Eq. (2) shows how the cycles change with a change in makeup-water TDS for a given level of circulating-water TDS. Rewriting:

$$dC = -(b/m^2)dm \quad [\text{for constant } b] \quad (5)$$

Using algebra, one can write:

$$dC/C = -dm/m$$

Integrating the above relation between the initial and final makeup-water TDS (m_1 and m_2), an equation is obtained that allows calculation of the final cycles, C_2, in terms of m_1, m_2 and the initial cycles C_1:

$$C_2 = (m_1/m_2)C_1 \quad (6)$$

This change in cycles, from C_1 to C_2, as will be shown, is the second source of unnecessary blowdown and, unless the cycles are held constant such that Eq. (5) equals 0, the cycles will vary due to changing makeup-water TDS, and Eq. (6) will be in effect.

The blowdown requirement

If the cooling-tower system is to be in equilibrium, the total mass of minerals dissolved in the circulating water must be maintained at some constant level, Q, such that there is no net change in Q in the long run. This condition can be stated mathematically as:

$$dQ/dt = mM - bB = 0 \quad (7)$$

This is the cooling-tower mineral-balance equation, in which dQ/dt is the time rate of change of Q, M is the rate of makeup, B is the blowdown rate, and m and b are, as before, the makeup-water TDS and the circulating-water TDS. Also, the constant-volume constraint requires that:

$$M = E + B \quad (8)$$

the makeup rate, M, equals the evaporation rate, E, plus the blowdown rate, B. Note that B includes any source of blowdown such as leaks, drift, etc.

Using algebra, and Eq. (7) and (8), the blowdown requirement is defined as:

$$(B/E) = \frac{1}{(b/m) - 1} = \frac{1}{C - 1} \quad (9)$$

This familiar result is in units of gallons of blowdown per gallon of evaporation, over the same period of time where B and E are average values. Eq. (9) shows explicitly that the blowdown requirement of the system is a function of cycles only, and that the blowdown requirement will change with cycles for all the reasons discussed earlier.

Obviously, determining the blowdown requirement for a system with varying cycles requires further calculation. The following analysis is appropriate: Assume that a system is operating in a cycle range of C_1 to C_2, i.e., $C_1 \leqslant C \leqslant C_2$. In the sense of Eq. (7), then, the cycles oscillate between C_1 and C_2 at some arbitrary frequency (depending on the reasons for the change), and all levels within this range are equally probable. The blowdown requirement for this system can be calculated by averaging Eq. (9) over the cycle range

Nomenclature

- b Total dissolved solids (TDS) in circulating water
- B Rate of blowdown
- C Cycles (cycles of concentration)
- E Rate of evaporation
- m Total dissolved solids (TDS) in makeup water
- M Rate of makeup
- Q Constant level of minerals dissolved in circulating water

Subscripts
- 1   Lower total dissolved solids (TDS) level
- 2   Upper total dissolved solids (TDS) level

$C_1 \leqslant C \leqslant C_2$. By incrementing C in sufficiently small steps for arbitrary accuracy, and evaluating B/E at each step, the average blowdown requirement is:

$$[B/E] \simeq \frac{(B/E)_1 + (B/E)_2 + \cdots + (B/E)_N}{N} = \frac{\sum_{i=1}^{N}(B/E)_i}{\sum_{i=1}^{N} i} \quad (10)$$

Greater accuracy is attained as integer N increases—as N approaches infinity, the average blowdown requirement can be written in closed form by integrating Eq. (9). This is:

$$[B/E] = \frac{\int_{C_1}^{C_2}(B/E)dC}{\int_{C_1}^{C_2}dC} = \frac{\ln[(C_2-1)/(C_1-1)]}{C_2 - C_1} \quad (11)$$

Eq. (11) is the exact blowdown requirement of a system varying between two cycle levels in the way described above, whereas Eq. (9) is the exact blowdown requirement of a system where the cycles are kept constant.

By comparing Eq. (9) with (11), it is quickly seen that the blowdown requirement of the varying system will always be greater than that of a system being held at C_2, i.e.:

$$[B/E]_{C_1 \leqslant c \leqslant c_2} > (B/E)_{C_2} \quad (12)$$

and it is easy to take the limit in Eq. (11) to obtain:

$$\lim_{C_1 \to C_2}[B/E] = 1/(C_2 - 1) \quad (13)$$

Eq. (12) and (13) together state that the blowdown requirement for a given cycle level, C_2, is minimized when the cycles are held constant at C_2.

Practical calculations and conclusion

Varying of cycles (the inability to maintain cycles at a constant level) is responsible for excessive blowdown and equally excessive chemical usage in cooling-tower systems. This is qualitatively intuitive. The amount of waste generated by varying cycles is not intuitive, however; in fact, it can be quite staggering.

Without the use of cycle control, the maximum allowable cycles of a given system are calculated, and control is set such that the system is always functioning at less than maximum cycles. The resulting blowdown requirement, $[B/E]$, is calculated with Eq. (11). With cycle control, the maximum allowable cycles of the system are calculated, and control is set such that the system is always equal to the maximum cycles. The resulting blowdown requirement, (B/E), is calculated with Eq. (9). The reduction of the blowdown and chemical feed requirements resulting from this change in control is given by:

$$\% \text{ Reduction} = \frac{[B/E] - (B/E)}{[B/E]} \times 100 \quad (14)$$

The notions of cycle control and blowdown requirement provide the basis for another calculation that could not previously be made accurately. This is, since

$$B = (B/E)E \quad (15)$$

and (B/E) is a constant of the system, it follows directly that:

% change in the evaporation rate (E) =
% change in the blowdown rate (B) =
% change in the chemical feedrate.   (16)

Therefore, the blowdown rate and the chemical feedrate are predictable functions of the evaporation rate, i.e., the load.

As an example, take a system that is known to vary between 2.0 and 3.0 cycles. According to Eq. (11), the blowdown requirement is 0.963. Actually, it will vary between 0.500 and 1.000 at a frequency depending on the varying factors. If the frequency of one of these factors is not known, as is the case with changing makeup-water sources, the blowdown and treatment consumptions will not be linear functions of the cooling load, and will vary unpredictably with the cycles.

By maintaining the cycles at a constant 3.0, the blowdown requirement will be 0.500 at all times. According to Eq. (14), this is a 27.8% lower requirement of blowdown and treatment. Further, Eq. (9), (15) and (16) state that since the cycles are constant, the blowdown and treatment consumptions can now be predicted as exclusive functions of the system load.

The author

Thomas Haupt is the founder of Logic Control Systems, P.O. Box 934, Redondo Beach, CA 90277. He was involved in cooling- and boiler-water treatment sales and service from 1971 to 1976. He holds a B.S. in applied physics from West Coast University at Los Angeles, and has done graduate work in physics at the University of California (Los Angeles).

LIQUID heat-transfer system (center background) for pilot plant.

Heat-Transfer Agents for High-Temperature Systems

How to select the transport system, its associated equipment and a suitable heat-transfer agent in order to obtain the most efficient system for carrying heat to high-temperature process operations.

JOEL R. FRIED, General Electric Co.

Water in all of its phases has long been a favorite agent for heat transfer in the chemical process industries. It is nontoxic, inexpensive, nonflammable, and available in great quantities. Its technology is well understood.

These qualities, in addition to extremely high heat-transfer coefficients, contribute to the importance of water in heat exchange. Although using water for heating or cooling may be determined by availability, process temperature requirements are most often the controlling factor. Below the normal freezing point of water (32 F.) and above about 350 F., where rapidly rising steam pressures greatly increase equipment cost and decrease handling safety, substitute heat-transfer agents are sought.

In the low-temperature range,[1] the technology is well developed. Typical refrigerants include the Freons and ammonia. Other low-temperature heat-transfer agents commonly used include air, brine, and solutions of glycol and water. Air and the other so-called permanent gases, although enjoying many of the advantages of water in terms of cost and safety, have very low coefficients of heat transfer. Heat-transport capability (density × heat capacity) and film-transfer coefficient at constant velocity and temperature for the various fluids should be compared.

Above 350 F. (where steam pressure is 135 psia.) but below 600 F. (1,543 psia.), the heat-transfer agents most commonly used are petroleum-derived heat-transfer oils.[2] Certain synthesized organic fluids[3] can extend this range to about 750 F., well above the critical temperature of water (705.34 F. at 3,206.2 psia.). Silicon-containing

Originally published May 28, 1973

CYCLE CONTROL CUTS COOLING-TOWER COSTS

Selection Guide for Heat-Transfer Agents—Table I

Name	Composition	Temperature Range, Deg. F.	Principal Usage
Dowtherm E[12]	o-Dichlorobenzene	0 to 500	Vapor 356 to 500 F.
Dowtherm H[13]	Aromatic oil	15 to 550	Liquid
Dowtherm J[14]	Alkylated aromatic	−100 to 575	Vapor 358 to 575 F.
Dowtherm G[15]	Di- and tri-aryl ethers	12 to 650	Liquid
Dowtherm A[12]	Eutectic mixture of diphenyl and diphenyl oxide	60 to 750	Vapor 495 to 750 F.
Humble therm 500[16]	Aliphatic oil	−5 to 600	Liquid
Hitec Salt[5,17]	40% $NaNO_2$, 7% $NaNO_3$, and 53% KNO_3	400 to 850	Liquid
Mobiltherm 600[18]	Alkylated aromatic	−5 to 600	Liquid
Therminol 44[21]	Modified ester	−60 to 425	Liquid
Therminol 55[19]	Alkylated aromatic	0 to 600	Liquid
Therminol 60[21]	Aromatic hydrocarbon	−60 to 600	Liquid
Therminol 66[20]	Modified terphenyl	20 to 650	Liquid
Therminol 88[22]	Mixed terphenyl	293 to 800	Liquid
Ucon 50-HB-280 X[23,24]	Ether of poly-alkylene oxide	0 to 500	Liquid

compounds[4] have been used up to a limit of about 800 F. For temperatures above 500 F. but below 1,000 F., mixtures of inorganic salts[5,6] in the melt state demonstrate relatively high coefficients of heat transfer and good stability. In fact, the high thermal efficiencies of such inorganic systems, in addition to their low cost and high degree of safety, recommend them well for temperatures from 500 through 750 F.[7]

For processing in the temperature range from 1,000 to about 2,000 F., mercury and certain liquefied metals[1,8] such as sodium, potassium, and mixtures of the two are used. These metals are good heat-transfer agents, but the extreme costs of such systems and the high operating risks indicate that they be avoided if possible. Liquid metals may be used in such special applications as the cooling of nuclear reactors, and in certain areas of the petroleum and petrochemical industries for extreme-high-temperature reactions.

In this article, the temperature range from 350 to 1,000 F. will be treated in detail with respect to both selecting the proper heat-transfer agent and designing the transport system. Applications of such systems in this temperature range are numerous. In plastics processing, relatively small electrically heated units are used to provide precise temperature control for extruders, calender rolls, presses, platens, and dies. Identical systems may be used in the pilot plant for heating (or cooling) jacketed reactors or for providing boiler heat to distillation columns and to more-specialized applications such as wiped-film evaporators. In scaleup, large direct-fired units capable of delivering up to 150 million Btu./hr. and larger have been used to meet most industrial heat-transfer requirements.[9]

System Specification

The first step in the design of the heat-transfer system is to establish the process operating temperature, including consideration of whether multiple operations may be required at various temperatures. Once these requirements are known, a decision can be made as to whether liquid-phase or vapor-phase heating can meet processing needs.

In the temperature range from 750 to 1,000 F., only liquid heating with inorganic melts is available. From about 360 to 750 F., the choice of phase can be important. It is possible that a combination of liquid-phase and vapor-phase heating may be superior to either alone. The decision should be based upon all factors, including the product heat tolerances, the equipment, and the associated economics.

Vapor-phase systems[10] allow transfer of greater quantities of heat per pound of heat-transfer medium than do liquid-phase ones that rely upon exchange of only sensible heat, and hence have smaller coefficients of heat transfer. Since only vapor must fill heating pipelines, vapor-phase systems require less working inventory. These factors contribute to savings in both inventory and operating costs for large units over liquid systems. Vapor systems are particularly attractive when processing conditions require uniform heating of heat-sensitive products. This is especially so with difficult-to-control liquid-flow patterns and velocity, as for example in unbaffled kettle jackets. Examples for tight temperature control in the vapor phase can be found in the spinning of synthetic fibers and in deodorization of vegetable oils. The high mass-flowrates required to achieve such precision temperature control in a liquid-phase system would in many cases be economically unjustified.

Liquid-phase systems are specified when a wide range of temperatures may be required, and when temperature control is important without system upset. It is possible to achieve variable heat transfer over the surface of the process unit by appropriately adjusting the circulation rate. Liquid systems also enable heating and cooling within the same system.[11] In addition, these systems achieve rapid startup and attainment of design temperatures, and rapid response to load changes, and require no purging or venting after shutdown. Vapor-phase systems do not share these features. The use of smaller-diameter piping and valves in the liquid system may also lower capital costs below those for vapor systems, although the operating and maintenance costs for the liquid systems are generally higher.

Selection of Heat-Transfer Medium

Once design temperature and system phase have been specified, it is then possible to select the proper heat-transfer agent. The availability of different heat-transfer media by many manufacturers has made the choice of the proper medium increasingly difficult. Most commercially available heat-transfer fluids find principal use in the liquid phase. A few are available for use in the vapor state. A list of some of the most important heat-transfer agents (including their composition, temperature range, and phase applicability) is given in Table I. Some of their physical properties are presented in Table II.

The primary consideration is to match the process-temperature requirements with the recommended temperature range for the fluids. For choosing among several fluids in the correct range, other factors such as inventory cost, toxicity and ecology, flammability, thermal stability, freezing point, corrosive and fouling characteristics, and pumping costs may become deciding considerations. These factors will be reviewed in the following sections.

Inventory costs—Initial investment for heat-transfer fluids differs widely. Generally, the petroleum oils and the inorganic salts are the least expensive. Some specialty organic fluids are quite expensive, particularly certain ethers. Among these organic fluids, the terphenyls are generally the least costly.

In large liquid-phase systems, inventory costs can represent a significant part of the total capital investment. A final decision should be based upon a comparison of inventory costs (with consideration of expected fluid life and regeneration costs) with utility costs for fluid heating and pumping.

Toxicity and ecology—All commercial heat-transfer agents are at most moderately toxic and, with normal precautions, present little or no danger to operating personnel. Several fluids have characteristically penetrating odors that give warning of vapor leaks far below dangerous levels.

In recent years, ecological considerations have become important. In the past year, all polychlorinated biphenyl compounds were removed from the market for concern of environmental contamination. The disposal or replacement of such fluids in existing facilities should be done through the manufacturer. Detailed information on toxicity and ecology for individual heat-transfer fluids is available through most suppliers.

Flammability—All heat-transfer fluids in the temperature range from 350 to 1,000 F. are flammable, provided the fluid is at sufficiently high temperature, and a source of ignition is present. Inorganic salts are the single exception. At the flash point, all other fluids will momentarily ignite on application of a flame or spark, and at the fire point, vapor is given off at a rate sufficient to sustain combustion. At the autoignition temperature, no source of ignition is required, but typically this temperature is well above the fluid's recommended range. The complete line of "fire resistant" fluids (Aroclor and the FR series) that was previously available consisted of the chlorinated biphenyl compounds that have been withdrawn for the reason already given. Halogenation retards combustion; but at sufficiently high temperatures, halogenated compounds show a fire point and even an explosive range.

Thermal stability—Deterioration of petroleum and organic heat-transfer fluids occurs through either cracking or oxidation. In the cracking process, carbon-hydrogen bonds are broken, forming new compounds. Some may be volatile products that lower the flash point.

Testing of the fluid's flash point at intervals is one method of determining fluid life. Cracking can also produce less-volatile materials (polymer formation), increasing fluid viscosity. This higher viscosity will in turn decrease fluid flow, allowing film temperatures to increase at heater surfaces. These higher temperatures lead to acceleration of thermal cracking, buildup of coke formation at heater surfaces, and ultimately to heater failure. In a similar fashion, oxidation products can lead to fluid thickening and the formation of insoluble materials that may deposit on heater surfaces. Film temperatures again will increase until heater failure occurs.

In well-designed systems with adequate fluid circulation and proper heat fluxes, fluid deterioration is seldom a problem when heat-transfer agents are used within their recommended range. Stability studies[25] have indicated that in general aromatic materials are superior to aliphatic compounds, but in time all fluids may deteriorate. By exceeding the maximum recommended fluid temperature by 50 F., a 10% sample loss (attributable to venting of volatile products and carbon deposition) has been shown to result from 2 to 4 weeks of continuous operation. Many of the fluid manufacturers provide sample-analysis services and will either regenerate deteriorated fluids themselves for a small charge or recommend a company equipped for such reclamation. At the process site, it is often common practice to partially regenerate many of the fluids. This involves the regeneration of small amounts on a continuous basis. This bypassed fluid, once regenerated, is then refed to the heating system. Some fluid may be drained off and new liquid admitted. By partial regeneration, system downtime for periodic fluid draining and replacement is eliminated.

Inorganic salts also undergo thermal decomposition.[26] In the range from 850 to 1,000 F., the salts experience a slow thermal breakdown of the nitrite to nitrate, alkali metal oxide, and nitrogen. This is apparently an endo-

Physical Properties of Heat-Transfer Agents—Table II

Name	Freeze Point, Deg. F.	Boiling Point, Deg. F.	Flash Point,[1] Deg. F.	Fire Point,[1] Deg. F.	Autoignition Temperature,[2] Deg. F.
Dowtherm E	<0	356.4	155	285	>932
Dowtherm H	0*	695	370	410	860
Dowtherm J	−100	358	145	155	806
Dowtherm G	−18*	575	305	315	>1,030
Dowtherm A	53.6	494.8	255	275	1,150
Humbletherm 500	15	—	425	475	—
Hitec Salt	285	none	none	none	none
Mobiltherm 600	0*	—	350	390	—
Therminol 44	−80 to −90*	640	405	438	705
Therminol 55	−20*	—	355	410	670
Therminol 60	−80 to −90*	600	310	320	835
Therminol 66	−18*	675 est.	340 est.	380 est.	705
Therminol 88	140	700 est.	375	460	>1,000
Ucon 50-HB-280 X	−35*	—	500	600	743

[1] Cleveland open cup, ASTM D92-56
[2] ASTM D2155-66
*Pour point

Nomenclature

A	Area of heat transfer, sq.ft.
C_p	Specific heat, Btu./(lb.)(°F.)
D	Diameter, ft.
E	Energy loss/unit surface area, (ft.)(lb.$_\text{force}$)/(hr.)(sq.ft.)
f	Friction factor, dimensionless
G	Mass velocity, lb./(hr.)(sq.ft.)
g_c	Conversion factor in Newton's law of motion, 4.169×10^8 (ft.)(lb.)/(sq.hr.)(lb.$_\text{force}$)
h	Heat-transfer coefficient, Btu./(hr.)(sq.ft.)(°F.)
k	Thermal conductivity, Btu./(hr.)(sq.ft.)(°F./ft.)
N_{Pr}	Prandtl number, dimensionless
N_{Re}	Reynolds number, dimensionless.
P	Friction power, (ft.)(lb.$_\text{force}$)/hr.
ΔP	Pressure drop, lb.$_\text{force}$/sq.ft.
Q	Volumetric flowrate, cu.ft./hr.
S	Cross-sectional area, sq.ft.
V	Velocity, lb./hr.
μ	Viscosity, lb./(hr.)(ft.)
ρ	Density, lb./cu.ft.

Subscript

f	Friction

thermic reaction that contributes to a gradual rise in the freezing point of the salt mixture. In addition, the nitrite within this temperature range is subject to slow oxidation by atmospheric oxygen. Slow oxidation can be eliminated by blanketing the salt with an inert gas such as nitrogen. Such blanketing will also eliminate additional deterioration due to the absorption of carbon dioxide (carbonate precipitation) and absorption of water vapor (alkali-metal hydroxide formation).

Freezing point—If the heat-transfer system is to be exposed to seasonal variations in ambient temperature, the freezing point is important to consider. Fluids with high freezing points require steam tracing to prevent possible blocking of lines. Fortunately, all such fluids contract upon freezing, preventing possible damage to pumps and other costly parts of the system. The chief drawback to the use of inorganic salts in the past has been their high freezing points (285 F.). Current practices for such melts involve dilution with water at lower temperatures.* During heatup, the water is vaporized as the salt solution is pumped through the complete system.

Corrosive and fouling characteristics—Fouling of heat-transfer surfaces is nominal for well-designed systems. Fouling may result from the deposition of polymer, or of coke byproducts from the thermal degradation of organic media.

All heat-transfer fluids given in Table I are noncorrosive to mild steel. For inorganic salts above 850 F., low-alloy or stainless steels are recommended. Chlorinated fluids may liberate hydrogen chloride when they are heated above their recommended maximum temperature. This leads to the corrosion of mild steel, and to the stress corrosion of stainless steels if traces of water are present in the heating system. Therefore, proper venting of systems containing these chlorinated fluids becomes particularly important to remove all water vapor during heatup.

---
*Patented by American Hydrotherm Corp., New York.

Pumping Costs

To select from several heat-transfer fluids that are otherwise close in most properties, a decision is often based upon the relative pumping efficiencies of transferring heat. Several such studies have been reported in the literature, comparing many of the heat-transfer fluids listed in Table I. Among these fluids, water and Dowtherm A have been compared by Parsons and Gaffney,[27] who plotted the heat-transfer coefficient, h, against the frictional energy expended pumping the fluids per unit surface area of pipe, E, in a single-tube heat exchanger. The tube considered in this study was 5.0 ft. long and 0.05 ft. dia.

Carberry[28] has extended this comparison to include Hitec salt by plotting h versus the horsepower requirements for pumping the various heat-transfer fluids through 1,000 ft. of 3-in. pipe in a shell-and-tube exchanger.

More recently, Kasper[29] has also included Dowtherm E and Mobiltherm 600 in a cost study of several heat-transfer fluids for the case in which the heat-transfer coefficient controls on the process-fluid side (the shell), and the case in which the heat-transfer coefficient controls on the media side (the tube). Where the heat-transfer media are controlling, the total annual cost of exchanging heat in a single-tube exchanger is a function of exchanger area and the power required for pumping the heat-transfer medium through the tube. As expected, there is a greater difference in relative costs among the heat-transfer fluids as compared to the case where the process coefficient is controlling. Hence, the total annual cost becomes dependent on power requirements alone.

Seifert and Jackson[25] have plotted h versus temperature, and pumping horsepower versus temperature for both the process-controlling and media-controlling cases, using as a basis for their calculations a shell-and-tube heat exchanger with 70 tubes per pass of 0.05 ft. dia. Although the position of the horsepower curves varied slightly for one heat-transfer fluid relative to another in the two controlling cases, the absolute ratios of horsepower requirements for any two fluids at a given temperature were not significantly different. The study extended to Dowtherm G, Humbletherm 500, and Therminol 66 in addition to the media already reviewed.

The most convenient technique of comparing heat-transfer efficiencies without consideration of a particular exchanger geometry is to determine the ratio of the heat-transfer coefficient to the frictional energy expended pumping the fluid per unit surface area of the pipe, E. The most efficient fluid will show the largest ratio over a given temperature range.

The first step is to express the frictional energy and the heat-transfer coefficient in terms of the mass velocity, G, in the form:[30]

$$E = P/A = KG^{3-m} \qquad (1)$$
$$h = \alpha G^n \qquad (2)$$

All parameters m, n, α and K are determined. Eliminating G between Eq. (1) and (2) yields the appropriate expression for the heat-transfer coefficient as a function of the frictional energy expended per unit surface area:

$$h = \alpha K^{n/(m-3)} E^{n/(3-m)} \quad (3)$$

For transferring heat in a shell-and-tube heat exchanger where a liquid heat-transfer agent is tubeside, the following equations apply:

$$E = \frac{0.023}{g_c \rho^2} \left(\frac{\mu}{D}\right)^{0.2} G^{2.8} \quad (4)$$

$$h = 0.023 C_p (\mu/D)^{0.2} (N_{Pr})^{-0.667} G^{0.8} \quad (5)$$

Eliminating G from Eq. (4) and (5) yields:

$$h = 19.75 \frac{(C_p)^{0.333} k^{0.667} \rho^{0.572}}{\mu^{0.524} D^{0.143}} E^{0.286} \quad (6)$$

Eq. (6) is valid for turbulent flow in smooth pipes for Reynolds numbers (DG/μ) from 5,000 to 200,000, and for Prandtl numbers $(C_p \mu/k)$ from 0.6 to 120. The derivation is based upon the following three equations:

The Fanning equation:[28]

$$\frac{f}{2} = \frac{g_c \Delta P_f \rho S}{G^2 A} \quad (7)$$

The empirical equation of Koo for long smooth pipes:[31]

$$\frac{f}{2} = 0.023(N_{Re})^{-0.2} \quad (8)$$

The Chilton and Colburn analogy:[32]

$$j_H = \frac{f}{2} = \frac{h}{C_p G} (N_{Pr})^{0.667} \quad (9)$$

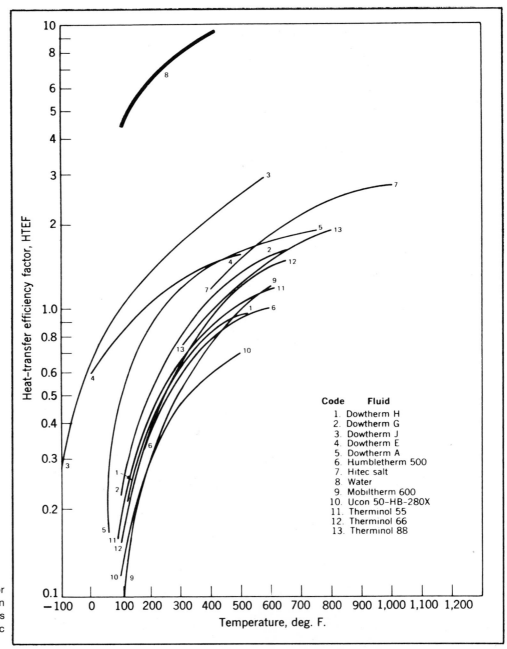

HEAT-TRANSFER efficiency factor provides a first-order approximation for selecting the heat-transfer agents most suitable for obtaining specific temperature requirements—Fig. 1

Code — Fluid
1. Dowtherm H
2. Dowtherm G
3. Dowtherm J
4. Dowtherm E
5. Dowtherm A
6. Humbletherm 500
7. Hitec salt
8. Water
9. Mobiltherm 600
10. Ucon 50-HB-280X
11. Therminol 55
12. Therminol 66
13. Therminol 88

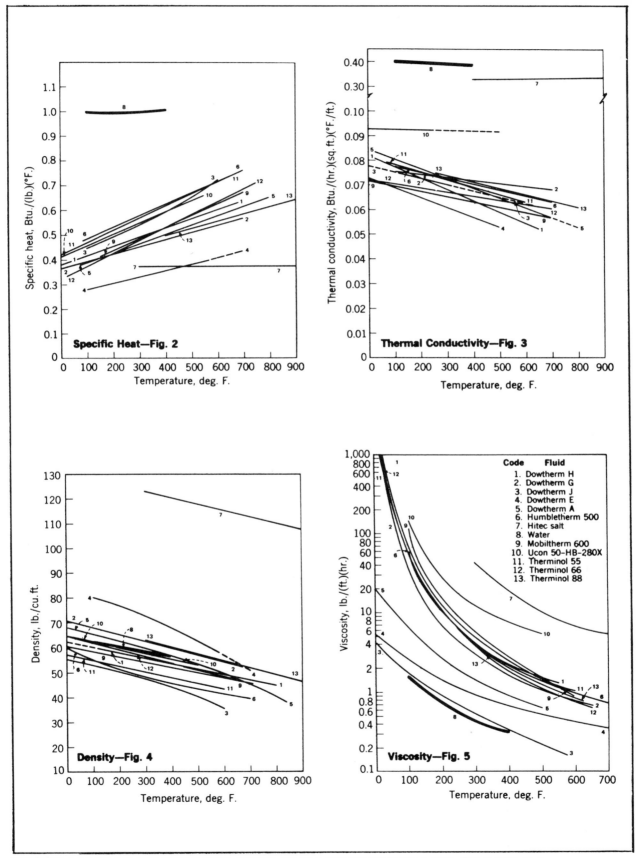

PHYSICAL properties of heat-transfer fluids establish the heat-transfer efficiency factors—Fig. 2 through 5

Eq. (4) may be expressed in an alternate form as:

$$E = \frac{0.023 V^{2.8}}{g_c D^{0.2}} \rho^{0.8} \mu^{0.2} \quad (10)$$

Eq. (10) demonstrates that in fluids having low densities and viscosities, frictional-energy losses are minimized.

From the preceding analysis, a heat-transfer efficiency factor (HTEF) can be determined as an expression of the ratio of the heat-transfer coefficient to the frictional energy expended pumping the fluid. By allowing HTEF to represent the coefficient of $E^{0.286}$ in Eq. (6), i.e.:

$$HTEF = 19.75 \frac{(C_p)^{0.333} k^{0.667} \rho^{0.572}}{\mu^{0.524}} \quad (11)$$

and by rearranging Eq. (6), the following is obtained:

$$HTEF = \frac{h}{E^{0.286}} D^{0.143} \quad (12)$$

Eq. (12) demonstrates that fluids with high values of HTEF, as determined by Eq. (11), are characterized by high heat coefficients for a given pipe diameter and frictional-energy loss.

Values of HTEF for the heat-transfer fluids listed in Table I and for water as a comparison are plotted against temperature over each fluid's recommended temperature range in Fig. 1.

The fluid properties (C_p, k, ρ and μ) that determine the value of HTEF are plotted in Fig. 2 through 5. The greatest variation among fluids is in their viscosity (see Fig. 5). Therefore, viscosity becomes a predominant influence on each fluid's HTEF. Fluids with low viscosities show high values of HTEF, as seen by comparing Fig. 5 with Fig. 1. The single exception is the salts, whose high viscosities are compensated for by high values of thermal conductivity and high densities. Combined, these lead to a high value of HTEF.

Salts appear superior in the higher temperature range (600 to 1,000 F.). From 100 to 400 F., water is the most efficient fluid. In general, the petroleum oils, including the aromatics and aliphatics, appear the least efficient. The terphenyls are midway by comparison, slightly less efficient on the HTEF scale than diphenyl and a few other specialty organics.

The HTEF values have been determined on the basis that the heat-transfer-media coefficient is controlling when transferring heat from shell-and-tube exchangers. It is clear from the study of Kasper,[29] for this media-controlling case, that the choice of the most efficient heat-transfer fluid may be an important economic decision. HTEF values can be used as a first-order approximation to fluid selection. This will narrow choice of a fluid from the many available to a few that appear the most economical for specific temperature requirements.

For example, it may be necessary in the course of a plant situation to heat a processing stream to 600 F. By looking at Fig. 1, we find that Dowtherm A and Hitec salt appear the most likely candidates on the basis of high values of HTEF at 600 F. A final choice between the two may involve comparative inventory costs, heating costs, and costs for pumping each fluid through the piping, valves, and fluid heater that comprise the remainder of the heat-transfer system.

Vapor-Phase Systems

For vapor-phase heating, the simplest and most economical system calls for gravity condensate return, eliminating the need for pumps. Such a system is possible if the static liquid head is sufficient to compensate for all friction losses in the system without flooding the heated equipment. Multiple process units can be accommodated by a single vaporizer if each unit is at the same temperature, and pressure drops are approximately equal for each. Individual units, with different pressure drops, also may be handled by increasing the static head to compensate for the unit with the highest pressure drop. Vapor bypass or the flooding of one unit by condensate from another can be prevented by using traps, or by joining the condensate return lines at the lowest elevation.

A Hartford loop[10,12,33] should be used when condensate is returned to the vaporizer at a level below the recommended liquid fill point of the heater tubes. The condensate return line is formed into an inverted "U" just before reaching the vaporizer. The height of this loop should equal the lowest permissible liquid level in the vaporizer, and the horizontal connection of the loop should not be more than two pipe diameters in length. The apex of the Hartford Loop is joined to the vapor line leaving the vaporizer, as illustrated in Fig. 6.

The purpose of the Hartford Loop is to ensure against extended operation of the vaporizer when the level of liquid heat-transfer fluid in the tubes is so low that the resulting high heat flux at the dry tube-surface may lead to carbonization, complete tube failure, and the possibility of extensive fire damage.[33] When throttling of the vapor line is required to control the supply of heat to the process unit and the temperature of the entering vapor, the resulting resistance to flow in the vapor line will cause a reduction in pressure within the unit. This

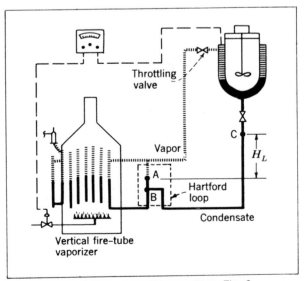

NATURAL circulation vapor-phase system—Fig. 6

HEAT-TRANSFER AGENTS FOR HIGH-TEMPERATURE SYSTEMS

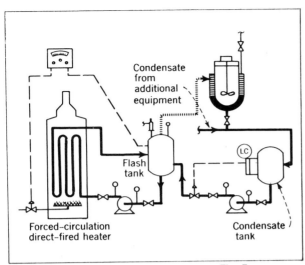

FORCED circulation vapor-phase system—Fig. 7

pressure void will cause the level of liquid to rise within the condensate line from its normal position at Point C (Fig. 6). The fluid above the Hartford Loop will correspondingly fall from Point A. (The initial distance between points A and C represents the head of the friction loss in the vapor line, heater, and condensate line.) If the liquid is drawn below the horizontal connection in the loop (at Point B), the liquid in the vaporizer is cut off from the condensate line by vapor filling the loop. This vapor being drawn back through the condensate line, and the sudden discontinuity in liquid flow, will result in a water hammer, giving warning that liquid level is too low. The throttling valve should then be appropriately opened to allow the liquid level to rise within the vaporizer.

Forced condensate return is used when the elevation of the process unit is insufficient to allow gravity return. A centrifugal pump then returns condensate to the vaporizer. For multiple units operating at varying pressures, several pumps may be used. Surging can be prevented by installing a small hand-operated bypass valve around the pumps or by pumping condensate directly from a holding tank (Fig. 7) that is gravity fed from the unit. For small systems with a single temperature requirement, positive-displacement pumps may be used.

Fired heaters for vapor-phase systems range in capacity from 500,000 Btu./hr. to over 300 million Btu./hr. Heat fluxes range from about 8,000 to 12,500 Btu./(sq.ft.)(hr.) of actual tube mean-circumferential area. Vaporizers are generally oil or gas fired and very often are the forced-circulation type. Forced circulation allows operation over a wide temperature range and rapid startup. For forced-circulation systems, great flexibility may be achieved in the location of the equipment. The vapor drum may be located adjacent to the process site to facilitate condensate return and to reduce the amount of large-diameter vapor piping, while the vaporizer may be remotely located. It is generally good practice to minimize the amount of brickwork and refractory to prevent damage caused by back radiation if flow through the heater should fail. Duplicate condensate pumps may also be included as an additional precaution.

Liquid-Phase Systems

A typical liquid-phase system is shown in Fig. 8. Liquid is forced-circulated through the heater at the discharge end of the pump and fed to the unit. A relief valve is installed to discharge back to the suction line of the pump.

The distinguishing feature of liquid systems is the presence of an expansion tank on the pump suction side. This piece of equipment is one of the most important in the liquid system but seldom receives adequate attention in the design phase. Its purpose is to accommodate thermal expansion of the liquid during heatup, and possible surging due to the sudden venting of line-trapped steam. For these purposes, tank volume should be at least twice the calculated expansion volume of the liquid capacity of the entire system at maximum operating temperature. In this manner, the expansion tank will be one-quarter full when cold but no greater than three-quarters full at the highest system temperature.

The expansion tank is positioned above the highest point of the heater at such an elevation to provide adequate net positive suction head (NPSH) for the pump. Oil in the expansion tank is heated only by convection currents. Practice calls for maximum tank temperature between 120 and 140 F., thereby providing a cold-oil seal to the atmosphere (minimizing fluid oxidation). To achieve these cool temperatures, the tank is not insulated and may in fact be water-jacketed. In addition, the diameter of the pipeline joining the tank to the pump suction should be smaller than the suction line itself (generally by a factor of one-half) to minimize heat convection but large enough to take care of possible surging. An orifice may be incorporated in the pipeline to further minimize convection heating.

Expansion tanks are generally built with a large height-to-diameter ratio (typically two or three to one). Tanks are vented to the atmosphere when the vapor

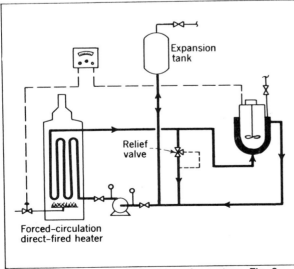

FORCED circulation liquid heat-transfer system—Fig. 8

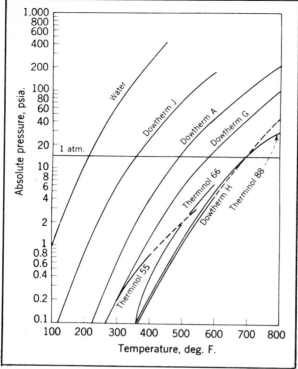

VAPOR pressures for heat-transfer liquid—Fig. 9

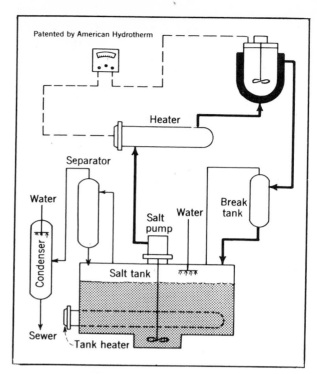

SALT dilution process for heat transport—Fig. 10

pressure of the heat-transfer fluid is low enough to allow for an unpressurized system, and the fluid is sufficiently stable to air to avoid the necessity of an inert-gas blanket. The large height-to-diameter ratio allows minimal contact of the cool fluid with the air, thereby extending fluid life. Some manufacturers use a float inside the tank to further minimize air contact. The float also serves as a level indicator. The expansion tank is often fitted with a sight glass to provide an indication of liquid inventory. An overflow connection is often included as an added precaution against system surge.

For proper venting, needle valves should be installed at all high points of the system. During startup of a new system, or after the heat-transfer fluid has been replaced in an existing one, it is important that system temperature be raised gradually, with frequent venting to remove water vapor. Typically, temperature should be raised about 100 F./hr. Adequate venting in this initial stage will ensure long life for the fluid, and minimize pump cavitation and fluid surging through the expansion tank.

An unpressurized liquid system is used when the vapor pressure of the heat-transfer fluid at the maximum operating temperature is low, and vapor flash in high-pressure-drop areas of the system is not a problem. Higher-vapor-pressure fluids require system pressurizing. This is accomplished by blanketing the expansion tank with nitrogen or another inert gas at about 25 to 150 psig. A graph of vapor pressures for some heat-transfer fluids is presented in Fig. 9.

Salt dilution (Fig. 10) requires as basic equipment, a salt tank, pump, process heater, and a separator. During startup, salt is dissolved in water and pumped through the heater to the process unit as system temperature is slowly raised. At 500 F., all water has been evaporated and the salt has reached its anhydrous molten state. When shutting down, water is again introduced at a controlled rate, eliminating any danger of solidification or uneven thermal contraction and vapor losses.

Heaters for large systems may be vertical up-fired types, floor-fired, or cabin-type end arrangements. The heating surfaces may be all-radiant, all-convective, or radiant/convective types. Overall efficiencies of 80% (up to 95% with supplemental air preheaters) are typically achieved. Heating coils may be of the serpentine, helical-coil, or horizontal-tube variety.

For smaller units, heaters are electric-immersion types. Such electric heaters are generally rated in the range of 6 to 12 w./sq.in. of heater surface. The limit to heater size for a given liquid system depends upon the maximum allowable fluid-film temperature. The film temperature for the system in turn depends upon heater design, fluid velocity, and desired bulk-temperature of the fluid. For well-designed units, film temperature will be about 30 to 50 deg. (F.) higher than the bulk fluid temperature. Therefore, it is important that the heat-transfer fluid be selected so that the film temperature (and not the bulk temperature) is below the maximum recommended temperature of the fluid.

System Equipment

In all cases but one, carbon steel is the recommended construction material for all parts of the heat-transfer system. For inorganic salts above 850 F., stainless steels are specified. Copper and copper-containing alloys should be avoided because of certain code restrictions against their use at elevated temperatures. Cast iron is subject to thermal fracture and should not be used. Aluminum and stainless steels may be attacked by some of the chlorinated fluids, especially in the presence of traces of water, and should be avoided.

Centrifugal or positive-displacement (rotary vane) pumps built for high-temperature operation may be used. Positive-displacement pumps should be protected by line strainers on the suction side. For salt systems, the pump is the submerged vertical-centrifugal type. This pump should be specified to permit no contact of the liquid with the packing gland. Pump manufacturers can assist in the selection of the most suitable seals for operation at specified process temperatures. Mechanical seals have gained much favor and can enjoy extended life in properly maintained service. New bellows-type seals can be used well above 500 F. without the need of water-cooled jackets. Canned pumps available at higher costs than either centrifugal or positive-displacement pumps offer the advantage of leakfree operation at temperatures up to 1,000 F. Gear pumps have seen service in particular operations.

Piping should be made of ASTM A-53 or ASTM A-106 seamless carbon steel. For piping less than 1 in. dia., schedule 80 pipe with clean-cut screw threads may be used. For piping larger than 1 in. dia., schedule 40 pipe with welded connections is recommended.

Flanges should be either 150 lb. or 300 lb. raised-face with a smooth finish. For salts at the highest temperature ranges, tongue-and-groove-type fittings are recommended. It is good practice to minimize the number of flanges in the system and rely chiefly on welded connections wherever possible. Suitable gaskets include spiral-wound stainless steel and asbestos.

Valves are either the gate or globe variety. Gate valves allow smaller pressure drops and have seen the greatest service in heat-transfer systems. Stuffing boxes for high-temperature application should be deep, with metallic packing. Above about 850 F., valves should be fitted with radiation fins in order to protect the packing glands.

Relief valves for high-temperature operation should be selected with care. Above 450 F., only tungsten-steel springs should be used, and O-ring type seats should be avoided in favor of flat metallic seating surfaces. Bodies should be constructed of cast steel with stainless-steel trim.

Instrumentation and Controls

Bourdon or diaphragm-type pressure gages equipped with siphons to seal the Bourdon element or diaphragm from high-temperature liquid or vapor may be used—provided all parts are steel. For outdoor installations, the diaphragm type is preferred because of the ease with which the diaphragm housing can be heated to prevent possible freezing of the heat-transfer fluid within the gage.

Control valves should be made of steel bodies with stainless-steel trim, and contain fin-cooled stuffing boxes equipped with chevron-type packing of graphited asbestos or TFE composition. For precise control, a valve positioner is recommended.

Level controllers should be either float or displacer types without packing glands. Movement is monitored by means of a torque tube, electrical inductance, or permanent magnet. These operate a pneumatic pilot or electrical contact.

Safety controls should be given adequate attention. These should be able to shut down the heater or vaporizer in case of excessive fluid temperature or pump failure. Heaters or vaporizers and pumps should be shut off when there is excessive pressure. Additional controls may include a level controller for the expansion tank in the liquid-phase system to cause unit shutdown when liquid inventory is too low for adequate pump NPSH. ■

References

1. Danziger, W. J., Heat-Transfer Media Other Than Water, in "Encyclopedia of Chemical Technology," 2nd ed., Vol. 10, p. 846, Interscience, New York, 1963.
2. Purdy, R. B., Balow, R., Shaffer, J. J. and Fanaritis, J. P., *Chem. Eng. Progr.,* May 1963, p. 43.
3. Conant, A. R. and Seifert, W. F., *Chem. Eng. Progr.,* May 1963, p. 46.
4. Geiringer, P. L. and Beanland, E., *Chem. Eng. Progr.,* May 1963, p. 50.
5. Uhl, V. W. and Voznick, H. P., *Chem. Eng. Progr.,* May 1963, p. 33.
6. Kirst, W. E., Nagle, W. M. and Castner, J. B., *Trans. AIChE,* **36,** 371 (1940).
7. Davies, D. J. I., *Chem. Proc. Eng.,* **44,** 473 (1963).
8. Brunn, F. M., Thermal-Liquid Systems, in "Chemical Engineers' Handbook," 4th ed., p. 9–52, McGraw-Hill, New York, 1963.
9. "Fired Heaters and Heat Energy Systems," American Schack Co., Pittsburgh, Pa.
10. Dean, D. K., *Ind. Eng. Chem.,* **31,** 797 (1939).
11. Stack, T. G. and Friden, J. E., *Chem. Eng. Progr.,* **48,** 409 (1952).
12. Form No. 176-276-71, Dow Chemical Co., Midland, Mich.
13. Form No. 176-1244-72, Dow Chemical Co., Midland, Mich.
14. Form No. 176-1240-72, Dow Chemical Co., Midland, Mich.
15. Form No. 176-1213-72, Dow Chemical Co., Midland, Mich.
16. Humble-Therm 500, Humble Oil and Refining Co., New York, 1964.
17. Du Pont Hitec Heat Transfer Salt, Du Pont, Wilmington, Del.
18. Mobil Technical Bulletin: "Heating With Mobiltherm," Mobil Oil Co., New York.
19. Technical Bulletin T-55, Monsanto Co., St. Louis, Mo.
20. Technical Bulletin T-66, Monsanto Co., St. Louis, Mo.
21. Interim Product Bulletins on Therminol 44 and Therminol 60, Monsanto Co., St. Louis, Mo.
22. Technical Bulletin T-88, Monsanto Co., St. Louis, Mo.
23. Petersen, D. E. and Bedell, R., *Chem. Eng. Progr.,* May 1963, p. 36.
24. Form Nos. F-6500 and F-7490F, Union Carbide Co., New York.
25. Seifert, W. F. and Jackson, L. L., *Chem. Eng.,* Oct. 30, 1972, p. 96.
26. Hydrotherm Molten-Salt Heat-Transfer System, American Hydrotherm Corp., New York.
27. Parsons, P. W. and Gaffney, B. J., *Trans. AIChE,* **40,** 655 (1944).
28. Carberry, J. J., *Chem. Eng.,* June 1953, p. 225.
29. Kasper, S., *Chem. Eng.,* Dec. 2, 1968, p. 117.
30. McAdams, W. H., "Heat Transmission," 3rd ed., pp. 438–439, McGraw-Hill, New York, 1954.
31. McCabe, W. L. and Smith, J. C., "Unit Operations of Chemical Engineering," 2nd ed., p. 345, McGraw-Hill, New York, 1967.
32. Kern, D. Q., "Process Heat Transfer," p. 574, McGraw-Hill, New York, 1950.
33. Badger, W. L., *Chem. Eng.,* May 1955, p. 192.

Meet the Author

Joel R. Fried is a chemical engineer in the chemical laboratory at the Corporate Research and Development Center, General Electric Co., P. O. Box 8, Schenectady, NY 12301. His work involves scaleup of polymer reactions, distillation operations, and design and operation of heat-transfer equipment for pilot plants. He has a B.S. in biology, B.S. in chemical engineering, and an M.Ch.E. from Rensselaer Polytechnic Institute. He is a member of AIChE and the Soc. of Plastics Engineers.

Low-toxicity cooling-water inhibitors—how they stack up

The author reports on seven formulations that have been used to curb corrosion in cooling-water circuits.

A.S. Krisher, Monsanto Co.

☐ In recent years, tighter effluent controls have hindered the use of conventional chromate-type inhibitors to control corrosion in cooling-water systems.

The first alternative to chromate formulations is to substitute treatments that will not violate effluent regulations. Such systems generally employ nonmetallic (i.e., non-heavy-metal) inhibitors.

A second alternative is to use alloy exhangers that do not corrode in the water, making necessary only a small amount of inhibitor to prevent excessive corrosion on the distribution piping and exchanger heads.

A third alternative is to retain zinc chromate inhibitors but treat blowdown so as to remove the metallic ions to a compliance level. This is being done commercially by two procedures: (1) chemical reduction of chromate to the trivalent chromium form, followed by an increase in pH to precipitate chromium and zinc as hydroxides, and (2) treating with ion-exchange resins to remove the offending ions. In some cases, ion exchange permits recovery of the metallic ions for recycling to the cooling water system.

These three alternatives show promise for commercial applications and have been tried in many systems. Two other alternatives have been investigated much less extensively. One, zero discharge with blowdown side-stream treatment, uses conventional lime soda ash to soften and precipitate out the hardness ions—primarily calcium and magnesium—which are objectionable. The remaining liquid, which bears the chromate ions, is then returned to the tower. Another alternative method treats the makeup water so that no deliberate blowdown is needed. This approach requires that the makeup be softened.

These last two alternatives require the recirculation of water having much greater dissolved-solids content than currently prescribed. The effectiveness of zinc chromate type inhibitor is not very well defined in such regions.

This article reports on some test work done by the Monsanto Co. to evaluate a total of seven low-toxicity inhibitor systems at three plant locations. All of these tests used a standard test exchanger to evaluate corrosion and fouling on heated carbon-steel exchanger tubes. The test exchanger was installed either as a pilot unit, or in parallel with plant equipment on a cooling tower.

Test procedure

Water analyses for the three test locations are shown in the key to Fig 1. Also shown is a proposed index for the waters' corrosivity [1]. This index is equal to the chloride content plus sulfate content, divided by the alkalinity, all expressed in equivalents per million. The values for the index at the three sites illustrate that the three cooling waters are significantly different in their inherent corrosiveness, a factor which has been substantiated by plant experience.

The shell-and-tube heat exchanger used for the tests follows a standard design [2, 3] outlined before the Natl. Assn. of Corrosion Engineers; the unit contains twelve ³/₄-in. tubes, 4 ft long. Water flowrate was controlled at 2.5 ft/s on the tube side, and steam pressure on the shell side at 10 psig (250°F). These conditions are believed to represent a difficult but fair test for the inhibitors.

(A survey of exchangers in some of our plants confirmed that many operate at flowrates above 2.5 ft/s, and process-side temperatures below 250°F. Those few that operate under conditions more severe—lower velocity, higher temperature—will not be protected by normal inhibitor treatments and, therefore, must be handled by other means; for example, by redesign, or by alloy substitution.)

Test measurements were taken over a period of three to six months. After the tubes were pulled, they were marked for identification, cleaned by either inhibited-acid immersion or sandblasting with a fine abrasive, and then examined for corrosion. The depth of the corroded areas was measured directly, and the maximum depth on a given tube was taken to be a measure of the severity of corrosion on that tube. Over a period of time we collected a series of up to twelve data points, which were plotted vs. the time of exposure. The slope of the resulting regression line (or the rate of increase), was taken to be the long-term corrosion rate.

Fouling measurements

Evaluation of fouling was done using data logged by operators once a day or once a shift. Conventional heat-transfer analysis was used to calculate the actual overall heat-transfer coefficient, and to estimate what the coefficient would be if the exchanger surfaces were clean.

The difference between the reciprocals of these two coefficients is the fouling factor. Although this factor represents the total fouling in the exchanger, it was assumed to be a reasonable approximation of water-side fouling since the steam source on the shell side was relatively clean. For the conditions of each run, we have established a range of corrosion rates and fouling factors.

Comparisons between low-toxicity inhibitors

These data were then arranged on a single plot of corrosion rate vs. fouling factor, as shown in Fig. 1. This plot shows the current limits to the state-of-the-art for low-

Originally published February 13, 1978

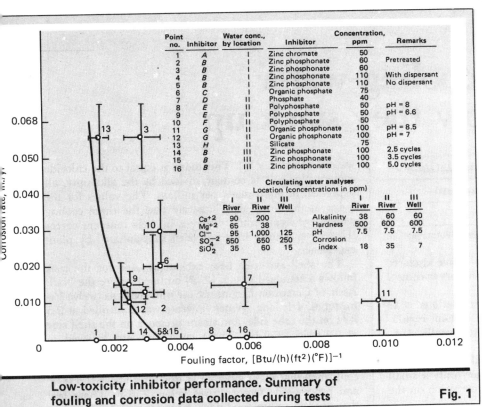

Low-toxicity inhibitor performance. Summary of fouling and corrosion data collected during tests Fig. 1

not present fouling data in meaningful quantitive form. Breske [4] has reported some lab- and plant-scale work that indicates a trade-off between corrosion effectiveness and fouling. Kumar and Fairfax [5], reporting on some very short-term lab work, gave a "fouling index" that also allows a comparison of fouling vs. corrosion for various treatments.

From available data we can conclude the following:

- Low toxicity inhibitors, as currently marketed, do not achieve the same degree of overall effectiveness as the zinc chromate formulations.

- The currently available low-toxicity inhibitors can achieve either corrosion control or fouling control equivalent to that achieved by zinc chromate systems, but they can not achieve both simultaneously.

- Corrosion control plus a design allowance for increased fouling appears to be a feasible approach for a new plant, if adequate allowances can be incorporated.

toxicity inhibitors. Note that the zinc chromate inhibitors have a long-term corrosion rate of essentially zero and an equilibrium fouling factor of 0.0015 $[Btu/(h)(ft^2)(°F)]^{-1}$ under these test conditions. Data for the low-toxicity inhibitors tested here all fall to the right of the line for chromate inhibitors. In all cases there was a significant amount of scattering of both fouling and corrosion data, as indicated by the bands on each point in Fig. 1.

For several inhibitors, tests were run using the same base water concentration, but with varying pH and inhibitor concentrations. For example, a phosphonate-type treatment was used at two different pH levels. At pH 8 (point 8), we observed a very low corrosion rate but a fairly high fouling rate. When the pH dropped to about 6.6 (point 9), the fouling factor fell, but the corrosion rate increased. Tests were also made at different concentrations for a zinc phosphonate blend, which was not totally nonmetallic. Runs at 60 ppm (point 3) and 110 ppm (point 4) demonstrated the benefits of higher concentration for this inhibitor.

During these runs with phosphonate treatments, some of the exchanger tubes were pretreated, whereas others were installed as received. As shown by points 2 and 3 in Fig. 1, at the 60 ppm level improved corrosion protection was achieved by pretreating. When the concentration level was raised to 110 ppm, however, there was no detectable benefit from pretreatment. Although these results do not disprove the well-established benefits of pretreatment, they do suggest that the same protection can be attained by increasing the inhibitor level.

Other workers have generated data that show a similar relationship between fouling and corrosion for low-toxicity inhibitors, although many of the recent papers do

Recommendations for users of inhibitors

Although available low toxicity treatments are useful, several operating aspects must not be overlooked.

First, control over water chemistry must be excellent. This necessitates more-elaborate instrumentation, more-frequent operator intervention, and greater service attention than is usually accorded cooling-water circuits.

Second, the control of biological growth in towers using a low toxicity inhibitor is more difficult. Conventional chromate materials offer a degree of biocidal activity. In contrast, nonmetallic materials not only diminish this toxic component, but also nourish biological growth.

Third, the low toxicity materials are often more expensive to use. They generally cost more per pound and are used at higher concentration levels.

References

1. "Standard Methods for the Examination of Water and Wastewater," 13th ed., American Public Health Assn., 1971.
2. Standard Heat Exchanger for Cooling Water Tests, *Materials Protection*, Vol. 4, Aug. 1965, pp. 70–73.
3. Krisher, A.S., "NACE Standard Heat Exchanger Monitors Cooling Tower Water Corrosion," *Materials Protection*, Vol. 4, Aug. 1965, pp. 73–79.
4. Breske, T.C., Testing and Field Experience with Non-Heavy Metal Corrosion Inhibitors, Paper No. 92, Corrosion/76 meeting Houston, Tex., March 22–26, 1976.
5. Kumar, J., and Fairfax, J.P., Rating Alternates to Chromate in Cooling Water Treatment, *Chem. Eng.*, April 26, 1976, pp. 111–112.

The author

A. S. Krisher, an Engineering Fellow in Monsanto Co.'s Corporate Engineering Dept., 800 N. Lindbergh Blvd., St. Louis, MO 63166, has 26 years of experience as a specialist in materials of construction and corrosion prevention. He holds a B.S. in chemical engineering from Oklahoma State University. A member of the AIChE, the Natl. Assn. of Corrosion Engineers and the American Soc. for Metals, he has authored a number of articles on corrosion.

Organic Fluids for High-Temperature Heat-Transfer Systems

For heating materials to temperatures between 350 F. and 750 F., organic heat-transfer fluids are usually used. Here is how to select the best one for your process, as well as information on designing or selecting the entire heat-transfer system.

W. F. SEIFERT, AND L. L. JACKSON, Dow Chemical U.S.A. and C. E. SECH, Charles E. Sech Associates

The most common heat-transfer fluids used are steam and water, and if the temperature is above the freezing point of water (32 F.) and below about 350 F., the choice is usually between these two fluids. On the other hand, if the temperature of application is below the freezing point of water or above about 350 F., it is necessary (or at least desirable) to consider other fluids.

For temperatures below the freezing point of water, the most common heat-transfer fluids are air, refrigerants such as halogenated hydrocarbons and ammonia, and brines or solutions of glycol and water.

As temperatures increase above 350 F., the vapor pressure of water increases rapidly (120 psig. at 350 F., 233 psig. at 400 F., and 665 psig. at 500 F.), and the problems of structural strength for processing equipment become more and more severe. Thus, with high-temperature systems, it becomes increasingly important to consider fluids having vapor pressures lower than those of water.

This article is adapted from a paper presented at the Process Heat Transfer Symposium of the 71st National AIChE Meeting, Dallas, Tex., Feb. 22-24, 1972.

Originally published October 30, 1972

Some of the more frequently used organic high-temperature heat-transfer fluids available today are shown in Table 1. Included in the table is the manufacturer's recommended maximum-use temperature. Their physical properties are shown in Table II.

In the design of an organic high-temperature heat-transfer system, the engineer has two key problem areas to evaluate. These are: selection of the heat-transfer medium, and system design (including selection of process equipment).

Heat-Transfer-Media Evaluation

Once the decision has been made to use an organic high-temperature heat-transfer fluid, the engineer needs to select a material that will perform satisfactorily and safely at the process temperatures required. To do this, he can draw on his past experience or make relative comparisons of the existing fluids by compiling the data available from fluid manufacturers. The important factors he must evaluate in selecting a high-temperature heat-transfer fluid can be categorized into four areas:

Toxicity and Environmental Ecology—Toxicity and the effect on the ecology are, of course, extremely important from both an operating standpoint and a process standpoint. There is always a chance that a heat-transfer fluid may find its way through packing glands on valves, pumps, heat exchangers, etc., and if this happens, operators, maintenance men, and the environment in general will be exposed to the fluid. More ecological information for evaluating this subject is being made available by many fluid manufacturers.

Corrosiveness to Materials of Construction—In general, a heat-transfer fluid should be noncorrosive to mild steel. Otherwise, the first cost of the equipment can become prohibitively high. It should be noted that all of the chlorinated compounds recognized as heat-transfer fluids are essentially noncorrosive to mild steel as long as all traces of water are kept out of the system and as long as the fluid is not overheated. If halogenated materials are overheated, either by a bulk temperature higher than the recommended maximum limit or by localized hot spots in a furnace, hydrogen chloride gas will be evolved. This gas will remain relatively noncorrosive to mild steel as long as the system is kept absolutely dry; but if traces of water are present, the hydrochloric acid formed will be extremely corrosive, particularly at elevated temperatures. Chlorides can also cause stress corrosion of stainless steels if water is present.

Flammability—Lack of flammability is always vital whenever there is a chance that a fluid may not be completely separated from all sources of ignition. Some of the chlorinated compounds such as chlorinated biphenyls are fire resistant because they will not support combustion (due to the chlorination). However, if they are heated to a sufficiently high temperature, they exhibit a flash point and an explosive range. They will burn if subjected to the ignition conditions encountered in the firebox of a fired heater. Thus, organic fluids must not be exposed to a source of ignition.

Although nonchlorinated heat-transfer fluids can burn, this factor presents no problems if they are contained properly. If, due to some unusual occurrence, they leak from the system into a space other than the firebox of a furnace, they will almost invariably be below their autoignition temperatures before they come in contact with air. Thus there must be a source of ignition before a leak outside a firebox can be serious. Moreover, combustion requires a mixture of air and vapor having a concentration within the flammability limits of the fluid. For continued burning, the liquid must be at temperatures higher than its fire point.

Thermal Stability and Engineering Properties—These last factors will be reviewed in depth in the next two sections of the article. Included in the evaluation are some of the major heat-transfer fluids used today.

Thermal Stability and Chemical Structure

This study covers the relative thermal stability of existing, commercial, heat-transfer fluids. In addition, data on experimental materials are included to broaden the spectrum of comparison and to better define the relationship between chemical structure and thermal stability.

Several generalizations can be made about thermal stability and degradation of organic heat-transfer media:

1. In comparing classes of compounds, aromatic materials are generally superior in thermal stability to aliphatic compounds.

2. For commercial products, the recommended maximum operating temperature is a rough measure of relative thermal stability.

3. Polymer formation is detrimental, particularly if the polymerization is exothermic. Polymers increase the viscosity of a fluid and promote carbonization, leading to

Frequently Used Organic Heat-Transfer Fluids — Table I

Heat-Transfer Fluid Composition	Trade name	Producer	Usable Temp. Range,* °F. Low	High
Aliphatic petroleum oil	Humbletherm 500	Humble Oil	−5	600
Alkyl-aromatic petroleum oil	Mobiltherm 600	Mobil Oil	−5	600
o-Dichlorobenzene	Dowtherm E	Dow Chemical	0†	500
Diphenyl-diphenyl oxide eutectic	Dowtherm A	Dow Chemical	55†	750
Di- and tri-aryl ethers	Dowtherm G	Dow Chemical	12	650
Hydrogenated terphenyls	Therminol 66	Monsanto	25	650
Polychlorinated biphenyl	Therminol FR-1	Monsanto	25	600
Polyphenyl ether	Therminol 77	Monsanto	60	700

*The low-temperature limit was estimated for each fluid from its minimum pumpability characteristic. This pumping factor has been generally accepted by centrifugal pump manufacturers. It is defined as the temperature where the fluid exhibits a 2,000-cp. viscosity.
†This fluid exhibits a true freezing point below the temperature shown. The viscosity at this temperature is less than 10 cp.

Physical Properties — Table II

Compound	Freezing Point, °F.	Boiling Point, °F.	Flash Point, °F.	Fire Point, °F.
1,2,4-trichlorobenzene	63	417	210	†
Tetrachlorobenzene (isomer mixture)	170	480	None	†
Chlorinated biphenyl	7‡	515-680	330	>500
Dichlorodiphenyl ether (isomer mixture)	−4	590	335	530
Trichlorodiphenyl ether (isomer mixture)	130	650	400	>600
Octachlorostyrene	210	—	None	None
Diphenyl ether–diphenyl eutectic	54	495	255	275
Biphenylyl phenyl ether (isomer mixture)	99	680	370	410
o-Biphenylyl phenyl ether	122	670	370	410
Di- and triaryl ethers	<0	572	305	315
Dimethyl-diphenyl ether (isomer mixture)	−40†	554	—	—
Tetramethyl diphenyl ether (isomer mixture)	—	590	—	—
Di-sec-butyl diphenyl ether (isomer mixture)	—	705	380	400
Dicyclohexyldiphenyl ether (isomer mixture)	—	785	—	—
Dodecyldiphenyl ether (isomer mixture)	45‡	>800	410	440
Ethyldiphenyl (isomer mixture)	<−60‡	536	—	—
Partially hydrogenated terphenyl	−15‡	690	335	375
Aliphatic oil	15	720-950	425	475
Alkylaromatic oil	20	~650	350	390

*Cleveland Open Cup method
†None to boiling point
‡Pour point

inefficiency and the potential failure of the heater.

4. Fluid degradation should produce a minimum of volatile materials such as hydrogen, ethylene and other light hydrocarbons. These decomposition products will increase operating losses and are a safety hazard (fire and toxicity) in a vented heating loop.

5. Degradation should not produce reactive or corrosive compounds. Acids, such as HCl, are corrosive, toxic, and accelerate fluid breakdown at high temperatures. Cracking products such as olefins will polymerize under operating conditions.

6. Oxidative stability can be an important factor if air is present at high temperatures.

Data on the static thermal stability of the fluids tested are shown in Table III.

In the case of the commercial fluids, the recommended maximum bulk temperature has been exceeded in certain tests. All available evidence indicates that such accelerated tests give sound data on fluid stability, while reducing the test time significantly (one week at 750 F. = approximately 40 weeks at 650 F.).

The degradation products are the same in tests at normal operating temperatures or in accelerated tests at elevated temperatures. The formation rate of degradation products was the primary variation as the test temperatures increased (approximately doubling for each 18 deg. temperature rise).

The sample losses reported in Table III result from either venting of volatile decomposition products or from carbon deposits on the walls of the pressure vessel. Normal handling loss was less than 4%.

A guideline to fluid stability and maximum operating temperature can be derived from the data in Table III as follows: When 10% sample loss was observed in a 2 to 3 wk. test, it can be said that the recommended maximum temperature has been exceeded by 50° F. This general

Static Thermal-Stability Comparisons—Table III

Compound	Length of Test, Wk.	Temp. °F	Sample Lost, %	Physical State*	Color	Residual Pressure†
1,2,4-trichlorobenzene	0	—	—	L	Colorless	—
	3	650	7.0	L	Black	No
	7	650	21	L&S	Black	Yes
Tetrachlorobenzene	0	—	—	S	White	—
	3	650	12	S	Black	No
	7	650	23	S	Black	Yes
Chlorinated biphenyl "Therminol FR-1"	0	—	—	L	Colorless	—
	3	650	8.0	L	Black	No
	4	650	5.5	VL	Black	No
Dichlorodiphenyl ether	0	—	—	L	Colorless	—
Dichlorodiphenyl ether A‡	3	650	41	S	Black	Yes
B‖	3	650	2.1	L	Brown	No
B	4	650	14.9	L	Black	No
Trichlorodiphenyl ether (isomer mixture)	0	—	—	S	White	—
	3	650	48	VL	Black	No
Octachlorostyrene	0	—	—	S	Yellow	—
	4	650	4.1	S	Brown	No
	10	650	7.9	S	Brown	No
Diphenyl ether–diphenyl eutectic "Dowtherm A"	0	—	—	L	Lt. Yellow	—
	3	775	1.9	L	Yellow	No
Biphenylyl phenyl ether (isomer mixture)	0	—	—	S	White	—
	4	650	2.1	VL	Brown	No
	6	750	5.7	VL	Black	No
	2	775	3.1	VL	Brown	No
	4	775	2.6	VL	Brown	No
o-biphenylyl phenyl ether	0	—	—	S	White	—
	3	750	1.0	VL	Tan	No
	4	775	1.3	VL	Brown	No
m-biphenylyl phenyl ether	0	—	—	L	Yellow	—
	2	775	2.5	L	Brown	No
	4	775	2.3	L	Brown	No
di- and triaryl ethers "Dowtherm G"	0	—	—	L	Yellow	—
	8	650	2-4 ¶	L	Brown	No
	4	700	2-4 ¶	L	Brown	No
	3	750	2-4 ¶	L	Brown	No
	3	775	2-4 ¶	L	Brown	No
Dimethyl diphenyl ether (isomer mixture) "Diphyl DT" (Bayer, Germany)	0	—	—	L	Colorless	—
	4	650	3.2	L	Brown	No
	3	700	17	L	Black	Yes
	3	750	24	VL	Black	Yes
Tetramethyl diphenyl ether (isomer mixture)	0	—	—	L	Yellow	—
	2	650	1.5	L	Yellow	No
	4	650	1.7	L	Brown	No
	2	700	4.5	L	Black	No
	4	700	8.5	L	Black	No
4,4'-diethyldiphenyl ether	0	—	—	L	Yellow	—
	4	650	3.7	L	Yellow	No
	2	700	12.0	L	Black	Yes
Triethyldiphenyl ether (isomer mixture)	0	—	—	L	Yellow	—
	2	700	13.6	L&S	Black	Yes
	4	700	36.9	L&S	Black	Yes
4,4'-di-sec-butyl diphenyl ether	0	—	—	L	Yellow	—
	2	650	4.3	L	Brown	Yes
	3	650	7.0	L	Brown	Yes
	4	650	32	L&S	Black	Yes
Dicyclohexyl diphenyl ether	0	—	—	L	Yellow	—
	2	650	7.8	L	Brown	Yes
	4	650	13.0	L	Black	Yes
Dodecyldiphenyl ether	0	—	—	L	Yellow	—
	2	650	15.0	L&S	Brown	Yes
	4	650	36	L&S	Black	Yes
Ethylbiphenyl "Therm-S-600" (Yawata Chemical Co., Japan)	0	—	—	L	Colorless	—
	2	650	1.5	L	Yellow	No
	4	650	3.9	L	Yellow	No
	3	700	4.0	L	Brown	Yes
	2	750	9.5	VL	Brown	Yes
Partially Hydrogenated terphenyl "Therminol 66"	0	—	—	L	Yellow	—
	8	650	2.5	L	Brown	No
	2	700	10.6	L	Brown	No
	4	700	10.3	L	Brown	Yes
	2	725	9.4	L	Brown	Yes
	4	725	11.9	L	Black	Yes
	1	750	42.7	L&S	Black	Yes
Aliphatic oil "Humbletherm 500"	0	—	—	L	Yellow	—
	2	650	8.3	L	Brown	No
	4	650	57.6	L&S	Brown	Yes
Alkylaromatic oil "Mobiltherm 600"	0	—	—	L	Brown	—
	2	650	8.5	L	Brown	Yes
	4	650	77.2	L&S	Brown	Yes

*Physical state L = Liquid, S = Solid, V = Viscous
† Residual pressure in reactor at dry ice temperature (-70 C.)
‡ Mixture of 2,4'-, 2,4'-, and 4,4'-isomers
‖ Mixture of 2,4'- and 4,4'-isomers. 2,4-isomer absent
¶ Range of losses for several samples

Physical and Thermal Properties — Table IV

Temp. °F.	DTG	DTA	TFR1	HBL500	MBL600	T66	DTE	Hitec Salt	Water
Specific heat, Btu./(lb.)(ft.)									
100	0.3950	0.3880	0.2830	0.4850	0.3900	0.3750	0.281	solid	1.00
200	0.4200	0.4260	0.3047	0.5333	0.4380	0.4300	0.303	solid	1.00
300	0.4540	0.4630	0.3255	0.5791	0.4850	0.4700	0.328	0.373	1.03
400	0.4780	0.5000	0.3464	0.6249	0.5320	0.5300	0.357	0.373	1.08
500	0.5100	0.5370	0.3672	0.6707	0.5780	0.5700	0.383	0.373	—
600	0.5410	0.5790	0.3880	0.7165	0.6250	0.6300	0.410	0.373	—
700	0.5670	0.6110	0.4089	0.7623	0.6710	0.6800	0.440 est.	0.373	—
Viscosity, lb./(ft.)(hr.)									
100	36.2000	6.2900	72.5000	66.4000	136.0000	72.0000	2.500	solid	1.620
200	7.0200	2.5700	8.2340	12.4000	15.2500	10.1500	1.500	solid	0.725
300	2.9000	1.4000	3.4160	4.5400	4.8400	3.7600	0.970	43.6	0.485
400	1.6500	0.9000	1.8900	2.6200	2.8300	1.8800	0.655	18.1	0.330
500	1.0600	0.6500	1.2350	1.4100	1.4600	1.0800	0.510	10.2	—
600	0.7700	0.4600	0.9000	1.0100	0.9510	0.7490	0.433	7.25	—
700	0.6000	0.3600	0.7070	0.7000	0.5350	0.5490	0.350 est.	5.32	—
Thermal conductivity, Btu./(hr.)(sq.ft.)(°F./ft.)									
100	0.0755	0.0805	0.0608	0.0775	0.0695	0.0703	0.0685	solid	0.397
200	0.0742	0.0765	0.0597	0.0750	0.0675	0.0685	0.0645	solid	0.392
300	0.0731	0.0725	0.0587	0.0725	0.0653	0.0670	0.0607	0.33	0.387
400	0.0720	0.0685	0.0576	0.0705	0.0631	0.0657	0.0567	0.33	0.382
500	0.0706	0.0645	0.0563	0.0675	0.0611	0.0640	0.0528	0.33	—
600	0.0693	0.0607	0.0549	0.0655	0.0589	0.0620	0.0490	0.33	—
700	0.0681	0.0568	0.0533	0.0638	0.0566	0.0605	0.0450 est.	0.33	—
Density, lb./cu. ft.									
100	68.0000	65.2700	85.5000	53.0000	58.4000	61.7000	80.30	solid	62.0
200	65.2400	62.4600	82.2000	50.7900	56.2700	59.3000	76.40	solid	60.4
300	62.4900	59.5000	78.7600	48.6100	54.2800	56.7000	72.20	123.0	57.3
400	59.6200	56.6700	75.2400	46.4200	52.1600	53.7000	67.70	121.0	53.7
500	57.1200	53.0000	71.8000	44.2400	50.1700	50.5000	62.60	118.0	—
600	54.2500	49.2900	68.3600	41.3100	48.1100	48.1000	56.60	115.0	—
700	51.4000	45.0300	64.9200	39.1200	45.9800	45.6000	51.00 est.	113.0	—

statement conforms to recommendations in the commercial-fluid product literature.

Comparison of Heat-Transfer Fluids

We made an engineering comparison of some widely used high-temperature heat-transfer fluids. This study shows how the pertinent physical and thermal properties of each fluid affect heat-transfer coefficients and pumping horsepower requirements. The fluids studied in this comparison were Dowtherm A, Dowtherm E, Dowtherm G, Hitec Salt, Humbletherm 500, Mobiltherm 600, Therminol FR-1, Therminol 66, and water. The fluids selected have had many years of use in high-temperature applications.

There are, of course, other available products but they were not evaluated in this study because of:

1. Lack of significant industrial usage.
2. The physical and thermal property data for the fluids were not available.
3. Many of the fluids have closely related physical and thermal properties to those included in this study.

The method used in making this comparison can be used for all other fluids except liquid metals. This study limits its comparison to only liquid-phase forced-circulation systems.

It is recognized that some of these fluids cannot be used over the temperature range studied in this comparison. For example, water is generally not used above 400 F. because of its high vapor-pressure. Also, all organic fluids have maximum temperature limitations. However, it was considered desirable to evaluate the fluids throughout the entire temperature range.

This was done in an assumed counterflow heat exchanger at bulk temperatures ranging from 100 F. to 700 F. Heat-transfer coefficients and power requirements were determined for all fluids. The heat duty, temperature change for the heating fluid, and temperature difference across the heat exchanger were all preestablished. The mass flow of each liquid was determined from its specific heat at the average bulk temperature. The pertinent physical and thermal properties of these fluids are included in Table IV. These data were obtained from literature supplied by the manufacturers of the fluids.

508 HEAT TRANSFER MEDIA

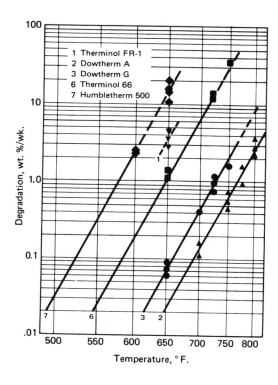

THERMAL STABILITY of heat-transfer fluids at constant temperature—Fig. 1

Two case studies were made in this evaluation. The first comparison was made for the systems where the film coefficient of the heat-transfer fluid is controlling (film-coefficient approximates the overall coefficient). This comparison involves variable coefficients, areas, and horsepower requirements. The second comparison is made for the case when the process-fluid film coefficient is controlling (process fluid coefficient approximates the overall coefficient). This comparison involves essentially a constant coefficient and thus a constant heat-transfer surface area, with a variation only in horsepower.

Reynolds numbers were calculated for all organic fluids (inside a tube) from 100 F. to 700 F., for Hitec Salt from 400 F. to 700 F., and for water from 100 F. to 400 F. When the calculated Reynolds number fell below 4,000—a restriction for the correlation—no additional calculations were made.

In the first comparison, the heat-transfer film coefficient was calculated for turbulent flow conditions. The exchanger area was then determined and used to find the horsepower requirement needed just to pump the fluid through the heat exchanger. In the second case, the process-fluid film coefficient was set to control. Because this restriction establishes a constant surface area, the fluids were compared only on horsepower requirements. For these cases, small differences in overall coefficients would actually result with each fluid. However, compensation for such differences could be made by small adjustments of the heat-transfer fluid's temperature.

In all the cases, the power consumed in the heat ex-changer was only a fraction of that required for the total heat-transport system. The total system would include exchangers, piping, valves, fluid heater, and other related equipment. No attempt was made in this evaluation to determine pumping horsepower requirements for an entire system.

Heat-Exchanger Sizing

Heat-transfer film coefficients and horsepower requirements were determined for the individual fluids at tube inside diameters ranging from 0.5 to 1.4 in. over the entire temperature range. The calculations showed that the ratio of film coefficients and horsepower for one fluid compared to another remained the same when evaluated at the same temperature and tube diameter over the entire range of tubes studied. After observing that an optimum diameter, temperature, fluid relationship did not exist, a single diameter was selected and the data developed for this article.

The tube selected for this study had an inside dia. of 0.6 in. and yielded Reynolds numbers in the turbulent range for the cases evaluated, and heat-transfer coefficients over 100 Btu./(hr.)(sq.ft.)(°F.). Tubes smaller than 0.6 in. were not selected for comparison because the fluid operating conditions were recognized as not being representative of typical industrial systems.

Calculation Basis

The fluids are compared on the following basis:
1. Equal heat duty.
2. Constant temperature change for the heating fluid.
3. Calculations based on the physical and thermal properties of the fluids at average bulk-temperature increments of 100 F., from 100 F. to 700 F. (except as noted).
4. Assumption of a constant temperature difference between the heat-transfer fluid and process fluid for determining exchanger surface area.

The total fluid flowrate in gal./min. was calculated from the heat balance. The Reynolds number, heat-transfer coefficient, heat-exchanger area and length were determined. The pressure drop was then calculated based on the Reynolds number, velocity and length per tube. Knowing the pressure drop, total flow and specific gravity, the horsepower was calculated for each fluid. Results of these calculations are shown in graphical form. Fig. 2 presents the variation of heat-transfer film coefficients of the fluids studied with temperature. Fig. 3 shows the variation of horsepower with temperature for the case when the heat-transfer fluid is controlling. Fig. 4 gives the variation of horsepower with temperatures for the fluids when the process fluid is controlling.

The relative use of these charts can be seen from the following illustration:

A plant has a high-temperature heat-transfer system using Mobiltherm 600 fluid. The circulation rate and heat-transfer-surface areas are fixed for this particular plant process. Consideration is being made to increase the bulk temperature of the heating fluid to 650 F. and replace the fluid with Dowtherm A liquid. The engineers

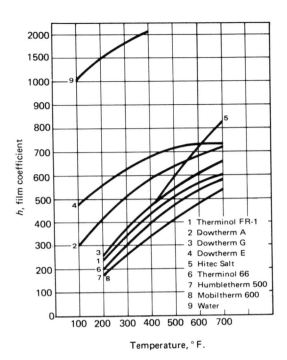

RELATIVE HEAT-TRANSFER film coefficient vs. temperature*—Fig. 2

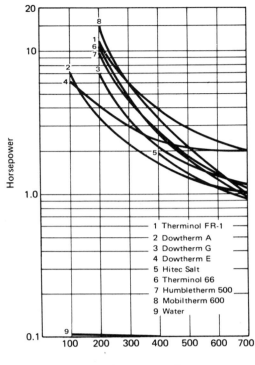

RELATIVE HORSEPOWER vs. temperature* (heat-transfer fluid film controlling)—Fig. 3

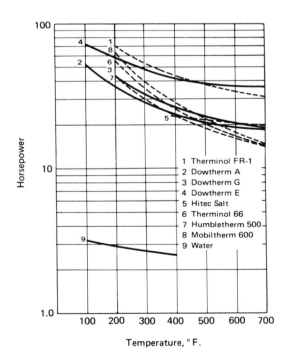

RELATIVE HORSEPOWER vs. temperature* (process-fluid film controlling—Fig. 4

are interested in the effect these changes will have on the heat-transfer-surface area and horsepower requirements in the system.

The relative comparison between the two fluids is given in Fig. 2, 3 and 4.

From these relative ratios, the engineer can visualize that his existing heat-transfer-surface area will not limit his existing heat duty for his process. In fact, he may be able to increase the heat delivered to his process because of the greater overall heat-transfer coefficient and the potentially higher temperature difference between the heating medium and process fluid. In the case of the horsepower requirement, the engineer observes that he will require from 0.86 to 1.16 times his existing horsepower. This factor will depend on how significantly the heating medium or the process fluid controls the overall heat-transfer rate.

These relative comparisons enable the engineer to make a quick evaluation of the effects of the engineering properties of the fluids under consideration. He should remember when using these charts, however, that the absolute values of the factors are not necessarily design figures. They represent numbers generated by a standard base case, defined earlier in this article. In this case, the heat duty, tube diameter, fluid-temperature differences, and equipment configuration are fixed for all fluids. Any

---

*Absolute ratios between fluids will not be affected significantly by changing the physical dynamics of the system (i.e., tube diameter, fluid velocity, temperature differential between fluids, etc.). See calculation basis page 508.

Nomenclature

A	Cross-sectional area, sq.ft.
A'	Surface area, sq.ft.
C_p	Specific heat, Btu./(lb.)(°F.)
D	Dia., ft.
E	Pump efficiency, dimensionless
F	Flowrate, gal./min.
G	Mass velocity, lb./(hr.)(sq.ft.)
h	Film heat-transfer coefficient, Btu./(hr.)(sq.ft.)(°F.)
H	Horsepower
K	Thermal conductivity, Btu./(hr.)(sq.ft.)(°F./ft.)
L	Length, ft.
n	Number of tubes per pass
N_{Re}	Reynolds number, dimensionless
Q	Heat duty, Btu./hr.
S	Specific gravity, dimensionless
V	Velocity, ft./sec.
W	Flowrate, lb./hr.
ΔP	Pressure drop, ft.
ΔT	Temperature change, °F.
μ	Viscosity, lb./(hr.)(ft.)
ρ	Density, lb./cu.ft.

Subscripts

f	Friction
F	Heat-transfer liquid
H	Heat exchanger
i	Inside
lm	Logarithmic mean
P	Process fluid
t	Tube
T	Total
w	Wall

variance of these factors on all fluids will have a minimal effect on the relative comparison ratios of the horsepower and film coefficient.

Case I: Heat-Transfer Fluid Controlling

In establishing the relative comparisons of heat-transfer surface area and horsepower requirements between fluids, the following basis was used:
$A = 0.00196$ sq.ft., $D_i = 0.05$ ft., $Q = 10,000,000$ Btu./hr., $n = 70$ tubes/pass,* $\Delta T_F = 100°$F., $\Delta T_{lm} = 50°$F., $E = 0.70$.

With the above data, W_T, h_F, A_i, N_{Re_t}, P, and H were calculated using the following equations:

Overall heat balance:

$$Q = (W_T)(C_p)(\Delta T_F) = (h_F)(A'_i)(\Delta T_{lm}) \quad (1)$$

or

$$W_T = Q/(C_p)(\Delta T_F) = 10,000,000/(C_p)(100) = 100,000/C_p$$

$$A'_i = Q/(h_F)(\Delta T_{lm}) = 10,000,000/(50)(h_F) = 200,000/h_F$$

Reynolds number:

$$N_{Re_t} = \frac{(D_i)(G_t)}{\mu} = \frac{(D_i)(W_T)}{(\mu)(A)(n)}$$

$$= \frac{(0.05)(100,000)}{(0.00196)(70)(\mu)(C_p)} = \frac{36,443}{(\mu)(C_p)} \quad (2)$$

*In obtaining a reasonable exchanger length for the heat duty and temperature range studied, 70 tubes per pass was selected. This configuration gave a good design basis for the relative comparison of the heat-transfer coefficients and pumping horsepower for all fluids.

Heat-transfer film coefficient:†

$$h_F = \left(\frac{D_i G_t}{\mu}\right)^{0.8} \left(\frac{C_p \mu}{K}\right) \left(\frac{K}{D_i}\right)^{0.33} \left(\frac{\mu}{\mu_w}\right)^{0.14} \quad (3)$$

Assume $(\mu/\mu_w) = 1$

Pressure drop:‡

$$\frac{(\Delta P)(D_i)}{(L)(V_t)^2} = 0.0001906 + \frac{0.01171}{(N_{Re_t})^{0.38}} \quad (4)$$

Horsepower:

$$H = \frac{(F)(\Delta P)(S)}{3,960(E)} \quad (5)$$

Case II: Process Fluid Controlling

When the heat-transfer coefficient controls on the process-fluid side, the heating medium used will have minimal effect on the surface-area requirements for the exchanger. Therefore, the relative comparison for the heat-transfer fluids is restricted to the horsepower evaluation.

Basis: $Q = 10,000,000$ Btu./hr., $h_P = 40$ Btu./(hr.)(sq.ft.)(°F.), $T_{lm} = 50°$F., $n = 70$ tubes/pass,** $D_i = 0.05$ ft., $A = 0.00196$ sq.ft.

Overall heat balance, Eq. (1):

$$A'_i = \frac{Q}{(h_P)(\Delta T_{lm})} = \frac{10,000,000}{(40)(50)} = 5,000 \text{ sq.ft.}$$

$$L_T = \frac{A_i}{\pi D_i} = \frac{5,000}{(\pi)(0.05)} = 31,831 \text{ ft.}$$

$$W_T = \frac{Q}{(C_p)(\Delta T_F)} = \frac{10,000,000}{(C_p)(100)} = \frac{100,000}{(C_p)}$$

Reynolds number, Eq. (2)
Pressure drop, Eq. (3)
Horsepower, Eq. (4)

Equipment Design and Operation

Design of Fired or Electrical Heaters—Heaters for organic heat-transfer fluids generally are designed with lower heat fluxes than heaters for water or steam boilers. Fired steam boilers and water heaters are frequently designed with heat fluxes as high as 40,000 to 50,000 Btu./(hr.)(sq.ft.). Fired heaters for organic heat-transfer fluids are usually designed with average heat fluxes ranging from 5,000 to 12,000 Btu./(hr.)(sq.ft.). The actual allowable heat flux is usually limited by a maximum film temperature, and this in turn is dependent upon factors such as maximum bulk-temperatures, velocity of the fluid across the heat-transfer surface, uniformity of heat distribution in the furnace, and heat-transfer properties of the fluid in question.

Precautions must be taken to guard against excessive

†Kern, D. Q., "Process Heat Transfer," McGraw-Hill, New York, 1950, p. 103.
‡Badger, W. L., McCabe, W. L., "Elements of Chemical Engineering," McGraw-Hill, New York, 1936, p. 36.

**The same tube configuration and fluid velocity were selected as in Case I. Using this basis, the surface area and tube lengths calculated were significantly greater than in the previous case. A review of the calculations showed, however, that the effect on reducing the velocity in the tube, or increasing the number of tubes per pass, will reduce the horsepower requirements for all of the fluids by the same factor. Thus the relative comparisons between fluids remain the same.

accumulations of high-molecular-weight decomposition products, corrosive gases, etc. In the case of fluids that can vaporize within their respective recommended operating temperature ranges, attention must also be paid to the percentage of liquid that will be vaporized. On one hand, vaporization can be beneficial in that it tends to increase the film heat-transfer coefficient and also can vaporize at a constant temperature, which tends to limit the maximum film-temperature for a given set of conditions.

On the other hand, if too much fluid is vaporized so that the heat-transfer surface is, for all practical purposes, blanketed with vapor rather than liquid, the film heat-transfer coefficient will be reduced very rapidly, and dangerously high surface-temperatures can develop, resulting in severe fluid degradation and mechanical failure.

All other things being equal, any organic heat-transfer fluid degrades in proportion to its temperature. When considering operating a fluid at temperatures higher than the manufacturer's recommended maximums, it is extremely important to guard against approaching the critical temperatures that may cause the fluid to carbonize and form hard carbon scale on the heat-transfer surface. When this happens, the hard carbon tends to insulate the surface, decreasing the rate of heat transfer at this point. This results in an increased metal temperature under and around the edges of the carbon, accelerating the rate of hard-carbon-scale formation. If this condition is not discovered and corrected, it will lead to overheating and ultimate destruction and rupture of the heat-transfer surface.

The formation of soft or particulate carbon as well as high-molecular-weight polymers need not be serious as long as the condition is recognized and kept under control. It is important, however, that the fluid be sampled periodically in accordance with the recommendations of both the fluid and heater manufacturers. With fluids having an atmospheric boiling point within their recommended operating temperature ranges, the quantity of soluble degradation products is usually determined by measuring the viscosity of the sample at a given temperature.

In addition to these high-molecular-weight products, which are soluble in the fluid, there will frequently be an accumulation of fine particulate organic insolubles as well as some particles of mill scale in the sample. The percentage of this sediment in the fluid must also be kept within limits recommended by the manufacturer.

Whenever the decomposition products, the insolubles, or the acidity of the heat-transfer medium exceeds the established recommendations, it is important that the fluid in the system be replaced or reclaimed. Some of the manufacturers of high-temperature organic heat-transfer fluids offer a reclamation service for their products. Onstream purification units are available to reclaim some of the fluids semicontinuously in the field without returning the fluid to the manufacturer. However, it is important that the cost of reclaiming the fluid be included in evaluating the overall operating cost. If any given fluid is overstressed severely above the recommended maximum temperature, the decomposition rate will become quite high. If this happens, the cost of removing the high-molecular-weight polymers, the cost of replacement fluid, and, especially, the cost of lost production can be appreciable.

The most frequent operating difficulties that cause excessive fluid degradation in a fired heater are:

Flame Impingement—Flame impingement on a heat-transfer surface will invariably cause trouble if not corrected. The results will be lower heat efficiencies, loss of capacity and, ultimately, tube failures. Flame impingement can be caused by using improper burners, improper adjustment of a burner, or by poor furnace design.

Low Circulation Rate (Forced-Circulation Heater)—Poor circulation rates through the heater can be the result of a power failure, instrumentation failure, or pump cavitation due to either low liquid level or system contamination with a low-vapor-pressure material.

Low Circulation Rate (Natural-Circulation Heater)—Poor flow in a natural-circulation heater can be attributed to low fluid level in the heater, insolubles restricting the flow in the tubes, or uneven firing of the burners in the furnace.

High Heat Fluxes—When a heater is operated above its rated design capacity, the film temperatures that the fluid is exposed to will be in excess of the recommended design value. To ensure that excessive degradation of the fluid does not take place, the heater manufacturer should be consulted to determine its recommendations for circulation-rate and burner modifications.

Heat-Transfer-Fluid Contamination—Contamination can be caused by a process material leaking into (or unintentionally being charged into) the high-temperature heating system. This situation can cause significant problems, depending on the thermal stability and quantity of process fluid added.

Almost any organic contaminants are potential hazards because of their lower thermal stability. If the contamination is excessive, hard carbon scale can deposit on the heat-transfer surface of the heater; once this hard carbon deposit starts, degradation of the heat-transfer fluid will be accelerated due to the excessive film temperatures. Even if a hard carbon scale is not formed on a heat-transfer surface, organic contaminants can accelerate the decomposition rates for a heat-transfer fluid.

If the organic contaminant is acid in nature, such as a fatty acid, severe corrosion of mild steel (in the presence of moisture) can result. Likewise, severe corrosion of mild steel can occur with inorganic acids or strong alkalis. The corrosion is usually most severe in the hottest part of the system, the tubes of the heater, unless the corrosive material is quite volatile. In the case of HCl, for example, regardless of whether the acid has leaked into the heat-transfer fluid or whether it is formed by thermal decomposition, the corrosion always appears to be most severe in the vapor or gas space *above* the liquid. This is probably because these are the areas where traces of moisture are most likely to collect, and the moisture causes the HCl to become very corrosive.

Heater and Vaporizer

As is the case with any refractory-lined furnace, boiler, etc., heaters for organic heat-transfer fluids should be started and operated at low heat input when they are started for the first time or whenever they have been shut down for an extended period. New refractory, or refractory in a furnace that has been standing idle, will always contain considerable quantities of absorbed water. Unless the refractory is heated slowly so the water can escape slowly as it is vaporized, internal pressure will develop, causing the refractory to spall.

Additionally, fluid manufacturers recommend maximum heatup rates for cold systems to protect their fluids and heaters. The viscosities of all of the organic fluids increase as their temperatures decrease (until they become solid, or so viscous that they are difficult to pump). The increase in viscosity causes a decrease in the film heat-transfer coefficient in two different ways. First, it reduces the amount of flow through the heater for a given pump head, which decreases the film heat-transfer coefficient because of a decrease in velocity. Second, the increased viscosity causes a reduction in the film heat-transfer coefficient for a given flowrate or velocity. These two effects are additive and, unless care is exercised in starting up a cold system, some extremely high film-temperature rises can be experienced, resulting in damage to the fluid or equipment. For a given heat flux, the film temperature rise is, of course, inversely proportional to the film heat-transfer coefficient.

Pump Seals and Piping

It is common practice to use welded joints wherever possible and to keep flanged joints at a minimum. Prior to the development of the spiral-wound stainless-steel-and-asbestos gaskets, it was standard practice to use Series-30 steel raised-face flanges at all flanged joints. The reason for the heavy flanges, in many cases, was not to withstand high pressures but to be able to achieve the gasket pressures required for a leakproof joint without warping the flanges. Although many users still prefer the Series 30 flanges in all instances because of their ability to not yield under cycling thermal stresses and to therefore maintain tighter joints, the spiral-wound gaskets have made it possible to maintain tight joints in many instances with Series 15 flanges. A new material known as Grafoil gasketing material, made by Union Carbide, has also proven to be satisfactory in high-temperature systems.

Pump glands can be sealed either with special pump packings for high-temperature service or with mechanical seals. Such seals for high-temperature service are becoming more and more popular. When the system is clean, free of abrasive sediment and contaminants, mechanical seals have been known to function effectively without leakage and without attention for as long as five years.

Some of the pump and pump-seal manufacturers recommend that mechanical seals be flushed with a stream of clean cooled fluid taken from the pump discharge. Filters, or small cyclone separators, may be used to clean the fluid. In some cases, a small water-cooled heat exchanger cools the fluid before it is put into the flushing connection on the seal chamber.

Until recently, seal and pump manufacturers recommended water-jacketed seal chambers when pumping fluids at temperatures above about 450 F. The purpose of this was to protect the seal rings made of Teflon resin or Viton fluoro-elastomer, which formed a seal between the pump shaft or shaft sleeve and the portion of the seal that was free to slide on the shaft. Many of the seal manufacturers are now offering a bellows-type seal for use without cooling at temperatures considerably above 450 F.

Seal manufacturers are also offering seals with graphite rings in place of Teflon rings to seal the sliding porton of the mechanical seal to the pump shaft. It is claimed that these seals can also be operated at temperatures above 350 F. without external cooling. Most of the mechanical seals used in pumps for handling high-temperature organic heat-transfer fluids have a dense carbon or tungsten carbide stationary ring that mates with a rotating ring having either a Stellite or a tungsten carbide face. Experience indicates there is a limiting temperature at which these materials can be used without external cooling or without flushing them with a cooled stream of fluid.

In the final analysis, thermal degradation of an organic heat-transfer fluid is important to the extent that it influences the functionality of the fluid. Decomposition is manifested by the appearance of low-boiling components and/or high-boiling materials (including polymeric tars). Low-boiling materials in a fluid may cause excessive venting, and high makeup rates may be encountered. In contrast, high-boiling materials and polymers in a fluid will result in higher viscosities. Increasing viscosity will accelerate the degradation process due to less efficient heat transfer and higher film temperature. Excellent fluid stability is a critical ingredient in arriving at a reliable, efficient, and safe high-temperature heat-transfer system. ■

Meet the Authors

W. F. Seifert   L. L. Jackson   C. E. Sech

Walter F. Seifert is technical service and development manager for Dow Chemical U.S.A., Midland, MI 48640, where he is also a specialist on heat-transfer fluids. He has worked for Dow on secondary oil recovery, and has also worked with Continental Oil. He received his B.S. in chemical engineering from the University of Colorado.

Larry L. Jackson works in the organic chemical production research laboratory for Dow Chemical U.S.A., Midland, MI 48640. He holds a B.S. in chemistry from Virginia Military Institute and a Ph.D. in organic chemistry from Ohio State University.

Charles E. Sech is President of Charles E. Sech Associates, Inc., 206 So. Main, Ann Arbor, MI 48108, where he provides consulting, design, construction and operating services to a wide variety of projects in the chemical process industries and for the federal government. Previously, he was president of W.L. Badger Associates, Inc. He holds a B.S. in chemical engineering from the University of Michigan and is a registered professional engineer.

Understanding Vapor-Phase Heat-Transfer Media

Heat-transfer systems based on a mixture of diphenyl and diphenyl oxide are used at temperatures higher than are practical with steam. The engineer should understand the subtle but critical differences between the two types of systems.

D. R. FRIKKEN, K. S. ROSENBERG, and D. E. STEINMEYER, Monsanto Co.

The eutectic mixture of diphenyl and diphenyl oxide (EDIPDO)* is an excellent vapor medium for precise temperature control at temperatures higher than practical with steam. Unfortunately, EDIPDO systems can behave rather strangely at times. In the following discussion, we explain this behavior and give guidelines on how to design an EDIPDO system.

EDIPDO vapor heating systems differ from steam systems in apparently insignificant ways:

1. Molecular weight is higher.
2. Vapor pressure is much lower for a given temperature, e.g. at 500°F, EDIPDO vapor pressure = 10.95 psig and steam 666 psig; at 600°F, EDIPDO vapor pressure = 30.64 psig and steam 1,538 psig.
3. Common noncondensables (such as water, phenol and benzene) are lighter than EDIPDO vapor; in steam systems, the common noncondensables (air) are heavier than steam and thus tend to flush out with the condensate.
4. Condensate returns by gravity in most systems.
5. Normal loads are frequently small compared with startup loads.

These differences probably would not worry an engineer familiar with steam systems when he designed his first EDIPDO vapor system. However, when he started it up, he might encounter some rather baffling phenomena. For example, the vaporizer would mysteriously lose level-refill and then proceed to repeat the cycle for many hours. After the liquid level finally stopped cycling, there might be widely varying temperatures within the system, despite only a 1-2-psi pressure drop.

To correct the temperature-variation problem, the engineer might enlarge the piping to reduce the pressure drop to 0.1 psi. Some temperature problems would go away—some would not. He might put in special vent lines and vent accumulators. Again, this would solve some problems, but other lines would have to be vented almost hourly to maintain the temperature where the vapor-pressure curve said it should be.

Other solutions might include bringing in additional feed lines or straightening out the condensate drain lines to assure a downward slope on all runs. Again, eliminating a problem in one place would cause a problem elsewhere.

The engineer might come upon an earlier draft of this paper and install supply, drain and vent piping, as suggested, on one of the parallel loops. He would find that this trial loop required venting much more frequently than any of the others. At this point, he would seriously consider going back to school for an MBA.

Loss of Level

The basic problem here is the high ratio of startup load to steady-state load, which results from the high-temperature operation. This means that piping large enough for normal operation can be grossly undersized for startup loads. The vaporizer gains pressure quickly

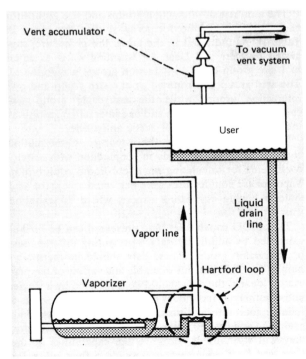

HARTFORD LOOP stops back-up in drain line—Fig. 1

*Commercially available as Dowtherm A (Dow), Thermex (ICI), Therm-S (Yawade), Diphyl (Bayer), and Therminol VP-1 (Monsanto).

Originally published June 9, 1975

514 HEAT TRANSFER MEDIA

as it is heated, but the rest of the system stays cold and the liquid is forced back up the liquid line. If a Hartford loop (Fig. 1) is present, vapor cannot leave that way, so it simply moves through the vapor line, condenses in the cold user, and remains. Either way, the vaporizer loses level. A sequence like this is illustrated in Fig. 2:

A. The system starts under vacuum, with the whole system cold.

B. The vaporizer heats up, but stays at high vacuum until it approaches 200°C. As the reboiler heats, the rest of the system stays much cooler because the combination of low pressure and very-low vapor density prevents much vapor from flowing out. (It is "choked" by piping pressure drop.)

C. At 210°C, the pressure in the vaporizer has risen to 4.6 psi, while the rest of the system is still very low. Assuming that all inerts are purged and the balance of the system is at 50°C, its pressure will be only 0.003 psi.

The difference in pressure is enough to push (or suck) liquid up the liquid return line to an elevation of 12 ft, high enough to reach the large-volume jackets. There is enough volume in these jackets to drop the liquid level in the vaporizer below the level of the bundle.

D. Liquid level and vaporizer pressure will then oscillate with just enough flow to maintain 4.6-psi pressure drop in the supply piping.

E. Finally, the system heats up and gains pressure to a point where there is less than 4.6-psi drop in the supply piping. We are then out of trouble, until the next startup.

To avoid this type of problem, we can either:*

Reduce the system pressure differential under startup conditions by enlarging the piping; or increase the vertical distance between the vaporizer and the user. Note that a Hartford loop does not avoid the loss-of-level problem. It simply increases the time for the problem's evolution and somewhat smooths out its effects.

If the heating time is not too critical, we may choose to just wait until the system stops bouncing.

Temperature Variation

To find a solution to more-complex problems, we must understand an EDIPDO system having parallel users operating at the same conditions. There can be three explanations for temperature variation: noncondensables (NCs) in the EDIPDO; varying pressure drops between users; or choked condensate drains.

NCs are a mixture of air from startup, degradation products, and water. The more troublesome NCs are probably dissolved in the condensate and simply circulate round and round through the system. There are two key elements that explain why NCs build up at odd places, causing low temperatures:

■ They separate from the EDIPDO vapor only when there is condensation.

■ Dissolved NCs will be stripped out of returning EDIPDO liquid only when condensate returns countercurrent to an EDIPDO vapor flow.

NCs in a condenser build up until their concentration

---

* Electrically heated vaporizers are often provided with a low-level power cutoff to prevent overheating of the electrical elements. A system that has such a vaporizer might have a reduced heat-up time if the heat input to the system were reduced, avoiding low-level shutdowns caused by loss of level.

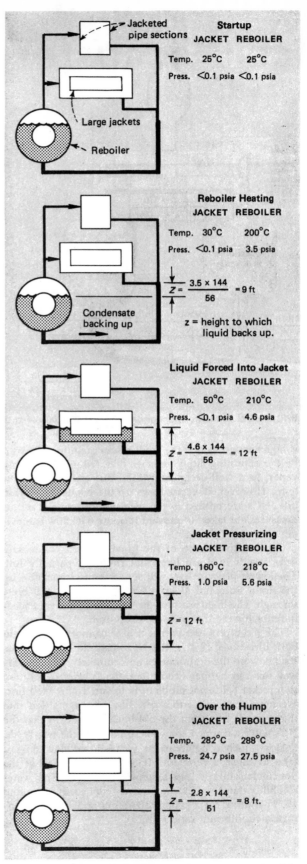

STARTUP LOAD leads to liquid back-up and low level in reboiler until system comes up to temperature—Fig. 2

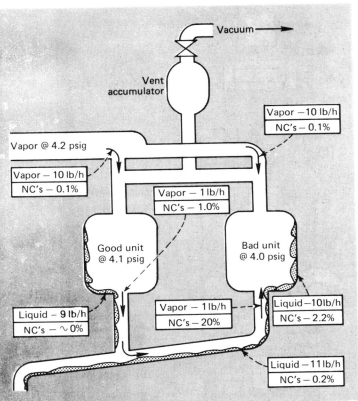

BACKFLOW can cause noncondensables to build up—Fig. 3

is high enough so that the equilibrium quantity dissolved in the returning liquid equals that fed in the supply vapor. In a well-designed system, this creates no problems. However, if vapor flows up the condensate drain line, NCs are bottled in the user. Unfortunately, unless measures are taken to prevent it, some backflow is inevitable.

Conservative sizing of the gravity-return condensate lines results in drain lines that run only partially full. The lowest-pressure user will always receive backflow up the drain line, and at least one of the users will blow through. The high-pressure-drop user acts as a vent accumulator for the low-pressure-drop user.

The basic problem is that a user cannot be fed from both directions. If it is, the NCs have no escape route. Backflow up the drain line is, unfortunately, not the only way this can happen. Too often, the "obvious" solution to a jacket cold-spot problem is to run a new feed line. We have found that this new line often gets tied into the opposite end from the old line. This means that the NCs will be trapped somewhere between the two feeds.

How much countercurrent vapor-liquid flow does it take to cause a problem? It all depends on what the noncondensable is, but assume it is water. The water solubility data* indicate an activity coefficient of around 10. Therefore, for a system operating at 550°F, the estimated equilibrium ratio is:

$$K = \frac{Y}{X} = \frac{\text{Water vapor pressure}}{\text{EDIPDO vapor pressure}} \times 10 = \frac{1045}{27.5} = 380$$

*Dowtherm Heat Transfer Fluid, The Dow Chemical Co., 1969.

This means that for a stripping factor of 5, all we would need would be 1 part of vapor moving up the drain line for 76 parts of liquid moving down. For perspective, with a stripping factor of 5 in a drain line, the following fractions of water would be stripped from the condensate:

Equilibrium stage	Fraction stripped
1	0.833
2	0.967
3	0.993

The point is that with a noncondensable as volatile as water, virtually any backflow up the line will keep it bottled in the user. Fig. 3 illustrates this situation.

Water is a troublesome noncondensable. But even in a poorly designed installation, it will eventually be purged from the system. There are two other NCs that are always present in EDIPDO systems: phenol and benzene, which are formed by EDIPDO degradation. Neither is nearly as volatile as water, and a much higher ratio of vapor flow is required to do significant stripping.

	Approximate K value in 500°F EDIPDO	Fraction stripped by 1 part vapor/10 parts liquid		
		1 stage	2 stages	3 stages
Benzene	26	.72	.90	.96
Phenol	6	.38	.49	.54

Experience tells us that NCs do strip out. We have found that the optimum way to make stripping occur in the right place is to arrange piping in the following manner:

■ Vapor inlets should be at or near the bottom of the user.

■ Condensate should be drawn off at the bottom of the user.

■ Vent piping should come from the top of the user and make a vertical run to the vent accumulator.

Thus, as the EDIPDO vapor enters the users, the NCs separate and are concentrated near the top. The line to the vent accumulator acts as a stripping column to further concentrate the NCs. The vent accumulator does not need a separate return to the condensate header. However, if this is omitted, the vapor line to the vent accumulator should be short, vertical and at least 1 in. in dia. We encounter systems where the vapor velocity in this line is so high that countercurrent draining of condensate cannot occur. This results in a low temperature in the vent accumulator, and is easy to misdiagnose as a bad case of NCs.

Seals and Venting

For a system that has many users, it is often difficult and expensive to supply vapor to the bottom of all of them. For systems such as these, we must provide seals in the drain lines to prevent upward flow of vapor. The easiest way to do this is to install a loop, as shown in Fig. 4. Of course, inverted bucket traps or other devices could also be used.

In a system with a liquid seal, it is desirable to have at least one condenser that is efficient in concentrating

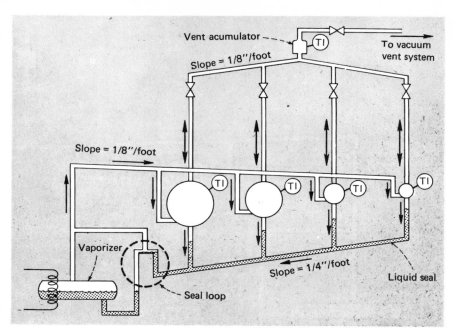

SEAL LOOP in the drain line is designed to prevent upward flow of vapor in the users—Fig. 4

noncondensables. If the concentration of water in the vapor supplied to the condensers could be kept as low as 0.01%, for example, the vapor saturation temperatures in the other condensers could only be 4°F lower than the vapor supplied to the users (assuming the system is operating at 550°F).

What about venting of parallel systems? The care with which the venting system is designed depends on the amount of temperature variation the designer is willing to accept. Suppose we use a single-vent accumulator, as shown in Fig. 5. The vapor flow through the vent lines from the "good unit" to the "bad unit" will keep the

VAPOR FLOW to "bad" unit traps noncondensables—Fig. 5

noncondensables bottled up in the "bad unit." The only way the noncondensables can leave is with the condensate.

One obvious answer is to provide a vent accumulator for each user. For a long jacketed line having many spools, this is expensive, and one is tempted to tie several users into a common vent accumulator, as indicated by Fig. 4. However, with this arrangement, only one user at a time can be properly vented. To escape this problem, temperature indicators (TIs) and valves are required, so that the operator can force-vent the proper users. The TIs and valves for this approach require almost as much capital as a vent accumulator for each user. Obviously, they require much more operator attention.

There is an attractive alternative to providing vent accumulators: put the users in series, as shown in Fig. 6. When this is done, the vapor should be made to leave the opposite end from which it enters.

Should the temperature in the vent accumulator drop? Yes, a falling vent-accumulator temperature is normally a good sign, in that it means the NCs are collecting where they are supposed to. As a corollary, if one vent accumulator is collecting NCs and another is not, the one collecting them probably is the one that is properly installed.

Pressure Drop

Another problem peculiar to parallel EDIPDO vapor systems is the penalty for high pressure drop in normal operation. The low-pressure/high-temperature operation means that the $\Delta T/\Delta P$ ratio at saturation is quite high:

Temperature, °F	$\Delta T/\Delta P$ ratio (°F/psi)	
	EDIPDO	Steam
500	5.3	0.16
550	3.5	0.12
600	2.4	0.09

This, in turn, means that a difference in pressure drop

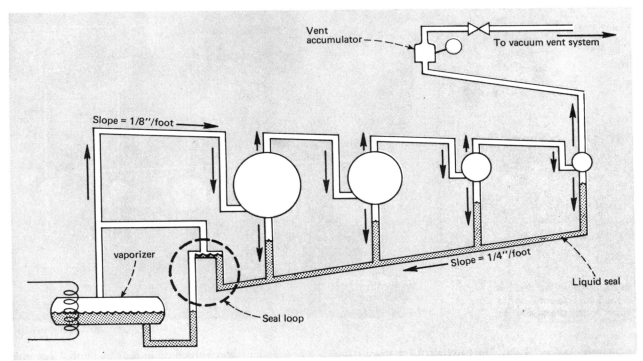

SERIES OPERATION is alternative to numerous vent accumulators. Vapor must enter and exit at opposite ends—Fig. 6

between parallel loops of 1 to 2 psi is not tolerable in systems where a relatively constant temperature is desirable. In order to approximate zero ΔT between parallel equal users, the vapor-supply piping must either be of equal ΔP or negligible ΔP.

Condensate choking as a cause of temperature variation is simply the backing of condensate up into the user. It will occur in a system whenever the combined pressure drop (in feet of liquid) in vapor and condensate lines exceeds the vertical distance available for draining.

Good Piping Practices

Our experience suggests that piping for EDIPDO vapor systems is customarily sized for unusually low pressure drops. There are basically three reasons for using these larger pipe sizes:

1. Total pressure drop in the system must be less than or equal to the vertical distance between the bottom of the user and the top of the liquid level in the vaporizer or reboiler. Minimum pressure drops thus will result in both a lower vertical requirement and more-compact layouts.

2. A more important reason is to minimize the temperature variations caused by the pressure variations. As noted earlier, this is much more likely to give problems in EDIPDO systems than in a steam system. The vapor-supply lines should be sized so the system pressure drop is 1 to 3 psi or less for the maximum flow condition. To avoid temperature variation problems, the system should be designed so that most of this pressure drop is in the common header(s).

3. It is normally advisable to size nothing less than 1 in. The increased cost of the 1-in pipe over the smaller sizes will be more than compensated for by fewer required supports and the engineering time saved in selecting the smaller sizes.

Vapor lines should slope downward to the user to prevent restriction due to condensate buildup. Condensate lines should slope to the vaporizer to facilitate draining.

Are EDIPDO systems really that complicated? The answer is no—but it's a bit qualified. The guidelines for good design are fairly clear and simple, but there are a surprising number of ways to get in trouble if the guidelines are ignored. #

Meet the Authors

D. R. Frikken **K. S. Rosenberg** **D. E. Steinmeyer**

Donald R. Frikken is a member of Monsanto Co.'s Corporate Engineering Dept. He holds a B.S. in mechanical engineering from Kansas State University and an M.S. in civil engineering from the University of Missouri at Rolla.

Kenneth S. Rosenberg is a product specialist for Monsanto's heat-transfer fluids. He also has been a senior process design engineer at Monsanto, and a production engineer for Shell Oil Co. Mr. Rosenberg has a B.S. in chemical engineering from Purdue University and an MBA degree from Southern Illinois University.

Daniel E. Steinmeyer is a principal engineering specialist, Monsanto Co., Corporate Engineering Dept., 800 N. Lindbergh Blvd., St. Louis, MO 63166. He has been responsible for design-method development and consultation in heat transfer at the company since 1966. Previously, Mr. Steinmeyer was a process engineer at Monsanto's Chocolate Bayou ethylene plant.

Using wastewater as cooling-system makeup water

No magical skill is required to adapt wastewater streams for cooling-tower makeup. What is normally required is an in-depth investigation, and the application of good water-treatment technology.

Edwin W. James, William F. Maguire and *William L. Harpel*, Betz Laboratories, Inc.

☐ Water is called wastewater when it has received some sort of contaminant that makes it undesirable for conventional uses. However, the water may still be useful, and some applications may actually serve as an effective treatment for the contaminant. Proper definition of the contaminants is thus the key factor in determining the water's suitability for cooling-tower makeup.

The microorganisms in a cooling tower, for example, may remove 90% or more of the phenols and reduce effluent organic carbon and oxygen demand, but this ability decreases with rising amounts of sulfide because sulfides apparently kill the better phenol-oxidizer microorganisms [2]. Also, hydrogen sulfide is corrosive to cooling-water equipment, reduces chromate inhibitors to ineffective chromic oxide sludge, and causes an odor. Therefore, when sulfide levels exceed 2 ppm in the wastewater, the water should be diverted from the cooling system.

The first step in evaluating a waste stream for cooling-water makeup is a thorough analysis of that stream. If the stream passes the analysis test, the next

Secondary sewage effluent*	Table I
Total hardness as $CaCO_3$, ppm	192
Calcium as $CaCO_3$, ppm	128
M-Alkalinity as $CaCO_3$, ppm	210
Silica as SiO_2, ppm	22
Orthophosphate as PO_4, ppm	8
pH	7.3
Chloride as Cl, ppm	172
Sulfate as SO_4, ppm	180
Specific conductance, micromhos at 18°C	900
Ammonium as N, ppm	10
Iron as Fe, ppm	0.5
Zinc as Zn, ppm	0.2
Copper as Cu, ppm	0.04
Suspended solids, ppm	8
BOD, mg/l	15
COD, mg/l	40
Nickel as Ni, ppm	0.18
Turbidity	5
Nitrate as N, ppm	0.3
Nitrite as N, ppm	3.1
Chloroform extractable, ppm	2

*Representative of water used in laboratory studies

Recirculator study (7 days)	Table II
Analysis of cycled water	
Calcium ($CaCO_3$)	615 ppm
Magnesium ($CaCO_3$)	240 ppm
pH	7.2
Chromate (CrO_4)	13.5 ppm
Chloride (Cl)	920 ppm
Orthophosphate (PO_4)	20 ppm
Treatment	
Chromate-zinc-phosphonate-polymer	
Corrosion data	mpy
Unpretreated and pretreated high carbon steel	1
Unpretreated and pretreated low carbon steel	1
304 Stainless steel	0
Admiralty	2
Welded high carbon steel	1

Comments
1. Admiralty corrosion not localized.
2. Some evidence of biological growth.
3. Heat transfer coefficient decreased by only 10%.
4. Minor foaming occurred and was eliminated by use of antifoam.

Originally published August 30, 1976

Foam potential studies — Table III
(Secondary sewage effluent at 5 cycles)

Chromate-zinc-phosphonate-polymer-biocide program

Antifoam	Antifoam level (ppm)	Foam height (mm)
None	0	175
A	1	0
B	1	0
C	1	0
D	1	55
D	5	15
D	7	0

Phosphate-phosphonate-polymer-organic-biocide program

Antifoam	Antifoam level (ppm)	Foam height (mm)
None	0	175
A	1	0
B	1	0
C	1	0
D	1	0

Comments: foaming tendency was determined using an inverted cone into which the solution under study was added. Foam was generated by bubbling air at a constant rate through the solution.

Ion exchange unit effluent — Table IV

Total hardness as $CaCO_3$, ppm	2
M-Alkalinity as $CaCO_3$, ppm	60
Sulfate as SO_4, ppm	10
Chloride as Cl, ppm	5
Silica as SiO_2, ppm	18
Total phosphate as PO_4, ppm	0.5
Orthophosphate as PO_4, ppm	0.5
pH	7.0
Specific conductance, micromhos at 18°C	200
Nitrate as NO_3, ppm	60
Ammonia as N, ppm	20

Kellogg ammonia plant boiler blowdown — Table V

M-Alkalinity as $CaCO_3$, ppm	32
Silica as SiO_2, ppm	0.1
pH	9.7
Iron as Fe, ppm	0.05
Sodium as Na, ppm	17.5
Specific conductance, micromhos at 18°C	80
Phosphate as PO_4, ppm	15

step is usually a feasibility study that would, via various tests, determine the stream's potential for corrosion, general fouling, scale formation, biological fouling and foaming, and what treatment would be needed to control such problems. Obviously, the nature of the contamination is indicated by the plant process generating the wastewater, but sophisticated organic analyses may be required to pinpoint unusual contaminants and concentrations.

Also, the variability of the composition of the wastewater is important in determining the treatment methods, and analyses should be extended over as long a period as possible. This will define the consistency of the effluent composition, and at the same time provide information helpful toward planning additional treatment. If water reuse is being considered part of a new plant, the cost of improved water quality may be compared to the savings it allows in the cost of heat exchangers or other equipment.

The problems tend to be magnified as cycles of concentration are increased in the cooling tower. In some cases, increased cycles mean the addition of an external treatment system, such as cold lime softening for phosphate removal and hardness reduction. In all cases, increased cycles call for an application of proper internal water-treatment methods such as application of biocides, deposit-control agents, antifoams, as well as blowdown and pH control.

A good understanding of the problems and practical answers may be achieved by actual in-plant runs or laboratory evaluations. Any in-plant runs should follow a gradual approach, with slowly increasing proportions of wastewater in the makeup water and with close monitoring, so as to avoid disruption of plant operations. If laboratory evaluation is the method, a test cooling-tower might be involved. The size of the test tower varies, but is normally about 15 gpm. Cooling-water conditions can be altered quite readily in these towers, and detrimental effects monitored in test heat exchangers and other monitoring devices.

While the foregoing principles apply to almost any instance of wastewater application as cooling-tower makeup, the variables inherent in water, plant operating conditions and metallurgy, and the cooling-tower environment, combine to make extensive generalization impossible. Instead, it will be more instructive to give some examples of waste streams used as cooling-tower makeup: domestic-sewage effluent, nitrogen-fertilizer-plant effluent, boiler blowdown and process condensate, chemical-plant effluent, bottoms from refinery sour-water strippers, and effluent from dissolved-air flotation systems.

Domestic-sewage effluent

Among wastewater streams, sewage-plant effluent typically provides large volumes of water that may serve as the sole source of makeup for a cooling tower. Such reuse offers two benefits: ecological, in that fresh water is conserved; and economic, in that the sewage effluent may be available at no charge, or at a greatly reduced cost, compared with fresh water.

Sewage-plant effluent has been used by a number of plants over the years, and first-hand experiences are documented in the literature [10, 11, 12, 13, 14, 17]. A literature review will give a general idea of what to expect with this type of water; however, a more specific

Ammonia plant process condensate		Table VI
	Gulf coast plant	Alaskan plant
Ammonia as N, ppm	350	824
Carbon dioxide as CO_2, ppm	360	2,200
pH	6.7	7.1
Specific conductance, micromhos at 18°C	1,600	5,000
Iron as Fe, ppm	0.0	0.7

Southern chemical plant waste stream	Table VII
Total hardness as $CaCO_3$, ppm	32
Calcium as $CaCO_3$, ppm	24
Magnesium as $CaCO_3$, ppm	8
M-Alkalinity as $CaCO_3$, ppm	24
Sulfate as SO_4, ppm	3
Chloride as Cl, ppm	8
Silica as SiO_2, ppm	3
Total phosphate as PO_4, ppm	0.4
Orthophosphate as PO_4, ppm	0.2
pH	7.2
Specific conductance, micromhos at 18°C	80
Copper as Cu, ppm	0.05
Iron as Fe, ppm	0.05
COD, mg/l	450
BOD, mg/l	150
TOC, ppm	135
Aromatics, ppm	125
Total hydrocarbons, ppm	200

Southern chemical plant waste stream corrosion study results				Table VIII
Cycles	Mild steel (mpy)	Copper (mpy)	Admiralty (mpy)	304 SS (mpy)
3.0	0.5	0.4	0.4	0.0
6.0	0.8	0.4	0.5	0.0

Experimental conditions: 120°F, 2 ft/sec., pH 6.5, spinner studies with CrO_4-Zn-Phosphonate-Polymer treatment. Chromate level was 30 ppm.

Spinner study results*		Table IX
	Corrosion rate, mpy	
Treatment	Steel	Admiralty
Chromate-phosphate-Zn (pH 6.5)	1	5
Phosphate-phosphonate (pH 7.5)	17	10
Cycled water analysis		
Calcium as $CaCO_3$, ppm	480	
Magnesium as $CaCO_3$, ppm	430	
Silica as SiO_2, ppm	130	
Chloride as Cl, ppm	135	
Sulfate as SO_4, ppm	20	
Organic carbon as C, ppm	150	

*All studies at 120°F with process condensate and simulated cycled water mixed at 1:6 ratio.

characterization is usually necessary. Knowledge of the type of sewage-plant treatment—primary, secondary, tertiary—is required, along with analyses.

Water quality from a sewage treatment plant may vary, depending on the percentage of industrial waste, the percentage of domestic waste (which may vary according to traditional washdays), and the runoff that is handled by the given plant. Thus the analyses should cover as long a period as possible.

Corrosion is usually less severe with sewage effluent than with fresh water. Although such effluent typically contains high orthophosphate concentrations, its tendency for calcium phosphate scaling is inhibited by the stabilizing effect of organic material that is also typically present in high concentrations. So, proper control of cycles of concentration and pH, along with the addition of deposit-control agents, usually permits higher calcium phosphate loadings than may be carried in fresh-water systems.

The nature of sewage effluents presents a severe microbiological fouling problem when such water is used as cooling-tower makeup. This can be controlled, but at increased biocide cost. A combined program of chlorine and nonoxidizing biocide addition is typically used to maintain good control.

A representative analysis of secondary sewage effluent is presented in Table I. Based on prior experience with water of this type, a review of the analysis indicates a tendency for: biological fouling (indicated by the BOD level), calcium phosphate deposition (indicated by the calcium and orthophosphate concentrations), and foaming (indicated by the suspended solids and MBAS concentrations).

A laboratory evaluation of sewage effluent is presented in Table II. This indicates a need for corrosion control with a chromate/zinc/phosphonate/polymer program. A well-chosen nonchromate approach may also be used in this instance. Although there are reports of some very severe, hard-to-contain foaming problems, most systems can be controlled with a small amount of antifoam. The data in Table III support this.

Fertilizer-plant effluents

Many plants have waste streams that are treated for quality improvement and subsequent discharge—but why throw this treated water away?

In fertilizer plants, process waters typically contaminated with high levels of ammonia and nitrate are treated by ion exchange to reduce this type of contamination. Some ammonia and nitrate may remain along with some urea, but, overall, the quality of this treated water is good. A sample analysis is presented in Table IV. There is no reason to discard this water; it may be

Waste treatment system effluent	Table X
Ammonia as N, ppm	10
Total hardness as CaCO$_3$, ppm	212
Calcium as CaCO3, ppm	160
Magnesium as CaCO$_3$, ppm	52
M-Alkalinity as CaCO$_3$, ppm	1540
Sulfate as SO$_4$, ppm	112
Chloride as Cl, ppm	152
Silica as SiO$_2$, ppm	17
Total phosphate as PO$_4$, ppm	29
Ortho phosphate as PO$_4$, ppm	28
pH	8.2
Copper as Cu, ppm	0.03
Iron as Fe, ppm	0.05
Suspended solids, ppm	18
COD, mg/l	306
Chlorine demand, ppm	108
BOD, mg/l	17
Triethylamine, ppm	1
Sodium as Na, ppm	790
Chromium as Cr, ppm	0.0

used as part of the cooling-tower makeup. Depending on the plant, this water may amount to only about 30% of the makeup stream; otherwise, the percentage may depend on the quality of the treated water. Nutrients in the form of ammonia, nitrate and urea present a substantial potential for microbiological growth, requiring closer monitoring of biological fouling, and a more intensive biocide program to minimize slime outbreaks.

Boiler blowdown and process condensates

Table V presents the characteristics of a boiler blowdown stream that was added to an ammonia plant's cooling-tower water circulating at 45,000 gpm. In this plant, deionized water had been used for boiler makeup, along with a coordinated phosphate/pH treatment program. The major concern was the phosphate concentration, because of its potential for phosphate deposition. The reconciling factor was the small size of the boiler blowdown—only 1% of the required cooling-tower makeup. Thus the blowdown is diluted, and the use of this water is appropriate.

Condensate streams are prime candidates for water reuse. The typical mineral salts are not present to create problems, although condensate streams usually do contain contaminants of some sort and must be evaluated. An ammonia plant condensate is shown in Table VI. The substantial concentrations of ammonia and carbon dioxide in this condensate provide nutrients for biological growth and create an increased biocide demand. In addition, the ammonia concentration would present a corrosion potential for copper alloys.

The increased biocide requirement will raise costs approximately 25–75% over the cost of fresh water, but microbiological fouling can be controlled. Although a substantial portion of the ammonia will be volatilized over the tower, the remaining concentration will still be enough to attack copper and admiralty-copper metals. While the exact nature and degree of this attack remain to be defined, system metallurgy should be thoroughly reviewed before proceeding with water of this quality.

Chemical-plant effluents

The possibility of reusing a particular water stream may not be easy to establish, even with extensive laboratory work. Table VII presents an analysis of a plant waste stream that has good mineral quality but is high in aromatic content. It was theorized that much of the aromatic contamination would be volatilized over the cooling tower. A laboratory study

Analyses of sour water stripper bottoms						Table XI
	1	2	3	4	5	6
Chlorine demand, ppm	—	—	—	400	—	—
Ammonia as N, ppm	28	30	30	12	50	50
BOD, ppm	—	—	288	121	—	—
COD, ppm	—	—	387	203	—	—
Total hardness as CaCO$_3$, ppm	14	1.2	92	5	6	2
Calcium as CaCO$_3$, ppm	12	0.8	32	5	2	2
Magnesium as CaCO$_3$, ppm	2	0.4	60	0	0	0
Phenolphthalein alkalinity as CaCO$_3$, ppm	0	0	16	51	0	42
Methyl orange, alkalinity as CaCO$_3$, ppm	88	140	68	100	46	94
Sulfate as SO$_4$, ppm	16	37	125		32	34
Chloride as Cl, ppm	39	1.5	260	28	142	93
Silica as SiO$_2$, ppm	<2.0	1.3	4.4	—	—	—
Total phosphate as PO$_4$, ppm	3.0	0.4	2.1	—	—	—
Total inorganic phosphate as PO$_4$, ppm	—	0.2	0.5	—	—	—
Ortho phosphate as PO$_4$, ppm	—	0.1	0.5	0.06	—	—
pH	6.8	7.7	8.7	9.6	6.8	9.
Specific conductance, micromhos at 18°C	250	500	1,200	280	—	425
Total carbon, ppm	—	—	130	—	—	—
Inorganic carbon, ppm	—	—	11	—	—	—
Organic carbon, ppm	—	—	119	32	—	—
Chloroform extractable, ppm	13	—	71	—	—	—
Phenol, ppb	110,000	33,000	79,800	50,000	—	—
Total copper as Cu, ppm	.05	0.02	0.12	—	0.1	—
Total iron as Fe, ppm	1.6	0.6	0.5	—	9.4	32.
Soluble zinc as Zn, ppm	0.0	0.0	0.0	—	0.0	—
Nitrite as NO$_2$, ppm	0	253	276	—	—	—
Nitrite as NO$_3$, ppm	<1	<1	<1	—	—	—
Hydrogen sulfide as H$_2$S, ppm	.01	0.1	0.1	0.1	0	0.6
Suspended solids, ppm	28	—	—	6	—	—

using wastewater cycled both three and six times confirmed volatilization of the organics and further indicated that a chromate-based treatment would provide satisfactory corrosion protection. These results are illustrated in Table VIII.

However, those laboratory tests also indicated problems from a gummy type of deposit, and this required further studies on a test tower to establish that the deposits would not be a serious problem.

Table IX presents the results of another study conducted for a plant that contained significant amounts of admiralty tubing. The corrosion rate indicated by this study was prohibitively high, due (it was thought) to the organic content of the water, principally formic acid. However, additional studies indicated that a substantial portion of the organics would be volatilized over the cooling tower. The preliminary study actually presented the worst case possible and not the average condition that might be experienced in the cooling tower.

The next step was to experiment reusing the water in the plant cooling tower, starting at a low percentage of the makeup, and building toward the final desired percentage, with constant monitoring of corrosion and fouling. In this actual operation, the predicted corrosiveness of the water did not materialize, and the water was successfully reused in its entirety, which amounted to 10–20% of the cooling-tower makeup.

Table X presents the analysis of the effluent from a bio-oxidation pond. Significant features of this analysis include high alkalinity, high chemical oxygen demand and high phosphate levels. The latter occurred when phosphoric acid was added as a nutrient to the activated-sludge basins. A brief review of this analysis did not indicate any insuperable problem.

Chromate inhibitors provide the best protection against corrosion. However, preliminary testing indicated that in this case the water had a high chromate demand, which precluded chromate treatment. Additional studies would be required to define the corrosion, fouling and biological growth potentials for the system. These studies would include laboratory evaluations and a gradual approach to the use of this water in the actual system by increasing the proportion of reused water to makeup water, with close monitoring.

Sour-water stripper bottoms

Refineries commonly use the sour-water stripper to remove dissolved gases from process waters. Typically, these strippers will remove 99% of the sulfides, 80–95% of the ammonia, and 30–60% of the phenol. Except for the phenol and ammonia contaminants, with their resultant contributions to COD, BPD and TOC, the bottoms from these towers are essentially distilled water. Analyses for six such bottoms are shown in Table XI.

Dissolved air flotation effluent			Table XII
	DAF inlet	DAF effluent without polymer	DAF effluent with 5 ppm polymer
Oil	80	14	4
Suspended solids	95	28	8
Dissolved solids	700	690	695
Total hardness, as $CaCO_3$	210	205	210
Calcium, as $CaCO_3$	140	138	142
Magnesium, as $CaCO_3$	70	67	68
"P" Alkalinity	0	0	0
"M" Alkalinity	60	62	60
Chloride	220	214	215
Sulfate	230	216	222
Sulfide	trace	None	None
pH	6.3	6.3	6.3
Iron	6.5	1.4	0.3
Ammonia	15	13	12
Phenol	12	10	10
COD	400	265	240
BOD	120	77	73
TOC	110	68	60

This wastewater is commonly reused to conserve fresh water and to remove up to 98% of the phenol from the refinery effluent. Alternate applications for reuse are: crude-oil desalting, process wash-water, and cooling-tower makeup. Since the cooling tower provides very effective bio-oxidation of the phenol, but is hampered in this by the presence of sulfides, the efficiency to which sulfides, ammonia and phenol are removed in the stripper determines if the stripper bottoms should be used as cooling-tower makeup.

As mentioned earlier, this water should be diverted from the cooling system whenever sulfide levels exceed 2 ppm. Also, ammonia has three important effects: Its

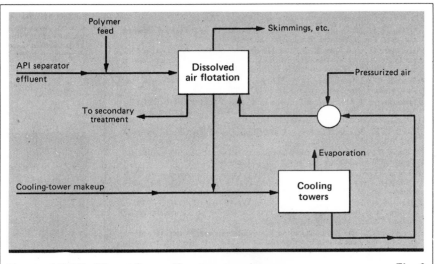

Reuse of DAF effluent for cooling tower makeup   Fig. 1

fluctuation causes pH changes; it is quite corrosive to copper and copper alloys; it is a nutrient for fouling organisms. During upset conditions, therefore, the sour-water stripper bottoms should be diverted away from the cooling tower and to the API separator or other waste-treatment facilities.

Microbiological fouling will occur as a result of the bio-oxidation of phenols, so that biocides are necessary for preventing massive growth of slime. Any tower biocide program must be carefully controlled, however, to achieve the proper phenol bio-oxidation. Chlorine can be injected intermittently to prevent slime growth and still permit efficient oxidation of the phenols, but this can result in chlorinated phenols and chloramines. Consequently, non-oxidizing biocides may be preferred over chlorine to prevent the formation of toxic compounds.

Dissolved-air-flotation effluent

The dissolved-air-flotation (DAF) unit, which normally follows the API separator in a refinery waste-treatment plant, serves primarily to remove oils and suspended solids from refinery wastes. Typically, it removes 50–90% of these contaminants, but that removal can be improved with low concentrations of polyelectrolytes.

Since the cooling tower will remove phenol and reduce oxygen demand of the DAF effluent, and since this effluent is usually a large volume of water that can significantly reduce demand for fresh water, some refineries are now using it as cooling-tower makeup. The major disadvantage of this use is fouling. Oil remaining in the water tends to agglomerate other solids into masses that restrict flow and retard heat transfer. This can be minimized by oil detectors to divert this water during upset conditions. Deposit control agents and wetting agents should be used continuously in the cooling water.

In the case of the system shown in Fig. 1, cooling-tower blowdown also goes to the DAF unit, so that a corresponding amount of DAF effluent must be diverted away from the tower. Towers should be controlled, limiting the levels for hardness, dissolved solids, suspended solids, and oil, in order to minimize fouling and corrosion.

In summary, industry is finding that many waste streams provide very satisfactory cooling-tower makeup water, if good water-treatment policies are used to compensate for the potentially deleterious features of such waters. An ideal example is the increased use of secondary sewage effluent as makeup water. With the proper approach, more and more plants will find themselves using wastewater streams for tower makeup, and obtaining very satisfactory results.

References

1. Hart, James A., Waste Water Recycle for Use in Refinery Cooling Towers, *Oil & Gas Jour.*, June 11, 1973, pp. 92-96.
2. Mohler, E. F., and Clere, L. T., Development of Extensive Water Reuse and Bio-oxidation in a Large Oil Refinery, National Conference on Complete Water Reuse (AIChE), Washington, D.C., April, 1973.
3. Willenbrink, Ron, Waste Water Reuse and Inplant Treatment, Annual Petrochemical and Refining Conference (AIChE), New Orleans, Mar., 1974.
4. Petrey, E. Q., Waste Water and Pollution Control, 12th Annual Liberty Bell Corrosion Course, Philadelphia, 1974.
5. Maguire, W. F., Harpel, W. L., and Carter, D. A., Aspects of Wastewater Reuse as Makeup for Boiler and Cooling Systems, National Conference on Complete Water Reuse (AIChE), Chicago, May, 1975.
6. Maguire, W. F., Minimize Plant Effluent Through Proper Water Engineering, Annual Petrochemical and Refining Conference (AIChE), Houston, March, 1975.
7. Maguire, W. F., Reuse Sour Water Stripper Bottoms, *Hyd. Proc.*, Sept., 1975.
8. Harpel, W. L., and James, E. W., Waste Water Reuse as Cooling Tower Makeup, International Water Conference, Pittsburgh, Oct. 1973.
9. Harpel, W. L., Waste Water Treatment, Regulation and Reuse, Industrial Fuel Conference, Purdue University, Oct., 1973.
10. Gray, H. J., McGuigan, C. V., and Rowland, H. W., Treated Sewage Serves as Tower Makeup, *Power*, May, 1973.
11. Hofstein, H., and Kim, K. B., Treated Municipal Waste Water as a Major Water Source for Industry, National Conference on Complete Water Reuse (AIChE), Washington, DC, April, 1973.
12. Weddle, C. L., and Masri, H. N., Reuse of Municipal Waste Water by Industry, *Ind. Water Engrg.*, June/July, 1972.
13. Humphreys, F. C., Sewage Effluent in Use as Power Plant Circulating Water, Proceedings of the 14th Industrial Waste Conference, Purdue (1959), reprinted in *Betz Indicator*, 29, No. 3.
14. Cecil, L. K., Sewage Treatment Plant Effluent for Water Reuse, *Water & Swg. Wks.*, 111, 421 (1964).
15. Carnes, B. A., Eller, J. M., and Martin, J. C., Reuse of Refinery and Petrochemical Waste Waters, *Ind. Water Engrg.*, June/July, 1972.
16. Osborn, D. W., Nitrified Sewage Effluents: Their Corrosiveness and Suitability for Use as Power Station Cooling Water, *Jour. Inst. Swge. Purif.*, 243 (1964).
17. Terry, S. L., and Ladd, K., City Waste Water Reused for Power Plant Cooling and Boiler Makeup, National Conference on Complete Water Reuse (AIChE), Washington, DC, April, 1973.
18. Petrey, E. Q., The Role of Cooling Water Systems and Water Treatment in Achieving Zero Discharge, Cooling Tower Institute Meeting, Houston, Jan., 1973.
19. Weisberg, E., and Stockton, D. L., Water Reuse in a Petroleum Refinery, National Conference on Complete Water Reuse (AIChE), Washington, DC, Apr., 1973.

The authors

William L. Harpel is Assistant Director, Product Development, for Betz Laboratories, Inc., Somerton Road, Trevose, PA 19047. A member of the AIChE, the National Association of Manufacturing Engineers, the Air Pollution Control Assoc., and the National Catalysis Soc., he has presented technical papers at a number of national conferences. He holds a B.A. in chemistry and a B.S. in chemical engineering from Lehigh University and did additional graduate work at Stanford University and the University of Pennsylvania.

Edwin W. James is Product Manager, Cooling Water Services for Betz Laboratories, where he has worked since 1968, first as engineer and Assistant Product Manager, Cooling Water Services. An author of several papers dealing with cooling water systems, he holds a B.S. in chemical engineering from Drexel University.

William F. Maguire is Assistant Market Manager, Hydrocarbon Processing Industry for Betz Laboratories. A member of the AIChE and the API, he has prepared and presented papers at several conferences. He holds a B.S. in chemical engineering and a masters degree in business administration.

Water that cools but does not pollute

Although recirculated-cooling-water systems avoid thermal pollution, they often require anticorrosion, antiscaling and antifouling additives, which are then discharged into natural waters in the cooling-tower blowdown. Recent studies show that salty or brackish makeup waters may not need as many additives and may be environmentally safer.

Haia K. Roffman and *Amiram Roffman,* Westinghouse Environmental Systems Dept.

☐ The thermal pollution caused by cooling-water discharges was not considered important until about 10 years ago, but during the mid-1960s research into the effects of such discharges on aquatic life resulted in stringent regulations and a trend toward alternatives to once-through cooling-water systems.

The principal federal law related to thermal discharges is the Federal Water Pollution Control Act (FWPCA) of 1972, for which the Environmental Protection Agency is charged with primary responsibility for enforcement. In addition to the regulations of FWPCA, individual states have promulgated ambient thermal standards that also require compliance.

To meet these state and federal thermal criteria, cooling towers and air coolers are being used more extensively by many industries. This has in turn liberated many industries from a prior dependence on locating facilities in water-rich regions, and permitted reducing the costs of transportation and labor.

However, these possibilities have in turn focused attention on the effects that cooling-tower blowdown might have on a local aquatic environment, particularly with respect to chemical contaminants. The evaluation of those effects is complex, because the chemicals in cooling-tower blowdown are both concentrated there as a result of water losses to evaporation, and are put there to control scaling, fouling and corrosion within the plant. A careful analysis of the environmental effects of cooling tower blowdown must be done prior to designing and citing such systems.

The characteristics of blowdown

The blowdown of a cooling system is that portion of the water removed to prevent the buildup of harmful concentrations of dissolved solids. The amount of blowdown depends on the quality of the makeup water and the concentrations of chemicals to be maintained in the circulating water. Determining the relationships is a classic calculation for chemical engineers, and has been **reduced to nomograph form (see box, p. 525).**

Originally published June 21, 1976

The potential hazards of cooling-water additives must be monitored to protect the enviroment.

From the environmental point of view, the efficiency of cooling towers is measured by the cycles of concentration. The number of cycles can be increased, for higher efficiency, by reducing the blowdown requirements, i.e., by controlling the buildup of harmful concentrations of material. However, hazards to the environment may be presented by the chemicals added to the circulating water to control corrosion, scaling and

Quick calculation of cooling tower blowdown and makeup

When the system is at equilibrium, the makeup must equal the losses, and for cooling towers:

$$M = E + B + W \quad (1)$$

where:

- M = makeup, % of circulation
- E = evaporation loss, % of circulation
- B = blowdown, % of circulation
- W = windage loss, % of circulation

Since the evaporation water will be essentially free of dissolved solids, all those introduced with the makeup water must be removed by the blowdown plus windage loss, or:

$$Mp_m = (B + W)p_c \quad (2)$$

where:

- p_m = concentration in the makeup, ppm.
- p_c = concentration in circulating water, ppm.

For cooling towers, the concentration in the recirculating water is arbitrarily defined as cycles of concentration, C, as:

$$C = \frac{\text{concentration in cooling water}}{\text{concentration in makeup water}}$$

so that:

$$M = (B + W)p_c/p_m = (B + W)C \quad (3)$$

Now, experience has shown that windage losses range: 1.0–5.0% for spray ponds, 0.3–1.0% for atmospheric cooling towers, and 0.1–0.3% for forced draft cooling towers, and that 0.1% can be assumed for modern forced-draft towers. Also, a heat balance plus experience shows that evaporation losses are 0.85–1.25% of the circulation for each 10°F drop in temperature across the tower, and that an evaporation loss of 1.0% for each 10°F drop can be assumed for most calculations, so that: $E = \Delta T/10$.

Thus:

$$M = \Delta T/10 + B + 0.1 \quad (4)$$

Combining Eq. (3) and (4):

$$\Delta T/10 + B + 0.1 = (B + 0.1)C$$
$$\Delta T/10 - 0.1(C - 1) = B(C - 1)$$

$$B = \frac{\Delta T}{10(C - 1)} \quad (5)$$

The nomograph permits a rapid, simultaneous solution of Eq. (1) and (5). *Example:* a cooling tower handles 1,000 gpm of circulating water, which is cooled from 110°F to 80°F. What blowdown and makeup are required if the concentration of dissolved solids is allowed to reach 3 times that in the makeup? On the nomograph, align $C = 3.0$ with $\Delta T = 30$, and read $B = 1.4\%$ or 14 gpm, and $M = 4.5\%$ or 45 gpm. If that same tower were operated at 4.0 cycles of concentration ($C = 4.0$), then the required blowdown and makeup would be read as 40 gpm and 9 gpm, respectively; and if the C were reduced to 2.5, blowdown and makeup would likewise become about 50 gpm and 19 gpm.

Source: Caplan, F., *Chem. Eng.*, July 7, 1975, p. 110.

---

fouling; and the concentration of these chemicals in the blowdown should be monitored and controlled to avoid the potential damage.

A number of methods are available for controlling cooling-tower blowdown. While the application of these techniques is subject to economic and environmental evaluations on an individual basis, they may be generally considered as influent treatments, sidestream treatments, or effluent treatments:

Influent treatments

These may be classified as H_2SO_4 additions, clarification and/or filtration, biocide additions, and desalting.

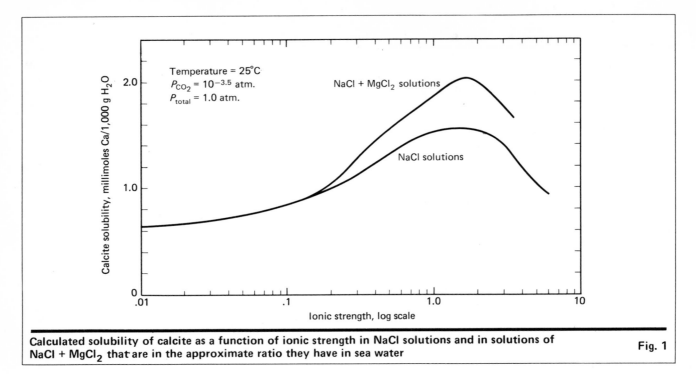

Calculated solubility of calcite as a function of ionic strength in NaCl solutions and in solutions of NaCl + MgCl$_2$ that are in the approximate ratio they have in sea water

Fig. 1

H$_2$SO$_4$ additions: Whenever necessary, lowering the pH to the point where the water has a negative Langelier index can prevent scale deposits on the heat-exchanger surfaces. This action permits increasing the cycles of concentration, but it will also increase the tendency of the cooling water to corrode.

Clarification and/or filtration: To reduce the amount of suspended solids, the makeup water is passed through a clarification and/or filtration system. Granular-bed filters or precoat filters can be used. This treatment will also reduce the amount of larvae and other organisms; it permits higher cycles of concentration by improving the makeup water.

Biocide additions: Additions of chlorine or other biocides may reduce the tendency toward biological fouling on the warm exchanger surfaces and in the cooling tower, permitting higher concentrations of biological substances and higher cycles of concentration in the circulating water.

Desalting: Although none of the many desalting techniques have yet been proven economical for cooling-tower makeup waters, the potential advantage of their application there (to eliminate salt drift, reduce chemical treatment, increase cycles of concentration) indicates that desalting should be evaluated for new installations.

Sidestream treatment

A fraction of the water in the circulating system can be removed, treated by filtration or contact-stabilization, and then returned to the system. Filtration of the sidestream, as in the influent treatment, will remove suspended material, especially organisms that may have developed in the system.

The contact-stabilization method is based on the principle that the crystallization rate, and the fixation of salts from supersaturated solutions, can be controlled. For example, a supersaturated solution of calcium carbonate can pass through a layer of limestone granules, deposit a portion of the dissolved calcium carbonate, and then return to the cooling system. The efficiency of this technique depends on the balance between the rate at which supersaturation occurs in the cooling system and the rate at which deposition occurs in the contact-stabilization unit. The technique is reported to have been successfully applied to seawaters. [1]

Effluent treatment

Hot and chemically rich blowdown should be controlled to reduce adverse ecological effects, before it is released to natural waters. The thermal problems are reduced, compared to a once-through cooling system, since the cooling tower releases most of the plant's heat to the atmosphere via evaporation. A blowdown's thermal plume, and its mixing zone in natural waters, can be calculated for both surface and submerged discharges, using various models, most of which take into consideration the higher density of the blowdown. [2-7] In general, the mixing and diffusion process in estuaries is affected by a more widely varying group of parameters than in oceans, making the problem of estuary discharge more complicated.

The blowdown can be diluted with a calculated quantity of available water in order to reduce the impact of its contained heat and dissolved solids. When the permissible concentration of total dissolved solids (TDS) in the effluent is given, and the TDS in the available dilution water is known, these can be related to the TDS in the blowdown, with allowance made for any evaporation from a dilution basin. Thus:

$$C_E = \frac{DC_D + BC_B}{D + B}$$

Principal dissolved solids in river water			Table I
Chemical	World average (ppm)	Minimum (ppm)	Maximum (ppm)
HCO_3^-	58.4	17.9	183.0
$SO_4^=$	11.2	0.8	289.0
F^-	–	–	0.2
Cl^-	7.8	2.6	113.0
NO_3^-	1.0	–	1.0
Ca^{++}	15.0	5.4	94.0
Mg^{++}	4.1	0.5	30.0
Na^+	6.3	1.6	124.0
K^+	2.3	1.8	4.4
Fe	0.7	1.9	–
Al	–	–	0.012
SiO_2	13.1	10.6	14.0
Total dissolved solids	120.0	43.1	853.0

where: C_E = concentration in the effluent or in a dilution basin, %
C_D = concentration in the dilution water, %
C_B = concentration in the blowdown, %
D = flow of dilution water, consistent units with B
B = flow of blowdown, consistent units with D

Also, C_E is related to C_D by:

$$\frac{C_E}{C_D} = \frac{D(R-1) + ER}{D(R-1) + E}$$

where: E = evaporation loss from a dilution basin, % of effluent
$R = C_B/C_D$

The cooling-water treatment

Since the permissible TDS and heat in the effluent waters are related to those in the diluent water through the blowdown, the choice of an optimal cooling-water treatment system becomes a balance between (1) the effects of the available water on equipment, and (2) the environmental effects of the chemical treatments necessary to control corrosion, scaling and fouling.

Principal dissolved solids in sea water		Table II
Chemical	Concentration (ppm)	Percent of total salt
Cl^-	18,980	55.05
Na^+	10,556	30.61
$SO_4^=$	2,649	7.68
Mg^{++}	1,272	3.69
Ca^{++}	400	1.16
K^+	380	1.10
HCO_3^-	140	0.41
Br^-	65	0.19
H_3BO_3	8	0.03

Treatment can mitigate the chemically and biologically aggressive attack of water constituents, as well as afford additional protection that cannot be designed into the materials of construction. In general, the chemical attack is caused by mineral salts that produce scale, corrosion, wood deterioration and fouling. Biological attack is caused by the rapid growth of algae, bacteria and fungi that produce fouling, corrosion and wood deterioration.

The introduction of the chemicals is a very important part of chemical treatment. The desired concentrations should be established in advance, and regulated through a suitable feeding system. Automatic systems are preferred, because they avoid frequent manual adjustments, and the exposure of personnel to potentially hazardous chemicals.

Continuous indicating and recording analyzers can be used to monitor critical variables, such as pH and oxidation-reduction potential, and the continuous analyzers can be used to regulate automatic feeders. For example, oxidation-reduction potential can be used to regulate Cl_2 feed, and the pH analysis can be used to control the acid.

The chemicals added through these systems can be evaluated according to their utility in controlling scale, corrosion, wood deterioration, and fouling, as follows:

Scale

Scale is usually formed on the heat-transfer surfaces of coolers by salts of calcium, magnesium and silica, which show a decreasing concentration in the cooling water with increasing temperatures. The most common scale-forming salts are $CaCO_3$ (solubility product, $K_{sp} = 5 \times 10^{-9}$), $MgCO_3 \cdot 3H_2O$ ($K_{sp} = 1 \times 10^{-5}$), $Mg(OH)_2$ ($K_{sp} = 8.2 \times 10^{-2}$) and $CaSO_4$ ($K_{sp} = 2.4 \times 10^{-5}$).

The solubility of cations such as calcium and magnesium ions is increased by the presence of anions such as chlorine and fluorine. Thus the solubility of calcium carbonate in distilled water at 20°C is 5.0×10^{-9}, whereas it is 2.66×10^{-6}, or 532 times greater, at the same temperature in seawater having 19% chlorinity. Accordingly, seawaters contain much more calcium carbonate than do fresh waters (Table I, II). The influence of NaCl and (NaCl + $MgCl_2$) on the solubility of calcite is shown in Fig. 1.

These solubilities determine the TDS at which scale formation begins in the plant—thus the cycles of concentration at which a cooling tower may be operated, and the required blowdown. For example, six cooling towers operated on seawater were maintained at TDS concentrations of 55,000–60,000 ppb in two towers and at 100,000 in the other four. Since seawater contains about 35,000 ppm TDS, that means the cycles of concentration for those towers were 1.57–1.71 and 2.85, respectively. [8]

It has been common practice in fresh-water cooling systems to add acid, usually H_2SO_4, to the circulating water to lower the pH and thus increase solubility. However, the effect of chlorine ions in inland brackish and seawaters sometimes makes acid addition unnecessary. Thus, of the six seawater systems mentioned above, two reported no maintenance problems while not using additives of any kind.

In any event, acid additions should be made for each specific situation. Although lowering the pH permits higher cycles of concentration, it also increases the tendency of the cooling water to corrode. Two indexes useful in evaluating the individual situations are the Langelier index and the Ryzner stability index (see box, p. 529).

Corrosion

Chemical corrosion-inhibitors interfere with either the anodic or cathodic reactions of the electrochemical corrosion cell (Fig. 2). Because of the nature of the reactions, it is essential to use enough corrosion inhibitor to inactivate the entire surface. Otherwise, severe pitting may occur on the unprotected parts.

Chromate-based inhibitors are best known, most effective and most widely used. When used for both cathodic and anodic protection, their concentrations can run as high as 3,500 ppm. For both economic and environmental reasons, zinc salts and inorganic phosphates are blended with the chromates to reduce the chromate concentrations to levels considerably below 600 ppm. Corrosion is believed to be prevented by an adherent film of iron oxide and chromic oxide on the anodic surfaces and a zinc hydroxide film on the cathodic surfaces.

The chromate-zinc treatment is usually excellent. However, if the metal surfaces are not clean, if the chromate-zinc concentrations are too low, or if the pH is too low, the protective zinc hydroxide film does not develop. Instead, a loose, unprotective $Zn(OH)_2$ product is formed, and because the low chromate concentrations are inadequate to protect the anodic areas, severe pitting can result. It has been observed that pretreatment of the metal surfaces to remove grease and foreign material, followed by initial high-level inhibitor dosage, generally results in better protection.

Chromates (mainly Cr^{+6}) are toxic to aquatic life and tend to concentrate in the body tissues of marine organisms. Thus, attempts have been made to replace chromates with chromate-free inhibitors. These substitute inhibitors can be classed as inorganic or organic.

Inorganic nonchromate inhibitors consist of various combinations of polyphosphates, silicates, ferrocyanides, nitrites, and metal ions such as Zn^{+2} and Cu^{+2}. Examples include straight polyphosphates, zinc polyphosphates, ferrocyanide-polyphosphate and zinc-ferrocyanide-polyphosphate.

When using the phosphates, the pH must be carefully controlled to prevent calcium *ortho*phosphate precipitation. The pH must also be regulated when silicates are used, since they are most effective at pH levels ranging from 6.5–7.5. The use of nitrite inhibitors requires high concentrations of 200–500 ppm, and careful pH control to 7.0–9.0, since the nitrites can attack metals at pH values below 6.5. Also, nitrite inhibition is limited to iron and steel, because it is inefficient with copper and aluminum alloys. Copper ions are detrimental to aluminum and probably also to steel; they are inferior to chromates, and cause pollution problems.

Organic corrosion-inhibitors consist of a variety of starch derivatives, lignosulfonates, tanning agents and modified tannings, glucosates, glyceride derivatives and many proprietary formulations. Soluble oils that have been buffered and emulsified inhibit corrosion on ferrous metals, copper and aluminum alloys. However, natural rubber cannot be used in the presence of these oils, since it softens and swells, and synthetic rubber is substituted.

Organic compounds can serve as nutrients for undesirable organisms that cause fouling problems in the cooling system and result in an increased biocide demand. Many of the organic compounds are not compatible with the chlorination used, and the cooling system requires nonoxidizing biocides. Thus, they present an inherent pollution problem.

In a relatively new approach to cooling-tower water treatment, pH levels are maintained at 7.5–9.0, instead of at the conventional lower values. This minimizes corrosion; and scaling is controlled with a variety of chemical chelates such as organophosphates and polyacrylates. [*9, 10*]

Salt concentration is the major contributor to corrosion problems in saltwater cooling towers. In general, a higher TDS concentration increases the corrosivity of the solution via increased conductivity in the electrochemical corrosive cell. The chloride ions increase corrosion rates to a high degree; however, at very high concentrations (brine), the corrosion rate drops. Also, high chloride concentrations have been observed to hinder the formation of protective coatings by inhibitors. Thus, corrosion problems with salty water are best resolved on an individual basis of specific water quality, operating conditions and materials of construction.

Wood deterioration

Cooling-tower wood is subject to both chemical and biological deterioration. Chemical deterioration results from destruction of the lignin holding the cellulose fibers together. In biological attack, the cellulose fibers are digested by microorganisms. At points of contact between iron and wood, "iron rot" occurs, i.e. iron causes the wood to become susceptible to fungus attack

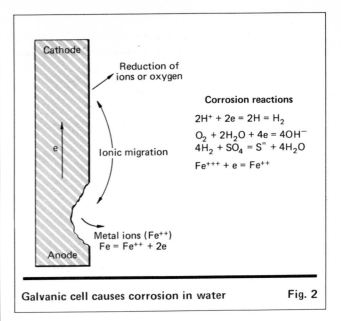

Corrosion reactions

$2H^+ + 2e = 2H = H_2$
$O_2 + 2H_2O + 4e = 4OH^-$
$4H_2 + SO_4 = S^= + 4H_2O$
$Fe^{+++} + e = Fe^{++}$

Galvanic cell causes corrosion in water — Fig. 2

Is your water scaling or corrosive?

☐ The commonly used indicators of the potential of a water to scale or corrode are the Langelier index, L, and the Ryzner stability index, R. L is somewhat qualitative, i.e., positive values indicate scaling tendencies, and negative values corrosive tendencies. R is somewhat quantitative, i.e., decreasing values below 6 indicate increasing scaling, and increasing values above 7 indicate increasing corrosion. A stable water is neither scale-forming nor corrosive.

Both indexes relate to water pH, alkalinity, calcium hardness, total dissolved solids and temperature. Water analyses usually report constituents as calcium carbonate equivalents ($CaCO_3$) in grains per gallon (gpg), or parts per million by weight:

$$\text{gpg} = (0.05838)(\text{ppm})$$
$$\text{ppm} = (17.13)(\text{gpg})$$

Also, Ca expressed as $CaCO_3$ equivalents is 2.5 times the Ca ion, and bicarbonate alkalinity expressed as $CaCO_3$ equivalents is 0.82 times the bicarbonate ion. The nomograph solves for both indices.

Example: Find L and R for 70°F water with: pH = 6.9, total dissolved solids (TDS) = 72, calcium hardness as $CaCO_3$ = 34 ppm, and alkalinity as $CaCO_3$ (methyl orange) = 47. Reading at the bottom of the left-hand scale, find TDS = 72 and note the intersection of this reading with the curved 70°F line. Carry this intersection horizontally to pivot line 2; connect that point with Ca hardness = 34 on the right-hand scale; note the intersection with pivot line 3; connect that point with alkalinity = 47 on the left-hand scale; and note the intersection on pivot line 4. This intersection is then connected to pH = 6.9, and the Langelier index and Ryzner index are read as −1.8 and 10.5 respectively. This water should be very corrosive.

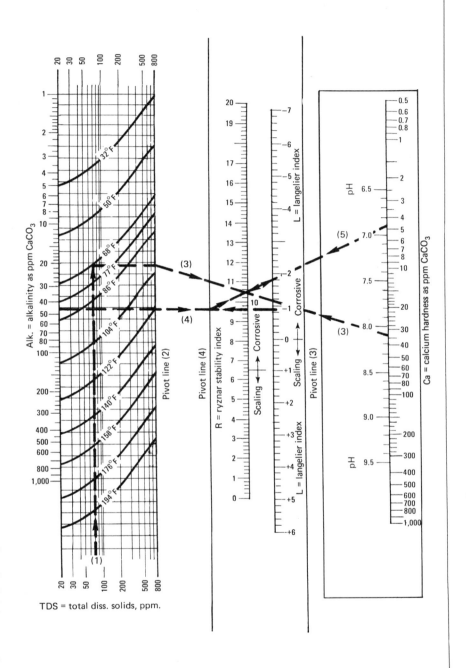

Source: Caplan, F., *Chem. Eng.*, Sept. 1, 1975, p. 129.

in the immediate area of contact. The best method for controlling iron rot is by insulating the iron and wood from each other by plastic.

Currently, wood is pressure-treated prior to construction. Preservatives are creosote, Celcure, Erdalith and Chemonite. A study by the Cooling Tower Institute found Celcure the most effective treatment, followed by creosote and Erdalith. However, creosote is not recommended because of its toxic effect on human skin, and the consequent special handling required.

In general, pressure pretreatment is most important in the damp area of the cooling tower, where biological attack is most likely to occur. Additional precautions include maintaining the pH between 6 and 7, adding background concentrations of 1 ppm or less of Cl_2, and adding chromate to the water.

Fouling

Whereas scaling results from precipitation of soluble salts, fouling results from the accumulation of insoluble

The use of cooling towers has liberated many industries from a prior dependence on water-rich regions.

material. Fouling may be either inert or biological. Inert fouling results from silt and other materials suspended in the makeup water, from airborne particles carried into the tower, from leaks and spills in the plant, and from the products of corrosion and wood deterioration. Biological fouling results from aquatic organisms.

Antifouling compounds are being used increasingly to reduce the effects of oils, dirt and other suspended foulants. The chemicals include flocculants, dispersants and chelates.

Flocculants, which are polymers charged like cations, perform their function through two mechanisms, adsorption and neutralization. Adsorption of the flocculant polymer occurs on the surfaces of some particles, so that the polymer chain links smaller particles together into larger ones. Neutralization occurs between the cationically charged polymer chain and negatively charged fouling particles, with a resultant formation of a floc that remains suspended because of a decrease in surface area per unit volume. Flocculants are usually added to the cooling-tower basin; and the suspended flocs can be removed by bleedoff.

Dispersants, which are anionic or nonionic natural organics and new synthetic polyelectrolytes, emulsify oily compounds so that they can be removed by bleedoff. Anionic dispersants coat the fouling particles and charge them negatively, causing them to repel each other, thereby preventing their coagulation and settling. These anionic polyelectrolytes can also interfere with scale by entering the crystal lattice as impurities and preventing crystal growth.

Chelates, which are used in scale prevention wherever possible, include natural and synthetic organic compounds that react with metal ions, such as calcium, magnesium, iron, aluminum and manganese, to form heat-stable chelated complexes. A variety of chelating agents is available for specific metal ions.

A combination of sunlight, warm temperatures, suspended oils and other bacterial foods, plus airborne contaminants, renders the cooling-tower environment beneficial for the growth of algae, fungi and bacteria. The algae are found mainly on those parts exposed to sunlight, the fungi and aerobic bacteria in the darker aerated areas, and the anaerobic bacteria on the inundated surfaces of wood.

The algae synthesize their own food from sunlight, air and minerals in the water. The fungi, with aerobic bacteria, yeasts, and in some cases protozoa, form a slime which, while unable to synthesize its own food, extracts its nourishment from the water. The anaerobic bacteria feed on the wood, causing serious deterioration problems.

Seawater fouling organisms, which are generally larger than those in fresh water, include seaweed, sea acorns, mussels and similar forms of sea life. These marine organisms can block pumps, cooling-water channels and pipes. Furthermore, leakage and spills from chemical plants containing organic compounds can nourish the bacteria to an enormous acceleration of growth. In a New Jersey petrochemical plant, for example, spills of alcohol and ketones accelerated the growth of bacteria to the point where it formed a 4-in-thick layer of dead matter on the cooling-tower basin, with consequent strong odors and accelerated fouling. [11]

A variety of biocides, including inorganic chemicals such as chlorine and bromine and a wide range of organic compounds, are employed to control biological growth in recirculated cooling water. Many of these compounds kill bacteria at high concentrations and inhibit their growth at lower concentrations, but they stimulate growth at dilute concentrations.

Proper concentrations of chlorinated phenols, acroleins and quaternary ammonium compounds are efficient biocides, but chlorine is the most widely used because it is usually the most economical method of treatment. However, chlorine concentrations in the circulating water should be kept below 0.5 ppm, in order to prevent cooling-tower wood deterioration and adverse effects on aquatic life near the effluent discharge. In cases where chlorination in low concentrations is unsuccessful in reducing or eliminating bacteriological fouling, a combination of chlorination and other biocide treatment can be employed.

Chlorination can be applied constantly in low concentrations, or intermittently in higher concentrations, as a shock treatment, although the latter is not effective with organisms such as mussels and barnacles that can close themselves when conditions are unfavorable and then open up when conditions improve. Where wood is used in the cooling tower, shock treatment is recommended to preserve the wood.

The use of chlorine poses a potential threat to aquatic life through chloramines, which can be formed from free chlorine and ammonia, according to the reactions:

$$NH_4^+ + HClO \rightleftharpoons H^+ + NH_2Cl$$
$$NH_2Cl + HClO \rightleftharpoons H_2O + NHCl_2$$
$$NHCl_2 + HClO \rightleftharpoons H_2O + NCl_3$$

The source of ammonium ions for these reactions could be any ammonium salts or ammonia in water. The source of hypochlorite ions could be hypochlorite or chlorine as it reacts in water.

The chloramines are relatively more stable than chlorine, have about half its oxidation potential, and are toxic to aquatic life. Current findings about this toxicity are inconclusive, however, and extensive research is underway. Long-term exposure of an amphipod and several minnow species of 15 and 21 weeks at concentrations greater than 0.0034 and 0.0165 mg/l, respectively, have shown some sublethal effects in terms of a reduction in the number of young amphipods and minnow eggs produced. [12] To date, there are no water-quality standards for these substances. However, EPA has suggested that continuous exposure of the aquatic community to chlorine compounds, including chloramines, should not exceed 0.003 mg/l. [13]

When chlorine-containing blowdown is combined with ammonia-containing effluents, such as sewage and hydrazine decomposition products, these could produce chloramines even prior to discharge. Thus, the combination of blowdown with other plant effluents should be treated with caution.

Recent research has also indicated that free chlorine could react with a variety of organic compounds to form carcinogenic compounds, which are of great concern if they enter public drinking-water systems. [14] To protect both the public and aquatic life, EPA has established allowable concentrations of free chlorine in new-plant effluents as an average of 0.2 mg/l and a maximum of 0.5 mg/l. [15]

The toxicity of high salinity to marine organisms has caused the Federal Water Pollution Control Administration to limit the permanent changes caused in estuaries to ±10% of the natural variation. [16] Within the cooling-tower basin, however, high salinities can be toxic to some marine organisms.

In summary, the choice of a proper fouling treatment requires an individual biological survey to determine qualitatively and quantitatively the biological population in the makeup water. The selection of suitable biocides in proper concentrations should take into consideration the effect of fouling on cooling systems, the effect of biocides on construction materials, the impact of biocides on natural aquatic organisms when discharged in untreated blowdown, and the handling properties of the biocides.

References

1. Langelier, W. F., Caldwell, D. H., Lawrence, W. B., and Spaulding, C. H., Scale Control in Sea Water Distillation Equipment, Contact Stabilization, *Ind. Eng. Chem.*, Vol. 42, 1, 1950, p. 126.
2. Stolzenbach, K. D., and Harleman, D. R. F., An Analytical and Experimental Investigation of Surface Discharges of Heated Water, Water Pollution Control Research Series, No. 16130 DJU, Environmental Protection Agency.
3. Motz, L., and Benedict, B., Heated Surface Jet Into a Flowing Ambient Stream, National Center for Research and Training in the Hydrologic and Hydraulic Aspects of Water Pollution Control, Report No. 4, Dept. of Environmental and Water Resources Eng., Vanderbilt University, Nashville, Tenn., 1970.
4. Carter, H. H., A Preliminary Report on the Characteristics of a Heated Jet Discharged Horizontally Into a Transverse Current, Part I, Constant Depth Technical Report No. 61, Chesapeake Bay Institute, Johns Hopkins University, Baltimore, Md., 1969, p. 30.
5. Fan, Loh-Nien, and Brooks, N. H., Numerical Solution of Turbulent Buoyant Jet Problems, W. M. Keck Laboratory of Hydraulics and Water Resources, California Institute of Technology, Pasadena, Calif., Publication No. KH-R-18, 1969, p. 23.
6. Harleman, D. R. F., Submerged Diffusers in Shallow Coastal Waters, Coastal Zone Pollution Management Symposium, Clemson University, Charleston, S.C., 1972.
7. Policastro, A. J., and Tokar, J. V., Heated Effluent Dispersion in Large Lakes, State-of-the-Art of Analytical Modelling, Part I, Critique of Model Formulation, Center of Environmental Studies, Argonne National Laboratories, 1972.
8. DeFlon, J. G., Design of Cooling Towers Circulating Brackish Waters, *Ind. Process Design for Water Pollution Control*, AIChE, Vol. 2, New York, 1969, p. 69.
9. Betz Handbook of Industrial Water Conditioning, Betz Laboratories, Trevose, Pa., 1962.
10. Millard, R. E., Operating Experience with Alkaline Cooling Waters, presented at the 34th Annual Meeting of the American Power Conference, Chicago, Ill., 1972.
11. Nester, D. M., Salt Water Cooling Tower, Cooling Towers, CEP Technical Manual, AIChE, New York, 1972, p. 115.
12. Arthur, T. W., and Eaton, T. G., Chloramine Toxicity to the Amphipod (*Gammarus pseudolimnaeus*) and the Fathead Minnow (*Pimephales promelas*), Jour. Fish. Res. Bd. of Canada, Vol. 28, 12, 1971.
13. Environmental Studies Board, National Academy of Sciences, Water Quality Criteria 1972, prepared for the Environmental Protection Agency, EPA-R3-73-033, U.S. Govt. Printing Office, Washington, D.C., March 1973.
14. Brungs, W. A., Effects of Residual Chlorine on Aquatic Life, *Jour. of Water Pollution Control Fedtn.*, Oct., 1974, p. 2180.
15. Permissable Chlorine Concentrations in Effluents from New Sources, *Federal Register*, Vol. 39, 196, Oct. 8, 1974.
16. Federal Water Pollution Control Administration, Water Quality Criteria, U.S. Govt. Office, Washington, D.C., April 1968.

The authors

Dr. Haia Roffman is a member of the Westinghouse Environmental Systems Dept., Westinghouse Bldg., Gateway Center, Pittsburgh, PA 15222, where she analyses reports for transmission lines and power plants, conducts geochemical studies of trace elements in power plant effluents, evaluates the impact of mining operations on land use and water, and performs studies on runoff from highways, parking areas, etc. She has more than 30 publications in the technical literature. She holds a Ph.D. in Geochemistry (1971) from the New Mexico Institute of Mining and Technology.

Dr. Amiram Roffman is manager of the Air Quality Group in the Westinghouse Environmental Systems Dept., where he has major responsibilities for air quality studies associated with steam electric generating plants, nuclear power plants, mines, etc. With a Ph.D. in physics (1971) from the New Mexico Institute of Mining and Technology, he joined the Westinghouse Environmental Systems Dept. in 1972. He has more than 50 publications in the technical literature.

Section X
WASTE-HEAT RECOVERY

Heat recovery in process plants
How to avoid problems of waste-heat boilers
Mystery leaks in a waste-heat boiler
Rankine-cycle systems for waste heat recovery
Useful energy from unwanted heat

Heat Recovery in Process Plants

Rising costs and shortages of fossil fuels emphasize the urgent need for lessening the demand of primary fuels through the optimum recovery and use of heat from elevated-temperature process streams.

J. P. FANARITIS and H. J. STREICH, Struthers Wells Corp.

The process designer involved with recovering heat to improve the economics of the process has an unlimited range of applications and techniques to consider. The increasing cost of primary fuel has broadened the range of heat-recovery applications that can be economically justified.

Several factors will influence the recovery of heat from a process:

1. Competitive market conditions on most products make it essential to reduce processing costs.

2. The cost of fuels keeps rising.

3. Limited fuel availability is already causing plant interruptions.

4. There are restrictions on using some of the lower-cost fuels because of environmental pollution.

5. Increasing emphasis is being placed on minimizing thermal pollution.

6. Increasing amounts of elevated-temperature flue-gas streams are becoming available from gas turbines, incinerators, etc.

7. New, sophisticated heat-recovery equipment is providing increased recovery efficiencies, and converting the recovered heat to high-pressure steam in the 600 to 1,500-psig. range, where the economic value of the steam is significantly higher.

The process designer has a number of alternatives to consider for recovering energy and/or heat. The specific method selected will depend on (a) type of energy available for recovery, (b) pressure or temperature level of the available energy, and (c) specific requirements of the process under consideration. This article will devote itself

Originally published May 28, 1973

only to the recovery of heat from elevated-temperature streams. Some of the methods used to recover heat are listed in Table I. These will be individually discussed as to advantages and disadvantages plus typical applications.

Steam Generation

Recovery of heat from elevated-temperature streams by the generation of steam is one of the oldest, easiest to engineer, and most widely used methods. This is probably the reason for the term "waste-heat boiler" evolving into a standardized term of reference for any type of heat recovery, including that in which no change of phase occurs in the coolant.

The purpose for which the generated steam is to be used will in large measure dictate the economics of the system, the type of heat-recovery equipment to be used, and whether or not steam generation is the proper heat-recovery technique.

The principal uses of steam generated by a heat-recovery system are:

- For process heating. In this application, the steam will probably be generated at pressures of 125 to 650 psig.
- For power generation. In this application, the steam will probably be generated at pressures of 650 to 1,500 psig., or higher, and will likely require superheating.
- For use as a diluent or stripping medium in a process. This is a low-volume use.

The use of steam generation as a means of recovering heat from high-temperature streams provides a number of advantages:

1. It generally results in a relatively compact heat-recovery installation because of the high rate of heat transfer associated with the boiling of water.

2. It usually will result in the lowest initial installation cost of any type of heat-recovery system.

3. It generally will incur fewer operating problems when applied to the cooling of high-temperature streams, since the high heat-transfer rates secured with boiling water will maintain metal temperatures close to the boiling-water temperature.

4. It will provide a rapid response rate.

5. It will permit some adjustability in heat-removal capacity, by raising or lowering the steam-side operating pressure within the design limitations of the equipment.

6. It does not require the close coordination between

Methods for Recovering Heat—Table I

☐ Generating steam.
☐ Preheating boiler feedwater.
☐ Preheating combustion air.
☐ Superheating steam.
☐ Preheating a process feedstream, or heating a process stream at some intermediate point in the process.
☐ Heating circulating heat-transfer media, which are then used to provide process heat.
☐ Preheating air for use in applications such as direct-contact dryers.
☐ Using flue-gas streams at elevated temperatures for process applications such as direct-contact dryers.
☐ Evaporating process streams.
☐ Providing space heating and utility steam.

Guidelines for Fire-Tube Steam Generators—Table II

1. Usually limited to steam pressure of under 1,000 psig., although technology is extending this to 1,850 psig.

2. Elevated-temperature streams being cooled may be liquid or gas.

3. Adaptable to cooling of elevated-temperature gas streams operating under pressure. Cooling of streams at pressures of 500 psig. and higher is not uncommon.

4. Most efficient when process-gas streams having a reasonable heat-transfer film coefficient are being cooled.

5. Generally limited to process-stream flowrates that can be handled by shop-assembled units.

6. Can handle clean or highly fouling elevated-temperature streams. Generally less susceptible to fouling and easier to clean on the high-temperature side than a water-tube design.

7. A fire-tube unit will generally be less expensive in applications where either fire-tube or water-tube design may be used.

8. In high-temperature service (1,000 to 1,800 F. inlet-gas temperatures), the inlet or hot tubesheet of a fire-tube unit is highly vulnerable.

9. Generally requires a higher pressure drop on the high-temperature streamside than a water-tube design.

10. Natural-circulation and forced-circulation designs are available.

Heat-recovery steam generators or waste-heat boilers may be either of the fire-tube or water-tube design, depending on a variety of factors. The process designer considering a steam-generator heat-recovery system must be familiar with the advantages and disadvantages of both designs. Improper application of either will likely result in a more-expensive-than-necessary installation and/or operating problems. Application guidelines for the two types of steam generators are given in Tables II and III.

The process designer must first establish whether heat recovery by means of steam generation is the most desirable for his application, and must then evaluate the available steam-generation arrangements. With guidelines previously outlined, the process engineer should be able to select the design that will result in the highest economic return and have minimum operating problems.

Preheating Boiler Feedwater

A preheating system for heating a large volume of feedwater through a reasonable temperature range will have a low initial cost and prove highly efficient in the recovery of heat. Here, heat may be recovered from flue gas or from process liquids and gases. Flue gas is the usual source of heat for this application.

Either water-tube or fire-tube heat-recovery designs may be used. Most of the guidelines for evaluation previously described are also applicable. Since heat is usually recovered from flue gases, an extended-surface water-tube type of coil is the normal design. In installations where small heat loads are involved and where the heat is to be recovered from a process stream, shell-and-tube heat exchangers are frequently used.

In feedwater preheating, the process designer should consider the following:

1. Flue-gas streams can become extremely corrosive when they are cooled below the dewpoint, particularly when traces of sulfur are present in the fuel. If the tube-wall temperature is below the dewpoint, condensation will occur even though the main-body gas temperature is still above the dewpoint. On occasion, semiconcurrent flow may be used to prevent condensation, even at the sacrifice of some of the log-mean temperature difference, and with the need for additional heat-transfer surface.

2. When heat is recovered from a process stream in a shell-and-tube exchanger, the stream is generally cooled over a long temperature range. With the process stream flowing through the tubes, a severe temperature differential across the multipass tubesheet can develop and result in operating problems.

3. It is generally not economical to use a high-temperature gas stream above about 800 F. for feedwater preheating. Greater economic value can be obtained by using higher-level heat for steam generation or the heating of process streams.

Preheating Combustion Air

Preheating combustion air is one of the oldest and most widely used means of recovering heat from elevated-temperature gas streams. This technique has

---

Guidelines for Water-Tube Steam Generators—Table III

1. Can be designed for any steam pressure, including supercritical.

2. Wider range of designs is available than in fire-tube units, including the use of extended-surface tubes as well as bare tubes.

3. Ability to use extended heat-transfer surface makes this design more efficient than a fire-tube unit when cooling gas streams that have poor heat-transfer characteristics or low allowable pressure drop.

4. Designs are available for handling low-pressure elevated-temperature gas streams at much higher flowrates, since water-tube designs may be field assembled in large sizes.

5. Water-tube designs do not lend themselves to cooling elevated-temperature gas streams with highly fouling characteristics. Even high-density bare-tube designs do not lend themselves to efficient soot blowing.

6. Water-tube designs lend themselves more readily to supplementary firing, where such additional heat input is desired, than fire-tube units.

7. Water-tube units can be designed for low friction losses (2 to 10-in. water column) on the high-temperature side while handling high flue-gas flows, and still provide efficient heat transfer.

8. Water-tube units are generally less susceptible to mechanical failure as a result of malfunction than fire-tube units.

9. Natural-circulation and forced-circulation designs available.

10. Certain water-tube designs may incorporate steam superheater and other auxiliary service coils beyond the primary function of steam generation. Included in a single housing can be superheater, generator and economizer coils to provide a high-efficiency heat-recovery unit.

---

process-stream flows and temperatures and feedwater flowrate, as required by other heat-recovery techniques.

Disadvantages of steam generation as a means of recovering heat:

1. A steam-generation system must operate at fairly high pressure to ensure economic justification. Current trend is for operation at the 650 to 1,500-psig. range, with higher operating pressures being evaluated.

2. Steam generation cannot cool elevated-temperature streams through as wide a range as other heat-recovery techniques because most of the heat is recovered primarily by vaporization of water at constant temperature corresponding to the system operating pressure.

3. A high-pressure waste-heat boiler generating steam requires high-quality feedwater, comparable to that required for a fired boiler. The cost of water treatment may significantly reduce the economic advantages steam generation might have over alternate heat-recovery techniques.

4. Steam generation has limited flexibility in utilizing the recovered energy. In certain types of installations, there is simply no use for the steam generated through heat recovery.

primarily been used in central-station power plants where large volumes of flue gas are available, and where plant thermal efficiency has always been of primary importance. Air preheaters for central stations are of specialized design because of the requirement to exchange heat between two streams having equally low heat-transfer characteristics. Rotating-element metallic preheaters are the principal type, but there are some shell-and-tube air preheaters used on smaller boilers. Some special extended-surface exchangers are also in this service.

Air preheaters have been used in conjunction with process heaters on a relatively small number of installations where high temperatures at the process inlet have not permitted use of convection sections.

Preheating of combustion air is an effective means of recovering heat from high-temperature flue-gas streams, but it can only be justified in a limited range of applications. Some of the disadvantages are:

1. It generally will require a relatively high energy expenditure in the form of horsepower for the induced-draft and/or forced-draft fans.
2. There is a relatively high initial cost because of the proportionately high heat-transfer surfaces.
3. The recovered energy must be used in conjunction with a combustion system.

Steam Superheating

Adding 50 deg. (F.) or more of superheat to high-pressure steam significantly adds to the dollar value of the steam when it is to be used for power generation. Elevated-temperature streams can frequently be used to superheat steam, thereby appreciably increasing the efficiency of a process.

Both water-tube and fire-tube waste-heat boilers can be provided with auxiliary superheaters to handle the steam that they generate. The water-tube design, which recovers heat from the exhaust of gas turbines, is particularly adaptable to using a steam superheating coil. Process furnaces also lend themselves to the superheating of steam in the convection section.

The one precaution to be exercised in the design of steam superheater coils in heat-recovery installations is that the steam must have a relatively low solids content, preferably under 1 ppm. Deposit of solids on the superheater tubes can lead to rapid tube failure when the elevated-temperature stream is over about 1,000 F.

Preheating Process Streams

Preheating of a process stream by an elevated-temperature flue gas or process stream generally will result in the most favorable economics of any form of heat recovery. This technique directly reduces the primary energy input required by the process, as well as eliminates or minimizes the size of the equipment required to transfer the primary energy to the process.

Feed-to-effluent heat exchangers have been used by refineries for many years to conserve heat. Any catalytic process involving high exothermic-reaction heat can very effectively conserve energy by exchanging heat between the feed and effluent.

Process-stream preheating is generally performed in water-tube coils or in shell-and-tube exchangers. Water-tube coils, usually of the extended-surface type, are used when heat is being recovered from flue gases. Shell-and-tube heat exchangers are generally used when heat is being exchanged between two process streams.

Heat transferred to a process is invariably more valuable than heat converted into steam. The process designer must explore all possibilities for heating process streams by recovering heat from elevated-temperature streams before considering other types of heat recovery. Among the considerations involved in this technique:

1. There is some hazard in tieing in an elevated-temperature gas stream from one processing unit into a process stream in a different unit. A shutdown in the one unit for any reason can force a shutdown of the second. From this standpoint, feed-to-effluent heat exchange within a single processing unit is ideal.
2. The source of the elevated-temperature stream and the point of use of the coolant stream must be relatively close to permit effective utilization of this system.
3. Preheating process streams from gas-turbine exhaust, especially with auxiliary firing, offers a means of heat recovery that can produce excellent economic returns.
4. The system must have a way to control the exit temperature of the process stream being heated.

Circulating Heating System

An effective method of recovering heat from elevated-temperature streams is to use a circulating heating medium as the coolant in an intermediate step, and then to transfer heat from the medium to other process units such as tower reboilers, evaporators, etc. This permits distribution of the heat recovered from a single elevated-temperature stream to multiple units that may be widely separated.

A heat-transfer medium other than steam has these advantages:

- Heat can be transferred to the process at high temperature levels without the high pressure associated with steam. This can sharply reduce the cost of equipment.
- Organic heat-transfer fluids can transfer heat at temperatures up to 750 F., a significantly higher temperature than can be attained with saturated steam.

A circulating heating medium to recover heat provides the designer with a highly versatile tool for optimizing the design of a process unit. Vapor-phase or liquid-phase heat transfer can be used, or a combination heating system can be provided with certain heat-transfer media.

Heat-Recovery Equipment

The primary equipment for heat recovery is broadly classified as either fire-tube or water-tube type, depending on whether the elevated-temperature stream is flowing on the inside or the outside of the tubes. There are some specialized designs of heat-recovery equipment that do not fit into these two categories. However, such equipment is outside the scope of this article. *(Cont'd.)*

HEAT RECOVERY IN PROCESS PLANTS

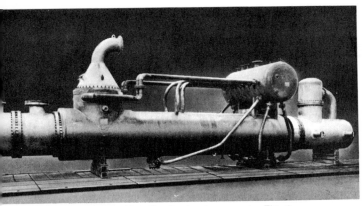

WASTE-HEAT boiler for heat recovery—Fig. 1

A typical fire-tube waste-heat boiler with separate steam drum, and with interconnecting risers and downcomers, is shown in Fig. 1. This design is widely used in process-plant heat-recovery applications. The separate steam drum permits the generation of higher-purity steam and better circulation through the bundle than can be achieved with a design incorporating integral steam-separation space above the tubes.

A fire-tube unit presents a highly versatile and low-initial-cost design for heat recovery. Some of the advantages and disadvantages of this design have been outlined under steam generation. In any fire-tube unit, certain features must be incorporated in order to minimize potential operating problems:

1. The inlet channel should generally be refractory lined, since most process applications involve inlet temperatures of 1,200 to 2,150 F.

2. The inlet or hot-end tubesheet is the single most vulnerable area of a fire-tube boiler. Maximum success has been achieved with relatively thin, $5/8$ to $1\frac{1}{4}$-in., flat flanged tubesheets having a generous corner radius to provide some expansion flexibility. Other features that have contributed to greater reliability of the hot tubesheet are: (a) strength welding of the tubes to the tubesheet, (b) refractory lining of the tubesheet, (c) insulated inlet ferrules to reduce the transfer of heat to the tubesheet ligaments, and (d) a wide tube spacing to ensure adequate tubesheet cooling area.

3. Heat flux in the inlet portion of the unit must be carefully analyzed to ensure that film boiling conditions leading to rapid tube burnout do not occur. This is especially important when cooling streams that have a high hydrogen content, because such streams have excellent heat-transfer characteristics. The combination of high heat-transfer rates and a high temperature differential can produce dangerously high heat fluxes in the inlet portion of a fire-tube steam generator. With carefully engineered fire-tube units, heat fluxes on the order of 140,000 to 160,000 Btu./(sq.ft.)(hr.) have been successfully achieved.

4. Design of the risers and downcomers is important for successful operation of a high-temperature fire-tube process cooler. The design must meet two criteria: (a) risers and downcomers must be sized and located to compensate for the much heavier steam generation that

WATER-TUBE heat-recovery coil for direct-fired heater will be used as a natural-circulation steam generator—Fig. 2

occurs in the hot inlet end of the unit, and (b) design must ensure high-velocity circulation of water across the hot tubesheet, and adequate steam removal so that there is no possibility of steam blanketing the tube sheet.

5. An internal bypass arrangement is necessary when outlet temperature of the process stream must be closely controlled.

The least complex of the water-tube designs used in heat recovery consists of a rectangular hairpin coil having tubes connected by 180-deg. return bends. This design is invariably used with gaseous elevated-temperature streams operating at low pressures. The tubes are frequently of the extended-surface type in order to compensate for the low heat-transfer rates characteristic of flue gases. A wide range of materials can be heated in this type of coil including steam, water, process liquids and gases.

The basic water-tube design of a heat-recovery unit, as shown in Fig. 2, is a typical direct-fired-heater convection section. This can serve as a complete heat-recovery unit or be incorporated into other units. It is limited to use with low-pressure, gaseous, elevated-temperature streams.

Another water-tube design frequently used in heat recovery is the two-drum boiler, as shown in Fig. 3. This may be used with elevated-temperature streams at any temperature level, but these streams must generally be at low pressures, usually under 2 psig. Extended surface can reduce the number of tubes. The design can be used over wide capacity ranges but must be in vaporizing service, with water generally, but occasionally with hydrocarbon fluids. It can be installed as a separate unit or as part of an integrated heat-recovery train.

A water-tube heat-recovery boiler is widely used in ammonia plants. It features conventional shell-and-tube construction, with modifications required to handle the high process-stream temperature and pressure, plus high generated-steam pressure. Special features include:

- The shell is internally refractory lined for protec-

TWO-DRUM boiler with extended surface tubes—Fig. 3

tion against inlet-gas temperatures of 1,500 to 1,800 F. Some designs include a water jacket to protect the shell from overheating.

- The bundle is of the U-tube, bayonet tube, or other construction, to provide maximum differential thermal expansion between the shell and tubes.

- High velocity is maintained in the tubes with either natural or forced-water circulation, so as to attain high heat fluxes without tube damage.

- Control of the process-gas outlet temperature within narrow limits can be achieved by bypassing a

WATER-TUBE steam generators recover heat from exhaust of gas turbines. These units include steam superheater, steam generator and economizer coils to produce steam at 650 psig. and 750 F. —Fig. 4

MODEL of cascaded water-tube unit generating steam and preheating process streams by recovering heat from high-temperature flue gas leaving a reforming furnace. Steam drum is at top of structure and one of the two fire-tube steam generators, recovering heat from the process stream, is shown at the grade level—Fig. 5

portion of the partially cooled process gas from an intermediate point in the shell.

• The units are normally designed for steam pressures in the 1,500 to 1,650-psig. range, and can be designed for even higher steam pressures.

This water-tube heat-recovery unit has primarily been used on hydrogen units in an ammonia plant but has the versatility for other applications where process gases in the 500-psig. range at elevated temperatures must be cooled. The principal limitations are that the unit must be kept within shop-fabricated sizes, and that it is noncompetitive economically unless steam pressures over 1,200 psig. are desired.

The dramatic increase in gas turbines as prime movers for electrical generators, compressors and pumps has accelerated development of a high-capacity heat-recovery unit. Fig. 4 shows a typical installation.

A gas turbine produces high volumes of exhaust gases at temperatures ranging from 750 to 975 F., and with an oxygen content of about 17%. Thus, there are not only large volumes of gas available for cooling but the high oxygen content permits auxiliary firing at high thermal efficiency to further increase the available heat. In order to handle this combination of conditions with efficient heat recovery, the multiple-coil water-tube heat recovery unit was developed. This design may also find other applications where large volumes of low-pressure flue gas are available.

Features of the highly versatile design include unlimited size and capacity, and very low gas-side friction losses, from 2 to 12 in. water column.

These units can handle any type of coolant stream. Many have multiple-duty coils. A typical unit might preheat a process stream, superheat steam, generate steam, preheat boiler feedwater, and produce low-pressure steam for deaeration.

The design lends itself to either natural- or forced-circulation-type steam-generator coils using extended-surface or bare tubes, or any combination of the two. Furthermore, the ability to provide either single-stage or dual-stage auxiliary firing provides added versatility and control adaptability.

Steam can be developed at high pressure and superheated for power generation. Alternately, lower-pressure steam can be generated for use in process heating, eliminating the need for a separate fired boiler.

Applications for Heat Recovery

Any flue-gas stream available at elevated temperatures (600 F. or more) is a prime candidate for heat recovery. Typical applications include high-temperature process furnaces, incinerators, glass furnaces, high-temperature direct-fired dryers, and steel-mill furnaces. Fig. 5 shows extensive multiple-coil heat recovery as applied to a high-temperature process furnace. In this instance, a portion of the heat is recovered by the process feedstream, and the remaining heat is recovered by generation of steam. The expanding use of incinerators has opened up a completely new field for the application of heat recovery. With exit flue-gas temperatures of 1,200 to 1,800 F. from typical incinerators, heat recovery can reduce the costs of waste disposal.

Any process that requires the feed material to be heated to elevated temperatures in order for catalytic or thermal conversion to occur has the problem of cooling the effluent gas. A typical example is a plant where natural gas or liquid hydrocarbons are cracked to produce hydrogen. Cooling of the effluent stream is necessary in order to prepare it for further processing, and the magnitude of available heat is such that efficient heat recovery is essential in order to improve the economics of the process. A similar but more critical application is the recovery of heat from ethylene-furnace effluent, where the process stream must be cooled rapidly through a critical range in order to prevent continuation of the reaction.

Many catalytic reactions are highly exothermic, and the temperature of the effluent stream is substantially higher than that of the feedstream. The exothermic heat may have to be totally or partially removed in the reactor, and the effluent stream will also generally require cooling. The exothermic reaction heat can prove to be a valuable source of additional energy when it is properly converted by a heat-recovery system. A special extended-surface tube, developed to remove heat from the catalyst bed, uses heat-transfer salt as the coolant, and subsequently generates steam while cooling the molten salt.

Economics of Heat Recovery

The broad range of applications for heat recovery makes it virtually impossible to provide specific guidelines for economic evaluations. Such evaluations must be based on the cost of primary energy, depreciation rates, tax benefits and all the other factors that enter into an economic analysis. However, certain basic guidelines for heat-recovery applications are almost universally applicable:

1. Economic value of the recovered heat should exceed the value of the primary energy required to produce the equivalent heat at the same temperature and/or pressure level. An efficiency factor must be applied to the primary fuel in determining its value compared to that obtained from heat recovery.

2. An economic evaluation of a heat-recovery system must be based on a projection of fuel costs over the average life of the heat-recovery equipment.

3. Environmental pollution restrictions may force the use of a more costly fuel.

(Continued)

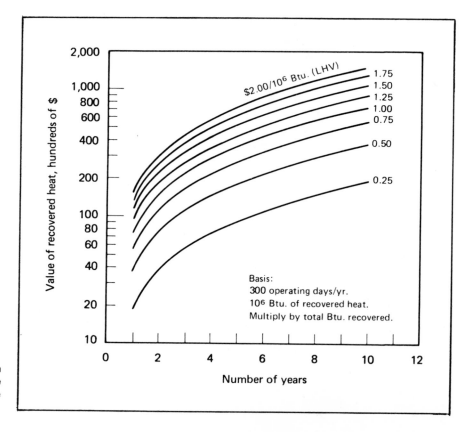

CHART yields value of one-million Btu. of recovered heat. This value is based on the projected average costs for primary fuel—Fig. 6

Equipment Cost for Heat-Recovery Units—Table IV

System Type	Approximate Cost $/10⁶ Btu.
1. Fire-tube steam generator, 650-psig. design, with external steam drum and interconnecting piping.	1,670
2. Water-tube shell-and-tube steam generator, 1,650-psig. design, with external steam drum and interconnecting piping.	2,170
3. Gas turbine waste-heat boiler with superheater, generator and economizer coils, including field erection	1,500
4. Water-tube steam generator for incinerator.	1,400

4. Many elevated-temperature process streams require cooling over a long temperature range, regardless of whether or not heat recovery is used. In such instances, the economic analysis should credit the heat-recovery installation with the saving that results from eliminating the nonheat-recovery equipment that would normally have been provided.

5. Fuel availability might have an impact on the ability of a plant to maintain operations. It is possible that extensive heat recovery in a plant may reduce primary-fuel requirements to the point that plant operation can be maintained, where it otherwise might be faced with interruptions during periods of fuel shortage.

Fig. 6 represents a plot of the value of the recovered heat when evaluated against a range of fuel costs. This value has been developed for a heat-recovery rate of 10^6 Btu./hr. over 300 operating days per year. The value for any other heat-recovery rate may be determined by direct ratio. Similarly, the value of the recovered heat may be determined directly for any period from 1 to 10 yr., and by direct ratio for any other time interval. Fuel values have been plotted in $/$10^6$ Btu., based on the lower heating value (LHV) of the fuel. Conversion efficiency from the lower heating value of the fuel (which represents available energy) to the useful heat actually recovered from the fuel must be applied.

An example will illustrate how to use Fig. 6. Heat is recovered at the rate of 150 million Btu./hr. in the form of 650-psig. steam superheated to 750 F. The projected average primary-fuel cost over a 12-yr. evaluation period is $0.75/10^6$ Btu. LHV. Expected thermal efficiency of a conventional power boiler to produce steam at the equivalent pressure and temperature is 86%, based on the LHV of the fuel.

From Fig. 6, the value of recovered heat for one year at a recovery rate of 1-million Btu./hr. and a fuel cost of $0.75/10^6$ Btu. LHV is $5,400.

The total value of the recovered heat would be:

$$\$5,400 \times \frac{150 \times 10^6 \text{ Btu./hr.}}{1 \times 10^6 \text{ Btu./hr.}} \times \frac{12 \text{ yr.}}{1 \text{ yr.}} = \$9,720,000.$$

Since the power boiler has a thermal efficiency of 86%, the equivalent cost of the primary fuel would be:

$$\$0.75(1/0.86) = \$0.873/10^6 \text{ Btu.}$$

Hence, the total value of the recovered heat if purchased as primary fuel would be:

$$\$9,720,000 \times \frac{\$0.873}{\$0.75} = \$11,314,000.$$

This is a startling amount to anyone not thoroughly familiar with the large savings available through heat-recovery systems. For a plant producing 1,000 tons/day of product, the use of heat recovery per this example could reduce cost of the product by about $3.14/ton. While not all installations will produce results as spectacular as the example, it must be recognized that properly applied heat recovery can produce significant savings.

Any economic evaluation of heat-recovery equipment will require a reasonable estimate of the cost of the equipment. Here again, there are so many different types of heat-recovery equipment and so many alternate designs that it is difficult to provide cost guidelines. The cost figures in Table IV are given for some of the principal types of heat-recovery units, but the user must recognize that they are very approximate, and are based on relatively large-sized units: 50,000 lb./hr. steam or more.

This article has attempted to cover the field of heat recovery from process and flue-gas streams so as to provide some broad guidelines. Increased interest in energy conservation because of rising fuel costs and serious fuel shortages already has forced, or will force, the process designer to become heat-recovery conscious. ■

Meet the Authors

J. P. Fanaritis is executive vice president of Struthers Wells Corp., Warren, PA 16365. Mr. Fanaritis joined Struthers Wells in 1941 upon graduating from the University of Pittsburgh with a B.S. in chemical engineering. He has been actively involved in the design of distillation equipment, direct-fired heaters, and heat-recovery equipment.

H. J. Streich is manager of the special products department at Struthers Wells Corp., Warren, PA 16365. He has been involved in the design of direct-fired heaters and heat-recovery equipment, including high-temperature applications involving liquid metals. He has a B.S. in chemical engineering from Case Institute of Technology.

How to avoid problems of waste-heat

*Peter Hinchley, Imperial Chemical Industries Ltd.**

Substantial reductions in the running costs of many process plants can be obtained by using their waste heat to raise steam. This heat is removed in economizers/feedwater-heaters, in boilers and in superheaters. The range of size and duty is wide, but on single-stream plants for certain processes, the steam-raising capacity is comparable to that of a large, industrial, power-station boiler. For example, modern plants for making ammonia raise approximately 250,000 kg/h (550,000 lb/h) of steam at pressures of 100 to 120 bar (1,450 to 1,740 psi) and temperatures of 460 to 520°C (860 to 970°F).

In nitric and sulfuric acid plants, the steam is raised by cooling the process gas. This is also the case with ammonia, methanol, hydrogen and town-gas plants. Further some of the steam in these plants is raised by cooling the flue gases from furnaces within the plant. Frequently, these flue gases are supplemented by auxiliary firing, particularly during plant startup.

Recovering heat from flue gases

Typically, flue gas under a slight draft, at 900 to 1,200°C (1,600 to 2,200°F) is cooled down to between 150 and 200°C (300 to 400°F). In modern plants, the flue gases, almost without exception, are drawn across banks of tubes containing the fluids being heated. Depending on the flowsheet adopted, there may be banks of tubes for superheating steam, heating process fluids, generating steam, vaporizing feedstocks, heating boiler feedwater, and preheating combustion air for the furnace.

Steam is generated in watertube boilers that are connected to steam drums. The boilers are usually of the horizontal forced-circulation type (F/1) or the vertical natural-circulation type (F/2), the choice being dictated largely by plant layout. For natural circulation, the drum should be some distance above the boiler to give sufficient driving force to the water and steam. The cost of the additional structure needed for elevating the steam drum to the required height may outweigh the savings in circulating pumps.

The actual disengagement of the steam from the boiling water takes place in the steam drum, which normally has a water level at or near the center line. The steam usually has to pass through a series of internal separators that remove entrained water before the steam reaches the outlet pipe. The complexity of the separators depends upon the required steam purity, which in turn depends on the type of plant. If the steam is passed through a superheater or into a turbine, very high purity is required to prevent the deposition of solids.

Recovering heat from process gases

Some process-gas boilers handle gases varying in pressure between 1 and 350 bar (15 and 5,000 psi) and in temperature between 400 and 1,200°C (750 and 2,200°F), and raise steam at pressures up to 120 bar (1,740 psi). They fall into two categories, firetube and watertube. The watertube type are generally for the higher outputs and pressures because of mechanical-design difficulties encountered with large, high-pressure, firetube boilers.

The principal features of a typical natural-circulation firetube boiler on a reformer are shown in F/3. Several companies design and make such boilers. However, there are significant differences between boilers with respect to maximum heat flux, maximum steam pressure, method of attaching the tubesheet to the shell and channel and of attaching tubes to the tubesheet, and method of protection of the tubesheet.

The manufacturers also differ in their ability to carry out a sufficiently rigorous stress analysis, bearing in mind the very high stresses that occur due to the high pressures involved and the differential expansion between tubes and shell.

---

* This article is adapted from a paper titled "Waste-Heat Boilers in the Chemical Industry," presented at a meeting on Energy Recovery in Process Plants, held by The Inst. of Mech. Engrs., 1 Birdcage Walk, Westminster, London SW1 H933. The paper will be published, along with others from the meeting, by IME.

Originally published September 1, 1975

boilers

Recovering heat from process gases, and from flue gases of associated furnaces, saves energy and money. Here is a review of the problems occurring in the various types of boilers, along with cases and solutions for actual plants.

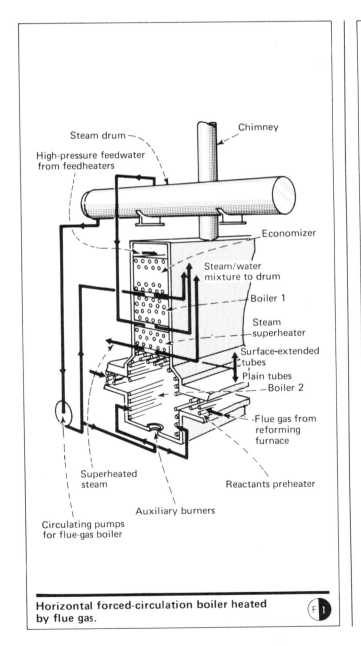

Horizontal forced-circulation boiler heated by flue gas. F1

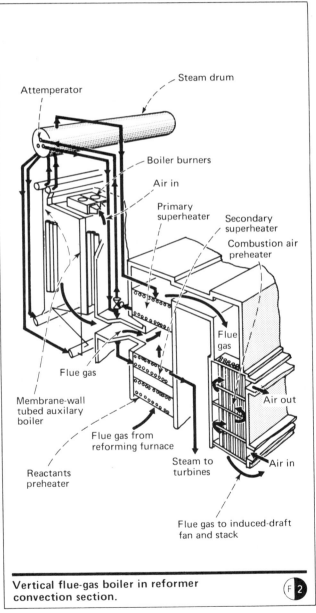

Vertical flue-gas boiler in reformer convection section. F2

The situation for watertube boilers heated by process gases is completely different, since a large number of companies have developed and patented their own designs. Some of these are dependent on pumped circulation, but others work on natural circulation. F/4 shows the principal features of one of the simpler designs, the vertical U-tube, which has been progressively developed for reformed-gas boiler duties by a U.S. company.

Whether the boilers are of the firetube or watertube type, they are generally connected to steam drums in the same way as flue-gas boilers, and commonly share the same drums. There are, however, occasional deviations from this arrangement. Some firetube boilers have their own integral steam spaces, and some watertube boilers are of the once-through type, where the water is converted directly into superheated steam in the tubes and there is no drum.

Failures in flue-gas boilers

In flue-gas-heated boilers, onstream corrosion has occurred due to excessively caustic water conditions together with a combination of heat flux, mass velocity and steam quality, which caused "dryout," i.e., inadequate wetting of the top of the tube [1]. The corrosion takes the form of deep gouging of the top of the tube. The problem has been solved by improving the control of water quality and by increasing the water circulation rate.

Several superheaters associated with flue-gas boilers have failed due to creep-rupture as a result of carrying over of boiler solids. The reasons include inadequate means of separating the steam from the water within the steam drum, and mechanical failure due to overload of good-quality primary and secondary separators. Another failure was caused by a badly made joint in a drum-type attemperator, which let boiler solids pass into the secondary superheater.

Some economizers associated with flue-gas boilers have failed as a result of oxygen pitting. Careful and thorough checking using stainless-steel sampling lines has shown that many proprietary deaerators fail to achieve their specified duties.

At least one steam economizer failed as a result of overheating caused by water starvation due to "excursive instability" or the so-called Ledinegg effect [2]. The problem was overcome by fitting distribution orifices at the inlets of the economizer tubes [4].

Failures in process-gas boilers

In firetube boilers, severe corrosion on the water side of the tubes has occurred just beyond the end of the heat-resisting ferrules in the tube inlets. This has happened in several plants, and has usually been caused by caustic or acid breakthrough into the boiler water. Naturally, boilers having the highest heat flux or poorest water distribution, or both, are most prone to failure should boiler-water quality deviate from the required optimum value.

Crevice cracking of the tubes within the tubesheet adjacent to the welds has been caused by failure to seal the crevices by post-weld expansion, and has occurred because the boiler-water quality deteriorated as a result of one of the dosing pumps being out of commission for one or two days.

Failure of the tubes and tubesheet as a result of overheating caused by the buildup of boiler solids behind the tubesheet has occurred on several boilers, as a result of either inadequate blowdown provision in the design

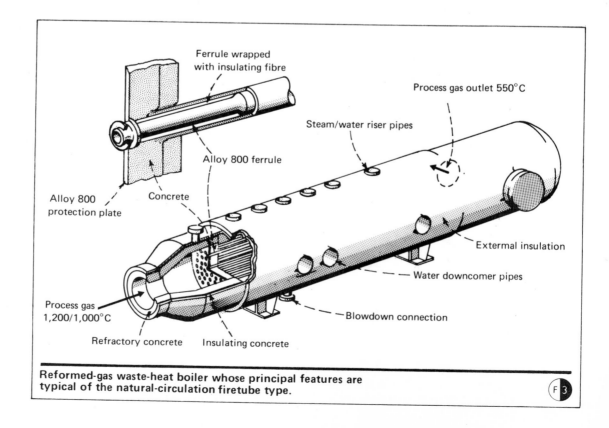

Reformed-gas waste-heat boiler whose principal features are typical of the natural-circulation firetube type.

or the failure to use the blowdown device provided.

Initial leakage of tubes on a sulfuric acid plant was a failure due to a combination of inadequate specification, poor workmanship and deficient quality control. The heat capacity of the sulfuric acid plant was such that even to repair one leaking tube a shutdown of 10 days was necessary. Furthermore, even a small leak could not be tolerated because of the immediate formation of corrosive acid.

The gas channels of several boilers have swollen as a result of gross overheating when external insulation on the shell has been extended on to a refractory-lined channel, or when there has been failure of the internal refractory lining.

Cracking and overheating of tubeplates may be due to their exposure to hot gases following the failure of their refractory protection.

Mechanical/thermal failure of tubes can occur when circulation stops following sudden depressurization of the steam system. In one case, this was caused by failure of the blowdown line.

Overheating of tubes within a tubesheet can result from using a thick tubesheet, inadequately protected by heat-resisting ferrules in the tube ends. The gaps between the ferrules and the tubes should be increased and filled with ceramic fiber paper and the inner ends of the ferrules should be swaged out to prevent gas tracking. (See detail on F/3)

Failure of the main tubesheet-to-shell welds occurred in one case during hydrostatic test, and in another after several months of operation. In both, there was a combination of unsatisfactory design, poor workmanship and inadequate quality control.

Failures in watertube boilers

It is not possible to go into the same amount of detail on watertube-boiler failures because the relevant details are for proprietary designs. Nevertheless, in order to present a balanced picture, an attempt will be made to indicate the reasons for the many failures that have occurred.

Horizontal-watertube boilers

One U-tube boiler handling gas at 980°C (1,800°F) had frequent failures at the top of the tubes as a result of dryout. The problem was eventually solved by increasing the water rate, changing to a higher grade of tubing, and fitting twisted tapes inside the tubes to ensure that the upper surfaces of the tubes were constantly wetted.

A similar U-tube boiler operating at much lower temperature had corrosion/erosion failures at the top of the tubes, caused by dryout—leading to a concentration of harmful chemicals in the boiler water. The problem was overcome by increasing the water-circulation rate.

A boiler for a nitric acid plant failed by overheating as a result of dryout, finally attributed to a low water velocity (which was made even lower than the design velocity as a result of blockage of some of the inlet distribution nozzles by excess hardness in the boiler feedwater). The hardness of the feedwater was improved and an additive was introduced to keep the solids in

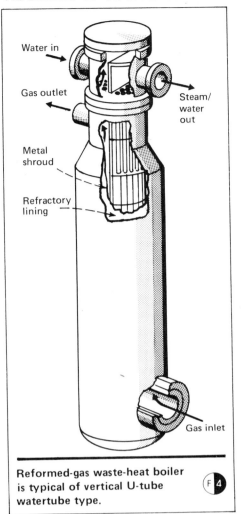

Reformed-gas waste-heat boiler is typical of vertical U-tube watertube type. F 4

suspension. This appeared to have overcome the problem until two years later, when another spate of failures occurred. This time, it was shown that these failures were due to the additive, causing the boiler solids to be laid down in the tubes. Another additive is now in use and so far there have been no failures.

Another boiler for a nitric acid plant, with a circular inlet manifold, failed as a result of the buildup of solids at the remote ends of the manifold, thus blocking off the end-tubes in this region. The problem was solved by putting blowdown connections at the ends of the manifold and by improved control of the boiler-water quality.

Vertical-watertube boilers

Creep-rupture failures can be caused by the blocking of individual tubes by construction/fabrication debris, or by magnetite scale dislodged during an upset in water treatment.

On-load corrosion failures can be caused by the concentration of harmful chemicals (in the boiler water) on any deposits existing in the regions of the tubes subject to the highest heat flux.

There can be creep-rupture failures of the gas pressure shell because of the failure of the refractory lining—the latter usually being due to a combination of poor design, workmanship and quality control.

Water quality for all types of boilers

Many of the above failures have occurred as a result of breakthrough of acid or alkali from the water treatment, or by contamination of return condensate. Now it is becoming common practice to monitor continuously the conductivity of water entering deaerators and to dump automatically any water that is out of specification. The supply is then taken from a standby tank of treated water.

Choice of waste-heat boiler

It is not possible to give a general answer to the question, "Which is the best boiler?" However, the following points can be made for flue-gas boilers and process-gas boilers.

Flue-gas boilers—Vertical, natural-circulation boilers are intrinsically more reliable than horizontal, forced-circulation boilers. True, the vast majority of horizontal boilers are reliable, but there is always a potential risk of dryout that can cause overheating or corrosion of the tubes and other components if harmful chemicals are present in the boiler water [1].

Insufficient data exist on the conditions that cause dryout or departure from nucleate boiling (DNB), so that we cannot be certain at the design stage that this will not occur in horizontal boilers. With vertical waste-heat boilers, there is little risk of DNB because the critical heat flux is many times greater than that in horizontal boilers [1].

Process-gas boilers—If their mechanical design and fabrication are to high enough standards, horizontal, natural-circulation, firetube boilers are basically more reliable than forced-circulation, watertube boilers. The latter are prone to failure if the dirt or debris gets into the system, particularly with vertical boilers that have their low point at the point of highest temperature and heat flux. However, faulty water treatment can give rise to failures of either type of boiler; and if the tubes fail, then a spare watertube bundle can be fitted in a short time, whereas a firetube boiler requires a very long time to retube or replace. For some processes such as nitric acid manufacture, there are well-proven designs of horizontal-watertube boilers.

Cost and consequences of boiler failures

There is an obvious reluctance in industry to report the cost of specific failures. Nevertheless, it is vital to indicate some idea of the order of costs, so as to illustrate the magnitude of losses that have occurred and do occur.

In the past, plant outputs were much smaller, the plants frequently had multiple units of equipment, and any steam that was raised was at low pressure. Thus, failures were less frequent, and the financial losses were not great.

It is now common to have single-stream plants, having outputs ten times those of a decade ago, which contain boilers raising steam at 120 bar (1,740 psi) and 500°C (930°F). A recent survey of 27 U.S. ammonia plants analyzed the causes of plant shutdowns [3]. Boilers were the second most frequent cause of shutdowns. One failure could be expected every three years, and on average this failure has caused the loss of five days' output. Thus, the average annual loss of profit due to boiler failures on an ammonia plant is $180,000.

The above statistics deal with normal plants, but there have been a number of plants where the financial losses were much greater. One pair of nitric acid plants had such frequent boiler troubles that in a three-year period there was a loss of over $2.4 million. Several ammonia plants have had single failures that have shut them down for up to 50 days, and one methanol plant was shut down for 106 days. These failures have caused millions of dollars in production losses because the plants were shut down while repairs were being made.

Conclusion

It will be apparent from this review of problems on waste-heat boilers that they are potentially a major cause of unreliability. As plants have become larger, the financial consequences of failures on single-stream plants have become more serious. One failure of a waste-heat boiler can cause a complete plant shutdown of between 5 and 100 days, and loss of millions of dollars.

It is not often appreciated by the boiler vendor that a single failure of his boiler can cause a loss of profit at least equal to the cost of that boiler, and frequently very much more.

From this, it will be clear that waste-heat boilers warrant very careful attention at all stages, from selection, design, fabrication, erection, through to commissioning, and also during normal operation.

Acknowledgements

The author acknowledges the helpful comments offered by his colleagues, and in particular the assistance given by one of his design engineers, J. W. Watson. Thanks are also due to the following who permitted the inclusion of illustrations based on equipment they have designed: Foster Wheeler; Humphreys & Glasgow; and Struthers Wells.

References

1. Collier, J. G., "Convective Boiling and Condensation," McGraw-Hill, New York, 1973.
2. Ledinegg, M., Instability of Flow During Natural and Forced Circulation, *Die Wärme*, Vol. 61.8 (AEC-TR-1861), 1954.
3. Sawyer, J. G. and Williams, G. P., Papers presented at AIChE Symposium on Safety in Air and Ammonia Plants, Atlantic City, N.J., 1971, and Vancouver, B.C., 1973.
4. Margetts, R. J., Excursive in Feedwater Coil, presented at Thirteenth National Heat Transfer Conference, AIChE/ASME, Denver, Colo., 1972.

The author

Peter Hinchley is furnace and boiler section manager of the agricultural div. of Imperial Chemical Industries Ltd., Billingham, England. He has worked for ICI since 1956 on equipment development, plant maintenance, engineering design and project engineering. He is a graduate of Sheffield University with a first class honors degree in chemical engineering, and is a member of the Institution of Mechanical Engineers.

Mystery leaks in a waste-heat boiler

After nine years of almost trouble-free operation, reported leaks led to three tube-bundle changeouts in six months. Inspection found leaks in only one of the three bundles. What happened?

W. J. Salot, Allied Chemical Corp.

☐ Tours of large ammonia plants feature several obvious points of interest:
- The primary reformer firebox inferno, which consumes more fuel than some entire cities.
- The roaring turbocompressor trains, with turbine-blade tips moving faster than the speed of sound.
- The tall, silent ammonia synthesis converter with its half-foot thick steel shell and quarter-million pound catalyst charge.

Top conversation piece

In our plant, tucked inside the structure supporting the steam drum and the high-vent silencer, are two 1,500 psig, natural-circulation, bayonet-tube, primary waste-heat boilers. Although not as eye-catching as the primary reformer, the turbocompressors or the ammonia converter, they make up in renown what they lack in conspicuousness. They have two claims to fame: One is their tremendous steam-generating capacity per square foot of heating surface. Each of them produces more than 137,000 lb/h with only 3,700 ft of 2-in.-O.D. tubes. The other is the huge cumulative total of production losses that their failures have wrought on the ammonia industry. Continuing failures have been blamed on a surprising variety of factors and have fueled evergrowing literature on the subject.

General surveys

A fitting beginning was made by a paper [2] presented in 1967. It summarized insurance losses in ammonia plants over a four-year period and suggested, "Perhaps the most troublesome problem concerns the large natural-circulation boilers."

Trouble continued. A survey [3] on ammonia-plant shutdown-causes during 1969-1970 reported, "Waste-heat boiler failures appear to be the current number-one industry problem on basis of total downtime."

This dubious honor was lost in succeeding surveys covering 1971-1972 [4] and 1973-1976 [5]. Waste-heat-boiler downtime had decreased as a result of fewer flanged-joint leaks, fewer shell ruptures, more spare tube-bundles, etc. However, the number of shutdowns caused by waste-heat-boiler tube leaks "has remained fairly constant" [5].

Bayonet-tube boiler surveys

Although not singled out for special study, bayonet-tube waste-heat boilers strongly influenced the results of the above surveys.

Owing to this, an unpublished 1972 survey [6] was made of bayonet-tube waste-heat-boiler tube-bundles in eleven centrifugal ammonia plants. It covered twenty-seven tube-bundles with an average service time of 1.9 years. Of these, three had been removed for retubing before any tubes failed, and seven were still in service. The remaining seventeen had suffered tube leaks in an average of less than 1.5 years.

The same survey [6] also referred to an "AIChE Average" tube-bundle life of 32,000 hours (3.7 years). This average was repeated in a later paper [13]. I have been unable to find its source, and question whether it applies to bayonet-tube boilers.

To compare the years before and after 1972, pertinent, unpublished, raw data were examined from the surveys [3–5] on ammonia-plant shutdown causes. Table I shows that the frequency of tube-bundle failures in bayonet-tube boilers causing ammonia-plant shutdowns has increased markedly since 1972.

These figures should not be used to estimate tube-bundle life because some tube-bundle leaks are allowed to continue, or are not discovered, until a shutdown is taken for other reasons, and some non-leaking tube-bundles are retubed one or more times without ever causing a shutdown.

None of the above surveys included data from one company that has already experienced more than twenty bayonet-tube-bundle failures [7].

Frequency of tube-bundle failures			Table I
Survey years	Tube-bundle leaks causing shutdowns	Waste-heat boilers surveyed	Percentage
1969–70	4	29	13.8
1971–72	3	31	9.8
1973–74	8	33	24.2
1975–76	6	33	18.2

Originally published September 11, 1978

Causes of tube failures	Table II

Overheating
  A. Fouling
    1. Dissolved solids and silica
      a. Feedwater contaminants [9-11, 15]
      b. Condensate return contaminants [9-11, 15]
      c. Solid water-treatment chemicals [6, 9, 10]
    2. Magnetite corrosion product [9, 10, 14]; Fig. 4
  B. Dryout
    1. Flow restriction
      a. Debris [11]
      b. Scale dislodged during plant upsets [11]
      c. Insufficient end-cap clearance
      d. Untrimmed gasket at inlet tubesheet [6]; Fig. 2
    2. Insufficient driving force
      a. Low water level [2, 6]
      b. High feedwater temperature
      c. Internal bypassing
  C. Reverse circulation [13]

Corrosion
  A. General
    1. Massive acid breakthrough [11]
    2. Caustic or strong alkali attack [6, 9]
    3. Hydrogen damage [9]
    4. Chemical cleaning damage [9]
  B. Localized
    1. Underdeposit corrosion [11]; Fig. 4
    2. Concentration-cell corrosion at nail-spacers [12]
    3. Erosion-corrosion at nail-spacers [8]

Mechanical
  A. Freezing water during turnaround popped off end-caps [8]
  B. Water-hammer after pressure swings cracked end-caps; Fig. 4
  C. Vibrating bayonet contributes to split at nail-spacer [6]; Fig. 4

Causes of tube leaks in bayonet-tube boilers

Various factors contributing to the mounting number of tube-bundle failures in bayonet-tube waste-heat boilers are covered in the literature. Other possible factors have only been talked about. Table II summarizes most of the factors.

The first mystery leak

The preceding review of ammonia-industry experience sets the stage for our mystery leaks.

We were reasonably aware of these problems in other bayonet-tube waste-heat-boiler tube-bundles, but not in our own. Until mid-1976, our tube-bundles had operated almost trouble-free for more than nine years. They were not in continuous service because we occasionally changed them out for inspection and external cleaning. Except in one freak case, they never leaked. That case is described in Fig. 1. It was easily diagnosed and repaired, and it should never happen again.

When asked the secret of our success with waste-heat boiler tubes, I used to say, "It's good water and superb operation."

My complacence was shattered in early June 1976, when the new ammonia production supervisor broke up a meeting with the statement, "We have a small leak in a waste-heat boiler."

There was instant mobilization! Other ammonia plant operators were called for advice. The quench line was disconnected. Thermocouples were replaced and re-replaced. Recorder charts and logs were checked and rechecked. Temperatures and temperature differentials were plotted and replotted. Calculations were made and re-made. The upshot of it all was this:

After six months of normal operation, a short shutdown had been taken in mid-June 1976, to weld-repair a pinhole-leak in the boiler feedwater preheat coil. The ensuing startup seemed normal except the outlet-gas temperature of the B Primary Waste-Heat Boiler had lined out at a level more than 100°F below where it had been running before the shutdown. On July 1, 1976, the B boiler outlet-gas temperature began to drop and continued downward even after the boiler inlet-gas temperature was increased. Steam generation was correspondingly affected (see Fig. 2).

We concluded that either the B Primary Waste-Heat Boiler tube-bundle was leaking or the Secondary Waste-Heat Boiler was leaking in such a way as to affect only the B-boiler outlet-gas-thermocouple reading.

The plant was shut down on July 12, 1976. The bottom connection of the B Primary Waste-Heat Boiler shell was opened, and the bottom head of the Secondary Waste Heat Boiler was lowered. The waterside was pressurized to 1,600 psig, using a boiler feedwater pump, and held for about an hour. No water leakage was observed in either unit.

We took the conservative approach. The B Primary Waste-Heat-Boiler tube-bundle was replaced with the spare. The questionable tube-bundle was mounted on the test gantry and hydrotested to 2,500 psig for several hours. There was no leak.

I proposed a theory that tight cracks in the tube-to-tubesheet welds might leak in service, but close up completely on a cold hydrotest. The tube-bundle was shipped to an outside shop for disassembled inspection. Our inspector witnessed dye-penetrant testing of the tube-to-tubesheet welds and reported that there was no sign of cracks or fouling.

We took a conservative approach again. The tube-bundle was retubed, and all physical evidence of a leak was thereby lost. Thus, a classic confrontation ended in a draw. Production data indicated there was a leak and maintenance inspection indicated there was not.

The second mystery leak

Meanwhile, back at the plant, the spare tube-bundle in B position had been started up on July 18, 1976. Referring again to Fig. 2, imagine the frustration when the B-boiler outlet-gas temperature lined out about where the previous bundle left off.

At first, this was explained as improved heat transfer

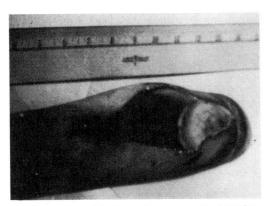

Rupture, 2 in. wide x 9 in. long, of an outermost tube, only 30 h after light-off; tube-bundle had been retubed and assembled with tube covered by untrimmed inlet tubesheet gasket. Repair was made by welding on a 22-in.-long replacement section. 1970. The repair was in service 2 yr when it was retired without failure.

The only leaking tube before mid-1976 — Fig. 1

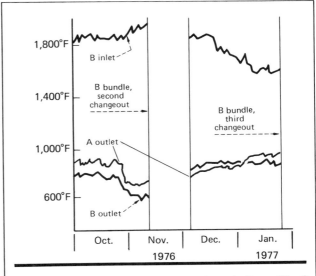

Gas temperatures of A and B waste-heat boilers — Fig. 3

from internal acid cleaning and external sandblasting of the spare tube-bundle. But then the B boiler outlet-gas temperature drifted slowly downward again after plant rate changes in late August and early September. Surely, the clean tubes were not getting cleaner! No, we must have had another leaking B tube-bundle.

In late October, we had one of those bad days. Failure of a valve controller led to MEA carryover, which forced shutdown of the ammonia loop, which caused a low level in the steam drum, which tripped the burners, which incited wild fluctuations in steam pressure. As a further complication, there was about a minute delay in starting the electric boiler-feedwater pump. Fig. 3 shows that, when order was restored two days later, both A and B boiler outlet-gas temperatures had plummeted and would not return to normal, even when the boiler inlet-gas temperature was raised to 1,940°F.

After a short struggle, the plant was shut down on November 13, 1976, with a concern that perhaps both tube-bundles were leaking. A second replacement tube-bundle was borrowed.

The first tube-bundle removed was from the B position. It was the spare that had been installed in July 1976. Immediately apparent were ruptures in several tubes on one side of the tube-bundle. That side could be variously described as roughly opposite the gas inlet nozzle, or roughly under the outboard steam riser, or roughly near the bottom when the tube-bundle was acid-cleaned in the horizontal position.

Fig. 4, upper right, shows two ruptures in one of the tubes in this bundle. It was one of many tubes that were cut off and split for inspection. The following observations were made:

1. The 2-in.-O.D. x 0.134-in. minimum wall, SA213-

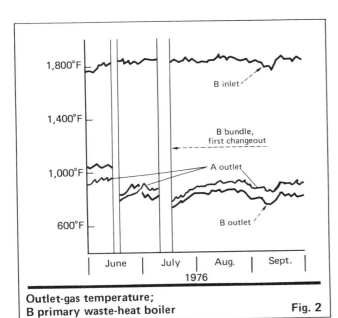

Outlet-gas temperature; B primary waste-heat boiler — Fig. 2

Left: ¼ in. thick, hard, tenacious magnetic deposit with under-deposit corrosion in nonleaking end-cap. 1968
Right: Circumferential leak caused by water hammer; also longitudinal leak in slightly bulged area at nail-spacer location 4-5/8 in. above inside bottom of end-cap. Tube was not fouled. 1976.

Water-hammer leak, and a corrosion-causing deposit — Fig. 4

T11, 1¼ Cr - ½ Mo outer tubes were clean and not corroded. At least two of their ruptures coincided with nail-spacer locations.

2. The 7/8-in.-O.D. x 0.065-in. minimum wall, SA214, carbon-steel inner tubes, had a thin, hard deposit in a stripe up to 150 deg. wide along the outside surface. It was bounded by two streaks of much heavier, soft deposit. The remainder was rusty. This was all probably the result of acid cleaning in the horizontal position.

3. Many of the inner tubes were bulged and longitudinally ruptured outward approximately 4 to 6 feet from the bottom. Some of the ruptured inner tubes were located in outer tubes that had not ruptured, but all were on the same side of the tube-bundle. Our calculations indicate that the inner tube ruptures must have occurred after one or more outer tube ruptures. However, an inner tube rupture could be present when an old inner tube-bundle is installed in a freshly retubed outer tube-bundle. If so, water could bypass the bottom of that tube, allowing it to overheat.

When A tube-bundle was removed, no tube leaks were evident. This was the only other tube-bundle of its type that we had ever acid-cleaned. Acid cleaning had been done by the same procedure as used in the ruptured tube-bundle. Therefore, if acid cleaning was a factor contributing to the rupture, the A tube-bundle could be in trouble, too.

The intention was to replace both tube-bundles, but the borrowed replacement was belatedly found to have a joint design that would not fit our shell. Therefore, we had to either reinstall the A tube-bundle or delay our startup pending delivery of another tube-bundle.

We decided to reinstall the A tube-bundle if we could prove that there were no leaks in the inner tubes. The ammonia production engineer devised a sensitive method of locating such leaks without disassembling the tube-bundle [1]. It was used and four leaks were found. The leaking inner tubes were replaced without disassembling the tube-bundle. The leaks turned out to be pinholes where nails were welded on, and probably had been there from the beginning.

The A tube-bundle was then installed in the B position, and a retubed spare tube-bundle was installed in the A position.

The third mystery leak

The startup in early December 1976 appeared back to normal for a change. As shown in Fig. 3, the A boiler outlet-gas temperature was lower because the A tube-bundle had new tubes and superior heat transfer. It climbed gradually, as would be expected for a slowly fouling tube-bundle.

The first unusual event on this run occurred in mid-December 1976. As shown in Fig. 3, the boiler inlet-gas-temperature began to drift downward. Since firing had not been decreased, it was obvious the thermocouple was failing. This unusually slow type of failure is typical of unpurged, chromel-alumel thermocouples in high-temperature hydrogen service.

After the Christmas holidays, the New Year greeted us with the realization that the A and B boiler outlet-gas temperature plots had crossed each other. For the third successive time, we presumed the tube-bundle in the B position was leaking. Apparently we should not have transferred that old tube-bundle from the A position to the B position in November.

But things were not as they seemed. We should not have been discouraged. As Ellery Queen might have said, "You now have all the information you need to explain all of the leaks."

CHALLENGE TO THE READER

You are where we were in January, 1977. Do not hesitate to review the facts before reading the solution that follows.

Mystery leaks solved

The key clue is in Fig. 2 and 3. It is the fact that after June 1976, the B boiler outlet-gas-temperature never returned to normal.

This means the three tube bundles involved were abnormally cooling the gas throughout the entire period. It also means the problem, once initiated, was continuous and independent of which tube-bundle was in service.

The trick is recognizing that the continuously low outlet-gas temperature can be caused by something other than continuously leaking tube-bundles.

That something is low gas flowrate. It produces low outlet-gas temperature via long residence time in the cooling environment of the tube-bundle.

With this in mind, we need only predict some sort of restriction in the gas flowpath through the B boiler. The most likely location is on the inlet side where temperatures are highest and the metal and refractory liners are known to be periodic maintenance items.

Restricted flow through the B boiler of course means increased flow in the parallel path through the A boiler. Notice in Fig. 2 and 3 that the cross-over points of A and B boiler outlet gas temperatures, in early July 1976 and in January 1977, were both partly the result of increasing temperature in the A outlet gas.

In contrast, the known B boiler leaks found in November 1976 were evidenced by a severe drop in both A and B boiler outlet-gas temperatures, as shown in Fig. 3. These temperatures are measured near where the two outlet gas streams mix. Apparently, the extreme cooling effect of the leaking water in one stream is sufficient to affect the outlet temperature of the other.

We, therefore, conclude that the tube-bundles removed in July 1976, and presumed to be leaking in January 1977, were not leakers.

That leaves us with the November 1976 failure, which assuredly did involve leaks in the B tube-bundle. The leaks no doubt resulted from the low water level and fluctuating steam pressures experienced in late October 1976, but this does not explain why the leaks appeared only in the B tube-bundle.

One might feel that the leaks should have been in the A tube-bundle because it was running hotter. The opposite is true for two reasons:

1. The cooler B tube-bundle was generating less steam and was, therefore, more susceptible to loss of circulation in the event of low water level and fluctuating steam pressure.

2. There was probably reverse circulation in the outboard riser, causing higher tube temperatures in that half of the tube-bundle. That is where the tubes failed.

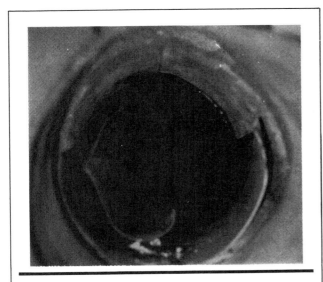

Flow restriction in B-boiler inlet piping   Fig. 5

joint and folded back over the inlet gas thermowell, thus blocking over 60% of the inlet flow area.

From these observations and a reappraisal of the temperature records, the explanation became painfully clear. The final ray of light was the very recent receipt of the paper [13] on reverse circulation.

By hindsight, we also discovered:

1. An early stage of our B boiler inlet-shroud failure had been reported in Nov. 1975, but was not considered serious enough to correct at that time. Seven months later, our troubles began.

2. When the leaking tube-bundle, removed from B position in Nov. 1976, was disassembled for retubing, the inlet tubesheet gasket was missing. This undoubtedly allowed some internal bypassing and contributed to the failure. However, the inlet tubesheet gasket was also missing from the non-leaking, 5-year-old tube-bundle that was in A position until November 1976, and subsequently in B position until January 1977. Apparently, lack of this gasket is not of itself sufficient to cause failure.

The facts match reverse-circulation theory [13] in every respect, from maldistribution on the gas side to the locations of the reverse circulating riser and overheated tubes.

Epilogue

You may find it hard to believe, but we failed to make the above diagnosis until after natural gas curtailment forced the ammonia to be shut down in late January 1977.

At that time, we removed the tube-bundle from B position and could see no evidence of leaks, but there was a surprising amount of refractory dust on the tubes.

Inspection of the B boiler inlet piping then explained the refractory dust on the tubes and made us aware of the flow restriction. As shown in Fig 5 and 6, the metal shroud had peeled down from the upstream expansion

References

(1) Rudenstein, C. S., Test the Bayonet of a Bayonet-in-Scabbard Tube, unpublished (1976).
(2) Badger, F. W., The Alternatives to Financial Loss—A Boiler and Machinery Insurer Looks at the New Ammonia Plants, "CEP Technical Manual, Ammonia Plant Safety," Vol. 10, 1968.
(3) Sawyer, J. G., Williams, G. P., and Clegg, J. W., Causes of Shutdowns in Ammonia Plants, "CEP Technical Manual, Ammonia Plant Safety," Vol. 14, pp. 62-66, 1972.
(4) Williams, G. P., and Sawyer, J. G., What Causes Ammonia Plant Shutdowns? "CEP Technical Manual, Ammonia Plant Safety," Vol. 16, pp. 4-9, 1974.
(5) Williams, G. P., Causes of Ammonia Plant Shutdowns, AIChE Paper No. A5a, presented at Ammonia Symposium, Denver, Colo., 1977.
(6) Demand, L., Centrifugal Ammonia Plants—Primary Waste Heat Boiler Survey, unpublished, 1972.
(7) Private communication, 1975.
(8) Private communications, 1977.
(9) Lux, J. A., Water Treatment Corrosion, and Cleaning of Stream Systems, "CEP Technical Manual, Ammonia Plant Safety," Vol. 15, pp. 6-11, 1973.
(10) Webb, L. C., Development of Correct Feedwater Treatment, "CEP Technical Manual, Ammonia Plant Safety," Vol. 16, pp. 92-99, 1974.
(11) Hinchley, P., A Review of Waste Heat Boiler Problems and Their Solutions, "CEP Technical Manual, Ammonia Plant Safety," Vol. 19, 1977.
(12) Akitsune, K., Imai, T., and Tange, M., Trouble with Primary Waste Heat Boiler of Ammonia Plant, presented at the 11th Pullman Kellogg Ammonia Club Meeting, Atlantic City, N. J., 1976.
(13) Livingstone, J. G., and Brook, P. H., Reverse Circulation in a Bayonet Tube Primary Waste Heat Boiler, presented at the 12th Pullman Kellogg Ammonia Club Meeting, Denver, Colo. 1977.
(14) Kusha, A., Problems Faced with the Primary Waste Heat Boiler in Shahpur Ammonia Plant and Remedial Measures Undertaken, presented at Ammonia Symposium, Denver, Colo., 1977
(15) Castorina, E., Current Water Treating Practices in an Ammonia Plant, presented at the 12th Pullman Kellogg Ammonia Club Meeting, Denver, Colo., 1977.

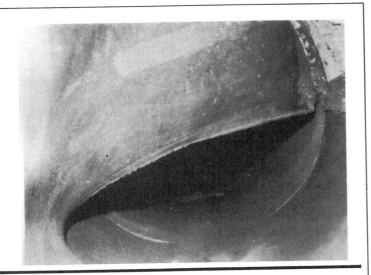

Another view of the B-boiler inlet   Fig. 6

The author

W. J. Salot is senior reliability engineer, Hopewell Chemical Plant, Fibers Div., Allied Chemical Corp., P.O. Box 761, Hopewell, VA 23860. He holds a B.S. in mechanical engineering from the University of Michigan, is a registered professional engineer in Virginia, an NACE corrosion specialist, and an American Welding Soc. certified welding inspector. He is a member of American Petroleum Institute, American Soc. for Metals, American Soc. of Mechanical Engineers, American Soc. for Testing and Materials, American Welding soc., and National Assn. of Corrosion Engineers.

Rankine-Cycle Systems For Waste Heat Recovery

Rising fuel costs are improving the economic feasibility of recovering waste heat by means of systems that employ organic working fluids.

ROBERT E. BARBER, Barber-Nichols Engineering Co.

In the past, major industries have foregone the most efficient use of excess process heat and instead purchased supplemental power and fuel, which used to be the lower-cost alternative. Much process heat above 200°F was (and still is) rejected to cooling streams. But as power and fuel costs rise, attitudes are changing. Efficient use of all energy resources is becoming more important, on both sociological and economic grounds.

Waste heat can be utilized by an organic Rankine-cycle system to produce useful electrical or shaft power without any fuel expense. The technology is already available, but until now cost has deterred development.

Such a recovery system is shown in Fig. 1. In its simplest form, it starts with the heat recovery boiler, which provides saturated or superheated vapor to the expander by transferring heat from the waste-heat fluid to the Rankine-cycle working fluid. Power is extracted in the expander, and the fluid passes on to a condenser, which provides liquid to the feed pump. The feed pump raises the pressure and resupplies fluid to the boiler, thereby completing the cycle. The working-fluid condenser heat is rejected to a cooling fluid in the condenser. This latter fluid could be air, or water from a cooling tower, perhaps circulating through a building heating system.

The expander shaft work is transmitted through a speed reducing gearbox, if necessary, and is ultimately used as shaft power to drive compressors or pumps, or to drive a generator to produce electrical power.

Depending on the particular application and working fluid selected, it may be economical to add a regenerator to improve cycle efficiency. This is a heat exchanger that transfers heat from the vapor leaving the expander exhaust to the liquid at the pump exit.

Estimating a Waste Stream's Potential

The waste heat from any cooling system and any hot-gas stack is a source for a possible heat-recovery application. To evaluate the potential for one of these streams, the maximum exhaust temperature under current conditions, the flowrate, and the required final temperature must all be obtained. Then the total amount of heat duty that should be available to a heat recovery system can be calculated.

The first step in the analysis is to plot the temperature profiles of a countercurrent heat exchanger for the hot- and working-fluid streams on the same graph, as in Fig. 2 (see footnote pg. 102). In this case, the hot stream is to cool from 450°F to 250°F, while the working fluid, Re-

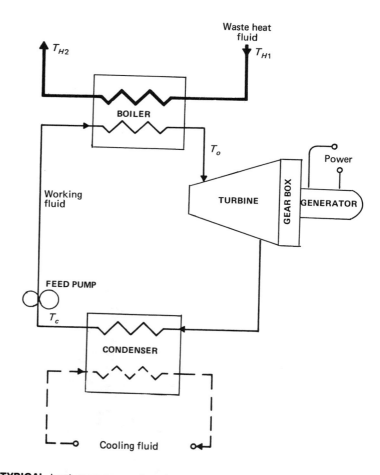

TYPICAL heat recovery system involves the flow of three streams: hot, working and cooling fluids—Fig. 1.

Based on a paper presented at the Ninth Intersociety Energy Conversion Engineering Conference, San Francisco, Aug. 1974.

Originally published November 25, 1974

RANKINE-CYCLE SYSTEMS FOR WASTE HEAT RECOVERY

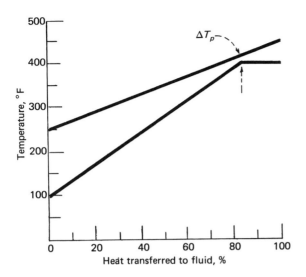

PINCH POINT is the closest distance between hot and cold streams on temperature-enthalpy chart—Fig. 2.

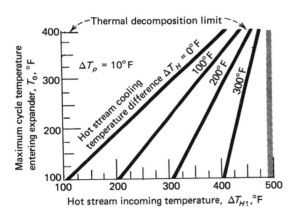

MAXIMUM Rankine cycle temperature depends on the degree of hot stream cooling for a given pinch point—Fig. 3.

frigerant-113, heats up from 100°F to its boiling point at 400°F, whereupon the temperature stays constant until complete saturation is achieved. The boiling point in this example is also the "pinch point" at which the temperature difference between both fluids is at a minimum (about 10°F).

Knowing the pinch point, Fig. 3* can be used as a guide to determining the maximum temperature of the Rankine-cycle fluid (at the inlet to the expander), which in general will be approximately 50°F less than the maximum temperature of the waste-heat stream. The boundary conditions are as follows: (a) the decomposition temperature of the working fluid, (b) the zero stream-cooling parameter at which the working-fluid temperature is equal to the hot-stream temperature minus the pinch temperature, (c) the hot-stream inlet temperature for

*Fig. 2 and 3 are specific to Refrigerant-113, the Rankine-cycle fluid chosen to demonstrate this method. For other fluids, the data for these graphs would have to be recalculated. All other curves are applicable in their present form.

which the pinch point occurs at the hot-stream exit from the boiler (480°F for Refrigerant-113).

Having determined the amount of heat available and the maximum Rankine-cycle temperature, the potential horsepower output can be obtained from Fig. 4. This graph has parameters of cycle efficiency for a water-cooled condenser, which can be determined from Fig. 5. If an air-cooled condenser is desired, the ratio of water-cooled to air-cooled cycle efficiencies from Fig. 5 can be applied to adjust the values of Fig. 4 accordingly.

The cycle efficiencies shown in Fig. 5 are appropriate for Rankine cycles having saturated vapor entering the expander; the working fluid chosen in this example is R-113. By way of comparison, data for other working fluids are shown by the data points scattered above and below the curves [1-4]. (Automotive Rankine-cycle systems [2] fall below the air-cooled condenser curve because condensing temperatures are above the 160°F shown here, and because the working fluid is highly superheated, which reduces cycle efficiency. The maximum cycle temperature–range of most interest for a heat recovery system is probably in the neighborhood of 200 to 600°F; above this range, valuable process steam can be generated instead.)

If electrical power is preferred instead of shaft power, the horsepower can be converted to electrical units by multiplying the value obtained in Fig. 4 by 0.95 to account for the generator efficiency, and then multiplying by 0.746 to convert to kilowatts.

A shaft-power utilization approach improves the economics and usefulness of the waste heat because generator losses are excluded. In some cases, a substantial improvement may be obtained if a high-speed compressor or pump could be driven with a high-speed expander. If this is possible, a reduction in the system's size and cost can be achieved.

A hot liquid stream has a potential for exceptional amounts of power generation. For example, a water flowrate of 100 gal/min at approximately 450°F would provide the potential for about 400 hp if the water were cooled by only 100°F. If this cooling differential were increased to 300°F, the output would jump to 1,200 hp.

Other potential sources for waste-heat power augmentation are the conventional gas-turbine, diesel and gasoline engines [5, 6, 7]. It has been shown (in the references cited) that the output of a simple-cycle gas turbine can be improved from 50 to 130% if augmented by an exhaust-gas-heated Rankine cycle. The output of a regenerative gas turbine can be increased 25 to 33%, and the output of diesel and gas engines can be upped 12 to 25%. Thus, virtually every major industrial installation probably could utilize a waste-heat-recovery power system.

Fluid Selection

Considerable work has been done in selecting optimum working fluids [8, 9]. A large number of variables come into play. Sometimes, flammability and toxicity are a major consideration, thereby ruling out many candidate fluids. Limitations imposed on the selection of a working fluid may cause a reduction in actual cycle efficiency below its maximum potential shown in Fig. 5.

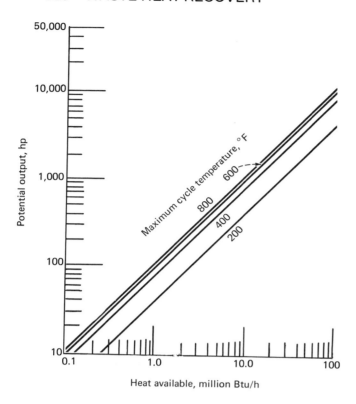

POTENTIAL POWER from waste heat depends on the heat available and highest cycle temperature—Fig. 4.

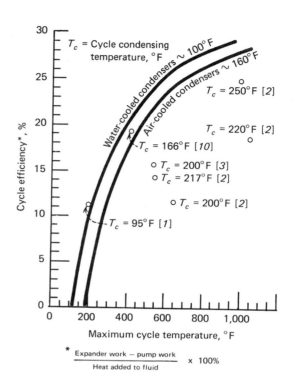

CYCLE EFFICIENCY is determined from maximum cycle temperature in air-and water-cooled units— Fig. 5.

Water is not a good working fluid for a Rankine cycle in low-temperature applications, because its large latent heat of vaporization makes it necessary to employ low boiling pressures; hence, cycle efficiency is low. For example, in cooling a hot waste-heat stream from 500°F to 200°F while at the same time heating a Rankine-cycle working fluid, the ideal cycle efficiency for a stream system would be about 18%, but an organic fluid, such as Refrigerant-113, would probably yield greater than 24% efficiency.

Many fluids have been used or are being used in Rankine-cycle systems. These include trifluoroethanol (made by Halocarbon Co.), FC-75 (3M Co.), and Refrigerants 12 and 113 (made by a number of manufacturers, led by Allied Chemical and Du Pont). The list also includes monochlorobenzene, toluene, and of course water, to name a few others.

Expander Selection

Usually, the best choice for an expander is a high-speed turbine or low-speed piston unit. Nichols [10] offers tips for selecting a turboexpander. In general, a piston expander has the potential for slightly higher efficiency and does not require a speed-reducing gearbox as does the turbine. The turbine, on the other hand, can be expected to have fewer development problems and be quieter and lighter than the piston engine. A 250-hp turbine/gearbox unit, for example, has a 22-lb turbine and a 50-lb gearbox [4, 11].

Heat Exchangers and Other Components

Two and possibly three heat exchangers are used in a heat recovery system. These are the boiler and condenser, and, if needed, the regenerator. The boiler and regenerator usually have a cross-counter flow design and can be analyzed and fabricated by conventional techniques. The condenser would have either a conventional tube-and-shell design if water-cooled, or a finned-tube one if air-cooled. The only unusual feature of the regenerator and condenser is that a larger than typical volume flows on the vapor sides and lower pressure-drop requirements probably will be experienced.

The boiler is the important heat exchanger. The highest saturated cycle-temperature possible for a given heat source must be obtained in order to get the maximum cycle efficiency. In a typical heat-recovery situation, there will be a maximum temperature available to the boiler from the waste heat source, and a particular required temperature drop for the power desired. Reasonable expander inlet-temperatures that can be expected as a function of these two variables are plotted in Fig. 3.

The boiler heat-transfer area is greatly influenced by the required temperature drop of the heat source, as shown in Fig. 6. As the drop becomes larger, the boiler area increases rapidly, because the required heat-exchanger efficiency must increase. Thus, the boiler costs can range from 5 to 20% of the installed cost of a particular installation.

The controls for a waste heat system can vary from the

RANKINE-CYCLE SYSTEMS FOR WASTE HEAT RECOVERY

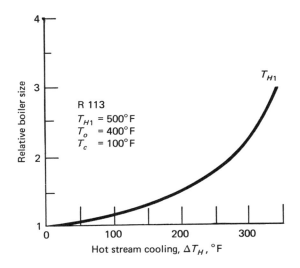

BOILER SIZE increases with cooling—Fig. 6.

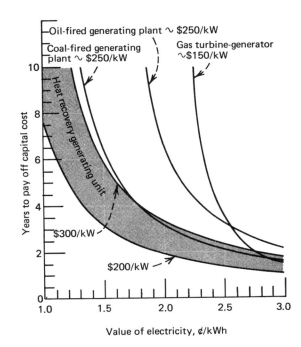

PAYOFF PERIOD for waste heat recovery compares favorably with other energy methods—Fig. 7.

extremely simple for a system operating at constant load and at temperatures well below thermal decomposition limits of the working fluid, to the very complex for systems operating at variable load and temperatures near the decomposition limit.

A typical heat-recovery system, however, should operate under nearly steady-state conditions and should not have rapid changes in load, heat-source temperature, or flowrate. Nor will the temperature be too high. Since process steam can be generated more economically from high-temperature waste heat, it is quite likely that most heat-recovery applications for organic Rankine cycles will have heat source temperatures that will not cause rapid thermal decomposition of the working fluid, even during upset conditions. Working fluids such as the refrigerants that are good up to 400°F, and trifluoroethanol fluids that withstand temperatures to approximately 650°F are applicable. Both these fluid types are nonflammable or only moderately flammable.

Benzene and toluene type fluids are usable up to temperatures of 700-750°F but are very flammable. Highly fluorinated fluids, such as FC-75, are stable to 750°F, are nonflammable and nontoxic, but require a large regenerator and are costly. Consequently, no ideal fluid exists and each application should be considered in detail in order to select the superior fluid.

Economics

Installed costs of waste-heat-recovery systems producing electrical power are estimated to be $150-300/kW, assuming low-quantity factory production of 100 units/yr. The cost range is primarily a function of the cycle efficiency, which reflects heat-exchanger requirements, and the temperature range available. Thermo Electron Corp. has manufactured a number of prototype Rankine-cycle systems and is in the process of developing systems for low quantity production, with an installed price goal of less than $200/kW. Turbodyne Corp.'s steam-heat-recovery system, which attaches to a gas-turbine combined cycle, has an installed cost of $133/kW for one system and between $155 and $215/kW for another, both of which were recently contracted for delivery in 1975 [12]. To be conservative, in the calculations that follow, the economics are based on a $200-300/kW installed cost range.

Table I presents installed cost, fixed cost, operation and maintenance costs, and fuel cost for base-power oil and coal steam-generating plants [13], gas-turbine generating plants [13] and the Rankine-cycle heat recovery system. The fixed costs include: carrying costs on original plant investment; cost of equity, common and preferred stocks; cost of indebtedness, interest on bonds and notes; depreciation; federal-income, state gross-revenue, and local property taxes. These figures approximate an annual cost of 14.7% of the initial installed cost [13]. The operation and maintenance costs include: maintenance

Costs of Alternative Methods of Power Generation*

	Base Power		Gas Turbine	Rankine Cycle
	Coal	Oil		
Installed cost, $/kW	250†	250†	105†	200 to 300
Fixed cost @ 14.7%, ¢/kWh	0.466†	0.466†	0.196†	0.373 to 0.559
Operation & Maintenance, ¢/kWh	0.091†	0.091†	0.370†	0.300
Fuel, ¢/kWh	0.45**	0.964†	1.557†	0
Total generating costs, ¢/kWh	1.007	1.521	2.123	0.673 to 0.859

*90% capacity factor
†Data from reference 12, fuel costs based on 1973 cost of fuel
**Coal cost at 50¢/million Btu

and overhaul; salary for operating personnel and supervision; administrative and general expenses; insurance; as well as lubricants, additives, and miscellaneous expenses. Unfortunately, a cost figure for operation and maintenance that is based on experience for heat recovery systems is not available; therefore, a value of 0.3¢/kWh has been assumed, which is slightly less than gas turbine operation-and-maintenance costs, and appreciably higher than nuclear or base-load systems. The fuel costs [13] are based on the costs experienced in 1973 (for example, 14¢/gal for gas-turbine #2 fuel oil). Since current prices are approximately twice as high, the costs shown in Table I are certainly conservative.

The installed cost of a nuclear system is estimated to be $350/kW [13]. Putting this figure against the base-power system estimate of $250 and the gas-turbine cost of $105, the $200-300 estimate for the Rankine-cycle system seems in the ballpark. Furthermore, Rankine-cycle cost could dip as low as $150/kW if greater quantity of production is possible (all of the preceding figures based on 1973 dollars).

The years to pay off the capital costs were calculated as a function of the value of electricity generated—assuming a 90% utilization factor—and are presented in Fig. 7. It is interesting to note that the payback period for the heat recovery unit is generally less than that of a coal-fired generating plant, and, at today's value of electricity (approximately 2.2¢/kWh), a payback can be obtained in 1.5 to 3.0 yr. As the value of electricity increases, the payback time is greatly reduced.

The average electricity cost for large industrial users has increased from 1.7¢/kWh in 1970 to 2.2¢/kWh in 1973 [14]. The accelerated fuel cost of the past year is likely to raise electrical costs to 2.5¢/kWh in the near future. The states of New York, Massachusetts, Connecticut, Rhode Island, Pennsylvania, Maryland, Delaware, Michigan, Wisconsin and New Mexico have hefty industrial electricity costs of from 2.2¢/kWh (New Mexico) to 3.85¢/kWh (New York), and are the highest cost areas in the U.S. [14].

The payback period can be reduced further if the waste heat power can be utilized as shaft power, thus avoiding the cost and efficiency losses of the generator. This savings can amount to $25/kW in installed cost. The payback time is, in general, reduced about one year by the utilization of shaft power and makes periods in the range of one to two years possible at current electrical costs.

If the payback is this quick, why aren't systems of this nature in wide use today? There are undoubtedly several reasons why. One probably is that until recently the value of electricity was in the neighborhood of 1.5 to 2.0¢/kWh, with payback periods of greater than 3 yr, which, all things being considered, has not been enough to warrant widespread development.

Current Rankine-Cycle Activities

Considerable effort to develop a Rankine-cycle automobile has been carried out over the past four or five years by companies including Thermo Electron, Lear Motors, Steam Engine Systems, Steam Power Systems, and Aerojet General, to name a few in the U.S. Thermo Electron has also come up with a small engine for generators and vehicular power in military uses. The Israeli firm of Ormat Turbines has developed and sold several hundred long-life Rankine-cycle systems. Honeywell and Barber-Nichols Engineering are working on a solar-heated Rankine-cycle generator and air conditioning system. Magmamax Industries is developing a geothermally heated organic Rankine-cycle system, and Du Pont is offering to license for commercial use an advanced Rankine-cycle in which the boiler and condenser are rotated, making use of increased heat transfer due to centrifugal forces. As a result, smaller heat-exchanger components are possible. [15].

Thermo Electron and Barber-Nichols are developing a 400-kW generating set that uses the exhaust heat from an 8,000-hp marine diesel engine. Thermo Electron and Ormat Turbines are the only firms to have Rankine-cycle systems for sale in prototype quantities. It has been reported that the Japanese installed three large heat-recovery units operating on Refrigerant-11 [16]. #

References

1. Prigmore, D. R. and Barber, R. E. "A Prototype Solar Powered Rankine Cycle System Providing Residential Air Conditioning and Electricity," Proceedings of the 9th Intersociety Energy Conversion Engineering Conference, Aug. 1974.
2. "Fourth Summary Report, Automotive Power System Contractors Coordination Meeting," EPA report, Ann Arbor, Mich., Dec. 1972.
3. Teagan, W. P. and Morgan, D. T., "3-kW Closed Rankine Cycle Powerplant," Report No. TE 5092-99-72, Thermo Electron Corp., June 1973.
4. Lodwig, E., "Performance of a 35 Hp Organic Rankine Cycle Exhaust Gas Powered System," Report No. 700160, Soc. of Automotive Engineers, Jan. 1970.
5. Angelino, G. and Moroni, V., "Perspectives for Waste Heat Recovery by Means of Organic Fluid Cycles," Paper No. 72-WA/Pwr-2, ASME, 1973.
6. Mortan, D. T. and Davis, J. P., "High Efficiency Gas Turbine/Organic Rankine Cycle Combined Power Plant," ASME 74-GT-35, Apr. 1974.
7. Davis, J. P., unpublished data to be presented in forthcoming NSF study report.
8. Miller, D. R., others, "Optimum Working Fluids for Automotive Rankine Engines," Report No. APTD-1564, EPA, June 1973.
9. Bjerklie, J. and Luchter, S., "Rankine Cycle Working Fluid Selection and Specification Rationale," ASME Paper No. 67-GT-6, Mar. 1967.
10. Nichols, K. E., "How to Select Turbomachinery for Your Application," Barber-Nichols Eng. Co., Arvada, Colo., 1971
11. Barber, R. E. others, "The Design and Development of a Turbine-Gearbox for Use in an Automotive Organic Rankine Cycle System," Paper No. 710564, SAE, 1971.
12. "Turbine Technology and Marketing News," Gas Turbine Publications, Jan. 15, 1974.
13. Shortt, J. H., "Power Generation Economics," Gas Turbine International, Jan.-Feb. 1974, pp. 32-36.
14. "Typical Electrical Bills—1973," Federal Power Commission Report, No. FPCR-82, Jan. 1974.
15. Doerner, W., others, "Rankine Cycle Engine with Rotary Exchangers," SEA Report No. 720053, 1972.
16. "Use of Fluorocarbon Turbine in Chemical Plants," Ishikawajima-Harima Heavy Industries, Co., (brochure) 1971.

Meet the Author

Robert E. Barber is vice-president of Barber-Nichols Engineering Co., 6325 West 55th Ave., Arvada, Colo. Previously he worked at Sundstrand Corp. and United Aircraft Research Laboratories. He holds a B.S.M.E. from Oregon State University and an M.S. from Rensselaer Polytechnic Institute. Mr. Barber is a registered professional engineer in Colorado and California.

Useful Energy From Unwanted Heat

Federal restrictions on thermal pollution threaten to push companies without plans headlong into adopting uneconomical control measures. Costs now favor the wet cooling tower, but other promising technologies loom on the horizon.

J. I. BREGMAN, WAPORA Inc.

It makes no sense for one industry to spend a great deal of money to get rid of energy when other industries are spending money to acquire it.

Granted there are some difficult technical problems, e.g., the waste heat can be at an undesirable temperature and pressure. Nevertheless, there is an unusual economic situation: This energy is not competing with conventional energy in terms of cost. Rather, we are trying to convert a very substantial outlay of many millions of dollars for minimizing or eliminating thermal pollution into no outlay or perhaps even into a return asset.

These are the new economics of the situation, and the reason I believe that we will begin to see beneficial uses of waste heat becoming more commonplace in the latter half of the 1970's.

Reasons for Controls

Thermal pollution control must now be a part of the planning and operation of electric-power systems in order to comply with the Federal Water Pollution Control Act. The philosophy of this Act is to assure that our waters will be protected equally for all legitimate uses. Propagation of fish and wildlife is one of these. Use by industry is another.

Industry must not be allowed to pollute a river and destroy the fish and wildlife that exist. Similarly, standards set for the propagation of fish and wildlife must not be such as to destroy or impose intolerable burdens upon the industries that use the river.

Basically, there are four types of temperature standards. The first, universally applied, is the maximum-temperature value. This is set for each body of water within the state, and differs from stream to stream. Generally, the temperature is a function of the natural history of the stream plus an addition of a small amount of artificial heat.

A second highly critical temperature standard, also universally found in various state requirements, is temperature increase, ΔT. This term indicates the allowable change in the temperature of the stream from what it would be without the artificial addition of heat.

A third standard that is not used too frequently relates ΔT to time, such as an increase of 1 F./hr. over a 12-hr. time period, but with the total increase in temperature to be not more than perhaps 5 F. or 7 F. This approach makes sense because aquatic life is affected by the rate of temperature increase. However, this standard is not used very often because it is exceedingly difficult to monitor. As sophisticated checking techniques are developed, it will probably become a fairly common standard.

Finally, as a fourth standard, there is the "mixing zone," which is inherent in every set of temperature standards. In this zone, the hot water is allowed to discharge directly from the cooling system. It is generally recognized that there must be a small stretch of water with no temperature limitations in order to dilute and disperse the heat. Temperature standards then go into effect at the border of this mixing zone.

In almost every state, the definition of the mixing zone is determined on a case-by-case basis by the state regulatory agency, because the mixing zone must be adjusted to the volume of the effluent, size of the

---

This article was originally presented at the 25th Annual Technical Meeting, Nov. 5, 1970, of the South Texas Section of the American Institute of Chemical Engineers, Houston, Tex.

Originally published January 25, 1971

Temperature Standards for Water—Table I

1. Allowable increase for open rivers and streams shall be 5° (F.). Maximum temperature is 90 F.
2. Allowable increase for lakes shall be 3° (F.) to a maximum of 90 F. A current exception to this policy is Lake Michigan.
3. A 1.5° (F.) rise is allowable in June, July and August, with a 4° (F.) rise during the rest of the year for estuaries and coastal waters. Maximum temperature in all cases is 90 F.
4. Mixing zones will no longer be recognized.

stream, velocity of the current, etc. There are certain general guidelines, e.g., a mixing zone must not stretch all the way across the stream to form a "thermal block." In a river, the mixing zone usually consists of a long, narrow stretch hugging the river bank and rarely extending out to more than one-third of the river's width. In a lake, the mixing zone will generally be a circle with a defined radius, whose center is at the point of discharge.

The temperature standards required for federal approval are becoming more restrictive. The Federal Water Quality Administration now has adopted the guidelines shown in Table I.

PRESENT CONTROL SYSTEMS

In a good many cases, no control measures are presently being taken. This may be for several reasons. For one, the industry or utility may be fortunate enough so that the hot effluent it discharges into the stream does not violate water-quality standards.

Perhaps standards have been set but no action has yet been taken toward meeting them. Those situations are coming to an end. Firms that have delayed in installing controls are going to find that they have a relatively short time left to do so, and they will have to go with whatever is the quickest approach regardless of whether it may be the most efficient on a cost basis or whether it uses the best technology.

Once-Through Cooling

Once-through cooling is certainly the most economical way of accomplishing heat dissipation. With the increase in size of stream-electric plants, and particularly with the advent of nuclear-fuel units, the number of sites available where this technique may be used has become quite limited because of the impact on the receiving-water bodies. Sources of cooling water for once-through systems include flowing streams, ponds, lakes, reservoirs, estuaries and the ocean. Rivers are naturally used as a primary source of cooling water but the number of those large enough for this purpose in the U.S. is limited.

Thus, for example, a 4,000-megawatt (Mw.) plant with a designed rise of 15 F. through the condenser, would require a condenser flow and a withdrawal rate of 4,600 to 7,000 cu.ft./sec. In order to limit the temperature rise in the river to 5 F., a minimum flow of the order of 14,000 to 21,000 cu.ft./sec. would be required. Few rivers in this country can meet that requirement.

Discharge of heated water to a receiving-water body is accomplished either as a layer (surface or subsurface), or as a mixing jet. The purpose of the jet is to minimize the temperature rise at any point by mixing the hot water rapidly with as much of the available receiving water as possible. The design of discharge outfalls is still not too well understood, and is based on empiricism to a substantial extent. Physical and hydraulic characteristics of the effluent and of the receiving water largely determine the rate of dispersion and the areal extent of the dilution.

Floating the warm water on the surface of the receiving stream or reservoir is common. In this manner, heat dissipation to the atmosphere is maximized, and the water areas and the volume affected by the effluent are minimized.

Impoundments

Impoundments such as cooling ponds, reservoirs or lakes for cooling heated water are frequently used. *Power Engineering* conducted a survey in early 1970 of all U.S. utilities with 8 Mw., or more, of steam-electric capacity. Table II shows the large number of times that cooling ponds and cooling lakes were used.

The Federal Power Commission has conducted a study of the possible capacities of reservoirs (existing and under construction) in the U.S. to supply projected cooling-water needs. It was assumed that a minimum surface area of 5 acres/Mw. would be required for fossil-fuel plants, except that 10 acres/Mw. was assumed for reservoirs and the Columbia River Basin. The results show that about 71 stations in the 1,000 to 4,000-Mw. range are capable of being supported by reservoirs, 24 in the 4,000 to 12,000 Mw. range, and 8 at 12,000 Mw. or greater. There is an imbalance in the location of the larger stations in terms of geography. While reservoirs or impoundments

Methods for Achieving Compliance With Water Temperature Standards—Table II

	Existing Units	Future Units			
		Next Unit	Second	Third	Fourth
Once-through cooling tower	27	7	5	1	1
Closed-cycle and cooling tower	84	36	18	9	2
Cooling pond	24	14	12	7	3
Cooling lake	7	6	4	—	—
Direct discharge to receiving body	—	38	21	12	6
Dry air-cooled condenser	2	1	—	—	—
Limit temperature rise of cooling water	7	4	—	—	—

will be helpful, there is a limit on the extent to which they can be used.

Within those limits and from an economic point of view, a company-owned cooling pond or reservoir is a very logical answer to the requirement for cooling the heated water. The water in question belongs to the user. It is cooled by nature in the storage system in which it is kept. Basic costs to the user consist of the land, the construction of the pond or reservoir, and the necessary conveyance equipment for getting the water to the plant and back. The obvious disadvantage is the land area needed as the generating system grows and requires more cooling water.

Efficiency of a cooling pond may be increased markedly by introducing a spray into the system. It has been estimated that the adoption of this procedure will reduce the required pond surface-area by a factor of 20. A spray pond is constructed by the installation of a spray system located 6 to 8 ft. above the water surface. Nozzles break the heated water into fine droplets. These provide a maximum water surface and thereby facilitate very efficient heat dissipation. It is estimated that spray ponds will cool 15 to 20 gal. water/(hr.)(sq. ft. surface area) when the heated water is cooled from about 110 F. to 120 F., down to 70 F.

Major advantage of a spray pond is the sharp reduction in costs of land. Disadvantages include the following: (1) time of contact between the air and sprayed water is limited, resulting in limited heat dissipation, and (2) there are added costs due to the spray-system pumping costs, and increased water loss.

Lakes and manmade reservoirs are also used as impoundments for cooling purposes. They become stratified throughout the summer months, during which time, of course, their use as cooling waters is especially critical. There are two distinct layers: an upper, warm portion of fairly uniform temperature called the epilimnion and a cooled, lower portion called the hypolimnion. Steam-power plants benefit from use of the cold water trapped in the hypolimnion of a reservoir during the winter months. A steam plant located on or near a deep reservoir not only has the cold water available for condenser operation but also a stored volume of cold water that may be adequate for the entire summer period.

Cooling Towers

Recirculation of cooling water through towers is becoming a most-common cooling method planned for future high-capacity electrical-generating units. Cooling towers have been used in this country for many years to recirculate water for other industrial applications, such as in petroleum refineries and steel mills. Their application to the electrical-utility industry started within the last decade and involves a major escalation in tower size and costs.

The two basic types of cooling towers are wet and dry. In the wet type, warm water from the condensers is passed through the cooling towers, where the temperature is lowered by evaporation, and cooled water is recirculated to the condensers. In the dry towers, the hot water inside pipes is cooled by using moving air as a heat-exchange medium. Wet cooling towers are more efficient and less expensive.

Although used extensively in Europe, natural-draft hyperbolic cooling towers have only come into use in the U.S. during the past decade. The first such tower was the Big Sandy installation in Kentucky, which was completed in 1962. Since that time, over 20 of these towers are now in various stages of operation or construction in the U.S. for heat discharge from large power plants.

In a natural-draft tower, the heated water is pumped through the top of the tower and is allowed to flow downward through a distribution system. Air movement through the tower is created by the wind velocity outside and by the temperature difference between the top and bottom of the tower. Louvers provide a horizontal flow of entering air.

Until 1963, practically all cooling towers used in the U.S. were of the mechanical-draft type because they are more economical for small heat loads. The advent of the huge new nuclear and fossil-fuel electrical-generating requirements changed the picture.

A mechanical-draft tower uses fans to move air through the tower. There are three types of mechanical-draft towers. In the first, the forced-draft tower, one or more fans are located at the air intake of the tower and push the air up through the water that is being distributed over the packing. The other two types are of the induced variety. Air is pulled through the towers by fans at the top. These towers may be either crossflow or counterflow, depending on whether the air flows through the transfer section of the tower in a horizontal or vertical manner.

Dry cooling towers are merely huge air-cooled heat exchangers. Air flows across the fins and absorbs heat from the hot water. Therefore, the heat from the hot water is dissipated by conduction and radiation, instead of by evaporation as in wet cooling towers.

Dry towers tend to be considerably more expensive than wet towers because of the large surface area required for heat transfer, the larger volumes required for cooling, and the consequent larger shell sizes. Nevertheless, certain cooling conditions, such as very high water temperatures, insufficient water, and blowdown problems, may make the air towers preferable in certain situations.

There are a number of problems common to large cooling towers. Several arise in connection with the use of water. These include wood deterioration, biological fouling, the formation of deposits, corrosion and scaling. These problems can all be eliminated or prevented, but the ways of doing so cost money and add impurities to the water that is then discharged to the stream, thereby causing water-quality problems.

Another major problem associated with cooling towers is that of water drift and fogging. Large amounts of water vapor may be carried out of the top of the tower with the air, become condensed and

Costs of Cooling-Water Systems for Steam-Electric Plants—Table III

	Investment Cost	
System Type	Fossil-Fueled* $/kw.	Nuclear-Fueled* $/kw.
Once-through†	2 to 3	3 to 5
Cooling ponds‡	4 to 6	6 to 9
Wet cooling towers:		
Mechanical draft	5 to 8	8 to 11
Natural draft	6 to 9	9 to 13

* Based on unit sizes of 600 Mw., and larger.
† Circulation from lake, stream or sea, and involving no investment in pond or reservoir.
‡ Artificial impoundments designed to dissipate entire heat load to environment. Cost data are for ponds capable of handling 1,200 to 2,000 Mw. of generating capacity.

cause ground fog or icing. Alternately, the hot contaminated water that is being recirculated in the tower may be blown out of the tower and settle on nearby structures and transmission lines. This causes a hazard in the immediate area. It is particularly troublesome with saltwater towers. Thus a typical tower with a flow of 250,000 gal./min. operating on seawater with a salinity of 35,000 ppm., and a drift loss of 0.1%, will emit about 4,400 lb./hr. of sodium chloride. This amount of salt thrown out on the surrounding area may be catastrophic in terms of its effect on crops and on metals.

The major problem connected with the gigantic cooling towers that are now coming into use is their cost. The Federal Power Commission has developed a table of comparative costs of cooling-water systems for steam-electric plants based on data supplied by a number of electric utilities. Table III lists these costs.

For each type of system, the costs of condensers and auxiliaries have been excluded. The installation costs cover items such as pumps, piping, canals, ducts, intake and discharge structures, dams and dikes, reservoirs, cooling towers and associated equipment.

The difference in economics between wet and dry cooling towers is shown by the approximate costs for the dry towers, which run from $25 to $30/kw., compared to costs for wet towers of $8 to $13/kw.

Capital Costs of Cooling Towers—Table IV

Type of Tower	Approximate Cost, $ Millions	Ratio of Total Investment Cost, %
Wet, mechanical draft	5 to 8	3 to 7
Wet, natural draft	6 to 11	4 to 10
Dry, mechanical draft	25 to 28	16 to 22
Dry, natural draft	25 to 30	16 to 23

Typical costs for the installation of cooling towers at utilities are given in Table IV. It shows the very large costs involved both in terms of actual dollars and percent of total investment. The values are based on a range of $100 million to $130 million as a cost of a 1-million-kw. plant.

The values in Table IV are for the total costs of towers. Individual costs today for wet towers range from $1.5 to $4 million, with at least two towers required at each installation.

THE FUTURE

We have examined today's techniques for thermal pollution control. The logical question now is what happens next. Without any doubt, almost every future major nuclear utility, and many of the fossil-fuel ones, will require cooling towers. We may safely apply this generalization to generating units that are under construction today, on the drawing boards, or actively being planned. There just is no practical alternative. Expensive or not, the cooling tower appears to be the only way at present to meet temperature standards.

The switch from cooling towers to more economically desirable approaches to thermal-pollution control will begin with the planning for new capacity that gets underway within the next year or two. This means that by the time these plans come to fruition, in terms of actual hardware, it will be the middle or latter part of the 1970's. Then we will begin to see beneficial uses of waste heat. What are these likely to be?

Perhaps the first one is the combination of nuclear electrical power with giant desalting units. This is not a new idea—it has been studied for a considerable period of time. But thus far, there has been much planning and no construction. This is because the cost of the desalted water was underestimated substantially. Further, the costs that have been calculated thus far have been primarily based on the need for desalted water only. The correct economic approach is to consider a situation where power will be generated anyhow. Then the logical question is how to tie in desalting units to power generation, so that advantage is taken of the excess available energy. This is now being done at a number of locations throughout the world.

We are still talking in terms of small scale desalting units, i.e., of the order of 5 million to 8 million gal./day. For nuclear power generation, desalting capacities will be on the order of 50 million to 100 million gal./day. By the time that this process is properly thought out and executed in the late 1970's, the cost of desalted water will probably be down to about 35¢ to 40¢/1,000 gal. as compared to the 85¢ to $1 today.

Satellite Locations

As future nuclear-generating plants are planned, considerable attention will be paid to developing satellite industrial and residential areas around them. The excess energy will then be made available to these industries at reasonably low cost. In addition, the hot water or the energy taken from it may be used to heat homes in nearby residential areas.

We are all familiar with snowfree and icefree walks in shopping centers that use this technique. I envision the same sort of thing being done on a giant scale. Here again, this is not a novel idea. Researchers at Oak Ridge have been planning a nuclear-complex approach for a number of years, in which the nuclear utility will be the center of an industrial, residential, commercial and perhaps even an agricultural complex. The government of Puerto Rico has a detailed study underway right now as to how to accomplish this. On the Hudson River near Indian Point, N.Y., eleven power plants planned or operating could heat 800,000 housing units while supplying 5,500 Mw. of electricity.

Cooling water from a nuclear power plant may be particularly suited to aquaculture. Desirable types of aquatic life could be cultivated at temperatures suitable for them in locations where they cannot presently exist. For example, in Bridgewater, England, the farming of shrimp is obtaining three times "normal" production. Near the Hinkley Point Power Station, in water 7 F. warmer than nearby Bristol Channel, 25,000 shrimp are reaching maturity in 18 mo. instead of the normal 3 to 5 yr.

The Tennessee Valley Authority is conducting an interesting experiment at its Gallatin Steam Plant. A research project is being carried out on commercial catfish production. Because the fish do not feed actively in cold weather, heated water may be a way of lengthening the growing season for catfish fingerlings.

A study at Par Pond in Aiken, S.C., by Du Pont, which operates the Savannah River Plant of the U.S. Atomic Energy Commission, indicates that 1 billion gal. of hot water—up to 115 F.—which are pumped into Par Pond every day may actually be very helpful to aquaculture. This hot water appears to make turtles and fish grow faster and in much greater abundance than they do under normal conditions. One interesting feature is that one vanishing American species of wildlife is coming back as a result. The alligators that thrive on the turtle and the fish in the pond are rapidly increasing in numbers.

A San Diego Gas and Electric cooling-water outfall is being successfully used to hatch and raise fish, lobster and shrimp. Commercial applications will begin soon. Some of the hot effluent of the utility will be converted to a 50-acre salt pond stocked with 600,000 shrimp. The shrimp will be harvested in about a year. It is also believed that lobsters, which require a normal 4 to 7-yr. maturity, will be matured in 2 yr.

A number of people have speculated on the possibility of using hot air from nuclear units for gigantic greenhouses in which huge quantities of vegetables could be grown. For example, studies carried out in Mexico show that plants raised under desert conditions, but with a hot, moist climate, will require much less water. For example, red kidney beans grown in 100% humidity used only a third of the water required by those grown in a 35% or 70% humidity situation. The crops are planted directly in desert sand but in a greenhouse where the air is kept hot and humid. Large quantities of a number of different vegetables such as beans, beets, broccoli, cabbage, cantaloupe, watermelon, etc., have been grown and have matured much more rapidly and with greater yield than they would in open fields. For example, cucumber and lettuce production have been at least six times more than would be expected from open fields. Results of the studies in Puerto Penasco, Mexico, and Abu Dhabi on the Arabian Gulf have indicated that about 20,000 miles of desert coastline existing around the world could be converted into likely locations for integrated power-water-food packages. Personnel at Oak Ridge state that plants, poultry or swine could be grown inside 200 acres of greenhouses next to a 300-Mw. (electrical) gas-cooled nuclear reactor.

The possibility of placing a chemical plant next to a nuclear power plant appears attractive. Dow Chemical is planning this now with Consumers Power in Midland, Mich. Dow will get about 40% of the heat as live steam from the boilers, while the rest is low-grade heat from the generator's cooling system. Du Pont's Chambers Works in New Jersey, and Union Carbide in South Charleston, S.C., also use heat from power plants. This is all on a small scale but indicates the great potential that does exist.

The Eugene (Ore.) Water and Electric Board has a rather interesting warm-water irrigation project underway on a 170-acre tract lying within a bend of the McKenzie River. The goal is to demonstrate that warm water, which in this case is obtained from a Weyerhauser paper mill, can be used to stimulate and enhance plant growth and to protect fruit from the killing frosts. In a similar manner, Pacific Power and Electric Co. is supporting a small project to see whether growing seasons cannot be lengthened and crop yields increased by warming the soil with electric cables. The cables simulate a network of pipes carrying heated water from power plants. Promoters of warm-water irrigation feel that it could be particularly useful in the Northwest, where millions of acres of potentially irrigable land exist.

All the developments I have mentioned indicate that the potential does exist for turning waste heat into a useful commodity. This is being done now on an experimental scale. In the next five to ten years, we will begin to see it occur on a practical and useful scale. I also believe that the major contributions of this development will almost certainly be made by chemical engineers. ∎

Meet the Author

J. I. Bregman, president of WAPORA Inc., 1725 DeSales St., N.W., Washington, D.C. 20036, combines a technical background in research and development with administrative experience in pollution control. He has served as Deputy Assistant Secretary for Water Quality and Research. His earlier positions include service with IIT Research Institute, Nalco Chemical Co., Illinois Air Pollution Control Board and Ohio River Valley Water Sanitation Commission. He is a graduate of Brooklyn Polytechnic Institute with a Ph.D in physical chemistry, and is the author of three books on environmental management.

Section XI
MISCELLANEOUS

Conserving fuel by heating with hot water instead of steam
Guide to trouble-free evaporators
Hot water for process use—storage vs. instantaneous heaters
Sizing vacuum equipment for evaporative coolers
Spot heating with portable heating systems
Low-cost evaporation method saves energy by reusing heat
New directions in heat transfer

Conserving fuel by heating with hot water instead of steam

Hot-water systems for unit processes can effect fuel savings of 20% over systems using steam. Such reduced costs, plus increased reliability and less maintenance, will accelerate the use of hot-water heating during the coming years.

William M. Teller, William Diskant and Louis Malfitani, American Hydrotherm Corp.

☐ Heating with high-temperature water under pressure was in use in Germany in the 1920s in the 250–420°F range. Apparently, engineers at I. G. Farben had trouble obtaining uniform temperatures on presses that were heated with steam. The presses were therefore heated with the hot water from the drum of the boiler, by circulating it at fairly high velocity through the presses. The water was returned to the boiler, which also produced steam for other applications.

While such changes were made to obtain a better product, it was quickly realized that a heat saving of up to 50% could also be obtained. In contrast to the U.S., fuel costs were a sizable manufacturing item in Europe. Within a very few years, high-temperature water (HTW) was adopted by many plants as a substitute for steam, with fuel savings amounting to 20–50%. HTW refers to water at about 250–420°F (300 psig). Mechanical problems caused by high water pressures above 420°F make this temperature the practical upper level.

At present, conditions in the U.S. are not unlike those in Europe in the twenties. Conversion from steam heat to HTW is quite attractive. Conversion costs from steam to hot water can usually be paid off in not more than two years for systems rated at about 20 MM Btu/h.

Economic studies made 25 years ago in the U.S. proved that, despite cheap fuel, substantial savings could be realized if HTW heating systems were used in multibuilding complexes.

Originally published June 21, 1976

Since 1950, most American air bases have been required to install HTW district heating in preference to steam systems. University campuses, hospitals and industrial-building complexes also started to follow this trend. Large segments of the chemical process industries have also benefited from such conversions, by having hundreds of HTW systems supplied for process heating.

The smaller systems, ranging from 5–15 MM Btu/h are only marginally more economical to operate than steam, but they are still favored because they provide much more accurate and uniform temperature control. Because all these factors have become increasingly critical in the U.S., it is timely to review the advantages of installing HTW systems or of converting existing steam-heating systems to hot water.

Disadvantages of steam heating

Apart from fuel savings, the following situations contribute to heat losses in steam-heating systems: open vents on condensate receivers; flashout losses and leaks in steam traps; and boiler blowdown. These losses involve at least an additional 5% in fuel costs. Another disadvantage is boiler deterioration, caused by instantaneous changes from low fire to peak loads. This is practically eliminated by HTW systems, because their "flywheel" effects dampen those of peak loads.

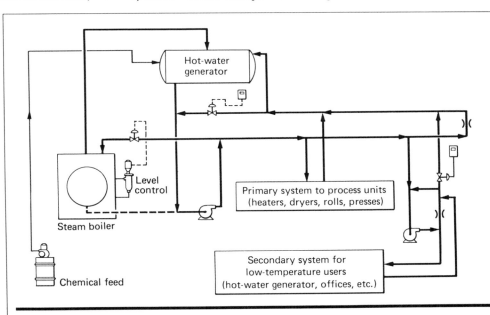

Hot-water generator produces water from steam Fig. 1

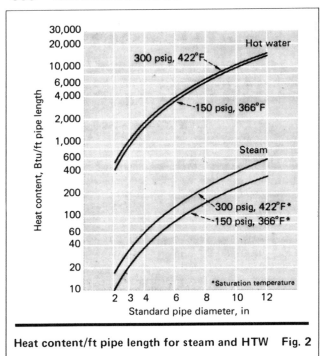

Heat content/ft pipe length for steam and HTW Fig. 2

Plants that have steam boilers can convert to hot-water heating simply and quickly by installing direct-contact water heaters in, or adjacent to, the boiler room. When boilers must be replaced because of age, hot-water generators can be installed as indicated in Fig. 1.

Direct-contact, hot-water heat exchangers (cascade heaters) not only can convert steam into hot water but can also serve as heat reservoirs. Such cascade heaters absorb sudden peak loads and make it possible to operate the boilers at fairly constant load levels. (Continuous changing of loads is one of the main factors that shorten boiler life.)

A central HTW system that supplies the higher temperature levels for process equipment, as well as low temperatures for space heating, is simple and uncomplicated, and has high distribution efficiency. A continuously circulating closed loop discharges HTW at constant temperature from the cascade heater (at only a few degrees below the saturation steam temperature), returning to the heater after releasing any required heat into separate secondary loops.

Each loop is individually controlled, either manually or automatically, to maintain the various temperatures at which the process equipment, the space heating, and the utility hot-water system operate. Cooled water from the secondary loops discharges into the return side of the primary loop and from there back to the cascade heater.

Since the only heat extracted is that required for the various services, distribution efficiency of the overall system approaches 95%. A typical system is shown schematically in Fig. 1.

The film heat-transfer coefficient, $Btu/[(h)(ft^2)(°F)]$, of hot water flowing through a pipe is constant at all points around its periphery for any given cross-section, varying only with velocity and temperature. For process applications requiring extremely close temperature control, the circulating rate through a secondary loop can be designed to limit the difference between inlet and outlet temperatures to as little as $\pm 2°F$.

On the other hand, the film heat-transfer coefficient of condensing vapors flowing through a horizontal pipe is higher in the top region and lower in the bottom, due to the forming of a condensate film along the bottom. The increasing accumulation of condensate between the inlet and the outlet of the pipe further reduces the overall inside-film heat-transfer coefficient. It has been demonstrated by actual tests that the condensate film thickness on the bottom of a pipe is five times thicker than that at the top.

The greater heat capacity of hot water over steam at equivalent saturation temperatures and the narrower pipelines required are other advantages. A steam line must be many times larger than an equivalent hot-water line, which need only be one or two sizes larger than the condensate line required in the steam system.

As to heat capacity, steam yields up about 800–960 Btu/lb when it condenses, while HTW can transfer only 100–150 Btu/lb. This, however, is no basis for comparison. Consider, for example, a 1-ft-long, standard 6-in-dia. pipe filled with 150-psig steam. This pipe will hold 0.073 lb of steam with a heat capacity of roughly 90 Btu. This same pipe with water at 366°F (saturation temperature corresponding to 150 psig) will hold 11

A hot-water system, on the other hand, is a closed circulating loop, with only very minor losses from leakage at valve stems and at pump stuffing boxes. Inasmuch as the amount of makeup water required is only a small fraction of that required for a steam system, this eliminates the need for blowdown, thereby doing away with a source of considerable heat loss.

Condensate lines are usually subject to corrosion, because mineral-free water, combined with atmospheric oxygen, becomes very corrosive. Theoretically, a properly maintained hot-water-cascade system, derived from condensing steam, will last indefinitely. The water volume increase in a hot-water-cascade system supplies makeup water to the boilers without flashing, thereby eliminating steam-condensate flashout losses.

Condensate collection systems usually flow by gravity, so all lines are pitched in the direction of the receiver. Forced-circulation hot-water lines are entirely independent of the building layout. They may, for instance, frame plant windows or doors to prevent cold air leaks.

Hot-water system advantages

For new installations, total capital investment is about the same for both steam and hot-water systems. However, the savings involved in fuel costs and maintenance make the payout period for a new hot-water system shorter than for conversion of an existing steam system.

Many plants use their steam boilers for both process and space heating. Cascade heaters can generate up to 350°F water from 150-psi steam (or 400°F from a 250-psi boiler). This temperature is adequate for rolls, presses, extruders, evaporators, conveyors and reactors. Steam-pressure reducing valves are not necessary to maintain the different temperature levels required by each machine.

CONSERVING FUEL BY HEATING WITH HOT WATER INSTEAD OF STEAM

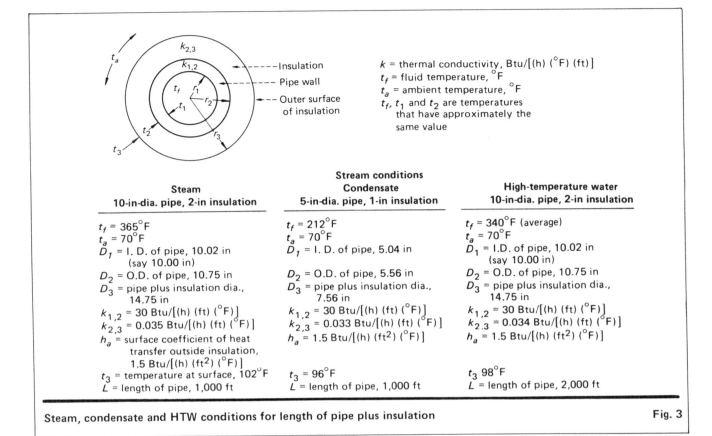

Steam, condensate and HTW conditions for length of pipe plus insulation — Fig. 3

lb/ft, which contains about 3,800 Btu. Hence, the ratio of absolute heat-storing capacity is 42 to 1 in favor of water. A graphical comparison of heat contents for different pipe diameters is provided in Fig. 2.

An HTW-distribution system acts as a heat accumulator due to its capacity to store heat. This may be likened to an energy reservoir, which can accommodate sudden heat demands without loss in temperature. Steam systems often suffer a temperature drop when shock or peak loads occur, which cause a drop in the boiler pressure. But because of the heat storage, HTW generators need not be sized for maximum peak loads. Steam boilers require such sizing to prevent pressure losses and accompanying temperature losses. Such smaller pipe diameters for HTW systems are economically important when heat is transported over long distances. Smaller pipes also reduce insulation costs.

One argument given for preventing many installations from switching to hot water is that steam lines are needed for certain production methods, and that such steam would be difficult to obtain from hot water. This is a fallacy, because obtaining steam from high-temperature water can be done inexpensively.

Comparison example

To compare the advantages of high-temperature water versus steam process heating, assume that 50,000 lb/h of steam at 150 psig are to be produced to deliver heat to equipment 1,000 ft away. Assume further that the saturation temperature of the 150-psig steam = 360°F; specific volume = 2.75; enthalpy of evaporation = 857 Btu/lb; enthalpy of saturated vapor = 1,195.6 Btu/lb; ambient temperature = 70°F; velocity = 5,000 ft/min; and density of water at 340°F = 56 lb/ft^3. (See Fig. 3 for further information.)

Steam system

Area required = $(50{,}000 \times 2.75)/(5{,}000 \times 60)$
= 0.46 ft^2, or 66 in^2

For a 5-lb pressure drop, a 10-in pipeline would be required, plus a 5-in condensate line. Assume also that the 10-in line would have a calcium silicate insulation 2 in thick, and that the 5-in line would have a 1-in insulation of the same material.

Pipeline heat losses:

To calculate the heat losses, this equation* is used:

$$q = \frac{\pi(t_f - t_a)L}{\left(\dfrac{1}{h_f D_1}\right) + \left[\left(\dfrac{2.3}{2k_{1,2}}\right)\log\left(\dfrac{D_2}{D_1}\right)\right] + \left[\left(\dfrac{2.3}{2k_{2,3}}\right)\log\left(\dfrac{D_3}{D_2}\right)\right] + \left(\dfrac{1}{h_a D_3}\right)}$$

where h_f = surface coefficient of heat transfer on the inside wall of the pipe; h_a = surface coefficient of heat transfer on the outside of the insulation. Usually, the first term of the denominator can be neglected because its effect is minimal.

The surface temperature, t_3, and the surface coefficient, h_a, must be computed by trial and error. Here, only the final computation for the heat losses are provided because the intermediate steps would be too long.

*Kern, Donald Q., "Process Heat Transfer," Chapter 2, p. 19, McGraw-Hill, New York (1950).

Therefore:

Loss in 10-in line =

$$\frac{\pi(365-70)(1{,}000)}{\left(\frac{2.3}{2\times 30}\right)\log\left(\frac{10.75}{10.00}\right)+\left(\frac{2.3}{2\times 0.035}\right)\log\left(\frac{14.75}{10.75}\right)+\left(\frac{1}{1.5(14.75/12)}\right)}$$

$$= 183{,}200 \text{ Btu}$$

Loss in 5-in line =

$$\frac{\pi(212-70)(1{,}000)}{\left(\frac{2.3}{2\times 30}\right)\log\left(\frac{5.56}{5.04}\right)+\left(\frac{2.3}{2\times 0.033}\right)\log\left(\frac{7.56}{5.56}\right)+\left(\frac{1}{1.5(7.56/12)}\right)}$$

$$= 78{,}100 \text{ Btu}$$

Total heat loss for steam system = 261,300 Btu
Amount of condensate = 183,200/857 = 214 lb
Amount of heat delivered = (50,000 − 214)857
$$= 42{,}666{,}600 \text{ Btu/h}$$

Flashout heat losses:

If the produced flash vapor is not used when the condensate is flashed out to atmospheric pressure in the return line and condensate receiver, the losses due to flashout will equal the enthalpy of the saturated water at 365°F minus the enthalpy of the saturated water at 212°F, or 338.5 − 180 = 158.5 Btu/lb.

To produce 857 Btu of latent heat per pound of steam, the boiler must supply 1,195.6 − 180 = 1,015.6 Btu/lb (assuming the condensate is returned to the boiler at 212°F). Therefore, condensate losses due to flashout = (158.5/1,015.6)100 = 15.6%. In addition, an approximate 5% loss occurs due to leakage of steam and condensate, plus blowdown losses, bringing the total losses up to 20%.

Total heat required:

Inasmuch as 80% of the condensate (40,000 lb/h) would be returned to the boiler, the enthalpy of the feedwater to the boiler, including makeup water, is:

40,000 lb condensate at 212°F (180 Btu/lb)
$$= 7{,}200{,}000 \text{ Btu}$$
10,000 lb makeup H$_2$O at 50°F (18 Btu/lb)
$$= 180{,}000 \text{ Btu}$$
Total heat to boiler = 7,380,000 Btu

The boiler must therefore produce (50,000 × 1,195.6) − 7,380,000 = 52,400,000 Btu/h.

Assuming 75% boiler efficiency, the adjusted total amount of energy needed for steam heat is 52,400,000/0.75 = 69,867,000 Btu.

Water system

To deliver 42,666,600 Btu to the equipment, assume a 40°F temperature drop between flow and return. The amount of water needed would then be 42,666,600/40 = 1,066,700 lb/h. If we assume a velocity of 10 ft/s, the pipe area needed would be 1,066,700/[(3,600)(10)(56)] = 0.529 ft², or 76.2 in².

To arrive at this pipe area, the equation $Q = AV$ is used, where Q = flow, ft³/h; A = area, ft²; and V = velocity, ft/h. Thus:

$$\frac{1{,}066{,}700 \text{ lb/h}}{(3{,}600 \text{ s/h})(10 \text{ ft/s})(56 \text{ lb/ft}^3)} = 0.529 \text{ ft}^2$$

This area requires a 10-in pipe.

The flow and return lines would require 2,000 ft of 10-in line, with a 2-in calcium silicate insulation. If the flow temperature = 360°F, and the return temperature = 320°F, the mean temperature would be 340°F. Then the heat loss would be:

$$\frac{\pi(340-70)(2{,}000)}{\left(\frac{2.3}{2\times 30}\right)\log\left(\frac{10.75}{10.00}\right)+\left(\frac{2.3}{2\times 0.034}\right)\log\left(\frac{14.75}{10.75}\right)+\left(\frac{1}{1.5(14.75/12)}\right)}$$

$$= 326{,}800 \text{ Btu}$$

Heat delivered to equipment = 42,666,600 Btu
Total heat required = 42,993,400 Btu

Assuming a boiler efficiency of 77%, the total adjusted amount of energy required from the hot-water system would then be 42,993,400/0.77 = 55,835,600 Btu/h.

Steam versus hot-water system

Heat required by steam system = 69,867,000 Btu
Heat required by hot-water system = 55,835,600 Btu
Difference = 14,031,400 Btu

Saving of high-temperature-water system over steam system = (14,031,400/69,867,000)100 = 20%.

The authors

William M. Teller has been associated with American Hydrotherm Corp., 470 Park Ave. South, New York, NY 10016 for over 28 years. Formerly, he was vice-president and partner of the corporation; now he only works part-time. He is a graduate in textile engineering from the University of Florence, Italy, and worked for three years in Austria for Caliqua. Thereafter he held various positions as a textile-machinery and photogrammetric-equipment designer.

William Diskant is Executive Vice-President, American Hydrotherm Corp., 470 Park Ave. South, New York, NY 10016, with which he has been associated since 1956. Previously, he worked for Bechtel Corp. and was an instructor at the College of the City of New York, from where he received a B.M.E. degree. He also holds an M.M.E. degree from New York University. A registered professional engineer in five states, he belongs to the American Soc. of Mechanical Engineers, the American Soc. of Heating, Refrigerating and Air Conditioning Engineers, and the International District Heating Assn.

Louis Malfitani is Vice-President, American Hydrotherm Corp., 470 Park Ave. South, New York, NY 10016, with which he has been associated since 1952. Since 1965 he has been chief of the High Temperature Water District Heating and Cooling Dept. He has a degree in mechanical engineering from the College of the City of New York, and is a member of the American Soc. of Mechanical Engineers and the International District Heating Assn.

GUIDE TO TROUBLE-FREE Evaporators

When evaporators fail, it usually means that the whole plant must be shut down. Here are practical tips on how evaporators should be designed and operated.

DAVID WETHERHORN, Continental Can Co.

The design of evaporators is a well-defined science according to the handbooks and textbooks. This is true as far as the determination of the required heating surface, method of feeding (forward, backward or intermediate), forced or natural circulation, long tube or short, etc. The characteristics of the liquid to be processed, and the product or products desired, pretty well determine the type of evaporator required; and the economics will be a factor in the number of effects, stages of flashing, materials of construction, available space, condenser design, and type of energy used.

But the design and operation of evaporators involves the "art" of engineering as well. Where does the "art" come in? There have been many cases where, even though the evaporators were correctly designed, good operation was not obtained because some practical aspects were overlooked. In many processes, pulping being a typical example, the cost of evaporators is a small percentage of the total cost of the mill; but all production ceases if the evaporators are taken out of service for a period of time that exceeds the liquor storage capacity. It is impractical to have a spare set of evaporators sitting around, so the proper design of the required evaporator is a key item in the process.

The primary aspects to be discussed in this article are:
- Design (process and mechanical).
- Specification and selection of evaporators.
- Troubleshooting and debugging.

DESIGN (PROCESS AND MECHANICAL)

The design of any evaporator should be done by, or in cooperation with, the personnel of the company supplying the basic equipment. Not only does this assure the employment of expert knowledge, it places the design responsibility on the supplier, where it belongs. If the purchaser specifies the design parameters, whether worked out by his own personnel or by a consulting engineer, the supplier becomes nothing more than a fabricator, and thus bears only limited responsibility for proper evaporator functioning.

If the process is well known, any reputable supplier of evaporators will be aware of the design requirements, so all you have to specify is the amount of evaporation required in your plant. If it is a new process, a confidential agreement can be worked out; and sufficient samples of the liquor to be evaporated should be furnished, to allow the necessary pilot-plant studies to determine the parameters required for the proper design. Every effort should be made to ensure that the samples represent the typical variations that may occur in day-to-day plant operations.

Choice of Design

There are cases where the supplier can offer a choice of designs, the one chosen depending on the overall economics of the entire process. For instance, a natural-circulation set of evaporators will have to be cleaned once a week, but if forced circulation is used, cleaning will be required only every two weeks. Both the initial and operating cost will be higher for the forced-circulation unit; but if having to clean the evaporators once a week were to reduce the production rate of the plant, then the forced-circulation set might indeed be the proper choice. After the decision is made, the responsibility for ensuring the two weeks between "boilouts" rests upon the supplier of the evaporators.

It is poor policy to put restrictions on the designer that may force him to limit the flexibility of the evaporators, at least in the initial design. For instance, space limitations may require that the vapor heads

Originally published June 1, 1970

be made quite small in diameter, with severe erosion resulting due to the excessive vapor velocity. Such a situation actually occurred when the original design used mesh-pad separators to reduce entrainment. Since the pressure drop was greater with the pads, the vapor heads were smaller in diameter. However, after a few months of operation, it was found necessary to remove the pads and use another type of separator. Ultimately, the vapor heads had to be replaced because of erosion, but because of installing the original evaporators to fit the smaller vapor heads, the diameter could not be increased. The final solution was to add another effect and thus reduce the amount of evaporation in each effect.

Never Buy on Price Alone

The science of designing evaporators is well established, as is proved by the number of units successfully operating in industry today, but little information is available concerning the problems that had to be overcome in order to make them operable. Not all evaporators have problems, but there have been many stories from the people who operate them that indicate there were problems in overcoming some of the minor difficulties before satisfactory performance was obtained. For this reason, if for no other, it is strongly recommended that management people who will be responsible for the operation of the evaporators be parties to decisions about purchasing this equipment.

EVAPORATOR set, in a typical installation.

The policy in some companies of allowing the purchase of process equipment by purchasing agents or consulting engineers is a "penny-wise and pound-foolish" approach. This leads to buying because of the lowest price, and does a disservice to both the designer and the operator. There have been instances where designers of two sets of equipment have differed by as much as 20% in the amount of heating surface offered for the same service. While both sets would have met the guarantee, the larger set would allow the operator greater time between boilouts, and thus enable him to keep up with greater production by the mill.

In this case, buying on price alone would have created problems for the man who must stay up day and night when the evaporators are not able to handle the required load. However, the operator should know that the "turndown" rate will be less (the minimum throughput will be greater) with the larger set, because the danger of plugging tubes becomes greater if the feed is reduced too low to maintain the minimum velocity in the tubes. Since it is much easier to take a set of evaporators off the line than to try to keep them on after they are fouled, the operator will pick the greater heating surface if given a choice. The other facet to be considered here is the designer's need to protect both his company and the purchaser from unforeseen possibilities. Even though similar evaporators may be in service in other plants throughout the world that make the same products, there are many variables that may affect the evaporators in different localities. Such things as water hardness, process variables ahead of the evaporators, product requirements, etc., might make the liquor react differently during the evaporation step and thus require some design alterations after installation. The designer must leave as much flexibility as possible, since his guarantee must be met and great pressure can arise when a $100-million plant is idled because a $1-million piece of equipment does not perform. Some designers prefer to build more insurance into their evaporators than others do! If given the opportunity to discuss this with the operating people, they would win their point in many cases. It is impractical to think of keeping a large plant down while waiting for another set of evaporators to be built and installed.

The closer the designer and operator work together, the closer the science and art approach the optimum evaporator.

SPECIFYING AND SELECTING EQUIPMENT

The final decision on selecting the proper equipment calls upon much more practical experience—art, if you will—than science. The best way to illustrate this is to take an example of how the final evaporator installation can be arrived at after considering alternatives.

A new kraft pulp mill is to be built, and the suppliers of evaporators are given the following data on which to base their bids:

Feed—428,000 lb./hr., 15% solids, 190 F.
Product—128,000 lb./hr., 50% solids.
Evaporation—300,000 lb./hr.
Steam available at 80 psig. and 150 psig.
Water temperature—85 F.
Quintuple set required—sixth effect may be added later.

Since this is a standard process, the designer uses data from his files to develop the basic design—actually, the "art" data are much more critical in this case than the "science." He knows that the viscosity of this liquor increases with the concentration, but that higher temperature lowers the viscosity. The liquor is very foamy and contains alkaline compounds that are somewhat corrosive to mild steel. In other words, since it is a well-known fact that long-tube-vertical (LTV) evaporators are widely used for kraft black-liquor service, this eliminates the need to search for all the alternative designs and, thus saves the time required to complete all the calculations normally involved in the scientific approach.

However, many other decisions must be made before the final installation. The buyer will need to know the cost of stainless steel vs. mild steel for internals and tubes, and what effect each will have on the heat-transfer coefficient. What type of entrainment separator is recommended, and what are the alternatives. All of the auxiliaries, such as pumps, condensers, instruments, etc., must be decided upon. In many cases, this is not done on a purely scientific basis.

Why Not All Science

With today's scientific development, it does seem difficult to understand why so much art should be acceptable in the design of evaporators. But there is a logical explanation. In the case given above, the designer uses the heat-transfer equation, $Q = UA\Delta T$, and by running trial heat-balances determines the heating surface required in each effect for the quintuple set, as requested. He selects the steam pressure and vacuum for the normal operating conditions, and uses values for U based on "commercially clean" condition of the tubes and the metal from which they are made. This calculation gives the complete picture of the basic design—size of each effect, number of tubes in each, evaporation in each, amount of liquor to and from each, condensate from each, temperature distribution throughout, and pressure distribution. This is a science, and if the conditions upon which the calculations were based remain constant, there is no doubt about the successful operation of the evaporator.

But anyone who has experience in kraft pulp mills will quickly point out that maintaining the average conditions used for the design is the exception rather than the rule. How will the evaporators react when the solids in the feed liquor drop to 12%?

Since the liquor is an alkaline solution, the alkali content affects the foaming characteristics and therefore how will the evaporators react to an increase in foaminess? The product concentration is critical and is the controlling factor in the proper operation of the evaporators. When the pressure in the steam chest of Number I effect reaches maximum, the concentration of the product will fall if the feed is maintained, and further fouling will continue. Unless curtailed production can be tolerated, this condition requires taking the set off the line for cleaning, and the frequency at which this occurs must fit into the overall operating schedule of the mill.

None of the variations mentioned above would change the heat balance of the design for 300,000 lb./hr. of evaporation. The drop in feed solids would reduce the feedrate to maintain the total evaporation rate but not the heating surface required. What could be done to minimize the problems that could arise from variations in the day-to-day operation? The designer should discuss possible variations with operating people and obtain agreement on the degree of flexibility to be designed into the evaporators. This will explain many cost questions before they reach the decisive stage. Since this is the "art" portion of the design, it should be a simple matter to demonstrate why flexibility is offered as a part of the design and what it will mean in the operation of the evaporators.

Specifying Auxiliary Equipment

Specification of pumps and motors should be made by the designer of the evaporators, but the choice of manufacturer may be left to the purchaser. This decision will be influenced by other pumps in service, spare-parts inventories, total-plant purchasing contract, and other extraneous factors. As long as the requirements are met to fill the design duty, this should be of no concern.

The greatest concern in the selection of equipment for better evaporator operation is, by far, instrumentation. Obtaining accurate data from the multipoint instruments installed on many sets of evaporators is difficult, if not impossible. The reason this situation is allowed to exist is that the evaporators will run well merely with control of the feed volume plus a steam-meter controller; but when trouble develops, the operating data available from a record of temperatures will be of great value in correcting the problem. As a matter of fact, if the designer can recommend an instrumentation system that can be kept in calibration by plant personnel, a great service could be done for the users of evaporators. Since this problem does exist, the designer should recommend installation of pressure-gage taps on all effects, and also sample taps for obtaining liquor samples from each effect. The sampling of vacuum effects may cause leaks that could cause upsets, and is not desirable; but in many instances, it is necessary to determine the temperature that should be recorded by the multipoint.

574 MISCELLANEOUS

Using computers to control evaporators will probably become common practice in the future, but only if the proper instrumentation can be developed.

TROUBLE SHOOTING AND DEBUGGING

The operation of a set of multiple-effect evaporators is an amazing phenomenon. Once they are started, by introducing the feed and turning on the steam, they reach their equilibrium and appear to "think" for themselves. When clean, the steam pressure in the No. I chest will be well below the design; and as they foul, it goes up just enough to increase the ΔT to the level necessary to produce the heat transfer required. When the maximum allowable pressure is reached, the evaporators must be cleaned, or the evaporation rate will fall below the design. The time for this cycle might be days or weeks, depending on the service, but during that time they operate pretty much on their own. Even when a change is made that causes an upset, they quickly establish an equilibrium and stabilize to the new condition. Considering that most evaporators are 4, 5 or 6 effects per set, good

Schematic Diagram of Evaporators, Showing Data from Operating Log

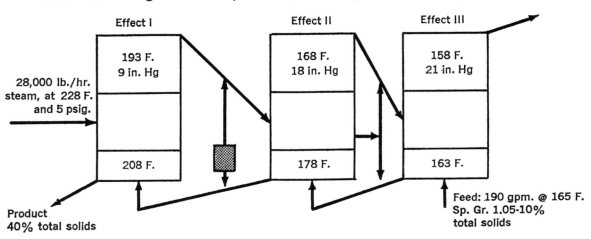

Specific heat of solids: 0.5 Btu./(lb.) (°F.)

$190 \times 500 \times 1.05 = 100{,}000$ lb./hr.

$\dfrac{100{,}000 \times .10}{0.40} = 25{,}000$ lb./hr. product

$100{,}000 - 25{,}000 = 75{,}000$ lb./hr. evaporated

Uncorrected economy $= \dfrac{75{,}000}{28{,}000} = 2.6$ lb. evaporated/lb. steam

Heat Balance for Triple-Effect Evaporation

Item	ΔT	C_p	Thousand Btu./Hr.	h_{fg}	Lb. Evap./Hr.
Steam to I, 28,000 × 0.97			26,074	960	
Liquor heat, (25,000 + 25,170)	30	0.90	1,355		
Heat from I			24,719	982	25,170
Condensate flash, 28,000	35	1.0	980		(1,000)*
Heat to II			25,699		
Liquor heat, (50,170 + 24,700)	15	0.93	1,048		
Heat from II			24,651	998	24,700
Condensate flash, (25,170 + 1,000*)	25	1.0	654		
Heat to III			25,305		
Liquor flash, 100,000	2	0.95	190	1,004	25,390
Total evaporation, lb./hr.					75,260

* Condensate flash, not evaporated from liquor.

operation is so simple that it is obvious the design has been based on sound principles. However, there are times after cleaning when, instead of stabilizing as before, the liquor carries over to the steam chest of the next effect and goes to the sewer. This might occur in any of the effects, or in two or more at the same time. It is situations like this that convince operators that evaporators are the worst equipment to analyze and debug of any ever made.

Troubleshooting is primarily an art, developed by experience in operating evaporators, and the approach will vary from process to process. Such simple things as checking for leaks, and checking pumps, noncondensables removal, steam and water flows, etc., are second nature to an operator. But when a set of evaporators will not operate after those things have been checked that, over the years, experience has taught the operator to look for, what can be done? This has happened more times than we care to admit. The next step is to call the designer and request an immediate visit. Science is required to augment the art.

When the engineer arrives, he will ask that the evaporators be started and will try to obtain temperature and pressure readings, so that a heat balance can be run; usually this will pinpoint the problem area. This is the big step, after which it may be necessary to go into the effect to find the exact trouble; but at least the search is narrowed so that all of the effects do not have to be opened. The ability of the engineer to run the heat balance in a comparatively short time, by trial and error, is the key to solving the problem. Solving any problem by trial and error can be very time-consuming unless the skill has been developed by continuous practice, and evaporator heat balances can be among the toughest to solve by this method. Look at the time, and money, that would have been saved had someone at the plant been able to run a heat balance.

Running an Evaporator Heat Balance

This raises a question that has been on my mind for many years: Why do our engineering schools spend many hours training young chemical engineers how to design evaporators, but offer nothing about how to run a heat balance on a set that is operating? Only a small percentage of chemical engineers ever design evaporators, but a much larger percentage would utilize the ability to run a heat balance from operating data. A paper was published on this subject in 1964* using an LVT, sextuple-effect black-liquor evaporator as an example. For this article a simple, triple-effect, backward-feed evaporator will illustrate the method.

The figure shows a schematic diagram of the set, with the temperatures and pressures in each effect, as well as the other operating data normally recorded by the operator at regular intervals. The calculated evaporation and economy may be in error if the data are not accurate, and this will show up in the heat balance. The table presents the complete heat balance based on the operating data.

To show how this calculation is accomplished, a step-by-step explanation follows:

1. Radiation losses are assumed to be 3%, therefore 97% of the initial heat is used for evaporation.

2. From the steam tables, the latent heat of evaporation (h_{fg}) for 5-psig. steam is 960 Btu./lb.

3. Here we reach the point where the trial-and-error method, which comes with experience, is normally required. The heat needed to raise the liquor to boiling will not be available for evaporation; so if you know approximately how much evaporation will be accomplished in No. I effect, you can put the trial-and-error method to good use. But if you know nothing about the evaporation, you can calculate it as follows:

$$26{,}074{,}000 - (25{,}000 + x)30 \times 0.90 = 982x$$
$$26{,}074{,}000 - 675{,}000 - 27x = 982x$$
$$25{,}399{,}000 = 1{,}009x$$
$$x = 25{,}170 \text{ lb. evaporated/hr., I}$$

The only assumption required was that 20% T.S. (total solids) was the content of the liquor from II—$(0.5 \times 0.20) + (1.0 \times 0.80) = 0.90$. However, if we had used 30%, the answer would have proved wrong and the approximate range would be apparent—i. e., $(0.5 \times 30 + 1.0 \times 0.70 = 0.85)$

$$26{,}074{,}000 - (25{,}000 + x)30 \times 0.85 = 982x$$
$$26{,}074{,}000 - 637{,}500 - 25.5x = 982x$$
$$25{,}436{,}500 = 1{,}007.5x$$
$$x = 25{,}250$$

$$\frac{25{,}000 \times 0.40}{25{,}000 + 25{,}250} = 19.9\% \text{ T.S.}$$

Thus, it is obvious that the original assumption of 20% was right.

4. After the liquor heat is subtracted from the total heat through the tubes, a check on the evaporation calculated in Step 3 is immediately available simply by dividing the latent heat of evaporation in I into the heat from I. If the two answers do not agree, an error has been made and should be corrected before proceeding further.

5. The flashing of the condensate from I is just one of the many ways of increasing the economy and is only included to illustrate how flashing is handled in calculating the heat balance. It is necessary to know all sources of all heat addition or loss to make this calculation worthwhile.

6. This step is a repeat of Step 3 for II effect.

$$25{,}699{,}000 - (50{,}170 + x)15 \times 0.93 = 998x$$
$$25{,}699{,}000 - 702{,}000 - 14x = 998x$$
$$24{,}997{,}000 = 1{,}012x$$

$$x = 24{,}700 \text{ lb. evaporated/hr., II}$$

7. Again, a check is made to confirm the accuracy of the evaporation.

8. The condensate from II flashes to the steam chest of III to increase the economy. Note that the 1,000 lb.

---

* Wetherhorn, David, The Calculation of Evaporator Heat Balances From Operating Data, *Tappi,* 47, No. 2, p. 168A (1964), or *Southern Pulp and Paper Mfr.,* May 10, 1964.

of flash from I must be included since the flash remains in the condensate of later effects.

9. Since the feed liquor is at a higher temperature than the liquor in III, a flashing occurs, and thus there is no heat required to raise the liquor to boiling.

10. The total evaporation is very close to the uncorrected evaporation in this example, but due to the inaccuracies in the instruments a check within 5% is considered acceptable when mill data are used.

This example has been simplified to illustrate the principles involved in running a heat balance; but regardless of the number of effects or the complexity, if a diagram can be drawn that accurately shows the flows, anyone can calculate a heat balance in a reasonable length of time. After its completion, if the surface in each effect is known, the heat-transfer coefficients can be calculated, and sources of most troubles can be pinpointed.

A computer program is easily developed using this approach, and all of the calculations can be quickly completed once the data are entered. As computer control of evaporators becomes more common, this program will enable continuous, or at least frequent, running of heat balances.

Even when there is no trouble in operation of the evaporators, this heat balance is a valuable aid in proving the errors in the instrument readings. In many cases, serious arguments have occurred between the operating people and the instrument department concerning the accuracy of the data, and a heat balance has been useful. Obviously, errors in the feed flow or steam flow will have a greater effect on the heat-balance-calculated evaporation rate, but errors in the temperature readings will change the coefficients of heat transfer in each effect, and thus cause false indications of trouble.

So much for the "art" of running heat balances even though you are not an expert on evaporator design. Many tricks-of-the-trade are used by operators in various industries, and obviously I cannot cover them all, even if they were all known to me, but a couple will be mentioned as illustrations.

1. As mentioned earlier, evaporators are designed to handle a feed that contains a specified solids content. In practical application, it is not uncommon to have variations from the norm and thus create operating problems. This can be minimized by circulating some of the product liquor back to the feed at a controlled rate and thereby maintaining the solids at a uniform level. This is called "sweetening" and is widely used by operators.

2. Another variable that creates operating problems is changes in the feed-liquor characteristics. This may take different forms in different industries, but a good example is available in the kraft pulping industry where the alkali content of the black liquor affects foaming characteristics during evaporation. When the alkali content drops below a certain level, carryover increases, so many operators will add caustic or caustic-containing liquor available from the process (white liquor). With only a small addition of caustic, the evaporators settle down to a smooth level. When the feed liquor returns to the normal level, the caustic addition is stopped.

Conclusions

Operators of evaporators in most industries today earned their jobs by learning how to keep equipment running efficiently. They do not have to know why a certain action helps to correct a problem, as long as the result is obtained. If the level in one effect goes up, and by switching pumps the level returns to normal, the operator doesn't have to be an engineer to conclude that the first pump needs repairs. If, when the pressure in the front end starts to climb and the ΔT in a middle effect is abnormally high, the operator opens the noncondensable line in the middle effect and the evaporator returns to normal, his primary task—to keep the evaporator running—has been accomplished. Then the cause can be looked for without the pressure of an imminent shutdown.

Operators of today have a greater knowledge of their art (just as designers have a greater knowledge of their science). They want to learn more of the science involved, and the scientists should be just as anxious to learn the art of operating evaporators.

Computers will no doubt control the evaporators in the future, but having an "artist" around will be a valuable asset.

Meet the Author

David Wetherhorn is Director of Research and Development of the Paperboard and Kraft Paper Div. of Continental Can Co., P. O. Box 1425, Augusta, Ga. 30903. He received his bachelor's degree in chemical engineering from Vanderbilt University, and studied advanced unit operations at Johns Hopkins University. He is a past chairman of the Alkaline Pulping Committee of Technical Assn. of the Pulp and Paper Industry (TAPPI), and chairman of the Technical Div. of Fourdrinier Kraft Board Inst.

Hot water for process use—storage vs. instantaneous heaters

When hot water of a given temperature is needed for a process, the engineer has several options on the best way of obtaining it. Here are the pros and cons of each.

Walter J. Schweitzer, Leslie Co.

☐ Which system—instantaneous or storage type—is best for maintaining process water at desired elevated temperatures? In what areas will the choice of system impact on your total operation? What are the alternatives, and what are the key factors to consider?

Good selection of hot-water-tempering systems will have a direct impact on water bills, space and structural costs, capital costs, installation, and even on the time it takes to get online. The choice can make a difference of more than 100% in terms of total owning and operating costs of the tempering system over its service life.

Alternatives to consider

Basically, there are three alternatives: (1) Storage-type, (2) instantaneous feedback and (3) instantaneous feedforward hot-water heaters. Depending upon the type of installation, one of these, or perhaps a combination, may offer the best choice.

An examination of alternatives may help you to identify and isolate possible performance factors and cost reduction areas in your system:

Storage heater—In its simplest form, a storage-type system (Fig. 1) consists of a large storage tank, a heat source, thermostatic controls, and piping to connect the tank with the process. The entire tankful is kept at a desired elevated temperature so that makeup hot water can be fed. The heat may be supplied by direct firing, electricity, steam coils, or direct steam-injection.

The storage heater has two major advantages over instantaneous types. First, it can be heated by gas, oil, or electricity, as well as by a steam coil. If you have no steam available, or not enough, you have no choice but to use a storage-type heater.

Second, the storage heater can meet steady-state requirements or a series of large-volume peak demands quite adequately—provided the time between peaks is

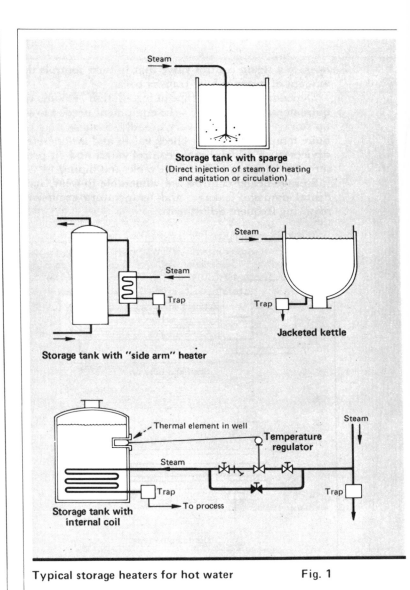

Typical storage heaters for hot water   Fig. 1

long enough for the tank to recover. (Of course, the more hot water needed, the larger the tank should be.)

The steam-heated storage heater maintains temperature by responding to changes in outlet temperature as sensed by a thermal element. As can be seen in Fig. 2, water temperature, sensed at the outlet, is fed back to

Originally published November 8, 1976

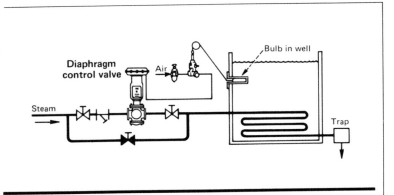

Typical control system for storage heater Fig. 2

operate a steam control valve that in turn controls the amount of steam in heat-transfer coils.

Depending upon the type of installation—volume requirements, demand, etc.—the equipment needed to set up storage heaters may vary broadly. Systems may require temperature pilots, check valves and air-operated devices (such as diaphragm control valves and air pressure regulators) as well as large tanks and piping.

Storage-heater controls are vulnerable to wear, accidental damage, leakage, and temperature excursions requiring frequent adjustment.

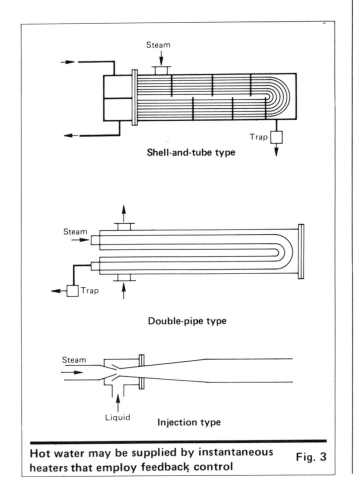

Hot water may be supplied by instantaneous heaters that employ feedback control Fig. 3

Also, because storage heating systems occupy large areas of valuable space due to tank size, and component and piping requirements, they are expensive to install. Structural support must be adequate for the weight of the full tank. It is not unusual to have to remove walls, piping and wiring to replace an old tank.

Instantaneous feedback heater—The instantaneous heater (Fig. 3), as the name implies, heats only a small amount of water, and only when needed. There is no large tank, and no large mass of water to support and keep at temperature (even when it is just in reserve). The heating medium is always steam, with either indirect or direct injection.

Instantaneous feedback, or closed-loop temperature-control heaters offer some fuel-saving economies. Such heaters are generally used when the demand for hot water does not fluctuate widely and when temperature control within a ±10°F range is satisfactory.

Like the storage heater, instantaneous feedback heaters respond to "after the fact" changes in water temperature. Feedback signals are transmitted from liquid- or gas-filled vapor-pressure measuring devices, or bimetallic thermostatic arrangements, that actuate a valve that controls steam flow or pressure.

The problems here are also similar to storage-tank heaters in operation. Often too slow, or too fast, in responding to sensed temperature, they cannot provide the complete and absolute reliability needed for process operations where strict temperature control or varying load conditions are the case.

In on/off services, instantaneous feedback systems are prone to temperature override, or buildup. Quick changes in demand cause a time (or response) lag in the system control operation before temperature-corrected water can reach the outlet.

This means a certain amount of water of the wrong temperature will be delivered until the system has had time to react and correct itself—a possible costly hazard to final product yield. Failures in feedback temperature-sensing components are common occurrences and also could result in dangerous overheating.

Instantaneous feedforward heater—The instantaneous feedforward, (or open-loop temperature control) heater satisfies small-space requirements and large swings in demand (as in batching and other on/off situations) to tight temperature ranges—usually to within ±4°F.

Feedforward control responds to demand changes that anticipate temperature changes in the delivered water. Instead of a conventional, closed-loop, feedback temperature-sensing system, the heat exchanger is controlled by an open-loop pressure system. This means it does not wait for an outlet temperature change. Rather, it anticipates that change by sensing a change in water flow demand at the feedforward unit (Fig. 4). Differential water pressure is sensed at a control valve to move a blender that correctly proportions hot and cold water to achieve a preset temperature.

Since the unit is actuated by water flow, the blender mixes hot and cold water to produce the desired outlet temperatures immediately, regardless of demand. Fluid flow changes are measured instantaneously, providing an immediate and automatic adjustment (Fig. 4).

Combination systems—Occasionally, demand require-

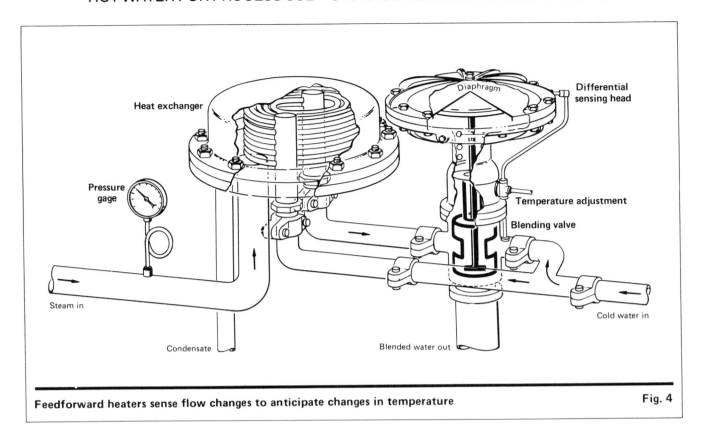

Feedforward heaters sense flow changes to anticipate changes in temperature Fig. 4

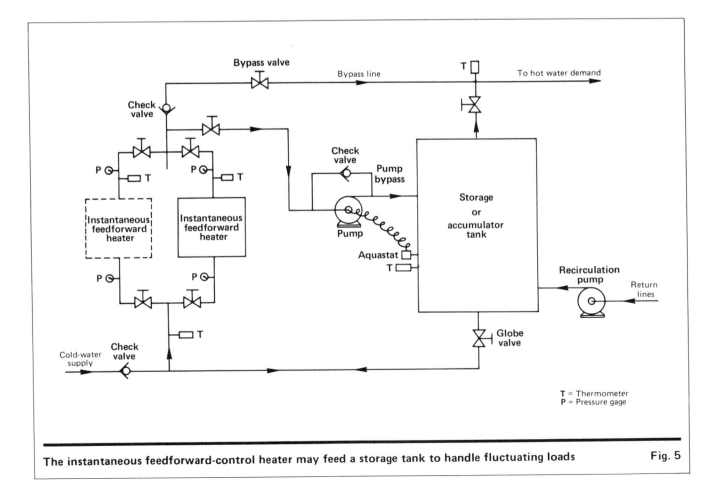

The instantaneous feedforward-control heater may feed a storage tank to handle fluctuating loads Fig. 5

Typical costs to consider when selecting steam water-heaters

Space, cost/ft^2
Labor (installation)
Accessories:
    Steam valve, strainers
    Gage and pigtail
    Steam traps, strainers
    Thermometer
    Steam relief valve
    Other
Materials:
    Foundations
    Crane
    Construction
    Mounting brackets
    Pipestands
Maintenance:
    Inspections (downtime)
    Tank repair
    Cleaning
    Thermal-system replacement
Initial purchase cost

ments favor neither the storage nor the instantaneous heater—for example, when peak volume is greater than an instantaneous heater can handle, yet demand for peak volume occurs frequently enough to prevent a storage tank from recovering. Another possible situation would be one characterized by high steady-state demand topped by a fluctuating load.

The solution then may be a combination instantaneous heater feeding a storage tank, with the storage tank serving as an accumulator (Fig. 5). The storage tank (or accumulator) can handle the extreme volume of peak-load demand, while the instantaneous heater supplies hot water to the storage tank at the required temperature.

Key cost factors to look for

Total owning and operating costs play as large a role in selecting instantaneous or storage heaters as does principle of operation. Cost factors that may not have been considered significant ten years ago may now be important. Key factors are shown in the table.

Nature of the demand—Is demand steady-state or intermittent? Does the operation run continuously or just for a few shifts? What temperature tolerance range must be held? What is the proportion of steady-state vs. peak demand as a function of time? What response time is needed in getting to temperature, or getting back to temperature after an excursion?

The best place to start is to actually plot water requirements against time for a whole day, or better yet, an entire week if the operation shuts down over the weekend. This will isolate the worst-case requirement which the equipment selection should accommodate.

Space requirements—The fewer square feet wasted as nonrevenue-producing space, the better.

Storage heaters occupy huge areas, nearly nine times the space of instantaneous types of equal capacity. And they always require heavy support structures.

Labor costs—Labor rates vary across the nation, but time estimates for two men to install either a storage or instantaneous feedbackheater run approximately 16h. Time estimates to install an instantaneous feedforward heater are about half that.

Accessories—What is an accessory to one manufacturer is standard with another. Costs vary, but standard items to look for will be steam valves and strainers, gages and pigtails, steam traps and strainers, thermometers and steam relief valves. Depending upon the alternative you choose, other items included might be pilot controllers and air-pressure valves. It is important to know these to determine total installation cost.

Material-handling costs (including installation materials)—Installation might require rigging services to move a unit into place. Rental costs for a crane can run into thousands of dollars. If you can, eliminate the need for special material-handling equipment in the first place.

Some instantaneous heater systems are light and small enough for two men to carry to the installation site. Additionally, these may be mounted on a wall or suspended from a ceiling, thereby offering a further space savings.

Watch for unanticipated outlays, such as for foundations, pipestands or mounting brackets, additional piping, insulation, or even additional construction costs if structural considerations are a problem.

Remember, too, the cost of removing the existing system, or better yet, see if you can avoid having to remove it. Instantaneous systems often make this possible because of their small size and light weight. Another alternative is to install an instantaneous unit, leaving the old storage tank in place for the time being, then have your own maintenance men remove it during slack times. Where space is really tight, instantaneous heaters have been installed inside the idled storage tank.

Maintenance costs—Factors to consider here are labor, inspection downtime, tank repair and cleaning downtime, product loss due to downtime, thermal system replacement (if required) and component maintenance.

Maintenance varies drastically from one alternative to the next. Storage-type heaters typically take days for complete maintenance, while instantaneous feedback heaters can take half that time. Feedforward heaters take only hours.

Depending upon the complexity, capacity and demand of the system, a fresh appraisal of alternatives may produce surprising results.

The author

Walter J. Schweitzer is manager, application engineering, Leslie Co., 409 Jefferson Rd., Parsippany, NJ 07054, where he is directly associated with solving fluid control problems in the chemical process industries. He serves on the ASHRAE (American Soc. of Heating, Refrigerating and Air-Conditioning Engineers) technical committee on service water heating and on the American Soc. of Mechanical Engineers research committee on water requirements for buildings. He majored in industrial engineering at New York University and business administration at Rutgers University, and is a member of Instrument Soc. of America.

SIZING VACUUM EQUIPMENT FOR EVAPORATIVE COOLERS

PAUL F. WALTRICH, Stokes Div. of Pennwalt Corp.

Evaporative cooling is widely used. The product to be cooled is put under vacuum, causing water vapor to flash off adiabatically, so that the latent heat of vaporization is drawn from the product, thus lowering its temperature. Once the initial air has been evacuated from such systems, the pressure decreases as a function of the product temperature, following essentially the vapor-pressure curve for water.

Since the capacity of the equipment evacuating the system varies with the system pressure, such equipment cannot be sized directly. Instead, vacuum pump capacity must be determined on the basis of achieving a given product temperature within a specified time. This calls for graphical integration techniques, which are developed as follows.

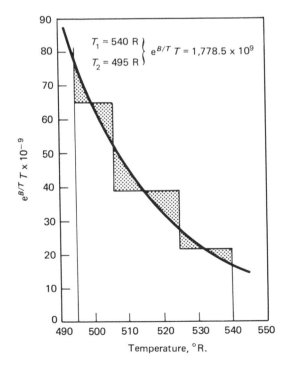

Vapor removal is given by the following:

$$\frac{dw}{dt} = S \times \frac{18 \text{ lb.}}{360 \text{ ft.}^3} \times \frac{P}{760} \times \frac{492}{T} = \frac{SP}{31.4\,T} \quad (1)$$

where:
S = pump speed, cfm.
P = absolute pressure, TORR
T = absolute temperature, R.
dw/dt = mass flow of vapor removed

A heat balance (assuming adiabatic conditions) gives the following equation:

$$\lambda \frac{dw}{dt} = \left[W - \frac{dw}{dt}(dt) \right] C_p \frac{dT}{dt} \quad (2)$$

where:
W = original product mass, lb.
$\frac{dw}{dt}(dt)$ = total mass of water removed
C_p = specific heat of mass W, Btu./lb.-F.
dT = temperature change in mass W caused by adiabatic evaporation, F.
λ = latent heat of evaporation, Btu./lb.

Expressing pressure as a function of temperature using the basic vapor pressure equation:

$$\ln P = A - \frac{B}{T} \quad \text{or} \quad P = e^{A-B/T} \quad (3)$$

Substituting Eq. (3) into Eq. (1)

$$\frac{dw}{dt} = \frac{PS}{31.4\,T} = \frac{e^{A-B/T}\,S}{31.4\,T} \quad (4)$$

Combining Eq. (4) and (2)

$$\frac{W C_p}{\lambda} \frac{dT}{dt} = \frac{e^{A-B/T}\,S}{31.4\,T} \quad (5)$$

The quantity dw has been assumed negligible, compared to the total mass W.

Rearranging terms gives an expression for temperature of the mass, W, in terms of time.

$$\int_0^t dt = \frac{31.4\, W C_p}{\lambda S e^A} \int e^{B/T}\, T\, dT$$

Originally published March 1, 1976

For the simple case where the vapor is water,

$$\lambda = 1{,}000 \text{ Btu./lb.}$$
$$A = 20.5648, \; e^A = 8.55 \times 10^8$$
$$B = 9{,}340$$

and the equation reduces to:

$$t \text{ (minutes)} = 3.67 \times 10^{-11} \frac{WC_p}{S} \int_{T_1}^{T_2} e^{B/T} T \, dT$$

The integral must be evaluated graphically.

Example: A ton of lettuce is to be cooled evaporatively from 80 F. to 35 F. Cooling time is to be 20 minutes. Specific heat is given as 0.9 Btu./lb.-F.

From a plot of $e^{B/T} T$ vs. T, the value of the integral between 80 and 35 F. is $1{,}778.5 \times 10^9$.

Then:

$$S = \frac{3.67 \times 10^{-11} \, (2{,}000) \, (0.9)}{20} \, 1{,}778.5 \times 10^9$$

$$S = 5{,}860 \; cfm.$$

If desired, a refrigerated condenser may be used in order to reduce the size of the vacuum pump.

The heat given up by the product is

$$WC_p \frac{dT}{dt}$$

and is equal to the heat picked up in the condenser:

$$UA \Delta T$$

Where
 U = overall condensing coefficient, Btu./hr.-ft.^{2}-F.
 A = condensing area, ft.^{2}
 ΔT = differential between product temperature, T, at time t, and the brine temperature (assumed as constant value, C)

$$\frac{W C_p \, dT}{dt} = UA \, (T - C)$$

$$dt = \frac{WC_p}{UA} \int_{T_1}^{T_2} \frac{dT}{T - C}$$

$$t = \frac{WC_p}{UA} \ln \frac{T_1 - C}{T_2 - C}$$

Note that here, the time is in hours. ∎

Spot heating with portable heating systems

Matching portable-heating methods to the heating job to be done (welding, heat to reactors, etc.) can save industry money, time and manpower.

Darrell W. Maukonen, Exomet, Inc.

☐ How do you provide spot heat economically to a process unit? Usually, the need is so urgent that no thought is given to what alternatives are available, or which one will provide the greatest savings. Examples range anywhere from heat-treating and emergency repair welding to curing a reactor lining, or from straightening a pipe to starting a process reaction. The costs of providing the required heat can amount to more than $1,000 on a seemingly small job, even without considering process downtime.

Heating alternatives can be exothermic, electrical-resistance or gas-fired, applied either alone or combined. To perform the job, equipment and materials can be purchased, leased or subcontracted out. The time to consider what is available is when non-emergency heating jobs are scheduled; choosing the wrong method can skyrocket energy, labor and downtime costs. The benefits from matching the correct heating alternative to the job can range from a reduction of $200 on a fuel bill to $60,000/d downtime.

At a Delaware City refinery, repair heat-treating saved millions of dollars by getting the plant back on line in half the time after a big fire; and at an East Coast refinery (Fig. 1), custom exothermic kits keep schedule turnarounds on schedule, and reduce labor and equipment costs. In reactor-vessel construction, gas-fired-curing operations can cut thousands of dollars in energy costs alone.

Portable heat can maintain process temperature if a processing unit goes down temporarily. It can also batch heat-treat on the production line, provide a short burst of heat to bring a reactor up to operating temperature, and provide help in shrink-fitting, brazing, soldering and other fabrication processes.

Considering the alternatives

An examination of the alternatives available provides a wide range of positive or negative features for specific jobs. Such features vary in accuracy of temperature control; personnel skill required; degree of portability; labor, material, energy and equipment costs; ability to meet existing codes; ease of control, etc. To select the correct heating means, engineers must forsake old hunches and make a thorough evaluation of the specific advantages and limitations of each alternative for the specific problem.

Exothermic kit heats metal rapidly, holds it for needed 'soaking' time, and allows it to cool slowly — Fig. 1

Temperatures in the range of 200–2,500°F (93.3–1,232°C) can be delivered by applying a portable heating unit alone, or in combination with others. Such heaters are ideal for intermittent or one-time process heating, or where capital investment of permanent heating units cannot be justified.

Exothermic heating is a fast, inexpensive portable-heat system that requires no outside source of power. Because the thermal cycle is designed into the system, monitoring functions can be minimized or omitted.

If slow controlled heating or extended "soaking" times are required, the electrical-resistance system is usually recommended. One exception would be when the cost of energy can be reduced by switching to gas. Electrical-resistance systems consist of heating elements, a controller, a recorder, a power source, insulation, thermocouples and connecting leads. The system can supply heat to one point or to multiple points at the same time, control them automatically, and document each point with a permanent record.

Originally published August 19, 1974

Applications of electrical, exothermic and gas-fired portable heating systems — Table I

Type of application	Electrical	Exothermic	High-Btu gas
Stress-relieving welds	Almost all cases	Primarily in piping applications	Large pieces of work
Normalizing welds	Almost all cases	Primarily in piping applications	Large pieces of work
Annealing welds	Almost all cases	Primarily in piping applications	Large pieces of work
Weld preheating	Almost all cases	When welding time is below 15 min	Usually not efficient
Heat-treating entire vessels	Only if vessels are small	Too costly	Most economical
Heat-treating spools of small-bore piping	If furnace is not available	If there are not too many welds	Only with a temporary furnace setup
Straightening lines	Almost all cases	Most cases	Never
Drying refractory linings	Only if gas is forbidden	Never	Most cases
Starting of reactions	Never	Never	Possible in most instances
Curing epoxy linings	Only if gas is forbidden	Never	Most cases
Maintaining temperature in processes	Most cases	Never	Unlikely

Gas is often the answer to heat-treating entire vessels or large sections of a vessel that would be uneconomical to heat-treat with other methods. Refractory lining in vessels and reactors can also be cured with a gas system.

Most portable-heat applications can be handled with either the electrical-resistance or the exothermic method; but gas has a few advantages that put it in the running for some jobs. Electrical-resistance methods are advantageous when these conditions exist: (1) a controlled rate of heating is needed; (2) a permanent record of the heating process is necessary; (3) configuration of shapes requiring heat is complex (tees, valves, etc.); (4) sections have diameters over 14 in, and wall thicknesses above 1 in (Fig. 2); (5) no more than five or six welds per day require heat; (6) adequate power is available; (7) all welds requiring heat are relatively close together.

Exothermic methods are generally convenient if these conditions prevail: (1) hardness is a criterion for acceptability; (2) joints are mainly butt welds, ells, flanges and reducers; (3) average area requiring heat is relatively small, and wall thickness of structure is under 1 in; (4) only a few welds are involved; (5) jobs are spread over a large area; (6) time schedule for completion is critical; (7) skilled labor and electrical power are in short supply.

Gas-fired methods are ordinarily preferable if these conditions exist: (1) electrical-power requirements for the job exceed the 150–200-kW range; (2) the area requiring heat is very large, i.e., a reactor chamber; (3) very high heat is required for extended periods of time (one week or longer); (4) precise but varying control of heat is needed.

For many jobs, the most efficient and economical solution becomes a combination of systems. Matching the correct portable-heating method to a specific requirement can maximize time, labor and fuel-cost savings. Table I lists applications where each method is best suited.

Matching heaters to the application

When comparing heating alternatives, a clear picture of the total job is needed at the start. A miscalculation, such as cost and availability of manpower, or cost and availability of energy source, can eat up dollars. The following checklist can help determine which portable heating system to use:

1. What is the code situation? Code requirements may leave you with only one choice.
2. What is the labor situation? Each area of the country has local rules governing the cost of labor for the portable-heating method selected.
3. What is the energy situation? Can you get enough gas or electric power to the job site to move ahead as

Ribbon heaters heat-treat weld between valves — Fig. 2

Relative cost of types of portable heating systems — Table II

Heating type	Relative cost					
	Labor	Material	Energy	Portability	Ease of control	Equipment
Electrical resistance	Low to medium	Medium	Medium	High	High	Medium
Electrical induction	Medium to high	High	High	Medium	High	High
Electrical, infrared	Medium to high	High	High	Low	Medium	Medium
Exothermic premolded systems	Low to medium	High	None	High	Limited	Nil
Exothermic flexible systems	Low to medium	High	None	High	Limited	Nil
Gas, natural	Low	Low	Low	Medium	Medium	High
Gas, propane	Low	Low	Medium	Medium	Medium	High
Gas and infrared	Low to medium	High	Low	Medium	Medium	Medium

fast as you would like? In remote jobs, you may not be able to do this without running extensive lengths of cable or pipe. Remember, too, that energy costs have risen sharply, and are likely to become more severe in the years to come.

4. What are the job logistics? For instance, what is the number and location of the heating points, the area to be heated and the type of material? Are roads accessible, and is the area to be worked on above ground level? (Cables and hose are difficult to handle if they must be carried up several stories.)

5. What is the timing schedule? Generally, exothermic-heating kits provide more scheduling freedom because these kits are self-sufficient energy sources.

6. Is equipment available? This should be anticipated early; equipment unavailability can limit the options that must be performed.

7. What degree of control (time, temperature) is required? The closer the control requirements, the more electrical-resistance heating is favored.

8. How long is the heating cycle? Gas is usually favored when longer heating or "soaking" cycles are required on large sections.

Recent advances in portable-heating methods include electrical-resistance systems that will monitor and control up to six points at the same time,* and flexible, exothermic materials that can be cut to size at the job site.* Adaptability of field-fitted, flexible exothermic materials have made portable heating much simpler.

Total cost (labor, material, power consumption and downtime) must be considered. Although materials for the exothermic system tend to be more expensive, labor costs usually are lower. If thickness of the pipe wall is more than 1 in, material cost may exceed labor savings. A complex configuration also increases the cost of exothermic materials needed. Table II compares types of portable-heating methods.

Portability is a significant advantage of the exothermic system, as is the speed with which exothermic kits can be applied. The operator simply sets the kit up and walks away. Yet, heating with gas (natural or propane) may be the only means for handling some situations, such as curing interior linings of large vessels.

A combination of heating systems for a single job may be more economical than either method alone. For example, heat-treating joints or sections with walls so thick that heating time for the large mass of metal becomes excessive can be handled by installing both exothermic material and resistance heaters at the joint. Also, stress-relieving of welds up to 7 in thick can be done more economically by combining two methods; even the time needed to do the job might be shortened.

If the heating method selected is either electric or gas, you generally have the choice of contracting the job, or of leasing or buying the necessary equipment. Although cost considerations are usually the basic factor, some of the points mentioned may affect your decision. If, for example, the degree of skill required is not available at your plant, contracting the job may be the only feasible means, just as it would be if it is a one-time-only job. But if the work is to be done often, buying the equipment may be more economical. Budget considerations alone may dictate the way to go.

* Exo-Lec and Flex-Anneal systems, Exomet, Inc., Conneaut, Ohio.

The author

Darrell W. Maukonen is product manager of the Heat Treating Div., Exomet, Inc. (subsidiary of Air Products and Chemicals, Inc.), Conneaut, OH 44030. His career started 16 years ago during the development of the exothermic heat-treating method. He holds an associate degree in metallurgical engineering technology from Tri-State Engineering College, Angola, Ind., and has written several articles on heat treating. He is a member of the American Welding Soc., the American Soc. of Mechanical Engineers, and the American Railway Engineers Assn.

Low-cost evaporation method

A unique multiple-stage process liquor or waste chemical or to incinerate

William G. Farin,

☐ The direct-contact multiple-effect system combines the best features of a direct-contact evaporator and a multiple-effect evaporator.

By itself, the direct-contact avaporator (DCE) is a high consumer of energy because the liquor to be concentrated is directly contacted by hot combustion gases. The resulting latent heat in the evaporated water dissipates to the atmosphere. However, the DCE is a low capital-cost unit capable of handling difficult liquors of high solids content. In directly firing the fuel into an evaporator, the operating cost per million Btu can be as low as half that of a million Btu of steam from a boiler.

The multiple-effect evaporator, on the other hand, is a high-capital-cost combination, because a boiler to produce steam, as well as considerable heat-transfer surfaces in several stages (effects), are required. The steam from boiler passes into the first stage to evaporate the liquor; vapor from the first stage then passes to a second stage to evaporate liquor at a lower pressure and temperature; and so on. With this system, energy savings are appreciable.

The direct-contact multiple-effect systems now being used [1-4], an example of which is shown in Fig. 1, include a:

"First Effect" of direct-contact evaporation, in which the combustion gases from burning a fuel come into contact with the liquor directly in a venturi scrubber to evaporate water, which saturates the flue gases in the conventional manner.

"Second Effect" provides heat recovery by vacuum evaporation. Condensate heated by countercurrent scrubbing contacts the saturated flue gas to remove its latent heat. The heated condensate passes through a heat exchanger to heat liquor that is circulated and flash-evaporated in a vacuum system. This procedure uses the latent heat for one or more effects of vacuum evaporation.

"Third Effect" of air evaporation heats liquor in the vacuum-evaporator's surface condenser. Additional heat is obtained by further cooling the flue gas with condensate and passing the heated condensate through a heat exchanger to heat the liquor. The heated liquor is circulated to an air evaporator, where it is contacted with air—thereby heating and saturating the air for added liquor evaporation and cooling.

The techniques can be applied to evaporation systems to use:
- Any fuel economically.
- Heat generated in burning waste liquors.
- Flue heat from recovery or waste-heat boilers.
- Waste heat contained in a plume.

Originally published May 14, 1973

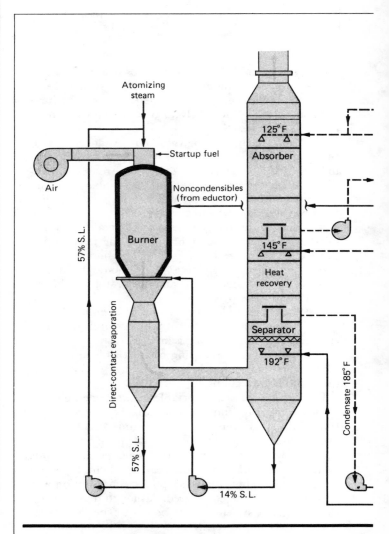

Direct-contact triple-effect system concentrates and incinerates

saves energy by reusing heat

evaporation system concentrates any stream to effect the recovery of a and dispose of a waste product.

Marathon Engineering Inc.

■ Heat from a condensing system for vacuum or air evaporation.

Furthermore, the system eliminates the steam normally needed for heating, and the water for condensing, by providing basic methods for heat-recovery evaporation that can be effective for energy conservation and environmental control in the chemical process industries.

Direct-contact evaporation

The direct-contact evaporator is particularly useful in handling liquors high in soluble or precipitated solids. It avoids the problems of low heat transfer due to high viscosity, scaling of heat-exchanger surfaces, and high boiling-point rise (where liquor temperatures exceed their vapor temperatures and result in a reduction of the available temperature differential and capacity in multiple-stage steam evaporators).

The design of the DCE is quite critical in order to provide complete saturation at minimum pressure drop, to avoid wet and dry buildups, and to control scaling, foaming and entrained carryover to the heat-recovery stages.

When incinerating a product, separation of the ash may require cooling the flue gas below 1,600° F to solidify the ash, facilitate separation and prevent carbonization of the product being direct-contact evaporated. This is achieved in one mill [7] by quenching the 5% liquor being concentrated in the combustion chamber with the 5% solids also being incinerated. At another mill, flue gas is recycled after scrubbing, to reduce its temperature to 900° F before it goes to the DCE [8]. Ash from the system is a molten smelt that discharges to a flaker for chemical recovery. Flue-gas temperatures have also been controlled with excess air. However, this reduces the potential for heat recovery.

Heat-recovery vacuum evaporation

Heat-recovery rates for vacuum evaporation often exceed direct-contact evaporation rates, due to the utilization of latent heat. Vacuum evaporation is particularly useful following a recovery boiler, fluidized-bed reactor, or incinerator burning wastes high in water content where it can be two to three times as high as that for direct-contact evaporation.

The amount of latent heat can be determined from Fig. 2, where the water-vapor content along with the heat content is shown per pound of dry flue gas at various saturation temperatures. The molecular weight of flue gas (about 31) compared to that for air (about 29) proportionally reduces the water-vapor content at a

combustible solids from feedstream   Fig. 1

587

588 MISCELLENAOUS

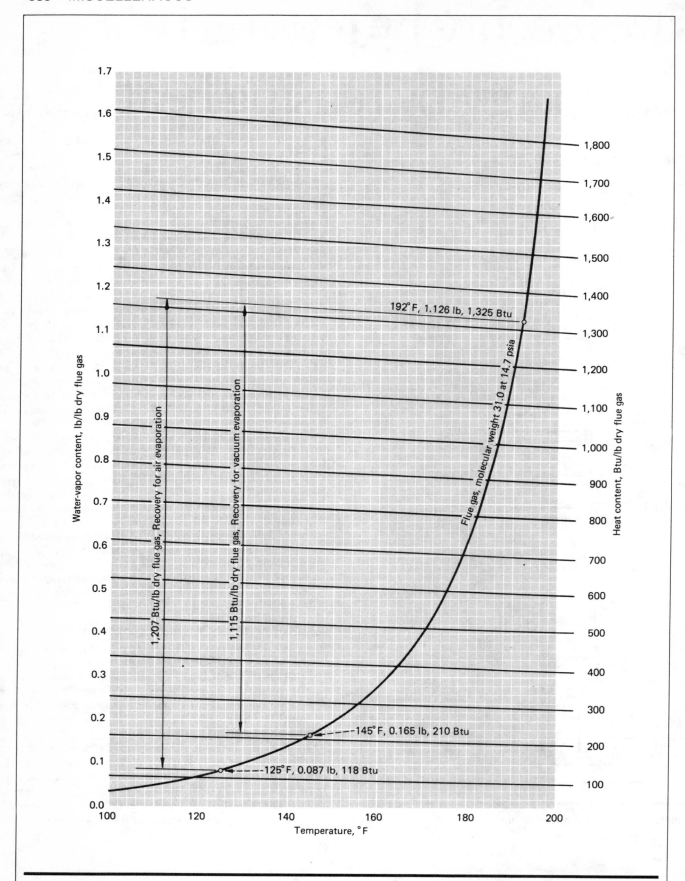

Flue-gas saturation determines amount of direct-contact and heat-recovery evaporation Fig. 2

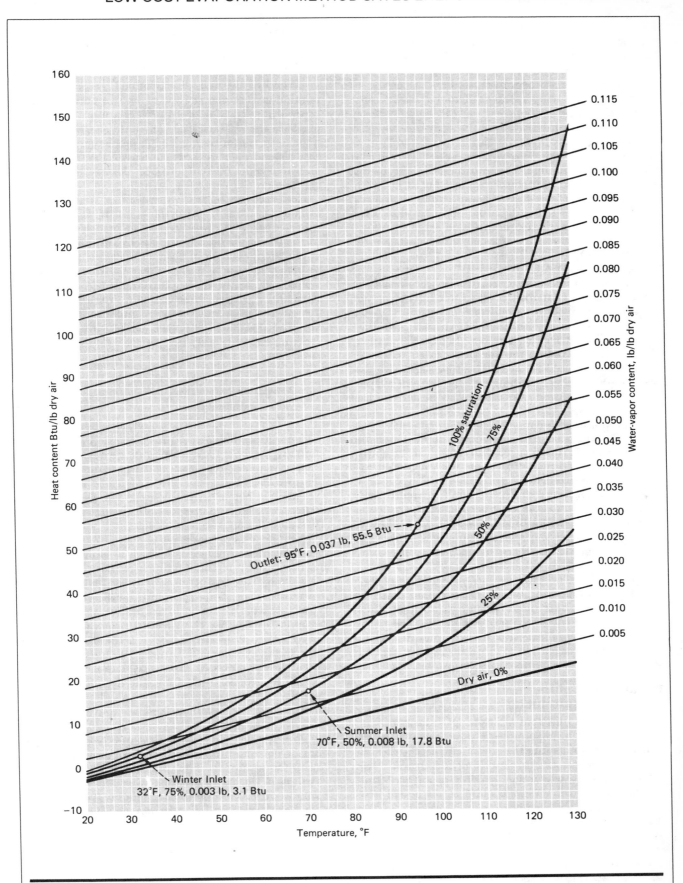

Air saturation establishes rates for air evaporation and concentration of waste feedstream   Fig. 3

given temperature, and thereby increases the saturation temperature. Increased pressure can also reduce water-vapor content and thereby increase saturation temperature. Both factors aid in providing an increased temperature differential for heat recovery.

If a sufficient temperature differential exists, two stages of vacuum evaporation can be considered. A multiple-effect flash-type evaporator may also be used for greater economy [1].

The type of evaporator can be varied to suit the product. When the final concentration is done in the DCE, the vacuum evaporator can operate at low concentrations and temperatures. This provides good heat-transfer rates with reduced scaling and corrosion. This unit can also be a final concentration stage or a vacuum evaporator-crystallizer for chemical recovery.

Heat-recovery air evaporation

The air evaporator is an effective tool for using low-quality heat for evaporation and for eliminating the need of cooling water for condensation. It can provide an added effect on any multistage evaporator by using heat from the surface condesnser. It can also utilize heat removed in cooling the flue gas to lower temperatures, and can exceed the evaporation possible with the DCE and sometimes that obtained by vacuum evaporation. Air-evaporation rates can readily be determined from Fig. 3.

The air evaporator cannot be used for concentrating liquors containing volatile chemicals that will cause air pollution. It has been successfully applied to pulp-mill sulfite-waste liquor. Provisions for vacuum stripping and preneutralization (originally provided as a safeguard) have proven unnecessary in this application. Concentrations up to 20% have been successfully maintained. The air evaporator is a good stripper [5], and will oxidize liquors such as Kraft-pulp waste [6]. This oxidation controls the release of odorous sulfur compounds during direct-contact evaporation and burning.

For the air evaporator, we can use standard cooling-tower equipment built of corrosion-resistant materials, exercising special care when providing additional mist-collection facilities. The design is also modified to prevent foam and gasification of circulated liquor, and to handle any solids encountered. In certain applications, a scrubber may also be necessary.

Scrubbing requirements

All scrubbing functions after the DCE are ordinarily carried out in a single multipurpose tower. In addition to recovery of heat in this tower, chemicals may be added to the condensate for scrubbing and chemical recovery if condensate dilution is not a problem. More often, scrubbing will follow cooling in an added absorption stage.

The heat-recovery scrubber should have low pressure drop, and should prevent condensate gasification, avoid loss of condensate back to the DCE, and avoid entrainment.

Particulates can normally be removed in the DCE and heat-recovery scrubber. However, when submicron particulates are present, a high-energy venturi can be added. Wet-wall precipitators or wet fiber-glass filters can also be used.

Waste-heat recovery

Whether the full direct-contact, triple-effect evaporator or just one or two of its components can be used for a particular heat-recovery application will depend on the characteristics of the flue gas and the liquor to be concentrated. The most economical method will depend on total heat content of the flue gas, and on water-vapor content and heat content per pound of dry flue gas.

Heat recovery with a waste-heat or recovery boiler, producing steam from incineration of wastes high in water-vapor, will ordinarily result in the discharge of 25 to 50% of the heat to the stack. Most of this heat is in the latent form from product moisture, steam atomization, combustion moisture, soot blowing, etc.

For example, in one application, in which discharge is at 450° F, the stack gas contains 365 Btu/lb of dry flue gas due to a water-vapor content of 0.21 lb/lb of dry flue gas. Or, for every million Btu ordinarily vented, this low-quality heat can still evaporate the following amounts of water:

By direct-contact evaporation	233 lb
By vacuum evaporation	616 lb
By air evaporation	580 lb
Total evaporation	1,429 lb

In addition, the flue gas is cooled and prepared for chemical scrubbing and air-pollution control.

Direct-contact evaporation depends on a high-temperature flue gas (at least 350° F) because latent heat is created but not used in this step. For the same heat content, dry flue gas provides the greatest potential for DCE since more of its sensible heat can be transformed to latent heat by evaporation.

On the other hand, flue gas high in moisture content has the greatest potential for heat-recovery evaporation even though the temperature may be well below 200° F. In these cases, the DCE would be eliminated and the type of evaporation determined by the saturation temperature (see Fig. 2). For vacuum evaporation, a saturation temperature above 150° F is required; a second stage of air evaporation provides an added potential. For air evaporation, a saturation temperature above 135° F is necessary, the amount depending on the quantity of flue gas available and evaporation requirements.

Heat-recovery evaporation provides an energy-saving, low-cost method for any process liquor or waste liquor being concentrated for chemical recovery, incineration or disposal. The oxidizing and stripping potential of the air evaporator as well as its ability for evaporation without producing condensate provides us with a variety of additional methods for evaporation, treatment and disposal that may be effectively used to meet present and future environmental requirements.

Example illustrates techniques

A direct-contact triple-effect evaporator will concentrate a feed liquor containing about 5% solids, having a dry-solids heat value of 7,500 Btu/lb, to a final

LOW-COST EVAPORATION METHOD SAVES ENERGY BY REUSING HEAT

Typical heat and material balances in a direct-contact triple-effect evaporator system

Item	Balance	Quantity*
1.00	Solids burned	
.01	Feed to burner, 57% solids	1.754 lb/lb
.02	Atomizing steam	0.3 lb/lb
.03	Air required	5.697 lb/lb
.04	Dry flue gas	6.165 lb/lb
2.00	Heat generated in burning	
.01	Combustion heat	7,500 Btu/lb
.02	Heat in liquor feed	208 Btu/lb
.03	Heat in atomizing steam	358 Btu/lb
.04	Heat in air	176 Btu/lb
.05	Radiation loss	−346 Btu/lb
.06	Heat in flue to direct-contact evaporator	7,896 Btu/lb
3.00	Water vapor from burning	
.01	Water from steam	0.300 lb/lb
.02	Water in air	0.088 lb/lb
.03	Water in liquor	0.754 lb/lb
.04	Combustion moisture	0.432 lb/lb
.05	Total moisture content	1.574 lb/lb
4.00	Direct-contact evaporation	
.01	Heat in feed	483 Btu/lb
.02	Heat from flue	7,896 Btu/lb
.03	Heat loss in product	−208 Btu/lb
.04	Heat for direct-contact evaporation	8,171 Btu/lb
.05	Heat in dry flue gas	1,325 Btu/lb dry flue gas
.06	Saturation temperature	192° F
.07	Water-vapor content of dry flue gas (Fig. 2)	1.126 lb/lb dry flue gas
.08	Water vapor	6.942 lb/lb
.09	Original moisture content	1.574 lb/lb
.10	Direct-contact evaporation	5.368 lb/lb
.11	Feed at 105° F, 14% solids	7.122 lb/lb
5.00	Heat recovery for vacuum evaporation	
.01	Flue heat, temperature out	145° F
.02	Heat content, (Fig. 2)	210 Btu/lb dry flue gas
.03	Heat content	1,294.7 Btu/lb
.04	Water-vapor content (Fig. 2)	0.165 lb/lb dry flue gas
.05	Water-vapor content	1.017 lb/lb
.06	Original water content	6.942 lb/lb
.07	Condensation	5.925 lb/lb
.08	Heat lost in condensate discharged	−610 Btu/lb
.09	Heat from direct-contact evaporator	8,171 Btu/lb
.10	Flue heat discharged	−1,294.7 Btu/lb
.11	Heat to heat exchanger	6,266.3 Btu/lb
6.00	Vacuum evaporation	
.01	Heat in feed at 170° F	2,704.8 Btu/lb
.02	Heat in product at 120° F	−1,109.7 Btu/lb
.03	Heat from heat exchanger	6,266.3 Btu/lb
.04	Total heat for vacuum evaporator	7,867.5 Btu/lb
.05	Heat for evaporation	1,113.3 Btu/lb water
.06	Vacuum evaporation	7.060 lb/lb
7.00	Heat recovery for air evaporation	
.01	Flue heat, temperature out	125° F
.02	Heat content (Fig. 2)	118 Btu/lb dry flue gas
.03	Heat content	727.5 Btu/lb
.04	Water-vapor content (Fig. 2)	0.087 lb/lb dry flue gas
.05	Water vapor	0.536 lb/lb
.06	Original water content	1.017 lb/lb
.07	Condensation	0.481 lb/lb
.08	Heat lost in condensate discharged	−39.9 Btu/lb
.09	Heat of inlet flue gas	1,294.7 Btu/lb
.10	Heat in discharged flue gas	−727.5 Btu/lb
.11	Heat from surface condenser	7,244.4 Btu/lb
.12	Recovered heat for air evaporation	7,771.7 Btu/lb
8.00	Heat-recovery air evaporation	
.01	Heat in feed	1,103.7 Btu/lb
.02	Heat in product	−483.0 Btu/lb
.03	Recovered heat for air evaporation	7,771.7 Btu/lb
.04	Total heat for air evaporation	8,392.4 Btu/lb
.05	Heat needed for summer evaporation (Fig. 3)	1,300 Btu/lb water
.06	Summer evaporation	6.45 lb/lb
.07	Summer feed to air evaporator, 7.4% solids	13.572 lb/lb
.08	Summer feed to vacuum evaporator, 4.8% solids	20.632 lb/lb
.09	Heat needed for winter evaporation (Fig. 3)	1,541 Btu/lb water
.10	Winter evaporation	5.44 lb/lb
.11	Winter feed to air evaporator, 7.9% solids	12.562 lb/lb
.12	Winter feed to vacuum evaporator, 5.1% solids	19.622 lb/lb

*lb/lb indicates lb/lb of solids burned.

content of 57% solids. The system whose flowsheet is shown in Fig. 1 provides for air evaporation, vacuum evaporation, scrubbing, combustion and direct-contact evaporation, along with the necessary heat recovery.

We will begin our step-by-step analysis of this system by starting with the incineration of the 57%-solids liquor in the burner at the conditions shown in Table I under Item 1.00. Heat is generated and water vapor formed (per Items 2.00 and 3.00). Flue gas discharges to the venturi, directly contacts the circulating liquor, and concentrates it to the 57% burning consistency from the 14% feed concentration (Item 4.00).

Flue gas saturated at 192° F is cooled to 145° F (Item 5.00) by countercurrent scrubbing with 135° F condensate, which is then heated to 185° F by the flue gas. The 185° condensate is sent through a heat exchanger to heat liquor that is circulated and flash-evaporated in a vacuum evaporator. This provides a second stage by vacuum evaporation at 120° F (Item 6.00 in Table I). The feed liquor is concentrated from the 5% to 7½% range.

The condensate, now cooled to 135° F, is recycled to the scrubber. Excess condensate (from condensed water vapor) is discharged. Noncondensibles from the vacuum evaporator are injected into the burner for combustion in order to enable recovery of chemicals and utilize the latent heat.

The flue gas is cooled further to 125° F in the absorption section by countercurrent scrubbing (Item 7.00) with 115°-F condensate. The condensate, in turn, is heated to 135° F by the flue gas, and also absorbs sulfur dioxide. The heated condensate is circulated through a heat exchanger, where cool (85° F) liquor is heated to 105° F. This liquor is now circulated through the air evaporator and cooled to 85° F, and it also becomes the coolant for the surface condenser of the vacuum evaporator. Heat from both exchangers serves to concentrate the liquor from the 7½% to 14% range by air evaporation (Item 8.00).

Calculation procedures

To find evaporation rates for the several stages of this system, we will use the data plotted as charts for flue gases and for air. From Fig. 2, we can quickly determine the rates for direct-contact evaporation and for heat-recovery evaporation; and from Fig. 3, the rates for air evaporation.

Direct-contact evaporation is calculated by totaling the heat input per pound of dry flue gas. The heat generated in burning is listed in Table I as Item 2.00. Direct-contact evaporation rates are shown in Item 4.00. Each pound of solids burned provides 8,171 Btu, or 1,325 Btu/lb of dry flue gas.

The flue gas following direct-contact evaporation will have a temperature of 192° F at 14.7 psia, and a saturated-water content of 1.126 lb/lb dry flue gas (Fig. 2), or 6.942 lb/lb of dry solids. Subtracting the 1.574 lb of water originally contained in the flue gas (Items 3.00 and 4.00), a direct-contact evaporation rate of 5.368 lb/lb dry solids is achieved. Feed to the direct-contact evaporator will have an added 1.754 lb./lb of the product sent to the burner, or 7.122 lb/lb of solids at 14% concentration.

For heat recovery during vacuum evaporation, flue gas is cooled to 145° F. As shown in Fig. 2, this reduces the water content to 0.165 lb/lb of dry flue gas, or 1.017 lb/lb of solids. The heat content is reduced to 210 Btu/lb of dry flue gas, or 1,294.7 Btu/lb of solids. This recovers 6,876.3 Btu/lb of solids, with the condensate heated from 135° F to 185° F. Of this total, 610 Btu/lb of solids is lost in the condensate (5.925 lb/lb of dry solids) discharged at 135°F. The balance (6,266.3 Btu/lb of solids) is available for heat transfer to the vacuum evaporator (Item 5.00).

The vacuum evaporator receives additional heat from the feed liquor (see Item 6.00 in Table I). This provides 7,867.5 Btu/lb of solids for the vacuum evaporation of 7.06 lb of water/lb of solids. Feed liquor is thus concentrated from 5% to 7½% solids.

For heat recovery during air evaporation, the flue gas is further cooled to 125° F (Item 7.00), thereby reducing the water-vapor content to 0.087 lb/lb of dry flue gas, or 0.536 lb/lb of solids. The heat content is reduced to 118 Btu/lb of dry flue gas, or 727.5 Btu/lb of solids. This recovers 567.2 Btu/lb of solids and condenses 0.481 lb of water/lb of solids. A loss of 39.9 Btu/lb solids occurs in the condensate discharged to acid makeup at 115° F. Added to 7,244.4 Btu/lb solids recovered in the surface condenser, a total of 7,771.7 Btu/lb solids is available for air evaporation.

Air-evaporation rates are obtained by determining the relative changes in heat content and water-vapor content of the air going in and out of the air evaporator.

Referring to Fig. 3, we find that, under summer conditions, air at 70° F and 50% saturated has a heat content of 17.8 Btu/lb of air and a water content of 0.008 lb/lb of air. If this air is heated to 95° F and saturated, its heat content rises to 55.5 Btu/lb and water content to 0.037 lb/lb air. The 0.029 lb of water evaporated into the heated air requires 37.7 Btu, or 1,300 Btu/lb water evaporated.

Under winter conditions at 32° F and 75% saturated, the initial heat content of air is 3.1 Btu/lb air, with 0.003 lb water vapor. The 0.034 lb of water evaporated requires 52.4 Btu, or 1,541 Btu/lb water evaporated.

The heat in the product raises the heat for air evaporation to 8,392.4 Btu/lb solids (Item 8.00). This provides a summer evaporation rate of 6.45 lb/lb of solids, and a winter rate of 5.44 lb/lb of solids. Thus, the concentration of the liquor stream increases from 7.4 to 14% in the summer, and from 7.9 to 14% in the winter.

Direct-contact double- and triple-effect evaporation systems are already operating for processing pulpmill waste liquors for evaporation and disposal [7, 8]. These systems provide for double and triple heat utilization and have very low capital and operating costs.

References

1. Farin, W. G., U.S. Patent 3,425,477 (Feb. 4, 1969).
2. Farin, W. G., U.S. Patent 3,638,708 (Feb. 1, 1972).
3. Farin, W. G., Canadian Patent 906,394 (Aug. 1, 1972).
4. Farin, W. G., *Tappi*, Sept. 1973, p. 69.
5. Estridge, B. G., Turner, B. G., Smathers, R. L. and Thibodeaux, L. J., *Tappi*, Jan. 1971, p. 53.
6. Sarkanen, K. V., Hrutfiord, B. J., Johanson, L. N and Gardner, H. S., *Tappi*, May 1970, p. 766.
7. MacLeod, M., *Pulp & Paper*, Oct. 1974, p. 58.
8. Evans, J. C., *Pulp & Paper*, Oct. 1975, p. 63.

The author

William G. Farin is President of Marathon Engineering Inc., a consulting and engineering firm, P. O. Box 335, Menasha, WI 54952. He obtained his experience in evaporation with E. I. du Pont de Nemours & Co., as chief engineer of the Krystal Div. of Struthers Wells, and with Marathon Div. of American Can Co. His experience on evaporation and burning includes systems for many paper companies. He has written extensively on the processing of spent liquors and other chemicals for the pulp and paper industry.

New Directions in Heat Transfer

The melting-point inversion process, which recovers work by means of a flow-work exchanger, and the indirect freezing process, which reuses heat, offer significant reduction in energy requirements and accomplish a high degree of separation for both aqueous and nonaqueous mixtures.

CHEN-YEN CHENG, University of Denver

The melting-point inversion process uses a unique way of upgrading heat energy, in which there is no gas phase involved. This method takes advantage of the abnormal melting-point curve of water. Water melts at a lower temperature under a higher applied pressure, while an ordinary substance melts at a higher temperature under a higher applied pressure.

Due to this difference, a substance that melts at a temperature lower than the freezing point of an aqueous solution may melt at a temperature higher than the melting point of water at a sufficiently high applied pressure. Thus, a suitable working medium can be used to form a cyclic auxiliary system that can be incorporated within the main system to (a) remove the heat of crystallization of water in the partial freezing of an aqueous solution by melting the working medium at a low pressure, and (b) supply the heat of melting the ice by solidifying the working medium at a sufficiently high applied pressure.

The process is distinct from the conventional freezing process in that it deals only with condensed (liquid and solid) phases. This has considerable effect on the energy requirements of the process and promotes the control of ice crystallization. The advantages of the present process are:

1. Low energy consumption—Since heat exchanges between the aqueous system and the medium system are accomplished by direct contact operations, Δt's for heat transfer may have very small values, less than 0.5°F. This contributes to a high thermodynamic efficiency of the process. The major energy input is made during the high-pressure ice-melting step. Fortunately, a flow-work exchanger (which simultaneously pressurizes a condensed stream and depressurizes a substantially equivalent volume of another condensed stream) has been developed by C. Y. Cheng and S. W. Cheng. The energy-recovery efficiency of a flow-work exchanger may be higher than 95% of the theoretically recoverable work.

2. Low-cost working media—Working media may be selected from nonvolatile paraffinic hydrocarbons that can be supplied by the petroleum industry at a cost of 5¢/lb, or less. It is worth noting that the latent heat of fusion of a straight-chain hydrocarbon in the range of C_{12} to C_{16} is particularly high (40 to 60 cal/g).

3. Low-cost equipment—Contributing to this low cost are: nonvolatility of the working medium, operation of most equipment at near-atmospheric pressures, and small equipment size due to the absence of a gas phase.

Energy Recovery

A flow-work exchanger can accomplish excellent energy recovery in a simultaneous pressurization and depressurization of two condensed streams.

In a continuous high-pressure process, a feed stream has to be pressurized and introduced into the high-pressure system, and a product stream depressurized and discharged from the high-pressure system. When the two streams are condensed streams (not gas streams), a flow-work exchanger can extract work from the product stream and add it to the feed stream. Work recovery can be better than 95% of the theoretically recoverable work; in contrast, the efficiency of the conventional combination of a hydraulic turbine with a high-pressure pump is less than 60%.

A flow-work exchanger can be used whenever there is a simultaneous pressurization and depressurization of two condensed streams. It can exchange flow-work (PV) between substantially equivalent volumes of the two fluid streams. Flow-work exchangers play an important role in the high-pressure ice-melting step of the melting-point inversion process, and contribute in reducing the energy consumption of the process.

When the volume of the feed stream (pressurizing stream) is in excess of the product stream (depressurizing stream), a part of the feed stream has to be introduced into the system by a high-pressure pump. In the reverse situation, a part of the product stream has to be depressurized in the conventional way.

Originally published August 19, 1974

Theory of Flow-Work Exchange

The theory and operation [1, 2] of a flow-work exchanger are described as follows:

Fig. 1 shows a high-pressure processing system to which a feed fluid and a product fluid are continually introduced and discharged, respectively, in a conventional manner by a pump, J_1, and a turbine, J_2. The feed fluid is pressurized from pressure P_{L1} to P_{H1} by the pump, and the product fluid is depressurized from P_{H2} to P_{L2} by the turbine.

When a condensed fluid is pressurized without phase change to a high pressure, the reversible shaft work, $-w_f$, received by the fluid in a flow process is:

$$-w_f = \int_{P_L}^{P_H} VdP$$

This shaft work is on the order of 200 times the corresponding value for a nonflow process:

$$-w_{nf} = -\int_{P_L}^{P_H} PdV$$

This difference is due to the noncompressibility of a condensed fluid—a liquid shrinks by about 1% on the application of 100-atm pressure. Similar statements can be made for the depressurization operation.

The reversible shaft work for the pump and turbine (Fig. 1) can be represented by:

$$(-w_1)_f = \int_{P_{L1}}^{P_{H1}} VdP = (P_{H1}V_{H1} - P_{L1}V_{L1}) - \int_{P_{L1}}^{P_{H1}} PdV \quad (1)$$

$$(+w_2)_f = \int_{P_{L2}}^{P_{H2}} VdP = (P_{H2}V_{H2} - P_{L2}V_{L2}) - \int_{P_{L2}}^{P_{H2}} PdV \quad (2)$$

Eq. (1) and (2) show that the shaft work for a flow process under high or low pressures is the sum of the shaft work for a corresponding nonflow process and the difference in the flow-work terms. The equations also show that the large values of the shaft work for the reversible flow pressurization and flow depressurization are to be attributed to the large values of the differences in flow-work terms, $\Delta(PV)$'s.

A flow pressurization may be considered as a superposition of two operations: (1) a nonflow pressurization and (2) a movement of fluid. The $|\Delta(PV)|$ term is large because the movement of fluid takes place across a large pressure differential between P_{L1} and P_{H1}. Similarly, a flow depressurization may be considered as a superposition of two operations: (1) a nonflow depressurization and (2) a movement of fluid. Again, the $|\Delta(PV)|$ term is large because of the movement across a large pressure differential between P_{L2} and P_{H2}.

For simultaneously flow-pressurizing a condensed fluid and flow-depressurizing another fluid, as shown in Fig. 1, it is possible to arrange the flow system so that movements of fluids take place across small pressure differentials, i.e. between P_{H2} and P_{H1}, and between P_{L1} and P_{L2}. Then, the $|\Delta(PV)|$ terms become very small, and the shaft work becomes small. The shaft work for each change can approach the values of the corresponding nonflow processes.

The simultaneous flow-pressurization and flow-depres-

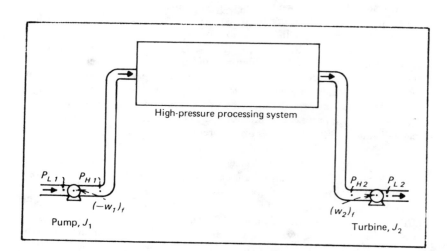

CONVENTIONAL method for pressurizing and depressurizing a high-pressure process uses a pump and turbine for handling the fluid—Fig. 1

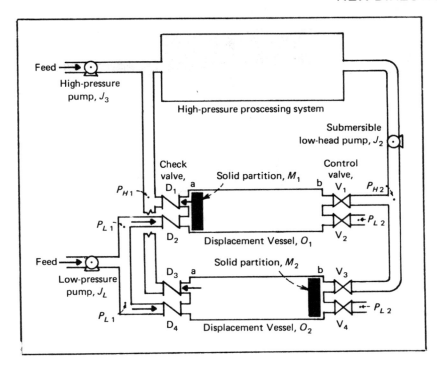

FLOW-WORK exchangers for a high-pressure processing system provide the means for heat reuse in the indirect contact-freezing process—Fig. 2

surization operations then involve the following steps:
1. Low-pressure and small-pressure-differential, $(P_{L1} - P_{L2})$, displacement operation.
2. Substantially nonflow pressurization of feed.
3. High-pressure and small-pressure-differential, $(P_{H2} - P_{H1})$, displacement operation.
4. Substantially nonflow depressurization of the product.

In the following discussion, the first and third steps will simply be referred to as the low-pressure displacement operation and the high-pressure displacement operation, respectively. The whole operation will be called flow-work exchange operation. A fluid to be pressurized in a process may exchange flow work with another fluid to be depressurized in the same process or other processes. The whole set of equipment will be called the flow-work exchanger.

Referring to Fig. 2, the flow-work-exchanger set consists of one or more displacement vessels (O_1 and O_2), check values (D_1, D_2, D_3 and D_4), control valves (V_1, V_2, V_3 and V_4), a low-pressure, low-head pump (J_L), and a submersible low-head pump (J_2). Pump J_2 is used to recover the pressure drop of fluid during its passage through the processing system, and to maintain P_{H2} higher than P_{H1} by an amount sufficient for a high-pressure displacement operation. Pump J_L is used to maintain P_{L1} somewhat higher than P_{L2} in order to carry out a low-pressure displacement operation. A high-pressure pump, J_H, is to pressurize the excess part of the feed. The feed-end and product-end of a displacement vessel will be called a-end and b-end, respectively.

Each displacement vessel is operated in the following four steps:

Step 1—Substantially nonflow depressurization. As shown in Fig. 2, displacement vessel O_1 is filled with high-pressure product. By closing valve V_1 and opening valve V_2, the content in the displacement vessel is depressurized, and some product fluid in the amount corresponding to the volume expansion due to depressurization flows out of the vessel through valve V_2. This operation takes a very short time. Check valves D_1 and D_2 are closed during this operation.

Step 2—Low-pressure displacement operation. When the pressure in the vessel drops below P_{L1}, check valve D_2 opens, low-pressure feed flows in through it, and the depressurized product flows out of the vessel through valve V_2. The solid partition, M_1, moves from the a-end to the b-end. The valves D_1 and V_1 are in closed positions. At the end of this operation, the vessel is filled with low-pressure feed. This is the situation shown in the displacement vessel, O_2, in Fig. 2.

Step 3—Substantially nonflow pressurization. Displacement vessel O_2 is filled with low-pressure feed. When valve V_4 is closed and valve V_3 is opened, some high-pressure product flows into the vessel to pressurize the contents. This takes a very short time, because only a small amount of fluid (to compensate for the volume shrinkage) has to be introduced. During this operation, check valves D_3 and D_4 are closed.

Step 4—High-pressure displacement operation. When the pressure in the vessel exceeds P_{H1}, valve D_3 opens. High-pressure product flows continually into the vessel through V_3, and the pressurized feed is transferred to the high-pressure system through valve D_3. The solid partition, M_2, moves from the b-end to the a-end. At the end of this operation, the vessel is filled with high-pressure product. This is the situation shown in displacement vessel O_1. Then, the whole operation returns to Step 1, and starts over.

The displacement operations, Steps 2 and 4, occupy most of the time in an operating cycle. The nonflow processes, Steps 1 and 3, take rather short periods of time.

Thus, when two displacement vessels are run with proper timing, fluid flow through the processing system will be continuous except for the short periods during Step 1, Step 3, and the time taken in operating the valves. These disturbances may be remedied by a small accumulator in the system.

Development of Flow-Work Exchangers

Several units of the flow-work exchanger have been built and tested. Early ones were constructed from commercially available component parts for the purpose of demonstrating the principle [2]. Recent units are designed for commercial operations, and an extensive integration of component parts has been made.

During the last few years, compact and reliable flow-work exchangers have been developed. At this time, at least two models of flow-work exchangers are available for commercial use:

A rotary-type energy-exchange engine (U.S. Patent 3,431,747) employs special seal structures (described in U.S. Patent 3,582,092). Several units have been built and successfully used in the pilot plant for the melting-point inversion process.

A recently introduced model (entitled: Integrated Flow-Work Exchanger With Pressure-Balanced Valve Pistons [3]) features a displacement vessel and its associated valves, integrated into a unit. Two valve pistons, one at each end, are used with the displacement vessel, and are tied together by a tie rod to form a movable valve assembly so that the entire assembly becomes pressure balanced. The assembly can be shifted between two desired positions by exerting a small force.

Indirect Freezing Process

A precisely controlled indirect contact-freezing process with heat reuse by alternative pressurization and depressurization incorporates the following features:

1. A uniquely designed unified freezer-melter comprising a longitudinal multivoid metal body. The body contains two sets of small longitudinal conduits that simultaneously and alternately serve as freezers or melters.

2. The heat released in an indirect freezing operation (conducted in one set of conduits) is used in supplying the heat needed in the *in situ* melting operation (simultaneously conducted in the other set of conduits) by alternately maintaining the two sets of conduits under appropriate pressures. This method of accomplishing heat reuse will be referred to as an alternate pressurization-depressurization technique.

3. An indirect freezing of a feed is done in each of the freezer conduits, and is followed by an *in situ* washing operation and an *in situ* melting operation.

The success of the process depends on a proper design of the freezer-melter, and controlling the freezing and melting operations within narrowly defined conditions. When properly conducted, the process can accomplish a high degree of separation of both an aqueous and nonaqueous solution at low energy consumption. Furthermore, because of the *in situ* freezing, washing and melting operations, the solids-handling problems that cause trouble in a conventional process have been eliminated. In addition, by using a freezer-melter with a small conduit diameter, a high heat-transfer rate has been obtained [overall heat-transfer coefficient is in excess of 100 $Btu/(h)(ft^2)(°F)$]. All these features have contributed to cutting equipment cost.

This process is currently under intensive development at the University of Denver under the support of the Office of Saline Water, U.S. Dept. of Interior.

Construction of Freezer-Melter

A multivoid unified freezer-melter [4, 5] consists of a metal body that has two sets of small and longitudinal conduits, denoted as *A*-conduits and *B*-conduits, as

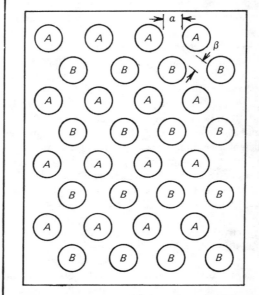
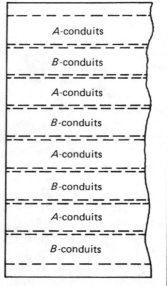

FREEZER-MELTER is a multivoid metal block containing two sets of small, longitudinal conduits that alternately and simultaneously are the freezing and melting units—Fig. 3

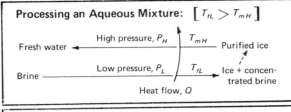
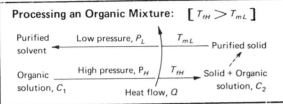

PRESSURE change is the basis for heat reuse—Fig. 4

shown in Fig. 3. These conduits are to be alternately and simultaneously used as a freezer and melter. Diameter of the conduits is less than ¼ in.

In Fig. 3, we may distinguish two types of separating walls, denoted, respectively, as α-walls and β-walls. A wall separating an A-conduit from an adjacent A-conduit or a B-conduit from an adjacent B-conduit is called an α-wall; and a wall separating an A-conduit from an adjacent B-conduit is called a β-wall. A β-wall may be half as thick as an α-wall. By taking advantage of this, a great saving in the amount of metal used in building a freezer-melter can be made. A multivoid metal block can be built at a low cost by first placing thin-wall steel tubes into a desired array, and then casting a heat-treatable aluminum alloy around the array.

Heat Reuse Is the Key to Success

An essential requirement of any successful fractional-solidification process is to use the heat given off in the freezing step to supply the heat required in the melting step, in order to achieve a low production cost. Conventional fractional-solidification processes either neglect heat reuse completely or recover heat by some kind of heat pump.

In the alternate pressurization-depressurization technique, the freezing and melting steps are conducted under sufficiently different pressures so that the prevailing temperature of the melting step becomes lower than that of the freezing step, as shown diagrammatically in Fig. 4. The desired heat reuse is accomplished simply by establishing heat exchange between the two steps [4]. In this process, a heat pump is not necessary; and for processing an aqueous mixture, the melting step is done under high pressure, while in processing an organic mixture, the freezing step is done under high pressure.

The shape and direction of the melting curve of a substance can be represented by the Clapeyron-Clausius equation. According to Bridgman [6], the melting point drops by 1°C for an increase in the applied pressure of about 100 atm. The general tendency of all systems is to have a positive slope of the fusion curve whose average value (as observed by Bridgman) is 50–60 atm/°C. Table I shows $\Delta P/\Delta T$ values for some solid-liquid systems [7].

Fig. 5 illustrates a pressure-temperature diagram for water. Referring to the diagram, we find that in general the pressures in the freezer and the melter, P_f and P_m, respectively, differ by an amount calculated by:

$$(P_f - P_m)/(dP/dT)_{S/L} = \Delta T_f + \Delta t \tag{3}$$

where ΔT_f is the freezing-point depression of the solution in the freezer and is defined as freezing temperature of pure solvent minus freezing temperature of the solution (both evaluated at 1 atm); Δt is the temperature differential between the freezer and melter allowed for heat transfer, and $(dP/dT)_{S/L}$ is the slope of the melting curve of the pure solvent.

Since ΔT_f is generally a positive value and Δt is always positive, P_f is greater than P_m when (dP/dT) is positive, and P_f is less than P_m when (dP/dT) is negative. The latter is the case for an aqueous system, the former for a nonaqueous system.

By using Eq. (3), the pressure to be applied to melter conduits for seawater and brackish-water desalting are about 8,000 psi and 3,000 psi, respectively.

Since the volume of water decrease about 9.5% as ice is melted, an amount of water equivalent to the volume shrinkage must be pumped into the melter conduits to maintain melting pressure. This is the major work needed in the process. For example, in this operation, the theoretical work needed to pump in 95 gal against a pressure of 8,000 psi is equivalent to 5.5 kWh/1,000 gal of fresh water produced from seawater. The corresponding work in brackish-water desalting requires pumping in 95 gal against a head of 3,000 psi, which is equivalent to 2.6 kWh/1,000 gal of fresh water.

Of course pumping efficiency, loss of product water, heat leakage, and other irreversibilities have to be allowed for. The results of a careful process analysis [8] based on experimentally observed heat- and mass-transfer rates, and allowing for all other losses, show that the power requirements for seawater and brackish-water

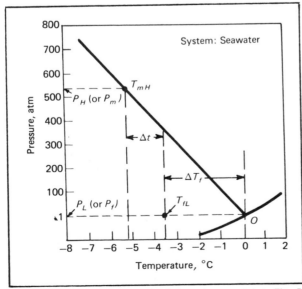

PRESSURE-TEMPERATURE relations for seawater—Fig. 5

Slopes of the Melting Curves—Table I

Substance	Normal Melting Point, °C	Slope, $\Delta P/\Delta T$ Atm/°C
Water	0	−100
Methane	−182.49	+ 39
Ethylene	−169.5	+ 70
Benzene	5.50	+ 37
p-Xylene	13.2	+ 29
Bibenzyl	51.8	+ 34
Cyclohexane	6.55	+ 19
Carbon tetrachloride	−22.95	+ 28
Ethylene dibromide	9.95	+ 40
o-Dichlorobenzene	53.15	+ 28
p-Dibromobenzene	87.3	+ 27.5
Cetyl alcohol	49.10	+ 43
Acetic acid	16.55	+ 48.5
Bromobenzene	−5.50	+ 53
n-Caproic acid	−3.9	+ 55

Source: Ref. [7]

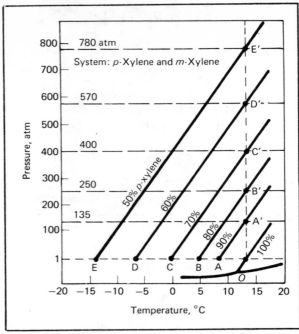

TEMPERATURES of p- and m-xylene mixtures—Fig. 6

desalting are 20 to 25 kWh/1,000 gal and 10 to 12 kWh/1,000 gal, respectively.

The energy requirement of the process is low because (1) heat released in a freezing step is used immediately in a melting step without first being transferred to an intermediate working medium, and (2) the amount of high-pressure pumping is limited to about 100 gal. in producing 1,000 gal or product water.

The phase diagram for the system p-xylene and m-xylene shows that the freezing temperatures of 100%, 90%, 80%, 70%, 60% and 50% solutions at 1 atm are 13.26°C, 9°C, 5°C, 0°C, −6°C and −13°C, respectively. Since we may consider that the freezing temperatures of these solutions will increase by 1°C as the applied pressure is increased by about 29 atm, we may obtain the freezing-temperature versus applied-pressure relations shown in Fig. 6. From Fig. 6, we find that the freezing temperatures of 90% solution at 135 atm, 80% solution at 250 atm, 70% solution at 400 atm, 60% solution at 570 atm, and 50% solution at 780 atm are all equal to 13.26°C, the melting point of pure p-xylene at 1 atm.

By adding 58 atm to these pressures, the freezing temperatures of these solutions would become 15.26°C, which is 2°C higher than the melting point of pure p-xylene at 1 atm. When the condition prevailing in the freezing step can be represented by any of these conditions, the desired heat reuse can be accomplished with an available ΔT for heat transfer of 2°C.

How the Multivoid Unit Works

The unified freezer-melter has two sets of conduits, as shown in Fig. 3. These are simultaneously and alternately used as freezer conduits and melter conduits. The unit is operated in a cyclic fashion, and each cycle consists of four steps. Using the desalting of an aqueous solution as an example, these four steps will be explained in detail:

Step 1—Throughout this step, an aqueous solution is frozen under a low pressure, P_L, in the A-conduits, and ice is melted under a high pressure, P_H, in the B-conduits. The pressure to be applied to the melter (or B-conduits) can be evaluated from Eq. (3). Oscillatory motion in the fluid enhances mass transfer, reduces concentration polarization in the freezer conduits, and promotes heat transfer in the melter conduits during this step [9].

The brine rejected from the freezer (A-conduits) is heat exchanged with the incoming brine before being removed from the system. This operation is not shown in the diagram.

Step 2—At the end of Step 1, the remaining void space in the A-conduits is filled with brine, and that in the B-conduits with fresh water. As will be described (in Step 3), the A-conduits will become a melter, and B-conduits a freezer.

Step 2 is a transition period between Steps 1 and 3, during which the brine in the A-conduits and the fresh water in the B-conduits have to be replaced by fresh water and brine, respectively. These displacement operations may be achieved by pumping fresh water and brine, respectively, in the A-conduits and B-conduits. Intermixing at the fluid boundaries should be prevented.

The displacement of fresh water by the feed brine may also be conducted in the following manner to include an air-displacement step:

1. Displace fresh water from B-conduits by introducing air into the conduits.
2. Introduce a quantity of feed brine into the B-conduits, and displace air out of the conduits.

The net water-production is equal to the water displaced in B-conduits minus the water used in the displacement of A-conduits and minus the water pumped in by the high-pressure pump in Step 1. This product is heat exchanged with feed brine before being removed from the system.

Step 3—During this step, the *A*-conduits are used as a melter and the *B*-conduits as a freezer. Heat of freezing, liberated in the *B*-conduits, is used in the melting of ice in the *A*-conduits. The discussions made in connection with Step 1 apply to this step; one simply interchanges *A*-conduits with *B*-conduits, and low-pressure brine with high-pressure fresh water.

Step 4—This step is similar to Step 2, and the discussions given in connection with Step 2 apply with appropriate changes.

The four steps are repeated cyclically. Since deposited solids are, washed and melted *in situ*, there is no need for mechanically scraping the heat-transfer surfaces and transferring a slurry stream.

Solids Separation

The degree of purification that can be attained in the programmed indirect-freezing process depends on how effectively the mother liquor in the interstitial spaces can be displaced by fluid.

The dimension of the central unfilled spaces and the interdendritic spaces should be such as to minimize dispersion and preferential channeling of fluid either through the central unfilled spaces or through a part of the interdendritic spaces during the *in situ* washing step. Therefore, the degree of solidification, defined as the fraction of the freezer volume occupied by solid phase at the end of the freezing step, should be within an appropriate range that is to be determined by experiment. Generally, it is between 45 to 60% for a dendritic deposit.

In conventional processes, a thick layer of deposit, 1 cm or more, is formed under a rather large temperature differential, 10°C or more. Under these operating conditions, the interdendritic spaces become extremely narrow, and enclosing spaces containing mother liquor are formed.

In order to form a solid bed of the desired characteristics, to have a low resistance to heat transfer, and to accomplish an efficient heat reuse, it is desirable to limit ΔT for heat transfer to less than 5°F, or preferably less than 3°F, and thickness of deposit to less than 3 mm.

Initiation of Freezing

A freezing/melting operation conducted simultaneously and alternately in two sets of conduits in a multi-void freezer-melter has a Δt for heat transfer between the two sets of conduits limited to a rather low value, say less than 3°F. With such a small Δt, spontaneous nucleation does not take place. Therefore, there is a need to bring or form seed crystals, or bring effective nucleating agents uniformly into the freezer conduits. Two satisfactory methods have been found, and these will be referred to as a seeded-feed method [10] and a seeded-film method [11].

The success of these approaches represents a major breakthrough in developing the programmed indirect-freezing process. We are now able to freeze more uniformly, avoid local plugging, attain a higher degree of solidification, and reduce conduit diameter. By being able to reduce conduit diameter to $3/16$ in, freezing time can be reduced to 6 to 8 min, and the overall heat-transfer coefficient increased to about 120 Btu/(h)(ft^2)(°F).

Summary

A flow-work exchanger can accomplish a very high energy recovery—in the range of 90 to 95% or better—in a simultaneous pressurization and depressurization of condensed streams. The integrated flow-work exchanger with pressure-balanced valve pistons can be made from honed hydraulic cylinder tubes, and commercially available piston seals.

The melting-point inversion process and the programmed indirect-freezing process are both thermodynamically efficient. Properly conducted, the energy consumptions of both processes in seawater desalting are within 30 to 35 kWH/1,000 gal of fresh water produced. For brackish water, the energy consumption is considerably less.

The programmed indirect-freezing process is very versatile, and can be used in separating both aqueous and nonaqueous mixtures. It is important to note that there is no need for a working medium in this process. These two processes are complementary and do not compete.

Acknowledgment

Some of the work reported herein has resulted from studies made under grants from the Office of Saline Water, U.S. Dept. of Interior. #

References

1. Cheng, C. Y., Cheng, S. W. and Fan, L. T., *AIChE J.*, **13**, 438–442 (1967).
2. Dynatech R/D Corp., Research and Development Report No. 680, Office of Saline Water, Washington, 1971.
3. Cheng, C. Y. and Cheng, S. W., patent pending.
4. Cheng, C. Y. and Cheng, S. W., U.S. Patent 3,678,696; British Patent 1,265,733; German Patent 1,912,019; Canadian Patent 897,584.
5. Cheng, C. Y., Van Riper, G. and Fox, G. V., Research and Development, Report No. 802, Office of Saline Water, Washington, 1973.
6. Ricci, J. E., "Phase Rule and Heterogeneous Equilibrium," p. 34, Van Nostrand, New York, 1951.
7. Timmermans, J., "Physico-Chemical Constants of Binary Systems," Interscience, New York, 1959.
8. Yang, P. T., M.S. Thesis, University of Denver, 1972.
9. Lin, K. S., M.S. Thesis, University of Denver, 1973.
10. Chang, E. W. Y., M.S. Thesis to be submitted to University of Denver, 1974.
11. Ford, T., Ph.D. Thesis to be submitted to University of Denver, 1974.

Meet the Author

Chen-yen Cheng is professor of chemical engineering at the University of Denver, Denver, CO 80210. Among his fields of interest are desalting, wastewater treatment, energy conversion, phase equilibria, and thermodynamics. He has published extensively in these fields and holds a number of patents on processes and equipment dealing with fractional solidification and heat reuse. He has a B.S. from National Taiwan University, Formosa, an M.S. from the University of Michigan and a Dr. Eng. from Kyoto University, Japan. He is also a member of AIChE and the Japanese Institute of Chemical Engineers.

INDEX

Acid aqueous solutions, 107
Agitated vessels, 36
Agitators
    mechanical, 239–245
Air-cooled fans, 427–431
    basic variables, 428–429
    rating, 429–431
Air-cooled heat exchangers, 29, 35
    design, 413–431
    heat transfer coefficients, 416
    procedure for preliminary estimates, 413–418
    simulation, 111–112
    temperature difference calculation, 416–418
Air-cooled steam condensers, *see* Steam condensers
Air coolers
    calculations, 423–426
    comparison with water cooling, 419–426
    costs, 404
    design, 413–431
    energy savings, 406–407
    fans, 427–431
    forced-draft air-fin, 421–423
    gear box vs. V-belt drives, 410
    induced-draft air-fin, 422
    rating, 401–402
    renewable energy sources, 402–404
    selection for heat disposal, 401–410
    space and power allocation, 415–418
    water power, 402
    wind power, 402
    winter operation, 404–406
Air evaporation
    heat recovery, 592
Air-to-air heat exchangers, 152
Algae, in cooling water, 488–490
Alkaline aqueous solutions, 107
Annealing, in-place
    of high-temperature furnace tubes, 209–210
Austenitic stainless steels, 183–184
Axial-flow exchangers, 140

Bacteria in cooling water, 488–490
Baffle-tray columns, 121–122
Batch processes
    computer-aided calculations, 247–251

    heating and cooling, 246–251
    use of inflated-plate heat exchangers in, 153
Bayonet-type heat exchangers, 30–31, 174
Blowdown, automatic, 366
Blowers, soot, 366
Boilers
    auxiliary equipment, 350–351
    balancing against plant loads, 339–342
    bayonet-type, 551
    burner, 348–349, 362, 364–365, 378–379
    calculation of generated steam, 345
    conversion from gas to oil and coal firing, 346–354
    economizers, 366
    efficiency, 342, 377–379
    emission control, 351
    feedwater preheating, 539
    fuel systems, 349–350, 379
    gas-fired, 346–354
    heat-loss minimization, 377–378
    heat recovery, 541, 546–550
    leaks, 551–555
    optimization of combustion, 378
    package, 359–367
    requirements for maximum economy, 339
    short-cut calculations, 392
    stack requirements, 352
    vertical, 347, 549
    waste-heat, 34, 539, 546–555
    water quality, 379, 550
    water-tube, 363, 549–550
Brass heat transfer, 183
Buried pipelines
    calculation of heat transfer, 261–262
Burners, 298–301
    of boilers, 348–349

Calandrias
    forced-circulation, 49–50
Cartridge-block exchangers, 173–174
Cascade coolers, 172–174
Coal firing of boilers, 346–354
Columns
    baffle-tray, 121–122, 125
    crossflow-tray, 123
    packed, 123, 125–126

[Columns]
    spray, 122-123, 125
Combustion air preheating, 539-540
Computer programs
    for calculation of enthalpy, 441-448
Condensers and condensation, 35
    cleaning, 201-203
    column-mounted, 48
    constant-pressure vent systems, 48
    design, 44-51, 113
    direct-contact, 117-126
    Dukler plot, 45-46
    fluted tubes, 46-47
    for vacuum steam fractionator, 118
    horizontal tubes, 47
    tubing, 200-203
    updraft vs downdraft, 46
    vapors, 129
    water velocity effect, 200-201
    See also Steam condensers
Contractors
    pipeline, 123-124, 126
Coolers
    cascade, 172-174
    direct-contact, 117-126
        evaporative, 34-35, 583-584
    See also Aircoolers, Fans, Gas cooling, Open recirculating cooling systems, Unitized tower and air coolers
Cooling
    in batch processes, 246-251
    use of agitators, 239-245
Cooling systems, Sec. VII
    open recirculating, calculation, 477-487
Cooling towers, 403-404
    basins, 411-412
    blowdown, 491-493
    cold-weather operation, 432-434
    cost reduction, 491-493
    flumes, 412
    fouling, 531-533
    nonpolluting water, 526-533
    operation and maintenance, 432-434
    plumes, 403, 407
    recirculating systems, 435-437, 563-564
    shutdown, 434
    startup, 432, 435-437
    testing, 435-437
    troubleshooting, 433
    wastewater utilization, 520-525
    water treatment, 434, 437
    wet type, 561
    wood, 436-437
    wood deterioration, 530-531
Cooling water
    calculations, 477-487
    corrosion, 488-490, 530
    fouling, 504, 531-533
    low-toxicity inhibitors, 504-505
Corrosion
    avoidance, 187-188
    control in cooling-water systems, 488-490
    of stainless steel, 185-188

CPI (chemical process industries) heat exchangers, 99-108, 180-182
    fluted tubes, 104-105
    fouling, 105
    roped tubes, 103-104
    tube selection, 99-108
Cryogenic systems, 155
Cubic (or rectangular) heat exchangers, 172
Cyclohexane
    liquid isobaric heat capacity, 457
Cylindrical exchangers, 173

Direct-contact condensers, 117-126
Direct-contact coolers, 117-126
    countercurrent, 118
Distillation columns, 72-73
Double-pipe and cascade exchangers, 30
Double-tubesheet heat exchangers, 56-59
    conventional, 57
    design problems, 57-58
    integral double tubesheets, 59
    leakage prevention, 56
    shell-tube leakage, 56-59
    special tubesheets, 59

EDIPDO (eutectic mixture of diphenyl and diphenyl oxide) vapor system, 515-519
Effluents
    chemical-plant, 523-524
    dissolved-air-flotation, 525
    domestic-sewage, 521-522
    fertilizer-plant, 522-523
    treatment, 528-529
EHT (enhanced-heat-transfer) tubes, 99, 102
Electric heating of pipelines, 282-284
Electric immersion heaters, 235-238
    corrosion and watt density, 236-238
Electric pipe tracing, 270-273
Enthalpy
    calculation with pocket calculator, 441-448
Evaporative coolers, 35
    sizing of vacuum equipment, 583-584
Evaporators and evaporation
    design and operation, 573-578
    direct contact, 589
    energy saving, 588-594
    heat balance, 577-578
    plate, 148-149
    troubleshooting, 576-577
    See also Air evaporation, EDIPDO vapor system, Plate evaporators, Vaporizers

Falling-film heat exchangers, 35
Fans, air-cooler, 427-431
Ferritic stainless steel, 183-184
Finned-tube calculations
    factors for selection, 454-455
    use of charts, 450-455
Fired heaters, Sec. V
    air supply, 292

[Fired heaters]
  auxiliary equipment, 331
  burners, 298-301
  combustion, 303-314
  combustion-air preheating, 317-318, 330
  construction materials, 293-302, 335-336
  convection section, 301-302, 309-310, 322-326, 334-335
  conversion from gas to liquid firing, 319
  cost reduction, 315-319, 328-336
  design, 287-292, 303-314, 320-336
  excess-air reduction, 315-316
  flue-gas removal, 292
  fluid pressure drop, 310
  heat balance, 323
  heat coils, 305-306
  horizontal, 329
  horizontal vs. vertical, 289-292
  mechanical features, 292-302
  overall furnace rating, 327
  performance, 293-302, 320-327
  radiant section, 307-309, 320-321, 325-326, 333-334
  refractories, 293-294
  sample calculation, 312-314
  stack design, 326-327
  thermal efficiency, 306-307
  tube coils, 294-297
  tube-wall thickness, 311
  vertical-tube, 329
Fixed-tubesheet exchangers, 82
Floating-head exchangers, 82
Floating-tubesheet exchangers, 28
Flowrate calculation
  through steam traps, 368-372
Flow-work exchangers, 595-601
Fluid-fluid heat exchangers, 14
Fluids, organic
  for high-temperature heat-transfer system, 506-514
Flumes for cooling tower, 412
Froth-contact heat exchangers, 34-35
Fuel conservation
  by heating with hot water instead of steam, 569-572
Fungi in cooling water, 488-490
Furnace tubes
  in-place annealing, 209-210

Gas boilers, 548-549
  conversion to oil and coal, 346-354
Gas cooling, 118-121
  hot, 124
  line quencher, 124
Gases
  heat recovery, 546-548
  hot pyrolysis, 118, 124-125
Gas heat-transfer coefficient
  estimation by nomograms, 467-470
Gas-to-coal conversion, 353
Gas-to-gas exchangers
  use of heat pipes, 152
Glass-lined reactors
  heat transfer system, 231-234
  temperature control, 233-234

Graphite heat exchangers, 170-175
  applications, 170-171, 175
  heat transfer, 174-175
  shell-and-tube type, 170

Heat disposal
  air cooler vs. water tower, 401-410
Heat emission
  relating to surface temperature, 472
Heaters
  direct-fired process, 320-327
  feedforward, 581
  fired, 287-336
  immersion, 174
  matching to applicaton, 586-587
  operating and maintenance records, 215-218
  room, 472
Heat exchangers
  axial-load effect, 65
  bibliographies, 34-36
  cartridge-block, 173
  classification, 26-34
  computer-aided design, 109-113
  computer programs, 89-91
  construction materials, Sec. I-C
  cylindrical, 173
  design, Sec. I-A, B, C, 14-36, 37-44, 60-67
  EHT (enhanced-heat-transfer) tubes, 102
  equipment, Sec. II
  extended surfaces, 29-30
  failures, 63-64
  flow-induced vibration, 65
  flow-work type, 595-601
  fluid flow, 36
  fluid-fluid, 14
  fouling, 60-61, 105, 134, 471
  handbooks and textbooks, 34
  heat recovery, 558
  heat transfer, 88, 99-100
  heat-transfer rate, 38-41
  insulation and tracing, 36
  leakage, 61-62
  for liquids in laminar flow, 37-44
  materials of constructon, Sec. I-C
  materials selection, 106-108, 176-179
  mechanical design, 36
  mechanical modification, 83-84
  modular-block, rectangular, 173
  operating temperature calculation, 88
  operation and maintenance, Sec. II, 36, 215-218
  optimum layout, 79-87
  performance prediction, 88-91
  piping, 79-87
  problems and cures, 66
  rating, 18-22
  record keeping, 215-217
  repair, 62
  salt water, 107
  specification, Sec. I-A, B, C
  spiral-plate, 127-144
  stainless-steel, 180-188
  steel tubing selection, 176-179

604 INDEX

[Heat exchangers]
   successive summation method, 88–89
   transversely finned tubes, 100–102
   trapping, 380–385
   trouble-free design, 60–67
   troubleshooting, 36
   tube selection, 99–108
   tubing, 176–179, 200–203, 208–209
   vibrations, 36, 208–209
   See also: Air-cooled heat exchangers, Air-to-air heat exchangers, Axial-flow exchangers, Bayonet-type heat exchangers, Boilers, Cartridge-block exchangers, CPI heat exchangers, Cubic (or rectangular) heat exchangers, Cylindrical exchangers, Double-pipe and cascade exchangers, Double-tubesheet heat exchangers, Falling-film heat exchangers, Fixed-tubesheet exchangers, Floating-head exchangers, Floating-tubesheet exchangers, Flow-work exchangers, Fluid-fluid heat exchangers, Froth-contact heat exchangers, Gas-to-gas exchangers, Graphite heat exchangers, Inflated-plate heat exchangers, Kettle-type heat exchangers, Modular-block rectangular exchangers, Nonmetallic exchangers, Plate heat exchangers, Plate-type exchangers, Rectangular exchangers, Scraped-surface exchangers, Shell-and-tube heat exchangers, Spiral-plate exchangers, Spiral-tube exchangers, Stainless-steel heat exchangers, Thin-film heat exchangers, Tubular heat exchangers, U-tube exchangers

Heating
   circulating system, 540
   electric, of pipelines, 282–284
   hot water, 569–587
   in batch process, 246–251
   of metals, 585
   portable systems, 585–587
   steam, 569–572
   use of agitators, 239–245
   See also Electric immersion heaters, Fired Heaters, Room heaters, Spot heating, Storage heaters, Superheaters, Superheating

Heat pipes, 150–152
   future applications, 152

Heat recovery
   by air evaporation, 592
   by low cost evaporation, 588–594
   control systems, 562–564
   economic aspects, 544–545
   flow-work exchanger, 595–601
   from flue gases, 546
   from process gases, 546–548
   in boilers, 546–550
   in process plants, 537–545
   Rankine-cycle systems, 556–560
   useful energy, 561–565

Heat transfer
   calculations, Sec. VIII, 510–512
   calculation from buried pipeline, 261–262
   enthalpy, 441–448
   fundamentals, 99–100
   gas heat-transfer coefficient, 467–470
   high-temperature systems, 494–503
   in mechanically agitated units, 239–245

[Heat transfer]
   in piping systems, Sec. IV
   in reactor units, Sec. III
   liquid heat capacities, 456–464
   radiant heat transfer calculation, 449
   relating fouling factors to conductivity coefficient, 469
   relating heat emission to surface temperature, 472
   spiral finned-tube, 450–455
   steam from flashing condensate, 473
   thermal conductivity, 465–466

Heat-transfer agents
   for high-temperature systems, 494–503
   instrumentation and controls, 503
   liquid-phase systems, 501–502
   pumping costs, 497–500
   vapor-phase systems, 500–501

Heat transfer media, Sec. IX
   high-temperature systems, 494–503, 506–514
   selection, 495–496
   vapor phase, 515–519

Heat transfer systems
   organic fluids for high-temperature applications, 506–514
   piping, 514, 519
   pump seals, 514
   seals and venting, 517

High-temperature applications
   heat-transfer agents, 494–503
   steel tubing, 176–179

High-temperature fluids, 36
Horizontal-thermosiphon reboilers, 68–71
Hot water heating, 569–587
   fuel conservation, 569–572
   storage vs. instantaneous heaters, 579–582
Hydraulics in horizontal reboilers, 76

Immersion heaters, 174
   electric, 235–238
Improved boiling surfaces, 36
Inflated-plate heat exchangers, 153–155
Inhibitors, in cooling water, 504–505

Jackets and jacketing
   choice of medium, 256–258
   for vessels, 252–258
   heat transfer, 254–255
   pressure drop, 255–256
   types, 252–254

Kettle-type heat exchangers, 82

Laminar flow
   heat-transfer coefficient, 37–38
Leakage
   in waste-heat boiler, 551–555
   of steam, 358
Liquid heat capacitors
   definition, 456–459
   estimation, 456–464

[Liquid heat capacitors]
    thermodynamic cycle calculation, 462–463
Liquid *n*-heptane
    heat capacities, 456
Liquid in laminar flow
    heat exchangers, 37–43
Low-toxicity inhibitors
    in cooling water, 504–505

Makeup water, 520–533
Mechanically agitated units
    heat transfer, 239–245
Medicine
    application of heat pipes, 151
Melting-point inversion process, 595–601
Metal heating, 585
Microorganisms control
    in cooling water, 488–490
Miscellaneous topics, Sec. XI
Modular-block rectangular exchangers, 173

Natural frequency of vibration
    exchanger tubes, 208–209
Nonmetallic exchangers, 36

Oil firing
    of boiler, 346–354
Open recirculating cooling systems
    calculations, 477–487
Organic fluids
    for high-temperature heat transfer systems, 506–514
    heater and vaporizer, 513–514
Organic liquids
    predicting thermal conductivity, 465–466
Orifice traps
    sizing, 368–372

Package boilers
    advantages, 361–362
    air heaters, 364
    design, 360–361
    economizers, 363–364
    energy-saving accessories, 363
    safety equipment, 363
    selection, 359–367
Pipe coils, 36
Pipeline contactors, 123–124, 126
Pipelines
    calculation of heat transfer, 261–262
    electric heating, 282–284
    steam tracings, 275–281
Piping
    discharge, 382–383
    heat transfer, Sec. IV
    steam traps, 382–385
Plate evaporators, 148–149
Plate heat exchangers, 35
Plate-type exchangers, 31–32, 134, 145–149, 156–162
    biological materials, 213

[Plate-type exchangers]
    comparison with tubular heat exchanger, 160–161
    corrosion, 168–169, 214
    costs, 145–146, 159–160
    crystallization, 212
    design, 165, 167
    erosion-corrosion, 166–169
    fouling, 160, 211–214
    frames, 161
    gaskets, 161, 168–169
    heat degradation, 214
    limitation, 161–162
    materials of construction, 161, 165–169
    plate configuration, 147–148
    plate performance, 146–147
    pressure drop, 158–159
    scaling, 211–212
    suspended solids, 212–213
Plumes, 403, 407–408
Pocket calculators
    use for computing enthalpy, 441–448
Pollution control, 562–564
Portable heating systems
    spot heating, 585–587
Process plants
    heat recovery, 537–545

Radiant heat transfer calculation, 449
Rankine-cycle systems
    for waste heat recovery, 556–560
Reaction units
    heat transfer, Sec. III
Reactors, glass-lined, 231–234
Reboilers
    design, 44–51, 75–76
    distillation units, 72–73
    elevation of drawoff nozzle, 77
    elevations, 68, 74–75
    flow relations, 78
    fouling, 52, 204–207
    friction losses, 76–77
    heat transfer, 204–206
    horizontal kettle-type, 50
    horizontal-thermosiphon, 68–71
    hydraulics in horizontal type, 76
    piping, 68, 72–87
    shell and tube, 70
    steam-heated, 219–222
    thermosyphon vs. forced circulation, 54–55
    vertical thermosyphon, 52–55, 204–207
Recirculating water
    conductivity, 482–483
    dissolved materials, 483–485
    electroneutrality, 479
    Langelier and Ryznar indexes, 483
    predicting pH, 480–483
    TDS, 482–483, 485
Rectangular exchangers
    modular-block, 173
Room heaters
    relating heat emission to surface temperature, 472
Rubber-ball cleaning, 201–203

INDEX

Salt waters, 107–108
Scraped-surface exchangers, 32, 35
Shell-and-tube heat exchangers, Sec. I-A, 14, 34, 80
   axial load, 94
   baffles, 11, 18, 80–82
   classification, 26–29
   computer-aided design, 110–111
   corrosion, 187–188
   costs, 12–13
   damping, 93
   design, 5–13, 80, 97–98
   flow distribution, 18
   flow-induced vibration, 17, 92–93, 97
   fluid allocation, 10
   fluid-elastic excitation, 96
   fluid properties, 17
   fouling, 6
   graphite type, 171–172
   heat conduction, 26–27
   mean temperature difference, 16
   natural frequencies, 93–95
   optimization, 5–13
   pressure drop, 6, 15, 27–28
   shellside condensing, 25
   shellside heat transfer, 24–25
   shortcut rating techniques, 14, 18–19
   stainless steel, 180–188
   TEMA nomenclature, 6–9
   tube pitch, 10
   tubeside heat transfer, 23–24
   tube size, 10, 17
   turbulent buffeting, 97
   U-bend tubes, 95
   vibration mechanism, 97
   vibration prevention, 92–98
   virtual mass, 94
   vortex shedding, 95–96
Skin current, electric
   heating of pipelines, 282–284
Solar power plants, 402
Soot blowers, 366
Spiral-plate exchangers
   design, 127–144
   fabrication, 127, 137
   flow arrangements, 127–129
   liquid-liquid, 135–136
   pressure drop, 135–136, 140–141, 143
   shortcut rating method, 129–132, 138–143
Spiral-tube exchangers, 31–32
   pressure drop, 140–144
Spot heating with portable heating systems, 585–587
Spray quenchers
   for pyrolisis gases, 118
Stainless steel
   austenitic, 183–184
   corrosion, 185–188
   ferritic, 183–184
   heat-transfer properties, 182–183
   precipitation-hardening, 184
   types and characteristics, 183
Stainless-steel heat exchangers, 180–188
   bulk temperature in pipe, 266–267
   heat transfer, 182–183

Steam condensers
   air-cooled, 191–199
   backflow, 195
   cold-climate considerations, 198–199
   design, 191–199
   thermal specificaions, 197–198
Steam from flashing condensate
   calculation by nomograph, 473
Steam generation, Sec. VI
   heat recovery, 538–539
   package boiler selection, 359–367
Steam generators
   basic data, at a glance, 343–344
   calculation by nomograph, 345
   heat recovery, 540–543
   package boiler selection, 359–367
   parameter charts, 343–344
Steam-heated reboilers
   condensate removal, 220, 227–228
   fouling, 220–222
   heat exchange, 219
   performance, 219–228
   steam-inlet control, 223–226
Steam heaters and heating
   disadvantages, 569–570
   short-cut calculations, 392
Steam leaks
   estimation of costs, 358
Steam superheating, 540
Steam tracing
   design, 263–269
   of pipelines, 278–281
Steam transmission, Sec. VI
Steam transmission lines
   without steam traps, 355–357
Steam traps, 355–357, 393–398
   basic function, 393
   costs, 391
   flowrate calculation, 368–372
   installation, 380–385
   mechanical, 394–395
   piping, 382–385
   selection, 386–391
   sizing of orifices, 368–372
   thermostatic, 395–398
Steel
   stainless, 180–188
   tubing for high-temperature service, 176–179, 182
Storage heaters
   for hot water, 580
Superheaters, 366
Superheating of steam, 540

Thermal conductivity
   new correlation, 465–466
Thin-film heat exchangers, 33–35
Tracing
   electric pipe, 270–273
   steam, 263–269
Trapless steam transmission lines
   design, 355–357
Traps and trapping, 380–385, 393–398

[Traps and trapping]
   bimetallic, 388, 396
   costs, 391
   discharge piping, 382-383
   disk, 389-390
   double, 380
   float, 386
   impulse-type, 389, 396-397
   mechanical, 394-395
   metal-expansion, 395-396
   multiple, 380-381
   open-bucket, 386-387
   piston, 396-397
   sizing, 390
   testing, 373-376
   thermodynamic, 397
   thermostatic, 387, 395-398

Tubing
   cleaning, 200-203
   high-temperature furnace, 209-210
   steel, 176-179, 182

Tubular heat exchangers, 117, 221

Unitized tower and air coolers, 408-409
U-tube exchangers, 82, 95

Vacuum equipment
   for evaporative coolers, 583-584
Vaporizers, 35
   falling-film, 50
   long-tube vertical, 51
Vapor-phase heat-transfer media, 515-319

Vertical-thermosyphon reboilers, 52-55, 204-207
Vessel jackets
   optimal selection, 252-258
Vibrations
   of exchanger tubes, 208-209

Waste-heat recovery, Sec. X
   economic aspects, 559-560
   problems in boilers, 546-550
   Rankine-cycle systems, 556-560
   useful energy, 561-565
Waste-gas streams
   heat transfer coefficients, 468-471
Wastewater
   use as cooling-system makeup water, 520-525
Water
   absence of pollution, 526-533
   chemical composition calculation, 477-487
   corrosion, 530-531
   desalting, 99
   pH, alkalinity and hardness, 478-479
   rivers, 529
   scaling, 529-531
   sea, 529
   sour-water stripper bottoms, 524-525
   wastewater utilization, 520-525
Water towers
   rating, 401-410
   renewable energy sources, 402-404
   selection for heat disposal, 401-410
   water power, 402
   wind power, 402